Periodic Table of the Elements

Key:

1 — Atomic number
H — Symbol
1.0079 — Atomic mass

Period	1 I IA	2 II IIA	3 IIIB	4 IVB	5 VB	6 VIB	7 VIIB	8	9 VIIIB	10	11 IB	12 IIB	13 III IIIA	14 IV IVA	15 V VA	16 VI VIA	17 VII VIIA	18 VIII VIIIA
1	1 H 1.0079																	2 He 4.00
2	3 Li 6.94	4 Be 9.01											5 B 10.81	6 C 12.01	7 N 14.01	8 O 16.00	9 F 19.00	10 Ne 20.18
3	11 Na 22.99	12 Mg 24.31											13 Al 26.98	14 Si 28.09	15 P 30.97	16 S 32.06	17 Cl 35.45	18 Ar 39.95
4	19 K 39.10	20 Ca 40.08	21 Sc 44.96	22 Ti 47.88	23 V 50.94	24 Cr 52.00	25 Mn 54.94	26 Fe 55.85	27 Co 58.93	28 Ni 58.71	29 Cu 63.54	30 Zn 65.37	31 Ga 69.72	32 Ge 72.59	33 As 74.92	34 Se 78.96	35 Br 79.91	36 Kr 83.80
5	37 Rb 85.47	38 Sr 87.62	39 y 88.91	40 Zr 91.22	41 Nb 92.91	42 Mo 95.94	43 Tc 98.91	44 Ru 101.07	45 Rh 102.91	46 Pd 106.40	47 Ag 107.87	48 Cd 112.40	49 In 114.82	50 Sn 118.69	51 Sb 121.75	52 Te 127.60	53 I 126.90	54 Xe 131.30
6	55 Cs 132.91	56 Ba 137.34	71 Lu 174.97	72 Hf 178.49	73 Ta 180.95	74 W 183.85	75 Re 186.2	76 Os 190.2	77 Ir 192.2	78 Pt 195.09	79 Au 196.97	80 Hg 200.59	81 Tl 204.37	82 Pb 207.19	83 Bi 208.98	84 Po 210	85 At 210	86 Rn 222
7	87 Fr 223	88 Ra 226.03	103 Lr 262.1	104 Rf	105 Db	106 Sg	107 Nh	108 Hs	109 Mt	110 Uun	111 Uuu	112 Uub	113 UUt					

Metals
Nonmetals
Metalloids

Lanthanide series	57 La 138.91	58 Ce 140.12	59 Pr 140.91	60 Nd 144.24	61 Pm 146.92	62 Sm 150.35	63 Eu 151.96	64 Gd 157.25	65 Tb 158.92	66 Dy 162.50	67 Ho 164.93	68 Er 167.26	69 Tm 168.93	70 Yb 173.04
Actinide series	89 Ac 227.03	90 Th 232.04	91 Pa 231.04	92 U 238.03	93 Np 237.05	94 Pu 239.05	95 Am 241.06	96 Cm 247.07	97 Bk 249.08	98 Cf 251.08	99 Es 254.09	100 Fm 257.10	101 Md 258.10	102 No 255

General, Organic, and Biochemistry

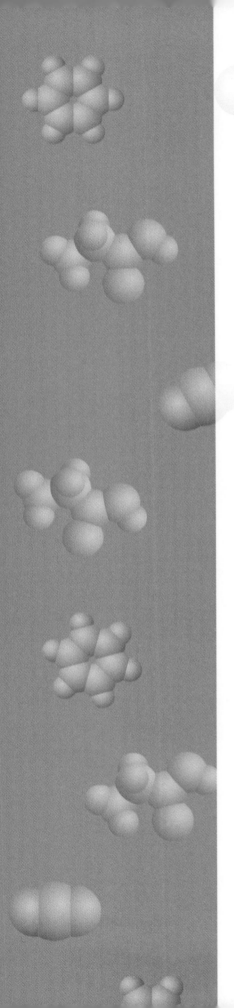

ABOUT THE AUTHORS

Ira Blei was born and raised in Brooklyn, New York, where he attended public schools and graduated from Brooklyn College with B.S. and M.A. degrees in chemistry. After receiving a Ph.D. degree in physical biochemistry from Rutgers University, he worked for Lever Brothers Company in New Jersey, studying the effects of surface active agents on skin. His next position was at Melpar Incorporated, in Virginia, where he founded a biophysics group that researched methods for the detection of terrestrial and extraterrestrial microorganisms. In 1967, Ira joined the faculty of the College of Staten Island, City University of New York, and taught chemistry and biology there for three decades. His research has appeared in the *Journal of Colloid Science,* the *Journal of Physical Chemistry,* and the *Archives of Biophysical and Biochemical Science.* He has two sons, one an engineer working in Berkeley, California, and the other a musician who lives and works in San Francisco. Ira is outdoors whenever possible, overturning dead branches to see what lurks beneath or scanning the trees with binoculars in search of new bird life, and he recently served as president of Staten Island's local Natural History Club.

George Odian is a tried and true New Yorker, born in Manhattan and educated in its public schools, including Stuyvesant High School. He graduated from The City College with a B.S. in chemistry. After a brief work interlude, George entered Columbia University for graduate studies in organic chemistry, earning M.S. and Ph.D. degrees. He then worked as a research chemist for 5 years, first at the Thiokol Chemical Company in New Jersey, where he synthesized solid rocket propellants, and subsequently at Radiation Applications Incorporated in Long Island City, where he studied the use of radiation to modify the properties of plastics for use as components of space satellites and in water desalination processes. George returned to Columbia University in 1964 to teach and conduct research in polymer and radiation chemistry. In 1968, he joined the chemistry faculty at the College of Staten Island, City University of New York, and has been engaged in undergraduate and graduate education there for three decades. He is the author of more than 60 research papers in the area of polymer chemistry and of a textbook titled *Principles of Polymerization,* now in its third edition, with translations in Chinese, French, Korean, and Russian. George has a son, Michael, who is an equine veterinarian practicing in Ohio. Along with chemistry and photography, one of George's greatest passions is baseball. He has been an avid New York Yankee fan for more than five decades and has been especially pleased with their performance in the past few seasons.

Ira Blei and George Odian arrived within a year of each other at the College of Staten Island, where circumstances eventually conspired to launch their collaboration on a textbook. Both had been teaching the one-year chemistry course for nursing and other health science majors for many years and, during that time, they became close friends and colleagues. It was their habit to have intense, ongoing discussions with each other about how to teach different aspects of the chemistry course, each continually pressing the other to enhance the clarity of his presentation. One of those conversations developed their ideas for this textbook.

General, Organic, and Biochemistry

Connecting Chemistry to Your Life

Ira Blei
George Odian
College of Staten Island
City University of New York

W. H. FREEMAN AND COMPANY
NEW YORK

Publisher: Michelle Russel Julet
Associate Editor: Jessica Fiorillo
Development Editor: Moira Lerner
Marketing Manager: Kimberly Manzi
Project Editor: Jane O'Neill
Text Designer: Diana Blume
Cover Designer: Patricia McDermond
Illustration Coordinators: Bill Page, Diana Blume
Illustrations: Network Graphics, Janet Hamlin
Photo Researcher: Kathy Bendo
Production Coordinator: Julia DeRosa
Supplements and Multimedia Editors: Matthew P. Fitzpatrick and Patrick Shriner
Composition: Progressive Information Technologies
Layout: Eileen Burke
Manufacturing: RR Donnelly & Sons

Library of Congress Cataloging-in-Publication Data

Blei, I (Ira) 1931–
General, organic, and biochemistry : connecting chemistry to your life / by Ira Blei, George Odian.
p. cm.
Includes Index
ISBN 0-7167-2872-9
1. Chemistry I. Odian, George G., 1933– . II. Title.
QD33.B663 1999 99-29954
540—dc21 CIP

Printed in the United States of America

Second printing 2003

Cover Images:
Human red blood cell: Bill Loncore/Photo Researchers
Groups of red blood cells: OmikronScience source/ Photo Researchers
Hemoglobin molecule: Ken Eward/BioGrafx, Mount Vernon, Ohio
Vein: From *Biology Fundamentals* by Gil Brum, Larry McKane, and Gerry Karp. (New York: Wiley, 1995). Used with permission of the authors.

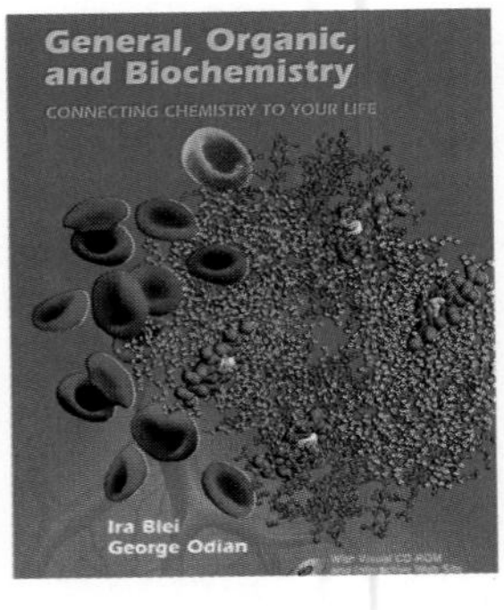

CONTENTS IN BRIEF

CONTENTS

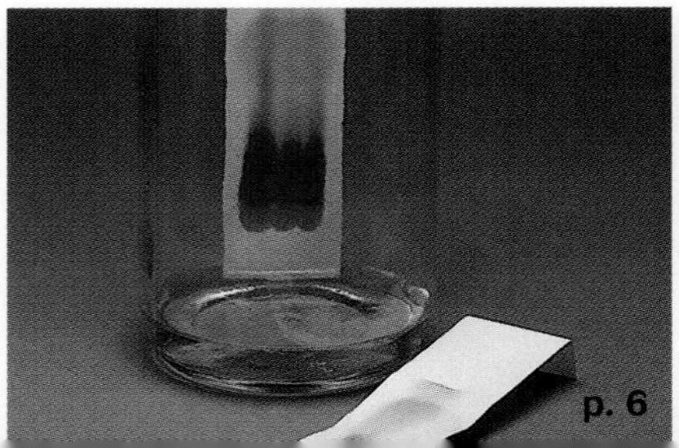

p. 6

p. 7

p. 26

p. 65

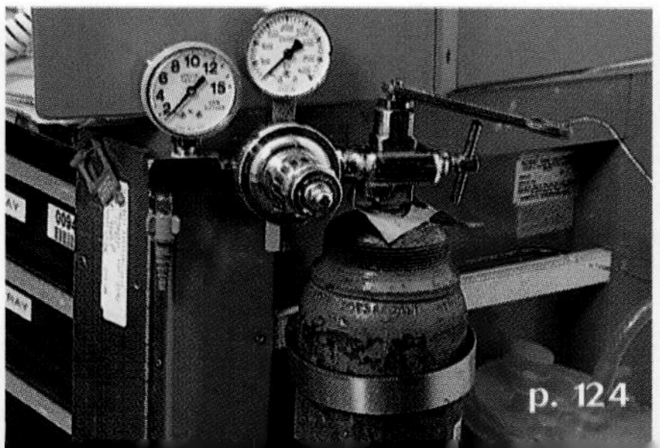
p. 124

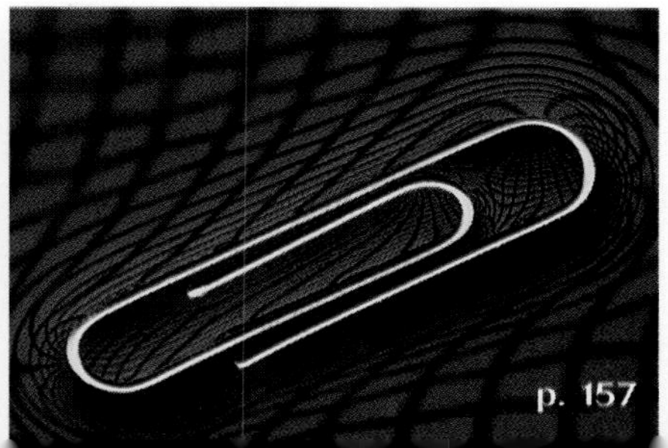
p. 157

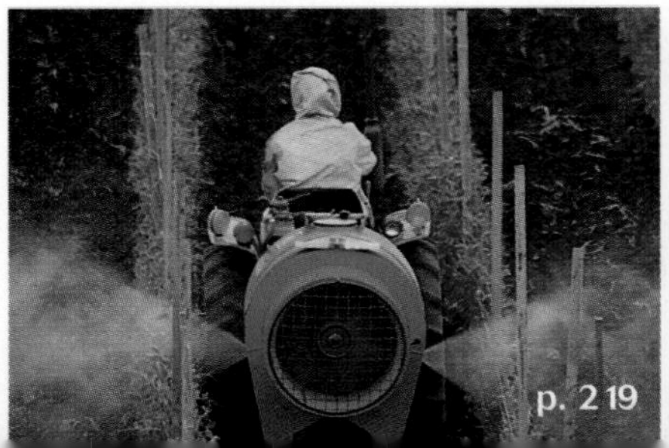
p. 219

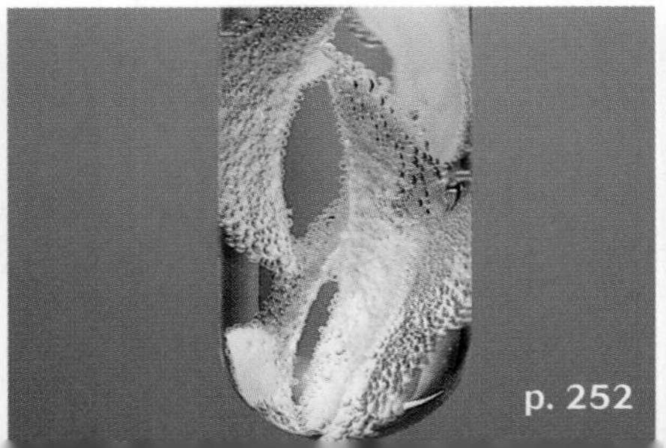
p. 252

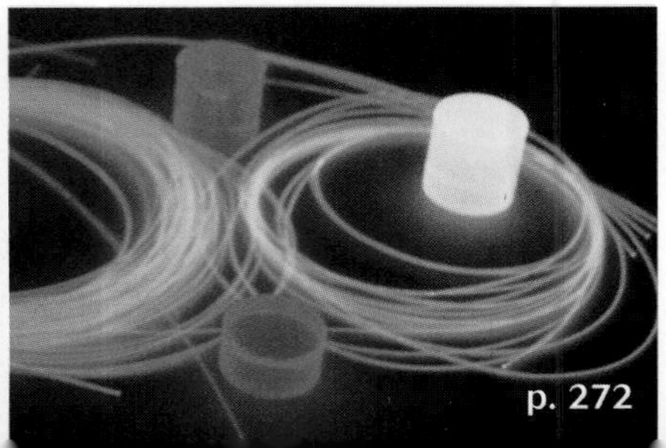
p. 272

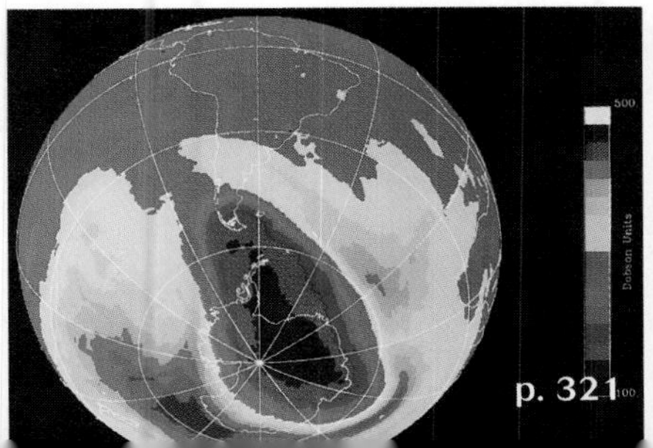
p. 321

329

p. 332

p. 379

p. 406

433

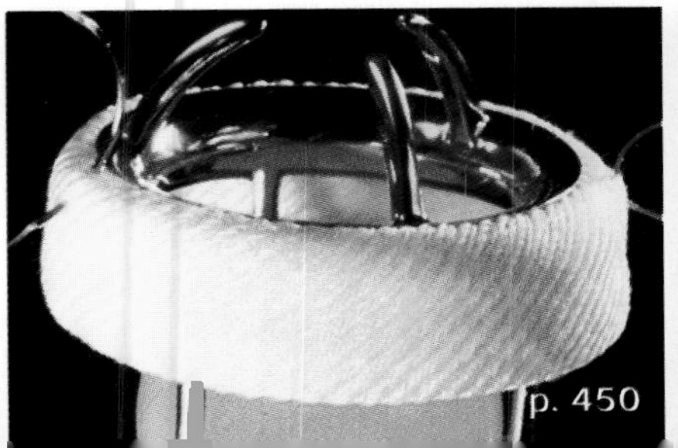
p. 450

p. 471

p. 498

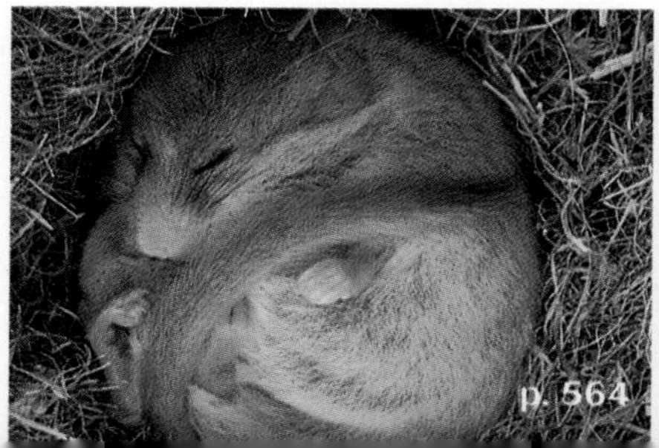
p. 564

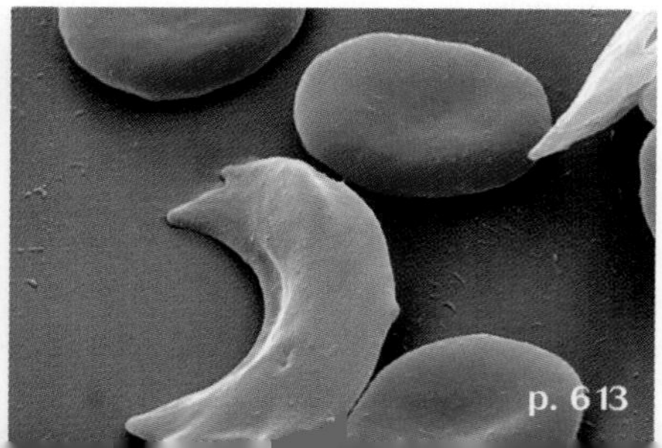
p. 613

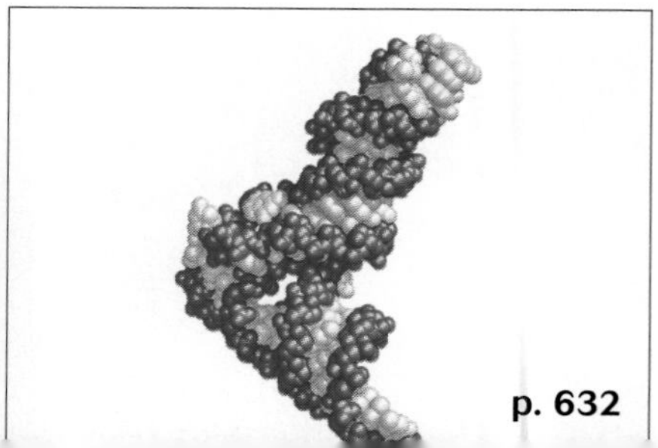
p. 632

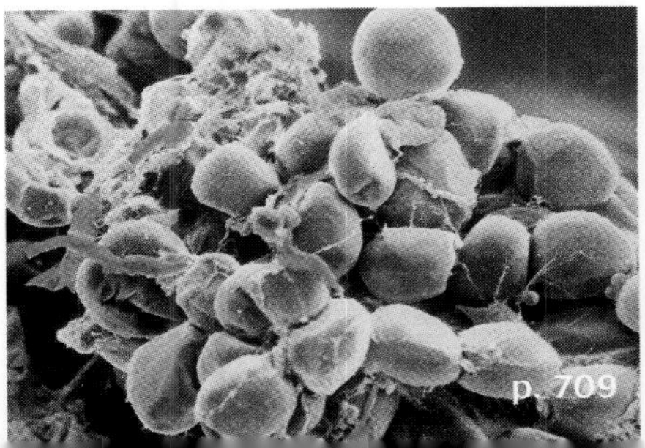
p. 709

p. 722

p. 767

PREFACE

General, Organic, and Biochemistry: Connecting Chemistry to Your Life is designed to be used in a one-year course presenting general, organic, and biochemistry. Our text was written for students who intend to pursue careers as nurses, dieticians, physician's assistants, physical therapists, or environmental scientists.

GOALS OF THIS BOOK

Our chief objective in writing this book has been to emphasize chemical *principles*—the comprehensive laws that help explain how matter behaves—because an introductory textbook that offers little more than a series of facts with no strong supporting explanation is of limited value to the student. New scientific information is discovered every day, and technological development is continuous. Students who merely memorize today's scientific information without understanding the basic underlying principles will not be prepared for the demands of the future. On the other hand, students who have a clear understanding of basic physical and chemical phenomena will have the tools to understand new facts and ideas and will be able to incorporate new knowledge into their professional practice in appropriate and meaningful ways.

The other central goal of our book is to introduce students to how the human body works at the level of molecules and ions—that is, to the chemistry underlying physiological function. In pursuit of this objective, our focus in Part 1, General Chemistry, and Part 2, Organic Chemistry, is on providing a clear explication of the chemical principles that are used in Part 3, Biochemistry. In the process of exploring and using these principles, we emphasize two major themes throughout: (1) the ways in which molecules interact, and how that explains the nature of substances, and (2) the relations between molecular structures within the body and their physiological functions. Throughout the book, we illustrate chemical principles with specific examples of biomolecules and, in many chapters, with problems having a biological or medical context.

PEDAGOGICAL FEATURES

The features of this book are **applications, problem-solving strategies, visualization,** and **learning tools,** using a real world context to connect chemistry to students' lives.

Making Connections with Applications

Students are more motivated to learn a subject if they are convinced of its fundamental importance and personal relevance. Examples of the relevance of chemical concepts are woven throughout the text and emphasized through several key features.

Chemistry in Your Future

A scenario at the beginning of each chapter describes a typical workplace situation that illustrates a practical, and usually professional, application of the contents of that chapter.

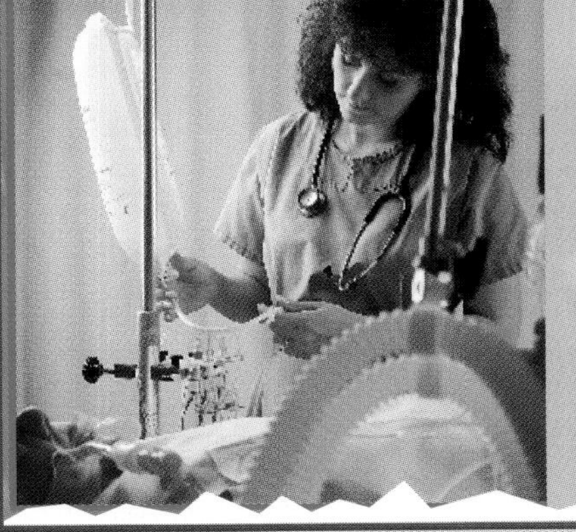

CHEMISTRY IN YOUR FUTURE

When a head-injury patient arrives at the hospital, one of the key concerns of the health-care team is to prevent or reduce excess fluid collection around the brain. Such swelling, or edema, is a natural response to injury, but when the injured organ is the brain, the added fluid pressure could cause severe damage. One way to deal with this problem is to administer an intravenous solution of mannitol, a water-soluble compound having no biological activity. Its only physiological effect is to increase the osmotic pressure of the filtrate in the kidneys' tubules, thus increasing the amount of fluid disposed of in the urine. What is osmotic pressure, and how does it affect body fluids? Aft reading this chapter

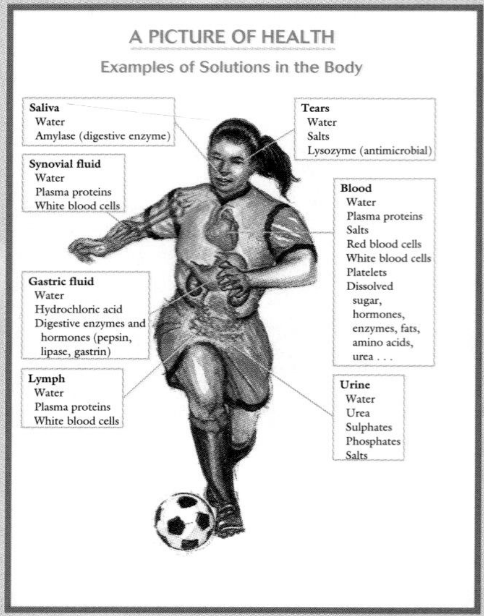

A Picture of Health

This series of richly rendered drawings shows how chapter topics apply to human physiology and health.

Three Categories of Boxes

A total of 80 boxed essays, divided into three categories, broaden and deepen the reader's understanding of basic ideas:

Chemistry Within Us These boxes describe applications of chemistry to human health and well-being.

Chemistry Around Us These boxes describe applications of chemistry to our everyday life (including commercial products) and to biological processes in organisms other than humans.

Chemistry in Depth These boxes provide a more detailed description of selected topics, ranging from chromatography to the mechanisms of key organic reactions.

Professional Connections

In these interviews, seven health-science professionals describe their work and educational backgrounds, talk about the relevance of chemistry to their daily routines, and give advice to students planning careers like theirs.

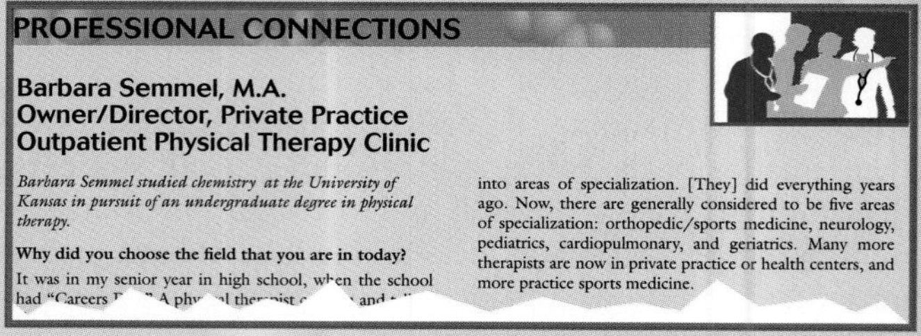

PROFESSIONAL CONNECTIONS

Barbara Semmel, M.A.
Owner/Director, Private Practice
Outpatient Physical Therapy Clinic

Barbara Semmel studied chemistry at the University of Kansas in pursuit of an undergraduate degree in physical therapy.

Why did you choose the field that you are in today?

It was in my senior year in high school, when the school

into areas of specialization. [They] did everything years ago. Now, there are generally considered to be five areas of specialization: orthopedic/sports medicine, neurology, pediatrics, cardiopulmonary, and geriatrics. Many more therapists are now in private practice or health centers, and more practice sports medicine.

Making Connections Through Problem Solving

Learning to work with chemical concepts and developing problem-solving skills is integral to understanding chemistry. We help students develop these skills.

In-Chapter Examples Nearly 250 in-chapter examples with step-by-step solutions, each followed by a similar in-chapter problem, allow students to verify and practice their skills.

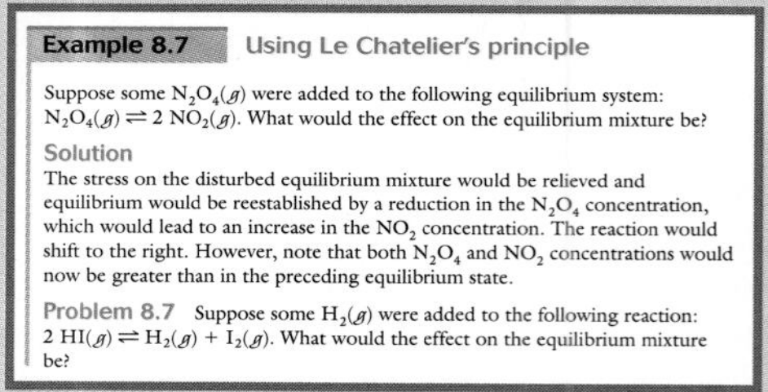

Example 8.7 Using Le Chatelier's principle

Suppose some $N_2O_4(g)$ were added to the following equilibrium system: $N_2O_4(g) \rightleftharpoons 2\ NO_2(g)$. What would the effect on the equilibrium mixture be?

Solution

The stress on the disturbed equilibrium mixture would be relieved and equilibrium would be reestablished by a reduction in the N_2O_4 concentration, which would lead to an increase in the NO_2 concentration. The reaction would shift to the right. However, note that both N_2O_4 and NO_2 concentrations would now be greater than in the preceding equilibrium state.

Problem 8.7 Suppose some $H_2(g)$ were added to the following reaction: $2\ HI(g) \rightleftharpoons H_2(g) + I_2(g)$. What would the effect on the equilibrium mixture be?

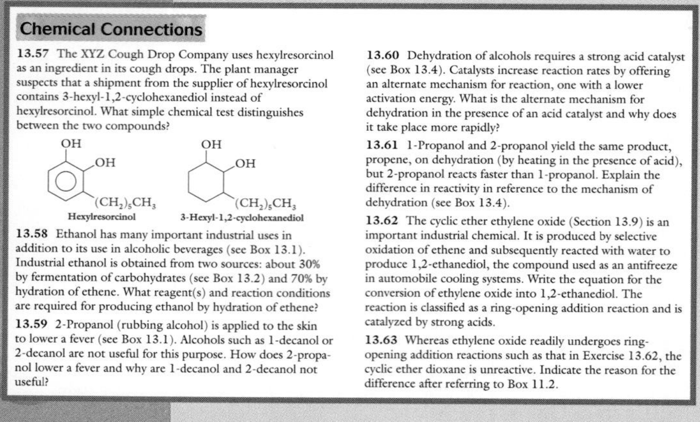

Chemical Connections

13.57 The XYZ Cough Drop Company uses hexylresorcinol as an ingredient in its cough drops. The plant manager suspects that a shipment from the supplier of hexylresorcinol contains 3-hexyl-1,2-cyclohexanediol instead of hexylresorcinol. What simple chemical test distinguishes between the two compounds?

Hexylresorcinol 3-Hexyl-1,2-cyclohexanediol

13.58 Ethanol has many important industrial uses in addition to its use in alcoholic beverages (see Box 13.1). Industrial ethanol is obtained from two sources: about 30% by fermentation of carbohydrates (see Box 13.2) and 70% by hydration of ethene. What reagent(s) and reaction conditions are required for producing ethanol by hydration of ethene?

13.59 2-Propanol (rubbing alcohol) is applied to the skin to lower a fever (see Box 13.1). Alcohols such as 1-decanol or 2-decanol are not useful for this purpose. How does 2-propanol lower a fever and why are 1-decanol and 2-decanol not useful?

13.60 Dehydration of alcohols requires a strong acid catalyst (see Box 13.4). Catalysts increase reaction rates by offering an alternate mechanism for reaction, one with a lower activation energy. What is the alternate mechanism for dehydration in the presence of an acid catalyst and why does it take place more rapidly?

13.61 1-Propanol and 2-propanol yield the same product, propene, on dehydration (by heating in the presence of acid), but 2-propanol reacts faster than 1-propanol. Explain the difference in reactivity in reference to the mechanism of dehydration (see Box 13.4).

13.62 The cyclic ether ethylene oxide (Section 13.9) is an important industrial chemical. It is produced by selective oxidation of ethene and subsequently reacted with water to produce 1,2-ethanediol, the compound used as an antifreeze in automobile cooling systems. Write the equation for the conversion of ethylene oxide into 1,2-ethanediol. The reaction is classified as a ring-opening addition reaction and is catalyzed by strong acids.

13.63 Whereas ethylene oxide readily undergoes ring-opening addition reactions such as that in Exercise 13.62, the cyclic ether dioxane is unreactive. Indicate the reason for the difference after referring to Box 11.2.

End-of-Chapter Exercises The nearly 1700 end-of-chapter exercises are divided into three categories:

- **Paired Exercises** are arranged according to chapter sections; each odd-numbered paired exercise is followed by an even-numbered exercise of the same type.
- **Unclassified Exercises** do not reference specific chapter sections but test the student's overview of chapter concepts.
- **Chemical Connections** are exercises that challenge students to expand their problem-solving skills by applying them to more complex questions or to questions that require the integration of material from different chapters.

Answers to Odd-Numbered Exercises are supplied at the end of the book. Step-by-step solutions to the odd-numbered exercises are supplied in the *Solutions Manual*. Step-by-step solutions to even-numbered exercises are supplied in the *Instructor's Manual*.

Making Connections Through Visualization

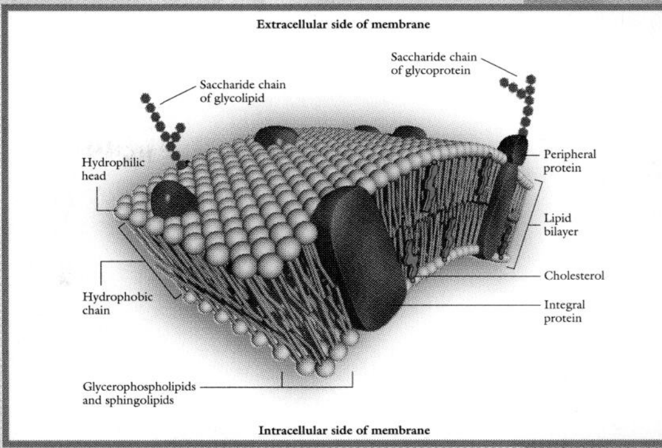

Illustrations Figures, tables, and other illustrations have been carefully chosen or designed to support the text and are carefully captioned and labeled for clarity. Special titles on certain illustrations—Insight into Properties, Insight into Function, and A Look Ahead—emphasize the use of secondary attractive forces and molecular structure as unifying themes throughout the book and remind readers that the concepts learned in Parts 1 and 2 will be applied to the biochemistry in Part 3.

Ball-and-Stick and Space-Filling Molecular Models Molecular structures of compounds, especially organic compounds, offer students considerable interpretive challenge. Throughout the text, two-dimensional molecular structures are supported by generous use of ball-and-stick and space-filling molecular models to aid in the visualization of three-dimensional structures of molecules.

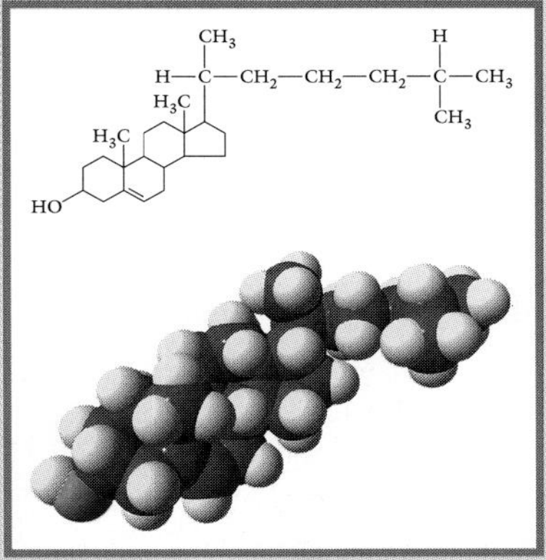

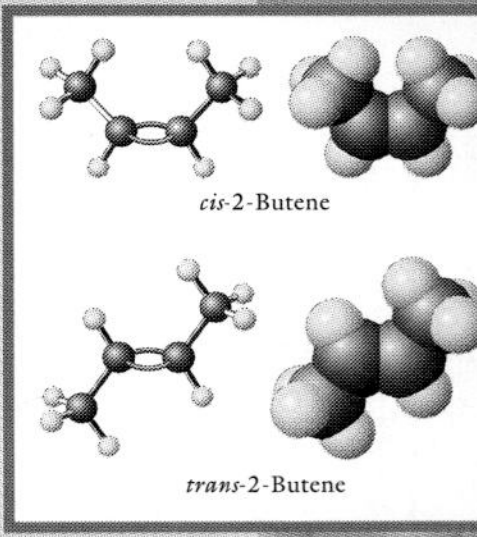

Functional Use of Color Color is used functionally and systematically in schematic illustrations and equations to draw attention to key changes or components and to differentiate one key component from another. For example, in molecular models, the carbon, hydrogen, oxygen, and nitrogen atoms are consistently illustrated in black, white, red, and blue, respectively. In structural representations of chemical reactions, color is used to highlight the parts of the molecule undergoing change.

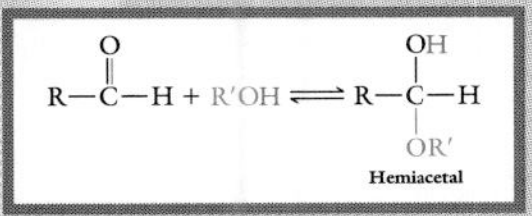

The strategic use of color makes diagrams of complex biochemical pathways less daunting and easier to understand.

Making Connections Using Learning Tools

Learning Objectives Each chapter begins with a list of learning objectives that preview the skills and concepts that students will master by studying the chapter. Students can use the list to gauge their progress in preparing for exams.

Concept Checklists The narrative is punctuated with short lists serving to highlight or summarize important concepts. They provide a periodic test of comprehension in a first reading of the chapter, as well as an efficient means of reviewing the chapter's key points.

> **Concept checklist**
>
> ✔ When the attractive forces between different molecules are similar, solutions will form.
>
> ✔ When the attractive forces between different molecules are different, solutions will not form.

Rules Rules for nomenclature, balancing reaction equations, and other important procedures are highlighted so that students can find them easily when studying or doing homework.

> ▶▶ Fatty acids, the components of fats and oils, are the subject of Chapter 19.

Cross References Cross-referencing in the text and margins alerts students to upcoming topics, suggests topics to review, and draws connections between material in different parts of the book.

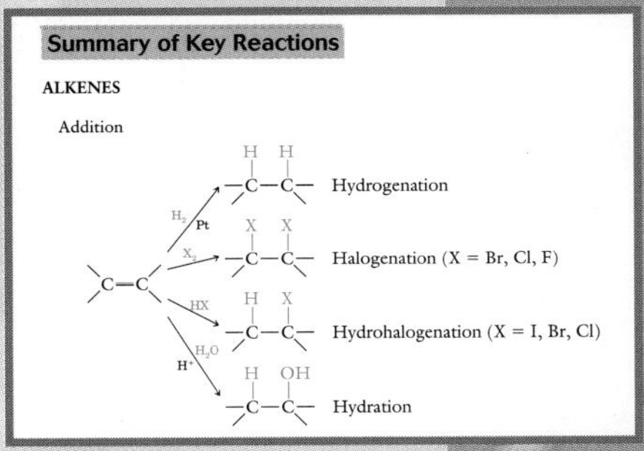

Chapter Summaries Serving as a brief study guide, the Summary at the end of each chapter points out the major concepts presented in each section of the chapter.

Summaries of Key Reactions At the end of most organic chemistry chapters, this feature summarizes the important reactions of a given functional group.

Key Words Important terms are listed at the end of each chapter and keyed to the pages on which their definitions appear.

ORGANIZATION

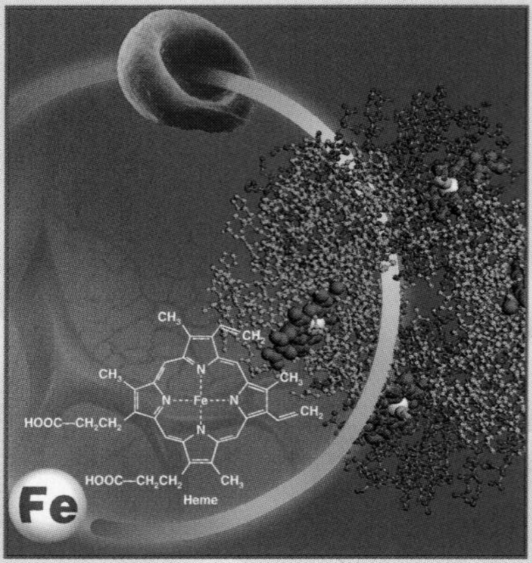

Part 1: General Chemistry (Chapters 1 Through 10)

To appreciate the molecular basis of physiological functioning, students must have a thorough grounding in the fundamental concepts of general chemistry. Part I emphasizes the structure and properties of atoms, ions, and molecules. Chapter 1 describes the qualitative and quantitative tools of chemistry. It is followed by a consideration of atomic and molecular structure and chemical bonding in Chapters 2 and 3. In Chapter 4, the major types of chemical reactions are presented, along with the quantitative methods for describing the mass relations in those reactions. Chapters 5 and 6 consider the physical properties of molecules and the nature of the interactions between them. Chapter 7 examines the properties of solutions, particularly diffusion and osmotic phenomena. A study of chemical kinetics and equilibria, in Chapter 8, paves the way for a later consideration of enzyme function. Chapter 9 treats acids and bases, critical for an understanding of physiological function. Chapter 10 deals with the effects of the interaction of radiation with biological systems and with the use of radiation in medical diagnosis and therapy.

Part 2: Organic Chemistry (Chapters 11 Through 17)

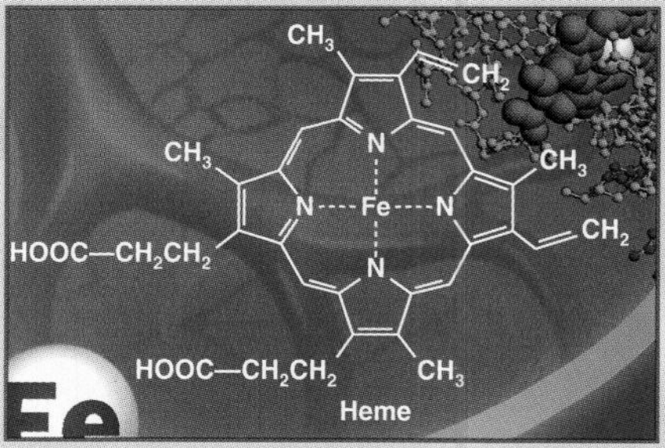

Having completed a study of the basic structure and properties of atoms and molecules, we proceed in Part 2 to a study of organic compounds. Chapter 11 presents a foundation for the study of organic chemistry and then examines saturated hydrocarbons. Unsaturated hydrocarbons are the subject of Chapter 12. Chapter 13 begins the study of oxygen-containing organic compounds by examining alcohols, phenols, ethers, and related compounds; together with Chapter 14, on aldehydes and ketones, it lays the foundation for the subsequent study of carbohydrates. Chapter 15 examines carboxylic acids and esters, preparing students for the subsequent study of lipids and nucleic acids. Amines and amides are considered in Chapter 16, a prelude to the subsequent examination of amino acids, polypeptides, proteins, and nucleic acids. Chapter 17 describes the concepts of stereochemistry and their importance in biological systems.

Part 3: Biochemistry (Chapters 18 Through 26)

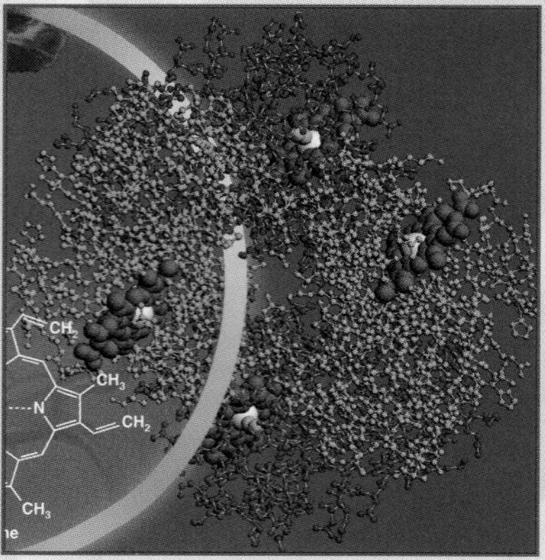

Biochemistry is the study of the chemical processes of the biomolecules that govern life functions. Chapters 18 through 21 present the principal biomolecules: carbohydrates, proteins, lipids, and nucleic acids. The structural features of these biomolecules are described in regard to the relations between their chemical structures and their physiological functions. Chapters 22 through 26 focus on those functions—specifically, on metabolism, the extraction of energy from the environment, and the use of energy to synthesize biomolecules. Chapter 22 provides a general survey of cell structures, metabolic systems, and enzymes, whereas Chapters 23 through 25 describe the key features of carbohydrate, lipid, and amino acid metabolism, respectively. Chapter 26 demonstrates how these principal metabolic pathways are integrated into the overall functions of the body. It does so by examining nutrition and digestive processes and then comparing the responses of the body under moderate and severe physiological stress.

SUPPLEMENTS

In the margins of the textbook you will notice this media icon. Each one indicates that there is an item on the CD-ROM that accompanies that section of the textbook. This *Multimedia CD-ROM,* developed by W. H. Freeman and Company, Sumanas, Inc., and Wendy Gloffke, of Cedar Crest College, complements and augments the material covered in the textbook. Its animations, simulations, and video demonstrations bring the book to life, and its practice tools such as interactive quizzes help students review for exams. The CD contains a customized calculator and dynamic periodic table to facilitate problem solving. Animations on the student CD-ROM were developed by Dr. Roy Tasker and Stefan Markworth, University of Western Sydney, Nepean, on behalf of VisChem, a joint project of the University of Western Sydney, Nepean, University of Technology, Sydney, and the University of Western Australia.

A *Student Solutions Manual,* by Mark D. Dadmun, of the University of Tennessee, Knoxville, provides complete solutions to the odd-numbered end-of-chapter exercises.

A *Study Guide,* by Wendy Gloffke, of Cedar Crest College, provides reader-friendly reinforcement of the concepts covered in the textbook. It includes detailed chapter outlines, summary pages of important equations, helpful hints, concept maps, and additional practice exercises with answers. Objective lists are keyed to the exercises, illustrations, and tables in the book, and all in-chapter problems are worked out for the student.

The *General, Organic, and Biochemistry Laboratory Manual,* by Sara Selfe, University of Washington, offers a choice of classic chemistry experiments and innovative ones. All of them place special emphasis on the biological implications of chemical concepts.

An *Instructor's Manual,* by Mark D. Dadmun, University of Tennessee, Knoxville, contains complete solutions to the even-numbered end-of-chapter exercises.

An *Instructor's Resource CD-ROM* provides all the textbook's illustrations for use in W. H. Freeman's Presentation Manager Software or other commercially available software.

A *Test Bank,* by Philip J. Wenzel, of Monterey Peninsula College, contains two sets of 50 multiple-choice, fill-in-the-blank, and short-answer questions per chapter. The test bank is available in both printed and electronic formats.

200 *Overhead Transparencies* with large-type labels illustrate the key figures and tables.

The *Web site* for this text offers a number of features for students and instructors including online study aids such as practice chapter quizzes for students and an online instructor's guide that provides ideas for teaching, using the media in combination with this book.

ACKNOWLEDGMENTS

We are especially grateful to the many educators who reviewed the manuscript and offered helpful suggestions for improvement: Brad P. Bammel, Boise State University; George C. Bandik, University of Pittsburgh; Bruce Banks, University of North Carolina, Greensboro; Lorraine C. Brewer, University of Arkansas; Martin L. Brock, Eastern Kentucky University; Steven W. Carper, University of Nevada, Las Vegas; John E. Davidson, Eastern Kentucky University; Geoffrey Davies, Northeastern University; Marie E. Dunstan, York College of Pennsylvania; James I. Durham, Blinn College; Wes Fritz, College of DuPage; Patrick M. Garvey, Des Moines Area Community College; Wendy Gloffke, Cedar Crest Community College; T. Daniel Griffiths, Northern Illinois University; William T. Haley, Jr., San Antonio College; Edwin F. Hilinski,

Florida State University; Vincent Hoagland, Sonoma State University; Sylvia T. Horowitz, California State University, Los Angeles; Larry L. Jackson, Montana State University; Mary A. James, Florida Community College, Jacksonville; James Johnson, Sinclair Community College; Morris A. Johnson, Fox Valley Technical College; Lidija Kampa, Kean College; Paul Kline, Middle Tennessee State University; Robert Loeschen, California State College, Long Beach; Margaret R. R. Manatt, California State University, Los Angeles; John Meisenheimer, Eastern Kentucky University; Michael J. Millam, Phoenix College; Frank R. Milio, Towson University; Renee Muro, Oakland Community College; Deborah M. Nycz, Broward Community College; R. D. O'Brien, University of Massachusetts; Roger Penn, Sinclair Community College; Charles B. Rose, University of Nevada, Reno; Richard Schwenz, University of Northern Colorado; William Schloman, University of Akron; Michael Serra, Youngstown State College; David W. Seybert, Duquesne University; Jerry P. Suits, McNeese State University; Tamar Y. Susskind, Oakland Community College; Arrel D. Toews, University of North Carolina, Chapel Hill; Steven P. Wathen, Ohio University; Garth L. Welch, Weber State University; Philip J. Wenzel, Monterey Peninsula College; Thomas J. Wiese, Fort Hays State University; Donald H. Williams, Hope College; Kathryn R. Williams, University of Florida; William F. Wood, Humboldt State College; Les Wynston, California State College, Long Beach.

We also wish to thank the students of George C. Bandik, University of Pittsburgh; Sharmaine Cady, East Stroudsburg University; Wes Fritz, College of DuPage; Wendy Gloffke, Cedar Crest Community College; Paul Kline, Middle Tennessee State University; Sara Selfe, University of Washington; Jerry P. Suits, McNeese State University, and Arrel D. Toews, University of North Carolina, Chapel Hill, whose comments on the text and exercises provided invaluable guidance in the book's development.

Special thanks are due to Irene Kung, University of Washington, Stan Manatt, California Institute of Technology, and Mark Wathen, University of Northern Colorado, who checked all calculations for accuracy.

We also thank Molecular Simulations, Inc. (96825 Scranton Road, San Diego, CA 92121, www.msi.com) for providing the molecular viewing program found on the CD-ROM.

Finally, we thank the people of W. H. Freeman and Company for their wonderful encouragement, suggestions, and conscientious efforts in bringing this book to fruition. Although most of these people are listed on the copyright page, we would like to add some who are not and single out some who are listed but who deserve special mention. We want to express our deepest thanks to Deborah Allen who initiated this project; to Michelle Russel Julet for the opportunity and encouragement to see it through; to Jane O'Neill and Patricia Zimmerman for their painstaking professionalism in producing a final manuscript and published book in which all can feel pride; and to Moira Lerner whose creativity, cheerful encouragement, and tireless energy were key factors in the book's evolution and production.

PART 1

GENERAL CHEMISTRY

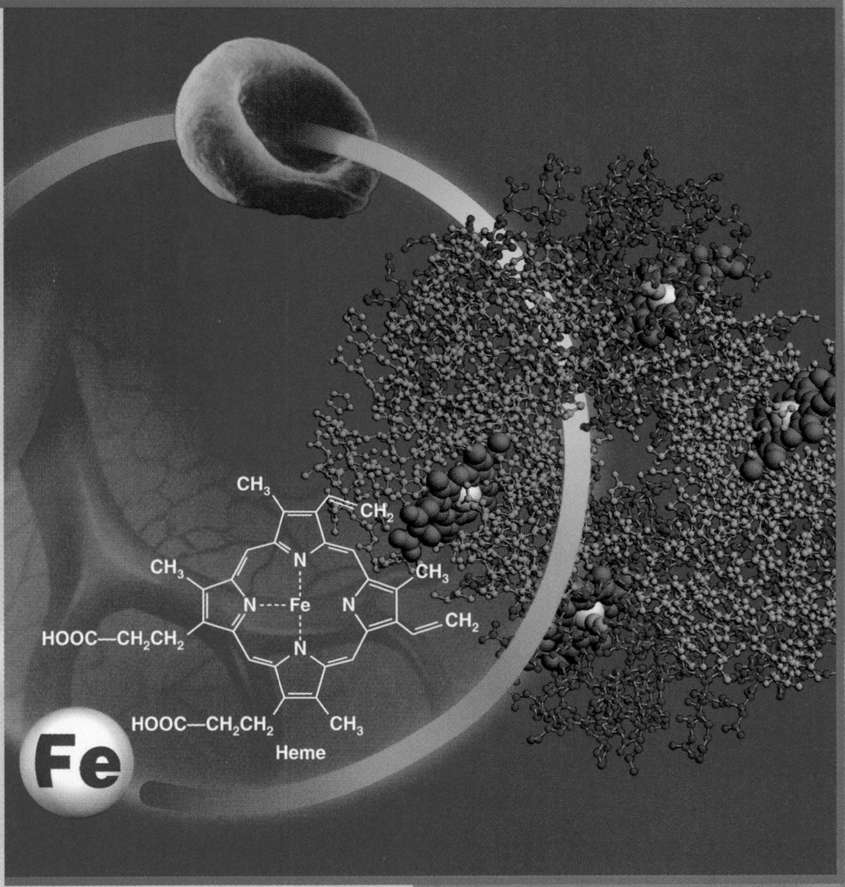

Living organisms are highly organized, with each successive level of organization more complex than the last. Atoms and small molecules are bonded together into molecules of great size, which are then organized into microscopic structures and cells. Cells are organized into macroscopic tissues and organs, organs into organ systems and organisms. The book's cover illustration, repeated here, provides a case in point. Red blood cells (top), which carry oxygen to all parts of our bodies, are able to do so because of the special structure of the protein called hemoglobin (center right) that they contain; and the key components of these large proteins are smaller molecules called heme, which contain a form of iron (Fe), to which oxygen becomes attached. Part 1 begins the story of how the properties of simple atoms and molecules lead to the construction of this complex machinery of life.

CHAPTER 1

THE LANGUAGE OF CHEMISTRY

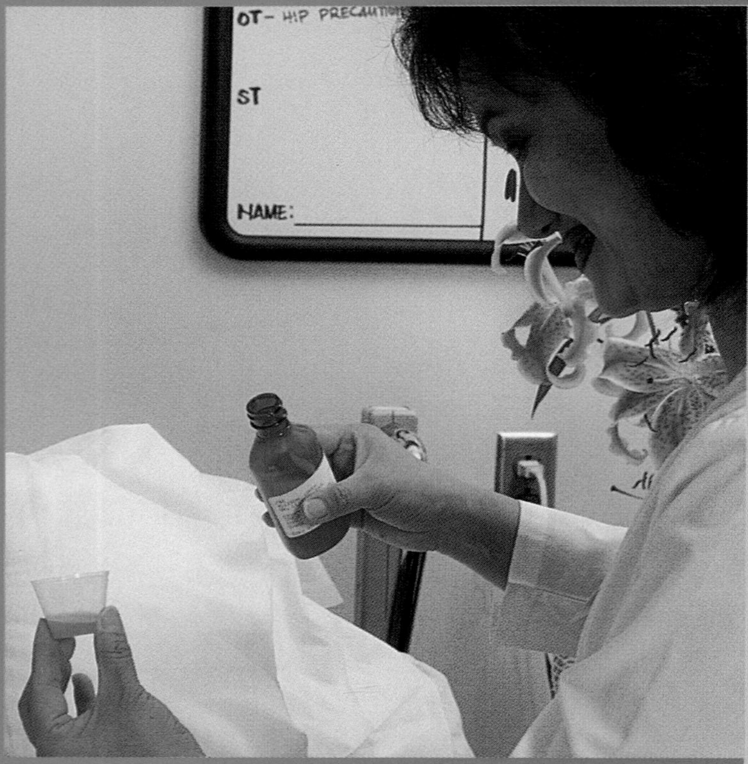

CHEMISTRY IN YOUR FUTURE

You arrive for your shift at the skilled nursing facility and read on a patient's chart that the doctor has prescribed a 100-mg dose of Colace. The pharmacy sends up a bottle of the medication in syrup form, containing 20 mg of medicine in each 5 mL of syrup. How many milliliters of the syrup do you give to your patient? A simple calculating technique that you learned in Chemistry helps you find the answer.

LEARNING OBJECTIVES

- Describe the characteristics of elements, compounds, and mixtures.
- Name the units of the metric system and convert them into the units of other systems.
- Describe the relation between uncertainty and significant figures.
- Use scientific notation in expressing numbers and doing calculations.
- Use the unit-conversion method in solving problems.
- Define mass, volume, density, temperature, and heat, and describe how they are measured.

Chemistry is the study of matter and its transformations, and no aspect of human activity is untouched by it. The discoveries of chemistry have transformed the food that we eat, the homes that we live in, and the manufactured objects that we use in our daily lives. In addition to explaining and transforming the chemical world outside of our bodies, chemists have developed a detailed understanding of the chemistry within us, the underlying physiological function. Today, students preparing for careers in any of the life sciences must learn the basic principles of chemistry to acquire a meaningful understanding of biology. If you are one of those students, the purpose of this book is to provide you, first, with a firm grounding in chemical science and, second, with a broad understanding of the physiological processes underlying the lives of cells and organisms.

The practical results of chemical research have greatly changed the practice of medicine. As recently as 60 to 70 years ago, families were regularly devastated when children and young adults died from bacterial infections such as diphtheria and scarlet fever. Entire hospitals were once dedicated to the care of patients with tuberculosis, and mental wards were filled with patients suffering from tertiary syphilis. That our experience is so different today is a result of the development of antibacterial drugs such as the sulfonamides, streptomycin, and penicillin. Medical professionals are no longer forced to stand by as disease takes its toll. Armed with a powerful pharmacological arsenal, they have some confidence in their ability to cure those formerly deadly infections.

Since the early 1950s, when the chemical structure of deoxyribonucleic acid (DNA) was described by James Watson and Francis Crick, the pace of accomplishment in the understanding of life processes has been truly phenomenal. The Watson and Crick model of DNA structure was rapidly followed by further developments that allowed biologists and chemists to treat chromosomes (the molecules of inheritance, which dictate the development of living things) literally as chemical compounds. In one of the more interesting and promising of these new approaches, a childhood disease caused by a defective gene has been treated by implanting a healthy gene into an affected child's chromosomes. In addition to direct medical applications, basic research into the chemistry and biology of DNA has led to the development of new pharmacological products, such as human insulin produced in bacteria.

▶▶ Chapter 21 describes the chemistry of DNA.

Parts 1 and 2 of this book, "General Chemistry" and "Organic Chemistry," will provide you with the tools that you need to understand and enjoy Part 3, "Biological Chemistry." At times you may feel impatient with the pace of the work. Your impatience is understandable because it is difficult to see an immediate connection between elementary chemical concepts and the biochemistry of DNA, but a good beginning will get us there. The present chapter launches our exploration of the chemistry underlying physiological processes with introductory remarks about the composition of matter, conventions for reporting measurements and doing calculations in chemistry, and descriptions of basic physical and chemical properties commonly studied in the laboratory.

1.1 THE COMPOSITION OF MATTER

Humans have been practicing chemistry for hundreds of thousands of years, probably since the first use of fire. Chemical processes—processes that literally transform the identity of substances—are at the heart of cooking, pottery making, metallurgy, the concoction of herbal remedies, and countless other long-time human pursuits. But these early methods were basically recipes developed in a hit-or-miss fashion over periods of thousands of years. The science of chemistry is only about 300 years old. Its accomplishments are the result of

1.1 Chemistry in Depth

The Scientific Method

The scientific method is basically a common sense approach to establishing knowledge about the natural world. The elements of the scientific method are the observation of demonstrable facts, the creation of hypotheses to explain or account for those facts, and experimental testing of hypotheses. As more tests validate a hypothesis, more confidence is placed in it until, finally, it may become a theory. An important aspect of this method is the willingness to discard or modify a hypothesis when it is not supported by experiment. A hypothesis is only as good as its last exposure to a rigorous test.

Repeated observations of natural processes can also lead to the development of what are called **laws**—concise statements of how nature behaves with no explanation of that behavior, to which there is no exception. For example, Newton's law of gravity says nothing about the mechanism underlying the law but merely asserts its universality. These ideas are illustrated by a flow diagram in Figure 1.1. Let's see how the scientific method worked in a real situation.

In 1928, it was discovered that a nonpathogenic strain of pneumococcus could be transformed into a virulent strain by exposure to chemical extracts of the virulent strain. Call this discovery a **fact** or an **observation.** The bacteria is *Diplococcus pneumoniae,* and the virulent strain causes pneumonia. The biological process was called transformation. The material in these extracts responsible for the transmittance of inheritance was called "transforming principle," but its chemical nature was unknown.

To uncover the chemical identity of the transforming principle, scientists required a **hypothesis,** a guess or hunch regarding what that transforming principle might be. Most biochemists at that time believed that inheritance was carried by proteins, and that became the first hypothesis proposed. It could be readily tested because proteins could be inactivated by heat and destroyed by enzymes such as trypsin and pepsin (the stomach enzyme that degrades proteins). The transforming principle survived all experiments devised to inactivate or destroy proteins in the transforming cell extracts. This fact established that the transforming principle could not be a protein, and that hypothesis had to be discarded. An alternative testable hypothesis was proposed—that the transforming substance could be DNA. The transforming principle was exposed to an enzyme that could degrade only DNA and no other substance. The result was the complete inactivation of the transforming principle. This result was the first indication that the transforming principle was DNA and that DNA was probably the universal carrier of genetic information. Since that time, many other experimental discoveries have supported the original hypothesis. Because of all the subsequent experimental support of the idea that DNA is the molecule that carries genetic information, it now has the status of a **theory,** a hypothesis in which scientists have a high degree of confidence.

quantitative methods of investigation and experimentation—that is, of systematic measurement and calculation. The general approach, called the scientific method, is discussed in Box 1.1 and diagrammed in Figure 1.1.

The science of chemistry began when it became recognized that, to develop an understanding of chemical processes, one must first study the proper-

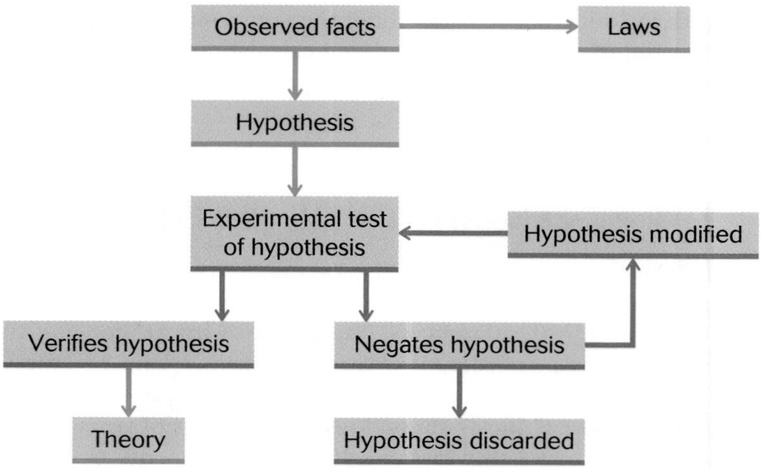

Figure 1.1 A flowchart illustrating the scientific method. A hypothesis is only as good as its last exposure to a rigorous test. The elements of the scientific method are facts, hypotheses, experimental tests, and theories (hypotheses in which scientists have developed a high degree of confidence). An important aspect of this method is the willingness of scientists to discard or modify a hypothesis when it is not supported by experiment.

Figure 1.2 Filtration is used to separate liquids from solids. The filter paper retains the solid because the particles of the solid are too large to pass through the pores, or openings, in the paper. Micropore filters, which have pore sizes small enough to retain bacteria, are used to produce sterile water, sterile pharmaceutical preparations, and bacteria-free bottled beer.

ties of **pure substances.** The notion of purity is not a simple one and requires more than a simple definition. At this point, however, let us simply say that the early chemists were familiar with certain substances—mercury, for example—which appeared to be neither adulterated by nor mixed with anything else. These substances were therefore called pure, and their characteristics served as a model for determining the purity of other, more complicated substances.

These pure materials could be obtained by filtration (Figure 1.2), distillation (Figure 1.3), and other "separation methods" (Box 1.2, Figure 1.4), both on the following page. They were found to have unique and consistent physical and chemical properties. **Physical properties** include the temperature at which a substance melts (changes from a solid to a liquid) or freezes

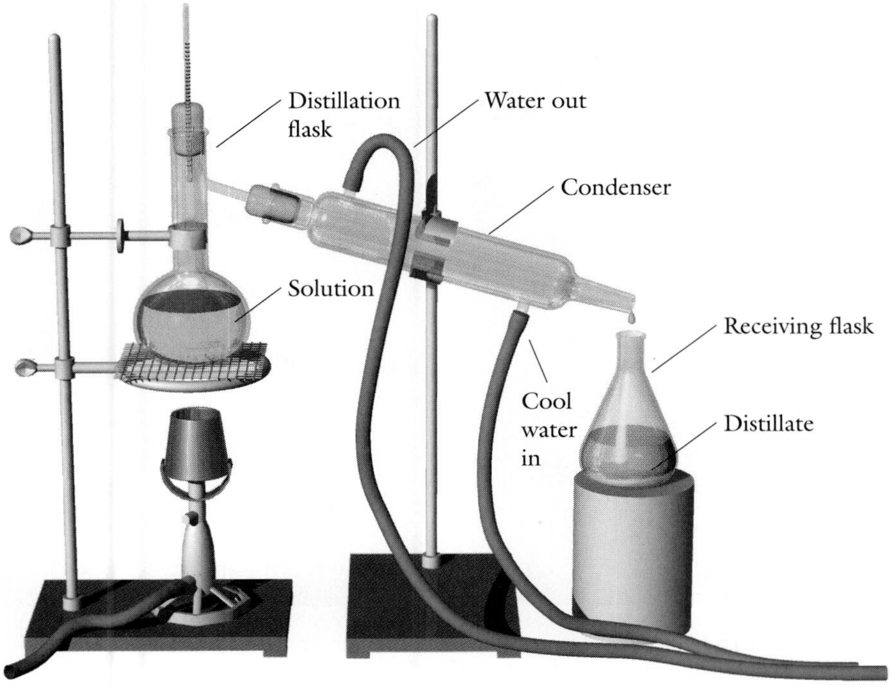

Figure 1.3 A distillation apparatus. If two liquids are to be separated, the liquid with the lower boiling temperature will vaporize at a lower temperature and leave the distillation flask before the higher-boiling liquid. The vaporized liquid leaves the flask and enters the condenser, a long glass tube with a glass jacket surrounding it through which cold water is circulated. There the cooled vapor condenses to a liquid and is collected in the receiving flask. A solution of a solid also can be separated by this technique, in which case the solid remains in the distillation flask.

1.2 Chemistry in Depth

Chromatography

Chromatography is a separation technique in which a mixture of substances in a pure liquid called the developer moves past another substance that remains stationary. Each component of the mixture interacts to a different extent with the stationary substance and therefore moves at a different rate. The developer does not interact with the stationary substance and acts as a neutral medium, allowing the components of the mixture to interact with the stationary solid. Just as athletes running at different rates will become separated from each other, the different components also become separated. In the earliest application of this method, the green pigments of plants (the chlorophylls) were separated as their liquid mixture flowed down a column packed with solid calcium carbonate. The colored (chroma means color) components moved down the column at different rates. Those that interacted most strongly with the solid lagged behind those that interacted weakly. Eventually, the various components cleanly separated. This method is called column chromatography.

Paper chromatography and thin-layer chromatography (TLC) are two related methods for separating substances in solution. In both, a drop of the mixture is placed on a strip of filter paper or on a thin layer of solid (such as silica gel or aluminum oxide deposited on a plastic strip) and allowed to dry. The strip is placed upright in a small pool of developer and acts as a wick, drawing the liquid along with the mixture of substances along the solid. After sufficient time, the strip is removed, and the solvent is allowed to evaporate. The components interacting least strongly with the solid will have moved farthest along the solid strip, leaving behind those interacting most strongly with the solid. If the compounds possess color, they will appear as a series of spots at different positions along the strip. If the compounds are colorless, additional treatment is necessary to locate them. Some of these treatments use radioactivity and are discussed in Chapter 10. Chromatography is used not only to separate homogeneous mixtures of substances, but also to identify unknown substances by comparison of their chromatographic characteristics with those of known pure compounds under identical conditions.

Paper chromatography and thin-layer chromatography have proved invaluable in separating the products of biochemical reaction products and identifying complex substances with very similar chemical properties. In particular, TLC (Figure 1.4) is used extensively in the pharmaceutical industry as a quality control check in the manufacture of complex substances such as penicillin and steroid hormones.

Figure 1.4 Thin-layer chromatography can separate complex mixtures and allow the identification of each compound.

(changes from liquid to solid), color, and densities (Figure 1.5). A pure substance undergoes physical changes (freezing, melting, evaporation, and condensation), illustrated in Figure 1.6, without losing its identity.

A **chemical property** is the ability of a pure substance to chemically react with other pure substances. In a chemical reaction (Figure 1.7 on the following page), substances lose their chemical identities and form new substances with new physical and chemical properties.

When chemists applied various separation methods to the materials around them and studied the physical and chemical properties of the resulting substances, they discovered that most familiar materials were **mixtures;** that is, they consisted of two or more pure substances in varying proportions. Some mixtures—salt and pepper, for example—are visibly discontinuous; the different components are easy to distinguish. Such a mixture is called **heterogeneous.** Other mixtures—sugar and water, for example—have a uniform appearance throughout. The eye cannot distinguish one component from another, even under the strongest microscope. They are called **solutions** and are described as **homogeneous.**

The pepper–salt mixture can be separated into its components by, first, the addition of water: the salt will dissolve in the water, whereas the pepper will remain a solid. Next, the mixture is poured through a filter as illustrated in Figure 1.2: the pepper remains on the filter, and the dissolved salt passes through. Finally, the salt–water solution is separated into its components by allowing the water to evaporate, which leaves the salt behind.

In contrast with pure substances, whose properties are consistent and predictable, mixtures have properties that are variable and depend on the proportions of the components. Consider the mixture of sugar in water. You can dissolve one, two, or more teaspoonsful in a cup of water, and the appearance of the mixture remains the same (in other words, the mixture is a solution and homogeneous). Yet you know from experience that the property known as sweetness increases as the sugar content of the mixture increases.

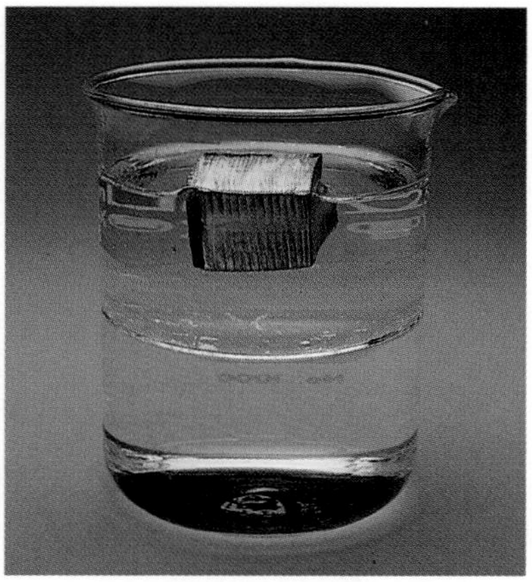

Figure 1.5 The element lithium is less dense than water or oil, and oil is less dense than water. The oil floats on water, and lithium floats on the oil.

▶▶ Much of chemistry is concerned with solutions as you'll see in Chapters 7 and 9.

Concept check

✔ A mixture is composed of at least two pure substances and is either homogeneous (visibly continuous) or heterogeneous (visibly discontinuous).

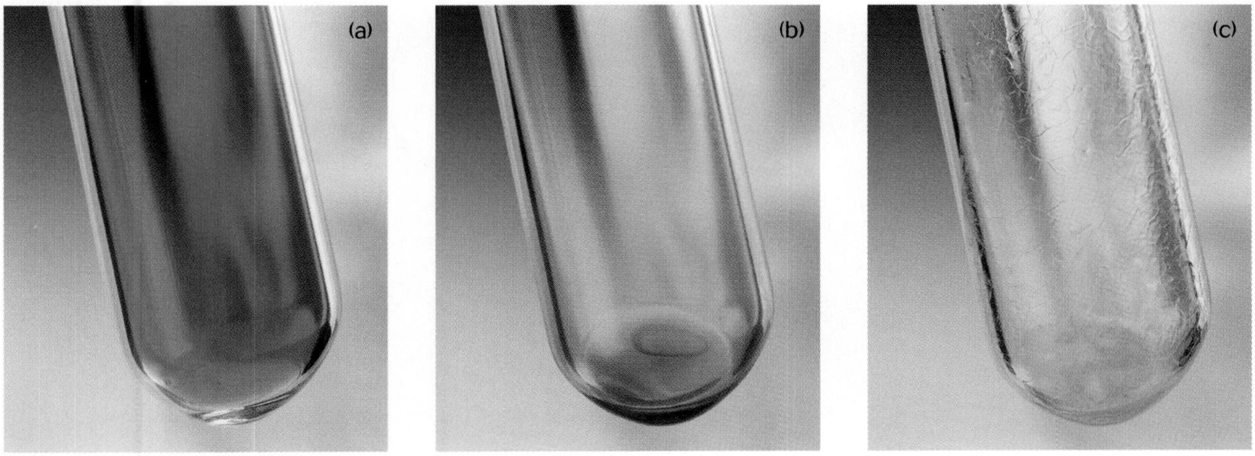

Figure 1.6 The substance nitric oxide (NO_2) can exist in three states, (a) gas, (b) liquid, and (c) solid, and can be reversibly transformed from one state into another without losing its chemical identity.

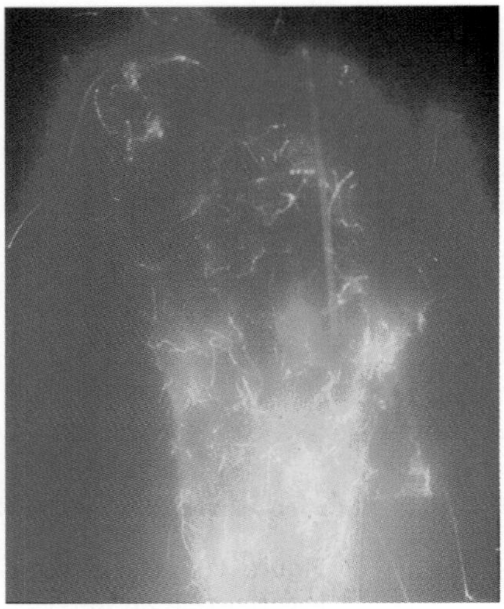

Figure 1.7 The solid metallic element iron reacts vigorously with the gaseous element chlorine to form the new solid substance iron chloride.

Figure 1.8 Chemical compounds found in the kitchen.

Chemists studying the chemical properties of pure substances found that some of the substances could be decomposed, by chemical means, into simpler pure substances. Furthermore, they found that those simpler substances could not be further decomposed. Decomposable pure substances are called **compounds** (Figure 1.8), and those that cannot be further decomposed are called **elements** (Figure 1.9). An element is a substance that cannot be separated chemically into simpler substances, nor can it be created by combining simpler substances.

When elements combine to form compounds, they always do so in fixed proportions. For example, glucose, also called dextrose, is a chemical combination of carbon, oxygen, and hydrogen. One hundred grams of glucose will always contain 40.00 grams of carbon, 53.33 grams of oxygen, and 6.67 grams

Figure 1.9 Some common elements. *Clockwise from left:* the red-brown liquid bromine, the silvery liquid mercury, and the solids iodine, cadmium, red phosphorus, and copper.

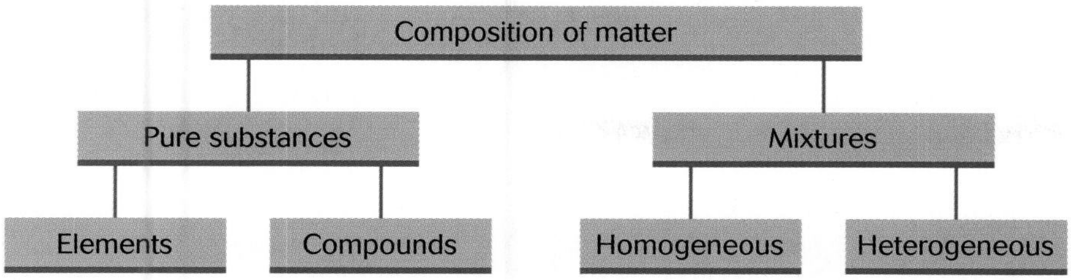

Figure 1.10 Analysis of the composition of matter.

of hydrogen, whether the glucose is extracted from rose hips or synthesized in the laboratory. The relations between the various categories of matter are illustrated in Figure 1.10.

The millions of pure compounds known today are built from the elements whose names and symbols can be found in the table inside the back cover of this book. Most of the symbols that chemists use to represent elements are derived from the first letter of the capitalized name of the element. However, when the names of two or more elements begin with the same first letter, the symbol for the more recently discovered element is usually formed by adding the second letter of the name, in lower case. For example, the symbol for carbon is C; for calcium, Ca; for cerium, Ce. Other symbols are formed from the elements' Latin, Arabic, or German names; for example, the symbol for potassium is K, after *kalium,* the element's Latin name of Arabic origin. Other examples are W for tungsten, whose German name is *Wolfram,* and Fe for iron, whose Latin name is *ferrum.* These and additional examples are listed in Table 1.1.

Concept checklist

- ✔ There are only two kinds of pure substances: elements and compounds.
- ✔ An element cannot be decomposed into simpler pure substances, nor can it be created by combining simpler substances.
- ✔ Elements combine to form compounds, which are substances containing fixed proportions of their constituent elements. The composition of a given compound is always the same, regardless of where or how the substance may have formed.
- ✔ A compound can be decomposed, by chemical means, into simpler pure substances.

TABLE 1.1 Common Names of the Elements Whose Symbols Are Derived from Latin or German Names

Common Name	Symbol	Symbol Origin
silver	Ag	*argentum*
gold	Au	*aurum*
iron	Fe	*ferrum*
mercury	Hg	*hydrargyrum*
potassium	K	*kalium*
sodium	Na	*natrium*
lead	Pb	*plumbum*
tin	Sn	*stannum*
antimony	Sb	*stibium*
tungsten	W	*Wolfram*

- ✔ The physical and chemical properties of compounds are always different from those of the elements from which they were formed.
- ✔ Elements are identified by symbols derived from their English, Latin, Arabic, Greek, or German names.

1.2 MEASUREMENT AND THE METRIC SYSTEM

It is far more common to use quantitative rather than qualitative language to describe properties of matter. After sulfur is described as a yellow powder, there are virtually no other descriptive qualities that can help differentiate pure sulfur from all other elements. On the other hand, by carefully measuring sulfur's quantitative properties—its melting point, density, specific heat, and coefficient of expansion, as well as the exact composition of its compounds with oxygen, chlorine, and so forth—we soon compile a profile that is unique. The key concept of the preceding sentence is **measure,** and the meaning of that word is best expressed by a common dictionary definition:

> the size, capacity, extent, volume, or quantity of anything, especially as determined by comparison with some standard or **unit.**

As this definition suggests, to measure anything, we need a standard system of units. We also need a device designed to allow comparison of the object being measured with the standard or reference unit. The story describing Noah building his ark illustrates two ways of using numbers: (1) Noah *counts* the animals that he is going to take and (2) he *measures* the dimensions of the ark (in cubits, a unit no longer in use but a unit nonetheless). A number resulting from counting is considered exact, but a number resulting from a measurement will always have a degree of uncertainty, depending on the device used for making the measurement. This idea will be considered more fully in Section 1.3.

The measurement system used in science and technology is called the **metric system.** The newest version of this system is called the **Système International d'Unités,** abbreviated as the **SI system.** The units defined by this system are found in Table 1.2. All other units are derived from these fundamental units. For example:

$$\text{Area} = \text{m}^2$$
$$\text{Volume} = \text{m}^3$$
$$\text{Density} = \text{kg/m}^3$$
$$\text{Velocity} = \text{m/s}$$

TABLE 1.2 Fundamental Units of the Modern Metric System

Fundamental Quantity	Unit Name	Symbol
length	meter	m
mass	kilogram	kg
temperature	kelvin	K
time	second	s
amount of substance*	mole	mol
electric current	ampere	A
luminous intensity	candela	cd

* The mole is a chemical quantity that will be considered in Chapter 4.

TABLE 1.3 Non-SI Units in Common Use

Quantity	Unit	Symbol	SI Definition	SI Equivalent
length	Ångström	Å	10^{-10} m	0.1 nanometer
volume	liter	L	10^{-3} m^3	1 $decimeter^3$
energy	calorie	cal	$(kg \cdot m^2 \cdot s^{-2})$*	4.184 joules

* A centered dot (·) is used to denote multiplication in derived units.

The first five units in Table 1.2 are those with which we will be concerned in chemistry. The SI system is widely used in the physical sciences because it greatly simplifies the kinds of calculations that are most common in those fields. Because certain older, non-SI units continue to be used in clinical and chemical laboratories, we will also use them in many of our quantitative calculations. A few of them are given in Table 1.3.

The great convenience of the metric system is that all basic units are multiplied or divided by multiples of ten, which makes mathematical manipulation very simple, often as simple as moving a decimal point. It also simplifies the calibration of measuring instruments: all basic units are subdivided into tenth, hundredth, or thousandth parts of those units. The multiples of ten are denoted by prefixes, all of Greek or Latin origin, and are listed in Table 1.4. They are combined with any of the basic metric units to denote quantity or size.

TABLE 1.4 Names Used to Express Metric Units in Multiples or Parts of Ten

Multiple or Part of Ten	Prefix	Symbol
1,000,000,000	1 giga-	G
1,000,000	1 mega-	M
1,000	1 kilo-	k
100	1 hecto-	h
10	1 deka-	da
0.1	1 deci-	d
0.01	1 centi-	c
0.001	1 milli-	m
0.000001	1 micro-	μ
0.000000001	1 nano-	n
0.000000000001	1 pico-	p

Example 1.1 Using metric system prefixes

Express (a) 0.005 second (s) in milliseconds (ms); (b) 0.02 meter (m) in centimeters (cm); (c) 0.007 liter (L) in milliliters (mL).

Solution

(a) Use Table 1.4 to find the relation between the prefix and the base unit. Milli represents 0.001 of a unit, so

$$0.001 \text{ s} = 1 \text{ ms}$$

therefore

$$0.005 \text{ s} = 5 \text{ ms}$$

(b) In Table 1.4,

$$0.01 \text{ m} = 1 \text{ cm}$$

therefore

$$0.02 \text{ m} = 2 \text{ cm}$$

(c) In Table 1.4,

$$0.001 \text{ L} = 1 \text{ mL}$$

therefore

$$0.007 \text{ L} = 7 \text{ mL}$$

Problem 1.1 Express (a) 2 ms in second (s); (b) 5 cm in meters (m); (c) 100 mL in liters (L).

Many of you are familiar with the English system of weights and measures—pounds (lb), inches (in.), yards (yd), and so forth. Section 1.5 will illustrate a formal mathematical procedure for converting units from that or any system of units into any other. This procedure, called the unit-conversion method, also forms the basis for the general method of problem solving that we will use throughout this book. It relies on the use of equivalences—so-called conversion factors—such as those found in Table 1.5 (see Section 1.7). First, however, let us look at some of the practical aspects of taking a measurement.

1.3 MEASUREMENT, UNCERTAINTY, AND SIGNIFICANT FIGURES

It is unlikely that a series of measurements of the same property of the same object made by one or more persons will all result in precisely the same value. This inevitable variability is not the result of mistakes or negligence. No matter how carefully each measurement is made, there is no way to avoid small differences between measurements. These differences arise because, no matter how fine the divisions of a measuring device may be, when a measure falls between two such divisions, an estimate, or "best guess," must be made. This unavoidable estimate is called the **uncertainty** or **variability.** All measurements are made with the assumption that there is a correct, or true, value for the quantity being measured. The difference between that true value and the measured value is called the **error.**

You may already have encountered this difficulty yourself in your chemistry laboratory, which is no doubt equipped with several types of balances for measuring mass (a property related to how much an object weighs; see Section 1.8). Let us assume that a balance has a variability of about 1 gram. This means that, every time a mass of, say, 4 g is placed on this balance, the reading will be slightly different but will probably fall within 1 g of the actual mass (no higher than 5 g and no lower than 3 g). Thus, if we decide to measure 4 g of a substance with this balance, we must take account of its variability and report the mass as 4 ± 1 g (4 plus or minus 1 gram). If, instead, we used a balance with a variability of 0.1 g, the measured value of the mass would be written 4 ± 0.1 g. For a third balance, with a variability of 0.001 g, we would report the mass as 4 ± 0.001 g. Finally, we could use an analytical balance with a variability of 0.0001 g and report the mass as 4 ± 0.0001 g.

Although masses are often reported with accompanying variabilities, as just illustrated, scientists also use a simpler system that takes advantage of a concept called the **significant figure.** This system eliminates the need for a $\pm$ notation. It indicates the uncertainty by means of the number of digits instead.

$4 \pm 1 \text{ g} = \underline{4} \text{ g}$	1 significant figure
$4 \pm 0.1 \text{ g} = \underline{4.0} \text{ g}$	2 significant figures
$4 \pm 0.001 \text{ g} = \underline{4.000} \text{ g}$	4 significant figures
$4 \pm 0.0001 \text{ g} = \underline{4.0000} \text{ g}$	5 significant figures

In this way, degrees of uncertainty are communicated through the numbers of significant figures—here one, two, four, and five significant figures, respectively. For the purpose of counting significant figures, zero can have different meanings, depending on its location within a number:

We have seen that the last digit in a reported value is an estimate. Therefore, a reported measurement of 4.130 g indicates an uncertainty of ±0.001 g and thus contains four significant figures.

- A trailing zero, as in 4.130 is significant.

In a report recording a measured value of 35.06 cm, the last digit is assumed to be an estimate, but the zero after the decimal and before the last digit is considered to be an accurate part of the measurement and is therefore significant. There are four significant figures in the number.

- A zero within a number, as in 35.06 cm, is significant.

A report lists a liquid volume of 0.082 L. In this case, the zeroes are acting as decimal place holders, and the measurement contains only two significant figures. The insignificance of the zeroes becomes clear when you realize that 0.082 L can also be written as 82 mL.

- A zero before a digit, as in 0.082, is not significant.

A report such as 20 cm is ambiguous. It could be interpreted as meaning "approximately 20" (say, 20 ± 10) or it might be understood as 20 ± 1. It might also mean 20 cm exactly. The number of significant figures in 20 cm is unclear.

- A number ending in zero with no decimal point, as in 20, is ambiguous.

Ambiguities of this last type can be prevented by the use of scientific, or exponential notation.

1.4 SCIENTIFIC NOTATION

Scientific notation, or **exponential notation,** is a convenient method for preventing ambiguity in the reporting of measurements and for simplifying the manipulation of very large and very small numbers. To express a number such as 233 in scientific notation, we write it as a number between 1 and 10 multiplied by 10 raised to a whole-number power: 2.33×10^2. The number between 1 and 10 (in our example, the number 2.33) is called the **coefficient,** and the whole-number exponent of 10 (in our example, 10^2) is called the **exponential factor.** A key rule to remember in using scientific notation is that any number raised to the zero power is equal to 1. Thus, $10^0 = 1$.

The following examples illustrate numbers rewritten in scientific notation:

3 is written:	3×10^0
24 is written:	2.4×10^1
346 is written:	3.46×10^2
2537 is written:	2.537×10^3

A PICTURE OF HEALTH

Ranges of Measurement in the Body

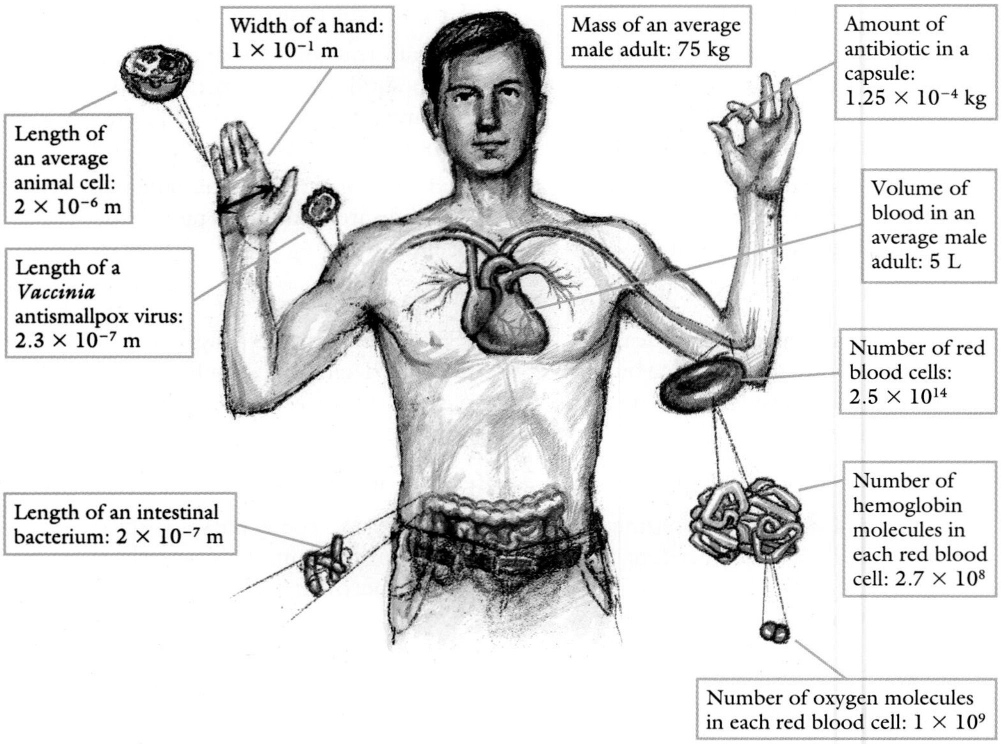

- For any number greater than 1, the decimal is moved to the left to create a coefficient between 1 and 10.
- Next, an exponential factor is created whose positive power is equal to the number of places that the decimal point was moved to the left.

To express numbers smaller than 1 (decimal numbers) in scientific notation, use the following algebraic rule describing the reciprocal of any quantity (units are also expressed in this way):

$$1/X = X^{-1}$$

Therefore, $1/8 = 8^{-1}$, and $1/\text{kg} = \text{kg}^{-1}$.

To express the number 0.2 in scientific notation, we transform it into a whole-number coefficient between 1 and 10, multiplied by an exponential factor that is decreased by the same power of ten:

$$0.2 = 2 \times 0.1 = 2 \times \frac{1}{10^1} = 2 \times 10^{-1}$$

The number 0.365 is therefore written 3.65×10^{-1}. The numbers 0.046 and 0.00753 are written 4.6×10^{-2} and 7.53×10^{-3}, respectively.

- For any number smaller than 1, the decimal is moved to the right to create a coefficient between 1 and 10.
- Next, an exponential factor is created whose negative power is equal to the number of places that the decimal point was moved to the right.

Example 1.2 Expressing a number in scientific notation

Express the number 0.00964 in scientific notation.

Solution

The first task is to create a coefficient between 1 and 10. This task is accomplished by moving the decimal point three places to the right. The result is 9.64. Moving the decimal to the right three places is equivalent to multiplying the decimal number by ten three times, as in the following three steps. To retain the value of the number, each time the decimal number on the left-hand side is multiplied by 10, the result on the right-hand side is reduced by a power of 10.

$$0.00964 = 0.0964 \times 10^{-1}$$
$$0.0964 = 0.964 \times 10^{-1}$$
$$0.964 = 9.64 \times 10^{-1}$$

Instead of taking three separate steps, the transformation into scientific notation is done in one step:

$$0.00964 \times 10^{3} = 9.64 \times 10^{-3}$$

The coefficient (a number between 1 and 10) was created by moving the decimal point three places to the right. To retain the actual numerical value, the coefficient must be multiplied by 10 raised to the negative number of places that the decimal was moved to the right.

Problem 1.2 Express the number 0.0007068 in scientific notation.

In the discussion of significant figures, it was pointed out that a number ending in zero with no decimal point in the number (21,600, for example) is ambiguous. Scientific notation provides a way to express the number without ambiguity. If the last zero in 21,600 is significant, the number has five significant figures and should be written 2.1600×10^4. If both zeroes are not significant, the number has three significant figures and should be written 2.16×10^4. If a number containing zeroes loses those zeros when the number is expressed in scientific notation, they were not significant.

1.5 CALCULATIONS USING SCIENTIFIC NOTATION

Although the rules of standard scientific notation require the coefficient to be a number between 1 and 10, for calculations requiring addition or subtraction, it is useful to write numbers by using nonstandard coefficients. For example, 4573 can be written as

$$4573 = 4573 \times 10^{0}$$
$$4573 = 457.3 \times 10^{1}$$
$$4573 = 45.73 \times 10^{2}$$
$$4573 = 4.573 \times 10^{3}$$

Each of these forms expresses the same numerical value. Every time the value of the coefficient was changed, the value of the exponential factor was adjusted to preserve the original numerical value of 4573. Because we can vary the coefficient and exponential factor of a number without changing the number's value, scientific notation simplifies additions and subtractions of numbers having different exponential factors.

Example 1.3 Adding numbers written in scientific notation

Perform the addition

$$(3.63 \times 10^{-2}) + (4.85 \times 10^{-3})$$

Solution

The more arduous solution would be to convert these expressions into full decimal notation, then add,

$$\begin{array}{r} 0.0363 \\ +\ \underline{0.00485} \\ 0.04115 \end{array}$$

and, after that, convert back into scientific notation: 4.12×10^{-2} (the "rounding" of such numbers will be explained in the next section). But most scientists use the more convenient approach of modifying the expressions so that the exponents are equal:

$$(3.63 \times 10^{-2}) + (0.485 \times 10^{-2})$$

Now, with each coefficient multiplied by the same exponential factor, the addition (or subtraction) takes the form

$$(3.63 + 0.485) \times 10^{-2} = 4.12 \times 10^{-2}$$

Problem 1.3 What is the result of subtracting the following numbers?

$$7.953 \times 10^{-4} - 6.42 \times 10^{-5}$$

To multiply numbers written in scientific notation, we multiply the coefficients and add the exponents.

Example 1.4 Multiplying numbers written in scientific notation

Multiply 3.4×10^3 by 2.8×10^{-2}.

Solution

$$(3.40 \times 2.80) \times 10^{[3+(-2)]} = 9.52 \times 10^1 = 95.2$$

Problem 1.4 Multiply 4.2×10^5 by 0.64×10^{-4}.

To divide numbers written in scientific notation, divide the coefficients and subtract the exponent of 10 in the denominator from the exponent of 10 in the numerator.

Example 1.5 Dividing numbers written in scientific notation

Divide 2.8×10^5 by 4.0×10^2.

Solution

$$\frac{2.8 \times 10^5}{4.0 \times 10^2} = 0.70 \times 10^{[5-(+2)]} = 0.70 \times 10^3 = 7.0 \times 10^2$$

Problem 1.5 Divide 3.45×10^4 by 7.2×10^{-2}.

Concept check

- ✔ To multiply in scientific notation, multiply the coefficients and add the exponents.
- ✔ To divide in scientific notation, divide the coefficients and subtract the exponent of 10 in the denominator from the exponent of 10 in the numerator.

1.6 CALCULATIONS AND SIGNIFICANT FIGURES

Calculations that are numerically correct can sometimes lead to unrealistic results. For example, how should we report the area of a square whose dimensions have been measured as 8.5 in. on a side? Mathematically,

$$\text{Area of a square} = \text{side} \times \text{side} = 8.5 \text{ in.} \times 8.5 \text{ in.}$$
$$= 72.25 \text{ in.}^2$$

Multiplication of 2 two-digit numbers always yields a number with more than two digits. However, information regarding the size of an object can be obtained only by measurement, not by an arithmetic operation. The results of multiplications and divisions using measured quantities are reported according to the following rule:

> The number of significant figures in a number resulting from multiplication or division may not exceed the number of significant figures in the least well known value used in the calculation.

In the earlier example of the area of the 8.5 in. × 8.5 in. square, the length of a side is known to two significant figures, and therefore the area of the square (length × length) cannot be known with any greater accuracy. Should we report it as 72 or 73 in.^{2}?

To reduce the number of significant figures and determine the value of the final significant digit, we commonly use a practice called **rounding.** The rules of rounding stipulate that, if the digit after the one that we want to retain is 5 or greater, we increase the value of the digit that we want to retain by 1 and drop the trailing digits. If its value is 4 or less, we leave unchanged the value of the digit that we want to retain and drop the trailing digits.

Note that rounding takes place after the calculation has been completed. That is, the calculation is done by using as many digits as possible. Only the final result is rounded. In determining the area of the 8.5-in. square, because the least well known measurement has only two significant figures, we should round the calculated result of 72.25 and report an area of 72 in.^{2}.

A more perplexing situation might be encountered if we needed to know the area of a rug required to fit a room 74 in. by 173 in. The calculated area is 12,802 in.^{2}. The least accurately known measurement possesses two significant figures, so the area must be expressed with that number of significant figures as well. The value of the area is reported by first converting the value into scientific notation and then rounding to two significant figures:

$$12{,}802 \text{ in.}^2 = 1.2802 \times 10^4 \text{ in.}^2 = 1.3 \times 10^4 \text{ in.}^2$$

The same considerations hold for division.

Example 1.6 Multiplying and dividing measured quantities

Velocity is defined as $\frac{\text{distance}}{\text{time}}$. What velocity must an automobile be driven to cover 639 km in 9.5 hours (h)?

Solution

$$\text{Velocity} = \frac{\text{distance}}{\text{time}} = \left(\frac{639\text{ km}}{9.5\text{ h}}\right) = 67.3\,\frac{\text{km}}{\text{h}} = 67\,\frac{\text{km}}{\text{h}}$$

Because the least well known quantity, 9.5 h, has two significant figures, it is necessary to round the resultant calculated value to two significant figures.

Problem 1.6 Calculate the volume of a cube that is 8.5 cm on a side. (Volume of a cube 1 cm on a side = 1 cm × 1 cm × 1 cm.)

A somewhat different approach is required for addition and subtraction. In both these situations, the number of figures after the decimal point decides the final answer. The final sum or difference cannot have any more figures after the decimal point than are contained in the least well known quantity in the calculation. All significant figures are retained while doing the calculation, and the final result is rounded.

Example 1.7 Adding measured quantities

Add the following measured quantities: 24.62 g, 3.7 g, 93.835 g.

Solution

The least well known of these quantities has only one significant figure after the decimal point, so the final sum cannot contain any more than that. We add all the values, and round off after the sum is calculated, as follows:

$$\begin{array}{rl} 24.62 & \text{grams} \\ 3.7 & \text{grams} \\ 93.835 & \text{grams} \\ \hline 122.155 & \text{grams} = 122.2 \text{ grams} \end{array}$$

Problem 1.7 Add the following quantities: 1.9375, 34.23, 4.184.

The same considerations apply to subtractions.

Example 1.8 Subtracting measured quantities

Calculate the result of the following subtraction:

$$5.753 \text{ grams} - 2.32 \text{ grams}$$

Solution

The least well known quantity has two significant figures after the decimal point, so the result cannot contain any more than that. As in addition, we round off after having done the subtraction.

$$\begin{array}{rl} 5.753 & \text{grams} \\ -2.32 & \text{grams} \\ \hline 3.433 & \text{grams} = 3.43 \text{ grams} \end{array}$$

Problem 1.8 What is the result of the following subtraction?

$$94.935 \text{ m} - 7.6 \text{ m}$$

Rules for determining significant figures

- The number of significant figures in the result of a multiplication or division may not exceed the number of significant figures found in the least well known value used in the calculation.
- The number of figures after the decimal point in the result of an addition or subtraction may not exceed the number of significant figures after the decimal point in the least well known quantity being used.

Rules for rounding numbers

- If the digit after the one to be retained is 5 or greater, increase the value of the digit to be retained by 1 and drop the trailing digits.
- If the digit after the one to be retained is 4 or less, leave unchanged the value of the digit to be retained and drop the trailing digits.
- Only a final result is rounded. All digits are retained until a calculation is complete.

1.7 THE USE OF UNITS IN CALCULATIONS: THE UNIT-CONVERSION METHOD

All of the quantities that you will be working with when you do chemical calculations will have units—for example, mL, cal, and so forth. The method used in solving problems with quantities having units is called the **unit-conversion method.** It is also referred to variously as the factor-label method, the unit-factor method, or dimensional analysis.

▶▶ We begin using unit conversions in the very next chapter to understand the nature of matter.

The underlying principle in this problem-solving strategy is the conversion of one type of unit into another by the use of a **conversion factor.**

$$\text{Unit}_1 \times \text{conversion factor} = \text{unit}_2$$

The conversion factor has the form of a ratio that allows cancellation of unit_1 and its replacement with unit_2. The units are quantities that are treated according to the rules of algebra.

Earlier in this chapter, Example 1.1 asked us to convert 0.001 s into milliseconds, which we accomplished by using the definition 0.001 second (s) = 1 millisecond (ms). Let us now see how the unit conversion method takes this kind of information and uses it to solve problems.

Suppose we wish to add 0.0230 s to 156 ms. To add these numbers, we must express them in the same units. Rather than immediately concerning ourselves with the given numbers, we will first consider only the units in the problem. This initial focus on units is the strength of the unit-conversion method. Let us decide now that the units of the answer will be in milliseconds. The heart of the problem, then, is to convert the units given in seconds into the desired units, milliseconds. We accomplish this task through the use of a conversion factor.

The required conversion factor is obtained by expressing the relation between seconds and milliseconds in the form of an equality:

$$1 \text{ s} = 1000 \text{ ms}$$

This relation is contained within a single system of measurement and is exact by definition. Therefore the number of significant figures in the answer is not determined by this relation but only by the measured values. However, relation between two different systems of measurement—for example, the relation between pounds and kilograms—are not necessarily exact and will affect the number of significant figures in an answer.

Dividing both sides of the equation by 1 s produces

$$1 = \frac{1000 \text{ ms}}{1 \text{ s}}$$

The expression to the right of the equals sign is a unit-conversion factor. It will be used to convert the number given in units of seconds into a number in units of milliseconds. Because the value of the conversion factor is unity, or 1, its use does not change the intrinsic value of any numerical quantity, merely its name.

We convert the number given in units of seconds into its value in units of milliseconds by multiplying it with the conversion factor just derived:

$$0.0230 \text{ \sout{second}} \times \frac{1000 \text{ milliseconds}}{\text{\sout{second}}} = 23.0 \text{ milliseconds}$$

The conversion factor allowed the cancellation of the old unit, so the result of multiplication is a numerical answer in the new units.

Because 0.0230 s = 23.0 ms, the sum of 0.0230 s and 156 ms is

$$156 \text{ milliseconds} + 0.0230 \text{ \sout{second}} \left(\frac{1000 \text{ milliseconds}}{\text{\sout{second}}}\right) = 179 \text{ milliseconds}$$

Because the reciprocal of the conversion factor also is equal to unity, we can also solve the problem by converting milliseconds into seconds.

$$0.0230 \text{ second} + 156 \text{ \sout{milliseconds}} \left(\frac{1 \text{ second}}{1000 \text{ \sout{milliseconds}}}\right) = 0.179 \text{ second}$$

Both results have the same number of significant figures.

The usefulness of this method of problem solving is that it allows you to check whether your approach to obtaining an answer is correct before any calculations have been done. Determine what form the unit-conversion factor must have if the units of the answer are to be derived from the units of the data provided.

Example 1.9 Using the unit-conversion method: I

Convert 0.164 liters (L) into milliliters (mL).

Solution

The conversion factor for this unit conversion is based on the equivalence

$$1 \text{ L} = 1000 \text{ mL}$$

Because L must be canceled, the conversion factor must be

$$1 = \frac{1000 \text{ mL}}{1 \text{ L}}$$

The solution is

$$0.164 \text{ L} \left(\frac{1000 \text{ mL}}{1 \text{ L}}\right) = 164 \text{ mL}$$

Problem 1.9 Convert 74.1 mL into liters.

As long as a measurement with its units is multiplied by the correct form of the conversion factor, all that really changes is the name of the units, not the intrinsic value of the measurement. However, you must be sure to use an appropriate conversion factor. To do so, you should become familiar with the conversion factors listed in Table 1.5.

TABLE 1.5 Relations between English and Metric Units

Length	Volume	Mass
1 in = 2.54 cm	1 ft^3 = 28.32 L	1 lb = 453.59 g
39.37 in = 1 m	1.057 qt = 1 L	1 oz = 28.35 g
1 mile = 1.609 km	1 gal = 3.7853 L	1 dram = 1.772 g

Example 1.10 Using the unit-conversion method: II

How many grams are in 1.81 lb?

Solution

First identify the unit to be converted and then the correct form of the conversion factor (but do not worry about the numbers yet). Pounds must be converted into grams. Table 1.5 shows that the conversion factor will be based on the relation

$$1 \text{ lb} = 453.6 \text{ g}$$

Then substitute the appropriate ratio:

$$1.81 \text{ lb} \times \frac{453.6 \text{ g}}{1 \text{ lb}} = 821 \text{ g}$$

Problem 1.10 How many pounds are in 752.4 g?

There are situations where it becomes necessary to use two or more conversion steps to achieve a solution, as in the following example.

Example 1.11 Using two or more unit-conversion factors

How many milliliters are in 2.35 gal?

Solution

Converting gallons into milliliters in one step would require a conversion factor with the dimensions mL/gal. You will not find this conversion factor in Table 1.5. However, the same thing can be accomplished by multiplying a series of factors that will result in the desired conversion.

The solution to the problem of converting gallons into milliliters is first to convert gallons into liters and then to convert the liters into milliliters.

We can find the first conversion factor in Table 1.5 and deduce the second from the relations in Table 1.4. The calculation takes the following form:

$$2.35 \text{ gal}\left(\frac{3.7853 \text{ L}}{1 \text{ gal}}\right)\left(\frac{1000 \text{ mL}}{1 \text{ L}}\right) = 889.5455 \text{ mL} = 8.90 \times 10^2 \text{ mL}$$

There should be only three significant figures in the answer because the measured value, 2.35 gal, has only three significant figures. Note that the last digit rounded to zero and was significant, so the answer is written in scientific notation.

Problem 1.11 How many seconds are in seven days? (Hint: There are exactly 60 s in a minute, 60 min in an hour, and 24 h in a day.)

1.8 TWO FUNDAMENTAL PROPERTIES OF MATTER: MASS AND VOLUME

Having established the basis for quantitative description, we are ready to examine two of the basic properties of matter and some of the ways in which they are measured in the laboratory.

Mass

Mass is a measure of the quantity of matter. It is a useful property in the study of matter because it remains constant in the presence of environmental changes such as fluctuations in temperature and pressure. Because an object's

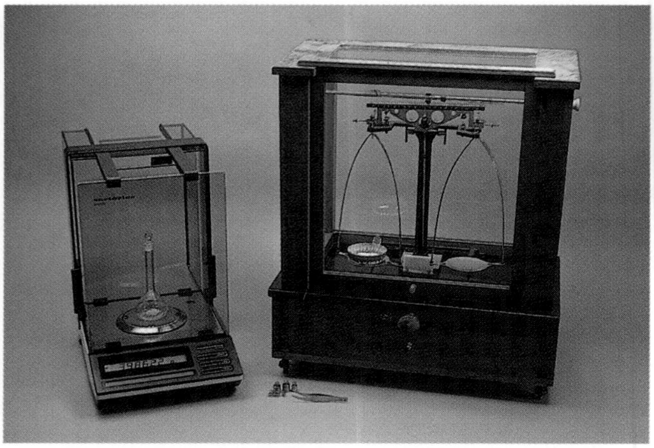

Figure 1.11 A modern automatic analytical balance (left) and an older laboratory beam balance (right). The mass of a sample on one pan of the beam balance is determined by balancing it with a known reference weight on the other pan.

mass is determined by weighing, it is common to speak of mass and weight interchangeably; however, they are not the same thing. A mass has weight because it is under the influence of a gravitational field. The gravitational field at the top of Mt. Everest is weaker than that at sea level by about 0.2%, so a 150-lb astronaut at sea level weighs 149.8 lb at the mountain's summit. Astronauts aboard the space shuttle in outer space are nearly weightless because there is little effect of gravity on a mass so far from Earth. Nevertheless, an astronaut's mass at the summit of Mt. Everest, at sea level on Earth, and in outer space is the same. The SI unit of mass is the kilogram.

Mass is measured relative to that of a standard mass. In weighing, an unknown mass is balanced against a known, or reference, mass. Thus, mass measuring devices, both mechanical ones and those that are a combination of mechanical and electronic devices, have come to be known as "balances."

A balance operates on the same principle as the playground seesaw, or teetertotter—a plank supported at its center by a wedge-shaped fulcrum. If equal weights are placed equal distances from the fulcrum, the plank is held in balance, with each end equidistant from the ground. In a laboratory balance, weights are added or removed until they equal that of the unknown mass, and balance is achieved. Figure 1.11 illustrates an older beam type and the newer electronic type of balance used in the chemical laboratory.

Example 1.12 Converting measures of mass

Express a mass of 76 g in units of kilograms.

Solution

Using the unit-conversion method, first write the given quantity, and then multiply by the appropriate conversion factor:

$$76 \text{ g}\left(\frac{1 \text{ kg}}{1000 \text{ g}}\right) = 0.076 \text{ kg}$$

Problem 1.12 Express a mass of 2.87 kg in grams.

Volume

Volume is the amount of space that a sample occupies. Figure 1.12 illustrates the devices commonly used in the laboratory to measure liquid volume. The one most commonly used in the beginning chemistry laboratory is the graduated cylinder. A graduated cylinder has an error of about 1%, which means plus or minus about 0.1 mL for a 10-mL graduated cylinder and plus or minus about 1 mL for a 100-mL graduated cylinder. When less error is desired, volu-

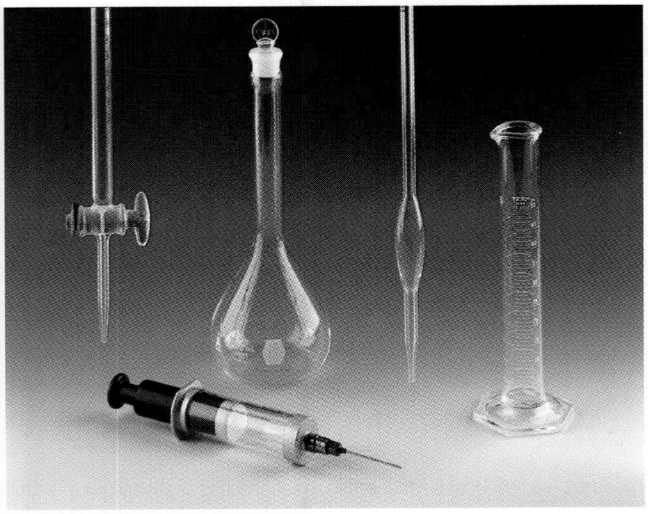

Figure 1.12 Devices for measuring liquid volumes: *(back row, left to right)* a buret, a volumetric flask, a pipet, and a graduated cylinder; *(in front)* a hypodermic syringe.

metric flasks and delivery pipets are used for fixed volumes, and burets are used for variable quantities. These devices will contain or measure out their stated volumes repeatedly to within 0.1%. The hypodermic syringe, also shown in Figure 1.12, has an error of about 5% to 6%, so a 5-mL drug injection may be in error of plus or minus about 0.3 mL. That size of error is acceptable because of the convenience and versatility of the delivery system.

All volumetric (volume-measuring) containers are calibrated in milliliters. One milliliter is equal to 1 cm^3 (a cubic centimeter, or cc), and there are 1000 mL in 1.0 L. There are 10 cm × 10 cm × 10 cm = 1000 cm^3 (mL) in 1.0 L; so, in SI units, 1 L = 1×10^{-3} m^3. To get an idea of the dimensions of a liter container, consider Figure 1.13, which shows an adult who is 2 meters tall, a child who is 1 meter tall, and a 1-liter cube, which is 10 centimeters "tall."

Example 1.13 Converting measures of volume

Express a volume of 364 mL in liters.

Solution

Using the unit-conversion method, first write the given quantity, and then multiply by the appropriate conversion factor:

$$364 \text{ mL} \left(\frac{1 \text{ L}}{1000 \text{ mL}} \right) = 0.364 \text{ L}$$

Problem 1.13 Express the quantity 3.97 L in milliliters.

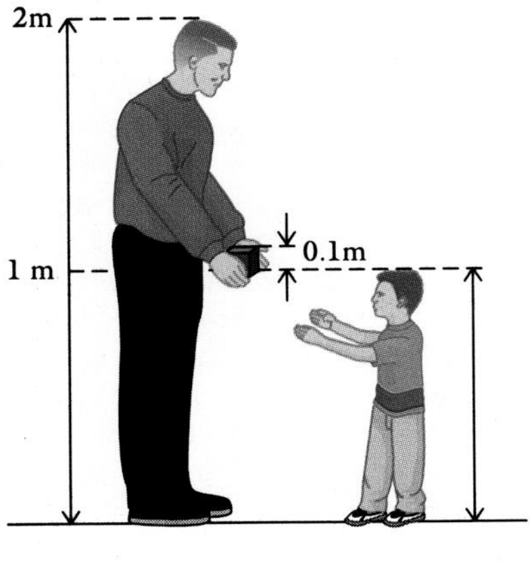

Figure 1.13 The dimensions of a liter container in the shape of a cube are 0.1 meter, or 10 centimeters, on a side. The adult is 2 meters tall and the child is 1 meter.

1.9 DENSITY

The characterization of substances requires measurement of their physical properties; that is, properties that can be determined without changing the substance's nature. One of the most useful of these properties is **density,** a derived unit defined as mass per unit volume:

$$\text{Density} = \frac{\text{mass (g)}}{\text{volume (cm}^3\text{)}}$$

Density is a property of substances in any physical state—gaseous, liquid, or solid. It is used to evaluate the purity of solids and liquids (pure gold in the solid state, for example, has a density of 19.32 g/cm^3). Density is also used to estimate the amount of dissolved solids in solutions, as in the clinical analysis of urine samples to evaluate kidney function. The density of a solution rises as the amount of a dissolved solid increases. Failure of the kidneys to concentrate the urine can be detected by the fact that the urine density is abnormally low. Density also serves as a conversion factor for translating mass into volume or volume into mass.

▶▶ Temperature and its measurement are discussed in Section 1.10.

It is important to note that mass does not change when the temperature changes, but volume does. The volumes of most liquids increase slightly as the temperature increases; as a result, they undergo a slight decrease in their densities. For this reason, any measurement of density must always be accompanied by the value of the temperature at which the measurement was made. An interesting case of the temperature dependence of density serving a biological need is discussed in Box 1.3.

Example 1.14 Calculating mass from density

Calculate the mass of 96.0 mL of a solution that has a density of 1.09 g/mL at 20°C.

Solution

Use the unit-conversion method, first writing the quantity to be converted and then multiplying it by the appropriate conversion factor—in this case, the density:

$$96.0\ \text{mL}\left(\frac{1.09\ \text{g}}{\text{mL}}\right) = 104.64 = 105\ \text{g}$$

Problem 1.14 Calculate the mass of 135.0 mL of a liquid that has a density of 0.8758 g/mL at 20°C.

To measure a substance's density, one must first determine the mass of a small volume of it. The mass of a solid is easily obtained on a balance. Its volume is then ascertained by filling a graduated cylinder about halfway with a liquid in which the solid will neither float nor dissolve. The solid is placed in the cylinder and the resulting rise in the liquid's level is equal to the volume of the solid.

1.3 Chemistry Around Us

Temperature, Density, and the Buoyancy of the Sperm Whale

The head of a sperm whale is huge, making up about one-third of the animal's total weight. Most of the head is composed of the whale's unique spermaceti organ, which plays a key role in the feeding behavior of the sperm whale. This organ contains about 4 tons of spermaceti oil, a mixture of triacylglycerols and waxes.

The sperm whale's diet consists almost exclusively of squid found in very deep waters, at depths of almost 1 mile or more. Squid are plentiful at these depths, and there are no competitors. The only problem for the sperm whale is staying at these depths to wait for the squid. It would take a tremendous amount of energy for a marine animal to stay at a given depth by swimming. The spermaceti organ is an energy-conserving solution to this problem.

The sperm whale's normal body temperature of about 37°C keeps the spermaceti oil in a liquid state as long as the whale is resting on the surface. As the whale dives to feeding depths, however, it pumps cold seawater through chambers in the spermaceti organ. The spermaceti oil cools and becomes partly solid. The density of the oil increases because its solid form is more dense than its liquid form. The increased density allows the whale to descend to the ocean depths without much swimming effort. The whale controls the depth to which it dives by controlling the temperature and hence the density of the spermaceti oil. After the whale feeds, an increase in the circulation of blood to the spermaceti organ warms the spermaceti oil, decreasing its density and thereby lifting the whale to the ocean surface.

The mass of a liquid can be obtained by using a special container whose volume and mass are already known. One then fills the container with a test liquid and determines the liquid's mass (the difference in mass between the filled and unfilled container). That mass divided by the known volume yields the test liquid's density. However, there is a simpler way to measure liquid density, and that is through the use of a hydrometer.

Objects placed in liquids either float on the surface, sink to the bottom, or remain at some intermediate position in which they have been placed and neither float nor sink. If an object floats, its density is less than that of the liquid. If it sinks, its density is greater than that of the liquid. If it neither floats nor sinks, its density is equal to that of the liquid. A **hydrometer** is a solid, vertical floating object that rises or falls to a level where its density is equal to the density of the liquid in which it is placed.

A hydrometer is a hollow, sealed glass vessel in the form of a narrow, graduated tube with a bulb at the lower end. The hollow tube floats in an upright, vertical orientation because the bulb at the lower end is filled with weights. Graduations on the tube are calibrated so that the mark coinciding with the surface of the liquid indicates the liquid's density.

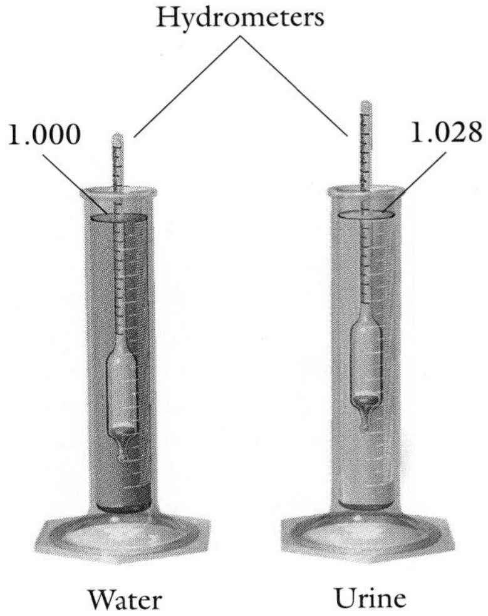

Hydrometers in common use in the clinical laboratory are graduated not in density units (such as grams per milliliter) but in values of **specific gravity,** a property defined as the ratio of the density of the test liquid to the density of a reference liquid:

$$\text{Specific gravity} = \frac{\text{density of test liquid}}{\text{density of reference liquid}}$$

Note that, because specific gravity is a ratio of densities, the units cancel, and specific gravity has no units. The standard reference liquid for measuring the specific gravity of aqueous solutions is pure water at 4°C. (Aqueous solutions are solutions containing water.) The density of water is at its maximum, $1.000\ g/cm^3$, at that temperature (actually 3.98°C). The temperature dependence of water and its importance to living organisms is considered in Box 1.4 on the following page. Specific gravities of blood or urine (both are aqueous solutions) may be reported as 1.028, for example. This means that the sample's density was 1.028 times the density of pure water.

1.10 TEMPERATURE

If a hot object makes contact with a cold object, the hot object will cool and the cool object will warm. This effect is explained by proposing that heat always flows from a hot body to a cold body in much the same way that water flows downhill under the influence of gravity. Substances can either gain heat or lose it, depending on whether they are cooler or hotter than their environments. The measurement of heat itself will be described in the next section but, to measure heat, we must first have a way to measure how hot or cold an object is. In other words, we need a device capable of indicating the object's position on a scale of "hotness." This position on the scale is called a **temperature,** and the device is called a **thermometer.** It is important to keep in mind that the flow of heat and the concept of temperature are inextricably related but are not the same thing.

▶▶ The role of heat in chemistry is introduced in chapters 6 and 8.

Many thermometers consist of a very small diameter glass tube partly filled with a fluid. The fluid, often mercury, expands when heated and contracts when cooled, thereby rising or falling to different levels within the tube. The scale on the thermometer is created by determining the locations of two reference positions on the tube. The lower one shows the fluid's level at the temperature at which pure water freezes. The upper one shows the fluid's level at

1.4 Chemistry Around Us

Density and the "Fitness" of Water

In 1912, the American physiologist Lawrence J. Henderson introduced a new idea into the study of ecology and ecosystems—the idea of the "fitness" of the environment to support life. He meant by this idea that evolution is a reciprocal process, requiring not only that living things adapt to their environments, but also that an environment must have certain unique characteristics that enable it to support life.

You are probably familiar with the fact that water is the principal component of living organisms and, furthermore, that life probably evolved in water. Water's density is one of the physical properties that explains this liquid's central role in living systems and in their environments. Specifically, water is the one notable exception to the rule that the density of ordinary solids is greater than the density of their corresponding liquids. If water behaved like other liquids, life on Earth would not exist in the form that we know it today.

In general, when a liquid cools, its density increases until the substance solidifies. If the solid and liquid forms of the substance are present simultaneously, the solid, which is denser, will rest at the bottom of its container with the liquid above it. As with other substances, the density of water also increases as the temperature decreases—but only until the temperature reaches 4°C. At that point, unlike the densities of other liquids, its density begins to decrease, and, at water's freezing point, the ice that forms floats on the remaining liquid water instead of sinking as another substance would. The density of water reaches its maximum at 4°C and is lower above and below that temperature. That fact is illustrated by the adjoining graph, in which the density of water is plotted as a function of Celsius temperature. This unusual behavior will be explained in Chapter 6.

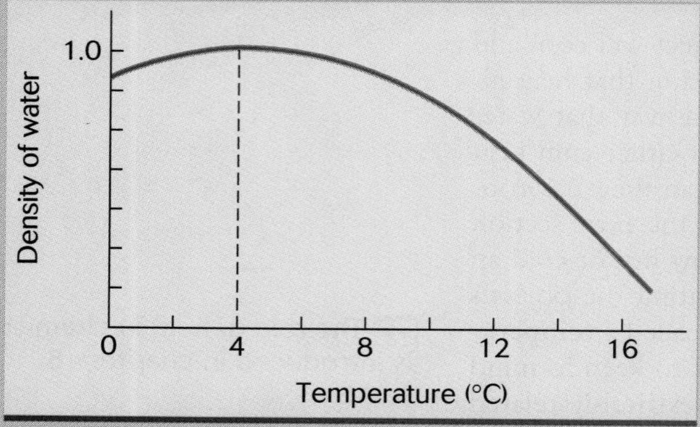

The density of water is plotted as a function of increasing temperature. Water melts at 0°C and, as it is warmed, its density increases to a maximum value of 1.0 g/cm^3 at 4°C. Above that value, its density decreases as the temperature is raised.

When natural bodies of water are gradually cooled from above-freezing temperatures to below-freezing ones, the surface water eventually reaches 4°C. Because the density is highest at that temperature, the surface water descends to the bottom, where it remains at 4°C (close to, but not quite, freezing). As the environmental temperature continues to drop, however, the density of the water that is still near the surface decreases, and that water rises, to remain at the surface until it freezes. Thus, the ice floats on the surface of liquid water. The final result of this anomalous density curve is a stratification, or layering, of temperature zones, with the "heaviest" water at 4°C at the bottom, surmounted by "lighter" water at lower temperatures and, finally, ice at the surface. The period during which the water below the ice can persist in the liquid state depends on the depth of the water.

Life is abundant under the antarctic ice.

If ice sank in liquid water, Earth's first winters would have seen large quantities of solid water sinking to the bottom of rivers and lakes. The ice formed during those winters would not have melted because of the time required to transfer sufficient heat through the water above it. In addition, the densities would have increased uniformly with depth, so no convection or mixing could have occurred.

This process would have been repeated yearly until eventually most or perhaps all of the planet's liquid water would have been transformed into ice. It is because water has its maximum density at a temperature above its freezing point that most of Earth's water exists in the liquid form and allows the continued existence of life as we know it.

the temperature at which pure water boils. Then, a uniform scale of equal divisions is established between these two positions.

The freezing and boiling points of pure water have different numerical values in the three temperature scales commonly encountered in chemistry. On the **Fahrenheit scale** (on which the units are degrees), the freezing point is 32°F and the boiling point is 212°F (both are exact numbers). On the **Celsius scale** (also in degrees), the freezing point of water is 0°C and the boiling point is 100°C (both exact numbers). The SI temperature scale is the **Kelvin scale,** and its units are called kelvins, symbol K (no degree symbol is used with the symbol for kelvins). The size of the kelvin is identical with the size of the Celsius degree. However, there is a difference between the two scales in that the Kelvin scale recognizes a low temperature limit called absolute zero, the lowest theoretically attainable temperature, and gives it the value of 0 K. On the Kelvin scale, the freezing point of water is 273.15 K (Celsius scale, 0°C). For convenience, this value will be rounded to 273, so that Celsius degrees can be converted into kelvins by simply adding 273 to the Celsius value:

$$\text{K} = {}^\circ\text{C} + 273$$

Conversion between Celsius and Fahrenheit scales is a bit more complicated. Figure 1.14 emphasizes that there are two problems in conversion:

1. The sizes of the degrees are different.
2. The numerical values of the reference points for the freezing and boiling points of water are not the same.

Let us consider these two issues one at a time.

A Celsius degree is larger than a Fahrenheit degree. There are 100 Celsius degrees between the freezing and the boiling points on that scale and 180 degrees between the freezing and the boiling points on the Fahrenheit scale. In other words, there are 180/100, or 9/5, Fahrenheit degrees per 1 Celsius degree. In this text, we will write the equations comparing the sizes of Fahrenheit and Celsius degrees in either of two ways:

$$\text{Fahrenheit degrees} = {}^\circ\text{C} \times (9/5)({}^\circ\text{F}/{}^\circ\text{C})$$

or

$$\text{Celsius degrees} = {}^\circ\text{F} \times (5/9)({}^\circ\text{C}/{}^\circ\text{F})$$

Because these equations are written in the form of unit-conversion calculations, they allow us to convert from one type of degree into another.

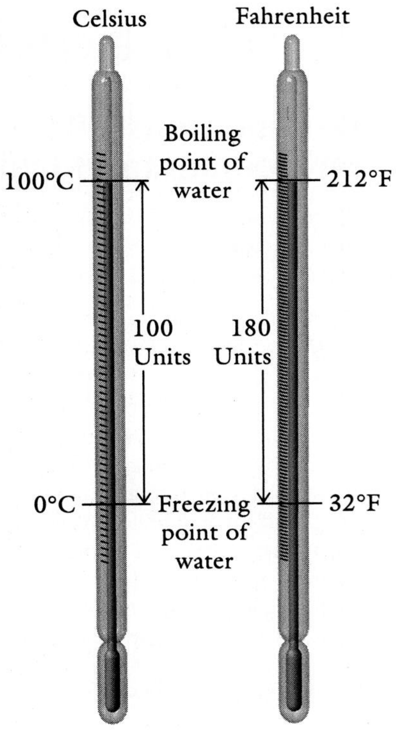

Figure 1.14 This drawing of two thermometers emphasizes the two pieces of information necessary for converting temperatures from one scale into another: (1) the relation between the sizes of degrees and (2) the difference in the numerical values of the freezing and boiling points in the two systems.

Example 1.15 Converting number of Celsius degrees into the equivalent number of Fahrenheit degrees

If a temperature is seen to rise 30 degrees on the Celsius scale, how many degrees will it have risen on the Fahrenheit scale?

Solution

We use the first of two conversion equations, $^\circ\text{F} = {}^\circ\text{C} \times (9/5)({}^\circ\text{F}/{}^\circ\text{C})$, and insert the given value:

$$^\circ\text{F} = 30^\circ\text{C} \times (9/5)({}^\circ\text{F}/{}^\circ\text{C}) = 54^\circ\text{F}$$

The temperature will have risen 54 Fahrenheit degrees.

Problem 1.15 If a temperature is seen to rise 65 degrees on the Fahrenheit scale, how many degrees will it have risen on the Celsius scale?

The preceding examples dealt only with the problem of equating numbers of Celsius degrees with numbers of Fahrenheit degrees or vice versa, not with the conversion of one thermometer reading into another. To interconvert tem-

PROFESSIONAL CONNECTIONS

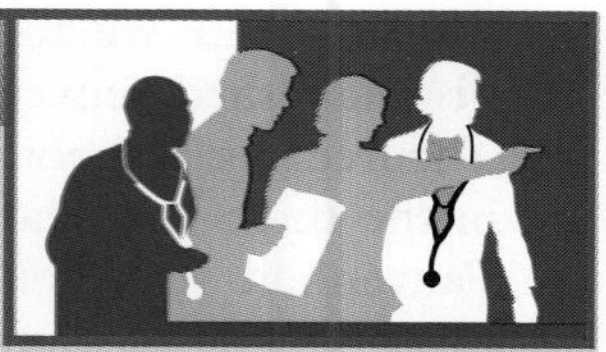

Darlene Reuss, BS
Environmental Health Specialist

"Professional Connections" are brief interviews with women and men who have chosen careers for which a chemistry course such as this one is a requirement.

Darlene Reuss studied undergraduate chemistry at California State University, Sacramento.

Why did you choose the field that you are in today?

I liked the science classes that I took at Sacramento State, particularly the chemistry and microbiology classes, so I knew that I wanted to go into a related field but then needed to know if I would prefer work in a lab or out with the public. I interviewed and then interned in the San Joaquin County Health Department and decided that I really liked the public service aspects of that job, being out with the public and being able to use the science that I've learned.

What positions did you have before this one?

I worked in the San Joaquin County Health Department for 2 years, beginning in 1988, and then moved to Mississippi and worked for Aerojet, ASRM, as an environmental analyst. It was one of three companies at the Advanced Solid Rocket Motor facility (ASRM) in Iuka, Mississippi, and was the company that would build the solid rocket motors for the space shuttles. I checked equipment designs for compliance with various laws, such as the Clean Air Act. When Congress canceled the program, I worked for the Mississippi Health Department for 3 years before returning to my native state of California and obtaining the position that I have today.

Please describe a typical day.

From 7:00 A.M. to 9:00 A.M., I do paperwork and answer phone calls; then I'm traveling and out in the field until 3:00 P.M. A typical day includes inspecting about six restaurants: checking for cleanliness, safe temperatures of food, proper storage (for example, chemicals shouldn't be stored next to food!), condition of the garbage areas, and so forth. By 3:30, I'm back in the office and return any phone calls that have come in during the day. A nontypical day is one in which some crisis occurrs in my territory, such as a sewage spill or an incident of a food-borne illness.

What advances have been made in your field? Have they changed your everyday life?

In the past, we had to use probe thermometers to check the internal temperatures of food, which would mean inserting the probe and cleaning it off thoroughly every time a different sample was tested. In San Mateo County, we now have ultraviolet temperature guns: we simply scan the food and get a readout! This is much easier and faster. We also use sensor tapes for food and dishes; these tapes can change color with the correct pH or temperature. In addition, we're about to get a new digital probe thermometer for testing water; the probe can be placed under water and can stand up to the temperatures and motion of dishwashers.

What was the most valuable thing that you learned in school?

That I was smart, and good in science and math! I was a returning student to college in the 1980s and had been out of school since 1963. I started at American River College, a community college in the Sacramento area, and made use of all of the resources available to me there. I started with remedial and self-paced classes and learned that, in contrast with my prior beliefs, I really could do well in math and science. The faculty at American River was great.

How do you use chemistry in your job?

When I was an environmental analyst at Aerojet, I used it to calculate potential stack emissions by using the density of a chemical and the total quantity of each product to be used in construction of the rockets. In my current job, it primarily provides me with the proper vocabulary to competently understand the situations that I encounter. I also find that the skills that I learned in the chemistry labs, keeping good records and documenting everything carefully, are very valuable in my current job. They taught me to investigate and be precise.

What advice do you have for students who are studying chemistry today?

Start at the beginning. The most important class that I took in chemistry was the very first one, the introduction to chemistry. The early classes are your building blocks—you start from there and go on. Also, labs are very important: they show you the "whys" behind what you are doing and make it easier to understand.

perature readings, we must take account of the fact that the numerical values of the reference points for freezing and boiling points of water on the two scales are not the same. This concept can best be understood through the following examples.

Example 1.16 Converting Celsius temperature into Fahrenheit temperature

Convert a temperature reading of 50°C into the equivalent temperature on the Fahrenheit scale.

Solution

To do this conversion, let us consult the temperature-scale diagram in Figure 1.14. As the diagram indicates, 50°C is exactly halfway between the freezing and the boiling points of water on the Celsius scale. On the Fahrenheit scale, the equivalent halfway point is 90 Fahrenheit degrees higher than the Fahrenheit freezing point of 32°F. The reading on the Fahrenheit scale directly opposite the Celsius scale must then be 122°F, the result of adding 32°F to the 90 Fahrenheit degrees above the freezing point. Therefore. to convert a Celsius temperature into a Fahrenheit temperature, we first multiply °C by 9/5 and then add 32 to the result. In the form of an equation,

$$
\begin{aligned}
{}^\circ F &= {}^\circ C \times (9/5)({}^\circ F/{}^\circ C) + 32{}^\circ F \\
&= 50{}^\circ C \times (9/5)({}^\circ F/{}^\circ C) + 32{}^\circ F = 122{}^\circ F
\end{aligned}
$$

Problem 1.16 What Fahrenheit temperature is equivalent to 100°C?

Now let us consider the conversion of a Fahrenheit reading on the thermometer into the equivalent Celsius reading.

Example 1.17 Converting Fahrenheit temperature into Celsius and Kelvin

A thermometer reads 122° on the Fahrenheit scale. What is its equivalent reading on both the Celsius and the Kelvin scales?

Solution

This Fahrenheit value was chosen so that we can simply reverse the steps that we took in Example 1.16. The value 122°F is exactly 90 Fahrenheit degrees above the freezing point of water. Therefore, to convert Fahrenheit temperature into Celsius temperature, first, subtract 32 and, second, multiply by 5/9. The result is, not surprisingly, 50°C. In the form of an equation,

$$
\begin{aligned}
{}^\circ C &= ({}^\circ F - 32{}^\circ F)(5/9)({}^\circ C/{}^\circ F) \\
&= (122{}^\circ F - 32{}^\circ F)(5/9)({}^\circ C/{}^\circ F) \\
&= 50{}^\circ C
\end{aligned}
$$

To calculate the equivalent Kelvin temperature, simply add 273 to the Celsius temperature:

$$K = {}^\circ C + 273 = 50 + 273 = 323\ K$$

Problem 1.17 Convert 98.6°F into both Celsius and Kelvin temperatures.

1.11 HEAT AND CALORIMETRY

Heat is a form of energy. In the section on temperature, heat was described as flowing from a region of higher temperature to one of lower temperature. By focusing on the loss or gain of heat, scientists have enlarged the list of properties that are useful for characterizing different substances.

Each substance has a different capacity to absorb heat. This capacity is measured by noting the rise in temperature of a fixed mass of a substance that has absorbed a known amount of heat. We know how to measure mass and temperature, but how do we measure heat? A unit of heat is defined by its effect (the rise in temperature) on a fixed mass of a reference substance.

A widely used non-SI unit of heat is called the **calorie,** abbreviated cal, and the official SI unit is called the **joule,** abbreviated J. The relation between them (a useful conversion factor) is 4.184 J = 1 cal. The temperature of 1.0 g of water rises 1 Celsius degree when 1 cal of heat is absorbed. Compare this with the effect of that same calorie on 1 g of aluminum: a temperature rise of almost 5 Celsius degrees.

The characteristic response (increase in temperature) of a given mass of a given substance exposed to a given amount of heat is expressed by a quantity called the **specific heat** (symbol, C_p). If the mass is in grams, the temperature on the Celsius scale, and the heat in joules, the units of the specific heat are:

$$C_p = \frac{\text{joules}}{\text{grams} \times \Delta°\text{C}}$$

This equation reads as follows: the specific heat is equal to the heat absorbed or lost per Celsius degree change in temperature per gram of substance. (The change in temperature is the difference, Δ, between final and initial temperatures, or $\Delta t = t_{\text{final}} - t_{\text{initial}}$.) The higher a substance's specific heat, the more slowly its temperature rises in response to heating. The specific heat is a unique property of a substance, and Table 1.6 lists various materials along with their specific heats. Liquid water has one of the highest values of specific heats of liquids. Water's role in temperature control in mammals is discussed in Box 1.5.

In summary, specific heat describes the capacity of 1 gram of a substance to absorb heat. It should not be confused with a related term, heat capacity, which varies according to how much of the substance is under study. **Heat capacity** is the capacity of a given sample to absorb heat and therefore depends

TABLE 1.6 Specific Heats of Some Common Substances

Substance	Specific Heat [J/(g × °C)]
METALLIC SOLIDS	
aluminum	0.89
copper	0.39
iron	0.46
platinum	0.13
NONMETALLIC SOLIDS	
coal	1.26
concrete	0.67
glass	0.68
rubber	2.01
LIQUIDS	
water	4.184
ethanol	2.51
gasoline	2.12
olive oil	1.97

1.5 Chemistry Around Us

Specific Heat and the "Fitness" of Water

In describing the "fitness" of water to support life, Lawrence J. Henderson also brought attention to the fact that water has the highest specific heat of any common liquid, a property that is central in the life of every cell. It enables all living organisms to keep their internal temperatures within certain bounds; below a certain temperature, chemical reactions take place too slowly to support life, and, above a certain temperature, the structures of proteins are destroyed. Species that derive their body temperatures from internal metabolic activity, such as mammals and birds, are called endotherms. They have adapted to far more stressful environmental conditions than ectotherms, species that absorb heat from their environment, such as reptiles. A constant, elevated temperature leads to a higher and therefore a more efficient rate of metabolism: endotherms are able to extract more energy per unit time from the nutrients in their diets than ectotherms can. An endotherm's internal temperature is produced by chemical reactions that generate heat. It is maintained at a consistent level—neither too high nor too low—through regulated avenues of heat loss.

Because of water's high specific heat, temperature increases due to chemical reactions that take place within water are kept to a minimum compared with the same reactions taking place in liquids with low specific heats. Let us compare the temperature rise of living tissue (which contains water) with the rise that might occur in another liquid, one having a typical specific heat of about 2.0 J/g × °C, when we add the same number of joules of heat (say, 440 kJ, the heat generated by a 75-kg male in 1 h) to each. To simplify the exercise, we will assume that the tissue has the same specific heat as water, because water is tissue's chief component. We will calculate the temperature rises per hour by solving for Δt in the equation for specific heat:

$$\text{Heat (joules)} = C_p \times \text{mass of sample (grams)} \times \Delta t$$

$$\Delta t = \frac{\text{heat (J)}}{C_p \times \text{mass (g)}}$$

In living tissue,

$$\Delta t_{\text{water}} = \left\{ \frac{440{,}000\ \text{J}}{4.184\ \frac{\text{J}}{\text{g} \times {}^\circ\text{C}} \times 75\ \text{kg}} \right\} = 1.4^\circ\text{C/h}$$
$$= 2.5^\circ\text{F/h}$$

In the other liquid,

$$\Delta t_{\text{“other”}} = \left\{ \frac{440{,}000\ \text{J}}{2.0\ \frac{\text{J}}{\text{g} \times {}^\circ\text{C}} \times 75\ \text{kg}} \right\} = 2.9^\circ\text{C/h}$$
$$= 5.2^\circ\text{F/h}$$

Body temperature-regulating mechanisms are able to handle a load of 1 to 3 Fahrenheit degrees per hour but would be overwhelmed at 5 to 6 Fahrenheit degrees per hour.

The problem of temperature control in living organisms, then, is minimized by the fact that water, with its peculiarly high specific heat, is the chief component of living cells. In Chapter 6, we will continue this consideration of temperature control in organisms when we examine another of water's biologically significant physical properties, the energy required for causing its evaporation.

not only on the substance's innate characteristics, but also on its mass. The greater the mass, the greater the amount of heat that the sample can absorb. The heat capacity of 100 g of water is 418.4 J/°C, and that of 200 g of water is 836.8 J/°C. Conversely, if we add the same amount of heat to two samples of the same substance, one of which has twice the mass of the other, the smaller mass will reach a higher temperature.

Because the amounts of heat encountered in chemical processes are in the range of thousands of joules or calories, the values are reported in kilojoules (kJ) or kilocalories (kcal). In discussions of nutrition, however, caloric values are instead measured and reported in Calories (capital C). A **Calorie** is equal to a kilocalorie, or 1000 "small" calories.

Example 1.18 Calculating specific heat

What is the specific heat of a substance if 6.00 joules of heat added to 16.0 grams of that substance cause its temperature to rise from 20°C to 38°C?

Solution

Insert the given values into the formula for specific heat given previously:

$$C_p = \frac{6.00 \text{ J}}{16.0 \text{ g} \times (38°\text{C} - 20°\text{C})} = 0.021 \text{ J}/(\text{g} \times °\text{C})$$

Problem 1.18 What is the specific heat of a substance if 334 J of heat added to 52 g of that substance cause its temperature to rise from 16°C to 48°C?

Example 1.19 Using specific heat in calculations

How much heat must be added to 45.0 g of a substance that has a specific heat of 0.151 J/(g × °C) to cause its temperature to rise from 21.0°C to 47.0°C?

Solution

The relation between heat and temperature rise is obtained by solving the equation for specific heat:

$$\text{Heat} = C_p \times \text{mass of sample (g)} \times \Delta t$$

Substituting the appropriate values:

$$\text{Heat} = \left(0.151 \frac{\text{J}}{\text{g} \times °\text{C}}\right)\left(45.0 \text{ g}\right)\left(47.0°\text{C} - 21.0°\text{C}\right) = 177 \text{ J}$$

Problem 1.19 How much heat must be added to 37 g of a substance that has a specific heat of 0.12 J/g × °C to cause its temperature to rise from 11°C to 22°C?

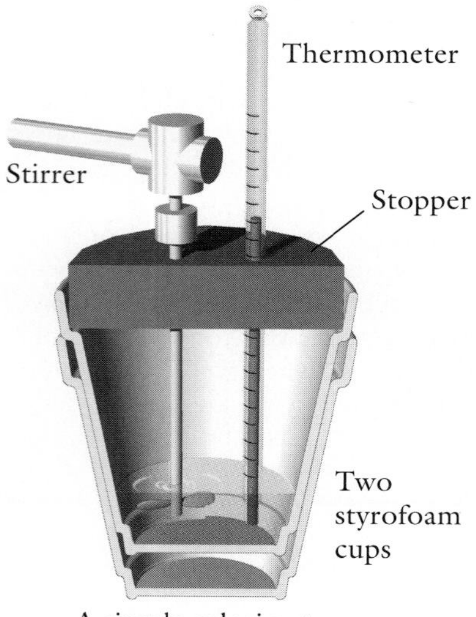

A simple calorimeter

Heat is measured by the rise in temperature of a given mass of water. The equation for calculating the specific heat of a substance was rearranged in Example 1.19 into the form

$$\text{Heat} = C_p \times \text{mass of sample (g)} \times \Delta t$$

When the heat produced by some physical or chemical process is absorbed into a given mass of water, the water's temperature rise will allow us to calculate the heat produced by the process. The same approach can be used if the process absorbs heat rather than producing it. In that case, the surrounding water's temperature will decrease. The laboratory device in which such an experiment is performed is called a **calorimeter.**

A calorimeter can be as simple as a styrofoam cup outfitted with a thermometer. The styrofoam is an excellent insulator and prevents loss of heat. If 50 mL of hydrochloric acid is added to 50 mL of potassium hydroxide in water in the calorimeter, the temperature of the water will rise and the heat of the reaction can be determined.

▶▶ Metabolism is the subject of Chapters 22 through 26.

A much more complex calorimeter is necessary to measure the heat produced as a result of human basal metabolism. The **basal metabolic rate (BMR)** is the minimal metabolic activity of a human at rest and with an empty gastrointestinal tract. When considering quantities of heat in the context of nutrition and metabolism, those quantities are commonly given in calories (1 cal = 4.184 J). Therefore in discussions about heat in the context of nutrition and metabolism, we will use specific heat in units of calories. The calorimeter for this measurement consists of a small room in which a patient reclines. The room is surrounded by a hollow jacket through which water flows at a fixed rate. The water temperature is measured at the inflow and at the outflow, which determines Δt—say, 2.0 Celsius degrees. The specific heat of water is known, and it is necessary to measure only the mass of water. If the water flow rate is 0.6 L per minute and the time for the experiment is 125 min, the volume of water whose temperature was raised by 2.0 Celsius degrees is 75 L. If we assume that the density of water is 1.0, then the mass of water is 75 kg. The heat produced in 125 min is:

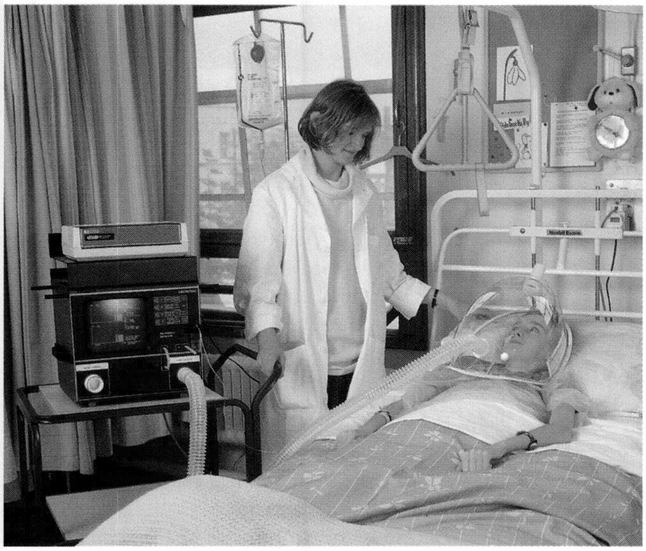

This patient is attached to a device that measures oxygen, carbon dioxide, and energy consumption in order to determine the basal metabolic rate.

$$\text{Heat} = 1.0\ \text{cal/g}\cdot{}^\circ\text{C} \times 75{,}000\ \text{g} \times 2.0^\circ = 150\ \text{kcal}$$

$$\text{Basal metabolic rate} = 150\ \text{kcal}/125\ \text{min} = 1.2\ \text{kcal}/\text{min}$$

Because this kind of calorimeter is quite elaborate and can be found at only a few research centers, basal metabolic rates are usually determined indirectly by measuring the oxygen uptake of a patient at rest. As the body consumes nutrients such as carbohydrates, fats, and proteins, it uses the same amount of oxygen as would be consumed by burning these nutrients in a test tube. An adult human male produces about 4.8 kcal of heat per liter of oxygen consumed. Therefore, measuring oxygen uptake can be equated with basal metabolic rate. An adult human male consumes about 250 mL of oxygen per minute at rest. At 4.8 kcal/L of oxygen, his basal metabolic rate is

$$\left(0.25\frac{\text{L O}_2}{\text{min}}\right)\left(4.8\ \frac{\text{kcal}}{\text{L O}_2}\right) = 1.2\ \frac{\text{kcal}}{\text{min}}$$

which corresponds quite well with the BMR obtained with the direct calorimetric method.

Example 1.20 Estimating minimal human caloric requirements

An equation for the estimation of minimal daily caloric needs (24-hour basal metabolic rate) for human males is:

$$\text{BMR} = 66 + [6.22 \times \text{weight (lb)}] + [12.7 \times \text{height (in.)}] - [6.8 \times \text{age (yr)}]$$

What is the daily minimal Calorie (kcal) requirement of a 25-year-old man, 70 inches tall and weighing 160 pounds?

Solution

$$\text{BMR} = 66 + (6.22 \times 160) + (12.7 \times 70) - (6.8 \times 25)$$
$$= 1.8 \times 10^3\ \text{kcal/day}$$

Problem 1.20 An equation for the estimation of minimal daily Caloric needs (24-hour basal metabolic rate) for human females is:

$$\text{BMR} = 655 + [4.36 \times \text{weight (lb)}] + [4.32 \times \text{height (in.)}] - [4.7 \times \text{age (yr)}]$$

What is the daily minimal Calorie (kcal) requirement of a 22-year-old woman, 66 inches tall and weighing 132 pounds?

Summary

The Composition of Matter Chemists classify substances on the basis of their composition. They first establish the purity of a substance by verifying that one and only one substance is present. If a pure substance can be decomposed into simpler substances, it is called a compound. If it cannot be decomposed further into simpler substances, it is called an element.

Mixtures are composed of two or more pure substances, whose proportions may vary. If the mixture appears uniform to the eye, it is called homogeneous; if nonuniform, heterogeneous.

Measurement and the Metric System The modern metric system is called the SI system. Its base units are the kilogram, meter, and second. The more commonly encountered older metric or non-SI system units are the gram, centimeter, and second. The measurement of very large and very small quantities is simplified by the use of prefixes denoting multiples of ten.

Measurement, Uncertainty and Significant Figures The last digit in any measurement must be considered an estimate and is therefore an unavoidable uncertainty or variability. The degree of uncertainty is expressed by the use of the concept of significant figures.

Scientific Notation Very large and very small numbers are encountered in chemistry. The difficulty of calculating with such numbers is reduced by the use of scientific notation. In scientific notation, a number is expressed as a coefficient (a number between 1 and 10) multiplied by an exponential factor (10 raised to a whole-number power).

Calculations Using Scientific Notation Addition and subtraction are done by converting all exponents into the same value. Exponents are added in multiplication and subtracted in division.

Calculations and Significant Figures The number of significant figures in a measured quantity controls the number of significant figures in any calculation done with that measured quantity. The number of significant figures in the calculated quantity cannot be greater than the smallest number of significant figures in any of the measured quantities.

Use of Units in Calculations Problems in chemistry are solved by the unit-conversion method. The first step is to identify the kind of units required in the answer and the kind of units given in the problem. Then relate the two kinds of units in an equation that defines the conversion factor. Multiplication of the given units by the conversion factor must yield the units required by the answer.

Two Fundamental Properties of Matter The mass of an object is the quantity of matter in that object. When an object is weighed on a chemical balance, its mass is being compared with a standard mass.

Volume is the amount of space that the object occupies. It is the product of three linear dimensions and is therefore a derived quantity.

Density Density is defined as mass divided by volume and so is determined by measuring the mass of a given volume of a substance. The density of a liquid or a solution can also be measured by its effect on the buoyancy of a float called a hydrometer.

Temperature The concept of temperature allows us to quantitatively specify how hot or cold an object is. There are three thermometric scales: the Celsius, Kelvin, and Fahrenheit scales.

Heat and Calorimetry Heat is measured by the change in temperature undergone by a given mass in response to the addition or loss of heat. The instrument by which heat is measured is a calorimeter, in which the heat produced by some physical or chemical process is captured by allowing it to be absorbed into a given mass of water. Human basal metabolic rates are determined by calorimetry or indirectly by oxygen consumption.

Key Words

basal metabolic rate (BMR), p. 32
calorie, p. 30
Celsius scale, p. 28
chemical property, p. 7
compound, p. 8
conversion factor, p. 19
density, p. 22
element, p. 8
error, p. 12
Fahrenheit scale, p. 28
heat, p. 29
Kelvin scale, p. 28
mass, p. 21
metric system, p. 10
mixture, p. 7
physical property, p. 5
scientific notation, p. 13
significant figure, p. 12
temperature, p. 25
unit conversion, p. 19
volume, p. 22

Exercises

The Composition of Matter

1.1 Classify the following processes as either chemical or physical properties of oxygen: (a) oxygen reacts with carbon to form carbon dioxide; (b) oxygen boils at $-183°C$; (c) oxygen dissolves in water; (d) oxygen reacts with hydrogen to form water.

1.2 Classify the following processes as either chemical or physical properties of magnesium: (a) magnesium dissolves in hydrochloric acid, with the production of bubbles of hydrogen gas; (b) magnesium melts at 649°C; (c) magnesium is shiny and conducts electricity; (d) when heated in oxygen, magnesium changes from a shiny metal to a white solid that no longer conducts electricity.

1.3 Classify the following as pure substances or mixtures: (a) air; (b) mercury; (c) aluminum foil; (d) table salt.

1.4 Classify the following as pure substances or mixtures: (a) seawater; (b) table sugar; (c) smoke; (d) blood.

Measurement and the Metric System

1.5 What are the fundamental quantities and their units in the SI system that are encountered in chemistry?

1.6 Give examples of units that can be derived from the fundamental units of the SI system that we use in this book.

1.7 Write the following quantities in more convenient form by changing the units: (a) 630 m (meters); (b) 1440 ms (milliseconds); (c) 0.000065 kg (kilograms); (d) 1300 μg (micrograms).

1.8 Write the following quantities in more convenient form by changing the units: (a) 0.813 kilometers; (b) 0.367 seconds; (c) 7140 mg (milligrams); (d) 0.096 mg (milligrams).

Significant Figures

1.9 How many significant figures are in the following numbers? (a) 0.0945; (b) 83.22; (c) 106; (d) 0.000130

1.10 Deduce the number of significant figures contained in the following quantities: (a) 16.0 cm; (b) 0.0063 m; (c) 100 km; (d) 2.9374 g; (e) 1.07 lb/in.^{2}.

1.11 How many significant figures are in the following measurements? (a) 25.9000 g; (b) 102 cm; (c) 0.002 m; (d) 2001 kg; (e) 0.0605 s

1.12 How many significant figures are in the following measurements? (a) 21.2 m; (b) 0.023 kg; (c) 46.94 cm; (d) 453.59 g; (e) 1.6030 km

Scientific Notation

1.13 Express the following numbers in scientific notation: (a) 0.00839; (b) 83,264; (c) 372; (d) 0.0000208.

1.14 Convert the following numbers into nonexponential form: (a) 2.3×10^{-3}; (b) 2.9×10^{2}; (c) 3.92×10^{-4}; (d) 1.73×10^{4}.

1.15 Write the following numbers in scientific notation: (a) 936,800; (b) 1638; (c) 0.0000568; (d) 0.00917.

1.16 Write the following quantities in scientific notation, along with correct metric abbreviations: (a) 275,000 kilograms; (b) 87,000 years; (c) 0.097 second.

Calculations Using Significant Figures

1.17 With careful concern for the correct number of significant figures in the answers, calculate (a) 3.21 cm $\times$ 15.091 cm; (b) $3.82 \times 1.1 \times 2.003$.

1.18 Report the result of dividing (a) 13.87 by 1.23 and (b) 0.095 by 1.427, with strict attention to the correct number of significant figures in the answers.

1.19 What is the result of adding (a) 12.786 to 1.23 and (b) 3.961 to 24.6543?

1.20 What is the result of subtracting (a) 2.763 from 3.91 and (b) 54.832 from 89.2?

Unit-Conversion Method

1.21 How many (a) kilometers, (b) meters, (c) centimeters, and (d) millimeters are there in 0.800 miles calculated to the correct number of significant digits? The conversion factor is 1 mile = 1.6093 km.

1.22 How long would it take to drive 375 miles at 55 miles per hour? Calculate your answer in hours.

1.23 How many milliliters are in 3.75 gallons?

1.24 How many milligrams are in 2.95 ounces?

1.25 How many miles are run in a 10-K (10 kilometer) foot race? (Use 10-K as an exact number.)

1.26 World class times in the 100-meter foot race are about 9.10 s. How fast is that in miles per hour? (Use 100 meters as an exact number.)

Mass, Volume, Density, Temperature, and Heat

1.27 How many grams are in 1 oz?

1.28 How many milligrams are in 1 oz?

1.29 How many (a) kilograms, (b) grams, and (c) milligrams are there in 0.60 lb calculated to the correct number of significant digits? Use the conversion factor 453.59 g = 1 lb.

1.30 How many (a) kilometers, (b) meters, and (c) centimeters are there in 0.820 mi. (miles) calculated to the correct number of significant digits? Use the conversion factor 1 mile = 1.609 km.

1.31 Calculate the length in centimeters of the side of a cube whose volume is 1.0 L.

1.32 Calculate the SI equivalent of 1.0 L.

1.33 Calculate the mass of a piece of glass of square cross section, of density 2.2 g/cm^3, and dimensions 3.25 cm $\times$ 6.50 cm $\times$ 17.00 mm.

1.34 What is the volume of 250.0 grams of acetone whose density is 0.7899 g/cm^3?

1.35 Convert 68°F into °C.

1.36 Convert 45°F into °C.

1.37 Express 27°C in kelvins.

1.38 Express 125°C in °F.

1.39 Calculate the specific heat of 32 g of a substance that required 22.0 J to raise its temperature from 18°C to 28°C.

1.40 How many joules are required to raise the temperature of 0.750 kg of a substance whose specific heat is 0.220 J/(g $\times$ °C) from 13.0°C to 68.0°C?

1.41 A glass vessel calibrated to contain 19.84 mL of water at 4°C was found to weigh 31.962 g when empty and dry. Filled with a sodium chloride solution at the same temperature, it was found to weigh 54.381 g. Calculate the solution's density.

1.42 A sample of urine weighed 25.853 g at 4°C. An equal volume of water at the same temperature weighed 23.718 g.

At 4°C, the density of water is exactly 1.000 g/cm^3. Calculate the density of the urine sample.

1.43 How many pounds of carbohydrate yielding 16.74 kJ/g must be consumed to obtain the 8.4×10^3 kJ of energy required by an average woman in a 24-h period?

1.44 The specific heat of the glass used in thermometer manufacture is 0.84 J/(g × °C). Calculate the heat in joules required to raise the temperature of a 100-gram thermometer 2.0 Celsius degrees.

1.45 A glass vessel that can contain 12.3 mL of water at 4°C was found to weigh 28.463 g when empty and dry. Filled with a sodium chloride solution at the same temperature, it was found to weigh 41.242 g. Calculate the solution's density.

1.46 A volume of glucose solution weighed 22.842 g at 4°C. An equal volume of water at the same temperature weighed 22.394 g. At 4°C, the density of water is exactly 1.000 g/cm^3. Calculate the density of the glucose solution.

1.47 The density of gold is 19.32 g/cm^3. Calculate the volume of a sample of gold that weighs 2.416 kg.

1.48 A urine sample has a density of 1.09 g/cm^3. What volume will 15.0 g of the sample occupy?

Unclassified Exercises

1.49 Classify the following properties as either chemical or physical: (a) an electric current sent through water results in the production of bubbles of hydrogen gas; (b) lead melts at 327°C; (c) lead is a metal, has a silvery luster, and conducts electricity; (d) when lead is heated in oxygen, its appearance changes from a silvery luster to a white solid that no longer conducts electricity.

1.50 Write the following quantities in simpler metric forms: (a) 0.0045 L; (b) 2.87×10^{-8} s; (c) 0.0057 km; (d) 0.0000036 kg.

1.51 Write the following quantities in scientific notation, along with correct metric abbreviations: (a) 9,620,000 kilograms; (b) 54,870 days; (c) 253 milliseconds; (d) 0.000274 kilometers.

1.52 Count the significant figures in the following numbers: (a) 0.00256; (b) 128.009; (c) 2.00730; (d) 201; (e) 0.09864.

1.53 How many ounces are in 45.8 mg?

1.54 What are the results (a) of multiplying 3.2×10^3 by 3.1×10^{-5}; (b) of dividing 9.47×10^{-3} by 2.32×10^{-2}; (c) of adding 5.6×10^{-3} and 2.3×10^{-2}; and (d) of subtracting 1.4×10^{-4} from 3.6×10^{-3}?

1.55 Perform the following calculations. Answers must have the correct number of significant figures.

(a) $(8.20 \times 10^2) + (3.75 \times 10^4)$

(b) $(5.21 \times 10^{-2}) + (2.74 \times 10^{-3})$

(c) $(1.01 \times 10^{-4}) + (7.23 \times 10^{-3})$

1.56 Perform the following calculations. Answers must have the correct number of significant figures.

(a) (2.7 cm + 1.08 cm) × 22.47 cm

(b) (2.54 cm − 0.541 cm) ÷ 2.2 cm

(c) (21.63 g + 4.284 g) × 0.0372 g

(d) (183.7 mL − 38.57 mL) ÷ 21.3 mL

1.57 What is an alternative SI equivalent for 1.0 L?

1.58 Calculate the specific heat of 12.7 g of an unknown metal whose temperature increased 25.0 Celsius degrees when 80.0 J of heat energy was absorbed by the metal.

Chemical Connections

1.59 A 175-lb man was placed in a tank filled with 1000 gallons of water. The temperature of the water increased by 2.49 Celsius degrees over a 1.5-h period. Calculate the heat lost by the man during this experiment. Using that data, calculate the patient's basal metabolic rate; that is, the energy required over a 24-h period. (Hint: The heat lost is equal to the heat gained by the water.)

1.60 A prescription for nifedipine calls for 0.2 mg/kg of body weight, four times per day. The drug is packaged in capsules of 5 mg. How many capsules per dose should be given to a patient who weighs 75 kg?

1.61 You need 5.00 g of sodium chloride to add to water to prepare a mixture containing 5.00 g of NaCl per 100 mL of mixture. There is a solution of NaCl in water in the stockroom, but the label on the container specifies only the substances in the mixture and not its quantitative composition. How could you obtain the required amount of NaCl to prepare your mixture?

1.62 Two students were required to determine the area of a sheet of paper in units of square centimeters. One student had a centimeter ruler marked in centimeters, and the other an inch ruler marked in half inches. The results of their measurements were:

	Length	Width	Area
Student A	28 cm	22 cm	6.2×10^2 cm^2
Student B	11 in.	8.5 in.	6.0×10^2 cm^2

Which student had the correct answer, and what mistake did the other student make?

1.63 On the basis of some previous experimental results, you design a new experiment and predict its outcome. The results of the new experiment do not agree with the results that you predicted. What is your next move?

1.64 What is the difference between a law and a theory?

CHAPTER 2

ATOMIC STRUCTURE

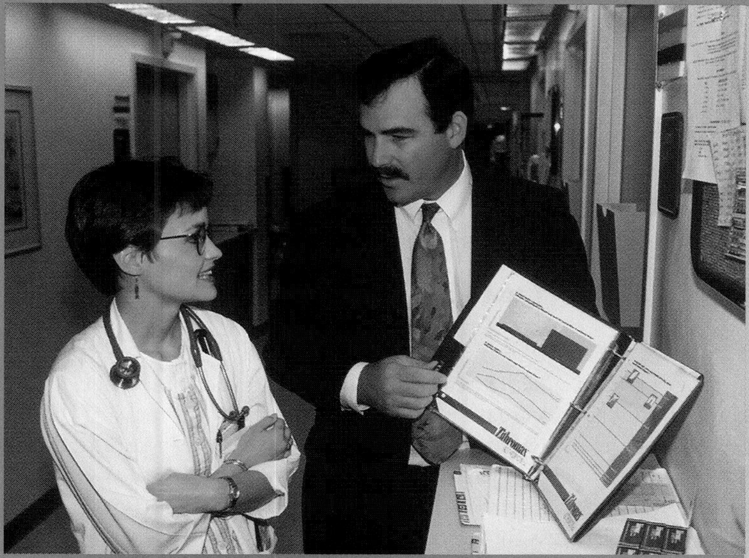

CHEMISTRY IN YOUR FUTURE

When you were training as a sales representative for a pharmaceutical company, it surprised you to learn that there are many important medications whose mode of action is not at all well understood. A case in point is the drug lithium carbonate, which has long been used to control the psychiatric dysfunction known as bipolar disorder, or manic depression. Some scientists think its efficacy stems from lithium's similarity to the elements sodium and potassium, which are known to play major roles in the transmission of nerve impulses. Chapter 2 presents the fundamental concepts that explain the chemical relatedness of these elements.

LEARNING OBJECTIVES

- Use Dalton's atomic theory to explain the constant composition of matter and the conservation of mass in chemical reactions.
- Define atomic mass.
- Describe the structure of the atom in terms of its principal subatomic particles.
- Identify main-group and transition elements, metals, nonmetals, and metalloids.
- Correlate the arrangement of the periodic table with the electron configurations of the valence shells of the elements.
- Describe the octet rule.

For thousands of years, people have speculated about the fundamental properties of matter. The basic question to be answered was, Why are there so many different substances in the world? Many explanations were proposed but the most enduring one—the one that has proved to be correct—was that all matter is composed of particles, or atoms. The atoms were assumed to combine in various ways to produce all the existing and future kinds of matter. This idea, first proposed by Greek philosophers in the fourth century B.C., remained more a philosophical argument than a scientific theory until the early nineteenth century. At that time, John Dalton, an English schoolmaster, proposed an atomic model that convincingly explained experimental observations on which chemists of the day had come to rely.

In this chapter, we examine the nature of atomic structure and correlate it with the chemical reactivity of the elements. By the chapter's end, we will know the answer to such questions as, Why does one atom of calcium react with one atom of oxygen but with two atoms of chlorine?

2.1 CHEMICAL BACKGROUND FOR THE EARLY ATOMIC THEORY

By the early 1800s, two experimental facts became firmly established:

1. Mass is neither created nor destroyed in a chemical reaction. When substances react chemically to create new substances, the total mass of the reacting substances is the same as the total mass of the resulting products. This is called the **Law of Conservation of Mass.**
2. The elements present in a compound are present in fixed and exact proportion by mass, regardless of the compound's source or method of preparation. This is called the **Law of Constant Composition.**

▶▶ Vitamins receive special consideration in Chapters 19 and 26.

Consider vitamin C, or ascorbic acid, which can be obtained from citrus fruits or synthesized in the chemical laboratory. No matter where vitamin C comes from, the relative amounts of carbon (symbol C), hydrogen (symbol H), and oxygen (symbol O) of which it is composed are the same. They are specified as the mass percent of each element, which taken together is called the **percent composition** of the compound.

The mass percentages of C, H, and O in vitamin C are:

$$\text{Mass percent of carbon in vitamin C} = \frac{\text{mass of carbon}}{\text{mass of vitamin C}} \times 100\%$$

$$\text{Mass percent of hydrogen in vitamin C} = \frac{\text{mass of hydrogen}}{\text{mass of vitamin C}} \times 100\%$$

$$\text{Mass percent of oxygen in vitamin C} = \frac{\text{mass of oxygen}}{\text{mass of vitamin C}} \times 100\%$$

Multiplication of the mass ratios by 100% converts them into percentages.

Example 2.1 Calculating the mass percent of a compound

Analysis of 4.200 g of vitamin C (ascorbic acid) yields 1.720 g of carbon, 0.190 g of hydrogen, and 2.290 g of oxygen. Calculate the mass percentage of each element in the compound.

Solution

$$\text{Mass percent of carbon in vitamin C} = \frac{1.720\text{ g}}{4.200\text{ g}} \times 100\% = 40.95\%$$

$$\text{Mass percent of hydrogen in vitamin C} = \frac{0.190\ \text{g}}{4.200\ \text{g}} \times 100\% = 4.52\%$$

$$\text{Mass percent of oxygen in vitamin C} = \frac{2.290\ \text{g}}{4.200\ \text{g}} \times 100\% = 54.52\%$$

Problem 2.1 Analysis of 4.800 g of niacin (nicotinic acid), one of the B-complex vitamins, yields 2.810 g of carbon, 0.1954 g of hydrogen, 0.5462 g of nitrogen (symbol N), and 1.249 g of oxygen. Calculate the mass percentage of each element in the compound.

2.2 DALTON'S ATOMIC THEORY

In the years from 1803 to 1808, John Dalton showed how the atomic theory could be used to explain the quantitative aspects of chemistry considered in Section 2.1. Since that time, our knowledge of the nature of the atom has changed significantly; Dalton was not correct in every particular. Nevertheless, he succeeded in creating, on the basis of a relatively small number of specific observations, the first workable general theory of the structure of matter.

Dalton's model consisted of the following proposals:

- All matter is composed of infinitesimally small particles called **atoms** (believed in Dalton's time to be indestructible but now known to be composed of even smaller parts).
- The atoms of any one element are identical.
- Atoms of one element are distinguished from those of a different element by the fact that the atoms of the two elements have different masses.
- Compounds are combinations of atoms of different elements and possess properties different from those of their component elements.
- In chemical reactions, atoms are exchanged between starting compounds to form new compounds. Atoms can neither be created nor destroyed.

Later in this chapter, we will examine some of the ways in which these proposals have undergone modification, but the key ideas have survived to this day.

Dalton's idea of chemical reactions is illustrated in Figure 2.1. Atoms (called **elementary particles** by Dalton) that are combined in compounds (called **compound particles** by Dalton) rearrange to form new compounds. (You can see where the modern names element and compound, Section 1.1, come from.) This model shows that, in a chemical process, mass is conserved. The new compounds possess the mass of their constituent atoms. No mass has been gained or lost. The atoms have simply become rearranged.

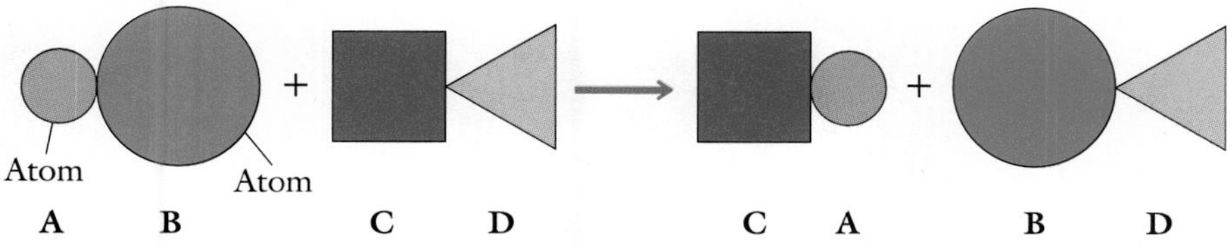

Figure 2.1 A chemical reaction as John Dalton might have described it. Atoms of elements are represented by different shapes and colors and labeled A, B, C, and D. The elements have combined to form two compounds, AB and CD, which react to produce two new compounds CA and BD. Note that no mass has been gained or lost and that the composition of all compounds is constant.

Dalton's model explained why the elements in a compound are present in fixed and exact proportion by mass. If the atoms of each element have a definite and unique mass and if compounds consist of a fixed ratio of given atoms—say, one atom of sulfur to two atoms of hydrogen—then the ratio of their masses also must be constant.

2.3 ATOMIC MASSES

In Dalton's definition, the principal difference between different elements is the different masses of their atoms. Because he could not weigh individual atoms to measure their mass, he proposed a relative mass scale instead. He identified the lightest element then known—hydrogen—and proposed that it be assigned a relative mass of 1.0. He then assigned masses to other elements by comparing them with hydrogen.

Dalton also examined compounds in which hydrogen combined with other elements in what Dalton believed were 1:1 ratios of the different atoms. By determining the percent compositions of these compounds, he could compare the masses of other elements with the mass of hydrogen in each compound and thus establish their relative atomic masses.

For example, the mass percentages of hydrogen and chlorine in the compound hydrogen chloride are 2.76% and 97.24%, respectively. If it is true that there is a 1:1 ratio of hydrogen and chlorine atoms in this compound, their relative masses in 100 grams of hydrogen chloride are

$$\frac{\text{Mass of a chlorine atom}}{\text{Mass of a hydrogen atom}} = \frac{97.24}{2.76} = \frac{35.2}{1}$$

or mass of a chlorine atom = 35.2 × mass of a hydrogen atom. Today the relative masses of atoms are determined by a technique called mass spectroscopy. A mass spectroscope can be seen in Figure 2.2.

The relative masses of atoms are called the **atomic masses** of the elements, although chemists often call them **atomic weights,** because mass is determined by weighing. In this book, we will use the term atomic mass. Atomic masses were first assigned by assuming hydrogen's mass to be 1. Atomic masses are now defined relative to the mass of the most common isotope of the element carbon, whose mass is specified as exactly 12 **atomic mass units,**

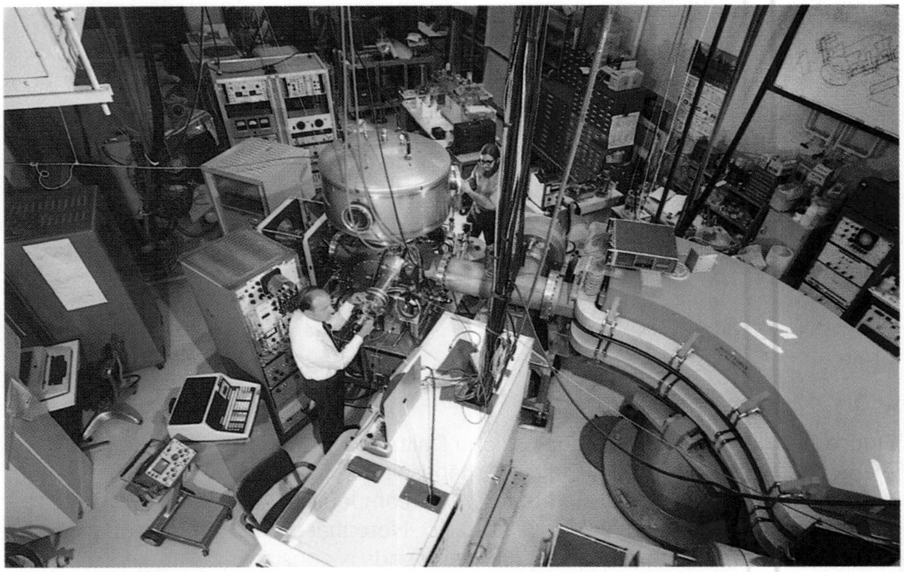

Figure 2.2 A mass spectroscope in action.

which now makes hydrogen's mass 1.008. (Isotopes are considered in Section 2.5). The precise atomic masses of the elements are given in the table inside the back cover. The symbol for atomic mass unit before 1962 was **amu.** Since that time, a new symbol, **u,** has been slowly coming into use. We will continue to use the amu abbreviation because it is a more explicit reminder of its meaning.

It is important to note here that atoms are much too small to be handled individually, much less be seen. Chemists had to learn about them indirectly. In the laboratory, mass is measured, not numbers of atoms. However, if the relative masses of atoms are known, it is possible to count the atoms. For example, suppose there are 19 g of fluorine and 1.0 g of hydrogen in a compound containing only hydrogen and fluorine. If the weighed masses of the two elements produce the same ratio as those elements' relative masses, there must be equal numbers of atoms of the two elements in the compound.

The relation between numbers of objects and their masses is easier to understand when familiar objects are used to illustrate it, as in Example 2.2.

Example 2.2 Relating mass to numbers of objects

The head chef at a large banquet requires 1000 oranges and 1000 cantaloupes for a dessert. Your job depends on being able to supply the correct numbers of each in the next 15 min, too short a time in which to count them. Luckily, you do have a balance large enough to accommodate such large numbers of items. The average mass per orange is 150 g and that per cantaloupe is 450 g. Calculate the total mass of each kind of fruit equivalent to the numbers required by the chef.

Solution

What you must do is weigh an amount of each fruit equivalent to the required numbers of items.

$$1000 \text{ oranges}\left(\frac{150 \text{ g orange}}{1 \text{ orange}}\right) = 150{,}000 \text{ g oranges} = 150 \text{ kg oranges}$$

$$1000 \text{ cantaloupes}\left(\frac{450 \text{ g cantaloupe}}{1 \text{ cantaloupe}}\right) = 450{,}000 \text{ g cantaloupes}$$

$$= 450 \text{ kg cantaloupes}$$

Slight differences in mass from orange to orange and cantaloupe to cantaloupe aside, weighing out 150 kg of oranges and 450 kg of cantaloupes should give you the required numbers.

Problem 2.2 Suppose, in one crate, there were 4500 g of oranges weighing 150 g each and, in another crate, 4500 g of cantaloupes weighing 450 g each. Calculate the numbers of oranges and cantaloupes in each crate.

The relation between numbers of atoms and their masses will be described in detail in Chapter 4.

2.4 THE STRUCTURE OF ATOMS

Since Dalton's time, scientists have learned that the atom is neither featureless nor indestructible. It is composed of subatomic particles, some that have electrical charges and some that do not. We will be concerned with three subatomic particles: protons, neutrons, and electrons.

Of these three types of particles, two are charged: the positively charged **proton** and the much lighter, negatively charged **electron.** The third is an

TABLE 2.1 Properties of Subatomic Particles

Subatomic Particle	Electrical Charge	Mass (amu)	Location
proton	1+	1.00728	nucleus
neutron	0	1.00894	nucleus
electron	1−	0.0005414	outside nucleus

electrically uncharged (neutral) particle, the **neutron,** that has about the same mass as the proton. These particles and their properties are listed in Table 2.1.

The protons and neutrons make up most of an atom's mass and are located together in a structure that lies at the center of the atom and is called the **nucleus.** The nucleus occupies only a small part, about $1 \times 10^{-13}\%$ of the atom's volume; therefore, the electrons occupy most of the volume of the atom.

Electrically charged particles repel one another if their charges are the same and attract one another if their charges are opposite. Despite the fact that electrons and protons have very different masses, the magnitudes of their respective charges are the same. That the atom is electrically neutral indicates that the oppositely charged particles are present in equal numbers: the number of protons in an atom is balanced by exactly the same number of electrons. The number of protons in the nucleus is called the **atomic number** and is unique for each element. Atomic numbers are listed on the inside back cover, along with the atomic masses.

Example 2.3 Calculating the charge of an atom

What are the sign and magnitude of the charge of an atom containing 12 protons, 11 neutrons, and 12 electrons?

Solution

Each proton has a single positive charge, neutrons have no charge, and each electron has a single negative charge; so, in this atom, the overall charge is the sum of all charges.

$$12 \text{ protons} \times [+1 \text{ (charge per proton)}] = +12$$
$$11 \text{ neutrons} \times [0 \text{ (charge per neutron)}] = 0$$
$$12 \text{ electrons} \times [-1 \text{ (charge per electron)}] = -12$$
$$\text{Charge of atom} = +12 + 0 + (-12) = 0$$

Problem 2.3 What are the sign and magnitude of the charge of an atom containing 9 protons, 10 neutrons, and 9 electrons?

Atoms of elements may acquire a charge by gaining or losing electrons in reactions with other compounds or elements. An atom bearing a net electrical charge is called an **ion.** A positively charged ion is called a **cation,** and a negatively charged ion is called an **anion.**

Example 2.4 Calculating the charge of an ion

What are the sign and magnitude of the charge of an atom containing 14 protons, 15 neutrons, and 12 electrons?

Solution

$$14 \times (1+) + 15 \times (0) + 12 \times (1-) = 2+$$

Problem 2.4 What are the sign and magnitude of the charge of an atom containing 13 protons, 14 neutrons, and 10 electrons?

Chemists assume that the electrons' contribution to the mass of an atom is negligible, because the mass of an electron is only about 1/2000 the mass of either a proton or a neutron (see Table 2.1). This means that, for all practical purposes, all the mass of the atom is located in the nucleus. The following example demonstrates why chemists feel justified in considering the mass of an atom to be the sum of the masses of its protons and neutrons.

Example 2.5 Adding the masses of subatomic particles

Calculate and compare the masses of an atom of 6 protons, 6 neutrons, and 6 electrons with the mass of the same atom minus the mass contribution of the 6 electrons.

Solution

The mass of each kind of particle can be obtained from:

$$\begin{aligned} \text{Mass of protons} &= 6 \times 1.007275 \text{ amu} = 6.04365 \text{ amu} \\ \text{Mass of neutrons} &= 6 \times 1.008665 \text{ amu} = 6.05199 \text{ amu} \\ \text{Mass of electrons} &= 6 \times 0.0005486 \text{ amu} = 0.003292 \text{ amu} \\ \text{Total mass} &= 12.09893 \text{ amu} \end{aligned}$$

Then:

$$\begin{aligned} \text{Total mass} - \text{electron mass} &= 12.09564 \text{ amu} \\ \text{Percent difference} &= (0.003292/12.09893) \times 100\% = 0.02721\% \end{aligned}$$

Problem 2.5 What are the name and approximate mass of an element that has 22 protons and 26 neutrons?

2.5 ISOTOPES

Although all the atoms of a particular element have the same number of protons—hence the same atomic number—the number of neutrons in the atoms of that element can vary. The result is a family of atoms of the same element that have the same chemical properties but have slightly different masses. They are known collectively as the **isotopes** of the element. Isotopes are therefore atoms of the same element that all have the same atomic number but have different mass numbers. (This is one of the ways in which Dalton's theory required modification.) Isotopes are routinely used in medicine and biology to follow the movement of molecules through the body, as well as in therapeutic applications.

▶▶ The medical use of isotopes is described in Chapter 10.

An isotope is identified in chemical notation by its **mass number,** which is the sum of the protons and neutrons that it contains. We can distinguish one isotope from another in two ways. One approach is to write the mass number as a superscript in front of the symbol for the element and write the atomic, or proton, number as a subscript. An example is $^{17}_{8}O$. The other is to simply write the name of the element followed by its mass number. The atomic number is then implicitly stated in the element's name—for example, oxygen-17.

Example 2.6 Identifying isotopes by symbolic notation

Explain the difference between two isotopes of carbon, carbon-12 and carbon-14, and express that difference with a symbolic notation that uses superscripts and subscripts.

Solution
The two isotopes of carbon have mass numbers of 12 and 14, respectively, and the same atomic number—that is, the same number of protons, 6. The symbolic notation is

$$^{12}_{6}C \quad \text{and} \quad ^{14}_{6}C$$

Problem 2.6 What is the symbolic notation for the two isotopes of nitrogen that contain 7 and 8 neutrons, respectively?

The number of neutrons in an isotope can be calculated from the isotope's mass number and atomic number.

Example 2.7 Calculating an isotope's protons and neutrons

Calculate the number of protons and neutrons in the following two isotopes of carbon: (a) $^{12}_{6}C$; (b) $^{14}_{6}C$.

Solution
An atom's mass number is the sum of its protons and neutrons; so, for (a), protons + neutrons = 12. The atomic number of an element is the number of protons = 6. Therefore,

$$\text{Neutrons} = 12 - 6 = 6$$

For (b), protons + neutrons = 14. The atomic number of carbon is the number of protons = 6. Therefore,

$$\text{Neutrons} = 14 - 6 = 8$$

Problem 2.7 Calculate the number of protons and neutrons in the two isotopes of nitrogen: (a) $^{14}_{7}N$; (b) $^{15}_{7}N$.

Any sample of an element found in nature will consist of a number of isotopes. Each atomic mass found in the table inside the back cover is an average of the atomic masses of the element's isotopes. The average atomic mass of an element is based on the **natural abundances** of its isotopes—the relative amounts in which they are found in nature.

Example 2.8 Calculating an element's average atomic mass

What is the average mass of neon, which is composed of three naturally occurring isotopes having mass numbers (to the closest atomic mass unit) of 20, 21, and 22? The natural abundance of each isotope (that is, the percentage of each in any sample of neon found in nature) is 90.92%, 0.257%, and 8.82%, respectively.

Solution
The average can be found by multiplying each isotope's mass number by its natural abundance expressed as a fraction and adding the results:

$$(20\ \text{amu} \times 0.9092) + (21\ \text{amu} \times 0.00257) + (22\ \text{amu} \times 0.0882) = 20.18\ \text{amu}$$

Problem 2.8 Magnesium consists of three isotopes of masses 24.0 amu, 25.0 amu, and 26.0 amu with abundances of 78.70%, 10.13%, and 11.17%, respectively. Calculate the average atomic mass of magnesium.

2.6 THE PERIODIC TABLE

After the atomic masses of about 60 elements had been determined, Dmitri I. Mendeleev, in Russia, and Lothar Meyer, in Germany, independently of each other studied the chemical and physical properties of the known elements as a function of increasing atomic mass. They found that the properties of the elements did not change smoothly and continuously as atomic mass increased but, instead, repeated periodically. This idea has since become established as the **Periodic Law,** which states that the properties of the elements repeat periodically as the elements are arranged in order of increasing atomic number (not of their atomic masses as these early chemists thought).

The Periodic Law is embodied in the periodic table. This table, reproduced inside the front cover of this book, hangs on the wall in most instructional chemistry laboratories. Currently, the columns are numbered in several ways, as in the version inside the front cover. We have chosen to use an abbreviated version, Figure 2.3, which we believe will be most useful in discussing the relation of position in the table to an element's reactivity and in focusing attention on the elements that we are most likely to meet within a biological context.

The elements consist of three major categories: metals, nonmetals, and transition elements. The boxes in all tables contain the symbols of the elements and their corresponding atomic numbers (for example, hydrogen has atomic number 1 and nitrogen has atomic number 7). Note that in the complete periodic table, shown inside the front cover, elements 57 through 70 and 89 through 102 are removed from the sequential numerical representation and placed at the bottom of the table. The reasons for placing them there will be presented soon. In all versions of the periodic table, the elements are arranged so that those with similar chemical properties (to be discussed shortly) are aligned in the same vertical column, called a **group** or a family. Each horizontal row is called a **period.**

Let us look briefly at some of the groups in the table. For example, the elements in the last column on the right-hand side are called the **noble gases.**

	I	II											III	IV	V	VI	VII	VIII
1	1 H																	2 He
2	3 Li	4 Be											5 B	6 C	7 N	8 O	9 F	10 Ne
3	11 Na	12 Mg	Transition elements										13 Al	14 Si	15 P	16 S	17 Cl	18 Ar
4	19 K	20 Ca	21 Sc	22 Ti	23 V	24 Cr	25 Mn	26 Fe	27 Co	28 Ni	29 Cu	30 Zn	31 Ga	32 Ge	33 As	34 Se	35 Br	36 Kr
5	37 Rb	38 Sr	39 y	40 Zr	41 Nb	42 Mo	43 Tc	44 Ru	45 Rh	46 Pd	47 Ag	48 Cd	49 In	50 Sn	51 Sb	52 Te	53 I	54 Xe
6	55 Cs																85 At	86 Rn

Metals
Metalloids
Nonmetals

Figure 2.3 An abbreviated form of the periodic table. The metals are separated from the nonmetals by a thick zigzag line. The elements shown in green on either side of that line have properties common to both groups and are called metalloids. The main group elements and the noble gases are designated with Roman numerals, and the periods are labeled at the left, 1 through 6. This table contains only the first two transition series so as to focus attention on the elements that we are most likely to meet within a biological context. An additional simplification is that only the atomic numbers are given with the elemental symbols.

A PICTURE OF HEALTH

Percentage of Atoms of Different Elements in the Body

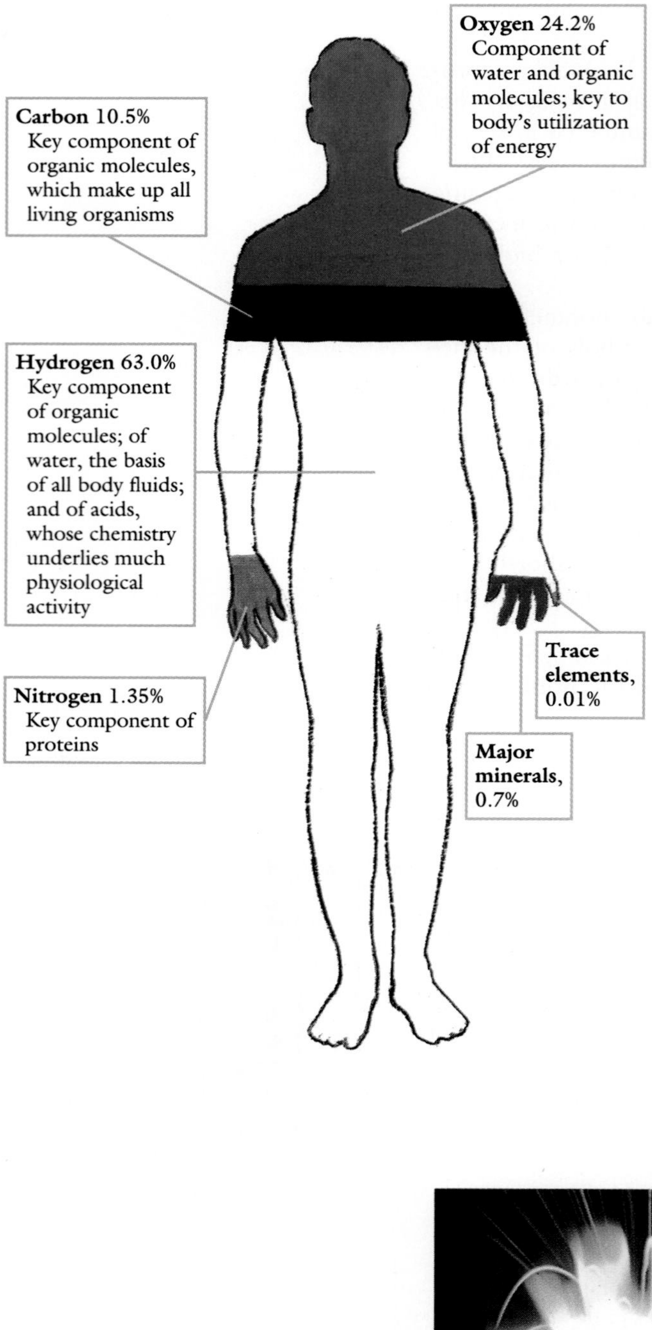

They are helium, neon, argon, krypton, xenon, and radon. These elements are called noble because they are virtually inert (nonreactive) and do not easily form compounds as most of the other elements do. Before 1962, they were called inert gases, but, at that time, it was discovered that xenon could form compounds with fluorine and oxygen, and krypton formed compounds with fluorine. No compounds of helium, neon, or argon are known.

The elements lithium, sodium, potassium, rubidium, and cesium, in the first column on the far left, also are characterized by their similar chemical properties. They are **metals**—shiny, malleable substances that can be melted and cast into desirable shapes and are excellent conductors of heat and electricity. We will see later that, when metals react chemically, they tend to lose electrons to form cations.

The elements in the first column react vigorously with water (Figure 2.4), producing hydrogen and chemical products called bases—for example, potassium hydroxide. Bases added to water form caustic, or alkaline, solutions. The production of caustic, or basic, solutions is the reason that these elements are called **alkali metals** (potassium hydroxide is an alkali metal hydroxide). Drano and Easy-Off (Figure 2.5) are extremely caustic products found in many kitchens.

Basic solutions react with chemicals called acids to produce salts. All salts produced by reaction of the alkali metal hydroxides with hydrochloric acid and isolated by removal of water are white solids and very soluble in water. Acids and bases will be considered briefly in Chapter 3 and more fully in Chapters 8 and 9. The acid that you are most familiar with is acetic acid, the principal component of vinegar.

The elements—fluorine, chlorine, bromine, iodine, and astatine—in the vertical column, or group, that directly precedes that of the noble gases are called the **halogens.** Three members of this group can be seen in Figure 2.6. They are **nonmetals**—elements that cannot be cast into shapes and do not conduct electricity. When nonmetals react, they tend to gain electrons to form anions. All the halogens react with hydrogen to form compounds that are very soluble in water and whose so-

Figure 2.4 The violent reaction of the alkali metal potassium with water. The products of the reaction are hydrogen gas and potassium hydroxide, and lots of heat. The flames are a result of the hydrogen reacting with oxygen from the air; the sparks are pieces of the molten potassium driven into the air by the force of the explosion.

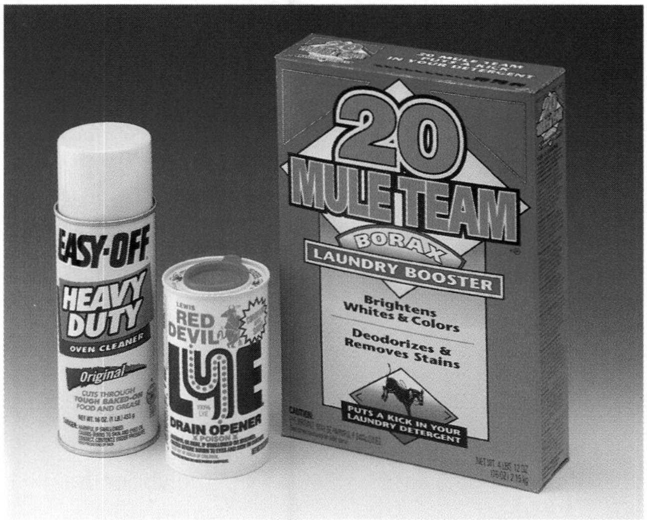

Figure 2.5 Many home cleaning products consist of caustic, or basic, chemical compounds.

lutions are acidic. These solutions can be neutralized with potassium hydroxide; if the water is then removed, the remaining product is a white potassium salt, which is quite soluble in water and, in solution, an excellent conductor of electricity.

Other groups of elements that have similar chemical properties can be identified in the same way. For example, the group to the right of the alkali metals and consisting of beryllium, magnesium, calcium, strontium, and barium comprises the **alkaline earth metals.**

The first period, or row, has only two members: hydrogen and helium. Although hydrogen is not an alkali metal, it does have some of the properties of that family and is often placed in the first position in Group I. Helium is inert, so it is placed in the first position in Group VIII. The second and third periods each contain eight elements, and, if there were no other kinds of elements, the periodic table would consist of eight columns. However, in the fourth period, a new feature emerges.

In period 4, a new set of ten elements appears. The properties of these elements are intermediate between Groups II and III of the preceding period 3. These intervening elements are called the **transition elements.** An important property of these elements is that they can form cations of more than a single electrical charge. For example, iron can form cations of charge 2+ and 3+, and copper can form cations of 1+ and 2+. Transition metals often form highly colored salts. The transition metals are found in biological systems where electrons are transferred in the course of metabolic processes. Iron is a transition element of great biological importance, the source of the deep red color of oxygenated blood. Molybdenum is a key factor in the biological process of atmospheric nitrogen fixation. Cobalt is the central feature of the structure of vitamin B_{12}.

Four series of transition elements of increasing complexity appear in periods 4 through 7. For example, in periods 6 and 7, larger transition groups appear, each consisting of 14 elements, called **inner transition elements.** The elements numbered 57 through 70 are called the lanthanide inner transition elements, and the elements numbered 89 through 102 are called the actinide inner transition elements, after the first element in each series. The chemical properties of the elements in each inner transition series resemble one another so closely that each series is traditionally assigned a single position in the periodic table. Each series in its entirety is placed below the table, resulting in the table's compact arrangement.

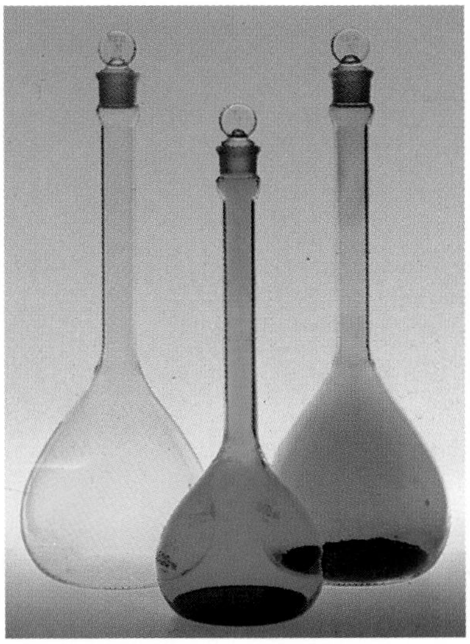

Figure 2.6 Members of the halogen family at room temperature. Left to right: chlorine (Cl_2), a pale yellow gas; bromine (Br_2), a reddish liquid with a reddish vapor; and iodine (I_2), a dark purple solid with a purple vapor.

TABLE 2.2 Essential Elements in the Human Body Listed by Their Relative Abundance

ELEMENTS COMPRISING 99.3% OF TOTAL ATOMS (MAJOR ELEMENTS)	
hydrogen	carbon
oxygen	nitrogen
ELEMENTS COMPRISING 0.7% OF TOTAL ATOMS (MAJOR MINERALS)	
calcium	sulfur
phosphorous	magnesium
potassium	chlorine
sodium	
ELEMENTS COMPRISING LESS THAN 0.01% OF TOTAL ATOMS (TRACE ELEMENTS)	
iron	selenium
iodine	molybdenum
copper	fluorine
zinc	tin
manganese	silicon
cobalt	vanadium
chromium	

In the version of the periodic table used in this book (Figure 2.3), the columns of transition elements are not numbered. The elements belonging to the first two and last five groups in the table denoted by Roman numerals, are called the **main-group elements.** The noble gases are designated by the symbol VIII.

Figure 2.3 indicates the general locations of the metals, which occupy most of the left-hand side of the table, and the nonmetals, which occupy the extreme right-hand side of the table. An intermediate category of elements, called **metalloids** or **semimetals,** have properties between those of metals and nonmetals and lie between the two larger classes. Table 2.2 is a list of elements essential in the nutrition of humans. The physiological consequences of nutritional deficiencies of those elements are listed in Table 2.3.

▶▶ Nutrition is a special focus of Chapter 26.

The reasons for this discussion of the periodic table are twofold: first, to point out that, once you are familiar with the chemistry of one member of a group, you will also have a good idea of the chemistry of the other members of the group; second, to set the stage for considering the next great advance in chemical knowledge—the relation between the atomic structure of an element and the element's chemical reactivity.

2.7 ELECTRON ORGANIZATION WITHIN THE ATOM

The picture of the atom, as we last left it, was of a tiny nucleus containing most of the atom's mass, surrounded by electrons equal in number to the atom's atomic number. Until the early twentieth century, the arrangement of those electrons was a mystery. Were they distributed in some sort of order or would we discover that they had no order at all? Today we know that an ordered arrangement of the electrons within atoms explains the structure of the periodic table and the chemical reactivity of the elements. Before we can profitably look into this arrangement, however, we must become familiar with

TABLE 2.3 Elements Required for Human Nutrition and the Consequences of Their Deficiencies

Element	Result of Nutritional Deficiency
calcium	bone weakness, osteoporosis, muscle cramps
magnesium	calcium loss, bowel disorders
potassium	muscle weakness
sodium	muscle cramps
phosphorus	muscle and bone weakness
iron	anemia
copper	anemia
iodine	goiter
fluorine	tooth decay
zinc	poor growth rate
chromium	hyperglycemia
selenium	pernicious anemia
molybdenum	poor growth rate
tin	poor growth rate
nickel	poor growth rate
vanadium	poor growth rate

some of the scientific developments that led to its discovery. The modern view of the atom's structure began with what scientists had observed about the colors emitted from elements made incandescent by flames or sparks.

Atomic Spectra

Light from the sun, a lit candle, or an incandescent bulb is called white light. It can be separated, or resolved, into its component colors, or frequencies, called its **visible spectrum,** by means of a prism, as shown in Figure 2.7. The same phenomenon causes a rainbow after rain, around the edges of a piece of window glass in the sun, or in the reflection of light from the surface of a phonograph record.

The visible spectrum from incandescent sources such as the sun or a heated filament in a light bulb is a series of colors that continuously change and merge into one another, as we see in Figure 2.7. In contrast, when light is emitted by excited, or energized, atoms or individual elements that have been

Figure 2.7 A glass prism resolves white light into a continuous spectrum.

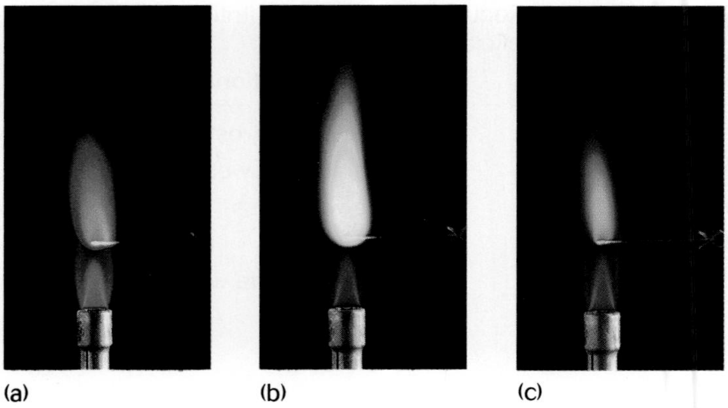

Figure 2.8 Colors of light emitted by Group I metals in flame tests. (a) Sodium, Na; (b) potassium, K; (c) rubidium, Rb.

vaporized in a flame, it produces not a continuous spectrum but a series of sharply defined and separated lines of different color. Figure 2.8 shows the appearance of flames into which various compounds of Group I elements have been injected. The flame's colors are distinctive in regard to the identity of the elements. These emissions are then sent through an instrument called an atomic spectrometer (Figure 2.9) to be resolved into their component colors. The spectral lines produced by the spectrometer, called the element's **atomic emission spectrum,** are recorded on a photographic plate as in Figure 2.10. Every element has its own characteristic emission spectrum, and these spectra are used to identify and quantitatively analyze unknown materials.

Radiation is absorbed by an atom when that radiation corresponds to the same specific frequency of light emitted from the atom when it is excited in a flame. In this way, the emission of light of a specific frequency by excited atoms of an element is complemented by the absorption of radiation of that same frequency when the atoms of the element in an unexcited state are exposed to that frequency of light. The absorption of light of specific frequencies is called an absorption spectrum. Box 2.1 describes how atomic absorption

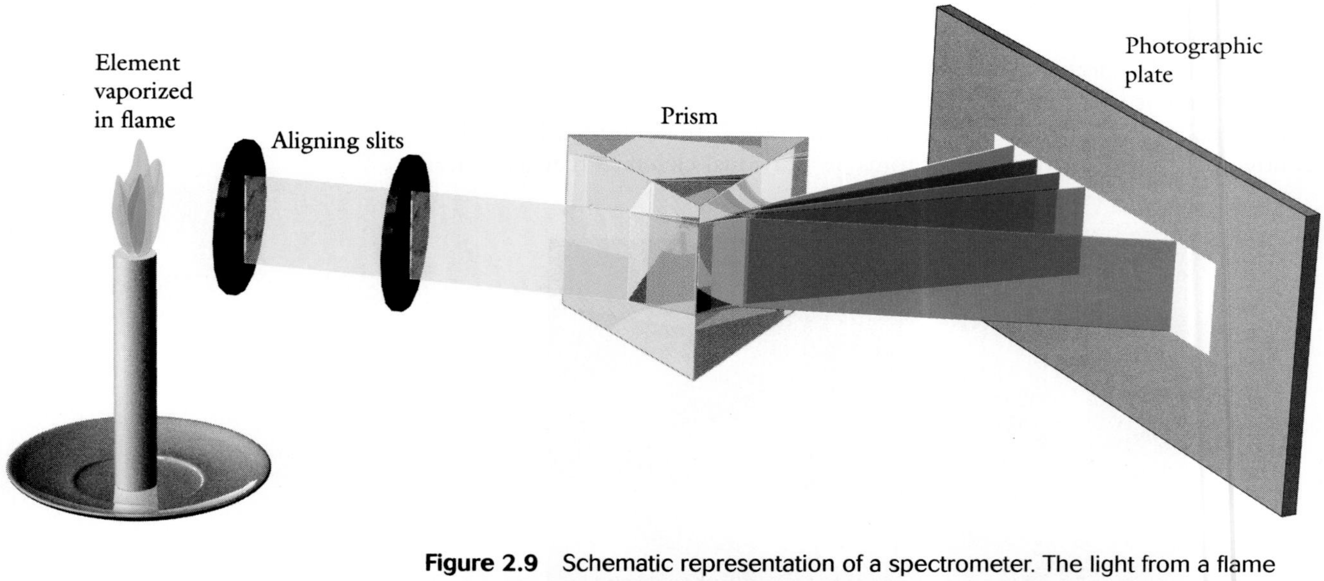

Figure 2.9 Schematic representation of a spectrometer. The light from a flame such as those in Figure 2.8 is shaped by a series of slits and passed through a prism that resolves the light into its constituent frequencies (colors). The resulting line spectrum is projected onto a photographic plate for recording.

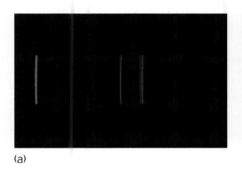

Figure 2.10 Atomic emission spectra of (a) hydrogen and (b) helium.

spectra are obtained and what they are used for. Other types of absorption spectra are used to detect and characterize molecules and will be presented in succeeding chapters.

The meaning of these line spectra puzzled physicists for more than 100 years. Until early in the twentieth century, chemists had no idea that atomic spectra could be the key to an understanding of chemical reactivity. That key lies in the relation between the color (frequency) of light and its energy, which is considered next.

Electromagnetic Radiation and Energy

Radiation describes the transfer of energy from one point in space to another. For example, after an iron bar has been heated to incandescence, you do not

2.1 Chemistry in Depth

Absorption Spectra and Chemical Analysis

Absorption spectra are obtained by generating light of a specific frequency and passing it through a sample. Substances that absorb that frequency will remove energy from the incoming light. The specific frequency absorbed identifies the material, and the amount of energy that it removes from the incoming light depends on the amount of a particular substance in the sample. An absorption spectrum is therefore simultaneously a qualitative and a quantitative tool for chemical analysis.

The important components for obtaining absorption spectra are a device for producing light of a particular frequency and another device for determining the loss of energy in the incident light beam after it emerges from the sample. The first can be accomplished with the same kind of prism (Figure 2.7) used to resolve light into its components. By rotating a prism irradiated with white light, one can select particular frequencies and direct them through a sample. A photocell placed after the sample can measure the light energy before and after the sample is placed in the beam and thus determine the energy loss of the beam due to absorption by the sample. The photograph is of a modern atomic absorption spectrometer. It is used in the clinical laboratory to determine, in the blood, the presence and the amounts of ions such as sodium, potassium, and calcium.

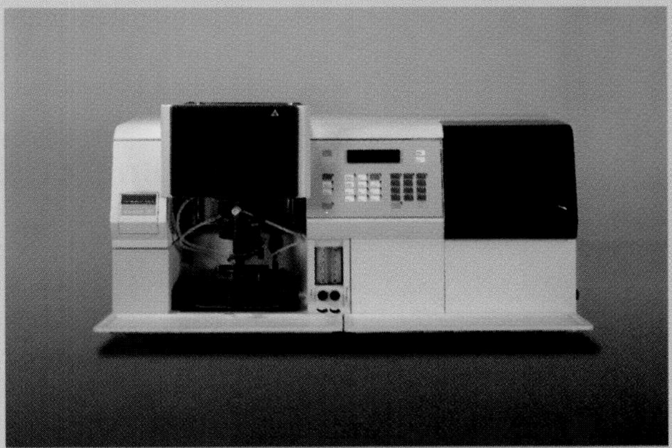

TABLE 2.4 Frequency and Energy of Electromagnetic Radiation

Type of Radiation	Frequency (Hz)	Energy Relative to Visible Light
cosmic rays	10^{22}	100,000,000
γ-rays	10^{20}	1,000,000
X-rays	10^{18}	10,000
ultraviolet light	10^{16}	10
visible light	10^{15}	**1**
infrared light	10^{14}	0.1
microwave radiation (radar)	10^{10}	0.0001
FM radio	10^{8}	0.0000001
AM radio	10^{6}	0.000000001

need to touch it to feel its heat. We say that the heat "radiates" from the incandescent metal. Heat and visible light are only small parts of the full spectrum of **electromagnetic radiation,** which has its origin in the oscillation, or vibration, of charged particles. For this reason, electromagnetic radiation is quantitatively described by its frequency; that is, by how many times per second a vibration is completed.

Frequency is calculated in cycles per second, or cps. The SI unit of frequency is the hertz, abbreviated Hz, named after Heinrich Rudolf Hertz, a German physicist who contributed a great deal to this branch of knowledge. Table 2.4 shows the frequency ranges for the entire electromagnetic spectrum. Cosmic radiation, at the top of the list, has the highest frequency; radio waves, at the bottom, have the lowest. The colors of visible light have frequencies ranging from 8×10^{14} to 4×10^{14} Hz, which is a very small part of the entire spectrum.

The energy of electromagnetic radiation is related to its frequency: the higher the frequency, the greater the energy. Ultraviolet (UV) radiation, produced by the sun, has an energy great enough to cause chemical damage to human skin (see Box 10.2). Therefore the use of UV-blocking preparations is strongly recommended by dermatologists. Medical practitioners keep close watch on their patients' exposure to diagnostic X-rays to prevent possible damage to deep tissue, and gamma-radiation (γ-rays) is used to destroy cancers. The last column in Table 2.4 compares the energies of the various radiation frequencies with that of visible light. Frequencies corresponding to, or less than, that of ultraviolet light are used to obtain absorption spectra of molecules, to be discussed in Chapter 3.

Atomic Energy States

The underlying reason for the line structure of atomic emission spectra was discovered by Niels Bohr, a Danish physicist. Several years before Bohr's discovery, the physicists Max Planck and Albert Einstein demonstrated that light consists of small packages that they called photons. They further showed that the energy of these photons depends on the frequency of the light. Bohr knew that, for light energy to be emitted, energy must be absorbed. He proposed that the discrete lines of atomic spectra must represent unique energies of emitted light and that, therefore, only certain energies can be absorbed. The absorption of energy raises the atom from a stable, low-energy state, or **ground state,** to a higher-energy **excited state.** Consequently, the lines of atomic spectra are the result of light emitted when an atom returns from an excited state to its ground state. In other words, the atom first absorbs and then emits a unique quantity of energy corresponding to the frequency of that spectral line.

Electromagnetic radiation energies that can be absorbed by very small systems such as atoms and electrons come in small individual packages called **quanta** (singular, quantum). The size of a quantum depends directly on the frequency of the radiation. This means that the energy of atoms can be increased only in discrete units, or small jumps. If the size of a quantum of energy striking an atom is equal to the energy difference between two of that atom's energy states, then the atom will absorb the energy and enter an excited state. If the quantum is larger or smaller than the energy difference between the two states, there will be no absorption and the atom will remain in its ground state. We might count our money down to the last penny, for example, but the Internal Revenue Service prefers not to recognize anything smaller than dollars.

Concept checklist

- ✔ The discrete lines of atomic emission spectra represent energy states of the atom. (Remember that each color of light represents electromagnetic radiation of a specific energy.)
- ✔ The existence of discrete lines of specific frequencies means that only certain energy states, and no others, are allowed.

The first model to incorporate the relation between line spectra and atomic energy levels pictured the atom as a miniature solar system, with electrons moving about the nucleus much as planets revolve about the sun. Because electrons in the atom are limited to certain permissible values of energy, it was argued that they must remain at fixed distances in paths about the nucleus and cannot occupy any position intermediate between these paths.

This "solar system" model was attractive because it used familiar images to describe the unseen; however, although the model worked well for hydrogen, it failed to predict the behavior of the other elements. The model treated the electron as a discrete particle whose speed and trajectory, like those of a rocket ship or bullet, could be known at any time. However, the electron is such a light and small particle that, when it is observed, its position is altered by the observation. The consequent uncertainty about position and motion was incorporated into a new model of the atom called quantum mechanics.

2.8 THE QUANTUM MECHANICAL ATOM

The modern theory that describes the properties of atoms and subatomic particles is called **quantum mechanics.** It is a complex mathematical theory that has had great success in predicting a wide variety of atomic properties. Because it is a mathematical theory, its results are very difficult to describe in terms of everyday experience. The most notable of these difficulties is that the ability to locate an electron with any precision had to be abandoned. The best that scientists can do is to estimate the probability of finding an electron in a given region of space.

The chief goal of this chapter is to show how the detailed structure of the atom explains the periodicity of the chemical properties of the elements. To do so, we will construct the periodic table by building elements from electrons and protons. The elements will be built by adding electrons one at a time to nuclei that increase in proton number in the same way. This cannot be done without a set of rules to guide us in the way in which the incoming electrons

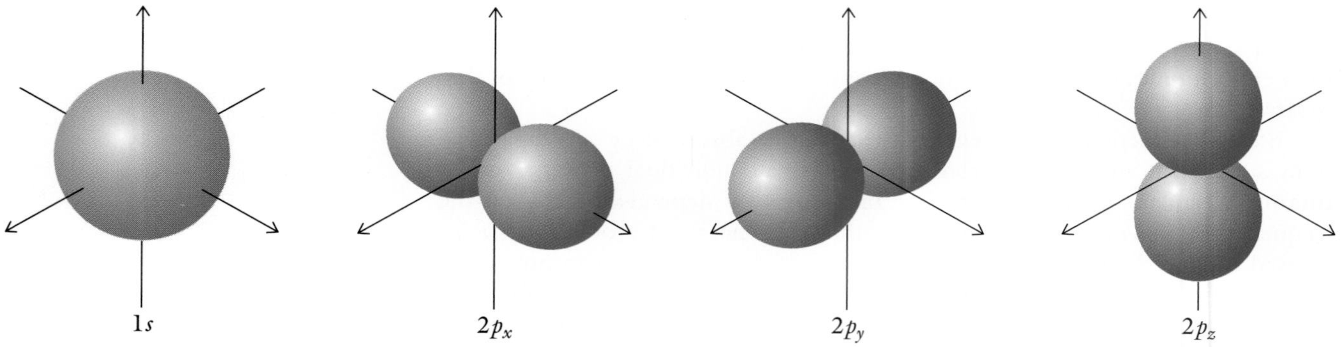

Figure 2.11 Atomic *s* and *p* orbitals. The surfaces enclose a region of space in which there is a greater than 90% probability of finding an electron. (See Section 2.9 for significance of subscripts *x*, *y*, and *z*.)

are to be organized within the atoms constructed in this way. Quantum mechanics has provided those rules and has shown that electrons are organized within the atom into shells, subshells, and orbitals, whose relation to each other depends on the energy state of the atom, which is defined by its principal quantum number.

Shell: Identified by a principal quantum number (1, 2, 3 . . . *n*) that specifies the energy level, or energy state, of the shell. The higher the quantum number, the greater the energy and the farther the shell electrons are from the nucleus.

Subshells: Locations within a shell, identified by the lowercase letters *s, p, d,* and *f.*

Orbital: The region of space within a subshell that has the highest probability of containing an electron.

The region of space in which an electron is most likely to be found is called an **atomic orbital.** Atomic orbitals are best visualized as clouds surrounding the nucleus. Their size and shape depend on the atom's energy state: the greater the energy, the larger and more complex the orbitals. The *s* and *p* types of atomic orbitals are illustrated in Figure 2.11. Remember that the atom's electrons cannot be located with any precision, but the surface of an orbital encloses a space in which there is a high likelihood, or high probability, of finding a given electron. As the principal quantum number increases, electrons are farther from the nucleus and their orbitals occupy a larger volume. Figure 2.12 shows how *s* orbitals retain their overall shape but increase in volume. The same holds true for *p, d,* and *f* orbitals.

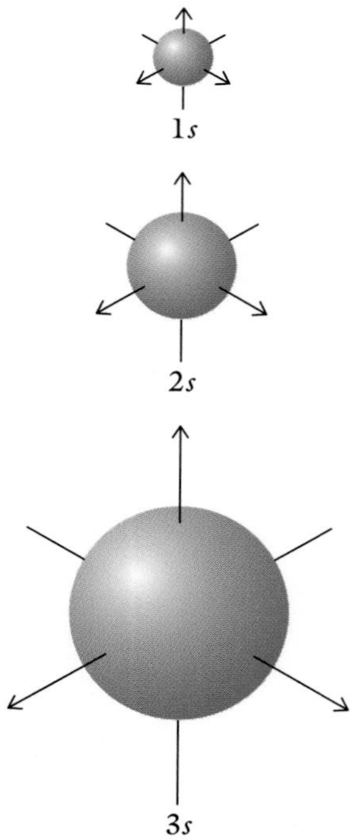

Figure 2.12 The 1*s*, 2*s*, and 3*s* atomic orbitals. The *s* orbitals all have the same shape but grow larger as their distance from the nucleus increases.

In Chapter 3, we will see that the formation of one kind of chemical bond is the connection (the bonding) of atoms by electrons contained in these orbitals. Because the orbitals are restricted to certain positions in space around the nucleus, the connections that they create between atoms give rise to the unique shapes and therefore the characteristic properties of molecules of different substances.

The relation between principal quantum number, number of subshells, and subshell names is tabulated in Table 2.5. You can see that, as the principal quantum number increases, the number of subshells increases. The number of orbital types within each subshell also increases. There is an *s* orbital in every shell; *s* and *p* orbitals are in the second shell; *s, p,* and *d* orbitals are in the third shell; and *s, p, d,* and *f* orbitals are in the fourth shell.

The numbers of orbitals within each subshell are listed in Table 2.6. The data in Tables 2.5 and 2.6 tell us that an atom in principal quantum state 1 has

TABLE 2.5 Relation Between Principal Quantum Number and Number and Type of Orbital Within a Subshell

Principal Quantum Number (Shell)	Number of Subshells	Subshell Names (Orbital Types)
1	1	*s*
2	2	*s*, *p*
3	3	*s*, *p*, *d*
4	4	*s*, *p*, *d*, *f*

one *s* orbital available for electrons, an atom in principal quantum state 2 has one *s* and three *p* orbitals available for electrons, and an atom in principal quantum state 3 has one *s*, three *p*, and five *d* orbitals available for electrons.

The final feature of electron behavior is that electrons possess a property called **spin,** analogous to the rotation of a planet. A single electron within an orbital is called an **unpaired electron** or a **lone electron.** For an orbital to accommodate two electrons, their spins must be opposite. The electrons are then said to be **spin-paired.** The spin-pairing will be represented symbolically in this book by a pair of arrows pointing in opposite directions (Section 2.9). The significance of spin-pairing is that one orbital can contain no more than two electrons, and, on that basis, the total number of electrons that can be accommodated in each subshell is shown in Table 2.7.

2.9 ATOMIC STRUCTURE AND PERIODICITY

To see how the modern quantum mechanical theory explains the arrangement of elements in the periodic table, we will imagine creating the elements starting with hydrogen by adding electrons one at a time around an atomic nucleus. To keep the resulting atom neutral, it is necessary that the atomic number of the nucleus must increase simultaneously every time that we add a new electron. This imaginary exercise has been given the German name ***Aufbau,*** which means "buildup." The result of the aufbau procedure is the complete description of the electron organization of the atom, also called the atom's **electron configuration.**

We will use two methods for representing electron configuration. Method 1 identifies electrons, first, by the principal quantum number (1, 2, and so forth); next, by the subshell type (*s*, *p*, and so forth); and, finally, by the number, written as a superscript, of electrons in each orbital ($1s^2$ and $2p^3$, for example). Method 2 uses boxes to represent orbitals at different energy levels and arrows in the boxes to represent electrons—for example, [↑].

TABLE 2.6 Relation Between Subshell Type and Number of Orbitals

Subshell Type	Number of Orbitals
s	1
p	3
d	5
f	7

TABLE 2.7 Number of Orbitals and Electrons in Each Subshell Type

Subshell	Number of Orbitals	Electrons per Orbital	Total Electrons per Subshell
s	1	2	2
p	3	2	6
d	5	2	10
f	7	2	14

Here is a list of rules for constructing an atom. We will begin to put them into practice in Example 2.10.

Aufbau rules

1. The larger the principal quantum number, the greater the number of subshells (see Table 2.5). An electron described as 2*s* is in the second shell and in an *s* subshell. A 3*d* electron is in the third shell and in a *d* subshell.
2. Each subshell has a unique number of orbitals (see Table 2.6).
3. The order in which electrons enter the available shells and subshells depends on the orbital energy levels. The orbitals fill according to the diagram in Figure 2.13, with the lowest energy orbitals (closest to the nucleus) filling before those more distant. Boxes at each level correspond to the number of orbitals in the subshell.
4. There can be no more than two electrons per orbital, and the two electrons can occupy the same orbital only if they are spin-paired (their spins are opposite). This rule is known as the **Pauli exclusion principle.** Method 2 emphasizes the exclusion principle by showing that orbitals can contain no more than two electrons.
5. In a subshell with more than one orbital (for example, *p* or *d* orbitals), electrons entering that subshell will not spin-pair until every orbital in the subshell contains one electron. This rule is known as **Hund's rule.** All electrons have the same electrical charge and tend to repel one another. This repulsive force can be reduced if the electrons stay as far apart as possible (by occupying different orbitals whenever possible).

Example 2.9 Interpreting an atom's principal quantum number

If an atom is in an energy state of principal quantum number 3, how many and what kinds of subshells are allowed?

Solution

We see in Table 2.5 that the principal quantum number 3 is characterized by three subshells, the *s*, *p*, and *d* subshells.

Problem 2.9 How many and what kinds of subshells are allowed in principal quantum state 4?

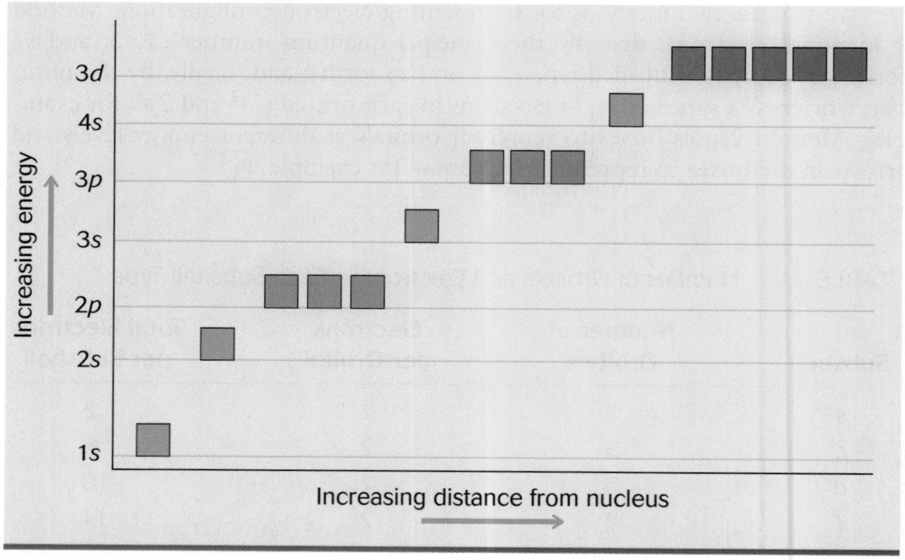

Figure 2.13 The quantized energy levels of atomic subshells. Atomic orbitals are represented by boxes.

Figure 2.13 emphasizes two important features of electron behavior. First, the allowed energies for electrons do not represent a gradually increasing continuum but a series of jumps from one level to another, with no intermediate energy levels allowed. This feature is what physicists mean when they say that energy is **quantized.** Second, beginning with principal quantum number 4, the *s* subshell is lower in energy than the *d* subshell of the preceding period. Therefore the 4*s* level fills before the 3*d* level.

Example 2.10 Diagramming electron configurations by using method 1

Diagram the electron configuration of elements 1, 2, 3, 4, 5, and 6. (Remember, the atomic number and the number of electrons are identical.)

Solution

Use Table 2.7 on page 55 to determine the principal quantum number ("shell"), available subshells, and number of electrons per subshell, and construct a table as shown here for each of these elements. An *s* subshell can contain a maximum of two electrons. Because we fill the orbitals one electron at a time, the first two elements will exhaust the first *s* subshell, 1*s*. In the third and fourth elements, electrons will fill the second *s* subshell, 2*s*; electrons in the fifth and sixth elements must go into the next available (empty) subshell, 2*p*, which can accommodate a maximum of six electrons in its three orbitals.

Element	Shell	Subshell	Number of Electrons	Electron Configuration
hydrogen	1	*s*	1	$1s^1$
helium	1	*s*	2	$1s^2$
lithium	1	*s*	2	
	2	*s*	1	$1s^22s^1$
beryllium	1	*s*	2	
	2	*s*	2	$1s^22s^2$
boron	1	*s*	2	
	2	*s*	2	
	2	*p*	1	$1s^22s^22p^1$
carbon	1	*s*	2	
	2	*s*	2	
	2	*p*	2	$1s^22s^22p^2$

Hydrogen's electron configuration is $1s^1$; that is, there is one electron (indicated by the superscript) in the *s* orbital of the first shell (indicated by 1). Helium's configuration is $1s^2$; helium's two electrons fill and complete the *s* orbital of the first shell. Lithium, atomic number 3, with three electrons, follows helium. Its third electron must enter the second shell, where the first available empty subshell is the 2*s* orbital. Thus, the configuration of lithium is $1s^22s^1$. With the fourth element, beryllium, the 2*s* subshell is complete. Beryllium's electron configuration is $1s^22s^2$. Boron, atomic number 5, follows beryllium. Its fifth electron must enter the next available empty orbital. The electron configuration of boron is $1s^22s^22p^1$. Carbon's electron configuration is $1s^22s^22p^2$.

Problem 2.10 Diagram the electron configuration of element 16 by using method 1.

The electron configuration of carbon in Example 2.10 must be looked at more closely. The three *p* orbitals lie perpendicularly to one another, so chemists label them according to the Cartesian axes on which they lie: x, y, and z (Figure 2.11). This designation is added as a subscript. Thus, the three p orbitals are known as p_x, p_y, and p_z. According to rule 5, these orbitals are occupied singly until each contains one electron, after which an electron of opposite spin may enter each one. Therefore, the complete electron configuration of carbon is $1s^2 2s^2 2p_x^{\,1} 2p_y^{\,1}$.

Example 2.11 Diagramming electron configuration by using method 2

Use method 2 to diagram the electronic configuration of the first four elements.

Solution

Represent each orbital by a box at an appropriate energy level (see Figure 2.13), and indicate the presence of electrons by arrows pointing up or down. Remember, rule 4 states that an orbital can contain no more than two spin-paired electrons. The atomic number is equal to the number of protons, and the number of protons is equal to the number of electrons.

		Orbital	
Element	**Atomic Number**	**1s**	**2s**
hydrogen	1	↑	
helium	2	↑↓	
lithium	3	↑↓	↑
beryllium	4	↑↓	↑↓

Problem 2.11 Using method 2, diagram the electronic configuration of element 5. (Hint: Because the 1s and 2s orbitals of element 5 are filled, the fifth electron must go into a new orbital.)

Example 2.12 Diagramming electron configurations by using method 2

Diagram the electronic configuration of elements 6 and 8 by using method 2.

Solution

Elements 6 and 8, carbon and oxygen, possess six and eight electrons, respectively. Therefore, because the 1s and 2s subshells hold a maximum of four electrons, both those subshells will be filled, and electrons 5 through 8 must enter the p subshell. In diagramming these structures, we must pay close attention to rule 5 (Hund's rule).

		Orbital				
Element	**Atomic Number**	**1s**	**2s**	**$2p_x$**	**$2p_y$**	**$2p_z$**
carbon	6	↑↓	↑↓	↑	↑	
oxygen	8	↑↓	↑↓	↑↓	↑	↑

Problem 2.12 Using method 2, diagram the electronic configuration of element 7.

▶▶ **Carbon, hydrogen, and oxygen are the elements of interest throughout Part 2, "Organic Chemistry."**

TABLE 2.8 Electron Configurations of the First 11 Elements of the Periodic Table

Element	Atomic Number	Symbol	Electron Configuration
hydrogen	1	H	$1s^1$
helium	2	He	$1s^2$
lithium	3	Li	$1s^22s^1$
beryllium	4	Be	$1s^22s^2$
boron	5	B	$1s^22s^22p^1$
carbon	6	C	$1s^22s^22p_x{}^12p_y{}^1$
nitrogen	7	N	$1s^22s^22p_x{}^12p_y{}^12p_z{}^1$
oxygen	8	O	$1s^22s^22p_x{}^22p_y{}^12p_z{}^1$
fluorine	9	F	$1s^22s^22p_x{}^22p_y{}^22p_z{}^1$
neon	10	Ne	$1s^22s^22p^6$
sodium	11	Na	$1s^22s^22p^63s^1$

Table 2.8 is a list of the first 11 elements in the periodic table along with their electron configurations represented in the notation of method 1. You can see that, by the time we reach elements with atomic numbers of 20 or so, this notation can take up quite a bit of space. A more compact method for writing electron configurations uses the fact that every period in the table ends with a noble gas, an element whose outer shell contains the maximum number of electrons that the shell can hold. We can use the bracketed name of a noble-gas element to represent its full electron configuration, thus shortening the notation for all the elements in the period below. This notation is demonstrated in the following example.

Example 2.13 Diagramming electron configurations by using the compact method

Show the compact method of writing the electronic configuration of element 19, potassium.

Solution

The electron configuration of element 19, potassium, is $[1s^22s^22p^63s^23p^6]4s^1$. Brackets have been placed around the part of the electron configuration that matches the configuration of argon, (symbol Ar), the preceding noble gas. This configuration is called the argon **core** of the potassium atom. The argon core of the potassium atom is denoted as [Ar], and the potassium electron configuration is written in compact notation as $[\mathrm{Ar}]4s^1$.

Problem 2.13 Write the electron configuration of element 16, sulfur, by using the compact method.

In compact notation, lithium is $[\mathrm{He}]2s^1$, sodium is $[\mathrm{Ne}]3s^1$, and calcium is $[\mathrm{Ar}]4s^2$. (You should be able to locate these elements in Figure 2.3.)

At the beginning of this discussion, we selected three families of elements to illustrate the periodicity of chemical properties: the noble gases, the halogens, and the alkali metals. Table 2.9 on the following page presents the electron configurations of the elements in these families for periods 2 through 6. The electrons of the outermost shell are printed in color.

TABLE 2.9 Electron Configurations of Some of the Alkali Metals, the Halogens, and the Noble Gases

Alkali Metals (Group I)		Halogens (Group VII)		Noble Gases (Group VIII)	
Li	[He] $2s^1$	F	[He] $2s^22p^5$	Ne	[He] $2s^22p^6$
Na	[Ne] $3s^1$	Cl	[Ne] $3s^23p^5$	Ar	[Ne] $3s^23p^6$
K	[Ar] $4s^1$	Br	[Ar] $3d^{10}4s^24p^5$	Kr	[Ar] $3d^{10}4s^24p^6$
Rb	[Kr] $5s^1$	I	[Kr] $4d^{10}5s^25p^5$	Xe	[Kr] $4d^{10}5s^25p^6$
Cs	[Xe] $6s^1$	At	[Xe] $4f^{14}5d^{10}6s^26p^5$	Rn	[Xe] $4f^{14}5d^{10}6s^26p^6$

Notice that (1) each noble gas has eight electrons in its outer shell, (2) each alkali metal has a single electron in its outermost shell, and (3) each halogen has seven electrons in its outermost shell. In addition to the fact that the elements in a family (group) have the same number of outer-shell electrons, it is most important to note that the number of outer-shell electrons is the same as the main-group elements' group number in Figure 2.3. This relation is further emphasized in Table 2.10.

In the first three periods, we easily see that the number of electrons in the outer electronic shell (electrons outside the inner noble-gas shell) is identical with the element's group number in Figure 2.3. Things become more complex in period 4, however, when the inner $3d$ subshell begins to fill to form the transition elements after the outer $4s$ subshell has filled.

The numbers of outer-shell electrons in elements of the transition series do not correlate simply with the main-group elements' group numbers; so, in Figure 2.3, group numbers are not assigned to these elements. We will examine the properties of specific members of the transition series—iron or copper, for example—in the next chapter and in Chapter 8.

2.10 ATOMIC STRUCTURE, PERIODICITY, AND CHEMICAL REACTIVITY

When an alkali metal atom, such as sodium, reacts with a halogen atom, such as chlorine, the sodium atom loses its single electron and the chlorine atom gains a single electron. Method 1 can be used to show the changes in the electron configurations of these elements as a result of their reaction:

$$\begin{array}{ccccccc} \text{Na} & + & \text{Cl} & \longrightarrow & \text{Na}^+ & + & \text{Cl}^- \\ 1s^22s^22p^63s^1 & + & 1s^22s^22p^63s^23p^5 & \longrightarrow & 1s^22s^22p^6 & + & 1s^22s^22p^63s^23p^6 \end{array}$$

TABLE 2.10 Electron Configurations of the Main-Group Elements in the Second and Third Periods

	Group							
	I	II	III	IV	V	VI	VII	VIII
Period 2	Li $1s^22s^1$	Be $1s^22s^2$	B $1s^22s^22p^1$	C $1s^22s^22p^2$	N $1s^22s^22p^3$	O $1s^22s^22p^4$	F $1s^22s^22p^5$	Ne $1s^22s^22p^6$
Period 3	Na [Ne]$3s^1$	Mg [Ne]$3s^2$	Al [Ne]$3s^23p^1$	Si [Ne]$3s^23p^2$	P [Ne]$3s^23p^3$	S [Ne]$3s^23p^4$	Cl [Ne]$3s^23p^5$	Ar [Ne]$3s^23p^6$

The electron configuration of the resulting positively charged sodium atom (sodium ion) is identical with that of the noble gas preceding sodium in the periodic table—neon. The electron configuration of the resulting negatively charged chlorine atom (chloride ion) is that of the noble gas immediately following chloride in the table—argon. By reacting with one another, both atoms have achieved a noble-gas electron configuration. In general, chemical reactions of the elements achieve the same results: they lead to a noble-gas electron configuration in the atoms' outer electron shells.

Because the noble gases are inert and chemically stable, chemists have concluded that a filled outermost electron shell is the most stable configuration that an atom can have. Because a noble gas has eight electrons in its outer shell, this conclusion is summarized as the **octet rule.** The outermost electron shell (in a given period) is called the **valence shell.** The electrons in the outer shell are called the **valence electrons.**

Chemists use a symbolic representation first proposed by the American chemist Gilbert N. Lewis and called **Lewis symbols.** The nucleus and inner-shell electrons are represented by the element's symbol, and dots around it represent the valence-shell electrons. The Lewis symbols for elements in periods (rows) 1, 2, and 3 of the periodic table are:

H· :He

Li· ·Be· :B· ·C· ·N· ·O· :F· :Ne:

Na· ·Mg· :Al· ·Si· ·P· ·S· :Cl· :Ar:

With the exception of helium, the number of valence-shell electrons is equal to the group number of the main-group element.

▶▶ The names of anions, such as lithium ions and fluoride ions, will be discussed in Chapter 3.

Example 2.14 Portraying a chemical reaction with Lewis symbols

Use Lewis symbols to illustrate the reaction of lithium and fluorine; that is, the formation of lithium and fluoride ions.

Solution

Lithium's Lewis symbol suggests that lithium atoms have a tendency to lose one electron, resulting in the formation of a cation with a single charge. The formation of a lithium ion can be represented as:

$$\mathrm{Li}\cdot - \mathrm{e} \longrightarrow \mathrm{Li}^+$$

in which "e" stands for electron. Fluoride's Lewis symbol suggests that fluoride atoms are driven to gain an electron, resulting in the formation of an anion with a single charge:

$$:\ddot{\underset{\cdot\cdot}{\mathrm{F}}}\cdot + \mathrm{e} \longrightarrow :\ddot{\underset{\cdot\cdot}{\mathrm{F}}}:^-$$

An atom's loss of an electron cannot occur without another atom's gain of that electron. Therefore, we will show that the two processes take place at the same time

$$\mathrm{Li}\cdot \curvearrowright \cdot\ddot{\underset{\cdot\cdot}{\mathrm{F}}}: \longrightarrow \mathrm{Li}^+ + :\ddot{\underset{\cdot\cdot}{\mathrm{F}}}:^-$$

Problem 2.14 Use Lewis symbols to illustrate the formation of sodium and chloride ions.

Concept checklist

✔ All members of the same group of the main group elements have the same outer-shell electron configuration. Periodicity in chemical properties is the result of these identical outer-shell electron configurations.

- ✔ The chemical properties of elements within the same group are almost identical, an observation indicating that it is the outer-shell electron configuration only that dictates the chemical reactivity of the elements.
- ✔ The octet rule states that elements react to attain an outer-shell electron configuration of eight valence electrons, or, in other words, the electron configuration of the nearest noble gas.
- ✔ Lewis symbols serve to highlight the electrons of the valence shells of the elements and to predict the loss or gain of electrons in chemical reactions.

In the next chapter, we will further explore the tendency of atoms to attain the noble-gas outer-shell electron configuration and its role as a driving force behind chemical reactions between the elements.

Summary

Dalton's Atomic Theory John Dalton promoted the idea that matter consists of ultimate particles called atoms and used it to explain two key experimental facts of chemistry: (1) the conservation of mass in chemical reactions and (2) the constant composition of matter. His theory was that all matter is composed of infinitesimally small particles called atoms, which are indestructible, and that the atoms of any one element are identical. Atoms of one element are distinguished from those of a different element by the fact that the two types of atoms have different masses. Compounds consist of combinations of atoms of different elements. In chemical reactions, atoms trade partners to form new compounds.

Atomic Masses The relative mass of an element is called its atomic mass. Atomic masses are calculated in relation to the mass of the most common form of the element carbon, whose mass is defined to be exactly 12 atomic mass units (amu).

The Structure of Atoms Atoms contain three kinds of particles: positively charged protons; much lighter, negatively charged electrons; and electrically uncharged (neutral) particles called neutrons, of about the same mass as protons. The protons and neutrons, making up most of the atom's mass, are located together in a structure at the center of the atom, the nucleus. The number of protons in an electrically neutral atom is balanced by exactly the same number of electrons. An atom bearing a net electrical charge is called an ion. A positively charged ion is called a cation, and a negatively charged ion is called an anion.

Isotopes Naturally occurring elements consist of isotopes. All isotopes of a given element have the same atomic number but differ slightly in mass because they have different numbers of neutrons. The sum of the numbers of protons and neutrons in its atom is an isotope's mass number; chemical notation differentiates between isotopes by providing both mass number and atomic number for each isotope.

The Periodic Table The periodic law states that the properties of the elements repeat periodically if the elements are arranged in order of increasing atomic number. In the periodic table, elements of similar chemical properties are aligned in vertical columns called groups. Horizontal arrangements, called periods, end on the right with a noble gas. Each of the second and third periods contains eight elements. But, in the fourth period, a new group of ten transition elements appears.

Elements are designated as either main-group or transition elements. The metals occupy most of the left-hand side of the periodic table, and the nonmetals occupy the extreme right-hand side of the table. Elements called metalloids, or semimetals, have properties between those of metals and nonmetals and lie between the two larger classes.

Electron Organization Within the Atom Electrons in atoms are confined to a series of shells around the nucleus. The distance from the nucleus increases in proportion to the energy state of the electrons: the farther away the electrons are from the nucleus, the greater their energy. Electron energy states are characterized by an integer called the principal quantum number. Shells consist of subshells, and electrons in these subshells reside in orbitals. Only two electrons are allowed in any orbital, and, if two electrons are to occupy the same orbital, they must have opposite spins, in which case they are said to be spin-paired. The atom's orbitals define the probability of finding an electron in a given region of space around the nucleus. With an increase in the number of shells, the number of subshells increases, and the orbitals assume increasingly more complex shapes.

Atomic Structure, Periodicity, and Chemical Reactivity The electron structure of an atom is described by its electron configuration. This is a symbolic notation that indicates the distribution of electrons in an atom. An alternative method of representation, Lewis symbols, emphasizes the number of electrons in an atom's valence shell.

The valence electrons dictate the chemical reactivity of each element and, in turn, the periodicity of chemical properties (the result of identical outer-shell electron configurations within each group of elements). The tendency of atoms to attain noble-gas outer-shell electron configurations, expressed in the octet rule, is an important driving force for chemical reactivity.

Key Words

anion, p. 42
atom, p. 39
atomic mass, p. 40
atomic number, p. 42
aufbau, p. 55
cation, p. 42
electron, p. 41
emission spectrum, p. 50
isotope, p. 43
Lewis symbol, p. 61
mass number, p. 43
neutron, p. 42
nucleus, p. 42
octet rule, p. 61
periodic table, p. 44
proton, p. 41
quantum mechanics, p. 53
valence electron, p. 61

Exercises

Atomic Theory

2.1 Explain the concept of conservation of mass.

2.2 Explain the concept of the constant composition of chemical compounds.

2.3 Analysis of 3.800 g of glucose yields 1.520 g of carbon, 0.2535 g of hydrogen, and 2.027 g of oxygen. Calculate the mass percent of each element in the compound.

2.4 Analysis of 5.150 g of ethanol (ethyl alcohol) yields 2.685 g of carbon, 0.6759 g of hydrogen, and 1.789 g of oxygen. Calculate the mass percent of each element in the compound.

2.5 If carbon is assigned a mass of 12.0000 amu and oxygen weighs four-thirds as much as carbon, what is the atomic mass of oxygen?

2.6 If sulfur weighs twice as much as oxygen, what is its atomic mass?

2.7 If nitrogen weighs 0.875 the mass of oxygen, what is its atomic mass?

2.8 If beryllium is 0.75 the mass of carbon, what is its atomic mass?

Atomic Structure

2.9 Describe the mass and electrical charges of the three principal subatomic particles making up the atom.

2.10 If an electrically neutral atom can be shown to possess 16 electrons, what must its atomic number be? Identify the element.

2.11 Explain the fact that the mass of a newly discovered element is about twice its atomic number.

2.12 Complete the following chart by filling in the blanks.

Protons	Neutrons	Electrons	Mass (amu)	Element
19	20	____	____	____
34	____	____	79	____
____	20	____	40	____
____	____	11	23	____

2.13 What would be an important feature of an atom consisting of 23 protons, 28 neutrons, and 20 electrons?

2.14 What would be an important feature of an atom consisting of 9 protons, 10 neutrons, and 10 electrons?

Isotopes

2.15 Calculate the atomic mass of strontium, which has four naturally occurring isotopes:

Natural Abundance (%)	Atomic Mass (amu)
0.56	83.91
9.86	85.91
7.02	86.91
82.56	87.91

2.16 Element X consists of two isotopes, masses 25 amu and 26 amu, and its mass is reported as 25.6 in the table of atomic masses. What are the percent abundances of its two isotopes?

2.17 Write the symbolic isotopic notation, with superscripts and subscripts, for the following pairs of isotopes: oxygen 16 and 17; magnesium 24 and 25; silicon 28 and 29.

2.18 Write the symbolic isotopic notation, with superscripts and subscripts, for the following pairs of isotopes: sulfur 32 and 33; argon 36 and 40; chromium 52 and 54.

Periodic Table

2.19 What are the names of the elements having the following symbols? (a) Li, (b) Ca, (c) Na, (d) P, (e) Cl

2.20 What are the names of the elements having the following symbols? (a) He, (b) Mg, (c) K, (d) N, (e) F

2.21 Identify the period and group of the following elements: Li, Na, K, Rb, Cs

2.22 Identify the period and group of the following elements: Si, Ge, As, Sb

2.23 Can the elements in Exercise 21 be described as metals or nonmetals? Explain your answer.

2.24 Can the elements in Exercise 22 be described as metals or nonmetals? Explain your answer.

2.25 For which elements (main-group or transition) do the numbers of electrons in the outer, or valence, shell correspond to the group number?

2.26 Which group of the periodic table is known as the alkali metals?

2.27 Which group of the periodic table is known as the halogens?

2.28 Which group of the periodic table is known as the alkaline earth metals?

Atomic Structure and Periodicity

2.29 Can one observe the emission spectrum of an atom in the ground state? Explain your answer.

2.30 True or false: Atomic spectra indicate the changes in energy states of electrons. Explain your answer.

2.31 On the basis of the quantum mechanical model of the atom, state the main features of the organization of an atom's electrons.

2.32 State the number and names of the subshells of the first three electron shells—principal quantum numbers 1, 2, and 3.

2.33 What is the definition of an orbital?

2.34 What is the maximum number of electrons that can occupy any orbital?

2.35 How many orbitals are there in an *s* subshell?

2.36 What is the maximum number of electrons that a *p* subshell can accommodate?

2.37 Is there any spatial orientation of an *s* subshell?

2.38 Is there any spatial orientation of orbitals in *p* subshells? If so, describe it.

2.39 Which elements are described by the following electron configurations? (a) $1s^22s^22p^63s^2$; (b) $1s^22s^22p^63s^23p^4$; (c) $1s^22s^22p^63s^23p^64s^1$

2.40 Which elements are described by the following electron configurations? (a) $[He]2s^22p^1$; (b) $[He]2s^22p^5$; (c) $[Ar]4s^2$; (d) $[Kr]5s^2$

2.41 What positively charged ion (cation) is represented by the following electron configuration? $1s^22s^22p^6$

2.42 What negatively charged ion (anion) is represented by the following electron configuration? $1s^22s^22p^6$

2.43 Which group of the periodic table is known as the alkali metals, and what electron configuration do they have in common?

2.44 Which group of the periodic table is known as the halogens, and what electronic configuration do they have in common?

2.45 Which group of the periodic table is known as the alkaline earth metals, and what electronic configuration do they have in common?

2.46 Which group of the periodic table is known as the noble gases, and what electronic configuration do they have in common?

Unclassified Exercises

2.47 True or false: The number of protons in an atom is called the atomic number.

2.48 What is the name of the light emitted by the atoms of an element that have been injected into a flame?

2.49 How does the energy of an electron change as its distance from the nucleus increases?

2.50 What is the maximum number of electrons that can occupy an electron shell with principal quantum number $n = 3$?

2.51 What is the principal characteristic of the elements in the same group of the periodic table?

2.52 Is sulfur a metal or a nonmetal?

2.53 Is calcium a metal or a nonmetal?

2.54 What is the chief property of the elements at the right-hand end of every period of the periodic table?

2.55 What is the symbolic notation of the two isotopes of oxygen whose masses are 16 and 18?

2.56 What is the octet rule?

2.57 The element neptunium possesses 93 electrons and weighs 237 amu. Calculate the contribution (in %) of its electrons to its total mass.

2.58 What are the similarities and differences between the two principal isotopes of carbon, atomic number 6, whose atomic masses are 12 and 13, respectively?

2.59 If the frequency of blue light is greater than that of red light, which has the greater energy?

2.60 Will the emission spectrum of an atom contain all frequencies of light? Explain your answer.

2.61 What is the name of the rule stating that, when the outermost *s* and *p* subshells of elements in any period are filled (that is, have a total of eight electrons), the elements are in the greatest state of stability?

Chemical Connections

2.62 In Example 2.1, the mass percentages of carbon, hydrogen, and oxygen in vitamin C were determined. Calculate the mass in grams of each of these elements in an 8.200-g sample of pure vitamin C.

2.63 Titanium has five naturally occurring isotopes of atomic masses 45.95, 46.95, 47.95, 48.95, and 49.95, respectively. What do you need to know to account for the fact that titanium's mass in the table of atomic masses is listed as 47.88?

2.64 If necessary, correct the following statement: Most of the elements listed in the periodic table have equal numbers of proton, electrons, and neutrons.

2.65 Criticize the following statement: All of Dalton's atomic theory is considered valid today.

2.66 Can you think of a method to detect sodium, potassium, or calcium in body fluids? Explain your answer.

CHAPTER 3

MOLECULES AND CHEMICAL BONDS

CHEMISTRY IN YOUR FUTURE

One of your tasks as a nutritionist is to instruct people placed on special diets by their physicians. For example, people with chronically high blood pressure are often put on a low-sodium diet. Most of these clients understand that this means limiting their use of table salt (sodium chloride), but many are not aware that it also means reducing their use of foods preserved with sodium nitrite, flavored with monosodium glutamate, and so forth. The ingestion of these substances contributes to the body's total concentration of sodium ions—but how? The answer to this and similar questions can be found in Chapter 3.

LEARNING OBJECTIVES

- Describe the difference between ionic and covalent bonds.
- Describe the relation between the octet rule and the formation of ions.
- Predict the formulas of binary ionic compounds.
- Construct the names of ionic compounds.
- Construct the names of covalent compounds.
- Create Lewis structures.
- Describe the difference between polar and nonpolar covalent bonds.
- Use VSEPR theory to predict molecular shape.

How many different substances are there? The number of substances known is estimated to be about 1×10^7, but the precise number can never be known, because new substances are created daily by chemists throughout the world. Most of the synthetic fibers of your clothes, the preservatives in your food, the antibiotics prescribed by your doctor, and the human insulin now produced by bacteria did not exist 50 years ago. Biological structures such as membranes, chromosomes, and muscle can now be described in chemical detail. Medicine has acquired a more comprehensive scientific basis because of the many chemical laboratory tests now available for the diagnosis of pathological conditions. All these remarkable accomplishments have resulted from advances in the understanding of the nature of chemical bonds.

To explain bonding, we begin with the octet rule presented in Chapter 2. All elements, with the exception of the noble gases, are unstable and react chemically to achieve the outer electron configuration of the noble gases. The result of that chemical process is the formation of compounds in which the atoms are connected to each other by chemical bonds.

There are two major types of chemical bonds, ionic and covalent. An **ionic bond** is a strong electrostatic attraction between a positive ion and a negative ion. It results when electrons are transferred from the valence shell of one atom into the valence shell of another. A **covalent bond** is a bond formed when electrons are shared between atoms. In general, ionic bonds are formed between metals and nonmetals, and covalent bonds are formed between nonmetals. As we proceed in our discussion of chemical bonding, however, you will find that the differences between ionic and covalent bonds are more of degree than of kind.

3.1 IONIC VERSUS COVALENT BONDS

The kind of bond—ionic or covalent—that will form as a result of a chemical reaction depends on the relative abilities of the reacting atoms to attract electrons. When one atom has a much greater ability to attract electrons than the other atom does, electron transfer will occur, and the resulting bond will be ionic. This is the case when metals react with nonmetals. When two atoms are similar in their ability to attract electrons, electron transfer is no longer possible. In those cases, octet formation is accomplished through electron sharing, and the resulting bond is covalent. This kind of bonding generally takes place in reactions between the nonmetals.

The ease or difficulty of removing electrons from the outer, or valence, shell of an element depends on how strongly the positively charged nucleus interacts with the outer electrons. It takes work to overcome the attraction of the positively charged nucleus for the negatively charged electron, and the energy for that process is called the **ionization energy.** If the ionization energy of metals were prohibitively high, so that electrons could not be removed from the valence shell of metals, then iron would never rust and silver would never tarnish.

Ionization energy is one of the chemical properties that shows periodicity when elements are arranged by increasing atomic number. In period 1, for example (Figure 3.1), the ionization energy increases across the period from Li through Be, B, C, N, O, F, and Ne. In general, the ionization energies increase across any given period, proceeding from Group I to Group VIII. The reason for this trend is that, within a period, the nuclei grow in positive charge, and, at the same time, the distance between each nucleus and the electrons going into that outer shell remains nearly the same. As a result, the intensity of electrical attraction between nuclei and outer electrons increases with atomic number across each period, and it becomes more and more difficult to

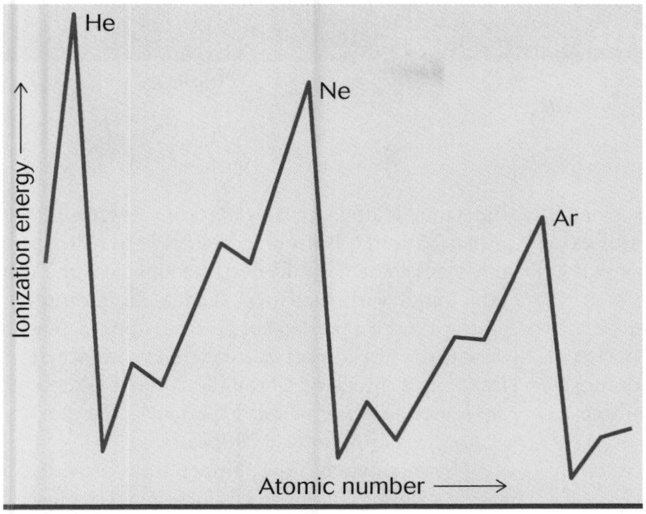

Figure 3.1 The ionization energies of the elements through the first four periods to scandium (atomic number 21). Ionization energy is the work required to remove an electron from an element's valence shell. It is greatest for elements at the end of a period (the noble gases) and smallest for elements at the beginning of a period (the Group I metals).

remove electrons from the valence shell. (See Box 3.1 on the following page, On the Intensity of Electrical Fields.)

On the other hand, ionization energy decreases from the top to the bottom of any group of the periodic table. Within Group I, for example, the energy needed to form cations is greatest for lithium and decreases in the order lithium > sodium > potassium > rubidium > cesium. It is easier to form a cesium ion than a lithium ion, because the number of electron shells increases down each group. Thus the outer electrons become increasingly insulated from the influence of the positively charged nucleus (see Box 3.1).

Concept checklist

- ✔ An ionic bond will form if one of the bonding partners has a much greater electron-attracting power than the other.
- ✔ A covalent bond will form if the electron-attracting powers of the bonding partners are similar in strength.

3.2 IONIC BONDS

The chemical reactivity of the elements depends, then, on their different tendencies to gain, lose, or share the electrons in their valence shells. After reaction, the valence shells have the configuration characteristic of a noble gas—they have acquired a complete octet.

Lewis Symbols and Formulas of Ionic Compounds

Lewis symbols were used to illustrate these changes in Example 2.14, where we observed that octet formation in the reaction between a metal, lithium, and a nonmetal, fluorine, created electrically charged atoms called ions, specifically lithium ions and fluoride ions. Let's extend that approach to metals in Groups II and III of the periodic table. We will use Lewis symbols to analyze the reactions between the metals lithium, magnesium, and aluminum and the nonmetal fluorine.

The group numbers of the representative elements tell us the number of electrons in the valence shell. Lithium, in Group I, has one electron in its valence shell; magnesium, in Group II, has two electrons in its valence shell; and aluminum, in Group III, has three electrons in its valence shell. Fluorine, a

3.1 Chemistry in Depth

On the Intensity of Electrical Fields

We have seen that ionization energy increases from left to right across the periodic table and decreases from top to bottom. The key to understanding this pattern is the variation, from element to element, in the attraction that the nucleus exercises over the valence electrons. Many of the properties of molecules depend on the intensity of the electrical field surrounding their component atoms, so it is worthwhile to consider this phenomenon in more detail. To simplify the concept, we will use the intensity of light as a stand-in or a model for the intensity of the field surrounding an electrical charge.

Imagine a library containing a circular reading table with a seating capacity of eight. The reading lamp in the center of the table is fitted with a bulb whose wattage can be varied and depends (magically) on the number of readers at the table. Each time a reader sits at the table, the lamp's intensity miraculously increases by 25 watts. When the first reader sits at the table, the bulb lights with an intensity of 25 watts. With the seating of the next reader, the lamp's intensity automatically increases by an additional 25 watts. Both readers experience the increase in intensity of light. Six additional readers join the others at the table, and, with each new reader, 25 additional watts are added to the reading lamp. Each time a new reader sits down, the already seated readers experience an increase in the lamp's intensity. The table is filled when eight readers are seated, each of whom experiences the maximum intensity of the lamp, now at 200 watts.

Unfortunately, the library has provided no other reading light. Therefore any new readers must sit in a circle outside the first. The light's intensity continues to increase by 25 watts for each new reader, but they must sit at a greater distance from the source than the first eight readers.

The intensity of light from a source such as an electric light bulb decreases rapidly with distance. We have all experienced what happens as we move away from or move toward a light source. As we move toward the light, it be-

nonmetal, is in Group VII and has seven electrons in its valence shell. The convention used to describe ionic charge is to write the number first and the sign of the charge second.

Lewis symbols (Section 2.10) depict the reaction between lithium and fluorine, with arrows indicating the transfer of electrons, as

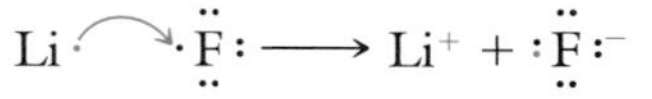

The reaction between magnesium and fluorine is

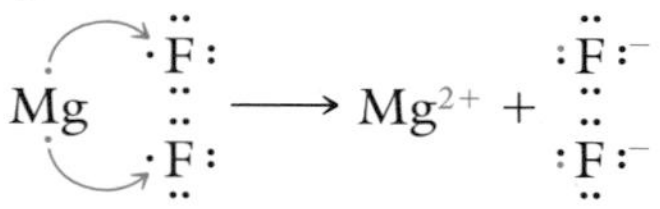

The reaction between aluminum and fluorine is

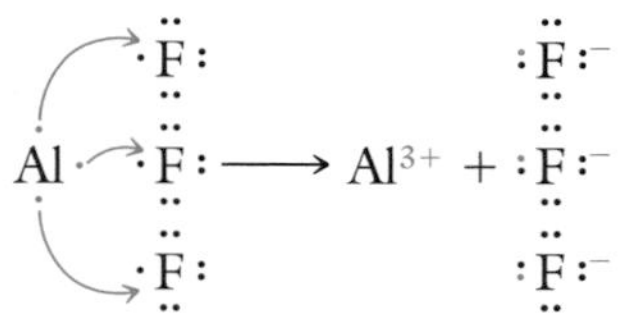

In the reaction between lithium and fluorine, lithium needed to lose only one electron to achieve the helium electron configuration. (Helium's electron configuration is a duet rather than the octets of succeeding periods, but helium is a noble gas nonetheless.) Remember, the electron loss can occur only if the electron can be accepted by another atom. For that reason, the process is more properly described as **electron transfer** rather than electron loss. Fluorine required only one electron for octet formation in its valence shell. The electron transfer therefore required 1:1 ratio of reacting atoms, 1 Li to 1 F.

comes more intense. As we move away, the intensity of the light decreases. We experience the same effect with respect to a source of heat. The closer we get, the more intensely we feel the heat; and the farther away, the less we feel the heat. These effects can be precisely expressed mathematically by a relation that shows the intensity of radiation such as light or heat (electromagnetic radiation) decreases with the square of the distance from the source of radiation:

$$\text{Intensity} \propto \frac{1}{(\text{distance})^2}$$

This relation tells us that, if the distance from the light source doubles, the intensity decreases by a factor of 4. The same relation holds for the intensity of electrical attraction. We will see it again in Chapter 10.

Because of their distance from the source, the intensity of light is less for those in the second circle than for those in the first, even though the intensity of the light source has increased. In fact, the light intensity experienced by the readers in the second circle is even lower than distance would predict because, in addition to the distance factor, the readers in the first circle obscure some of the light simply by getting in the way.

This decrease in intensity is precisely what electrons experience as they are added to a growing nucleus in the Aufbau process. Instead of light intensity, however, it is the positive charge of the nucleus that reaches out to them, with diminishing effect. The electron that enters the fluorine atom to form a fluoride ion is the eighth "reader" sitting down at the table, and it experiences close to the maximum electrical attraction that any nucleus in the periodic table could exert over an additional electron. This is why fluorine's ionization energy is the greatest of all the elements (except the noble gases). At the other extreme is cesium, whose outer electron is very weakly held by its nucleus despite the fact that the nuclear charge is 55+. There are four "circles of readers" intervening between cesium's outer electron and its "reading lamp."

In the reaction between magnesium and fluorine, magnesium needed to lose two electrons to achieve the neon electron configuration. Fluorine again required only one electron for octet formation in its valence shell. The transfer of two electrons from magnesium therefore required acceptance of those two electrons by two fluorine atoms. This resulted in a 1:2 ratio of reacting atoms, 1 Mg to 2 F.

In the reaction between aluminum and fluorine, aluminum needed to lose three electrons to achieve the neon electron configuration. Fluorine again required only one electron for octet formation in its valence shell. The transfer of three electrons from aluminum therefore required acceptance of those three electrons by three fluorine atoms. This resulted in a 1:3 ratio of reacting atoms, 1 Al to 3 F.

Although no oxygen is involved, chemists consider the loss of electrons to be an oxidation, and the gain in electrons a reduction.

A chemical compound is identified by its **formula,** which is a statement about the combining ratio of its constituent elements. In the formula, the symbols of all the elements in the compound are combined with subscripts that tell us the relative numbers of atoms of each element. When the formula of any ionic compound contains only one cation or anion, its subscript is left out, and the 1 is understood. Thus, when lithium reacts with fluorine, the 1:1 ratio of atoms in the resulting product is expressed as the formula of the newly formed compound, LiF. The 1:2 ratio of ions in the product of the magnesium–fluorine reaction is described in the formula MgF_2. Similarly, the 1:3 ratio of reacting aluminum and fluorine atoms leads to a product of formula AlF_3. Two-element compounds such as these are called **binary compounds.** The first column in the following table lists the relative numbers of atoms of each element as a ratio, and the second column lists the corresponding formulas of each of the three compounds.

Ratio of Atoms in Each Compound	Formulas of Each Compound
1 Li:1 F	LiF
1 Mg:2 F	MgF_2
1 Al:3 F	AlF_3

When we look at the reaction of Group III metals with Group VII, Group VI, and Group V nonmetals, a similar pattern emerges. The first reaction is a repeat of the aluminum–fluorine reaction:

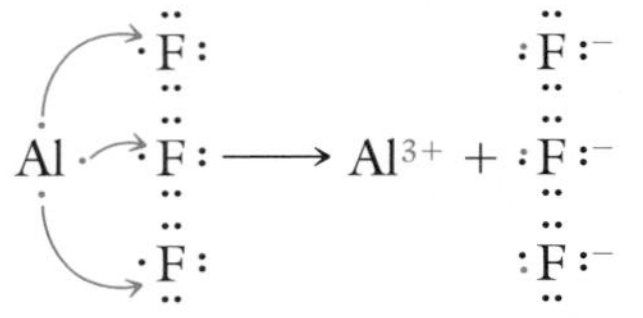

Look next at the reaction between aluminum and oxygen (Group VI):

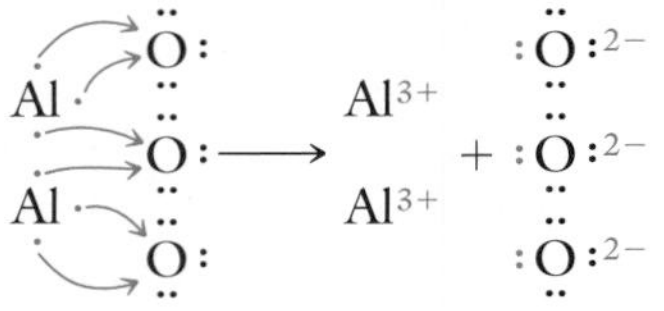

Finally, examine the reaction between aluminum and nitrogen (Group V):

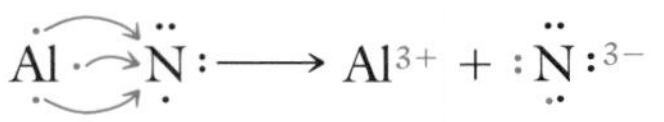

Aluminum must lose three electrons to achieve the neon electron configuration. It therefore requires partners that can accept those three electrons. We have already seen how the transfer takes place in the aluminum–fluorine reaction. In the aluminum–oxygen reaction, oxygen can accept only two electrons to complete its octet, and so the electron exchange requires that three oxygen atoms react with two aluminum atoms. Nitrogen (Group V), however, requires three electrons to complete its octet, and aluminum needs to lose three electrons to achieve noble-gas configuration; so, when these two elements react, a 1:1 exchange takes place. The first column in the following table lists the relative numbers of atoms of each element as a ratio, and the second column lists the corresponding formulas of each of the three compounds.

Ratio of Atoms in Each Compound	Formulas of Each Compound
1 Al:3 F	AlF_3
2 Al:3 O	Al_2O_3
1 Al:1 N	AlN

Formulas of Ionic Compounds

To predict the formula of a binary ionic compound, one first needs to know the ionic charges of all component elements.

Example 3.1 Predicting the electrical charges of ions

What are the electrical charges of ions formed from elements of Groups I, II, III, V, VI, and VII?

Solution

Remember that the elements in Groups I, II, and III are metals. They lose electrons from their valence shells to form cations, whose positive charge is equal to their group number. In contrast, Groups V, VI, and VII gain electrons to form anions, whose charge is equal to the group number minus 8:

Group number	I	II	III	V	VI	VII
Ion charge	1+	2+	3+	3−	2−	1−

Problem 3.1 Predict the charge of ions formed from the elements (a) potassium; (b) calcium; (c) indium; (d) phosphorus; (e) sulfur; (f) bromine.

The next step in determining the formulas of binary ionic compounds is to satisfy the requirement that the net electrical charge of an ionic compound must be zero. In other words, the magnitude of the total positive charge must be equal to the magnitude of the total negative charge. Therefore, once we know the charges of the individual ions, we use that information to calculate how many of each ion are needed to achieve a net, or overall, charge of zero. We call this the net-charge approach to determine the formulas of ionic compounds. Dot diagrams allowed us to find the answers by visual examination, but there are less-cumbersome ways to do it as well. Take another look at the aluminum–oxygen reaction.

In Example 3.1, we determined that aluminum forms ions of 3+, and oxygen of 2−. To find a combination of aluminum and oxygen ions in which the net charge is zero, we can multiply those charge numbers by factors that will make the total positive and total negative charges equal. If we multiply the charge on the aluminum ion by 2, and the charge on the oxygen ion by 3, the positive and negative charges are the same, and the net charge is zero.

$$Al^{3+} \times 2 = 6+$$
$$O^{2-} \times 3 = 6-$$

We take the factors that were used to multiply each elemental ion and insert them into the formula as subscripts of those same elements. The formula for the ionic compound formed by aluminum and oxygen is therefore Al_2O_3.

You can take the guesswork out of finding the right multipliers by simply using the number of the cation charge as the anion subscript and using the number of the anion charge as the cation subscript. This shortcut has been called the “cross-over approach.” For the aluminum–oxygen compound,

$$Al^{3+} \times O^{2-} \longrightarrow Al_2O_3$$

When the ions have the same charge coefficient (for example, Ca^{2+} and O^{2-}), a literal use of the cross-over approach would yield the formula Ca_2O_2. Therefore, when the charge values are the same, assume a 1:1 ratio of ions, and the formula for calcium oxide is CaO.

Example 3.2 Predicting the formulas of ionic compounds by using the net-charge approach

Use the net-charge approach to predict the formulas for the ionic compounds formed from (a) Li and Cl; (b) Al and I; (c) Mg and N.

Solution

Use the elements' group numbers to discover their ionic charges. Multiply these ionic charges by factors that result in a net charge of zero for each compound.

(a) $Li^{+} \times 1 = 1+$
$Cl^{-} \times 1 = 1-$
Formula: LiCl

(b) $Al^{3+} \times 1 = 3+$
$I^- \times 3 = 3-$
Formula: AlI_3

(c) $Mg^{2+} \times 3 = 6+$
$N^{3-} \times 2 = 6-$
Formula: Mg_3N_2

Problem 3.2 Use the net-charge approach to predict the formulas for the ionic compounds formed from (a) K and Br; (b) Ga and F; (c) Ca and P.

Example 3.3 Predicting the formulas of ionic compounds by using the cross-over approach

Use the cross-over approach to predict the formulas of the ionic compounds formed from (a) Na and I; (b) Ca and N; (c) Al and S.

Solution

Use the elements' group numbers to discover their ionic charges. Use the number of the charge on one element as the other element's subscript.

(a) $Na^{+} \; I^{-} \longrightarrow Na_1I_1 \longrightarrow NaI$

(b) $Ca^{2+} \; N^{3-} \longrightarrow Ca_3N_2$

(c) $Al^{3+} \; S^{2-} \longrightarrow Al_2S_3$

Problem 3.3 Use the cross-over approach to predict the formulas of the ionic compounds formed from (a) K and Cl; (b) Mg and P; (c) Ga and O.

Figure 3.2 summarizes the preceding discussion about the formulas of binary ionic compounds. It shows the possible combining ratios for anions (abbreviated N, for nonmetal) and cations (abbreviated M, for metal) of various charges. At the juncture of column and row in Figure 3.2 is the formula for the compound formed between a metal ion of a given charge (1+, 2+, or 3+) and a nonmetal ion of a given charge (1−, 2−, or 3−). Each formula results in a net charge of zero.

We have seen that, when a single metal and a single nonmetal react, the result is the formation of a binary ionic compound. Other ions and their compounds are somewhat more complex, but the binary ionic compounds are a good place to start to learn how ionic compounds are named.

		Nonmetal ion		
		1−	2−	3−
Metal ion	1+	MN	M_2N	M_3N
	2+	MN_2	MN	M_3N_2
	3+	MN_3	M_2N_3	MN

Figure 3.2 Possible combining ratios for ionic compounds. The formula of the binary compound produced when a metal (M) reacts with a nonmetal (N) can be found where the row representing the charge on a metal ion meets the column representing the charge on a nonmetal ion.

TABLE 3.1 Names and Formulas of Ions with Variable Charge

Stock System Name	Ion Formula	Common Name
copper(I)	Cu^{+}	cuprous
copper(II)	Cu^{2+}	cupric
iron(II)	Fe^{2+}	ferrous
iron(III)	Fe^{3+}	ferric
lead(II)	Pb^{2+}	plumbous
lead(IV)	Pb^{4+}	plumbic
mercury(I)	Hg_2^{2+}	mercurous
mercury(II)	Hg^{2+}	mercuric
tin(II)	Sn^{2+}	stannous
tin(IV)	Sn^{4+}	stannic
titanium(III)	Ti^{3+}	titanous
titanium(IV)	Ti^{4+}	titanic

3.3 NAMING BINARY IONIC COMPOUNDS

In the names of binary ionic compounds, the name of the metal element is included unchanged, followed by the first part of the name of the nonmetal element combined with the suffix **-ide.** Examples include potassium chloride (KCl), aluminum oxide (Al_2O_3), and calcium fluoride (CaF_2).

This simple rule suffices for elements contained within the first three periods of the periodic table. However, more elaboration is required for the elements of period 4 and beyond, because many of those elements, among them the transition elements, can form ions of more than one charge type. For example, iron can form ions with charges of 2+ and 3+, and copper can form ions with charges of 1+ and 2+. Therefore, the chlorides of these metals can have the compositions $FeCl_2$, $FeCl_3$, CuCl, and $CuCl_2$. The older, and sometimes commonly used, names of these compounds are ferrous chloride, ferric chloride, cuprous chloride, and cupric chloride, respectively; but a modern, more systematic method for naming them is becoming more widely used.

The modern method of nomenclature for ions with more than one charge type is called the **Stock system.** In this system, the charge of the cation is indicated with a roman numeral: iron(II) chloride, iron(III) chloride, copper(I) chloride, and copper(II) chloride. The combining ratios (the formulas) of such compounds can be deduced from their names and the charge of the anion. Table 3.1 lists the commonly used names, the Stock system names, and the charges of several common metal ions of variable charge.

Example 3.4 Writing the Stock system names of ionic compounds

Write the Stock system names of the following compounds: (a) $TiCl_3$; (b) $FeCl_3$; (c) CuCl; (d) $CoCl_2$.

Solution

We determine the charge of the cation in each compound by examining the combining ratios and the charge of the anion. In each of these cases, the chloride ion possesses a stable electrical charge of 1−. Therefore, Ti must be Ti^{3+}, Fe must be Fe^{3+}, Cu must be Cu^{+}, and Co must be Co^{2+}. The compound names are therefore (a) titanium(III) chloride; (b) iron(III) chloride; (c) copper(I) chloride; and (d) cobalt(II) chloride.

Problem 3.4 Referring to Table 3.1, write the common names for the following compounds: (a) $TiCl_3$; (b) $HgCl_2$, (c) $FeCl_2$; (d) PbO_2.

3.4 POLYATOMIC IONS

Some ions consist of two or more atoms combined into a single charged unit. These ions are called **polyatomic ions.** The names and formulas of some of the more important ones are listed in Table 3.2. In Chapter 9, we shall consider the chemical origin of polyatomic ions. For now, we confine ourselves to the formulas and names of compounds resulting when polyatomic ions combine with ions of opposite charge.

As with other ions, when a polyatomic ion is present in a compound, the sum of positive and negative charges must be zero. If two or more of a given polyatomic ion are present in a formula, the formula is written with the polyatomic ion enclosed in parentheses followed by a subscript indicating the number of those ions in the formula. For example, aluminum carbonate is $Al_2(CO_3)_3$, and calcium phosphate is $Ca_3(PO_4)_2$. If only one polyatomc ion is present, no parentheses (or subscript) are necessary. For example, potassium carbonate is K_2CO_3, and sodium phosphate is Na_3PO_4. Finally, by convention, in every ionic compound, the cation is indicated first, followed by the anion.

Example 3.5 Writing formulas of compounds containing polyatomic ions

Write the formulas of the following compounds: (a) sodium acetate; (b) ammonium phosphate; (c) potassium hydrogen phosphate; (d) calcium dihydrogen phosphate.

Solution

Find the polyatomic ion from each compound in Table 3.2 and note its formula and ionic charge:

Acetate	$C_2H_3O_2$	1−
Ammonium	NH_4	1+
Phosphate	PO_4	3−
Hydrogen phosphate	HPO_4	2−
Dihydrogen phosphate	H_2PO_4	1−

Remember that sodium and potassium, both in Group I, form cations of charge 1+; and calcium, in Group II, forms a cation of charge 2+. Using Figure 3.2 or the methods illustrated in Examples 3.1 and 3.2, we therefore determine the formulas of these compounds to be (a) $NaC_2H_3O_2$; (b) $(NH_4)_3PO_4$; (c) K_2HPO_4; (d) $Ca(H_2PO_4)_2$.

Problem 3.5 Write the names of the following compounds: (a) NH_4HSO_3; (b) $CaCO_3$; (c) $Mg(CN)_2$; (d) $KHCO_3$; (e) $(NH_4)_2SO_4$.

3.5 DOES THE FORMULA OF AN IONIC COMPOUND DESCRIBE ITS STRUCTURE?

Figure 3.3 represents an experimental arrangement called a conduction cell. It consists of a vessel containing a medium that will not conduct electricity—for example, pure water; a pair of electrodes, one negative and the other positive, attached to an electrical power source; and a device that can reveal the flow of

TABLE 3.2 Some Important Polyatomic Ions

Ion	Name	Ion	Name
NH_4^+	ammonium	CO_3^{2-}	carbonate
NO_3^-	nitrate	HCO_3^-	hydrogen carbonate
NO_2^-	nitrite	PO_4^{3-}	phosphate
OH^-	hydroxide	HPO_4^{2-}	hydrogen phosphate
$C_2H_3O_2^-$	acetate	$H_2PO_4^-$	dihydrogen phosphate
CN^-	cyanide	SO_4^{2-}	sulfate
HSO_4^-	hydrogen sulfate	HSO_3^-	hydrogen sulfite

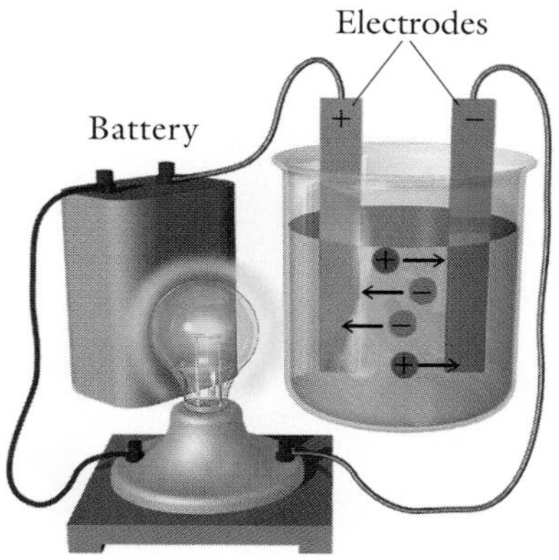
Electrodes
Battery
+
−

Figure 3.3 Diagram of an electrical conductivity cell. It consists of a vessel containing pure water (to which experimental substances can be added); a pair of electrodes (one negative, one positive) attached to an electric power source; and an electric light bulb that lights up when an electric current is able to flow through the medium between the electrodes.

an electric current through the cell. This device could be as simple as an electric light bulb that will light up whenever an electric current flows through the medium between the electrodes.

Ionic solids do not conduct electricity; neither does pure water. However, when ionic solids, such as sodium chloride, are dissolved in water or melted, the solutions and the melted solids do conduct electricity. For this reason, ionic compounds are called **electrolytes.** Water—and compounds such as glucose, which when dissolved in water do not conduct electricity—are called **nonelectrolytes.**

In their solid form, the ions of an ionic compound are locked into place and all their charges are neutralized. It is only when the solid melts or is dissolved in water, thereby freeing the individual ions, that the ions can carry

A PICTURE OF HEALTH

Examples of Molecules and Ions in the Body

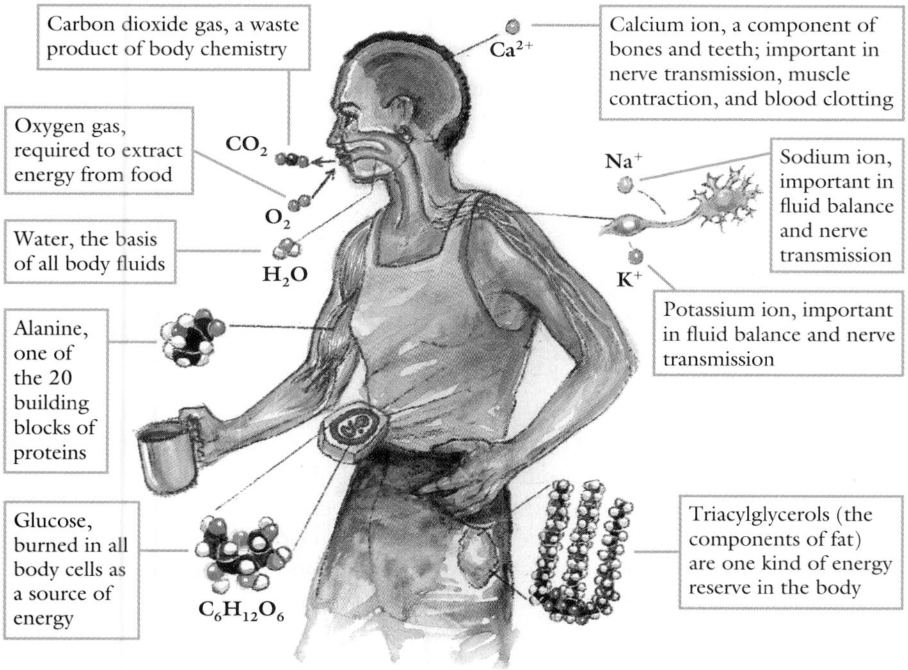
Carbon dioxide gas, a waste product of body chemistry
Oxygen gas, required to extract energy from food
Water, the basis of all body fluids
Alanine, one of the 20 building blocks of proteins
Glucose, burned in all body cells as a source of energy
CO_2
O_2
H_2O
$C_6H_{12}O_6$
Ca^{2+}
Calcium ion, a component of bones and teeth; important in nerve transmission, muscle contraction, and blood clotting
Na^+
Sodium ion, important in fluid balance and nerve transmission
K^+
Potassium ion, important in fluid balance and nerve transmission
Triacylglycerols (the components of fat) are one kind of energy reserve in the body

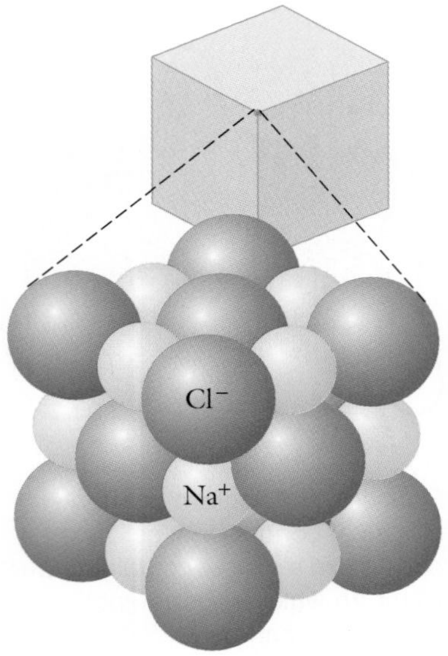

Figure 3.4 A drawing of the crystal structure of sodium chloride. Because each sodium ion is surrounded by six chloride ions and each chloride ion is surrounded by six sodium ions, it is not possible to identify a unique structural unit.

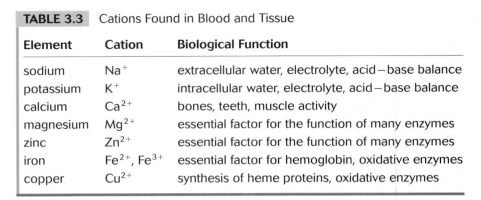

TABLE 3.3 Cations Found in Blood and Tissue

Element	Cation	Biological Function
sodium	Na^+	extracellular water, electrolyte, acid–base balance
potassium	K^+	intracellular water, electrolyte, acid–base balance
calcium	Ca^{2+}	bones, teeth, muscle activity
magnesium	Mg^{2+}	essential factor for the function of many enzymes
zinc	Zn^{2+}	essential factor for the function of many enzymes
iron	Fe^{2+}, Fe^{3+}	essential factor for hemoglobin, oxidative enzymes
copper	Cu^{2+}	synthesis of heme proteins, oxidative enzymes

electrical current by moving to an electrode of opposite charge. These phenomena reveal that ionic bonds exist only in the solid state. Once the solid is melted or dissolved in water, the constituent ions become virtually independent of one another.

That independence explains why we can speak of sodium or potassium cations in blood or tissue without referring to a corresponding anion. The sodium and potassium ions, not their compounds, are of primary importance in their biological function. Table 3.3 lists several of the cations found in blood. Their cationic charges are balanced by the presence of chloride ions, bicarbonate ions, and proteins, which normally have a negative ionic charge in blood and tissue.

Ionic solids consist of three-dimensional arrays of positive ions surrounded by negative ions and negative ions surrounded by positive ions. This arrangement makes it impossible to identify a discrete structural unit that can be represented by the formula of the solid (Figure 3.4). There is no such structural unit as a molecule for an ionic compound.

When we consider the nature of covalent bonds, you will see that this distinction becomes very important. The formulas of compounds such as water (H_2O), carbon dioxide (CO_2), and oxygen (O_2), whose chemical bonds are covalent, represent not only the combining ratios of component elements, but also the actual, ultimate structural units that we can call molecules.

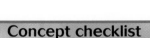

Concept checklist

- ✔ The formula of an ionic compound represents only the combining ratios of the component ions, not a unique structural unit.
- ✔ In contrast with ionic solids, the chemical identities of molecules, such as water, are retained whether they are in the solid, liquid, or gaseous state.

3.6 COVALENT COMPOUNDS AND THEIR NOMENCLATURE

One of the simplest covalent bonds unites two hydrogen atoms to form the hydrogen molecule, H_2. Hydrogen atoms possess only one electron, and they can achieve the noble-gas configuration of helium by gaining one electron more. The helium configuration calls for a 1*s* orbital containing two electrons. Hund's rule tells us that two electrons can share an orbital only if the electrons have opposite spins. Therefore, a single covalent bond uniting two atoms consists of two spin-paired electrons shared by both atoms.

Before proceeding to a detailed discussion of covalent bonds, let's first consider some aspects of the system used for naming covalent compounds. As

TABLE 3.4 Systematic Names of Covalent Compounds

Nitrogen and Oxygen Compound	Systematic Name	Carbon and Oxygen Compound	Systematic Name
N_2O	dinitrogen monoxide	CO	carbon monoxide
NO	nitrogen monoxide	C_2O_3	dicarbon trioxide
NO_2	nitrogen dioxide	CO_2	carbon dioxide
N_2O_3	dinitrogen trioxide		
N_2O_4	dinitrogen tetroxide		
N_2O_5	dinitrogen pentoxide		

with ionic compounds, we will find that the names themselves provide considerable information on which to base a first acquaintance.

Most of the compounds that we shall refer to in this chapter—such as ammonia, carbon dioxide, and water—have common names with which you are already familiar, names adopted before the development of systematic methods of nomenclature, and those are the names that we shall be using for them. However, should an unfamiliar substance be referred to, a systematic method is available to us.

The systematic method for naming simple, binary molecular compounds is based on the fact that some pairs of elements can form more than one covalently bonded compound. Two such pairs, for example, are nitrogen and oxygen as well as carbon and oxygen. The system adopted for naming their compounds uses Greek prefixes (see the table in the margin) and is illustrated in Table 3.4.

You can take your first clue concerning a covalent compound's name from the order in which its elements appear in the chemical formula: the first element in the formula is named first; the second element is named second, but the name is altered to accommodate the suffix **-ide** (as in the naming of simple ionic compounds). The Greek prefixes are added to the elements' names to indicate the numbers of different kinds of atoms in each molecule of the compound. Note, however, that NO is called nitrogen monoxide, not mononitrogen monoxide, and CO is called carbon monoxide, not monocarbon monoxide. The prefix **mono-** is never used for naming the first element.

GREEK PREFIXES IN CHEMICAL NOMENCLATURE

1	mono-
2	di-
3	tri-
4	tetra-
5	penta-
6	hexa-
7	hepta-
8	octa-
9	nona-
10	deca-

Example 3.6 Naming covalent compounds

Name the following compounds: (a) CCl_4; (b) PCl_3; (c) PCl_5; (d) SO_2.

Solution

(a) Carbon tetrachloride; (b) phosphorus trichloride; (c) phosphorus pentachloride; (d) sulfur dioxide.

Problem 3.6 Name the following compounds: (a) NI_3; (b) P_2O_5; (c) S_2Cl_2; (d) SO_3.

3.7 REPRESENTATION OF COVALENT BONDS

Chemists use a number of different notation methods to represent covalent bonds and the ways in which they are organized within molecules. One such method is electron-dot notation. It was used in Chapter 2 to represent the structure of the valence shells of the elements, but it can be effectively adapted for representing covalent bonds as well.

When electron-dot notation is used to demonstrate the structure of molecules, the resultant diagrams are called **Lewis structures.** These diagrams are named after their originator, Gilbert N. Lewis, who used them to predict how atoms are connected to each other within a given molecule. The key concept in building Lewis structures lies in satisfying the octet rule, the rule stating that an element achieves stability by filling its outer shell with eight electrons. However, recall that hydrogen forms a duet to achieve stability.

The **combining power** of an element is the number of covalent bonds it forms to complete its octet. For example, carbon, in Group IV, requires four additional electrons to fill its octet. It can do this by forming four covalent bonds with hydrogen to form methane, CH_4, or with chlorine to form carbon tetrachloride, CCl_4. Nitrogen and other elements in Group V require three additional electrons and can form three covalent bonds. The Group VI elements, such as oxygen and sulfur, require two additional electrons and can form two bonds to complete their octets. Group IV elements are called tetravalent; Group V elements, trivalent; Group VI elements, divalent; and Group VII elements, monovalent (the Greek prefixes again). Table 3.5 contains nonmetals from Groups IV through VII and indicates their bonding requirements for octet formation.

Example 3.7 Predicting formulas of simple molecular compounds

For each of the following pairs of elements, predict the formula of the product of the reaction between the two elements. The order of the elements in each formula is that given for each pair of elements in the example. (a) carbon and bromine; (b) nitrogen and iodine; (c) carbon and sulfur.

Solution

Refer to Table 3.5 for the bonding requirements (combining power) of nonmetals.

(a) Carbon is tetravalent and bromine is monovalent; therefore, carbon will react with four bromine atoms, and the compound formed is CBr_4.

(b) Nitrogen is trivalent and iodine is monovalent; therefore, nitrogen will react with three iodine atoms, and the compound formed is NI_3.

(c) Carbon is tetravalent and sulfur is divalent; therefore, carbon will react with two sulfur atoms, and the compound formed is CS_2.

Problem 3.7 For each of the following pairs of elements, predict the formula for the product of the reaction between the two elements. The order of the elements in each formula is that given for each pair of elements in the problem. (a) carbon and hydrogen; (b) phosphorus and chlorine; (c) carbon and oxygen.

Let's try out the Lewis method of representing molecular structure by using it to examine how two fluorine atoms react to form a diatomic molecule. If each fluorine atom shares one electron with the other, then, in effect, they will both have eight. In this way, each fluorine atom can achieve the stable electron configuration of neon. Note that the result of their bonding is the sharing of a pair of electrons between them:

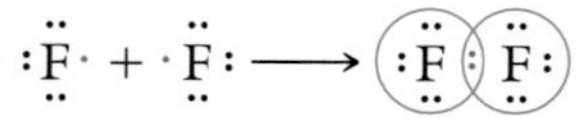

On the left side of the arrow, the reacting atoms are represented with numbers of valence electrons equal to their group number. On the right side of the arrow, we see the fluorine molecule, with each atom surrounded by a completed

TABLE 3.5 Bonding Requirements for Octet Formation of Some Nonmetals*

Group			
IV	V	VI	VII
$-\overset{\vert}{\underset{\vert}{C}}-$	$-\overset{\vert}{N}-$	$-O-$	$F-$
	$-\overset{\vert}{P}-$	$-S-$	$Cl-$
			$Br-$
			$I-$

* The lines indicate the number of bonds needed for octet formation.

octet. (Overlapping circles have been drawn around the bonded fluorine atoms to indicate that the shared pair of electrons is part of each atom's octet.)

The Lewis structure (the molecule drawn with Lewis symbols) is simplified by representing the bonding pair of electrons as a line between the atoms and the other pairs as dots surrounding the atoms: $:\ddot{\underset{\cdot\cdot}{F}}-\ddot{\underset{\cdot\cdot}{F}}:$. The pairs of electrons not shared in the covalent bond are called **nonbonded electrons** or **lone pairs.** Lewis structures can be further simplified by leaving out all nonbonded electrons, in which case they are called **structural formulas,** F—F.

It is important to recognize that hydrogen is one of several exceptions to the octet rule. Hydrogen requires only two electrons, or a **duet,** to achieve the nearest noble-gas configuration, helium. For this reason, hydrogen will always form a covalent bond to only one other atom and will therefore never be found between two other atoms in Lewis structures. In the formation of the hydrogen molecule, each hydrogen atom achieves the helium configuration by sharing the other's single electron.

Other exceptions to the octet rule are encountered with boron and beryllium, which form compounds, such as BeH_2 and BF_3, that contain four and six electrons, respectively, in the central atoms' completed valence shells. However, the octet rule applies to all the other elements in the first and second periods of the periodic table, the principal focus of our interest.

Drawing Lewis Structures

The derivation of a Lewis structure requires (1) knowledge of a compound's molecular formula and (2) a set of simple rules (illustrated in Example 3.8). The **molecular formula** (Chapter 4) tells us the number of atoms for each element in a molecule of the compound. It does not tell us, as the Lewis structure does, how they are connected to one another.

Example 3.8 Drawing Lewis structures: I

Construct the Lewis structure for ammonia, NH_3.

Solution

Step 1. Place the symbols for the bonded atoms into an arrangement that will allow you to begin distributing electrons. In ammonia (as well as in many other compounds), there is only one atom of one type (here, N) and several of another (here, three H atoms); so as a first approximation choose N as the central atom to which the rest are bonded.

H
H N H

Step 2. Determine the total number of valence electrons in the molecule. To do so, add up the valence electrons contributed by each one of the atoms in the molecule.

Element	Valence Electrons	Atoms per Molecule	Number of Electrons
N	5	1	5
H	1	3	3
		Total valence electrons available	= 8

Step 3. Represent shared pairs of electrons by drawing a line between the bonded atoms. For ammonia,

H
|
H—N—H

Each line (bond) represents two electrons, accounting for six of the eight available valence electrons.

Step 4. Position the remaining valence electrons (two in this case) as lone pairs to satisfy the octet rule for each atom in the compound (remember that hydrogen's "octet" consists of only two electrons).

H
|
H—N̤—H

Problem 3.8 Construct the Lewis structure for phosphorus trichloride, PCl_3.

Example 3.9 Drawing Lewis structures: II

Construct the Lewis structure for water, H_2O.

Solution

Step 1. Position the bonded atoms in a likely arrangement. In water, there is only one atom of oxygen and two of hydrogen, so a good guess is that the oxygen is the central atom to which the others are bonded:

H O H

Step 2. Determine the total number of valence electrons in the molecule by adding up the valence electrons contributed by each of the atoms in the molecule.

Element	Valence Electrons	Atoms per Molecule	Number of Electrons
O	6	1	6
H	1	2	2
		Total valence electrons available	= 8

Step 3. Represent electron pair bonds by drawing a line between bonded atoms. For water:

H—O—H

This structure accounts for four of the eight available valence electrons.

Step 4. Position the remaining four valence electrons as lone pairs around the atoms to satisfy the octet rule for each:

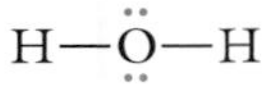

Problem 3.9 Construct the Lewis structure for sulfur dichloride, SCl_2.

The next step in learning to use Lewis structures is to diagram a compound in which there is no unique central atom.

Example 3.10 Writing Lewis structures for more complex compounds: I

Write the Lewis structure for ethane, C_2H_6. Ethane is one of a large family of compounds called **alkanes,** also called hydrocarbons because they contain only the elements hydrogen and carbon. Remember that hydrogen cannot be inserted between the two carbon atoms, because it is not capable of participating in more than one bond at a time.

Alkanes are the simplest organic compounds and the subject of Chapter 11.

Solution

Step 1. Place the carbon atoms so that they are joined together, with the hydrogens in terminal positions:

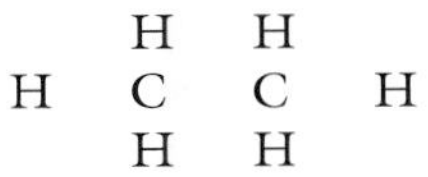

Step 2. Calculate the total number of valence electrons.

Element	Valence Electrons	Atoms per Molecule	Number of Electrons
Carbon	4	2	8
Hydrogen	1	6	6
		Total valence electrons available =	14

Step 3. All these electrons are accommodated by drawing the seven bonds required to connect all the atoms of the molecule.

Step 4. Because there are no lone pairs, the final structure is:

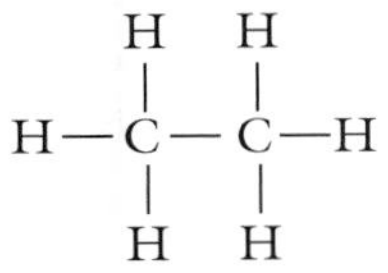

Problem 3.10 Write the Lewis structure for C_3H_8.

Lewis Structures Containing Multiple Bonds

Some interesting cases arise when there appear to be too few electrons to satisfy the octet rule. This apparent deficiency occurs when we try to write the Lewis structures for compounds such as ethylene (C_2H_4), acetylene (C_2H_2), nitrogen (N_2), carbon dioxide (CO_2), and hydrogen cyanide (HCN).

Example 3.11 Writing Lewis structures for compounds with multiple bonds: I

Write the Lewis structure for ethylene, C_2H_4.

Solution

Step 1. As we guessed with ethane, we can assume that the carbon atoms are joined and the hydrogen atoms occupy terminal positions:

Step 2. Calculate the total number of valence electrons.

Element	Valence Electrons	Atoms per Molecule	Number of Electrons
Carbon	4	2	8
Hydrogen	1	4	4
		Total valence electrons available =	12

Step 3. We can insert 10 of the 12 electrons by drawing five bonds to connect all the atoms, with the result that 2 electrons are left over:

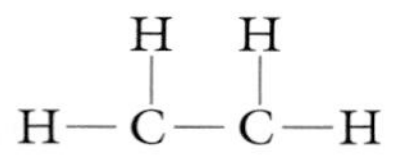

Step 4. The hydrogen atom duets are satisfied. However, if we are to satisfy the octet rule, two pairs of electrons are needed for each carbon atom. That would seem to mean that 4 more electrons are required, but only 2 electrons are available. This apparent dilemma can be solved by adding the 2 remaining electrons as an additional bonding pair between the carbon atoms, to form a **double bond.** The shared double bond provides an octet around each carbon atom:

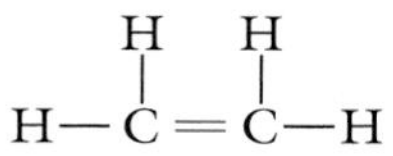

▶▶ **Ethylene and other hydrocarbons with double bonds are discussed in Chapter 12.**

Problem 3.11 Write the Lewis structure for carbon dioxide, CO_2.

Example 3.12 Writing Lewis structures for compounds with multiple bonds: II

Write the Lewis structure for nitrogen, N_2.

Solution

Step 1. The gaseous element nitrogen is a diatomic molecule whose structure consists of two nitrogen atoms joined by a covalent bond.

N—N

Step 2. Calculate the total number of valence electrons.

Element	Valence Electrons	Atoms per Molecule	Number of Electrons
Nitrogen	5	2	10
		Total valence electrons available =	10

Step 3: When we attempt to place one bond between the atoms and add the remaining electrons as lone pairs on each nitrogen atom,

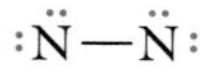

we find that we are short two pairs of electrons for the necessary octets.

Step 4. However, we can move two of the lone pairs to a position between the nitrogen atoms to form two more bonds; then we add the remaining four electrons as lone pairs on each of the atoms to obtain

:N≡N:

Three pairs of electrons are shared to complete the octets, and the bond between the nitrogen atoms is called a triple bond.

Problem 3.12 Write the Lewis structure for the rocket fuel hydrazine, N_2H_4.

When you begin your study of organic chemistry, you will find that many important organic molecules contain single, double, or triple bonds. Your vision depends on the properties of a double bond, and its story is told in Box 12.2.

3.8 LEWIS STRUCTURES OF POLYATOMIC IONS

Lewis structures are also used to depict polyatomic ions, in much the same way as we used them in the preceding worked examples. However, in calculating the total number of valence electrons, the charge on the ion must be taken into account. If it is a negative ion, electrons must be added; if it is a positive ion, electrons must be subtracted. For example, the ammonium ion possesses a charge of 1+, and the carbonate ion has a charge of 2−, so the total number of valence electrons is calculated:

Ion	Total Number of Valence Electrons
NH_4^+	N (1×5) + H $(4 \times 1) - 1 = 8$
CO_3^{2-}	C (1×4) + O $(3 \times 6) + 2 = 24$

With reference to preceding examples for details, the Lewis structures of the ammonium ion and the carbonate ion are:

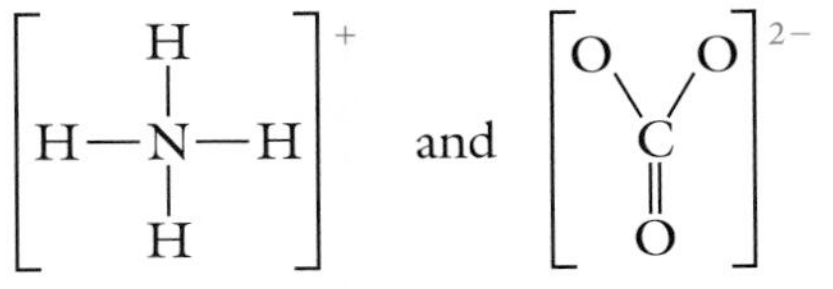

Ammonium ion   Carbonate ion

Polyatomic ions do not have their origin in the direct transfer of electrons from element to element. It is therefore important to recognize that ions can be formed in other types of chemical reactions.

Most of the polyatomic ions of interest to us arise from reactions taking place in aqueous solutions—mixtures consisting mostly of water, with other substances dissolved in it. All the polyatomic ions listed in Table 3.2 were created by reactions in water, when a dissolved substance reacted either with the water itself or with some other dissolved substance. For example, two substances, HCl (hydrogen chloride) and NH_3 (ammonia), react with water in the following ways:

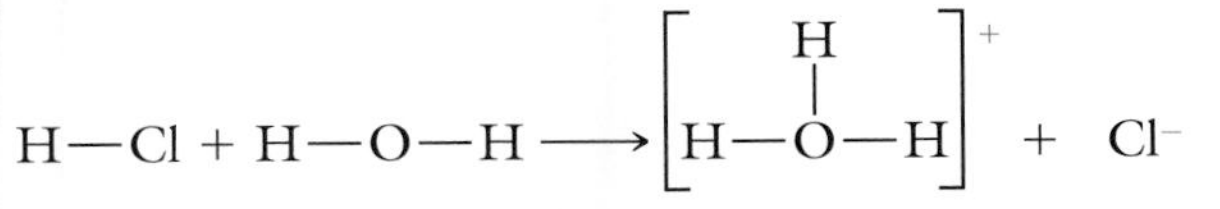

Hydronium ion   Chloride ion

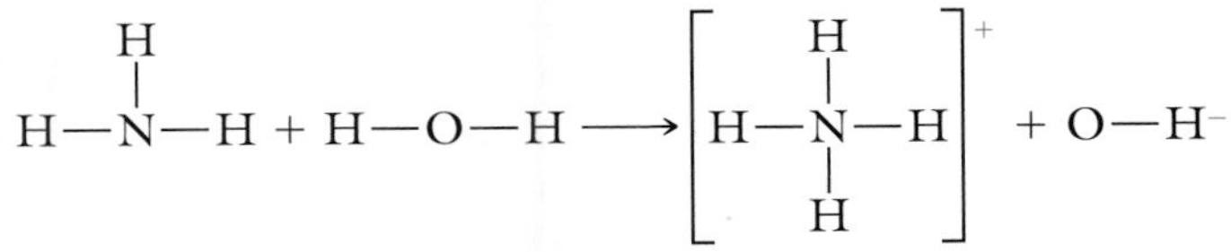

Ammonium ion   Hydroxide ion

In the first reaction, the product H_3O^+ ion, called the hydronium ion, is the result of the transfer of a hydrogen ion (H^+, not an electron) to water. This reaction always takes place when compounds called acids react with water. In the second reaction, a hydrogen ion (not an electron) is transferred from water to ammonia with the resultant formation of the hydroxide ion, which, in water, forms a basic solution. In each case, the newly formed covalent bond is created from a lone pair contributed by the central atom and is called a **coordinate covalent bond.** Once formed, such a bond cannot be differentiated from any of the other covalent bonds. Note carefully that the bonds uniting all atoms in these polyatomic ions are covalent.

We will explore the details of the chemistry of acids and bases in Chapter 9.

3.9 POLAR AND NONPOLAR COVALENT BONDS

When an electron pair is shared by two atoms of the same element, as in molecules such as H_2 and F_2, the electrons are attracted equally to both atomic nuclei and are found midway between them. Other examples are O_2, N_2, and Cl_2. This type of covalent bond is called a **nonpolar covalent bond.** When molecules consist of different elements, however, such as the molecules NO and HCl, one atom participating in the bond is likely to exert a greater attraction for the electrons than the other atom. In these cases, the electrons tend to reside closer to the nucleus with the greater ability to attract them. A bond of this type is called a **polar covalent bond.**

Actually, polar bonds are distributed along a continuum. At one extreme of this continuum is the ionic bond, in which both electrons reside totally on one of the ions of a cation–anion pair. At the other extreme is the pure covalent bond, in which the electron pair spends most of its time halfway between the bonded atoms. Most covalent chemical bonds are polar, however, and lie somewhere between these two extremes.

The ability of an atom within a molecule to draw electrons toward itself is called its **electronegativity.** Figure 3.5 provides an abbreviated list of electronegativities of various elements. Electronegativity is used to determine the

H 2.2							He
Li 0.98	Be 1.6	B 2.0	C 2.6	N 3.0	O 3.4	F 4.0	Ne
Na 0.93	Mg 1.2	Al 1.6	Si 1.9	P 2.2	S 2.6	Cl 3.2	Ar
K 0.88	Ca 1.0	Ga 1.8	Ge 2.0	As 2.2	Se 2.6	Br 2.8	Kr
Rb 0.82	Sr 0.95	In 1.8	Sn 2.0	Sb 1.9	Te 2.1	I 2.7	Xe
Cs 0.79	Ba 0.89	Tl 1.8	Pb 1.9	Bi 1.9	Po 2.0	At 2.2	Rn

Figure 3.5 An abbreviated table of electronegativities.

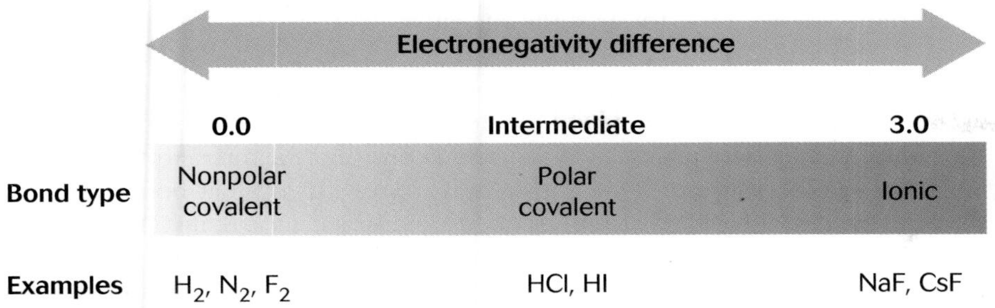

Figure 3.6 A continuum showing the effect of electronegativity difference on the type of bond formed between elements. At one extreme is the nonpolar covalent bond, which can form only between identical atoms (electronegativities are therefore identical). At the other extreme is the ionic bond, which forms when the difference between the elements' electronegativities exceeds about 2.0.

order of elements in the name of a binary compound. That is why the name of the binary compounds formed between oxygen and nitrogen and between carbon and oxygen are called NO and CO. The less-electronegative element comes first in the compound's name. Notice that the electronegativities of fluorine, oxygen, and nitrogen are among the largest of all the elements. This fact will become important when we consider weak interactions between molecules in Chapter 6.

A useful (though not infallible) rule of thumb is that, if the difference between the electronegativities of two elements is greater than 2.0, the bond between their atoms will probably be ionic; if the difference is less than 1.5, the bond will probably be covalent. This rule of thumb is demonstrated schematically in Figure 3.6. For example, the electronegativity difference between lithium and fluorine is 3.0, and the LiF bond is probably ionic. The electronegativity difference between boron and carbon is 0.5; therefore a bond between the atoms of those two elements will be covalent. When the difference in electronegativities is in the range of 2.0 to 1.5, the bond will be ionic if the compound is a metal–nonmetal combination and will be covalent if it is a nonmetal–nonmetal compound.

When the two bonded atoms of a binary molecule (a two-atom molecule) have different electronegativities, the electrical assymmetry that is produced causes one end of the molecule to possess a negative charge and the other end a positive charge. The electrical charges are not full charges; rather they are partial charges and are indicated by lowercase Greek symbols $\delta+$ or $\delta-$. The covalent bond of such a molecule is electrically polarized, and, for convenience, chemists call that kind of bond a polar covalent bond. The H–F molecule is a good example. When we want to show the existence of partial charges on its atoms we write the formula $\overset{\delta+}{H}—\overset{\delta-}{F}$. Because it possess only two atoms, the polar covalent bond of HF makes it a **polar molecule** as well. It is important to note that, although the HF molecule is electrically assymetric, it is still electrically neutral overall, because the collective positive charges of the combined nuclei are exactly balanced by the collective negative charges of the combined electrons.

A polar molecule is attracted to a neighboring polar molecule, positive end to negative end. The electrical force causing the attraction is measured by a quantity called the **dipole moment.** The magnitude of the dipole moment depends on the distance separating the partial charges and on the difference in electronegativities of the combined atoms. The greater the distance of separation and the greater the difference in electronegativities, the larger the dipole moment. It is customarily represented by an arrow pointing along the bond from the positive end toward the negative end, +⟶; from the less-electronegative element to the more-electronegative element.

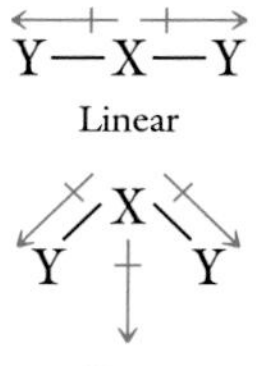

A molecule may contain two or more polar bonds and not be polar. Its overall shape may be such that the dipole moments cancel each other out. As an example, imagine a generalized molecule consisting of three atoms, of formula XY_2. In this molecule, X is the central atom with covalent bonds to each Y atom. Because the joined atoms X and Y are different, they must have different electronegativities. Therefore each individual bond in the molecule must be polar and must contribute to the overall polarity of the molecule. There are two possibilities for the molecular shape: linear and bent, as shown in the margin.

In the linear molecule, the two internal dipoles cancel each other's effect because the bonds are identical, and therefore they are equal in magnitude and act in directly opposite directions. Neither end of the molecule is more positive or negative than the other. Therefore the linear molecule will possess no overall dipole moment—is not polar. An example is carbon dioxide, O=C=O. If the polar bonds in a linear molecule are different (for example, Cl—Be—F), the molecule will be polar.

In the bent molecule, the internal dipoles act in concert to produce an overall dipole moment, so that the X "end" of the molecule has a partial positive charge and the Y "end" a partial negative one. An experiment that tested the hypothetical molecule for the presence of a dipole moment would therefore reveal whether the molecule was bent or straight. The absence of a dipole moment would mean that the molecule was linear, like CO_2. The presence of a dipole moment in a molecule with two identical bonds would mean the molecule was bent, like H_2O.

The consequences of molecular polarity can be seen primarily in physical properties. Molecules with large dipole moments tend to interact strongly with one another, as well as with other molecules possessing dipole moments. As a result, substances composed of polar molecules have higher melting and boiling points than do those composed of nonpolar molecules. At room temperature, methane, CH_4, which has no dipole moment, is a gas, but water, H_2O, which has a large dipole moment, is a liquid at the same temperature. Moreover, water, because of its polarity, is an excellent solvent for polar substances. Many of the physical and chemical properties of organic compounds depend on their overall molecular polarity, which is the result of the polarity of covalent chemical bonds within their molecules. Figure 3.7 shows how the polarity of a covalent bond can can be manifested in a substance's properties. A more thorough discussion of the effects of structure on physical properties will be found in Chapter 6.

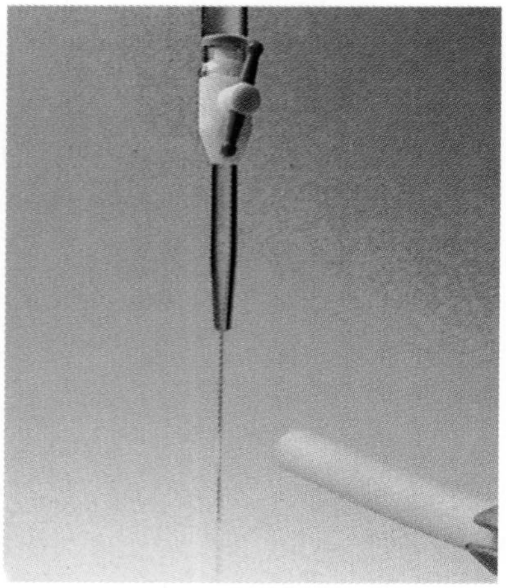
(a)

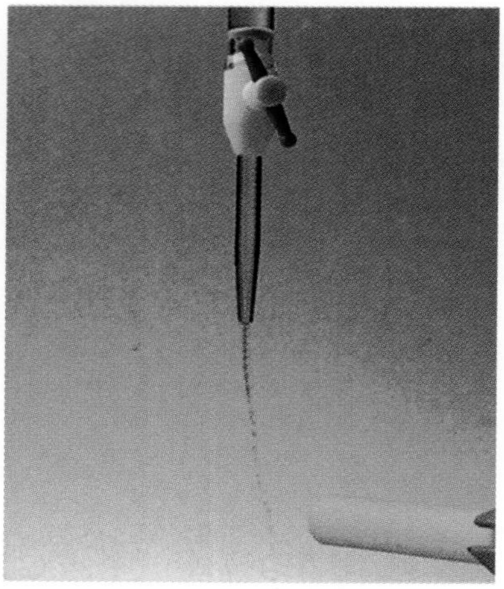
(b)

Figure 3.7 (a) The buret on the top contains hexane, a nonpolar substance. (b) The buret on the bottom contains water, a polar substance. When both are allowed to trickle past an electrically charged rod, only the polar substance responds to the electrical field.

3.10 THREE-DIMENSIONAL MOLECULAR STRUCTURES

Lewis structures provide us with two kinds of information about molecular structure:

- How atoms within the molecule are connected to one another
- Whether the connecting bonds are single, double, or triple

However, Lewis structures cannot describe the actual spatial arrangements of atoms within molecules. The important point to remember is that, although Lewis structures allow us to decide which atoms are connected and tell us about the nature of the bonds holding them together (single, double, or triple; polar or nonpolar), Lewis structures cannot reveal the three-dimensional arrangements of the atoms. It is the three-dimensional, or spatial, arrangement of atoms in a molecule that provides a molecule's unique characteristics. Those spatial arrangements are of central importance in all biological functions, such as enzymatic activity, membrane transport, nerve transmission, and antigen–antibody interactions. See Box 17.1 on smell and taste, Box 17.2 on chiral drugs, Section 17.5 on chiral recognition, Box 12.2 on vision, Box 12.3 on pheromones, and Section 22.3 on enzymes.

▶▶ **Enzymes mediate almost all the reactions described in Chapters 22–26.**

VSEPR Theory

A simple and very useful theory enables us to predict the shapes of both simple and complex molecules. The theory states that the shape of a molecule is affected by each of the valence electron pairs surrounding a central atom, not just the pairs participating in bonds.

The central idea of the theory is that, because electrons tend to repel one another, the mutual repulsion of all the electron pairs forces the atoms within the molecule to take positions in space that minimize those mutual repulsions. In other words, the electrons strive to get as far away from one another as possible. The name of the method is the **valence-shell electron-pair repulsion theory,** or the **VSEPR theory.**

Because the most stable arrangement for the pairs of electrons in a molecule, is to be as far away from one another as possible, they must occupy points in space that define symmetric geometrical figures called regular polyhedra. A number of these figures are shown in Figure 3.8. A linear molecule such as $BeCl_2$ (beryllium chloride) takes its linear shape so that the two pairs of electrons may be as far apart as possible. Boron trifluoride, BF_3, takes the shape of an **equilateral triangle** for the same reason. The **tetrahedron** is the result of the symmetric arrangement of four pairs of electrons around a central atom. Each of the four pairs is as far from the others as possible (while still remaining attached to the central atom). Six groups surrounding a central atom will be located at the corners of an octahedron.

To illustrate how these shapes emerge from considerations of the repulsion of the electron pairs of covalent bonds, let's apply VSEPR theory to predict the three-dimensional structure of methane, CH_4.

(a) Linear

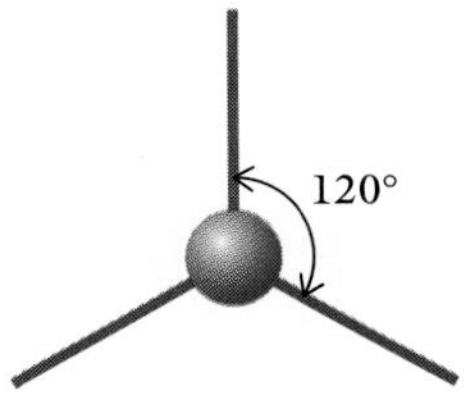

(b) Trigonal planar

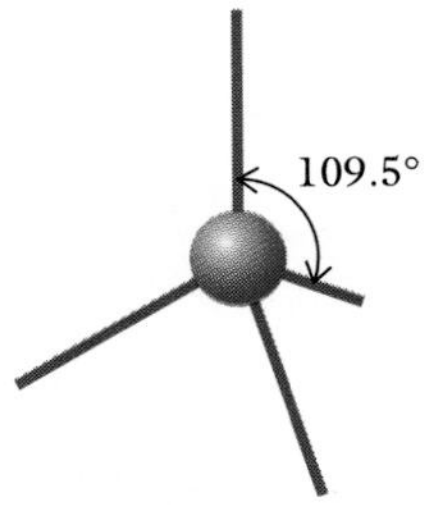

(c) Tetrahedral

Figure 3.8 The positions assumed by pairs of electrons around a central atom are represented by straight lines emanating from the central atom. (a) Two pairs form a linear molecule; (b) three pairs, an equilateral triangle; and (c) four pairs form a regular tetrahedron.

Example 3.13 Using VSEPR theory to predict molecular structure: I

Use VSEPR theory to predict the three-dimensional structure of methane, CH_4.

Solution

The most direct way to go about this is to first determine the Lewis structure for methane.

Step 1. It is most likely that the carbon atom is central, with the hydrogens terminal:

```
   H
H  C  H
   H
```

Step 2. Calculate the total number of valence electrons.

Element	Valence Electrons	Atoms per Molecule	Number of Electrons
Carbon	4	1	4
Hydrogen	1	4	4
		Total valence electrons available =	8

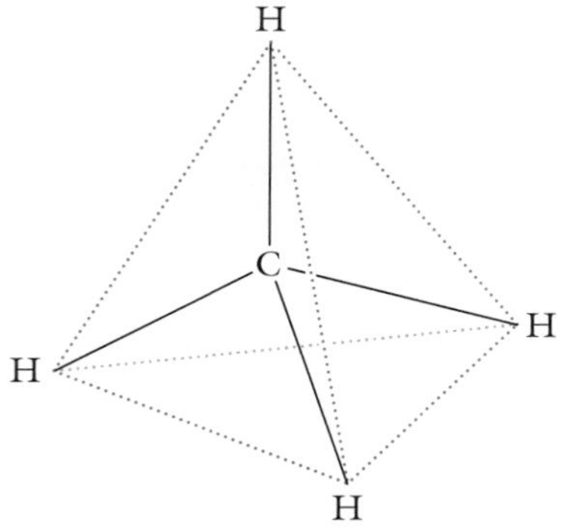

Figure 3.9 Methane has a tetrahedral molecular shape. The dotted lines indicate the outline of a regular tetrahedron, and the solid lines represent the C–H bonds.

Step 3. These electrons are all accommodated by drawing four single bonds between all atoms of the molecule.

Step 4. Because there are no lone pairs, the final structure is

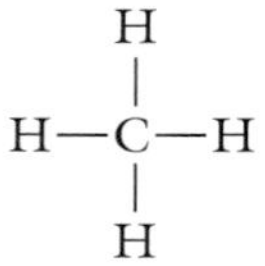

The Lewis structure shows four pairs of electrons surrounding the central carbon atom. The three-dimensional figure that allows maximum separation of four similar electric charges is the tetrahedron (Figure 3.8). Therefore, VSEPR theory predicts that methane will have a tetrahedral structure, as illustrated in Figure 3.9.

Problem 3.13 Use VSEPR theory to predict the three-dimensional structure of carbon tetrachloride, CCl_4.

Figure 3.9 shows the carbon atom of CH_4 in its central location, with single bonds, separated by angles of 109.5°, connecting it to each of four hydrogen atoms. The dotted lines indicate the outline of the hypothetical solid figure enclosing the carbon atom, and the solid lines represent the carbon–hydrogen single bonds. Although each C–H bond is very slightly polar, the four identical bonds are arranged symmetrically; therefore there can be no overall molecular polarity. Experiments have confirmed that the structure of methane is tetrahedral and that the molecule has no dipole moment. X-ray diffraction is used to determine the structure of molecules. In this technique, X-rays are passed through crystals of a compound, and the pattern of emerging X-rays is recorded as in Figure 3.10. Mathematical treatment of the data locates the positions of all atoms within the structure—hence the molecule's shape.

Molecular Absorption Spectra and Molecular Structure

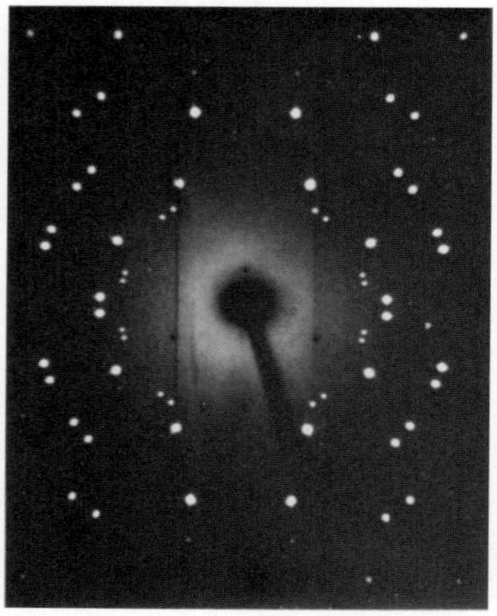

Figure 3.10 The X-ray diffraction pattern produced by passing X-rays through a crystal of NaCl. The pattern of spots provides information regarding the arrangement of the ions within the crystal.

Other aspects of molecular structure can be clarified by using radiation of lower energy than X-rays. In Section 2.7, we learned that the absorption of particular frequencies of electromagnetic energy causes changes in atomic energy states. Molecules as well as atoms possess unique energy states, and these energy states generally correspond to particular molecular structural features. Just as elements can be detected and quantified by atomic absorption spectra, molecules and parts of molecules can be characterized by molecular absorption spectra.

In your study of organic chemistry, you will learn about molecules that possess a number of multiple bonds. The electrons in those types of molecules possess energy states corresponding to frequencies in the ultraviolet and visible spectrum (see Table 2.4). Other molecular energy states correspond to the motions of atoms attached to other atoms—that is, to vibrations within molecules. These energy states are excited by infrared frequencies. A technique called nuclear magnetic resonance depends on the response of nuclear energy states to radiofrequencies. Nuclear magnetic resonance is of primary importance in the elucidation of complex molecular structures and has become an important tool in medical diagnosis, where it goes by the name of MRI (see Chapter 10).

The application of these techniques requires the ability to generate the frequencies of interest and the ability to detect changes in the energy of the beam emerging from a sample. The absorption of ultraviolet, visible, infrared, and radiofrequencies is in daily use in all modern laboratories engaged in the characterization and chemical analysis of molecular structure. Box 14.3 describes the use of some of these techniques in the analysis of organic compounds.

Nonbonding Electrons

In contrast with the structure of methane, there are situations in which the arrangement of four electron pairs is not symmetric. Specifically, when one or more of an atom's electron pairs are not participating in a bond, those nonbonding pairs take up considerably more space, causing the bonding electron pairs to squeeze closer together.

VSEPR theory explains the symmetry-altering effect of nonbonding electron pairs by pointing out that electron pairs in bonds are localized because they are fixed between two positively charged nuclei. Nonbonded pairs are not subjected to a similar localizing influence, because they are not fixed between two positively charged nuclei. As a result, they occupy a greater volume of space than a bonded pair does. In doing so, they push the bonded pairs closer to one another than the tetrahedral geometry would predict. As an example, let's see what VSEPR theory predicts for the structure of ammonia.

Example 3.14 Using VSEPR theory to predict molecular structure: II

Use VSEPR theory to predict the structure of the ammonia molecule.

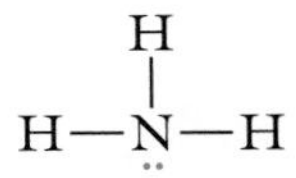

Ammonia

Solution

The Lewis structure for ammonia, NH_3, was worked out in Example 3.8. As in methane, four pairs of electrons surround the central nitrogen atom. VSEPR theory therefore predicts that the structure will be tetrahedral; but, in this case, one of the pairs is nonbonded, which alters the perfect symmetry exhibited by methane. Experiments have shown that the bonds between the central nitrogen atom and the hydrogens are separated by angles of 107°, rather than the 109.5° angles of a perfect tetrahedron.

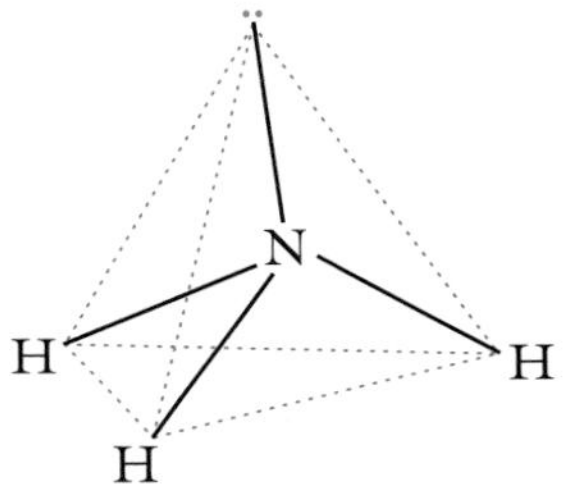

Ammonia

We consider the shape of a molecule to be defined by the positions of the bonded atoms. Therefore, the ammonia molecule forms a pyramid (outlined by the blue dotted lines) with three hydrogen atoms forming the base. This shape is called **trigonal pyramidal** because the pyramid has a triangle for its base. The asymmetric arrangement of bonds and one lone pair of electrons causes this molecule to have a strong dipole moment.

Problem 3.14 Use VSEPR theory to predict the structure of nitrogen triiodide, NI_3.

Example 3.15 Using VSEPR theory to predict molecular structure: III

Use VSEPR theory to predict the three-dimensional structure of water, H_2O.

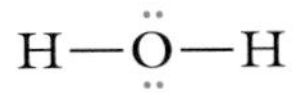

Water

Solution

The Lewis structure for water, with its two lone pairs on the oxygen atom, was worked out in Example 3.9. From VSEPR theory, we know that the total of four electron pairs, two bonded and two nonbonded, surrounding the oxygen atom must be accommodated by a roughly tetrahedral configuration (outlined by blue dotted lines). However, because two of the pairs are nonbonding, they will occupy even more space around the central atom than did the one lone pair in ammonia. The result is that the two hydrogen atoms of water are squeezed even closer together than the hydrogen atoms in ammonia, with the two lone electron pairs directed at the other two corners of a distorted tetrahedron. The bond angle between the oxygen atom and the two hydrogen atoms has been

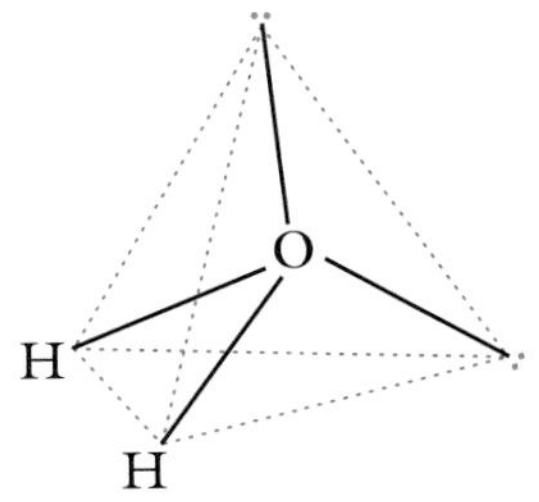

Water

determined experimentally to be 104.5°. The water molecule has a bent appearance because the two lone pairs cannot be "seen." This molecule has an even greater overall polarity than that of ammonia.

Problem 3.15 Use VSEPR theory to predict the three-dimensional structure of oxygen difluoride, OF_2.

For a molecule with no central atom, such as ethane (see Example 3.10), we can assume that each of the carbon atoms possesses tetrahedral symmetry; so the shape of the molecule must be defined by two tetrahedrons joined at one apex.

VSEPR theory treats double and triple bonds as though they were single bonds. The Lewis structure for acetylene, C_2H_2, for example, is $H—C≡C—H$ and the molecule possesses a triple bond. Because VSEPR theory treats a triple bond as though it were a single bond, we proceed as though there were only two bonding pairs of electrons to be accommodated around each carbon atom. The geometrical arrangement that allows maximal separation of two bonding pairs is a straight line separating the bonding pairs by 180°. Acetylene must therefore be a linear molecule with all atoms in a straight line and an appearance similar to that depicted in its Lewis structure. Experiments have shown that acetylene has no dipole moment, thus verifying the VSEPR prediction.

The steps for determining the shapes of molecules by VSEPR theory are

Rules for predicting molecular shape by VSEPR theory

- Determine the Lewis structure.
- Determine the number of electron pairs (bonded and nonbonded) surrounding a central atom.
- Use VSEPR theory to decide what symmetric three-dimensional shape will accommodate all bonded and nonbonded electron pairs.
- Classify the shape of the molecule by considering the arrangement of the joined atoms only, ignoring the nonbonded electron pairs. That is why, although the CH_4, NH_3, and H_2O molecules each have four pairs of electrons around the central atom, the shape of methane is tetrahedral, that of ammonia is trigonal pyramidal, and that of water is bent.

Three-Dimensional Representations of Molecules

VSEPR theory allows us to deduce the overall three-dimensional shapes of molecules, but we must use a different approach to illustrate three-dimensional internal molecular structure on a two-dimensional page. Although there are variations associated with each approach, there are two fundamental ways to illustrate three-dimensional molecular structure. One is called the ball-and-stick model, and the other is the space-filling model. Both are illustrated in Figure 3.11.

In the ball-and-stick model of methanol, the atoms are the "balls" and are differentiated by color; the bonds are the "sticks" and are shown about equal in length. The bond angles are correct, but the atom size and bond length do not represent the actual atom size to bond length ratio. Actual bond lengths are much shorter than represented by the ball-and-stick models. In the space-filling models, the atoms and bond lengths are drawn to precise relative sizes and are disposed in space at the correct angles with respect to one another. The surfaces of the space-filling "atoms" in the models represent, in relative terms, the closest the atoms can approach one another. However, using space-filling models for instruction presents difficulties because the bonds cannot be seen (the atoms hide them).

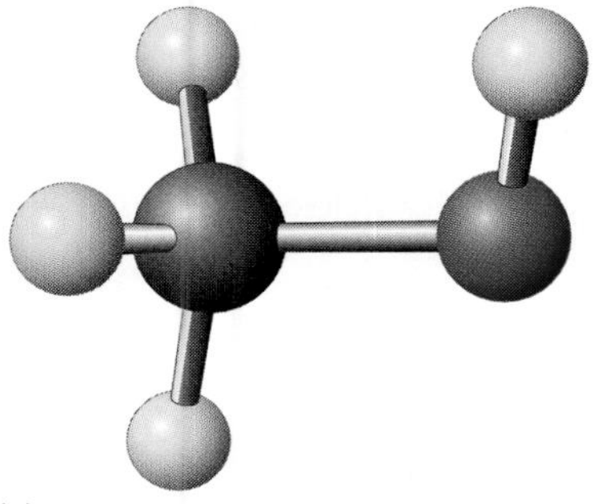

(a)

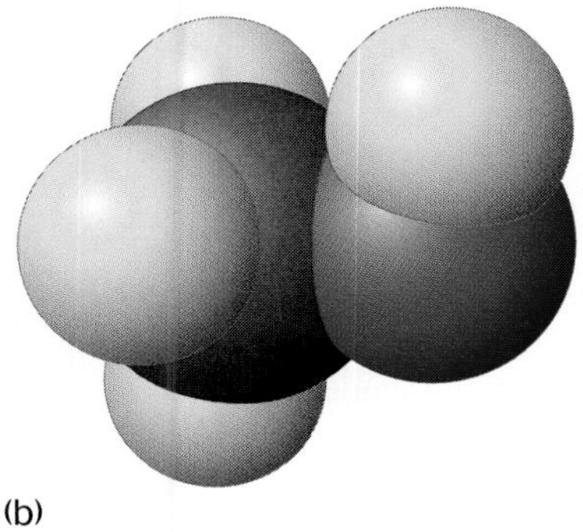

(b)

Figure 3.11 (a) Ball-and-stick and (b) space-filling models of methanol, CH_3OH.

LOOKING AHEAD

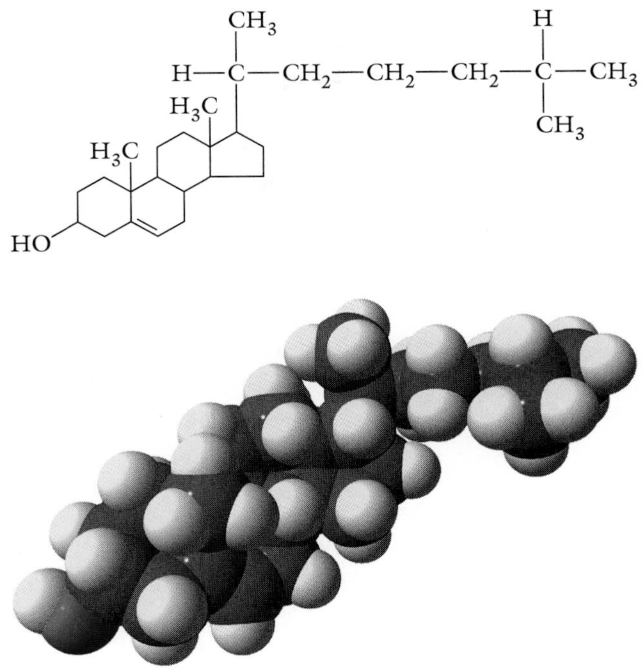

Figure 3.12 A space-filling molecular model of cholesterol, along with its planar molecular representation. We will have more to say about the structure and function of cholesterol in Chapters 19 and 26.

When it seems important to illustrate a particular molecular structure in three dimensions, one of these two models will be used. For example, the space-filling model of cholesterol shown in Figure 3.12 helps us to comprehend the molecular shape of an important biochemical. However, most of the molecular structures you will encounter in this book will be two-dimensional Lewis structures minus the nonbonded electrons. We have chosen this style because the Lewis structures are easy to draw and well within the capability of those of us who are not artists.

Summary

Compounds are the result of the stable association between atoms of the elements. This close association is called chemical bonding. There are two kinds of chemical bonds: ionic bonds and covalent bonds.

Ionic Bonds When metals react with nonmetals, electrons are transferred from the valence shell of the metal into the valence shell of the nonmetal so that each outermost valence shell has a complete octet. In consequence, charged atoms called ions are formed. The force holding ions together in a chemical bond is the strong attraction of oppositely charged particles, and it is called an ionic bond. The compounds thus formed are called ionic compounds.

Naming Binary Ionic Compounds The result of reaction between a metal and a nonmetal is a binary ionic compound. In binary compounds, the metals take the name of the element, and the anion's name begins with that of the element but takes the ending -ide. The modern systematized method of nomenclature of compounds with ions of variable valence is called the Stock system. Some ions consist of combinations of atoms and are called polyatomic ions.

Covalent Compounds In reactions between nonmetals, the participating elements acquire an octet of valence electrons by sharing electrons between them. The result is the formation of one or more covalent bonds, each consisting of two spin-paired electrons.

Electron pairs shared by atoms of the same elements, as in H_2 and F_2, form nonpolar covalent bonds. Electron pairs shared by atoms of different elements form polar covalent bonds. The ability of an atom within a molecule to draw electrons toward itself is defined as its electronegativity.

Three-Dimensional Molecular Structures Lewis formulas reveal the connections between atoms in a molecule but cannot be used to suggest the three-dimensional arrangements of those connections. However, the shapes of molecules can be predicted by the use of the valence-shell electron-pair repulsion theory, or the VSEPR theory. Molecular shape depends on the total number of electron pairs surrounding a central atom. The mutual repulsion of all the electron pairs causes them to assume positions that define symmetrical geometric figures, called regular polyhedra.

Key Words

binary compound, p. 69
covalent bond, p. 66
dipole moment, p. 85
double bond, p. 82
electronegativity, p. 84
-ide, pp. 73, 77
ionic bond, p. 66
Lewis structure, p. 78
lone pair of electrons, p. 79
molecular formula, p. 79
Stock system, p. 73
valence-shell electron-pair repulsion (VSEPR) theory, p. 87

Exercises

3.1 What is the basis for the formation of chemical bonds?

3.2 What is the octet rule?

3.3 What are the ways in which elements can satisfy the octet rule?

3.4 What are the results of the two ways of satisfying the octet rule?

Nomenclature

3.5 Write the systematic name of each of the following compounds: (a) K_2SO_4; (b) $Mn(OH)_2$; (c) $Fe(NO_3)_2$; (d) KH_2PO_4; (e) $Ca(C_2H_3O_2)_2$; (f) Na_2CO_3.

3.6 Write the formulas of the following compounds: (a) ferric chloride; (b) stannic oxide; (c) cupric nitrate; (d) mercurous chloride; (e) plumbous chloride.

3.7 In the blank space to the right of the ion given, write the name of the ion.

Ion	Name
(a) NH_4^+	________
(b) NO_3^-	________
(c) SO_4^{2-}	________
(d) PO_4^{3-}	________
(e) $C_2H_3O_2^-$	________

3.8 In the left-hand column, write the formula of the ion named in the right-hand column.

Ion	Name
(a) ________	Carbonate
(b) ________	Hydrogen carbonate
(c) ________	Hydroxide
(d) ________	Hydrogen phosphate
(e) ________	Dihydrogen phosphate

3.9 Write the formulas of the compounds formed in the reactions between aluminum and (a) chlorine; (b) sulfur; (c) nitrogen.

3.10 Write the names of the compounds formed in the reactions between aluminum and (a) chlorine; (b) sulfur; (c) nitrogen.

3.11 What are the formulas of the compounds formed from the following elements? (a) lithium and oxygen; (b) calcium and bromine; (c) aluminum and oxygen; (d) sodium and sulfur.

3.12 What are the names of the compounds formed from the following elements? (a) lithium and oxygen; (b) calcium and bromine; (c) aluminum and oxygen; (d) sodium and sulfur.

Ionic Bonds

3.13 What are the combining ratios of elements in Group I with elements in Group VII of the periodic table?

3.14 What are the combining ratios of elements in Group I with elements of Group VI of the periodic table?

3.15 What are the combining ratios of elements in Group II with elements in Group VII of the periodic table?

3.16 What are the combining ratios of elements in Group III with elements in Group VI of the periodic table?

3.17 Write the formulas of the compounds formed when calcium reacts with each of the first three elements of Group VI of the periodic table.

3.18 Write the names of the compounds formed when calcium reacts with each of the first three elements of Group VI of the periodic table.

3.19 Write the formulas of the compounds formed when aluminum reacts with each of the first three elements of Group VI of the periodic table.

3.20 Write the names of the compounds formed when aluminum reacts with each of the first three elements of Group VI of the periodic table.

3.21 Write the formulas of the compounds formed when aluminum reacts with each of the first four elements of Group VII of the periodic table.

3.22 Write the names of the compounds formed when aluminum reacts with each of the first four elements of Group VII of the periodic table.

Covalent Bonds and Lewis Structures

3.23 What is the Lewis structure of carbon tetrachloride, CCl_4?

3.24 Write the Lewis structure for chloroform, $CHCl_3$.

3.25 What is the Lewis structure of OF_2?

3.26 What is the Lewis structure of NCl_3?

3.27 Write the Lewis structure of carbon dioxide, CO_2.

3.28 Alkanes are compounds that contain only carbon and hydrogen. In most alkanes, the carbon atoms are connected to one another to form a chain. Write the Lewis structure for butane, C_4H_{10}. Does this Lewis structure have another name?

Three-Dimensional Molecular Structure

3.29 Use VSEPR theory to predict the angular orientation of fluorine atoms around the central boron atom in the polyatomic ion BF_4^-.

3.30 Use VSEPR theory to predict the three-dimensional structure of OF_2.

3.31 For each of the following molecules, (1) describe the molecular shape and (2) predict whether it has a dipole moment. (a) CCl_4; (b) BF_4^-; (c) CO_2.

3.32 For each of the following molecules, (1) describe the molecular shape and (2) predict whether it has a dipole moment. (a) $CHCl_3$; (b) NCl_3; (c) H_2O.

Unclassified Exercises

3.33 Describe a covalent chemical bond.

3.34 Describe an ionic bond.

3.35 Write the systematic name of each of the following compounds: (a) $AgNO_3$; (b) $HgCl_2$; (c) Na_2CO_3; (d) $Ca_3(PO_4)_2$.

3.36 Write the formulas of the compounds formed in the reactions between magnesium and (a) chlorine; (b) sulfur; (c) nitrogen.

3.37 Write the names of the compounds formed in the reactions between magnesium and (a) chlorine; (b) sulfur; (c) nitrogen.

3.38 Write the formulas of the compounds formed when barium reacts with the first three elements of Group VI of the periodic table.

3.39 Write the names of the compounds formed when barium reacts with the first three elements of Group VI of the periodic table.

3.40 Write the formulas of the compounds formed when potassium reacts with the first three elements of Group V of the periodic table.

3.41 Write the names of the compounds formed when potassium reacts with the first three elements of Group V of the periodic table.

3.42 Write the Lewis structure of carbon monoxide.

3.43 Use VSEPR theory to predict the angular orientation of the atoms in $TeBr_2$.

3.44 Write the formulas of the compounds formed by the cations and the anions in the following table.

	chloride	hydroxide	sulfate
Sodium	(a)	(b)	(c)
Calcium	(d)	(e)	(f)
Iron(II)	(g)	(h)	(i)

3.45 Indicate what kinds of bonds will be formed by the reaction of the following pairs of elements: (a) Si and O; (b) Be and C; (c) C and N; (d) Ca and Br; (e) B and N.

3.46 An organic compound containing on OH group is called an alcohol. Write the Lewis structure for the alcohol, CH_3OH (methanol). (Hint: The OH group is connected to a carbon atom and is not a hydroxide ion.)

3.47 Write the formulas of the compounds formed when lithium reacts with the first four elements of Group VII of the periodic table.

3.48 Write the names of the compounds formed when lithium reacts with the first four elements of Group VII of the periodic table.

3.49 Are there any exceptions to the octet rule?

3.50 Write the formulas of the compounds formed in the reactions between magnesium and (a) iodine; (b) selenium; (c) arsenic.

Chemical Connections

3.51 Your instructor has given you a pure compound and has asked you to determine whether it is a molecular or ionic substance. Suggest an experiment to answer that question.

3.52 Is it possible for a substance to contain both ionic and covalent bonds? Explain your answer.

3.53 Write the Lewis structure of hydrogen peroxide, H_2O_2.

3.54 The Ag^+ ion reacts with cyanide ion, CN^-, to form a polyatomic ion $Ag(CN)_2^-$. Predict the three-dimensional structure of that ion.

3.55 Write the Lewis structure of nitrogen tetroxide, N_2O_4.

CHAPTER 4

CHEMICAL CALCULATIONS

CHEMISTRY IN YOUR FUTURE

One of your customers at the pharmacy is a strict vegetarian as well as lactose intolerant. He wants to take vitamin B_{12} supplements to avoid coming down with pernicious anemia from the lack of animal products in his diet. He reads the label on the bottle of tablets you hand him and remarks that 125 micrograms per tablet seems a tiny amount. (He says his vitamin C tablets contain vitamin C in milligram amounts.) You assure him that minute amount of vitamin B_{12} is enough for the body to function healthily. To reassure him further, you look up the compound's molecular formula, $C_{63}H_{88}CoN_{14}O_{14}P$, punch some buttons on your calculator, and tell him that 125 micrograms provides 5.6×10^{16} molecules of the vitamin, a very large number. Chapter 4 will show you how to do this calculation.

LEARNING OBJECTIVES

- Calculate the formula mass of a compound.
- Define the mole.
- Use the mole as a unit-conversion factor for converting mass into moles and moles into mass.
- State the magnitude of Avogadro's number and what it implies about the size of atoms and molecules and the numbers of them in a weighable sample.
- Calculate the empirical and molecular formulas of a compound.
- Write a chemical reaction as an equation.
- Balance a chemical equation.
- Use a balanced chemical equation to predict the masses of compounds produced and used up in chemical processes.

In this chapter, we will use our knowledge of atoms and molecules to answer some practical questions about the conservation of mass in chemical processes. For example, the technician who measures a patient's oxygen uptake to determine his or her basal metabolic rate can compare the amount of oxygen being used with the amount of carbon dioxide produced and discover the composition of the patient's diet. When a chemist weighs out 83 grams of sodium carbonate in the laboratory, she can calculate the number of atoms in the sample; and she knows that, if 111 g of calcium chloride is added to the 83 g of sodium carbonate, 100 g of calcium carbonate will form. Both the economics of the manufacture of drugs and the control of pollution depend on a detailed understanding of the quantitative aspects of chemical reactions. These are the kinds of connections that we will explore in the following pages.

The ability to predict the quantitative outcome of chemical processes is based on one of the most fundamental ideas in chemistry: the concept of the mole. To define the mole, we must first explain how a compound's formula is used to calculate its relative mass. We will then use the mole concept to calculate the number of formula units, given the mass of a compound, and to determine empirical and molecular formulas. Lastly, we will explain how to balance chemical equations and to use the mole concept to predict the quantitative outcomes of reactions.

▶▶ The skills developed here will be useful in Chapter 23, when we quantify the energy obtained from carbohydrates.

4.1 CHEMICAL FORMULAS AND FORMULA MASSES

The **formula** of a chemical compound states how many atoms of each element are in a fundamental unit of the compound. The fundamental unit is called a **formula unit.** Examples of formula units that you have seen before are NaCl, $CaBr_2$, and Al_2O_3.

We have seen that relative masses have been assigned to the different elements (Section 2.3). A relative mass may also be assigned to a compound by simply adding up the relative masses of all the atoms in the formula unit. This sum is called the **formula mass.** To determine the formula mass of water, H_2O, we add the relative masses (taken from the inside back cover of this book) of its two hydrogen atoms and one oxygen atom: 1.008 amu + 1.008 amu + 16.00 amu = 18.02 amu.

The formula unit represents the compound's composition only. It is not intended to describe the compound's physical structure. Recall that, in Chapter 3, a distinction was made between ionic compounds, such as NaCl, and compounds that exist as molecules, such as H_2O. The definition of formula mass using the idea of a formula unit permits us to ignore those distinctions when we are concerned primarily with mass relations. All we need to know from a formula unit is what elements are present in the compound and how many atoms of each element are present.

A compound's formula mass can be calculated only if its formula is known. The best way to illustrate the procedure is to calculate the formula masses of a few representative compounds. The atomic and molecular masses in this chapter will be expressed to four significant figures, unless otherwise noted.

Example 4.1 Calculating the formula mass of a compound

Calculate the formula masses of the following compounds: NaCl; $CaCl_2$; Al_2O_3; and $C_6H_{12}O_6$ (glucose, a sugar).

Solution

Formula mass is defined as the sum of the atomic masses of a compound's constituent atoms. Therefore,

$$\text{Formula mass of NaCl} = \text{atomic mass of Na} + \text{atomic mass of Cl}$$
$$= 22.99 \text{ amu} + 35.45 \text{ amu} = 58.44 \text{ amu}$$

$$\text{Formula mass of } CaCl_2 = \text{atomic mass of Ca} + 2 \times \text{atomic mass of Cl}$$
$$= 40.08 \text{ amu} + (2 \times 35.45 \text{ amu}) = 111.0 \text{ amu}$$

$$\text{Formula mass of } Al_2O_3 = 2 \times \text{atomic mass of Al} + 3 \times \text{atomic mass of O}$$
$$= (2 \times 26.98 \text{ amu}) + (3 \times 16.00 \text{ amu}) = 102.0 \text{ amu}$$

$$\text{Formula mass of } C_6H_{12}O_6 = 6 \times \text{atomic mass of C}$$
$$+ 12 \times \text{atomic mass of H}$$
$$+ 6 \times \text{atomic mass of O}$$
$$= (6 \times 12.01 \text{ amu}) + (12 \times 1.008 \text{ amu})$$
$$+ (6 \times 16.00 \text{ amu})$$
$$= 180.2 \text{ amu}$$

Problem 4.1 Calculate the formula masses of the following compounds: (a) $Zn(HPO_4)_2$; (b) Fe_3O_4; (c) H_2O; (d) H_2SO_4; (e) $C_3H_6O_2N$ (the amino acid alanine, a component of proteins).

In Chapter 2, we saw that samples of different elements must contain equal numbers of atoms when the mass of each sample, in grams, is identical with the element's relative mass in atomic mass units. This means that there are equal numbers of atoms in 55.85 g of iron, 22.99 g of sodium, 16.00 g of oxygen, and so forth. The definition of formula mass allows us to make the same statement about the formula mass of a compound. That is, the mass of any compound in grams that is equal to its formula mass in atomic mass units contains the same number of formula units as the mass of any other compound in grams that is equal to its formula mass in atomic mass units. Therefore, the number of formula units in 58.44 g of NaCl is equal to the number of formula units in 180.2 g of glucose. Figure 4.1 shows the masses of a number of compounds in which there are equal numbers of formula units.

Figure 4.1 Molar quantities of some molecular compounds. The mass of each substance is equal to its formula mass in grams, or 1 mole. The substances from left to right are water, H_2O (molar mass, 18.02 g); ethanol, C_2H_6O (molar mass, 46.07 g); glucose, $C_6H_{12}O_6$, (molar mass, 180.2 g), sucrose, $C_{12}H_{22}O_{11}$ (molar mass, 342.3 g).

Example 4.2 Converting formula units into mass

Calculate the masses, in grams, of NaCl, $CaCl_2$, Al_2O_3, and $C_6H_{12}O_6$ that will contain equal numbers of formula units of each compound.

Solution

The formula masses of these compounds were calculated in Example 4.1, and those masses in grams will all contain the same numbers of formula units. The masses are 58.44 g of NaCl, 111.0 g of $CaCl_2$, 102.0 g of Al_2O_3, and 180.2 g of $C_6H_{12}O_6$.

Problem 4.2 Calculate the masses, in grams, of (a) Na_2SO_4, (b) Kbr, (c) MgO, and (d) C_6H_{14} that will contain equal numbers of formula units of each compound.

4.2 THE MOLE

As noted in Chapter 2, it is impossible to weigh out one atom, but we can weigh out the mass of an element in grams that is equal to the relative mass of an atom, and we can do the same with a compound. The number of atoms contained in the atomic mass (in grams) of an element and the number of formula units contained in the formula mass (in grams) of a compound is given a name of its own. This quantity is called the **mole.** When numerical amounts are specified, the mole is abbreviated **mol.** The mole allows us to know the number of formula units in a sample of a given mass.

A mole of any element or compound contains the same number of atoms or formula units as a mole of any other element or compound. The mass of 1 mol of a compound is called its **molar mass.** The molar mass of sodium chloride weighs 58.44 g. It has the same number of formula units as 1 mol of glucose, which weighs 180.2 g. Table 4.1 illustrates the relations between formula mass (atomic mass units), molar mass (grams per mole), particles per mole, and number of moles for hydrogen atoms, hydrogen molecules, water, and calcium chloride.

TABLE 4.1 Relations Between Molar Quantities

Substance	Formula	Formula mass (amu)	Molar mass (g/mol)	Particles per mole	Moles
atomic hydrogen	H	1.008	1.008	6.022×10^{23} hydrogen atoms	1 mol H atoms
molecular hydrogen	H_2	2.016	2.016	6.022×10^{23} hydrogen molecules 12.044×10^{23} hydrogen atoms	1 mol H_2 molecules 2 mol H atoms
water	H_2O	18.02	18.02	6.022×10^{23} water molecules 6.022×10^{23} oxygen atoms 12.044×10^{23} hydrogen atoms	1 mol H_2O molecules 1 mol O atoms 2 mol H atoms
calcium chloride	$CaCl_2$	111.0	111.0	6.022×10^{23} $CaCl_2$ formula units 6.022×10^{23} Ca^{2+} ions 12.044×10^{23} Cl^- ions	1 mol $CaCl_2$ formula units 1 mol Ca^{2+} ions 2 mol Cl^- ions

The definition of molar mass,

1 formula unit of a compound in grams = 1 mole of that compound,

allows us to write the following conversion factors:

$$\frac{\text{1 formula unit of any compound in grams}}{\text{1 mol of that compound}} = 1$$

or

$$\frac{\text{1 mol of any compound}}{\text{1 formula unit of that compound in grams}} = 1$$

By using these conversion factors, we can calculate the number of formula units in any sample of a compound if we know both the formula mass and the mass of the sample in grams.

Concept check

✔ The mole permits the conversion of mass into number of formula units and the conversion of number of formula units into mass.

Example 4.3 Converting mass into moles

Calculate the number of moles in 270.0 g of glucose.

Solution

The conversion factor required for this calculation uses the fact that 1 formula mass of a compound is equal to 1 mol of that substance. The formula mass of glucose was calculated in Example 4.1 and is 180.2 amu, so we know that, by definition, 1 mol of glucose weighs 180.2 g. To convert grams of glucose into moles of glucose, let's first look at the conversion from the point of view of units only:

$$\text{g } \cancel{\text{glucose}} \left(\frac{\text{mol glucose}}{\text{g } \cancel{\text{glucose}}}\right) = \text{mol glucose}$$

Next, we insert the relevant data and calculate the answer:

$$\text{mol glucose} = 270.0 \text{ g glucose} \left(\frac{1 \text{ mol glucose}}{180.2 \text{ g glucose}}\right) = 1.498 \text{ mol glucose}$$

It is interesting to note that, although it is not possible to weigh out 1.498 molecules of glucose (because any fraction of a molecule of glucose is no longer glucose), 1.498 mol of glucose can be weighed out with no difficulty. Remember, a mole is not a molecule.

Problem 4.3 Calculate the number of moles of $CaCl_2$ in 138.8 g of $CaCl_2$.

Example 4.4 Converting moles into mass

How many grams are in 0.3300 mol of Al_2O_3?

Solution

To calculate the mass of a given number of moles of Al_2O_3 (or any compound or element), we derive a conversion factor from the fact that 1 formula mass of Al_2O_3 contains 1 mol of Al_2O_3 and weighs 102.0 g.

$$\text{g } Al_2O_3 = 0.3300 \text{ mol} \left(\frac{102.0 \text{ g } Al_2O_3}{1 \text{ mol } Al_2O_3}\right) = 33.66 \text{ g } Al_2O_3$$

Problem 4.4 How many moles are in 0.3600 g of glucose?

▶▶ The mole concept will be important in Chapter 7, when we learn to describe the concentration of solutions.

4.3 AVOGADRO'S NUMBER

The actual number of fundamental units in a mole has been determined by experiment and is called **Avogadro's number.** (Amadeo Avogadro was the first to clarify the difference between atoms and molecules.) It is a very large number, 6.022×10^{23}. It is defined as the number of carbon atoms in exactly 12 g of the pure isotope of carbon of mass 12 amu. One mole of anything—pencils, books, students, molecules, atoms—consists of 6.022×10^{23} of that thing.

Example 4.5 Calculating the number of moles of elements in a compound

How many moles of ions are there in 1 mol of each of the following ionic compounds? (a) NaCl; (b) $CaCl_2$; (c) $AlCl_3$; (d) Mg_3N_2.

Solution

Moles of Each Kind of Ion	Number of Moles of All Ions
(a) 1 mol of Na^+ ions and 1 mol of Cl^- ions	2 mol
(b) 1 mol of Ca^{2+} ions and 2 mol of Cl^- ions	3 mol
(c) 1 mol of Al^{3+} ions and 3 mol of Cl^- ions	4 mol
(d) 3 mol of Mg^{2+} ions and 2 mol of N^{3-} ions	5 mol

Problem 4.5 How many moles of atoms of each kind are there in 1 mol of each of the following substances? (a) $ZnHPO_4$; (b) H_2SO_4; (c) Al_2O_3; (d) Ca_3P_2.

It is difficult to grasp the magnitude of Avogadro's number, but perhaps the following example will help. Imagine a flat area the size of the United States. One mole of dimes spread out uniformly on this area would reach a height of about 10 miles. Mt. Everest, in comparison, is only about 6 miles high, and the highest altitude of a transcontinental airplane trip is about 8 miles. Because a mole of carbon occupies a volume of about a teaspoon, the size of Avogadro's number tells us two other things: (1) atoms and molecules are very small and (2) individually they do not weigh very much.

The relation between the atomic mass unit and the physical mass unit—that is, the gram—can now be stated numerically. That is, we are now in a position to calculate the mass of 1 amu in grams.

Because an Avogadro number of C-12 atoms has a mass of exactly 12 g, we can set up the following expression:

$$6.022 \times 10^{23}\ \text{C}\ \cancel{\text{atoms}} \left(\frac{12\ \text{amu}}{\text{C}\ \cancel{\text{atom}}}\right) = 12\ \text{g}$$

$$6.022 \times 10^{23}\ \text{amu} = 1\ \text{g}$$

$$1\ \text{amu} = \frac{1\ \text{g}}{6.022 \times 10^{23}} = 1.661 \times 10^{-24}\ \text{g}$$

Now we can calculate the masses of individual atoms.

Example 4.6 Calculating atomic masses in grams

Calculate the mass, in grams, of the following atoms: hydrogen, oxygen, sulfur, sodium, phosphorus.

Solution

This problem is solved by listing the atoms and their relative atomic masses in atomic mass units and then multiplying each atomic mass by the mass, in grams, of 1 amu: 1.661×10^{-24} g.

Atom	Atomic Mass (amu)	Grams per Atomic Mass Unit	Atomic Mass (grams per atom)
H	1.008	1.661×10^{-24}	1.674×10^{-24}
O	16.00	1.661×10^{-24}	2.658×10^{-23}
S	32.06	1.661×10^{-24}	5.325×10^{-23}
Na	22.99	1.661×10^{-24}	3.819×10^{-23}
P	30.97	1.661×10^{-24}	5.144×10^{-23}

Problem 4.6 Calculate the mass, in grams, of the following atoms: lithium, nitrogen, fluorine, calcium, carbon.

From the results calculated in Example 4.6, we can see that it is very difficult to weigh out a small number of atoms of any element. The weighing done in chemistry laboratories takes place on a much larger scale, with the use of a balance calibrated in units of 0.1000 g at the least. The following example shows how the weighing of many atoms of an element becomes a practical matter.

Example 4.7 **Relating the numbers of oxygen atoms to mass**

What number of oxygen atoms constitutes a practical amount of oxygen to weigh?

Solution

This number can best be illustrated by arranging numbers of atoms of oxygen and corresponding masses in grams in tabular form. (The mass of an oxygen atom was given in Example 4.6.)

Number of Atoms of Oxygen	Mass in Grams
10	2.658×10^{-22}
100	2.658×10^{-21}
100,000	2.658×10^{-18}
1.0×10^{15}	2.658×10^{-8}
6.022×10^{20}	0.01600
6.022×10^{23}	16.00

The tabular arrangement shows that a very large number of atoms is required to obtain a practical mass of any substance. Only the last two numbers of atoms, 0.0010 mol and 1.000 mol, are practical for weighing.

Problem 4.7 Using the numbers of atoms in Example 4.7, calculate the corresponding masses, in grams, of calcium.

4.4 EMPIRICAL FORMULAS

Chemists did not use the periodic table to determine that the formula for calcium chloride was $CaCl_2$. Instead, it was the other way around. The formulas

4.1 Chemistry in Depth

Determining the Composition of a Compound

To determine the formula of a compound requires decomposing it into its component elements or converting each of the elements in it into compounds whose identity and mass are readily determined. The latter is the approach used for compounds consisting of carbon and hydrogen or of carbon, hydrogen, and nitrogen. When such a compound is fully combusted or burned in excess oxygen, the products will always be gaseous carbon dioxide and water, and, if the compound contained nitrogen, nitrogen dioxide. The gases are collected by absorption and the increase in weight of the absorbent is measured, as illustrated in the accompanying diagram.

As oxygen flows through the combustion chamber (the furnace), all of the substance's carbon is converted into CO_2, its hydrogen into H_2O, and its nitrogen into NO_2. The CO_2 and H_2O are collected by absorbent materials whose increase in weight is then measured. After the mass of carbon and hydrogen in the original sample is accounted for, the mass of nitrogen is determined by difference. The percent composition of the compound is determined by the methods described in Sections 2.1 and 4.4.

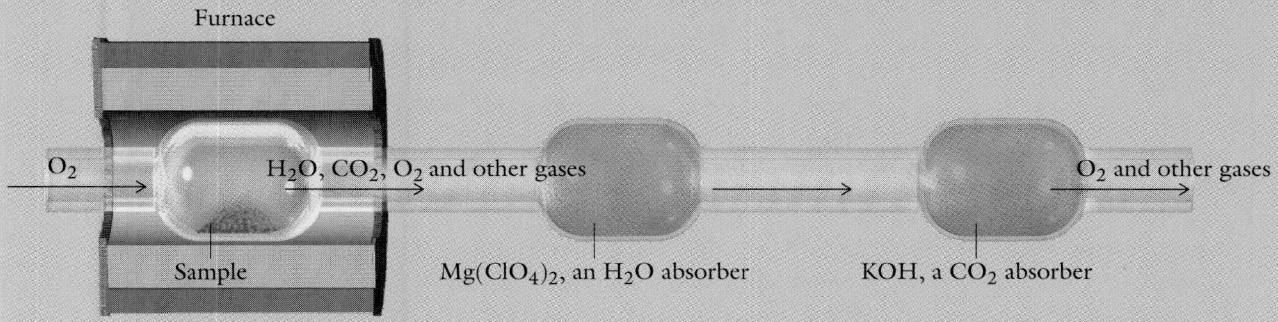

A combustion device used to analyze substances for carbon, hydrogen, and nitrogen.

for compounds are first determined by quantitative analysis, specific identification methods that determine the identity and percent composition of each element in a sample of the compound. Box 4.1 describes an experimental method for the analysis of compounds containing carbon, hydrogen, and nitrogen. The calculation of percent composition of a compound was presented in Section 2.1. The relative mass of each atom is then used to determine the number of moles of each atom in the sample. From that, we derive the simplest whole-number ratio of the constituents, which is called the **empirical formula.** The following example illustrates this approach.

Example 4.8 Finding the empirical formula: I

Quantitative analysis of calcium chloride reveals that composition of the compound is 36.04% calcium and 63.96% chlorine. From those data, determine its empirical formula.

Solution

Because the data are expressed in percentages, it is convenient to assume that we have a 100.0 g sample of calcium chloride. This assumption means that, of the

100.0 g of calcium chloride, 36.04 g is calcium and 63.96 g is chlorine. The atomic mass of each kind of atom is used to calculate the number of moles of each:

$$36.04 \text{ g Ca} \left(\frac{1 \text{ mol Ca}}{40.08 \text{ g Ca}}\right) = 0.8992 \text{ mol Ca}$$

$$63.96 \text{ g Cl} \left(\frac{1 \text{ mol Cl}}{35.45 \text{ g Cl}}\right) = 1.804 \text{ mol Cl}$$

It is possible to have 0.8992 mol of calcium and 1.804 mol of chlorine but impossible to have 0.8992 atoms of calcium and 1.804 atoms of chlorine, because atoms are indivisible, cannot be present fractionally, and must be represented in the formula by small whole numbers. This apparent discrepancy is reconciled by dividing each of the experimentally derived values by the smallest numerical value of moles present. This procedure converts all the experimental values of moles into small whole numbers representing the relative numbers of atoms within the formula unit.

$$\frac{0.8992 \text{ mol Ca}}{0.8992} = 1.000 \text{ mol Ca}, \quad \text{which rounds to 1}$$

$$\frac{1.804 \text{ mol Cl}}{0.8992} = 2.006 \text{ mol Cl}, \quad \text{which rounds to 2}$$

Because these molar quantities are based on experimental values, it is unlikely that they will be exact whole numbers. In this type of analysis, it is acceptable to round 2.006 to 2. Therefore, the empirical formula of calcium chloride is $CaCl_2$. Example 4.10 presents another aspect of this procedure.

Problem 4.8 Quantitative analysis of magnesium oxide reveals that the composition of the compound is 60.31% magnesium and 39.69% oxygen. From those data, determine its empirical formula.

Exercise 4.9 Finding the empirical formula: II

Chemical analysis shows the composition of a compound containing potassium, hydrogen, phosphorus, and oxygen, to be 44.66% potassium, 18.13% phosphorous, 36.63% oxygen, and 0.5785% hydrogen. Determine its empirical formula.

Solution

Convert percentages into grams by assuming that the sample contains 100.0 g of the compound; then use the elements' atomic masses (as conversion factors) to determine the number of moles of each element in the sample:

$$44.66 \text{ g K} \left(\frac{1 \text{ mol K}}{39.10 \text{ g K}}\right) = 1.142 \text{ mol K}$$

$$18.13 \text{ g P} \left(\frac{1 \text{ mol P}}{30.97 \text{ g P}}\right) = 0.5854 \text{ mol P}$$

$$36.63 \text{ g O} \left(\frac{1 \text{ mol O}}{16.00 \text{ g O}}\right) = 2.289 \text{ mol O}$$

$$0.5785 \text{ g H} \left(\frac{1 \text{ mol H}}{1.008 \text{ g H}}\right) = 0.5739 \text{ mol H}$$

Again, the numbers of moles of each element must be represented by small whole numbers. This task is accomplished by dividing each of the preceding experimentally derived values by the smallest value.

$$\frac{1.142 \text{ mol K}}{0.5739 \text{ mol H}} = 1.990, \quad \text{which rounds to 2}$$

$$\frac{0.5854 \text{ mol P}}{0.5739 \text{ mol H}} = 1.020, \quad \text{which rounds to 1}$$

$$\frac{2.289 \text{ mol O}}{0.5739 \text{ mol H}} = 3.988, \quad \text{which rounds to 4}$$

$$\frac{0.5739 \text{ mol H}}{0.5739 \text{ mol H}} = 1.000, \quad \text{which rounds to 1}$$

These amounts now reveal the smallest whole-number ratios of the different elements in the compound. The reason that the calculated numbers of moles are not quite whole numbers is that the percentages were determined experimentally and were therefore subject to variability. The rounded values show the empirical formula of potassium monohydrogen phosphate to be K_2HPO_4.

Problem 4.9 Chemical analysis shows the composition of magnesium sulfate to be 20.20% magnesium, 26.60% sulfur, and 53.20% oxygen. Determine its empirical formula.

Example 4.10 Finding the empirical formula: III

Chemical analysis shows the composition of potassium dichromate to be 26.43% potassium, 38.09% oxygen, and 35.48% chromium. Determine its empirical formula.

Solution

Convert percentages into grams by assuming a sample of 100.0 g of the compound; then determine the number of moles of each element by using the atomic mass of each element as a unit-conversion factor:

$$26.43 \text{ g K}\left(\frac{1.000 \text{ mol K}}{39.10 \text{ g K}}\right) = 0.6760 \text{ mol K}$$

$$38.09 \text{ g O}\left(\frac{1.000 \text{ mol O}}{16.00 \text{ g O}}\right) = 2.381 \text{ mol O}$$

$$35.48 \text{ g Cr}\left(\frac{1.000 \text{ mol Cr}}{52.00 \text{ g Cr}}\right) = 0.6823 \text{ mol Cr}$$

Dividing each value by the smallest one, we get

$$\frac{0.6760 \text{ mol K}}{0.6760 \text{ mol K}} = 1.000, \quad \text{which rounds to 1}$$

$$\frac{2.381 \text{ mol O}}{0.6760 \text{ mol K}} = 3.522, \quad \text{which rounds to 3.5}$$

$$\frac{0.6823 \text{ mol Cr}}{0.6760 \text{ mol K}} = 1.009, \quad \text{which rounds to 1}$$

The numbers just calculated are relative numbers of atoms. They are based on experimental fact. Therefore, the 3.522 atoms of oxygen should be rounded to 3.5. At the same time, because atoms are indivisible, only whole numbers of atoms are possible. A formula unit of this compound cannot have 3.5 atoms of oxygen. To solve the dilemma, we double the calculated values, which produces the smallest whole-number values and shows that the formula of potassium dichromate must be $K_2Cr_2O_7$.

Problem 4.10 By chemical analysis, the composition of an oxide of phosphorus was 43.66% phosphorus and 56.34% oxygen. Determine its empirical formula.

4.5 MOLECULAR FORMULAS

The empirical formula represents the basic combining ratios of a compound's elements. It does not reveal the actual numbers of each kind of atom in a molecule. That information is contained in the **molecular formula** of a compound. For example, the percent composition of fructose is: carbon, 40.00%; hydrogen, 6.716%; and oxygen, 53.29%. From these data, the empirical formula can be calculated.

$$\text{Carbon:} \quad 40.00 \text{ g C} \left(\frac{1 \text{ mol C}}{12.00 \text{ g C}}\right) = 3.331 \text{ mol C}$$

$$\text{Hydrogen:} \quad 6.716 \text{ g H} \left(\frac{1 \text{ mol H}}{1.000 \text{ g H}}\right) = 6.663 \text{ mol H}$$

$$\text{Oxygen:} \quad 53.29 \text{ g O} \left(\frac{1 \text{ mol O}}{16.00 \text{ g O}}\right) = 3.331 \text{ mol O}$$

Dividing all numbers by 3.331, we find the smallest ratio of whole numbers to be 1:2:1 and the empirical formula of fructose to be CH_2O. The molecular mass based on this empirical formula is approximately 30.

If the molecular mass were 30, then CH_2O would have to be accepted as the molecular formula, the presumed actual numbers of atoms of all kinds making up the molecule. However, it is known that the molecular mass of fructose, determined by an independent method, is approximately 180 amu. This information in combination with the empirical formula allows us to determine the molecular formula. We know that the ratio of elements cannot be changed, only the actual numbers. Thus, if 1:2:1 is the correct ratio of C:H:O, the molecular mass of 180 must be some whole-number multiple of the empirical formula mass of 30.

▶▶ **Beginning in Chapter 11, we'll learn to describe organic compounds that have the same molecular formulas but different molecular shapes.**

The clearest way to derive the molecular formula is to construct a table of multiples of the mass of CH_2O, along with the corresponding molecular formulas. We can then select by inspection the correct molecular formula corresponding to the appropriate molecular mass.

Molecular Formula	Molecular Mass (amu)
CH_2O	30
$C_2H_4O_2$	60
$C_4H_8O_4$	120
$C_6H_{12}O_6$	180
$C_8H_{16}O_8$	240

The correct molecular formula corresponding to the molecular mass of 180 is $C_6H_{12}O_6$.

4.6 BALANCING CHEMICAL EQUATIONS

The atomic theory allows us to restate the Law of Conservation of Mass (Section 2.1) as the **Law of Conservation of Atoms:** In any chemical process, whatever atoms are present at the beginning of the reaction must be present at the end, albeit in new combinations. This idea is put into practice in the writing of balanced chemical equations. When you have mastered the skill of writing a chemical reaction as a balanced equation, you will also be able to predict the mass quantities of the products emerging from the reaction on the basis of the masses of the starting materials.

Chemists express the Law of Conservation of Mass in chemical shorthand by writing reactions as equations. An equal sign can be used, but most chemists use an arrow to separate components that react, called **reactants,** from components that result from the reaction, called **products.** Consider the reaction between sodium and chlorine. Remember that chlorine exists in elemental form as a diatomic molecule. A first attempt at writing an equation for that reaction might look like

$$Na + Cl_2 \longrightarrow NaCl$$

The components on the left-hand side of the equation, Na and Cl_2, are the reactants, and the component on the right-hand side is the product. The equation is not written correctly, however, because conservation of mass (or of atoms) requires that equal numbers of each kind of atom appear on both sides of the equation. The number of sodium atoms on the reactant side is the same as the number of sodium atoms on the product side, which means that the equation is **balanced** with respect to the sodium atoms. However, there are two chlorine atoms on the left but only one on the right, and the equation is therefore **unbalanced** with respect to the chlorine atoms. The chemical equation cannot be considered correct until the numbers of atoms of the elements on both sides of the equation are equal or balanced.

Balancing a reaction equation is accomplished by inserting numbers in front of individual molecular or ionic formulas. These numbers serve as multipliers. The numbers of chlorine atoms, for example, can be balanced by placing a 2 in front of the NaCl.

$$Na + Cl_2 \longrightarrow 2\ NaCl$$

The number 2 in front of the NaCl is called a balancing **coefficient.** Just as in an algebraic equation, the absence of a number in front of a formula is understood to indicate a balancing coefficient of 1: Na and Cl_2 can be read as 1 Na and 1 Cl_2.

Inserting a coefficient of 2 in front of NaCl increases the number of sodium atoms and chlorine atoms on the right to two each, but now sodium is unbalanced because there is only one sodium atom on the left. This imbalance is remedied by placing a balancing coefficient of 2 in front of the sodium on the left.

$$2\ Na + Cl_2 \longrightarrow 2\ NaCl$$

There are now equal numbers of each kind of atom on both sides of the equation, and the relation is called a **balanced chemical equation.**

Balancing can be accomplished only by inserting appropriate numbers before formulas, as multipliers or balancing coefficients. Changing subscripts within a compound's formula is not permissible, because it changes the chemical identity of the substance. It might be mathematically correct to balance the equation by writing

$$Na + Cl_2 \longrightarrow NaCl_2$$

but it is not chemically correct, because $NaCl_2$ is not the same substance as NaCl and, in fact, does not exist. In NaCl, the element chlorine is no longer in the form of a gas; rather, it is in the form of chloride ions, and these ions combine with sodium ions in a 1:1 ratio.

Concept check

✔ To balance an equation, the correct formulas of reactants and products must be known.

Consider another reaction in which two elements react:

$$Al + O_2 \longrightarrow Al_2O_3$$

The key issue here is to recognize that there are two oxygen atoms on the left and three oxygen atoms on the right. By multiplying the left-hand side by 3 and the right-hand side by 2, the numbers of oxygen atoms on the left and right are made equal. Therefore, the first step in balancing this equation is:

$$Al + 3\ O_2 \longrightarrow 2\ Al_2O_3$$

The result is six oxygen atoms on the left and six oxygen atoms on the right. However, there are now four aluminum atoms on the right, so a coefficient of 4 must be placed before the Al on the left. The final balanced equation is:

$$4\ Al + 3\ O_2 \longrightarrow 2\ Al_2O_3$$

In chemical notation, a polyatomic ion is enclosed in parentheses when more than one of the ion is present in a formula. A subscript after the closing parenthesis indicates how many of that ion the formula unit contains. The compound $Ca_3(PO_4)_2$, for example, contains two phosphate ions in each formula unit. It is important to remember that the coefficients in a balanced chemical equation multiply the entire formula unit, and therefore the numbers of each atom in the formula—including each atom in each polyatomic ion in the formula.

Now consider the reaction between aluminum chloride, $AlCl_3$, and sodium phosphate, Na_3PO_4. Note that the PO_4^{3-} group is a polyatomic ion, which, as stated earlier, behaves as an individual ionic unit.

$$AlCl_3 + Na_3PO_4 \longrightarrow AlPO_4 + NaCl \quad \text{(unbalanced)}$$

At this point, the numbers of aluminum atoms and phosphate ions are the same on both sides of the equation, but only one-third the number of reacting sodium and chlorine atoms appear as products. To correct this imbalance, a 3 is placed before the formula for NaCl, as follows:

$$AlCl_3 + Na_3PO_4 \longrightarrow AlPO_4 + 3\ NaCl$$

There is no single, consistently successful method for balancing equations. It is an artful procedure, and each of us goes about it in his or her own way, developing effective methods as a result of diligent repetition. However, the following approach may prove helpful. This method of balancing equations focuses on one type of atom (or polyatomic ion) at a time, ignoring its partners in the compounds in which it exists. With that in mind, let's return to the reaction between aluminum chloride and sodium phosphate.

The most effective first step is to examine the fate of the cation having the largest charge—in this case, aluminum:

$$Al(X) \longrightarrow Al(Y)$$

X and Y are used to indicate that the compound containing aluminum on the reactant side is different from the compound containing aluminum on the product side. There are identical numbers of aluminum atoms on each side of the equation; therefore no multipliers or coefficients are required.

Phosphate is a polyatomic ion and can be considered a unit for balancing purposes. Examination of the preceding unbalanced equation reveals equal numbers of phosphate on both sides, so no further modification is suggested by the phosphate ion.

The next step is to examine the fate of sodium:

$$Na_3(X) \longrightarrow Na(Y)$$

In this case, a coefficient of 3 before the product, Na(Y), is required for balance.

$$Na_3(X) \longrightarrow 3\ Na(Y)$$

Now, consider the other consequence of inserting that coefficient. It balanced the sodium atoms, but it also multiplied the chloride in NaCl, resulting in three chlorides on the right-hand side. Because there are also three chloride atoms on the left-hand side of the arrow, the equation is now fully balanced:

$$AlCl_3 + Na_3PO_4 \longrightarrow AlPO_4 + 3\ NaCl$$

Let's try a more complex reaction, that of aluminum hydroxide with sulfuric acid, which results in the formation of the salt aluminum sulfate and water:

$$Al(OH)_3 + H_2SO_4 \longrightarrow Al_2(SO_4)_3 + H_2O$$

Notice that, in these compounds, the parentheses emphasize that there are three polyatomic OH^- ions per unit of aluminum hydroxide and three polyatomic SO_4^{2-} ions per unit of aluminum sulfate.

Again, we will consider one type of atom at a time, ignoring its partners in the compounds in which it exists. We focus first on aluminum, the cation of greatest charge, Al^{3+}:

$$Al(X) \longrightarrow Al_2(Y)$$

Because two atoms of Al appear in the product, there must be two atoms in the reactant. This imbalance requires a coefficient of 2 before the reactant:

$$2\ Al(X) \longrightarrow Al_2(Y)$$

Now we turn to the polyatomic ion of greatest charge, SO_4^{2-}, counting its numbers in product and reactant before balancing:

$$(X)SO_4 \longrightarrow (Y)(SO_4)_3$$

This ion can be balanced by inserting a coefficient of 3 in front of the reactant that contains the sulfate ion:

$$3\ (X)SO_4 \longrightarrow (Y)(SO_4)_3$$

The equation is now balanced with respect to Al and SO_4, but it is not balanced with respect to H_2O:

$$2\ Al(OH)_3 + 3\ H_2SO_4 \longrightarrow Al_2(SO_4)_3 + H_2O$$

The formula for water, H_2O, may also be written as HOH. The formula for water written in this way makes it easier to visualize the balancing coefficients needed to complete the procedure. There are 6 OH^- ions and 6 H atoms on the left-hand side of the arrow. A coefficient of 6 inserted in front of H_2O on the product side balances them out:

$$6\ (OH) + 6\ H \longrightarrow 6\ HOH \longrightarrow 6\ H_2O$$

The final, balanced equation is

$$2\ Al(OH)_3 + 3\ H_2SO_4 \longrightarrow Al_2(SO_4)_3 + 6\ H_2O$$

Example 4.11 Balancing a chemical equation

Balance the following equation:

$$K_3PO_4 + BaCl_2 \longrightarrow Ba_3(PO_4)_2 + KCl$$

Solution

Focus on the cation of greatest charge, Ba^{2+}:

$$Ba(X) \longrightarrow Ba_3(Y)$$

The barium atoms can be balanced, provisionally, by placing a 3 before the Ba on the left-hand side.

$$3\ Ba(X) \longrightarrow Ba_3(Y)$$

The consequence of inserting this balancing coefficient is that there are now six chlorines on the left-hand side. Inserting a coefficient of 6 in front of the Cl-containing compound on the right-hand side increases the number of potassium atoms in the product to six. To balance those 6 K, we must now place a coefficient of 2 in front of the reactant K_3PO_4. The final balanced equation is

$$2\ K_3PO_4 + 3\ BaCl_2 \longrightarrow Ba_3(PO_4)_2 + 6\ KCl$$

Problem 4.11 Balance the following equation:

$$Ca(OH)_2 + H_3PO_4 \longrightarrow Ca_3(PO_4)_2 + H_2O$$

4.7 OXIDATION–REDUCTION REACTIONS

Chemical reactions in which oxygen or hydrogen, or both, is gained or lost are called oxidation–reduction or **redox** reactions. An increase in a compound's oxygen content or a decrease in its hydrogen content, or both, is considered to be an **oxidation.** Conversely, a decrease in a compound's oxygen content or an increase in its hydrogen content, or both, is considered to be a **reduction.** Remember that there are reactions in which oxygen has no role (Section 3.2), and chemists consider the loss of electrons in these reactions to be an oxidation and the gain in electrons a reduction.

Direct Reaction with Oxygen

Metals react directly with oxygen to form oxides: for example,

$$2\ Mg + O_2 \longrightarrow 2\ MgO$$
$$4\ Fe + 3\ O_2 \longrightarrow 2\ Fe_2O_3$$
$$4\ Al + 3\ O_2 \longrightarrow 2\ Al_2O_3$$

Each of these reactions is a redox reaction in which the metals are oxidized and oxygen is reduced.

A PICTURE OF HEALTH

Examples of Stoichiometry in Nutritional Intake

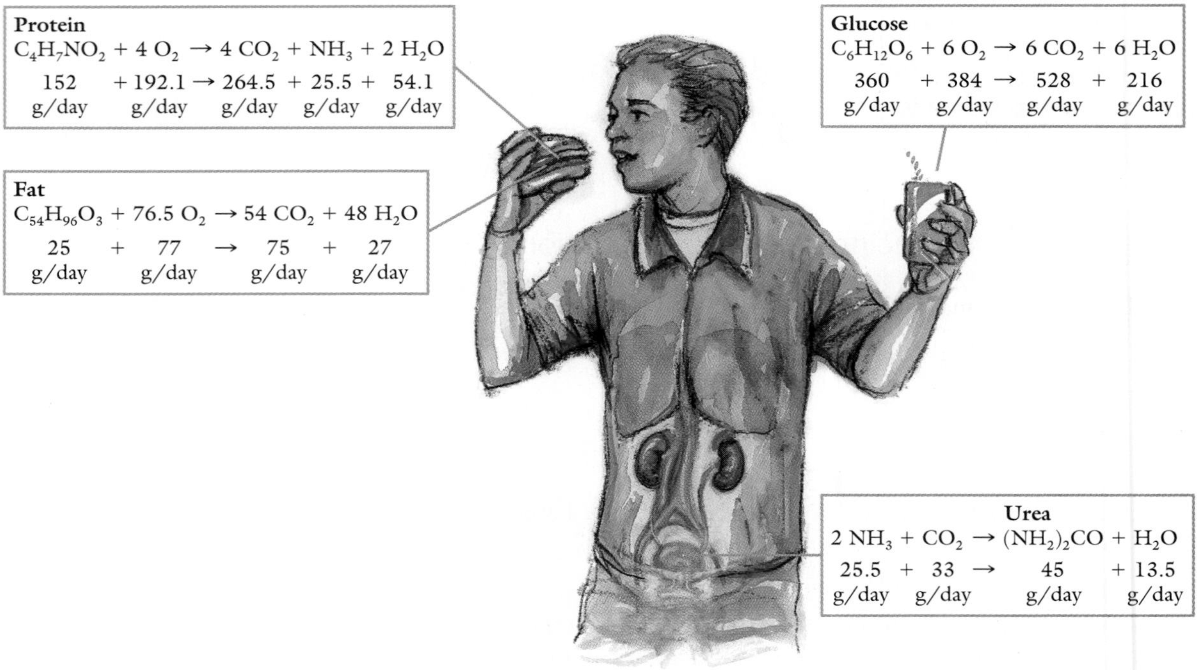
Protein
$C_4H_7NO_2 + 4\ O_2 \rightarrow 4\ CO_2 + NH_3 + 2\ H_2O$
152 g/day + 192.1 g/day → 264.5 g/day + 25.5 g/day + 54.1 g/day
Fat
$C_{54}H_{96}O_3 + 76.5\ O_2 \rightarrow 54\ CO_2 + 48\ H_2O$
25 g/day + 77 g/day → 75 g/day + 27 g/day
Glucose
$C_6H_{12}O_6 + 6\ O_2 \rightarrow 6\ CO_2 + 6\ H_2O$
360 g/day + 384 g/day → 528 g/day + 216 g/day
Urea
$2\ NH_3 + CO_2 \rightarrow (NH_2)_2CO + H_2O$
25.5 g/day + 33 g/day → 45 g/day + 13.5 g/day

An example of a reaction in which a compound loses oxygen is the reaction of an ore (in this case, iron ore) with a reducing agent (in this case, carbon):

$$2\ Fe_2O_3 + 3\ C \longrightarrow 4\ Fe + 3\ CO_2$$

Most metals are obtained in elemental form by the reduction of ores that, in many cases, are oxides of metals.

When the direct reaction of an element or compound with oxygen is accompanied by the familiar sight of burning, the reaction is called a **combustion reaction.** Hydrogen, sulfur, and nitrogen, for example, each burn in excess oxygen to produce H_2O, SO_2, and NO_2, respectively.

Compounds containing carbon react with oxygen to produce CO_2 if sufficient oxygen is present. For example, if an alkane (Chapter 11) such as C_3H_8 is burned, the products will be CO_2 and H_2O. In Chapters 22 through 24, the energy-yielding reactions of metabolism will be shown to be equivalent to slow, controlled combustion reactions.

If less than sufficient O_2 is available, the combustion product of an alkane will be CO (carbon monoxide). A significant number of deaths occur every year from the carbon monoxide generated when fossil fuels are burned in heating systems that have insufficient oxygen intake. For our purposes, assume that there is sufficient oxygen to produce CO_2 unless otherwise indicated.

Example 4.12 Balancing a combustion equation: I

Write the balanced equation for the combustion of the alkane C_5H_{12}.

Solution

The products of the alkane's combustion will be water and carbon dioxide:

$$C_5H_{12} + O_2 \longrightarrow CO_2 + H_2O \quad \text{(unbalanced)}$$

All the carbon is contained in one product, and all the hydrogen in the other, so the coefficients of the products are quickly determined:

$$C_5H_{12} + O_2 \longrightarrow 5\ CO_2 + 6\ H_2O \quad \text{(unbalanced)}$$

The sum of O atoms on the right is 16; therefore, the coefficient of O_2 is 8.

$$C_5H_{12} + 8\ O_2 \longrightarrow 5\ CO_2 + 6\ H_2O$$

Problem 4.12 Write the balanced equation for the combustion of the alkane C_4H_{10}. (Hint: To avoid fractional balancing coefficients, multiply all coefficients by an appropriate whole number—for example, 2 or 3—to change the fractions to whole numbers.)

Example 4.13 Balancing a combustion equation: II

Write the balanced equation for the combustion of the carbohydrate glucose ($C_6H_{12}O_6$).

Solution

The elements of which glucose (and any other carbohydrate) is composed are carbon, hydrogen, and oxygen. Therefore the products of the combustion of glucose will be water and carbon dioxide:

$$C_6H_{12}O_6 + O_2 \longrightarrow CO_2 + H_2O \quad \text{(unbalanced)}$$

All the carbon is contained in one product and all the hydrogen in the other, so the coefficients of the products are quickly determined:

$$C_6H_{12}O_6 + O_2 \longrightarrow 6\ CO_2 + 6\ H_2O \quad \text{(unbalanced)}$$

The sum of O atoms on the right is 18, but, in this case, both reactants contribute O atoms. Six oxygen atoms are contributed by glucose; therefore oxygen need contribute only 12, and its coefficient is 6.

$$C_6H_{12}O_6 + 6\ O_2 \longrightarrow 6\ CO_2 + 6\ H_2O$$

Although the metabolic oxidation of glucose in living cells is far more complex, it, too, is represented by this balanced equation.

Problem 4.13 Write the balanced equation for the combustion of lactic acid ($C_3H_6O_3$), a product of carbohydrate metabolism.

Dehydrogenation

One of the reactions through which the cell extracts energy from food is the oxidation of malic acid to oxaloacetic acid. This reaction is an example of oxidation by hydrogen loss, also called a **dehydrogenation** reaction. The hydrogen is transferred as a hydride ion ($H:^-$) to an intermediary compound called nicotinamide adenine dinucleotide, abbreviated NAD^+ (see Figure 14.6), which is reduced to form NADH in an enzyme-catalyzed reaction (Chapters 22 and 23). The hydrogen eventually emerges in the form of water. The oxidation can be written as

$$C_4H_6O_5 + NAD^+ \longrightarrow C_4H_4O_5 + NADH + H^+$$

In this reaction, one hydrogen atom in the form of a hydride ion is transferred to NAD^+ along with two electrons, and the other hydrogen atom enters solution as a proton.

By using structural formulas, we write the reaction as follows:

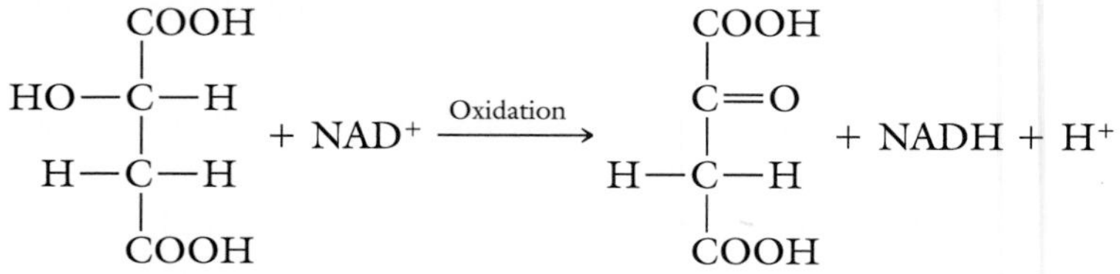

Malic acid   Oxaloacetic acid

The reverse of this reaction, a reduction, in which the oxaloacetic acid gains hydrogen to form malic acid also takes place in biological systems.

Concept checklist

- ✔ Oxidation takes place when a compound's oxygen content increases or its hydrogen content decreases or both.
- ✔ Reduction takes place when a compound's oxygen content decreases or its hydrogen content increases or both.

Loss and Gain of Electrons

In Chapter 3, the reaction of sodium with chlorine was written as two separate processes called half-reactions in which electrons lost by one of the components were gained by the other. Although oxygen has no role in such reactions, chemists consider the loss of electrons to be an oxidation and the gain in electrons a reduction, as mentioned earlier. The practice of separating the complete reaction equation into equations showing only reduction or oxidation allows balancing with respect to electrons as well as atoms.

$$2\ Na \longrightarrow 2\ Na^+ + 2\ e^-$$
$$Cl_2 + 2\ e^- \longrightarrow 2\ Cl^-$$

The reactions between sodium and chlorine or aluminum and chlorine can be written in the usual manner and balanced by inspection:

$$2\ Na + Cl_2 \longrightarrow 2\ NaCl$$
$$2\ Al + 3\ Cl_2 \longrightarrow 2\ AlCl_3$$

However, finding the balancing coefficients for more complex oxidation–reduction reactions that take place in solution is more difficult. Fortunately, a systematic procedure called the **ion-electron method** can be used to balance any oxidation–reduction reaction. The method is illustrated in Box 4.2 on the next page.

The list below summarizes the process of balancing chemical equations.

Balancing chemical equations

- Write a chemical reaction as an equation with an arrow separating the reacting components, or reactants, from the resulting components, or products.
- To balance an equation, the correct formulas of reactants and products must be known.
- It is not permissible to change the subscripts in a compound's formula in an effort to balance an equation, because altering subscripts changes the chemical identity of the substance.
- An equation can be balanced only by inserting whole numbers, called balancing coefficients, in front of the formulas of both reactants and products. The coefficient multiplies the entire formula unit and therefore the numbers of all the atoms in it.
- When a final check shows that there are equal numbers of each kind of atom on both sides of the equation, it is balanced.

4.8 STOICHIOMETRY

The coefficients, or multipliers, in balanced equations can now be used to connect the Law of Conservation of Atoms to the Law of Conservation of Mass. In Section 4.7, the emphasis was on the balancing of equations with respect to number of atoms. Note, however, that, if we multiply a balancing coefficient by Avogadro's number, we are in effect converting it into Avogadro numbers of formula units, or into moles. With this conversion, we can relate the balancing coefficients to mass. The quantitative mass relations of chemical reactions define an aspect of chemistry called **stoichiometry.**

Stoichiometric calculations in chemistry are analogous to multiplying or dividing the amounts of ingredients in a bread recipe in order to produce a smaller or larger number of loaves.

4.2 Chemistry in Depth

Balancing Oxidation–Reduction Reactions by the Ion-Electron Method

The ion-electron method of balancing redox reactions requires the balance of mass (atoms) and electrons. This process is accomplished by the following series of steps.

- **Step 1.** Separate the oxidation and reduction half-reactions.
- **Step 2.** Make certain that both half-reaction equations are balanced with respect to all elements other than oxygen and hydrogen.
- **Step 3.** Balance the oxygen atoms by adding water.
- **Step 4.** Balance the equations with respect to hydrogen by using protons.
- **Step 5.** Balance the equations with respect to charge.
- **Step 6.** Obtain the complete and balanced equation by adding the two half-reactions and canceling any terms that appear on both sides.

Balancing the reaction equation used for the roadside detection of alcohol on the breath is an illustration of the ion-electron method. In this reaction, ethanol, C_2H_5OH, is oxidized to acetaldehyde, C_2H_4O, by the yellow dichromate ion, $Cr_2O_7^{2-}$, which is reduced to the green chromium(III) ion, Cr^{3+}. The unbalanced net ionic equation is:

$$H^+ + Cr_2O_7^{2-} + C_2H_5OH \longrightarrow Cr^{3+} + C_2H_4O + H_2O$$

The first step in the solution of this problem is to separate the oxidation half-reaction from the reduction half-reaction:

Step 1. The two half-reactions are:

$$Cr_2O_7^{2-} \longrightarrow Cr^{3+}$$

$$C_2H_5OH \longrightarrow C_2H_4O$$

Step 2. Make certain that both equations are balanced with respect to all elements other than oxygen and hydrogen. In this case, the oxidation half-reaction is balanced, but the dichromate reduction requires a coefficient of 2 before the product, chromium(III) ion.

$$Cr_2O_7^{2-} \longrightarrow 2\ Cr^{3+}$$

$$C_2H_5OH \longrightarrow C_2H_4O$$

Step 3. Balance the oxygen atoms. This balancing is done by noting that the oxygen of $Cr_2O_7^{2-}$ is lost on its conversion into Cr^{3+}. The lost oxygen is assumed to have reacted with H^+ to form H_2O. In this case, only the reduction half-reaction includes changes in oxygen. There are seven oxygen atoms on the left and zero on the right, so the oxygen atoms are balanced by adding 14 H^+ ions to the left-hand side of the dichromate half-reaction, which results in the formation of 7 H_2O:

$$Cr_2O_7^{2-} + 14\ H^+ \longrightarrow 2\ Cr^{3+} + 7\ H_2O$$

$$C_2H_5OH \longrightarrow C_2H_4O$$

Step 4. Balance with respect to hydrogen. There are six hydrogen atoms in ethanol on the left-hand side of the oxidation half-reaction and four hydrogen atoms in

When we use the mole (the Avogadro number of formula units), the balanced equation

$$2\ Al(OH)_3 + 3\ H_2SO_4 \longrightarrow Al_2(SO_4)_3 + 3\ H_2O$$

can be interpreted to mean that 2 mol of $Al(OH)_3$ will react with 3 mol of H_2SO_4 to produce 1 mol of $Al_2(SO_4)_3$ and 3 mol of H_2O. These relations can be expressed as a series of equivalences:

$$2\ Al(OH)_3 \Leftrightarrow 1\ Al_2(SO_4)_3$$
$$2\ Al(OH)_3 \Leftrightarrow 3\ H_2SO_4$$
$$2\ Al(OH)_3 \Leftrightarrow 3\ H_2O$$
$$3\ H_2SO_4 \Leftrightarrow 1\ Al_2(SO_4)_3$$
$$3\ H_2SO_4 \Leftrightarrow 3\ H_2O$$
$$1\ Al_2(SO_4)_3 \Leftrightarrow 3\ H_2O$$

acetaldehyde on the right-hand side. Therefore two hydrogen atoms in the form of H^+ ions must be added to the right-hand side of that half-reaction:

$$Cr_2O_7^{2-} + 14\,H^+ \longrightarrow 2\,Cr^{3+} + 7\,H_2O$$

$$C_2H_5OH \longrightarrow C_2H_4O + 2\,H^+$$

The two half-reactions are now individually balanced with respect to atoms. Next, they must be balanced with respect to charge.

Step 5. Balancing with respect to charge means that the charges on both sides of the arrow must be equal to each other but not necessarily 0. Each half-reaction must be charge balanced by adding electrons to the side with excess positive charge. The dichromate half-reaction has a total charge of +12 on the left and +6 on the right, so six electrons must be added to the left-hand side. This addition of electrons makes the charge of +6 on the left equal to a +6 charge on the right:

$$Cr_2O_7^{2-} + 14\,H^+ + 6\,e^- \longrightarrow 2\,Cr^{3+} + 7\,H_2O$$

The alcohol half-reaction has a charge of 0 on the left-hand side and +2 on the right-hand side, so two electrons must be added to the right-hand side. This addition of electrons makes the charge of 0 on the left equal to the charge of 0 on the right.

$$C_2H_5OH \longrightarrow C_2H_4O + 2\,H^+ + 2\,e^-$$

It is now necessary to recognize that, in a redox reaction, the number of electrons consumed in the reduction step must be equal to the number of electrons produced in the oxidation step. The reduction of 1 mol of dichromate requires 6 mol of electrons. The oxidation of 1 mol of ethanol produces 2 mol of electrons. In this case, the oxidation half-reaction must be multiplied by 3 to provide the 6 mol of electrons required by the reduction reaction. When that is done, the equations are balanced with respect to both atoms and electrons.

Step 6. The complete and balanced equation is obtained by adding the two half-reactions and canceling any terms that appear on both sides:

$$Cr_2O_7^{2-} + 14\,H^+ + 6\,e^- \longrightarrow 2\,Cr^{3+} + 7\,H_2O$$

$$3\,C_2H_5OH \longrightarrow 3\,C_2H_4O + 6\,H^+ + 6\,e^-$$

$$Cr_2O_7^{2-} + 3\,C_2H_5OH + 8\,H^+ \longrightarrow 2\,Cr^{3+} + 3\,C_2H_4O + 7\,H_2O$$

In the complete equation, the electrons do not appear. They were equal on both sides of the equation and therefore canceled by summation. Electrons should never appear in the completely balanced equation.

The notation $\approx$ is an equivalence sign. In this context, equivalence means, for example, that 2 mol of $Al(OH)_3$ in this reaction will produce 1 mol of Al_2SO_4. It does not mean that 2 mol of $Al(OH)_3$ is equal to 1 mol of Al_2SO_4.

Relations among these balancing coefficients provide conversion factors that allow us to calculate how much product will be formed by the reaction of given amounts of starting materials. In the more explicit example that follows, a balanced chemical equation is used to construct unit-conversion factors relating moles of one component to moles of any other component.

Example 4.14 Using stoichiometry to create conversion factors: I

Derive all possible unit-conversion factors from the following balanced equation:

$$2\,Al(OH)_3 + 3\,H_2SO_4 \longrightarrow Al_2(SO_4)_3 + 3\,H_2O$$

Solution

The following 6 unit-conversion factors and their reciprocals add up to a total of 12 different unit-conversion factors from this chemical equation.

$$\frac{2\text{ mol Al(OH)}_3}{1\text{ mol Al}_2(\text{SO}_4)_3} \qquad \frac{2\text{ mol Al(OH)}_3}{3\text{ mol H}_2\text{SO}_4} \qquad \frac{2\text{ mol Al(OH)}_3}{3\text{ mol H}_2\text{O}}$$

$$\frac{3\text{ mol H}_2\text{SO}_4}{1\text{ mol Al}_2(\text{SO}_4)_3} \qquad \frac{3\text{ mol H}_2\text{SO}_4}{3\text{ mol H}_2\text{O}} \qquad \frac{1\text{ mol Al}_2(\text{SO}_4)_3}{3\text{ mol H}_2\text{O}}$$

Problem 4.14 Derive all possible unit-conversion factors from the following balanced equation:

$$3\text{ Mg(OH)}_2 + 2\text{ FeCl}_3 \longrightarrow 2\text{ Fe(OH)}_3 + 3\text{ MgCl}_2$$

Example 4.15 Using stoichiometry to create conversion factors: II

In the reaction of aluminum hydroxide with sulfuric acid, how many moles of aluminum sulfate can be produced if the reaction is begun with 1.500 mol of $Al(OH)_3$?

Solution

This problem can be solved with the unit-conversion method, by first deciding on the required conversion factor. Starting with some number of moles of $Al(OH)_3$, how many moles of $Al_2(SO_4)_3$ can we prepare?

$$\text{moles of Al(OH)}_3 \times \text{conversion factor} = \text{moles of Al}_2(\text{SO}_4)_3$$

$$\text{Conversion factor} = \frac{\text{moles of Al}_2(\text{SO}_4)_3}{\text{moles of Al(OH)}_3}$$

Substituting the appropriate numerical values from the balanced equation, we have

$$1.500\text{ mol Al(OH)}_3\left(\frac{1\text{ mol Al}_2(\text{SO}_4)_3}{2\text{ mol Al(OH)}_3}\right) = 0.7500\text{ mol Al}_2(\text{SO}_4)_3$$

Problem 4.15 In the reaction of aluminum hydroxide with sulfuric acid, how many moles of aluminum hydroxide are required to produce 3.200 mol of $Al_2(SO_4)_3$?

We must again point out that the mathematical conventions followed in Example 4.15 might seem to suggest that, in the reaction of $Al(OH)_3$ with H_2SO_4, 1.500 mol of $Al(OH)_3$ is equal to 0.7500 mol of $Al_2(SO_4)_3$. That is not quite correct; $Al(OH)_3$ is not $Al_2(SO_4)_3$. A more precise meaning is that 1.500 mol of $Al(OH)_3$ is equivalent to 0.7500 mol of $Al_2(SO_4)_3$ in that particular reaction. It is important to note that conversion factors of the kind used in Example 4.15 allow us to express the numerical equivalences between moles of different materials. Such numerical values are one thing, but the identity of the substances is a separate issue entirely. To clarify this idea, a few more examples will be presented.

Example 4.16 Using stoichiometry to create conversion factors: III

In the reaction

$$\text{N}_2 + 3\text{ H}_2 \longrightarrow 2\text{ NH}_3$$

how many moles of hydrogen are equivalent to 0.5000 mol of nitrogen?

Solution

We require a factor that will convert moles of nitrogen into their equivalence in moles of hydrogen, in terms of this particular balanced equation.

$$0.5000 \text{ mol } N_2 \left(\frac{3 \text{ mol } H_2}{1 \text{ mol } N_2}\right) = 1.500 \text{ mol } H_2$$

This calculation shows that 0.5000 mol N_2 is equivalent to, but certainly not equal to, 1.500 mol H_2 in the reaction of nitrogen with hydrogen to form ammonia.

Problem 4.16 How many moles of ammonia will be formed from 0.5000 mol of N_2 and 1.500 mol of H_2?

In the laboratory, chemists weigh grams of substances, not moles of substances. But balanced chemical equations relate moles of reactant to moles of products. To predict grams of product from grams of reactant, we therefore must convert grams of reactant into moles, then use the balanced equation to predict the number of moles of product, and then, finally, convert the number of moles of product into grams of product. The detailed procedure is as follows:

- First, balance the equation describing the reaction.
- Using the formula mass as a unit-conversion factor, convert nitrogen's mass in grams into its equivalent number of moles.
- Using the balancing coefficients as unit-conversion factors, calculate the number of moles of hydrogen required to react with the calculated number of moles of nitrogen.
- Using the formula mass as a unit-conversion factor, convert the calculated number of moles of hydrogen into its mass in grams.

Figure 4.2 is a flow diagram of the procedure for calculating either mass or moles from a balanced chemical equation. The procedure is based on the use of the balancing coefficients to convert moles of reactants into moles of products or vice versa. The methods of stoichiometry are based on the fact that reactions can be interpreted on a molecular, a molar, or a mass basis, and these interpretations are summarized in Figure 4.3 on the following page. Example 4.17 illustrates the method.

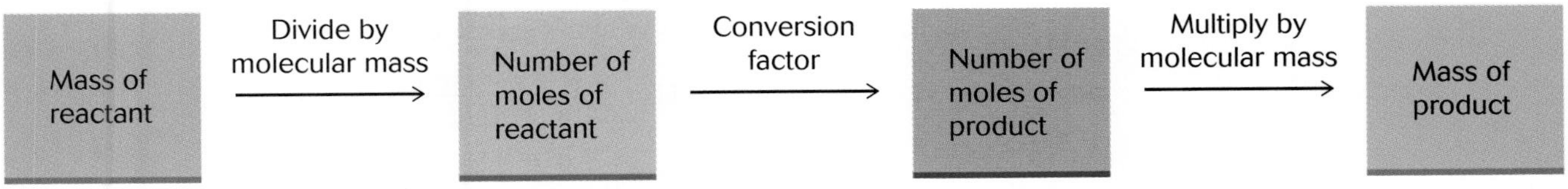

Figure 4.2 The procedure for calculating either mass or moles from balanced chemical equations. The central feature of the procedure is the use of balancing coefficients to convert moles of reactants into moles of product or vice versa.

REACTION 1	$2\ H_2$	+	O_2	⟶	$2\ H_2O$
Molecular	2 molecules	+	1 molecule	⟶	2 molecules
Molar	2 mol	+	1 mol	⟶	2 mol
Mass	4.032 g	+	32.00 g	⟶	36.03 g
REACTION 2	Mg	+	Cl_2	⟶	$MgCl_2$
Molecular	1 atom	+	1 molecule	⟶	1 formula unit
Molar	1 mol	+	1 mol	⟶	1 mol
Mass	24.31 g	+	70.90 g	⟶	95.21 g

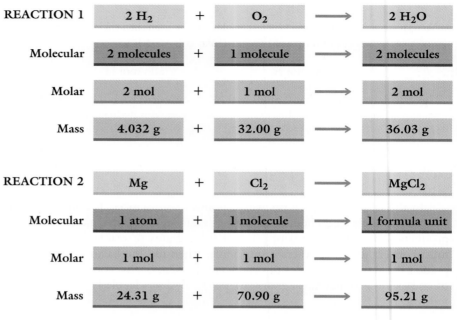

Figure 4.3 Three interpretations—molecular, molar, and mass—of the stoichiometric relations in two chemical reactions.

Example 4.17 Calculating the mass relations in a chemical reaction

How many grams of hydrogen are required for the complete conversion of 42 g of nitrogen into ammonia, and how many grams of ammonia can be formed from that quantity of nitrogen?

Solution

Step 1. Balance the equation:

$$N_2 + 3\ H_2 \longrightarrow 2\ NH_3$$

Step 2. Convert the mass of nitrogen into moles of nitrogen:

$$42.00\ \text{g}\ N_2 \left(\frac{1\ \text{mol}\ N_2}{28.02\ \text{g}\ N_2}\right) = 1.499\ \text{mol}\ N_2$$

Step 3. Calculate the numbers of moles of hydrogen and ammonia equivalent to 1.499 mol of N_2 by using the balancing coefficients as unit-conversion factors:

$$1.499\ \text{mol}\ N_2 \left(\frac{2\ \text{mol}\ NH_3}{1\ \text{mol}\ N_2}\right) = 2.998\ \text{mol}\ NH_3$$

$$1.499\ \text{mol}\ N_2 \left(\frac{3\ \text{mol}\ H_2}{1\ \text{mol}\ N_2}\right) = 4.497\ \text{mol}\ H_2$$

Step 4. Convert moles of H_2 and NH_3 into their corresponding mass in grams by using their formula masses as unit-conversion factors:

$$2.998\ \text{mol}\ NH_3 \left(\frac{17.03\ \text{g}\ NH_3}{1\ \text{mol}\ NH_3}\right) = 51.06\ \text{g}\ NH_3$$

$$4.497\ \text{mol}\ H_2 \left(\frac{2.016\ \text{g}\ H_2}{1\ \text{mol}\ H_2}\right) = 9.066\ \text{g}\ H_2$$

Add up the masses of the reactants and see if they equal the mass(es) of the product(s) to check for conservation of mass:

$$42.00\ \text{g}\ N_2 + 9.066\ \text{g}\ H_2 = 51.07\ \text{g}\ NH_3$$

Problem 4.17 Given the reaction

$$K_3PO_4 + BaCl_2 \longrightarrow Ba_3(PO_4)_2 + KCl,$$

calculate the number of grams of $BaCl_2$ needed to react with 42.4 g of K_3PO_4, and calculate the number of grams of $Ba_3(PO_4)_2$ produced.

The mass relations developed in this chapter are independent of the state of the chemical substances reacting. They apply equally well to substances in the solid, liquid, or gaseous states. Nevertheless, there are several specific uses of stoichiometry that are best discussed in a particular context. Therefore, the determination of molecular mass will be presented in Chapter 5, "The Physical Properties of Gases," and more detailed aspects of the stoichiometry of reactions in solution will be delayed until Chapters 7 and 9.

Summary

Chemical Formulas and Formula Masses The formula of a chemical compound tells us how many atoms of each element are in a fundamental unit of the compound. The fundamental unit is called a formula unit. The sum of the relative masses of the atoms of the formula unit is called the formula mass. The formula unit represents the compound's composition, not its structure. A compound's formula mass can be calculated only if its formula is known.

The Mole The atomic mass (in grams) of an element, or the formula mass (in grams) of a compound, is called the mole. A mole of any element or compound contains the same number of atoms or formula units as a mole of any other element or compound. The mole is the factor for converting mass into number of formula units, as well as number of formula units into mass.

Avogadro's Number The number of fundamental units in a mole has been determined by experiment to be 6.022×10^{23}. This number is called Avogadro's number. It is defined as the number of carbon atoms in exactly 12 g of the pure isotope of carbon of mass 12 amu. One mole of any substance consists of 6.022×10^{23} formula units of that substance.

Empirical and Molecular Formulas The formulas for compounds are first determined by quantitative analysis, which yields the number of moles of each atom in a sample of the compound. The numbers of moles of atoms in the sample can be fractional—for example 0.932 or 2.64—but the atoms themselves must be present in whole numbers. One cannot have 0.5 or 0.33 atoms in a compound. If the numbers of moles obtained in a quantitative analysis are fractional, transforming them into whole numbers, by dividing each number of moles present by the smallest numerical value of moles present, produces the compound's empirical formula, a representation of the relative number of atoms of each element in a formula unit of the compound. The molecular formula represents the actual number of atoms of each element in a molecule.

Balancing Chemical Equations The Law of Conservation of Mass is expressed by writing chemical reactions as equations in which an arrow separates components that react, called reactants, from components resulting from the reaction, called products. Conservation of mass (atoms) requires equal numbers of each kind of atom on both sides of the equation. When an equation has equal numbers of each kind of atom on both sides of the equation, it is called a balanced chemical equation.

To balance an equation, the correct formulas of reactants and products must be known. Changing subscripts in a compound's formula is not permissible, because that changes the compound's chemical identity. Therefore, an equation can be balanced only by inserting whole numbers, called balancing coefficients, in front of the formulas of both reactants and products. The coefficients in a balanced chemical equation multiply the entire formula unit that they precede and therefore the numbers of all the atoms in it.

Oxidation–Reduction Reactions Chemical reactions in which oxygen or hydrogen, or both, is gained or lost are called oxidation–reduction or redox reactions. An increase in a compound's oxygen content or a decrease in its hydrogen content, or both, is considered to be an oxidation. Conversely, a decrease in a compound's oxygen content or an increase in its hydrogen content, or both, is considered to be a reduction. In reactions in which oxygen has no role, the loss of electrons is an oxidation and the gain in electrons is a reduction.

Stoichiometry The study of the quantitative mass relations in chemical reactions is called stoichiometry. These relations stem from the consequences of the Law of Conservation of Mass. Specifically, a balanced chemical equation expresses the quantitative relations between moles of components in the reaction. By using these relations and the formula masses of the reaction components, we can convert moles into mass and mass into moles. We can determine the number of moles of product given the number of moles of reactant, as well as the mass of product in grams given the mass of reactant in grams.

Key Words

Avogadro's number, p. 99
balanced chemical equation, p. 105
combustion reaction, p. 109
dehydrogenation, p. 110
empirical formula, p. 100
formula, p. 95
formula mass, p. 95
formula unit, p. 95
mole, p. 97
oxidation, p. 108
product, p. 105
reactant, p. 105
redox reaction, p. 108
reduction, p. 108
stoichiometry, p. 111

Exercises

Chemical Formulas and Formula Masses

4.1 Calculate the formula masses, in grams, of the following compounds: (a) $CaCrO_4$; (b) $Mg(OH)_2$; (c) $TiCl_4$; (d) $Na_2Cr_2O_7$; (e) C_3H_7OH (propyl alcohol).

4.2 Calculate the formula masses, in grams, of the following compounds: (a) $Zn_3(PO_4)_2$; (b) $Ba(C_2H_3O_2)_2$; (c) Hg_2Cl_2; (d) $Sr(NO_3)_2$; (e) BF_3.

4.3 Calculate the formula masses, in grams, of the following compounds: (a) $Zn_3(AsO_4)_2$; (b) $Al(C_2H_3O_2)_3$; (c) $HgCl_2$; (d) $Sr(ClO_3)_2$; (e) BI_3.

4.4 Calculate the formula masses, in grams, of the following compounds: (a) Ag_2CO_3; (b) Cr_2O_3; (c) Hg_2Br_2; (d) $ZnCl_2$; (e) B_3N_2.

The Mole

4.5 How many moles are there in 24.67 g of sodium chloride?

4.6 How many moles are there in 20.27 g of sodium thiocyanate, NaSCN?

4.7 How many grams of sodium tripolyphosphate, $Na_5P_3O_{10}$, are there in 0.275 mol of that compound?

4.8 How many grams are there in 0.8750 mol of nickel iodate, $Ni(IO_3)_2$?

Avogadro's Number

4.9 How many zinc ions are present in 30.5 g of zinc pyrophosphate, $Zn_2P_2O_7$?

4.10 How many H atoms are there in 5.40 g of water?

4.11 What is the mass, in grams, of 10.54×10^{24} molecules of CH_4?

4.12 How many moles of water are there in 2.3×10^{24} molecules of water?

Empirical Formulas

4.13 By chemical analysis, the composition of zinc chromate is 36.05% zinc, 28.67% chromium, and 35.28% oxygen. Calculate its empirical formula.

4.14 By chemical analysis, the composition of calcium phosphate is 38.71% calcium, 20.00% phosphorous, and 41.29% oxygen. Determine its empirical formula.

Balancing Chemical Equations

4.15 Balance the following equations:

(a) $N_2 + Br_2 \longrightarrow NBr_3$

(b) $HNO_3 + Ba(OH)_2 \longrightarrow Ba(NO_3)_2 + H_2O$

(c) $HgCl_2 + H_2S \longrightarrow HgS + HCl$

4.16 Balance the following equations:

(a) $Mg(OH)_2 + FeCl_2 \longrightarrow Fe(OH)_2 + MgCl_2$

(b) $KBr + Cl_2 \longrightarrow KCl + Br_2$

(c) $Li + H_2O \longrightarrow LiOH + H_2$

4.17 Balance the following equations:

(a) $P + O_2 \longrightarrow P_2O_5$

(b) $FeCl_2 + K_2SO_4 \longrightarrow FeSO_4 + KCl$

(c) $HgCl + NaOH + NH_4Cl \longrightarrow Hg(NH_3)_2Cl + NaCl + H_2O$

4.18 Balance the following equations:

(a) $BrCl + NH_3 \longrightarrow NBr_3 + NH_4Cl$

(b) $Cu(NO_3)_2 + NaI \longrightarrow I_2 + CuI + NaNO_3$

(c) $TiCl_4 + NaH \longrightarrow Ti + NaCl + H_2$

Oxidation–Reduction Reactions

4.19 Complete and balance the following combustion reaction:

$$C_6H_6 + O_2 \longrightarrow$$

4.20 Complete and balance the following combustion reaction:

$$C_6H_{12}O_6 + O_2 \longrightarrow$$

4.21 Complete and balance the following equations:

$$C_5H_{12} + O_2 \longrightarrow$$
$$C_8H_{18} + O_2 \longrightarrow$$
$$C_{10}H_{20} + O_2 \longrightarrow$$

4.22 Complete and balance the following equations:

$$C_2H_6O + O_2 \longrightarrow$$
$$C_4H_8O_2 + O_2 \longrightarrow$$
$$C_{12}H_{22}O_{11} + O_2 \longrightarrow$$

4.23 Write the balanced equation for the combustion of methane, CH_4.

4.24 Complete and balance the following equation:

$$C_3H_8O_3 + O_2 \longrightarrow$$

Stoichiometry

4.25 What unit-conversion factors can be derived from the following balanced equation?

$$2\ Al(OH)_3 + 3\ H_2SO_4 \longrightarrow Al_2(SO_4)_3 + 6\ H_2O$$

4.26 What unit-conversion factors can be derived from the following balanced equation?

$$Ca(NO_3)_2 + Na_2SO_4 \longrightarrow CaSO_4 + 2\ NaNO_3$$

4.27 How many moles of $Al_2(SO_4)_3$ can be prepared from 2.5 mol of H_2SO_4 by using sufficient $Al(OH)_3$?

4.28 How many moles of $Al(OH)_3$ are required to completely react with 2.75 mol of H_2SO_4?

4.29 How many moles of H_2SO_4 must be used to produce 6.5 mol of water?

4.30 How many moles of $Al_2(SO_4)_2$ can be prepared from 245 g of H_2SO_4 by using sufficient $Al(OH)_3$?

4.31 How many grams of $Al_2(SO_4)_3$ can be prepared from 46.0 g of H_2SO_4 by using sufficient $Al(OH)_3$?

4.32 How many grams of $Al(OH)_3$ are required to completely convert 144 g of H_2SO_4 into $Al_2(SO_4)_3$?

4.33 Given the following equation,

$$2\ Cu(NO_3)_2 + 4\ NaI \longrightarrow I_2 + 2\ CuI + 4\ NaNO_3$$

(a) How many moles of $Cu(NO_3)_2$ will react with 0.87 mol of NaI?
(b) How many moles of NaI must react to produce 1.45 mol of CuI?
(c) How many grams of $Cu(NO_3)_2$ will react with 15.0 g of NaI?
(d) How many grams of $NaNO_3$ will be produced from 15.00 g of NaI?

4.34 Given the following equation,

$$3\ BaCl_2 + 2\ Na_3PO_4 \longrightarrow Ba_3(PO_4)_2 + 6\ NaCl$$

(a) How many moles of Na_3PO_4 will react with 0.45 mol of $BaCl_2$?
(b) How many moles of $Ba_3(PO_4)_2$ can be produced from 33.3 g of $BaCl_2$?
(c) How many grams of NaCl can be produced from 59.0 g of Na_3PO_4?
(d) How many grams of $Ba_3(PO_4)_2$ can be produced from 0.550 mol of Na_3PO_4?

Unclassified Exercises

4.35 Balance the following equations:

(a) $Ca_3(PO_4)_2 + H_3PO_4 \longrightarrow Ca(H_2PO_4)_2$
(b) $FeCl_2 + (NH_4)_2S \longrightarrow FeS + NH_4Cl$
(c) $KClO_3 \longrightarrow KCl + O_2$
(d) $O_2 \longrightarrow O_3$
(e) $C_5H_6 + O_2 \longrightarrow CO_2 + H_2O$

4.36 How many grams are there in 1.50 mol of Na_3PO_4?

4.37 Calculate the number of moles contained in the given amounts of the following compounds: (a) 102.6 g of $BaCO_3$; (b) 60.75 g of HBr; (c) 148.5 g of CuCl; (d) 50.4 g of HNO_3; (e) 65 g of $C_6H_{12}O_6$.

4.38 Calculate the number of moles contained in the given amounts of the following compounds: (a) 45 g of H_2SO_4; (b) 133 g of $Ca(OH)_2$; (c) 35 g of C_2H_6O; (d) 110 g of C_4H_{10}; (e) 210 g of $FeCl_2$.

4.39 How many formula units are there in 133 g of $Ca(OH)_2$?

4.40 How many molecules are there in 35 g of C_2H_6O (ethyl alcohol)?

4.41 How many electrons are lost by 42 g of sodium when its atoms are oxidized to ions?

4.42 How many molecules are there in 1 kg of water?

4.43 By chemical analysis, silver chloride was found to be 75.26% silver and 24.74% chlorine. Determine its empirical formula.

4.44 By chemical analysis, an oxide of copper was found to be 88.81% copper and 11.19% oxygen. Determine its empirical formula.

4.45 Chemical analysis of methyl ether showed it to be composed of 52.17% carbon, 13.05% hydrogen, and 34.78% oxygen. Determine its empirical formula.

4.46 The empirical formula of glycerin is $C_3H_8O_3$. Determine its percentage composition.

4.47 Complete and balance the following equations:

(a) $AgNO_3 + KCl \longrightarrow AgCl +$
(b) $Ba(NO_3)_2 + Na_2SO_4 \longrightarrow BaSO_4 +$
(c) $(NH_4)_3PO_4 + Ca(NO_3)_2 \longrightarrow Ca_3(PO_4)_2 +$
(d) $Mg(OH)_2 + H_3PO_4 \longrightarrow Mg_3(PO_4)_2 +$

4.48 Calculate the formula masses of the following compounds: (a) $Hg_3(PO_4)_2$; (b) $BaSO_4$; (c) $C_{12}H_{22}O_{11}$; (d) $MgC_8H_4O_4$; (e) CuCl.

4.49 What is the mass, in grams, of 2.000 mol of $Al_2(SO_4)_2$?

4.50 Balance the following equations:

(a) $P + O_2 \longrightarrow P_2O_5$
(b) $Ti + O_2 \longrightarrow Ti_2O_3$

Chemical Connections

4.51 The chemical formula for cocaine is $C_{17}H_{21}NO_4$. Calculate its percent composition.

4.52 How many electrons are lost by 60 g of calcium when its atoms are oxidized to ions?

4.53 Calculate the number of grams contained in the given amounts of the following compounds: (a) 0.46 mol H_2SO_4; (b) 1.80 mol $Ca(OH)_2$; (c) 0.76 mol C_2H_6O; (d) 1.89 mol C_4H_{10}; (e) 1.66 mol $FeCl_2$.

4.54 When aluminum is heated with an element from Group VI of the periodic table, it forms an ionic compound. On one occasion, the Group VI element was unknown, but the compound was 35.93% Al by weight. Identify the unknown element.

4.55 Analysis of an oxide of phosphorus showed it be 43.64% phosphorus and 56.36% oxygen. Its molecular mass was shown to be 282. Calculate its molecular formula.

CHAPTER 5

THE PHYSICAL PROPERTIES OF GASES

CHEMISTRY IN YOUR FUTURE

While scuba-diving with some friends, you surface to find that one of the party is barely conscious and appears to be in pain. The group had been exploring a reef 150 feet below the surface and didn't notice when this person surfaced much earlier than the rest. Three of you immediately force this diver back down 100 feet and slowly bring him up in stages of 50 feet, staying at each depth for about 30 minutes. At the end of the process, he is fully recovered. The certified divers in the group knew how to handle the emergency, but you are the only one who studied chemistry and can explain what occurred. It has to do with the material in Chapter 5.

LEARNING OBJECTIVES

- Define gas pressure and its units, and describe how it is measured.
- Summarize the gas laws' quantitative descriptions of the physical behavior of gases.
- Apply the appropriate gas laws to particular experimental conditions.
- Describe the properties of mixtures of gases.
- Use the gas laws to determine molecular mass.
- Determine the amount of gas dissolved in a liquid.

About 60% of a person's body mass is skeletal muscle, a tissue that often functions quite effectively for significant periods of time in the absence of oxygen. However, the heart and brain cannot be deprived of oxygen for even a few minutes without serious long-term consequences. To maintain homeostasis—that is, a healthy physiological steady state—the body's blood oxygen concentration must remain high and the carbon dioxide concentration low. A general understanding of the behavior of gases is vital in this regard. The respirators, "iron lung" machines, and barometric chambers used in inhalation therapy are only a few of the lifesaving tools that are found in every hospital. These machines function according to a short list of basic principles that chemists have discovered about gases. We will be exploring these principles in this chapter.

▶▶ Oxygen's central role in the body is a key theme of Chapter 22.

Experiments have shown that all gases exhibit virtually the same behavior in response to temperature and pressure changes, no matter what the gaseous compound's molecular mass or shape. To illustrate this universal behavior, Table 5.1 lists several gases, differing in number and type of atoms per molecule and in formula mass. At the same temperature and pressure (defined in Section 5.1), each of these gases occupies about the same amount of space—close to 22.4 L.

The space around each individual particle in these samples of gas may be a thousand or more times the actual volume of the gas molecule itself. In other words, the volume of a gas molecule is insignificant compared with the space surrounding it. That is the basic reason why both the quantitative and the qualitative ideas presented in this chapter apply to all gases without regard to their chemical nature.

5.1 GAS PRESSURE

This chapter explores the physical properties of gases. These properties include two that we have studied before—volume and temperature—and a new one called pressure, which is one of the most important of the physical properties with which a gas can be characterized. Differences in pressure cause the movement of gases from one place to another. This phenomenon explains the transport of oxygen from the lung air-space into the blood, the transport of carbon dioxide from the blood into the lung air-space, and the transport of oxygen from a respirator into the lung air-space.

Pressure is defined as a force applied over a unit area, and it is measured in units such as pounds per square inch, kilograms per square meter, or grams per square centimeter. In mathematical terms:

$$\text{Pressure} = \frac{\text{force}}{\text{area}}$$

TABLE 5.1 Volume Occupied by 1 mol of Several Different Gases at 0°C and 1 atm Pressure

Gas	Formula	Formula Mass (amu)	Volume (L)*
hydrogen	H_2	1.008	22.43
helium	He	4.003	22.42
nitrogen	N_2	14.01	22.38
carbon monoxide	CO	28.01	22.38
oxygen	O_2	32.00	22.40

* The volumes are expressed to four significant figures to show the variability that accompanied these experimentally determined values.

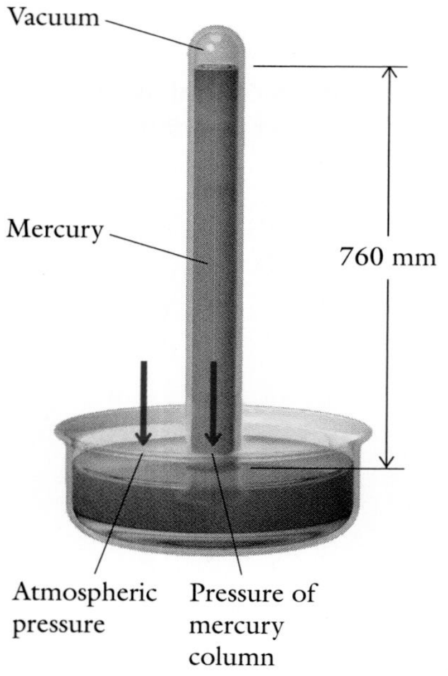

Figure 5.1 A mercury barometer. The weight of the mercury column is balanced by the pressure of the atmosphere. The atmospheric pressure is measured by the height of the mercury column. The small space at the top of the column is a vacuum and does not contribute to any pressure inside the barometer tube.

This means that pressure will increase if the force increases but the area remains constant, or if the force stays the same but the area decreases. For example, the same force applied over a smaller and smaller area will produce a larger and larger pressure. Compare how it would feel to have your toe stepped on by an infant or by a pro-football linebacker. It is not difficult to imagine the pressure exerted by someone stepping on your toe. But a gas cannot step on your toe, so how does it exert pressure? The first clue can be found by examining the meaning of atmospheric pressure, a gas pressure that is routinely mentioned in weather reports. Low pressure usually means bad weather on the way, and high pressure signals a pleasant day. The atmosphere extends to heights in excess of 20 miles and is a mixture of about 20% oxygen, 79% nitrogen, and 1% other gases, including variable amounts of water vapor. These gases have mass, and **atmospheric pressure** is the result of their weight pressing against Earth's surface. The variations in atmospheric pressure are caused by variation in the mass of the atmospheric gas column over a particular location. Cold air is more dense than hot air; therefore it is heavier and leads to a higher local atmospheric pressure. The higher the elevation, the shorter the air column above that location and the lower the average atmospheric pressure.

Atmospheric pressure is often measured by means of a **barometer** (Figure 5.1). The barometer consists of a mercury-filled, closed-end tube inverted in a bath of mercury, whose surface is exposed to the atmosphere. The weight of the atmosphere presses down against the exposed mercury surface and supports the column of mercury in the inverted tube to about 760 mm high. The small space above the mercury column contains no gas. It is a vacuum that is created as the mercury recedes slightly from the end of the tube because of its own weight.

Because atmospheric pressure varies from place to place and from time to time, a standardized unit of pressure independent of location and time has been created. It is called the **atmosphere** (abbreviated **atm**) and is defined as the pressure that will support a vertical column of mercury to a height of exactly 760 mm at 0°C. (Remember, an exact number has an infinite number of significant figures.)

Pressure can be expressed either in atmospheres, millimeters of mercury (mm Hg), or torr. The unit called **torr,** which is equal to **1 mm Hg,** is named in honor of Evangelista Torricelli, who invented the first barometer. In English units, 760 mm Hg is equivalent to 14.7 lb per square foot. In this book, pressure will be expressed in units of torr or atmospheres. To summarize,

$$1 \text{ atm} = 760 \text{ torr} = 760 \text{ mm Hg}$$

The effect of gas pressure on a physiological state is described in Box 5.1.

Example 5.1 Converting units of pressure

Calculate (a) the pressure, in atmospheres, that a gas exerts if it supports a 385-mm column of mercury and (b) the equivalent of 0.750 atm in millimeters of mercury and in torr.

Solution

To solve these problems, we will use conversion factors based on the equivalence 1 atm = 760 mm Hg = 760 torr.

$$\text{(a)}\quad 385 \text{ mm Hg}\left(\frac{1 \text{ atm}}{760 \text{ mm Hg}}\right) = 0.507 \text{ atm}$$

$$\text{(b)}\quad 0.750 \text{ atm}\left(\frac{760 \text{ mm Hg}}{1 \text{ atm}}\right) = 570 \text{ mm Hg} = 570 \text{ torr}$$

5.1 Chemistry Within Us

Diving Time and Gas Pressure

In a manometer using water rather than mercury as the column of fluid, 1 atmosphere of pressure would raise the liquid's level to a height of more than 33 feet. Compare this height with that of a mercury column at that pressure, about 30 inches. The reason for the difference is that mercury has a density of 13.6 g/cm^3 compared with water's density of 1.0 g/cm^3. Nevertheless, water's density is so much greater than that of the air that marine divers are subjected to one additional atmosphere of pressure for every 33 feet that they descend. This effect translates to 4 atm at a depth of about 100 feet.

Novice divers describe the experience of breathing at that depth and pressure as quite unsettling. Some say that the sensation is almost like "drinking" air. They must make a significant physical effort to maintain gas flow into and out of their lungs. This effort is required because an increase in gas pressure is accompanied by a corresponding increase in gas viscosity. (Viscosity is a property of gases and liquids that describes their resistance to flow. For example, glycerol, a liquid whose viscosity is about the same as that of maple syrup, is more than 1000 times as viscous as water.) The divers' problem is alleviated somewhat by breathing-tube design, but the increased physical effort nevertheless leads rapidly to fatigue and requires close attention to diving time. Divers who forget to check their wristwatches pay a heavy price.

Problem 5.1 Calculate (a) the pressure, in atmospheres, that a gas exerts if it supports a 563-mm column of mercury and (b) the equivalent of 0.930 atm in millimeters of mercury and in torr.

To measure the pressure of a gas in a closed container, a device called a **manometer** is used. Figure 5.2 shows a common type of manometer known as a differential manometer, so called because it measures pressure by revealing the difference between two pressures. It consists of a glass U-tube that is open at both ends and filled with a fluid, typically mercury; however, other fluids will serve as well (for example, as described in Box 5.1), as long as the gas does not dissolve in the fluid. In Figure 5.2*a*, the liquid reaches the same height in

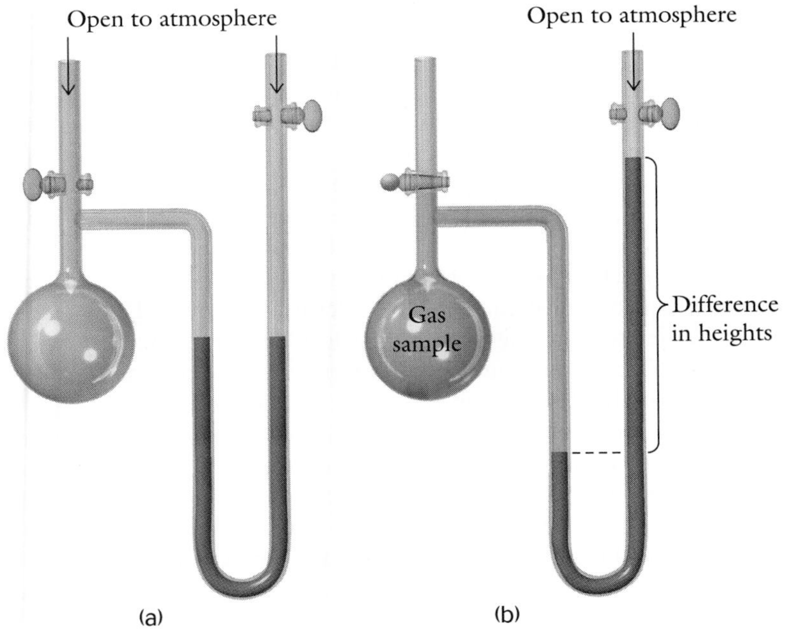

Figure 5.2 Differential manometers. (a) Both columns are at the same height because both sides are exposed to the atmosphere; (b) the stopcock on the left is closed, and the stopcock on the right is open to the atmosphere. The difference in heights is a direct measure of the difference in pressure between the flask on the left and the atmospheric pressure.

Figure 5.3 A gas-pressure gauge on a tank of oxygen used in respiratory therapy.

each arm of the tube because each surface of the liquid is in contact with the atmosphere, which exerts the same force on each surface.

In Figure 5.2*b,* the left arm of the manometer is connected to a gas source, and the other arm is open to the atmosphere. The difference in the levels of the liquids on each side is caused by a difference in pressure. The pressure of the gas is reported as the difference between the heights of the mercury columns in each arm in millimeters of mercury, or torr.

Three possible conditions affect the positions of the fluid in a differential manometer. When gas pressure is greater than atmospheric pressure, the mercury column will be higher on the atmospheric side of the U-tube, as illustrated in Figure 5.2*b*. When gas pressure is less than atmospheric pressure, the mercury column will be higher on the side of the gas source. When gas pressure is equal to atmospheric pressure, the mercury column will be the same on both sides of the manometer.

The differential manometer is not a sturdy device, and a more practical meter must be used in a setting such as a hospital. The gas gauge on an oxygen tank (Figure 5.3) employs a membrane to separate the gas, not from the atmosphere but from a springlike mechanical device. The device exerts a force opposing the gas pressure that is calibrated to read in pressure units of pounds per square inch or millimeters of mercury.

The pressure of a gas in a container is the result of the gas molecules colliding with the walls of their container. The greater the collision rate—that is, the more collisions per unit time—the greater the pressure. The rate of collisions can be increased in two ways. The first is to increase the number of molecules in a fixed-volume container. The second is to increase the temperature of the gas. The latter causes the molecules to move faster and therefore collide more often with the container walls. Imagine sitting in a shed with a metal roof at the beginning of a rainstorm. The individual drops on the roof can be heard as the rain begins. But, as the intensity of the storm increases, the sounds of the individual drops begin to merge and, finally, become a constant roar. That "roar" of a gas as it hits the walls of a container is what we detect when its pressure is measured. A practical use of manometry in medicine is described in Box 5.2.

▶▶ The influence of pressure on physical states and on the properties of solutions is developed in Chapters 6 and 7.

5.2 THE GAS LAWS

The physical properties of gases depend on only four variables: pressure (P), volume (V), Kelvin temperature (T), and number of moles (n). All four variables can be changed simultaneously, but experimenting with only two at a time while keeping the other two constant results in some simple but important mathematical descriptions of gas behavior.

The effects of physical changes on the properties of gases can be visualized by imagining a system consisting of a cylinder with a movable piston (see Figures 5.4 to 5.7). The gas cannot escape, and so the number of moles in the cylinder must remain constant. If the cylinder is not insulated, any temperature change in the gas will be followed by the loss or gain of heat to or from the environment, effectively keeping the gas temperature constant. If the cylinder were insulated, such a loss or gain of heat would be prevented.

This model allows us to predict qualitative changes in gas properties when its pressure, temperature, volume, and mass are varied. Remember that a gas takes the shape of its container and occupies all of it. This means that 1 L of a gas is a sample of that gas enclosed in a 1-L container. Furthermore, the pressure of the gas within the cylinder is defined by the external force on the piston. If the piston does not move, the pressure of the gas within the cylinder must be equal to that external pressure. For example, a 10-kg weight on a pis-

5.2 Chemistry Around Us

Manometry and Blood Pressure

Blood pressure is measured by means of a mercury-filled manometer called a sphygmomanometer. The sphygmomanometer is a manometer in which the column of mercury is connected to both a hand-held air pump and a balloon in the form of a cuff on the other side. The cuff is designed to be wrapped around the upper arm above the elbow. It is then inflated with air until its pressure is great enough to stop the blood flow through the artery to the lower arm. This pressure is transmitted to the manometer and can be read as the height of the mercury column in millimeters. A stethoscope is placed just below the cuff, and the air in the cuff is slowly released. When the cuff pressure falls to a point just below what is known as the systolic pressure (the peak pressure developed by the pumping of the heart), the artery begins to open under the pressure developed by the heart's contractions. At that point, the blood begins to flow with great velocity through a narrower than normal opening, which causes turbulence and the production of vibrations indicated by the appearance of soft tapping sounds (heard through the stethoscope).

As the cuff pressure falls further, the tapping sounds are separated by longer intervals, because the arteries remain open longer during each cycle of cardiac contractions. The tapping sounds also become louder. At the point at which the cuff pressure equals what is known as diastolic pressure, the artery remains open, allowing continuous turbulent flow. The sounds heard through the stethoscope become dull and muffled. Just below diastolic pressure, the blood flow becomes continuous, nonturbulent, and silent because the artery is now completely open.

Systolic pressure is recorded as the cuff pressure at which sounds first appear, and diastolic pressure is recorded as the cuff pressure at which sounds disappear. Normal systolic pressures are in the range from 120 torr to 150 torr. Normal diastolic pressures are in the range from 70 torr to 90 torr. Diastolic pressures above that range indicate difficulty in pumping blood through the capillary bed because of either vasoconstriction or vascular obstruction. This capillary narrowing may cause systolic pressures to rise high enough to cause small blood vessels to burst. Such events in the central nervous system are called strokes.

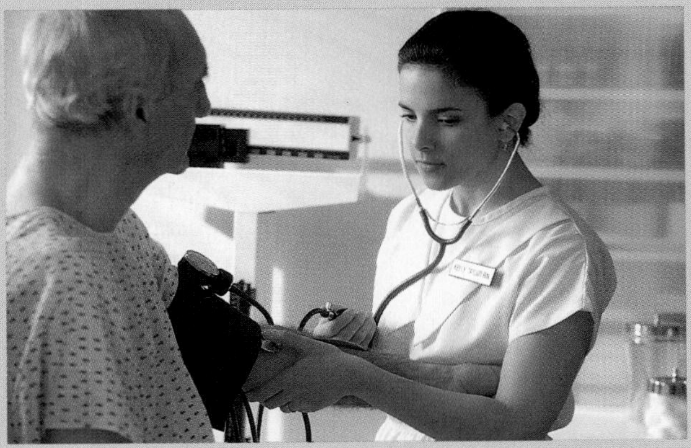

ton compressing a gas in a cylinder of 10-cm^2 area exerts an external pressure of 1 kg/cm^2, and that is also the pressure of the gas contained in the cylinder.

On the following pages, we will use our cylinder–piston system to perform four "thought" experiments that illustrate the four fundamental principles formulated in the laws of Boyle, Charles, Gay-Lussac, and Avogadro:

- Imagine a downward force is exerted on the piston handle to increase the gas pressure (P). Provided that temperature and number of moles of gas remain constant, the downward force will cause the gas to occupy a smaller volume (V), as in Figure 5.4 (see next page). This universal behavior of a gas is described mathematically by Boyle's law:

$$P \times V = \text{constant}$$

- Imagine that the temperature (T) of the gas is increased or decreased. Provided that the force on the piston (the gas pressure) and the number of moles of gas remain constant, when the temperature increases the piston will rise as the gas expands (volume, or V, increases) and fall when the temperature decreases and the gas contracts (volume, or V, decreases). This behavior is described mathematically by Charles's law:

$$V = \text{constant} \times T$$

- Imagine that the temperature (T) of a fixed amount of gas is raised. The only way in which the volume can be kept constant is to increase the force on the piston, a change that results in an increase in gas pressure (P). This behavior is described by Gay-Lussac's law:

$$P = \text{constant} \times T$$

- Imagine adding n moles of gas to the cylinder while keeping both the temperature and the pressure constant. The piston will rise (gas volume will increase) to accommodate the increased amount of gas. This behavior is described by Avogadro's law:

$$V = \text{constant} \times n$$

5.3 BOYLE'S LAW

Figure 5.4 shows three "snapshots" of a gas confined in our model of a cylinder with a movable piston. We can increase the pressure of the gas in the system by increasing the external force on the piston. In Figure 5.4, this force is represented by a weight sitting on the piston's handle. Note that, as the pressure increases, the gas volume decreases.

Figure 5.4 also shows that, as the volume decreases, the gas particles become more and more crowded. This crowding causes more collisions with the container walls per unit time and accounts for the increase in gas pressure.

Robert Boyle

The relation between gas pressure and volume is described by **Boyle's law,** which states that the product of a gas's pressure multiplied by its volume is a constant. (This means that multiplying the pressure of any sample of gas by its volume will always yield the same numerical value). Increasing the pressure on a gas will decrease its volume, and vice versa. This principle underlies the mechanics of breathing presented in Box 5.3, on page 128.

The data listed in Table 5.2 illustrate the results of an experiment in which a gas's pressure or volume were varied while the amount of gas and the temperature (0°C) were held constant. The table reveals a number of things:

- As the pressure decreases, the volume increases; or, as the pressure increases, the volume decreases.
- The $P \times V$ product is a constant.

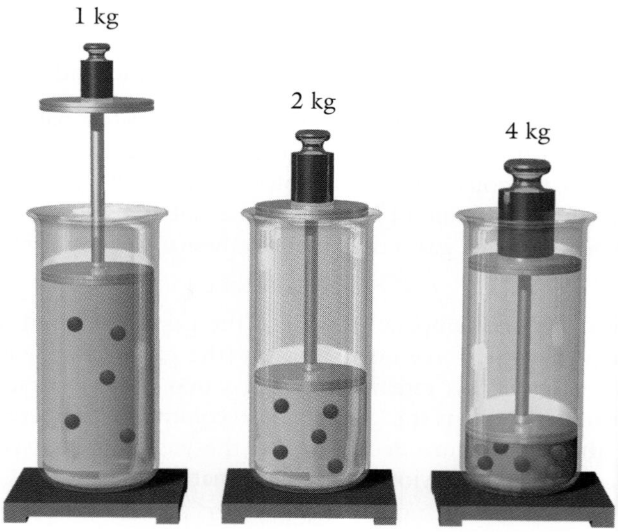

Figure 5.4 A gas confined in a cylinder at three different pressures illustrates Boyle's law. Doubling the pressure on the gas halves its volume. The diagram shows a doubling of the force on the piston from 1 kg to 2 kg and again from 2 kg to 4 kg.

TABLE 5.2 Variation in Gas Pressure and Volume When Temperature and Number of Moles of Gas Are Held Constant

State	Pressure (atm)	Volume (L)	$P \times V$ (L × atm)
1	5.31	0.377	2.00
2	2.67	0.75	2.00
3	1.20	1.67	2.00
4	0.663	3.02	2.00

If the condition of the gas at state 1 is compared with that of the gas after it has been changed to state 2, then

$$P_1 \times V_1 = 2.00 \text{ L} \times \text{atm} = P_2 \times V_2$$

or, because both are equal to the same constant, Boyle's law at constant temperature and number of moles becomes:

$$P_1 \times V_1 = P_2 \times V_2$$

This equality can be solved in two ways, representing two experimental situations. It can be used to tell us (1) what happens to the volume when pressure is varied and (2) what happens to the pressure when volume is varied.

The equation can be solved directly by substitution. If you know any three of the variables, the fourth is obtained algebraically. It can also be solved as a unit-conversion problem.

For situation 1, when pressure is varied,

$$V_2 = V_1 \left(\frac{P_1}{P_2} \right)$$

This is a unit-conversion calculation in which the conversion factor is a ratio of pressures. If the pressure on the gas is increased [that is, $P_2 > P_1$ (P_2 is greater than P_1)], then the conversion factor must be smaller than 1.0, and the gas volume must decrease, as predicted by Boyle's law. If, instead, P_2 is less than P_1, the reverse will occur.

For situation 2, when volume is varied,

$$P_2 = P_1 \left(\frac{V_1}{V_2} \right)$$

This also is a unit-conversion calculation, but now the conversion factor is a ratio of volumes. If the volume of the gas increases—that is, $V_2 > V_1$—the conversion factor must be smaller than 1.0, and the pressure must decrease, as Boyle's law predicts. If, instead, V_2 is less than V_1, the conversion factor will be greater than 1.0, and, as Boyle's law predicts, the pressure must increase.

Example 5.2 Using Boyle's law: I

A 712-mL sample of gas at 505 torr is compressed at constant temperature until its final pressure is 825 torr. What is its final volume?

Solution

First, tabulate the conditions:

	State 1	State 2
Pressure	505 torr	825 torr
Volume	712 mL	?
Temperature	constant	constant
Number of moles	constant	constant

5.3 Chemistry Within Us

Breathing and the Gas Laws

All fluids, whether gas or liquid, will flow when subjected to a gradient in pressure. In other words, a gas will flow from a region of higher pressure to a region of lower pressure. Breathing, a process in which air must flow into and then out of two elastic sacs called lungs, takes advantage of this universal principle. Air will flow into the lungs if the pressure within them is lower than the pressure outside the body, and it will flow out of the lungs if the pressure within them is higher than the pressure outside. The necessary pressure changes are brought about by changes in lung volume caused by the action of a number of powerful muscles within the chest. Breathing is Boyle's law in action.

The figure below shows the lungs located in the airtight thoracic cavity, a space formed by the rib cage and its muscles and a muscle at the base of the cavity called the diaphragm. When relaxed, the diaphragm muscle assumes an upwardly convex shape. The increase in lung volume is caused by expansion of the rib cage and a flattening of the diaphragm. Lung volume decreases on relaxation of the diaphragm upward accompanied by relaxation of the muscles of the rib cage. Boyle's law describes the interaction: As lung volume increases, gas pressure within the lung will decrease and external air will flow into the lungs. As lung volume decreases, gas pressure within the lung will increase and air within the lungs will flow out.

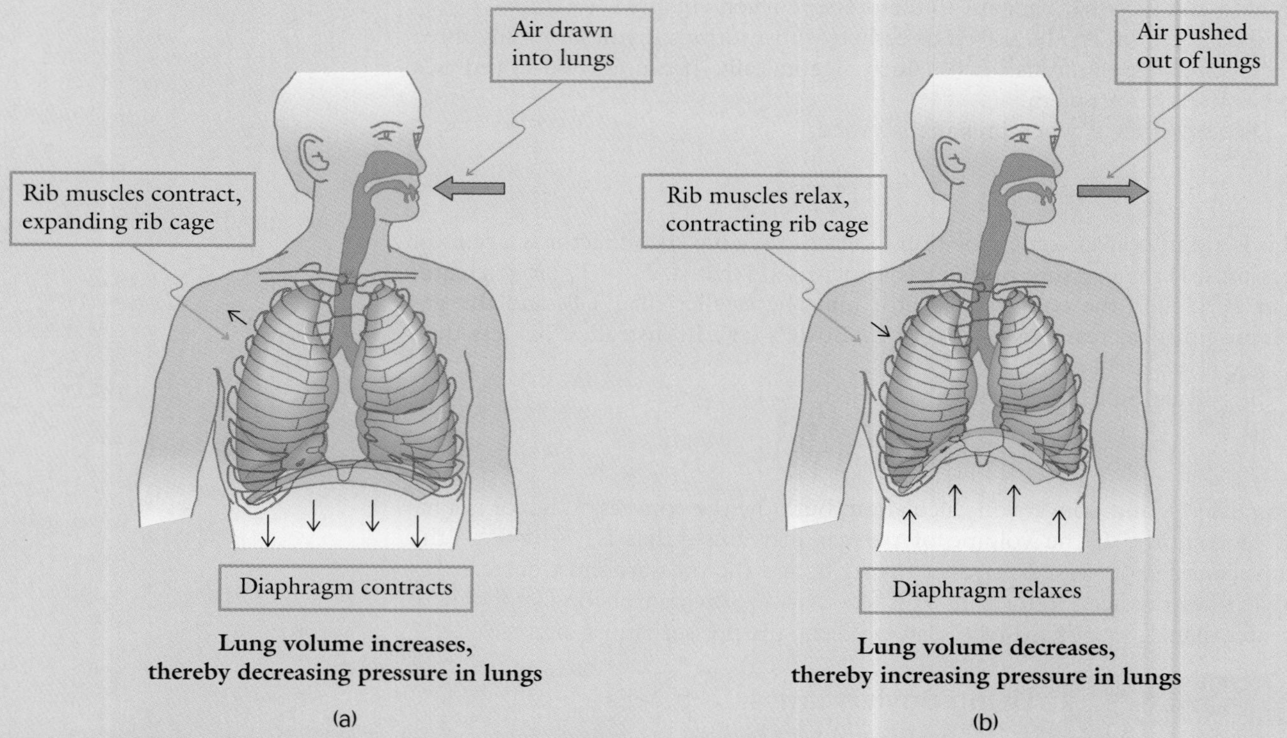

Movement of the diaphragm causes the lungs to expand and contract, allowing Boyle's law to take effect. Remember, the $(P \times V)$ product remains constant at constant temperature. This means that changes in the volume of the lungs will change the air pressure within the lungs. Internal air pressure above atmospheric pressure will force air out of the lungs, and air pressure below atmospheric pressure will force air at atmospheric pressure to enter the lungs. The movement of air into and out of the lungs is controlled by the movement of the muscular diaphragm. When it relaxes and moves upward, the space within the thoracic cavity is reduced, the pressure within the lungs increases, and air is expelled. When the diaphragm contracts and moves downward, the space within the thoracic cavity increases, the pressure within the lungs decreases, and air enters the lungs.

Because the amount of gas and its temperature remain constant, Boyle's law applies and is solved for V_2, the new volume:

$$V_2 = V_1\left(\frac{P_1}{P_2}\right)$$

Because the pressure on the gas increased, the volume must have decreased, and the conversion factor must be smaller than 1.0. Substituting the tabulated values gives

$$V_2 = 712\ \text{mL}\left(\frac{505\ \text{torr}}{825\ \text{torr}}\right) = 436\ \text{mL}$$

The smaller value for the final volume makes sense because the gas was compressed. The pressure units cancel because they are in a ratio, and the final volume units are the same as the original units. Always make certain that the pressure and volume units are clearly stated.

Problem 5.2 An 862-mL sample of gas at 425 torr is compressed at constant temperature until its final pressure is 901 torr. What is its final volume?

Example 5.3 Using Boyle's law: II

A 2.00-L sample of gas at 0.800 atm must be compressed to 1.60 L at constant temperature. What pressure in atmospheres must be exerted to bring it to that volume?

Solution

First, tabulate the conditions:

	State 1	State 2
Pressure	0.800 atm	?
Volume	2.00 L	1.60 L
Temperature	constant	constant
Number of moles	constant	constant

Because the amount of gas and its temperature remain constant, Boyle's law applies and is solved for P_2.

$$P_2 = P_1\left(\frac{V_1}{V_2}\right)$$

The volume of the gas decreased from state 1 to state 2, so the pressure must have increased. Therefore, the ratio of volumes must be greater than 1.0. Substituting the tabulated values gives:

$$P_2 = 0.800\ \text{atm}\left(\frac{2.00\ \text{L}}{1.60\ \text{L}}\right) = 1.00\ \text{atm}$$

Here, again, the volume units in the ratio cancel, and the final pressure unit is the same as the original unit. If asked to calculate the final pressure in torr, use the conversion factor 1 atm = 760 torr to change the final answer to torr.

Problem 5.3 A 1.80-L sample of gas at 0.739 atm must be compressed to 1.40 L at constant temperature. What pressure in atmospheres must be exerted to bring it to that volume?

5.4 CHARLES'S LAW

If the temperature of a gas is increased and the external pressure (the weight on the piston) and amount of gas are kept constant, then the volume of the

Figure 5.5 Equal quantities of a gas confined in insulated cylinders at two temperatures illustrate Charles's law. When the Kelvin temperature is doubled and the pressure is held constant, the gas volume doubles.

gas will increase. Raising the temperature of a gas increases the velocity of the gas molecules. They undergo more collisions with the container walls per unit time, creating a consequent rise in pressure. Because the external pressure on the gas is not increased, the gas expands. This effect is illustrated in Figure 5.5.

The quantitative relation between gas volume and temperature, called **Charles's law,** relates the volume of a gas to the Kelvin temperature. Mathematically, Charles's law is:

$$V = \text{constant} \times T$$

This form of the law tells us, for example, that, if the Kelvin temperature is doubled, the volume doubles, and when it is halved, the volume also is halved.

Table 5.3 lists data from an experiment in which temperature changes led to volume changes for a fixed amount of gas kept at a constant pressure. The table indicates that

- As the temperature increases, the volume increases.
- The ratio V/T is a constant, but the ratio V/t (t representing the Celsius temperature) is not constant.

When the temperature is expressed on the Kelvin scale, the ratio of volume to temperature is a constant that takes the same value at all temperatures if the quantity of gas and its pressure remain constant.

The volumes of a sample of gas at two different temperatures can be compared by noting that

$$\frac{V_1}{T_1} = \text{constant} = \frac{V_2}{T_2}$$

Because the two ratios of volume to Kelvin temperature are both equal to the same constant, they are also equal to each other. The final form of Charles's law is:

$$\frac{V_1}{T_1} = \frac{V_2}{T_2}$$

Two experimental situations can be solved with Charles's law: one situation determines what happens to the volume of a gas when the temperature is changed, and the other determines what happens to the temperature of a gas when the volume is changed.

For the first experimental situation, the equation is solved for V_2:

$$V_2 = V_1\left(\frac{T_2}{T_1}\right)$$

This is a unit-conversion calculation, in which the conversion factor is a ratio of Kelvin temperatures. Charles's law predicts that the volume of the gas will increase if the temperature increases. Because $T_2 > T_1$, the ratio of temperatures is greater than 1.0, and the volume must increase.

TABLE 5.3 Variation in Gas Volume with Change in Temperature at Constant Pressure

State	t (°C)	T (K)	Volume (L)	V/T (L/K)	V/t (L/°C)
1	−23.0	250	5.50	0.0220	−0.239
2	27.0	300	6.60	0.0220	0.244
3	77.0	350	7.70	0.0220	0.100
4	152	425	9.35	0.0220	0.0620

Example 5.4 Using Charles's law: I

A 512-mL sample of a gas, in a cylinder with a movable piston, at 0.000°C is heated at a constant pressure of 0.800 atm to 41.0°C. What is its final volume?

Solution

To use Charles's law, it is necessary to convert Celsius temperature into kelvins. The conversion uses the relation

$$\text{Kelvin temperature} = \text{Celsius temperature} + 273$$

Tabulate the conditions.

	State 1	State 2
Pressure	0.800 atm	0.800 atm
Volume	512 mL	? mL
Number of moles	constant	constant
Celsius temperature	0.000°C	41.0°C
Kelvin temperature	273 K	314 K

The pressure and number of moles are constant, so Charles's law applies. Solving for V_2 and introducing the Kelvin temperatures into Charles's law gives

$$V_2 = 512\ \text{mL}\left(\frac{314\ \text{K}}{273\ \text{K}}\right) = 589\ \text{mL}$$

Charles's law informs us that, if the temperature increased, the volume had to increase. Therefore, it is necessary to multiply the initial volume by a fraction, formed by the initial and final temperatures, that is greater than 1.0, or 314/273.

Problem 5.4 A 755-mL sample of a gas, in a cylinder with a movable piston, at 3.00°C is heated at a constant pressure of 0.800 atm to 142°C. What is its final volume?

Example 5.5 Using Charles's law: II

A 1.00-L sample of gas at 27.0°C is heated so that it expands at constant pressure to a final volume of 1.50 L. Calculate its final temperature.

Solution

Tabulate the conditions:

	State 1	State 2
Pressure	constant	constant
Volume	1.00 L	1.50 L
Number of moles	constant	constant
Celsius temperature	27.0°C	? °C
Kelvin temperature	300 K	? K

The pressure and amount of gas are constant, so Charles's law applies. Again, it is necessary to convert Celsius temperature into kelvins. Solving for T_2 and making the appropriate substitutions gives

$$T_2 = 300\ \text{K}\left(\frac{1.50\ \text{L}}{1.00\ \text{L}}\right) = 450\ \text{K}$$

Charles's law specifies that, if the volume increased, the temperature had to increase. Therefore the initial temperature is multiplied by a fraction, formed by the initial and final volumes, greater than 1.0, or 1.5/1.0.

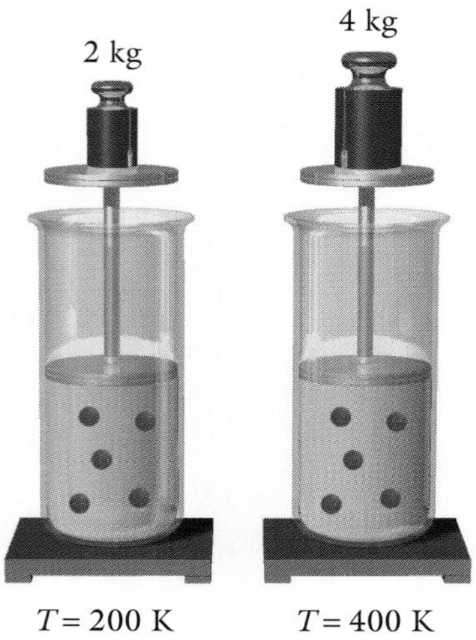

Figure 5.6 Illustration of Gay-Lussac's law. The volume of a gas confined in an insulated cylinder will remain the same when its Kelvin temperature is doubled if, at the same time, the pressure also is doubled.

Problem 5.5 A 0.630-L sample of gas at 16.0°C is heated so that it expands at constant pressure to a final volume of 1.35 L. Calculate its final temperature.

5.5 GAY-LUSSAC'S LAW

Gay-Lussac's law states that, if the Kelvin temperature of a fixed amount of gas kept at a constant volume is increased, the pressure will increase. In addition, if the pressure on a gas is increased at constant volume, the temperature will increase. Figure 5.6 indicates that, to keep the volume of a fixed amount of gas constant when the Kelvin temperature is doubled, the external pressure also must be doubled. Gay-Lussac's law is expressed mathematically as

$$P = \text{constant} \times T$$

The data from an experiment that measured the variation in pressure of a fixed amount of gas with change in temperature at a constant volume are found in Table 5.4. The table indicates that

- As the temperature increases, the pressure increases.
- The ratio P/T is a constant, but the ratio P/t is not constant.

When the temperature is expressed on the Kelvin scale, the ratio of pressure to temperature is a constant that takes the same numerical value at all temperatures if the quantity of gas and its volume remain constant.

The pressures of a sample of gas at two different temperatures can be compared by noting that

$$\frac{P_1}{T_1} = \text{constant} = \frac{P_2}{T_2}$$

Because the two ratios of pressure to Kelvin temperature are both equal to the same constant, they are also equal to each other, and the final form of Gay-Lussac's law is:

$$\frac{P_1}{T_1} = \frac{P_2}{T_2}$$

It is important to remember that both the amount of gas—that is, the number of moles of gas—and the volume remain constant.

Gay-Lussac's law is applicable in experimental situations when we want to know (1) what happens to the pressure of a gas when the temperature is changed and (2) what happens to the temperature of a gas when the pressure is changed.

For experimental situation 1, we solve the equation for P_2:

$$P_2 = P_1\left(\frac{T_2}{T_1}\right)$$

This is a unit-conversion calculation, with the conversion factor in the form of a ratio of kelvin temperatures. Gay-Lussac's law predicts that the pressure of the gas increases if the temperature increases. Accordingly, because $T_2 > T_1$,

TABLE 5.4 Variation in Gas Pressure and Change in Temperature at Constant Volume

State	t (°C)	T (K)	Pressure (atm)	P/T (atm/K)	P/t (atm/°C)
1	−23.0	250	1.50	0.00600	−0.0652
2	27.0	300	1.80	0.00600	0.0667
3	77.0	350	2.10	0.00600	0.0273
4	152	425	2.55	0.00600	0.0168

5.4 Chemistry Around Us

Sterilization and the Autoclave

A number of different approaches have been developed for using heat to inactivate microorganisms. The choice of method depends on the nature of the materials being sterilized. Items such as scalpels or glass syringes, characterized by hard surfaces, can withstand harsh treatment at high temperatures, but heat-sensitive materials must be sterilized by more gentle methods. The fact that steam can be heated to high temperatures is particularly useful because bacteria such as *Salmonella* are killed in only 7 to 12 min when exposed to steam at 125°C but require much longer exposures when subjected to dry heat at the same temperature (see Section 20.9).

At atmospheric pressures, liquid water and its accompanying vapor can never reach temperatures above its boiling point of 100°C, but the temperature of steam can be raised to much higher values. Because steam is gaseous water, it is subject to the gas laws. When heated but allowed to expand, its temperature will not rise; however, if it is heated at constant volume, its temperature will increase—the situation that is quantitatively described by the Gay-Lussac law. The pressure developed in converting steam at 100°C into steam at 125°C at a constant volume can be calculated as follows:

$$760 \text{ torr}\left(\frac{498 \text{ K}}{373 \text{ K}}\right) = 1015 \text{ torr} = 1.34 \text{ atm}$$

Hospital and clinical laboratories are equipped with a device called an autoclave for producing steam at high temperatures. An autoclave is shown in the photograph above, at right. Essentially a double-walled, airtight, constant-volume kettle, the autoclave resists pressures of well over 2 atm. Steam is produced in a boiler and is introduced at higher temperature, and consequently under high pressure, into the autoclave's hollow wall. When the temperature of the inner chamber reaches that within the hollow wall, the high-temperature steam is injected into the chamber and comes into contact with the instruments and other objects placed there for sterilization. From 15 to 20 min is sufficient for complete sterilization. Gay-Lussac's law indicates that higher temperatures could be attained if desired, but temperatures near 125°C are sufficient for all practical purposes.

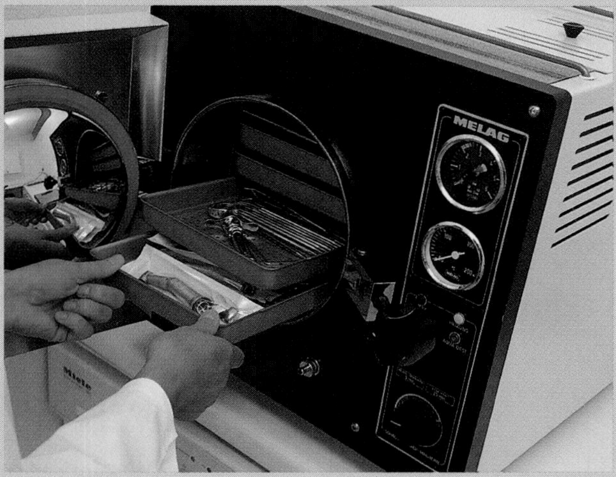

the ratio of temperatures is greater than 1.0, and the pressure must increase. Gay-Lussac's law is the basis for the operation of the autoclave, which is described in Box 5.4.

Example 5.6 Using Gay-Lussac's law: I

Calculate the final pressure of a sample of gas in a fixed-volume cylinder at 0.500 atm and 0.000°C when the gas is heated to 41.0°C.

Solution

Tabulate the conditions:

	State 1	State 2
Pressure	0.500 atm	? atm
Volume	constant	constant
Number of moles	constant	constant
Celsius temperature	0.000°C	41.0°C
Kelvin temperature	273 K	314 K

The volume and number of moles are constant, so Gay-Lussac's law applies. To use Gay-Lussac's law, Celsius temperature must be converted into kelvins. The conversion uses the relation

$$\text{Kelvin temperature} = \text{Celsius temperature} + 273$$

Solving for P_2 and introducing the Kelvin temperatures into Gay-Lussac's law gives

$$P_2 = 0.500 \text{ atm} \left(\frac{314 \text{ K}}{273 \text{ K}}\right) = 0.575 \text{ atm}$$

According to Gay-Lussac's law, if the gas temperature increased, the pressure also had to increase. Therefore, the initial pressure had to be multiplied by a fraction that is formed by the initial and final temperatures and is greater than 1.0, or 314/273.

Problem 5.6 Calculate the final pressure of a sample of gas in a fixed-volume cylinder at 1.25 atm and 14.0°C when the gas is heated to 125°C.

Example 5.7 Using Gay-Lussac's law: II

A constant-volume sample of gas at 27.0°C and 1.00 atm is heated so that its pressure increases to a final value of 1.50 atm. What is its final temperature?

Solution

Tabulate the conditions:

	State 1	State 2
Pressure	1.00 atm	1.50 atm
Volume	constant	constant
Number of moles	constant	constant
Celsius temperature	27.0°C	? °C
Kelvin temperature	300 K	? K

The volume and amount of gas are constant, so Gay-Lussac's law applies. Again, it is necessary to convert Celsius temperature into kelvins. Solving for T_2 and making the appropriate substitutions gives

$$T_2 = 300 \text{ K} \left(\frac{1.50 \text{ atm}}{1.00 \text{ atm}}\right) = 450 \text{ K}$$

Gay-Lussac's law informs us that, if the pressure increased, the temperature had to increase. Therefore the initial temperature is multiplied by a fraction that is formed by the initial and final pressures and is greater than 1.0, or 1.5/1.0.

Problem 5.7 A constant-volume sample of gas at 3.00°C and 0.650 atm is heated so that its pressure increases to a final value of 2.15 atm. What is its final temperature?

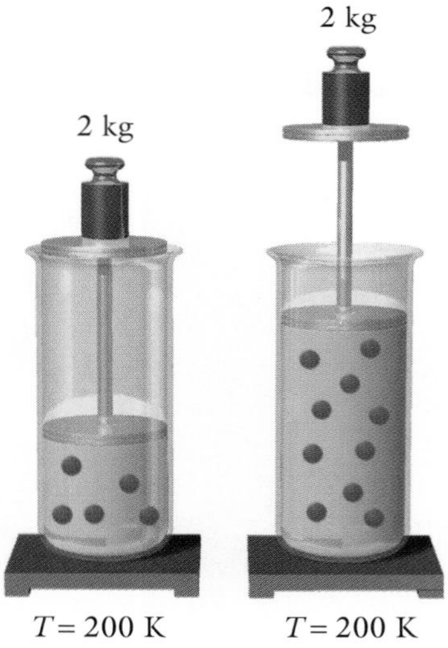

Figure 5.7 Illustration of Avogadro's law. Doubling the amount of a gas at the same Kelvin temperature and pressure will double its volume. Equal volumes of gases at the same Kelvin temperature and pressure contain the same number of molecules.

5.6 AVOGADRO'S LAW

Avogadro's law, illustrated by the data presented in Table 5.1 and in Figure 5.7, states that equal volumes (V) of gases at the same temperature and pressure contain equal numbers of molecules (n). The molecules may be monatomic like helium, diatomic like oxygen or chlorine, triatomic like carbon dioxide, or tetratomic like ammonia, or they can be of any size that can exist in the gaseous state. Expressed mathematically,

$$V = \text{constant} \times n$$

Therefore, the volumes of gas samples containing different numbers of moles can be compared by noting that

$$\frac{V_1}{n_1} = \text{constant} = \frac{V_2}{n_2}$$

Because the two ratios of volume to numbers of moles are both equal to the same constant, they are also equal to each other, and the final form of Avogadro's law is

$$\frac{V_1}{n_1} = \frac{V_2}{n_2}$$

It is important to remember that application of this law requires that both the temperature of the gas sample and its pressure remain constant.

Avogadro's law can be used to answer the questions, (1) What happens to the volume of a gas when the number of moles changes? and (2) What number of moles would be contained in a new and different volume?

The equation expressing Avogadro's law can be solved, for example, for V_2:

$$V_2 = V_1\left(\frac{n_2}{n_1}\right)$$

This is a unit-conversion calculation in which the conversion factor is a ratio of moles. Avogadro's law predicts that the volume of the gas increases if the number of moles increases. If $n_2 > n_1$, the ratio of moles is greater than 1.0, and the volume must increase.

In another form, when $n_1 = n_2$, then $V_1 = V_2$. This is a quantitative expression of the most important consequence of Avogadro's law:

- Equal volumes of gases at the same pressure and temperature contain the same number of molecules.

This idea is the basis for relating the measured mass of a gas to the relative mass of its molecules, thus establishing its molecular mass (see Example 5.11) and, in addition, for many of the relations considered in Section 4.6.

5.7 THE COMBINED GAS LAW

Suppose that we use our model cylinder for an experiment in which a fixed quantity of gas is compressed but a constant temperature is not maintained. According to Boyle's law, if the pressure on a gas is increased, its volume will decrease; but, according to Gay-Lussac's law, if the pressure goes up, the temperature will increase. The proposed experiment takes place in one step, but we can think of it as being composed of two separate but simultaneous events and describe it by a combination of the two laws:

$$\frac{PV}{T} = \text{constant}$$

The equation can be written to describe two different states of the same sample of gas:

$$\frac{P_1V_1}{T_1} = \text{constant} = \frac{P_2V_2}{T_2}$$

Because each expression is equal to the same constant, they are equal to each other:

$$\frac{P_1V_1}{T_1} = \frac{P_2V_2}{T_2}$$

This relation is called the **combined gas law.**

Concept Check

✔ The combined gas law enables us to calculate the new value of one of the variables when both of the others are changed simultaneously.

Example 5.8 Using the combined gas law

A 1.20-L sample of gas at 27.0°C and 1.00 atm pressure is heated to 177°C and 1.50 atm final pressure. What is its final volume?

Solution

Tabulate the conditions:

	State 1	State 2
Pressure	1.00 atm	1.50 atm
Volume	1.20 L	? L
Number of moles	constant	constant
Celsius temperature	27.0°C	177°C
Kelvin temperature	300 K	450 K

Use the combined gas law, $\frac{P_1V_1}{T_1} = \frac{P_2V_2}{T_2}$, and solve it for V_2:

$$V_2 = V_1\left(\frac{T_2}{T_1}\right)\left(\frac{P_1}{P_2}\right)$$

This expression has the form of a unit-conversion calculation in which two conversion factors multiply the initial volume. Because the volume increases as the temperature increases, the temperature ratio is T_2/T_1; and, because the volume decreases as the pressure increases, the pressure ratio is P_1/P_2.

Substituting appropriate values gives

$$V_2 = 1.20\ \text{L}\left(\frac{450\ \text{K}}{300\ \text{K}}\right)\left(\frac{1.00\ \text{atm}}{1.50\ \text{atm}}\right) = 1.20\ \text{L}$$

Notice that increasing the temperature should increase the volume, but the increase in pressure was just enough to keep the volume constant.

Problem 5.8 A 1.76-L sample of gas at 6.00°C and 0.920 atm pressure is heated to 254°C and 2.75 atm final pressure. What is its final volume?

The combined gas law can be used to calculate the answers to three kinds of questions about gases:

1. What is the new pressure of a gas when both the temperature and the volume are simultaneously changed?
2. What is the new temperature of a gas when both the pressure and the volume are simultaneously changed?
3. What is the new volume of a gas when both the temperature and the pressure are simultaneously changed?

5.8 THE UNIVERSAL GAS LAW

All four gas variables, P, V, T, and n, can be combined into the following single mathematical expression:

$$\frac{PV}{nT} = \text{constant}$$

The constant, given the symbol **R,** is called the **universal gas constant.** The equation is more commonly written as

$$PV = nRT$$

At room temperature and atmospheric pressure, all gases, regardless of type or molecular mass, follow this relation quite closely. The law is universally applicable, and its use is necessary to solve problems concerning moles or masses of gas.

The value of the universal gas constant R was determined by experimentally establishing that, at 1.00 atm and 273 K, 1.00 mol of any gas occupies 22.4 L. Avogadro predicted that all gases would occupy the same volume under the same conditions, but the precise value of the volume could not be predicted and had to be determined by experiment. Because the volumes of gases depend on temperature and pressure, the only way to compare volumes of different gases is to measure them at the same temperature and pressure. The specific conditions adopted by scientists were 1.00 atm and 273 K and were designated as the **standard temperature and pressure (STP)** of a gas.

Inserting the experimentally determined gas volume and STP values into the universal gas equation and solving for R, we have

$$R = \left(\frac{1.00 \text{ atm}}{1.00 \text{ mol}}\right)\left(\frac{22.4 \text{ L}}{273 \text{ K}}\right) = 0.0821 \,\frac{\text{L} \times \text{atm}}{\text{K} \times \text{mol}}$$

To avoid expressing the units of R as a fraction, in this chapter R will be written as follows: $R = 0.0821 \text{ L}\cdot\text{atm}\cdot\text{K}^{-1}\cdot\text{mol}^{-1}$. A centered dot ($\cdot$) is used to denote multiplication in R, which is a derived unit. Take note of these units: pressure is in atmospheres, temperature in kelvins, volume in liters, and number (n) in moles. To use this equation for solving gas-law problems, you must be careful to express all measurements in these units.

Example 5.9 Using the universal gas law

Inhaled air contains O_2 at a pressure of 160 torr. The capacity of a human lung is about 3.00 L. Body temperature is 37.0°C. Calculate the mass of O_2 in a lung after inspiration.

Solution

The variables given are P, T, and V, and the requirement is to find mass, so the universal gas law must be used. However, pressure must be converted from torr into atmospheres, and temperature from degrees Celsius into kelvins before the universal gas law can be used.

$$160 \text{ torr}\left(\frac{1.00 \text{ atm}}{760 \text{ torr}}\right) = 0.211 \text{ atm}$$

$$37°\text{C} + 273°\text{C} = 310 \text{ K}$$

Solving the universal gas law for n, we have

$$n = \frac{PV}{RT} = \frac{0.211 \text{ atm} \times 3.00 \text{ L}}{0.0821 \text{ L}\cdot\text{atm}\cdot\text{K}^{-1}\cdot\text{mol}^{-1} \times 310 \text{ K}} = 0.0249 \text{ mol}$$

Finally,

$$\text{grams } O_2 = 0.0249 \text{ mol}\left(\frac{32.0 \text{ g } O_2}{1.00 \text{ mol } O_2}\right) = 0.797 \text{ g}$$

Problem 5.9 Exhaled air contains CO_2 at a pressure of 28.0 torr. The capacity of a human lung is about 3.00 L. Body temperature is 37.0°C. Calculate the mass of CO_2 in a lung before expiration.

5.9 THE UNIVERSAL GAS LAW AND MOLECULAR MASS

Substances in the gaseous state consist of molecules. Application of the universal gas law allows us to determine the mass of a mole of molecules—called, appropriately, **molecular mass.**

The number of moles in any mass of a substance is calculated by dividing the mass in grams by the formula mass. In a sample of a gas, the number of moles, n, is equal to the mass of the sample in grams (g) divided by the molecular mass (M). That is,

$$n = \frac{\text{g}}{\text{M}}$$

This relation is combined with the universal gas law in the following equations:

$$n = \frac{\text{g}}{\text{M}} = \frac{PV}{RT}$$

Solving for M gives

$$\text{M} = \frac{\text{g}RT}{PV} \quad \text{or} \quad \text{M} = \left(\frac{\text{g}}{V}\right)\frac{RT}{P}$$

Note that the factor (g/V) is in fact the gas density in grams per liter. When correct units are used for R, T, and P, the units of molecular mass are grams per mole.

Example 5.10 Calculating molecular mass with the universal gas law

The density of CO_2 at 25.0°C and 1.00 atm is 1.80 g/L. Calculate its molecular mass.

Solution

The density of the gas is the factor g/V in the preceding equation

$$M = \left(\frac{1.80\text{ g}}{1.00\text{ L}}\right)\left(\frac{0.0821\text{ L}\cdot\text{atm}\cdot\text{K}^{-1}\cdot\text{mol}^{-1} \times 298\text{ K}}{1.00\text{ atm}}\right)$$

$$M = 44.0\text{ g/mol}$$

Problem 5.10 The density of O_2 at 15.0°C and 1.00 atm is 1.36 g/L. Calculate its molecular mass.

Avogadro's law provides another method for using gas densities to establish molecular mass. Because 1.00 mol of any gas occupies 22.4 L at STP, all we need to do is weigh some volume of a gas and, using the combined gas equation, calculate its equivalent mass at 22.4 L—the volume of 1.00 mol at STP.

Example 5.11 Using Avogadro's law and gas density to determine molecular mass

Determine the molecular mass of methane, 6.60 L of which, collected at 745 torr and 25.0°C, weighs 4.24 g.

Solution

Because 1.00 mol of a gas occupies 22.4 L at STP, the 6.60 L of methane at 25.0°C and 745 torr must be changed into its equivalent volume at STP.

$$6.60\text{ L}\left(\frac{273\text{ K}}{298\text{ K}}\right)\left(\frac{745}{760}\right) = 5.93\text{ L}$$

The next step is to calculate the mass of gas that would occupy 22.4 L. This calculation requires a conversion factor defined by the fact that 1.00 mol of any gas = 22.4 L.

$$\left(\frac{4.24\text{ g}}{5.93\text{ L}}\right)\left(\frac{22.4\text{ L}}{1.00\text{ mol}}\right) = 16.0\,\frac{\text{g}}{\text{mol}}$$

Problem 5.11 Determine the molecular mass of argon, 14.10 g of which occupies 7.890 L at STP.

5.10 DALTON'S LAW OF PARTIAL PRESSURES

Mixtures of gases are described by **Dalton's law of partial pressures.** This law states that the total pressure of a mixture of gases is equal to the sum of the individual pressures of all the individual gases in the mixture. These individual pressures are called **partial pressures,** defined as the pressure that each gas would exert if it were present alone instead of in a mixture. Expressed mathematically,

$$P_{\text{total}} = P_{\text{A}} + P_{\text{B}}$$

The chief characteristic of gases is their physical similarity to one another. Gaseous substances composed of widely differing molecules nevertheless display virtually identical physical characteristics. Dalton's law merely states this characteristic in a new way—namely, that 0.10 mol of oxygen and 0.10 mol of carbon dioxide in a 10-L container will have the same pressure as 0.20 mol of oxygen or 0.20 mol of any other gas in the same container under the same conditions of temperature and pressure. Each molecule of a gas is independent of every other molecule present, whether it is the same substance or a different one. Therefore, the total pressure of a gas mixture is the sum of the partial pressures of each member of the mixture.

In Example 5.9, the pressure of oxygen in air was specified as 160 torr. Because air is a mixture principally of O_2 and N_2 and the total pressure of air is 760 torr, 160 torr is the partial pressure of O_2. Medical oxygen from a tank at 760 torr therefore has a pressure that is five times the pressure of oxygen in air. The result of this difference is that oxygen from the tank is transported into the blood from the lung air-space at about five times the rate of transport from air.

Example 5.12 Calculating partial pressures: I

As just stated, the partial pressure of O_2 in air is 160 torr. What is the partial pressure of N_2 in air?

Solution

Dalton's law of partial pressures can be written as

$$P_{\text{air}} = P_{O_2} + P_{N_2}$$

Atmospheric pressure (pressure of air) is usually about 760 torr. Assuming that value and rearranging the equation to solve for P_{N_2} gives

$$P_{N_2} = 760\text{ torr} - 160\text{ torr} = 600\text{ torr}$$

Problem 5.12 Water can be decomposed by an electric current into oxygen and hydrogen whose collected volumes are in the ratio of 1 to 2, respectively. The total pressure of these two dry gases is 741 torr. What are their partial pressures?

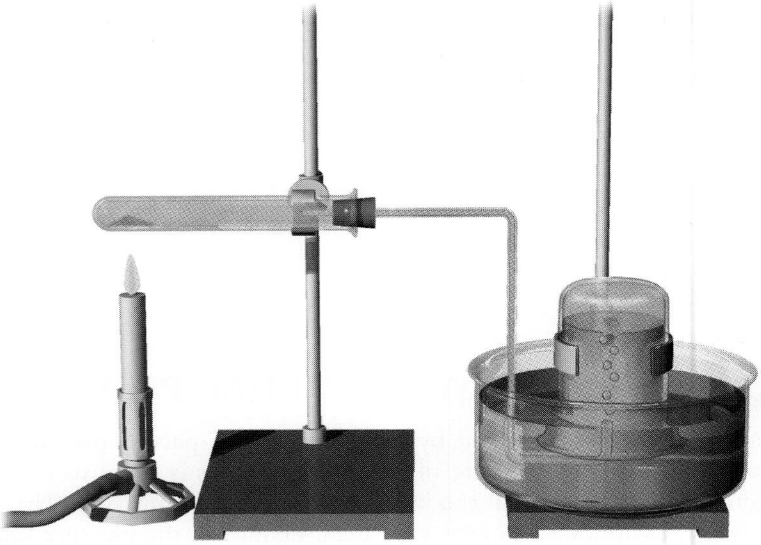

Figure 5.8 Collection of a gas over water. A gas produced by chemical reaction in the test tube is conducted by tubing under and into an inverted beaker filled with water. The gas displaces the water and is trapped above the water surface.

Example 5.13 Calculating partial pressures: II

Gases are collected in the laboratory by bubbling them into a bottle that is filled with water and then inverted under water that is contained in a vessel open to the atmosphere (Figure 5.8). Water in the bottle is then displaced by the entering gas. This experimental arrangement can keep the water's level inside the beaker even with the level of water surrounding it, ensuring that the trapped gas is present at atmospheric pressure. However, because the gas is in contact with the water, some water in the form of gas or vapor is also mixed in with the collected gas. If the atmospheric pressure is 755 torr and the partial pressure of water is 23.0 torr, what is the pressure of a gas collected under these conditions?

Solution

$$P_{\text{atmosphere}} = P_{\text{gas}} + P_{\text{water}}$$
$$755 \text{ mm} = P_{\text{gas}} + 23.0 \text{ torr}$$
$$P_{\text{gas}} = 755 \text{ torr} - 23.0 \text{ torr} = 732 \text{ torr}$$

Problem 5.13 Hydrogen was collected over water at 25.0°C and a total pressure of 755 torr. The vapor pressure of water at that temperature is 23.8 torr. The amount of hydrogen collected was 0.320 mol. What was the volume of the gas?

5.11 GASES DISSOLVE IN LIQUIDS

The amount of gas that will dissolve in a liquid such as water depends strongly on the gas pressure. As the gas pressure increases, more gas will dissolve.

- The general rule is, the greater a gas's partial pressure in a gas mixture, the greater the extent to which that gas will dissolve in any liquid present.

Gas solubility is described by **Henry's law,** a statement of the relation between the amount of gas dissolved in a liquid (at a fixed temperature) and the gas pressure (or partial pressure for a mixture of gases). Solubility is a quantitative measurement of one substance's ability to dissolve in another. We will explore the concept of solubility in detail in Chapters 6 and 7 but introduce it here to briefly consider the extent to which gases dissolve in liquids. Henry's law for a pure gas takes the form

$$\left(\frac{\text{Amount of gas dissolved}}{\text{Unit volume of solvent}}\right) = C_{\mathrm{H}} \times \text{Gas pressure}$$

and, for a mixture of gases,

$$\left(\frac{\text{Amount of gas dissolved}}{\text{Unit volume of solvent}}\right) = C_{\mathrm{H}} \times \text{Gas partial pressure}$$

C_{H} is called the **Henry's law constant.** It has a unique value for a particular gas at a particular temperature. The Henry's law constant for CO_2 at 25°C is different from that for O_2 at 25°C and from that for CO_2 at 30°C. A practical

PROFESSIONAL CONNECTIONS

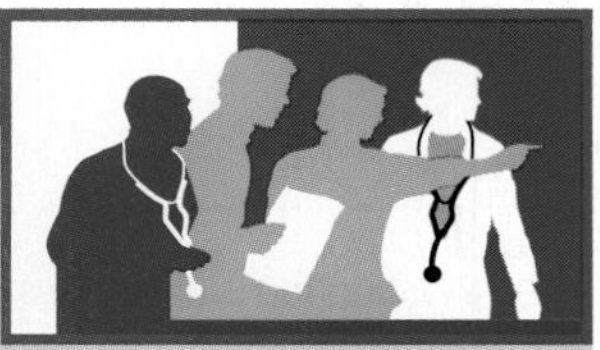

June G. Neal, R.N.
Emergency Department

At Polk Community College, in Winterhaven, Florida, June Neal studied chemistry in pursuit of a nursing degree.

What is your current occupation and why did you choose it?

I am currently employed as a registered nurse in an emergency department. I chose emergency nursing because of the diversity of the specialty. I enjoy the fact that I never know what situation I might encounter when I start my shift. It seems as if, just when you "thought you'd seen it all," something new comes in.

What health care jobs have you held in the past?

I've worked my way up the nursing ladder. I have been a medical assistant, and a licensed practical nurse in a physician's office, and now I am an RN in an emergency department. As a fun sideline, I also work as the on-site nurse for an international beauty pageant and have served as a consultant on a television series.

Please describe a typical day on the job.

There isn't usually a "typical day" in an emergency department. I work the night shift on the weekends, so we take care of patients who have suffered a trauma or have a medical problem. In my current position as clinical services coordinator, I ensure the flow of patients through the department. I also serve as a supervisor for the clinical staff, as a resource person and mentor for new employees.

Are any scientific or technological advances or other changes affecting the nature of your work?

Scientific and technological changes affect my work daily. One of the newest technological changes that we have employed in our department is the use of the bedside diagnostic testing tool. Licensed staff collect as little as two drops of blood and test it in a hand-held analyzer. We are able to have the results of a blood chemistry test in 2 minutes. This means to the patient a more rapid diagnosis and shorter stay in the emergency department.

Do you use chemistry in your job?

I use my chemistry studies in every aspect of my job. Chemical reactions underlie all physiological processes—digestion, respiration, and even the movement of our arms and legs. A quick example that I might see in the emergency department would be in the evaluation of a patient experiencing shortness of breath. Part of the patient's care would be the interpretation of an arterial blood gas. By checking the pH and the oxygen and carbon dioxide levels of the arterial blood, we could determine the efficiency of the gas exchange in the lungs. This would assist us in developing a plan of care for the patient.

What advice do you have for health science students who are studying chemistry today?

Take the time to really understand the theory that you will be learning. So much of the time, we study just to pass the test, then we can't recall much of what we've learned. You're going to find that this chemistry course will give you a better understanding of the information that you'll receive in your other health care couses.

definition for the units of the Henry's law constant is obtained by solving for C_H in Henry's law for a pure gas:

$$C_H = \frac{\left(\dfrac{\text{amount of gas dissolved}}{\text{within a unit volume of solvent}}\right)}{\text{gas pressure}}$$

$$C_H = \frac{\text{mL gas}}{\text{mL solvent} \times \text{atm}}$$

The units of gas dissolved and the volume of solvent are given in milliliters, and the gas pressure is given in atmospheres. Gas pressure may also be given in torr. To use Henry's law to calculate the solubility of a gas, we need to know the Henry's law constant for that gas or be able to calculate it from the data in hand. Values of Henry's law constants for many gases can be found in chemical and biological handbooks.

A PICTURE OF HEALTH

Solubilities of O_2 and CO_2 in Body Fluids

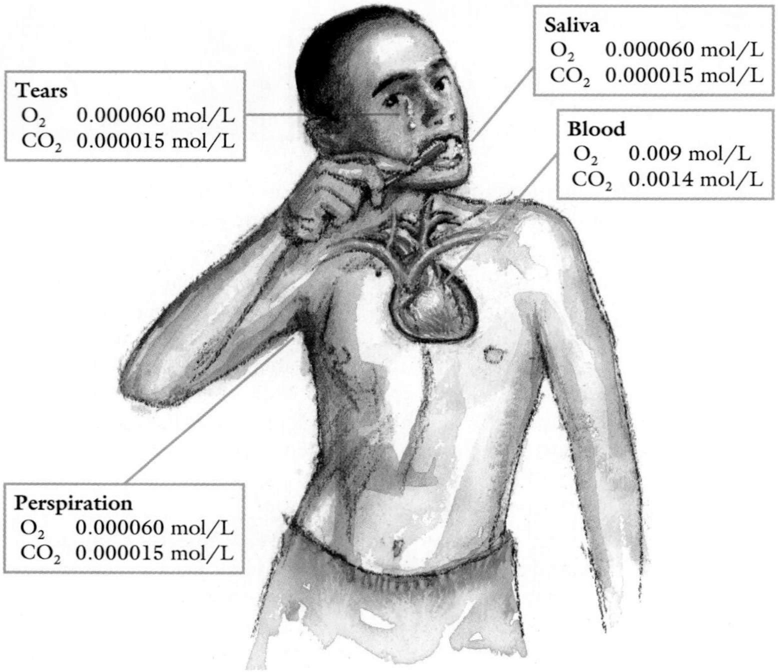

Because gas solubility depends on the pressure of the gas in contact with the liquid, the amounts dissolved can be described in terms of gas pressure alone. For example, the solubility of carbon dioxide in arterial blood is expressed as 46 torr. This means that the amount of carbon dioxide dissolved in arterial blood is the amount that dissolves when the partial pressure of carbon dioxide gas in contact with arterial blood is 46 torr. Physiologists and anesthesiologists define this method of expressing gas solubility as the **gas tension.** The amounts of gas dissolved in blood plasma are critical in anesthesiology and respiratory therapy. Remember, the general rule is, the greater a gas's partial pressure in the gas mixture, the greater the extent to which that gas will dissolve in any liquid present.

Air consists mostly of nitrogen. When blood comes into contact with air in the lungs, nitrogen, as well as oxygen, dissolves in it. Nitrogen is not used up in metabolism, and, for that reason, in contrast with oxygen, the blood is saturated with nitrogen at all times. Under special circumstances, such as under water at a significant depth, the total gas pressure of inhaled air is much greater than at Earth's surface. For every 10 m of depth in water, this pressure increases by 1 atm. Consequently, the amounts of all gases dissolving in the blood will also increase. If the external gas pressure is then suddenly decreased, as occurs when a person rises rapidly from deep immersion, the gases will quickly leave solution in the form of bubbles. When this formation of gas bubbles occurs within the body, it results in a potentially lethal condition called the bends (also called caisson disease), which is considered in more detail in Box 5.5.

Example 5.14 Using Henry's law for pure gases

The Henry's law constant for pure oxygen at 20.0°C is 0.0310 mL O_2/mL H_2O at 1.00 atm total pressure. Calculate the oxygen content of water at 20.0°C in contact with pure oxygen at 1.00 atm O_2 pressure.

Solution

The Henry's law expression for the solubility of O_2 in water under the stated conditions is:

$$1.00 \text{ atm} \left(\frac{0.0310 \text{ mL } O_2}{\text{mL } H_2O \times 1.00 \text{ atm}} \right) = 0.0310 \text{ mL } O_2/\text{mL } H_2O$$

Problem 5.14 The Henry's law constant for pure nitrogen in water at 20.0°C and 1.00 atm pressure of N_2 is 0.0152 mL N_2/mL H_2O. Calculate its solubility in water at 20.0°C and a pressure of 1.00 atm.

Example 5.15 Using Henry's law for a mixture of gases

Calculate the oxygen content of pure water in the presence of air at 20.0°C.

5.5 Chemistry Within Us

Gas Solubility and Caisson Disease

Caisson disease, more commonly known as "the bends," was first encountered in the United States in 1867, in the course of the construction of the Eads Bridge over the Mississippi River at St. Louis, Missouri. The Eads was the first steel bridge ever built, and its construction required piers to be sunk down to bedrock some 100 feet below the river's surface. The construction workers accomplished the necessary digging by descending to the river bottom inside giant upright steel tubes, or caissons. The caissons, resembling a wide-mouthed jar with open end down, were filled with air under high pressure that kept water and mud out of the working area.

As a result of working under these novel conditions, 119 men of a labor force of 600 suffered intense pains in their muscles and joints, often with significant damage to the tissues. Similar problems arose in the construction of the Brooklyn Bridge in New York City in 1886. Although the disease was still not fully understood, the builders recognized that it developed when the caisson workers left the high pressure within the caisson and returned to the surface.

The fundamental cause of caisson disease is the increase in gas solubility as the result of an increase in gas pressure. Because nitrogen is the major component of air, large amounts of nitrogen as well as oxygen dissolved in the construction workers' blood under the high-gas-pressure conditions that they experienced inside the caissons. Oxygen is used up in metabolism, but nitrogen is not and therefore accumulates in large amounts in the blood. As long as the workers remained under those conditions, they suffered no symptoms. However, when they were too rapidly elevated to the surface, the dissolved gases literally bubbled out of solution, just as CO_2 escapes from warm soda when the bottle cap is suddenly removed.

Caisson disease is virtually unknown among tunnel and bridge workers today because adequate precautions are enforced in those industries. Workers exposed to high air pressures must decompress very slowly. They are returned to surface pressures in gradual stages over time, allowing the dissolved gases to leave their blood and tissues at a rate that does no damage.

One of the by-products of the decompression technology developed to prevent or treat the bends has been the introduction of hyperbaric chambers. These chambers are airtight and can be large enough to hold several people at a time. The pressure inside the chamber can be regulated so that, for example, a diver with the bends can quickly be provided with a high-gas-pressure environment to reverse the bubbling of nitrogen from the blood and relieve the painful symptoms. Then the pressure can be lowered slowly and gradually until the danger is over.

In the increased total pressure within the hyperbaric chamber, the partial pressure of oxygen also increases. Therefore, hyperbaric chambers have proved useful in other situations, including the prevention and reversal of infections caused by anaerobic pathogens (such as gangrene) and the treatment of fire fighters who have been exposed to carbon monoxide.

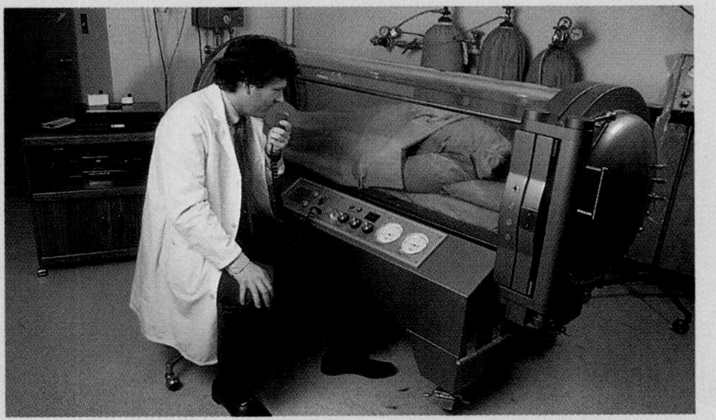

Solution

As given in Example 5.14, the Henry's law constant for pure oxygen at 20.0°C is 0.0310 mL O_2/mL H_2O at 1.00 atm total pressure. Air consists of about 20% O_2, which means that, at 1.00 atm air pressure, the O_2 content of air can be written as 0.200 atm. Therefore, the Henry's law expression for the solubility of O_2 in water in the presence of air is:

$$0.200 \text{ atm}\left(\frac{0.0310 \text{ mL } O_2}{1.00 \text{ mL } H_2O \times 1.00 \text{ atm}}\right) = 0.00620 \text{ mL } O_2/\text{mL } H_2O$$

Problem 5.15 The solubility of pure nitrogen in water at 20.0°C and 1.00 atm pressure is 0.0152 ml N_2/mL H_2O. It constitutes about 80% of air. Calculate its solubility when air is in contact with water at 20.0°C and a pressure of 1.00 atm.

The solubility of any gas decreases as temperature increases. Gas molecules dissolve in liquids by occupying spaces between the liquid's molecules. As the temperature increases, the liquid's molecules move about more and more chaotically, with the result that the number of spaces available to gas molecules decrease—and so does the gas's solubility. The loss of "zip" as a glass of cold carbonated beverage warms to room temperature is a good example of this behavior. Figure 5.9 shows how the solubility of gases in water decreases as the temperature increases.

The inverse of this phenomenon is that, as temperature decreases, more gas will dissolve. A fortunate consequence is the number of people—particularly children, with their low body masses—who have been able to survive near drowning in water at freezing temperatures. The amounts of oxygen dissolved in cold body fluids can be sufficient to support metabolism of the brain and other vital organs until the person is revived. This phenomenon also explains why cryogenic surgery is useful under certain conditions.

Quantitative relations to remember

- Boyle's law: pressure × volume = constant
- Charles's law: volume = temperature (K) × constant
- Gay-Lussac's law: pressure = temperature (K) × constant
- Avogadro's law: volume = number of moles × constant
- Combined gas law: pressure × volume = temperature (K) × constant

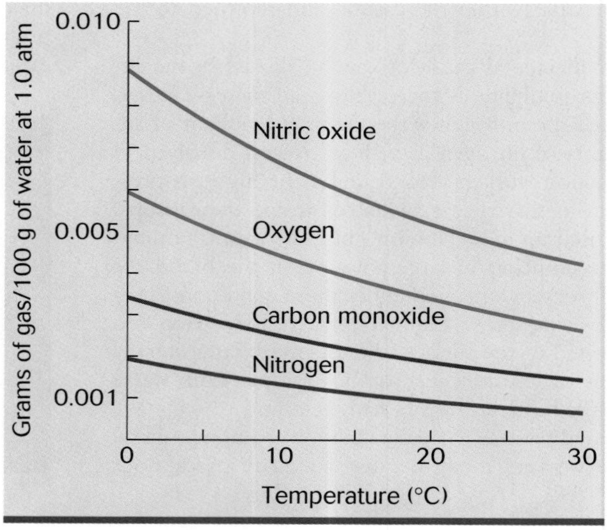

Figure 5.9 The solubility of gases in water decreases with increasing temperature.

- Universal gas law: pressure × volume = number of moles × temperature (K) × constant
- Henry's law: volume of gas dissolved/volume of liquid = C_H × gas pressure

Summary

Experiments have shown that all gases exhibit virtually the same behavior with respect to temperature and pressure, without regard to molecular mass or shape. This finding means that the physical properties of all gases, whatever their chemical nature, can be described by just a few simple mathematical laws. Moreover, these properties depend on only four variables: pressure (P), volume (V), temperature (T), and number of moles (n).

The Gas Laws Increasing the pressure on a gas will compress it so that it will occupy a smaller volume, provided the temperature and number of moles of gas remain constant. Gas behavior under these circumstances is described mathematically by Boyle's law.

When the temperature of a gas is raised, the gas will expand, and, if the gas is cooled, it will contract, provided that the gas pressure and the number of moles of gas remain constant. This behavior is described mathematically by Charles's law.

When the temperature of a fixed amount of gas is raised and the volume is kept constant, the gas pressure increases. This behavior is described by Gay-Lussac's law. Avogadro's law states that equal volumes of gases at the same temperature and pressure contain the same number of molecules.

These three gas laws can be combined to predict results when two variables are changed simultaneously.

The Universal Gas Law All four gas variables, P, V, T, and n, can be combined into a single mathematical expression called the universal gas law. It is used to predict the behavior of gases when mass is one of the variables. It is also used to determine the molecular mass of gaseous substances.

Dalton's Law of Partial Pressures Mixtures of gases are described by Dalton's law of partial pressures. This law states that the total pressure of a mixture of gases is equal to the sum of the individual pressures of all gases in the mixture. These individual pressure are called partial pressures.

Gases Dissolve in Liquids The amount of a gas that will dissolve in a liquid depends chiefly on its pressure and is described by Henry's law. As the gas pressure increases, more gas will dissolve.

Key Words

Exercises

Gas Pressure

5.1 Convert the following gas pressures into torr: (a) 364 mmHg; (b) 483 mmHg; (c) 675 mmHg; (d) 735 mmHg.

5.2 Convert the following gas pressures into atm: (a) 825 torr; (b) 375 torr; (c) 680 torr; (d) 745 torr.

5.3 Convert 735 mmHg into (a) torr and (b) atm.

5.4 Convert 160 mmHg into (a) torr and (b) atm.

5.5 Convert (a) 1.20 atm into torr; (b) 0.450 atm into mmHg; (c) 850 mmHg into atm.

5.6 Convert (a) 1421 torr into atm; (b) 0.675 atm into mmHg; (c) 927 mmHg into atm.

The Gas Laws

5.7 In applications of Boyle's law, which of the four gas variables, P, V, T, and n, are held constant?

5.8 In applications of Charles's law, which of the four gas variables, P, V, T, and n, are held constant?

5.9 In applications of Gay-Lussac's law, which of the four gas variables, P, V, T, and n, are held constant?

5.10 In applications of Avogadro's law, which of the four gas variables, P, V, T, and n, are held constant?

5.11 Calculate the new volume of 2.0 L of a gas when its pressure is changed from 1.0 atm to 0.80 atm at constant temperature.

5.12 While the pressure of 1.40 L of a gas is kept constant, its temperature is increased from 27.0°C to 177°C. Calculate the new volume.

5.13 The initial state of 1.00 L of a gas is $P = 0.800$ atm, $t = 25.0$°C. The conditions were changed to $P = 1.00$ atm, $t = 100$°C. Calculate the new volume.

5.14 The initial state of a gas is $P = 0.80$ atm, $t = 25°C$, $V = 1.2$ L. The conditions were changed to $P = 1.6$ atm, $t = 25°C$. What is the final volume?

5.15 The initial state of a gas is $P = 0.800$ atm, $t = 25°C$, $V = 1.20$ L. The conditions were changed to $P = 1.00$ atm, and $V = 2.00$ L. Find the new temperature.

5.16 The volume of a 3.60 L sample of gas at 27.0°C is reduced at constant pressure to 3.00 L. What is its final temperature?

5.17 The initial state of a gas is $P = 1.50$ atm, $t = 25.0°C$, $V = 1.20$ L. The conditions were changed to $V = 0.600$ L, $t = 50.0°C$. What is the final pressure?

5.18 The initial state of a gas is $P = 1.25$ atm, $t = 27.0°C$, $V = 1.50$ L. The conditions were changed to $V = 1.00$ L, $t = 77.0°C$. What is the final pressure?

5.19 The pressure on a 3.00-L sample of gas is doubled, and its absolute temperature (K) is increased by 50%. What is its final volume?

5.20 If the volume of a 4.00-L sample of gas is doubled, but its final absolute temperature (K) is then reduced to half of its original value, what is the final volume of the gas?

5.21 A sample of gas at 0.830 atm and 25.0°C has a volume of 2.00 L. What will its volume be at STP?

5.22 Ammonia gas occupies a volume of 5.00 L at 4.00°C and 760 torr. Find its volume at 77.0°C and 800 torr.

5.23 Calculate the value of R at STP when volume is in milliliters and pressure is in torr.

5.24 Calculate the value of R at STP when the volume is in milliliters and the pressure is in atm.

5.25 Calculate the new volume of 1.35 L of hydrogen gas at STP when 0.100 mol of hydrogen at STP is added to its expandable container.

5.26 How many moles of helium are required to fill a 3.50-L tank at 25.0°C to a final pressure of 8.74 atm?

5.27 Calculate the volume occupied by 8.00 g of oxygen at STP.

5.28 Compare the volumes occupied by 4.00 g of helium and 44.0 g of carbon dioxide at STP.

5.29 The volumes of hydrogen and ammonia samples at STP are 11.2 L and 5.60 L, respectively. What masses are contained in those volumes?

5.30 A 6.13-g sample of a hydrocarbon occupied a 5.00-L container at 25.0°C and 1.00 atm. Calculate its molecular mass.

Law of Partial Pressures

5.31 The total pressure in a vessel containing oxygen collected over water at 25.0°C was 723 torr. The vapor pressure of water at that temperature is 23.8 torr. What was the pressure of the gas in atmospheres?

5.32 The total pressure in a vessel containing hydrogen collected over water at 25.0°C was 0.8 atm. The vapor pressure of water at that temperature is 23.8 torr. What was the pressure of the gas in torr?

5.33 The volume of a sample of hydrogen collected over water at 25.0°C was 6.00 L. The total pressure was 752 torr. How many moles of hydrogen were collected?

5.34 The volume of a sample of methane (CH_4) collected over water at 25.0°C was 7.00 L. The total pressure was 688 torr. How many moles of methane were collected?

5.35 In different containers, you have 18.0 g of H_2O, 2.02 g of H_2, and 44.0 g of CO_2, all under standard conditions of temperature and pressure. Which occupies the largest volume?

5.36 In different containers, you have 9.0 g of H_2O, 16 g of CH_4, and 44.0 g of CO_2, all under standard conditions of temperature and pressure. Which occupies the smallest volume?

5.37 A mixture of 0.250 mol of oxygen and 0.750 mol of nitrogen is prepared in a container such that the total pressure is 800 torr. Calculate the partial pressure of oxygen in the mixture.

5.38 A mixture of 1.25 mol of argon and 3.75 mol of nitrogen is prepared in a container such that the total pressure is 760 torr. Calculate the partial pressure in torr of argon in the mixture.

Unclassified Exercises

5.39 What volume in liters does a 1.50 mol sample of H_2 occupy at 55.0°C and 1.00 atm pressure?

5.40 The pressure of a sample of N_2 was changed from 1.00 atm to 2.75 atm. As a result, its volume changed from 725 mL to 450 mL, and its final temperature was 235°C. Calculate its initial temperature in Celsius degrees.

5.41 The pressure of a sample of He at 25.0°C was changed from 565 torr to 760 torr. As a result, its volume changed from 800 mL to 850 mL. Calculate its final temperature.

5.42 The volume of a sample of O_2 at 0.75 atm was changed from 2.0 L to 3.5 L. As a result, its temperature changed from 25°C to 77°C. Calculate its final pressure.

5.43 The temperature of a 2.00-L sample of Ar at 600 torr was changed from 25.0°C to 125°C. As a result, its pressure changed from 600 torr to 783 torr. Calculate its final volume.

5.44 The temperature of a sample of CO_2 at 600 torr was changed from 25.0°C to 125°C. As a result, its volume changed from 1200 mL to 1440 mL. Calculate its final pressure.

5.45 The final volume of a sample of methane (CH_4) compressed from 550 torr to 850 torr was 1550 mL. In the process, its temperature changed from 50.0°C to 587°C. Calculate its initial volume.

5.46 How many moles of CO_2 are contained in a 1.5-L vessel at 50°C and 0.86 atm?

5.47 What is the density of H_2 contained in a 2.75-L vessel at 850 torr and 30.0°C?

5.48 The density of benzene vapor at 100°C and 650 torr is 2.18 g/L. Calculate its molecular mass.

5.49 In the reaction $Zn + 2\ HCl \rightarrow ZnCl_2 + H_2$, 13.1 g of Zn was consumed. At STP, how many milliliters of H_2 were formed?

5.50 The Henry's law constant for nitrogen at 20.0°C and 1.00 atm total pressure is 0.0152 mL N_2/mL H_2O. How many milliliters of nitrogen are dissolved in a milliliter of water at 20.0°C at a nitrogen partial pressure of 456 torr?

5.51 Convert the following gas pressures into torr: (a) 0.750 atm; (b) 1.23 atm; (c) 0.950 atm; (d) 1.750 atm.

5.52 What is the meaning of the expression standard temperature and pressure (STP)?

5.53 Fill in the correct values in the empty places in the table in the right hand column.

P (atm)	*V* (L)	*T* (K)	*n* (mol)
(a) 0.800	______	300	0.900
(b) ______	1.60	225	2.00
(c) 0.900	15.0	______	0.500
(d) 4.00	10.0	490	______

5.54 A 2.50-L sample of a gas at 1.00 atm is heated from 50.0°C to 375°C. Calculate the pressure required to keep the volume constant at 2.50 L.

Chemical Connections

5.55 Medical oxygen contained in a 20.0-L steel tank was removed at a constant temperature of 27.0°C for a period of 65 min, during which time the pressure in the tank decreased from 14.0 atm to 11.5 atm. How much oxygen, in moles, was removed from the tank?

5.56 If the oxygen removed from the tank in Exercise 5.55 were at STP, what would be the rate of delivery of the gas in liters per minute?

5.57 The partial pressure of CO_2 in contact with human arterial blood is 46 torr. The Henry's law constant for CO_2 in blood plasma at 37°C is 0.51 mL CO_2/mL plasma at 1.0 atm of CO_2. Calculate the amount of carbon dioxide, in milliliters, dissolved in a milliliter of arterial blood.

5.58 A helium weather balloon filled at ground level and atmospheric pressure of 1.0 atm contains 50 L of gas at 20°C. When it rises to the stratosphere, the external pressure will be 0.40 atm, and the temperature will be −50°C. What should the capacity of the balloon be at ground level? Hint: Have you noticed how limp stratospheric balloons appear as they are released from ground level?

CHAPTER 6

INTERACTIONS BETWEEN MOLECULES

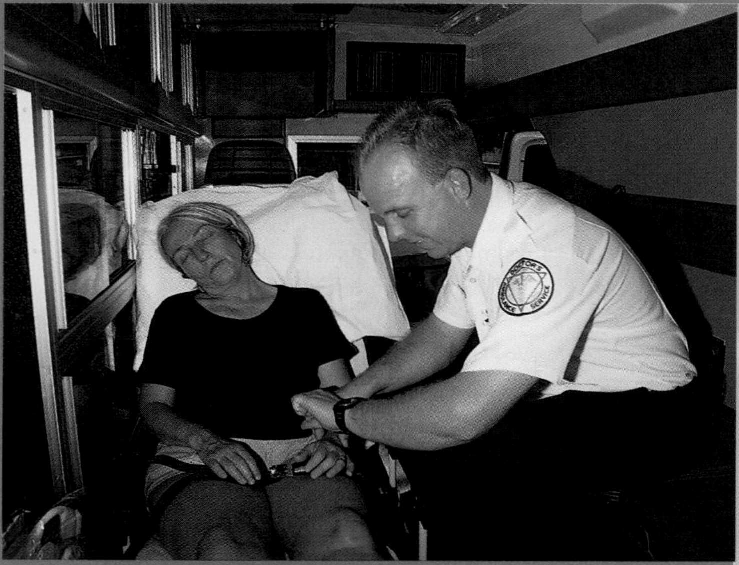

CHEMISTRY IN YOUR FUTURE

Having recently moved from Phoenix to Chicago, you start your new job at Northwestern Hospital's emergency room in the midst of a record-breaking heat wave. The temperatures are not any higher than the temperatures typical for this time of year in Phoenix, so you are perplexed when you encounter many more, and far more serious, cases of heat prostration than you ever saw back home. The phrase "it's not the heat, it's the humidity" runs through your mind. Could that explain the riddle? Chapter 6 reveals the physical basis of this phenomenon and of other physiological problems as well.

LEARNING OBJECTIVES

- Use molecular concepts to explain the properties of the three states of matter and transitions between them.
- Describe secondary forces, and correlate chemical structure with types of secondary forces.
- Correlate physical properties such as vapor pressure, normal boiling point, melting point, and surface tension with types of secondary forces.
- Understand the molecular characteristics necessary for solution formation.
- Define dynamic equilibrium.

So far, we have concentrated on the forces that hold atoms together to form chemical compounds—covalent and ionic bonds. In this chapter, we will study a new group of forces that operate not within molecules but between them. Walking on water—when it is frozen, of course—is a good reminder that it can exist as a solid (ice) or a liquid or a gas (steam). Water is not unique in this regard; all substances may exist in any of the three states of matter, depending on the conditions of temperature and pressure.

Substances exist as solids and liquids because of attractive forces between the fundamental particles. This concept is easy to explain when the substance is an ionic compound, such as sodium chloride or potassium nitrate, because the component particles are positive and negative ions that are strongly attracted to one another. But the molecules of water, carbon dioxide, ammonia, and many other molecules are electrically neutral, yet can exist as liquids or solids. What forces cause water and other neutral molecules to attract one another?

The existence of liquid and solid states in substances consisting of neutral molecules such as water or atoms such as helium is explained by a number of different types of attractive forces. Although these forces are much weaker than ionic or covalent chemical bonds, they underlie not only the formation of solids and liquids, but many physiological phenomena as well, including the structures and properties of biomolecules such as proteins, nucleic acids, carbohydrates, and lipids.

Many biological processes are based on brief weak interactions between different molecules. The conduction of nerve impulses, for example, depends on a small molecule released from one nerve cell interacting weakly with large molecules embedded in the surface of an adjacent nerve cell. The weak interaction guarantees that the interactions formed are very short lived. If the attractions were strong, a nerve impulse would never cease. You would feel pain, taste sourness, or see red endlessly.

▶▶ Other practical consequences of these interactions are explored throughout Part 2, Organic Chemistry.

Our primary interest in this chapter is in the weak interactions between neutral molecules. We begin with a discussion of the properties of the three states of matter and the transitions between them. Next, we consider the nature of the attractive forces between molecules. Then we explore some properties of liquids that are basic to many biochemical and physiological processes. A brief concluding section looks at some of the properties of solids.

6.1 THE THREE STATES OF MATTER AND TRANSITIONS BETWEEN THEM

A solid has a fixed volume and shape, both of which are independent of its container. A liquid also has a fixed volume but not a fixed shape. It can be poured from one container to another and takes the shape of its container. The volume of a gas depends on the volume of its container. A gas always occupies all of its container. The volumes of both liquids and solids change very little when pressure is applied. They are said to be incompressible. On the other hand, a gas is easily compressed into a smaller volume. Figure 6.1 is a sketch

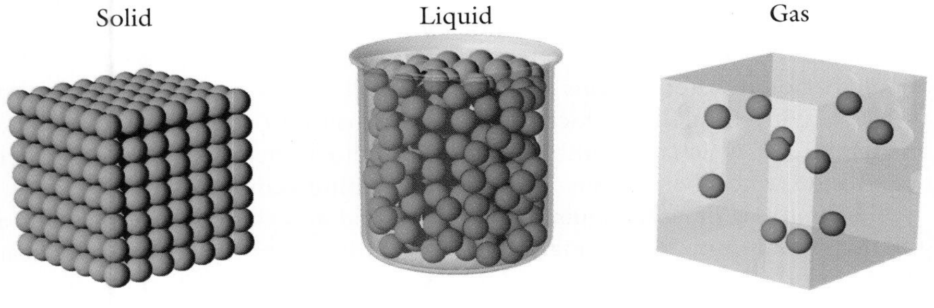

Figure 6.1 A molecular view of the three states of matter. The molecules in both the solid and the liquid states are in contact, but those in the liquid state are somewhat disordered. Molecules in the gaseous state are very far apart and totally disordered.

depicting the arrangements of molecules in each of the three states of matter. There is little increase in volume when a solid melts, but, when a liquid is vaporized, there is a large change in volume. The volume of 1 mol of liquid water at 100°C is 18.8 mL, and the volume of 1 mol of gaseous water (steam) at the same temperature and pressure is about 3.1×10^4 mL.

A solid cannot be compressed, because the constituent molecules are in contact, as close to one another as they can be. A solid holds its shape because strong attractive forces hold the molecules in fixed positions from which they cannot escape.

The compressibility of liquids is similar to that of solids, so the molecules are assumed to be as close to one another as they can be. The fact that a liquid flows and cannot hold its shape is explained by assuming that the attractive forces between molecules in the liquid state are weaker than those existing in solids. The internal structure of liquids is less orderly than that of a typical solid.

In a gas, the particles move at high speeds and change directions only when they collide with the walls of their container or with other particles. Because the molecules in gases are so far apart, they have virtually no opportunity to interact, so attractive forces between molecules play a small role in gases. Because the molecules of gases are not subject to such forces, their physical properties are independent of the nature of the individual molecules. For example, in the gaseous state, there is virtually no difference in the physical behavior of hydrogen, methane, or any other gaseous compound. As a consequence, all gases can be quantitatively described by the few simple mathematical laws presented in Chapter 5.

▶▶ Chapter 8 describes the role of kinetic energy in the mechanics of a chemical reaction.

The energy of a moving body, or energy of motion, is called **kinetic energy (KE).** It is equivalent to the work necessary to bring that body to rest or to impart motion to a resting body. Kinetic energy at the molecular level depends only on temperature. As the temperature increases, so does kinetic energy, and, as a result, molecules move faster. In hot systems, molecules move about much faster than in cold systems. Kinetic energy tends to oppose the attractive forces between molecules.

Most solids melt and are transformed into liquids on the addition of sufficient heat and can revert back to the solid state if the same amount of heat is removed. The transition from solid to liquid is called **melting.** The reverse process, transforming liquid into solid, is called **freezing.** The **melting point** and **freezing point** of a pure solid are identical. The transition from liquid to gas also requires heat and is called **vaporization.** The reverse of this process is called **condensation.** Continuous vaporization from the surface of a liquid into the atmosphere is called evaporation, which is considered further in Section 6.7.

The changes from one physical state into another as a result of the addition or removal of heat are called **phase transitions.** Phase transitions occur in either direction; that is, they are **reversible.** Heat added to a system in transition from solid to liquid or from liquid to gas does not cause the temperature to increase. Instead, this heat is used to overcome the attractive forces between molecules. Only when the phase transition is complete—that is, when all the molecules of a solid, for example, have been converted into the liquid state—will the added heat cause the temperature to increase. The temperature of a sample of liquid water remains at 100°C until all of it is converted into steam at 100°C. Any heat then added to the steam will raise its temperature. The temperature at which gas bubbles form throughout the liquid in a container open to the atmosphere is called the **normal boiling point.**

The amount of heat required to melt a solid and then to vaporize it is a unique physical property distinguishing one substance from another. The heat per mole of solid required for melting is called the **molar heat of fusion,** ΔH_f, in kcal/mol. The heat per mole of liquid required for vaporization is called the **molar heat of vaporization,** ΔH_v, in kcal/mol. The molar heat of vaporization is always considerably greater than the molar heat of fusion because, in the transition from liquid to gas, the attractive forces between molecules must be completely overcome. Melting requires only a loosening of the solid structure, and the attractive forces are largely still at work.

In the next section, we will examine the nature of the forces that operate between molecules.

6.2 ATTRACTIVE FORCES BETWEEN MOLECULES

The attractive forces between molecules are often called **intermolecular forces** or **van der Waals forces.** In this book, we will use the term **secondary forces** because it includes both intermolecular and intramolecular forces, and intramolecular forces are very important for biologically important molecules such as proteins, polysaccharides, and DNA. It is important to note that secondary forces are far weaker than chemical bonds—perhaps at most from 10 to 20% of the strength of a covalent or ionic bond—and must not be confused with a true chemical bond. Even though the energies of these secondary forces are much smaller than those of chemical bonds, they underlie not only the physical properties of solids and liquids, but many physiological phenomena as well. To better understand these physical properties, we must explore the origin and nature of secondary forces.

Dipole–Dipole Interactions

Secondary forces cause gases to condense into liquids and liquids to freeze into solids. These forces are electrical in nature and arise from molecular polarity. Permanent molecular polarity, characterized by a dipole moment, was described in Chapter 3. Molecules with permanent dipole moments, or **polar molecules,** attract each other and other, different molecules possessing dipole moments.

Figure 6.2 shows the interaction of polar molecules in the liquid state. The oppositely charged parts of the polar molecules attract, and the similarly charged parts repel each other. Attractive forces between polar molecules—for example, methyl chloride, CH_3Cl, or chloroform, $CHCl_3$, or mixtures of the two—are called **dipole–dipole forces.** Substances composed of polar molecules have higher melting and boiling points than do those composed of molecules possessing no permanent dipoles, or **nonpolar molecules.** Methane, CH_4, has no permanent dipole moment and, at room temperature, is a gas. However, water, H_2O, a molecule of about the same mass, has a large dipole moment and is a liquid at the same temperature.

▶▶ The polarity of lipids, described in Chapter 19, explains many of their important biological functions.

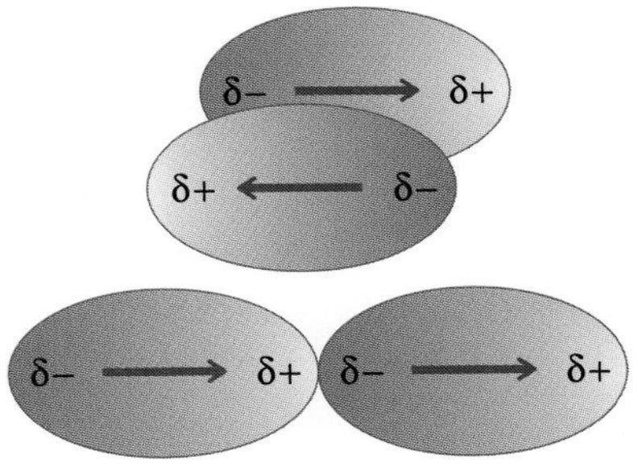

Figure 6.2 Polar molecules attract each other by the interaction of their electric dipoles (represented by the arrows). Two relative orientations are shown: end-to-end and side-by-side. (Adapted from P. Atkins and L. Jones, *Chemistry: Molecules, Matter, and Change*, 3d ed., New York, W. H. Freeman and Company, 1997.)

London Forces of Interaction

Even in molecules such as methane that have no permanent polarity, temporary dipoles can exist for brief moments. This is because the erratic motions of electrons result in uneven distributions of electrical charge within the molecule at any given time. For example, for a brief instant, perhaps 1×10^{-12} s, most of the electrons in a molecule may be at one end. The attractive force resulting from this temporary dipole is called a **London force.** This phenomenon occurs in all molecules. The strength of the London force can range from being many times as weak as dipole–dipole forces to being about equal to them. London forces exist between all molecules, but their contribution to physical properties is particularly significant in those molecules that do not possess permanent dipoles.

TABLE 6.1 Boiling Points Characteristic of the Three Types of Secondary Forces

Secondary Force	Substance	Boiling Point (°C)
London force	helium (He)	−269
	hydrogen (H_2)	−253
	methane (CH_4)	−164
	butane (C_4H_{10})	−0.5
	hexane (C_6H_{14})	69
dipole–dipole	methyl fluoride (CH_3F)	−78
	fluoroform (CHF_3)	−82
	formaldehyde (HCHO)	−21
hydrogen bond	methanol (CH_3OH)	65
	water (H_2O or HOH)	100
	ethylene glycol (HOC_2H_4OH)	198

For example, nonpolar substances such as methane or helium condense to liquids only because of London forces. Compare the boiling points of nonpolar substances with those possessing permanent dipoles in Table 6.1. Methane, a nonpolar molecule, condenses to a liquid at −164°C, compared with formaldehyde, a polar molecule that condenses at −21°C.

Because London forces arise from uneven distributions of electrons within molecules, molecules with large numbers of electrons exert larger London forces. Therefore nonpolar molecules of large molecular mass can exert greater attractive force than can nonpolar or even polar molecules of smaller molecular mass. Compare methane's normal boiling point with that of hydrogen and methanol's with that of hexane in Table 6.1.

Example 6.1 Using structural formulas to predict polarity

Which of the following substances are characterized by polar secondary forces?

F_2 CH_4 HF CH_3Cl

Solution

The structural formulas for these compounds were developed in Section 3.7 on Lewis formulas. They are:

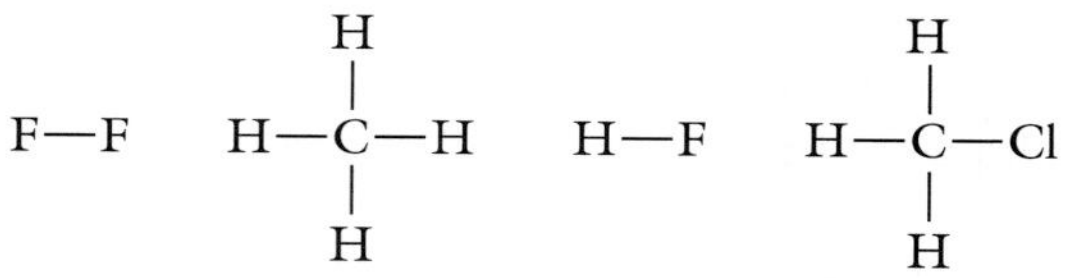

F_2 and CH_4 have electrically symmetrical structures and therefore possess no dipole moments. They cannot interact through polar secondary forces. HF and CH_3Cl have electrically asymmetric structures. Therefore they possess permanent dipole moments and interact through polar secondary forces.

Problem 6.1 Which of the following compounds condense to the liquid phase from the gas phase through nonpolar secondary forces?

(a) H—H (b) 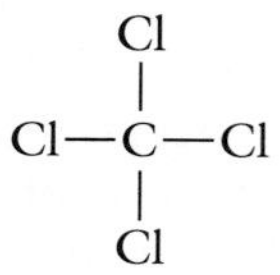 (c) H—C≡N (d) I—I

6.3 THE HYDROGEN BOND

A covalent bond between hydrogen and any one of the three elements oxygen, nitrogen, and fluorine possesses unique character. The differences in electronegativity between hydrogen and these elements, and therefore the polarity within such a bond, are very large. This polarity leads to a unique and very strong attractive force, called **hydrogen bonding,** between molecules such as H_2O, NH_3, and HF.

Figure 6.3 is a sketch of hydrogen bonding between two water molecules drawn with appropriate bond angles. The hydrogen bond between molecules appears as a dashed line to differentiate it from the covalent bonds within the water molecules. A hydrogen atom in one of the water molecules, with its intense partial positive charge (indicated by δ^+), is able to interact strongly with a lone electron pair of the oxygen atom in another water molecule. In Figure 6.3, the partial charges on all other atoms in the diagram are ignored so as to focus on this intermolecular hydrogen–oxygen interaction.

A hydrogen bond, sometimes abbreviated H-bond, is a special case of a dipole–dipole attraction because it is significantly stronger than any other dipolar secondary force. It forms not only between identical molecules, as occurs in liquid water between molecules of H_2O or in ammonia between molecules of NH_3, but also between different molecules, allowing mixtures, for example, of water and ammonia.

The hydrogen bond allows the hydrogen atom to act as "glue" between two different molecules; that is, the hydrogen of one molecule binds to the oxygen, nitrogen, or fluorine atom of another molecule. The bond, though strong, is weaker than a covalent bond and is therefore depicted as a longer bond than the covalent bonds.

Water molecules in the solid state, ice, are completely hydrogen bonded. Figure 6.4 shows that each molecule in ice forms four hydrogen bonds and

INSIGHT INTO PROPERTIES

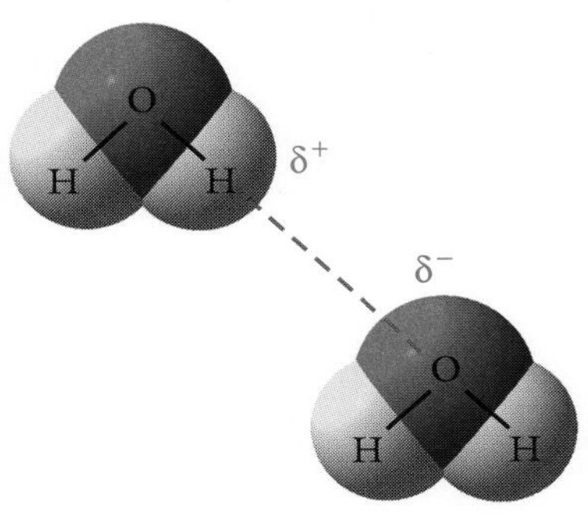

Figure 6.3 Sketch of hydrogen bonding between two water molecules. The hydrogen bond is a dashed line to differentiate it from the covalent bonds (solid lines) of the water molecules. The hydrogen bond explains why water is a liquid.

INSIGHT INTO PROPERTIES

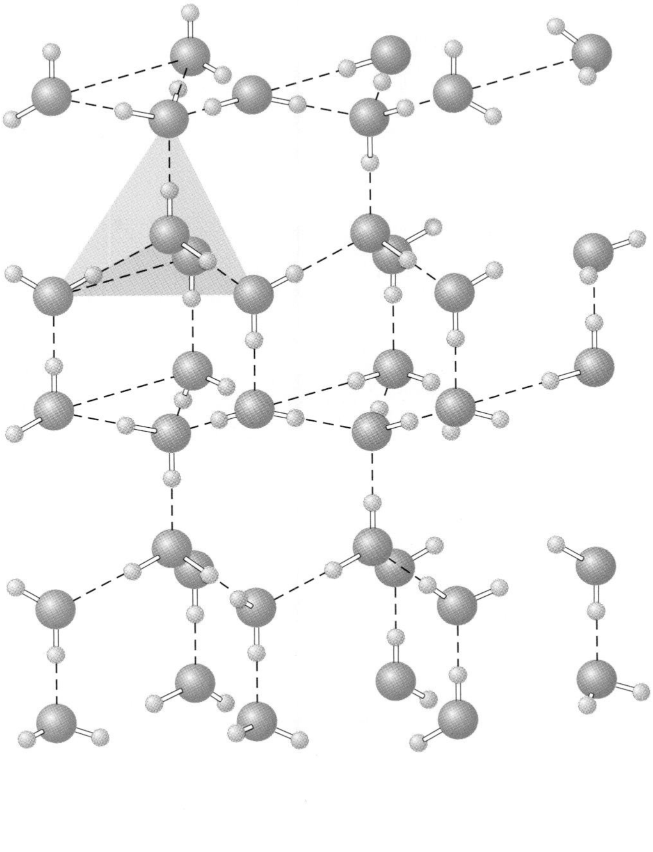

Figure 6.4 The crystal structure of ice. Every oxygen (shown in red) is located at the center of a tetrahedron formed by four other oxygen atoms. Hydrogen bonds (dashed lines) link all oxygen atoms together. To form this number of hydrogen bonds, the resulting structure requires a greater volume than does the liquid and is therefore of lower density than liquid water. That explains why ice floats in liquid water.

that every oxygen is located at the center of a tetrahedron formed by four other oxygen atoms. The entire solid structure is held together by hydrogen bonds, which forces the molecules in the solid to take up as much room as possible—to form an open structure. When ice melts, some 15 to 20% of the hydrogen bonds are broken, and the open structure collapses a bit. The result is that the density of liquid water is greater than that of ice, and solid water floats on its liquid form.

Binary (two-element) molecular compounds formed between hydrogen and one other element are called **hydrides.** Water, ammonia, and hydrogen fluoride are hydrides. In Figure 6.5, the boiling points of the binary molecular hydrides are plotted as a function of their row in the periodic table. The following table identifies these hydrides with respect to group and row of the periodic table.

	Group			
Row	**IV**	**V**	**VI**	**VII**
1	CH_4	NH_3	H_2O	HF
2	SiH_4	PH_3	H_2S	HCl
3	GeH_4	AsH_3	H_2Se	HBr
4	SnH_4	SbH_4	H_2Te	HI

For the most part, the boiling points increase with increasing mass, as would be expected for molecules interacting through London forces. The linear relation between boiling point and row of the periodic table implies that the boiling points of H_2O, NH_3, and HF should be below those of the hydrides in rows 2, 3, and 4. However, the behaviors of H_2O, NH_3, and HF

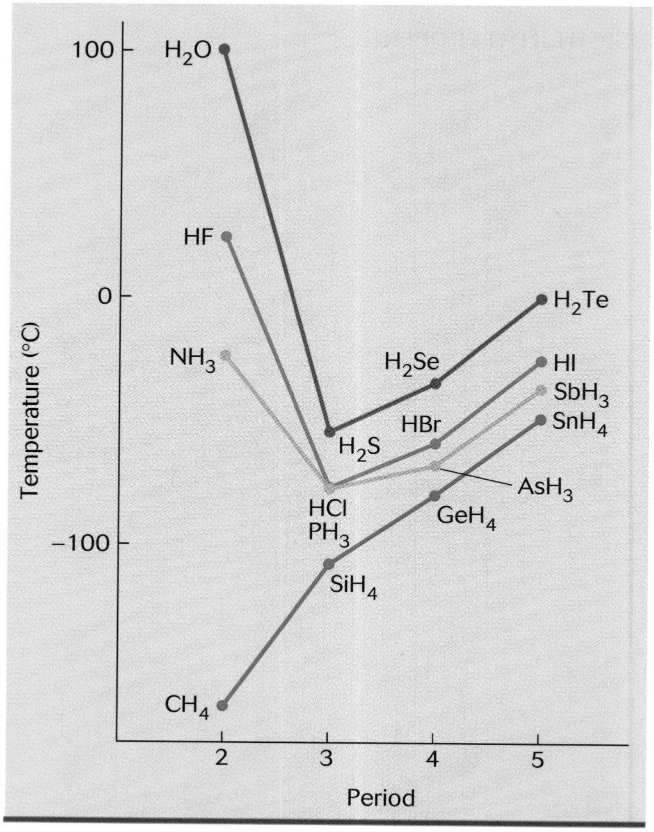

Figure 6.5 The boiling points of the hydrides of the Group IV, V, VI, and VII elements plotted as a function of their row in the periodic table. The anomalously high boiling points of the hydrides of the three elements nitrogen, oxygen, and fluorine are the consequences of hydrogen bonding.

break the pattern—their boiling points are actually higher than any of the others. The reason is that oxygen, nitrogen, and fluorine atoms are strongly electronegative, and this electronegativity causes extensive hydrogen bonding in H_2O, NH_3, and HF. The special behavior of the hydrides of only the three elements nitrogen, oxygen, and fluorine but not the other hydrides shows that hydrogen bonding cannot occur between any and all molecules that contain a hydrogen atom.

Concept check

✔ Look for hydrogen bonding when a molecule contains a hydrogen atom covalently bonded to either oxygen, nitrogen, or fluorine.

Example 6.2 Recognizing molecules that can form hydrogen bonds

Which of the following molecules can form hydrogen bonds?

H_2 CH_4 NH_3 HF CH_3OH

Solution

A hydrogen bond can form if compounds possess a hydrogen atom covalently bonded to either oxygen, nitrogen, or fluorine. Three of the preceding compounds fit that criterion: NH_3, ammonia; HF, hydrogen fluoride; and CH_3OH, methyl alcohol. Neither H_2 (hydrogen) nor CH_4 (methane) can form hydrogen bonds.

Problem 6.2 Which of the following molecules can form hydrogen bonds? (a) SiH_4; (b) CH_4; (c) CH_3NH_2; (d) H_2Te; (e) C_2H_5OH.

The importance of water in biological systems developed in large part because of its hydrogen-bonding properties. The anomalous behavior of water's density and its effect on aquatic life were presented in Box 1.4. Hydrogen bonding also has a key role in the regulation of body temperature, which will be considered in Section 6.8. The characteristics of many important biological substances—proteins, nucleic acids, and carbohydrates—strongly depend on their ability to form hydrogen bonds within their own molecules, as well as external hydrogen bonds with other molecules, particularly water. Hydrogen bonds are responsible in large part for the properties of biomolecules such as the protein hemoglobin (Section 20.7) and the double-helical DNA of chromosomes (Section 21.3).

6.4 SECONDARY FORCES AND PHYSICAL PROPERTIES

As mentioned earlier, Table 6.1 lists normal boiling points for compounds whose liquid states are dominated by different kinds of secondary forces: London forces, dipole–dipole forces, and hydrogen bonds. The boiling points are indicative of the relative strengths of those three types of interaction.

Nonpolar molecules can interact only by London forces, but polar molecules interact by both London and dipole–dipole forces, and hydrogen-bonding substances interact by all three forces. Table 6.2 on the following page lists substances that interact by one, two, or all three secondary forces, along with their normal boiling points.

The data of Tables 6.1 and 6.2 show that

- Hydrogen bonds are stronger than dipole–dipole forces, which are stronger than London forces.

TABLE 6.2 Boiling Points and Structures of Substances Whose Molecules Interact by One, Two, or Three Types of Secondary Forces

Substance	Structure	London	Dipole–Dipole	Hydrogen Bonds	Boiling Point (°C)
methane	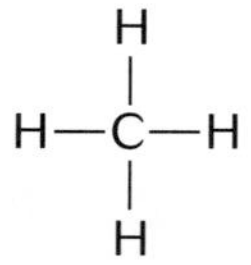	Yes	No	No	−164
trifluoromethane	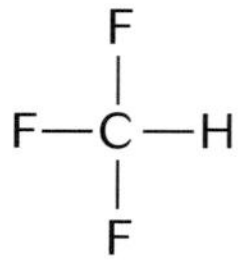	Yes	Yes	No	−82
methyl fluoride	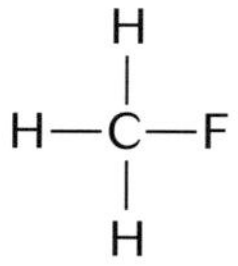	Yes	Yes	No	−78
formaldehyde	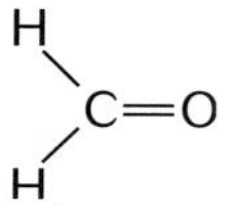	Yes	Yes	No	−21
methanol	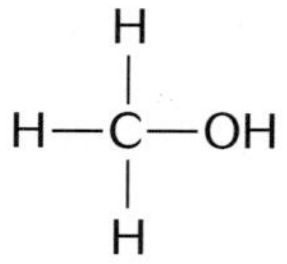	Yes	Yes	Yes	65
water	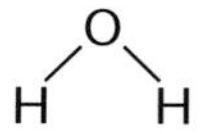	Yes	Yes	Yes	100
ethylene glycol	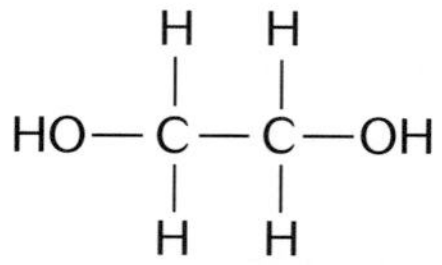	Yes	Yes	Yes	198

- Although both hydrogen bonds and dipole–dipole forces are stronger than London forces, that weak force still contributes to the overall interactions among polar molecules.

The answers to two questions allow us to predict trends in physical properties (such as melting points or boiling points) for a series of compounds:

- Is a compound polar or nonpolar?
- Can a compound form hydrogen bonds?

Example 6.3 Using molecular structure to predict physical properties: I

Arrange the normal boiling points of the following compounds from highest to lowest:

$$H_2 \quad CH_4 \quad NH_3 \quad HF$$

Solution

The two polar compounds, NH_3 and HF, must have higher boiling points than the two nonpolar molecules, H_2 and CH_4. The polar compounds can also form

H-bonds. Because fluorine is the element with the highest electronegativity (Section 3.9), HF forms stronger H-bonds than does NH_3. The order of boiling points for these two compounds, then, from higher to lower, is HF > NH_3. Methane and hydrogen interact by weak London forces, but methane is heavier, contains more electrons, and therefore interacts by stronger London forces than does hydrogen and will boil at a higher temperature. The overall order of boiling points, from highest to lowest, is:

$$HF > NH_3 > CH_4 > H_2$$

Problem 6.3 Arrange the normal boiling points of the following substances from highest to lowest: (a) He; (b) CH_4; (c) I_2; (d) HF.

Figure 6.6 Molecules in a liquid are attracted in all directions, but molecules at the surface are attracted only in an inward direction. The inward pull results in surface tension.

Surface Tension

Surface tension is an important property of liquids that depends on the type of attractive forces between molecules. Molecules behave differently when they are at a surface or at a boundary between phases, as between a liquid and its vapor. Molecules in the interior of a liquid undergo attractions uniformly in all directions around them. However, molecules at a surface are in contact with attracting molecules on one side but not on the other (Figure 6.6). They therefore undergo a net attraction inward toward the liquid's interior, reducing the number of molecules at the interface and consequently reducing the surface area. The result is a force, called the **surface tension,** that resists the expansion of the liquid's surface and acts, in effect, like an elastic skin. Figure 6.7 shows a metal paper clip floating on water despite the fact that its density is much greater than that of water. The surface tension of water is great enough and the mass of the paper clip is distributed in such a way as to to prevent the paper clip from penetrating the surface and sinking. Figure 6.8 illustrates how some insects take advantage of this effect to walk on the surface of water. As long as the weight of the insect does not exceed the force exerted by the surface tension, the creature remains on the water's surface.

The surface tension of water is reduced by the presence of dissolved substances called **surface-active agents** or **surfactants.** Examples of surfactants are soaps and detergents. The reduction of the surface tension of water by soaps typically leads to the formation of stable foams (suds). If a surfactant were added to the pond pictured in Figure 6.8, our water strider would have to learn to swim. The bile salts synthesized by the liver are surfactants crucial to the digestion of dietary fats (Box 6.1 on the following page). Surfactants are also produced by the cells that line our lungs (Box 6.2 on page 159).

All surface-active substances have similar characteristics: they consist of rather long molecules containing both polar and nonpolar segments. The

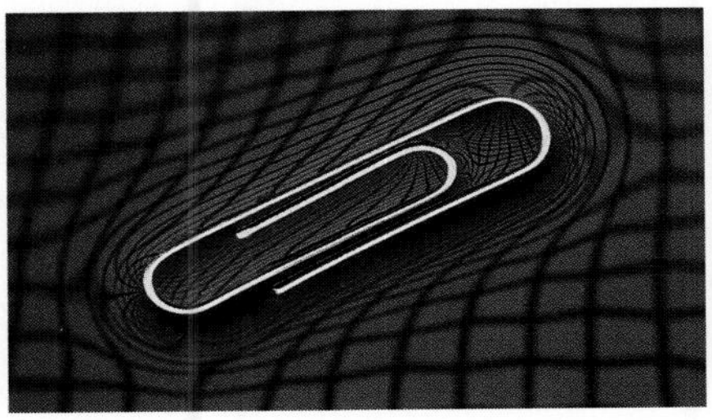

Figure 6.7 A metal paper clip floats on water. Because of the hydrogen bonds between the water molecules, the surface of the water behaves like an elastic membrane.

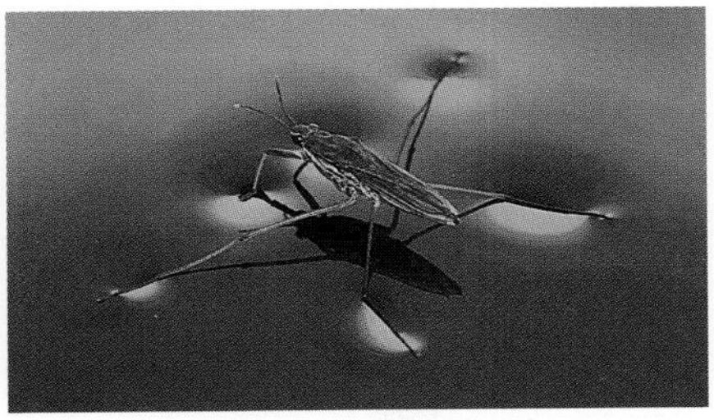

Figure 6.8 The water strider can walk on the surface of water even though it is more dense than the water.

6.1 Chemistry Within Us

Surface Tension and the Digestion of Dietary Fats

By the time that a dietary lipid (fat) has passed through the stomach and entered the small intestine, it has been separated into liquid droplets suspended in an aqueous medium. Digestive enzymes, whose function is to convert lipid molecules into soluble substances for absorption by intestinal cells, cannot enter these droplets and therefore can act on the lipid molecules only at the droplet surface. The rate of digestion of the droplets thus depends on the amount of lipid droplet surface area accessible to the enzymes. If the surface area of the droplets can be increased, the rate of enzyme action will be significantly enhanced. Not surprisingly, our intestines contain a compound to do just that.

When an oil—any salad oil, for example—is shaken with water, the mechanical agitation breaks the oil into tiny droplets and disperses them throughout the water, but, as the droplets collide with one another, they coalesce to form a continuous oil layer above a separate continuous water layer. However, if surface-active compounds are added to the oil and water mixture, they concentrate at the boundary between the oil and the water and prevent the oil droplets from coalescing. This process creates a stable oil–water mixture and is known as **emulsification.**

In the intestine, emulsification is accomplished by the action of biological surface-active compounds called bile salts (Section 19.7). They are synthesized in the liver from cholesterol, stored in the gall bladder, and released into the small intestine in response to the arrival of dietary fat. Their presence at the oil–water boundary combined with the mechanical action of the intestine emulsifies the lipid droplets, greatly increasing their total surface area. Specifically, bile salts cause dietary fats in the intestines to be reduced to fat droplets about 0.0001 cm in diameter. For example, when a teaspoon of olive oil enters the small intestine, its surface area will increase about 10,000 times. This increase in the surface area of lipid droplets results in a similar increase in the rate of their digestion.

sodium salt of palmitic, or hexadecanoic, acid [$CH_3(CH_2)_{14}COO^-Na^+$], a soap, is a good example (Figure 6.9) of a surface-active substance. The molecule consists of a nonpolar hydrocarbon chain (Example 3.10) 15 carbons long with a polar group called a carboxyl group at the end. The nonpolar hydrocarbon part of the molecule is described as **hydrophobic,** or "water hating." The polar end is called **hydrophilic,** or "water loving." Molecules simultaneously possessing both these properties are called **amphipathic.**

As a result of having hydrophobic and hydrophilic properties within the same molecule, surfactants tend to stay at interfaces (air–water or oil–water interfaces, for example) rather than remain in the interior of the water phase. At an interface, the hydrophobic end can escape the water, whereas the hydrophilic end remains nestled in the water's surface. The concentration of wa-

INSIGHT INTO FUNCTION

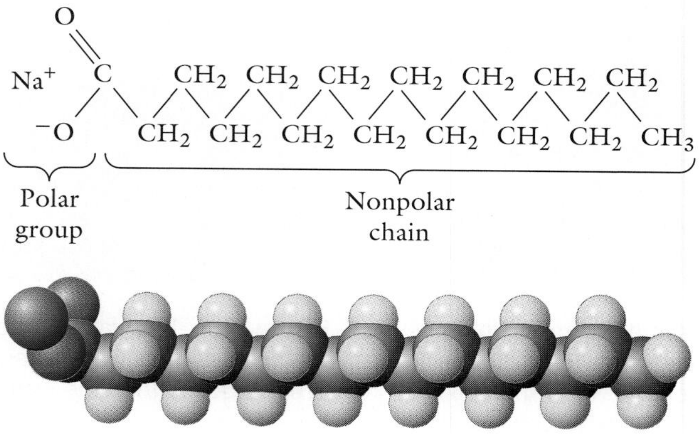

Figure 6.9 Space-filling and structural models of a surface-active agent, the sodium salt of palmitic acid. The molecule has two distinctly different ends: a nonpolar hydrocarbon chain that does not dissolve in water and, at the other end, a polar (ionized) group that does.

6.2 Chemistry Within Us

Respiratory Distress Syndrome

Breathing requires two types of work: (1) the work of stretching the thorax and diaphragm (see Box 5.3) and (2) the work needed to "stretch" the alveoli of the lungs. Gas exchange with the blood takes place in the air space within an alveolus, whose inner surface contacting the gas phase is covered with a thin layer of water. The surface of a liquid is like an elastic membrane. Its surface tension is a measure of the work that must be done to expand its surface, and the surface tension of water is among the largest known. The surface tension of the water lining the alveoli makes the water behave like a stretched rubber band, constantly contracting and resisting expansion. Considerable work must be expended to distend and expand the bubblelike alveolar sacs lined with water. If the alveoli were lined with pure water, it is likely that no amount of muscular effort could expand the lungs, and in fact they would tend to collapse. Fortunately, this does not happen in healthy people, because alveolar cells synthesize a surface-active agent called **lung surfactant,** which markedly reduces the surface tension of the fluid lining the alveoli.

A condition called respiratory distress syndrome of the newborn, also called hyaline membrane disease, is the best-known consequence of the absence of the lung's ability to synthesize lung surfactant. This condition is often present in premature newborns whose surfactant synthesizing cells have not had time to fully develop. Taking a breath requires very strenuous effort on the part of these infants and may cause exhaustion, inability to breathe, lung collapse, and death. The symptoms can be relieved by aspirating a synthetic lung surfactant directly into the lungs. As these infants mature and begin to produce their own surfactant, the problem is mitigated. Lung surfactant is dipalmitoylphosphatidylcholine, and you will learn more about it in Section 19.6.

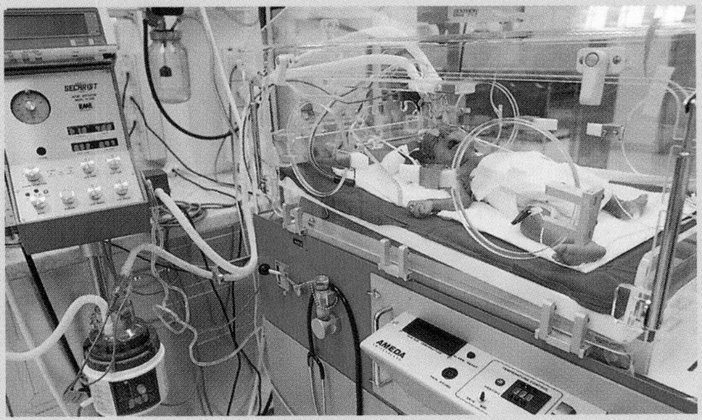

Life support equipment surrounds an incubator holding a premature baby born after a term of 24 weeks. The infant is suffering from respiratory distress syndrome, also called hyaline membrane disease.

ter at the interface is reduced; therefore the attractive forces between molecules at the interface are reduced and, as a consequence, the surface tension is reduced.

Molecular Mixtures

It is well known that oil and water do not mix. Other substances—ethyl alcohol and water, table sugar and water, or salt and water—combine so willingly that their individual components cannot be easily distinguished. A mixture of substances with different kinds of molecules (or ions) uniformly distributed visually throughout is called a **solution.** The key to the formation of this intimate mixture is that the secondary forces of each substance are similar. Solutions will not form between substances that each give rise to different kinds of secondary forces. An old rule of thumb expressive of that idea is, "Like dissolves like."

To explore this idea, let us represent the secondary forces between the molecules of liquid A by the symbol A---A, and those of liquid B by the symbol B---B. (The dashed lines represent the secondary forces operating between molecules). For liquids A and B to mix thoroughly (form a solution), the molecules must be able to change partners to form A---B and A---B. If the forces represented by A---A and B---B are about the same, then

INSIGHT INTO PROPERTIES

Figure 6.10 Schematic representation of the formation of a solution. A volume of liquid A (molecules colored red) is added to an equal volume of liquid B (molecules colored blue). In this case, the secondary forces indicated by the dashed lines are similar; therefore, A---B bonds form and a solution forms. We will see this principle in action throughout our study of organic and biochemistry.

molecules of A will be just as likely to interact with molecules of B as with each other, and a solution will probably form. As an equation we might write:

$$\text{A---A} + \text{B---B} \longrightarrow \text{A---B} + \text{A---B}$$

Figure 6.10 shows the process in diagrammatic form. In this sketch, a beakerful of A molecules (colored red) is added to a beakerful of B molecules (colored blue), and the two substances mix to form a solution (the forces between A molecules being about the same as those between B molecules) whose volume is equal to the sum of the volumes of the two individual components.

An understanding of the nature of secondary forces makes it possible to predict whether two molecular substances will form molecular mixtures, or solutions. You need only to examine the structure of the molecules in question and determine whether they are polar or nonpolar, or whether H-bonding is possible between them.

Example 6.4 Using molecular structures to predict whether solutions will form: I

Can water and ethanol form a molecular mixture? The structural formulas for ethanol and water are:

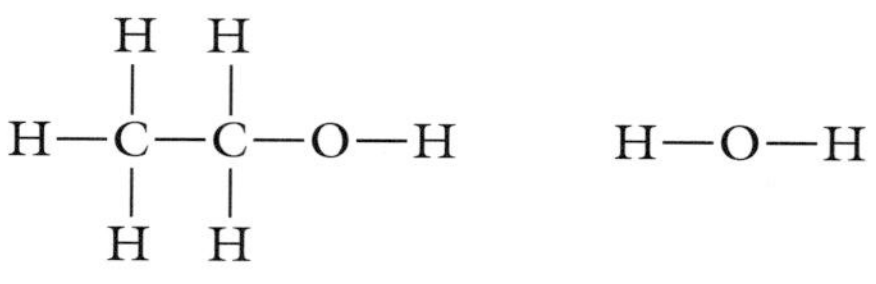

Ethanol Water

Solution

Both molecules are polar, and the presence of the hydroxyl (OH) group in both molecules indicates that both can form H-bonds not only between their own molecules, but also between each other's molecules. These observations suggest that the secondary forces within each individual substance and those that would form within a mixture of the substances are about the same; therefore a solution should form.

Problem 6.4 Can water and carbon tetrachloride

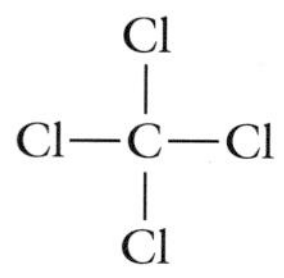

form a solution? Explain your answer.

Example 6.5 Using molecular structures to predict whether solutions will form: II

Will the hydrocarbon hexane, C_6H_{14}, and water, H_2O, form a solution? The structures for hexane and water are:

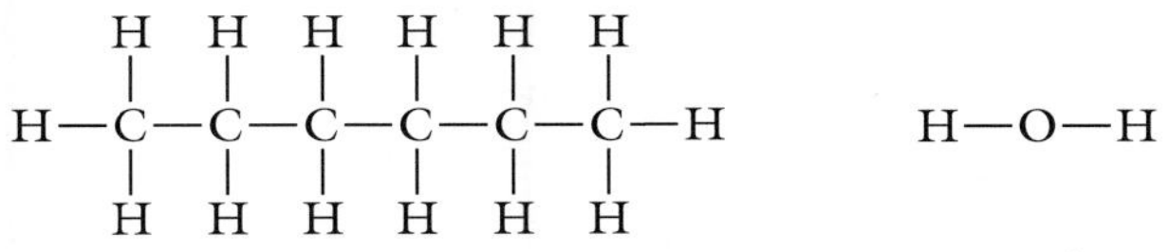

Hexane Water

Solution

The structural formulas must be examined to see how the secondary forces within each liquid compare with those of the other. Hexane is a hydrocarbon with no permanent dipole; so its liquid state results from molecular attraction by London forces. Water molecules interact with each other through hydrogen bonds. Therefore the attractions between hexane and water molecules would be significantly different from those that exist between water molecules and between hexane molecules. The result is that water will be attracted to water and hexane to hexane. Hexane will not mix with or dissolve in water.

Problem 6.5 Explain why the following pairs of substances form or do not form solutions.
(a) CCl_4 and H_2O; (b) C_5H_{12} and CCl_4.

Concept checklist

- ✔ When the attractive forces between different molecules are similar, solutions will form.
- ✔ When the attractive forces between different molecules are different, solutions will not form.

6.5 THE VAPORIZATION OF LIQUIDS

As already stated, vaporization is the process that transforms a liquid into a gas. From a molecular point of view, vaporization is the "escape" of a molecule from a liquid. The energy of a particular molecule in a liquid continuously changes as it collides with others and exchanges energy. It can in time acquire a kinetic energy greater than the average and, should it be at the liquid's surface, it will be able to free itself of its neighbors' attractive forces and enter the gas phase.

When a substance in a sample is present as a liquid and a gas simultaneously, the gaseous part of the sample is called a **vapor.** Oxygen is never called a vapor, because, at room temperature, it does not exist in the liquid state. However, at the temperature of liquid oxygen, about −200°C, when both liquid and gas are present, oxygen in the gaseous state can be called oxygen vapor. The familiar odor of gasoline at the filling station is a vapor because both liquid and gaseous states are present simultaneously. A liquid's vapor is a gas and exerts pressure just as any gas does.

6.6 VAPOR PRESSURE AND DYNAMIC EQUILIBRIUM

Vaporization of a liquid depends on its temperature. A molecule in a liquid must have sufficient kinetic energy (be moving fast enough) to escape the attractive forces of the other molecules and enter the gas phase. Vapor pressure therefore increases with increases in temperature. On the other hand, its condensation depends on the number of vapor molecules colliding with the liquid's surface per unit time. The more vapor molecules striking the liquid surface per unit time, the greater the rate of condensation.

Consider the following experiment, illustrated in Figure 6.11. A beaker of a liquid—say, water—is placed under a bell jar. The experimental setup illustrated is called a **closed system,** a system in which mass is held constant; that is, it is physically impossible for mass to escape or enter the container. The experiment begins when the gaseous atmosphere under the bell jar is removed by a vacuum pump. At first, only vaporization takes place because, in the absence of vapor, condensation cannot occur. Shortly, however, condensation also begins to take place. In time, the amount of liquid transformed into the gaseous state per unit time (its rate of vaporization) and the amount of gas condensing into the liquid state per unit time (its rate of condensation) will be the same, and the amounts of liquid and vapor in the bell jar will remain unchanged. This kind of balance is a condition known in chemistry as an **equilibrium state.** The vapor pressure of a liquid in such a closed system is called its **equilibrium vapor pressure,** and only in a closed system can an equilibrium vapor pressure of a liquid be established.

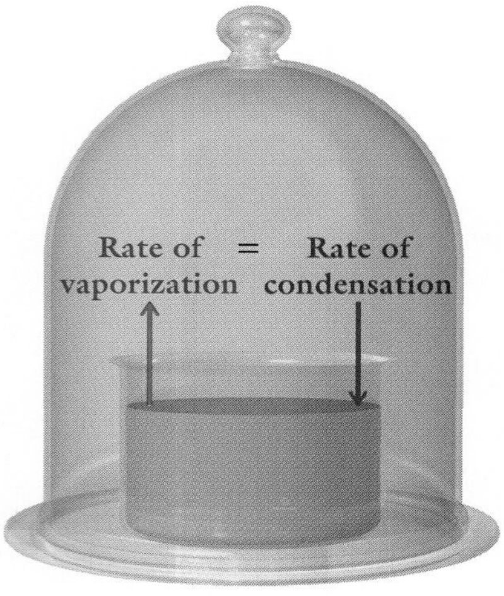

Figure 6.11 After all the gas has been removed by a vacuum pump from a bell jar containing a beaker of liquid, the vapor pressure above the liquid eventually reaches a constant value. That value depends on the specific liquid and the temperature.

When an equilibrium is the result of two opposing but active processes, it is called a **dynamic equilibrium.** Heat is required for the liquid-to-vapor-phase transition. That same quantity of heat is evolved when vapor condenses into a liquid.

$$\text{Liquid} + \text{heat} \xrightarrow{\text{vaporization}} \text{vapor}$$

$$\text{Vapor} \xrightarrow{\text{condensation}} \text{liquid} + \text{heat}$$

A dynamic equilibrium is represented by separating two states—such as liquid and vapor—by two arrows facing in opposite directions:

$$\text{Liquid} + \text{heat} \underset{\text{condensation}}{\overset{\text{vaporization}}{\rightleftharpoons}} \text{vapor}$$

The names of the processes are sometimes inserted above and below the arrows, as shown here. At other times, labels are placed on the arrows to indicate any special conditions (heat, pressure, the presence of a certain kind of chemical) without which the process will not take place.

If the bell jar in the experimental setup illustrated in Figure 6.11 is removed, mass can then escape or enter the container. That kind of situation is described as an **open system,** and vaporization from an open system is called **evaporation.** Evaporation is an imbalance in the vaporization–condensation equilibrium. Because vaporized molecules continue to diffuse away from the liquid's surface, little or no condensation can occur. Vaporization is the dominant process, and the liquid will be continuously transformed into the gas.

To evaporate, the liquid must continuously absorb heat from its surroundings. If the flow of heat from the surroundings is somehow restricted, the liquid's temperature will fall. That is why evaporation is called a cooling process and is used to reduce body temperature in the event of a high fever. This process is illustrated in Box 6.3 and considered in Section 6.8.

6.7 THE INFLUENCE OF SECONDARY FORCES ON VAPOR PRESSURE

Vapor pressure is a measure of the **escaping tendency** of a substance's molecules. The smaller the attractive forces within a liquid sample, the greater the tendency of molecules to escape from that liquid and, consequently, the

6.3 Chemistry Around Us

Topical Anesthesia

The crowd has finally settled down for the start of the fourth game of the World Series, and the first ball has been thrown. There is no way that the unfortunate batter can prevent the errant 92 mph fast ball from making contact with his elbow. The trainer rushes to the plate with an aerosol can and sprays the afflicted area. As the batter's contorted face relaxes, he aims a grateful smile at the trainer and then takes his painfully earned position at first base.

The substance sprayed on the traumatized zone is ethyl chloride (C_2H_5Cl), whose boiling point is 12.3°C. At temperatures above that limit, vapor forms throughout the liquid, which proceeds to evaporate at a speed dependant on the rate at which heat is transferred into it. When ethyl chloride is sprayed on human skin (temperature from approximately 30°C to 35°C), the rate at which heat transfers into the thin film of liquid is very rapid, and the liquid evaporates instantly. This extremely rapid evaporation of ethyl chloride occurs much faster than the rate at which the skin's heat can be replaced by the flow of blood; therefore there is a large temperature drop in the sprayed area. The cooling, in turn, significantly lowers the metabolic activity of pain receptors in the skin, and the sensation of pain diminishes as if by magic.

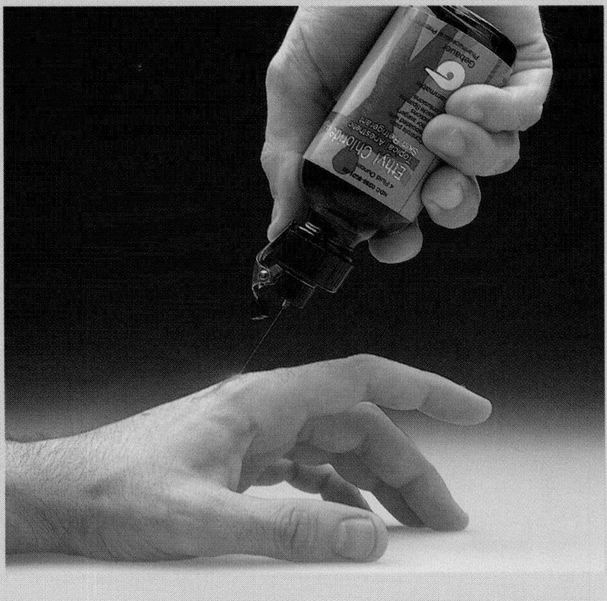

Ethyl chloride, a topical anesthetic, is sprayed on an injured joint.

greater the vapor pressure. The larger the attractive forces between molecules, the smaller the vapor pressure. Liquids that are called volatile have high vapor pressures because of their relatively small secondary forces. Examples are ethyl ether, formerly used as an anesthetic, or ethyl chloride, a topical anesthetic.

Boiling is a special case of vaporization in which bubbles of vapor form throughout the liquid in an open container. Atmospheric pressure acts like a piston pressing against the surface of a liquid open to the atmosphere. A liquid's molecules must overcome this opposing force to enter the gas space over the liquid. Bubbles of vapor cannot form within the bulk of the liquid when its vapor pressure is below atmospheric pressure, because the greater atmospheric pressure will cause any bubbles to collapse. Therefore, when the temperature of a liquid is such that its vapor pressure is less than atmospheric pressure, vapor will leave the liquid only at its surface. However, when the vapor pressure is equal to atmospheric pressure, stable bubbles of vapor can form throughout the bulk of the liquid, rather than only at its surface, and the liquid is said to **boil.**

There is no reason why boiling cannot occur at any temperature. All that is required is that the liquid's vapor pressure be equal to the external pressure. Atmospheric pressure decreases with altitude, so the higher one is above sea level, the lower the temperature of boiling. The boiling point of water in New York City is 100°C, but, at the top of Mt. Everest in the Himalaya Mountains, it is 70°C. Boxed cake mixes purchased in Denver, Colorado, will be culinary disasters when baked in Pensacola, Florida, because the recipe used in Denver must be designed for cooking at high altitude—in other words, at atmospheric pressures lower than that at sea level.

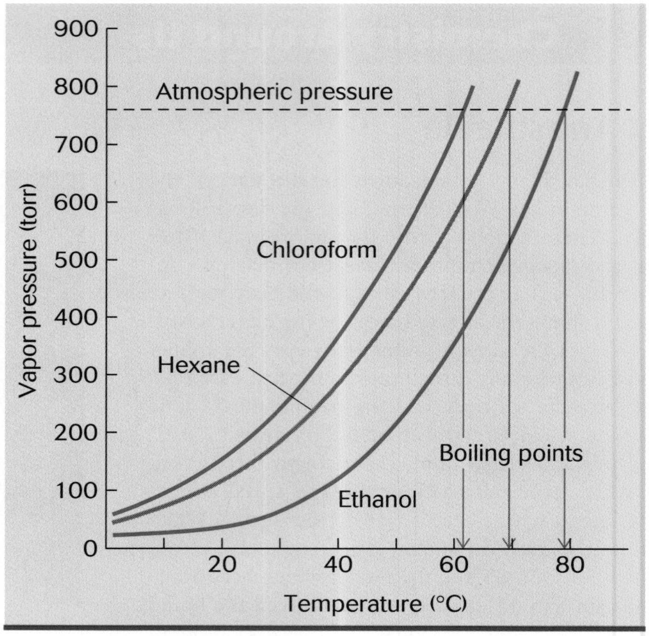

Figure 6.12 Vapor pressure of chloroform ($CHCl_3$), hexane (C_6H_{14}), and ethanol (C_2H_5OH) as a function of temperature. The horizontal dashed line represents atmospheric pressure of 1 atm. When the vapor pressure is equal to the atmospheric pressure, the liquids boil.

Because of the variations in atmospheric pressure and their effects on boiling points, chemists have found it helpful to define the **normal boiling point.** This point is the temperature at which boiling occurs under an external pressure of exactly 1 atm. Figure 6.12 shows changes in vapor pressure with changes in temperature for three compounds of varying volatility: chloroform ($CHCl_3$), hexane (C_6H_{14}), and ethanol (C_2H_5OH). As the temperature increases, the vapor pressures increase to a different extent for each compound. Of the three liquids, chloroform has the highest vapor pressure (is the most volatile), then hexane, and then ethanol. The dashed horizontal line represents atmospheric pressure. Boiling points are inversely related to vapor pressure: the more volatile a substance, the lower its boiling point.

The normal boiling point is a good qualitative indication of the secondary forces within a liquid. Differences in normal boiling points among liquids are due to differences in those forces within the liquids. The data in Table 6.2 illustrate this effect. The boiling points can be seen to increase with increasing strength of secondary forces.

Example 6.6 Reading a vapor-pressure-versus-temperature curve

Estimate the normal boiling points of the three liquids whose pressure-versus-temperature relations appear in Figure 6.12.

Solution

The normal boiling point is the temperature at which the vapor pressure of a liquid reaches atmospheric pressure. The temperatures in Figure 6.12 that correspond to that pressure for each liquid are: chloroform, approximately 62°C; hexane, approximately 69°C; and ethanol, approximately 78°C.

Problem 6.6 Estimate the vapor pressures at 50°C of the three liquids whose pressure-versus-temperature relations appear in Figure 6.12.

Concept check

✔ The normal boiling point is the temperature at which vapor forms throughout a liquid at an external pressure of exactly 1 atm.

Example 6.7 Relating vapor pressure to melting and boiling points

The following table lists the normal boiling points and melting points of two substances.

	Melting Point	Boiling Point
Substance A	−15°C	60°C
Substance B	−60°C	−10°C

(a) Compare the vapor pressure of substance A at 60°C with that of substance B at −10°C. (b) Then compare the vapor pressures of substances A and B when both substances are at −12°C.

Solution

(a) Because 60°C and −10°C are the respective normal boiling points of the two liquids, their vapor pressures must be equal to atmospheric pressure, and therefore equal to each other, at those temperatures.

(b) At −12°C, substance B is in the liquid state and close to its normal boiling point. Substance A also is in the liquid state but barely above its melting point and far from its normal boiling point. Therefore, at −12°C, the vapor pressure of liquid B must be greater than the vapor pressure of liquid A.

Problem 6.7 Methyl alcohol (CH_3OH) and methyl iodide (CH_3I) are both polar molecules, but methyl iodide's vapor pressure at room temperature is almost four times as great as that of methyl alcohol. What is the molecular basis for the difference?

Example 6.8 Using secondary forces to predict boiling points

Referring to the Lewis structures of H_2O (water) and H_2S (hydrogen sulfide), explain the much higher normal boiling point of water, 100°C, compared with −61°C for H_2S.

Solution

Hydrogen bonds can form between molecules when hydrogen is covalently bonded within the molecule to oxygen, nitrogen, or fluorine. Of these two compounds, only water can form H-bonds, whereas H_2S is limited to interacting through weaker secondary forces.

Problem 6.8 The structures for ammonia and phosphine are:

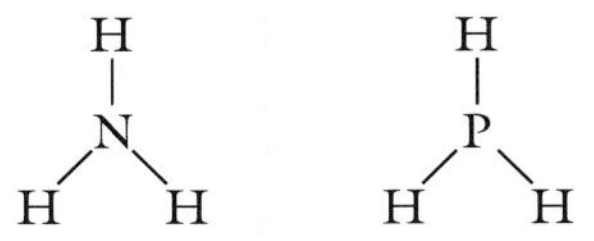

Ammonia Phosphine

Which has the higher boiling point?

6.8 VAPORIZATION AND THE REGULATION OF BODY TEMPERATURE

The regulation of body temperature is of vital importance to humans. It is achieved by a balance between the body's heat production and its ability to lose heat. Vaporization requires heat and therefore provides an important mechanism for heat loss. Water vaporizes from the body at the skin's surface

A PICTURE OF HEALTH

Temperature Regulation in the Body

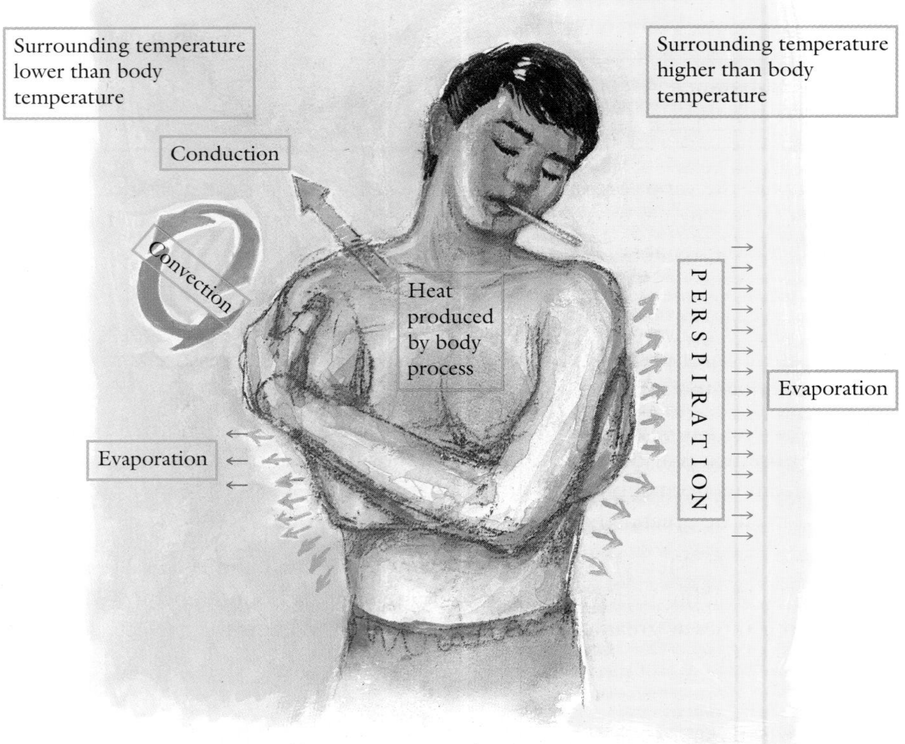

H_2O	9.73 kcal/mol
NH_3	5.59 kcal/mol
CH_4	1.96 kcal/mol

when we perspire, as well as from the lungs and mucous membranes of the mouth and nasal passages, removing heat from the body in the process.

The heat loss from the body through the vaporization of water is very efficient because of water's high heat of vaporization. That in turn is the result of hydrogen bonding among water molecules. When the molar heats of vaporization of water, ammonia, and methane are compared (see margin), the amount of heat required for the vaporization of water is about twice as much as that for another hydrogen-bonding molecule, ammonia, and five times that for nonpolar methane. At comfortable room temperatures, water loss from vaporization of a sedentary person might be only 600 mL/day, or about 350 kcal/day. Water loss due to sweating as a result of heavy work or running can be as high as 4 L/h, equivalent to as much as 2000 kcal. However, the rate of vaporization of body water can be seriously affected by the relative humidity, a factor discussed next.

Water vapor is one of the most important gases in Earth's atmosphere, but its concentration varies greatly from place to place. Water's atmospheric partial pressure may be as low as 0.5 torr in certain desert regions and as high as 55 torr in the tropics. That last figure is close to the equilibrium vapor pressure of water (that is, the vapor pressure of water in a closed system). Because the atmosphere is not a closed system, water's vapor pressure in the atmosphere must be expressed in relative terms. Thus we have the term **relative humidity,** defined as the ratio of the partial pressure of water vapor in the atmosphere to the equilibrium vapor pressure of water at the same temperature times 100. Mathematically:

$$\text{Relative humidity} = \left(\frac{\text{partial pressure of water}}{\text{equilibrium vapor pressure of water}}\right) \times 100\%$$

The equilibrium vapor pressure of water is 23.8 torr at 25°C. If the partial pressure of atmospheric water vapor is 12.8 torr, the relative humidity is

$$\text{Relative humidity} = \left(\frac{12.8}{23.8}\right) \times 100\% = 53.8\%$$

If the temperature goes down to 15°C, where water's equilibrium vapor pressure is 12.8 torr and if the partial pressure of atmospheric water vapor remains the same, the relative humidity is

$$\text{Relative humidity} = \left(\frac{12.8}{12.8}\right) \times 100\% = 100\%$$

At that temperature, water's partial pressure is equal to its equilibrium pressure. The equilibrium pressure represents the maximum concentration of water vapor at that temperature, and the air is said to be saturated with water vapor. The important point is that, under these conditions, the rate of condensation is equal to the rate of vaporization.

At environmental temperatures below body temperature, heat is transferred from our bodies to the environment in the same way that heat is transferred when a cup of coffee cools. Three mechanisms are responsible:

1. **Conduction.** Heat flows from a warm body to a cold one. If you are in contact with any substance (air, water, or a solid) that is at a lower temperature than your body, you will lose heat to that substance.
2. **Convection.** Air warmed by the body becomes less dense, rises, and is replaced by a flow of cooler, denser air.
3. **Evaporation.** Water from the body is vaporized and lost to the environment.

Conduction and convection are largely responsible for body-temperature control when the environmental temperature is below body temperature. However, when the environmental temperature approaches body temperature, convection and conduction begin to fail and we depend almost exclusively on the evaporation of water to control our body temperature.

We have all heard or used the phrase "It's not the heat, it's the humidity." This is an expression of the fact that, when environmental temperatures are near body temperature and water vapor partial pressures are near the equilibrium vapor pressure, we feel quite uncomfortable. For some people, these conditions are not only uncomfortable, but also dangerous. A recent short-lived heat wave in the Midwest caused the deaths of more than 100 elderly men and women. When the same temperatures are reached in the Southwest, very few are overcome by the heat; why, then, did so many die in the Midwest?

The answer is that the death-causing heat wave was accompanied by very high relative humidity, with the partial pressure of water vapor almost equal to the equilibrium vapor pressure of water. Under those conditions, the rate of condensation of water is close or equal to that for evaporation. The consequence is that any heat lost by evaporation is immediately gained back by condensation. Therefore no heat loss can be effected, and body temperature can rise to dangerous levels.

Air conditioning and electric fans would have allowed heat loss by conduction and convection, respectively. Not possessing either of these modern conveniences, the elderly poor succumbed rapidly to rising body temperatures.

6.9 ATTRACTIVE FORCES AND THE STRUCTURE OF SOLIDS

In contrast with the disorder of the gaseous state, solids are characterized by order. The atoms of a solid are arranged in regular patterns. Solids form crystals

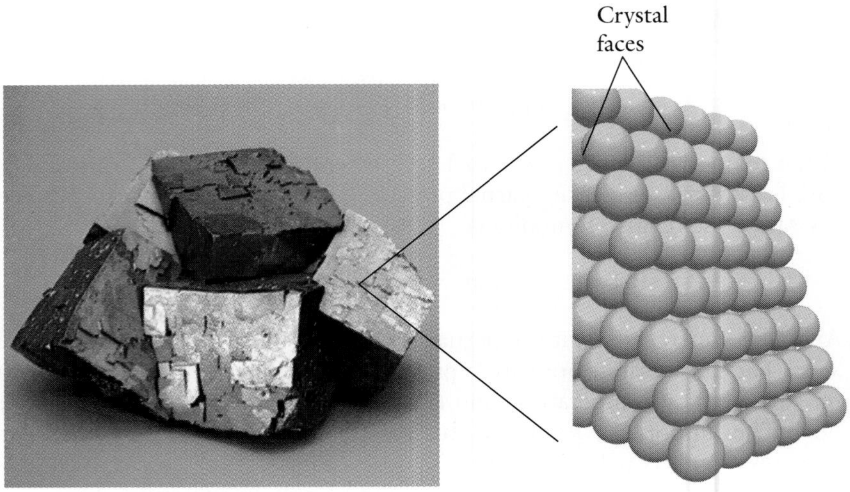

Figure 6.13 Crystalline solids have well-defined faces and edges. Each face is the exposed end of an orderly internal arrangement of atoms, ions, or molecules.

that have sharply defined melting points and can exist in geometrically well defined shapes, such as prisms, octahedrons, or cubes. The reason that crystalline solids have flat faces and well-defined edges is that each face is the exposed end of an orderly internal arrangement of atoms, ions, or molecules, as shown in Figure 6.13. Not all solids have an ordered crystalline form; those lacking it are called **amorphous.** For example, the noncrystalline form of SiO_2, called glass, is an amorphous solid. Figure 6.14a contains a photograph of the crystalline form of silica, SiO_2, or quartz, and a two-dimensional sketch showing how the atoms are arranged in an orderly network; Figure 6.14b shows a glass vessel formed when molten quartz is allowed to solidify, along with a sketch showing how the atoms now form a disorderly array.

The melting point is one of the most useful clues available for identifying the types of attractive forces operating in a solid. (Remember that the melting point and freezing point are identical, they are merely approached from opposite directions.) Table 6.3 lists several compounds along with their melting (freezing) points. The highest, by far, belongs to diamond, a form of elemental carbon in which the atoms are held in place in the crystal by primary covalent bonds. Quartz is a naturally occurring form of silicon dioxide also held to-

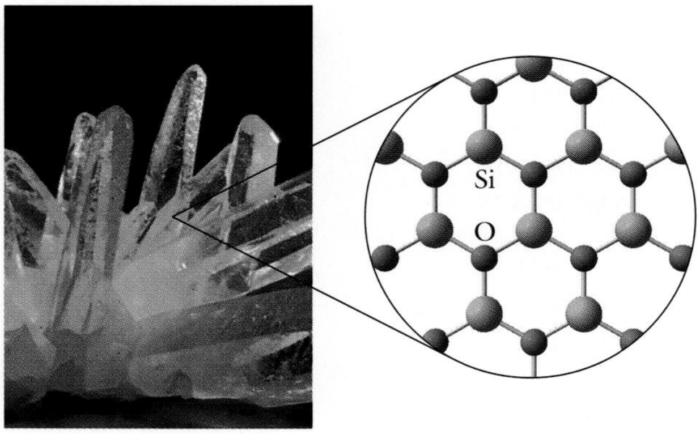

(a)

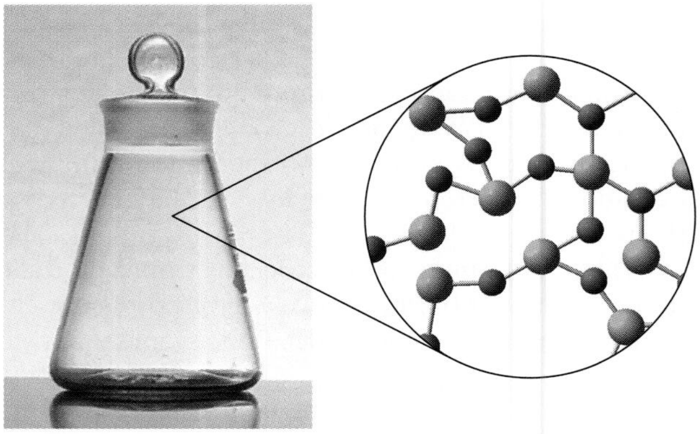

(b)

Figure 6.14 (a) The crystalline form of silica (quartz), and the orderly way in which its atoms are arranged. (b) The amorphous form of silica, called glass, and the disorderly array of its atoms.

TABLE 6.3 Melting Points of Covalent, Ionic, and Molecular Crystals

Substance	Solid State	Melting Point (°C)
diamond (carbon)	covalent crystal	3550
quartz (SiO_2)	covalent crystal	1610
sodium chloride	ionic crystal	800
calcium chloride	ionic crystal	782
citric acid	molecular crystal	153
cocaine (crack)	molecular crystal	98
chloroform	molecular crystal	−64
ethanol	molecular crystal	−117

gether by covalent bonds. Diamond and quartz are therefore called **covalent crystals.** Next in order are NaCl and $CaCl_2$, whose crystals are formed by primary ionic chemical bonds. Substances such as NaCl and $CaCl_2$ are therefore called **ionic crystals.**

Molecules such as ethanol, citric acid, cocaine, or chloroform are held together in the solid state by the much weaker secondary forces—that is, hydrogen bonds, dipole–dipole forces, and London forces—and, in the solid state, are called **molecular crystals.** The melting points in Table 6.3 illustrate the relative strengths of the attractive forces within these three different kinds of crystalline solids.

▶▶ Chapters 20 and 21 describe how secondary forces decide the shapes of proteins and DNA.

Example 6.9 Using molecular structure to predict physical properties: II

Arrange the order of melting points for the following solids:

$$CH_4 \quad C_6H_{14} \quad C_2H_6 \quad C_4H_{10}$$

Solution

All these compounds are hydrocarbons. They do not have permanent dipoles, nor can they form hydrogen bonds. They therefore can interact only by London forces. The greater the molecular mass, the more electrons, the larger the temporary dipoles, and the larger the secondary forces. The order of melting points from highest to lowest is therefore

$$C_6H_{14} > C_4H_{10} > C_2H_6 > CH_4$$

Problem 6.9 Arrange the order of melting points for the following solids, highest to lowest: (a) Br_2; (b) I_2; (c) F_2; (d) Cl_2.

Concept checklist

- ✔ Crystalline solids have an orderly internal structure.
- ✔ The internal structure of amorphous solids is disordered.

Under ordinary environmental conditions—for example, normal atmospheric pressure and room temperature—some solids can vaporize directly to the gaseous state without melting to an intermediate liquid state. Examples are solid CO_2 (dry ice); solid I_2, a solid that is always present in its container along with some purple vapor; and *p*-dichlorobenzene, a solid whose distinctive odor, as it vaporizes, is used to keep moths away from clothing. The direct transition from solid to gas is called **sublimation.** The energy required to transform the solid into a gas is about the same as it would be if the solid had gone through two phase transitions, melting and vaporization, instead of just one.

In preceding chapters, chemical bonds and the nature of molecules were the subjects of interest. In this chapter, the objective has been to broaden the view to include interactions between molecules. These interactions are far weaker than chemical bonds but are of central importance in defining the physical characteristics of pure substances. Of more interest to those who intend to work in the life sciences, these weak interactions are of paramount importance in all aspects of biochemical structure and function.

Summary

The Three States of Matter and Transitions Between Them The physical differences between the three states of matter are due to different degrees of molecular organization. The molecules themselves retain their sizes and shapes whether in the solid, liquid, or gaseous state. Molecules in the solid state are the most compact and highly organized. Liquids are almost as compact but are less organized, whereas both the amount of space between molecules and the degree of disorder are maximal in gases.

Phase transitions, conversions between physical states, occur when the balance between kinetic energy and secondary forces is changed by the addition or removal of heat. Melting, the conversion from solid into liquid, occurs when only a percentage of the secondary forces are overcome. Vaporization, the conversion from liquid into gas, occurs when all secondary forces are overcome. In contact with its liquid form, the resulting gas is called a vapor and exerts a pressure called the vapor pressure.

Vapor pressure increases with increase in temperature. Vaporization and condensation occur simultaneously. When the two processes proceed at the same rate, the amounts of liquid and gas (vapor) remain constant. A system that does not change over time as the result of two opposing and active processes, is said to be in dynamic equilibrium.

Attractive Forces Between Molecules Attractive forces between molecules are called secondary forces. Three principal secondary forces are associated with the formation of liquids and solids: London forces, dipole–dipole forces, and hydrogen bonds. The relative strengths of these forces are in the order hydrogen bonding > dipolar attractions > London force.

The ability of different substances to mix at the molecular level, forming solutions, depends on the nature of the secondary forces within each substance. If the secondary forces in one substance are of the same type as those in another substance, it is likely that the two substances will mix to form a solution. These considerations are the basis of the rule of thumb that "like dissolves like."

In contrast with the disorder of the gaseous state, the atoms of a solid are arranged in regular patterns. Solids form crystals that have sharply defined melting points and possess geometrically well-defined shapes.

Key Words

boiling point, p. 164
condensation, p. 150
crystalline solid, p. 168
dipole–dipole force, p. 151
dynamic equilibrium, p. 162
equilibrium state, p. 162
freezing, p. 150
hydrogen bonding, p. 153
kinetic energy, p. 150
liquid, p. 161
London force, p. 151
melting, p. 150
molar heat of fusion, p. 150
molar heat of vaporization, p. 151
secondary force, p. 151
solid, p. 167
solution, p. 159
surface tension, p. 157
vaporization, p. 150
vapor pressure, p. 161

Exercises

Solids, Liquids, and Gases

6.1 Describe the condition of matter if there were no secondary forces operating between molecules.

6.2 Describe the condition of matter if the secondary forces operating between molecules were as strong as covalent or ionic bonds.

6.3 Propose a molecular mechanism for the melting of a solid.

6.4 Propose a molecular mechanism for the evaporation of a liquid.

6.5 True or False: The vapor pressure of a liquid increases with temperature. Explain your answer.

6.6 True or False: The vapor pressure of a boiling liquid in an open beaker increases with temperature. Explain your answer.

6.7 Why are solids and liquids virtually incompressible?

6.8 Why are gases so much more compressible than liquids or solids?

6.9 What is the relation of the molar heat of fusion to the molar heat of vaporization?

6.10 What is the molecular explanation for your answer to Exercise 6.9?

6.11 True or False: The melting point of a solid is the same as its freezing point. Explain your answer.

6.12 True or False: The normal boiling point of a liquid is the temperature at which its vapor pressure is equal to the atmospheric pressure. Explain your answer.

6.13 If it takes 9.7 kcal of heat to vaporize 1.0 mol of water, how much heat will be released when 1.0 mol of water vapor condenses into liquid water?

6.14 If it takes 1.44 kcal of heat to melt 1.0 mol of solid water (ice), how much heat will be released when 1.0 mol of liquid water freezes and forms ice?

Attractive Forces Between Molecules

6.15 How can molecules in the gaseous state condense into a liquid if their molecules are not polar?

6.16 Compare the boiling points of liquids whose molecules interact by hydrogen bonds with those whose molecules interact by London forces.

6.17 Between which of the following pairs of molecules can hydrogen bonds be formed?

(a) $\text{H}-\overset{\overset{\text{H}}{|}}{\underset{\underset{\text{H}}{|}}{\text{C}}}-\text{H}$ and $\text{H}-\text{O}-\text{H}$ 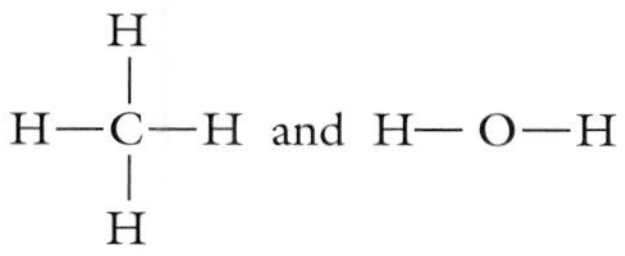

(b) $\text{H}-\text{O}-\text{H}$ and $\overset{\text{H}}{\underset{\text{H}\quad\text{H}}{\text{N}}}$ 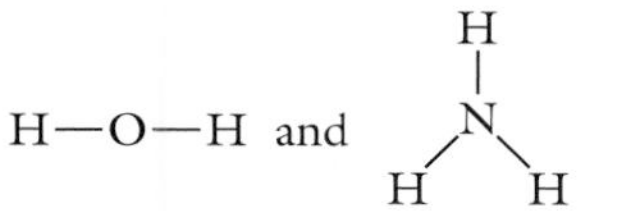

(c) $\text{Cl}-\overset{\overset{\text{Cl}}{|}}{\underset{\underset{\text{Cl}}{|}}{\text{C}}}-\text{Cl}$ and $\text{H}-\text{F}$ 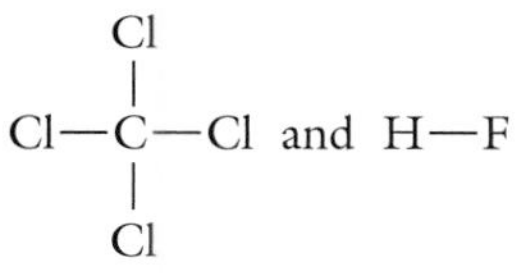

6.18 Between which of the following pairs of molecules can hydrogen bonds be formed?

(a) $\text{H}-\text{S}-\text{H}$ and $\text{Cl}-\overset{\overset{\text{H}}{|}}{\underset{\underset{\text{Cl}}{|}}{\text{C}}}-\text{Cl}$ 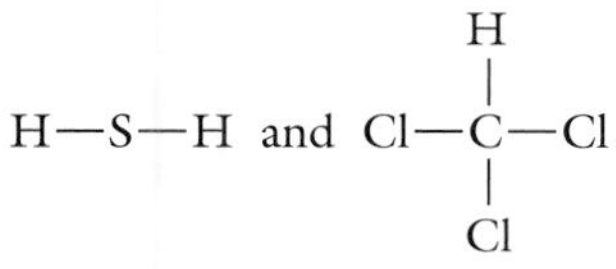

(b) $\text{H}-\overset{\overset{\text{H}}{|}}{\underset{\underset{\text{H}}{|}}{\text{C}}}-\text{H}$ and $\text{H}-\overset{\overset{\text{H}}{|}}{\underset{\underset{\text{H}}{|}}{\text{C}}}-\overset{\overset{\text{H}}{|}}{\underset{\underset{\text{H}}{|}}{\text{C}}}-\text{H}$ 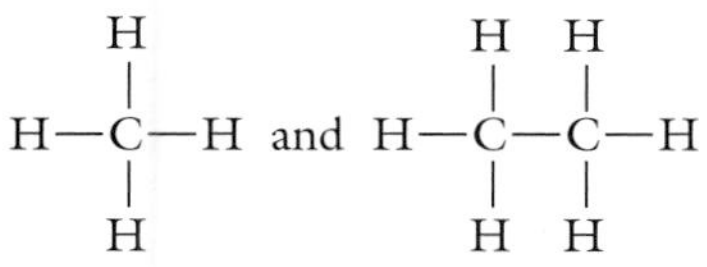

(c) $\text{H}-\text{F}$ and $\overset{\text{H}}{\underset{\text{H}\quad\text{H}}{\text{N}}}$ 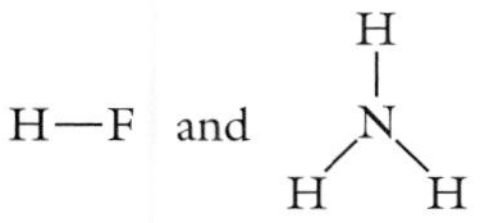

6.19 What is the relation between molecular mass and the strength of the London forces exerted between molecules?

6.20 Arrange the boiling points of the following compounds in the order from highest to lowest.

$$\text{C}_8\text{H}_{18} \qquad \text{C}_6\text{H}_{14} \qquad \text{C}_3\text{H}_8 \qquad \text{C}_4\text{H}_{10}$$

6.21 Why is it that carbon tetrachloride, CCl_4, a molecule with four polar covalent bonds, has no dipole moment?

6.22 Why is it that carbon dioxide, CO_2, a molecule with two polar covalent bonds, has no dipole moment?

6.23 Fill in the following table with a yes or no answer to the question, "Does this secondary force operate between the molecules of the given compound?"

	London Force	Dipole-Dipole	Hydrogen Bond
CH_4	_____	_____	_____
CHCl_3	_____	_____	_____
NH_3	_____	_____	_____

6.24 Fill in the following table with a yes or no answer to the question, "Does this secondary force operate between the molecule of the given compound?"

	London Force	Dipole-Dipole	Hydrogen Bond
CCl_4	_____	_____	_____
CHF_3	_____	_____	_____
H_2O	_____	_____	_____

6.25 Which of the following compounds has the greater heat of vaporization?

(a) $\text{H}-\overset{\overset{\text{H}}{|}}{\underset{\underset{\text{H}}{|}}{\text{C}}}-\overset{\overset{\text{H}}{|}}{\underset{\underset{\text{H}}{|}}{\text{C}}}-\text{O}-\text{H}$ or (b) $\text{H}-\overset{\overset{\text{H}}{|}}{\underset{\underset{\text{H}}{|}}{\text{C}}}-\text{O}-\overset{\overset{\text{H}}{|}}{\underset{\underset{\text{H}}{|}}{\text{C}}}-\text{H}$

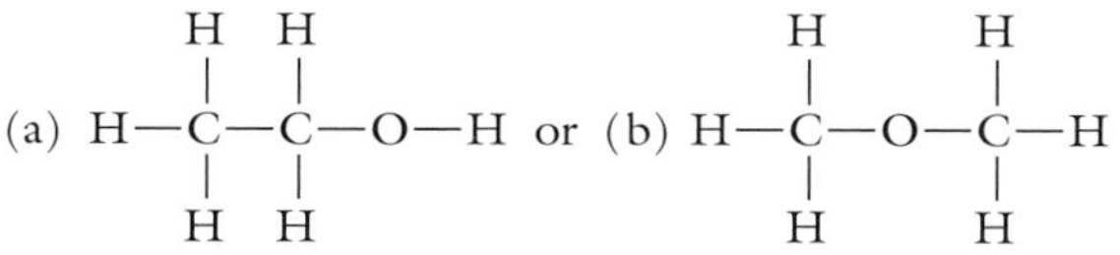

6.26 Which of the following compounds has the greater heat of vaporization?

(a) $\text{H}-\overset{\overset{\text{H}}{|}}{\underset{\underset{\text{H}}{|}}{\text{C}}}-\text{H}$ or (b) $\text{H}-\text{O}-\text{H}$

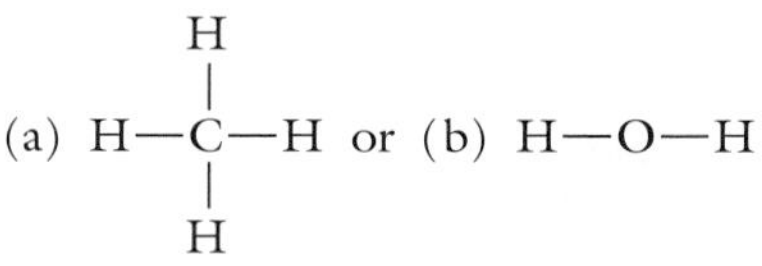

Liquids

6.27 Explain why a liquid can flow and assume the shape of its container.

6.28 Explain why a solid retains its shape and cannot assume the shape of its container.

6.29 Explain why water will continuously evaporate from an open container at room temperature.

6.30 If water continuously evaporates from an open container at room temperature, does it require energy? If it does, where does the energy come from?

6.31 From a molecular point of view, describe the relation between the volume of a solid and the volume of the resultant liquid.

6.32 From a molecular point of view, describe the relation between the volume of a liquid and its volume as a gas.

6.33 Why do we not use the word vapor to characterize oxygen or nitrogen?

6.34 Why do we use the word vapor to characterize water or ethanol?

6.35 Arrange the following compounds in order of increasing vapor pressure, lowest to highest: (a) H_2O; (b) KCl; (c) $CHCl_3$.

6.36 Arrange the following compounds in order of increasing vapor pressure, lowest to highest: (a) NaCl; (b) CH_4 (methane); (c) CH_3CH_2OH (ethanol).

6.37 Is it possible to measure the vapor pressure of a liquid in an open container? Explain your answer.

6.38 Is it possible to measure the vapor pressure of a liquid in a closed container? Explain your answer.

6.39 What is meant by equilibrium?

6.40 What is meant by a dynamic equilibrium?

6.41 What is the normal boiling point of a liquid?

6.42 Can a liquid boil at any pressure? Explain your answer.

6.43 Compare the surface tensions of water and chloroform, and explain any differences. The structures of water and chloroform are:

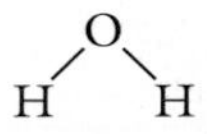
Water

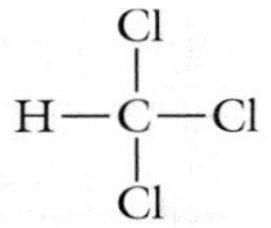
Chloroform

6.44 Compare the surface tensions of water and the hydrocarbon hexane, and explain any differences. The structures of water and hexane are:

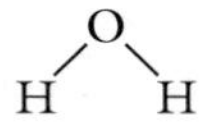
Water

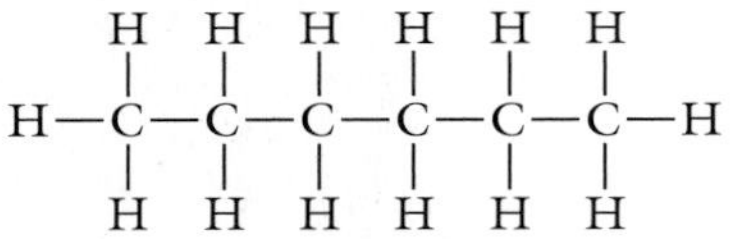
Hexane

6.45 Will either (a) acetic acid or (b) pentane form molecular mixtures with (dissolve in) water? Explain your answer. Their structures are:

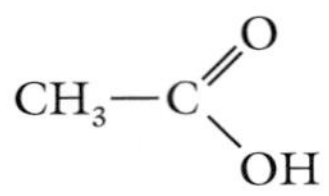
Acetic acid

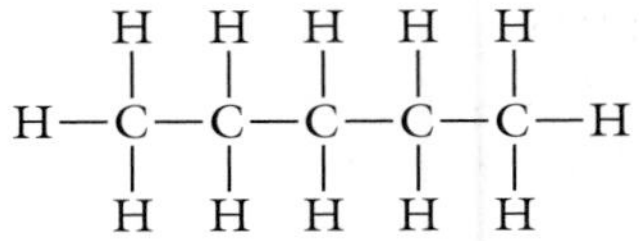
Pentane

Solids

6.46 What are the molecular characteristics and an important physical property of a crystalline solid?

6.47 What are the molecular characteristics and an important physical property of an amorphous solid?

6.48 Place the melting points of the following substances in the order from lowest to highest: KCl; H_2O; H_2S.

6.49 Place the melting points of the following substances in the order from lowest to highest: C_2H_5OH (ethanol); NaCl, CH_4 (methane).

6.50 The equilibrium vapor pressure of water at 75°F is 23.6 torr. The partial pressure of water in the atmosphere is 12.3 torr. Calculate the relative humidity.

6.51 The equilibrium vapor pressure of water at 86°F is 31.8 torr. The partial pressure of water in the atmosphere is 21.6 torr. Calculate the relative humidity.

Unclassified

6.52 Give three examples of molecules that interact exclusively by London forces.

6.53 Give three examples of molecules that have permanent dipole moments.

6.54 Why does it take so much longer to boil an egg in Denver than in San Francisco?

Chemical Connections

6.55 The boiling points of some hydrides of the first period are:

CH_4	NH_3	H_2O	HF
−161°C	−33°C	100°C	−19.5°C

Explain why water, not HF, has the highest boiling point, even though oxygen is less electronegative than fluorine.

6.56 Explain the use of alcohol rubs to lower body temperature.

6.57 The relation of strength of secondary force to vapor pressure is inverse. That is, the greater the secondary forces, the smaller the vapor pressure. Explain this phenomenon.

6.58 Why can you detect the presence of mothballs, composed of solid para-dichlorobenzene, without any sophisticated analytical instrumentation?

6.59 Explain why dimethyl ether, $CH_3—O—CH_3$, an apparently symmetrical molecule, has a dipole moment.

CHAPTER 7

SOLUTIONS

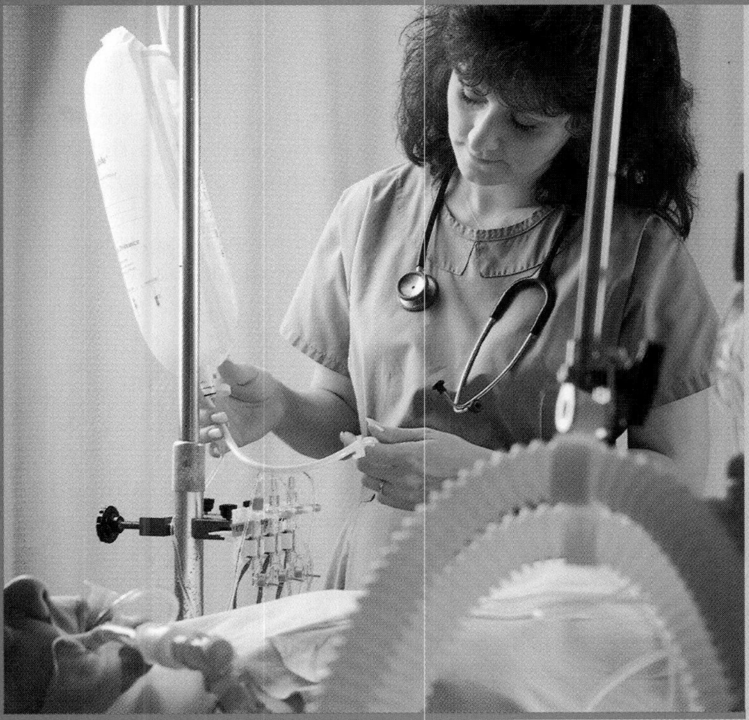

CHEMISTRY IN YOUR FUTURE

When a head-injury patient arrives at the hospital, one of the key concerns of the health-care team is to prevent or reduce excess fluid collection around the brain. Such swelling, or edema, is a natural response to injury, but when the injured organ is the brain, the added fluid pressure could cause severe damage. One way to deal with this problem is to administer an intravenous solution of mannitol, a water-soluble compound having no biological activity. Its only physiological effect is to increase the osmotic pressure of the filtrate in the kidneys' tubules, thus increasing the amount of fluid disposed of in the urine. What is osmotic pressure, and how does it affect body fluids? After reading this chapter on solutions, you will know.

LEARNING OBJECTIVES

- Describe how the formation of a solution depends on the molecular properties of the solute and solvent.
- Give the quantitative definitions of concentration and use them as conversion factors in calculations.
- Specify reasons and methods for preparing dilute solutions from concentrated solutions.
- Describe diffusion and the characteristics of semipermeable membranes from a molecular point of view.
- Describe the origin of osmotic pressure and how it is measured and used in calculations.
- Describe the properties of macromolecules and colloidal solutions.

Chapter 6 dealt with the properties of pure substances, but, in practice, most chemical reactions take place in solutions. A solution is a special kind of mixture in which substances are mingled so thoroughly that the separate components are no longer visible, even under the microscope. The composition of a solution is variable, and the components retain their chemical identities. However, the properties of a solution can be markedly different from the properties of the solution's components. This difference warrants special consideration of the properties of solutions.

It is difficult to find a biological process that does not take place in solution. The living cell itself consists of an aqueous medium containing a complex mixture of structures and dissolved substances. To the chemist, then, solutions offer a means of studying processes that could never be observed in the solid or gaseous states. But, to the biochemist, solutions can serve as experimental models that closely simulate conditions in the cell.

▶▶ Cell structures and components are presented in Chapter 22.

In this chapter, we will explore the qualitative and quantitative aspects of solution formation and some of the ways in which they are of central importance in biological systems.

7.1 GENERAL ASPECTS OF SOLUTION FORMATION

The simplest kind of solution is a mixture of two substances: the **solute,** the substance in smaller amount, dissolved in the **solvent,** the substance in larger amount. We have said that mixtures have no fixed recipe, so the components can be present in varying proportions. Nevertheless, there is generally a limit to the amount of solid solute that can be dissolved in a solvent. Gas mixtures are an exception to this general rule. Solvent and solute can be mixed in any proportions in all gas mixtures, as well as in certain liquid mixtures such as ethanol and water. When two or more substances can form solutions in all proportions, they are said to be **miscible.**

All sorts of substances can be made to mix with one another, but not all can be made to form solutions. If we categorize substances by physical state—gas, liquid, or solid—and mix the states two at a time, nine kinds of mixtures but only seven kinds of solutions are possible. The combinations are described in Table 7.1 as G/G (gas in gas), L/L (liquid in liquid), S/S (solid in solid), and so forth.

The pairs of states that cannot be combined to form true solutions are solids in gases (S/G) and liquids in gases (L/G). The S/G pairing describes smoke, and the L/G pairing describes mists or fogs; both these mixtures are called suspensions. A **suspension** consists of visible aggregates or particles suspended in a continuous medium. Sometimes the particles, although visible, are so small that gravity has a negligible effect on them and they remain suspended indefinitely. Particles that small are described as colloidal and will be considered separately later in this chapter.

7.2 MOLECULAR PROPERTIES AND SOLUTION FORMATION

Although many types of solutions are possible, liquid solutions—that is, solids in liquids or liquids in liquids—will be the primary objects of our interest. The physical properties of liquids and solids discussed in Chapter 6 were shown to depend on the secondary forces operating between molecules—that is, London forces, dipole–dipole forces, and hydrogen bonding. Solution formation also was shown to depend on the same secondary forces. The molecules of sol-

TABLE 7.1 The Nine Types of Mixtures Made from the Three States of Matter, with Examples

Mixture*	Example
G/G	air
G/L	carbonated water
G/S	hydrogen in steel
L/G	mist or fog
L/L	alcohol in water
L/S	dental amalgam, Hg in Ag
S/G	smoke
S/L	sugar in water
S/S	TV screen phosphors

* The letters G, L, and S stand for gas, liquid, and solid, respectively. The first letter denotes the solute, and the second is the solvent. For example, G/L means gas in liquid, and L/G means liquid in gas.

vent and solute must interact through secondary forces to form a solution. A solution will form when solute–solvent interactions are due to secondary forces closely similar to the solute–solute and solvent–solvent interactions. This idea is often stated simply as "like dissolves like." The interaction between any solute and solvent is described as **solvation.**

Water is a solvent in which solutions form because of strong, rather than weak, interaction between solvent and solute. Its polarity and its ability to form hydrogen bonds allow water to interact with, or solvate and dissolve, electrically charged substances, polar substances, or substances that can form hydrogen bonds (Section 6.2). The special case of solvation by water is called **hydration.** Solutions in which water is the solvent are called aqueous solutions. The hydration of ions or molecules is expressed by the notation *aq,* as in $Na^+(aq)$ and $Cl^-(aq)$. The hydration of Na^+ and Cl^- ions is illustrated schematically in Figure 7.1.

From a molecular point of view, what is the difference between a pure liquid such as water and a solution of some substance such as glucose in that same liquid? To answer this question, we must examine and compare the structures of glucose and water. The interactions between glucose molecules consist of hydrogen bonds between OH groups, the same type of bond as those within pure water. Therefore glucose will dissolve in water because similar attractive forces act within a mixture of the two substances.

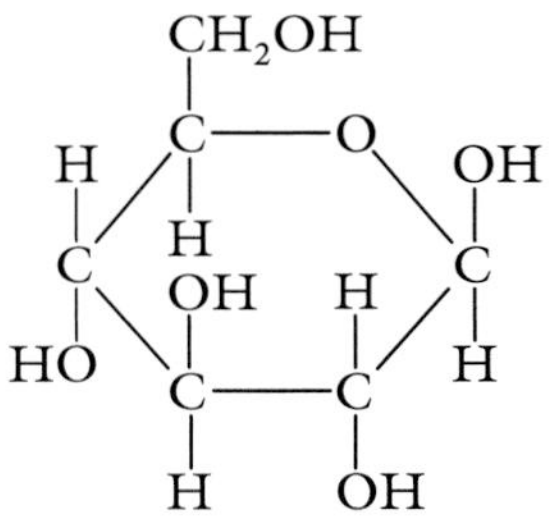

INSIGHT INTO PROPERTIES

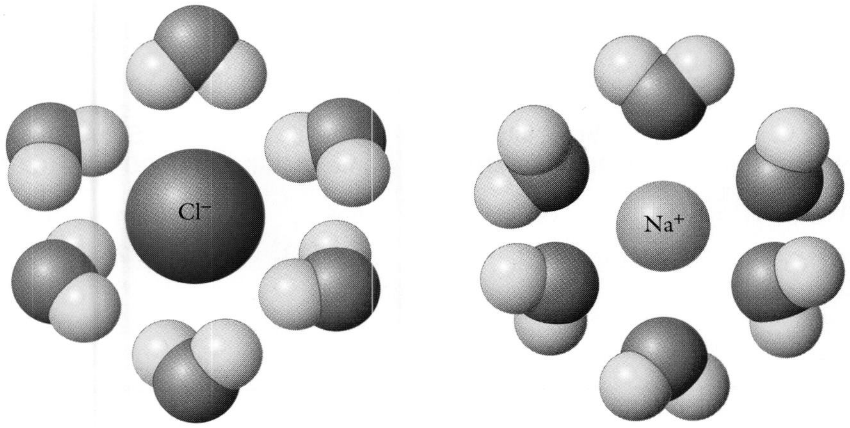

Figure 7.1 The solvation of sodium and chloride ions in water. The ions are surrounded by a shell of water molecules bound to the ions by secondary forces. This helps explain water's ability to dissolve so many ionic substances.

The difference between pure water and its glucose solution is that some of the water molecules have been replaced by glucose molecules. The glucose molecules are said to be hydrated—that is, to have water molecules closely attracted to them by secondary forces. But that situation is not significantly different from that in pure water, in which water molecules have water molecules attracted to them. Thus, in dilute aqueous solutions, polar solute molecules dissolve because some portion of the water originally present has been replaced by some other substance similar in structure or polarity or both.

▶▶ Chapter 13 considers the interactions between water molecules and organic molecules containing -OH groups.

7.3 SOLUBILITY

Solubility is a quantitative term that is often confused with the qualitative term soluble. The designation soluble is more properly reserved to describe substances that are able to dissolve in significant amounts in a given solvent, perhaps a few grams in 100 mL of solvent. For example, a tablespoon of sugar can be seen to dissolve rapidly in a beaker of water and so is considered to be water soluble.

For most substances, there is a limit to the amount that can be dissolved in a given amount of solvent. When less than that limit is dissolved, the solution is said to be **unsaturated.** When the limit is reached, the solution is said to be **saturated. Solubility** is a quantitative measure of the limiting amount of solute that will dissolve in a fixed amount of solvent at a specific temperature.

It is difficult to draw a line between soluble and insoluble, but some sense of the meaning and use of these descriptors can be gained from the data in Table 7.2. Although all the compounds listed in Table 7.2 dissolve in water, they clearly fall into two distinct categories with regard to the extent to which they are able to dissolve.

Chemists have developed a set of empirical rules that are quite useful for making qualitative predictions of whether an ionic substance will be soluble in water. They are encapsulated in Table 7.3; the general rule of solubility or insolubility for a given class of compounds is stated in the left-hand column, and important exceptions to that general rule are presented in the right-hand column.

7.4 CONCENTRATION

Concentration is a measure of the amount of solute dissolved in a given amount of solvent. Figure 7.2 demonstrates that concentration is a kind of

TABLE 7.2 The Quantitative Meaning of the Words Soluble and Insoluble

Soluble Salts		Insoluble Salts	
Name of Salt	Solubility (g/100 mL at 20°C)	Name of Salt	Solubility (g/100 mL at 20°C)
$BaCl_2$	35.7	$BaSO_4$	2.4×10^{-4}
BaI_2	203.1	$Ba(IO_3)_2$	2.2×10^{-2}
CaI_2	67.6	CaF_2	1.6×10^{-3}
$FeSO_4$	26.5	FeS	6.2×10^{-4}
$HgCl_2$	6.1	Hg_2Cl_2	2.0×10^{-4}

TABLE 7.3 Qualitative Solubility Rules for Ionic Solids in Water

Solubility	Important Exceptions
SOLUBLE	
Na, K, and NH_4 salts	
nitrates and acetates	
sulfates	Ca, Sr, and Ba
chlorides and bromides	Ag, Pb, and Hg(I) chlorides Ag, Pb, and Hg(I) bromides
INSOLUBLE	
hydroxides	Li, Na, K, Rb, Ca, Sr, and Ba
phosphates, carbonates, and sulfides	Li, Na, K, Rb, and NH_4^+

density, a measure of mass per unit volume. Figure 7.3 shows that, no matter how small or large an amount of a solution is observed, the amount of dissolved solute found in any fixed volume of it—the concentration—remains the same. It is defined mathematically as

$$\text{Concentration} = \frac{\text{amount of solute}}{\text{amount of solution}}$$

The amount of solute can be expressed either in physical mass units, such as grams, or in chemical mass units, such as moles. The amount of solvent can be expressed in units of either volume or mass—that is, liters or kilograms, respectively. For example,

$$\text{Concentration} = \frac{\text{grams of solute}}{\text{liter of solution}}$$

The effectiveness of drugs depends on their concentrations in the blood. If the concentration is too high, there will be toxic side effects; if it is too low, then the drug will be ineffective. The fluid balance of the body depends on the concentrations of blood components. In fact, there is literally no aspect of human physiology that can be adequately described without referring to concentration. In the next sections, we will more fully develop the quantitative concept of concentration.

Blood is depicted as "the mass transport system of the body" in Chapter 26.

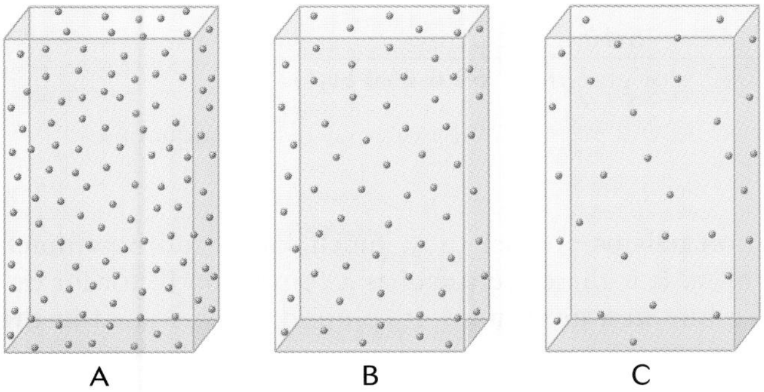

Figure 7.2 Visualizing molarity as a kind of molecular density. The concentrations and corresponding densities of these three solutions are in the order highest to lowest: A > B > C.

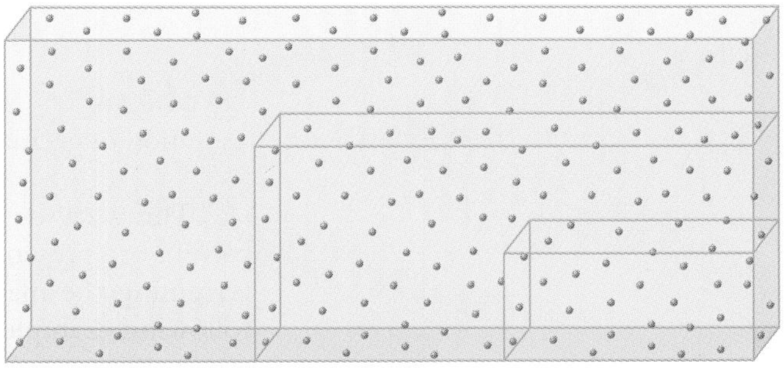

Figure 7.3 The concentration of a solution remains constant regardless of the volume of solution observed.

7.5 PERCENT COMPOSITION

The concept of percent composition of chemical compounds was introduced in Section 2.1. The concept can also be applied to the concentration of solutions. For that use, **percent concentration** is defined as 100% × concentration, or

$$\text{Percent concentration} = \frac{\text{amount of solute}}{\text{amount of solution}} \times 100\%$$

In this designation of concentration the amounts may be expressed either in mass or in volume or in both. It is important to note that the amount of solution in the definition consists of the sum of the amounts of solute plus that of the solvent.

w/w Percent Composition

When the compositions of both the solute and solvent are given in mass units, grams, the percent composition is designated as % w/w, meaning that the quantities of both the solute and solvent were determined by weighing on a balance. The concentration provides us with the unit-conversion factor for calculations aimed at the determination of solution composition.

Example 7.1 Preparing % w/w solutions

How would you prepare 75.0 g of a 7.20% w/w aqueous solution of glucose?

Solution

The problem is to calculate the mass of glucose required to add to water to make up 75.0 g of a 7.20% glucose solution. Mathematically, the percent composition of the solution is:

$$7.20\%\ \text{w/w} = \left(\frac{7.20\ \text{g of glucose}}{100\ \text{g of glucose solution}}\right) \times 100\%$$

The concentration of the glucose solution provides us with the unit-conversion factor in the form required to cancel out the glucose-solution units. To solve the problem, we multiply the given mass of solution by the conversion factor (the concentration):

$$75.0\ \cancel{\text{g of glucose solution}} \left(\frac{7.20\ \text{g of glucose}}{100\ \cancel{\text{g of glucose solution}}}\right) = 5.40\ \text{g of glucose}$$

The aqueous solution is prepared by adding 5.40 g of glucose to 69.6 g of water. Seventy-five grams of a 7.20% w/w aqueous glucose solution would therefore have the composition

$$\frac{5.40\ \text{g of glucose}}{5.40\ \text{g of glucose} + 69.6\ \text{g of } H_2O}$$

Problem 7.1 How would you prepare 200 g of a 5.50% w/w aqueous solution of gelatin?

The w/w designation tells us precisely how much solute and how much solvent are present by mass. It is therefore useful as a conversion factor for determining the mass of solute in a given mass of solution, as illustrated in the following example.

Example 7.2 Determining mass of solute in a % w/w solution

Calculate the number of grams of glucose present in 1.00 g of a 3.00% w/w aqueous solution of glucose.

Solution

First, write the 3.00% w/w concentration as a conversion factor:

$$\text{Concentration} = \frac{3.00 \text{ g of glucose}}{100 \text{ g of solution}}$$

The next step is to use the conversion factor to calculate the number of grams of glucose in 1.00 g of solution:

$$1.00 \cancel{\text{ g of solution}} \left(\frac{3.00 \text{ g of glucose}}{100 \cancel{\text{ g of solution}}}\right) = 0.0300 \text{ g of glucose}$$

Problem 7.2 Phenol is a strong germicide. Calculate the number of grams of phenol present in 1.0 g of a 0.20% w/w aqueous solution.

Suppose you need 5.0 g of NaCl for the preparation of an isotonic saline solution, and all that you have available is several liters of a 10% w/w NaCl solution. The next example illustrates how to determine what mass of a given % w/w solution contains a desired mass of solute.

Example 7.3 Determining the volume of solution containing a desired mass of solute

Calculate how many grams of a 4.50% w/w lactate solution are needed to provide 9.00 g of lactate, a product of glucose metabolism that is used to prepare isotonic saline solutions.

Solution

There are 4.50 g of lactate in 100 g of solution. Using that relation as a unit-conversion factor gives

$$9.00 \cancel{\text{ g of lactate}} \left(\frac{100 \text{ g of solution}}{4.50 \cancel{\text{ g of lactate}}}\right) = 200 \text{ g of solution}$$

Problem 7.3 How many grams of a 6.30% w/w solution of $CaCl_2$ must be used to obtain 27.6 g of $CaCl_2$?

v/v Percent Composition

The v/v designation means that the parts in the definition of percent composition are measured in units of volume. The v/v designation is the most practical way to characterize a solution of liquids in liquids. The following example describes the preparation of a solution of ethanol in water.

Example 7.4 Preparing % v/v solutions

How would you prepare an aqueous 8.0% v/v solution of ethanol?

Solution

The v/v designation means that the amounts of both solvent and solute are expressed in volume units—for example, in milliliters:

$$\% \text{ v/v} = \frac{\text{mL of liquid solute}}{100 \text{ mL of solution}} \times 100\%$$

An 8.0% v/v ethanol solution in water is therefore

$$8.0\% \text{ v/v} = \frac{8.0 \text{ mL of ethanol}}{100 \text{ mL of solution}} \times 100\%$$

The solution is prepared by adding 8.0 mL of ethanol to a small amount of water and then adding enough additional water to produce 100.0 mL of solution.

Problem 7.4 How would you prepare an aqueous 4.8% v/v solution of acetone?

It is important to note that the amount of water used in preparing the ethanol solution of Example 7.4 will not be 92 mL (even though 100 mL of aqueous solution was prepared by using 8 mL of ethanol). When liquids are mixed together, their final combined volume does not always equal the sum of the individual volumes. In regard to ethanol and water, 50 mL of ethanol added to 50 mL of water results in a solution of about 95-mL volume. A change in the volume of a mixture is caused by the nature of the interaction between the molecules of solvent and solute.

Concept check

✔ The v/v designation allows one to know the volume of solute but not the volume of solvent added to it.

w/v Percent Composition

The most common method of preparing a solid in liquid solution in the laboratory is to weigh a sample of a solid, place it in a container calibrated to contain a fixed volume (say, 100 mL), add just enough solvent to dissolve the solid, and then add sufficient additional solvent to fill the container to the calibration mark. This procedure is illustrated in Figure 7.4.

The w/v percent designation means grams of solid per 100 mL of solution. Calculations can be handled in the same way as % w/w or % v/v problems, but the units do not cancel and the % w/v is not a true fraction, as are % w/w and % v/v. It is more useful to think of a 5% w/v solution of KCl as 5 g of KCl per 100 mL of solution, and we will express the conversion factor in those terms.

(a) (b) (c)

Figure 7.4 The preparation of 1.00 L of a solution containing 0.100 mol of $K_2Cr_2O_7$. The 0.100 mol of $K_2Cr_2O_7$ (29.4 g) is (a) weighed out and (b) added to and dissolved in a 1.00-L volumetric flask partly filled with water. (c) More water is added to bring the total volume up to the 1.00-L mark on the neck of the flask. The solution is continually swirled to ensure complete mixing.

Example 7.5 Determining mass of solute in a given volume of % w/v solution

Calculate the number of grams contained in 45 mL of a 6.0% w/v aqueous glucose solution.

Solution

The w/v designation means that the amount of solute is expressed in grams and the amount of solution (not solvent) is expressed in milliliters; therefore the conversion factor is:

$$\frac{6.0 \text{ g of glucose}}{100 \text{ mL of glucose solution}}$$

and the answer is:

$$45 \text{ mL of glucose solution} \left(\frac{6.0 \text{ g of glucose}}{100 \text{ mL of glucose solution}}\right) = 2.7 \text{ g of glucose}$$

Problem 7.5 Calculate the number of grams of NaCl contained in 55 mL of a 12% w/v aqueous solution of NaCl.

7.6 MOLARITY

Molarity is a measure of concentration that is based on chemical mass units. It is defined as moles of solute per liter of solution. Concentrations described in this way are called **molar** and are denoted by the symbol ***M***. For example, a 2.0 *M* solution is a 2.0 molar solution, a solution whose molarity is 2.0 mol/L. Three important contexts for the use of molarity as a measure of concentration are:

- Preparation of a solution of a specific molarity
- Calculation of how many moles of solute are contained in a certain volume of solution
- Calculation of the molarity of a certain volume of solution containing a specific mass of a solute in grams

The following examples explore each of these cases.

Example 7.6 Preparing molar solutions

How would you prepare 1.00 L of a 0.650 *M* aqueous solution of $CaCl_2$, and how many grams of $CaCl_2$ would be required?

Solution

To prepare 1.00 L of a 0.650 *M* solution, 0.650 mol of $CaCl_2$ must be dissolved in 1.00 L of water. The mass of $CaCl_2$ in grams equivalent to 0.650 mol can be calculated by using the formula mass of $CaCl_2$ as a conversion factor (Section 4.1):

$$0.650 \text{ mol of } CaCl_2 \left(\frac{111 \text{ g of } CaCl_2}{1 \text{ mol of } CaCl_2}\right) = 72.2 \text{ g of } CaCl_2$$

Weigh out 72.2 g of $CaCl_2$ and dissolve it in sufficient water to make 1.00 L of solution.

Problem 7.6 How would you prepare 500 mL of a 0.275 *M* aqueous solution of KH_2PO_4, and how many grams of KH_2PO_4 would be needed?

Example 7.7 Using molarity as a conversion factor: I

How many moles of solute are contained in 1.75 L of a 0.950 *M* solution?

Solution

In this case, rewrite the concentration as a conversion factor, 0.950 *M* = 0.950 mol/L, and multiply the given volume by the conversion factor in the form that will convert liters into moles:

$$1.75\ \text{L}\left(\frac{0.950\ \text{mol}}{\text{L}}\right) = 1.66\ \text{mol}$$

Problem 7.7 How many moles of solute are contained in 0.680 L of a 1.35 *M* solution?

Example 7.8 Using molarity as a conversion factor: II

What volume of a 1.68 *M* solution will contain 0.250 mol of solute?

Solution

Write the concentration term as a unit-conversion factor: 1.68 *M* = 1.68 mol/L. Invert the conversion factor to obtain the form that will convert moles into liters:

$$0.250\ \text{mol}\left(\frac{1.00\ \text{L}}{1.68\ \text{mol}}\right) = 0.149\ \text{L}$$

Problem 7.8 What volume of a 0.750 *M* solution will contain 1.25 mol of solute?

Example 7.9 Calculating the molarity of a solution

What is the molarity of a $CaCl_2$ solution in which 72.15 g of $CaCl_2$ is dissolved in 200.0 mL of solution?

Solution

Because molarity is defined as moles per liter, first calculate the number of moles in 72.15 g of $CaCl_2$ (Section 4.2), then convert 200.0 mL into liters, and finally calculate *M*.

Step 1. $$72.15\ \text{g of CaCl}_2\left(\frac{1\ \text{mol}}{111.0\ \text{g of CaCl}_2}\right) = 0.6500\ \text{mol of CaCl}_2$$

Step 2. $$200.0\ \text{mL}\left(\frac{1\ \text{L}}{1000\ \text{mL}}\right) = 0.2000\ \text{L}$$

Step 3. $$\frac{0.6500\ \text{mol}}{0.2000\ \text{L}} = \frac{3.250\ \text{mol}}{1.000\ \text{L}} = 3.250\ M$$

Problem 7.9 What is the molarity of a KCl solution in which 20.6 g of KCl is dissolved in 920 mL of solution?

Solution concentration is often expressed in terms of **millimoles per milliliter,** abbreviated **mmol/mL.** Just as 1 mL is 1/1000th of a liter, 1 mmol is 1/1000th of a mole. Therefore, 1000 mmol = 1 mol and

$$\frac{1\ \text{mmol}}{\text{mL}} = \frac{1\ \text{mmol}\left(\dfrac{1\ \text{mol}}{1000\ \text{mmol}}\right)}{1\ \text{mL}\left(\dfrac{1\ \text{L}}{1000\ \text{mL}}\right)} = \frac{1\ \text{mol}}{\text{L}}$$

For example, the concentration of a 4.0 *M* solution of NaCl expressed in millimoles per milliliter is:

$$4.0\ M = \frac{4.0\ \text{mol}}{\text{L}} = \frac{4.0\ \text{mmol}}{\text{mL}}$$

and the concentration of a solution of KCl containing 7.4 mmol in 2 mL of solution is:

$$\frac{7.4\ \text{mmol}}{2.0\ \text{mL}} = \frac{3.7\ \text{mmol}}{\text{mL}} = \frac{3.7\ \text{mol}}{\text{L}} = 3.7\ M$$

The formula masses of compounds are often expressed in milligrams and millimoles. From the preceding discussion, you can see that the formula mass in grams per mole can also be expressed as milligrams per millimole:

$$\frac{1\ \text{g}}{\text{mol}} = \frac{1\ \text{mg}}{\text{mmol}}$$

Example 7.10 **Calculating concentrations in millimoles per milliliter**

Express the concentration in millimoles per milliliter of a solution in which 0.0746 g of KCl is dissolved in 100 mL of solution.

Solution

First calculate the mass of KCl in milligrams, then convert into millimoles, and finally calculate the molarity in millimoles per milliliter.

Step 1. $$0.0746\ \text{g of KCl}\left(\frac{1000\ \text{mg}}{\text{g}}\right) = 74.6\ \text{mg of KCl}$$

Step 2. $$74.6\ \text{mg of KCl}\left(\frac{1\ \text{mmol of KCl}}{74.6\ \text{mg of KCl}}\right) = 1.00\ \text{mmol of KCl}$$

Step 3. $$\frac{1.00\ \text{mmol of KCl}}{100\ \text{mL}} = 0.0100\ \frac{\text{mmol}}{\text{mL}} = 0.0100\ M$$

Problem 7.10 Express the concentration in millimoles per milliliter of a solution in which 0.259 g of $Ca(OH)_2$ is dissolved in 175 mL of solution (Section 4.2).

7.7 DILUTION

Suppose an analytical procedure requires a $1 \times 10^{-6}\ M$ solution of a reagent. You might attempt to prepare it by weighing out 1×10^{-6} mol of the reagent, with the intention of dissolving it in a liter of solvent, but a mass that tiny, perhaps from 0.2 to 0.3 mg, is difficult to weigh out accurately. A more practical approach is to first prepare a more concentrated solution and then dilute it to the desired concentration.

Suppose we wish to prepare a 1% w/w aqueous solution from a 2% w/w solution. Because the desired concentration is less than the solution's current concentration, we do not need to do a calculation to know that the desired concentration must be obtained by dilution. That is, water must be added to some quantity of the concentrated solution to reduce its concentration. The key point to remember is that the mass of solute that is present in the concentrated solution will still be present in the diluted solution, but it will occupy a larger volume. For example, after an 8-ounce can of frozen orange juice is made up to 1 quart, the mass of orange juice in the mixture remains the same; it is just distributed over a larger volume.

Let's say the original mass was measured in grams or moles. Because there is no loss in mass of solute on dilution, we can write:

$$\text{grams}_{\text{initial}} = \text{grams}_{\text{final}}$$

or

$$\text{moles}_{\text{initial}} = \text{moles}_{\text{final}}$$

If we replace the mass in moles with an equivalent expression that is the result of multiplying molar concentration by volume—namely,

$$\text{mol} = \text{L}\left(\frac{\text{mol}}{\text{L}}\right)$$

it follows that

$$\text{L}_{\text{initial}}\left(\frac{\text{mol}}{\text{L}}\right)_{\text{initial}} = \text{L}_{\text{final}}\left(\frac{\text{mol}}{\text{L}}\right)_{\text{final}}$$

or

$$\text{volume}_{\text{initial}} \times \text{concentration}_{\text{initial}} = \text{volume}_{\text{final}} \times \text{concentration}_{\text{final}}$$

This expression can be simplified to

$$V_{\text{I}} \times M_{\text{I}} = V_{\text{F}} \times M_{\text{F}}$$

The volumes and masses can be in any units, as long as they are the same on both sides of the equation. Returning to the dilution of the 2.0% solution to a concentration of 1.0% and choosing 10 mL as the starting volume, we can now write

$$10 \text{ mL} \times 2.0\% = x \text{ mL} \times 1.0\%$$

Solving the equation for x mL gives

$$x \text{ mL} = \frac{10 \text{ mL} \times 2.0\%}{1.0\%} = 20 \text{ mL}$$

Notice how the concentration units cancel; the final volume must be 20 mL. We obtain this final volume by adding sufficient water to the original 10-mL solution to obtain a final volume of 20 mL. The result is the desired 1.0% solution.

Example 7.11 Determining the concentration of a diluted solution

What is the final volume of a 0.0015 M solution prepared by dilution of 100 mL of a 0.090 M solution?

Solution

Use the expression $V_{\text{I}} \times M_{\text{I}} = V_{\text{F}} \times M_{\text{F}}$, where $V_{\text{I}} = 100$ mL, $M_{\text{I}} = 0.090\ M$, $M_{\text{F}} = 0.0015\ M$, and $V_{\text{F}} = ?$.

$$100 \text{ mL} \times 0.090\ M = V_{\text{F}} \times 0.0015\ M$$
$$V_{\text{F}} = 6.0 \times 10^3 \text{ mL} = 6.0 \text{ L}$$

Problem 7.11 What is the final volume of a 0.00750 M solution prepared by dilution of 75.0 mL of a 0.180 M solution?

Example 7.12 Preparing a dilute solution

What initial volume of a 1.25 M solution must be used to prepare a final volume of 2.00 L of a 0.500 M solution?

Solution

Use the expression $V_{\text{I}} \times M_{\text{I}} = V_{\text{F}} \times M_{\text{F}}$, where V_{I} = unknown, $M_{\text{I}} = 1.25\ M$, $V_{\text{F}} = 2.00$ L, and $M_{\text{F}} = 0.500\ M$.

$$V_{\text{I}} \times 1.25\ M = 2.00 \text{ L} \times 0.500\ M$$
$$V_{\text{I}} = 0.800 \text{ L}$$

Problem 7.12 What volume of a 0.730 *M* solution must be obtained to prepare 1.36 L of a 0.270 *M* solution?

7.8 CONCENTRATION EXPRESSIONS FOR VERY DILUTE SOLUTIONS

An expression of concentration often used in the clinical laboratory is mg %, which is a % w/v designation. It is defined as

$$\text{mg }\% = \frac{\text{mg of solute}}{100 \text{ mL of solution}} \times 100\%$$

For example, the concentration of glucose in a person's blood from 3 to 4 h after a meal ranges from 70 to 110 mg %, and normal concentrations of calcium in the blood range from 8.5 to 10.5 mg %. The SI equivalent of mg % is mg/dL, which is the preferred concentration notation.

Example 7.13 Calculating the concentration of solute in a mg % solution

Creatine is produced as a result of muscle metabolism. Normal creatine concentrations in blood are reported as 1.5 mg %. Calculate the concentration of blood creatine in units of milligrams per milliliter.

Solution

The conversion factor derived from the 1.5 mg % concentration is:

$$\frac{1.5 \text{ mg of creatine}}{100 \text{ mL of blood}} = \frac{0.015 \text{ mg of creatine}}{1.0 \text{ mL of blood}}$$

Creatine concentration = 0.015 mg/mL of blood.

Problem 7.13 Assuming that the normal serum calcium concentration is 9.0 mg %, calculate its concentration in grams per milliliter.

Another convenient expression for describing very dilute aqueous solutions is the designation **parts per million,** or **ppm.** This is a weight-per-volume designation for parts of solute per million parts of solution. For example, a 0.0001% w/v solution of some salt may be written as 0.0001 g of salt per 100 mL of solution. This ratio can be converted into whole numbers by multiplying both numerator and denominator by 1×10^4. The results will be clearer if we rewrite numerator and denominator in scientific notation:

$$\frac{0.0001 \text{ g}}{100 \text{ mL}} = \frac{1 \times 10^{-4} \text{ g}}{1 \times 10^{2} \text{ mL}}$$

and then multiply numerator and denominator by 1×10^4:

$$\left(\frac{1 \times 10^{-4} \text{ g}}{1 \times 10^{2} \text{ mL}}\right)\left(\frac{1 \times 10^{4}}{1 \times 10^{4}}\right) = \frac{1 \text{ g}}{1{,}000{,}000 \text{ mL}}$$

The solution can then be seen to contain 1 gram of solute per 1 million milliliters of solution—that is, to have a concentration of one part per million, or 1 ppm.

It is more common to express parts per million as the number of milligrams of solute contained in a liter of solution. For example, 1 mg per liter can be expressed as 0.001 g per 1000 mL. Then,

$$\frac{1 \text{ mg}}{\text{L}} = \frac{0.001 \text{ g}}{1000 \text{ mL}}$$

Multiplying numerator and denomonator by 1×10^3 gives

$$\left(\frac{0.001\ \text{g}}{1000\ \text{mL}}\right)\left(\frac{1 \times 10^3}{1 \times 10^3}\right) = \frac{1\ \text{g}}{1{,}000{,}000\ \text{mL}}$$

So,

$$\frac{1\ \text{mg}}{\text{L}} = 1\ \text{ppm}$$

In similar manner, one part per billion is expressed as 1 ppb and in terms of mass per liter is:

$$1\ \text{ppb} = \left(\frac{1\ \text{g}}{1 \times 10^9\ \text{mL}}\right)\left(\frac{1 \times 10^{-6}}{1 \times 10^{-6}}\right) = \frac{1 \times 10^{-6}\ \text{g}}{1 \times 10^3\ \text{mL}} = \frac{1\ \mu\text{g}}{\text{L}}$$

Concept checklist

✔ One part per million = 1 ppm = $\frac{1\ \text{mg}}{\text{L}}$

✔ One part per billion = 1 ppb = $\frac{1\ \mu\text{g}}{\text{L}}$

7.9 THE SOLUBILITY OF SOLIDS IN LIQUIDS

The solubility of a solid describes the maximum amount, or mass, of it that will dissolve in a fixed volume of solvent at a particular temperature. For example, no more than 24.99 g of the amino acid glycine will dissolve in 100 g of water at 25°C. Solubility must be determined by experiment, as was done for the compounds plotted in Figure 7.5. A solution containing the maximum amount of a dissolved solid is said to be saturated at that temperature.

A saturated solution is a system at equilibrium, a condition defined in Section 6.6 as being the result of two opposing processes whose rates are identical. In a saturated solution, the two opposing processes are (1) the dissolving of the solid, in which ions or molecules leave the solid and enter the liquid

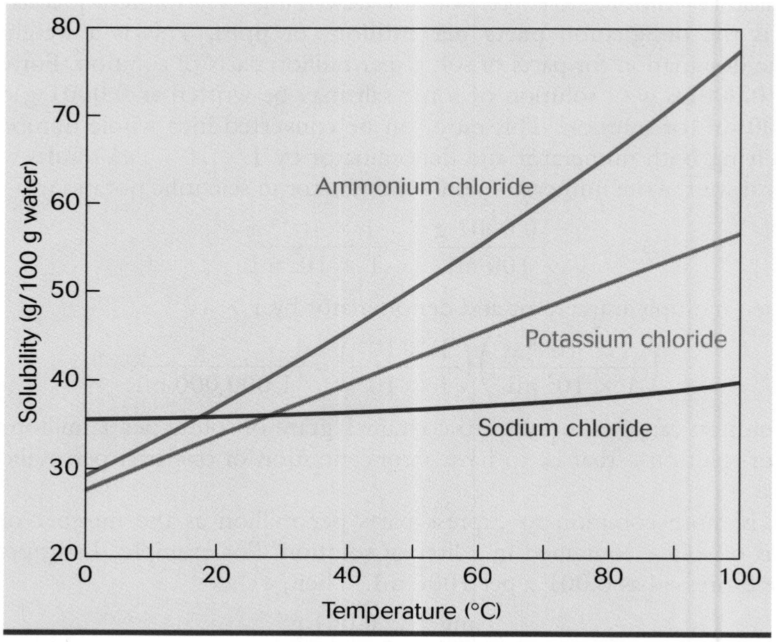

Figure 7.5 The solubilities of ammonium chloride, potassium chloride, and sodium chloride all increase with increase in temperature—however, each to a different extent.

phase, and (2) the crystallization of previously dissolved ions or molecules onto the surface of the solid. Let's think about what happens when enough of an ionic solid, such as potassium chloride, is mixed with water to form a saturated solution. The opposing processes can be represented by the following equation:

$$KCl\,(s) \underset{\text{crystallization}}{\overset{\text{solution}}{\rightleftharpoons}} K^+\,(aq) + Cl^-\,(aq)$$

where KCl(s) represents potassium chloride in the solid state and $K^+(aq)$ and $Cl^-(aq)$ represent hydrated ions.

At first, the rate of dissolving is greater than the rate of crystallization, but, as the concentration of dissolved ions increases, the rate of crystallization increases. When the two rates become equal, the solution is saturated. The maximum amount of dissolved solid (for that particular temperature) has been reached.

In the body, most nitrogen-containing compounds are transformed into urea, which becomes the principal solid component of urine. However, a certain group of nitrogen-containing compounds, the nucleotides (from which the genetic molecules RNA and DNA are synthesized; Chapter 21), undergo metabolic degradation to form uric acid. Under normal conditions, uric acid and its sodium salt are excreted in the urine. A small group of people suffer from a disorder in nucleotide metabolism that results in an overproduction of uric acid and leads to the clinical condition known as gout. Uric acid and its sodium salt are thousands of times less soluble than urea, and their overproduction rapidly saturates body fluids. In consequence, needlelike uric acid crystals are deposited in the cartilaginous structures of the joints, causing inflammation and great pain.

Figure 7.6 A solution of sodium chloride (NaCl) added to a solution of silver nitrate ($AgNO_3$) produces the white precipitate silver chloride (AgCl).

7.10 INSOLUBILITY CAN RESULT IN A CHEMICAL REACTION

When two ionic compounds are added to water, one of the reasons that new compounds can form is that one of the products is insoluble and forms a precipitate. Reactions of this type are shown in Figures 7.6 and 7.7. Reactions between $BaCl_2$ and K_2SO_4 or K_3PO_4 and $CaCl_2$, represented in Chapter 4 by

$$BaCl_2 + K_2SO_4 \longrightarrow BaSO_4 + 2\,KCl$$

and

$$3\,CaCl_2 + 2\,K_3PO_4 \longrightarrow Ca_3\,(PO_4)_2 + 6\,KCl,$$

are examples of this kind of reaction. They are stoichiometrically correct (balanced) but do not represent the actual process taking place in solution, because, in aqueous solution, all dissolved ionic compounds are present in the form of individual ions.

Let's examine the details of one of these reactions in which a precipitate is formed—the result of the addition of $BaCl_2$ to a solution of K_2SO_4. Tables 7.2 and 7.3 list soluble and insoluble salts and show (1) that both $BaCl_2$ and K_2SO_4 are ionic solids soluble in water and (2) that one of the products, $BaSO_4$, is an insoluble solid. We can represent the details of the process with the following equation, which is called the **complete ionic equation.**

$$Ba^{2+}(aq) + 2\,Cl^-(aq) + 2\,K^+(aq) + SO_4^{2-}(aq) \rightleftharpoons BaSO_4(s) + 2\,K^+(aq) + 2\,Cl^-(aq)$$

Again, the notations (aq) and (s) are used to denote the fact that the components of the system are in the form of hydrated ions and a precipitate.

Figure 7.7 A yellow precipitate of lead(II) iodide, PbI_2, forms immediately when colorless aqueous solutions of lead(II) nitrate, $Pb(NO_3)_2$, and potassium iodide, KI, are mixed.

Notice that the $K^+(aq)$ and $Cl^-(aq)$ ions appear unaltered on both sides of the equation. In this case, we can exclude such ions from the equation and consider only those ions that participate in the reaction:

$$Ba^{2+}(aq) + SO_4^{2-}(aq) \rightleftharpoons BaSO_4(s)$$

This is called the **net ionic equation.** The $K^+(aq)$ and $Cl^-(aq)$ ions omitted from the net ionic equation are called **spectator ions** because they do not take part in the reaction and would appear on both sides of the equation if included.

Example 7.14 Finding the net ionic equation

Write the net ionic equation that describes the result of adding silver nitrate, $AgNO_3$, to a solution of NaCl, a reaction in which a precipitate is formed. Refer to Table 7.3 for solubility data.

Solution

Table 7.3 shows that silver chloride is an insoluble salt. The reaction equation is:

$$Ag^+(aq) + NO_3^-(aq) + Na^+(aq) + Cl^-(aq) \rightleftharpoons AgCl(s) + Na^+(aq) + NO_3^-(aq)$$

The net ionic equation is:

$$Ag^+(aq) + Cl^-(aq) \rightleftharpoons AgCl(s)$$

Problem 7.14 Write (a) the complete ionic equation and (b) the net ionic equation for the reaction that takes place when calcium chloride, $CaCl_2$, is added to sodium phosphate, Na_3PO_4. (Refer to Table 7.3.)

7.11 DIFFUSION

Molecules free to move about a space will tend, in time, to distribute themselves uniformly throughout that space. To achieve such a uniform distribution, molecules move from regions of higher concentration to regions of lower concentration. This movement is called **diffusion.**

Diffusion occurs very rapidly in gases but much more slowly in liquids. In gases, comparatively few collisions between molecules occur, and progress is rapid. However, in liquids, diffusing molecules encounter and collide with large numbers of solvent molecules, and such collisions impede their progress. After adding sugar to your coffee or tea, you would never consider drinking it without first stirring. From experience, you know that it would take a long time for the sugar to become uniformly distributed throughout the drink. Would you be surprised to learn that it actually takes several months?

Figure 7.8 depicts events at the molecular level when a solute diffuses in a liquid. At first, the concentration of solute particles is greater on the left-hand side (in this representation) than on the right-hand side. The arrows are meant to indicate that the motion of individual solute particles is random; each particle is as likely to move in one direction as another. In this representation, only solute molecules are seen; the solvent is not shown.

In Figure 7.9, the molecular picture has been modified in two ways. First, the container of the solution has been divided into three imaginary compartments, labeled A, B, and C. Second, the arrows are redrawn so as to emphasize what, in this case, are the most relevant components of the particles' random motion—those movements that affect diffusion from left to right and cause particles to approach the compartments' imaginary boundaries directly from either side. There is no need to consider random motion in other directions.

Figure 7.9 represents our thought experiment at the start: more solute

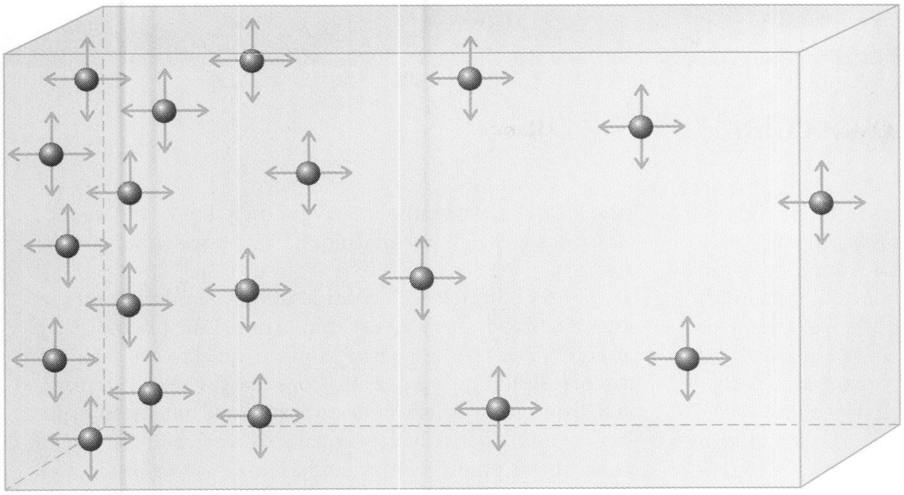

Figure 7.8 A molecular model illustrating random motion of solute molecules diffusing from left to right. Only back-and-forth and up-and-down motions are indicated.

molecules are in compartment A than in compartment B, which in turn contains more solute than compartment C. Because motion is entirely random, any solute molecule is just as likely to move to the right as to the left. Let us assume some convenient number of solute molecules in each compartment—say, 15 in A, 10 in B, and 5 in C. Some solute molecules in compartment A may leave A and enter B. The rate at which they leave is proportional to their number, so we may assign a rate at which they enter B as 15 per unit time. The rate at which solute leaves compartment B and enters A is proportional to their number, or 10 per unit time. If 10 solute molecules leave B but 15 molecules enter B, then, over time, solute will accumulate in compartment B. The same argument can be made for the movement of solute molecules between compartments B and C. The result is that, over time, solute molecules will appear to move from the left to the right, and, after sufficient time has passed, they will be uniformly distributed among the three compartments. It is important to recognize that molecules do not "know" that there is more room to their right into which to diffuse. Diffusion is a statistical phenomenon with its origin in the random motion of individual molecules. The rate of diffusion—that is, how fast diffusion occurs—depends on differences in concentration within a solution. The greater the concentration difference, the greater the

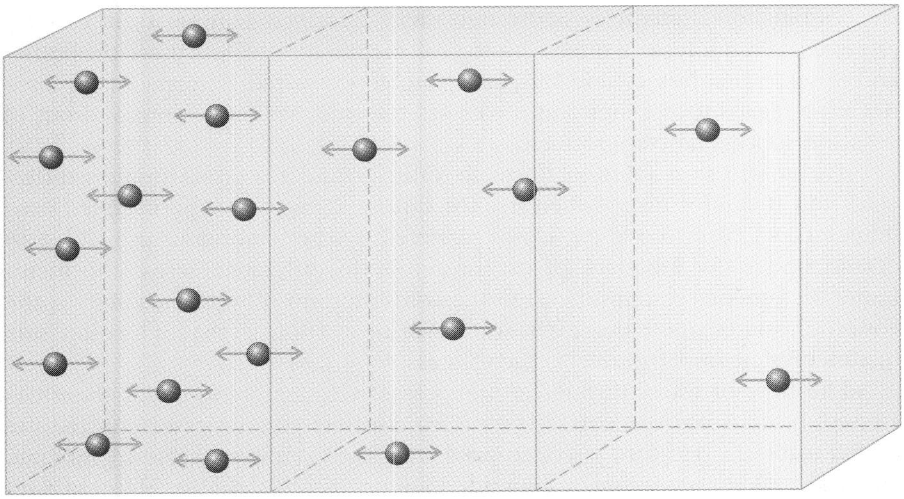

Figure 7.9 Random motion of solute molecules in a solution. The solute molecules are distributed in a nonuniform manner, and only the left-to-right random motions are shown.

7.1 Chemistry Within Us

Diffusion and the Cardiovascular System

As oxygen diffuses into cells, it is consumed by the numerous different metabolic processes taking place there. If all the oxygen is used up before some of it can diffuse to the most remote regions of the cell, certain aspects of metabolism will fail, and the cell will die. This puts a limit on the size to which a cell can safely grow, and in fact no cell of an oxygen-consuming organism, whether single cellular or multicellular, ever exceeds a thickness of about 0.5 mm. Some multicellular organisms, such as kelp and flatworms, transcend this limitation by growing in thin sheets so that all their cells are in contact with their oxygen-containing environment. In multicellular organisms such as humans, oxygen has a long way to go before it can even begin to diffuse into cells. This severe transport problem has been solved by the creation of a strong pump (the heart) that sends an oxygen-rich fluid (the blood) along a superhighway of "pipes," or vessels (the vascular system).

In the lungs, large quantities of oxygen diffuse rapidly into the blood across a very thin external membrane (lung alveoli). The blood is then pumped quickly through the vascular system, whose vessels constantly decrease in diameter from aorta to artery to arteriole and finally to capillary, a vessel approximately equal in diameter to the size of cells. At this point, the oxygen has only a short distance to diffuse to reach all parts of the metabolizing cells. Thanks to this elegant solution to the diffusion problem, oxygen travels from the outside world to all the cells of the body in a matter of seconds.

rate of diffusion. Gas and nutrient exchange between cells of the body and our environment constitutes a serious diffusion problem, which is discussed in Box 7.1.

7.12 OSMOSIS AND MEMBRANES

Membranes are sheetlike structures that can modify or control the molecular process of diffusion. Some possess holes, or pores, of molecular dimensions and regulate the passage of molecules on the basis of size: molecules larger than a certain diameter cannot pass, but those smaller than the limiting pore size will penetrate. Other membranes control diffusion by means of electrical charge: charge repulsions allow only negatively charged ions to pass through a positively charged membrane, and vice versa. Membranes that allow some substances but not others to pass through them are called **semipermeable.** They are essentially molecular filters. Biological membranes utilize these properties and others (Chapters 19 and 23), achieving an extraordinary array of specificities with regard to the ions and molecules that they will allow into and out of cells and subcellular compartments.

The solute in a solution normally diffuses under a concentration difference, but it cannot do so when it must diffuse across a semipermeable membrane. The solvent, however, is not hindered by the membrane and is free to diffuse under the influence of its concentration difference across the membrane. In aqueous systems, because the concentration of water increases as the concentration of solute decreases, water will always flow to the high-solute side of a membrane impermeable to the solute.

The flow of water through a semipermeable membrane under these circumstances is called **osmosis.** Figure 7.10 illustrates the process, with the use of a U-tube divided into two compartments by a semipermeable membrane. The right-hand compartment contains an aqueous 1.0% glucose solution, and the left-hand compartment contains only water (pure solvent).

PROFESSIONAL CONNECTIONS

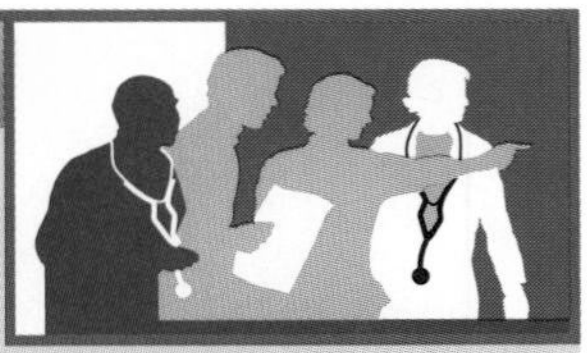

Beth Lerner Wilcox, R.N.
Long-Term Care for the Elderly

Beth Wilcox completed her first college-level chemistry course at Mount St. Mary's College in Los Angeles.

What is your current occupation?

I have been a registered nurse for 15 years. I chose nursing because I wanted to understand illness and support wellness, and I thought that I would enjoy the patients and their families. Health care is an environment in which both competence and compassion are important. It is rewarding because I can see the results of my efforts; we also have a lot of fun at work.

What are some of the jobs that you held in the past?

I currently work with the elderly whose needs require living in a long-term care facility. Previously, I worked in oncology (cancer treatment), including a bone marrow transplant unit.

Please describe a typical day on the job.

My day at St. Joseph Senior Community is spent assessing our clients, investigating any abnormal findings, and developing, as part of an interdisciplinary team, individualized care plans, which our caregivers will carry out. The data that I collect are sent to state and federal health agencies, which enables them to study long-term care trends.

Are any scientific or technological advances affecting the nature of your work?

Such advances affect the care of the elderly as they do all recipients of medical care; treatment options change as we learn more in the laboratory. Recent pharmacological developments come to mind, such as a drug for diabetics that enhances insulin acceptance across the cell membrane. Another drug, by slowing the degradation of acetylcholine, appears to slow the progression of Alzheimer disease.

Do you use chemistry in your job?

Because chemistry describes the building blocks of the human body, its concepts are inseparable from nursing assessment and care. For example, memory loss must never be assumed to be a "normal" part of the aging process, accepted without question; we may find that a treatable disease process or medication is responsible. The cause and treatment are based on principles of chemistry. In oncology, understanding the chemistry of treatment is essential to competent care and enables the nurse to help the patient understand his or her experience. One type of chemotherapy irritates the bladder wall during excretion, so another drug is added that binds with the irritating substance, enabling it to pass through the bladder without injury.

What advice do you have for health science students who are studying chemistry today?

I urge health science students to view chemistry as necessary to their careers as practitioners rather than merely a required course and to look for associations between what they are learning and "real life." Basic chemistry principles such as osmosis and electrolyte balance are evident throughout medical care and provide a solid foundation for assessment and problem-solving.

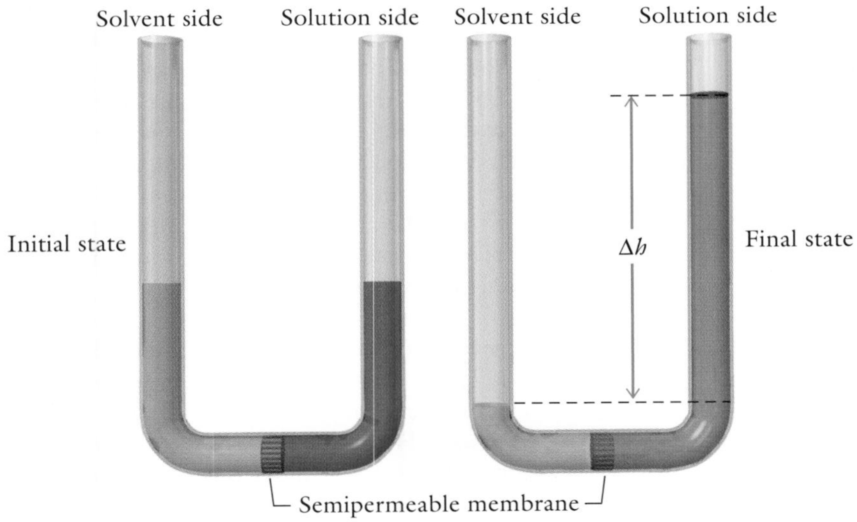

Figure 7.10 A U-tube manometer measures osmotic pressure. The right-hand compartment contains an aqueous 1.0% glucose solution, and the left-hand side contains only water (pure solvent). Osmosis causes water to cross the membrane into the solution compartment, which makes the liquid in the solution compartment rise to a higher level than the level in the solvent compartment. The difference in liquid levels results, in turn, in the generation of a force that causes water to flow back into the solvent compartment. When the liquid levels remain constant, the difference in the levels is a measure of the osmotic pressure.

Osmosis causes water to cross the membrane into the solution compartment, a movement of solvent that makes the liquid level in the solution compartment rise higher than the level in the solvent compartment. The difference in liquid levels results, in turn, in the generation of a force (the weight of the column of liquid) that causes water to flow back into the solvent compartment. Eventually, an equilibrium is established in which the inflow of solvent caused by the concentration difference becomes balanced by the outflow of solvent caused by the weight of the higher liquid level; at that point, the liquid levels no longer change.

The weight of the higher liquid column is measured as a pressure—the weight, or the force, per unit area. It is called a hydrostatic pressure because it is the result of the difference in water levels. In many cases, water flows from faucets at home because taps are at lower levels than the water at the reservoir. The difference in liquid levels leads to a pressure difference, hence flow.

7.13 OSMOTIC PRESSURE

The hydrostatic pressure is measured by the difference in height of the two liquid levels (Δh in Figure 7.10). Because the liquid levels do not change over time, the final pressures in both compartments must be equal. Therefore, the difference in liquid levels is also a measure of the opposing pressure generated by the concentration difference. That pressure is called the **osmotic pressure.**

The U-tube experimental system is a useful reminder that the flow of water through a semipermeable membrane can occur in two ways:

- Osmotic pressure, a pressure generated by a difference in concentration of dissolved substances separated by a semipermeable membrane
- Hydrostatic pressure, a pressure generated within a liquid by mechanical means.

A physiological example of hydrostatic pressure is the pressure generated by the muscular contraction of the heart, forcing blood into arterial circulation. The production of urine furnishes another example: Water can be squeezed out of a protein solution that is forced against a semipermeable membrane by hydrostatic pressure. The emerging protein-free solution is called an **ultrafiltrate.** The first step in urine formation is the formation, by similar means, of an ultrafiltrate from blood. This aspect of urine formation is discussed in Box 7.2.

An industrial example is a process, called **reverse osmosis,** used to desalinate water on a large scale in some of the Persian Gulf states and at a number of U.S. military bases. Figure 7.11 shows a commercial reverse-osmosis unit that uses hydrostatic pressure to force the water from saline (salt

A PICTURE OF HEALTH

Examples of Solutions in the Body

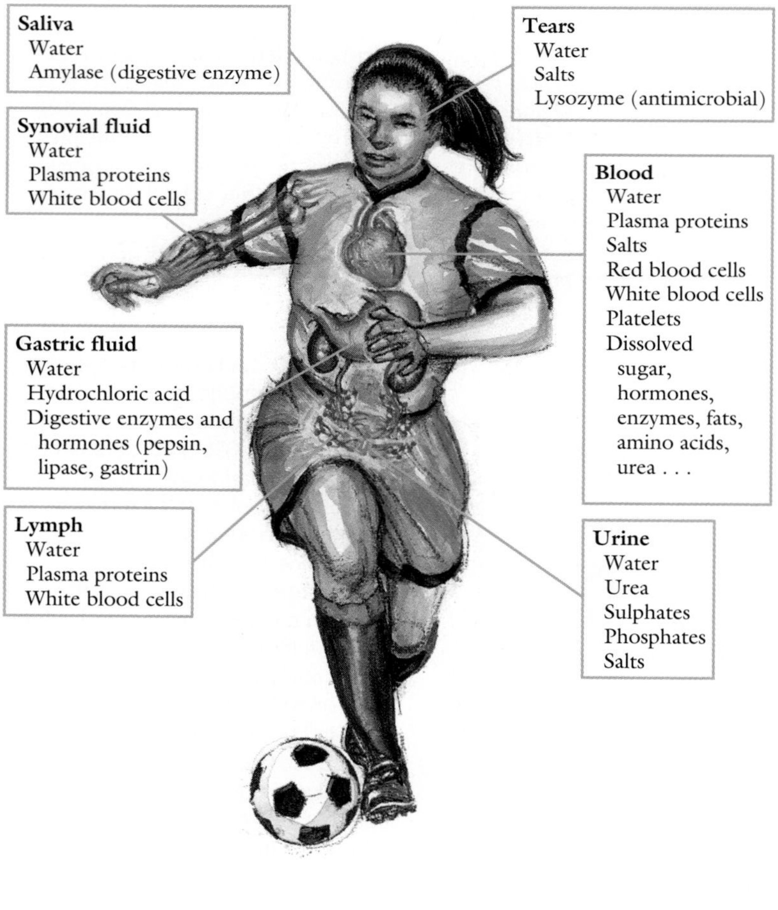

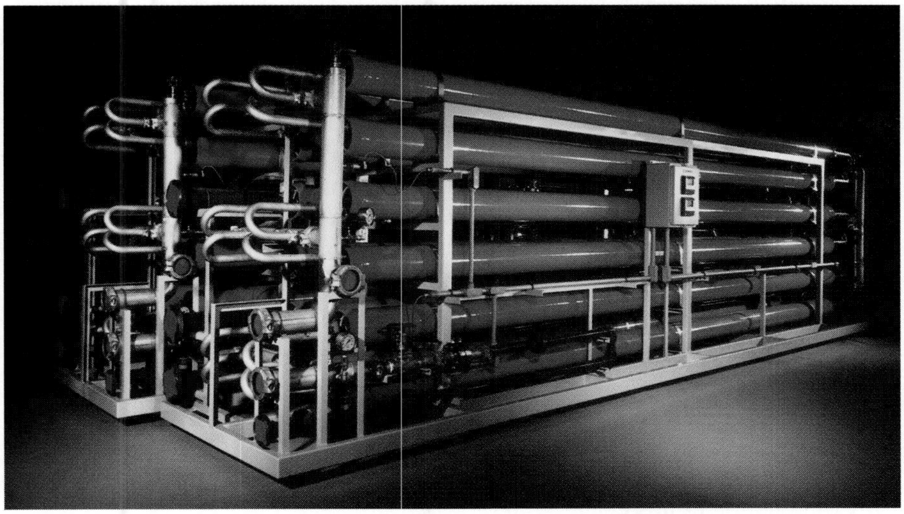

Figure 7.11 A commercial reverse-osmosis unit. The unit shown removes more than 95% of the salt initially present.

water) solutions through special semipermeable membranes that are able to retain salts, thus producing pure water. It is called reverse osmosis because the water is made to flow from a region of low water concentration to one of high water concentration.

7.14 OSMOLARITY

The osmotic pressure of a solution depends on the sum of molar concentrations of all independent particles in that solution, without regard to the particles' specific identities. One mole of chloride ions in solution exerts the same osmotic pressure as one mole of dissolved glucose molecules (which cannot ionize or dissociate). In certain ways, the behavior of a dilute solution is similar to that of an ideal gas, with the solute behaving like gas particles and the solvent representing empty space (just more viscous). Because of this similarity,

7.2 Chemistry Within Us

Semipermeability and Urine Formation

The blood supply to the kidneys, delivered by a major branch of the aorta, flows into those organs under very high pressures, about 100 torr. After blood enters the kidney, it is conducted by water-impermeable arteries to microscopic specialized tubular structures called glomeruli. The glomeruli contain capillaries whose walls are semipermeable. In consequence, under sufficient pressure, all molecules that are smaller than the membrane's smallest pore size will be forced through the capillary wall into the glomerulus tubule. Molecules larger than that minimum pore size will not pass through and will remain in the plasma.

In short, when whole blood enters the kidney glomerulus, hydrostatic pressure (Section 7.13) generated by the heart literally squeezes a protein-free solution through the semipermeable kidney glomerulus. This protein-free filtrate—called an ultrafiltrate—is the first step in urine formation. Not surprisingly, one of the earliest signs of kidney disease is the presence of protein in the urine, indicating some kind of pathology of the glomerular membranes. Unless the problem is remedied, the loss of protein from the blood plasma will cause a drop in plasma osmotic pressure, and generalized edema or swelling of the tissues will result.

the osmotic pressure can be calculated by using a law that is virtually identical with the ideal gas law:

$$P = \frac{nRT}{V} = \left(\frac{n}{V}\right) RT$$

However, in regard to osmotic pressure, an additional factor must be added to the equation to explicitly specify the number of particles contained in 1 mol of the solute. The reason for adding this factor is that ionic substances in solution provide more than one particle per mole, and n in the gas law, $PV = nRT$, must represent the actual number of moles present. The osmotic pressure relation is:

$$\Pi = i\left(\frac{n}{V}\right) RT$$

Π is the osmotic pressure in atmospheres
R is the gas constant, 0.0821 $L{\cdot}atm{\cdot}K^{-1}{\cdot}mol^{-1}$
T is the temperature in kelvins
V is the volume in liters
n is the number of moles present; (n/V) is the molar concentration, mol/L
i is a factor that corrects for the number of particles per mole of solute

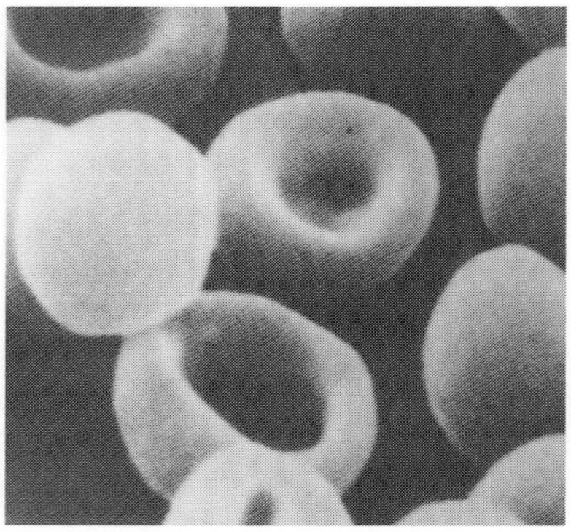

(a)

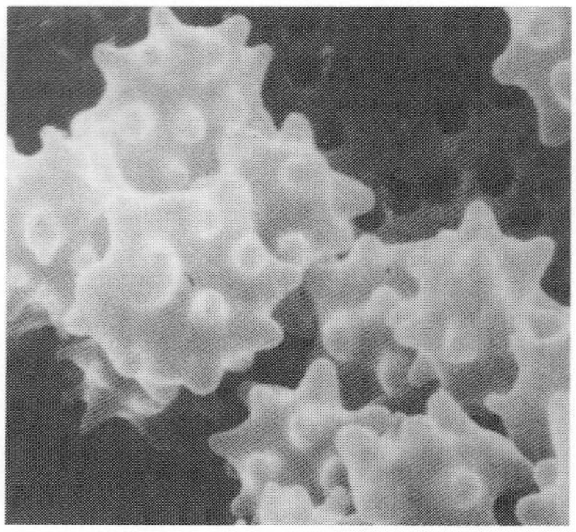

(b)

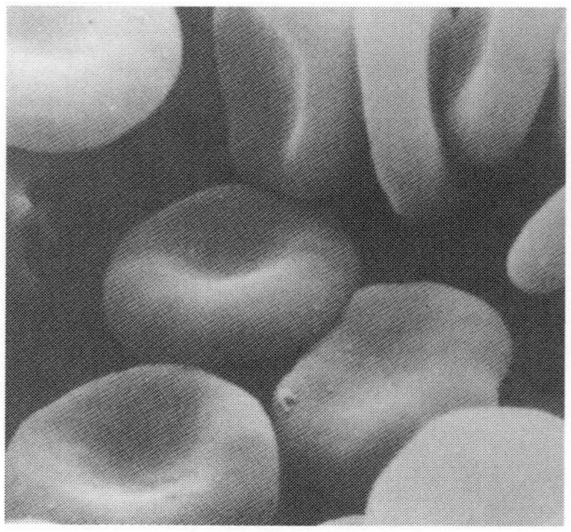

(c)

Figure 7.12 Osmosis in erythrocytes (red blood cells): (a) blood cells in an isotonic medium undergo no change in volume or shape; (b) blood cells in a hypertonic solution have contracted as a result of water loss; (c) blood cells in a hypotonic medium have swollen as a result of water entering the cells.

The quantity n/V is the molarity of the solution, but the quantity $[i \times (n/V)]$ is called the **osmolarity.** It is a measure of the total solute concentration of a solution. One osmole (1 osmol) is equal to one mole (1 mol) of an ideal nonionizing solute such as glucose. If n is the number of moles of particles that 1 mol of a compound will add to a solution when dissolved, then $[i \times n]$ is the number of osmoles.

Osmolarity recognizes that ionic solids such as NaCl dissolved in solution yield more than one particle per mole. A 1 molar (1 M) solution of NaCl contains 2 mol of dissolved particles per liter (1 mol of Na^+ ions and 1 mol of Cl^- ions), or 2 osmol/L. A solution that is 1 M in glucose and at the same time also 1 M in NaCl has an osmolarity of 3 osmol/L. The factor i accounts for dissociation. For glucose, $i = 1$ because glucose does not ionize in solution; for NaCl, $i = 2$, for $CaCl_2$, $i = 3$; for $AlCl_3$, $i = 4$; and so forth.

7.15 OSMOSIS AND THE LIVING CELL

The osmotic condition of a living cell bounded by a semipermeable membrane is the result of a balance between the rate of water entering the cell and the rate of water leaving the cell. The relations between a cell's inner and outer concentrations lead to three different osmotic conditions: hypotonic, hypertonic, and isotonic, illustrated in Figure 7.12.

When a solution external to a cell such as a red blood cell has a lower solute concentration than the internal solution does, the outer solution is called **hypotonic.** Under these circumstances, water will diffuse into the cell, thereby causing an increase in the hydrostatic pressure within the cell and consequent swelling. The flow may continue until the cell bursts. When this occurs in hypotonic solutions of red blood cells, the rupturing is called **hemolysis.**

When a solution external to a cell has a higher solute concentration than that of the solution inside the membrane, the outer solution is called **hypertonic.** Cells placed in a hypertonic solution begin to lose water and consequently shrink. Red blood cells in hypertonic solution shrink and appear crumpled and are described as **crenate.**

When the solute concentrations external and internal to a cell are identical, no net flow will occur. Therefore, no change in cell volume will be observed, and the external solution is said to be **isotonic.** In medical practice,

7.3 Chemistry Within Us

The Osmotic Pressure of Isotonic Solutions

In an isotonic medium, the rates of flow of water into and out of erythrocytes are the same. Why are physiological saline (0.90% w/v NaCl) and 5.4% w/v dextrose (glucose) solutions both called isotonic solutions, when their concentrations are not the same? The answer is that the osmotic equilibrium of a blood cell is regulated by the surrounding solution's osmolarity, not by its molarity. To understand why physiological saline and dextrose solutions are both isotonic, we will calculate their osmolarities.

The osmotic pressure of an aqueous solution is calculated with the following relation:

$$\Pi = i\left(\frac{n}{V}\right)RT$$

where $\left[i \times \left(\frac{n}{V}\right)\right]$ is the osmolarity.

To compare the osmotic characteristics of the two solutions, it is sufficient to calculate their osmolarities. The first step is to calculate their molarities.

For glucose:

$$5.4\%\ \text{w/v} = \frac{5.4\ \text{g}}{100\ \text{mL}} = \frac{54\ \text{g}}{\text{L}}$$

$$\frac{\text{mol glucose}}{\text{L}} = 54\ \text{g}\left(\frac{1\ \text{mol}}{180\ \text{g}}\right) = \frac{0.30\ \text{mol}}{\text{L}}$$

Because $i = 1$ for an aqueous glucose solution,

$$\left[i \times \left(\frac{n}{V}\right)\right] = \frac{0.30\ \text{osmol}}{\text{L}}$$

For a 0.90% w/v solution of NaCl (physiological saline),

$$0.90\%\ \text{w/v NaCl} = \frac{0.90\ \text{g}}{100\ \text{mL}}$$

$$\left(\frac{0.90\ \text{g}}{100\ \text{mL}}\right)\left(\frac{1000\ \text{mL}}{\text{L}}\right)\left(\frac{1\ \text{mol}}{58.45\ \text{g}}\right) = 0.15\ \text{mol/L}$$

Second, recognizing that each mole of NaCl provides 2 mol of ions in solution, we assign i a value of 2. The osmolarity of 0.90% w/v NaCl is:

$$\left[i \times \left(\frac{n}{V}\right)\right] = \frac{0.30\ \text{osmol}}{\text{L}}$$

The osmolarities of the glucose and saline solutions are the same; therefore they will have identical effects on osmotic equilibria.

any fluids added to the blood must be isotonic. If they are not isotonic, serious imbalances of fluid or electrolytes (ions in solution) will ensue, potentially leading to malfunction of all the major organs. Box 7.3 shows how different concentrations of sodium chloride (saline) and glucose (dextrose) solutions can be isotonically identical.

Concept checklist

- ✔ Under hypotonic conditions, the rate of water entering the cell is greater than the rate of water leaving.
  Result: swelling and possible bursting.
- ✔ Under hypertonic conditions, the rate of water leaving the cell is greater than the rate of water entering.
  Result: shrinkage and crumpling.
- ✔ Under isotonic conditions, the rates of water entering and leaving are the same.
  Result: no change in volume or shape.

7.16 MACROMOLECULES AND OSMOTIC PRESSURE IN CELLS

Proteins, polysaccharides, and nucleic acids are members of a class of high-molecular-mass molecules called **macromolecules** or **polymers.** They are formed by covalently linking together many small molecules. Proteins are

TABLE 7.4 Dimensions and Masses of Various Molecules, Macromolecules, and Biological Entities

Molecule	Largest Dimension (cm)	Molecular Mass: Atomic Mass Units	Molecular Mass: Grams
Na^+ (sodium ion)	0.4×10^{-7}	23	
H_2O	0.5×10^{-7}	18	
$C_6H_{12}O_6$ (glucose)	0.7×10^{-7}	180	
glutamic acid	0.7×10^{-7}	147	
oxytocin	1.5×10^{-7}	1,040	
myoglobin	3.5×10^{-7}	17,000	
serum albumin	5.5×10^{-7}	68,000	
γ-globulin	8.5×10^{-7}	250,000	
bacterial virus	25×10^{-7}	6,200,000	
plant virus	300×10^{-7}	40,000,000	
E. coli cell	$2{,}000 \times 10^{-7}$		2×10^{-12}
human liver cell	$20{,}000 \times 10^{-7}$		$2{,}000 \times 10^{-12}$

formed from amino acids such as glutamic acid; polysaccharides are formed from sugars such as glucose. These and other biological macromolecules will be described in detail in Chapters 18, 20, and 21. Table 7.4 lists a number of biologically significant molecules and macromolecules along with their dimensions and masses. The volume of the plasma protein molecule γ-globulin is more than 150 times as great as that of the glucose molecule. Several other biological entities are included in Table 7.4 to provide additional perspective.

Biological macromolecules are usually too large to penetrate cell membranes, and this characteristic is at the heart of a number of interesting biological phenomena. Some of these phenomena are caused by the fact that solvent but not solute will penetrate cell membranes. Consequently, an osmotic pressure will develop across such a barrier. Fluid balance in the body—that is, water in versus water out—depends on osmotic phenomena in the vascular system. Urine formation and the need for the digestive system are discussed in Boxes 7.2 (on page 193) and 7.4.

Particles ranging in size from about 1×10^{-7} cm to 1×10^{-4} cm are called **colloidal.** Although too small to be seen with the naked eye, colloidal particles can be identified by the fact that a beam of light passing through a colloidal solution can be seen at right angles to the beam's direction. (When a beam of light is passed through a solution containing molecules the size of glucose, the beam remains invisible). This quality of colloidal solutions is called the **Tyndall effect** and is illustrated in Figure 7.13.

Soap and detergent solutions form colloidal solutions. The physical basis underlying the formation of colloidal-sized particles (Figure 7.14) in soap and detergent solutions is discussed in Box 7.5 on page 198, explaining how globular proteins (Chapter 20) assume their characteristic shapes.

Figure 7.13 Colloidal particles are larger than molecules such as glucose but are too small to be seen even with a microscope. However, their size is great enough to intercept the passage of a light beam, scatter it, and reveal their presence.

7.4 Chemistry Within Us

Semipermeability and the Digestive System

Aside from their specific roles in functions such as circulatory water balance and urine formation, the semipermeable membranes of the body serve a crucial general purpose: they prevent key proteins and other macromolecules from leaking out of cells.

Cells contain a wide variety of macromolecules that are central for sustaining life (Sections 20.7 and 21.2). Reproduction and metabolism could not take place without them. If the membranes that surround the cells were permeable to such large molecules, the molecules would continually escape. Cells would have to synthesize new macromolecules at an impossibly rapid rate to keep up with the loss.

The evolution of semipermeable membranes was an excellent solution to two problems: (1) how to retain critical macromolecular components within the cells and (2) how to enable low molecular mass nutrients to enter the cell so that new macromolecules can be synthesized when necessary. However, this solution led to another physiological dilemma.

We obtain nutrition by ingesting the cells and tissues of other organisms, and these cells and tissues are composed primarily of biological macromolecules. How can our cells obtain the benefit of nutrients whose molecular dimensions are too large to penetrate the cells' semipermeable walls? This dilemma was solved by the evolution of our digestive system.

The digestive system is a tube about 15 feet long running through the body from mouth to anus. The lumen, or interior, of this tube is continuous with the external environment, in analogy with the hole in a doughnut. Because it is continuous with the external environment, its contents are technically "outside" of our bodies. Within this tube, macromolecular foodstuffs are chemically converted into molecules of small dimensions and mass. Nutrient molecules such as amino acids and sugars are small enough to penetrate the membranes of intestinal-wall cells, enter the bloodstream, and finally penetrate cell membranes to fuel the metabolic engine.

INSIGHT INTO FUNCTION

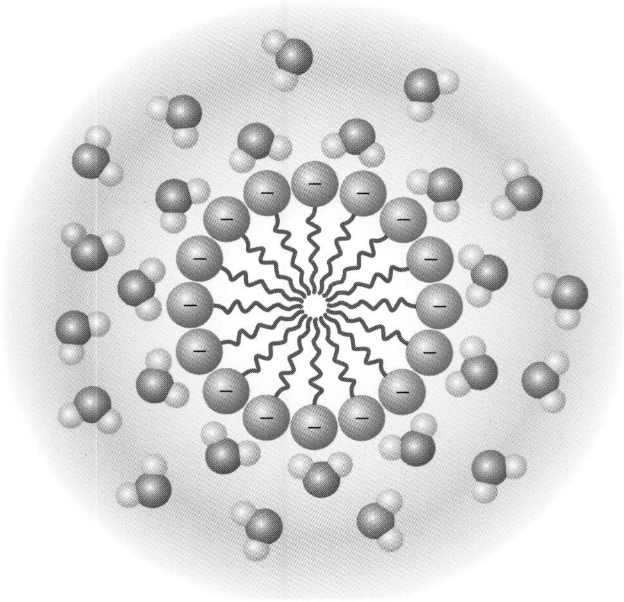

Figure 7.14 Sketch of a typical spherical micelle formed in water by an amphipathic substance such as a sodium soap. The water-soluble groups at one end of each amphipathic molecule (green circles) form the surface of the micelle and interact with the surrounding water molecules. The nonpolar hydrocarbon tails (zigzag lines) excluded from the water form the liquidlike core of the micelle.

7.5 Chemistry in Depth

Association Colloids, Micelles, and Protein Structure

In many ways, the properties of amphipathic substances in solution (Section 6.4) are very much like those of nonamphipathic substances such as glucose or sodium chloride. For example, at low concentrations, solution properties such as osmotic pressure and surface tension change gradually with increase in concentration, no matter what the solute. However, at some concentration unique to each amphipathic substance, a sudden and dramatic change occurs in a solution's properties. In soaps, for example, the solutions begin to form stable foams and exhibit their cleaning powers.

The sudden change in properties seen in amphipathic solutions is caused by the formation of large molecular assemblies called **micelles,** also called **association colloids.** A typical spherical micelle is sketched in Figure 7.14. The concentration at which micelles are created is called the **critical micelle concentration,** or **CMC.** The formula masses of micelles vary, but they usually exceed 50,000 and may be much higher, depending on molecular type. Each amphipathic substance has its own CMC.

The CMC of an amphipathic substance is the point at which an aqueous solution of the substance becomes "saturated" with respect to the solute molecules' hydrophobic tails. However, the molecules' hydrophilic head groups are still quite water soluble. Thus, one part of the molecule tends to leave the aqueous solution while the other part of the molecule is impelled to enter into solution. A compromise is reached in which groups of molecules assemble into a colloid-sized aggregate with the hydrophobic tails on the inside and the hydrophilic head groups on the outside. In this way, the hydrophobic groups escape from the aqueous environment, but, at the same time, the colloidal particle remains soluble by virtue of the large number of water-soluble groups on its surface. Micelles are called association colloids because they are of colloidal dimensions but are held together by secondary forces rather than by covalent bonds (Section 6.2).

Soluble proteins, such as those in blood plasma, are large enough so that we can also think of them as having an inside and an outside. The inside can, in fact, provide an environment that is less polar than water and can solvate molecules not normally soluble in water. For example, the protein serum albumin, an important component of blood plasma, transports water-insoluble nutrients such as palmitic acid (Section 19.2). The acid is able to dissolve in the interior of the protein, which, because of its water-soluble surface, is then able to carry the acid through the aqueous environment of the bloodstream.

The properties of micelles allow us to explain these structural characteristics of proteins, as well as the fact that each type of protein possesses a unique topography on its polar outside surface. A protein consists of a long covalently bonded chain of amino acids, and significant numbers of these amino acids contain structural parts that do not interact readily with the macromolecule's aqueous environment. Consequently, a number of amino acids seek an interior position in the protein, forcing the covalently bonded chain to organize itself in the way that a soap micelle does, with water-soluble groups on the outside and water-incompatible groups on the inside. Because each protein possesses a unique sequence of amino acids, the folding of each protein is unique—and so is the topography of its surface. These unique surface properties are at the heart of the ability of proteins such as enzymes and antibodies to recognize a given type of molecule out of the thousands of kinds of molecules within a cell.

Summary

General Aspects of Solution Formation The simplest kind of solution contains two substances: a solute (the substance present in smaller amount) dissolved in a solvent (the substance present in larger amount). Concentration is a measure of the amount of solute dissolved in a given amount of solvent. The amount of solute can be expressed in physical mass units (such as grams) or chemical mass units (such as moles), and the amount of solvent is expressed in units of either volume or mass.

Solubility When the strength of the interactions between solvent and solute molecules are about the same as those within a sample of pure solvent and a sample of pure solute, a solution will usually form. This general rule is sometimes stated as "like dissolves like." Solubility quantitatively describes the limiting amount of solute that will dissolve in a fixed amount of solvent at a specific temperature.

Percent Composition Percent composition is defined as the solution concentration $\times$ 100%. When both solvent and solute are given in mass units (grams), the percent composition is designated as % w/w. When both are given in units of volume, the percent composition is expressed as % v/v. For solutions prepared by placing a weighed sample of a solid in a container and adding solvent to a calibrated final volume, the percent composition is % w/v.

Common units for expressing the concentrations of very dilute solutions are mg %; ppm, meaning parts per million; and ppb, meaning parts per billion.

Molarity Molarity is an expression of concentration defined as moles of solute per liter of solution. Concentrations described in this way are called molar, a property identified by the symbol *M*. An alternative expression for the concentration of dilute solutions is millimoles (mmol) per unit volume.

Diffusion and Membranes Diffusion is a statistical phenomenon resulting from the random motion of molecules or ions. Membranes prevent or limit the process of diffusion. Membranes possessing the ability to select among penetrating ions and molecules are called semipermeable. Unless prevented by a mechanical force, water will always flow to the high-solute side of a semipermeable membrane. The flow of water under these circumstances is called osmosis, and the driving force for the flow is called the osmotic pressure. The osmotic pressure depends on the sum of molar concentrations of all independent particles in solution without regard to the particles' identity.

Macromolecules Macromolecules consist of long chains of small molecules linked to each other by covalent chemical bonds or assembled by the noncovalent association of amphipathic substances through weak secondary forces. Important classes of biological macromolecules are proteins, formed from amino acids such as glutamic acid; the carbohydrates, formed from sugars such as glucose; and nucleic acids, polymers of sugar phosphates and nitrogenous bases.

Key Words

colloidal particles, p. 196
concentration, p. 176
diffusion, p. 188
macromolecule, p. 195
membrane, p. 190
molarity, p. 181
osmosis, p. 190
solubility, p. 176
solute, p. 174
solvent, p. 174

Exercises

Percent Composition

7.1 How many grams of each solute are present in the following solutions? (a) 25.2 g of a 4.25% w/w solution of glucose; (b) 125 g of a 6.55% w/w solution of sodium sulfate.

7.2 Calculate the number of grams of each compound contained in (a) 60 g of a 4.5% w/w aqueous $CaCl_2$ solution; (b) 26.0 g of a 0.450% w/w solution of insulin.

7.3 How many grams of $KHCO_3$ must you add to 250 mL of water to prepare a 0.600% w/v solution?

7.4 What volume of 7.50% w/v KCl solution can be prepared from 30.0 g of KCl?

7.5 What volume of a 6.20% w/v NH_4Cl solution contains 5.40 g of salt?

7.6 What mass of glucose is required to prepare 210 g of a 8.50% w/w solution of glucose?

7.7 What weight of a 5.00% w/w solution of glucose must you take to obtain 6.25 g of glucose?

7.8 What volume of a 5.50% v/v solution of ethanol in water can be prepared with 22.6 mL of ethanol?

7.9 What volume of a 7.50% v/v solution of propylene glycol in water can be prepared with 33.6 mL of propylene glycol?

7.10 What is the concentration in mg % of a glucose solution that contains 0.049 g of glucose in 70 mL of solution?

7.11 What is the concentration in mg % of a uric acid solution that contains 0.230 mg of uric acid in 19.0 mL of solution?

7.12 What is the concentration in ppm of a solution that contains 0.147 g of glucose in 105 mL?

7.13 What is the concentration in % w/v of a NaCl solution whose concentration is 42 ppm?

Molarity

7.14 What is the volume of a 2.8 *M* solution that contains 0.70 mol of solute?

7.15 How many moles of solute are there in 2.5 L of a 0.40 *M* solution?

7.16 What are the molarities of solutions containing (a) 0.60 mol in 0.80 L; (b) 0.75 mol in 1.5 L?

7.17 What are the molarities of solutions containing (a) 2.6 mol in 1.3 L; (b) 0.810 mol in 0.240 L?

7.18 Calculate the number of moles contained in (a) 0.200 L of a 0.700 *M* solution; (b) 2.60 L of a 0.750 *M* solution.

7.19 Calculate the number of moles contained in (a) 1.05 L of a 2.65 *M* solution; (b) 452 mL of a 0.850 *M* solution.

7.20 Calculate the volumes containing the specified number of moles from the following solutions: (a) 0.750 mol of a 0.850 *M* solution; (b) 0.50 mol of a 2.5 *M* solution.

7.21 What are the volumes containing the specified number of moles from the following solutions? (a) 1.35 mol of a 0.900 *M* solution; (b) 2.80 mol of a 0.731 *M* solution.

7.22 What is the molarity of a solution containing 21.55 g of KCl in 643 mL?

7.23 What is the molarity of 275 mL of a solution containing 23.83 mg of $MgCl_2$?

7.24 How many grams of $MgCl_2$ are contained in 225 mL of a 0.650 *M* solution?

7.25 How many milligrams of $CaCl_2$ are contained in 75.0 mL of a 0.450 *M* solution?

7.26 What is the volume of a 0.750 *M* $MgCl_2$ solution containing 45.0 g of $MgCl_2$?

7.27 What is the volume of a 0.620 *M* KCl solution containing 25.0 g of KCl?

Dilution

7.28 Given 20.0 mL of a solution of 2.50 *M* HCl, to what volume must we dilute it to obtain a concentration of 0.200 *M*?

7.29 The concentration of sulfuric acid, H_2SO_4, is 36 *M*. What volume is required to prepare 6.0 L of 0.12 *M* acid?

7.30 What volumes of the following concentrated solutions are required to prepare solutions of the final concentrations indicated? (a) 12 *M* H_2SO_4 to prepare 2.0 L of 1.5 *M* H_2SO_4; (b) 2.0 *M* KCl to prepare 200 mL of 0.50 *M* KCl.

7.31 What volumes of the following concentrated solutions are required to prepare solutions of the final concentrations indicated? (a) 0.750 *M* glucose to prepare 4.50 L of 0.250 *M* glucose; (b) 0.360 *M* lactose to prepare 650 mL of 0.18 *M* lactose.

7.32 What volume of 0.900% w/v saline solution can be prepared from 0.300 L of a 3.00% w/v saline solution available in stock?

7.33 From 0.80 L of a glucose stock solution, 4.0 L of a 1.8% w/v solution of glucose was prepared. What was the concentration of the stock solution?

7.34 What is the molarity of a solution prepared by diluting 0.15 L of 0.80 *M* HCl to a final volume of 0.48 L?

7.35 Calculate the molarity of a solution prepared by diluting 0.100 L of 2.10 *M* KOH to a final volume of 0.420 L.

Solubility

7.36 An unknown amount of KCl was added to 250 mL of water at 35°C and formed a clear solution. The temperature was lowered to 25°C, and the solution remained clear. Was the solution at 35°C saturated? (See Figure 7.5.)

7.37 The solubility of strontium acetate is 43 g/100 mL at 10°C and 37.4 g/100 mL at 50°C. A clear solution of strontium acetate at 10°C was heated to 50°C and remained clear. Was the solution at 10°C saturated?

7.38 The solubility of aluminum sulfate in water at 20°C is 26.7 g/100 mL. Calculate its molarity.

7.39 The solubility of calcium chloride in water at 20°C is 74.5 g/100 mL. Calculate its molarity.

Diffusion and Membranes

7.40 Calculate the osmolarity of the following aqueous solutions: (a) 0.40 *M* glucose; (b) 0.10 *M* $CaCl_2$.

7.41 Calculate the osmolarity of the following aqueous solutions: (a) 0.10 *M* $AlCl_3$; (b) 0.01 *M* KCl.

7.42 What happens to erythrocytes (red blood cells) placed in a 0.05% w/v saline solution?

7.43 What happens to erythrocytes (red blood cells) placed in a 8.0% w/v dextrose (glucose) solution?

7.44 What happens to erythrocytes (red blood cells) placed in a 0.90% w/v sodium chloride solution?

Unclassified Exercises

7.45 Calculate the percent by weight of the following solutions: (a) 13.0 g of $CaCl_2$ in 133 g of H_2O; (b) 12.5 g of ethyl alcohol in 165 g of H_2O; (c) 2.5 g of procaine hydrochloride in 60 g of propylene glycol.

7.46 Given an aqueous stock solution that is 5.0% by weight in iron(II)chloride, how many milliliters must you take to obtain 7.5 g of $FeCl_2$?

7.47 How much of each compound is present in the following mixtures? (a) 55.0 g of a 4.00% w/w solution of glucose; (b) 175 g of a 7.50% w/w solution of ammonium sulfate.

7.48 Calculate the volume percent of the following mixtures: (a) 15.0 mL of ethyl alcohol made up to a final volume of 200 mL of solution with water; (b) 8.40 mL of chloral hydrate made up to 35.0 mL with ethyl alcohol; (c) 10.6 mL of propylene glycol made up to 50.0 mL with water.

7.49 Calculate the weight/volume percent of the following mixtures made up to final volume with water: (a) 6.40 g of LiCl in 66.0 mL; (b) 1.30 g of $NaHCO_3$ in 125 mL; (c) 3.60 g of $CaCl_2$ in 90.0 mL.

7.50 Calculate the molarity of the following aqueous solutions: (a) 0.320 mol in 1.25 L; (b) 0.220 mol of NaCl in 0.471 L; (c) 1.70 mol of $NaHCO_3$ in 2.26 L; (d) 0.325 mol HCl in 116 mL.

7.51 Calculate the number of moles in each of the following aqueous solutions: (a) 46 mL of 0.066 *M* $CaCl_2$ solution; (b) 1.2 L of 0.065 *M* KNO_3 solution; (c) 495 mL of 0.250 *M* LiCl solution; (d) 0.625 L of 1.75 *M* $NaHCO_3$ solution.

7.52 Calculate the number of grams of solute in each of the following aqueous solutions: (a) 75 mL of 1.75 *M* KCl; (b) 0.075 L of 0.82 *M* LiCl; (c) 2.75 L of 0.415 *M* NaSCN; (d) 62.5 mL of 18.0 *M* H_2SO_4.

7.53 How many moles of a 0.375 *M* solution of glucose will be contained in the following volumes of that solution? (a) 1.49 L; (b) 6.72 L; (c) 2.56 L; (d) 1.81 L.

7.54 How many grams of $KHCO_3$ must you add to 250 mL of water to prepare a 0.650 *M* solution?

7.55 How would you prepare 500.0 mL of a 0.160 *M* aqueous solution of KI?

7.56 What weight of a 5.00% w/w solution of glucose must you take to obtain 6.25 g of glucose?

7.57 Calculate the weight percent of the following solutions: (a) 15.0 g $CaCl_2$ in 250 g H_2O; (b) 16.5 g NaCl in 325 g H_2O; (c) 2.20 g C_2H_5OH in 122 g H_2O.

7.58 What volume of physiological saline (0.900% w/v NaCl) can be prepared from 90.0 mL of 2.75% w/v NaCl solution?

7.59 Calculate the grams of solute present in each of the following solutions: (a) 0.950 L of 1.25 *M* NaCl; (b) 625 mL of 0.750 *M* $Mg(NO_3)_2$; (c) 325 mL of 2.20 *M* K_2CO_3.

7.60 Calculate the molarity of the following aqueous solutions: (a) 62.0 g $K_2C_2O_4$ in 747 mL of solution; (b) 126 g $C_6H_{12}O_6$ in 350 mL of solution; (c) 26.0 g $BaCl_2$ in 0.825 L of solution.

7.61 What volume of 0.180 *M* HCl can be prepared from 1.50 L of 1.80 *M* HCl?

7.62 What volume of 0.1250 *M* KOH is required to prepare 750.0 mL of 0.06500 *M* KOH?

7.63 A 150-mL portion of a concentrated stock saline solution was used to prepare 1.50 L of 0.90% w/v saline solution. What was the concentration of the stock solution?

7.64 Calculate the molarity of a solution prepared by diluting 0.100 L of 2.10 *M* KOH to a final volume of 0.420 L.

7.65 Solutions of the following compounds are added together. Consult Table 7.3 to predict whether a reaction will take place. If a reaction will take place, write the net ionic equation that describes it. (a) NaCl + $AgNO_3$; (b) KOH + $MgCl_2$; (c) $Pb(NO_3)_2$ + KCl; (d) K_3PO_4 + $CaCl_2$.

7.66 Solutions of the following compounds are added together. Consult Table 7.3 to predict whether a reaction will take place. If a reaction will take place, write the net ionic equation that describes it. (a) K_2SO_4 + $BaCl_2$; (b) $AlCl_3$ + KOH; (c) NaBr + $AgNO_3$; (d) KCl + Na_2CO_3.

Chemical Connections

7.67 Seventy-five milliliters of propylene glycol, a pure liquid, is dissolved in 55 mL of water. Which component is the solute and which is the solvent?

7.68 A chemist has 75 mL of 0.25 *M* HCl available and requires 125 mL of 0.175 *M* HCl. Will she be able to obtain the required quantity? Explain your answer.

7.69 The hydrostatic pressure at the renal (kidney) artery is about 100 torr. The osmotic pressure of the blood is about 15 torr. Does water move into or out of the arterial blood?

7.70 How would you prepare an isotonic saline solution described by a medical dictionary as a 0.90% NaCl w/v solution?

7.71 What is the osmolarity of a solution that is a mixture of 0.45% w/v NaCl and 2.7% w/v glucose?

CHAPTER 8

CHEMICAL REACTIONS

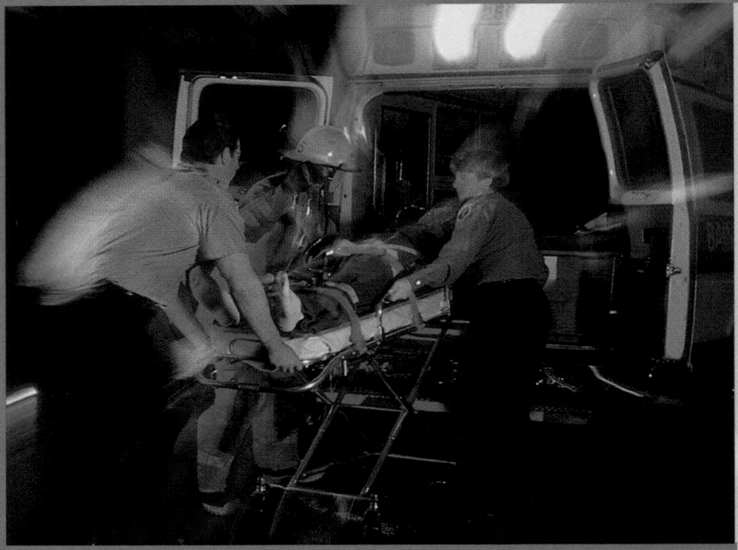

CHEMISTRY IN YOUR FUTURE

A house painter is brought to the emergency room in a state of collapse after complaining of visual problems. You and the other firefighter paramedics bring all his paints and solvents along to help the physicians find clues to his problem. One solvent is "mineral spirits" (methyl alcohol, also called wood alcohol), a toxic substance known to affect vision. The physicians immediately begin feeding small doses of ethyl alcohol (a nontoxic alcohol) to the patient, and overnight he completely recovers. You overhear one physician say that many biochemical processes are reversible. Chapter 8 explains this statement and how reactions can be reversed.

LEARNING OBJECTIVES

- Describe how the rates of chemical reactions are affected by the concentrations of reactants, the temperature, and the presence of catalysts.
- Describe how chemical reactions are the result of collisions that lead to the formation and decay of an activated complex.
- Explain how a chemical system reaches a state of dynamic equilibrium.
- Describe Le Chatelier's principle.
- Describe how an equilibrium constant provides a quantitative description of chemical equilibrium.

When the chemical reactions of metabolism (the subject of Chapters 22 through 26) are undertaken in test tubes, extraordinary conditions of temperature, concentration, and acidity are required to induce the reactions to take place. However, when these same reactions take place in cells, they proceed rapidly and efficiently under quite mild conditions. This is made possible by the presence of unique protein molecules called enzymes. When you reach the end of this chapter, you should have a clearer understanding of the reasons for the chemical ingenuity of cells. Our goals are to describe how chemical reactions take place, develop molecular mechanisms of how they occur, and learn how to characterize their outcome quantitatively.

8.1 REACTION RATES

In a chemical reaction, the concentrations of the starting components, the **reactants,** and those of the emerging chemical species, the **products,** change over time. The **rate** of a chemical reaction is the speed with which these changes take place. The study of the factors affecting the rates of chemical reactions is called **chemical kinetics.**

Consider reactions between substances such as nitrogen and hydrogen to form ammonia or between nitric oxide and bromine to form nitrosyl bromide. The balanced reaction equations for those processes are:

$$3\ H_2 + N_2 \longrightarrow 2\ NH_3 \qquad (1)$$

$$2\ NO + Br_2 \longrightarrow 2\ NOBr \qquad (2)$$

The equations imply that in reaction 1, the concentrations of hydrogen and nitrogen decrease and that of ammonia increases, and in reaction 2, the concentrations of nitric oxide and bromine decrease and that of nitrosyl bromide increases. Theoretical data representing these kinds of reactions are charted in Figure 8.1. As time passes, the molar concentrations of reactants are seen to decrease while the molar concentrations of products simultaneously increase.

The **reaction rate** is defined as the change in the number of moles of a reactant or a product per unit time. That is,

$$\text{Rate} = \frac{\text{moles of product appearing}}{\text{elapsed time}}, \text{ or } \left(\frac{\text{moles of reactant disappearing}}{\text{elapsed time}}\right)$$

The rate is positive for the appearance of product and for the disappearance of reactant. For example, if 2.5 mol of a reactant is used up in a 10-min period of time, the average rate of reaction is 0.25 mol/min, and, if 2.5 mol of a reactant is formed in a 10-min period of time, the average rate of reaction also is 0.25 mol/min.

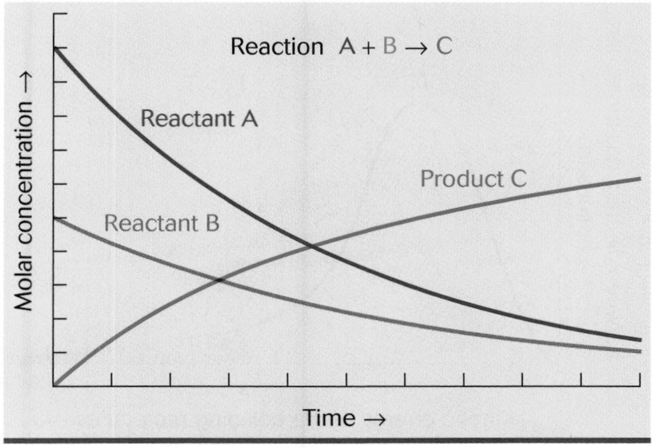

Figure 8.1 In a chemical reaction, product appears and reactant disappears with the passage of time.

Chemical reactions require that collisions occur between the reacting molecules. The rate of collisions (and hence of reaction) depends on four variables: (1) the concentration of reactants, (2) their spatial orientation when they collide, (3) the temperature at which the reaction takes place, and (4) the presence of catalysts.

8.2 REACTIVE COLLISIONS

For a chemical reaction to take place, a collision must result in the breaking or the rearrangement of bonds. This result can occur only if (1) the molecules collide with sufficient energy and (2) they collide in a favorable spatial orientation. A collision having both characteristics is called a **reactive collision.**

The kinetic energy (energy of motion) of molecules increases as the temperature increases. Molecules move about much faster in hot systems than in cold ones. It is important to note that, in any assembly of molecules, only a small fraction are moving about with velocities great enough to produce reactive collisions. This small fraction of fast-moving molecules can increase dramatically with an increase in temperature. Figure 8.2 plots the number of collisions between different gas molecules as a function of their kinetic energies at two different temperatures, T_1 and T_2. The minimum energy needed for a reactive collision—the activation energy, E_a—is noted on the *x*-axis. The graph shows that the number of collisions having kinetic energies above the activation energy increases rapidly with increase in temperature. The increase in the number of collisions in the high kinetic energy range is so rapid that, for many reactions, the reaction rate can be doubled by a mere 10-kelvin increase in temperature. An interesting application of this effect used by biological systems is discussed in Box 8.1 on page 206.

Figure 8.3 is a schematic representation of the reaction between a molecule of fluorine and a molecule of nitrogen dioxide to form nitrogen oxyfluoride. Three scenarios depict the collisions of the reactants. Figure 8.3a shows a nonreactive collision in which the kinetic energies of the molecules are not sufficient to cause bonds to break and reform. The molecules merely bounce off each other. Figure 8.3b shows a reactive collision in which both the kinetic energies and the spatial orientation are optimized for correct bond breakage and new bond formation. The new molecular structure formed on collision is called an **activated complex.** It is unstable and rapidly decays either into products or back into the original reactants. Figure 8.3c shows a nonreactive collision in which the kinetic energies were sufficient, but the orientation was not

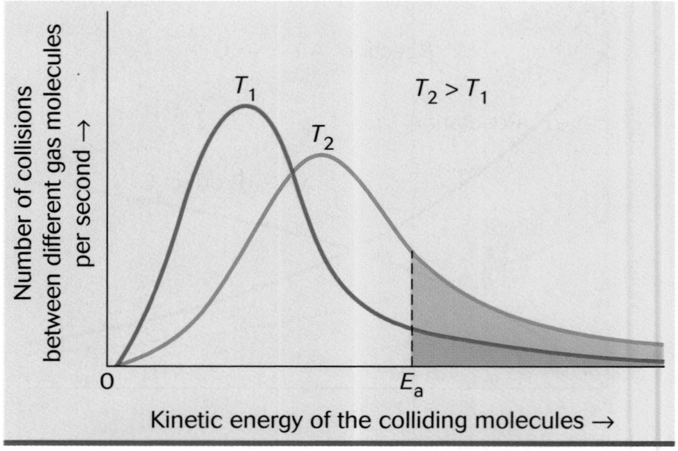

Figure 8.2 The number of collisions between different gas molecules is plotted versus their kinetic energies at two different temperatures (T_1 and T_2). The minimum energy needed for a reactive collision, the activation energy, is indicated on the *x*-axis.

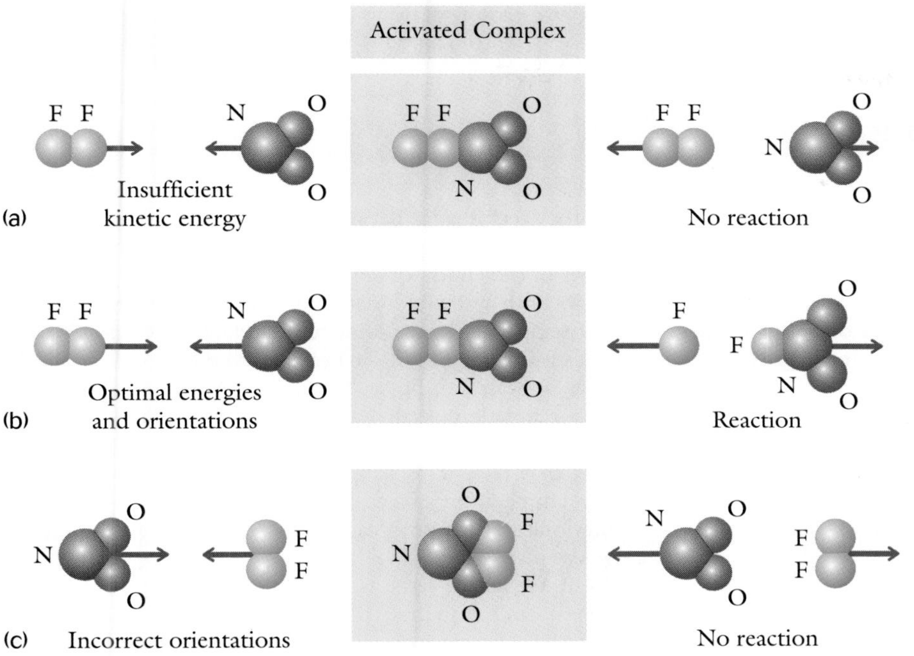

Figure 8.3 Three scenarios for the reaction between a molecule of fluorine and a molecule of nitrogen dioxide to form nitrogen oxyfluoride. (a) a nonreactive collision in which the kinetic energies of the molecules are insufficient to cause bonds to break; (b) a reactive collision in which both the kinetic energies and the spatial orientations are optimal for correct bond breakage and new bond formation; (c) a nonreactive collision with sufficient kinetic energy but incorrect orientations for bond breakage. The shorter arrows in (a) indicate that the kinetic energies of the reactants are insufficient for bond breakage.

correct. The fluorine molecule must strike the nitrogen atom of nitrogen dioxide for the complex to form and the reaction to take place.

The **activation energy,** then, is the minimum energy required to form an activated complex. The energy of motion of the colliding molecules is incorporated into the overall energy of the activated complex and is expressed by greater vibrations of the constituent atoms. In consequence, the bond lengths (the distances between atoms) become greater than normal and therefore unstable. (We use the word complex to describe any combination of molecules that has a short lifetime.)

Figure 8.4, called an activation-energy diagram, is a plot of the changes in energy (the *y*-axis) of a pair of colliding molecules as they form an activated complex; that is

$$A + B \longrightarrow \text{activated complex} \longrightarrow C + D$$

The *x*-axis in Figure 8.4 represents the progress of the reaction in which A approaches B to form the complex, which proceeds to form products C and D. The difference between the energy of the activated complex and that of the

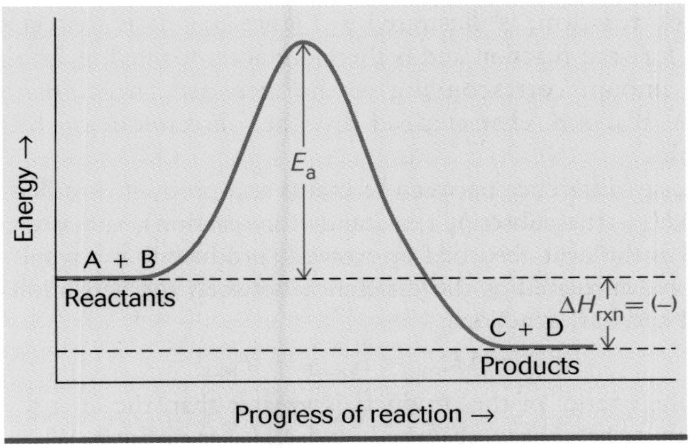

Figure 8.4 An activation-energy diagram plots the changes in the energy of an activated complex as an exothermic chemical reaction progresses from reactants to products. The heat of reaction for the process is designated by ΔH_{rxn}.

8.1 Chemistry Within Us

The Influence of Temperature on Biological Processes

One very important aspect of temperature regulation in warm-blooded organisms is the **febrile response,** the physiological changes that lead to an increase in body temperature in response to infection. Fever has been primarily considered a medical problem and thus a problem pertaining to humans. As such, it has not been the subject of experimental studies, because of the difficulties of conducting studies that use human subjects. More important, humans and other mammals are designed to maintain their body temperatures within a very narrow range, a fact that severely limits their potential usefulness in any study of how temperature can cause changes in biochemical processes.

Fortunately, there are organisms that do not possess physiological mechanisms for controlling their body temperatures. These organisms are the **ectothermic,** or "cold-blooded," animals, such as lizards, snakes, and insects. Ectothermic animals regulate their body temperatures behaviorally. For example, they warm their bodies in the sun to raise their metabolic rates to be able to pursue food or mates or to escape predation. If their body temperatures become too high because of extended exposure to the sun, they seek shade. Because their body temperatures are not fixed, ectothermic animals are excellent models for studies of the effects of temperature on selected processes.

In both warm-blooded and cold-blooded organisms, body temperatures are primarily regulated in response to signals from a region of the brain called the hypothalamus. There appears to be a "thermostatic setpoint" about which regulation occurs. If the body temperature drops below the setpoint, physiological or behavioral mechanisms that raise it to the desired value are instituted; if body temperature exceeds the setpoint, physiological or behavioral mechanisms leading to a lower body temperature are invoked.

In the febrile response in humans, the hypothalamus behaves as if the thermostatic setpoint has been raised to a higher level by approximately 2 to 3 Celsius degrees. The chills that humans experience at the onset of fever are a response to this new internal environment in which the hypothalamus is convinced that the body temperature should be much higher than it is. The person then shivers, experiencing violent muscle contractions, which are exothermic and therefore raise the body temperature. When body

reactants is the activation energy. In Figure 8.4, the difference in energy between the (A + B) level and the (C + D) level signifies that energy is lost when the reactants are chemically transformed into products. This difference in energy levels represents a loss of heat, and the reaction is characterized as **exothermic;** such reactions give off heat. The reaction as written, with reactants on the left and products on the right, is called a **forward reaction.**

There is no reason to suppose that the products of our hypothetical reaction (C + D), cannot react to form (A + B); that is, the reaction is reversible (see Section 8.5). That chemical reaction,

$$\mathrm{C + D \longrightarrow \text{activated complex} \longrightarrow A + B},$$

called a **back reaction,** is illustrated in Figure 8.5. It is the opposite of the exothermic forward reaction and is therefore accompanied by an absorption of heat in an amount corresponding to the increased energy of the products (A + B). A reaction characterized by the absorption of heat is called **endothermic.**

The energy difference between reactants and products is called the **heat of reaction,** ΔH_{rxn} (the subscript rxn stands for reaction), and is experimentally determined as the heat absorbed or evolved (produced) as a result of reaction. It can also be calculated as the difference between the activation energies of the forward and back reactions:

$$\Delta H_{rxn} = E_{forward} - E_{back}$$

When the energy of the products is greater than the energy of the reactants, heat must have been absorbed, and ΔH_{rxn} is positive. When the energy temperature reaches the new higher setpoint, shivering ceases and the person feels comfortable. When the fever breaks, the original setpoint is restored, and the person feels very warm because the body temperature is now a few degrees above the setpoint. The physiological response is to perspire copiously and to feel an urge to throw off the covers.

Research in the past 30 or 40 years has established that the febrile response occurs in virtually all animal phyla: fishes, amphibia, reptiles, and birds, as well as mammals. The phenomenon is so widespread, phylogenetically speaking, that it seems to be a fundamental, adaptive biological response of multicellular organisms to infection.

The most interesting research probing this possibility was performed with lizards, making use of the facts that they are prone to suffer from a particular bacterial infection and that they can regulate their body temperature behaviorally, actively moving about to find a location having the temperature that best suits their current metabolic needs. When researchers placed healthy lizards in an aquarium outfitted with a floor whose temperature varied along its length, low at one end, high at the other, the lizards regularly moved to the part of the aquarium floor where the temperature was 38°C. The researchers injected several groups of lizards with pathogenic bacteria and placed them in other aquaria whose temperatures also varied along the floor length. However, the high-end temperature of each aquarium was different and varied from 42°C to below 30°C. All infected lizards experienced a febrile response and tried to raise their body temperatures above the normal setpoint of 38°C, but only those in the aquaria with high-end temperatures above 38°C succeeded. The result was virtually 100% survival among those lizards who could raise their body temperatures to above 40°C. There was significant mortality among those who could not raise their body temperatures above the normal setpoint. The specific mechanisms leading to survival are not known.

These results indicate that the febrile state is an adaptive and useful response to infection. Although too high a fever for too long can be hazardous, researchers now question the generally held view that, at the first sign of fever, antipyretic, or fever-reducing, drugs such as aspirin should be taken to reduce the body temperature.

of the products is less than the energy of the reactants, heat must have been lost, and ΔH_{rxn} is negative.

Concept checklist

- ✔ In an exothermic reaction, heat is evolved (lost) and the heat of reaction is negative $[\Delta H_{rxn}(-)]$.
- ✔ In an endothermic reaction, heat is absorbed (gained) and the heat of reaction is positive $[\Delta H_{rxn} = (+)]$.

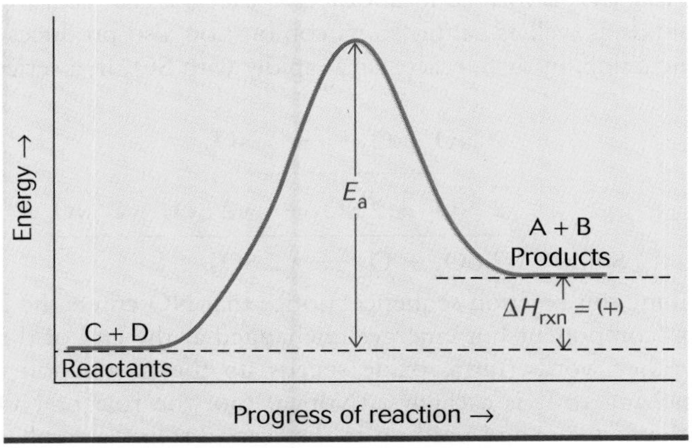

Figure 8.5 An activation-energy diagram plots the changes in the energy of an activated complex as an endothermic chemical reaction progresses from reactants to products.

Example 8.1 Using activation energy to calculate the heat of reaction

The activation energies characteristic of the reversible chemical reaction $A + B \rightleftharpoons C$ are $E_{forward} = 53$ kJ and $E_{back} = 82$ kJ. Calculate ΔH_{rxn}, the heat of the forward reaction. Is the forward reaction exothermic or endothermic?

Solution

$$\begin{aligned}\Delta H_{rxn} &= E_{forward} - E_{back} \\ &= 53 \text{ kJ} - 82 \text{ kJ} \\ &= -29 \text{ kJ}\end{aligned}$$

The heat of reaction is negative, which means that heat has been lost and the forward reaction is exothermic.

Problem 8.1 The activation energies characteristic of the reversible chemical reaction $A + B \rightleftharpoons C$ are $E_{forward} = 74$ kJ and $E_{back} = 68$ kJ. Calculate ΔH_{rxn}, the heat of the forward reaction. Is the forward reaction exothermic or endothermic?

8.3 CATALYSTS

A **catalyst** is a substance that increases the rate of a reaction but emerges unchanged after the reaction has ended. Catalysts in cells are called enzymes. For example, the enzyme carbonic anhydrase, found in red blood cells, is a catalyst that specifically increases the rate of formation of bicarbonate ion from carbon dioxide and hydroxide ion by a factor of 3.5×10^6. Without this rapid reaction, it would be impossible to rid the body of the carbon dioxide generated by cellular oxidative processes.

The serious damage to this statue reveals the presence of acid rain.

When sulfur burns, it forms two different oxides: sulfur dioxide and sulfur trioxide. The reactions are:

$$S + O_2 \longrightarrow SO_2 \qquad (1)$$

and

$$2\,SO_2 + O_2 \longrightarrow 2\,SO_3 \qquad (2)$$

Reaction 1 is much faster than reaction 2; so, when a sample of coal, which contains sulfur compounds, is burned, we might expect the bulk of the oxide produced to be SO_2. However, measurements have shown that in the course of coal combustion, the formation of large quantities of SO_3 contributes significantly to the acid rain descending on our cities and forests. Acid rain forms when SO_3 dissolves in water to form sulfuric acid:

$$SO_3 + H_2O \longrightarrow H_2SO_4$$

The reason that SO_3 is formed in unexpected excess is that coal contains nitrogen compounds as well as sulfur, so its combustion also produces nitric oxide (NO). Nitric oxide, in turn, reacts very rapidly with SO_2 in a series of steps to form SO_3:

$$2\,NO + O_2 \longrightarrow 2\,NO_2$$

$$2\,NO_2 + 2\,SO_2 \longrightarrow 2\,SO_3 + 2\,NO$$

$$\text{Sum:} \quad 2\,SO_2 + O_2 \longrightarrow 2\,SO_3$$

As you examine this reaction sequence, notice that NO enters the first reaction as a reaction component but emerges unchanged at the end of the second reaction. In other words, nitric oxide speeds up the production of SO_3 and emerges unchanged: it is a catalyst. Notice, too, the role of NO_2, which is formed in the first step and used up in the second step (as emphasized by the connecting line). The NO_2 is called a **reaction intermediate.** The two steps are connected in a sequence because the product of the first step—NO_2—becomes a reactant in the second step. The final stoichiometry of the production of SO_3 is derived from an algebraic summation of the two steps in the reaction sequence.

The catalytic action of nitric oxide is typical of all catalysts in that it provides a new pathway to the reaction products that has a lower activation energy.

Concept checklist

✔ A catalyst lowers the activation energy of a reaction by providing a different pathway leading to products.

✔ It does so by becoming an active participant in the chemical process, but it emerges unchanged.

✔ It does not alter the results of a reaction; it changes only the speed at which the reaction takes place.

8.4 BIOCHEMICAL CATALYSTS

▶▶ Chapter 20, on proteins, provides insight into enzymatic structure.

The protein molecules called **enzymes** are catalysts whose structures allow them to interact specifically with particular molecules or families of molecules. The specificity can be great enough that an enzyme can select one of thousands of different molecules in the cell for conversion into a product. Recall that carbonic anhydrase selects only CO_2 for its conversion into HCO_3^-.

The site, or location, where catalysis takes place on the enzyme—called an **active site**—is structurally related to the substance being converted—the **substrate**—as a lock is to a key. We will refer to this spatial relation as **complementarity.** The substrate is bound to the active site by secondary forces. The specifics of enzyme-catalyzed reactions will be considered in Chapters 22 through 26, but the important idea here is that their catalytic properties depend on the formation of an activated complex (see Section 8.2). In all cases, the formation of the complex of enzyme and substrate provides a reaction pathway of low overall activation energy and, consequently, a rapid reaction rate.

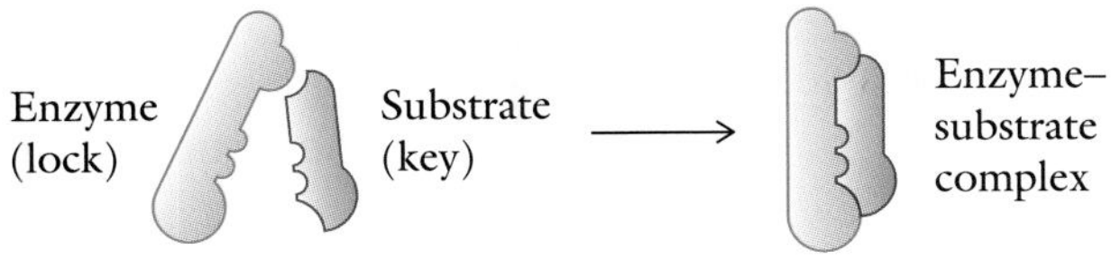

The activation energy controls the rate of a chemical reaction. The greater the activation energy, the smaller the percentage of colliding species that have the required activation energy and the slower the rate of reaction. The relation between activation energy, reaction rate, and temperature is an important feature of biological function and adaptation. In Figure 8.6, on the following page, the relative rates of chemical reactions possessing different activation energies are plotted as a function of temperature over a small temperature range. The temperature effects on a cricket's chirping rate are included in Figure 8.6. The temperature range corresponds to temperatures encountered by most organisms on our planet. Note that the greater the activation energy, the steeper the slope. In other words, in reactions with a large activation energy, the rate of reaction changes very rapidly with a small increment in temperature. Conversely, in reactions with a small activation energy, the rates of reaction are relatively insensitive to changes in temperature.

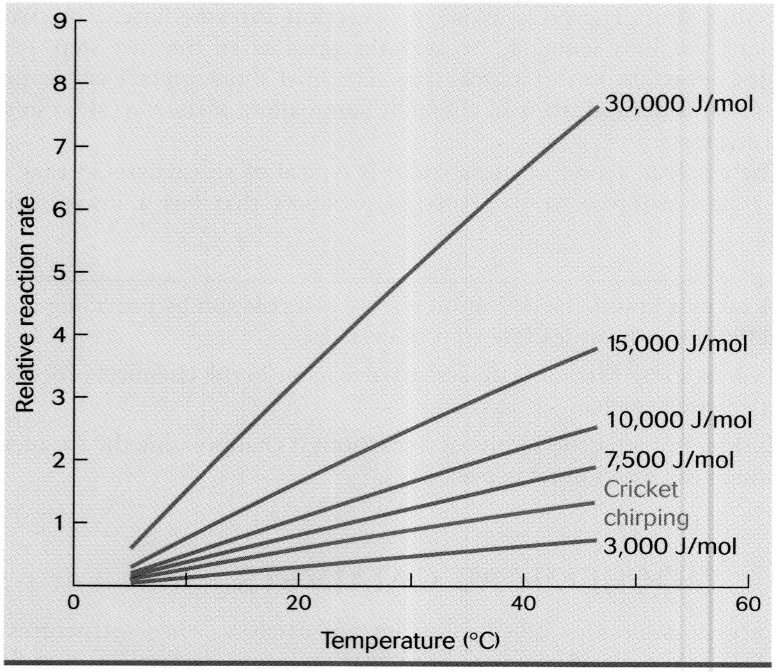

Figure 8.6 The effects of activation energy on the relative rates of chemical reactions as a function of temperature.

The survival of organisms depends in large part on the control of the chemical reactions that constitute metabolism. The body temperatures of warm-blooded animals are restricted to fairly constant values, and even so-called cold-blooded, or ectothermic, animals function within very narrow temperature ranges. This means that most animals are equipped with biochemical processes of low activation energies and therefore are not significantly dependent on changes in temperature to modify the rates of their metabolic reactions. Chemical reactions in cells are controlled by changes in enzyme activity and concentration (Chapter 22). Data on activation energies of uncatalyzed chemical reactions, enzyme-catalyzed reactions, and some typical biological processes are listed in Table 8.1. They show that enzyme-catalyzed reactions

TABLE 8.1 The Activation Energies of Some Chemical and Biological Processes

Process	Activation energy (cal)
DECOMPOSITIONS	
HI	45,600
N_2O	53,000
C_2H_5Br	54,800
Sucrose	25,600
ENZYME-CATALYZED REACTIONS	
sucrose hydrolysis by invertase	9,000
tributyrin hydrolysis by lipase	7,600
BIOLOGICAL PROCESSES	
cricket chirping	5,300
bacterial decomposition of urea	8,700
α-rhythm of human brain	8,000

and biological processes have significantly smaller activation energies than those of uncatalyzed chemical reactions.

8.5 CHEMICAL EQUILIBRIUM

The time course of a reaction such as that of hydrogen with nitrogen to produce ammonia is plotted in Figure 8.1. Figure 8.7 plots a similar reaction over a long period of time. After sufficient time, the concentrations of product and reactants no longer change. At that point in time, the system has reached a state of **chemical equilibrium.** The amounts of reactants left can vary from large quantities to none.

The idea of a dynamic equilibrium, a state or condition in which two opposing processes take place at the same rate, was introduced in Section 6.7. In the equation describing the liquid–vapor equilibrium, these processes were represented by arrows pointing in opposite directions:

$$\text{Liquid} \underset{\text{condensation}}{\overset{\text{vaporization}}{\rightleftharpoons}} \text{vapor}$$

As stated earlier in this chapter, chemical reactions also consist of simultaneous opposing processes—the forward and back reactions. Therefore, chemists have adopted the convention of representing equilibrium in chemical reactions by two arrows facing in opposite directions:

$$A + B \rightleftharpoons C + D$$

Whenever you see this representation, you may assume that:

- The chemical system is either at or is capable of reaching an equilibrium state.
- The equilibrium state may be reached from either direction; that is, by adding either A to B or C to D.

In other words, reactions are **reversible.** In theory, all reactions are reversible. However, there are reactions in which the amounts of reactants left at the point of equilibrium are much too small to be of practical significance. In other reactions, such as the combustion of a hydrocarbon, reaction of the products will not produce the original reactants. Such reactions are called **irreversible,** and, often but not always, equations are written with a single arrow pointing to products to denote this condition.

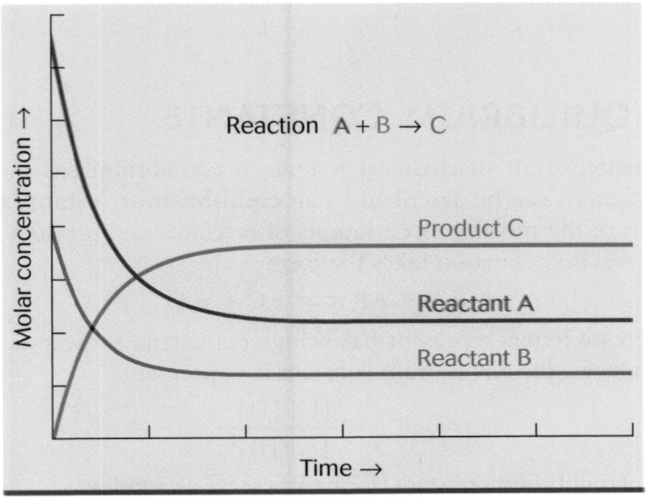

Figure 8.7 When the concentrations of products and reactants no longer change over time, a chemical reaction has reached equilibrium.

A PICTURE OF HEALTH

Examples of Enzymes in the Body

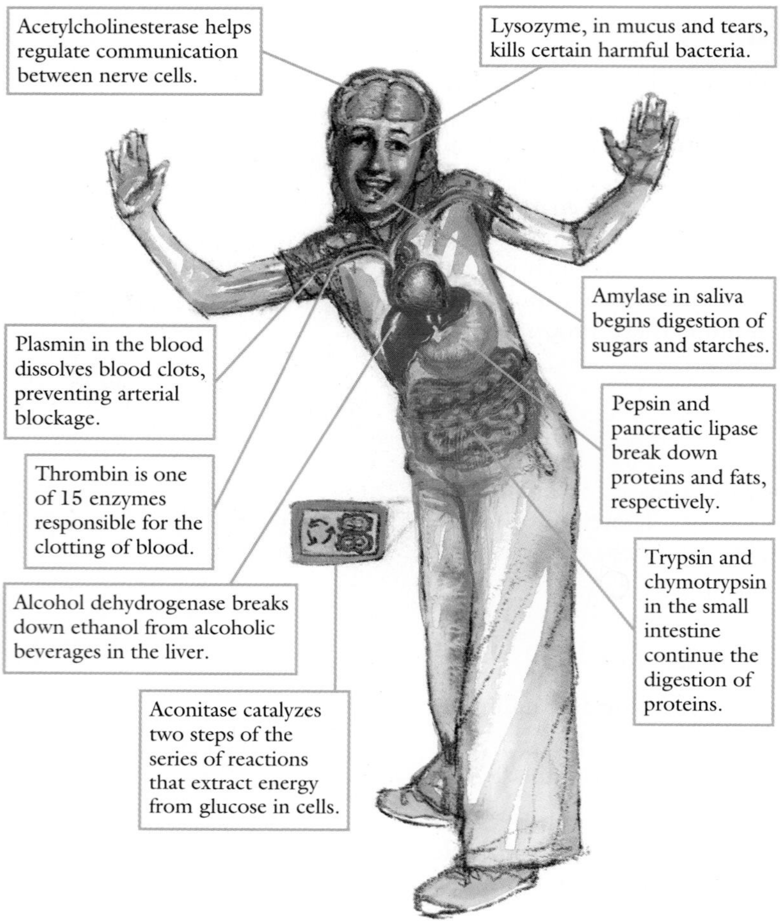
Acetylcholinesterase helps regulate communication between nerve cells.
Lysozyme, in mucus and tears, kills certain harmful bacteria.
Amylase in saliva begins digestion of sugars and starches.
Plasmin in the blood dissolves blood clots, preventing arterial blockage.
Pepsin and pancreatic lipase break down proteins and fats, respectively.
Thrombin is one of 15 enzymes responsible for the clotting of blood.
Trypsin and chymotrypsin in the small intestine continue the digestion of proteins.
Alcohol dehydrogenase breaks down ethanol from alcoholic beverages in the liver.
Aconitase catalyzes two steps of the series of reactions that extract energy from glucose in cells.

In many enzymatic reactions, a molecule closely resembling the normal substrate can bind to the active site. If such a molecule is present simultaneously with the normal substrate, both will compete for the active site. The most strongly bound molecule will win out if it is present in high enough concentrations. This is the basis for the action of drugs such as HIV protease inhibitors. If there is little difference in the degree of binding, the equilibrium can be shifted by increases in concentration of either the substrate or the substrate mimicker—that is, the competitor C. If we represent the enzyme–substrate complex as ES and the enzyme–competitor complex as EC, we can write the competition as

$$\mathrm{ES + C \rightleftharpoons EC + S}$$

This representation indicates that competitor can displace substrate, and vice versa. The enzyme–substrate complex concentration is favored when the substrate concentration, [S], is high, and the enzyme–competitor concentration is favored when the competitor concentration, [C], is high. The effects of certain kinds of toxins that bind to an enzyme's active site or to a transport protein can be reversed in this way. For example, carbon monoxide binds strongly to the iron in hemoglobin and displaces oxygen. This process can be reversed by increasing the oxygen concentration in the blood. To increase the blood's oxygen concentration, the patient is placed in a hyperbaric chamber in which the partial pressure of oxygen is increased by raising the atmospheric pressure to 3 or 4 atm.

Concept check

✔ Many biochemical reactions are reversible.

8.6 EQUILIBRIUM CONSTANTS

The quantitative study of chemical systems at equilibrium has shown that all chemical reactions can be described by an **equilibrium constant,** a mathematical function of the molar concentrations of reactants and products. If we propose that a reaction equation takes the form

$$a\,\mathrm{A} + b\,\mathrm{B} \rightleftharpoons c\,\mathrm{C} + d\,\mathrm{D}$$

where lowercase letters represent balancing coefficients and capital letters represent substances, the equilibrium constant is written as:

$$K_{\mathrm{eq}} = \frac{[\mathrm{C}]^c[\mathrm{D}]^d}{[\mathrm{A}]^a[\mathrm{B}]^b}$$

To write an equilibrium constant, four rules serve as a guide:

Rules for writing equilibrium constants

1. The products of the reaction always appear in the numerator, with reactants in the denominator.
2. Square brackets are used to indicate molar concentrations at equilibrium.
3. The molar concentration of each component at equilibrium is raised to a power equal to the component's coefficient in the balanced equation.
4. Pure solids or liquids to not appear in the equilibrium constant expression for a reaction in which they take part.

The first three of the four rules apply to equilibria of reactions in the gas phase or in solutions, defined as **homogeneous equilibria.** Many other equilibria consist of more than one phase and are called **heterogeneous equilibria.** These phases can be pure solid phases and pure liquid phases. Heterogeneous equilibria require rule 4. Because the equilibrium constant expression must take the physical states of the reaction components into account, these states must be specified for all components in the balanced equation. This specification is done by placing the following notations next to the components in the reaction equation: (*g*) for the gaseous state, (*l*) for the liquid state, (*s*) for the solid state, and (*aq*) for aqueous solution. You have already seen some of this notation in Sections 7.2 and 7.9.

The equilibrium constant changes with a change in temperature. The direction of change depends on whether the reaction is exo- or endothermic. If the reaction is exothermic—evolves heat—increasing the temperature will decrease the numerical value of the equilibrium constant, and vice versa, as discussed in more detail in Section 8.8. The temperatures of reactions specified in the following examples and problems are to be considered constant.

Example 8.2 Writing the equilibrium constant expression for a homogeneous equilibrium

Write the equilibrium constant for the following reaction in which all components are present as gases:

$$2\ N_2(g) + 3\ Br_2(g) \rightleftharpoons 2\ NBr_3(g)$$

Solution

The reaction equation specifies the balancing coefficients of all components, so the reaction's K_{eq} is expressed as

$$K_{eq} = \frac{[NBr_3]^2}{[N_2]^2[Br_2]^3}$$

Problem 8.2 Write the equilibrium constant for the following balanced equation:

$$H_2(g) + I_2(g) \rightleftharpoons 2\ HI(g)$$

Example 8.3 Writing the equilibrium constant expression for a heterogeneous equilibrium

Write the equilibrium constant expression for the reaction of solid carbon with gaseous oxygen to form gaseous carbon dioxide:

$$C(s) + O_2(g) \rightleftharpoons CO_2(g)$$

Solution

Because carbon is a pure solid, it will not appear in the equilibrium constant:

$$K_{eq} = \frac{[CO_2]}{[O_2]}$$

Problem 8.3 Write the equilibrium constant expression for the following reaction:

$$2\ C(s) + O_2(g) \rightleftharpoons 2\ CO(g)$$

The most important qualitative information that we get from the equilibrium constant is how far the reaction of the components proceeds toward products. If the equilibrium constant is much larger than 1.0—for example, 1×10^3—then the concentrations of product (components in the numerator) must be much larger than the concentrations of the reactants (components in the denominator). Such a reaction therefore yields significant amounts of product, and the equilibrium is called **favorable.** Conversely, if the equilibrium constant is much smaller than 1.0—say, 1×10^{-3}—small amounts of products are the result, and the reaction is called **unfavorable.**

Equilibrium constants tell us nothing about how fast a reaction proceeds. The equilibrium constant for the reaction between hydrogen and oxygen to form water is very large, but the reaction at room temperature takes many years to go to completion. This seems counterintuitive to the idea that a large equilibrium constant leads to large amounts of product. However, with a little help from a spark, which provides the required activation energy, the equilibrium state of large amounts of water can be rapidly attained.

Example 8.4 Estimating relative concentrations of reactants and products at equilibrium

Let us assume two possibilities for the equilibrium constant of the hypothetical reaction $A \rightleftharpoons B$: one where the equilibrium constant is much larger than 1.0, and the equilibrium concentration of B is far greater than that of A, and another where the equilibrium constant is much smaller than 1.0, and the equilibrium concentration of A is far greater than that of B.

Possibility 1. The equilibrium constant $K_{eq(1)} = \frac{[B]}{[A]} = 100$

Possibility 2. The equilibrium constant $K_{eq(2)} = \frac{[B]}{[A]} = 0.01$

For each possibility, estimate the concentration of product with respect to reactant at equilibrium.

Solution

Possibility 1. Rearrangement of the equilibrium constant expression gives

$$[B] = 100 \times [A]$$

In other words, the equilibrium concentration of B is 100 times the equilibrium concentration of A.

Possibility 2. Rearrangement of the equilibrium constant expression gives

$$[B] = 0.01 \times [A]$$

In other words, the equilibrium concentration of B is only 1/100th the equilibrium concentration of A.

Problem 8.4 Estimate the concentration of product with respect to reactant at equilibrium for the hypothetical reaction $A \rightleftharpoons B$, for which $K_{eq} = [B]/[A] = 1.0$.

Concept checklist

✔ The size of the equilibrium constant tells us about the yield of the reaction.

✔ When $K_{eq} > 1$, the reaction is favorable; and, at equilibrium, the product-to-reactant ratio is greater than one.

✔ When $K_{eq} < 1$, the reaction is unfavorable; and, at equilibrium, the product-to-reactant ratio is less than one.

Example 8.5 Calculating an equilibrium constant

For the following reaction at 25°C,

$$CO(g) + H_2O(g) \rightleftharpoons H_2(g) + CO_2(g)$$

the equilibrium concentrations are:

$$[CO] = [H_2O] = 0.09 \text{ mol/L}$$
$$[H_2] = [CO_2] = 0.91 \text{ mol/L}$$

Calculate the equilibrium constant.

Solution

The equilibrium constant expression is:

$$K_{eq} = \frac{[CO_2][H_2]}{[CO][H_2O]}$$

Substitute the equilibrium concentrations:

$$K_{eq} = \frac{[0.91][0.91]}{[0.09][0.09]} = 1 \times 10^2$$

Problem 8.5 In the reaction $2\ HI(g) \rightleftharpoons H_2(g) + I_2(g)$, the equilibrium concentrations at 25°C are: $[H_2] = [I_2] = 0.0011M$ and $[HI] = 0.0033\ M$. Calculate the value of the equilibrium constant.

Concept check

✔ If we know the equilibrium concentrations of reactants and products of a reaction, we can calculate the value of the equilibrium constant.

The equilibrium constant can be used to calculate the concentrations of components of a reaction at equilibrium if we know the starting concentrations of reactants. The general method for this calculation is shown in Box 8.2 on the following page.

8.7 BIOCHEMICAL REACTIONS ARE CONNECTED IN SEQUENCES

Cellular metabolism consists of sequences of reactions. Many of the individual reactions have unfavorable equilibrium constants; however, each sequence as a whole produces adequate quantities of the necessary end product by incorporating highly favorable reactions as intermediate steps. This idea is best illustrated by an example.

Example 8.6 Calculating the equilibrium constant for a reaction sequence

Consider a sequence of two separate reactions that have a common intermediate. Assume that the equilibrium constant for the first reaction is large and that for the second reaction is small. What will the overall equilibrium constant be for the total process?

Solution

To solve this problem, we must first construct a hypothetical two-step chemical process in which a reactant A leads to a product C in two steps connected by a common intermediate B. Each step written separately can be characterized by an equilibrium constant:

Reaction 1. A $\rightarrow$ B, equilibrium constant $K_1 = \dfrac{[B]}{[A]}$

Reaction 2. B $\rightarrow$ C, equilibrium constant $K_2 = \dfrac{[C]}{[B]}$

The overall reaction, A $\rightarrow$ C, is simply the result of adding reactions 1 and 2. This overall reaction is characterized by an equilibrium constant that also represents the overall process:

$$K_3 = \frac{[C]}{[A]}$$

The equilibrium constant for the reaction, K_3, is the product of the equilibrium constants for reactions 1 and 2:

$$K_3 = \frac{[B]}{[A]} \times \frac{[C]}{[B]} = \frac{[C]}{[A]} = K_1 \times K_2$$

Now assume a small value for K_1—say, 1×10^{-2}—and a large value for K_2—say, 1×10^4. The overall equilibrium constant would then be:

$$K_3 = K_1 \times K_2 = 1 \times 10^2$$

The overall equilibrium constant is quite large, showing that an unfavorable reaction can still produce significant amounts of product when it is one of a sequence of more favorable reactions. In Chapter 22, we will see that sequences of this kind are characteristic of metabolic processes.

Problem 8.6 Consider a set of three hypothetical reactions connected by common intermediates:

$$\begin{aligned} A &\longrightarrow B \quad K_1 = 1 \times 10^{-3} \\ B &\longrightarrow C \quad K_2 = 1 \times 10^{2} \\ C &\longrightarrow D \quad K_3 = 1 \times 10^{3} \end{aligned}$$

Calculate the overall equilibrium constant for the following reaction:

$$A \longrightarrow D$$

8.2 Chemistry in Depth

The Quantitative Description of Chemical Equilibrium

The distribution of mass among products and reactants of a reaction can be calculated by using the reaction's equilibrium constant. To illustrate the procedure, we will examine the reaction

$$CO(g) + H_2O(g) \rightleftharpoons H_2(g) + CO_2(g)$$

The equilibrium constant for this reaction is 1.00×10^2 at 25°C. Our approach will be to calculate how many moles of H_2 and CO_2 will be left at equilibrium if the reaction begins with 1.00 mol each of CO and H_2O in a 1.00-L container at 25°C. The stoichiometry of the reaction equation indicates that, for every molecule of CO and H_2O that disappears, one molecule each of CO_2 and H_2 will appear.

An accounting system to deduce the equilibrium concentrations for all components is constructed by defining the concentrations of all components of the equilibrium system for three phases of the reaction: at the start, in the course of the reaction when changes take place, and, finally, at the end, or at equilibrium. This accounting system takes the form of a table:

	[CO]	[H_2O]	[H_2]	[CO_2]
Start	1.00	1.00	0.0	0.0
Change	$-x$	$-x$	$+x$	$+x$
Final	$1.00 - x$	$1.00 - x$	x	x

The first line shows that, at the start of the reaction, 1 mol each of CO and H_2O are present and no products have yet formed. When the reaction has reached equilibrium, the stoichiometry tells us that, for every mole of CO that

Concept checklist

- ✔ In a reaction sequence, the products of one reaction become reactants in a second reaction; the products of that reaction may become reactants in a third, and so forth.
- ✔ In such cases, the equilibrium constants can be combined into one that characterizes the overall process.

8.8 LE CHATELIER'S PRINCIPLE

The equilibrium state of a liquid–vapor system is defined solely by its vapor pressure, which has a unique value at a fixed temperature (Section 6.6). Decreasing the volume of the gas phase of a liquid-water–water-vapor system (increasing its pressure) at a fixed temperature causes some of the gaseous water to condense to the liquid phase. This change reduces the temporarily increased gas pressure to its original equilibrium value. Conversely, an increase in the volume of the gas phase at a fixed temperature (decreasing its pressure) causes some vaporization of liquid to the gas phase, with the same result: the vapor pressure returns to its equilibrium value.

Just as a physical system in equilibrium can be temporarily displaced from equilibrium, a system in chemical equilibrium also can be temporarily displaced from equilibrium. However, a chemical system returns to its state of equilibrium by undergoing changes in the relative amounts of reactants and products. These amounts readjust so that the value of the equilibrium constant remains unchanged. In regard to the balanced chemical equation, there is a shift in mass from left to right or right to left, depending on the nature of the imposed change, or stress.

Qualitative predictions of how a chemical system will respond to a disturbance in its equilibrium conditions are described by **Le Chatelier's principle:**

- If a system in an equilibrium state is disturbed, the system will adjust so as to counteract the disturbance and restore the system to equilibrium.

has reacted, 1 mol of H_2O has disappeared and 1 mol each of H_2 and CO_2 has appeared. Because the actual amounts are unknown, we can insert the value of x in the table for the amounts that have disappeared and appeared, and the 1:1 ratio of moles disappearing and appearing is maintained.

The final concentrations, as they appear in the bottom line of the accounting table, are substituted into the equilibrium constant expression:

$$K_{eq} = \frac{[CO_2][H_2]}{[CO][H_2O]} = 1.00 \times 10^2$$

$$K_{eq} = \frac{[x][x]}{[1.0 - x][1.0 - x]} = \frac{[x]^2}{[1.0 - x]^2} = 1.00 \times 10^2$$

The relation is now in the form of a quadratic equation, which has a formally prescribed method of solution. However, because the expressions on both sides of the equal sign are perfect squares, their square roots can be taken directly and the simplified expression then solved for x:

$$\frac{[x]}{[1.0 - x]} = 10.0$$

$$[x] = 10.0 - 10.0[x]$$

$$11.0[x] = 10.0$$

$$[x] = 0.910$$

Therefore, at equilibrium,

$$[CO] = [H_2O] = (1.00\ \text{mol/L} - 0.910\ \text{mol/L}) = 0.090\ \text{mol/L}$$

$$[H_2] = [CO_2] = 0.910\ \text{mol/L}$$

This method will be used in Chapter 9 to quantitatively characterize solutions of acids and bases.

This principle applies to both chemical and physical systems. For example, consider the following reaction to be at equilibrium:

$$N_2(g) + 3\ H_2(g) \rightleftharpoons 2\ NH_3(g)$$

How would changes in concentration of any of the components of the system affect the distribution of all the chemical components? This question is best answered by focusing on the rates of the forward and back reactions of the dynamic equilibrium and noting that the rate of a chemical reaction increases with increases in the concentrations of the reactants.

Adding nitrogen or hydrogen or both to the reaction mixture at equilibrium will increase the rate of the forward reaction, resulting in an increase in the concentration of ammonia and a new equilibrium mixture of N_2, H_2, and NH_3. In other words, adding mass to the left-hand side of the reaction equation causes the equilibrium to shift to the right. Adding ammonia to an equilibrium mixture will increase the rate of the back reaction, resulting in an increase in the concentration of hydrogen and nitrogen and a new equilibrium mixture of H_2, N_2, and NH_3. As was the case with the forward reaction, adding mass to the right-hand side of the reaction equation causes the equilibrium to shift to the left. The application of Le Chatelier's principle in a particular biochemical context is described in Section 23.5.

▶▶ Biochemists use Le Chatelier's principle to isolate compounds called acetals, as described in Chapter 14.

Example 8.7 Using Le Chatelier's principle

Suppose some $N_2O_4(g)$ were added to the following equilibrium system: $N_2O_4(g) \rightleftharpoons 2\ NO_2(g)$. What would the effect on the equilibrium mixture be?

Solution

The stress on the disturbed equilibrium mixture would be relieved and equilibrium would be reestablished by a reduction in the N_2O_4 concentration, which would lead to an increase in the NO_2 concentration. The reaction would shift to the right. However, note that both N_2O_4 and NO_2 concentrations would now be greater than in the preceding equilibrium state.

Problem 8.7 Suppose some $H_2(g)$ were added to the following reaction: $2\ HI(g) \rightleftharpoons H_2(g) + I_2(g)$. What would the effect on the equilibrium mixture be?

Many reactions yield not only new substances, but heat as well. In these cases, the equilibrium state of the chemical system may also be displaced by the addition or removal of heat. The equilibrium will shift to either the product or the reactant side of the reaction, depending on the direction in which heat is evolved or absorbed. Thus, changes in temperature cause quantitative changes in the equilibrium constant.

Chemical equations can be written to show the exothermic or endothermic nature of a reaction. In exothermic reactions, heat is shown as a product; in endothermic reactions, it is listed as a reactant:

Exothermic:	$A + B \rightleftharpoons C + D + \text{heat}$	$\Delta H_{rxn} = (-)$
Endothermic:	$A + B + \text{heat} \rightleftharpoons C + D$	$\Delta H_{rxn} = (+)$

Example 8.8 Predicting the effect of heat addition or removal on a reaction

The reaction of N_2 with H_2 not only produces NH_3, but heat as well:

$$N_2 + 3\ H_2 \rightleftharpoons 2\ NH_3 + 98\ \text{kJ}$$

What changes would the system undergo if (a) heat were added to it and (b) heat were removed?

Solution

The balanced equation indicates that the reaction evolves heat—that is, it is exothermic—and the heat of reaction is designated as a negative quantity, $\Delta H_{rxn} = -98$ kJ (the system has lost heat).

(a) Because this equilibrium system evolves heat on reaction, adding heat to it is equivalent to increasing one of the components on the right-hand, or product, side. This stress is relieved by an equilibrium shift to the left, resulting in the absorption of heat with more reactants being formed.

(b) Removal of heat is equivalent to reducing one of the components on the left-hand, or product, side. This stress is relieved by an equilibrium shift to the right, resulting in the evolution of heat and more ammonia formed.

Problem 8.8 Consider the following equilibrium system:

$$CaCO_3(s) + 158 \text{ kJ} \rightleftharpoons CaO(s) + CO_2(g)$$

In which direction will the equilibrium shift if heat is (a) added or (b) removed from the system?

Box 8.3 describes the industrial synthesis of ammonia by the Haber process. This process is an excellent example of how chemists have learned to

8.3 Chemistry Around Us

Nitrogen Fixation: The Haber Process

Living organisms require nitrogen, but most of them cannot take it directly out of the air. Instead, metabolizable nitrogen is acquired in the form of nitrates, ammonia, or ammonia derivatives. Before 1915, the guano (shorebird droppings) beds of coastal Chile were the principal source of nitrogen-containing compounds used commercially for both life and death—fertilizer and explosives. During World War I, the Allies placed an embargo on these guano fields to keep Germany from acquiring the guano for the manufacture of explosives. Pressed for survival, the Germans developed a chemical process for fixing atmospheric nitrogen. The key to this synthetic process was a catalyst invented by Fritz Haber, who received a Nobel Prize in chemistry in 1918 for his discovery.

Modern agriculture has developed in large measure because of the invention of the Haber process for fixing nitrogen.

It is difficult to overemphasize the effect of this process on modern civilization. Nitrogen fixation by the Haber process is the foundation of both modern agriculture and the commercial synthesis of a wide variety of chemicals and chemical products.

The process uses the direct reaction between hydrogen and nitrogen gases to produce ammonia. This reaction is exothermic, so the percentage of ammonia in the equilibrium mixture decreases as the temperature is raised. However, the proportion of ammonia in the equilibrium mixture increases as the total pressure is increased. The way to get a good yield would therefore seem to be to raise the pressure and operate at low temperatures. Unfortunately, the rate of the reaction at low temperatures is extremely slow, and one must wait for commercially unfeasible lengths of time for a significant yield. Haber's solution to this problem was to invent a catalyst that significantly reduced the activation energy of the reaction and therefore reduced the time required to reach equilibrium. When the catalyst was employed at high pressures and moderate temperatures, it quickly led to a very favorable yield of ammonia.

use concentrations, temperature, and catalysts to obtain economically useful yields of chemical products.

In the next chapter, we will apply the general principles of chemical equilibrium to systems in which substances called acids and bases react with water to form new ionic species. The normal physiological function of cells and cellular systems depends strongly on these ionic equilibria.

Summary

Reaction Rates The rate at which a chemical reaction takes place is measured in moles of reactant disappearing or moles of product appearing per unit time. Experiments have shown that an increase in the concentration of reactants causes the rate of reaction to increase. The reaction rate also increases with a rise in temperature.

Reactive Collisions Chemical reactions take place as the result of reactive collisions between properly oriented molecules. In a reactive collision, two molecules combine into one complex molecule called an activated complex. The energy required to form the activated complex is called the activation energy. The activated complex can decay either into products or back into reactants. If the energy of the products is less than that of the reactants, heat is lost during the reaction and the reaction is called exothermic. If the opposite is true, the reaction absorbs heat and is called endothermic. The activation energy is a unique property of a given reaction system and controls the rate of reaction. The greater the activation energy, the slower the reaction.

Catalysts Catalysts are agents that increase the rate of a reaction but emerge unchanged after the reaction has ended. A catalyst lowers the activation energy of a reaction by providing a different pathway leading to products. It does so by becoming an active participant in the chemical process. It does not change the final result, only the rate of reaction.

Biochemical Catalysts Chemical processes within living cells are catalyzed by enzymes. The catalytic properties of an enzyme are based on the enzyme's ability to form an activated complex with reactants. The complex can take many different forms, but, in all cases, it is the formation of the complex that results in a reaction pathway with a low overall activation energy and, consequently, a more rapid reaction rate.

Chemical Equilibrium When the concentrations of reacting substances no longer change, the reaction system is considered to be at equilibrium. The relation among the concentrations of reactants and products at that point is quantitatively described by the equilibrium constant, which can be used to estimate the amount of products relative to reactants.

Le Chatelier's Principle A system in chemical equilibrium may be subjected to changes in conditions that will displace it from equilibrium, but the system will seek to reestablish its equilibrium through compensating changes in the relative amounts of reactants and products. This concept is called Le Chatelier's principle.

Key Words

activated complex, p. 204
catalyst, p. 208
chemical kinetics, p. 203
chemical equilibrium, p. 211
complementarity, p. 209
enzyme, p. 209
heat of reaction, p. 206
Le Chatelier's principle, p. 217
product, p. 203
reactant, p. 203
reaction rate, p. 203

Exercises

Reaction Rates

8.1 In the reaction $3\ O_2(g) \rightarrow 2\ O_3(g)$ (ozone), the rate of appearance of ozone was 1.50 mol/min. What was the rate of disappearance of oxygen?

8.2 In the reaction $3\ H_2(g) + N_2(g) \rightarrow 2\ NH_3(g)$, the rate of disappearance of hydrogen was 0.60 mol/min. What was the rate of disappearance of nitrogen?

8.3 In the following reaction of nitrogen and oxygen,

$$N_2(g) + O_2(g) \rightleftharpoons 2\ NO(g)$$

explain why the reaction rate doubled when the temperature was raised only 10 kelvins.

8.4 In the following reaction of nitrogen and oxygen,

$$N_2(g) + O_2(g) \rightleftharpoons 2\ NO(g)$$

explain why the reaction rate increased when the concentration of nitrogen was increased.

8.5 The activation energies for each of two reactions were found to be (a) 24 kJ and (b) 53 kJ. If the temperatures of both reactions are identical, which has the greater rate of reaction?

8.6 Hydrogen peroxide, H_2O_2, is unstable and slowly decomposes over time to form water and gaseous oxygen. However, in the presence of the enzyme catalase, its rate of decomposition was increased by a factor of more than 1×10^6, with no change in the H_2O_2 concentration. Explain.

8.7 The heat of a reaction, ΔH_{rxn}, is −21 kJ and the activation energy of the forward reaction, $E_{forward}$, is +37 kJ. Calculate the activation energy of the reverse reaction, E_{back}.

8.8 The heat of a reaction, ΔH_{rxn}, is +12 kJ, and the activation energy of the forward reaction, $E_{forward}$, is −46 kJ. Calculate the activation energy of the reverse reaction, E_{back}.

Chemical Equilibrium

8.9 Calcium carbonate is heated in a crucible open to the atmosphere. The reaction that takes place is a decomposition:

$$CaCO_3(s) \rightleftharpoons CaO(s) + CO_2(g)$$

Is the reaction reversible as performed? Explain.

8.10 The decomposition reaction

$$2\ HCl(aq) + CaCO_3(s) \longrightarrow H_2O(l) + CO_2(g) + CaCl_2(aq)$$

is used as a field test for the detection of limestone. When the reaction is conducted in the field in the open air, is it reversible? Explain.

8.11 For the reaction $CaCO_3(s) \rightleftharpoons CaO(s) + CO_2(g)$, which is the forward and which is the back reaction?

8.12 For the reaction $CaO(s) + CO_2(g) \rightleftharpoons CaCO_3(s)$, which is the forward reaction and which is the back reaction?

8.13 Write the equilibrium constant expression for the reaction in Exercise 8.11.

8.14 Write the equilibrium constant expression for the reaction in Exercise 8.12.

8.15 In the reaction $H_2(g) + I_2(g) \rightleftharpoons 2\ HI(g)$, the equilibrium concentrations were found to be $[H_2] = [I_2] = 0.86\ M$ and $[HI] = 0.27\ M$. Calculate the value of the equilibrium constant.

8.16 In the reaction $NH_4Cl(s) \rightleftharpoons NH_3(g) + HCl(g)$, the equilibrium concentrations were found to be $[NH_3] = [HCl] = 3.71 \times 10^{-3}\ M$. Calculate the value of the equilibrium constant.

Le Chatelier's Principle

8.17 For the reaction $N_2(g) + 3\ H_2(g) \rightleftharpoons 2\ NH_3(g)$, in what direction will the equilibrium shift if some NH_3 is removed from the equilibrium mixture?

8.18 For the reaction $N_2(g) + 3\ H_2(g) \rightleftharpoons 2\ NH_3(g)$ + 92 kJ, what will the effect of increasing the temperature be on the extent of the reaction?

8.19 In each of the following aqueous equilibria, a visual change can be observed when there is a shift in the equilibrium. Predict the visual change that occurs when the indicated stress is applied.

(a) Cool the following reaction:

$$\text{Heat} + Co^{2+}(\text{pink}) + 4\ Cl^-(\text{colorless}) \rightleftharpoons CoCl_4^{2-}(\text{blue})$$

(b) Add Cl^- to the following reaction:

$$\text{Heat} + Co^{2+}(\text{pink}) + 4\ Cl^-(\text{colorless}) \rightleftharpoons CoCl_4^{2-}(\text{blue})$$

8.20 In each of the following aqueous equilibria, a visual change can be observed when there is a shift in the equilibrium. Predict the visual change that occurs when the indicated stress is applied.

(a) Add Fe^{3+} to the following reaction:

$$Fe^{3+}(\text{brown}) + 6\ SCN^-(\text{colorless}) \rightleftharpoons Fe(SCN)_6^{3-}(\text{red})$$

(b) Heat the following reaction:

$$Pb^{2+}(\text{colorless}) + 2\ Cl^-(\text{colorless}) \rightleftharpoons PbCl_2(\text{white solid}) + \text{heat}$$

Unclassified Exercises

8.21 What is the meaning of the square brackets around the chemical components in equilibrium constant expressions?

8.22 Write the equilibrium constant expressions for each of the following reactions:

(a) $C(s) + O_2(g) \rightleftharpoons CO_2(g)$
(b) $COCl_2(g) \rightleftharpoons CO(g) + Cl_2(g)$

8.23 Write the equilibrium constant expressions for each of the following reactions:

(a) $CH_4(g) + H_2O(g) \rightleftharpoons CO(g) + 3\ H_2(g)$
(b) $C_8H_{16}(l) + 12\ O_2(g) \rightleftharpoons 8\ CO_2(g) + 8\ H_2O(g)$

8.24 Write the equilibrium constant expressions for each of the following reactions:

(a) $S_2(g) + C(s) \rightleftharpoons CS_2(g)$
(b) $ZnO(s) + CO(g) \rightleftharpoons Zn(s) + CO_2(g)$

8.25 Write the equilibrium constant expressions for each of the following reactions:

(a) $NH_2COONH_4(s) \rightleftharpoons 2\ NH_3(g) + CO_2(g)$
(b) $2\ Cu(s) + 0.5\ O_2(g) \rightleftharpoons Cu_2O(s)$

8.26 Given the equilibrium reaction

$$131\ \text{kJ} + C(s) + H_2O(g) \rightleftharpoons CO(g) + H_2(g)$$

complete the following table:

Change	Effect on Equilibrium
(a) Increase in temperature	________
(b) Increase in concentration of $H_2O(g)$	________
(c) Increase in concentration of products	________
(d) Decrease in temperature	________
(e) Adding a catalyst	________

8.27 Write the balanced equation corresponding to each of the following equilibrium constants:

(a) $K_{eq} = \dfrac{[CO][Cl_2]}{[COCl_2]}$

(b) $K_{eq} = \dfrac{[CO][H_2]^3}{[CH_4][H_2O]}$

8.28 Write the balanced equation corresponding to each of the following equilibrium constants:

(a) $K_{eq} = \dfrac{[PCl_5]}{[Cl_2][PCl_3]}$

(b) $K_{eq} = \dfrac{[O_3]^2}{[O_2]^3}$

8.29 Write the equilibrium constant expression for each of the following reactions:

(a) $N_2O_5(g) \rightleftharpoons NO_2(g) + NO_3(g)$

(b) $4\ NH_3(g) + 3\ O_2(g) \rightleftharpoons 2\ N_2(g) + 6\ H_2O(g)$

8.30 Write the equilibrium constant expression for each of the following reactions:

(a) $2\ NO(g) + 2\ H_2(g) \rightleftharpoons N_2(g) + 2\ H_2O(g)$

(b) $2\ Mg(s) + O_2(g) \rightleftharpoons 2\ MgO(s)$

8.31 The following equilibrium constants refer to the following reaction:

$$A(g) \rightleftharpoons B(g)$$

Predict the ratios of [A]/[B] at equilibrium in each of the following cases. (a) $K_{eq} = 1 \times 10^{-3}$; (b) $K_{eq} = 1 \times 10^{2}$; (c) $K_{eq} = 0.14$; (d) $K_{eq} = 0.0832$.

8.32 Write the equilibrium constants for each of the following reactions:

(a) $N_2O_4(g) \rightleftharpoons 2\ NO_2(g)$

(b) $2\ H_2(g) + O_2(g) \rightleftharpoons 2\ H_2O(g)$

8.33 When ammonium thiocyanate (NH_4CNS) dissolves in water, the solution becomes cold. If an aqueous solution of NH_4CNS together with additional solid NH_4CNS were heated, would some of the solid dissolve or would more solid precipitate?

8.34 When lithium iodide (LiI) is dissolved in water, the solution becomes hot. If an aqueous solution of LiI together with additional solid LiI were heated, would some of the solid dissolve or would more solid precipitate?

8.35 The ΔH_{rxn} of a reaction is +15 kJ, and the activation energy of the back reaction is $E_{back} = 57$ kJ. What is the activation energy of the forward reaction, $E_{forward}$?

Chemical Connections

8.36 One of the ways in which ozone is destroyed in the upper atmosphere is described by the following sequence of reactions:

$$O_3(g) + NO(g) \longrightarrow O_2(g) + NO_2(g)$$
$$NO_2(g) + O(g) \longrightarrow NO(g) + O_2(g)$$
$$\text{Sum:} \quad O_3(g) + O(g) \longrightarrow 2\ O_2(g)$$

Which chemical component is a catalyst and which is a reaction intermediate?

8.37 The heat of reaction of a chemical process was found to be +16 kJ/mol. Which energy of activation was larger, that of the forward reaction or that of the back reaction?

8.38 Magnesium hydroxide, $Mg(OH)_2$, is slightly soluble in water. The solubility can be described by the following equilibrium:

$$Mg(OH)_2(s) \rightleftharpoons Mg^{2+}(aq) + 2\ OH^-(aq)$$

What will happen to the equilibrium if hydrogen ion, H^+, is added to the solution?

8.39 State Le Chatelier's principle.

8.40 Many substances crystallize from aqueous solution in the form of a hydrate. The formula unit of a hydrate indicates the number of water molecules associated with it. Write the equilibrium constant for the following reaction:

$$CuSO_4 \cdot 5\ H_2O(s) \rightleftharpoons CuSO_4(s) + 5\ H_2O(g)$$

Deep blue solid copper(II) sulfate pentahydrate is heated, water vapor is driven off, and white solid copper(II) sulfate is formed.

CHAPTER 9

ACIDS, BASES, AND BUFFERS

CHEMISTRY IN YOUR FUTURE

Mr. Keenan, who was admitted to the medical-surgical ward at 4 this afternoon, is nervous about tomorrow's operation. His anxiety increases as the evening wears on, and he tries to breathe deeply to calm his rising fears. Instead, he begins to breathe more and more rapidly, until he cannot catch his breath. His room partner calls you (the night nurse), and you produce a paper bag and place it over Mr. Keenan's mouth and nose. His breathing becomes less labored and finally returns to normal. You give him a tranquilizer, and he rests comfortably. Apparently, you are familiar with the material presented in Chapter 9.

LEARNING OBJECTIVES

- Describe the central role of water in acid-base chemistry.
- Explain the difference between strong and weak acids and strong and weak bases.
- Use the ion-product of water to calculate hydronium ion and hydroxide ion concentrations.
- Define pH and describe its use as a measure of acidity.
- Describe the behavior of weak acids and bases in terms of chemical equilibria.
- Use the Brønsted-Lowry theory to explain the properties of salts of weak acids and bases and of buffers.
- Describe how the concentrations of acids and bases are determined by titration.

⏩ Our overview of cell structure and chemistry begins in Chapter 22.

In this chapter, we focus on an area of chemical equilibrium that is central to the life of the cell: the equilibria established when water reacts with two large groups of substances, acids and bases.

Foods, beverages, and household cleaning aids are familiar substances composed of or containing acids and bases. Acids impart the typical tang to carbonated drinks and the tartness of citrus fruits, and bases are responsible for the bitterness of, for example, soapy water. When we eat the "wrong" foods and complain of heartburn (in reality, excess acid produced in the stomach), relief is available in the form of antacids (compounds that react with the excess acid and neutralize its effects). The health of cells critically depends on a balance between the acids produced in metabolism and the body's ability to neutralize them.

Some of the properties of acids and bases were presented in earlier chapters: (Sections 2.6 and 4.6). A brief summary of some other properties of acids and bases is given in Table 9.1. One of the more important chemical properties listed is neutralization, the reaction of acids with bases to form salts, which is considered in detail in Section 9.10.

A property very useful to chemists is that acidic and basic solutions can change the colors of various dyes. Litmus is a naturally occurring dye that turns red in acidic solutions and blue in basic solutions. Figure 9.1 shows how paper strips impregnated with litmus can be used to characterize the acidity or basicity of a solution. Red litmus paper turns blue in contact with basic solution, and blue litmus paper turns red in contact with acid solution.

Although we encountered many of the chemical properties of acids and bases earlier, we will now enhance our practical understanding of their effects in biochemical and physiological processes by taking a quantitative look at these properties and by considering the central role of water in acid-base chemistry. Our study of acids and bases begins with a look at how water molecules react with each other to form hydronium and hydroxide ions.

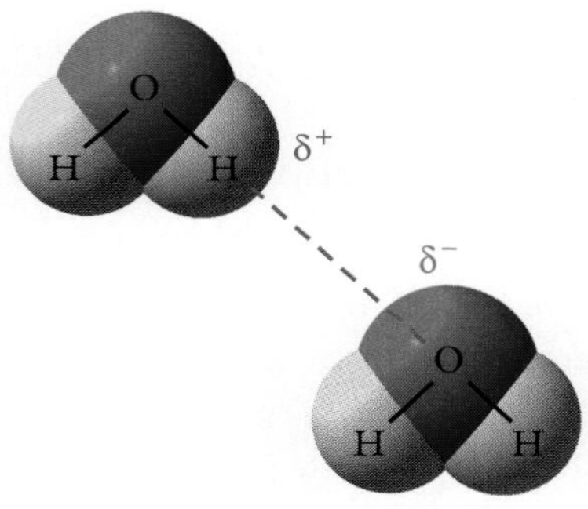

Hydrogen-bonded water molecules

9.1 WATER REACTS WITH WATER

Recall from the discussion in Section 6.3 that, in liquid water, water molecules interact with each other through hydrogen bonding. The δ^+ and δ^- in the drawing in the margin indicate partial electric charges that result from the strong electronegativity of the oxygen atom. (Only ions have full electric charges.) These partial charges and the molecule's bent geometry give rise to the dipole moment of the water molecule. The hydrogen bond, represented by a dashed line, is a weak secondary bond and is somewhat longer than the covalent bonds of the water molecule. A hydrogen bond in water is about 1.8 Å in length; the covalent bond in water is about 1.0 Å long (see Table 1.3).

TABLE 9.1 Properties of Acids and Bases

Acids	Bases
When dissolved in water, produce hydrogen ions	When dissolved in water, produce hydroxide ions
Neutralize bases to produce water and salts	Neutralize acids to produce water and salts
Solutions turn blue litmus paper red	Solutions turn red litmus paper blue

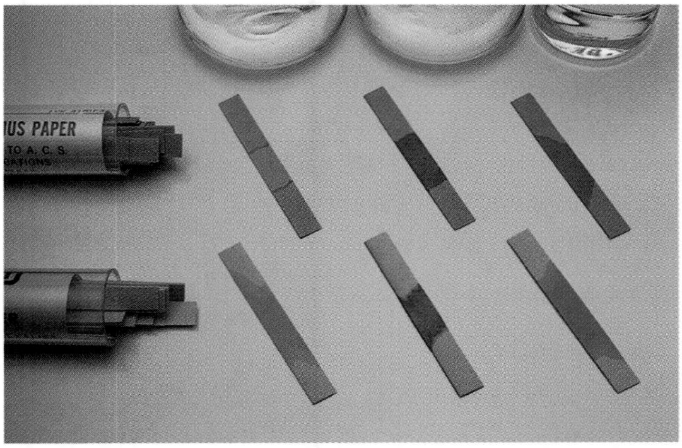

Figure 9.1 The use of litmus paper as a quick test for acidity or basicity. Litmus turns red with acid and blue with base. In the upper row, from left to right, blue litmus paper strips have been treated with acid, base, and pure water. In the lower row, from left to right, red litmus paper has been treated with acid, base, and pure water.

Part of the motion of atoms within molecules consists of vibrations, or back and forth movements, that alternately stretch and shorten the chemical bonds. At any instant in time, the hydrogen atom participating in a hydrogen bond may find itself much closer to the oxygen atom of the adjacent water molecule than to the oxygen atom of its own water molecule. The diagram in the margin shows that the former hydrogen bond has shortened to become a covalent bond and the former covalent bond has lengthened to become a hydrogen bond. The electrons of the hydrogen atom are left behind, and the resulting proton is covalently bound to the water molecule. The process creates two new molecular species that bear full electric charges: an H_3O^+, or **hydronium ion,** and an OH^-, or **hydroxide ion.** In subsequent discussions, it will become convenient to refer to "the protons of an acid," and to occasionally use the symbol H^+, rather than H_3O^+, in reaction equations. However, you must never forget that there can be no free protons in water, only hydronium ions.

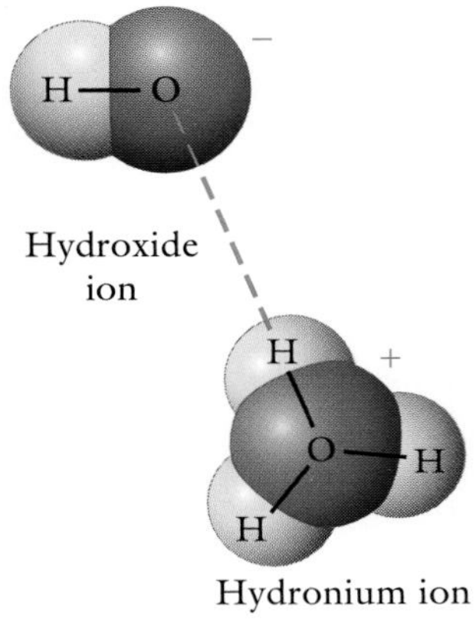

Formation of hydronium and hydroxide ions from hydrogen-bonded water molecules

In pure water, at any instant in time, only about one in a billion water molecules has undergone this reaction. The chemical equilibrium that is established can be represented by structural formulas:

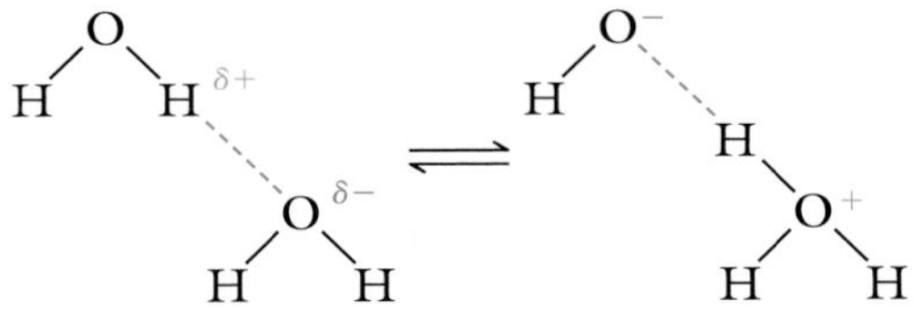

or by the more usual chemical equation:

$$H_2O(l) + H_2O(l) \longrightarrow H_3O^+(aq) + OH^-(aq)$$

This equilibrium reaction is of crucial importance to biological systems, and we need to look at it in detail.

A constant called the **ion-product of water (K_w)** is used to represent this equilibrium process. K_w is defined as the product of the molar concentrations of H_3O^+ and OH^- ions in pure water:

$$K_w = [H_3O^+][OH^-]$$

The concentrations of the H_3O^+ and OH^- ions in pure water are equal. They have been carefully measured and found to be 1.00×10^{-7} *M* at 25°C. Therefore,

$$K_w = [H_3O^+][OH^-] = 1.00 \times 10^{-14}$$

We can obtain two pieces of information from the expression for the ion-product of water. First, very little water exists in the form of its ion products; that is, K_w is very small. Second, given that the product of hydronium ion and hydroxide concentrations is a constant, we know that, when the concentration of H_3O^+ in water increases, the concentration of OH^- must decrease, and vice versa. Thus three situations can exist for the relative concentrations of hydronium and hydroxide ions in water:

- In a neutral solution, $[H_3O^+]$ is equal to $[OH^-]$.
- In acidic solutions, $[H_3O^+]$ is greater than $[OH^-]$.
- In basic solutions, $[OH^-]$ is greater than $[H_3O^+]$.

These definitions of acidic and basic solutions will become clearer in the next section.

9.2 STRONG ACIDS AND STRONG BASES

Experiments have shown that, when hydrogen chloride, HCl (*g*), is added to water, no HCl molecules are found in solution, only H_3O^+ ions and Cl^- ions. This observation means that each HCl molecule reacts with a water molecule to form a hydronium ion (H_3O^+) and a chloride ion. The aqueous HCl solution is called hydrochloric acid.

$$HCl(g) + H_2O(l) \longrightarrow H_3O^+(aq) + Cl^-(aq)$$

The process is called **dissociation** because it results in the formation of ions. Because HCl undergoes complete dissociation—every molecule added to water leads to the formation of a hydronium ion—it is called a **strong acid.** Hydrochloric acid is one of six strong acids that undergo complete dissociation in water (Table 9.2). It is the acid secreted by the parietal glands of your stomach as one of the first steps in the digestive process discussed in Section 26.3.

Similarly, when the base potassium hydroxide (KOH) is added to water, no KOH molecules are found in the solution, only hydrated K^+ and OH^- ions. The reason is that potassium hydroxide is already an ionic solid, composed of hydroxide anions and potassium cations. Its component ions simply separate when the ionic solid dissolves in water. Bases that are completely dissociated in water are called **strong bases** (see Table 9.2).

These substances are called "strong" acids and bases because there are many other acids and bases that do not fully dissociate in water. Such substances, called "weak" acids and bases, are discussed later in this chapter. Note

TABLE 9.2 Names and Formulas of All the Strong Acids and Bases

Strong Acids		Strong Bases	
Formula	**Name**	**Formula**	**Name**
$HClO_4$	perchloric acid	LiOH	lithium hydroxide
HNO_3	nitric acid	NaOH	sodium hydroxide
H_2SO_4	sulfuric acid*	KOH	potassium hydroxide
HCl	hydrochloric acid	RbOH	rubidium hydroxide
HBr	hydrobromic acid	CsOH	cesium hydroxide
HI	hydroiodic acid	TlOH	thallium(I) hydroxide
		$Ca(OH)_2$	calcium hydroxide
		$Sr(OH)_2$	strontium hydroxide
		$Ba(OH)_2$	barium hydroxide

* Only the first proton in sulfuric acid is completely dissociable. The first product of dissociation, HSO_4^-, is a weak acid.

that the terms strong and weak as used here do not denote the concentrations of acids and bases. The words strong and weak tell us only that the substance does or does not fully dissociate in solution. The listings in Table 9.2 include all the common strong acids and bases.

Concept checklist

- ✔ Strong acids are acidic compounds that undergo complete dissociation in water.
- ✔ Strong bases are basic compounds that undergo complete dissociation in water.

Many metals react with acids to produce gaseous hydrogen and metallic salts. Figure 9.2 shows the difference between the reactions of magnesium metal with a solution of a strong acid and with a solution of a weak acid of the same concentration. Because the concentration of hydronium ion in the strong acid solution is greater than that in the weak acid solution, the reaction of magnesium is more vigorous with a strong acid.

Because strong acids and strong bases are fully dissociated in water, we can easily calculate the hydronium or hydroxide ion concentrations in solutions of strong acids and bases as long as we know the solution concentrations. Example 9.3 will show, however, that it is also important to keep stoichiometry in mind.

Figure 9.2 The reactions of a strong acid (top) and a weak acid (bottom) at the same solution concentrations. The reaction is between hydronium ion and magnesium. The strong acid provides much more hydronium ion than does the weak acid at the same solution concentration and therefore causes a more vigorous reaction.

Example 9.1 Calculating ion concentrations in solutions of a strong acid

What is the concentration of hydronium ion in a 0.010 *M* aqueous solution of HCl? What is the total ionic concentration?

Solution

HCl, listed in Table 9.2 as a strong acid, is completely dissociated in aqueous solution; that is, 1 mol of HCl gives rise to 1 mol of H_3O^+ ion and 1 mol of Cl^- ion. In a 0.010 *M* solution, the concentrations of H_3O^+ and Cl^- are each 0.010 *M*. The total concentration of ions is therefore 0.020 *M*.

Problem 9.1 What is the concentration of hydronium ion in a 0.050 *M* aqueous solution of HCl?

Example 9.2 Calculating ion concentrations in solutions of a strong base

What is the concentration of hydroxide ion in a 0.0250 *M* aqueous solution of NaOH? What is the total ionic concentration?

Solution

NaOH is listed in Table 9.2 as a strong base and therefore consists completely of ions in aqueous solution: 1 mol of NaOH gives rise to 1 mol of Na^+ ion and 1 mol of OH^- ion. In a 0.0250 *M* solution, the concentrations of OH^- and Na^+ are each 0.0250 *M*, and the total ionic concentration is 0.0500 *M*.

Problem 9.2 What is the concentration of hydroxide ion in a 0.050 *M* aqueous solution of NaOH?

Example 9.3 Calculating ion concentrations in solutions of a strong base of a divalent cation

What is the concentration of hydroxide ion in a 0.020 *M* aqueous solution of $Ca(OH)_2$? What is the total ionic concentration?

Solution

Table 9.2 lists $Ca(OH)_2$ as a strong base. For this calculation, it is important to note the stoichiometry: 2 mol of OH^- ion per 1 mol of base. Thus the concentration of Ca^{2+} is 0.020 *M*, but the concentration of OH^- is twice that of Ca^{2+}, or 0.040 *M*. The total ionic concentration is 0.060 *M*.

Problem 9.3 What is the concentration of hydroxide ion in a 0.030 *M* aqueous solution of $Sr(OH)_2$?

The ion-product of water at 25°C, $K_w = 1.00 \times 10^{-14}$, provides us with a method for calculating either the hydronium ion concentration of an aqueous solution, given the hydroxide ion concentration, or the hydroxide ion concentration, given the hydronium ion concentration.

Example 9.4 Using the ion-product of water to calculate the H_3O^+ concentration in a solution of known OH^- concentration

Calculate the hydronium ion concentration in an aqueous 0.00100 *M* NaOH solution.

Solution

Because NaOH is a strong base, the hydroxide ion concentration of a 0.00100 *M* solution is 0.00100 *M*. We can substitute the value for hydroxide ion concentration into the ion-product expression and calculate the hydronium ion concentration:

$$K_w = [H_3O^+][OH^-] = [H_3O^+] \times 0.00100 = 1.00 \times 10^{-14}$$
$$[H_3O^+] = 1.00 \times 10^{-11}$$

Problem 9.4 Calculate the hydronium ion concentration in an aqueous 0.00500 *M* KOH solution.

Example 9.5 Using the ion-product of water to calculate the OH^- concentration in a solution of known H_3O^+ concentration

Calculate the hydroxide ion concentration in an aqueous 0.00200 *M* HCl solution.

Solution

Because HCl is a strong acid, the hydronium ion concentration of a 0.00200 *M* HCl solution is 0.00200 *M*. Substituting the value for hydroxide ion concentration into the ion-product expression yields:

$$K_w = [H_3O^+][OH^-] = 0.00200 \times [OH^-] = 1.00 \times 10^{-14}$$
$$[OH^-] = 5.00 \times 10^{-12}\ M$$

Problem 9.5 Calculate the hydroxide ion concentration in an aqueous 0.00500 *M* HNO_3 solution.

9.3 A MEASURE OF ACIDITY: pH

The acidity of a solution is measured in terms of the molar concentration of hydronium ions, a concentration that typically ranges from 1.0 *M* to 1.0×10^{-14} *M*. Because the use of scientific notation for these often very small concentrations can become quite cumbersome, a physiologist, S. P. L. Sørenson, devised the **pH** scale defined by the following relation:

$$[H_3O^+] = 10^{-pH}$$

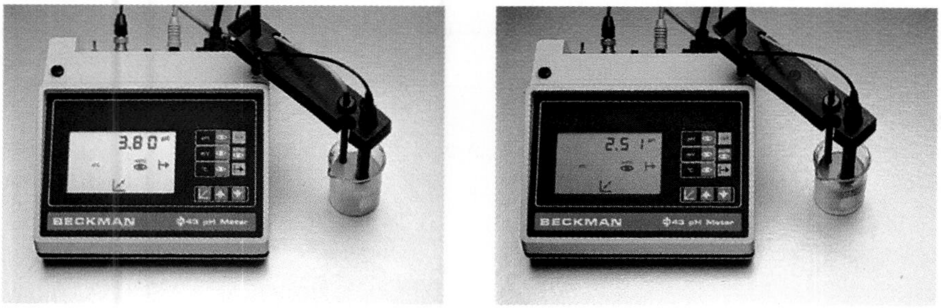

(a) (b)

Figure 9.3 A pH meter responds to the pH when a special electrode is immersed in a solution. This meter displays the measured pH digitally. The two samples are (a) orange juice and (b) lemon juice. The pH of lemon juice is lower than that of orange juice and therefore has a higher concentration of hydronium ion.

- This form of the definition can be used to calculate either the hydronium ion concentration, given the pH, or the pH, given the hydronium ion concentration.

Any number y may be expressed in the form $y = 10^x$. The number x to which 10 must be raised—that is, the exponent of 10—is called the **logarithm** or the **log** of y and is written as

$$x = \log y$$

Thus an alternative definition of pH is:

$$\text{pH} = -\log [H_3O^+]$$

- This alternative form of the definition also can be used to calculate the pH, given the hydronium ion concentration, or the hydronium ion concentration, given the pH.

The pH of a solution is measured with a pH meter (Figure 9.3), an electronic device that responds to changes in hydronium ion concentration. The pH scale is illustrated in Figure 9.4, and the pH values of various aqueous solutions are presented in Figure 9.5 on the following page.

Table 9.3 on the following page lists hydronium ion concentrations in decimal notation, scientific notation, and the corresponding values of pH. The first digit in a logarithm is a decimal place holder and is therefore insignificant. The number of significant figures in a logarithm is equal to the number of digits after the decimal point. For example, the log 7.1 contains one significant figure, the log 7.10 contains two significant figures, and so forth. Two things

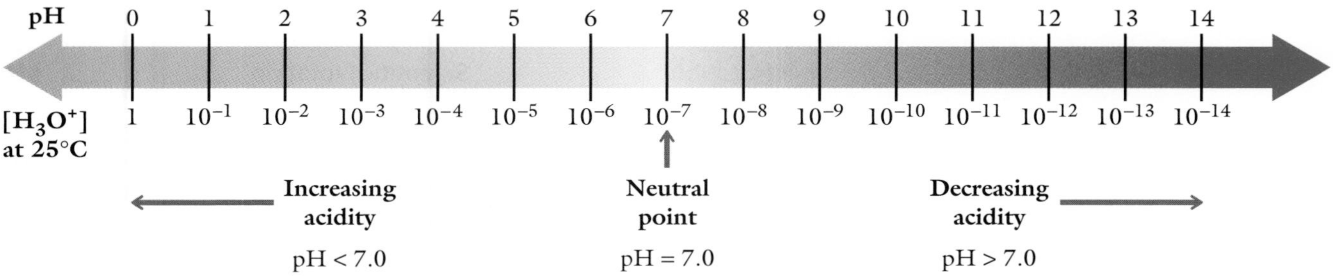

Figure 9.4 The pH scale, with pH values above and corresponding molar concentrations of hydronium ion below. Note how decreasing values of pH denote increasing acidity and how increasing values of pH denote increasing basicity.

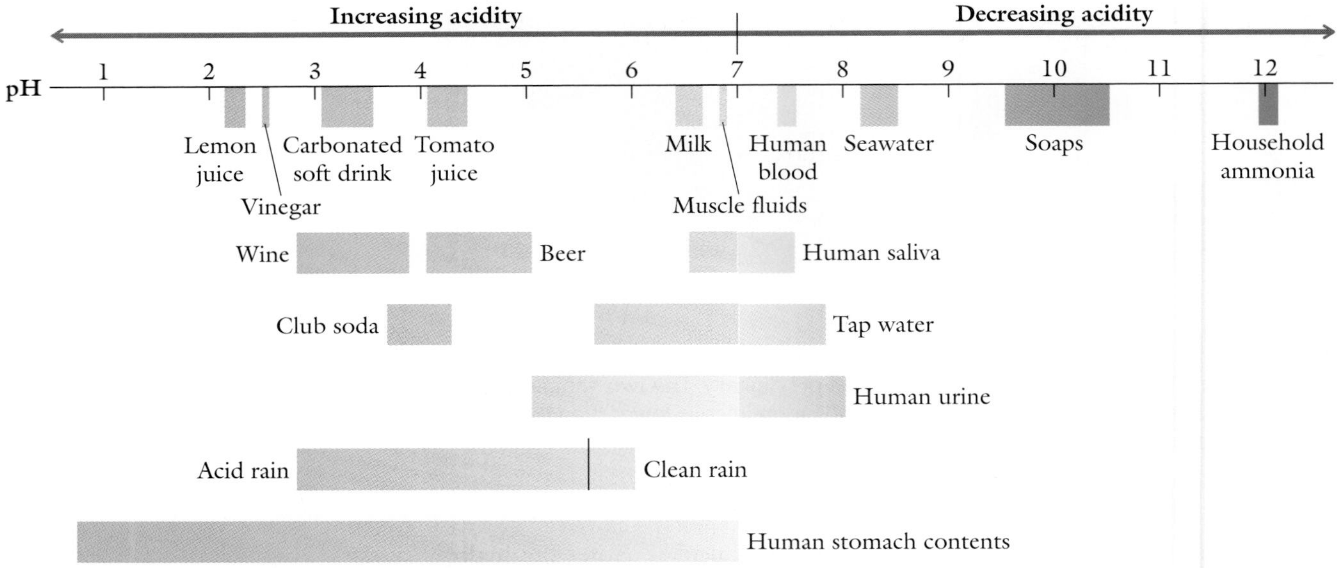

Figure 9.5 The ranges of pH found in some aqueous solutions.

important to note here are that, as Figure 9.4 also shows, the pH is equal to the negative exponent of the hydronium ion concentration and that a tenfold change in hydronium ion concentration results in a change in pH of one unit. Solutions that differ in acidity by one pH unit have a tenfold difference in hydronium ion concentration.

We ended Section 9.1 by noting three possible relations between the concentrations of hydronium ions and hydroxide ions in aqueous solutions. By referring to Table 9.3, we can now describe these conditions in terms of pH:

Concept checklist

- ✔ In a neutral solution, pH = 7, because $[H_3O^+] = [OH^-] = 1.00 \times 10^{-7}$.
- ✔ In an acidic solution, pH is less than 7, because $[H_3O^+]$ is greater than $[OH^-]$.
- ✔ In a basic solution, pH is greater than 7, because $[H_3O^+]$ is less than $[OH^-]$.

TABLE 9.3 Hydronium Ion Concentrations with Corresponding pH Values

$[H^+]$		
Decimal Notation	**Scientific Notation**	**pH**
1.0	1.0×10^{0}	0.00
0.10	1.0×10^{-1}	1.00
0.010	1.0×10^{-2}	2.00
0.0010	1.0×10^{-3}	3.00
0.000010	1.0×10^{-5}	5.00
0.00000010	1.0×10^{-7}	7.00
0.0000000010	1.0×10^{-9}	9.00

Example 9.6 Calculating pH, given OH^- concentration

Calculate the pH of an aqueous 0.0010 *M* NaOH solution.

Solution

Because NaOH is a strong base, the hydroxide ion concentration is equal to the NaOH concentration. The hydronium ion concentration is calculated by using the ion-product of water:

$$K_w = [H_3O^+] \times 0.0010 = 1.0 \times 10^{-14}$$
$$[H_3O^+] = 1.0 \times 10^{-11} M$$

The pH is most easily calculated by using the relation

$$[H_3O^+] = 10^{-pH}$$

Therefore, pH = 11.00.

Problem 9.6 Calculate the hydronium ion concentration and the pH of an aqueous 0.010 *M* KOH solution.

If the concentration of HCl in a solution is a value other than 0.00100 *M* —say, 0.0016—the $[H_3O^+]$ would be expressed as 1.6×10^{-3} *M*, and the pH could not be determined merely by inspection. Example 9.7 shows how to solve that problem.

Example 9.7 The general case of calculating pH from hydronium ion concentration

Calculate the pH of an HCl solution whose concentration is 0.0024 *M*.

Solution

To solve this problem, we must use the alternative definition of pH:

$$pH = -\log [H_3O^+]$$

The H_3O^+ ion concentration in scientific notation equivalent to 0.0024 *M* NaOH is 2.4×10^{-3} *M*. Substituting that value into the definition of pH gives:

$$pH = -\log [2.4 \times 10^{-3}]$$

The determination of pH is now a matter of obtaining the log of 2.4×10^{-3}. Logarithms can be obtained directly with a hand calculator fitted with a logarithm function key:

Step 1. Enter the molar concentration of hydronium ion, either 2.4×10^{-3} or 0.0024, and press the log key. The result is − 2.62.

Step 2. Press the change sign key, [+/−]. The answer is 2.62, which is the pH.

Problem 9.7 Calculate the pH of an aqueous 0.0037 *M* HCl solution.

Chemists have extended the use of the "p" notation (as used in pH) to simplify other constants that are expressed in scientific notation. Among them are hydroxide ion concentration, expressed as **pOH,** and a variety of equilibrium constants, expressed as **p*K*** with an identifying subscript. (The p*K* notation is used for acids and bases, as we will see at the end of Section 9.4.)

We can establish a useful relation between pH and pOH. Taking the negative logarithms of both sides of the equation for the ion-product of water, $K_w = [H_3O^+][OH^-]$, we get:

$$-\log K_w = -\log [H_3O^+] + (-\log [OH^-])$$

Using the "p" notation gives:

$$pK_w = pH + pOH$$

Substituting $pK_w = 14$ gives:

$$pH + pOH = 14$$

If the pH is known, the pOH can be calculated directly by difference, and vice versa.

Example 9.8 Calculating pH from pOH

What is the pH of an aqueous 0.0010 *M* NaOH solution?

Solution

NaOH is a strong base, so the concentration of OH^- ion is 0.0010 *M*, or 1.0×10^{-3} *M*.

$$pOH = -\log [OH^-] = -\log (1.0 \times 10^{-3}) = 3.00$$
$$pH = 14.00 - pOH = 14.00 - 3.00 = 11.00$$

Problem 9.8 What is the pH of a 0.00010 *M* aqueous solution of TlOH?

Example 9.9 Calculating pOH from pH

What is the pOH of an aqueous 0.00010 *M* HCl solution?

Solution

HCl is a strong acid, so the concentration of H_3O^+ ion is 0.00010 *M*, or 1.0×10^{-4} *M*.

$$pH = -\log [H_3O^+] = -\log (1.0 \times 10^{-4}) = 4.00$$
$$pOH = 14.00 - pH = 14.00 - 4.00 = 10.00$$

Problem 9.9 What is the pOH of an aqueous 0.0020 *M* HNO_3 solution?

As we have seen, for a strong acid or base, the pH of an aqueous solution is easily calculated if we know the concentration of acid or base in the solution. However, very often it is the pH itself that is experimentally determined in the laboratory. From the measured pH, the hydronium ion concentration and thus the concentration of the strong acid or base can be calculated. This calculation uses the definition

$$[H_3O^+] = 10^{-pH}$$

and requires determination of the **antilogarithm,** or **antilog,** of the measured pH value. The antilog can be found by using a calculator, as described in Example 9.10.

Example 9.10 Calculating the concentration of a strong acid solution from the measured pH

The pH of an aqueous HCl solution was found to be 3.50. Calculate the molar concentration of the acid.

Solution

We can substitute 3.50 in the definition of pH:

$$[H_3O^+] = 10^{-pH} = 10^{-3.50}$$

Your hand calculator can provide the antilog in a number of different ways. Here are three of the ways to calculate the antilog of − 3.50. If the first one does not work on your calculator, then try the next one.

- Enter 3.50, press the change sign key, press the $\mathbf{10^x}$ key. The answer is 0.00032.

- Enter 10, press the **Y**^{x} key, enter 3.50, press the change sign key, press the equal sign key. The answer is 0.00032.
- Enter 3.50, press the change sign key, press the **INV** (inverse) key, then press the log key. The answer is 0.00032.

Problem 9.10 The pH of an aqueous HI solution was found to be 2.46. Calculate the molar concentration of the acid.

9.4 WEAK ACIDS AND WEAK BASES

Although thus far we have focused on strong acids and strong bases, the vast majority of acids and bases (hundreds of thousands of substances) are **weak acids** and **weak bases.** Aqueous solutions of these substances react only partly with water; that is, only a small percentage of the acid and base molecules dissociate in water to form hydronium and hydroxide ions, respectively. The acids and bases found in biological systems are principally weak acids and bases, and we will encounter many such molecules in our study of biochemistry. In fact, any acid or base that you encounter in your study of inorganic and organic chemistry and biochemistry that is not listed in Table 9.2 is weak.

The weak inorganic acids include HCN (hydrocyanic acid), HNO_2 (nitrous acid), and HF (hydrofluoric acid). All organic compounds containing the carboxyl group (—COOH) are weak acids. When the carboxyl group is ionized (—COO^-), it is called the carboxylate group or anion. (Organic acids have the general formula RCOOH, where R stands for H, CH_3, or some other group. These acids are the subject of Chapter 15.)

▶▶ Amino acids, the building blocks of proteins, contain the carboxyl group. They are presented in Chapter 20.

Most acid-base problems in this chapter use acetic acid as an example of a weak acid. Vinegar is an aqueous 5% acetic acid solution. The structural formula for acetic acid is shown in the margin. The hydrogen of the OH in the carboxyl group is the dissociable hydrogen, or proton. Other examples of organic acids are formic acid, the irritant found in ant bites, and lactic acid, an important intermediate in carbohydrate metabolism (Chapter 23).

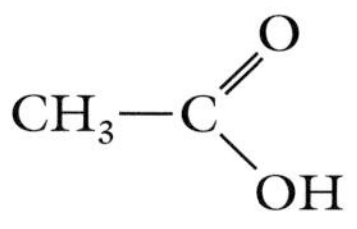
Acetic acid

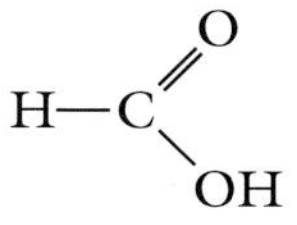
Formic acid

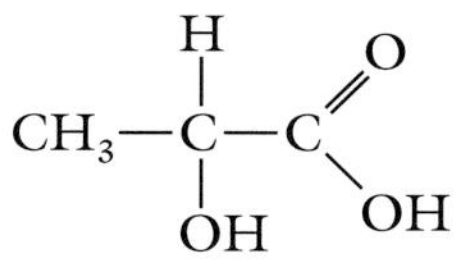
Lactic acid

Ammonia (NH_3) is an example of a weak base. Many organic compounds containing nitrogen are also weak bases. These organic bases can be considered to be derivatives of ammonia, with other groups substituting for one or more of the hydrogens of ammonia. Examples of organic bases are methylamine (CH_3NH_2), pyridine (C_5H_5N), and morphine ($C_{17}H_{19}O_3N$). The organic bases, including a number of physiologically active compounds such as morphine, are discussed in Chapter 16.

Reaction of Weak Acids and Bases with Water

Acetic acid and other weak acids react with water, in a way similar to the way that water reacts with water (Section 9.1), to produce a hydronium ion and a carboxylate anion:

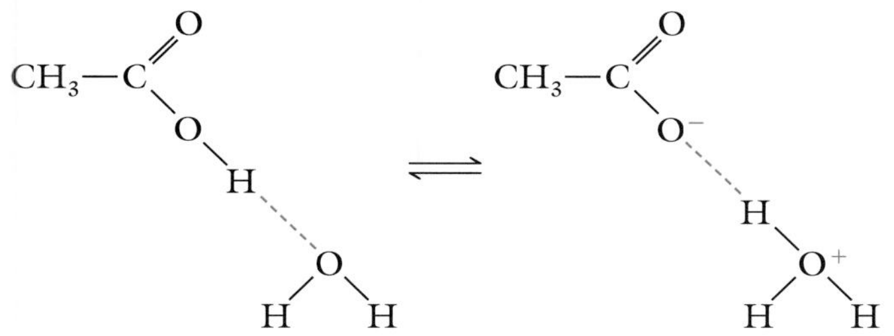

In pure water, one water molecule acts as an acid by donating a hydrogen ion, or proton, to a second water molecule, which acts as a base by accepting the proton. A weak acid simply takes the place of a water molecule, but its presence imposes a new set of equilibrium requirements on those that exist in pure water.

Acetic acid reacts with water to produce hydronium ions and acetate ions but in much smaller quantities than the amount of unreacted or undissociated acetic acid molecules.

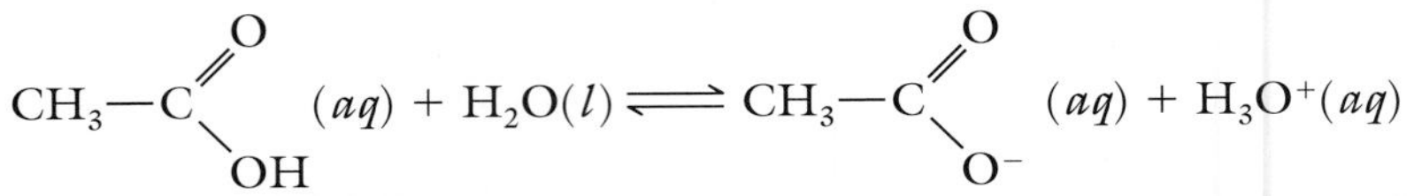

The structural formula for acetic acid is equivalent to the molecular formulas CH_3COOH and $HC_2H_3O_2$. We will use the molecular formula CH_3COOH in reaction equations describing dissociation, because it is a more simple representation.

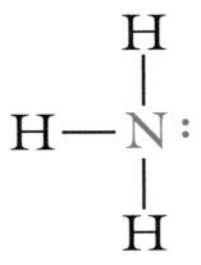

Ammonia

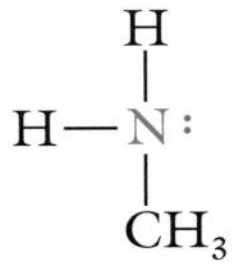

Methylamine

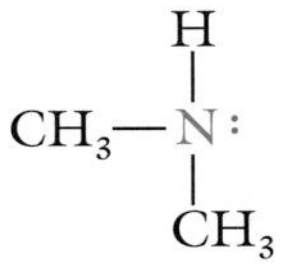

Dimethylamine

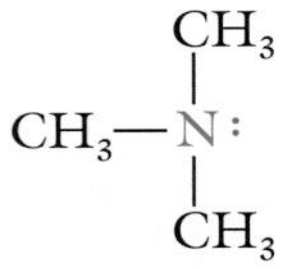

Trimethylamine

Ammonia (NH_3) is a weak base because it reacts with water to produce hydroxide ions and because the amount of hydroxide ion produced is much smaller than the total amount of ammonia added to water. The many nitrogen-containing organic compounds that are also weak bases react with water to produce hydroxide ions in the same way as ammonia does. The reaction is:

$$NH_3(aq) + H_2O(l) \rightleftharpoons NH_4^+(aq) + OH^-(aq)$$

with the equilibrium far to the left. The basic properties of ammonia and all other bases similar to it depend on the lone pair of electrons on its nitrogen, which bonds with the proton donated by water. This property is best illustrated by using Lewis structures to show the reaction with water:

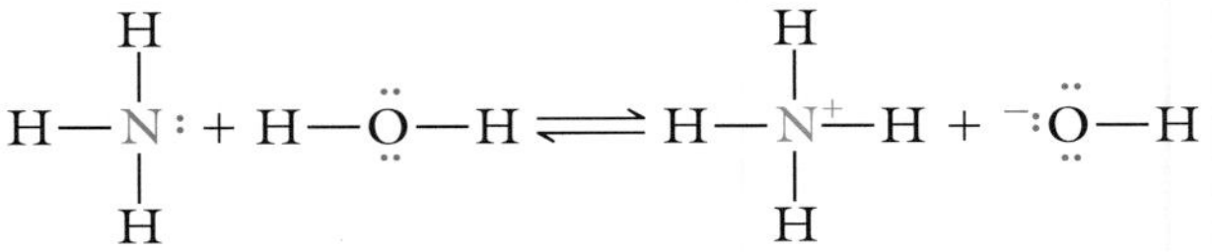

The structural feature responsible for basicity—the lone pair of electrons on nitrogen—is illustrated in the margin for ammonia as well as some organic bases.

Many organic bases with similar structural characteristics, such as morphine, codeine, and other physiologically active substances used as medicines, are insoluble in water. Before they can be used medicinally, they are converted into their soluble salts by reaction with HCl (see Section 16.5).

Quantitative Aspects of Acid-Base Equilibria

To consider weak acids and bases in quantitative terms, we must consider them to be systems in chemical equilibrium, characterized by equilibrium constants. [Remember that pure solids or pure liquids do not appear in the equilibrium constant expression (Section 8.3)]. For example, the equation for the reaction of acetic acid with water is:

$$CH_3COOH(aq) + H_2O(l) \rightleftharpoons H_3O^+(aq) + CH_3COO^-(aq)$$

and the equilibrium constant expression for its dissociation is:

$$K_a = \frac{[H_3O^+][CH_3COO^-]}{[CH_3COOH]}$$

This equilibrium constant is called an **acid dissociation constant, K_a** ("a" for acid), because the reaction that it describes results in the formation of a hydronium ion.

Similarly, the dissociation of a weak base that results in the production of a hydroxide ion is characterized by a **base dissociation constant,** K_b ("b" for base). Consider the reaction of ammonia with water:

$$NH_3(aq) + H_2O(l) \longrightarrow NH_4^+(aq) + OH^-(aq)$$

The dissociation constant is:

$$K_b = \frac{[NH_4^+][OH^-]}{[NH_3]}$$

The values of K_a and K_b are indicative of the relative amounts of products and reactants at equilibrium: the smaller the values of the dissociation constants, the smaller the amount of product relative to reactant, or, in terms of dissociation, the smaller the degree or extent of dissociation. In other words, the value of this constant provides us with a quantitative measure of just how "weak" a weak acid or weak base is.

Concept check

✔ The smaller the dissociation constant, the weaker the acid or base.

Acid and base dissociation constants are often expressed in the "p" notation mentioned earlier. For weak acids,

$$K_a = 10^{-pK_a} \quad \text{and} \quad -\log K_a = pK_a$$

For weak bases;

$$K_b = 10^{-pK_b} \quad \text{and} \quad -\log K_b = pK_b$$

Table 9.4 lists various weak acids and bases along with their dissociation constants and pK values. Note that, reading down each list, as the dissociation constants become smaller, the corresponding pK values become larger.

Box 9.1 on the following page shows how acid dissociation constants are used to calculate the pH of weak acids and bases.

9.5 BRØNSTED-LOWRY THEORY OF ACIDS AND BASES

At the end of Section 3.8, we described the reaction between hydrogen chloride and water that produces an H_3O^+ ion as the result of the transfer of a hydrogen ion to water, a reaction that always takes place when compounds called

TABLE 9.4 Values of K_a and pK_a for Some Weak Acids and Values of K_b and pK_b for Some Weak Bases*

Acids				Bases			
Name	**Formula**	**K_a**	**pK_a**	**Name**	**Formula**	**K_b**	**pK_b**
nitrous acid	HNO_2	4.47×10^{-4}	3.35	dimethylamine	$(CH_3)_2NH$	5.81×10^{-4}	3.24
cyanic acid	HCNO	2.19×10^{-4}	3.66	methylamine	CH_3NH_2	4.59×10^{-4}	3.34
formic acid	HCOOH	1.78×10^{-4}	3.75	ammonia	NH_3	1.75×10^{-5}	4.76
hydrazoic acid	HN_3	1.91×10^{-5}	4.72	cocaine	$C_{17}H_{21}O_4N$	3.90×10^{-9}	8.41
acetic acid	CH_3COOH	1.74×10^{-5}	4.76	aniline	$C_6H_5NH_2$	4.17×10^{-10}	9.38
carbonic acid	H_2CO_3	4.45×10^{-7}	6.35	morphine	$C_{17}H_{19}O_3N$	1.41×10^{-10}	9.85

* All the weak bases listed are derivatives of ammonia.

acids react with water. This description is essentially a statement of the **Brønsted-Lowry theory** of acids and bases. In this theory,

- Acids are described as **proton donors.**
- Bases are described as **proton acceptors.**

The reaction of HCl and H_2O is:

$$HCl(g) + H_2O(l) \longrightarrow H_3O^+(aq) + Cl^-(aq)$$

9.1 Chemistry in Depth

Acid Dissociation Constants and the Calculation of pH

In the discussion of chemical equilibrium in Chapter 8, we presented a general method (Box 8.2) for determining the concentrations of components of an equilibrium mixture. The same approach is used for quantitative analysis of solutions of weak acids and bases. The objective of the analysis is to determine the hydronium ion or hydroxide ion concentrations in these solutions and then, from these values, the pH. Here we illustrate the general approach by calculating the hydronium ion concentration of a 0.00500 *M* solution of acetic acid.

The equation for reaction of acetic acid with water is:

$$CH_3COOH + H_2O \rightleftharpoons H_3O^+ + CH_3COO^-$$

Because acetic acid is a weak acid, only a small amount of the acid dissociates to form an unknown number of moles of hydronium ion and an equal amount of acetate ion.

We can use the accounting method developed in Box 8.2 to describe the equilibrium concentrations of the reaction mixture. With the symbol x representing the unknown number of moles of hydronium and acetate ions per liter formed by dissociation, the equilibrium molar concentrations of the reaction mixture are:

	CH_3COOH	H_3O^+	CH_3COO^-
Start	0.00500	0	0
Change	$-[x]$	$+[x]$	$+[x]$
Final	$0.00500 - [x]$	$[x]$	$[x]$

The correct equilibrium constant to use here is the acid dissociation constant (see Table 9.4):

$$K_a = \frac{[H_3O^+][CH_3COO^-]}{[CH_3COOH]} = 1.74 \times 10^{-5}$$

Substituting appropriate values into the dissociation-constant expression from the accounting table gives:

$$1.74 \times 10^{-5} = \frac{[x][x]}{0.00500 - [x]} = \frac{[x^2]}{0.00500 - [x]}$$

This is a quadratic equation in one unknown and can be solved by using the standard solution for the general quadratic equation $ax^2 + bx + c = 0$, where

$$x = \frac{-b \pm \sqrt{b^2 - 4ac}}{2a}$$

Note that, although two values of x are obtained by solving this equation, for real substances only the positive value is possible.

We can simplify the calculation further. Because the equilibrium constant is quite small, the value of x is so small that $0.00500 - [x]$ is approximately equal to 0.00500. By using 0.00500 in place of $0.00500 - [x]$, the arithmetic is greatly simplified.

$$1.74 \times 10^{-5} = \frac{[x^2]}{0.00500}$$

$$x^2 = 8.70 \times 10^{-8}$$

$$x = [H_3O^+] = \sqrt{8.70 \times 10^{-8}} = 2.95 \times 10^{-4}\ M$$

$$pH = -\log [H_3O^+] = 3.53$$

The precise value for the hydronium ion concentration obtained by using the quadratic equation is:

$$[H_3O^+] = 2.87 \times 10^{-4}$$

$$pH = 3.54$$

The approximated value of $[H_3O^+]$ is about 3% greater than the precise value of $[H_3O^+]$, a discrepancy that is considered acceptable.

This simplified method of obtaining an approximate solution has its limitations. When the equilibrium constant is larger than 1×10^{-4}, the formal quadratic solution must be used because the error in calculating $[H_3O^+]$ becomes unacceptably large; that is, greater than 5% of the precisely calculated value.

The same quantitative analysis can also be used to calculate the pH of solutions of bases, such as ammonia, and solutions of the salts of weak acids or bases. The reaction equation that describes the equilibrium in an aqueous solution of sodium acetate is:

$$CH_3COO^-(aq) + H_2O(l) \rightleftharpoons CH_3COOH(aq) + OH^-(aq)$$

Because this reaction and those of bases such as ammonia produce hydroxide ions, the quantitative determination of pH requires the use of a base dissociation constant, K_b.

Note that both the hydronium ion and the chloride ion are hydrated. In every such reaction, the Brønsted-Lowry theory requires the presence of both a proton donor (the acid) and a proton acceptor (the base; in this case, water). Two facts are noteworthy here:

- Proton transfer requires a pair of substances, a donor and an acceptor.
- Substances other than water can serve as the proton acceptor (base).

This theory is a powerful tool for understanding the acid-base equilibria for weak acids and bases. Let's revisit these equilibria in the context of the Brønsted-Lowry theory.

Conjugate Acids and Bases

As you learned in Section 9.1, when water reacts with water, one water molecule acts as an acid, donating its proton to the other water molecule, which acts as a base. The result is one molecule of protonated water and one molecule of deprotonated water. The Brønsted-Lowry theory links the two pairs, H_3O^+/H_2O and H_2O/OH^-, and describes them as **conjugate acid-base pairs:**

Conjugate acid-base pair

$$H_2O(l) + H_2O(l) \rightleftharpoons H_3O^+(aq) + OH^-(aq)$$

Conjugate acid-base pair

The reactions between water and a weak acid, such as acetic acid, and between water and a weak base, such as ammonia, are described by the same scheme:

Conjugate acid-base pair

$$CH_3COOH(aq) + H_2O(l) \rightleftharpoons H_3O^+(aq) + CH_3COO^-(aq)$$

Conjugate acid-base pair

Conjugate acid-base pair

$$NH_3(aq) + H_2O(l) \rightleftharpoons OH^-(aq) + NH_4^+(aq)$$

Conjugate acid-base pair

The central idea is that, as a result of an acid donating its proton, it becomes a base; and, when a base accepts a proton, it becomes an acid. Therefore, the acetate ion is a base, as is the hydroxide ion; and the ammonium ion is an acid, as is the hydronium ion. In general, you will find that a conjugate base (a Brønsted-Lowry base) is the ion produced when its conjugate acid (a Brønsted-Lowry acid) donates a proton to water. For example,

$$HCN(aq) + H_2O(l) \rightleftharpoons CN^-(aq) + H_3O^+(aq)$$

CN^- is the conjugate base of the acid HCN

$$H_2PO_4^-(aq) + H_2O(l) \rightleftharpoons HPO_4^{2-}(aq) + H_3O^+(aq)$$

HPO_4^{2-} is the conjugate base of the acid $H_2PO_4^-$

$$NH_4^+(aq) + H_2O(l) \rightleftharpoons NH_3(aq) + H_3O^+(aq)$$

NH_3 is the conjugate base of the acid NH_4^+

When acetate ion is added to water, it disturbs the existing equilibrium by reacting with water (the acid):

$$CH_3COO^-(aq) + H_2O(l) \rightleftharpoons CH_3COOH(aq) + OH^-(aq)$$

One of the conjugate acid-base pairs is CH_3COOH/CH_3COO^- and the other is H_2O/OH^-. If you prepared an aqueous solution of sodium acetate, the result would be a solution with excess hydroxide ion and therefore a pH above 7.0. Similarly, if you prepared an aqueous solution of ammonium chloride, the result would be a solution with an excess of hydronium ion and therefore a pH below 7. The effects of salts on the pH of aqueous solutions through their hydrolysis reactions (reactions with water) are further considered in Section 9.7.

Relation Between Acid and Base Dissociation Constants

The dissociation constant describing what occurs when acetate ion is added to water,

$$CH_3COO^-(aq) + H_2O(l) \rightleftharpoons OH^-(aq) + CH_3COOH(aq),$$

is a base dissociation consant, K_b, because the product of the equilibrium is an OH^- ion, a base:

$$K_b = \frac{[CH_3COOH][OH^-]}{[CH_3COO^-]}$$

The dissociation constant describing the results of adding acetic acid to water is an acid dissociation constant, K_a, because the product of the reaction is an H_3O^+ ion:

$$K_a = \frac{[H_3O^+][CH_3COO^-]}{[CH_3COOH]}$$

When the two constants are multiplied,

$$K_a \times K_b = \frac{[H_3O^+][CH_3COO^-]}{[CH_3COOH]} \times \frac{[CH_3COOH][OH^-]}{[CH_3COO^-]},$$

TABLE 9.5 Values of K_a, pK_a, K_b, and pK_b for Selected Conjugate Acid-Base Pairs

Acids				Bases			
Name	**Formula**	K_a	pK_a	**Name**	**Formula**	K_b	pK_b
nitrous acid	HNO_2	4.47×10^{-4}	3.35	nitrite	NO_2^-	2.24×10^{-11}	10.65
cyanic acid	HCNO	2.19×10^{-4}	3.66	cyanate	CNO^-	4.57×10^{-11}	10.34
formic acid	HCOOH	1.78×10^{-4}	3.75	formate	$HCOO^-$	5.62×10^{-11}	10.25
acetic acid	CH_3COOH	1.74×10^{-5}	4.76	acetate	CH_3COO^-	5.75×10^{-10}	9.24
carbonic acid	H_2CO_3	4.45×10^{-7}	6.35	bicarbonate	HCO_3^-	2.24×10^{-8}	7.65
dihydrogen phosphate	$H_2PO_4^-$	6.32×10^{-8}	7.20	monohydrogen phosphate	HPO_4^{2-}	1.58×10^{-7}	6.80
ammonium	NH_4^+	5.71×10^{-10}	9.24	ammonia	NH_3	1.75×10^{-5}	4.76
hydrogen carbonate	HCO_3^-	4.72×10^{-11}	10.33	carbonate	CO_3^{2-}	2.19×10^{-4}	3.67
Hydrogen phosphate	HPO_4^{2-}	4.84×10^{-13}	12.32	phosphate	PO_4^{3-}	2.07×10^{-2}	1.68

and the common terms in the numerator and denominator are cancelled, we get

$$K_a \times K_b = [H_3O^+][OH^-]$$

or

$$K_a \times K_b = K_w$$

Although we used specific examples in this derivation, the result can be generalized to the behavior of all conjugate acids and bases in water:

Concept checklist

- ✔ The acid dissociation constant of a conjugate acid can be calculated from the base dissociation constant of its conjugate base, because their product is always K_w.
- ✔ The base dissociation constant of a conjugate base can be calculated from the acid dissociation constant of its conjugate acid, because their product is always K_w.

Table 9.5 lists the dissociation constants and pK values for some conjugate acid-base pairs.

Example 9.11 Deriving K_b from K_a for a conjugate acid-base pair

Using the values in Table 9.5, show how the K_b value for formate is derived from the K_a value of formic acid.

Solution

Table 9.5 shows that the K_a of formic acid is 1.78×10^{-4}. By substituting the known values in the expression

$$K_a \times K_b = [H_3O^+][OH^-] = K_w,$$

we can calculate K_b:

$$1.78 \times 10^{-4} \times K_b = 1.00 \times 10^{-14}$$
$$K_b = 5.62 \times 10^{-11}$$

Problem 9.11 Show how the K_b value for cyanate is derived from the K_a value of cyanic acid.

Relation Between pK_a and pK_b

Once again, the "p" notation can be applied to the dissociation constants of conjugate acid-base pairs, with the result that

$$pK_a + pK_b = pK_w = 14$$

Thus, we can calculate the pK_a of an acid from the pK_b of its conjugate base, and vice versa. For example, given the pK_a for acetic acid of 4.76 (Table 9.4), we can calculate the pK_b—and thus the K_b—for the reaction of acetate ion with water:

$$4.76 + pK_b = 14.00$$
$$pK_b = 14.00 - 4.76 = 9.24$$

Example 9.12 Deriving pK_b from pK_a for a conjugate acid-base pair

Using the values in Table 9.5, show how the pK_b value for formate is derived from the pK_a value of formic acid.

Solution

From Table 9.5, we find that the pK_a of formic acid is 3.75. Rearranging the expression $pK_a + pK_b = pK_w$ and substituting the known values gives:

$$pK_b = pK_w - pK_a = 14.00 - 3.75 = 10.25$$

Problem 9.12 Show how the pK_b value for cyanate is derived from the pK_a value of cyanic acid.

We will be applying the Brønsted-Lowry theory of acids and bases in the discussion of polyprotic acids, salt hydrolysis, and buffer systems in the next three sections.

9.6 DISSOCIATION OF POLYPROTIC ACIDS

Among the strong acids listed in Table 9.2 is H_2SO_4 (sulfuric acid) and among the weak acids listed in Table 9.4 is H_2CO_3 (carbonic acid). Both of these acids are **diprotic acids,** meaning that each molecule has two dissociable protons. In Section 4.7, we used two diprotic organic acids, malic and oxaloacetic acids, to illustrate a metabolic redox reaction. They are just two of a large group of similar metabolic products that we will encounter in Chapter 23. Phosphoric acid (H_3PO_4), which will be discussed in some detail here, is a **triprotic acid.** Citric acid, an important metabolic intermediate discussed in Chapters 22 and 23, also is a triprotic acid. All acids with two or more dissociable protons are called **polyprotic acids.** Carbonic acid, phosphoric acid, and their salts are of central importance in maintaining the proper acid-base balance of the blood and tissues of the body.

The Dissociation of Phosphoric Acid

The three sequential dissociations of phosphoric acid and the associated K_a values are:

$$H_3PO_4(aq) + H_2O(l) \rightleftharpoons H_3O^+(aq) + H_2PO_4^-(aq)$$

$$K_{a1} = \frac{[H_3O^+][H_2PO_4^-]}{[H_3PO_4]} = 5.93 \times 10^{-3}$$

$$H_2PO_4^-(aq) + H_2O(l) \rightleftharpoons H_3O^+(aq) + HPO_4^{2-}(aq)$$

$$K_{a2} = \frac{[H_3O^+][HPO_4^{2-}]}{[H_2PO_4^-]} = 6.32 \times 10^{-8}$$

$$HPO_4^{2-}(aq) + H_2O(l) \rightleftharpoons H_3O^+(aq) + PO_4^{3-}(aq)$$

$$K_{a3} = \frac{[H_3O^+][PO_4^{3-}]}{[HPO_4^{2-}]} = 4.84 \times 10^{-13}$$

The product of the first dissociation step is the dihydrogen phosphate ion, $H_2PO_4^-$; the product of the second is the monohydrogen phosphate ion, HPO_4^{2-}; and the product of the third is the phosphate ion, PO_4^{3-}.

Note that the dissociation constants become smaller for each successive dissociation. This trend occurs because, at each successive step, a positive ion (a proton) must be removed from an increasingly negatively charged anion; that is, after each step, the resulting anion exerts a greater force of attraction on the departing proton. Typically, the dissociation constant for each successive step in the dissociation of a polyprotic acid is smaller than the dissociation constant for the preceding step by a factor of about 1×10^{-5}. As a consequence, the hydronium ion in a solution of a polyprotic acid is virtually entirely derived from the first dissociation.

For example, when H_3PO_4 is added to water, the product of the first dissociation, $H_2PO_4^-$, contributes virtually nothing to the total H_3O^+ concentration. All H_3O^+ is derived from the dissociation of the H_3PO_4. When $H_2PO_4^-$ is added to water (in the form of a salt such as KH_2PO_4), the H_3O^+ is derived from the dissociation of the $H_2PO_4^-$, and virtually none comes from the dissociation of the product, HPO_4^{2-}.

Each dissociation step has a conjugate acid-base pair: H_3PO_4 can only donate a proton and is thus an acid; PO_4^{3-} can only accept a proton and is thus a base. But both $H_2PO_4^-$ and HPO_4^{2-} can behave either as an acid or as a base: each can both accept and donate a proton. All such substances are called **amphoteric.** You can see that water is also an amphoteric substance.

The $H_2PO_4^-/HPO_4^{2-}$ pair is an important system in cells because it helps to maintain the constant pH necessary for cellular functions (Section 9.8).

The Dissociation of Carbonic Acid

The dissociation steps for carbonic acid at 25°C are:

$$H_2CO_3(aq) + H_2O(l) \rightleftharpoons H_3O^+(aq) + HCO_3^-(aq)$$

$$K_{a1} = \frac{[H_3O^+][HCO_3^-]}{[H_2CO_3]} = 4.45 \times 10^{-7}$$

$$HCO_3^-(aq) + H_2O(l) \rightleftharpoons H_3O^+(aq) + CO_3^{2-}(aq)$$

$$K_{a2} = \frac{[H_3O^+][CO_3^{2-}]}{[HCO_3^-]} = 4.72 \times 10^{-11}$$

Note that HCO_3^- (hydrogen carbonate, also known as bicarbonate) is amphoteric and can behave as either acid or base.

The H_2CO_3/HCO_3^- pair is chiefly responsible for keeping the pH of the blood constant at pH = 7.4. The equilibria in this case are more complicated than indicated by the preceding equations because carbonic acid is also in equilibrium with H_2O and CO_2 in solution and with gaseous CO_2 in the lungs. This system is discussed in Section 9.8, and its physiological function is fully described in Chapter 26.

Concept check

✔ The protons of a polyprotic acid dissociate sequentially. Each step of the dissociation creates a new equilibrium system that is characterized by its own dissociation constant.

9.7 SALTS AND HYDROLYSIS

The conjugate bases of weak acids such as acetate, carbonate, and phosphate are the anions of salts such as sodium acetate and potassium dihydrogen phosphate. In aqueous solution they react with water—a reaction called **hydrolysis**—to produce hydroxide ion (OH^-) and basic solutions. These anions are referred to as basic anions. Similarly, the cation NH_4^+ is the conjugate acid of a weak base; it reacts with water, or **hydrolyzes,** to produce an acidic solution.

There are no basic cations. All Group I and II cations, such as Na^+ and Ca^{2+}, are neutral in aqueous solution. However, salts of Al^{3+}, Pb^{2+}, Sn^{2+}, and the transition metals such as Fe^{2+} and Fe^{3+} form acidic solutions. Water is covalently bound to the metal cation of the salt, and the dissociation of protons from the bound water makes the aqueous solutions of these salts acidic (Figure 9.6). Box 9.2 on the following page describes an environmental consequence of the acidity of an iron salt solution.

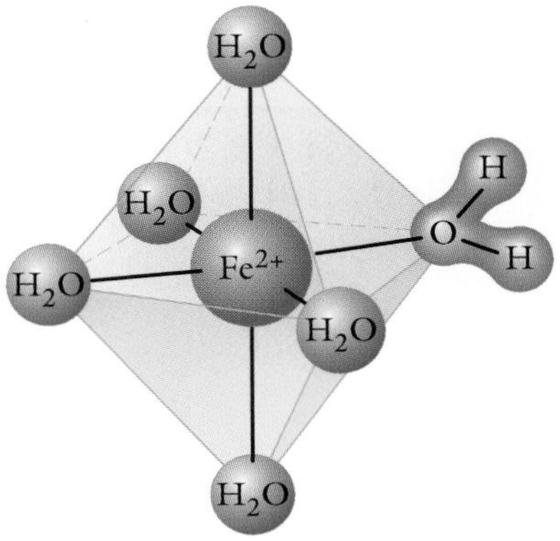

Figure 9.6 An acidic cation, $Fe(H_2O)_6^{2+}$. The structure of only one of the six bound water molecules is shown.

9.2 Chemistry Around Us

Acid-Mine Drainage

A biologist exploring a wooded area near an abandoned coal mine observes a box turtle moving toward a small pool of water. The turtle slips into the water and, in about 10 to 20 s, it is dead. The biologist notes that the water in the pool has a decided red color. She measures the pH with her portable pH meter, which records a pH of 3.3. She determines the density of the water with a hydrometer and finds it to be equivalent to about 4 g of solid per 100 mL of solution, an extraordinarily high solution density. The biologist follows the runoff from the pool down to a small creek, where she finds rocks bordering the stream that are covered with a bright yellow solid (a precipitate) and a small steel bridge that is severely corroded.

What is the explanation of these phenomena? Bituminous coal, the type of coal formerly obtained from this mine, contains significant quantities of an insoluble mineral called marcasite, which is iron(II) sulfide. On exposure to air, the sulfide is oxidized to sulfate, converting the insoluble iron(II) sulfide into water-soluble iron(II) sulfate. The hydrated Fe(II) ion is a weak acid, as described in Section 9.7, and produces a significant hydronium ion concentration in its aqueous solution:

$$Fe(H_2O)_6^{2+}\,(aq) + H_2O(l) \rightleftharpoons Fe(OH)(H_2O)_5^{+}\,(aq) + H_3O^{+}\,(aq)$$

The pH of the solution can be as low as 3.0, depending on the concentration of the Fe(II) salt.

The reddish pool of water that killed the box turtle consisted of groundwater that had penetrated the old bituminous coal mine and then drained out to the surface heavily laden with the acidic iron(II) sulfate solution. After entering the watershed, the acid-mine drainage eventually found its way into a larger stream. As the drainage entered the stream, it became diluted with water of somewhat higher pH. This pH change caused precipitation of the iron salt in a form called yellow-boy, which coated the rocks bordering the creek. Yellow-boy is a complex iron salt, mostly a mixture of iron(II) oxide and hydroxide. The corrosion of the steel bridge was caused by the acid content of the incoming stream.

Yellow-boy coats the rocks along a creek fed by acid-mine drainage from a local abandoned coal mine.

Rules for determining the acidity or basicity of a salt solution are:

Rules for determining the acidity or basicity of a salt solution

- Salts composed of cations and anions of strong acids and bases form neutral aqueous solutions.
- Salts having anions of weak acids form basic solutions.
- Salts having cations of weak bases form acidic solutions.

When a salt is formed from a strong acid and a weak base or a strong base and a weak acid, it is the weak base or weak acid that determines the pH of the salt solution.

Example 9.13 Predicting the acidity of salt solutions

Predict whether the following salts will produce an acidic, a basic, or a neutral aqueous solution: sodium chloride (NaCl), potassium acetate (CH_3COOK), ammonium chloride (NH_4Cl).

Solution
In each case, we must examine both the cation and the anion and determine whether it was formed by neutralization of a strong acid with a weak base or a strong base with a weak acid.

$NaCl$: derived from a strong acid, HCl, and a strong base, $NaOH$.
CH_3COOK: derived from a strong base, KOH, and a weak acid, CH_3COOH.
NH_4Cl: derived from a weak base, NH_3, and a strong acid, HCl.

For each salt, the acidity or basicity of the aqueous solution can be predicted from the rules given in the text.

Salt	Cation	Anion	Solution
NaCl	neutral	neutral	neutral
CH_3COOK	neutral	basic	basic
NH_4Cl	acidic	neutral	acidic

Problem 9.13 Predict whether the following salts will produce an acidic, basic, or neutral aqueous solution: $MgCl_2$, KNO_3, $SnCl_2$, NH_4NO_3.

Figure 9.7 The pH of the solutions of salts considered in Example 9.13 are measured with indicators. Ammonium chloride, on the left, is acidic; sodium chloride, in the center, is neutral; and sodium acetate, on the right, is basic.

Litmus paper tells us whether a solution is acidic or basic. However, many dyes change color over very narrow ranges of pH. In Figure 9.7, the pH of the solutions considered in Example 9.13 were determined by using such dye indicators.

▶▶ More details regarding indicators can be found in Section 9.10.

Example 9.13 illustrates how to predict the acidity of a salt solution by inspection and the application of simple rules. But how would we determine the acidity of a solution of a salt derived from the reaction of a weak base with a weak acid? The pH of such a salt solution depends on which dissociation constant, K_a or K_b, is larger. Recall that K_a describes a reaction in which hydronium ion is produced, and K_b describes a reaction in which hydroxide ion is produced. For example, for ammonium acetate, the K_a for NH_4^+ is 5.71×10^{-10} and the K_b for CH_3COO^- is 5.75×10^{-10} (Table 9.5). In this case, because K_a is about the same value as K_b, the solution will be approximately neutral.

9.8 BUFFERS AND BUFFERED SOLUTIONS

The normal metabolic activity of cells results in the production of acids. Muscle tissue, for example, produces lactic acid. Such acids are a potential danger to proteins, whose structures and biological functions depend on the maintenance of a proper pH. Even modest changes in pH can result in significant changes in the catalytic activity of enzymes, thereby causing cells to malfunction. The pH of blood must be maintained within very narrow limits, which, if exceeded, can result in death. Fortunately, the body's buffer systems can effectively counter changes in pH.

▶▶ The adverse effect of pH on proteins is considered futher in Section 20.9.

A **buffered solution** is a solution that resists changes in pH when hydronium or hydroxide ion is added. Such a solution contains a **buffer system** consisting of a conjugate acid-base pair. The $H_2PO_4^-/HPO_4^{2-}$ and H_2CO_3/HCO_3^- systems described in Section 9.6 are just two examples of conjugate acid-base pairs that can be used as buffer systems. Many of the commercially available antacids create a buffer system in the stomach by providing a Brønsted-Lowry base such as citrate. When these systems are added to the low-pH environment of the stomach, they react with hydronium ion to raise the pH.

For optimal buffering action to take place, the concentrations of conjugate acid and conjugate base in the buffer system must be approximately equal. We will see why this is so shortly.

If a small amount of hydroxide ion is introduced into a buffer solution, the conjugate acid will react with the hydroxide ion; and, if a small amount of hydronium ion is added, the conjugate base will react with the hydronium ion. The conjugate acid behaves as a hydroxide ion sponge, and the conjugate base as a hydronium ion sponge, both constantly ready to absorb the appropriate ion and thus maintain a constant pH.

The greater the concentrations of conjugate acid and base, the greater the capacity of the buffer to absorb changes in hydronium or hydroxide ion concentrations. We will return to this idea of buffering capacity after taking a quantitative look at the conjugate acid-base pair that makes up a buffer system.

Quantitative Aspects of Buffer Systems

Each conjugate acid-base pair has a specific pH that it is able to maintain through its buffering action. This pH can be calculated by using the **Henderson-Hasselbalch equation:**

$$\text{pH} = \text{p}K_a + \log\left(\frac{[\text{proton acceptor}]}{[\text{proton donor}]}\right)$$

The equation states that the pH of a buffered solution depends on the pK_a of the proton donor of the acid-base pair (not the pK_b of the proton acceptor) and on the logarithm of the proton acceptor to proton donor ratio. The derivation of the equation is presented in Box 9.3.

Table 9.6 illustrates how the acceptor-to-donor ratio affects resistance to change in pH in a buffer solution, as calculated from the Henderson-Hasselbalch equation. The effectiveness of a buffer can be seen by noting the change in pH for a wide range of acceptor-to-donor ratios. When the ratio changes from 1.6/1 to 0.6/1, the logarithm of the ratio and therefore the pH varies by only ±0.2 pH units. Even changes in acceptor-to-donor ratios ranging from 10/1 to 1/10—representing changes of about ±90% of the original concentrations—result in changes of only ±1.0 pH unit. However, as you can see, when the ratio is much larger or much smaller than 1.0, the pH changes rapidly.

TABLE 9.6 Effect on pH of the Ratio of Proton Acceptor to Proton Donor Forms of an $HPO_4^{2-}/H_2PO_4^-$ Buffer (pK_a = 7.2)

Ratio	Logarithm	pH
1000/1	3.00	10.20
100/1	2.00	9.20
10/1	1.00	8.20
5/1	0.70	7.90
1.6/1	0.20	7.40
1/1	0.00	7.20
0.6/1	−0.20	7.00
1/5	−0.70	6.50
1/10	−1.00	6.20
1/100	−2.00	5.20
1/1000	−3.00	4.20

9.3 Chemistry in Depth

The Henderson-Hasselbalch Equation

The Henderson-Hasselbalch equation is derived by considering the dissociation equilibrium of a generalized weak acid. The formula of the weak acid is represented as HA, the anion as A^-. The dissociation process is therefore:

$$HA + H_2O \rightleftharpoons H_3O^+ + A^-$$

where A^- is the conjugate base of the acid HA, and the acid dissociation constant expression is:

$$K_a = \frac{[H_3O^+][A^-]}{[HA]}$$

Now rewrite the dissociation constant to emphasize the ratio of conjugate base to conjugate acid:

$$K_a = [H_3O^+] \times \frac{[A^-]}{[HA]}$$

To introduce the "p" notation, we first take the negative logarithm of both sides, term by term:

$$-\log K_a = -\log [H_3O^+] - \log \frac{[A^-]}{[HA]}$$

Thus:

$$pK_a = pH - \log \frac{[A^-]}{[HA]}$$

The minus sign is eliminated by rearrangement:

$$pH = pK_a + \log \frac{[A^-]}{[HA]}$$

The relation can be expressed in the more generalized form

$$pH = pK_a + \log \frac{[\text{proton acceptor}]}{[\text{proton donor}]}$$

This form of the equation makes it clear that:

1. A buffer system can consist of any conjugate acid-base pair such as $H_2PO_4^-/HPO_4^{2-}$, HCO_3^-/CO_3^{2-}, or NH_4^+/NH_3.
2. The pH of a buffer system depends on two variables:
   - The acid dissociation constant (not the base dissociation constant) of the conjugate acid-base pair.
   - The ratio of molar concentrations of proton acceptor to proton donor of the conjugate acid-base pair.

Concept check

✔ For maximum buffering effect, the value of the acceptor-to-donor ratio in a buffer solution must be kept as close to 1.0 as possible.

Knowing the pK_a of the proton donor and the ratio of acceptor to donor, we can calculate the pH of any buffer solution.

Example 9.14 Calculating the pH of a buffer solution

Calculate the pH of an aqueous solution consisting of 0.0080 *M* acetic acid and 0.0060 *M* sodium acetate.

Solution

For this type of calculation, we use the Henderson-Hasselbalch equation, substituting the pK_a of acetic acid, 4.76 (Table 9.4), and the concentrations of acceptor and donor. (Note that it is the pK_a of acetic acid, not the pK_b of acetate, that is used.) The acceptor (numerator) is acetate, and the donor (denominator) is acetic acid.

$$pH = 4.76 + \log \frac{[0.0060]}{[0.0080]}$$

$$= 4.76 - 0.13$$

$$pH = 4.63$$

Problem 9.14 Calculate the pH of an aqueous solution consisting of 0.0060 *M* acetic acid and 0.0080 *M* sodium acetate.

The information provided by the Henderson-Hasselbalch equation has other uses besides the calculation of pH. It tells us how best to select a buffer system that will buffer at a specific pH—a common requirement in laboratory work. For example, a particular experiment on a biological reaction may require a buffer that maintains a solution at pH 7.0 ± 0.3.

The key to selecting a buffer system is to consider what the Henderson-Hasselbalch equation reveals when the concentrations of conjugate base and acid are equal; that is, when [acceptor] = [donor]. When that is the case,

$$pH = pK_a + \log 1.0$$

But log 1.0 = 0; so, when [acceptor] = [donor],

$$pH = pK_a$$

Concept check

✔ To prepare a buffer of specific pH, select a conjugate acid-base pair with a pK_a as close as possible to the desired pH.

For a solution that buffers at pH 7.0 ± 0.3, this acid-base pair is $H_2PO_4^-/HPO_4^{2-}$, with a pK_a of 7.2 (see Table 9.5). We can prepare the buffer by dissolving equal molar quantities of KH_2PO_4 and K_2HPO_4 in water.

Buffering Capacity

The ratio of the molar concentrations of a conjugate acid-base pair is independent of the concentrations themselves, but the buffering capacity of the system is not. The ratio of concentrations can have the value of 1.0 whether the concentrations of donor and acceptor are equal at 0.001, 0.01, or 0.1 *M*. If more acid is added to a buffer solution than the conjugate base can absorb, then the pH will drastically change. Because the conjugate acid and base behave like sponges, the amount of hydronium ion or hydroxide ion that can be buffered depends on the amount (concentration) of the conjugate acid-base pair present.

Concept check

✔ The buffering capacity of a buffer increases with increasing concentration of the buffer's conjugate acid-base pair.

9.9 BUFFER SYSTEM OF THE BLOOD

Carbon dioxide produced by cells as an end product of metabolism diffuses into the erythrocytes of the blood. There it is converted into bicarbonate ion in a reaction catalyzed by the enzyme carbonic anhydrase. This process can be represented by a two-step process:

$$CO_2(aq) + H_2O(l) \rightleftharpoons H_2CO_3(aq)$$
$$H_2CO_3(aq) + H_2O(l) \rightleftharpoons H_3O^+\ (aq) + HCO_3^-\ (aq)$$

You can see that each molecule of CO_2 added to the blood results in the formation of an H_3O^+ ion. The buffer system appears to consist of H_2CO_3/HCO_3^-, whose pK_a is 6.35, a value far lower than the blood pH of 7.4. That value is a full pH unit away from blood pH, so how can the H_2CO_3/HCO_3^- system act as the blood's buffer? This apparent discrepancy can be cleared up by noting that the concentration of H_2CO_3 depends on the concentration of dissolved CO_2.

In the Henderson-Hasselbalch equation, the proton donor (acid) component is the sum of the concentrations of dissolved carbon dioxide and carbonic acid ($CO_2 + H_2CO_3$), and the pH is given by:

$$pH = pK_a + \log\left(\frac{[HCO_3^-(aq)]}{[CO_2(aq) + H_2CO_3(aq)]}\right)$$

Substituting the values for blood pH (7.4) and pK_a (6.35) gives:

$$7.4 - 6.35 \approx 1 = \log\left(\frac{[HCO_3^-(aq)]}{[CO_2(aq) + H_2CO_3(aq)]}\right)$$

Because the log of the ratio is about 1, the ratio of proton acceptor to proton donor is approximately 10/1. It would appear that the acid component (the proton donor) is at such a low concentration that the introduction of a small quantity of base should drastically change the pH. What saves the day is the reservoir of dissolved CO_2.

Carbon dioxide is constantly renewed by metabolic processes, so the carbonic acid reservoir, $CO_2 + H_2CO_3$, resists depletion and the H_2CO_3/HCO_3^- system effectively buffers the blood.

Carbon Dioxide Maintains the Acid-Base Balance of the Blood

As CO_2 is transported from metabolizing cells to the lungs for excretion to the air, it is accompanied by the protons generated by metabolic activity. Bicarbonate and protons are formed from CO_2 in erythrocytes at the tissue level, and those same protons are eliminated at the lungs by recombination with bicarbonate:

Tissue level: $CO_2 + H_2O \rightleftharpoons H^+ + HCO_3^-$

At the lungs: $H^+ + HCO_3^- \rightleftharpoons CO_2 + H_2O$

This abbreviated scheme emphasizes the role of CO_2 in maintaining the acid-base balance of the blood. It is continually removed by diffusion because its concentration at the tissue level is greater than that at the lungs. Remember that metabolizing tissues and the lungs rapidly communicate through the transport of blood by the vascular system. The buffering system is an open system, not at equilibrium, and there are serious consequences if the CO_2 is removed too rapidly or too slowly at the lungs. The rate of CO_2 removal depends on the rate of breathing, which we call the **ventilation rate.**

A PICTURE OF HEALTH

Normal pH of Some Body Fluids

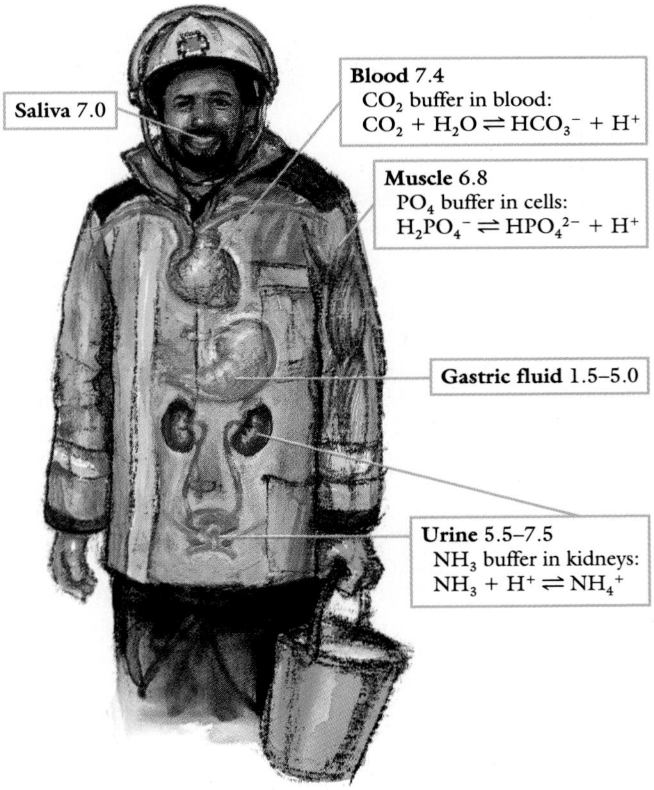

Acidosis and Alkalosis

In certain circumstances—for example, in the lung disease emphysema or when a person is under anesthesia—the ventilation rate is low, and CO_2 is not removed from the lungs rapidly enough. Consequently, the bicarbonate buffer system will "back up," hydrogen ion will not be removed by reaction with HCO_3^-, and the blood pH will fall. This condition is called **respiratory acidosis.** Immediate treatment consists of intravenous bicarbonate infusion. Conversely, if CO_2 is removed from the lungs faster than it arrives, the blood pH will rise. This condition, called **respiratory alkalosis,** occurs, for example, under conditions of great excitement (rapid breathing). First aid for this condition is to have the patient breathe into a paper bag, an action that increases the CO_2 content of the lungs.

Metabolic acidosis is a lowering of the blood pH as a result of a metabolic disorder, rather than as a result of a failure of the blood buffer system. For example, a large and serious decrease in pH can occur as a result of uncontrolled diabetes. The blood pH may fall from the normal 7.4 to as low as 6.8. The increased H^+ concentration is due to the large amounts of acidic compounds produced in the liver. The bicarbonate buffer system attempts to compensate for the excess H^+, producing excess CO_2, which must be eliminated at the lungs. However, so much CO_2 is lost by ventilation that the absolute concentration of the buffer system decreases. The capacity of the buffer system is severely compromised and cannot reduce the metabolically produced excess

PROFESSIONAL CONNECTIONS

Terri E. Weaver, R.N., Ph.D., Assistant Professor, University of Pennsylvania; Research on Sleep Apnea and Sleep Disorders

Terri Weaver studied chemistry as an undergraduate at the University of Pittsburgh.

Why did you choose the position that you are in today?

As long as I can remember, I wanted to be a nurse. My mother was a nurse, and so that influenced me. While I was a clinical specialist for 7 years at the University of Pennsylvania Medical Center, scientific questions arose in my clinical practice that made me realize that I needed further study in order to answer them. I obtained a Ph.D. in order to do research in the areas that intrigued me. My areas of interest are sleep disorders, sleep apnea, and functional impairment due to lack of sleep. I enjoy teaching, and being a professor allows me an opportunity to conduct research in my areas of interest and to share the results with students.

What positions did you have before your current one?

I practiced for 5 years as a staff nurse and supervisor in intensive care. Then I obtained my master's degree and was a pulmonary clinical nurse specialist at the University of Pennsylvania Medical Center.

Has technology affected your work life?

Computers certainly help us track data and make that part of the research job much easier. In sleep disorders, with the emphasis on CPAP (continuous positive airway pressure), new ways to deliver this technique to patients are constantly evolving.

What was the most valuable thing that you learned in school?

Organization and time management! At one point in my education, things clicked for me so that I realized the importance of organization. I realized for myself effective ways to process and seek information in order to complete an assignment—what goes where and how to find it.

Do you use chemistry in your job?

Chemistry is a big part of teaching about the physiology of the body: acid-base balance, renal, endocrine systems. In my research, it's important in a number of ways; for instance, the effect of hormones on the chemical process in cells and how that affects sleep disorders. You must know basic chemistry and apply it.

What advice do you have for students who are studying chemistry today?

Really learn it! Most students see chemistry as a rite of passage that they must get through—they don't see its importance until they get into a clinical setting. But, really, it's the building block for other sciences and the foundation for future course work. I can tell which students have had problems in their chemistry courses when they get into later courses, as well as in the clinical setting. Seniors tell us that they want to tell the freshmen that they'd wished they'd studied harder in chemistry—it's too late once they get to the advanced courses. You never know how you're going to need your building-block course work.

H^+. In such cases, buffer capacity can be temporarily restored by intravenous administration of sodium bicarbonate.

An increase in blood pH as a result of a metabolic disorder is called **metabolic alkalosis.** This condition can arise when excessive amounts of H^+ ion are lost—for example, during continuous vomiting. H^+ ion is then borrowed from the blood, and the blood pH will rise. Intake of excessive amounts of antacids also will cause the blood pH to rise. Immediate treatment consists of intravenous administration of ammonium chloride.

9.10 TITRATION

Acid-base chemistry as conducted in the laboratory, including the clinical laboratory, often requires the measurement of the concentration of an acid or a base in solution. This measurement is done with a titration using a neutralization reaction. Neutralization is the reaction between a strong acid and a strong base that results in the formation of a salt and water. The complete ionic equation for the reaction between hydrochloric acid and sodium hydroxide in water is:

$$H^+(aq) + Cl^-(aq) + Na^+(aq) + OH^-(aq) \rightleftharpoons H_2O(l) + Na^+(aq) + Cl^-(aq)$$

The net ionic equation for the neutralization of any strong acid by any strong base is:

$$H^+(aq) + OH^-(aq) \rightleftharpoons H_2O(l)$$

and emphasizes the formation of water. The ionic reaction takes place because the equilibrium constant for the reaction is 1×10^{14}. Nitric acid, HNO_3, or any strong acid added to a solution of calcium hydroxide, $Ca(OH)_2$, or any strong base is described by the same net ionic equation.

The first step in the **titration** procedure is to accurately measure a volume of the solution containing the acid. A basic solution of known concentration is then added to the acid solution in small measured amounts until the acid is neutralized. The experimental setup for titration can be seen in Figure 9.8.

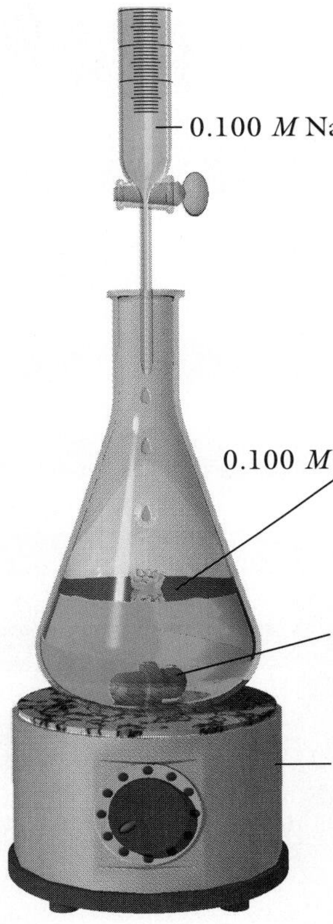

Figure 9.8 A titration setup. NaOH (aq) is added drop by drop from a buret into a constantly stirred solution of acid until the end point is neared. The end point is identified by a change in color of the indicator.

The success of this analytical technique depends on several factors:

- The reaction must go to completion (have a very large equilibrium constant).
- The reaction must be fast.
- There must be a way of detecting when the reaction is complete; that is, when neutralization has occurred.

A reaction that is stoichiometrically complete, or theoretically complete, is said to be at its **equivalence point.** In a titration, the experimentally determined point of completion is called the **end point.** End points are detected by using dyes called **indicators** (Table 9.7), which change color near the pH of the

TABLE 9.7 Indicator Dye pK_as and Titration End Points of Some Weak Acids and Bases

Indicator	pK_a	Weak Acid or Base	End-Point pH
methyl red	4.6	NH_3	4.5
β-naphtholphthalein	7.6	HCOOH	7.9
phenolphthalein	8.7	CH_3COOH	8.4
thymol blue	9.6	$H_2PO_4^-$	9.6

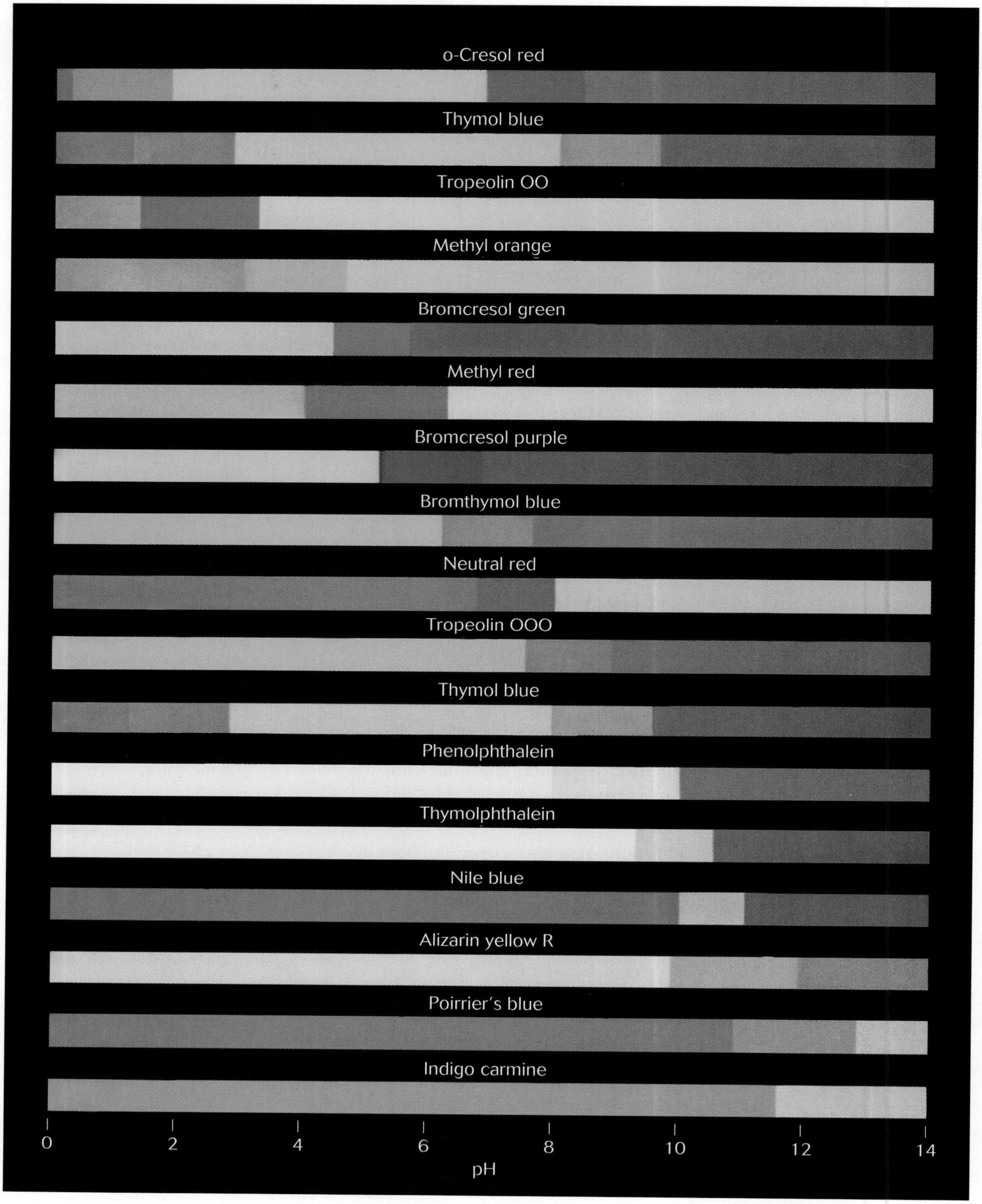

Figure 9.9 The colors of indicators used to determine end points of titrations over a wide range of pH values.

9.4 Chemistry in Depth

Use of Indicators in Determining pH

Not all titrations have end points at neutral pH—that is, pH 7—because, as we have seen, not all salt solutions are neutral. The salt of a weak acid forms a basic solution, and the salt of a weak base forms an acidic solution. For example, a solution of 0.1 *M* sodium acetate can have a pH of about 8.4; therefore, the end point of a titration of acetic acid with sodium hydroxide consists of a solution of sodium acetate at a pH of 8.4, not pH 7. To detect the end point of a titration, we must be able to determine when the pH of the solution coincides with the predicted pH. This measurement can be done with a pH meter, but most work of this type is accomplished with indicator dyes (Sections 9.7 and 9.10).

A typical indicator dye is itself a weak acid and, like any weak acid, it reacts with water. What makes the indicator so useful is that a solution of the undissociated acid and a solution of the dissociated acid, or anion, have different colors. The difference is usually so marked that it is a simple matter to determine the transition from one color to another. How can this color change be used to measure pH?

First, let's write the acid dissociation reaction, calling the acid form of the indicator HIn and its anion In^-:

$$HIn + H_2O \rightleftharpoons H_3O^+ + In^-$$

We can now express the equilibrium in terms of the Henderson-Hasselbalch equation (see Box 9.3):

$$pH = pK_a + \log \frac{[In^-]}{[HIn]}$$

From many years of experience, we know that the eye can detect a color change when 10% of an indicator has undergone dissociation. In other words, we can "see" the end point of a titration to well within ±1 pH unit of the pK_a of the indicator. However, for the accurate and rapid detection of the end point of a titration, we must choose an indicator with a pK_a very close to the calculated end point. Table 9.7 lists a number of indicator dyes, the pK_a of each dye, and the weak acid whose titration end point coincides with each pK_a; and Figure 9.8 shows the colors of many other indicators used to determine end points over a wide range of pH values.

Most indicator dyes used in the laboratory, are generally added directly to solutions, but they are also available as dye-impregnated paper strips that will determine pH to within about 0.5 pH unit. Litmus paper, by comparison, will reveal only whether the pH is above or below 7.

equivalence point (Figure 9.9 and Box 9.4) or by using a pH meter. The goal of titration is to get the theoretical equivalence point and the experimentally determined end point to coincide. The behavior of an indicator in solutions at the beginning, at the end point, and past the end point of a titration is illustrated in Figure 9.10.

When the titration is complete, we know the total volume of basic solution of known concentration required to neutralize a known volume of the acid. We can then calculate the concentration of the acid by a method very similar to the one used to quantitate dilution in Section 7.7. In Example 9.15, we will use this method to calculate the concentration of a solution of hydrochloric acid.

Figure 9.10 Aqueous solutions at pH 2 (red), pH 5 (orange), and pH 10 (yellow). The acid-base indicator is the dye methyl red.

Example 9.15 Calculating acid concentration from a neutralization reaction

The neutralization of 25.0 mL of a solution of HCl of unknown concentration required 16.7 mL of a 0.0101 *M* KOH standard solution. (A **standard solution** is one of precisely known concentration.) Calculate the concentration of the acid.

Solution

As in dilution calculations (Section 7.7), the number of moles of added base in a neutralization reaction must be equal to the number of moles of acid:

$$\frac{\text{moles}_{\text{base}}}{\text{liter}} \times \text{volume}_{\text{base}} = \frac{\text{moles}_{\text{acid}}}{\text{liter}} \times \text{volume}_{\text{acid}}$$

$$M_{\text{base}} \times \text{volume}_{\text{base}} = M_{\text{acid}} \times \text{volume}_{\text{acid}}$$

The only restriction on units is that they must be the same on both sides of the equation:

$$0.0101 \times 16.7 \text{ mL} = M_{\text{acid}} \times 25.0 \text{ mL}$$

$$M_{\text{acid}} = 0.00675\ M$$

Problem 9.15 What volume of a 0.0200 *M* KOH standard solution is required to neutralize 35.0 mL of a 0.0150 *M* solution of HNO_3?

In Section 7.10, we pointed out that reactions between ions can take place if one of the products is insoluble and forms a precipitate. We know that a neutralization reaction results in the formation of water, and now we can look at neutralization reactions in which not only water but a gas is also formed.

The reaction between sodium bicarbonate, taken orally to relieve the distress of excess stomach acid, and HCl results in the formation of a gas. The neutralization products are water and carbonic acid (H_2CO_3). Carbonic acid is unstable and rapidly decomposes to form CO_2 and H_2O. The reaction between HCl and $CaCO_3$ is shown in Figure 9.11. The equation for the formation of carbonic acid is:

$$NaHCO_3(aq) + HCl(aq) \rightleftharpoons H_2CO_3(aq) + NaCl(aq)$$

The net ionic equation is:

$$H^+(aq) + HCO_3^-(aq) \rightleftharpoons H_2CO_3(aq)$$

The gas-formation step consists of the decomposition of the unstable carbonic acid:

$$H_2CO_3(aq) \rightleftharpoons H_2O(l) + CO_2(g)$$

The net ionic equation showing the overall result is:

$$H^+(aq) + HCO_3^-(aq) \rightleftharpoons CO_2(g) + H_2O(l)$$

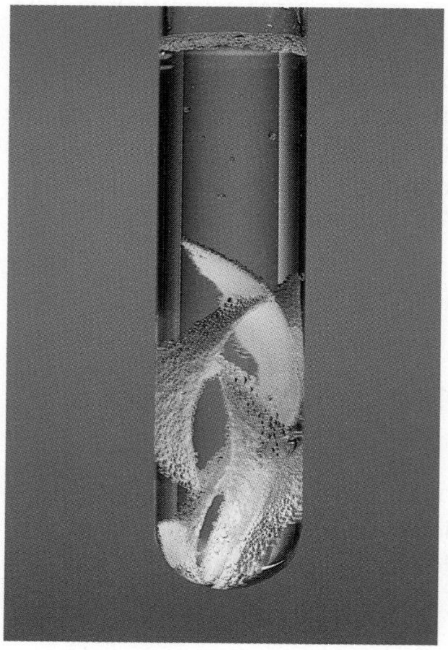

Figure 9.11 Carbon dioxide is one of the products of the reaction of metal carbonates with acids. Here, an eggshell, largely composed of calcium carbonate, reacts with hydrochloric acid. The bubbles on the eggshell surface are carbon dioxide.

9.11 NORMALITY

A unit of concentration that we have not yet introduced is **normality, *N*,** defined as **equivalents per liter.** Although less frequently used than molarity, this unit is often encountered in clinical chemistry. Its meaning rests on a quantity called equivalent mass.

Equivalent mass and **equivalents** are related to formula mass and moles. These units are used in acid-base chemistry to express the mass in grams of a base such as KOH that is equivalent to the mass in grams of an acid required for neutralization.

For example, consider the stoichiometry of the reaction between HCl and KOH: 1 mol of KOH is equivalent to 1 mol of HCl. Therefore, 56.1 g of KOH is equivalent to 36.5 g of HCl. We define 36.5 g of HCl as 1.00 equivalent of HCl, and 56.1 g of KOH as 1.00 equivalent of KOH. For either HCl or KOH, 1.00 equivalent contained in 1.00 L of solution is 1.00 *N* in concentration.

In the calculation of the normality for the solution of a polyprotic acid, the stoichiometry of its neutralization reaction is of key importance. Consider the reaction between H_2SO_4 and KOH: 1 mol of H_2SO_4 contains 2 mol of H^+, so $\frac{1}{2}$ mol of H_2SO_4 is equivalent to 1 mol of KOH. To obtain a mass of H_2SO_4 equivalent to 1 mol of KOH in this neutralization reaction, we must divide the molar mass of H_2SO_4 by 2: 98/2 = 49 g/mol. The equivalent mass of H_2SO_4 is 49 g/mol; that is, 49 g of H_2SO_4 is equivalent to 56 g of KOH. A 1.0 *N* solution of H_2SO_4 contains 49 g/L (see Example 4.11).

These concepts are best explained in the context of a typical laboratory problem.

Example 9.16 Preparing a solution of known normality

How would you prepare a 1.00 *N* solution of phosphoric acid (H_3PO_4)?

Solution

We first write a balanced equation for the neutralization of H_3PO_4 by a base such as KOH:

$$H_3PO_4(aq) + 3\ KOH(aq) \rightleftharpoons K_3PO_4(aq) + 3\ H_2O(l)$$

Then we rewrite the equation so that the acid reacts with only 1 mol of base.

$$\tfrac{1}{3}\ H_3PO_4(aq) + KOH(aq) \rightleftharpoons \tfrac{1}{3}\ K_3PO_4(aq) + H_2O(l)$$

The modified stoichiometry shows that $\frac{1}{3}$ mol of H_3PO_4 is equivalent to 1 mol of KOH in this neutralization reaction.

To prepare a 1.00 *N* solution of the acid, $\frac{1}{3}$ mol of H_3PO_4, 98.0/3 = 32.7 g (or 1.00 equivalent), should be dissolved in sufficient water to make 1.00 L of aqueous solution.

Problem 9.16 How would you prepare a 1.00 *N* solution of the diprotic acid $(CH_2)_2(COOH)_2$ (succinic acid)?

Concept check

✔ The equivalent mass of an acid is the formula mass of the acid divided by the number of reacting H^+ ions per mole of acid.

The normality of an acid solution used in neutralization is calculated by the following relation:

$$N = \frac{\text{equivalents}_{\text{acid}}}{\text{L}} = \left(\frac{\text{mol}_{\text{acid}}}{\text{L}}\right)\left(\frac{\text{moles dissociable H}^+}{\text{mol}_{\text{acid}}}\right)$$

The normality of a basic solution used for neutralization is calculated in the same way; simply replace H^+ with OH^-.

The meaning of equivalent mass or equivalents can be expanded to include the equivalent mass of charged species such as Na^+ or Mg^{2+}. In this case, an equivalent mass of a charged species is the mass that contains an Avogadro number of charges. For example, the equivalent mass of Na^+ ion is its atomic mass, that of Ca^{2+} is one-half its atomic mass, and that of Al^{3+} is one-third its atomic mass. The ion concentrations in blood plasma and other physiological fluids are frequently recorded in units of normality, specifically in **milliequivalents,** or **meq,** per liter, where 1000 meq = 1 equivalent. We will put these ideas to work in the next example.

Example 9.17 Calculating molar concentrations of ions from milliequivalents per liter (normality)

The concentrations of sodium and calcium ions in a sample of blood plasma are recorded as Na^+ = 142 meq/L and Ca^{2+} = 5.0 meq/L. Calculate the molar concentrations of these ions.

Solution

To convert equivalents into moles, we use the following conversion factor:

$$\frac{\text{mol}}{\text{L}} = \left(\frac{\text{equivalents}}{\text{L}}\right)\left(\frac{\text{mol}}{\text{equivalent of charge}}\right)$$

$$Na^+\text{:}\quad \left(\frac{142\ \text{meq}}{\text{L}}\right)\left(\frac{1.00\ \text{eq}}{1000\ \text{meq}}\right)\left(\frac{1.00\ \text{mol}}{1.00\ \text{eq of charge}}\right) = 0.142\ M$$

$$Ca^{2+}\text{:}\quad \left(\frac{5.0\ \text{meq}}{\text{L}}\right)\left(\frac{1.0\ \text{eq}}{1000\ \text{meq}}\right)\left(\frac{1.0\ \text{mol}}{2.0\ \text{eq of charge}}\right) = 0.0025\ M$$

Problem 9.17 The concentrations of potassium and magnesium ions in a sample of blood plasma are given as K^+ = 5.0 meq/L and Mg^{2+} = 3.0 meq/L. Calculate the molar concentrations of these ions.

Summary

Water Reacts with Water Water reacts with water to form two ions: H_3O^+, the hydronium ion, and OH^-, the hydroxide ion. The process is represented by a special equilibrium constant called the ion-product of water, K_w. In pure water and in neutral solutions, the concentration of hydronium ion is equal to the concentration of hydroxide ion. In acidic solutions, the concentration of hydronium ion is greater than the concentration of hydroxide ion. In basic solutions, the concentration of hydroxide ion is greater than the concentration of hydronium ion.

Strong and Weak Acids and Bases Acids and bases are divided into two broad classes: strong and weak. Strong acids and bases become completely dissociated on reaction with water. Because of this 100% dissociation, the concentration of hydronium or hydroxide ion in acidic or basic solutions is equal to the concentration of the acid or base at which the solution is prepared. Acids and bases that do not fully dissociate in water are called weak acids and bases.

A Measure of Acidity: pH Acidity, the molar concentration of hydronium ion, is expressed by an acidity scale called pH. The three possible conditions in aqueous solutions with respect to acidity are pH equal to 7 in a neutral solution, pH below 7 in an acidic solution, and pH greater than 7 in a basic solution.

Reaction of Weak Acids and Bases with Water There are six common strong acids and nine strong bases. Most other acids and bases (hundreds of thousands of substances) in aqueous solution react only partly with water. Only a small percentage of the molecules of a weak acid or a weak base dissociate in water to form hydronium or hydroxide ions, respectively.

Quantitative Aspects of Acid-Base Equilibria The quantitative treatment of weak acids and bases requires their consideration as equilibrium systems that are characterized by equilibrium constants. Equilibrium constants describing the reaction of weak acids or weak bases with water are called dissociation constants. The smaller the dissociation constant, the weaker the acid or base.

The Brønsted-Lowry Theory of Acids and Bases In the Brønsted-Lowry theory of acids and bases, acids are described as proton donors and bases as proton acceptors. The central idea of the theory is that, when an acid donates a proton, the acid becomes a base; and, when a base accepts a proton, the base becomes an acid. This theory links each acid with its deprotonated form, now a base, in a conjugate acid-base pair.

The Dissociation of Polyprotic Acids Acids with two or more dissociable protons are called polyprotic acids. The protons of a polyprotic acid dissociate sequentially, and each step of the dissociation process creates a new equilibrium system, characterized by its own dissociation constant. The intermediary products of the dissociation can behave as both proton donors and proton acceptors; that is, as both acids and bases. Such substances are called amphoteric.

Salts and Hydrolysis The anions of some salts are conjugate bases of weak acids (such as acetate). In aqueous solution, they hydrolyze—that is, react with water—to produce basic solutions. Cations that are weak acids (such as the ammonium ion or some hydrated metal ions) hydrolyze in water to produce acidic solutions.

Buffers and Buffered Solutions A buffered solution resists changes in pH on addition of hydronium or hydroxide ion. A buffered solution contains a buffer system consisting of a conjugate acid-base pair. The ratio of concentrations of proton acceptor (base) to proton donor (acid) controls the pH, which can be calculated with the Henderson-Hasselbalch equation, using the pK of the donor acid.

Titration The concentrations of acid and base in solutions are experimentally determined by titration. Titration is a neutralization reaction that is theoretically complete at the equivalence point, which should coincide closely with the experimentally determined end point. End points are detected by using indicator dyes that change color at the pH of the equivalence point or by using a pH meter.

Normality Normality, a unit of concentration used in clinical chemistry, is defined as concentration in equivalents per liter. In acid-base reactions, the equivalent mass of an acid is its formula mass divided by the number of its dissociable protons. An equivalent mass of a charged species such as Na^+ or Mg^{2+} is its atomic mass divided by its charge.

Key Words

acid dissociation constant, p. 234
amphoteric substance, p. 241
base dissociation constant, p. 234
Brønsted-Lowry theory, p. 235
buffer system, p. 243
dissociation, p. 226
Henderson-Hasselbalch equation, p. 244
hydronium ion, p. 225
hydroxide ion, p. 225
ion-product of water, p. 225
normality, p. 253
pH, p. 228
polyprotic acid, p. 240
strong acid, p. 226
strong base, p. 226
titration, p. 249
weak acid, p. 233
weak base, p. 233

Exercises

Water Reacts with Water

9.1 The concentration of hydronium ion and of hydroxide ion in pure water at 25°C is 1.00×10^{-7} *M*. What is the value of the ion product for the ionization of water at that temperature?

9.2 The ion-product of pure water, K_w, is 2.51×10^{-14} at body temperature (37°C). What are the molar H_3O^+ and OH^- concentrations in pure water at body temperature?

9.3 What is meant by an amphoteric substance?

9.4 Can water be considered to be an amphoteric substance?

Strong Acids and Bases

9.5 What is the definition of a strong base? Give two examples.

9.6 What is the definition of a strong acid? Give two examples.

9.7 Calculate the molar H_3O^+ and Cl^- concentrations in a 0.30 *M* aqueous solution of HCl.

9.8 Calculate the molar H_3O^+ and ClO_3^- concentrations of a 0.28 *M* solution of perchloric acid ($HClO_3$).

9.9 Calculate the $[OH^-]$ and $[Tl^+]$ concentrations in a 0.25 *M* aqueous solution of TlOH.

9.10 Calculate the $[OH^-]$ and $[Ca^{2+}]$ of a 0.025 *M* $Ca(OH)_2$ solution.

9.11 Calculate the $[OH^-]$ in a 0.350 *M* aqueous solution of HCl.

9.12 Calculate the $[H_3O^+]$ and $[K^+]$ in a 0.16 *M* aqueous solution of KOH.

9.13 Calculate the hydronium ion concentration of a 0.020 *M* solution of $Ca(OH)_2$.

9.14 Calculate the hydroxide ion concentration in a 0.285 *M* solution of HCl.

A Measure of Acidity: pH

9.15 What is the pH of pure water at 25°C?

9.16 What is the pH of pure water at 37°C?

9.17 What is the pH of a solution with $[H_3O^+] = 0.01$ *M*?

9.18 What is the pH of a solution with $[H_3O^+] =$ 0.0026 *M*?

9.19 What is the pH and the pOH of a solution with a hydroxide ion concentration of 0.015 *M*?

9.20 What is the pH and the pOH of a solution with a hydrogen ion concentration of 0.027 *M*?

9.21 Calculate the pH of the following solutions at 25°C: (a) 0.0033 *M* HCl; (b) 0.025 *M* HNO_3; (c) 0.00073 *M* HBr; (d) 0.074 *M* HI.

9.22 Calculate the pH of the following solutions at 25°C: (a) 0.0041 *M* KOH; (b) 0.00094 *M* $Ca(OH)_2$; (c) 0.035 *M* $Ba(OH)_2$; (d) 0.0084 *M* TlOH.

Weak Acids and Bases

9.23 What is the definition of a weak acid? Give two examples.

9.24 What is the definition of a weak base? Give two examples.

9.25 Is a solution prepared to contain equal concentrations of acetic acid (CH_3COOH) and ammonia (NH_3) acidic, basic, or neutral?

9.26 Is a solution containing equal concentrations of formic acid (HCOOH) and aniline ($C_6H_5NH_2$) acidic, basic, or neutral?

The Brønsted-Lowry Theory of Acids and Bases

9.27 Identify the conjugate acid-base pairs in the following equations:

(a) $HNO_2(aq) + H_2O(l) \rightleftharpoons NO_2^-(aq) + H_3O^+(aq)$

(b) $H_2PO_4^-(aq) + H_2O(l) \rightleftharpoons HPO_4^{2-}(aq) + H_3O^+(aq)$

9.28 Identify the conjugate acid-base pairs in the following equations:

(a) $HCOOH(aq) + NH_3(l) \rightleftharpoons HCOO^-(aq) + NH_4^+(aq)$

(b) $H_2O(l) + H_2O(l) \rightleftharpoons OH^-(aq) + H_3O^+(aq)$

9.29 Using Table 9.5, identify the equilibrium constant for the reaction of acetate ion with water (hydrolysis) at 25°C.

9.30 Using Table 9.5, identify the equilibrium constant for the reaction of nitrous acid with water at 25°C.

9.31 Compute the pK_a for the dissociation of acetic acid and the pK_b for the hydrolysis of acetate ion in water at 25°C.

9.32 Compute the pK_a for the dissociation of nitrous acid and the pK_b for the hydrolysis of nitrite ion in water at 25°C.

9.33 What is the sum of the pK_a for the dissociation of acetic acid and the pK_b for the hydrolysis of acetate ion at 25°C in water?

9.34 What is the sum of the pK_a for the dissociation of nitrous acid and the pK_b for the hydrolysis of nitrite ion at 25°C in water?

Polyprotic Acids

9.35 Write equations for the complete dissociation of carbonic acid (H_2CO_3) in water.

9.36 Write equations for the complete dissociation of phosphoric acid (H_3PO_4) in water.

9.37 Write equilibrium constant expressions for the complete dissociation of carbonic acid in water.

9.38 Write equilibrium constant expressions for the complete dissociation of phosphoric acid in water.

9.39 In an aqueous solution of H_2CO_3, is the proportion of the H_3O^+ contributed by HCO_3^- of major or minor significance?

9.40 In an aqueous solution of H_3PO_4, is the proportion of the H_3O^+ contributed by $H_2PO_4^-$ of major or minor significance?

Salts and Hydrolysis

9.41 Predict whether each of the following salts will produce an acidic, basic, or neutral aqueous solution: (a) $CaCl_2$ (calcium chloride); (b) Na_2CO_3 (sodium carbonate); (c) $FeCl_3$ (iron(III)chloride); (d) NH_4Cl (ammonium chloride); (e) $MgSO_4$ (magnesium sulfate).

9.42 Predict whether aqueous solutions of each of the following salts are acidic, neutral, or basic: (a) NH_4NO_3 (ammonium nitrate); (b) $Al_2(SO_4)_3$ (aluminum sulfate); (c) KI (potassium iodide); (d) $NaHCO_3$ (sodium hydrogen carbonate); (e) K_2HPO_4 (potassium hydrogen phosphate).

9.43 Predict whether aqueous solutions of the following salts are acidic, neutral, or basic: (a) $Ca(NO_3)_2$ (calcium nitrate); (b) $NaNO_2$ (sodium nitrite); (c) KCN (potassium cyanide); (d) CH_3COONa (sodium acetate); (e) $(HCOO)_2Mg$ (magnesium formate).

9.44 Predict whether aqueous solutions of the following salts are acidic, neutral, or basic: (a) $Fe(NO_3)_3$ (Iron(III) nitrate); (b) $Mg(NO_2)_2$ (magnesium nitrite); (c) KBr (potassium bromide); (d) $(CH_3COO)_2Ca$ (calcium acetate); (e) $MgCO_3$ (magnesium carbonate).

Buffers and Buffer Solutions

9.45 Why does a solution of a conjugate acid-base pair behave as a buffer solution?

9.46 Why must the concentrations of the conjugate acid and conjugate base in a buffer system be comparable?

9.47 What is the pH of an aqueous solution consisting of 0.0060 *M* acetic acid and 0.0080 *M* sodium acetate?

9.48 What is the pH of a solution consisting of 0.070 *M* formic acid (HCOOH) and 0.070 *M* sodium formate (HCOONa)?

9.49 Calculate the pH of a solution consisting of 0.075 *M* K_2HPO_4 and 0.050 *M* KH_2PO_4.

9.50 Calculate the pH of a solution consisting of 0.050 *M* K_2HPO_4 and 0.075 *M* KH_2PO_4.

Titration

9.51 The neutralization of a 20.0-mL solution of HCl of unknown concentration required 16.2 mL of a 0.021 *M* KOH standard solution. Calculate the concentration of the acid.

9.52 What was the molar concentration of 25.0 mL of a solution of HNO_3 that required 32.0 mL of a 0.0180 *M* standard solution of KOH for complete neutralization?

9.53 What was the molar concentration of 40.0 mL of an H_2SO_4 solution that required 33.2 mL of a 0.0410 *M* standard solution of KOH for complete neutralization?

9.54 What was the molar concentration of 20.0 mL of an H_3PO_4 solution that required 42.3 mL of a 0.0850 *M* standard solution of KOH for complete neutralization?

Normality

9.55 How would you prepare a 0.500 *N* solution of phosphoric acid (H_3PO_4)?

9.56 How would you prepare a 0.200 *N* solution of sulfuric acid (H_2SO_4)?

9.57 What is the normality of a 0.024 *M* solution of H_3PO_4?

9.58 What is the normality of a 0.016 *M* solution of H_2SO_4?

9.59 Calculate the molar concentrations of the following ions, whose concentrations are recorded in units of normality: (a) $Na^+ = 0.125$ eq/L; (b) $K^+ = 0.035$ eq/L; (c) $Ca^{2+} = 0.072$ eq/L; (d) $Mg^{2+} = 0.028$ eq/L.

9.60 Compute the normality of the following solution with respect to each ionic species. The solution contains (a) 27.4 mmol of K^+, (b) 7.81 mmol of Ca^{2+}, and (c) 6.08 mmol of Al^{3+} ions, all in 54.7 mL.

Unclassified Exercises

9.61 Calculate the pH of the following solutions: (a) 0.0031 *M* HNO_3; (b) 1.0 *M* HCl; (c) 0.0069 *M* HI; (d) 0.019 *M* HBr; (e) 0.023 *M* $HClO_3$.

9.62 Calculate the pOH of the solutions in Exercise 9.61.

9.63 Calculate the pOH of the following solutions: (a) 0.0062 *M* KOH; (b) 0.0041 *M* $Ca(OH)_2$; (c) 0.028 *M* $Ba(OH)_2$; (d) 1.0 *M* KOH: (e) 0.01 *M* NaOH.

9.64 Calculate the pH of the solutions in Exercise 9.63.

9.65 Are solutions of the following salts acidic, neutral, or basic? (a) $Fe_2(SO_4)_3$; (b) NaBr; (c) $NaNO_2$; (d) NH_4NO_3; (e) $Mg(CN)_2$.

9.66 Calculate the pH of a solution that is 0.050 *M* in H_2CO_3 and 0.075 *M* in $KHCO_3$.

9.67 What is the molar ratio of $HPO_4^{2-}/H_2PO_4^-$ in a buffered solution of pH 8.2?

9.68 Compute the molarity of 15.0 mL of nitric acid solution that required 35.9 mL of 0.0480 *M* KOH for complete neutralization.

9.69 Calculate the normality of 25.0 mL of a $Ca(OH)_2$ solution that required 62.0 mL of a 0.0250 *M* HCl solution for complete neutralization.

9.70 The pH of an HCl solution is 3.20. Calculate the concentration of the HCl solution.

9.71 Calculate the pOH and the hydroxide ion concentrations of the following solutions: (a) 0.0034 *M* HCl; (b) 0.025 *M* HNO_3; (c) 0.00073 *M* HBr; (d) 0.074 *M* HI.

9.72 What is the definition of a polyprotic acid?

9.73 Predict whether CH_3COONH_4 (ammonium acetate) will form an acidic or a basic solution.

9.74 Give some examples of buffer systems.

9.75 What are the components of a buffered solution?

Chemical Connections

9.76 The ion-product of water increases as temperature increases. Is the dissociation of water an exothermic or an endothermic reaction?

9.77 What is a conjugate acid-base pair?

9.78 What is the meaning of pH?

9.79 Suppose the capacity of your stomach to be 2.00 L and its pH to be 1.50. How many grams of the antacid sodium bicarbonate, $NaHCO_3$, would be necessary to bring the pH of your stomach to 7.00?

9.80 How would you prepare a buffer to maintain the pH of a solution as close to pH 5.0 as possible? (Consult Table 9.5.)

CHAPTER 10

CHEMICAL AND BIOLOGICAL EFFECTS OF RADIATION

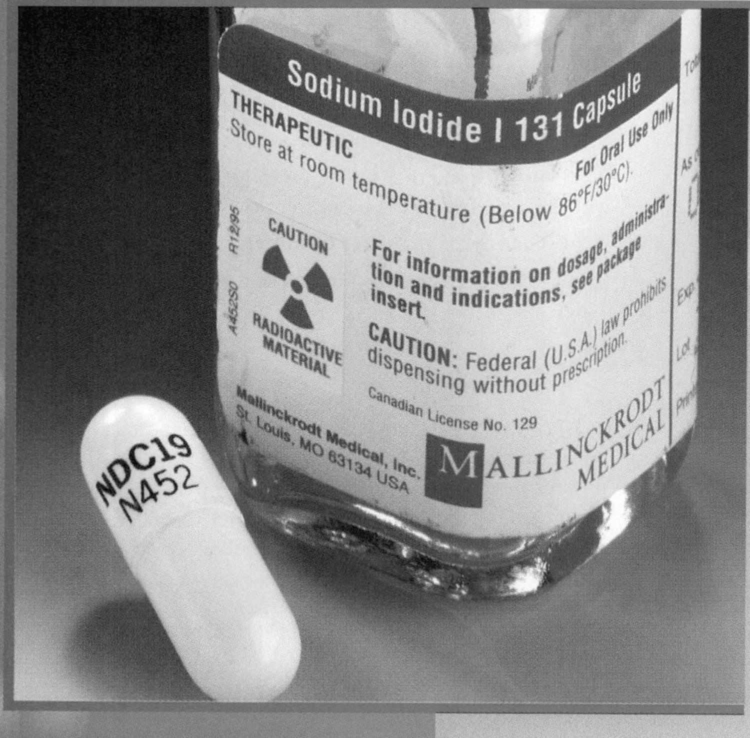

CHEMISTRY IN YOUR FUTURE

Less than a month ago, when Ms. Jensen came to the radiation laboratory where you work as a technician, she was experiencing extreme nervousness, insomnia, and weight loss. Her physician had discovered that Ms. Jensen's thyroid gland was enlarged and, after getting the results of some blood tests, determined that the problem was Graves disease, a condition characterized by an overactive thyroid gland. Now, after several visits during which you administered the prescribed series of radioactive iodine capsules, Ms. Jensen's symptoms are gone and she feels like herself again. This chapter identifies the radioactive isotope that was used and explains why it was chosen for her therapy.

LEARNING OBJECTIVES

- Define radioactivity and describe the principal kinds of nuclear emissions.
- Use nuclear equations to explain transmutation.
- Describe radioactive decay and the half-life concept.
- Describe the chemical effects of high-energy radiation.
- Describe the characteristics of radioisotopes used in diagnosis and in therapy.
- Explain how CAT scans, PET scans, and MRI are used in diagnosis.
- Describe the origin of nuclear energy and its effect on the biosphere.

We are immersed in a sea of electromagnetic radiation—cosmic rays, heat from the sun, radio frequencies. The energies of certain frequencies of radiation are great enough to break chemical bonds and to ionize atoms, with serious consequences for biological systems. It is therefore important for us to examine the origin and properties of radiation and its interaction with matter. We will find that radiation also plays a significant role in medical diagnosis and therapy and can be harnessed as an important source of energy.

10.1 RADIOACTIVITY

In 1895, Wilhelm Roentgen discovered that, by exposing metals to high electrical voltages, he could generate radiation that penetrated solid objects. He called this mysterious radiation X-rays. Within months of their discovery, they were being used in forensic applications and to explore pathologies of bone. Shortly after that, Henri Bequerel discovered penetrating radiation that emanated spontaneously from uranium salts. After Bequerel's discovery, Marie and Pierre Curie found that two new elements that they discovered—polonium and radium—also emitted penetrating radiation. Many other natural emitters of radiation were found in the ensuing years and came to be called radioactive substances.

By 1912, Ernest Rutherford had shown that radioactivity was the result of processes taking place in the atomic nucleus. He proposed that the emission of high-energy particles from the nucleus allowed unstable radioactive elements to achieve stability. Today, it is believed that the origin of the instability lies in the fact that an atom's nucleus contains protons in extraordinarily close contact, with the resulting strong repulsive forces reduced by the presence of neutrons. As atomic number increases, the number of neutrons increases to compensate for the increasing repulsive forces within the nucleus. Nevertheless, above atomic number 83, the atomic nuclei become so unstable that all elements above that atomic number are radioactive. Some of the isotopes of the lighter elements are radioactive as well. One of the radioactive isotopes of potassium is of particular importance because it is naturally present within our bodies and is a constant internal source of radiation.

Radioactive emissions often result in changes in mass number or atomic number or both, and these changes can be accounted for only by noting the precise number of **nucleons** (the name for either the protons or the neutrons in the nucleus) present before and after radioactive emissions. For this reason, in considering nuclear processes, we must always be careful to refer to a specific isotope. We will use the notation developed in Section 2.5 for characterizing isotopes, in which one isotope is distinguished from another in either of two ways. One method is to write the mass number as a superscript and the atomic or proton number as a subscript in front of the symbol for the element. An example is ${}^{17}_{8}O$. The alternate method is to write the name of the element followed by a hyphen and the atomic mass number. For example, oxygen-17 or potassium-38, with the atomic number implied by the element's name.

The word isotope is used to denote atoms of the same element possessing different mass numbers. The name **nuclide** is a more general term used for referring to isotopes of either the same or different elements. A radioactive isotope is called a **radionuclide.** For example, potassium-40 and oxygen-17 are both radionuclides.

10.2 RADIOACTIVE EMISSIONS

The major types of radioactive emissions are alpha(α)-particles, beta(β)-particles, gamma(γ)-rays, and positrons. **Alpha-particles** have been shown to

TABLE 10.1 Summary of the Properties and Results of Nuclear Emissions

		Change in Nucleus	
Emission	**Symbol**	**Mass Number**	**Atomic Number**
α	$^{4}_{2}He$	decreases by 4	decreases by 2
β	$^{0}_{-1}e$	no change	increases by 1
positron	$^{0}_{1}e$	no change	decreases by 1
γ	$^{0}_{0}\gamma$	no change	no change

consist of helium nuclei; that is, helium atoms with no electrons. **Beta-particles** are energetic electrons originating in the nucleus and emitted from it. **Positrons** have all the properties of electrons but with a positive rather than a negative charge. **Gamma-rays** are not particles; they are very high energy photons. Table 10.1 lists the emissions, their symbols, and the nuclear consequence of their loss.

Because the α-particle is a helium nucleus, its symbolic notation as a nuclear particle is $^{4}_{2}He$. Considered a helium nucleus—that is, a helium atom that has lost its electrons—it is also an ion and would appear in a chemical equation as He^{2+}. When an element loses an α-particle, $^{4}_{2}He$, its mass undergoes a change of 4 amu, and its atomic number is reduced by two units. It is therefore transmuted into a new element of lower atomic number with a reduced mass. **Transmutation** is a nuclear process in which a nuclide is transformed into a new element either spontaneously or artificially, the latter by interaction with a high-energy nuclear particle.

Transmutation is described by a **nuclear equation.** For example, to show that radium-226 emits an α-particle and forms radon-222, we write:

$$^{226}_{88}Ra \longrightarrow {}^{222}_{86}Rn + {}^{4}_{2}He$$

Notice that the sum of the subscripts (numbers of protons) on the right-hand side is equal to the subscript on the left-hand side and that the sum of the superscripts (protons + neutrons) on the right is equal to the superscript on the left.

As shown in Table 10.1, the symbol for a β-particle, $^{0}_{-1}e$, has a superscript of zero, denoting the infinitesimal mass of an electron relative to a nucleon. The subscript of -1 refers to the electron's charge. The loss of a β-particle from a nucleus causes a transmutation by increasing the positive charge of the nucleus (the atomic number) by one unit.

The symbol for a positron, $^{0}_{1}e$, indicates that it has no mass but does possess a positive charge. Its loss decreases the positive charge of the nucleus (the atomic number) and leads to a transmutation of the element into the next element that is one atomic number lower. The symbol for the γ-ray, $^{0}_{0}\gamma$, indicates no mass and no charge; thus γ-ray emission does not lead to transmutation.

Example 10.1 Writing a nuclear equation: I. Emission of an alpha particle

What is the effect of radioactive emission of an α-particle by an atom of uranium-238?

Solution

When a $^{238}_{92}U$ nucleus emits a helium nucleus, $^{4}_{2}He$, the mass number decreases by four, and the atomic number decreases by two. The change in atomic number reveals that a new element has been produced. The nuclear equation that describes the ejection of a helium nucleus from uranium-238 is:

$$^{238}_{92}\text{U} \longrightarrow {}^{234}_{90}\square + {}^{4}_{2}\text{He}$$

Check to be sure that the sum of the mass numbers of the product superscripts (protons plus neutrons) equals the mass number of the uranium-238 and that the sum of the product subscripts (protons) is equal to the atomic number of uranium-238. The new element, $^{234}_{90}\square$, is identified by referring to the periodic table to find the element with atomic number 90. It is thorium. The new nuclide is therefore thorium-234.

Problem 10.1 Radium-223 is a radioactive α-emitter. Write the nuclear equation for this emission event, and identify the product.

The following example shows how to predict the results of β-emission.

Example 10.2 Writing a nuclear equation: II. Emission of a β-particle

What is the effect that the emission of a β-particle has on an atom of thorium-234.

Solution

As in Example 10.1, we first write the nuclear equation for this emission event. Radioactive thorium-234 emits a β-particle as:

$$^{234}_{90}\text{Th} \longrightarrow {}^{234}_{91}\text{Pa} + {}^{0}_{-1}\text{e}$$

The mass number of the product is the same as that of thorium, because the β-particle has no mass. But notice that, because one negative nuclear charge has been lost, the positive nuclear charge, and consequently the atomic number, has been increased by one. Examination of the periodic table indicates that the element possessing atomic number 91 is protactinium.

Problem 10.2 Radium-230 is a radioactive β-emitter. Write the nuclear equation for the event, and identify the product.

We have said that a β-particle is equivalent to an electron. How, then, can it be emitted from a nucleus, which contains only protons and neutrons? The answer is that the β-particle is the result of the conversion within the nucleus of a neutron ($^{1}_{0}\text{n}$) into a proton:

$$^{1}_{0}\text{n} \longrightarrow {}^{1}_{1}\text{H} + {}^{0}_{-1}\text{e}$$

The energetic β-particle is ejected, whereas the proton, $^{1}_{1}\text{H}$, remains in the nucleus to increase the atomic number by one.

In an analogous process, the emission of a positron is the result of the conversion of a proton into a neutron:

$$^{1}_{1}\text{H} \longrightarrow {}^{1}_{0}\text{n} + {}^{0}_{1}\text{e}$$

If a positron comes into contact with an electron, it is rapidly annihilated to produce γ-rays as:

$$^{0}_{-1}\text{e} + {}^{0}_{1}\text{e} \longrightarrow 2\ {}^{0}_{0}\gamma$$

Note that the reaction of a positron with an electron produces two γ-rays. This process is the basis for positron emission tomography (PET), a medical imaging technique that will be discussed later in this chapter. Example 10.3 illustrates a nuclear reaction equation for positron emission.

Example 10.3 Writing a nuclear equation: III. Emission of a positron

A radioactive isotope of potassium, $^{38}_{19}\text{K}$, is a positron emitter. Describe the transmutation that takes place as a result.

Solution

The nuclear equation for the process is:

$$^{38}_{19}K \longrightarrow ^{38}_{18}Ar + ^{0}_{1}e$$

The product has the same mass number as that of potassium, but, because the positive nuclear charge has been reduced by one, its atomic number has decreased by one. The periodic table reveals the new element to be argon.

Problem 10.3 Sodium-21 is a radioactive positron emitter. Write the nuclear equation for the event, and identify the product.

The emission of an α- or β-particle often leaves the nucleus of the product in an excited state. Recall from Section 2.7 that, when an atom is in an excited state, it can return to its ground state by losing the energy of excitation as emitted light—that is, as a photon. The photons emitted by excited atomic nuclei, however, are far more energetic than the photons emitted by the return of excited electrons to the ground state. Moreover, they are invisible. These photons are the γ-rays, or γ-radiation, emitted by radioactive nuclei. Because γ-ray emission does not alter the mass of a radioactive nuclide, it is usually not included in nuclear equations.

Radioactivity can be induced in nonradioactive elements by bombarding them with high-energy nuclear particles. The first artificial conversion of one nucleus into another was performed by Ernest Rutherford in 1919. He bombarded nitrogen-14 (the target) with α-particles (the projectiles) emitted by radium and produced oxygen-17. The reaction is:

$$^{14}_{7}N + ^{4}_{2}He \longrightarrow ^{17}_{8}O + ^{1}_{1}H$$

Since that time, hundreds of radioisotopes have been produced in the laboratory.

In order for charged particles such as α-particles to penetrate a positively charged nucleus or even the electron cloud surrounding the nucleus, they must be moving at extremely high velocities. Such velocities are achieved in machines called particle accelerators such as the cyclotron, the Van de Graaf generator, and the betatron. All the elements above atomic number 92 have been created in this manner; none of them occur naturally.

Most radioisotopes used in medicine have been produced with the use of neutrons as subatomic projectiles. For example, cobalt-60, a γ-emitter used in radiation therapy for cancer, is prepared by bombarding iron-58 with neutrons (obtained from a nuclear reactor; Section 10.8). The successive reactions are:

$$^{58}_{26}Fe + ^{1}_{0}n \longrightarrow ^{59}_{26}Fe$$

$$^{59}_{26}Fe \longrightarrow ^{59}_{27}Co + ^{0}_{-1}e$$

$$^{59}_{27}Co + ^{1}_{0}n \longrightarrow ^{60}_{27}Co$$

10.3 RADIOACTIVE DECAY

Radioactive decay is the loss of radioactivity when a radioactive element emits nuclear radiation. The decay of radioisotopes found in nature results in the formation of products called **daughter nuclei,** which may or may not be radioactive. If the resulting products are nonradioactive (stable), the decay stops. However, if the products are radioactive, decay will continue to produce new elements until a stable state is achieved.

A series of nuclear reactions that begins with an unstable nucleus and ends with the formation of a stable one is called a **nuclear decay series** or **nuclear disintegration series.** Three such series are found in nature. One series begins with uranium-238, proceeds through 14 decay steps, and terminates with lead-206. One of the daughters of this decay series is radium-226, which decays to a number of radon radionuclides. Of these radionuclides, radon-222 is a major

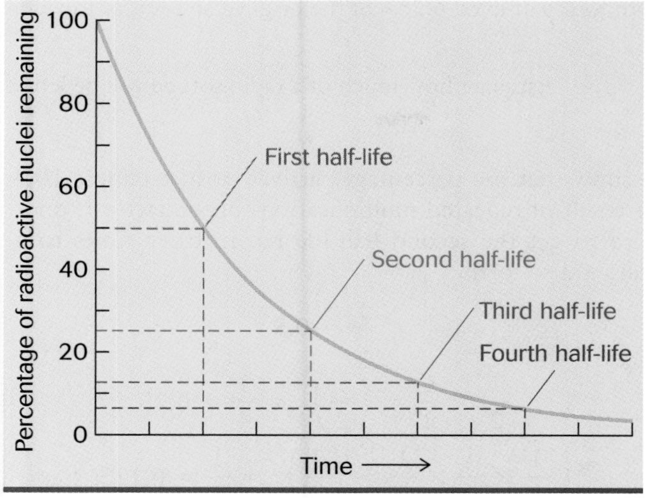

Figure 10.1 Relation between radioactive decay and half-life. Fifty percent of the radioactivity of a sample is lost at the end of every half-life.

health hazard and comprises almost 50% of the background radiation on Earth. Background radiation will be more fully discussed in Box 10.4. Another series begins with uranium-235 and terminates with lead-207. The third begins with thorium-232 and terminates with lead-208.

The rate at which a particular radionuclide decays is as characteristic of that nuclide as the melting point is of a pure compound. Some radionuclides decay in millionths of a second, others in millions of years. Decay is independent of the variables, such as temperature, pressure, and concentration, that affect chemical reactions. It is an intrinsic property of the unstable nucleus and is outside of human control. For this reason, disposal of nuclear wastes is very difficult and becomes one of the considerations in the selection of radioisotopes used for therapeutic purposes.

The rate of radioactive decay can be described as the loss of a constant percentage of the original element per unit time. Chemists measure it in terms of the time required for 50% of the original material to decay. This period of time is called the **half-life, $t_{1/2}$**. Because a constant percentage of the element decays in a given period of time, 50% of the original amount is lost during the first half-life, 50% of the remainder will decay over the second half-life, 50% of what remains after that will decay over the third half-life, and so forth. The relation between half-life and radioactive decay is illustrated in Figure 10.1.

Example 10.4 Determining the relation between half-life and percentage of decay

Estimate how much of a radioisotope will be left after four half-lives.

Solution

The simplest way to arrive at this estimate is to put the problem into tabular form.

End of Half-Life	Amount Left (%)
0	100.00
1	50.00 (half the original amount)
2	25.00 (half the preceding remainder)
3	12.50 (half the preceding remainder)
4	6.25 (half the preceding remainder)

At the end of four half-lives, 6.25% of the original amount of the radioisotope will be left.

Problem 10.4 Estimate how much of a radioisotope will be left after six half-lives.

We can show that the percentages arrived at in Example 10.4 and Figure 10.1 are the result of repeated multiplications of the factor $\frac{1}{2}$, once for the first half-life, twice to get the second half-life result, three times to get the third half-life result, and so forth.

$$\frac{1}{2} = 0.5$$

$$\frac{1}{2} \times \frac{1}{2} = \left(\frac{1}{2}\right)^2 = \frac{1}{4} = 0.25$$

$$\frac{1}{2} \times \frac{1}{2} \times \frac{1}{2} = \left(\frac{1}{2}\right)^3 = \frac{1}{8} = 0.125$$

$$\frac{1}{2} \times \frac{1}{2} \times \frac{1}{2} \times \frac{1}{2} = \left(\frac{1}{2}\right)^4 = \frac{1}{16} = 0.0625$$

These repeated multiplications can be summarized in a convenient equation. If the amount of radioisotope at time zero (before decay begins) is defined as N_0, and the amount remaining after n half-lives is N, then the fraction of isotope remaining after n half-lives is

$$\frac{N}{N_0} = \left(\frac{1}{2}\right)^n$$

and the percentage of isotope remaining after n half-lives is:

$$\frac{N}{N_0} \times 100\% = \left(\frac{1}{2}\right)^n \times 100\%$$

Conversion of the fractions remaining after one, two, three, and four half-lives into percentages gives:

$$0.5 \times 100\% = 50\%$$
$$0.25 \times 100\% = 25\%$$
$$0.125 \times 100\% = 12.5\%$$
$$0.0625 \times 100\% = 6.25\%$$

Example 10.5 Determining the amount of radioisotope remaining after an integral number of half-lives

Calculate the percentage of radioisotope remaining after 7.0 half-lives

Solution

The percentage of radioisotope left after any number of half-lives can be calculated by using the following relation:

$$\frac{N}{N_0} \times 100\% = \left(\frac{1}{2}\right)^n \times 100\%$$

where N = amount left after a number, n, of half-lives, and N_0 is the amount of isotope at the start of the process. You can calculate $(0.5)^n$ on your calculator in several ways. There may be a special function key labeled $\mathbf{Y}^x$. To use it, enter 0.5 (**Y** equals 0.5), press the $\mathbf{Y}^x$ key, enter n, press the enter key, and read the result. In the absence of a special key, enter 0.5, and multiply it n times by 0.5.

The percentage of a radioisotope left after seven half-lives is:

$$\frac{N}{N_0} \times 100\% = \left(\frac{1}{2}\right)^{7.0} \times 100\% = \frac{1}{128} \times 100\% = 0.781\%$$

Problem 10.5 Calculate the percentage of radioisotope remaining after 5.0 half-lives.

Example 10.6 Determining the number of half-lives from the amount of remaining radioisotope

A nurse locates the iodine-131 prescribed for measurement of thyroid activity and finds it was produced 21 days before. What percentage of the isotope is left in the sample?

Solution

This problem is solved by calculating how many half-lives have taken place since the isotope's manufacture. We find the half-life of the isotope in Table 10.2 to be 8.05 days. Therefore the number of half-lives is

$$21 \text{ days} \left(\frac{\text{half life}}{8.05 \text{ days}}\right) = 2.6 \text{ half-lives}$$

As in Example 10.5, the percent of radiosotope left after any number of half-lives can be calculated using the relation

$$\frac{N}{N_0} \times 100\% = \left(\frac{1}{2}\right)^{n} \times 100\%$$

Substituting the value for half-lives into this equation we get

$$\frac{N}{N_0} \times 100\% = \left(\frac{1}{2}\right)^{2.6} \times 100\%$$

We can solve this equation as we did in Example 10.5:

$$\left(\frac{1}{2}\right)^{2.6} \times 100\% = 17\%$$

This can also be solved by noting that $(0.5)^{2.6} = 2.6 \times \log 0.5$.

Problem 10.6 Calculate the percentage of radioisotope remaining after 3.8 half-lives.

Table 10.2 on the following page lists the half-lives of some biomedically useful radioisotopes. We will learn more about their practical applications in Section 10.5. In addition, radioisotopes such as carbon-14, hydrogen-3 (tritium), phosphorus-32, and sulfur-35 are important in biochemical research. They are incorporated into biochemical intermediates by synthesis. The resultant radioactivity, or "label," of the intermediates makes it possible to map their progress through the metabolic pathways in which they participate. An understanding of radioactive isotopes and their half-lives has also enabled scientists to assess the ages of ancient rocks, fossils, and human artifacts. Radiocarbon dating is discussed in Box 10.1 on page 267.

10.4 EFFECTS OF RADIATION

The relation between types of radiation and their energies in kilojoules per mole is illustrated in Table 10.3 on the following page. For our purposes, we can divide the table into two parts: examples of high-energy radiation

TABLE 10.2 Half-lives and Biomedical Applications of Some Radioisotopes

Nuclide	Half-life	Emission	Application
barium-131	11.6 days	gamma	detect bone tumors
carbon-11	20.3 min	positron	PET brain scan
chromium-51	27.8 days	gamma	determine blood volume and red blood cell lifetime
gold-198	64.8 h	beta	assess kidney activity
iodine-131	8.05 days	beta	measure thyroid uptake of iodine
iron-59	45 days	beta	assess blood iron metabolism
krypton-79	34.5 h	gamma	assess cardiovascular function
phosphorus-32	14.3 days	beta	detect breast carcinoma
selenium-75	120 days	beta	measure size and shape of pancreas
technetium-99	6.0 h	gamma	detect blood clots
indium-111	2.8 days	gamma	label blood platelets
gallium-67	78 h	gamma	diagnose lymphoma and Hodgkin disease
chromium-51	27.8 days	gamma	diagnose gastrointestinal disorders

(energies ranging from 2.0×10^8 kJ/mol to 1.0×10^3 kJ/mol) and examples of low-energy radiation (energies below 1.0×10^3 kJ/mol). Inserted in the table, as a reminder, is the typical energy of a covalent bond. Its placement there emphasizes the fact that radiation can initiate chemical processes.

Occasionally, you will see a non-SI unit of energy called the **electron-volt (eV).** Workers in the field of high-energy radiation use this unit of energy. One

TABLE 10.3 Energy Associated with Various Types of Radiation

Type of Radiation	Energy (kJ/mol)
HIGH-ENERGY RADIATION	
gamma-ray	2.0×10^8
beta-ray	2.0×10^8
alpha-ray	2.0×10^8
X-ray	1.0×10^7
UV in outer space	1.0×10^3
LOW-ENERGY RADIATION	
UVB at Earth's surface	420
UVA at Earth's surface	315
covalent bond energy	250 to 450
blue light	300
red light	250
infrared light	80
FM radio	10
AM radio	1×10^{-4}

10.1 Chemistry Around Us

Radiocarbon Dating

Many fossils and ancient human artifacts contain radioisotopes whose rate of decay allows chemists, archeologists, and other scientists to determine the article's age. To do so requires a way of measuring the amount of radioisotope in the sample, a knowledge of the radioisotope's half-life, and some way of determining the amount of radioisotope that was originally present. All these requirements are satisfied in an unstable radioisotopic form of carbon: carbon-14, or C-14.

Carbon-14 is produced at a constant rate in the upper atmosphere. The interaction of cosmic radiation with molecules there releases neutrons that in turn interact with nitrogen to produce a radioactive isotope of carbon:

$$^{14}_{7}N + ^{1}_{0}n \longrightarrow ^{14}_{6}C + ^{1}_{1}H$$

There is no chemical difference between C-14 and nonradioactive C-12, so they are both continuously incorporated into living systems through photosynthesis. While an organism is alive, the ratio of C-14 to C-12 remains constant because the organic substances of which all living things are composed are continuously replaced by new materials from the environment. However, once the organism is dead, the C-14 in its system continually decays but is not replaced. Therefore the ratio of the two isotopes in formerly living organisms continually changes through time.

A comparison of the ratio of C-14 to C-12 in archeological samples with the ratio in living systems reveals the fraction of C-14 remaining. The half-life of C-14 is very well established as 5730 ± 40 years. By using radiocarbon dating, researchers have determined the timing of many events over the past 50,000 years. A notable example has been its use in timing the prehistoric migrations of humans over the continents of North and South America.

electron-volt/particle or photon is equivalent to 96.5 kJ/mol. Nuclear emissions have energies in the millions of electron-volts (MeV) compared with the 100–200 KeV of X-rays and the 2 to 3 eV of visible light. The energy required to break a covalent bond ranges from about 250 kJ/mol to 450 kJ/mol, which is equivalent to about 3–5 eV.

The study of the effects of radiation on matter is called **radiochemistry** or **radiation chemistry.** Radiochemistry addresses two major questions: (1) In what circumstances will radiation cause chemical changes in matter? and (2) What specific chemical changes will result?

Penetrating Power of Radiation

In what circumstances will radiation cause matter to undergo chemical change? One major factor determining the effects of radiation is its penetrating power. Particles of the same energy but of different mass produce quite different results.

Consider an α-particle, which is massive and highly charged. Because of its size and charge, an α-particle undergoes atomic collisions and intense interactions with bound electrons all along its path. As it travels, it tears electrons away from the valence shells of the atoms that it encounters, leaving a trail of ions in its wake. However, these numerous encounters slow the α-particle down. It can penetrate ordinary matter only to a depth of a few millimeters before it discharges its kinetic energy and stops. However, it continues to produce extensive ionization in the place where it comes to rest.

Beta-particles also are charged but have the minute mass of the electron. Therefore, they undergo many fewer collisions and less-intense interaction with atoms along their paths. Consequently, β-particles have greater penetrating power than do α-particles. They create ions along a thin, much less dense track than do α-particles. Alpha-particles may penetrate human tissue to a depth of 0.05 mm, less than the thickness of human skin, and β-particles may penetrate from 5 to 10 mm.

Gamma-rays have no mass or charge and so have very large penetrating power. They cannot be stopped by human tissue, but their intensity is reduced (attenuated) by about 10 to 20% after passing through the body. A similar reduction in intensity is undergone by X-rays.

Chemical Effects of Radiation

High-energy radiation produces not only ions along its penetration track, but free radicals as well. A free radical is an uncharged fragment of a molecule resulting from the breakage of a covalent bond such that each of the resulting fragments bears an unpaired electron. Because of the unpaired electron, a free radical is highly reactive and has a very short lifetime. Free radicals interact rapidly with any matter that they encounter, forming new bonds or extracting electrons from other molecules to form ions that are more stable.

Water makes up from 60 to 70% of living tissue. For this reason, the effects of radiation on water have a strong bearing on the ways in which radiation affects health and life. The first step in the irradiation of water is ionization caused by the loss of an electron. The reaction is represented by the following equation:

$$H_2O + \text{radiation} \longrightarrow H_2O^+ + e^-(aq)$$

where $e^-(aq)$ is a free but hydrated electron. This reaction is called the **primary radiation event,** and it is rapidly followed by a number of **secondary chemical processes** resulting in free radical formation. (The single unpaired electron characteristic of a **free radical** is shown as a dot on the appropriate atom.)

$$H_2O^+ + H_2O \longrightarrow H_3O^{\cdot +} + \cdot OH$$
$$e^-(aq) + H_2O \longrightarrow \cdot H + OH^-$$

Potentially harmful reactions now can take place between biomolecules and the hydroxyl and hydrogen free radicals. Either of these free radicals may abstract a hydrogen atom from a donor biomolecule, which then produces a biomolecule free radical:

$$\cdot OH + H{-}R{-}NH_2 \longrightarrow HOH + \cdot R{-}NH_2$$

The new free radical may now combine with itself or another biomolecule. If that biomolecule is an enzyme or a nucleic acid, these events may severely alter its normal function and have a devastating effect on cellular activity.

Free radicals formed in the upper atmosphere protect us from ultraviolet (UV) radiation; this phenomenon is discussed in Box 10.2. The action of manufactured chemicals that reverse this beneficial effect is discussed in Box 11.5.

Biological Consequences of Radiation Damage

The effects of high-energy radiation on living tissue range from negligible to repairable damage to lethality. The damage occurs within cells; so organ or tissue damage is the result of the death of significant numbers of cells. The immediate chemical results of cellular irradiation are likely to be the same in different cell types, but the ultimate consequences of radiation exposure vary with the type of tissue and its biological role.

Exposure that does not result in death may produce a number of disorders known collectively as **radiation sickness.** Typical symptoms include nausea and a drop in white blood cell count. The earliest effects are seen in tissues that undergo rapid cell division, because any radiation-induced alterations in DNA or protein structure cause derangements in cell reproduction. Such rapidly dividing tissues include bone marrow (where white blood cells are

▶▶ The effects of radiation on DNA are described in greater detail in Section 21.8.

10.2 Chemistry Around Us

The Ozone Layer and Radiation from Space

In all likelihood, life did not evolve simultaneously in the sea and on land. Early in Earth's history, the land environments would have been too dangerous because the atmosphere contained no gases that could absorb the harmful high-energy UV radiation from the sun. As Table 10.3 indicates, UV radiation from the sun is energetic enough to break covalent bonds and severely damage biomolecules such as DNA. Neither biomolecules nor living organisms could survive on land under such conditions. The fact that water absorbs UV radiation allowed life to exist and evolve adaptively in the sea. But what changes took place in the atmosphere to enable life to move onto the land?

Interestingly, the necessary atmospheric changes resulted indirectly from a major advance in evolution—photosynthesis. Photosynthetic organisms produce oxygen gas as a by-product, and in fact injected vast amounts of this new gas into the primitive atmosphere. As the oxygen diffused throughout the air, it encountered the energetic radiation from space.

When oxygen molecules absorb high-energy UV radiation in the upper atmosphere, they break apart into oxygen free radicals:

$$O_2 \xrightarrow{\text{UV radiation}} 2\ O\cdot$$

Because the atmospheric density is very low in the upper atmosphere, these oxygen atoms rarely encounter other compounds with which to react and are thus very stable. As the free radical oxygen atoms diffuse toward Earth, however, and reach the lower atmosphere, they begin to meet and react with oxygen molecules to form ozone:

$$O\cdot + O_2 \longrightarrow O_3$$

The ozone produced at this lower level strongly absorbs the UV radiation and forms a powerful barrier to the high-energy radiation from space. It works in the following way. The UV radiation reaching this atmospheric level splits the bonds of the ozone molecule back into oxygen and a free radical:

$$O_3 \xrightarrow{\text{UV radiation}} O_2 + O\cdot$$

A cyclical process ensues in which ozone formed in the upper atmosphere is destroyed in the lower atmosphere, and high-energy UV radiation is prevented from reaching Earth's surface.

However, some UV radiation of lower energy does penetrate the atmosphere and reach the surface of Earth, where it produces sunburn—erythema (abnormal reddening) of the skin caused by capillary enlargement—and tanning—photooxidation of the skin pigment melanin.

Medical scientists consider sunlight that reaches Earth's surface to consist of two energy levels of ultraviolet radiation (see Table 10.3): UVB, with energies of about 420 kJ/mol, which causes sunburn and cancer; and UVA, with energies of about 315 kJ/mol, which penetrates the top layer of the skin and appears to be responsible for damaging the lower levels of the skin, causing roughening and premature aging.

Although the amount of UVA radiation in sunlight is about 10 to 100 times as great as the amount of UVB radiation, UVB radiation is 1000 times as effective in inducing cancer. The rates of appearance of skin cancer have increased in the past 20 years, and the increase has been correlated with deterioration in the ozone of the upper atmosphere. (The upper atmospheric depletion of ozone is described in Box 11.5). For every 1% loss of ozone, there is a 2% increase in the UVB radiation at Earth's surface. Animal studies and statistical analyses directly support the correlation between UVB radiation and skin cancer.

The most likely biochemical link between UVB and skin cancer is the response of DNA to that radiation. Thymine, one of the four nitrogen bases that are the building blocks of DNA, undergoes a reaction called dimerization when exposed to UVB radiation. If two thymine bases adjacent to one another along one strand of the double helix are irradiated with UVB, they can undergo a chemical reaction in which a new covalent bond forms between them. This dimerization can interfere with normal base pairing in the course of DNA replication (Section 21.8). The result is the introduction of a mutation that can either be immediately lethal to a cell or cause long-term problems such as cancer.

Dermatologists recommend avoiding exposure to the sun during the hottest 4-h period of the day, keeping covered at all other times, using a sunscreen preparation rated at 15+ or higher, wearing a broad-brimmed hat, and protecting the eyes with adequate sunglasses.

High-energy radiation can also be used to prevent the spoiling of food. Here we see two groups of 1-week old peaches. The rotting peaches on the left were not irradiated; those on the right were irradiated.

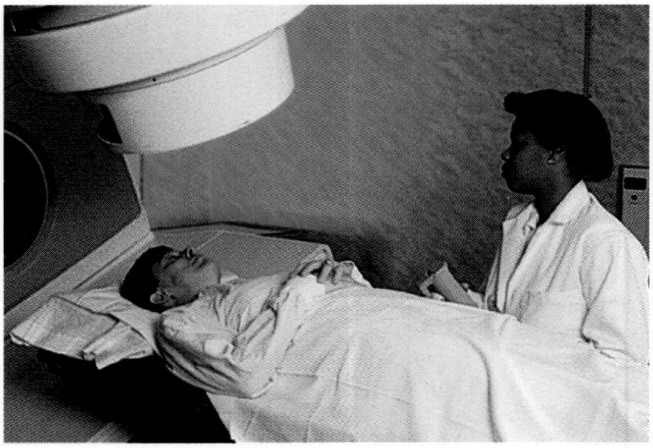

Figure 10.2 A particle accelerator uses high-energy radiation to treat a cancer patient.

produced) and the intestinal lining. Because children's tissues reproduce more rapidly than do those of adults, their susceptibility to radiation sickness is greater.

Cells that survive exposure to high-energy radiation may nevertheless have undergone subtle chemical changes that lead to chromosome damage or mutations. In somatic cells—that is, tissue other than testes and ovaries—these effects may cause cancer directly. For example, such changes may immediately induce cancer in bone tissue or result in the later development of leukemia due to radiation effects on bone marrow. However, if these changes take place in ovaries or testes—also known as germ tissue—the radiation effects may not show up until later generations.

Because radiation has its most immediate effects in tissue that undergoes rapid cell division and because cancer cells are among the most rapidly dividing cells, radiation has become an important mode of cancer therapy (Box 21.1). The technology here has concentrated on producing radiation beams that are extremely narrow and focusable so as to reduce damage in healthy surrounding tissue. Both high-energy X-rays and γ-radiation are employed in cancer therapy. Figure 10.2 shows a patient receiving radiation treatment.

Protection from Radiation

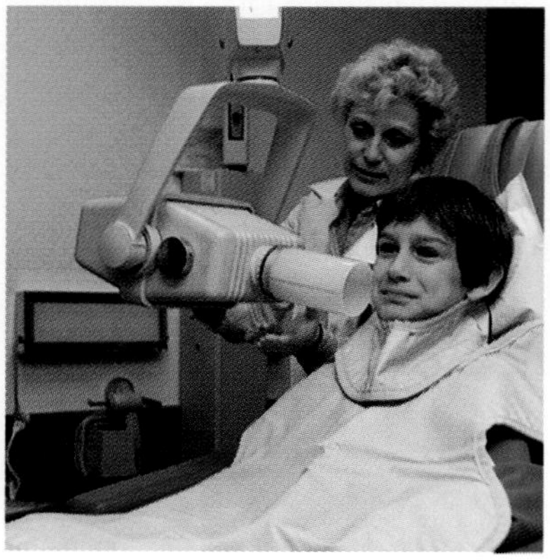

A technician positions a device to take an X-ray of a young man's teeth.

People can minimize their exposure to high-energy radiation in three ways: (1) by using a radiation-absorbing barrier of some kind, (2) by remaining as far away from the radiation source as is practicable, and (3) by limiting the time of exposure. All technicians working in research and the health sciences must take such measures seriously. For example, X-ray technicians wear aprons composed of layers of lead that serve as a radiation-absorbing barrier and often leave the room before activating an X-ray machine by remote control.

The reason for advising people to stay as far away as practicable from a radiation source is that radiation of all kinds follows an inverse-square law with respect to distance:

$$I \propto \frac{1}{d^2}$$

where I, the intensity of the radiation, is proportional to the reciprocal of the square of the distance d from the origin of the radiation. The radiation intensity falls off very rapidly as the distance from the source increases. The effect of this law is calculated in Example 10.7.

Example 10.7 Calculating the effect of increasing distance on radiation intensity

If the distance between a source and target is 2 m, how far should the source be moved from the target to decrease the radiation intensity to one-fourth (1/4) of its current value?

Solution

A ratio of two intensities can be constructed by using the equation introduced earlier:

$$\frac{I_{new}}{I_{old}} = \frac{\left(\dfrac{1}{d_{new}^2}\right)}{\left(\dfrac{1}{d_{old}^2}\right)}$$

This equation simplifies to:

$$\frac{I_{new}}{I_{old}} = \frac{d_{old}^2}{d_{new}^2}$$

Substituting appropriate values into these ratios gives:

$$\frac{1}{4} = \frac{4\ \text{m}^2}{d_{new}^2}$$

$$d_{new}^2 = 16\ \text{m}^2$$

$$d_{new} = 4\ \text{m}$$

Irrespective of the specific values of distance and intensity, a doubling of the distance causes the radiation intensity to decrease by a factor of four.

Problem 10.7 If the distance between a source and target is 6 m, how far should the source be moved from the target to decrease the radiation intensity to one-fourth of its current value?

10.5 DETECTION OF RADIOACTIVITY

Nuclear emissions, as we have seen, are quite energetic and leave a trail of ions behind them as they penetrate through matter. These ions serve as evidence that radiation is present. Many techniques have been devised for detecting them.

The most common and inexpensive method of detection is by the use of photographically sensitive film that becomes progressively darker as its exposure to radiation increases. Film badges are routinely worn by scientists and technicians who work with radioactive materials or around radiation sources. The badges can differentiate between different types of radiation. Gamma-rays penetrate much more matter than do β-particles, which in turn penetrate more matter than do α-particles before they are stopped, so covering photographic film with different thicknesses of material makes it possible to identify whatever type of radiation is present.

Other devices, such as the **Geiger counter,** use electronics to detect the ions created when radiation passes through matter. The counter, shown in Figure 10.3 on the following page, consists of a tube fitted with a thin window and filled with argon gas. A positively charged electrode runs down the center of the tube, and the tube itself serves as the negative electrode. When a photon or particle passes through the window into the tube's interior, it creates positive argon ions in the gas, which are collected by the negative electrode. The resulting current can be measured on a meter or heard audibly as a click.

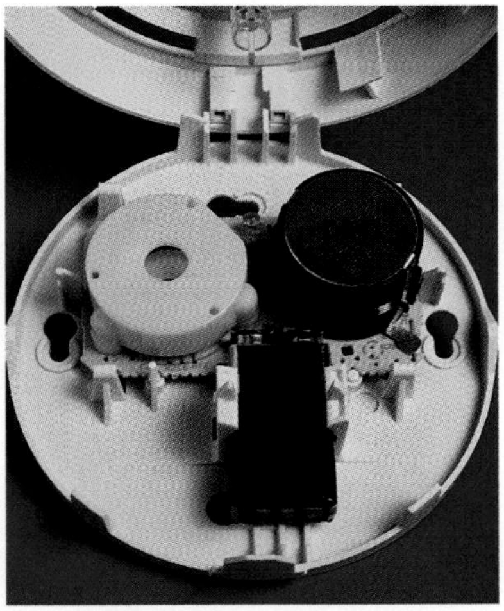

An ionizing smoke detector uses the basic principle of the Geiger counter. However, here, when smoke particles enter the chamber of an ionizing smoke detector, they attract the ionized air molecules already present and reduce the electrical current.

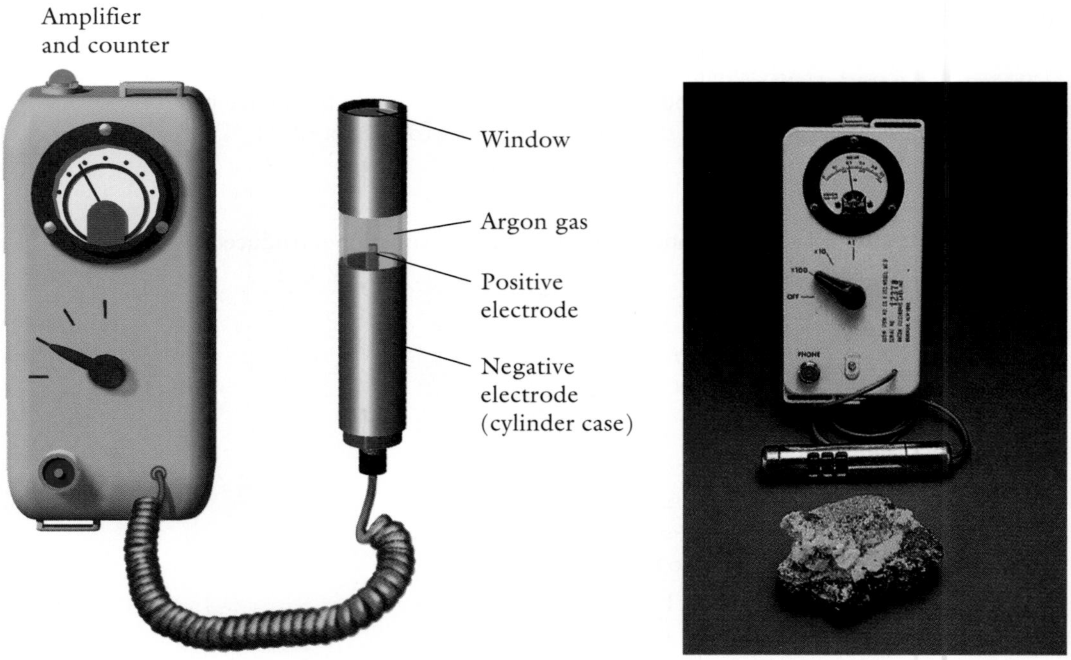

Figure 10.3 Diagram and photograph of a Geiger counter. Radiation enters through a thin window in the cylinder, ionizing the argon gas inside. The ions then carry electrical current between the cathode (negative electrode) and the anode (positive electrode), which is detected by the amplifer and counter.

Each burst of current signals the presence of an ionization event. The same ionization effect is used in one type of smoke detector.

In devices called **scintillation counters,** radiation generates flashes of light: the more flashes of light per unit time, the greater the dosage of radiation in the environment. A scintillation counter uses a device called a scintillator, which contains substances that emit a flash of light when struck by an energetic photon or particle (Figure 10.4). These substances—sodium and

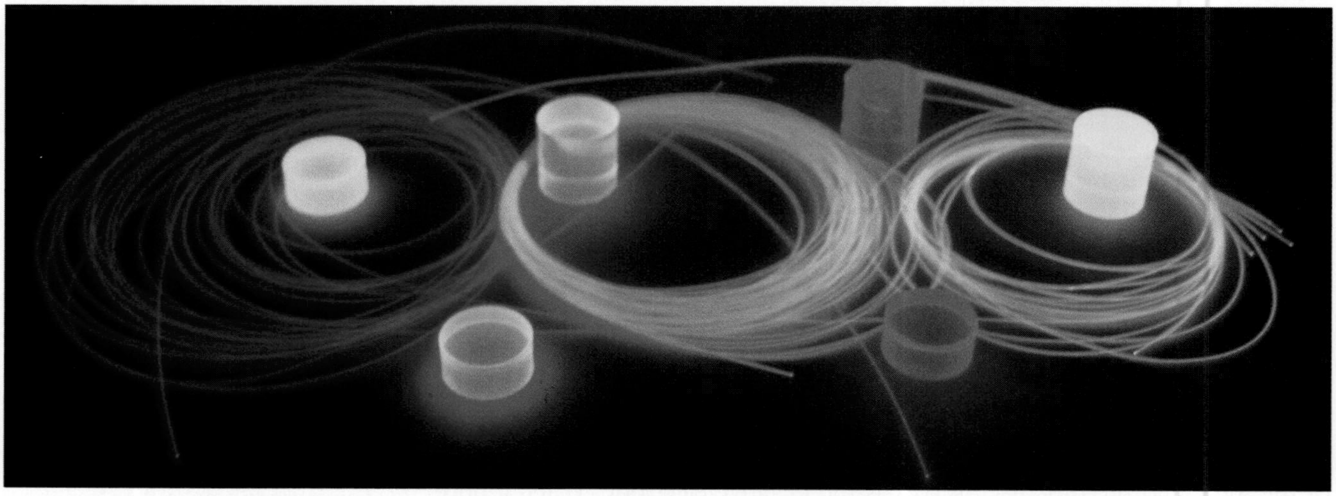

Figure 10.4 A scintillator, such as those in the photograph, emits a flash of light when a charged particle passes through it. The color of the emitted light depends on the type of material.

thallium iodides, zinc sulfide, and others—are similar to the substances that cover the inner surface of television tubes. The light is then converted into an electrical signal by a phototube.

10.6 MEASURING RADIOACTIVITY

The general approach to measuring the radioactivity of a sample is to measure the disintegrations per second (dps); that is, the number of atoms that decay per second. The most commonly used unit, the **curie (Ci)**, represents the number of radioactive disintegrations per second in a 1.0-g sample of pure radium. It is named after the discoverer of radium, Marie Sklodowska Curie.

$$1 \text{ Ci} = 3.7 \times 10^{10} \text{ dps}$$

This level of radioactivity is very intense, so it is usually more convenient to report a measurement in millicuries (mCi), microcuries (μCi), or picocuries (pCi). The SI unit of radioactivity is the becquerel (Bq), after the scientist credited with discovering radioactivity.

$$1 \text{ Bq} = 1\text{dps}$$

Radiation can also be quantified by the number of ions produced by the radiation passing through air. The unit describing this mode of detection is the roentgen (R), named after the discoverer of X-rays. It is used to measure the exposure to X-rays or γ-rays and provides an estimate of exposure to high-energy radiation.

$$1 \text{ R} = 2.1 \times 10^{9} \text{ units of charge/cm}^3 \text{ of dry air}$$

Table 10.4 on the following page lists these and other types of radiation units, as well as the situations in which they are appropriately applied.

What if we wish to predict the amount of biological damage likely to be caused by a given amount of radiation? The number of disintegrations per unit time of a radioactive sample or the number of ionizations in a volume of air will not tell us much in this regard. We saw earlier that biological damage depends to a great extent on the type of radiation to which the organism is exposed. Two units, the rad and the rem, have been devised to take this correlation into account.

We saw earlier that X-rays and γ-rays are particularly effective in penetrating human tissue, that α-particles are stopped by skin, and that β-particles may penetrate to a depth of 5 to 10 mm. Alpha-radiation and β-radiation from external sources are therefore not ordinarily considered dangerous. Nevertheless, when solid particles emitting these radiations enter the body and are not excreted quickly, they can do a great deal of damage in the tissue surrounding the emitting particle (see Box 10.3 on page 275). The experience of workers in the plutonium-purifying industry has shown that, if particles become lodged in the lungs, cancer is a very likely result.

The **rad** is a measure of the energy absorbed in matter as a result of exposure to any form of radiation. To convert this information into a measure of the biological damage likely to be induced by radiation, the rad is multiplied by a factor known as the **RBE (relative biological effectiveness).** The exact value of the RBE depends on the type of tissue irradiated, the dose rate, and the total dose, but it is approximately 1 for X-rays, γ-rays, and β-rays and 10 for α-particles, protons, and neutrons. The product is the **rem (roentgen equivalent for man).**

$$\text{rads} \times \text{RBE} = \text{rems}$$

Note that a rad of α-rays will produce more damage than a rad of β-particles. An SI unit called the **sievert (Sv)** measures a similar property—the effect of absorbed radiation on different kinds of tissues and under different situations.

TABLE 10.4 Units of Radiation, Their Symbols, and Definitions

Unit	Symbol	Definition
curie	Ci	An older non-SI unit that describes the activity of a radioactive sample. It is the number of disintegrations per second (dps) in 1.0 g of pure radium: 1 Ci = 3.7×10^{10} dps This is an extremely high activity; more commonly, activity is reported in milli- or microcuries: 1 μCi = 3.7×10^{4} dps
becquerel	Bq	An SI unit that describes the activity of a radioactive sample: 1 Bq = 1 dps 3.7×10^{10} Bq = 1 Ci
roentgen	R	A non-SI unit of exposure to γ- or X-radiation based on the amount of ionization produced in air. 1 R will produce 2.1×10^{9} ions in 1.0 cm^3 of dry air at 0°C and 1.0 atm pressure.
rad	rad	A measure of the energy absorbed in matter as a result of exposure to any form of radiation. It does not reveal the biological effects of the radiation. 1 Rad = 0.01 J absorbed per kilogram of matter
gray	Gy	An SI unit that describes the energy absorbed by tissue. Like the rad, it does not reveal the biological effects of radiation. 1 Gy = 1 J absorbed per kilogram of tissue 1 Gy = 100 rad
rem	rem	A measure of the energy absorbed in matter as a result of exposure to specific forms of radiation. This unit takes into account the fact that different forms of radiation produce different biological effects. 1 rem is equal to 1 rad times a factor dependent on the type of radiation.
sievert	Sv	An SI unit that, like the rem, takes into account the different biological effects caused by different forms of radiation. 1 Sv = 100 rem

One sievert = 100 rem. The clinical effects of short-term exposure to radiation are listed in Table 10.5.

When we weigh the relative hazards posed by different types of radiation, it is important to remember that all life forms on Earth are continuously exposed to low levels of **background radiation** from natural sources such as cosmic rays and radioactive elements in soil and rocks. A typical exposure to this radiation is about 350 mrem per year (1 millirem = 1×10^{-3} rem). About 50% of this background radiation is from radon, a naturally occurring radioac-

10.3 Chemistry Within Us

Radiation Dosimetry and Wristwatches

In the late 1930s, many cases of severe radiation sickness were discovered among workers in a factory that produced watch faces on which the numerals glowed in the dark. Workers painted numerals by hand with paint containing very low concentrations of a radium salt and a phosphorescent material. The radiation from radium caused the phosphor to glow in the dark. The work required precise location of the radioactive paint, so the workers used very thin brushes and kept the brush ends pointed by licking them with their tongues. In this way, α-emitting solids came to be embedded in the tissue of their mouths and tongues, causing radiation damage that eventually led to cancer.

In 1941, Robley Evans, a physician, completed a study of these workers and others whose jobs entailed exposure to radiation. It led to the development of standards for maximum allowable lifetime exposure that are still in use today. Such standards are crucial if scientists are to safely conduct experiments with materials such as the radioactive isotopes routinely used in medicine for diagnosis and therapy.

tive noble gas formed from the decay of radioactive radium-226 (see Box 10.4 on the following page). Additional background radiation from manufactured sources—X-rays, radiotherapy, radioisotopes—is about 65 mrem per year. Table 10.5 indicates that background radiation is too weak to be considered dangerous.

10.7 APPLICATIONS

Current knowledge of the properties of radiation and its interaction with matter has led to the development of a wide variety of applications, particularly in medical and biological science. Much of the recent research in molecular biology, metabolism, and cell biology would have been impossible without radioisotopes.

Radioisotopes

Radioisotopes introduced into the body are used for diagnosing pathological conditions as well as for therapy. The characteristic emission of a radioactive isotope is the decisive factor in its choice either as a diagnostic or as a therapeutic tool.

TABLE 10.5 Effects of Short-Term Exposure to Radiation

Dose (rem)	Clinical Effect
0–25	nondetectable
25–50	temporary decrease in white blood cell count
100–200	large decrease in white blood cell count; nausea; fatigue
200–300	immediate nausea, vomiting; delayed appearance of appetite loss, diarrhea; probable recovery in three months
500	within 30 days of exposure, 50% of the exposed population will die

10.4 Chemistry Within Us

Radon: A Major Health Hazard

The U.S. Environmental Health Agency (EPA) estimates that from 5000 to 20,000 deaths from lung cancer yearly are caused by radon-222 gas. Radon-222 is a naturally occurring radionuclide derived from the radioactive decay of uranium ores. It is one of the daughters arising in the 14-member decay series from uranium to lead. Because uranium ores are present in many types of rocks and soil, radon-222 and its decay products are also commonly found components of our environment.

The normal average background radon concentration is about 0.2 pCi (1 picocurie = 1×10^{-12} curie, see Table 1.4) per liter of air, constituting about 50% of natural background radiation. The EPA's maximum recommended level is 4 pCi/L of air. However, when radon-222 first came to the attention of authorities, it was because it was found present at concentrations of more than 2500 pCi/L of air in a home in Pennsylvania. It was detected there because it was the home of an engineer who worked at a nearby nuclear energy plant. He kept setting off the radiation alarms when he reported for work because his clothes were contaminated with radon-222 and its decay products, which had been picked up from the air in his house.

Because radon is a noble gas, most radon-222 is exhaled, unchanged, after spending only a short time in the lungs. However, two of its decay daughters—polonium-218 and polonium-214—are radioactive α-emitters like their parent, with the insidious difference that they are solids and can remain in contact with lung tissue. Thus polonium-218 and polonium-214 are able to remain in the lungs to irradiate tissue, damage cells, and possibly lead to cancer.

The radon-222 hazard is higher inside homes than outside because the ventilation indoors is restricted. The gas can be detected with inexpensive commercial test kits and then, as long as the home is not constructed with uranium-ore-containing granite, its concentration can be significantly reduced by adequate ventilation in the basement and careful sealing of any cracks and openings in the foundation.

- Diagnostic use requires that the radiation have significant penetrating power to be accurately detected; that is, it should be primarily a γ-emitter.
- Therapeutic use requires intentional damage to abnormal (cancerous) tissue; therefore the isotope should be an α- or β-emitter.

Another key factor in a physician's decision to make use of a radionuclide for either of these purposes is that the benefits of the radiation must outweigh the risks. Important considerations include the intensity of radiation—the shorter the half-life, the more intense the radiation—and the rate of excretion—if too low, unwanted radiation damage will occur; if too high, the isotope will not remain at the irradiation site long enough to have its desired effect. Moreover, any decay products should not be radioactive.

Some isotopes have a natural tendency to concentrate in particular cells of the body. In such cases, the isotope can be administered as a solution, whether orally or by injection. For example, the thyroid gland secretes thyroxin, an iodine derivative of the amino acid tyrosine. As a result, radioactive iodine,

TABLE 10.6 Radioactive Isotopes Used as Radiation Sources

Isotope	Half-life	Type of Emission	Energy (Mev)
polonium-210	138 days	α-particle	5.3
radium-226	1620 years	α-particle	4.8
cesium-137	30 years	β-particle	1.2
cobalt-60	5.27 years	γ-ray	1.33
phosphorous-32	14.3 days	β-particle	1.71

which is often used to treat thyroid disorders, will have a tendency to concentrate there. When an isotope is not one that selectively concentrates in the target organ, it can instead by inserted into the subject tissue as a package in a metal or plastic tube. Table 10.2 lists radionuclides used in diagnosis, and Table 10.6 lists some radionuclides used solely in therapy, along with their half-lives and the types of radiation emitted.

Iodine-131, a β-emitter and thus damaging to tissues, is the radionuclide often used to treat thyroid cancer and hyperthyroidism. Iodine-123, a γ-emitter, is used in diagnosis. Thyroid malfunction may result in serious metabolic disorders, with cardiovascular consequences. Direct observation of the radionuclide's distribution in the gland, by using a radiation detection apparatus, will reveal the presence and often the type of disorder.

Radiolabels and Metabolic Studies

There is virtually no chemical difference between a radioactive and a nonradioactive isotope of the same element. Both are able to participate in the same chemical reactions under the same physical and chemical conditions. For example, glucose containing a radioactive carbon will undergo the same metabolic reactions as nonradioactive glucose. A compound containing a radioactive atom is said to be **radiolabeled.** The metabolic fate of a radiolabeled compound can be followed because of the ease of detection of the label's radioactivity.

One technique used to follow the fate of a radiolabeled compound is called **autoradiography.** This technique recognizes radiolabeled substances by their effects on photographic film. Melvin Calvin made use of it in the research that led to his discovery of the chemical steps of glucose photosynthesis. Calvin fed radiolabeled bicarbonate ($H^{14}CO_3^-$) to actively growing algae. After short periods of growth, the cells were killed and broken, and chromatography was used (see Box 1.1) to separate all the different molecular components on a single sheet of paper (each component ended up in a unique position on the paper). When the paper was covered with photographic film, the radiolabeled compounds revealed their positions by darkening the film wherever they were located. The identity of each labeled compound was learned by comparing its position on the chromatogram with reference chromatograms, which had been made by using known compounds. Dr. Calvin received the Nobel Prize in 1953 for this work. The same techniques are used today to study the physiological distribution and metabolic fate of drugs.

▶▶ You will learn about photosynthesis in Section 18.7.

Autoradiography also uses photographic film to locate radiolabeled compounds in undisrupted cells and tissues. In a typical experiment, living tissue is exposed to a radiolabeled compound and then allowed to continue its normal metabolic activity for a while. When enough time has passed, a sample of tissue is removed, fixed, and mounted on slides. (To fix a tissue sample means to treat it with a preparation that stops all further chemical action and preserves the structures of the cell.) Next, the slides are covered with photographic film.

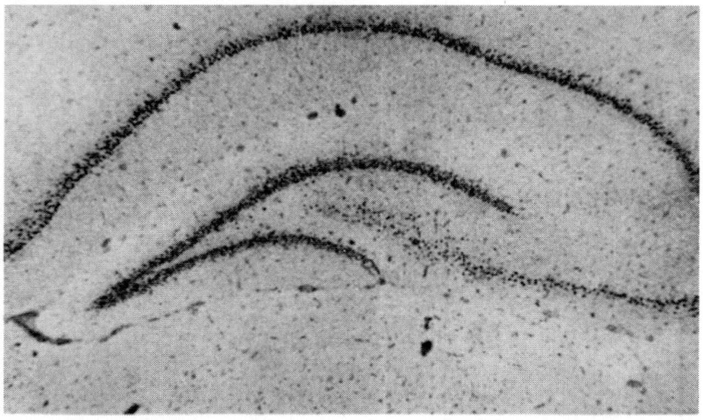

Figure 10.5 An autoradiograph showing the localization of steroid receptors in the brain of a rat injected with radioactive estradiol. The hormone was localized in the hippocampal region of the brain.

The location of the label within the tissue is revealed by the darkening of the corresponding location on the film. An image obtained from such an experiment can be seen in Figure 10.5.

Radiation and Computer Imaging Methods

Advances in radiochemistry and computer technology have combined to produce powerful new scanning methodologies for the diagnosis of injury and disease. The general approach common to many of these techniques is to direct radiation through the body and then detect the degree of attenuation of the radiation when it emerges. The information captured in this way is translated by a computer into an image of the body's inner structures.

In one of these methods, called a **CAT scan (computer-assisted tomography),** short bursts of X-rays are directed toward some part of the body. The X-rays are absorbed differentially, and the emerging radiation is detected by a circular array of detectors surrounding the body. The detected X-rays are then transformed by a computer program into an image of the internal structures.

In **magnetic resonance imaging,** or **MRI,** the body is placed in the field of a strong electromagnet and simultaneously exposed to a radiofrequency field. Protons, as well as many other atomic nuclei, behave as tiny magnets and will tend to become aligned in an intense magnetic field. These tiny magnets also interact with and absorb energy from the radio signals, to a degree that depends on the protons' local molecular environment. The signals detected during MRI (by detectors arrayed about the body, similar to the arrangement used in CAT scans) reflect this fact, and a computer is able to transform the differences in energy absorption into an image of soft tissue structures within the body. The energies needed for MRI imaging are much smaller than the energies need for a CAT scan and therefore allow much longer scan times.

In the technique known as a **PET scan (positron emission tomography),** a solution of positron-emitting isotope is injected into the body and produces γ-radiation that is detected outside the body by arrays of detectors similar to those used in X-ray tomography (the CAT scan). When an emitted positron encounters a nearby electron, both are annihilated, producing two γ-rays traveling in opposite directions. The most commonly used isotopes are O-13, N-13, C-11, and F-18. The half-lives of these positron emitters range from 2 to 110 min. Thus the key obstacle in the use of this technique has been the need to develop rapid chemical syntheses to incorporate the isotopes into metabolically useful compounds.

Positron emission tomography has been extensively used to study disorders in the brain. Glucose, labeled with the positron-emitting radionuclide carbon-11, is often used in such studies. After the glucose is fed intravenously

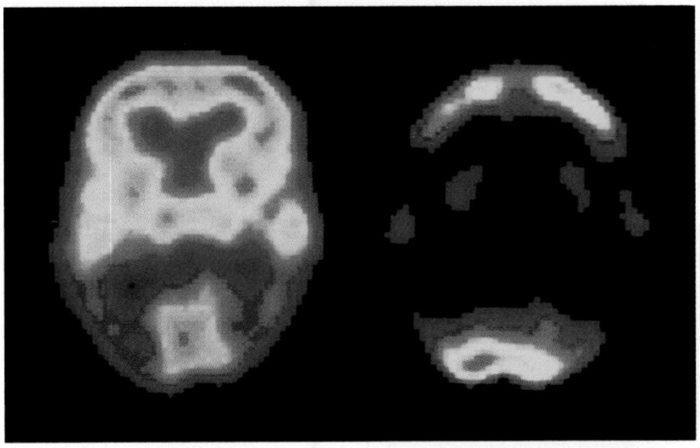

Figure 10.6 PET scan of a normal brain and that of a patient with Alzheimer disease. The differences in brain scans are easily detected.

to the subject (and enters the brain), the positrons that it emits quickly combine with electrons from any nearby molecules (for example, water) to create γ-rays. Because the brain uses glucose exclusively for energy, the internally generated γ-rays produce brain scans (like the one in Figure 10.6) that can reveal disorders in metabolism indicative of a variety of pathologies. Recent work has centered on the metabolism of neurotransmitters, compounds such as acetylcholine and norepinephrine that effect the transfer of nerve impulses between neurons. PET scans provided the first evidence that acetylcholine concentrations are significantly lower than normal in the brains of patients with Alzheimer disease.

10.8 NUCLEAR REACTIONS

In 1938, scientists found that, when uranium is bombarded with neutrons, its atoms fragment into smaller nuclei—such as barium, krypton, cerium, and lanthanum—and enormous amounts of energy are released. The general explanation is that the uranium nucleus absorbs the high-energy neutrons, becomes destabilized as a result, and responds by spontaneously decomposing into more stable products with the evolution of much heat. The new process was given the name **nuclear fission.**

There are many possible products of uranium fission. More than 200 isotopes of 35 different elements have been found as a result of the fission of uranium-235. Two possible outcomes are:

$$^{235}_{92}U + ^{1}_{0}n \longrightarrow ^{139}_{56}Ba + ^{94}_{36}Kr + 3\ ^{1}_{0}n$$

$$^{235}_{92}U + ^{1}_{0}n \longrightarrow ^{144}_{54}Xe + ^{90}_{38}Sr + 2\ ^{1}_{0}n$$

These equations provide a key to the power and danger of nuclear fission. Note that, when one neutron is absorbed by and destabilizes one uranium nucleus, at least two neutrons are released as a result. These fission-produced neutrons, in their turn, can each destabilize another uranium nucleus and result in the release of still more fission-produced neutrons. In other words, the multiple yield of neutrons results in an exothermic chain reaction that grows exponentially—explosively—and produces extremely large amounts of energy. This is the operating principle for the atomic bomb.

Nuclear reactors, which use fission to produce energy, keep the reaction to manageable levels with the use of control rods made of cadmium, an efficient absorber of neutrons. The control rods are arranged and deployed as shown in Figure 10.7. A nuclear power plant that produces electricity is diagrammed in

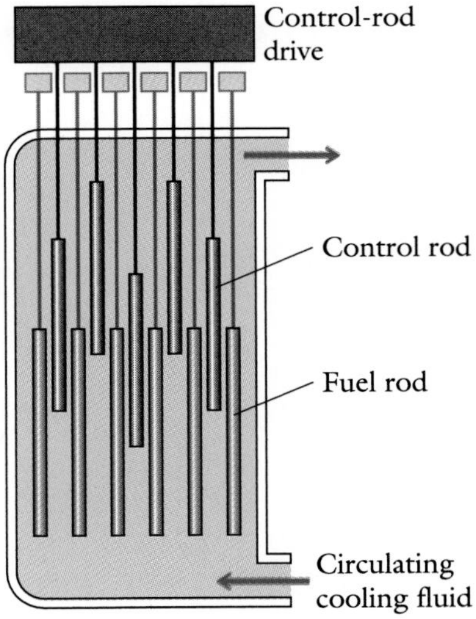

Figure 10.7 Diagram of the control mechanism in a nuclear reactor. The fuel rods contain fissionable material. The control rods are made from materials that are good absorbers of neutrons. The density of neutrons in the reactor core can be controlled by raising or lowering the control rods. Controlling the neutron density controls the rate of energy production.

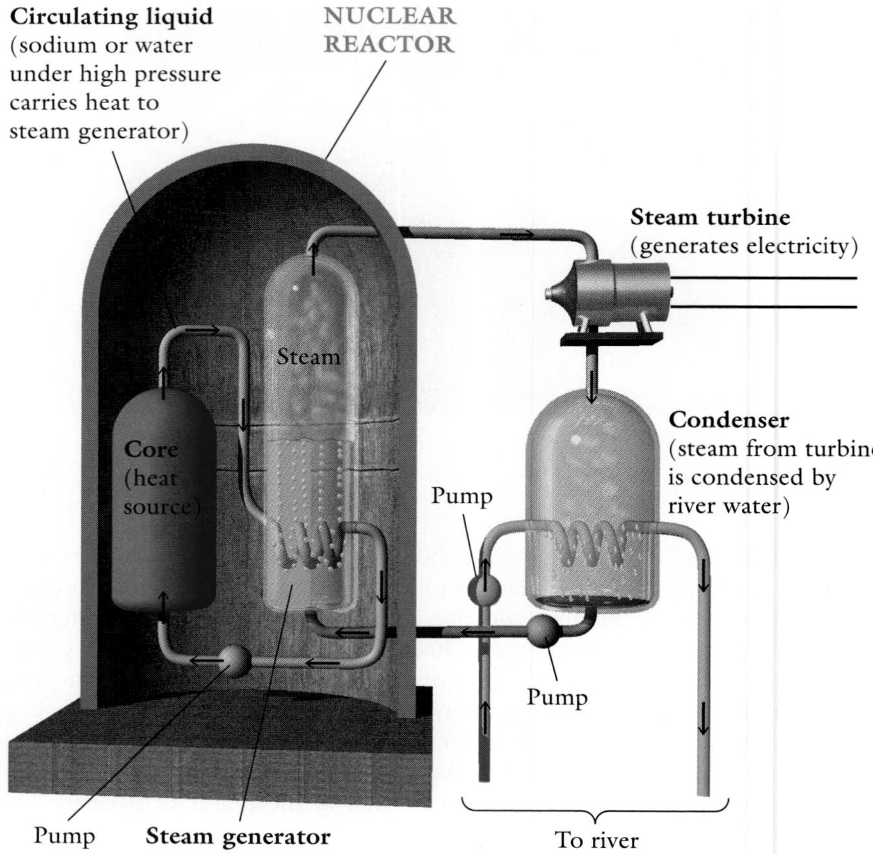

Figure 10.8 Heat transfer in a nuclear reactor. The heat produced in the core is transferred by a closed loop of liquid sodium metal or liquid water to a steam generator. The steam produced in the generator runs a steam turbine that produces electricity. The steam from the turbine must be cooled and pumped back into the generator. Cooling can be accomplished by using cooling towers (evaporative cooling) or by using nearby sources such as a river.

Figure 10.8. The heat released in the fission reaction is used to drive steam turbines that in turn generate electricity. Although we have stopped building nuclear plants in the United States, other countries—for example, France and Japan—use nuclear power extensively to generate electricity.

Nuclear fusion, a process that seems very much like the opposite of nuclear fission, also is able to produce large amounts of energy. In **nuclear fusion,** the nuclei of two light elements collide at velocities high enough to overcome the mutual repulsion of the nuclear protons and fuse to form a nucleus of higher atomic number. Nuclear fusion is the source of the energy continuously generated in stars such as our sun, where protons, deuterons ($^{2}_{1}H$), and helium nuclei collide successively as follows:

$$^{1}_{1}H + ^{1}_{1}H \longrightarrow ^{2}_{1}H + ^{0}_{1}e$$
$$^{1}_{1}H + ^{2}_{1}H \longrightarrow ^{3}_{2}He$$
$$^{3}_{2}He + ^{3}_{2}He \longrightarrow ^{4}_{2}He + 2\,^{1}_{1}H$$
$$^{3}_{2}He + ^{1}_{1}H \longrightarrow ^{4}_{2}He + ^{0}_{1}e$$

Nuclear fusion reactions require temperature of about 4×10^7 K to overcome the mutual repulsions of the various nuclei; but, when fusion occurs, it releases vast quantities of energy, more than enough to sustain stellar fusion processes (see Section 10.9).

Research on the use of nuclear fusion is in progress in the hope that it will be a major source of energy in the foreseeable future. The chief problem is to design a system that will produce and maintain the extraordinarily high temperatures required in a controlled manner.

The energy changes in nuclear reactions are millions of times as great as those associated with chemical reactions. The reason is that there are changes in mass in nuclear processes. Mass and energy are related by the Einstein equation:

$$E = mc^2$$

where m is the mass and c is the velocity of light, 3×10^8 m/s. As you can see, the value of the speed of light is so large that even a small change in mass produces an enormous change in energy.

The masses of nuclei are always less than the sum of the masses of the constituent nucleons. For example, the mass of a helium-4 nucleus is 4.00150 amu. The sum of two individual protons and two individual neutrons is 4.03190 amu. The difference in mass of 0.03040 amu represents the energy released when protons and neutrons fuse into a helium-4 nucleus. That change in energy can be calculated by using the Einstein relation in the form:

$$E = \Delta m \times c^2$$

where Δm is the change, or difference, in mass. Disintegration of 1 mol of helium-4 requires an energy input of 2.73×10^{12} J/mol. The fusion of protons and neutrons to form a helium nucleus will produce the same amount of energy. It is 1 million times the energy released in chemical reactions, which at the most may reach values of 1 to 2×10^6 J/mol.

10.9 NUCLEAR ENERGY AND THE BIOSPHERE

The dangers of uncontrolled exposure to high-energy radiation have been understood for many years. Because we use nuclear reactors to generate electricity and to manufacture weapons, we must confront the possibility of widespread damage that might occur through the accidental or deliberate release of radioactive substances.

The first concern about such possibilities was prompted by atmospheric testing of atomic weapons in the 1950s. Two of the more dangerous byproducts of nuclear fission released through bomb testing are strontium-90 and cesium-137. The properties of strontium are very similar to those of calcium; therefore, if radioactive strontium enters the body, it will be incorporated where calcium is found, in bone, particularly growing bone, such as the bones of children. When incorporated into bone, the radioisotope becomes a lifetime burden and can produce cancer of the bone or leukemia or both. Cesium, on the other hand, is an alkali metal, so its properties are similar to those of potassium, whose cation concentrates inside cells. The presence of radioactive cesium in cells can inflict damage in a wide variety of tissues. However, cesium cations are quite soluble and are rapidly eliminated from the body. There is no pharmacological treatment for exposure to these radioactive substances. This awareness led countries in possession of atomic weapons to enter into a treaty that banned atmospheric testing.

Today, a more acute concern is the possibility of accidental release of radioactivity from the nuclear reactors that are being used to produce electricity. In 1986, when a serious accident occurred in Russia at the plant in Chernobyl, the fallout of radioactivity was intense. One of the more dangerous of the radiation products was iodine-131, for which, fortunately, a pharmacological antidote is available.

Because iodine becomes concentrated in the thyroid gland, high concentrations of its I-131 isotope can produce serious radiation damage. For this reason, in the wake of the Chernobyl accident, anyone living in a region where iodine-131 was known to have been deposited was encouraged to use table salt enriched with nonradioactive iodine-127. The thyroid gland cannot discriminate between iodine-127 and iodine-131, so flooding the body with the nonradioactive iodine ensured that the body's concentration of iodine-131 would be significantly diluted and its rate of uptake sharply reduced compared with the uptake of iodine-127.

We can never fully escape exposure to high-energy radiation. We are constantly exposed to cosmic rays from space and to radiation from naturally occurring radioisotopes within our bodies (such as potassium-40) and in the surrounding rocks and soil (such as radon). Add to that the occasional medical or dental X-ray session. For residents of the United States, these sources of exposure add up to about 350 mrem per year, a small amount compared with the standard for safe maximum exposure, which is about 5000 mrem per year. Although this exposure varies from place to place, background radiation does not seem to pose a serious problem.

Summary

Radioactivity The process of spontaneous nuclear decomposition, in which high-energy subatomic particles are ejected from the nucleus, is called radioactivity. The major types of radioactive emissions are α-particles, β-particles, γ-rays, and positrons. Emission of α-, β-, and positron radiation results in changes in atomic number or mass or both to create a new element. Gamma-rays have no mass and no charge; therefore γ-ray emission does not lead to the formation of a new element.

Radioactivity can be induced in nonradioactive elements by interaction with high-energy nuclear particles. The first artificial conversion of one nucleus into another was performed by E. Rutherford in 1919. Since that time, hundreds of radioisotopes have been produced in the laboratory. All the elements above atomic number 92 have been created in this manner.

Radioactive Decay Radioactive decay occurs when a radioactive element is transformed into another element by emitting nuclear radiation. The rate of radioactive decay is characterized by the loss of a constant percentage of the original element per unit time. The decay rate is defined as the time required for 50% of the original material to decay. This period of time is called the half-life ($t_{1/2}$). Different radioisotopes have different half-lives.

Effects of Radiation The study of the effects of radiation on matter is called radiochemistry or radiation chemistry. High-energy radiation produces ions and free radicals along its penetration tracks and has various effects on living tissue, ranging from repairable damage to lethality. The immediate chemical results of cellular irradiation are likely to be the same in any kind of cell, but the consequences for the organism will depend on the kind of tissue that is damaged. People can protect themselves from the effects of high-energy radiation by using an absorbing barrier of some kind, remaining as far away from the radiation source as is practicable, and limiting their exposure time.

Detection of Radioactivity Nuclear radiation is detected by the use of photographically sensitive film, Geiger counters, or scintillation counters. Measurements of radiation intensity correlated with type of radiation can be used to predict the likelihood of biological damage. The likelihood of damage also depends on the type of tissue irradiated, the dose rate, and the total dose. With the same amount of energy absorbed, the damage is different for different types of radiation.

Applications Radioisotopes are employed extensively for the diagnosis of pathological conditions as well as for therapy. In diagnostic applications, radioisotopes are primarily γ-emitters, and, in therapeutic use, they are either α- or β-emitters.

Radiation originating inside or passing through the body can be detected externally and the information treated by computer to generate an image of the body's inner structures. In one of these methods, called the CAT scan, X-rays are sent into the body and detected after absorption. Another technique, known as a PET scan, is used primarily to produce images from the results of positron decay of a metabolite within the brain. In a third technique, called magnetic resonance imaging or MRI, differentially absorbed radio signals are detected externally and transformed by computer program into an image of soft tissue.

Nuclear Reactions Two nuclear processes that yield vast amounts of energy are nuclear fission and nuclear fusion. The energy changes in these nuclear reactions are the result of mass changes in the nucleus. These mass changes produce 1 million times the energy released in chemical reactions.

Key Words

α-particles, p. 259
background radiation, p. 274
β-particles, p. 260
computer assisted tomography (CAT scan), p. 278
free radical, p. 268
γ-rays, p. 260
Geiger counter, p. 271
half-life, p. 263
magnetic resonance imaging (MRI), p. 278
nuclear equation, p. 260
nuclear fission, p. 279
nuclear fusion, p. 280
nucleons, p. 259
positron emission tomography (PET scan), p. 278
positrons, p. 260
radioactivity, p. 259
radiation chemistry, p. 267
relative biological effectiveness (RBE), p. 273
transmutation, p. 260
X-rays, p. 259

Exercises

Radioactivity

10.1 What are the characteristics of α-radiation?

10.2 What are the properties of β-radiation?

10.3 Describe the important characteristics of γ-radiation?

10.4 How do positrons compare with other atomic particles?

10.5 Specify the missing component of the following nuclear equation:

$$^{240}_{95}\text{Am} + \square \longrightarrow {}^{243}_{97}\text{Bk} + {}^{1}_{0}\text{n}$$

10.6 Specify the missing component in the following nuclear equation:

$$^{238}_{92}\text{U} + {}^{12}_{6}\text{C} \longrightarrow {}^{244}_{98}\text{Cf} + \square$$

10.7 Complete the following nuclear equation:

$$^{40}_{19}\text{K} \longrightarrow {}^{40}_{20}\text{Ca} + \square$$

10.8 Complete the following nuclear reaction:

$$^{0}_{-1}\text{e} + \square \longrightarrow 2\,\gamma$$

10.9 What is induced radioactivity?

10.10 Give some examples of induced radioactivity?

10.11 What is a radioactive decay series?

10.12 What is a daughter nucleus?

10.13 The amount of C-14 remaining in a charcoal sample from an archeological site in New Mexico was determined to be 23.3% of the original quantity. The half-life of C-14 is 5730 years. How old was the campsite?

10.14 Paleontologists found that 2.75 half-lives of C-14 had passed since a bone knife had been carved at a site in Alaska. What was the percentage of C-14 left in that sample?

Effects of Radiation

10.15 Which is more effective in penetrating tissue, α-radiation or β-radiation?

10.16 Describe the pattern of radiation produced in tissue by β-radiation.

10.17 What happens to X-radiation as it passes through tissue?

10.18 What happens to γ-radiation as it passes through tissue?

10.19 What are the immediate results of the interaction of γ-radiation with water?

10.20 Do any chemical events follow the immediate results of γ-irradiation of water?

10.21 Which has the most long-term consequences, a primary radiation event or its secondary chemical processes?

10.22 Describe how biochemical radiation damage may have a deleterious biological consequence.

10.23 What is radiation sickness?

10.24 What is background radiation?

10.25 A radiation technician was standing 4 feet from a radiation source. She determined that she must move away to reduce the intensity of the radiation to 1/4 of its value. How much farther did she move?

10.26 If the technician in Exercise 10.25 could not work farther away from the radiation source, is there anything else that she could do to protect herself?

Detection

10.27 What is the basis of detection of radiation by the Geiger counter?

10.28 How does a scintillation counter detect radiation?

Units

10.29 What is a rad?

10.30 What is the difference between the rad and the rem?

10.31 Does 1 rad of α-radiation have the same effect on tissue as 1 rad of γ-radiation?

10.32 Does 1 rad of β-radiation have the same effect on tissue as 1 rad of γ-radiation?

Applications

10.33 Can α-emitting isotopes be used for diagnosis? Explain your answer.

10.34 Can γ-emitting isotopes be used for diagnosis? Explain your answer.

10.35 Can cobalt-60 be used for cancer therapy? Explain.

10.36 Can radium-226 be used for cancer therapy? Explain.

10.37 Can positrons be detected directly? Explain.

10.38 Can β-particles be detected directly? Explain.

10.39 Would iodine-131, a β-emitter, be used for scanning the thyroid for disorders?

10.40 Would technetium-99, a γ-emitter, be used for scanning for disorders of the vascular system?

10.41 What is a CAT scan used for?

10.42 Describe how a PET scan works.

Unclassified Exercises

10.43 Technetium-99m is a radioisotope used to assess heart damage. Its half-life is 6.0 h. How much of a 1.0-g sample of technetium-99 will be left after 30 h?

10.44 Iodine-126, a γ-emitter, has a half-life of 13.3 h. If a diagnostic dose is 10 ng, how much is left after 2.5 days?

10.45 A sample of $Na_3{}^{32}PO_4$ had an activity of 6.3 mCi. What does that mean in dps?

10.46 The radiation in rads absorbed by tissue from an α-emitter was the same as that absorbed from a β-emitter. The radiation absorbed from the β-emitter was determined to be 15.2 mrem. What was the absorbed dose in rems from the α-emitter?

10.47 The radioisotope fluorine-18 is used for medical imaging. It has a half-life of 110 min. What percentage of the original sample activity is left after 4 h?

10.48 The intensity of X-rays at 2 m from a source was 0.8 roentgens (R). How far from the source should the target be to reduce the intensity to 0.2 R?

10.49 What is radioactivity?

10.50 What are the primary chemical events when high-energy radiation penetrates tissue?

Chemical Connections

10.51 Does radioactive decay always result in transmutation?

10.52 Would you expect infrared light to initiate chemical events? Explain.

10.53 What kinds of tissue are most sensitive to radiation damage?

10.54 Is a radioisotope useful if its decay products are equally radioactive? Explain your answer.

10.55 What is the difference between nuclear fission and nuclear fusion?

PART 2

ORGANIC CHEMISTRY

Having completed our study of the basic structure and properties of atoms and molecules, we proceed to a consideration of organic chemistry, the study of compounds that contain the element carbon, usually together with hydrogen and often one or more other elements (oxygen, halogen, nitrogen, sulfur, or phosphorus). Both naturally occurring and synthetic organic compounds are important to our lives. An example of a naturally occurring organic molecule is heme—a component of the protein hemoglobin. Heme contains iron (Fe) at its center. That is the site to which oxygen attaches when hemoglobin takes on oxygen at the lungs and stores it for subsequent delivery to cells and tissues throughout the body.

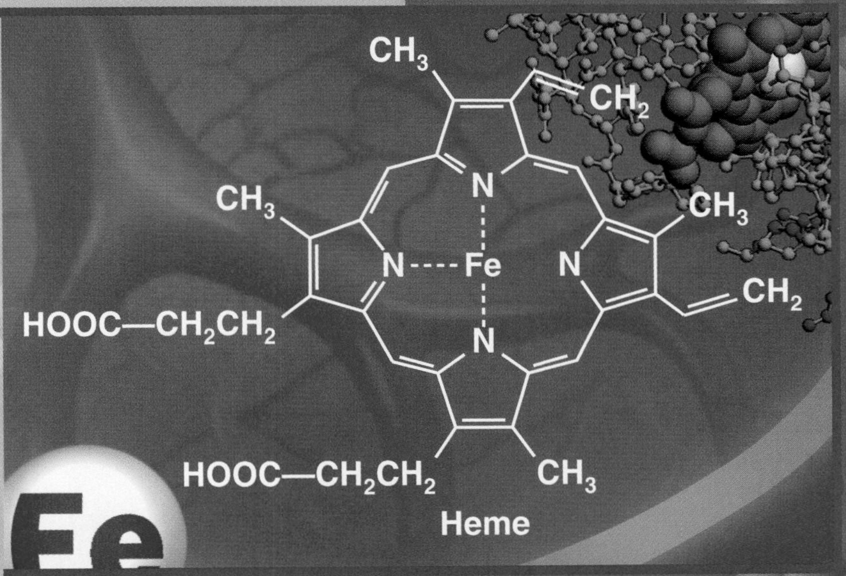

CHAPTER 11

SATURATED HYDROCARBONS

CHEMISTRY IN YOUR FUTURE

No sooner do you clock in at the poison control center than you receive a call from a frightened grandmother who thinks her tiny grandson may have drunk some lemon-scented furniture polish. The child appears well, but she believes she detects a smell of polish on his breath. You tell her to keep him warm and rush him to a hospital emergency room. You warn her above all not to induce vomiting, even though that is the recommended procedure when many other poisons are swallowed. The furniture polish contains petroleum distillates, which can do serious lung damage if aspirated during vomiting. This chapter describes the organic compounds called saturated hydrocarbons that are present in petroleum. Although they are used to make many important materials, including fuels, plastics, medicines, and furniture polish, they are poisons if significant amounts are swallowed.

LEARNING OBJECTIVES

- Define organic chemistry.
- Identify the families of organic compounds and their functional groups.
- Describe the bonding in alkanes.
- Draw condensed and expanded structural formulas of alkanes.
- Draw constitutional isomers of alkanes.
- Name alkyl groups and alkanes by using the IUPAC nomenclature system.
- Draw and name cycloalkanes.
- Draw geometric (cis-trans) stereoisomers of cycloalkanes.
- Describe and explain the physical properties of alkanes and cycloalkanes.
- Write equations for the halogenation of alkanes and cycloalkanes.
- Write equations for the combustion of alkanes and cycloalkanes.

The remainder of this text deals with organic chemistry and its extension into biochemistry. **Organic chemistry** is the study of **organic compounds,** compounds containing the element carbon. All other compounds are **inorganic** and are the subject matter of **inorganic chemistry.** However, chemists have traditionally assigned a small number of relatively simple carbon compounds to the category of inorganic compounds. Carbon monoxide (CO), carbon dioxide (CO_2), and various carbonates (for example, H_2CO_3 and Na_2CO_3), bicarbonates (for example, $NaHCO_3$), and cyanides (for example, HCN and NaCN) are usually classified and studied with the inorganic compounds.

With 109 other elements in the periodic table, why does the study of carbon and its compounds merit so much time and effort? The answer lies in the number, complexity, and importance of organic compounds. There are close to 10 million organic compounds, almost 50 times the number of inorganic compounds. Carbon atoms bond to one another (only a few elements in the periodic table have this ability) as well as to other elements and are therefore uniquely capable of creating molecules that range from very small to very large. There are organic compounds containing tens, hundreds, thousands, and more carbon atoms bonded one to another.

Many organic compounds contain only carbon and hydrogen. A larger number also contain one or more other elements, principally oxygen, halogen (fluorine, chlorine, bromine, iodine), nitrogen, sulfur, and phosphorus. The number of organic compounds is greatly enlarged because of the variety of structural arrangements in which carbon atoms bonds to each other and to other elements.

The term organic originally conveyed the idea of compounds produced by a so-called vital force that chemists thought was present in living organisms, both plant and animal. Examples of such compounds were sugar, starch, plant oils, animal fats, and proteins. Early chemists assumed that organic compounds could not be synthesized in the laboratory. That assumption proved incorrect, however, and, by the early 1800s, the definition of organic chemistry had to be revised. Since then, an enormous number of organic compounds have been created or recreated in the laboratory. We study organic compounds, both naturally occurring and synthetic, because of their great importance to our lives:

- Our bodies are composed of organic compounds—carbohydrates, lipids, nucleic acids, proteins, and other organic molecules. Carbohydrates provide energy. Lipids are used to build the membranes that surround all living cells. They are also used as a storage form of energy and for the synthesis of steroid hormones and various fat-soluble vitamins. Nucleic acids (DNA and RNA) direct and control the reproduction of an organism. Proteins, the key components of muscle and bone, also transport oxygen to tissues, catalyze

A PICTURE OF HEALTH

Examples of Functional Groups

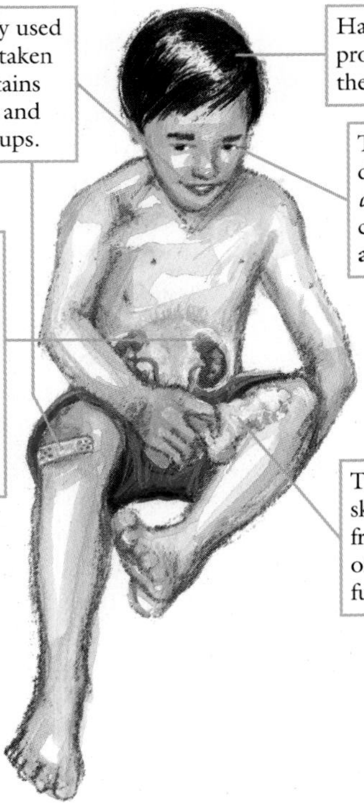

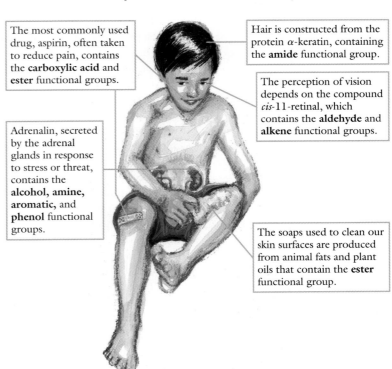

The most commonly used drug, aspirin, often taken to reduce pain, contains the **carboxylic acid** and **ester** functional groups.

Hair is constructed from the protein α-keratin, containing the **amide** functional group.

The perception of vision depends on the compound *cis*-11-retinal, which contains the **aldehyde** and **alkene** functional groups.

Adrenalin, secreted by the adrenal glands in response to stress or threat, contains the **alcohol, amine, aromatic,** and **phenol** functional groups.

The soaps used to clean our skin surfaces are produced from animal fats and plant oils that contain the **ester** functional group.

chemical reactions throughout the body, protect against viruses and bacteria, and regulate bodily functions in general.

- Most of the energy to run automobiles, trains, and aircraft, to heat homes, offices, and factories, and to operate electrical equipment is obtained by burning organic compounds: gasoline, diesel oil, and other fuel oils, heating and cooking gases, and other petroleum products.
- Organic compounds in naturally occurring and synthetic medicines and medicaments—aspirin, penicillin, anesthetics, rubbing alcohol, and so forth—relieve pain and illness.
- Synthetic plastics, textiles, and rubbers are organic compounds. In beverage and detergent containers, antistick cookware, toys, polyester and nylon clothing, synthetic grass, and automobile tires, they have transformed modern life and contribute heavily to our high standard of living.
- The active ingredients in soaps, detergents, polishes, cosmetics, deodorants, and shampoos are organic compounds.
- The materials for manufacturing artificial body parts, such as hip and knee prostheses, heart valves, and dentures, are organic compounds.

We begin the study of organic chemistry with an overview of the formulas and families of organic compounds, followed by a detailed look at the group of compounds called saturated hydrocarbons.

11.1 MOLECULAR AND STRUCTURAL FORMULAS

As stated in Section 3.7, the **molecular formula** of a compound indicates the number of atoms of each element present in a molecule of the compound, but it does not indicate how the various atoms are connected to one another. Thus, although the molecular formula for methane, CH_4, shows that a molecule of methane contains one carbon atom and four hydrogen atoms, it does not describe how those atoms are connected or arranged. The **structural formula,** however, does show how the various atoms in a molecule are bonded together, as shown in the margin for methane.

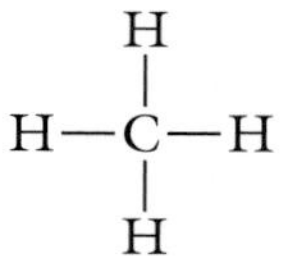

Methane

We can deduce the structural formula from the molecular formula by considering the **combining power**—the bonding requirements—of each element in the compound. The combining power of an element in a covalent compound is the number of covalent bonds that it typically forms to complete its octet of valence electrons (Section 3.7). The combining powers of the elements most commonly present in organic compounds are shown in Table 11.1. Carbon, with its four valence electrons, needs four more electrons to complete its octet, and so it is described as tetravalent; that is, it forms four covalent bonds. All carbon atoms in all compounds must have four bonds to other atoms, no more and no less. Nitrogen and phosphorus have five valence electrons, and so they need three electrons to complete their octets; they are trivalent. Oxygen and sulfur have six valence electrons, need two electrons to complete their octets, and are divalent. Hydrogen has one valence electron, needs one electron, and is monovalent. Each of the halogens (fluorine, chlorine, bromine, and iodine) has seven valence electrons, needs one electron, and is monovalent. Sulfur and phosphorus are unique among the elements found in organic compounds. They exhibit variable combining powers because their octets can expand. Thus, phosphorus and sulfur are also pentavalent and hexavalent, respectively, in some compounds. Examples are the phosphorus atoms

TABLE 11.1 Combining Powers of Elements Present in Organic Compounds

Element	Number of Bonds	Bonding Representation
C	4	—C— (with one bond above and one below)
N	3	—N— (with one bond above)
O	2	—O—
H, F, Cl, Br, I	1	H— F— Cl— Br— I—
P	3	—P— (with one bond above)
	5	—P— (with a double bond above and one bond below)
S	2	—S—
	6	—S— (with a double bond above and a double bond below)

in phosphoric acids and esters (Sections 15.13 and 21.2) and the sulfur atoms in sulfuric acid (Table 9.2) and benzenesulfonic acid (Section 12.12).

Concept checklist

- ✔ A structural formula cannot be correct—cannot represent a real compound—unless the combining power of each atom is represented correctly. A structural formula showing an incorrect number of bonds for any atom is not a correct structural formula.
- ✔ A molecular formula is correct only if it can be translated into a correct structural formula.

We should note that, for compounds containing only carbon and hydrogen, a structural formula is the same as a Lewis structure (Section 3.7). (For compounds that also contain oxygen or nitrogen, a structural formula is the Lewis structure without the nonbonded electrons shown on oxygen or nitrogen.)

Example 11.1 Determining correct molecular formulas

Which of the following molecular formulas are correct and which are incorrect? (a) CH_2F_2; (b) CH_3F_2; (c) CH_2F.

Solution

(a) CH_2F_2 is correct because we can draw a structural formula that shows the correct number of bonds for each of the atoms of the molecular formula. Carbon has four bonds, and hydrogen and fluorine each have one:

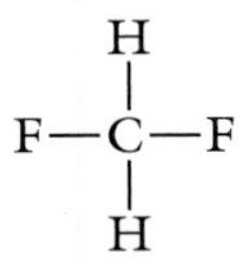

(b, c) We cannot draw a structure with the correct number of bonds for all atoms for either CH_3F_2 or CH_2F. We can draw incorrect structural formulas such as

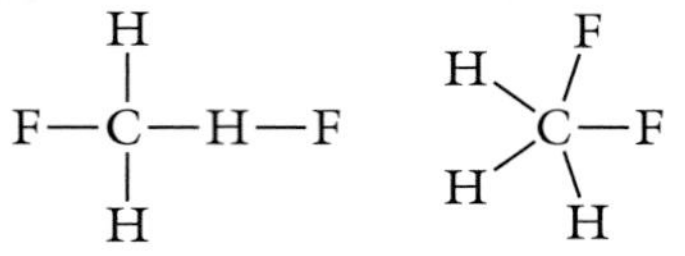

1 2

for CH_3F_2, but they violate the bonding requirements of one or another atom. Structure 1 shows one H atom with two bonds, but H is monovalent. Structure 2 shows carbon with five bonds, but carbon is tetravalent. Structures 1 and 2 and other variations do not represent real compounds.

A structure such as 3 drawn to represent CH_2F also does not represent a real compound, because carbon must have four bonds not three.

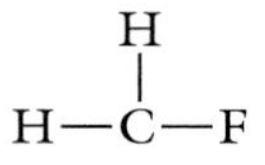

3

Problem 11.1 Which of the following molecular formulas are correct and which are incorrect? (a) CH_5N; (b) CH_5O; (c) C_2H_5Cl.

11.2 FAMILIES OF ORGANIC COMPOUNDS

Studying the chemistry of millions of organic compounds would be an impossible task were it not for the fact that very large groups of organic compounds have certain kinds of chemical behaviors in common. Thus organic compounds are organized into **families** (sometimes called **classes**), each consisting of a very large number of different compounds having a common characteristic pattern of behavior. For example, all alkenes react with bromine (Br_2), all carboxylic acids are acidic, and all amines are basic. We study organic chemistry by studying the characteristic chemical pattern that identifies the members of each family.

The common physical and chemical properties of all compounds in a family results from the presence in their molecular structures of a common **functional group**—a specific atom or bond or a specific group of atoms in a specific bonding arrangement. The functional group dictates the behavior of a compound.

Table 11.2 shows the major families of organic compounds, each with its functional group. Note, as you examine the table, that:

- The only bonds found in alkanes are carbon–carbon and carbon–hydrogen single bonds.
- Alkenes, alkynes, and aromatics contain multiple bonds, two adjacent carbon atoms that share more than one bond between them. An **alkene** has a carbon–carbon double bond, two bonds between a pair of adjacent carbon atoms. An **alkyne** has a carbon–carbon triple bond, three bonds between a pair of adjacent carbon atoms. An **aromatic** has six carbon atoms in a cyclic arrangement with alternating single and double bonds.
- An **alcohol** contains a **hydroxyl group,** an —OH group attached to a carbon atom.
- An **ether** contains an oxygen attached directly to two different carbons.

TABLE 11.2 Families of Organic Compounds

Family	Functional Group	Example
alkane	C—C and C—H single bonds	$CH_3{-}CH_3$ **Ethane**
alkene	$\rangle C{=}C\langle$	$CH_2{=}CH_2$ **Ethylene**
alkyne	$-C{\equiv}C-$	$CH{\equiv}CH$ **Acetylene**
aromatic	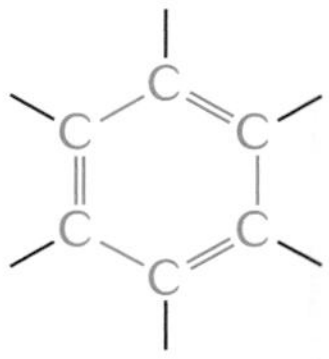	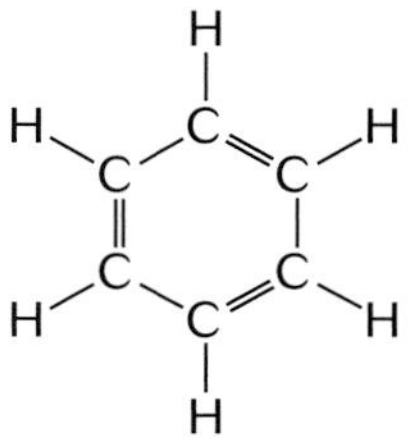 **Benzene**
alcohol	$-\overset{\vert}{\underset{\vert}{C}}-O-H$	$CH_3CH_2{-}O{-}H$ **Ethyl alcohol**
ether	$-\overset{\vert}{\underset{\vert}{C}}-O-\overset{\vert}{\underset{\vert}{C}}-$	$CH_3{-}O{-}CH_3$ **Dimethyl ether**
aldehyde	$-\overset{O}{\overset{\Vert}{C}}-H$	$CH_3{-}\overset{O}{\overset{\Vert}{C}}{-}H$ **Acetaldehyde**
ketone	$-\overset{\vert}{\underset{\vert}{C}}-\overset{O}{\overset{\Vert}{C}}-\overset{\vert}{\underset{\vert}{C}}-$	$CH_3{-}\overset{O}{\overset{\Vert}{C}}{-}CH_3$ **Acetone**
carboxylic acid	$-\overset{O}{\overset{\Vert}{C}}-OH$	$CH_3{-}\overset{O}{\overset{\Vert}{C}}{-}OH$ **Acetic acid**
ester	$-\overset{O}{\overset{\Vert}{C}}-O-\overset{\vert}{\underset{\vert}{C}}-$	$CH_3{-}\overset{O}{\overset{\Vert}{C}}{-}O{-}CH_3$ **Methyl acetate**
amine	$-\overset{\vert}{\underset{\vert}{C}}-\overset{H}{\overset{\vert}{N}}-H$	$CH_3{-}\overset{H}{\overset{\vert}{N}}{-}H$ **Methyl amine**
amide	$-\overset{O}{\overset{\Vert}{C}}-\overset{H}{\overset{\vert}{N}}-H$	$CH_3{-}\overset{O}{\overset{\Vert}{C}}{-}\overset{H}{\overset{\vert}{N}}{-}H$ **Acetamide**

- An **amine** contains an **amino group,** an $—NH_2$ group attached to a carbon atom. (In some amines, an NH or an N is attached to two or three carbon atoms, respectively.)
- Aldehydes, ketones, carboxylic acids, esters, and amides possess the **carbonyl group,** a carbon–oxygen double bond, but differ in the atom or group of atoms connected to the carbon of the carbonyl group. The carbonyl carbon of an **aldehyde** is directly connected to at least one hydrogen. The carbonyl carbon of a **ketone** is directly connected to carbon atoms, not hydrogen. The carbonyl carbon of carboxylic acids and esters is connected by a single bond to an oxygen, but these families differ in what is connected to that oxygen. In **carboxylic acids,** a hydrogen is connected to the oxygen; in **esters,** the oxygen is connected to a second carbon atom. An **amide** has a nitrogen connected to the carbonyl carbon.

Alkanes, alkenes, alkynes, and aromatics are also known as **hydrocarbons,** because they contain only carbon and hydrogen (Figure 11.1). Alkanes are **saturated hydrocarbons,** because they contain no carbon–carbon multiple bonds, only carbon–carbon single bonds. Alkenes, alkynes, and aromatics are **unsaturated hydrocarbons,** because they contain carbon–carbon multiple bonds.

Saturated and unsaturated hydrocarbons differ greatly in their ability to participate in chemical reactions. Saturated hydrocarbons undergo very few chemical reactions. Unsaturated hydrocarbons undergo many different chemical reactions.

Example 11.2 Identifying the family of a compound

Identify the family of each of the following compounds by referring to Table 11.2. Indicate how you arrived at the identification.

(a) $CH_3CH_2—O—CH_2CH_3$ (b) $CH_3CH_2CH_2—OH$

(c) $CH_3CH_2—\overset{\overset{\displaystyle O}{\|}}{C}—H$ (d) $CH_3CH_2—\overset{\overset{\displaystyle O}{\|}}{C}—OH$

Solution

(a) ether; (b) alcohol. Both an alcohol and an ether contains an oxygen atom, but its two single bonds are attached to different atoms. In an ether, oxygen is attached to two different carbon atoms. In an alcohol, one of oxygen's attachments is to a hydrogen.

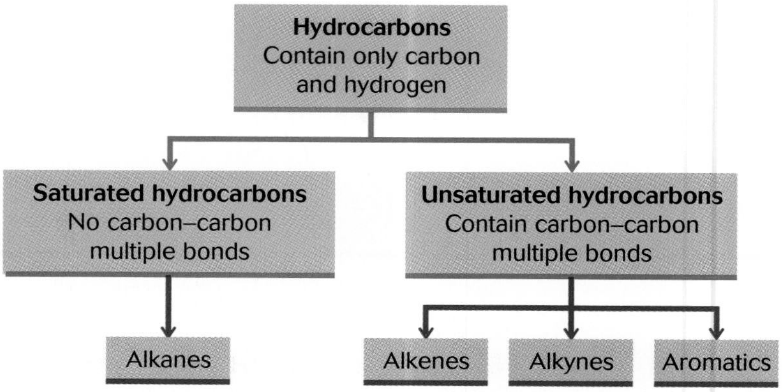

Figure 11.1 There are four families of hydrocarbons: alkanes, alkenes, alkynes, and aromatics.

(c) aldehyde; (d) carboxylic acid. Each compound has the carbonyl group, C=O, but the compounds differ in the atoms attached to the carbon of that group. In aldehydes, the carbonyl carbon is directly attached to an H; in carboxylic acids, the carbonyl carbon is attached to the O of an —OH group.

Problem 11.2 Identify the family of each of the following compounds by referring to Table 11.2:

(a) $CH_3CH_2-\overset{\overset{O}{\|}}{C}-CH_2CH_3$ (b) $CH_3CH_2-\overset{\overset{O}{\|}}{C}-O-CH_2CH_2CH_3$

(c) $CH_3-\overset{\overset{O}{\|}}{C}-NH_2$ (d) $CH_3CH_2-NH_2$

11.3 ALKANES

Alkanes, together with other hydrocarbons, are obtained from petroleum and natural gas. These materials are critical to our standard of living for two reasons: (1) they are used for cooking and heating and to generate power (Section 11.10); and (2) they are converted into other organic chemicals that are of commercial importance, such as plastics, textiles, rubbers, drugs, and detergents (see Box 11.1 on page 296).

Alkanes consist entirely of carbon–carbon and carbon–hydrogen single bonds. They have the general molecular formula C_nH_{2n+2}, where n, the number of carbon atoms, is an integer greater than 0. For example, for $n = 1$, $2n + 2 = 4$, and the molecular formula is CH_4; for $n = 2$, $2n + 2 = 6$, and the molecular formula is C_2H_6, and so forth. Each family of organic compounds can be described by a general molecular formula. In other words, all members of a given family contain the same ratios of the different kinds of atoms.

In all members of the alkane family, the C:H ratio is n:$(2n + 2)$. This ratio can be ascertained in Table 11.3, which lists the first ten unbranched alkanes, their molecular and structural formulas, names, and melting and boiling points. The first four alkanes are gases, and the others are liquid at normal

TABLE 11.3 Formulas and Properties of Normal Alkanes

n	Molecular Formula	Condensed Structural Formula	Name	Melting Point (°C)	Boiling Point (°C)
1	CH_4	CH_4	methane	−182	−162
2	C_2H_6	CH_3CH_3	ethane	−183	−89
3	C_3H_8	$CH_3CH_2CH_3$	propane	−190	−42
4	C_4H_{10}	$CH_3CH_2CH_2CH_3$	butane	−138	−1
5	C_5H_{12}	$CH_3CH_2CH_2CH_2CH_3$	pentane	−130	36
6	C_6H_{14}	$CH_3CH_2CH_2CH_2CH_2CH_3$	hexane	−95	69
7	C_7H_{16}	$CH_3CH_2CH_2CH_2CH_2CH_2CH_3$	heptane	−91	98
8	C_8H_{18}	$CH_3CH_2CH_2CH_2CH_2CH_2CH_2CH_3$	octane	−57	126
9	C_9H_{20}	$CH_3CH_2CH_2CH_2CH_2CH_2CH_2CH_2CH_3$	nonane	−51	151
10	$C_{10}H_{22}$	$CH_3CH_2CH_2CH_2CH_2CH_2CH_2CH_2CH_2CH_3$	decane	−30	174

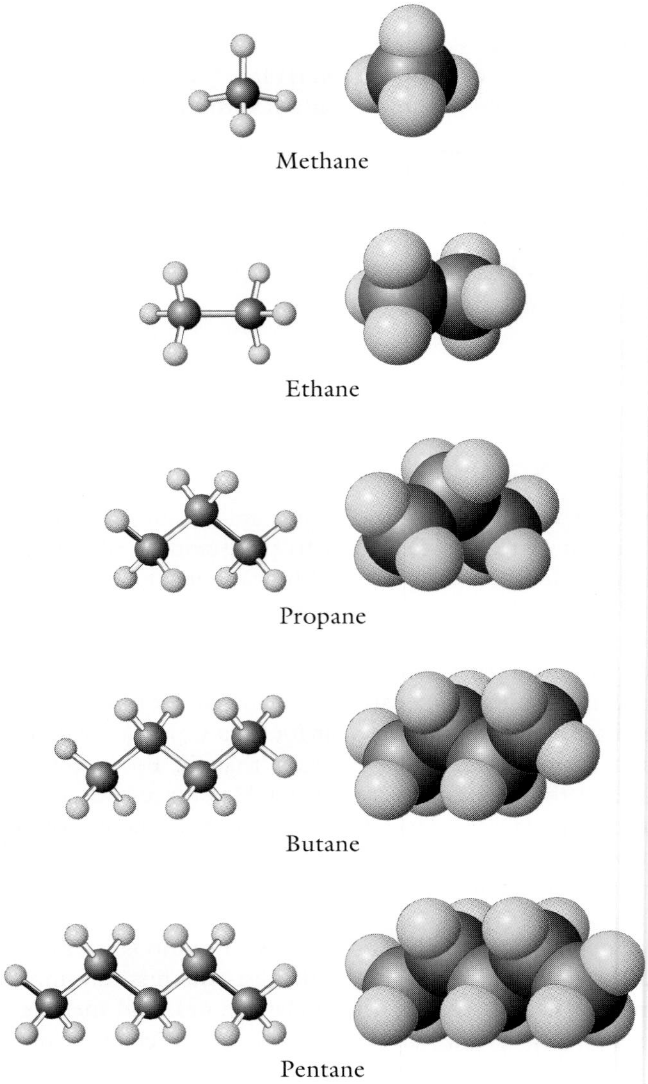

Figure 11.2 Ball-and-stick (left) and space-filling (right) models of alkanes.

ambient temperatures (near 20°C). (The term unbranched will be defined in Section 11.5.)

Both ball-and-stick and space-filling models of the first five alkanes are shown in Figure 11.2, to illustrate the three-dimensional structures of alkanes. Space-filling models more correctly represent the dimensions of atoms relative to bonds than ball-and-stick models do (that is, bonds are extremely short relative to atomic radii; Section 3.10). However, ball-and-stick models are more useful for the beginning student because they more clearly show the bonds between all the atoms in a molecule. We will generally use the ball-and-stick models.

Carbon Bonding in Alkanes

Carbon atoms in alkanes are distinguished by the following characteristics:

- Carbon is tetravalent: each carbon atom has four bonds.

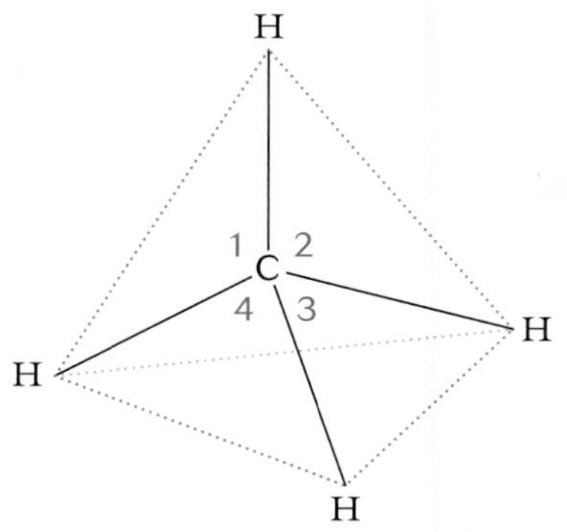

Figure 11.3 Tetrahedral sp^3-hybridized carbon. All angles (1 through 4) are 109.5°

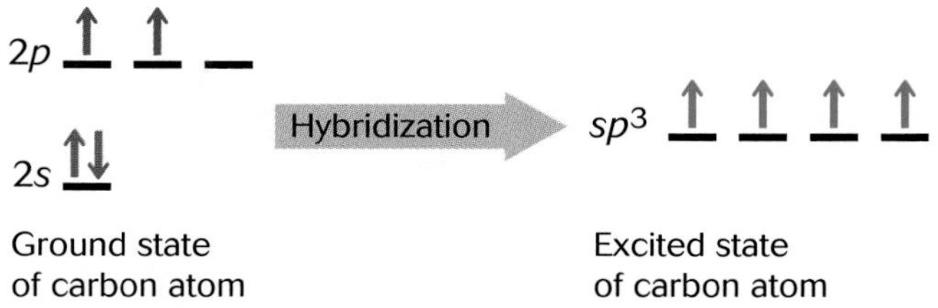

Figure 11.4 Formation of sp^3-hybridized carbon.

- The four bonds of carbon possess **tetrahedral geometry:** each carbon atom is at the center of a tetrahedron with its bonds directed to the four corners of the tetrahedron (Figure 11.3).
- The four bonds are equivalent and have similar properties.

The equivalence of the four bonds of carbon is not at all what we would expect to find on the basis of the electron configuration of isolated (ground-state) carbon atoms. The ground-state carbon atoms possesses four valence electrons distributed as follows: two paired electrons in the $2s$ orbital and one unpaired electron in each of two $2p$ orbitals. The third $2p$ orbital remains empty (left side of Figure 11.4). The ground-state electron configuration of carbon predicts that carbon is tetravalent but does not predict that the four bonds are equivalent.

A modified theory of bonding, called **orbital hybridization theory,** explains discrepancies such as the properties of carbon atoms in alkanes. The orbital hybridization theory proposes that an excitation of ground-state carbon, prior to bond formation, creates a very different electron configuration for the element. The $2s$ and three $2p$ orbitals of carbon, together with its four electrons, mix ("hybridize") to produce four equivalent tetrahedral orbitals called ***sp*³ orbitals,** each containing one electron (right side of Figure 11.4). The sp^3 orbitals are teardrop shaped (Figure 11.5).

Carbon uses its four sp^3 orbitals when forming bonds to other atoms. Each sp^3 orbital forms a covalent bond by overlapping with an orbital from another atom. Figure 11.5 shows the formation of CH_4 from one carbon and four hydrogen atoms. Each sp^3 orbital of the carbon atom (containing one electron) overlaps with the $1s$ orbital of a hydrogen (also containing one electron). In alkanes containing more than one carbon, the carbon atoms bond to one another by the overlapping of their sp^3 orbitals. The single bonds formed with the use of sp^3 orbitals, like other single bonds, are called **sigma bonds** (σ bonds).

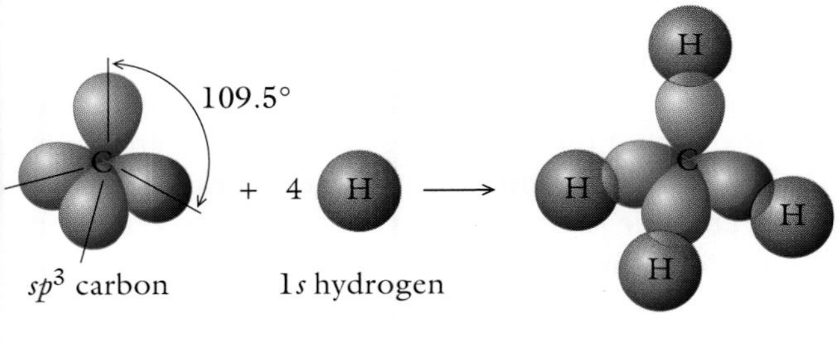

Figure 11.5 Formation of CH_4 from $1s$ orbitals of four hydrogens and sp^3 orbitals of one carbon.

11.1 Chemistry Around Us

Natural Gas and Petroleum

Petroleum (crude oil) and natural gas, the decay products of plant and animal remains, are dispersed together in porous rock formations below ground level throughout Earth.

Natural gas consists of alkanes of fewer than five carbons, mostly methane (typically 85%), with small amounts of ethane, propane, and butane. It is generally stored as a liquid in high-pressure tanks and separated into its various components for different uses. Methane, containing small amounts of ethane, is used to heat our homes (those that have a gas heating system) and for cooking purposes. In highly industrial nations such as the United States, extensive underground gas pipelines distribute the gas over long distances (as far as from Texas to Maine). Propane, available in low-pressure cylinders, is the main source of heating and cooking gas in many sparsely populated agricultural and rural areas, where gas pipeline distribution networks are too expensive to install. Propane is often used as an economical substitute for gasoline and diesel fuel in tractor and other internal combustion engines.

Petroleum—a thick, black liquid—contains hundreds of different hydrocarbons of five-carbon molecules and larger (mostly alkanes, branched and unbranched); but it also contains some cyclic compounds, alkenes, and aromatics. Petroleum is refined by fractional distillation, which separates the mixture into different products on the basis of differences in boiling points. The approximate compositions and boiling points of the different products are:

• Gasoline fuel	C_5–C_{10}	30–200°C
• Kerosene and jet fuel	C_{10}–C_{18}	180–275°C
• Diesel fuel and heating oil	C_{15}–C_{18}	200–300°C
• Lubricating and mineral oils	C_{17}–C_{25}	300–400°C
• Paraffin wax, asphalt	C_{20}–C_{40}	>400°C

The importance of the different products varies throughout the year. During the summer, the demand for gasoline is high, whereas more heating oil is consumed in the winter. Modern refineries can alter the relative amounts of the different fractions to meet demand by advanced chemical processes called cracking and alkylation. Cracking increases the amount of the lower-boiling-point fractions by thermally breaking larger molecules into smaller ones. Alkylation increases the amount of higher-boiling-point fractions by reacting smaller molecules to form larger ones. A full discussion of these reactions is beyond the scope of this text.

Petroleum is critical to life in modern, technological societies. Like natural gas, it is used to heat homes, schools, and factories, to power automobiles, trains, and aircraft, and to generate electricity. Various components of petroleum are used as lubricating oils and greases and as asphalt for roadways.

Just as important as the natural gas and petroleum used for energy are the approximately 5% and 20%, respectively, of the yearly yields of petroleum and natural gas that are converted into other organic chemicals. Although the percentage of the total amount of natural gas or petroleum being used is small, the volume of organic chemicals produced is enormous (trillions of pounds per year). Hydrocarbon compounds from natural gas and petroleum are

The angle between any two bonds of a tetrahedral (sp^3) carbon in any compound is 109.5°—the **tetrahedral bond angle** (see Figure 11.3). (A bond angle is the angle between two bonds that share a common atom.) Valence-shell electron-pair repulsion (VSEPR) theory (Section 3.10) identifies the tetrahedral bond angle as the most stable bond angle for an atom forming four equivalent orbitals. In this arrangement, the electrons of the four orbitals are as far away from one another as possible and thus experience the least amount of repulsion.

Conformations and Single-Bond Free Rotation

A single bond between any two atoms possesses **free rotation;** that is, the atoms in the bond are able to rotate freely relative to one another through 360°, and they do so continuously. There is little or no energy barrier to rotation about single bonds. (Exceptions to this generalization will be discussed as the need arises.) This free rotation makes it possible for a molecule to have different **conformations;** that is, different orientations of the atoms of a molecule that result only from rotations about its single bonds.

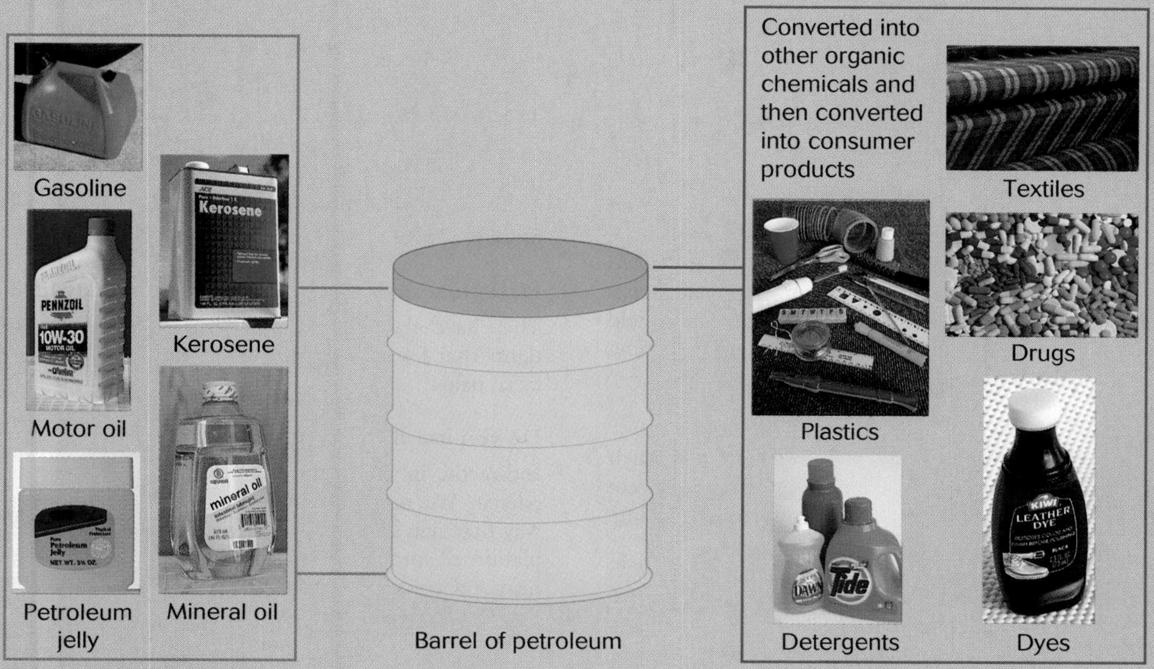

Utilization of a barrel of crude oil.

converted by chemical reactions into a variety of organic chemicals—members of other organic families such as alcohols, aldehydes and ketones, carboxylic acids and esters, amines, and amides. These organic chemicals are in turn used to produce more than 90% of all plastics, textiles, rubbers, drugs, detergents, dyes, and a host of other products critical to the high standard of living that we enjoy today.

Consider rotation about the C–C bond between the middle two carbons in butane (C_4H_{10}). This rotation occurs continuously, but we can visualize a butane molecule being frozen into different conformations at different points in time. Two conformations of butane are:

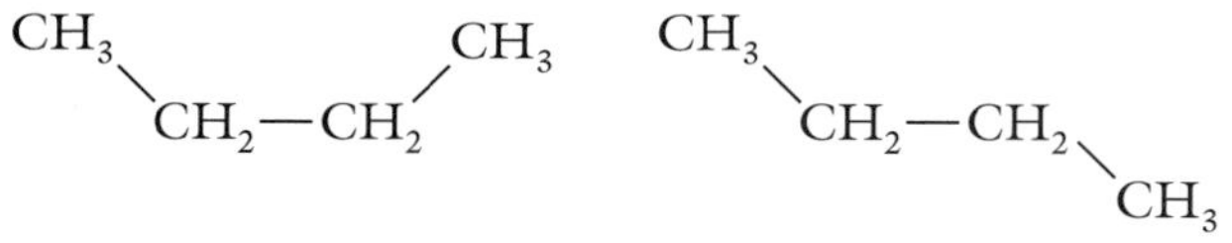

The first structure shows the conformation in which the CH_3 carbon atoms at the ends of the molecule are nearest each other. In the second structure, the middle C–C bond has rotated to the point where the CH_3 carbon atoms at the ends of the molecule are farthest from each other. In any sample of butane, the bonds of all the molecules are continuously undergoing rotation and changing their conformations. Never will they all have the same conformation at the same time.

The situation is even more complex. In a molecule such as butane, all the C–C and C–H single bonds are rotating simultaneously. The number of

PROFESSIONAL CONNECTIONS

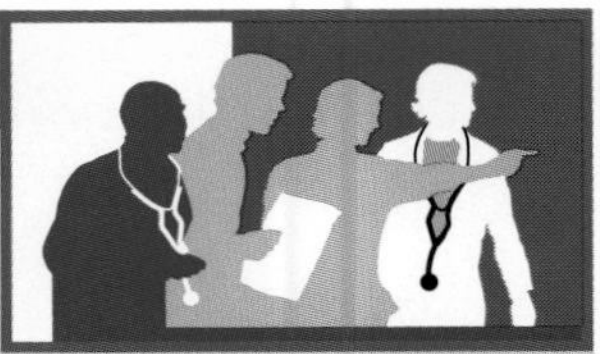

Barbara Semmel, M.A.
Owner/Director, Private Practice
Outpatient Physical Therapy Clinic

Barbara Semmel studied chemistry at the University of Kansas in pursuit of an undergraduate degree in physical therapy.

Why did you choose the field that you are in today?

It was in my senior year in high school, when the school had "Careers Day." A physical therapist came in and talked about her profession. At that time, physical therapists worked only in hospitals or rehabilitation centers. The field appealed to me because of the opportunity to be helping people and being in an active, nonrepetitive position.

What do you enjoy the most?

Working with people—helping people to improve their health or physical conditions, educating them to prevent future injuries. It's satisfying and fun—not at all boring.

What positions did you have before your current one?

I worked in many different facilities: an acute care hospital, chronic care rehab centers, and several years in a school system. This field allows you to move in a range of work, from pediatric to geriatric care, easily.

Please describe a typical day

We start at 9 A.M. and work 5 days a week, 2 evenings, and, about every half hour, we see patients. We evaluate or reevaluate their physical status and treat them—through exercising and the modalities of heat, massage, and mobilization. I specialize in orthopedic, postsurgical patients—particularly problems with TMJ, or temporal mandibular joints (jaw joints). I also take time for continuing education, usually taking one new course a year.

What advances have been made in your field?

The main change is that physical therapists have moved into areas of specialization. [They] did everything years ago. Now, there are generally considered to be five areas of specialization: orthopedic/sports medicine, neurology, pediatrics, cardiopulmonary, and geriatrics. Many more therapists are now in private practice or health centers, and more practice sports medicine.

Have these advances changed your everyday life?

They have changed the composition of the types of patients that I see. Most of my patients have head, neck, or facial pain.

Do you use chemistry in your job?

Indirectly, in many ways. We use pressure for reducing edema. We need to know how surface tension, and the way that skin conducts electricity affect the electrical stimulation of muscle. We need to know about the effects of heat and cold and similar physical chemistry principles on the transport of ions and medication into an inflamed joint. Paraffin wax, a hydrocarbon derived from petroleum, is used to produce heat for therapy. Soaps used for whirlpool treatment are chlorine-based detergents. We talk to people about chemistry regarding their diets: using the right combinations of nutrients to benefit the metabolism. In all of our therapy, we are basically making changes at the molecular level of structure below the observable surface of the body to correct a problem.

What advice to you have for students who are studying chemistry today?

It's more applicable than you think! Also, when I was a senior in high school, my teacher made chemistry interesting and dynamic. Good instructors and good books can make a big difference!

possible conformations for a molecule increases with the number of atoms. Ball-and-stick models of some of the conformations of butane are shown in Figure 11.6.

Conformation is not important for many of the small-sized molecules that we shall consider in Chapters 11 through 17. However, conformation is usually critical for large-sized biochemical molecules such as proteins, carbohy-

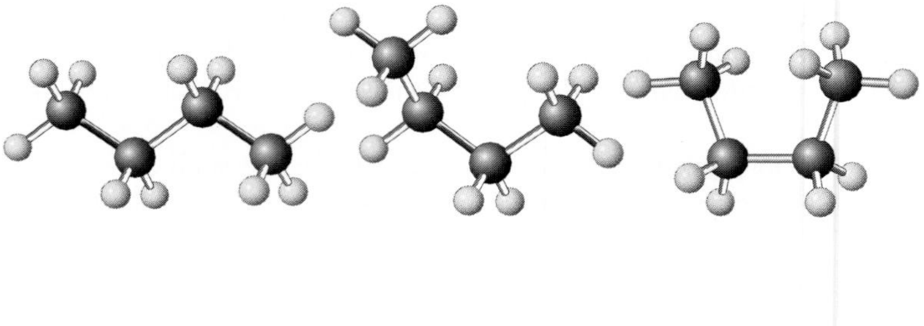

Figure 11.6 Ball-and-stick models of some conformations of butane.

drates, and nucleic acids (Chapters 18, 20, and 21). One conformation is favored over others for a large-sized molecule, and this specific conformation determines the molecule's overall shape, which in turn is critical to its physiological function.

11.4 TYPES OF STRUCTURAL FORMULAS

Writing and reading structural formulas present special difficulties for beginning students for several reasons:

- Different conformations of the same molecule can easily be mistaken for molecules of different compounds.
- Molecules are three dimensional but are usually represented by two-dimensional structural formulas.

Two-dimensional structural formulas of the same compound—even in the same conformation—can look quite different if drawn by different people or from different viewing angles. Consider what might happen if you asked several people to photograph a chair. One person might photograph the chair from the front, one from the side, one from the rear, one from the top, and so forth. The photographs would look different, but each would be a true representation of the same chair.

- Abbreviated structural formulas are often used to save time and space.

Consider the following structural formulas:

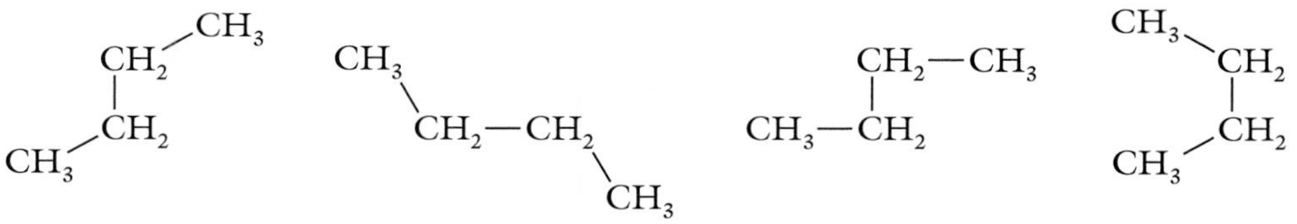

I II III IV

$CH_3-CH_2-CH_2-CH_3$

V

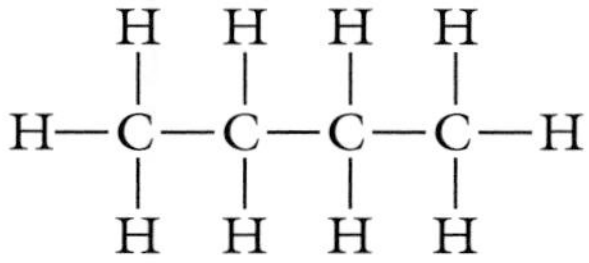

VI

$CH_3CH_2CH_2CH_3$

VII

Each of these structural formulas represents the same compound, butane. Formulas I, II, and III represent the same conformation as well—the conformation in which the end carbons are farthest away from each other. I and II are exactly the same drawing except that II is rotated 90° counterclockwise relative to I. III represents the same conformation as I and II except that it does not depict the C–C–C tetrahedral (109.5°) bond angles as accurately. IV represents a conformation different from that of I, II, and III; it represents the conformation in which the end carbons are nearest each other. V represents butane without specifying anything about the conformation. The first four representations all translate into V when observed at an appropriate viewing angle.

The only bonds seen in formulas I through V are the carbon–carbon bonds; no bonds at all are seen in formula VII. These structures are varieties of abbreviated, or shorthand, structural formulas, differing in the extent to which bonds are understood to be present without being explicitly drawn. Formula VII is a **condensed structural formula;** formula VI, in contrast, is a fully **expanded structural formula,** showing every carbon–hydrogen as well as every carbon–carbon bond in the molecule. Formulas I through V are partly expanded structural formulas, containing more detail than VII but less detail than VI. Formulas V and VII are shorthand versions of VI. When you have learned the combining powers of the various atoms in organic molecules (Table 11.1), you will not find it difficult to recognize that V, VI, and VII all represent the same molecule, butane.

C—C—C—C

Skeleton

Line

Skeleton and **line structural formulas** also are used. Only the carbon atoms in a molecule are shown in a skeleton structural formula. In a line structural formula, it is understood that carbon atoms are present at every intersection of two or more lines and wherever a line begins or ends.

Example 11.3 Drawing structural formulas

Draw different structural formulas representing pentane, C_5H_{12}. Include at least two different conformations, one expanded structural formula, one condensed structural formula, one skeleton structural formula, and one line structural formula.

Solution

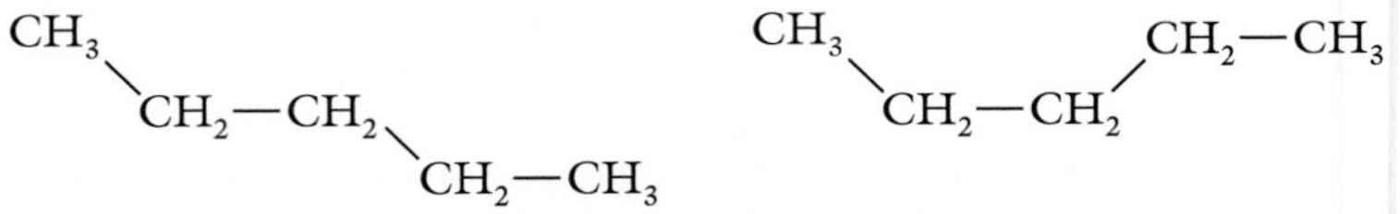

Different conformations

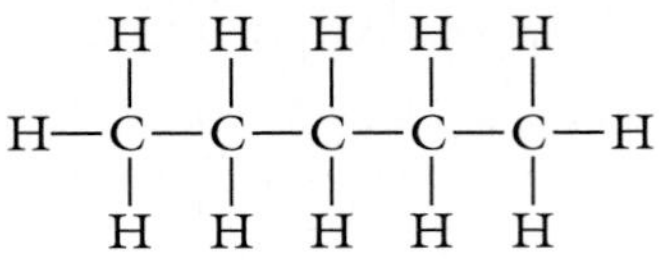

Expanded

$CH_3CH_2CH_2CH_2CH_3$

Condensed

C—C—C—C—C

Skeleton

Line

Problem 11.3 Draw different structural formulas representing hexane, C_6H_{14}. Include at least two different conformations, one expanded structural formula, one condensed structural formula, one skeleton structural formula, and one line structural formula.

11.5 CONSTITUTIONAL ISOMERS OF ALKANES

$CH_3—CH_2—CH_2—CH_3$

Butane

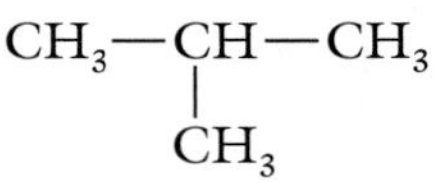

Isobutane

A compound called isobutane has the same molecular formula, C_4H_{10}, as that of butane. The two compounds have different structures as well as different properties.

Butane and isobutane are examples of **isomers,** different compounds that have the same molecular formula. There are several different types of isomers. Butane and isobutane illustrate **constitutional (structural) isomers**—compounds that differ from each other in **connectivity;** that is, in the order in which the atoms are attached to each other. Other kinds of isomers will be introduced in Section 11.8 and in Chapters 12 and 17. Ball-and-stick models of butane and isobutane are shown in Figures 11.6 and 11.7, respectively.

Constitutional isomers, like all types of isomers, are different compounds with different chemical and physical properties. For example, butane has a boiling point of −1°C and a melting point of −138°C, whereas the corresponding values for isobutane are −12°C and −159°C. In later chapters when we consider the molecules of nature, such as carbohydrates, lipids, proteins, and nucleic acids, we will see that different isomers have different chemi-

▶▶ Besides constitutional isomers, there are isomers called stereoisomers. Both types play key roles in the selectivity of biological processes, as described in Section 17.5.

cal and physiological behaviors. We often find that, although there are many possible isomers with the same molecular formula, only one isomer has physiological function. There is high selectivity in the molecules used in nature.

When we look closely at the structures of constitutional isomers, we see that their basic difference lies in the number of carbon atoms connected in a successive manner (one after another). This number is called the **longest continuous carbon chain** (or, simply, the **longest chain).** Whereas four carbons are bonded in a successive manner in butane, the longest chain in isobutane is three, not four, carbons long. In other words, three of isobutane's carbons are bonded one after another, and the fourth is a branch coming off the middle carbon of the longest chain.

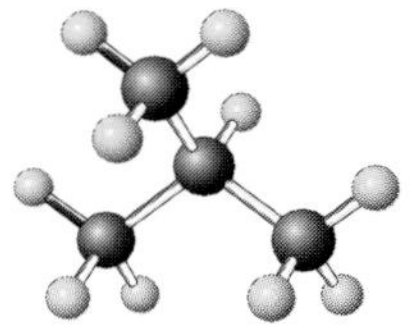

Figure 11.7 Ball-and-stick model of isobutane, the branched isomer of butane (see Figure 11.6).

Butane is called an **unbranched** (or **linear** or **straight-chain) alkane,** and isobutane is called a **branched alkane.** In an unbranched alkane, all the carbon atoms are contained in the longest chain. Such is not the case in a branched alkane, where the carbons not contained within the longest chain are found instead on branches attached to the longest chain.

The greater the total number of carbon atoms in an organic molecular formula, the higher the number of constitutional isomers represented by that formula. Thus, C_5H_{12} has more constitutional isomers than does C_4H_{10}; C_6H_{14} has more constitutional isomers than does C_5H_{12}; and so forth.

Example 11.4 Drawing constitutional isomers

Draw the different constitutional isomers of C_5H_{12}, taking care to draw only one structural formula for each isomer. It is incorrect to draw more than one structural formula for each isomer.

Solution

There are three constitutional isomers of C_5H_{12}:

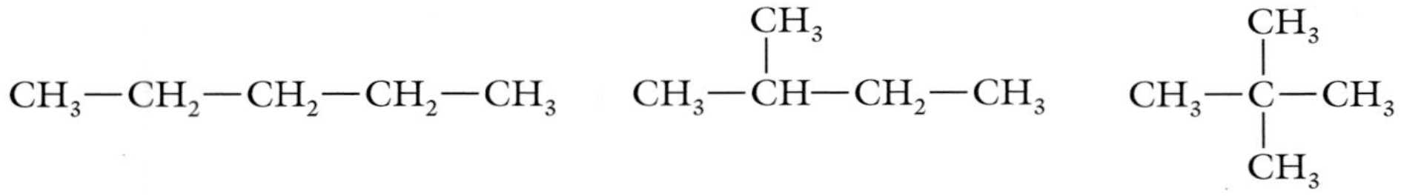

1 2 3

The various isomers can be represented by first drawing the one that has all five carbons in the longest chain (compound 1). Next, draw the branched isomers that have one carbon less in the longest chain. Compound 2 has four carbons in the longest chain, with the fifth carbon forming a branch on the next-to-last carbon from either end of the longest chain. Compound 3 has three carbons in the longest chain, with the fourth and fifth carbons forming separate branches on the middle carbon of the longest chain. Figure 11.8 shows ball-and-stick models of the three isomers of C_5H_{12} (the conventions for naming these compounds will be considered in Section 11.6). It is important in working out the answer to this problem not to incorrectly show two different structural formulas of the same isomer. For example, including structure 4 along with structures 1, 2, and 3 is a

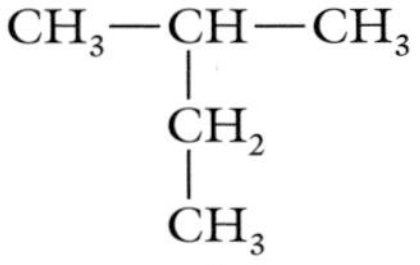

4

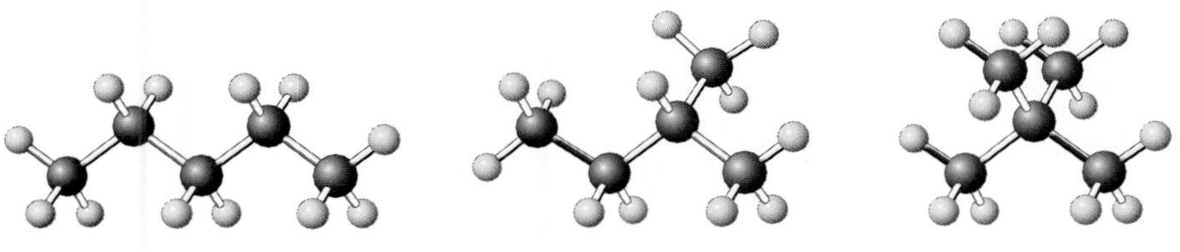

Pentane 2-Methylbutane 2,2-Dimethylpropane

Figure 11.8 Ball-and-stick models of isomers of C_5H_{12}.

mistake because structures 2 and 4 represent the same compound. Either structure is a correct representation of the isomer having four carbons in a continuous chain, with the fifth carbon forming a branch on the next-to-last carbon of the longest chain. If you include both structures, however, you convey the wrong idea—you imply that there are four constitutional isomers of C_5H_{12}. In fact, there are only three isomers, and your structural drawings must correctly express that fact. As you solve problems such as that in this example, you must compare each new structure with the previously drawn structures to eliminate duplicates, looking for the longest chain and analyzing the placement of branches along it. Use molecular models, if necessary.

Problem 11.4 Draw the different constitutional isomers of C_6H_{14}.

11.6 NAMING ALKANES

Compounds must be named by a generally accepted convention so that it is always clear what compound is being discussed.

Primary, Secondary, Tertiary, and Quaternary Classification

Chemists categorize a carbon atom in a compound on the basis of the number of other carbon atoms directly connected to it. This categorization is useful because it helps us to use the IUPAC nomenclature system and to understand the reactivity of a series of compounds. A carbon atom in a compound is categorized as a **primary** (1°), **secondary** (2°), **tertiary** (3°), or **quaternary** (4°) carbon, depending on whether it is directly bonded to a total of one, two, three, or four other carbon atoms, respectively.

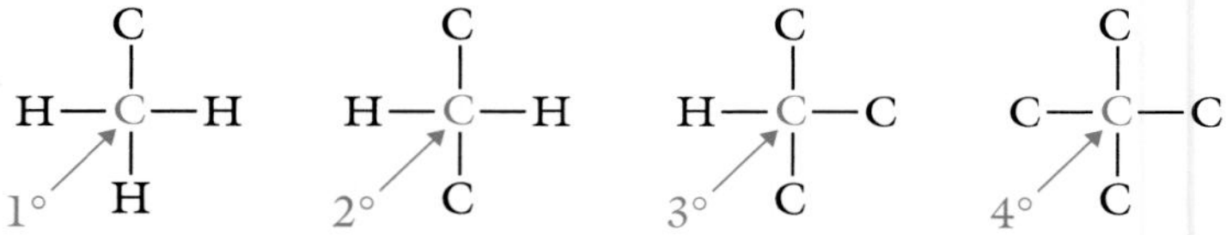

A functional group, such as —OH, attached to a primary, secondary, or tertiary carbon also is categorized as primary, secondary, or tertiary, receiving the same categorization as the carbon to which it is attached. Reactivity of most compounds changes in a regular manner, increasing or decreasing in the order primary, secondary, tertiary.

Note that there is no such thing as a quaternary functional group, because carbon is tetravalent. There is no way in which a functional group and four carbons can all bond at the same time to a single carbon atom.

Example 11.5 Identifying 1°, 2°, 3°, and 4° carbons

Indicate the 1°, 2°, 3°, and 4° carbons in the following compound:

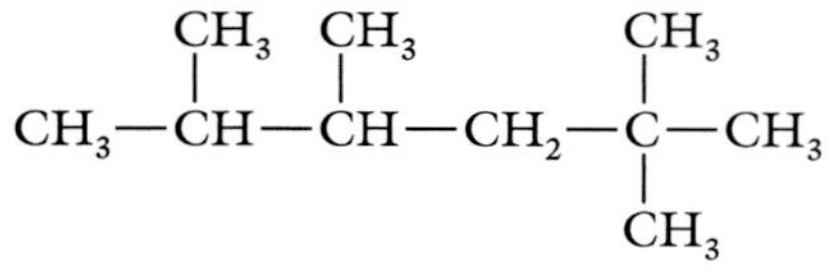

Solution

Count the number of carbons to which each carbon atom is directly bonded.

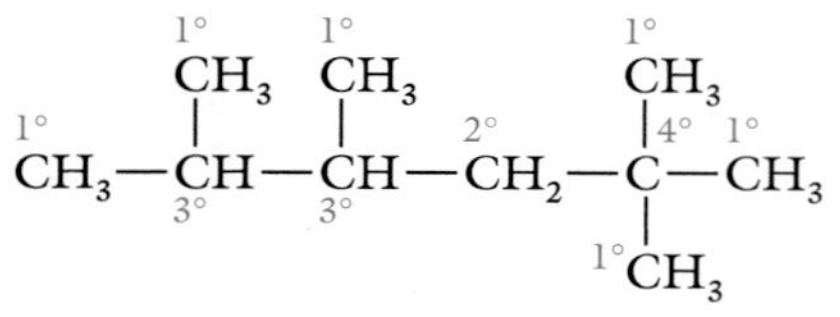

Problem 11.5 Indicate the 1°, 2°, 3°, and 4° carbons in the following compound:

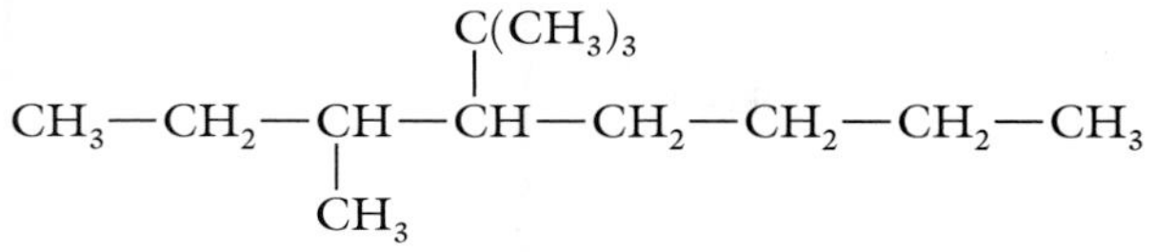

Common Nomenclature System

Two nomenclature systems—the common nomenclature system and the IUPAC nomenclature system—are used to name alkanes. Both nomenclature systems name straight-chain (unbranched) alkanes by attaching the suffix **-ane** to a prefix denoting the total number of carbons in the compound (see Table 11.3). The alkanes from C_1 to C_{10} are named methane, ethane, propane, butane, pentane, hexane, heptane, octane, nonane, and decane. The suffix of each name is the ending of the family name—in this case, -ane for alkane.

Alkanes with more than three carbons have constitutional isomers. In the common nomenclature system, prefixes are added to the names of straight-chain alkanes to form the names of their branched isomers. For example, we noted in Section 11.5 that the branched C_4 alkane is distinguished from the straight-chain isomer by the prefix **iso-**. The straight-chain C_4 alkane is called butane, with no prefix, whereas the branched C_4 alkane is called isobutane. Some authors use the prefix *n*- or the word **normal** for straight-chain alkanes, as in *n*-butane or normal butane. We will not use these terms, because they are redundant. In this textbook, an alkane name with no prefix always means the straight-chain isomer.

The three C_5H_{12} isomers of Example 11.4 can be named by using an additional prefix, **neo-**. The three isomers are called pentane, isopentane, and neopentane. However, the common nomenclature system becomes useless beyond three isomers. The number of constitutional isomers increases rapidly with increasing numbers of carbons. There are 5 constitutional isomers of C_6H_{14} and 75 constitutional isomers of $C_{10}H_{22}$. To use the common nomenclature system for $C_{10}H_{22}$, we would need to devise and use 75 prefixes and memorize which prefix designates which isomer.

Organic chemists developed an alternate nomenclature system—the IUPAC nomenclature system—nearly a century ago. IUPAC stands for the International Union of Pure and Applied Chemistry, an international organization of chemists and other scientists that devised the new nomenclature system. Nomenclature, like other scientific activities, is ever evolving as new chemical structures are synthesized by chemists.

Alkyl Groups

The IUPAC nomenclature system requires the use of the names of **alkyl groups** that are commonly encountered in organic molecules. These groups are found attached to other groups or to some atom in a molecule. We need to learn the alkyl groups and their names before we can learn the IUPAC nomenclature system. We can visualize the derivation of various alkyl groups from alkanes. We will consider a limited number of alkyl groups in this text.

One-carbon and two-carbon alkyl groups, called **methyl** and **ethyl,** respectively, are derived from methane and ethane by loss of a hydrogen atom:

$$\underset{\textbf{Methane}}{CH_4} \xrightarrow{-\,H} \underset{\textbf{Methyl}}{CH_3—}$$

$$\underset{\textbf{Ethane}}{CH_3—CH_3} \xrightarrow{-\,H} \underset{\textbf{Ethyl}}{CH_3—CH_2—}$$

One bond from the alkyl group is incomplete, and that is the bond through which the alkyl group becomes attached to some other group or atom in a molecule. Alkyl groups do not have an independent existence but are a hypothetical means by which we understand the structure of compounds and name them.

Whereas only one alkyl group each is derived from methane and ethane, two different alkyl groups, **propyl** and **isopropyl,** are derived from propane. Removal of a hydrogen from one of the end carbons of propane yields the propyl group. The isopropyl group is obtained by removal of a hydrogen from propane's middle carbon:

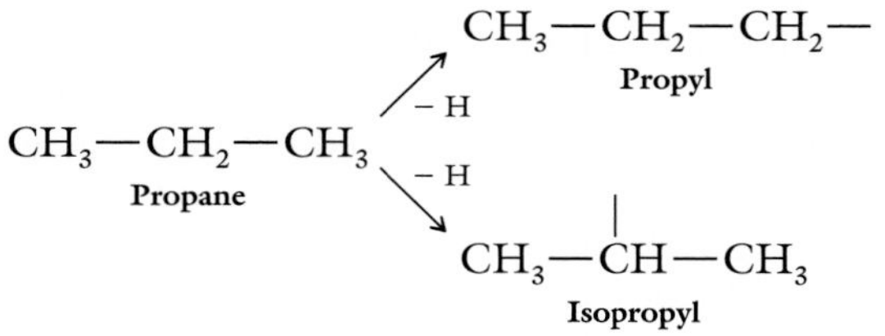

The **butyl** and ***s*-butyl groups** are derived from butane. Removal of a hydrogen from one of the end carbons of butane yields the butyl group; removal of a hydrogen from one of the inner carbons yields the *s*-butyl group:

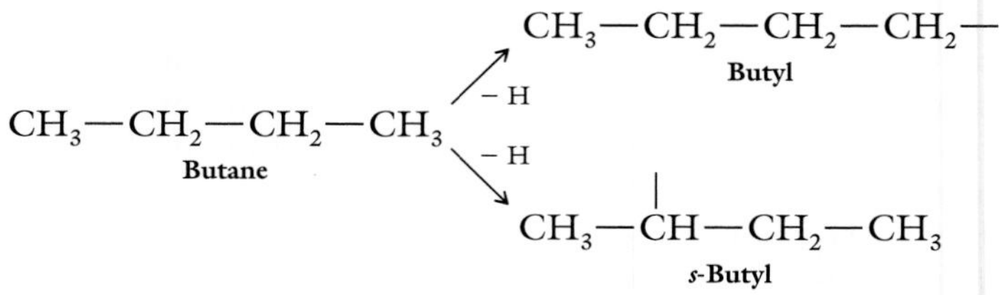

The **isobutyl** and ***t*-butyl groups** are derived from isobutane. Removal of a hydrogen from one of the CH_3 carbons yields the isobutyl group; removal of a hydrogen from the CH carbon yields the *t*-butyl group:

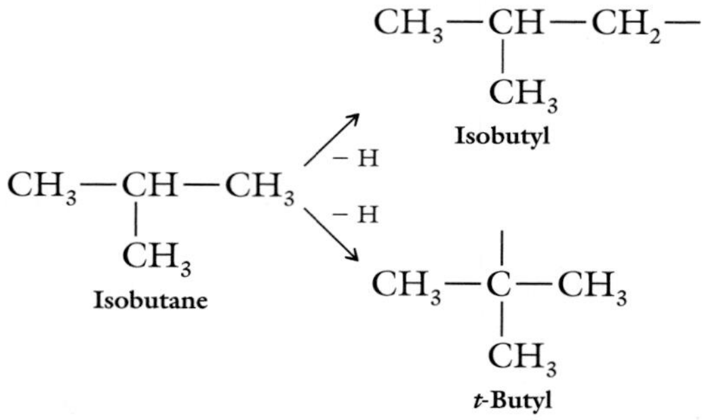

The prefixes *s*- or *sec*- and *t*- or *tert*- are abbreviations for secondary and tertiary, respectively. These prefixes are italicized when used. The prefix iso- is not italicized and is not abbreviated to *i*-. Recall that no prefix is used for the unbranched groups.

You need to memorize these alkyl groups, including their condensed formulas, until you can recognize them easily (Table 11.4). Here are some hints to help you remember the different alkyl groups:

Guide to recognizing the alkyl groups

- An alkyl group with no prefix is unbranched and has its incomplete bond at the end of the chain of carbons.
- The *s*- and *t*-butyl groups have the incomplete bond at a secondary and tertiary carbon, respectively.

TABLE 11.4 Alkyl Groups

Name	Expanded Structural Formula[a]	Condensed Structural Formula
methyl	CH_3-	CH_3-
ethyl	CH_3-CH_2-	CH_3CH_2- or C_2H_5-
propyl	$CH_3-CH_2-CH_2-$	$CH_3CH_2CH_2-$
isopropyl	$CH_3-\overset{\vert}{C}H-CH_3$	$(CH_3)_2CH-$
butyl	$CH_3-CH_2-CH_2-CH_2-$	$CH_3CH_2CH_2CH_2-$
s-butyl	$CH_3-\overset{\vert}{C}H-CH_2-CH_3$	$CH_3\overset{\vert}{C}HCH_2CH_3$
isobutyl	$CH_3-\underset{\underset{\displaystyle CH_3}{\vert}}{C}H-CH_2-$	$(CH_3)_2CHCH_2-$
t-butyl	$CH_3-\overset{\vert}{\underset{\underset{\displaystyle CH_3}{\vert}}{C}}-CH_3$	$(CH_3)_3C-$

[a]C–H bonds are not expanded.

- The isopropyl and isobutyl groups each possess two methyl groups and fit the general formula

$$CH_3-\underset{\underset{\displaystyle CH_3}{\vert}}{C}H(CH_2)_n-$$

with $n = 0$ and 1 for isopropyl and isobutyl, respectively.

We will occasionally refer to the **simple alkyl groups** when describing the nomenclature rules for organic compounds. The simple alkyl groups are the unbranched alkyl groups from methyl, ethyl, propyl, and butyl through decyl and the branched three- and four-carbon alkyl groups (isopropyl, isobutyl, *s*-butyl, *t*-butyl).

In organic chemistry, molecules are often written in a very generalized form that uses the letter R to represent any alkyl group. For example, R–OH represents an —OH functional group attached to any alkyl group.

IUPAC Nomenclature System

There are only a few rules for naming alkanes by the **IUPAC nomenclature system.** The utility of the IUPAC rules is that, if they are followed correctly, they will always result in the same name for the same compound. Here are those rules and the order in which they must be used:

Rules for naming alkanes

1. The name of the longest chain becomes the **base,** or **parent, name** of the compound. The suffix (ending) of the family name is added to the end of this base name. Note that, even if the subsequent rules are correctly applied, a failure to correctly identify the longest chain will result in an incorrect name for the compound.

2. The base name accounts only for the carbons in the longest chain. The carbons that are not part of the longest chain—those attached as branches to the longest chain and called **substituents** or **groups**—must also be included in the name. Substituents are included as follows:

a. The name(s) of any alkyl group(s) in the compound is placed in front of the base name.

b. Use the prefixes di-, tri-, tetra-, penta-, and hexa- before the name of the alkyl group when there are two, three, four, five, or six, respectively, of the same group.

c. Alphabetize the names of alkyl groups when there are two or more different types of groups. Ignore all prefixes (both the branching prefixes *s*- and *t*- and the multiplying prefixes such as di-, tri-, and tetra-) in alphabetizing, with one exception: iso- is not ignored in alphabetizing.

3. Number the carbons in the longest chain, starting from whichever end will result in the lowest number (or set of lowest numbers) for the alkyl group(s). An alternate rule is useful for most compounds: Number from the end nearest a branch.

4. In front of the name of each alkyl group, place the number of the carbon to which the group is attached.

5. Use hyphens to separate numbers from words; use commas to separate numbers.

In addition to naming organic compounds, the IUPAC nomenclature system helps us recognize whether two structural formulas represent the same compound or different compounds. As noted earlier, two different-looking structural formulas of the same compound will yield the same name if the nomenclature rules are correctly applied.

Example 11.6 Using the IUPAC nomenclature system to name alkanes: I

Give the IUPAC name for the following compound:

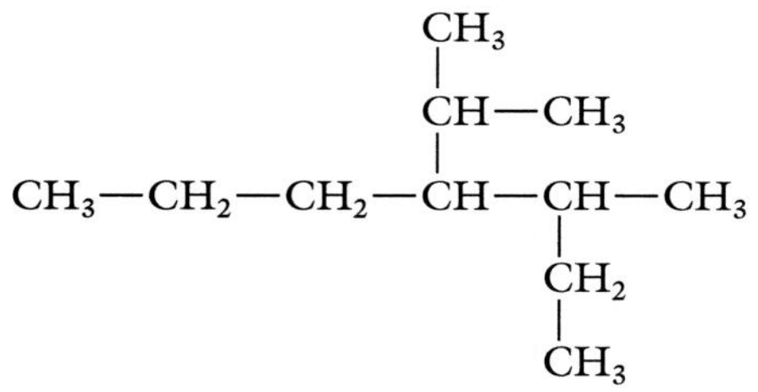

Solution

As often occurs in the depiction of organic compounds, the longest chain in this structural formula was not drawn in a horizontal line. Careful examination shows that the longest chain in the compound contains seven carbons. Table 11.3 tells us that the base name for a seven-carbon chain is heptane. The longest chain does not include the alkyl groups isopropyl (shown in blue in the following depiction) and methyl (shown in red). The names of these alkyl groups must be placed in front of the base name in alphabetical order, and each must be assigned a number to indicate its placement on the longest chain. The numbering of the longest chain begins with the terminal carbon nearest the methyl group.

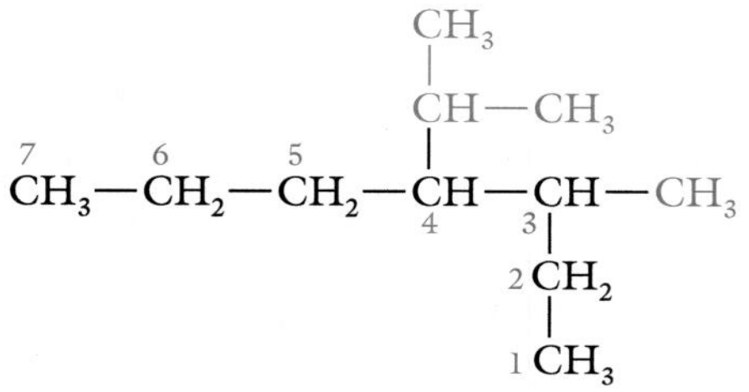

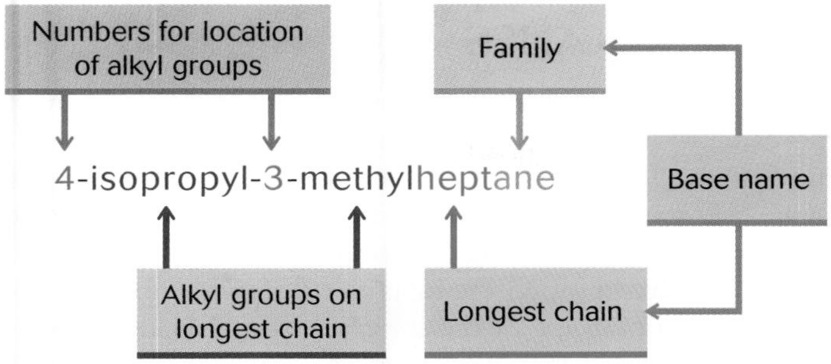

Figure 11.9 The IUPAC nomenclature system.

Numbering from the opposite end is incorrect, because that end is farther from an alkyl group (isopropyl). The IUPAC name of the compound is 4-isopropyl-3-methylheptane. Figure 11.9 shows the different segments of the IUPAC name.

Problem 11.6 Give the IUPAC name for each of the following compounds:

(a) 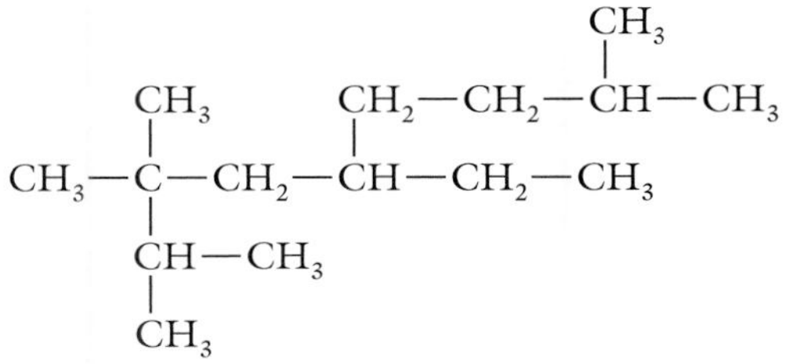

(b) 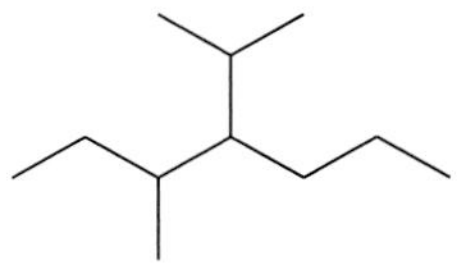

Example 11.7 Using the IUPAC nomenclature system to name alkanes: II

Give the IUPAC name for the following compound:

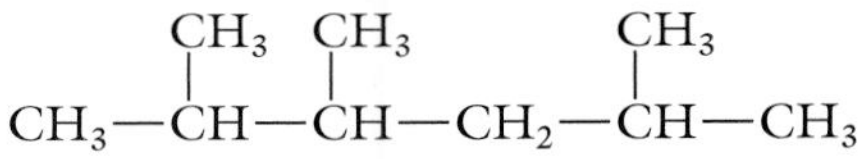

Solution

There are four longest chains of the same length, six carbons, in this compound. These are the chains that begin with one of the carbon atoms marked with an asterisk in the following structure and continue toward one of the asterisk-marked carbons at the other end of the molecule:

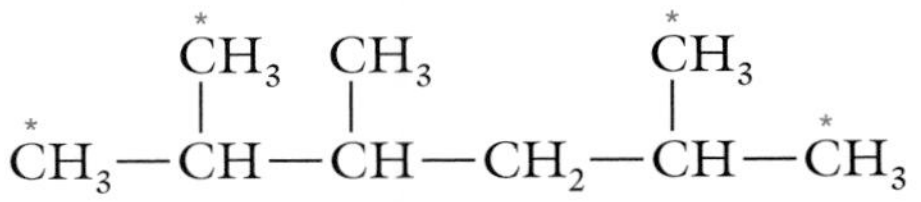

The correct name is the one that results in the set of lowest numbers for the alkyl groups. For many compounds, we can shorten this process because some or all parts of the molecule are equivalent. In this example, for instance, all four six-carbon longest chains are equivalent: the two asterisk-marked carbons on the left side are equivalent because they are each part of a methyl group; likewise, the two asterisk-marked carbons on the right side are equivalent because they, too, are each part of a methyl group. The compound's name can therefore be based on any one of those four longest chains.

The correct name is the one with the set of lowest numbers. Numbering from left to right, the name is 2,3,5-trimethylhexane. Numbering from right to left, the name is 2,4,5-trimethylhexane. The numbers in the set 2,3,5 are lower than the numbers in the set 2,4,5. The IUPAC name, therefore, is 2,3,5-trimethylhexane.

Problem 11.7 Draw the structural formula of each of the following compounds: (a) 4-ethyl-3-methylhexane; (b) 4-*t*-butyl-3-methyloctane.

11.7 CYCLOALKANES

We have seen that carbon atoms can join together to form short or long chains. They can also form rings. In **cyclic,** or **ring,** molecules, there is a bond between the first and last carbons of a chain. Cyclic structures are often found in nature. Glucose, starch, cholesterol, and the sex hormones testosterone and progesterone are examples of biologically important ring compounds.

Alkanes are **acyclic** or **open-chain** molecules, with no bond connecting the first and last carbon atoms of the longest chain. **Cycloalkanes,** the cyclic counterparts of the alkane family, possess the same characteristic functional groups as alkanes (C–H and C–C single bonds) but differ from alkanes in having a cyclic chain of carbon atoms. Starting at one of the carbon atoms in the cyclic chain and proceeding to successive carbons will return us to the starting carbon atom. Together, the alkanes and cycloalkanes constitute the saturated hydrocarbons, compounds that possess only C–C and C–H single bonds.

A cycloalkane is often represented by a condensed structural formula in the form of a geometrical shape—triangle, square, pentagon, hexagon, heptagon (Table 11.5). The geometrical shapes are the cyclic equivalent of the line structural formulas introduced in Section 11.4. Recall that the intersection of two lines defines a carbon atom. Like other line structural formulas, the geometrical shapes representing cycloalkanes do not show any of the hydrogens attached to the carbons of the ring. One always assumes, when reading these structures, that each carbon is bonded to whatever number of hydrogens is necessary to complete the carbon atom's tetravalent requirement.

TABLE 11.5   Cycloalkanes

| Name | Molecular Formula | Structural Formulas | | Boiling Point (°C) |
		Expanded[a]	Condensed	
cyclopropane	C_3H_6	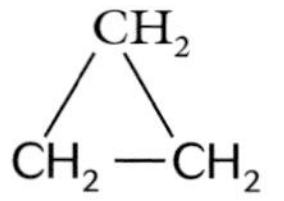 CH_2 / CH_2—CH_2	 △	−33
cyclobutane	C_4H_8	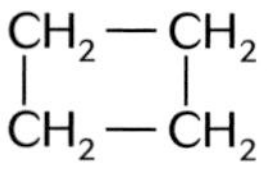 CH_2—CH_2 / CH_2—CH_2	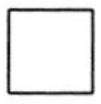 □	12
cyclopentane	C_5H_{10}	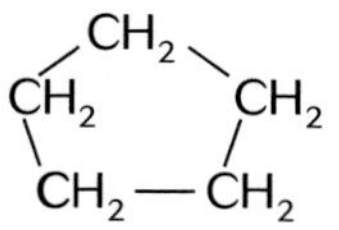 CH_2 / CH_2 CH_2 / CH_2—CH_2	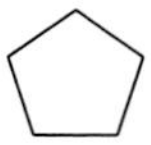 ⬠	49
cyclohexane	C_6H_{12}	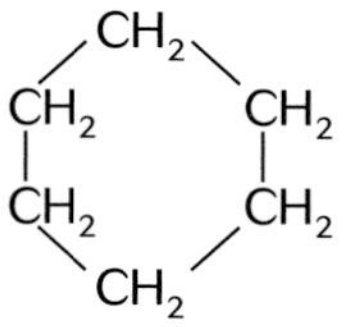 CH_2 / CH_2 CH_2 / CH_2 CH_2 / CH_2	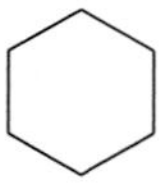 ⬡	81
cycloheptane	C_7H_{14}	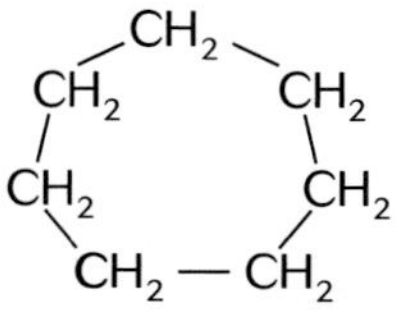 CH_2 / CH_2 CH_2 / CH_2 CH_2 / CH_2—CH_2	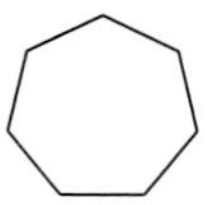 ⬡	119

[a]C–H bonds are not expanded.

The general molecular formula describing cycloalkanes, C_nH_{2n}, has two fewer hydrogens than the general formula for alkanes, C_nH_{2n+2}. Any cyclic compound has two fewer hydrogen atoms than does the corresponding acyclic compound (the acyclic compound with the same number of carbon atoms). This correspondence is easy to understand if we imagine an alkane turning into a cycloalkane by losing a hydrogen from each of its two terminal carbons, which then form a bond with one another.

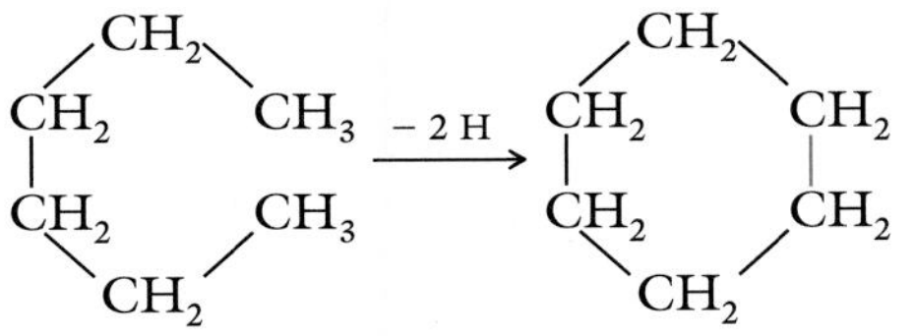

The IUPAC rules for naming cycloalkanes are straightforward variations on the rules for naming alkanes:

Rules for naming cycloalkanes

1. The base name of the ring structure consists of the prefix **cyclo-** followed, without hyphen or space, by the name of the straight-chain alkane that has the same number of carbons.
2. The names of substituents are placed in front of the base name. Numbers indicate the placement of substituents on the ring. The numbering of the ring carbons starts with a carbon holding a substituent and moves around the ring from there. When different sets of numbers can be obtained by starting at different carbons or counting in different directions, the correct name is the one with the set of lowest numbers.

Example 11.8 Using the IUPAC nomenclature system to name cycloalkanes

Give the IUPAC name for the following compound:

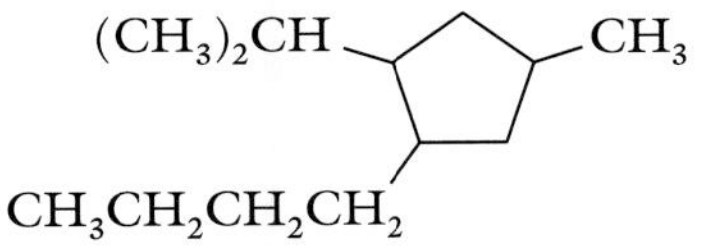

Solution

The ring structure is cyclopentane. Adding the names of substituents in alphabetical order gives butylisopropylmethylcyclopentane. Numbering from the carbon holding the methyl group yields 1,3,4 as the set of numbers designating the positions of the substituents on the ring, irrespective of whether we proceed clockwise or counterclockwise. Starting at the carbon holding the isopropyl group yields 1,3,5 when we proceed clockwise and 1,2,4 counterclockwise. Starting at the butyl group yields 1,2,4 or 1,3,5, depending on the direction. The set of lowest numbers among these possibilities is 1,2,4, but there are two such sets, one starting at the carbon with the isopropyl group as C1 and the other starting at the carbon with the butyl group as C1. Which is the correct set? Alphabetical order wins; the butyl group has precedence over the isopropyl group, so the carbon with the butyl group is C1. 1-Butyl-2-isopropyl-4-methylcyclopentane is the correct IUPAC name.

Problem 11.8 Give the IUPAC name for each of the following compounds:

(a)

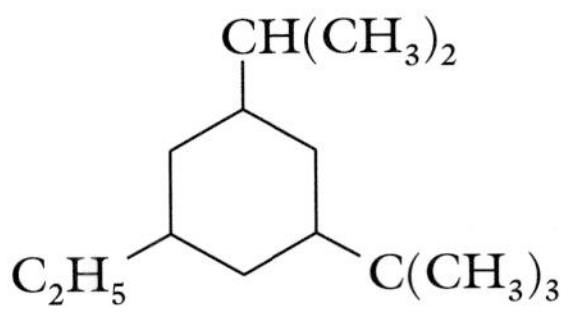

(b)

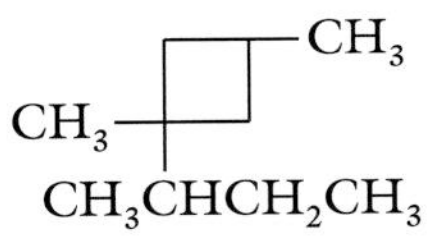

Cycloalkanes possess constitutional isomerism when the same atoms join in such a way as to form compounds with different-sized rings or when the placement of substituents around the ring is different or both. For example, there are five constitutional isomers of C_5H_{10}:

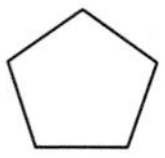

Cyclopentane

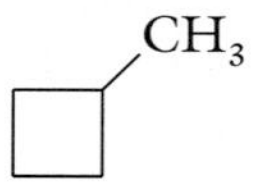

Methylcyclobutane

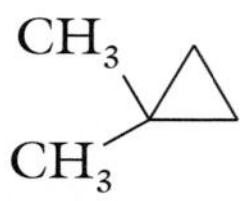

1,1-Dimethylcyclopropane

1,2-Dimethylcyclopropane

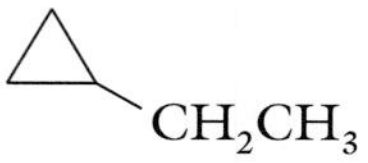

Ethylcyclopropane

Different isomers of five-carbon cycloalkanes can have either three, four, or five of the carbons in the ring. The two isomers containing the cyclopropane ring can have their two additional carbons attached either at a single location or at two different locations on the ring.

Note that there is no need to use 1- in the names methylcyclobutane and ethylcyclopropane. The methyl and ethyl substituents are understood to be at C1. The designation 1- is needed only when there are two or more substituents on a ring structure.

Example 11.9 Drawing constitutional isomers of cycloalkanes

How many constitutional isomers are possible for C_4H_8? Give one structural formula for each isomer. Give the IUPAC name of each isomer.

Solution

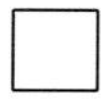

Cyclobutane

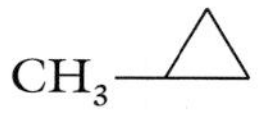

Methylcyclopropane

Problem 11.9 How many constitutional isomers are possible for C_6H_{12}? Draw one structural formula for each isomer. Give the IUPAC name of each isomer.

Special aspects of the shape and stability of different-sized ring compounds are presented in Box 11.2.

11.8 CIS-TRANS STEREOISOMERISM IN CYCLOALKANES

The carbon–carbon bonds of a ring, unlike those in acyclic structures, have **restricted rotation.** In other words, the carbon atoms of a ring are not free to rotate relative to each other. Certain cycloalkanes possess **geometric,** or **cis-trans, stereoisomerism,** a type of isomerism different from constitutional isomerism, because of this restricted rotation. For example, there are actually six cycloalkanes of molecular formula C_5H_{10}, one more than the five constitutional isomers described in Section 11.7. More specifically, there are two different 1,2-dimethylcyclopropane compounds, not one:

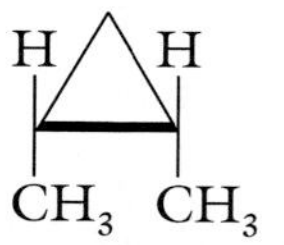

***cis*-1,2-Dimethylcyclopropane**

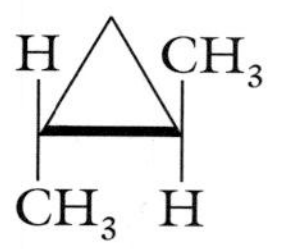

***trans*-1,2-Dimethylcyclopropane**

11.2 Chemistry in Depth

Stability and Shape of Cycloalkanes

Five- and six-membered ring structures are frequently found in nature in a variety of substances; examples include starch, glucose, sex hormones, cholesterol, vitamins A and D, DNA, and some of the amino acids. Three- and four-membered ring structures, in contrast, are found very infrequently, because carbon rings of those sizes are much less stable than the five- and six-membered rings. The difference in stability results from the difference in bond angles for the different-sized rings.

Ring structures are drawn as if the rings were flat; that is, as if all the carbon atoms of the ring lay in a single plane. Although this portrayal is approximately correct for the three-, four-, and five-membered rings, it is incorrect for the six-membered ring. Cyclohexane's "ring" is a puckered, nonplanar shape called the **chair conformation:**

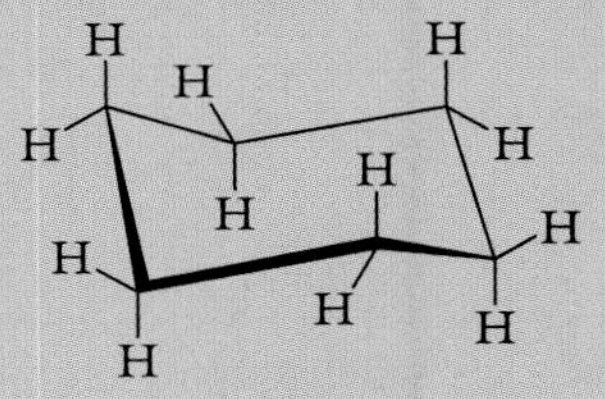

In the creation of three- and four-membered rings, the sp^3 orbitals of the participating carbon atoms must be forced out of their normal tetrahedral bond angles (109.5°) into angles of approximately 60° and 90° respectively, to overlap one another and bond. This distortion of the normal bond angle is the reason for the instability of such compounds. Five-membered rings are stable because their formation occurs with only a very small distortion from the tetrahedral bond angle. Six-membered rings would be unstable if they were flat, because flat geometry would result in 120° bond angles, a large distortion from the tetrahedral bond angle. However, the chair conformation allows the six atoms of the ring to maintain their tetrahedral bond angles. As a result, six-membered rings are highly stable.

The exact conformation of the six-membered carbon ring is not important at the level of this text; so, for the sake of simplicity, we will continue to use flat structures rather than the chair conformation in our drawings. Any exceptions to this generalization will be noted as appropriate.

The two compounds are **geometric** (or **cis-trans**) **stereoisomers** and are distinguished one from the other by writing one of the italicized prefixes *cis-* and *trans-* before the IUPAC name.

Whereas constitutional isomerism results from differences in connectivity, **stereoisomerism** results from differences in configuration. Stereoisomerism is more subtle than constitutional isomerism. Cis and trans stereoisomers have the same substituents attached to the same carbon atoms of the ring—the same connectivity—but the compounds differ in **configuration;** that is, in the spatial orientation in which the substituents extend from the ring carbons into the regions lying on opposite sides of the plane of the ring. The cis isomer is the isomer in which two similar substituents (either hydrogens or methyls in the case of 1,2-dimethylcyclopropane) on two different ring carbons are on the same side of the ring. The trans isomer has the two similar substituents on opposite sides of the ring. The difference between the cis and trans isomers can also be described by the proximity between substituents. Cis substituents are closer to each other than are trans substituents. (Cis-trans stereoisomers are one type of stereoisomers, often called **diastereomers** to distinguish them from a second type, called enantiomers, which will be considered in Chapter 17.)

Our convention for drawing and reading the two 1,2-dimethylcyclopropanes (and similar compounds) is that the ring is perpendicular to the plane of the paper with the H and CH_3 groups parallel to the plane of the paper. The front bonds of a ring are often drawn as thick lines to aid in visualizing this convention, especially when compounds are geometric stereoisomers.

Cis and trans isomers are different compounds with different chemical and physical properties. For example, the boiling points of the cis and trans isomers of 1,2-dimethylcyclopropane are 37°C and 29°C, respectively. The physiological behaviors of cis and trans isomers are often different, another instance of the high selectivity of the molecules used in nature.

A cis isomer cannot easily convert into a trans isomer, and vice versa, because the carbon–carbon bonds of a cyclic structure, unlike those of an acyclic structure, are not free to rotate. You can prove this statement by using molecular models to build a cyclic compound and then trying to rotate the carbon–carbon bonds of the ring.

Cis and trans isomers occur only when each of two ring carbons has two different substituents. Such ring carbons are stereocenters. A **stereocenter** is an atom in a compound bearing substituents of such identity that a hypothetical exchange of the positions of any two substituents would convert one stereoisomer into another stereoisomer. In the structural drawings of the two 1,2-dimethylcyclopropanes, the difference between the cis and trans isomers is reversal of the configurations (the positions of H and CH_3) at the stereocenter on the right side.

None of the following types of cycloalkanes possess geometric isomerism:

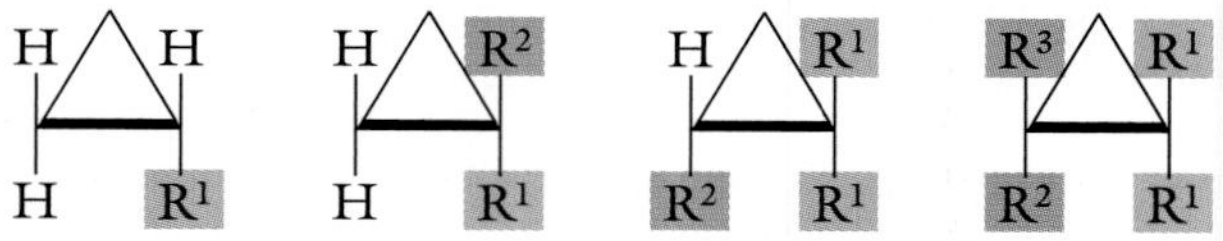

Concept check

✔ Geometric isomers are possible only when two of the ring atoms are stereocenters; that is, two ring atoms each bear two different substitutents.

Example 11.10 Identifying and drawing cis and trans isomers

Which of the following compounds exist as cis and trans isomers? If cis and trans isomers are possible, show one structural drawing for each isomer. (a) Methylcyclopentane; (b) 1,3-dimethylcyclobutane; (c) 1,1,2,2-tetramethylcyclohexane.

Solution

(a, c) Two stereocenters are required for cis-trans isomerism. Neither methylcyclopentane nor 1,1,2,2-tetramethylcyclohexane exists as a pair of cis-trans isomers because neither has two stereocenters.

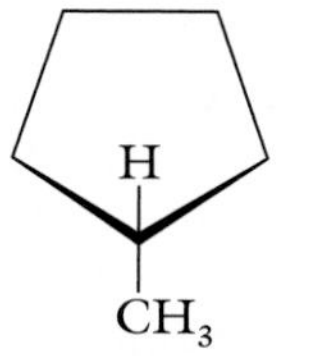

Methylcyclopentane

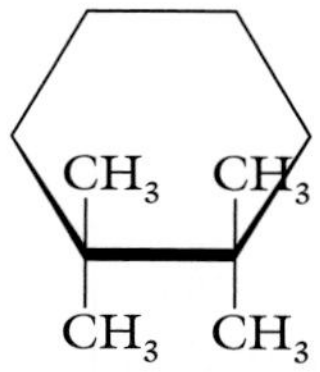

1,1,2,2,-Tetramethylcyclohexane

Methylcyclopentane has only one stereocenter, C1, which has H and CH_3 attached to it. Each of the other ring carbon atoms has two of the same substituents (H). 1,1,2,2-Tetramethylcyclohexane has no stereocenters. Each of two of the ring carbons has two CH_3 groups, whereas each of the other four ring carbons has two H atoms.

(b) 1,2-Dimethylcyclobutane has two stereocenters and thus exists as two different compounds. Ring carbons C1 and C3 each have two different substituents (one H and one CH_3).

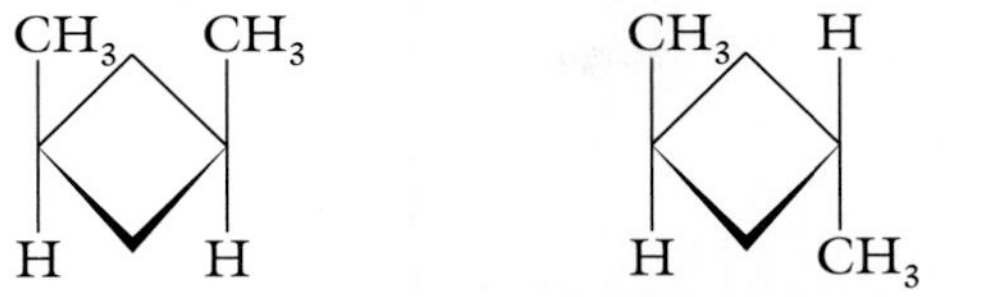

cis-1,3-Dimethylcyclobutane *trans*-1,3-Dimethylcyclobutane

Problem 11.10 Which of the following compounds exist as cis and trans isomers? If cis and trans isomers are possible, show one structural drawing for each isomer. (a) 1,3-Dimethylcyclopentane; (b) 1-ethyl-1,2,2-trimethylcyclopropane; (c) 1,4-dimethylcyclohexane.

11.9 PHYSICAL PROPERTIES OF ALKANES AND CYCLOALKANES

The physical properties of organic compounds—their melting and boiling points, physical strength, and solubility—are just as critical to their utilization, both in biological systems and in our everyday life, as are their chemical properties. A lawn chair must be constructed from a solid substance, not a liquid, and the chair must be of sufficient strength to bear the weight of anyone who might sit in the chair. In the body, the melting points of molecules used in constructing bones must be above body temperature, and bones must be of sufficient strength to withstand a wide range of physical stresses.

Solubility in water is another important physical property because our bodies are aqueous systems. The physiological functioning of bones, skin, blood vessels, the heart, the liver, the lungs, and other tissues and organs requires them to be water insoluble. Conversely, the functioning of digestive enzymes, hormones, hemoglobin, and neurotransmitters depends on their solubility in aqueous systems so that they can be transported throughout the body in the blood and other bodily fluids.

In this and subsequent chapters, we will correlate the physical properties of organic compounds with the strength of their secondary forces. We shall see that the level of the secondary forces varies considerably from one family to another, and this variation results in significant differences in physical properties. Substances become useful as materials of construction, both in the body and outside, when their molecular masses and secondary forces are sufficiently high that they develop the required physical strength.

Melting and Boiling Points and Density

Melting is the transition from the solid to the liquid state. Boiling is the transition from the liquid to the gas state. Whether a substance has a high or low melting or boiling point depends on the strength of the secondary forces holding its constituent particles together. The greater the secondary forces, the higher the temperature required to overcome them and bring about the melting or boiling transition.

◀◀ Secondary forces are the attractive forces between molecules, introduced in Section 6.2.

Greater secondary forces also result in greater densities. Density increases with an increase in the number of molecules packed into a unit volume. In compounds with large secondary forces, the molecules attract one another more strongly and pack more closely together, and thus the density is quite high.

Oil fires are difficult to fight with water because oil has a lower density than water. Water sinks below the oil and has limited effectiveness at shutting off the flames' access to oxygen.

In organic compounds, the strength of secondary forces is determined by three molecular features—family (and hence the type of secondary force), molecular mass, and molecular shape.

Family The type of secondary forces present in an organic compound is related to the compound's family because the family determines what chemical bonds are present and thus whether the bonds are polar or nonpolar and whether there is hydrogen bonding. The order of secondary forces, highest to lowest, is:

Hydrogen bonding > dipole–dipole > London

All molecules possess London forces. This secondary force is the only one present in nonpolar molecules. Polar molecules have dipole–dipole forces as well as London forces, so they possess greater secondary forces than nonpolar molecules do. Hydrogen-bonding forces, the strongest of all secondary forces, are present only in molecules containing O–H or N–H bonds.

Alkanes and cycloalkanes contain only nonpolar bonds (C–C and C–H). Thus, they are nonpolar compounds with only London forces acting between the molecules. (There is a very slight electronegativity difference between carbon and hydrogen, but it is negligible from a practical viewpoint.) Therefore, alkanes and cycloalkanes have the lowest secondary forces of all organic families and the lowest melting and boiling points.

Molecular mass For any series of compounds within the same organic family, secondary forces increase with increasing molecular mass. London forces arise from uneven distributions of electrons within molecules. Larger molecules exert larger London forces because they possess larger numbers of electrons (Section 6.2). In accordance with this generalization, boiling and melting points increase with increasing molecular mass for all families of organic compounds. Tables 11.3 and 11.5 show this trend for straight-chain alkanes and cycloalkanes, respectively.

Molecular shape For compounds within the same organic family and having the same or close to the same molecular mass, the order of boiling points, highest to lowest, is:

Cycloalkane > straight-chain alkane > branched alkane

This trend is seen in the boiling points of the C_4 compounds (see margin).

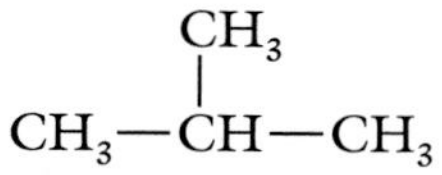

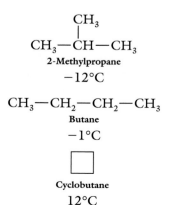

2-Methylpropane

−12°C

$CH_3—CH_2—CH_2—CH_3$

Butane

−1°C

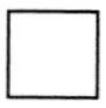

Cyclobutane

12°C

Molecular shape affects the strength of secondary forces because molecular shape affects the surface area over which molecules attract each other, and this surface area determines the effectiveness of attractions between molecules. Some simple geometrical analogies to straight-chain, branched, and cyclic molecules are useful for understanding the boiling-point differences between the differently shaped molecules (Figure 11.10). Straight-chain molecules can be visualized as flexible (jointed) cylinders, something like a string of connected sausage links. Parts of a cylinder can move with respect to one another because of the free rotation about the C–C single bonds. Branched molecules are compact in shape and can be thought of as spheres. Pairs of adjacent cylindrical molecules contact one another over a much larger surface area than pairs of adjacent spherical molecules do (a line compared with a point). Contact over a larger surface area results in greater secondary forces and higher boiling points for the straight-chain alkane than for the branched alkane.

The higher boiling point of the cycloalkane relative to the straight-chain alkane is due to the cycloalkane's rigid structure. The latter can be imagined as a flat, square doughnut. The rigidity of cyclic structures allows the molecules to pack closely together, one on top of the other, achieving contact over a comparatively large surface area. The flexibility of the straight-chain compounds' jointed cylinders prevents those molecules from packing together in a similarly efficient way.

The effect of molecular shape on melting points differs somewhat from its effect on boiling points. Cycloalkanes do have higher melting points than the corresponding straight-chain alkanes, as one might expect from their shape. However, not all branched alkanes have lower melting points than the corresponding straight-chain alkanes. The reasons for this behavior are complex and beyond the scope of this text.

INSIGHT INTO PROPERTIES

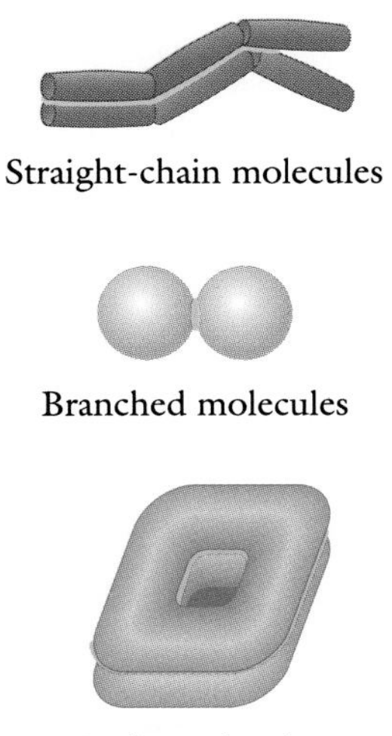

Figure 11.10 Straight-chain, branched, and cyclic molecules differ in their ability to pack closely together.

When compounds differ in more than one structural feature and the differences have opposing effects on secondary forces, we cannot reliably predict the order of increasing secondary forces. Consider compound A, a polar compound of molecular mass 60, and compound B, a nonpolar compound of molecular mass 128. Which has the higher boiling point? The molecular-mass factor favors compound B, but the family factor favors compound A. We are unable to make a reliable prediction.

Concept checklist

- ✔ Family, molecular mass, and molecular shape determine the strength of the secondary forces.
- ✔ The order of increasing secondary forces cannot be reliably predicted for a set of compounds that differ in more than one structural feature when those differences have opposing effects on the secondary forces.

Example 11.11 Predicting relative boiling points of various compounds

Which compound has the higher boiling point in each of the following pairs? Explain each of your answers in regard to the difference in secondary forces in the two compounds. (a) Heptane and hexane; (b) nonane and 3-methyloctane; (c) cyclopentane and pentane.

Solution

(a) Heptane has the higher boiling point because of its larger molecular mass. Larger molecular mass results in greater secondary forces because of the greater number of electrons.

(b) Nonane has the higher boiling point. Branched molecules attract one another over a smaller surface area relative to unbranched molecules of the same molecular mass.

(c) Cyclopentane has the higher boiling point. Its rigid structure results in tighter packing and greater secondary forces among molecules.

Problem 11.11 Which compound has the higher boiling point in each of the following pairs? Explain each of your answers in regard to the difference in secondary forces in the two compounds. (a) Cyclopentane and 2-methylbutane; (b) nonane and 3-methylnonane; (c) cyclobutane and cycloheptane.

Solubility

A solute is soluble in a solvent only if the secondary attractive forces between molecules of solute are similar to the secondary attractive forces between molecules of solvent. When this requirement is met, solute dissolves in solvent to form a solution because the solute molecules form sufficiently strong secondary attractive forces with solvent molecules (Section 6.4). As mentioned in Section 6.4, the requirement for solubility is often summarized by the phrase "like dissolves like."

Concept checklist

- ✔ Nonpolar compounds are soluble in nonpolar compounds.
- ✔ Polar or hydrogen-bonding compounds are soluble in polar or hydrogen-bonding compounds.
- ✔ Nonpolar compounds do not dissolve in polar or hydrogen-bonding compounds.

Example 11.12 Predicting the solubility of compounds

Explain why pentane is not soluble in water.

Solution

Pentane is a nonpolar compound and possesses only weak London forces. Water, in contrast, is highly polar, with very strong hydrogen-bonding forces. The energy needed to separate water molecules from one another is large. None of this energy can be retrieved by mixing, because water and pentane are so different in their types of secondary forces. The attractive secondary forces possible between the hydrogen-bonding water molecule and the nonpolar alkane are negligible.

Problem 11.12 Does sodium chloride (NaCl) dissolve in hexane? Why?

The physical properties considered so far are important both in the uses and in the health hazards associated with alkanes (Box 11.3).

11.10 CHEMICAL PROPERTIES OF ALKANES AND CYCLOALKANES

As we have seen, alkanes and cycloalkanes together constitute the larger organic category of saturated hydrocarbons, compounds possessing only C–C and C–H single bonds. Because of the stability of those bonds, the saturated compounds show low chemical reactivity relative to other families of organic compounds. They are inert to a wide range of reagents, including strong acids (for example, HCl or H_2SO_4), strong bases (for example, NaOH or KOH), reducing agents (such as H_2), and most oxidizing agents (for example, $KMnO_4$ or $K_2Cr_2O_7$). Alkanes and cycloalkanes undergo very few reactions. Each of the other families, in contrast, contains a functional group—specifically, some bond or set of bonds that are relatively weak—that gives rise to higher and more varied reactivity. For example, unsaturated hydrocarbons

11.3 Chemistry Around Us

Health Hazards and Medicinal Uses of Alkanes

Gaseous alkanes (C_1 to C_4) and the vapors from the volatile liquid alkanes (C_5 to C_7) can be harmful because alkanes are highly flammable and have an anesthetic effect. They can also kill by asphyxiation because their presence in the air reduces the concentration of oxygen.

Liquid alkanes also can be harmful if they are swallowed and then coughed or vomited up into the lungs. The alkanes, being nonpolar, dissolve the lipid molecules (also nonpolar) that make up the cell membranes of the air sacks in the lungs, and a condition called chemical pneumonia ensues because the lungs become less efficient at expelling gases and fluids. The term chemical pneumonia refers to a buildup of fluids in the lungs that is similar to that of bacterial or viral pneumonia. Hence, we are warned never to induce vomiting in someone who has swallowed gasoline or other liquid alkanes.

Another prudent warning in the handling of alkanes is to wear gloves when using liquid alkanes such as gasoline, paint thinner, and turpentine. These lower-molecular-mass compounds dissolve the natural body oils that protect our skin.

There are cases in which the similarities of alkanes to human body oils allow them to be put to beneficial use. The higher-molecular-mass alkanes (C_{18} to C_{25}) that make up mineral oil, for example, are quite similar to our natural body oils, enabling mineral oil to be used in hospitals as a laxative. It softens the stool because it passes through the digestive tract without being absorbed into the tissues. Prolonged use is not recommended, however, because the presence of mineral oil in the bowel prevents the absorption of the fat-soluble vitamins A, D, E, and K from foods during the digestive process. Hence, prolonged use of mineral oil depletes the body of these vitamins.

Petroleum jelly (petrolatum), a jellylike semisolid composed of C_{22} to C_{30} alkanes, is useful as both a lubricant and a solvent. It serves as the solvent in a variety of baby lotions, hand lotions, medicated salves, and cosmetics. Mineral oil is used for this purpose as well.

Spray painting an automobile. The worker uses a respirator to prevent inhalation of hydrocarbon solvents and other chemicals present in the paint.

(alkenes, alkynes, and aromatics) are highly reactive because of their carbon–carbon multiple bonds, and they undergo reactions with many of the reagents to which the alkanes and cycloalkanes are unreactive.

In spite of the limited scope of their reactions, alkanes and cycloalkanes are extremely important chemicals—they are the ultimate reason for our high standard of living. Petroleum and natural gas are the sources of these and other hydrocarbons (see Box 11.1). Hydrocarbons from petroleum and natural gas are important not only for heating and energy generation by combustion, but also for the manufacture of other chemicals that are in turn converted into the myriad synthetic materials—medicines, plastics, rubbers, and textiles—that enhance our lives.

Combustion

All hydrocarbons, indeed, all organic compounds, undergo **combustion** (burning in air) (Section 4.7). This reaction distinguishes organic compounds from most inorganic compounds, which do not undergo combustion.

Combustion requires oxygen and results in the breakage of all C–H and C–C bonds in the hydrocarbon. The reaction is exothermic (heat given off) and converts the carbon and hydrogen atoms of the hydrocarbon into carbon dioxide and water, respectively. The balanced equation for the combustion of propane is:

$$\underset{\textbf{Propane}}{CH_3{-}CH_2{-}CH_3} + 5\ O_2 \longrightarrow 3\ CO_2 + 4\ H_2O + \text{heat}$$

Combustion is an example of an **oxidation reaction.** Oxidation of an organic compound is generally defined as an increase in the oxygen content or a decrease in the hydrogen content of a compound or both (Section 4.7). More specifically, we describe oxidation of an organic compound by an increase in the oxidation state of carbon. The greater the **oxidation state** of a carbon atom in a compound, the more bonds there are from that carbon to oxygen atoms or the fewer bonds there are from that carbon to hydrogen and carbon atoms or both. Before combustion, each carbon in propane is in its lowest oxidation state—there are no oxygens bonded to the carbons. Only hydrogens and carbons are bonded to the carbons of propane. After combustion, no hydrogens or carbons are bonded to the carbons. Each carbon ends up in CO_2, where all four of its bonds are to oxygen—the highest oxidation state for a carbon atom.

The combustion of hydrocarbons is of enormous importance to human life. We depend on combustion for energy in the form of heat and for mechanical and electrical power. Heating our homes and cooking by combusting hydrocarbons is efficient because the reaction is highly exothermic, producing about 15,000 kJ per pound of fuel (1 cal = 4.184 J). In engines, the same large heat output results in a pressure buildup that is then converted into mechanical energy. At power plants, the heat of combustion is used to produce steam, which turns the blades of a turbine engine. The mechanical energy of the rotating turbine is converted into electrical energy because the turbine turns a coil positioned within a magnetic field. Some of the environmental consequences of the combustion reaction are discussed in Box 11.4.

The combustion reaction that we have described is **complete combustion.** When there is insufficient oxygen or when combustion conditions (temperature, mixing of hydrocarbon and oxygen) are not optimized, **incomplete combustion** occurs and various amounts of carbon monoxide (CO) are formed instead of carbon dioxide. Thus, some of the carbon atoms in the fuel are not oxidized to their highest oxidation state.

Incomplete combustion is detrimental in two ways. First, less heat is generated per pound of fuel, so heat, power, and electricity become more costly. Second, the resulting carbon monoxide is highly toxic and especially danger-

Petroleum is the source of energy for airplanes and other modes of transportation as well as for lighting our buildings and highways.

11.4 Chemistry Around Us

Greenhouse Effect and Global Warming

Many scientists worry about the potential greenhouse effect of increasing amounts of carbon dioxide in Earth's atmosphere. Earth's temperature directly depends on that blanket of air, which consists mostly of nitrogen, oxygen, and a mixture of several other gases, including carbon dioxide.

Most of the energy that warms our planet originates in the sun. The high-frequency solar energy, traveling through space, penetrates the atmosphere and reaches Earth's surface, where some is used by plants for photosynthesis and some is converted into heat. At the same time, Earth itself radiates low-frequency energy toward space. Not all of this low-frequency energy passes through the atmosphere, however. Much of it is absorbed by gases in the atmosphere and radiated back to Earth's surface.

The twentieth century is characterized by an enormous increase in the combustion of hydrocarbons, with a corresponding increase in the concentration of carbon dioxide (the end product of that combustion) in the atmosphere. Some scientists fear that the increased carbon dioxide will result in ever larger fractions of the low-frequency energy being radiated back to Earth (analogous to the way in which glass traps heat inside a greenhouse), with a variety of disastrous consequences. They warn that global warming could turn farmland into desert and flood coastal cities as a result of the melting of polar ice caps. Other scientists counter that there is no hard evidence of overall global warming, although there may be warming in local regions of Earth.

Another variable in the greenhouse effect is the role of plants. Plants absorb carbon dioxide in the course of photosynthesis. Some scientists suggest that plants are helping to control the extent of global warming, and this factor should cause us to be greatly concerned about the continuing loss of plant life in all regions of the planet.

It will take a considerable period of time to prove conclusively the extent of this problem. Whatever global warming is occurring is not large when measured in degrees per year but may well be large when measured in degrees per generation. Your life may not be significantly affected, but the situation may be very different for your grandchildren and theirs.

ous because it has no odor. Carbon monoxide is selectively and strongly absorbed by hemoglobin; and, even at low concentrations, it interferes with hemoglobin's ability to absorb oxygen in the lungs and deliver it to the tissues. Hence, we are warned never to run an automobile engine in a closed garage.

Throughout the remainder of the text, unless otherwise stated, the term combustion refers to complete combustion.

A reaction parallel to the combustion of saturated hydrocarbons is used by the body for energy generation. The body does not burn saturated hydrocarbons but uses instead the organic foodstuffs that we call carbohydrates, lipids, and proteins. These foodstuffs are oxidized in an indirect way, not by direct reaction with atmospheric oxygen, to produce energy needed by the body. The overall results are similar: the body takes in organic compounds and oxygen and produces energy, carbon dioxide, and water. There is one significant difference—in the body, the oxidation takes place by a complex multistep process that allows the energy to be captured and preserved in the form of new molecules that can perform useful work such as muscle contraction. This process is considered in Chapter 22.

Example 11.13 Writing equations for the combustion of hydrocarbons

Write balanced equations for the combustion of: (a) hexane; (b) 3-methylpentane; (c) 2,3-dimethylbutane.

Solution

Structural formulas are not needed to write combustion equations, because combustion results in complete dissection of the molecule. The molecular formula is sufficient. In this case, the same equation describes the combustion of all three compounds in the problem, because they are all constitutional isomers of C_6H_{14}. The unbalanced equation is:

$$C_6H_{14} + O_2 \longrightarrow CO_2 + H_2O$$

Balance the elements in the order C followed by H followed by O. The 6 C and 14 H on the left side are balanced by 6 CO_2 and 7 H_2O on the right side. This change leads to 19 O on the right side that require $9\frac{1}{2}$ O_2 on the left side. The balanced equation is:

$$C_6H_{14} + 9\tfrac{1}{2}\ O_2 \longrightarrow 6\ CO_2 + 7\ H_2O$$

All coefficients are multiplied by 2 to avoid fractional coefficients:

$$2\ C_6H_{14} + 19\ O_2 \longrightarrow 12\ CO_2 + 14\ H_2O$$

Problem 11.13 Write balanced equations for the combustion of: (a) cyclopentane; (b) 2,2,3-trimethylhexane.

Halogenation (Halogen Substitution)

Molecular halogens (F_2, Cl_2, Br_2) react with alkanes and cycloalkanes, but only in the presence of heat or ultraviolet (UV) light. The reaction, called **halogenation,** substitutes a halogen atom (—F, —Cl, —Br) for a hydrogen atom in the alkane or cycloalkane to form an **alkyl halide.** The term alkyl halide encompasses any organic compound containing one or more halogen atoms. A number of halogen-containing products are important materials of commerce and everyday life (Box 11.5).

Halogenation can be illustrated by the chlorination of methane:

$$CH_4 + Cl_2 \xrightarrow{\text{UV or heat}} CH_3Cl + HCl$$

Organic chemists often write chemical equations in an abbreviated form such as

$$CH_4 \xrightarrow[\text{UV or heat}]{Cl_2} CH_3Cl$$

to emphasize the transformation that takes place in the organic reactant. The reagent (Cl_2) and reaction conditions (UV or heat) required for reaction to take place are written above and below the arrow, respectively. The nonorganic product (HCl) is omitted because it is not important in this context. In studying the properties of organic reactants, we will almost always be interested in the organic product(s) only, not the inorganic product(s). (The combustion reaction is the rare exception.)

Equations can also be written in a generalized form, for the purpose of emphasizing family behavior. In such cases, R is used to represent any alkyl group (methyl, ethyl, propyl, and so forth):

$$R{-}H \xrightarrow[\text{UV or heat}]{Cl_2} R{-}Cl$$

Halogenation usually yields a mixture of products because each C–H bond in an alkane or cycloalkane is susceptible to reaction. Halogenation of methane can proceed sequentially, as follows:

$$CH_4 \longrightarrow CH_3Cl \longrightarrow CH_2Cl_2 \longrightarrow CHCl_3 \longrightarrow CCl_4$$

Replacement of one hydrogen by halogen is called **monohalogenation** and replacement of two or more hydrogens by halogens is called **polyhalogenation.** Monohalogenation predominates if a large amount of methane is present relative to the halogen; polyhalogenation predominates if the halogen is present in excess.

11.5 Chemistry Around Us

Applications of Alkyl Halides and Some of the Problems That They Create

Many kinds of organic compounds possess anesthetic properties, but all too often they have other properties that interfere with their potential use in the control of pain. Cyclopropane (C_3H_6) and chloroform ($CHCl_3$) are two examples. Although they were used as anesthetics in the past, neither is used today. Cyclopropane is quite explosive, and chloroform causes liver damage. One of the most commonly used anesthetics at present is the alkyl halide 2-bromo-2-chloro-1,1,1-trifluoroethane (Halothane). It is nonflammable and nonexplosive and presents minimal health hazards.

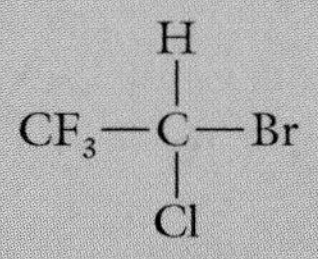

Halothane

Alkyl halides have many other medical applications. Chloroethane (ethyl chloride) serves as a topical cooling agent. Athletes use it during competitions, to temporarily numb the pain from a bruise (see Box 6.3). Physicians apply it before performing minor surgery.

The alkyl halides have everyday uses as well. Tetrachloromethane (carbon tetrachloride) was once used widely as the solvent in dry-cleaning processes, but it caused health problems and has since been replaced by 1,1,2,2-tetrachloroethene (a chlorinated alkene). 1,2-Dichloroethane is used as a spot cleaner (clothes, piano keys). 1,2,3,4,5,6-Hexachlorocyclohexane (Lindane) is used as an insecticide to prevent insect infestation in stored seeds.

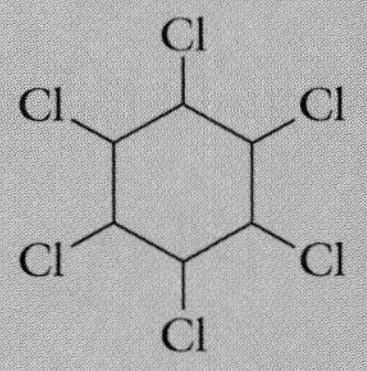

Lindane

The alkyl halide dichlorodifluoromethane (Freon) is one of several compounds that are known as chlorofluorohydrocarbons (CFCs) and used in refrigeration and air-conditioning equipment. CFCs have also been used extensively as the gaseous propellant in aerosol hair sprays and other similar products. Recently, however, worldwide concern about damage to the ozone layer of Earth's atmosphere has greatly curtailed the use of CFCs.

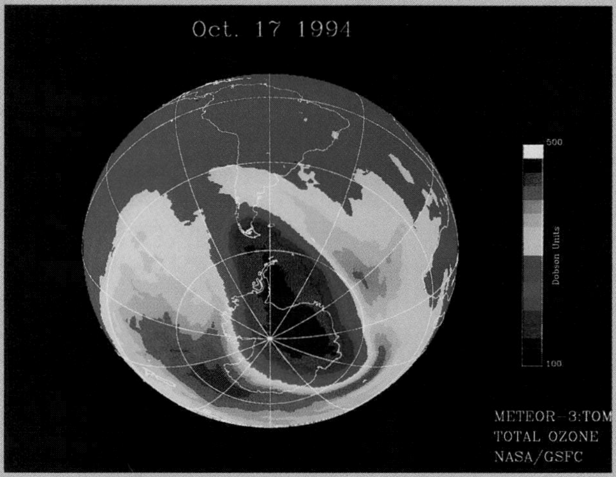

A view of the hole in the ozone layer over Antarctica in 1994. Ozone concentrations are expressed in Dobson Units (higher concentrations are indicated by larger values). The data come from an American Tiros weather satellite.

Ozone (O_3) in the stratosphere, a middle layer of the atmosphere, protects plant and animal life by absorbing much of the cancer-causing ultraviolet radiation from space and preventing it from reaching Earth (see Box 10.2). Scientists have noted significant depletion of the ozone concentration in some parts of the atmosphere, especially over Antarctica, and have identified CFCs as the agent responsible. CFCs escape from Earth and concentrate in the ozone layer, where they participate in a complex set of reactions that convert ozone into oxygen. Oxygen is much less effective than ozone in protecting Earth from ultraviolet radiation. Paul Crutzen, Mario Molina, and Sherwood Rowland received the 1995 Nobel Prize in chemistry for their work on the effect of CFCs on the ozone layer.

Most industrialized nations have adopted guidelines for restricting the use of CFCs. Certain alkanes, alcohols, and other compounds have been substituted as the propellant in aerosol cans. Substitutes are also being developed for use in refrigeration and air conditioning. In the stratosphere, these substitutes do not undergo reactions that harm the ozone layer.

Example 11.14 Writing equations for halogenation of hydrocarbons: I

Write an equation to show the bromination of cyclopentane—that is, the halogenation of cyclopentane with the use of bromine. (In this type of problem, the word halogenation refers to monohalogenation, not polyhalogenation, unless otherwise stated.)

Solution

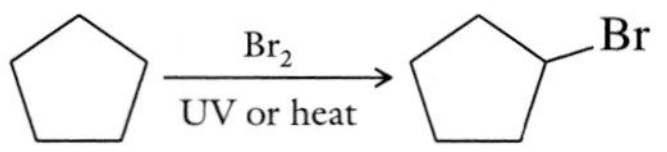

Problem 11.14 Write an equation for the chlorination of ethane.

Monohalogenation of some alkanes and cycloalkanes yields more than one monosubstituted product because substitutions at different C–H bonds yield constitutional isomers. For example, monochlorination of propane yields a mixture of 1-chloropropane and 2-chloropropane:

$$\underset{\textbf{Propane}}{CH_3{-}CH_2{-}CH_3} \xrightarrow[\text{UV or heat}]{Cl_2} \underset{\textbf{1-Chloropropane}}{CH_3{-}CH_2{-}CH_2{-}Cl} + \underset{\textbf{2-Chloropropane}}{CH_3{-}\overset{\overset{\large Cl}{|}}{C}H{-}CH_3}$$

1-Chloropropane is produced when reaction takes place at any one of the six hydrogens at the terminal carbons in propane. Substitution at either of the two hydrogens of the middle carbon yields 2-chloropropane.

Note that the preceding equation for the monochlorination of propane is unbalanced. The left side of the equation has half as many atoms of carbon, hydrogen, and chlorine as the right side. Unbalanced equations are often used to describe organic reactions, for two reasons. First, the major emphasis is on describing the organic reactant(s) used, the reaction conditions needed, and the organic product(s) formed. Second, most organic reactions give less than a 100% yield. For example, in a typical monochlorination of propane, the yields of 1-chloropropane and 2-chloropropane are 39% and 44%, respectively, with the remainder of the product being a mixture of more highly chlorinated compounds. Only the major products, the monochloro products, are usually shown in the equation. The minor (polychloro) products are omitted.

Example 11.15 Writing equations for halogenation of hydrocarbons: II

Write an equation to show the monochlorination of butane.

Solution

$$CH_3{-}CH_2{-}CH_2{-}CH_3 \xrightarrow[\text{UV or heat}]{Cl_2}$$

$$CH_3{-}CH_2{-}CH_2{-}CH_2{-}Cl + CH_3{-}\overset{\overset{\large Cl}{|}}{C}H{-}CH_2{-}CH_3$$

Problem 11.15 Write an equation to show the monobromination of 2-methylpropane.

Naming Alkyl Halides

The IUPAC rules for naming alkyl halides are the same as those for alkanes and cycloalkanes except that they also provide for the identification of the halogens. The presence of F, Cl, Br, and I is signaled by the prefixes **fluoro-,**

chloro-, **bromo-**, and **iodo-**, respectively. Thus, the two products of the monochlorination of propane are 1-chloropropane and 2-chloropropane.

In the common nomenclature system, alkyl halides are named by attaching the halogen prefix to the name of the alkyl group followed by the suffix **-ide** (for **halide,** Section 3.3). The common names of 1-bromopropane and 2-bromopropane are propyl bromide and isopropyl bromide, respectively. This naming procedure is applicable only to alkyl halides with simple alkyl groups.

Example 11.16 Naming alkyl halides

Give the IUPAC and common names for the following alkyl halides:

(a) $CH_3—CH_2—CH_2—CH_2—Cl$

(b) $CH_3—CH(CH_3)—CH(Cl)—CH_2—CH_2—CH_3$

Solution

(a) IUPAC: 1-chlorobutane; common: butyl chloride.

(b) IUPAC: 3-chloro-2-methylhexane. Note that the halogen is alphabetized along with the methyl group. There is no common name, because the alkyl group attached to Cl is not a simple alkyl group.

Problem 11.16 Give the IUPAC and common names for the following compounds:

(a) $CH_3—CH(CH_3)—CH_2—Br$

(b)

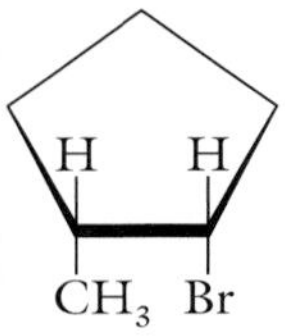

Summary

Molecular and Structural Formulas The molecular formula of a compound indicates the elements present in the compound and the number of atoms of each element. The structural formula shows the details of how various atoms are bonded together in a molecule and must be consistent with the combining powers of each of the atoms.

Families of Organic Compounds Organic chemistry is the chemistry of compounds of carbon—organic compounds. Its study is organized along family lines. A family contains compounds with a common set of physical and chemical properties, the result of possessing the same functional group. A functional group is a particular atom or type of bond or group of atoms in a particular bonding arrangement.

Alkanes Alkanes have the general formula C_nH_{2n+2} and contain only carbon–carbon and carbon–hydrogen single bonds. The carbon atoms in alkanes bond to hydrogens and other carbons through sp^3-hybrid orbitals, which are characterized by tetrahedral (109.5°) bond angles. The large number of organic compounds in each family is a consequence of two factors: (1) that carbon atoms bond to each other and (2) that there can be many more than one constitutional isomer for most organic molecular formulas. Constitutional isomers are compounds with the same molecular formula but different connectivity; that is, the order in which the atoms are attached to each other. Carbon atoms can bond to each other to form branched and unbranched acyclic molecules as well as cyclic molecules. The latter, called cycloalkanes, follow the general formula C_nH_{2n} and behave very much like the alkanes.

Naming Alkanes and Cycloalkanes By the IUPAC system, alkanes are named after the longest continuous carbon chain; prefixes and numbers indicate substituents attached to the longest chain. Cycloalkanes are named in a similar manner except for the addition of the prefix cyclo-.

Physical and Chemical Properties of Alkanes and Cycloalkanes Alkanes and cycloalkanes are nonpolar compounds with weak secondary forces. Their melting and boiling points are low relative to those of other organic families, and they are insoluble in water. Alkanes and cycloalkanes possess low chemical reactivity because their carbon–carbon and carbon–hydrogen single bonds are relatively strong. Halogenation and combustion are the only chemical reactions for alkanes and cycloalkanes. Halogenation is a substitution reaction in which a hydrogen in the alkane or cycloalkane is replaced by a halogen. The reaction takes place with molecular halogen such as Cl_2 or Br_2 and requires ultraviolet radiation or heat. In combustion (burning), an oxidation reaction, alkanes and cycloalkanes react with oxygen to yield carbon dioxide and water.

Summary of Key Reactions

COMBUSTION

$$C,H + O_2 \longrightarrow CO_2 + H_2O + \text{heat}$$

HALOGENATION (X = F, Cl, Br)

$$C{-}H + X_2 \xrightarrow[\text{heat}]{\text{UV or}} C{-}X + HX$$

Key Words

alkane, p. 293
alkyl group, p. 303
branched alkane, p. 301
cis-trans isomers, p. 310
combustion, p. 317
configuration, p. 311
conformation, p. 296
constitutional isomer, p. 300
cycloalkane, p. 308
family, p. 290
functional group, p.290
geometric stereoisomers, p. 310
halogenation, p. 320
hybridization, p. 290
hydrocarbon, p. 292
IUPAC nomenclature system, p. 305
longest continuous chain, p. 301
organic chemistry, p. 287
organic compound, p. 287
stereoisomer, p. 311
structural formula, pp. 288, 299
tetrahedral, p. 295

Exercises

Molecular and Structural Formulas

11.1 Define combining power.

11.2 How many bonds do each of the following atoms form? (a) C; (b) H; (c) N; (d) O; (e) S; (f) Cl; (g) Br; (h) F.

11.3 Which of the following molecular formulas are correct? (a) CH_4; (b) CH_5; (c) CH_3.

11.4 Which of the following molecular formulas are correct? (a) C_2H_5; (b) C_2H_6; (c) C_2H_7.

11.5 Which of the following molecular formulas are correct? (a) C_2H_4Cl; (b) CH_4O; (c) CH_6N.

11.6 Which of the following molecular formulas are correct? (a) $C_4H_{10}Cl$; (b) C_4H_8Cl; (c) C_4H_9Cl.

Organization of Organic Chemistry into Families

11.7 Draw the functional group that characterizes each of the following families (refer to Table 11.2): (a) alkane; (b) carboxylic acid; (c) aldehyde.

11.8 Draw the functional group that characterizes each of the following families (refer to Table 11.2): (a) ester; (b) amide; (c) alkene.

11.9 To what family does each of the following compounds belong? (Refer to Table 11.2.)

(a) $CH_3{-}CH(OH){-}CH_2{-}CH_3$

(b) $CH_2{=}CH{-}CH_3$

(c) $CH_3{-}C({=}O){-}CH_2CH_3$

(d) 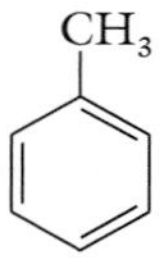

(e) $CH_3{-}C({=}O){-}CH_3$

(f) $CH_3{-}O{-}CH(CH_3){-}CH_3$

(g) $(CH_3)_2CH{-}C({=}O){-}OH$

(h) $CH_3CH_2{-}C({=}O){-}NH_2$

11.10 Some important compounds, especially those with physiological activity, belong simultaneously to more than one family. By referring to Table 11.2, identify the family for each of functional groups a–h in the following compound:

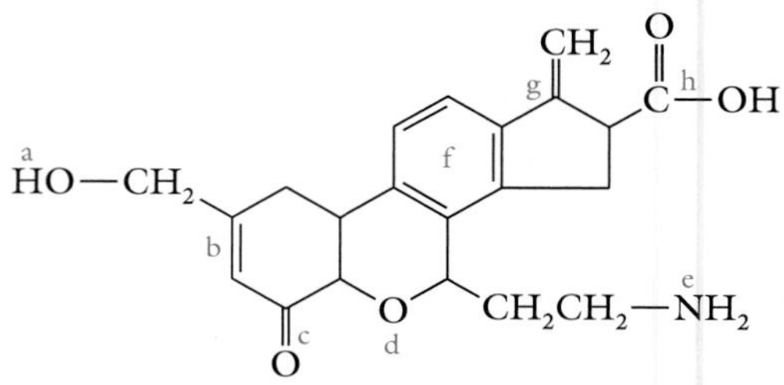

Alkanes

11.11 Why are sp^3 orbitals proposed for the carbon atoms in alkanes?

11.12 What are the values of bond angles A, B, and C in the following compound?

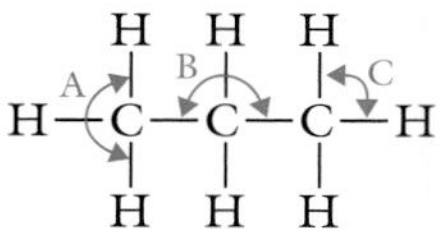

Types of Structural Formulas

11.13 Draw an expanded structural formula for each of the following compounds: (a) $(CH_3)_2CH_2$; (b) $CH_3CH_2CH_2CH_2CH_3$; (c) $CH_3CH_2CH_2CH(CH_3)_2$.

11.14 Draw a skeleton structural formula for each of the compounds in Exercise 11.13.

11.15 Draw a line structural formula for each of the following compounds:

(a) 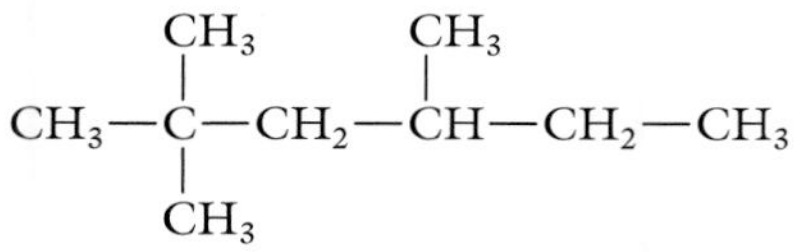

(b) 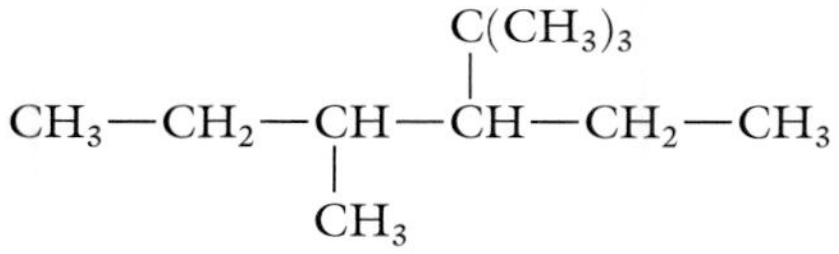

(c) 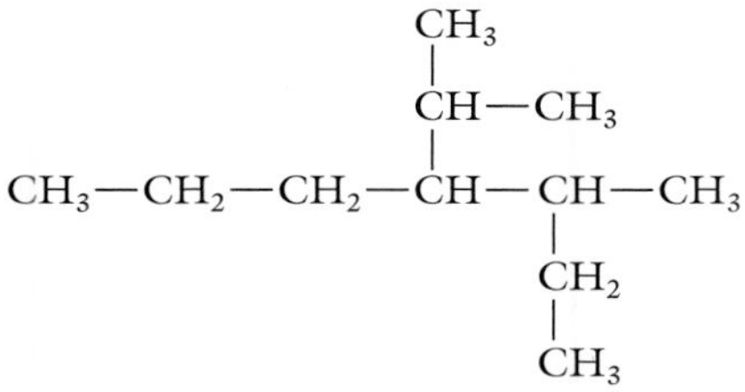

(d) 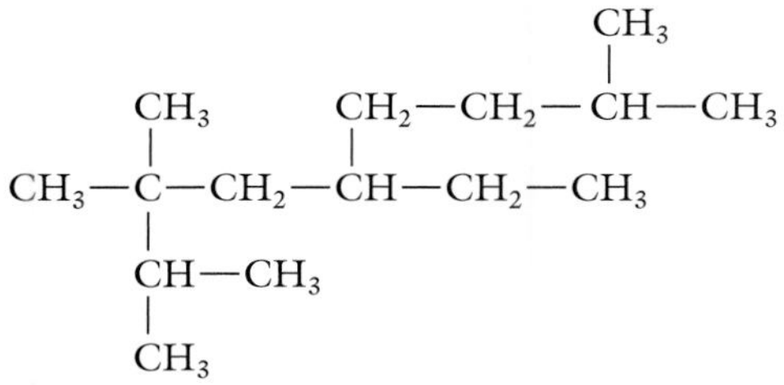

11.16 Draw a condensed structural formula for each of the following compounds:

(a) 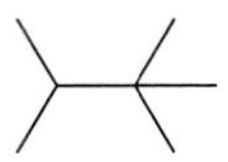 (b) 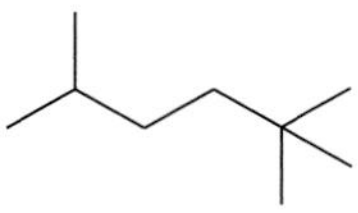

Constitutional Isomers

11.17 For each of the following pairs of structural formulas, indicate whether the pair represents (1) the same compound or (2) different compounds that are constitutional isomers or (3) different compounds that are not isomers.

(Note: Two structural formulas represent the same compound when they portray different conformations of a single compound, when they differ in the extent to which they are condensed or expanded, or when they differ in the extent to which they show the correct bond angles.)

(a) $CH_3CH_2CH(CH_3)_2$ and 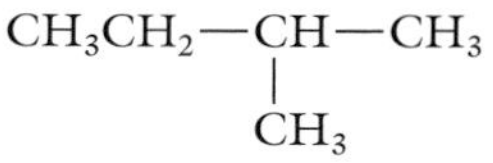

(b) $(CH_3)_3CCH_2{-}CH_3$ and 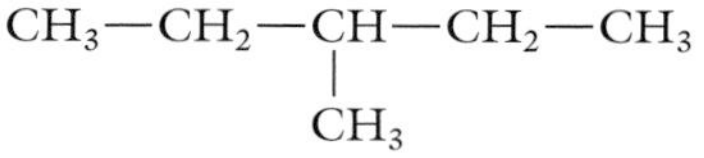

(c) 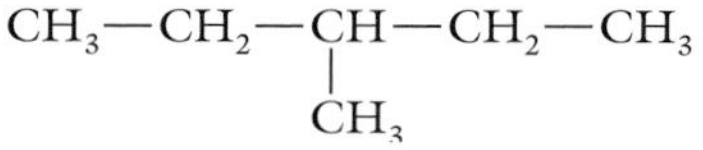 and 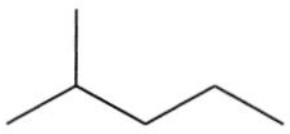

11.18 For each of the following pairs of structural formulas, indicate whether the pair represents (1) the same compound or (2) different compounds that are constitutional isomers or (3) different compounds that are not isomers.

(Note: Two structural formulas represent the same compound when they portray different conformations of a single compound, when they differ in the extent to which they are condensed or expanded, or when they differ in the extent to which they show the correct bond angles.)

(a) 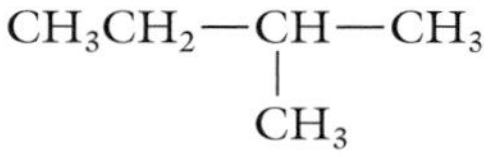 and $(CH_3)_2CH{-}CH_2{-}CH_3$

(b) 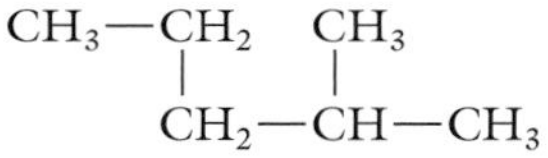 and $(CH_3)_2CHCH_2CH_2CH_3$

(c)  and 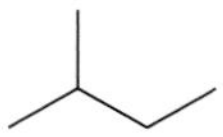

11.19 How many constitutional isomers are there for C_7H_{16}? Show one structural formula for each isomer.

11.20 Draw the structural formula for each of the following compounds: (a) an alkane, C_5H_{12}, that contains three methyl groups; (b) an alkane, C_5H_{12}, that contains four methyl groups; (c) an alkane, C_5H_{12}, that contains two methyl groups.

Naming Alkanes

11.21 Give the IUPAC names for each of the following compounds:

(a) $CH_3CH_2CH(CH_3)_2$ (b) 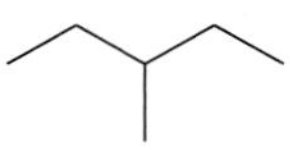

(c) 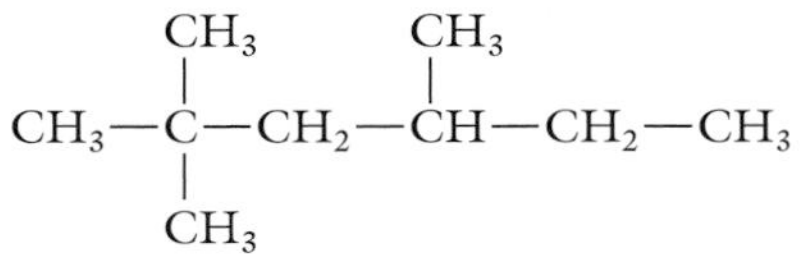

(d) 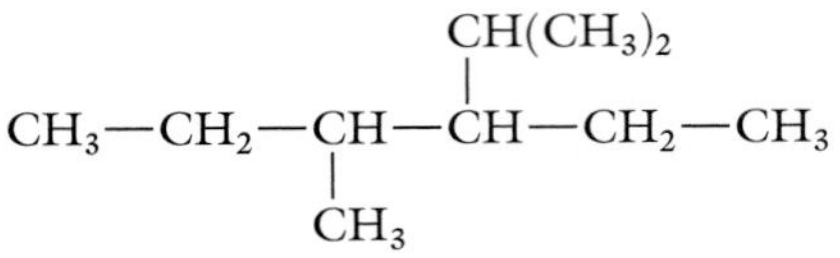

11.22 Give the IUPAC names for each of the following compounds:

(a) 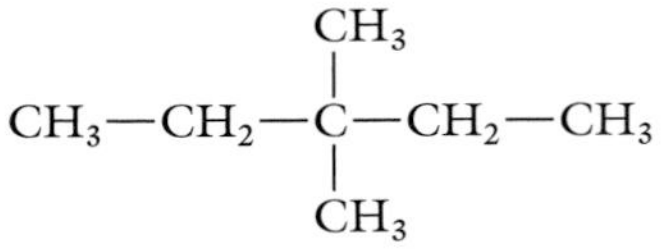

(b) 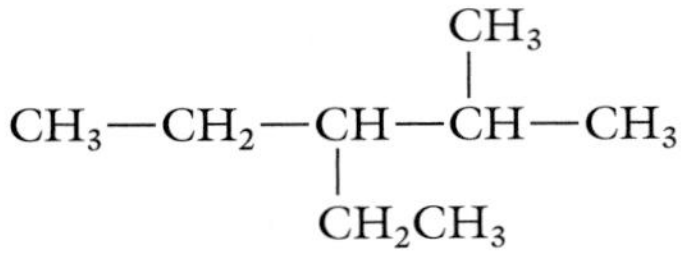

(c) 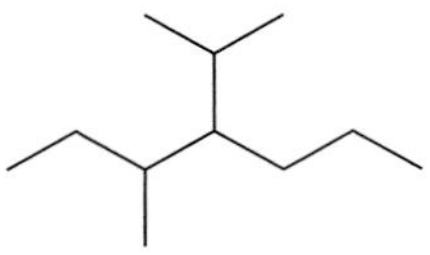

(d) 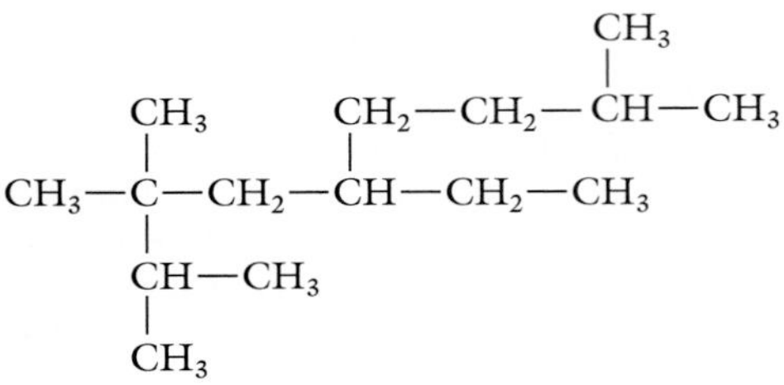

(e) 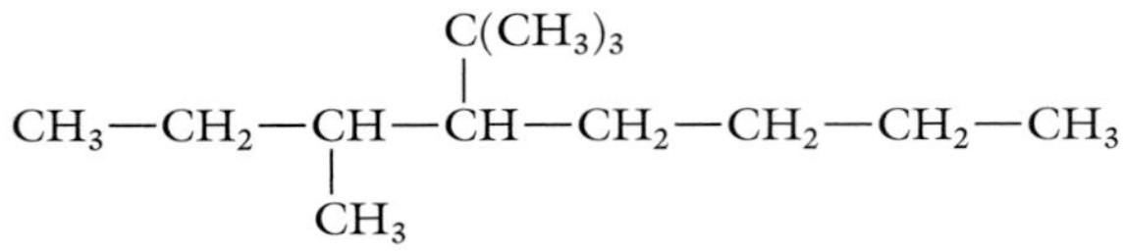

11.23 Draw the structural formula for each of the following compounds: (a) 2-methylbutane; (b) 2,2,4-trimethylhexane; (c) 3-isopropyl-2,3-dimethylhexane; (d) 4-ethyl-4-isopropyl-2,2-dimethyloctane; (e) 4-*s*-butyl-2-methylheptane.

11.24 Identify 1°, 2°, 3°, and 4° carbons in the compounds of Exercise 11.23.

Cycloalkanes

11.25 Draw the structural formula for each of the following compounds: (a) cyclooctane; (b) 4-isopropyl-1,2-dimethylcyclohexane; (c) 1-*t*-butyl-3-ethylcyclohexane.

11.26 Give the IUPAC names for each of the following compounds:

(a) 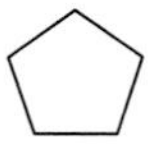 (b) 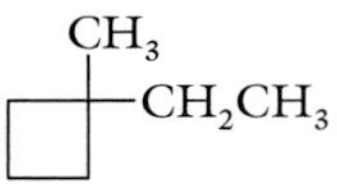 (c) 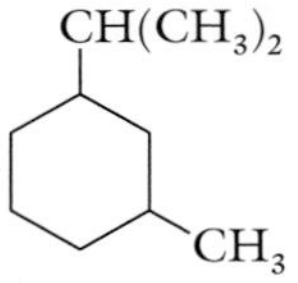

11.27 Identify 1°, 2°, 3°, and 4° carbons in the compounds of Exercise 11.26.

11.28 Draw a line structural formula for each of the following compounds: (a) *t*-butylcyclopropane; (b) 1,2-diethylcyclopentane.

11.29 Draw the structural formula and give the IUPAC name for each of the following compounds: (a) a cycloalkane, C_6H_{12}, that contains one methyl group; (b) a cycloalkane, C_6H_{12}, that contains one ethyl group.

11.30 Draw the structural formula and give the IUPAC names for all cycloalkanes, C_6H_{12}, that contains two methyl groups.

Cis-Trans Isomerism in Cycloalkanes

11.31 Which of the following compounds exist as cis and trans isomers? If cis and trans isomers are possible, show one structural formula for each isomer. (a) 1-Chloro-1-methylcyclobutane; (b) 1-chloro-2-methylcyclohexane; (c) 1-chloro-2,2,3-trimethylcyclohexane.

11.32 For each of the following pairs of structural formulas, indicate whether the pair represents (1) the same compound or (2) different compounds that are constitutional isomers or (3) different compounds that are not isomers or (4) different compounds that are cis-trans isomers.

(a) 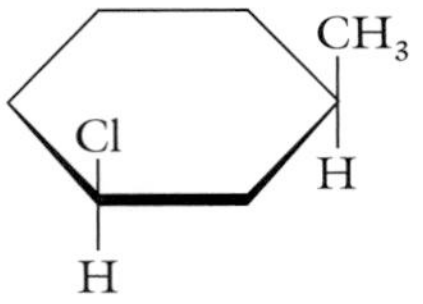 and 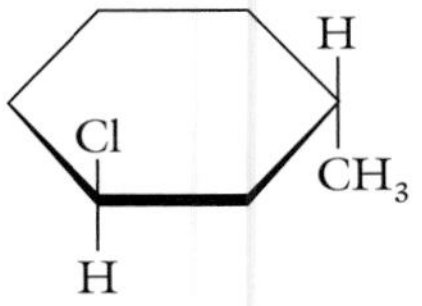

(b) 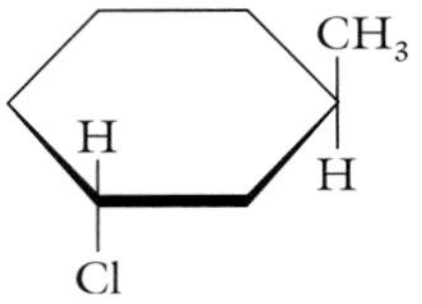 and 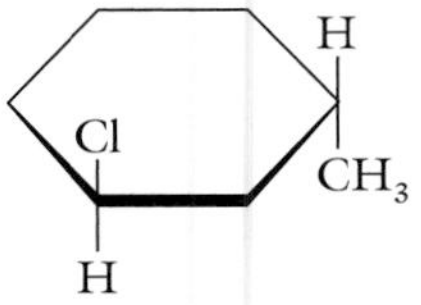

(c) 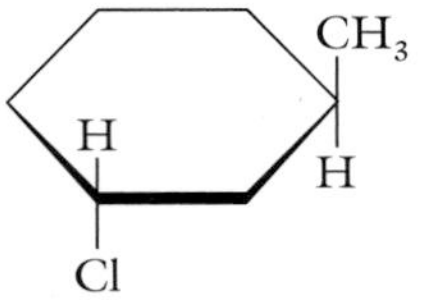 and 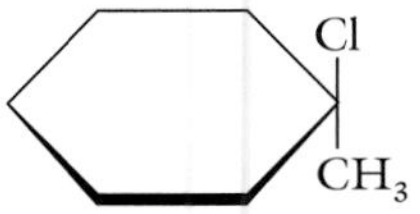

(d) 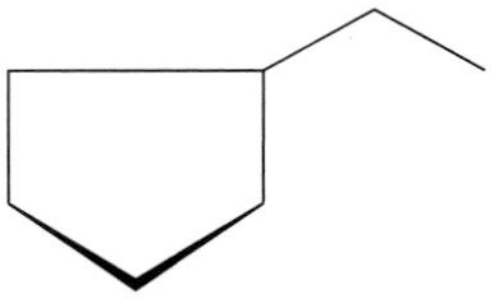 and 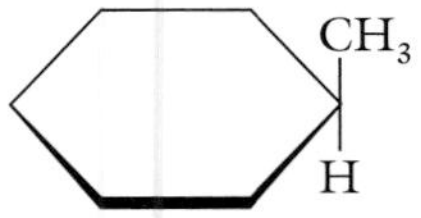

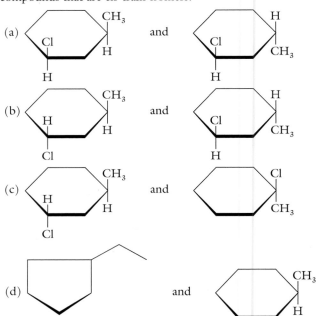

Physical Properties of Alkanes and Cycloalkanes

11.33 For each of the following pairs of compounds, indicate which compound in the pair has the higher boiling point. Explain each of your answers in regard to the difference in secondary forces in the two compounds. (a) Hexane and pentane; (b) cyclohexane and cyclopentane; (c) cyclohexane and methylcyclohexane; (d) hexane and 2-methylpentane; (e) cyclopentane and pentane.

11.34 For each of the following pairs of compounds, indicate whether a solution forms when the two compounds are mixed. Explain each of your answers in regard to the difference in secondary forces in the two compounds. (a) Cyclohexane and hexane; (b) hexane and water; (c) hexane and sodium chloride; (d) sodium chloride and water.

Chemical Properties of Alkanes and Cycloalkanes

11.35 Give the organic product(s) formed in each of the following reactions. If no reaction takes place, write "no reaction." If more than one product is formed, indicate only the major product.

(a) $CH_3—CH_2—CH_2—CH_3 \xrightarrow{H_2SO_4} ?$

(b) 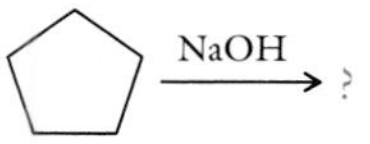

(c) 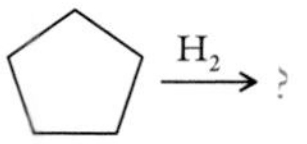

(d) $CH_3—CH_2—CH_2—CH_3 \xrightarrow{HCl} ?$

(e) $CH_3—CH_2—CH_2—CH_3 \xrightarrow[\text{UV or heat}]{Cl_2} ?$

11.36 Indicate whether reaction takes place when cyclohexane is treated with the following reagents. If reaction takes place, give the organic product(s) formed. If more than one product is formed, indicate which is the major product. If no reaction takes place, write "no reaction." (a) NaOH; (b) H_2SO_4; (c) H_2; (d) HCl; (e) Cl_2 (no heat, no UV); (f) Cl_2 (heat or UV).

11.37 Write the balanced equation for the complete combustion of each of the following compounds: (a) pentane; (b) 2-methylbutane; (c) cyclohexane; (d) methylcyclopentane.

11.38 Give the monochlorination product(s) of each of the following compounds: (a) pentane; (b) cyclopentane; (c) methylcyclopentane; (d) 2,2-dimethylpropane.

Unclassified Exercises

11.39 What property of carbon is responsible for the much larger number of organic compounds relative to the number of inorganic compounds?

11.40 Which of the following molecular formulas are correct? (a) C_5H_{10}; (b) C_5H_{14}; (c) C_5H_{12}

11.41 Draw structural formulas for the constitutional isomers of C_4H_9Cl, taking care to draw only one structural formula for each isomer.

11.42 Draw a condensed structural formula for each of the following compounds:

(a) pentane (b) 2-methylpentane

(c) 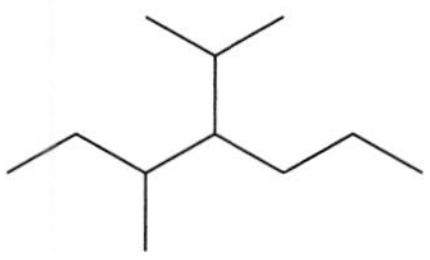 (d) 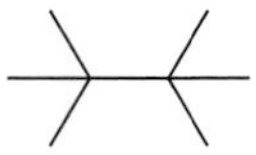

11.43 For each of the following pairs of structural formulas, indicate whether the pair represents (1) the same compound or (2) different compounds that are constitutional isomers or (3) different compounds that are cis-trans isomers or (4) different compounds that are not isomers.

(a) $ClCH_2CH(CH_3)_2$ and $ClCH_2—CH(CH_3)—CH_2Cl$

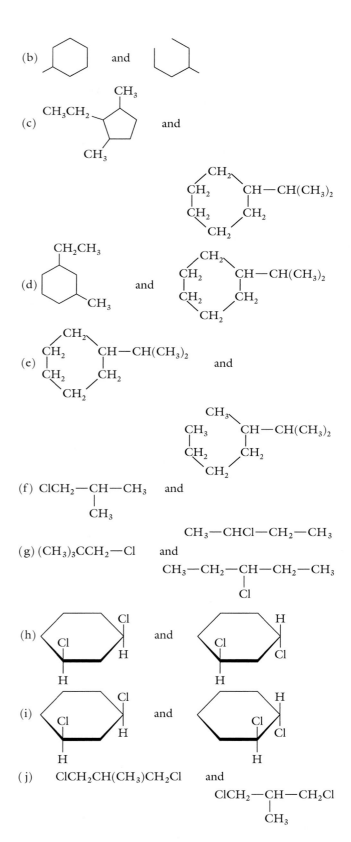

(b) 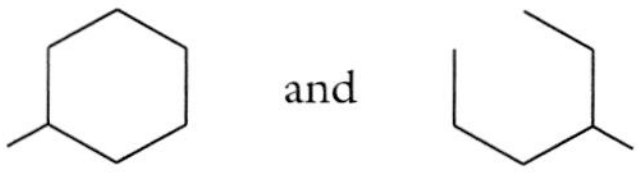

(c) 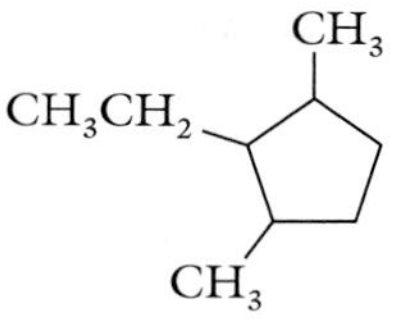

and 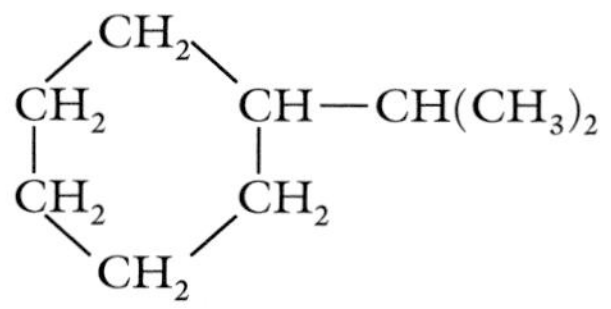

(d) 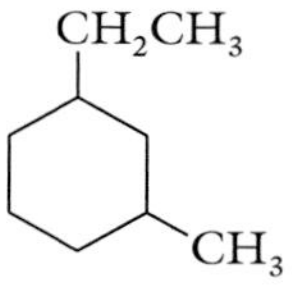

and 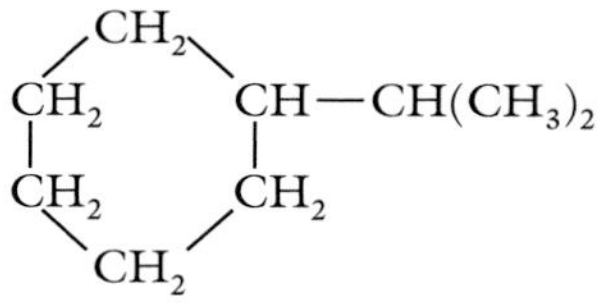

(e) 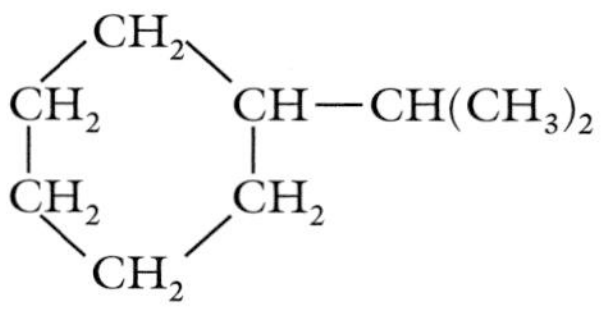

and 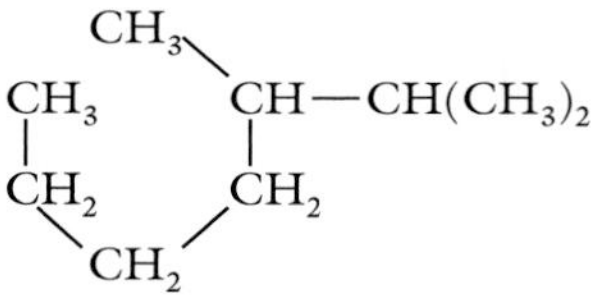

(f) $ClCH_2—CH(CH_3)—CH_3$ and $CH_3—CHCl—CH_2—CH_3$

(g) $(CH_3)_3CCH_2—Cl$ and $CH_3—CH_2—CH(Cl)—CH_2—CH_3$

(h) 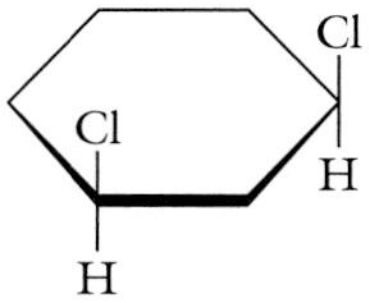

and 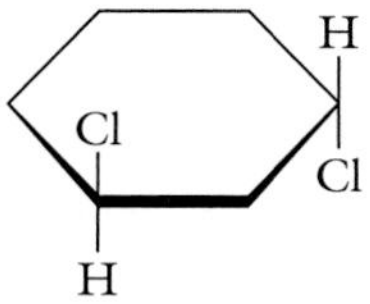

(i) 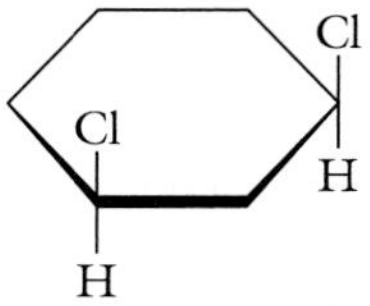

and 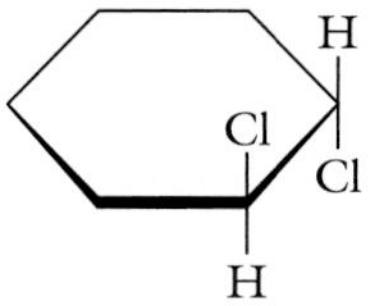

(j) $ClCH_2CH(CH_3)CH_2Cl$ and $ClCH_2—CH(CH_3)—CH_2Cl$

11.44 Each of the following names is an incorrect IUPAC name. State why the name is incorrect. Indicate, if possible, the correct name of the compound. (a) 4-Methylhexane; (b) 2-ethylpentane; (c) 1,3,4-trimethylcyclohexane;

(d) 1-propyl-5-methylcyclopentane;
(e) *cis*-1,1,2-trimethylcyclopentane.

11.45 An unknown compound has the formula C_5H_{10}. Is this a cyclic or acyclic compound?

11.46 An unknown compound has the formula $C_7H_{15}Br$. Is this a cyclic or acyclic compound?

11.47 What is the molecular formula of each of the following compounds?

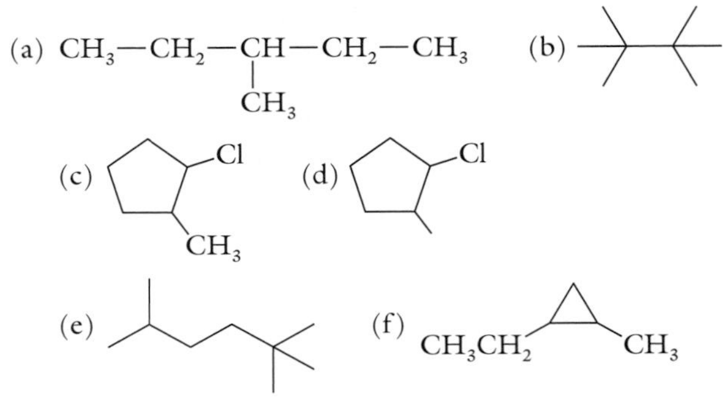

11.48 Indicate whether reaction takes place when ethane is treated with each of the following reagents. If reaction takes place, give the organic product(s) formed. If more than one product is formed, indicate which is the major product. If no reaction takes place, write "no reaction." (a) NaOH; (b) H_2SO_4; (c) H_2; (d) HCl; (e) Cl_2 (no heat, no UV); (f) Cl_2 (heat or UV).

11.49 Write an equation to show the product(s) of each of the following reactions with 2,3-dimethylbutane: (a) combustion; (b) monobromination.

11.50 Give the perchlorination product of each of the following compounds. (In perchlorination, chlorine substitutes for all hydrogens in an alkane or cycloalkane.) (a) Pentane; (b) methylcyclopentane.

11.51 Chlorination of butane yields several dichlorobutane isomers (in addition to monochloro-, trichloro-, and other products). Draw structural formulas of the different dichlorobutane isomers.

11.52 Chlorination of methylcyclopropane yields several dichloromethylcyclopropane isomers (in addition to monochloro-, trichloro-, and other products). Draw structural formulas of the different dichloromethylcyclopropane isomers.

Chemical Connections

11.53 Oil fires are difficult to fight with water because oil, having a lower density than that of water, floats on water. Why do the compounds in oil (mostly alkanes) have lower densities than water does?

11.54 2-Bromo-2-chloro-1,1,1-trifluoroethane (C_2HF_3BrCl) is the commonly used anesthetic Halothane (Box 11.5). How many constitutional isomers are possible for C_2HF_3BrCl?

11.55 Approximately one-half of all synthetic medicines have petroleum as their starting material. How is this possible when petroleum contains only hydrocarbons, whereas medicines contain a variety of oxygen- and nitrogen-containing functional groups such as —OH, $—NH_2$, and —COOH?

11.56 You are driving on an isolated back road when your car runs out of gas. Hours go by without help. It is now time to take your heart medication, for which you need some water. You have a gallon bottle of water in the trunk, but you also have a gallon bottle of gasoline in the trunk and neither bottle is labeled. How can you determine which bottle contains the water and which the gasoline? Gasoline has a distinctive smell but you have a cold and your sense of smell is completely absent this day. All that you have available to help you distinguish between the contents of the two bottles are some empty cups and some salt (NaCl). Describe two different tests that you can perform to determine which bottle contains the water.

11.57 The major components of lipsticks are wax and other materials that are mostly alkanes. Why is lipstick easily removed by petroleum jelly (Box 11.3) but not by water?

11.58 The commonly used topical cooling agent ethyl chloride (Boxes 6.3, 11.5) is produced from ethane. What other reagent(s) and reaction conditions are required to produce ethyl chloride from ethane?

11.59 Methane is the main component of the gas used to heat our homes and for cooking purposes (Box 11.1). The equation for the complete combustion of methane is:

$$CH_4 + 2\ O_2 \longrightarrow CO_2 + 2\ H_2O + 803\ kJ$$

Calculate the heat generated when 160.43 g of methane is burned in a gas heating system or in a gas cooking oven.

CHAPTER 12

UNSATURATED HYDROCARBONS

CHEMISTRY IN YOUR FUTURE

As a farmer and a consumer, you are interested in environmentally safe alternatives to pesticide use. You have studied a number of integrated pest-management programs whose purpose is to reduce pesticide use by combining pesticides with synthetic insect sex pheromones, or attractants. A major challenge to producing pheromones at reasonable costs is that the compounds usually have several isomeric configurations, and only one will work. Many pheromones belong to a family of unsaturated hydrocarbons called alkenes. Chapter 12 considers the chemistry of alkenes and their "configuration" problem.

LEARNING OBJECTIVES

- Describe the bonding in alkenes.
- Draw and name constitutional isomers of alkenes.
- Draw cis-trans (geometric) stereoisomers of alkenes.
- Describe and write equations for addition, polymerization, and oxidation reactions of alkenes.
- Draw and name alkynes.
- Describe the structure and stability of aromatics.
- Draw and name constitutional isomers of aromatics.
- Describe and write equations for substitution and other reactions of aromatics.

We began our exploration of organic chemistry in Chapter 11 with a look at the saturated hydrocarbons, compounds consisting of hydrogen and carbon in which all bonds are single bonds. Now we will examine the unsaturated hydrocarbons—the alkene, alkyne, and aromatic families—compounds consisting of hydrogen and carbon and containing carbon–carbon multiple bonds (double bonds or triple bonds).

The presence of a multiple bond greatly increases the chemical reactivity of an organic compound. As noted in Chapter 11, the reactivity of the saturated hydrocarbons is relatively low. Alkenes and alkynes, on the other hand, are highly reactive, and the aromatic compounds, although generally more stable than the alkenes and alkynes, are also more reactive under certain reaction conditions.

Many naturally occurring molecules contain carbon–carbon multiple bonds. Examples are animal fats and vegetable oils, natural rubber, and some proteins, nucleic acids (DNA, RNA), and vitamins. Molecules containing carbon–carbon double bonds also take part in visual perception and in the oxidation of fatty acids. The medical community's recommendation that we lower our dietary intake of saturated fats relative to unsaturated fats means that we should increase the proportion of fats containing carbon–carbon double bonds, which have less tendency to increase the cholesterol levels in blood. Unsaturated hydrocarbons are also the raw materials used to synthesize a wide variety of industrial and consumer products, including drugs, plastics, synthetic rubbers, dyes, soaps, and cosmetics.

A PICTURE OF HEALTH?

Examples of Alkenes

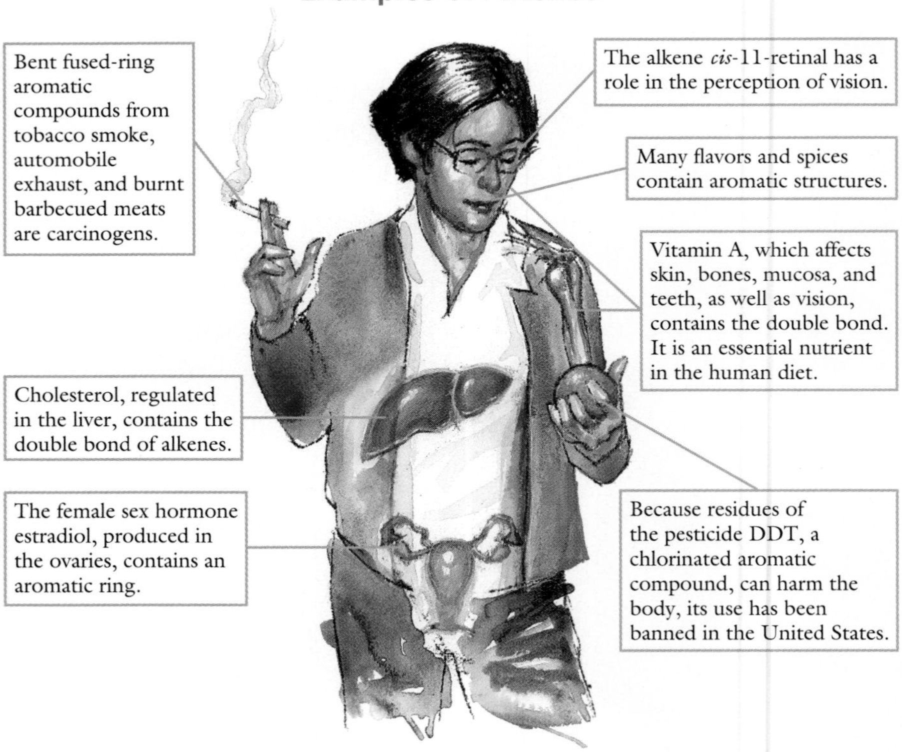

Bent fused-ring aromatic compounds from tobacco smoke, automobile exhaust, and burnt barbecued meats are carcinogens.
The alkene *cis*-11-retinal has a role in the perception of vision.
Many flavors and spices contain aromatic structures.
Vitamin A, which affects skin, bones, mucosa, and teeth, as well as vision, contains the double bond. It is an essential nutrient in the human diet.
Cholesterol, regulated in the liver, contains the double bond of alkenes.
The female sex hormone estradiol, produced in the ovaries, contains an aromatic ring.
Because residues of the pesticide DDT, a chlorinated aromatic compound, can harm the body, its use has been banned in the United States.

12.1 ALKENES

Alkenes are unsaturated hydrocarbons containing a carbon–carbon double bond; that is, two adjacent carbon atoms joined together with two bonds. The general formula for an alkene is C_nH_{2n}, where n is an integer greater than 1. The first (smallest) member of the alkene family is C_2H_4 (IUPAC name, ethene; common name, ethylene). Table 12.1 shows the first nine members of the alkene family and their melting and boiling points. The first three members are gases at ambient temperatures; the others are liquids.

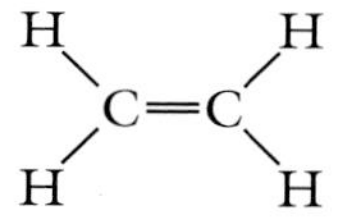

Ethene

The alkenes are nonpolar compounds with physical properties similar to those of the alkanes. They are insoluble in water and have boiling points close to those of the alkanes. There is more variation in melting points among the alkenes than among the alkanes; the reasons are complex and will be considered in Section 19.2, which describes the differences between saturated and unsaturated fatty acids.

▶▶ Fatty acids, the components of fats and oils, are the subject of Chapter 19.

The general formula for an alkene, C_nH_{2n}, is the same as that for a cycloalkene. Both alkenes and cycloalkanes contain two fewer hydrogens than do alkanes, whose general formula is C_nH_{2n+2}. We can visualize the formation of an alkene as the loss by an alkane of one hydrogen from each of two adjacent carbons, which then complete their octets by forming a carbon–carbon double bond. The formation of a cycloalkane is visualized in a similar manner, except that the hydrogens are lost from (and the new bond formed between) two carbons that were nonadjacent.

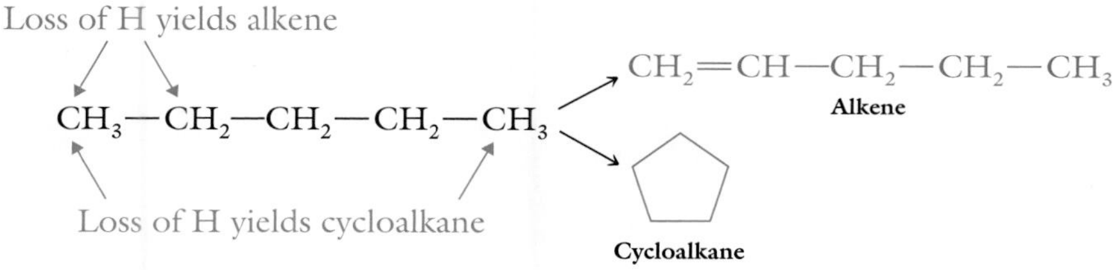

Example 12.1 Distinguishing among alkanes, cycloalkanes, and alkenes

An unknown compound has the molecular formula C_7H_{14}. Is the compound an alkane, cycloalkane, or alkene?

TABLE 12.1 Formulas and Properties of 1-Alkenes

n	Molecular Formula	Condensed Structural Formula	Name	Melting Point (°C)	Boiling Point (°C)
2	C_2H_4	$CH_2{=}CH_2$	ethene	−169	−104
3	C_3H_6	$CH_2{=}CHCH_3$	propene	−185	−47
4	C_4H_8	$CH_2{=}CHCH_2CH_3$	1-butene	−185	−6
5	C_5H_{10}	$CH_2{=}CHCH_2CH_2CH_3$	1-pentene	−138	30
6	C_6H_{12}	$CH_2{=}CHCH_2CH_2CH_2CH_3$	1-hexene	−140	63
7	C_7H_{14}	$CH_2{=}CHCH_2CH_2CH_2CH_2CH_3$	1-heptene	−119	94
8	C_8H_{16}	$CH_2{=}CHCH_2CH_2CH_2CH_2CH_2CH_3$	1-octene	−102	121
9	C_9H_{18}	$CH_2{=}CHCH_2CH_2CH_2CH_2CH_2CH_2CH_3$	1-nonene	−81	146
10	$C_{10}H_{20}$	$CH_2{=}CHCH_2CH_2CH_2CH_2CH_2CH_2CH_2CH_3$	1-decene	−66	171

Solution

This molecular formula does not fit the general formula for an alkane, C_nH_{2n+2}, but it does fit the general formula for both alkenes and cycloalkanes, C_nH_{2n}, with $n = 7$ in this case. The unknown compound is therefore either an alkene or a cycloalkane.

Problem 12.1 An unknown compound has the molecular formula C_7H_{16}. Is the compound an alkane, a cycloalkane, or an alkene?

Because alkenes and cycloalkanes follow the same general formula, the molecular formula alone does not allow us to distinguish between them. We need to know more about a compound, such as its chemical properties, to distinguish between the two possibilities.

Box 12.1 describes some of the alkenes found in nature.

12.2 BONDING IN ALKENES

Typical structural formulas of alkenes do not differentiate between the two bonds of a carbon–carbon double bond. The two lines of the double bond are

12.1 Chemistry Around Us

Alkenes in Nature

Ethene, the simplest alkene, is produced as a hormone by plants such as bananas and tomatoes. Its function is to trigger the ripening process of the fruits. In the commercial fruit industry, fruit is picked before it ripens because fruit that is unripe and hard is much less damaged during shipment. On arrival at its destination, the fruit is artificially ripened by exposure to ethene.

Geraniol is extracted from roses and used in perfume formulations.

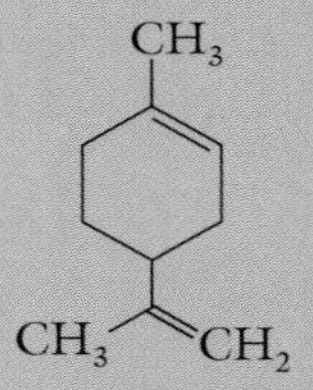

Limonene

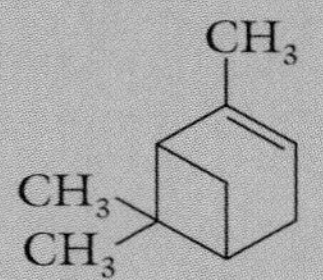

α-Pinene

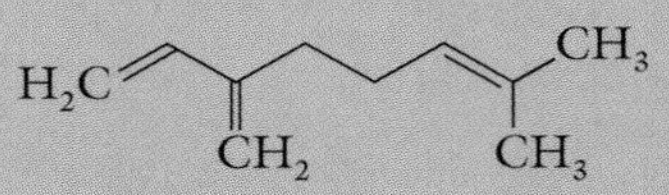

Myrcene

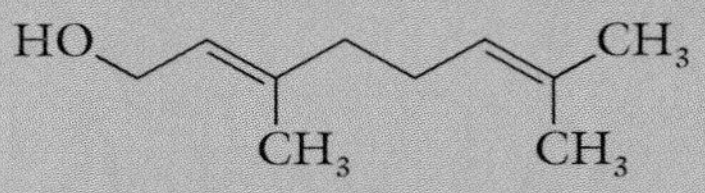

Geraniol

Other alkenes are responsible for the distinguishing odors of some plants. Limonene, myracene, and α-pinene, for example, are found in the oils of citrus fruits, bay leaves, and pine trees, respectively. Geraniol is extracted from oil of roses and used to produce perfumes.

Vitamin A, another alkene-containing compound, is required for good vision. Humans either obtain it directly from animal sources or produce it within the body from β-carotene. Nutritional sources of β-carotene include carrots, spinach, and sweet potatoes.

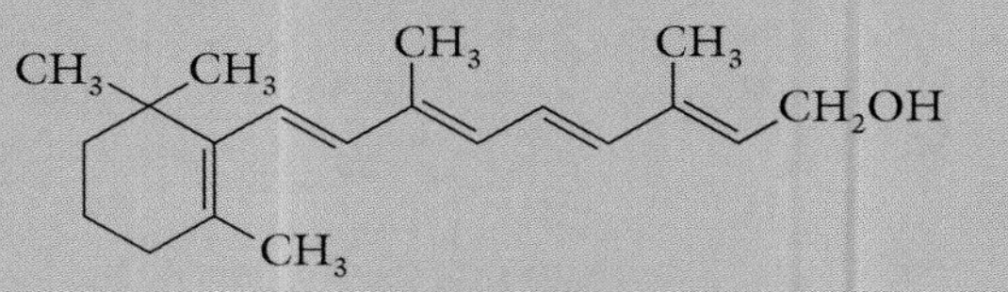

Vitamin A

Note that some of these molecules belong to more than one family: they contain two or more different functional groups. This multifunctionality is typical of many physiologically important molecules. Geraniol and vitamin A contain not only the carbon–carbon double bond of the alkene family, but also the —OH group of the alcohols.

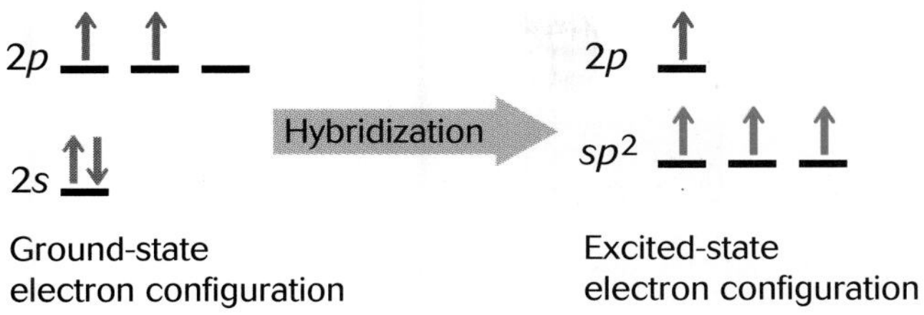

Figure 12.1 Formation of sp^2 orbitals for carbon–carbon double bonds.

drawn with equal weight and length. Note that this representation should not be taken to mean that the two bonds are equal in strength. In fact, the two bonds of the double bond in alkenes are not equivalent and have considerably different bond strengths. One bond is strong; about 355 kJ/mol (85 kcal/mol) are required to break it. The other bond is much weaker; its energy is about 250 kJ/mol (60 kcal/mol). The strong and weak bonds are called **sigma (σ)** and **pi (π) bonds,** respectively. The weak π bond is responsible for the characteristic behavior of alkenes—their ability to undergo chemical reactions with a range of chemical reagents that have no effect on alkanes. The strong σ bond is similar in strength to the bonds in alkanes, which also are called σ bonds.

The atomic orbitals used by carbon atoms to form the carbon–carbon double bond are not the sp^3 orbitals seen in the single bonds of alkanes. True, the carbon atoms of a double bond are tetravalent, as are the carbon atoms in alkanes; but the four orbitals of a double-bonded carbon are not all equivalent, so the bonds that they form have different strengths.

In a carbon–carbon double bond, only three of the four orbitals of each carbon (and the σ bonds that they form) are equivalent; and one of those three equivalent orbitals is used to join one carbon to another in the formation of the strong σ bond of the double bond. The fourth orbital is different from the other three and is used to form the weak π bond of the carbon–carbon double bond.

In Chapter 11, we saw how carbon uses hybridized sp^3 orbitals in the formation of alkanes. The orbitals used in the formation of carbon–carbon double bonds, in contrast, are known as sp^2 orbitals. The sp^2 orbitals, like the sp^3 orbitals, form through an excitation process in which the atom's ground-state orbitals become hybridized. To form a carbon–carbon double bond, each of carbon's $2s$ orbitals and two of its $2p$ orbitals hybridize (mix) to form three equivalent sp^2 orbitals, each containing one electron (Figure 12.1). One $2p$ orbital, containing one electron, is left unhybridized.

The spatial arrangement of the hybridized orbitals around the nucleus of the carbon atom is seen on the far left in Figure 12.2. The three sp^2 orbitals lie in a flat plane at close to 120° bond angles, called the **trigonal bond angles.** The unhybridized $2p$ orbital is perpendicular to the plane of the sp^2 orbitals. This spatial arrangement of one $2p$ and three sp^2 orbitals fulfills the requirements of VSEPR theory for minimizing repulsions between electrons in those orbitals.

Figure 12.2 shows the bonding in ethene, C_2H_4, formed from two sp^2-hybridized carbon atoms and four hydrogen atoms. The strong σ bond of the

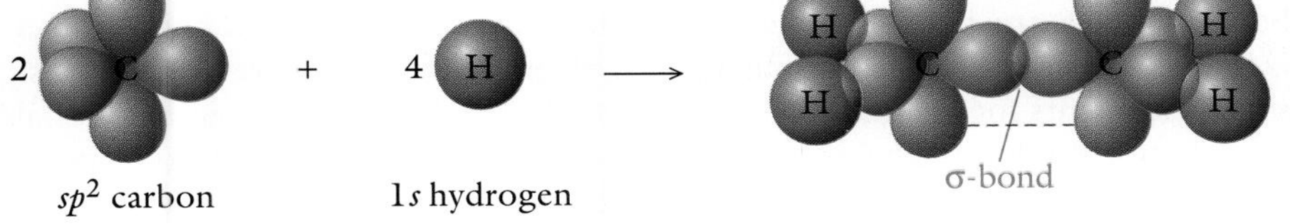

Figure 12.2 Double-bond formation in ethene.

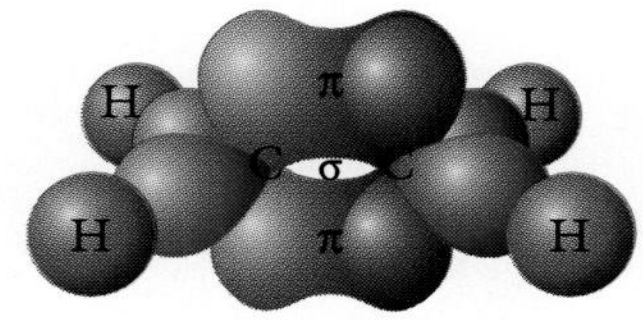

Figure 12.3 Pi-bond formation in ethene.

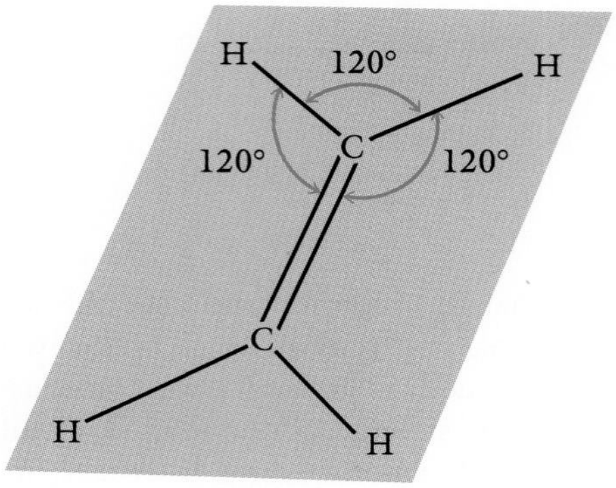

Figure 12.4 Trigonal bond angles of sp^2 carbons in the alkene double bond.

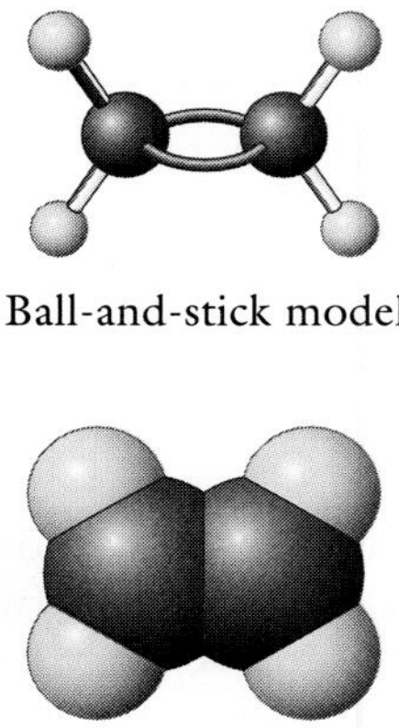

Figure 12.5 Ball-and-stick and space-filling models of ethene.

double bond is formed by overlap of one sp^2 orbital from each of the two carbons. Each of the remaining two sp^2 orbitals of each carbon forms a σ bond by overlap with the $1s$ orbital of a hydrogen atom.

Lateral (sideward) overlap between the $2p$ orbital from each of the carbons results in the π bond of the double bond. The π bond is weak relative to the σ bond because the location of the $2p$ orbitals on each atom does not allow a high degree of overlap between them. It is even difficult to show the lateral overlap of $2p$ orbitals pictorially. Dotted lines are used to indicate the π bond in Figure 12.2, and an alternate representation—one that deemphasizes the σ bond to show the π bond—is shown in Figure 12.3.

The two carbons of the double bond and the atoms to which they are attached all lie in the same plane (Figure 12.4). Note the 120° bond angles about the carbons of the double bond. The planar geometry of the atoms constituting and attached to the double bond is also evident in ball-and-stick and space-filling models (Figure 12.5).

12.3 CONSTITUTIONAL ISOMERS OF ALKENES

More constitutional isomers are possible for alkenes and other families than for alkanes because alkanes possess only one type of connectivity, whereas other families possess two:

- Different carbon skeletons are possible (the one option also open to alkanes).
- Different placements of the functional group (double bond for alkenes) along the carbon skeleton are possible (for all families except the alkane family).

Although there are only two C_4H_{10} alkanes, there are three C_4H_8 alkenes (Figure 12.6). In both the C_4 alkanes and the C_4 alkenes, the same two carbon skeletons are possible: one unbranched and one branched four-carbon skeleton. In regard to the unbranched C_4 alkene, however, there are two possible placements for the carbon–carbon double bond and, hence, an additional constitutional isomer. The double bond is between the first two carbons for 1-butene and between the middle two carbons for 2-butene. In alkenes with

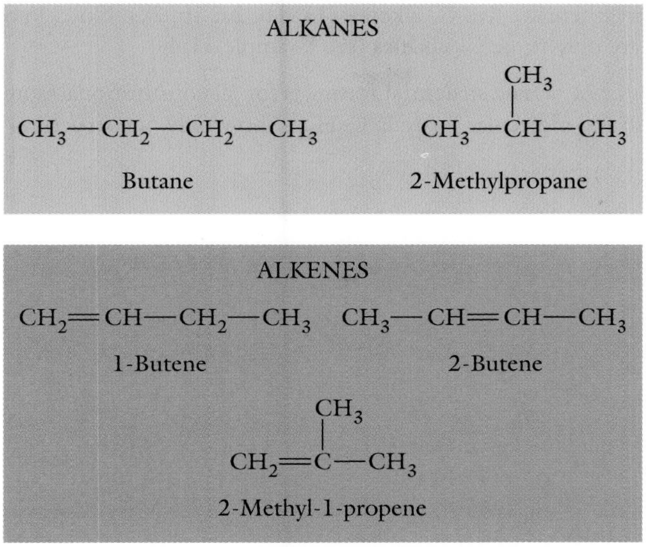

Figure 12.6 Structural isomers of C_4H_{10} and C_4H_8.

more than four carbons, both branched and unbranched, even more options exist for the location of the double bond.

Example 12.2 Drawing constitutional isomers

Draw structural formulas for all constitutional isomers of alkenes with the molecular formula C_5H_{10}. Be careful to show only one structural formula for each isomer.

Solution

There are two steps in solving this problem. First, draw the different carbon skeletons that are possible for branched and unbranched acyclic C_5 compounds:

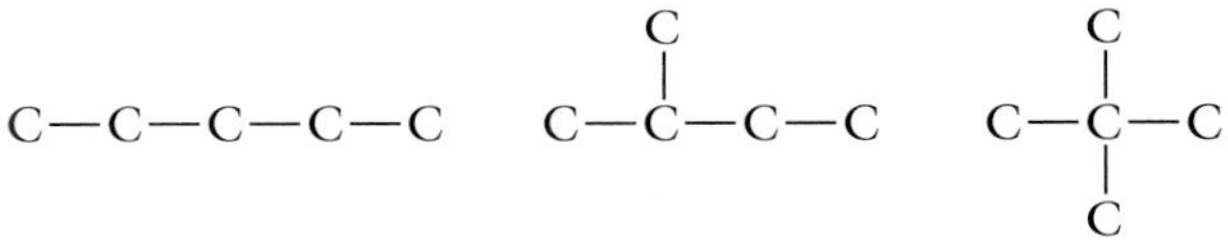

1 2 3

Second, consider each skeleton separately, looking for different possible locations for the double bond within that skeleton.

There are two C_5H_{10} alkenes with skeleton 1. The double bond can be located between the first two carbons of the chain (structure 1a) or between the second and third carbons (structure 1b):

$$\underset{\textbf{1a}}{CH_2{=}CH{-}CH_2{-}CH_2{-}CH_3} \qquad \underset{\textbf{1b}}{CH_3{-}CH{=}CH{-}CH_2{-}CH_3}$$

Do not draw duplicate structures. For example, placing the double bond between the first two carbons counting from the right side gives the same isomer as 1a. Placing the double bond between C2 and C3, counting from the right side, gives the same isomer as 1b.

Three constitutional isomers of C_5H_{10} have skeleton 2:

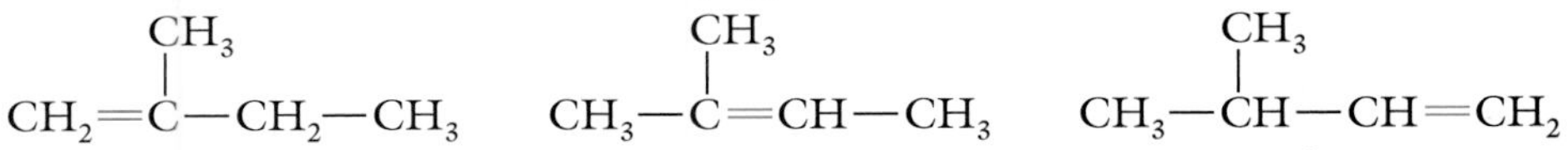

2a 2b 2c

No alkenes are possible for skeleton 3. Placing a double bond in skeleton 3 would result in a total of five bonds to the central carbon—an impossible arrangement because carbon is tetravalent.

In short, there are five C_5 alkenes (1a, 1b, 2a, 2b, 2c). Note, in comparison, that there are only three C_5 alkanes (see Example 11.4).

Problem 12.2 Draw structural formulas for all constitutional isomers of alkenes with the molecular formula C_6H_{12}. Show only one structural formula for each isomer.

12.4 NAMING ALKENES

In the IUPAC nomenclature system, alkenes are named according to the same rules used for naming alkanes, but with a few adjustments:

Rules for naming alkenes

- For nomenclature purposes, the longest continuous chain is redefined as the longest continuous chain containing both carbons of the double bond. For some alkenes, this chain may not be the longest continuous chain of carbons in the molecule.
- The family ending of the name is changed from **-ane** to **-ene.**
- The longest continuous chain is numbered from the end nearest the double bond. The position of the double bond is indicated by placing the lower of the two numbers for the double-bond carbons in front of the base name of the compound. (For ethene and propene, the number 1- is not required to indicate the position of the double bond because there is only one possible placement of the double bond.)
- The names of substituents are used as prefixes, preceded by numbers that indicate their positions on the longest chain.
- A compound with two double bonds, called a **diene,** is given the ending **-adiene** instead of **-ene.** Two numbers are then provided, specifying the position of each double bond.
- A cyclic compound with a double bond in the ring is named as a **cycloalkene.** Numbering of the ring starts at one of the carbons of the double bond and proceeds through the other carbon of the double bond to substituents on the ring in the direction that yields the lower numbers for the substituents.

Example 12.3 Using the IUPAC nomenclature system to name alkenes

What is the IUPAC name for each of the following compounds?

(a) $CH_3—CH_2—\underset{\underset{CH_2}{\|}}{C}—CH_2—CH_2—CH_3$

(b) 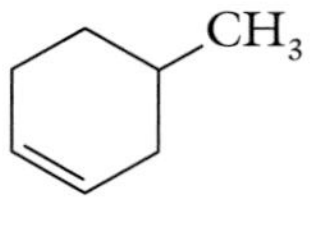

(c) $CH_2{=}CH—CH{=}CH—CH_3$

Solution

(a) The longest continuous chain of all possible chains of carbon atoms in this molecule contains six carbons:

$$\overset{1}{CH_3}—\overset{2}{CH_2}—\overset{3}{\underset{\underset{CH_2}{\|}}{C}}—\overset{4}{CH_2}—\overset{5}{CH_2}—\overset{6}{CH_3}$$

However, this is not the longest chain for nomenclature purposes, because it contains only one of the carbons of the double bond. The longest chain for nomenclature purposes is the five-carbon chain that contains both carbons of the double bond. This chain yields pentene as the base name of the compound.

There are two possible directions for numbering this chain:

$$\underset{}{CH_3-CH_2-\overset{2}{\underset{\underset{\underset{1}{CH_2}}{\|}}{C}}-\overset{3}{CH_2}-\overset{4}{CH_2}-\overset{5}{CH_3}} \qquad CH_3-\overset{4}{CH_2}-\overset{3}{\underset{\underset{\underset{5}{CH_2}}{\|}}{C}}-\overset{2}{CH_2}-\overset{1}{CH_2}-CH_3$$

The correct direction is the one shown on the left-hand side because it yields the lowest numbers (1,2 versus 4,5) for the carbons of the double bond. The lower of the two numbers is used in front of the base name: 1-pentene.

The complete name must also indicate the position of the ethyl group attached to the 1-pentene chain. Thus, the name of the compound is 2-ethyl-1-pentene.

(b) The numbering system for a cycloalkene always places the double bond between C1 and C2 of the ring. Thus, cycloalkene is automatically a 1-cycloalkene; we do not use the number "1." To find the position of the methyl group, we need to count through the two carbons of the double bond and toward the methyl group in the direction that assigns the lowest number to the methyl group. If we number in the counterclockwise direction, the methyl group is at C4. Numbering in the clockwise direction places the methyl group at C5. The lower alternative is the correct one, and the compound is named 4-methylcyclohexene.

(c) With two double bonds, this compound is a diene. Numbering from left to right yields the set of lower numbers for the positions of the double bonds (1,3- versus 2,4-). The compound is named 1,3-pentadiene.

Problem 12.3 What is the IUPAC name for each of the following compounds?

(a) $CH_3-\underset{}{\overset{\overset{CH_2CH_3}{|}}{CH}}-CH_2-CH{=}CH_2$ (b) $CH_3-\overset{\overset{CH_3}{|}}{CH}-CH{=}CH-CH_2Cl$

(c) 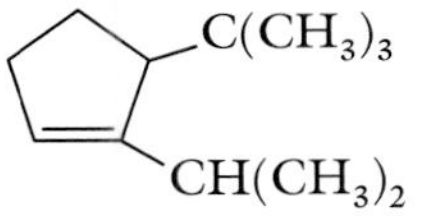

(d) $CH_2{=}CH-CH_2-\overset{\overset{CH_3}{|}}{C}{=}CH_2$

Alkenes containing as many as four carbons are more often known by common names instead of IUPAC names. Common names, in usage before development of the IUPAC system, employ the ending **-ylene** instead of **-ene,** as in ethylene and propylene instead of ethene and propene, respectively. Ethylene and propylene are important industrial chemicals. More than 40 billion pounds of ethylene and 20 billion pounds of propylene are used each year in the United States, mostly to produce the plastics polyethylene and polypropylene (see Section 12.7).

12.5 CIS-TRANS STEREOISOMERISM IN ALKENES

◀◀ **Single bonds in acyclic compounds have free rotation, as described in Section 11.3.**

The carbon–carbon double bond, unlike the single bond in acyclic compounds, has **restricted rotation** (Section 11.8). In other words, the carbon atoms of the double bond are not free to rotate relative to each other. Certain alkenes possess geometric, or cis-trans, stereoisomerism because of this restricted rotation. For example, there are two 2-butene compounds, not one as implied in Section 12.3; they are cis-trans **stereoisomers** that are distinguished

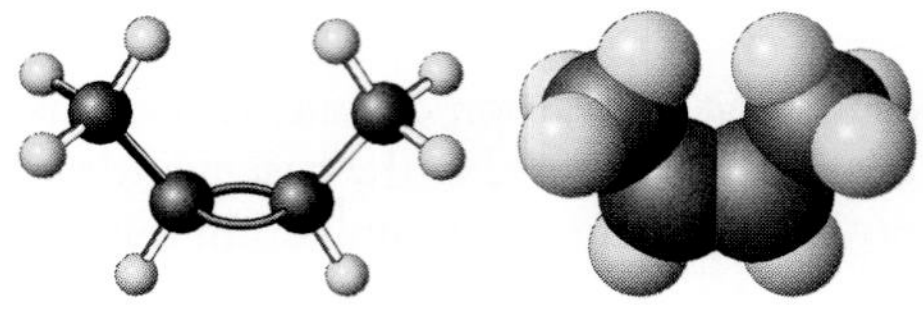

cis-2-Butene

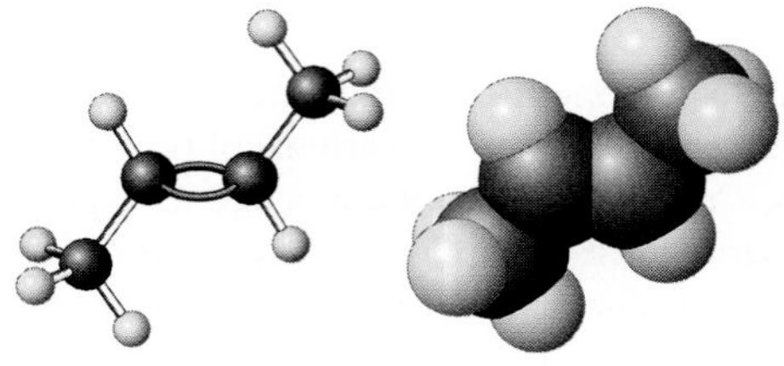

trans-2-Butene

Figure 12.7 Ball-and-stick and space-filling models of 2-butene. The cis and trans isomers are different compounds with different chemical and physical properties.

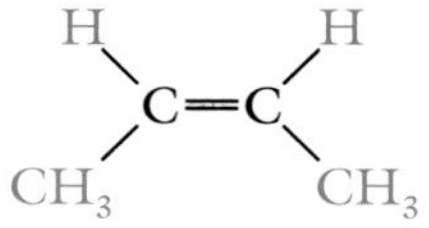

***cis*-2-Butene**

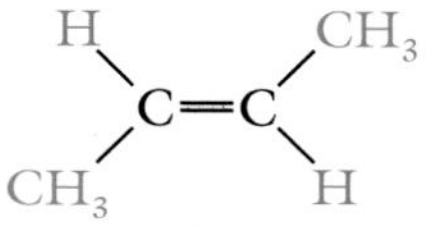

***trans*-2-Butene**

one from the other by the italicized prefixes *cis*- and *trans*- before the IUPAC name.

Cis-trans stereoisomerism in alkenes is analogous to cis-trans stereoisomerism in cycloalkanes (Section 11.8). In cis-trans steroisomers of cycloalkanes, the substituents on different carbons of a ring can be located either on the same or on opposite sides of the plane of the ring. In cis-trans stereoisomers of alkenes, substituents on the carbons of a double bond are located either on the same or on opposite sides of the double bond. Cis refers to the isomer that has the two similar substituents (hydrogens or methyls in 2-butene for example) on the same side of the double bond. The trans isomer has the two similar substituents on opposite sides of the double bond. The difference between the cis and trans isomers can also be described by the proximity between substituents. Cis substituents are closer to each other than are trans substituents. The structural differences are seen in Figure 12.7.

Cis and trans isomers are different compounds with differing chemical and physical properties. For example, the melting points of the cis and trans isomers of 2-butene are $-139°C$ and $-106°C$, respectively. Cis and trans isomers are different compounds because the π bond of the double bond presents a significant barrier to rotation of the two double-bonded carbons relative to each other. If rotation were free, as it is for acyclic carbon–carbon single bonds, cis and trans isomers would be easily converted one into the other, and separate compounds would not exist.

Cis and trans isomers have the same connectivity (the same order of attachment of the component atoms). Their difference resides in their **configurations,** the arrangements of their atoms in space, at the carbon atoms of the double bond. Two different configurations are possible when each of the carbon atoms of the double bond is a stereocenter. (Recall from Section 11.8 that a stereocenter is an atom in a compound bearing substituents of such identity that a hypothetical exchange of the positions of any two substituents would convert one stereoisomer into another stereoisomer.) A carbon of a double bond is a stereocenter when it bears two different substituents.

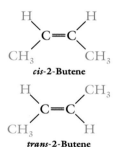

To better understand the relation between the cis and trans configurations, imagine the hypothetical exchange of the groups at one or the other stereocenter. The result is the conversion of one stereoisomer into the other by reversal of the configuration at a stereocenter. Thus, an exchange of the H and CH_3 groups at either carbon of the double bond converts *cis*-2-butene into *trans*-2-butene and vice versa.

None of the following types of alkenes possesses geometric isomerism:

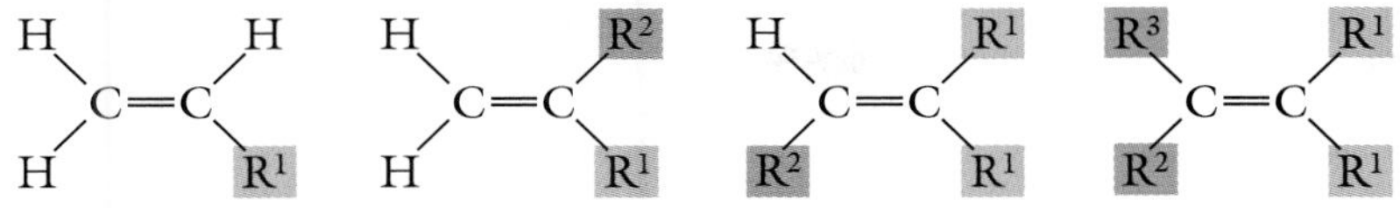

Concept check

✔ Geometric isomers are possible only when each carbon of the double bond is a stereocenter; that is, each bears two different groups.

The chemical and physiological behaviors of cis and trans isomers are often different—another instance of the high selectivity of the molecules used in nature. Typically, the two geometric isomers have very different physiological behaviors (Box 12.2 below and Box 12.3 on the following page).

12.2 Chemistry Within Us

Vision and Cis-Trans Isomerism

Human eyesight begins with a chemical event, and cis-trans isomerism—the mechanism that allows us to perceive light—is its basis. The chemical event in question takes place in the rod and cone cells—distinctively shaped cells in the retina of the eye.

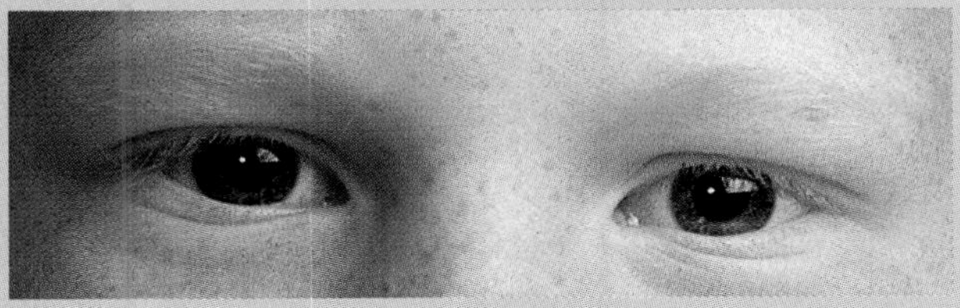

Cis-trans isomerism of 11-retinal is responsible for vision.

Rod and cone cells contain rhodopsin, which is a complex of two molecules: the protein opsin and *cis*-11-retinal. The shape of *cis*-11-retinal allows it to fit snugly into a cavity (a so-called receptor site) in opsin; and, in this associated form, the pair is ready to receive light. *cis*-11-Retinal is produced in the body from vitamin A (see Box 12.1).

When light strikes a molecule of *cis*-11-retinal, its energy is sufficient to momentarily break the π bond of the double bond between C11 and C12. The σ bond between C11 and C12 is then, for a brief period, free to rotate. By the time the π bond becomes reestablished, the molecule has converted (isomerized) into *trans*-11-retinal, the more stable of the two isomers. This change has an important consequence: *trans*-11-retinal dissociates from the cavity of opsin because its elongated shape, relative to that of *cis*-11-retinal, no longer fits the cavity. The loss of *trans*-11-retinal results in the transmission of an electrical signal to the optic nerve, and this signal is interpreted as sight by the brain. Subsequently, the enzyme retinal isomerase forces an extremely rapid reversal of the isomerization reaction, changing *trans*-11-retinal back into *cis*-11-retinal. Rhodopsin is reformed and ready to receive the next input of light energy.

The fit of *cis*-11-retinal and the nonfit of *trans*-11-retinal into opsin's cavity is an example of the **complementarity principle** commonly encountered in physiological processes (Sections 8.4, 22.5). A remarkable aspect of this mechanism is the subtleness with which it operates: it takes a change in the configuration of only one of the five double bonds in *cis*-11-retinal to cause dissociation from opsin and the subsequent transmission of a light signal to the brain.

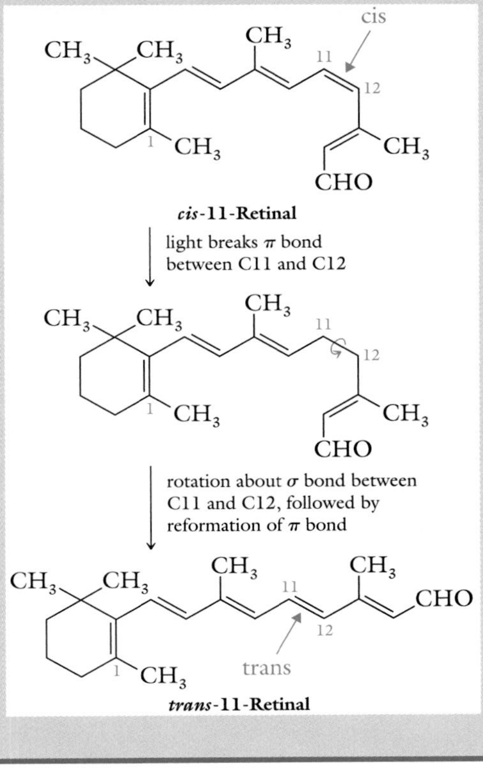

12.3 Chemistry Around Us

Cis-Trans Isomers and Pheromones

Pheromones are chemicals produced by insects and other animals for purposes of communication, such as attracting the opposite sex, sending a danger signal, or marking the location of food. Pheromones are specific for each species. Many pheromones contain carbon–carbon double bonds, and geometric isomerism is critical to their physiological activities.

The specificity of its geometry makes a pheromone effective in an exceedingly small amount. The sex attractant for the common housefly is muscalure (IUPAC: *cis*-9-tricosene), and it effectively lures male houseflies when secreted by female houseflies in an amount less than 10^{-10} g.

Pheromone action depends on the pheromone's ability to bind at protein-receptor sites in a manner analogous to the complementarity principle for vision (see Box 12.2). Here, again, we see that there is a huge difference in physiological response to cis and trans structures. The trans isomer of muscalure has been synthesized in the laboratory and found to be totally inactive as a sex attractant for houseflies.

Bombykol's pheromone action as a sex attractant for the silkworm moth depends on geometric isomerism.

Bombykol (IUPAC: *trans*-10-*cis*-12-hexadecadien-1-ol) is the sex attractant for the silkworm moth. Pheromone action in this case is, again, critically dependent on geometric isomerism. The physiologically active bombykol has a trans double bond between C10 and C11 and a cis double bond between C12 and C13 (carbons are

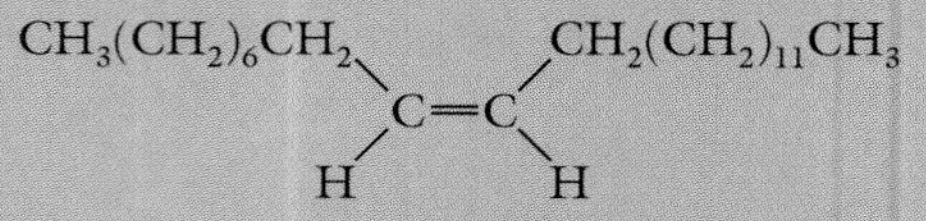

Muscalure

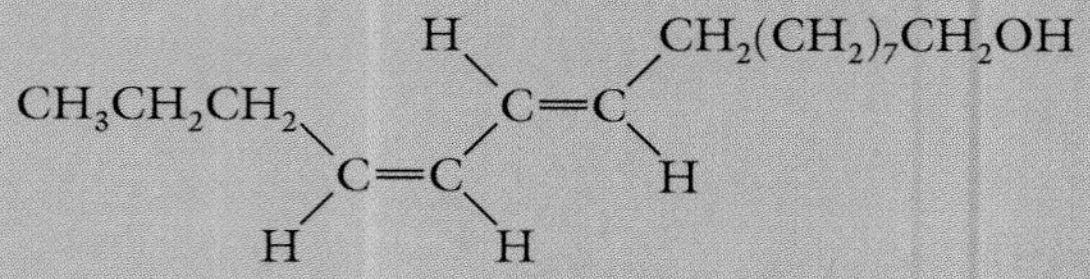

Bombykol

numbered from the end of the chain bearing the —OH group). Other geometric isomers of bombykol (for example, the *cis*-10-*trans*-12 isomer instead of *trans*-10-*cis*-12) are incapable of eliciting a sexual response from male silkworm moths.

Pheromone action is being studied as an alternative to pesticide use in the control of insect populations. The critical step comprises identification and industrial synthesis of the sex pheromone for a particular species. The pheromone is placed in traps to lure male insects and stop the reproductive cycle. An alternate strategy is to spray the pheromone over a large area, such as an apple orchid, to confuse male insects and make it difficult for them to locate mates.

Example 12.4 Identifying and drawing cis and trans isomers

Which of the following compounds exist as a pair of cis and trans isomers? Show structural formulas of the cis and trans isomers. (a) 1-Chloropropene; (b) 1-chloro-2-methylpropene; (c) 2-pentene.

Solution

Examine each carbon of the compound's double bond. Cis-trans isomerism is possible only when each carbon of the double bond has two different groups.

(a) 1-Chloropropene exists as a pair of cis-trans isomers. One carbon of the double bond has H and Cl; the other carbon has H and CH_3:

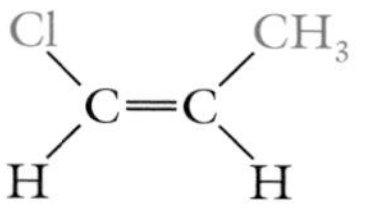

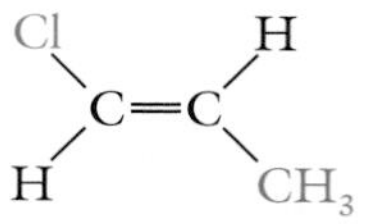

cis-1-Chloropropene *trans*-1-Chloropropene

(b) 1-Chloro-2-methylpropene cannot exist as a pair of cis-trans isomers. One of the carbons of the double bond bears two groups (CH_3) that are the same:

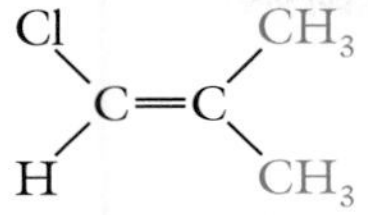

1-Chloro-2-methylpropene

(c) 2-Pentene exists as a pair of cis-trans isomers. One carbon of the double bond has H and CH_3, the other carbon has H and CH_3CH_2:

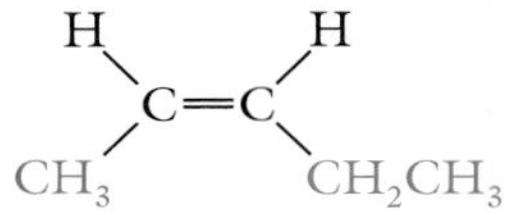

***cis*-2-Pentene**

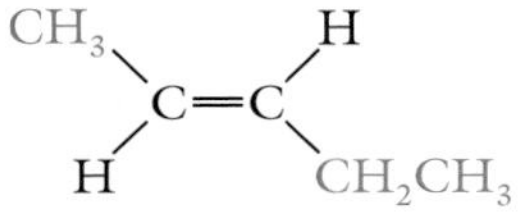

***trans*-2-Pentene**

Problem 12.4 Which of the following compounds exist as a pair of cis and trans isomers? Show structural drawings of the cis and trans isomers.
(a) 2-Methyl-2-pentene; (b) 3-methyl-2-pentene; (c) 4-methyl-2-pentene.

12.6 ADDITION REACTIONS OF ALKENES

Although alkenes and alkanes have similar physical properties because they are both nonpolar compounds, the two families are very different in their chemical properties. Alkenes are highly reactive, whereas alkanes are not. The presence of the weak π bond in alkenes is responsible for their high reactivity. Alkanes possess only the stronger σ bond. Alkenes undergo reaction with a wide range of chemical reagents to which alkanes are entirely unresponsive.

The π bond of an alkene is sufficiently weak to undergo reaction with several chemical reagents:

- Hydrogen (H_2)
- Halogen (F_2, Cl_2, Br_2)
- Hydrogen halide (HCl, HBr, HI)
- Water

With each of these reagents, alkenes undergo the same general type of reaction, an **addition reaction.** All the elements present in the two reactant molecules end up "added together" in the one product molecule, as described by the general equation

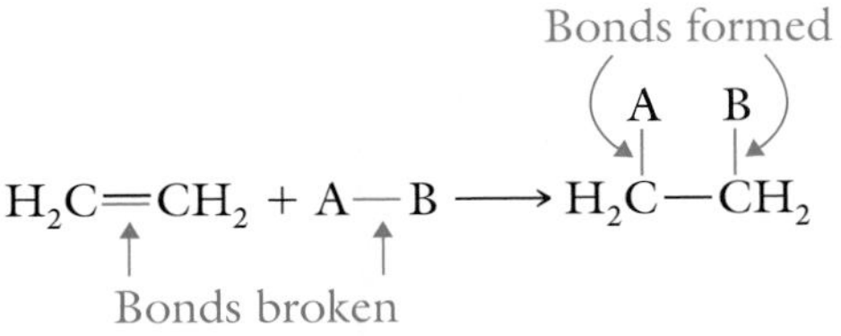

where A—B symbolizes the chemical reagent. The reagent A—B breaks into two reactive fragments A and B, whose presence forces the π bond to break. There is an interim loss of completed octets for the carbons of the double bond and for the A and B fragments. Formation of the product, however, through formation of a σ bond between each carbon and an A or B fragment, once again completes all the atoms' octets.

Addition of Symmetric Reagents (A = B)

Hydrogen and halogen are symmetric reagents; that is, the two fragments of the A—B reagent molecule are the same. The addition of hydrogen (H_2) adds one hydrogen to each carbon of the double bond and converts the alkene into an alkane:

$$H_2C{=}CH_2 + H{-}H \xrightarrow{\text{Pt or Ni}} H_2\overset{\displaystyle H}{\overset{|}{C}}{-}\overset{\displaystyle H}{\overset{|}{C}}H_2$$

The addition of hydrogen proceeds readily at ambient temperatures, but only in the presence of a metal catalyst such as platinum (Pt) or nickel (Ni). For that reason, the reaction is called **catalytic hydrogenation.** Note that the need for a Pt or Ni catalyst is indicated on the arrow in the preceding equation. Writing the equation without specifying the presence of Pt or Ni is incorrect, because it conveys incorrect information, wrongly suggesting that the reaction will take place as long as both reactants are present. The addition of hydrogen to the double bond does not take place in the absence of Pt or Ni. Recall that a catalyst accelerates the rate of a chemical reaction by providing an alternate reaction pathway (with a lower activation energy) for the reaction.

◀◀ A catalyst increases the rate of a reaction but emerges unchanged after the reaction has ended, as described in Section 8.3.

Hydrogenation is a **reduction reaction** (Section 4.7). Reduction of an organic compound, the opposite of oxidation, corresponds to a decrease in the number of oxygen atoms or an increase in the number of hydrogen atoms (or both) bonded to the carbon atoms of the compound (Section 11.10). Each carbon of ethene's double bond is bonded to two hydrogens, whereas, after hydrogenation, each carbon is bonded to three hydrogens. Hydrogenation is used in the manufacture of margarine from vegetable oils, as we will see in Section 19.4.

The addition of halogens to the alkene double bond entails the addition of a halogen atom to each of the carbons of the double bond:

$$H_2C{=}CH_2 + Br{-}Br \longrightarrow H_2\overset{\displaystyle Br}{\overset{|}{C}}{-}\overset{\displaystyle Br}{\overset{|}{C}}H_2$$

The reaction, called **halogenation,** proceeds readily at ambient temperatures.

◀◀ In the halogenation of alkenes, halogen adds to the double bond; in the halogenation of alkanes (Section 11.10), halogen substitutes for hydrogen.

The rapid reaction of alkenes with hydrogen and halogens at ambient temperatures emphasizes the large difference in reactivity between alkenes and alkanes. Alkanes do not react at all with hydrogen. Although alkanes react with halogens, the reaction is of the substitution type (Section 11.10), not the addition type as in alkenes, and takes place only in the presence of ultraviolet radiation or at high temperatures (200°C).

Example 12.5 Writing equations for addition reactions of alkenes: I

Complete each of the following reactions by writing the structure of the organic product. If no reaction takes place, write "no reaction."

(a) $CH_3{-}CH{=}CH{-}CH_3 \xrightarrow{Cl_2} ?$

(b) $CH_3{-}CH{=}CH_2 \xrightarrow{H_2} ?$

(c) $CH_3{-}CH{=}CH_2 \xrightarrow[Ni]{H_2} ?$

Solution

The first step in solving this problem is to determine whether the reagent is one that adds to the double bond. If it is, proceed to break the reagent into its component fragments and add them to the carbons of the double bond.

(a) $CH_3{-}CH{=}CH{-}CH_3 \xrightarrow{Cl_2} CH_3{-}\overset{\displaystyle Cl}{\overset{|}{C}}H{-}\overset{\displaystyle Cl}{\overset{|}{C}}H{-}CH_3$

(b) No reaction, because a catalyst (Pt or Ni) is not present.

(c) The required catalyst is present, and addition takes place:

$$CH_3—CH{=}CH_2 \xrightarrow[Ni]{H_2} CH_3—CH_2—CH_3$$

Problem 12.5 Complete each of the following reactions by writing the structure of the organic product. If no reaction takes place, write "no reaction."

(a) $CH_3—CH{=}CH—CH_3 \xrightarrow[Ni]{H_2}$?

(b) $CH_3—CH{=}CH_2 \xrightarrow{Br_2}$?

The ease with which bromine undergoes addition to alkenes provides chemists with a **simple chemical diagnostic test** for alkenes. A simple chemical diagnostic test is a reaction that takes place rapidly and gives a positive result that can be seen by the human eye. Most chemical reactions do not fit that description. Many chemical reactions proceed rapidly but give no visible indication of reaction. In most of them, colorless chemicals are mixed with other colorless chemicals to form colorless products. Bromine addition to alkenes is an exception. Bromine has a deep red color, but the product of bromine addition to an alkene is colorless. This color change affords us a ready means of differentiating between alkenes and alkanes. We add some bromine to an unknown compound and wait a few minutes (Figure 12.8):

- If the red color disappears, addition has taken place, and the unknown is an alkene.
- If the red color does not disappear, addition has not taken place, and the unknown is an alkane.

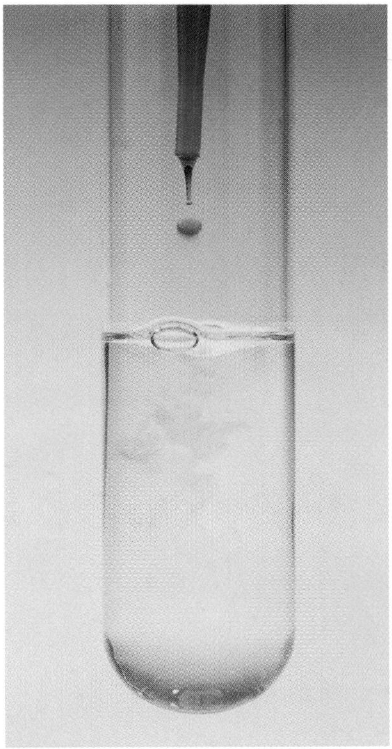

Figure 12.8 1-Hexene gives a positive test with Br_2. The deep red color of Br_2 is decolorized as it reacts with the double bond of 1-hexene.

Example 12.6 Identifying alkenes and alkanes

An unknown compound is either hexane or 1-hexene. When bromine is added to the test tube, the red color of the bromine persists. Identify the unknown.

Solution

The compound is hexane because hexane does not react with and decolorize bromine. If the unknown were 1-hexene, it would have reacted with and decolorized bromine.

Problem 12.6 An unknown compound is pentane, cyclopentane, or 2-pentene. The red color of bromine is decolorized when bromine is added to the unknown. Identify the unknown.

Addition of Asymmetric Reagents (A ≠ B)

Hydrogen halides (HCl, HBr, HI) and water (H_2O) are asymmetric reagents; that is, A—B reagents whose two fragments are not the same. In all of these reagents, H is one of the fragments, and the other fragment is either a halogen or —OH.

The addition of a hydrogen halide, called **hydrohalogenation,** to an alkene adds hydrogen to one carbon of the double bond and a halogen to the other carbon:

$$H_2C{=}CH_2 + H—Cl \longrightarrow \underset{}{H_2\overset{H}{\overset{|}{C}}—\overset{Cl}{\overset{|}{C}}H_2}$$

The addition of water to an alkene adds hydrogen to one carbon of the double bond and OH to the other carbon:

$$H_2C{=}CH_2 + H{-}OH \xrightarrow{H^+} H_2\overset{\displaystyle H}{\overset{|}{C}}{-}\overset{\displaystyle OH}{\overset{|}{C}}H_2$$

Note that the addition of water, called **hydration,** requires the presence of a strong acid catalyst, such as sulfuric acid (H_2SO_4), as denoted by H^+ over the reaction arrow. The product of hydration is an alcohol, R—OH. Hydration is used industrially to produce various alcohols and is an important reaction in the metabolism of carbohydrates, fats, and proteins.

These reactions, like those with hydrogen and halogen, show the high reactivity of alkenes relative to alkanes. Alkanes do not undergo reaction with hydrogen halide or water.

Symmetric alkenes are alkenes that have the same groups on each carbon of the double bond. Asymmetric alkenes carry different groups on the two carbons of the double bond.

- Symmetric alkenes: $CH_2{=}CH_2$, $RCH{=}CHR$, $R_2C{=}CR_2$
- Asymmetric alkenes: $CH_2{=}CHR$, $CH_2{=}CR_2$, $RCH{=}CR_2$

The addition of an asymmetric reagent (a hydrogen halide or water) to an asymmetric alkene is more complex than the addition of either a symmetric or an asymmetric reagent to a symmetric alkene. Only one product is possible for the last two cases. Two products are possible for the first case. Table 12.2 shows the different products formed from various combinations of symmetric and asymmetric reagents and alkenes.

As shown in Table 12.2, two products are possible for the addition of an asymmetric reagent to an asymmetric alkene because there are two ways for the addition to take place. Consider the addition of HCl to propene. One way is for HCl to add so that H adds to C1 and Cl adds to C2:

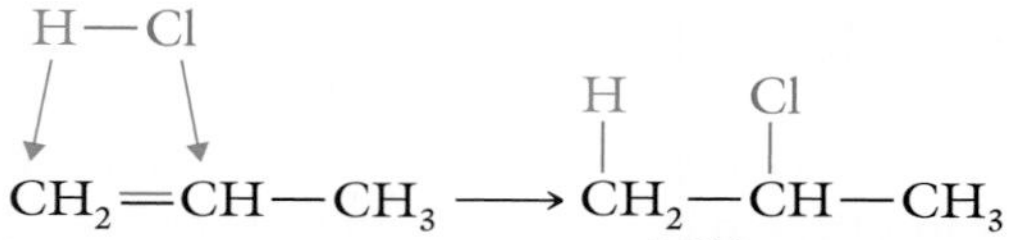

2-Chloropropane
Major product

The other way is the reverse; HCl adds so that Cl adds to C1 and H adds to C2:

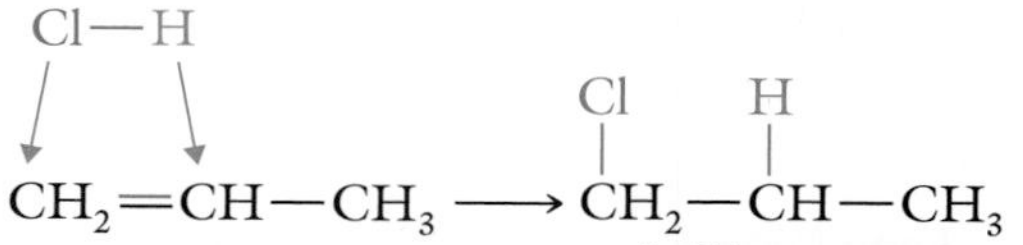

1-Chloropropane
Minor product

TABLE 12.2 Addition Reactions of Alkenes

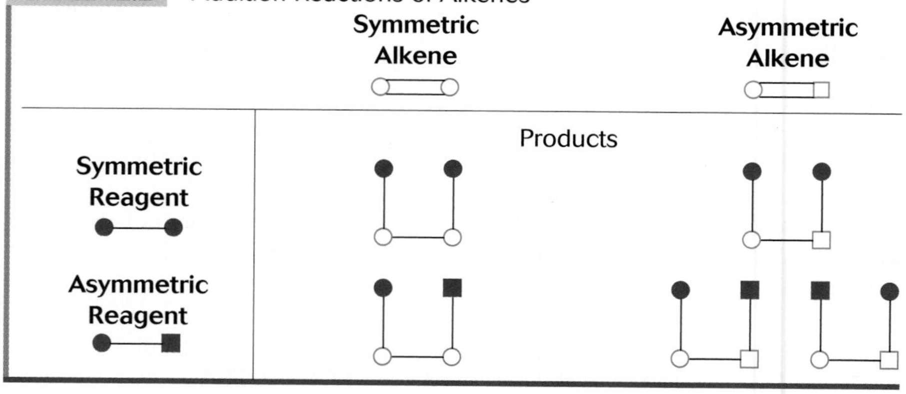

The two possible products are 2-chloropropane and 1-chloropropane, respectively. This reaction is an example of one that has two competing pathways.

When two products are possible, some reactions proceed in a **nonselective** manner and give more or less equal amounts of both. Other reactions are **selective,** with one product formed in greater amounts than the other. The two products are called **major** and **minor** products, respectively. Additions of asymmetric reagents to asymmetric alkenes are selective reactions. One of the two possible products is formed in much greater abundance. 2-Chloropropane is the major product ($>90\%$) in the addition of HCl to propene.

Additions of asymmetric reagents to asymmetric alkenes follow the **Markovnikov rule** proposed in 1869 by the Russian chemist Vladimir Markovnikov:

- The major product is the one formed when the hydrogen from the asymmetric reagent adds to the carbon of the double bond that already had more hydrogens prior to the reaction.

The rule is often paraphrased as "the rich get richer": Of the two carbons in the alkene double bond, the one already relatively "rich" in H gets even "richer" in H. For the reaction of propene with HCl, C1 of propene has more hydrogens than does C2; and as a result, the H of HCl is more likely to become attached to C1.

Only one product is possible when either the reagent or the alkene is symmetric, because there is only one way for the addition to take place.

The reason that reactions follow the Markovnikov rule has to do with the reactions' mechanism, described in Boxes 12.4 and 12.5 on pages 346 and 347.

Example 12.7 Writing equations for addition reactions of alkenes: II

Complete each of the following reactions by writing the structure of the organic product. If no reaction takes place, write "no reaction." If there is more than one product, indicate which is the major and which is the minor product.

(a) $CH_2{=}CH{-}CH_3 \xrightarrow{Br_2} ?$

(b) $CH_3{-}CH{=}CH{-}CH_3 \xrightarrow{HCl} ?$

(c) $CH_2{=}CH{-}CH_2{-}CH_3 \xrightarrow[H^+]{H_2O} ?$

Solution

(a) Only one product is possible, because the reagent is symmetric.

$$CH_2{=}CH{-}CH_3 \xrightarrow{Br_2} \overset{\displaystyle Br}{\overset{|}{C}}H_2{-}\overset{\displaystyle Br}{\overset{|}{C}}H{-}CH_3$$

(b) Only one product is possible, because the alkene is symmetric.

$$CH_3{-}CH{=}CH{-}CH_3 \xrightarrow{HCl} CH_3{-}\overset{\displaystyle H}{\overset{|}{C}}H{-}\overset{\displaystyle Cl}{\overset{|}{C}}H{-}CH_3$$

(c) Two products are possible, because both the reagent and the alkene are asymmetric. The Markovnikov rule tells you which product is the major product.

$$CH_2{=}CH{-}CH_2{-}CH_3 \xrightarrow[H^+]{HOH} \underset{\textbf{Major}}{\overset{\displaystyle H}{\overset{|}{C}}H_2{-}\overset{\displaystyle OH}{\overset{|}{C}}H{-}CH_2{-}CH_3} + \underset{\textbf{Minor}}{\overset{\displaystyle OH}{\overset{|}{C}}H_2{-}\overset{\displaystyle H}{\overset{|}{C}}H{-}CH_2{-}CH_3}$$

Problem 12.7 Complete each of the following reactions by writing the structure of the organic product. If no reaction takes place, write "no reaction."

12.4 Chemistry in Depth

Mechanism of Alkene Addition Reactions

Our considerations of chemical reactions have used equations to describe what chemicals (reactants) undergo change and what new chemicals (products) are formed. We have not described the molecular-level details of how atoms are exchanged, how bonds are broken and reformed—in short, how the reactants change into the products. The "how" of a reaction is called its **mechanism.** Chemists study reaction mechanisms so that they can better predict the outcome of chemical reactions and find ways of enhancing the efficiency with which important chemical products may be produced.

Most reaction mechanisms consist of more than one step. For example, addition reactions of alkenes, with the exception of hydrogenation, proceed by a common two-step mechanism in which ions play key roles. (The addition of hydrogen also proceeds by a multistep process, but neutral hydrogen atoms instead of ions have the key roles.) Consider the addition of HCl to ethene.

- **Step 1.** HCl interacts with the electron pair of the weak π bond of ethene. The H–Cl bond breaks to form H^+ and Cl^-, and the H^+ adds to ethene:

$$CH_2{=}CH_2 + H{-}Cl \longrightarrow \underset{\text{Carbocation}}{CH_3{-}CH_2^{+}} + Cl^-$$

Ions rather than neutral atoms are formed when the H–Cl bond breaks, a form of bond breakage described as **heterolytic** (uneven). Because chlorine is more electronegative than hydrogen, chlorine carries away both electrons when the HCl bond breaks, thus forming Cl^-. Hydrogen becomes H^+ because it carries away no electrons from the bond undergoing cleavage.

The process takes place by a "flow" of electron pairs, as shown by the curved arrows. the electron pair of the π bond of ethene moves away from one of the carbons of the double bond and forms a bond to H^+, and simultaneously the electron pair of the H–Cl bond moves away from H and onto Cl to form Cl^-. This flow results in neutralization of the positive charge on H^+ and the creation of a new carbon–hydrogen bond. The carbon from which the electrons flowed is now electron-deficient and carries a positive charge. An organic species carrying a positive charge on a carbon atom is called a **carbocation.**

- **Step 2.** Bond formation between the positive carbocation and the negative chloride ion yields the final product:

$$\underset{\text{Carbocation}}{CH_3{-}CH_2^{+}} + Cl^- \longrightarrow CH_3{-}CH_2{-}Cl$$

If there is more than one product, indicate which is the major and which is the minor product.

(a) $CH_2{=}CH{-}CH_2{-}CH_3 \xrightarrow{Cl_2} ?$

(b) $CH_2{=}CH{-}CH_3 \xrightarrow[H^+]{H_2O} ?$

(c) $CH_2{=}CH{-}CH_2{-}CH_3 \xrightarrow{HCl} ?$

12.7 ADDITION POLYMERIZATION

The twentieth century is often called the Age of Synthetic Polymers. This distinction results from the ability of alkenes as well as some other families of organic molecules to add to one another like links in a chain, thereby forming very large molecules (called **polymers**) consisting of hundreds or thousands of smaller units (Section 7.16). Box 12.6 on page 348 describes the many practical applications of synthetic polymers as plastics, fibers, and rubbers. In addition to the synthetic polymers produced in the laboratory (Figure 12.9), there are the natural polymers (carbohydrates, proteins, and nucleic acids) produced by living organisms.

12.5 Chemistry in Depth

Carbocation Stability and the Markovnikov Rule

We have seen that many organic reactions have more than one possible product. As in the addition of HCl (an asymmetric reagent) to propene (an asymmetric alkene), the two products are usually not formed in equal amounts. Each product is the result of a different mechanistic pathway, one of which will be favored over the other. Often the reason has to do with the presence of carbocation intermediates (see Box 12.4).

In the addition of HCl to propene, the two reaction pathways are:

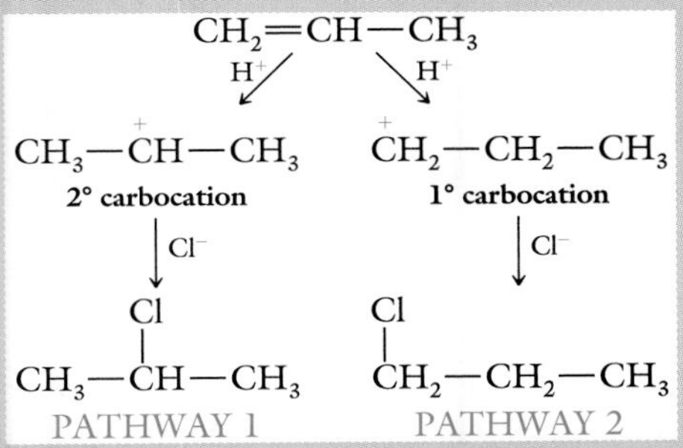

Each pathway begins with the formation of a different carbocation, and there are differences in stability between them. These differences determine the winner of the competition between the two possible pathways for this reaction. The pathway proceeding through the more stable carbocation is the winner.

The stability of carbocations follows the order

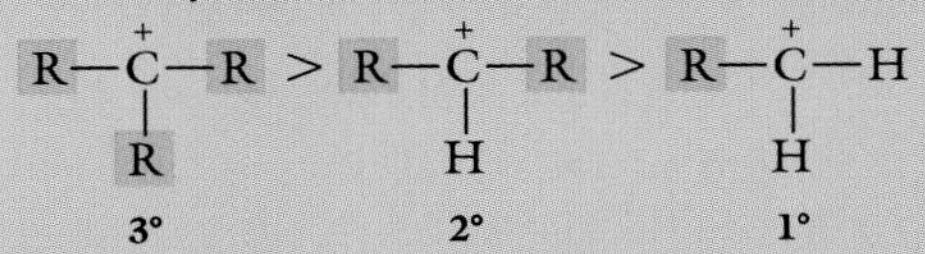

Tertiary, secondary, and **primary carbocations** are carbocations having three, two, and one nonhydrogen substituent, respectively, attached to the carbon carrying the positive (+) charge. The more stable a carbocation, the more easily and quickly it is formed. Reactions that proceed through more-stable carbocations take place at faster rates.

In the addition of HCl to propene, pathway 1 contains a secondary carbocation, whereas pathway 2 contains a primary carbocation. Because secondary carbocations are more stable and form faster than primary carbocations, more molecules of propene travel down pathway 1 than pathway 2. The overall result is that more 2-chloropropane than 1-chloropropane is formed. This is the result predicted by the Markovnikov rule.

Figure 12.9 Commercial products created through addition polymerization.

12.6 Chemistry Around Us

Synthetic Addition Polymers

Synthetic polymers are a key reason for the high standard of living in developed countries. The great utility of these materials is explained by a number of factors:

- Low cost
- Excellent combination of properties
- Ease of fabrication into products

The key property that makes polymers useful is their physical strength, the ability of the polymer in the form of a bottle, packaging film, a heart valve, or other product to carry a load without becoming distorted in shape or breaking. The physical strength of organic polymers is a consequence of their high molecular masses. In Section 11.9, we noted that secondary forces increase with increasing molecular mass, thereby causing corresponding increases in boiling and melting points. When molecular masses reach such magnitudes that a polymer's physical strength can be measured, we find that physical strength also increases with increasing molecular mass. Commercial polymers, ranging in molecular mass from 20,000 to 1,000,000, are chosen according to the strength needed for each particular application.

Synthetic polymers are inexpensive because the raw materials needed for making them, chiefly petroleum, are abundant and cheap. Aside from being a common source of fuel for generating heat and power, petroleum is converted by chemical reactions into the variety of monomers needed to produce useful polymers. Chemists have learned how to manage the reactions for producing and polymerizing monomers so that they run with maximum efficiency and economy to produce a polymer of the required molecular mass.

Another useful quality of synthetic polymers is their general inertness to broad ranges of conditions. Polymers utilized for food packaging are not affected by any moisture content in the food and therefore do not dissolve on contact with the food and contaminate it. Polymers used in medical devices (syringes, artificial joints) are inert to the physiological environment and therefore are less likely to trigger an allergic or other immune reaction; furthermore, they can be sterilized.

Low cost and excellent properties would not guarantee the usefulness of synthetic materials were it not for the ease with which the materials are fabricated. The products made from polymers fall into three categories—plastics, rubbers, and fibers—all easily and economically fabricated into a variety of shapes and products. End products include sheets, films, bottles and other containers, hoses, gaskets, appliance housings, foam cushioning, fast-food containers, textile fabrics for clothes and carpeting, vehicle dashboards and tires, and rubber bands (see Figure 12.9).

Large parts of buildings and their furnishings and appliances, automobiles, trucks, aircraft, and industrial equipment are made from synthetic polymers. More than 90 billion pounds of synthetic polymers are produced and used annually in the United States. The worldwide annual production is about 250 billion pounds.

Not represented among the addition polymers listed in Table 12.3 is an important class of addition polymers that are produced from dienes; that is, from monomers containing two double bonds. The dienes used for this purpose are 1,3-butadiene, isoprene (2-methyl-1,3-butadiene), and chloroprene (2-chloro-1,3-butadiene). Addition polymers formed from dienes are used as synthetic rubbers. Polybutadiene and polyisoprene are used for tires, footwear, gloves, and adhesives. Polychloroprene is used for adhesives, industrial belts and hoses, seals for building and highway joints, and roof coatings. The polymer structures obtained from diene monomers are more complicated than those from monomers containing only one double bond.

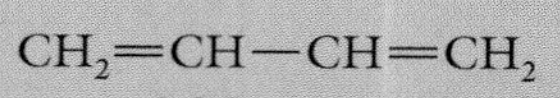

1,3-Butadiene

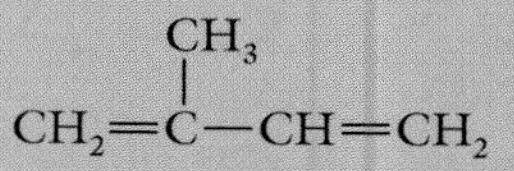

Isoprene

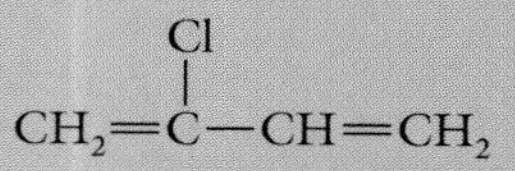

Chloroprene

Synthetic polymers contribute greatly to our life, but there are problems attendant on their use. Polymers are usually not biodegradable: the microorganisms in the ground do not attack and break them down. Instead, polymers accumulate in our landfills in ever increasing amounts. Two strategies have been used to resolve this problem. Many communities have banned or restricted certain uses of polymers (for example, in fast-food packaging). Other communities have instituted recycling programs.

Unfortunately, recycling of polymers is neither simple nor economical, because of the many different kinds of polymers being used. When all polymers are mixed together into a composite product, the result is a low-grade material of little practical use. Sorting the different polymers during recycling produces useful materials, but the process is expensive. Efforts to improve the economics of recycling must continue.

Another problem associated with the pervasiveness of polymers is the toxic gas produced by many of them when they burn; for example, when halogenated polymers such as poly(vinyl chloride) combust, hydrogen chloride is released. The toxic gases produced when a house burns make it more difficult for the occupants to escape and can cause permanent health problems in anyone who survives.

Addition polymerization is similar to the addition reactions described in Section 12.6 except that there is no reactant other than the alkene. Addition polymerization is a self-addition reaction in which many thousands of alkene molecules, individually called **monomers,** add to one another to form the polymer molecules. A representative part of such a process is shown here, illustrated by the polymerization of ethylene (IUPAC: ethene)

$$\underset{\textbf{Monomer}}{CH_2{=}CH_2 + CH_2{=}CH_2 + CH_2{=}CH_2} + \text{etc.} \longrightarrow$$

$$\underset{\textbf{Polymer}}{\sim CH_2{-}CH_2{-}CH_2{-}CH_2{-}CH_2{-}CH_2 \sim}$$

where $\sim$ is used to indicate that the polymer structure goes on and on, with large numbers of monomers linked together. The linkage process relies on the breaking of π bonds in the monomer molecules, which are then linked together by new σ bonds. The resulting linked CH_2CH_2 units are called **repeat units.** Addition polymerization is one of two types of polymerization reactions; the other type is called condensation polymerization.

▶▶ Condensation polymerization produces textile materials called polyesters and polyamides, as described in Sections 15.10 and 16.12.

The addition polymerization reaction is abbreviated as

$$\underset{\textbf{Ethylene}}{n\ CH_2{=}CH_2} \xrightarrow{\text{polymerization catalyst}} \underset{\textbf{Polyethylene}}{\left(CH_2{-}CH_2\right)_n}$$

where n is a large number, usually in the thousands. This notation emphasizes the structures of the monomer and repeat unit. Many alkenes, containing a range of substituents, undergo polymerization. Table 12.3 on the following page lists a few of the most important commercial addition polymers.

The polymerization reaction requires the presence of a catalyst. Moreover, polymerization will not take place in the presence of addition reagents (hydrogen, halogens, hydrogen halides, water, sulfuric acid), because they prevent polymerization by preferentially adding to the double bond. The type of catalyst required for polymerization is beyond the scope of this book and will be referred to simply as "polymerization catalyst."

Example 12.8 Writing equations for the polymerization of alkenes

Write the equation for the polymerization of 1-chloroethene (common name: vinyl chloride) to form poly(vinyl chloride).

Solution

$$n\ CH_2{=}\underset{\displaystyle Cl}{\underset{|}{CH}} \xrightarrow{\text{polymerization catalyst}} \left(CH_2{-}\underset{\displaystyle Cl}{\underset{|}{CH}}\right)_n$$

Problem 12.8 Write the equation for the polymerization of propene (common name: propylene) to form polypropylene.

12.8 OXIDATION OF ALKENES

Like other organic compounds, alkenes will combust, or burn in air (Section 11.10). Where sufficient oxygen is available, each carbon and hydrogen atom in the compound becomes oxidized to its highest oxidation state (CO_2 and H_2O, respectively). The balanced combustion equation for ethene is therefore

$$CH_2{=}CH_2 + 3\ O_2 \longrightarrow 2\ CO_2 + 2\ H_2O$$

Alkenes also undergo selective oxidations in which only the carbons of the double bond receive an oxygen atom. This reaction takes place with a variety of laboratory oxidizing agents, such as permanganate (MnO_4^-) and

TABLE 12.3 Structure and Uses of Addition Polymers

Monomer Name and Structure	Polymer Name and Structure	Uses
ethylene $CH_2{=}CH_2$	polyethylene $\text{+}CH_2{-}CH_2\text{+}_n$	food and detergent bottles; toys and housewares; electrical wire and cable insulation; plastic sheeting for agricultural use
propylene $CH_2{=}CH$ $\vert$ CH_3	polypropylene $\text{+}CH_2{-}CH\text{+}_n$ $\vert$ CH_3	outdoor carpeting for home, sports stadiums; food packaging; housings for appliances
styrene 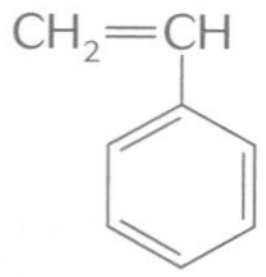	polystyrene 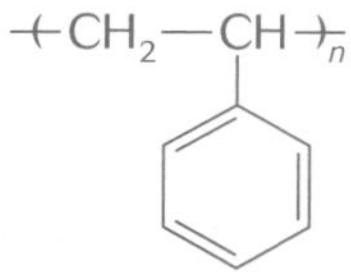	fast-food containers, hot-drink cups; food-packaging trays; hairbrush handles; toys
tetrafluoroethylene $CF_2{=}CF_2$	polytetrafluoroethylene $\text{+}CF_2{-}CF_2\text{+}_n$	nonstick cookware; high-performance mechanical parts; electrical insulation; chemical-resistant gaskets
vinyl chloride $CH_2{=}CH$ $\vert$ Cl	poly(vinyl chloride) $\text{+}CH_2{-}CH\text{+}_n$ $\vert$ Cl	home vinyl siding, rain gutters; flooring (sheet, tile); garden hose; surgical gloves; wire and cable insulation
acrylonitrile $CH_2{=}CH$ $\vert$ CN	polyacrylonitrile $\text{+}CH_2{-}CH\text{+}_n$ $\vert$ CN	acrylic textile fibers
methyl methacrylate $COOCH_3$ $\vert$ $CH_2{=}C$ $\vert$ CH_3	poly(methyl methacrylate) $COOCH_3$ $\vert$ $\text{+}CH_2{-}C\text{+}_n$ $\vert$ CH_3	factory and aircraft windows; bathtubs; contact lenses; dentures and dental fillings

dichromate ($Cr_2O_7^{2-}$) ions, as well as with oxygen and ozone (O_3) from the atmosphere. The products of selective oxidation are alcohols, aldehydes, and ketones or carboxylic acids, depending on the oxidizing agent and the reaction conditions (such as temperature and pH). The selective oxidation of lipids left in the open air is responsible for the disagreeable odor and taste of rancid butter (Section 19.4).

The selective oxidations by permanganate (MnO_4^-) and dichromate ($Cr_2O_7^{2-}$) are useful as simple chemical tests for alkene. A positive test is accompanied by a visible change:

- Permanganate oxidation: MnO_4^- (purple solution) is converted into MnO_2 (brown precipitate).
- Dichromate oxidation: $Cr_2O_7^{2-}$ (orange solution) is converted into Cr^{3+} (green solution).

Example 12.9 **Writing equations for the combustion of alkenes**

Give the balanced equation for the combustion of 1-pentene.

Solution

Proceed by the method described in Example 11.13.

$$2\ C_5H_{10} + 15\ O_2 \longrightarrow 10\ CO_2 + 10\ H_2O$$

Problem 12.9 Give the balanced equation for the combustion of 2-octene.

12.9 ALKYNES

Alkynes are unsaturated hydrocarbons containing the carbon–carbon triple bond, two adjacent carbon atoms bonded together with three bonds. Because of this triple bond, alkynes are more unsaturated than alkenes. Alkynes have two fewer hydrogens than the corresponding alkene or cycloalkane with the same number of carbon atoms. The general formula for alkynes is C_nH_{2n-2}, where n is an integer greater than 1. The first member of the alkyne family is C_2H_2 (IUPAC name: ethyne; common name: acetylene).

$$H-C\equiv C-H$$

Ethyne

We will spend little time on alkynes, because few biological molecules contain the triple bond. Furthermore, the chemistry of alkynes closely resembles that of alkenes. For our purposes, it will suffice to note the following points:

- The nomenclature rules for alkynes are the same as those for alkenes, except the ending is **-yne** instead of **ene.**
- Geometric isomerism is not possible for alkynes, because the bond angles at the carbons of the triple bond are 180° and only one group is attached to each carbon of the triple bond.
- The triple bond (formed from *sp*-hybridized carbon atoms) is composed of one strong σ bond and two weak π bonds. The π bonds are analogous to those in alkenes.

The triple bond reacts with the same reagents as the double bond does, but it does so with twice as much reagent because each π bond is reactive. Only the reactions with hydrogen, halogens, and hydrogen halides concern us here. Reactions with the other reagents give complex products.

Acetylene is an important industrial chemical, largely because of the oxyacetylene torch used in welding. The torch is supplied with acetylene and oxygen from separate high-pressure tanks. The combustion of acetylene produces a high-temperature flame capable of melting and vaporizing iron and steel. The high temperature achieved by the combustion of acetylene is a consequence of the relative instability of the two π bonds. This instability results in a much more exothermic combustion reaction than that of an alkane or alkene.

Example 12.10 Writing equations for addition reactions of alkynes

Show the sequential reaction of 1-butyne with excess HCl; that is, show the reaction of one molecule of 1-butyne with one HCl molecule and then with a second HCl molecule. Show only the major product for each addition.

Solution

Each addition of the asymmetric reagent HCl follows the Markovnikov rule.

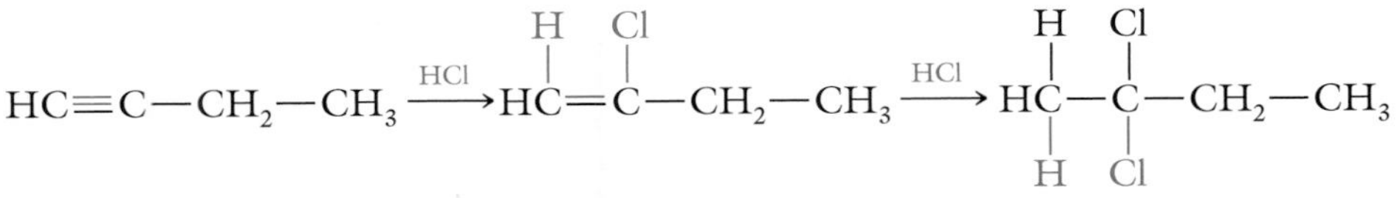

1-Butyne Major product Major product

Problem 12.10 Show the sequential reaction of 2-butyne with excess HCl. Show only the major product for each addition.

12.10 AROMATIC COMPOUNDS

Alkanes, alkenes, and alkynes are called **aliphatic hydrocarbons** to distinguish them from the family of hydrocarbons called **aromatic hydrocarbons.** The chemical properties of aromatic hydrocarbons are very different from those of the aliphatic hydrocarbons, although the physical properties are very similar for the two groups. Aromatic compounds are unsaturated hydrocarbons that do not behave like other unsaturated hydrocarbons. The term aromatic, originally used because many aromatic compounds have pleasant aromas, is now used to emphasize the property that distinguishes this family from other families—aromatic compounds have exceptional stability, far in excess of what one would expect of unsaturated compounds.

The simplest aromatic hydrocarbon is benzene, C_6H_6, whose six carbons form a ring:

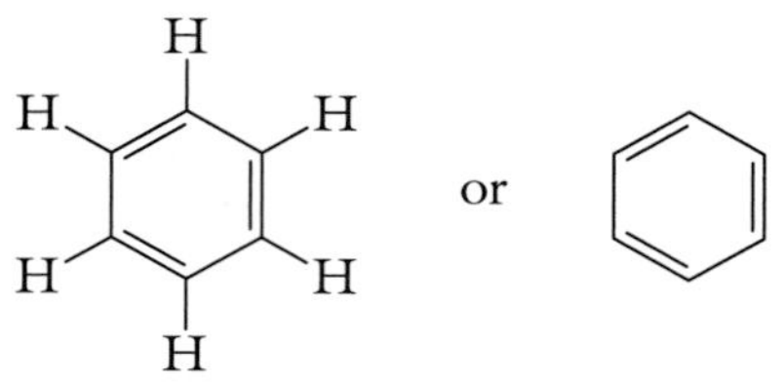

All carbons and hydrogens of benzene lie in one plane (Figure 12.10). The unsaturated ring system of benzene, called the **benzene ring** or **benzene system,** exists not only in benzene, but also in a wide variety of other compounds. Organic compounds whose structures contain a benzene ring are called **aromatic compounds.**

The benzene ring has exceptional stability, which it retains wherever it exists—whether as a constituent of a hydrocarbon or in some other family of organic compounds. The benzene ring, as depicted in the preceding structural formula, looks like a triene with alternating single and double bonds, and we might expect it to react with three times the number of reagent molecules that would react with an alkene double bond. But such reactions do not take place. Benzene and other aromatic compounds resist reactions that would break into the benzene ring.

- Halogens, hydrogen halides, water, and sulfuric acid do not add to the double bonds of benzene under any conditions. These reagents readily add to the double bonds of alkenes.
- The combustion of aromatic compounds is less exothermic than the combustion of aliphatic compounds. (More-stable compounds give off less heat in undergoing reaction than do less-stable compounds.)

This resistance of the benzene structure to change is called **aromaticity** or **aromatic behavior.**

When you see a compound containing a six-membered ring with three double bonds alternating with three single bonds, recognize that the compound does not behave as it looks—it is not like an alkene in behavior, be-

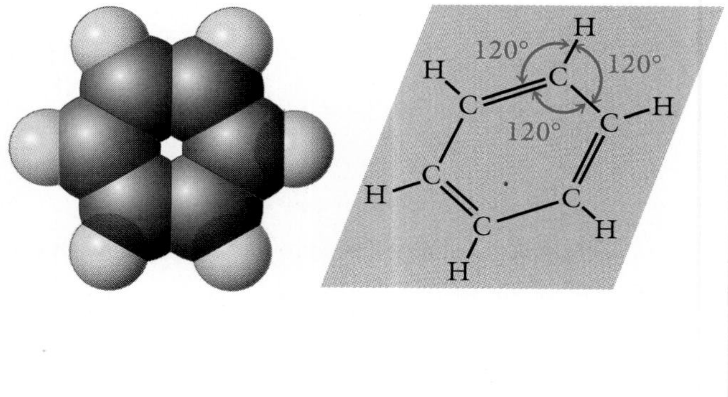

Figure 12.10 A space-filling model of benzene and a diagram showing the trigonal bond angles of its sp^2 carbons.

cause it has the special stability of aromaticity. Aromaticity resides in the unique bonding arrangement that results when three double bonds alternate with three single bonds in a six-membered ring (Box 12.7).

Chemists prefer an alternate representation of the benzene ring—a hexagon with a circle inside it, as shown in the margin. This representation of the benzene ring is preferred because it clearly shows the ring as a unique structure, not simply three alternating single and double bonds.

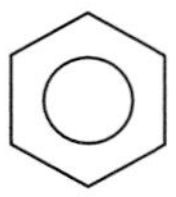

Benzene

Aromatic compounds are encountered throughout the biological world and in everyday life (Box 12.8 on the following page). The benzene ring is not the only aromatic structure in chemistry. Several kinds of heterocyclic aromatic structures containing nitrogen in the ring are of major importance in biological systems. Nucleic acids such as DNA and RNA, for example, the molecules responsible for heredity, contain the pyrimidine and purine aromatic structures (Section 21.1).

Pyrimidine

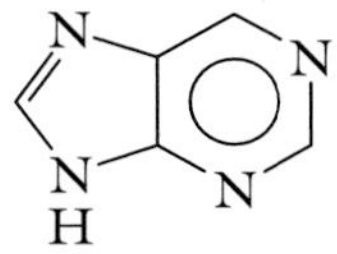

Purine

12.7 Chemistry in Depth

Bonding in Benzene

Each carbon atom of the benzene ring is sp^2 hybridized, as it is in a normal carbon–carbon double bond. That is, each carbon atom has three sp^2 orbitals and one $2p$ orbital; and each carbon atom uses its three sp^2 orbitals to form three σ bonds, one to each of the two adjacent carbons and one to a hydrogen atom (see the illustration at right). The key feature that distinguishes benzene from an alkene with three double bonds is the π bonding that results from sideward overlap of the six $2p$ orbitals. In an alkene double bond, $2p$ orbitals on two adjacent carbons overlap, thereby causing localization of the two electrons in the resulting π bond. In benzene, the $2p$ orbital on each carbon overlaps with two $2p$ orbitals, those from the adjacent carbon atoms on either side. This arrangement results in a continuous, circular overlap of six $2p$ orbitals (as shown in the illustration below). The overlapping $2p$ orbitals of the benzene ring prove to have a remarkable stability, a feature manifested in benzene's characteristic chemical behavior.

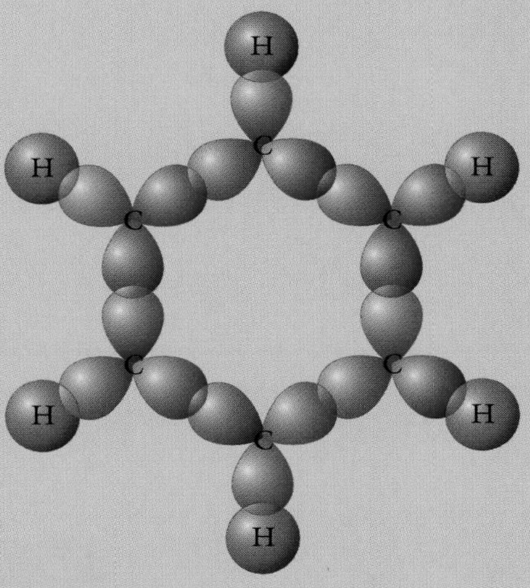

Sigma bonds in benzene.

INSIGHT INTO PROPERTIES

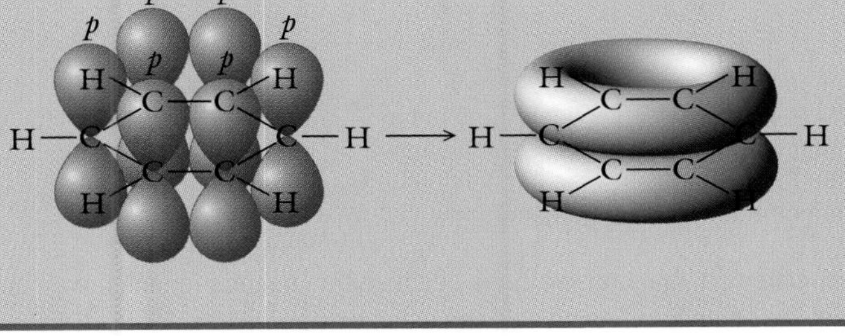

Overlap of $2p$ orbitals to form π bonding in benzene. The π bonding in benzene explains the benzene ring's great stability.

12.8 Chemistry Around Us

Aromatic Compounds in Everyday Life

Many compounds encountered in everyday life contain benzene rings. Prominent examples are the many aromatic compounds found in flavorings and spices, such as the benzaldehyde in almonds, the vanillin in vanilla beans, and the anethole in anise seeds.

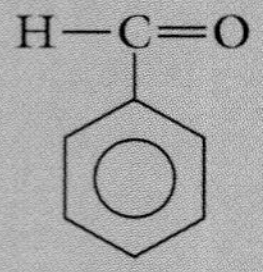

Benzaldehyde
(Almond)

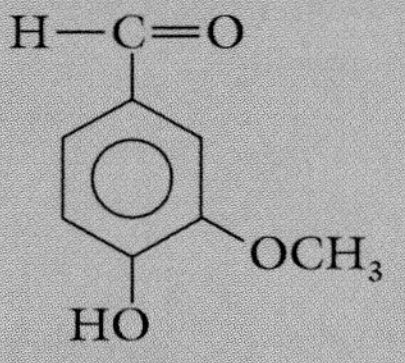

Vanillin
(Vanilla)

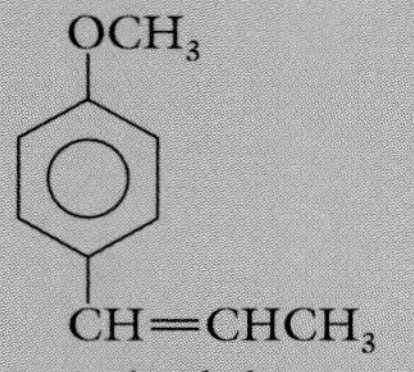

Anethole
(Anise)

Two other compounds containing the benzene ring are the female sex hormone estradiol and the synthetic pain killer aspirin.

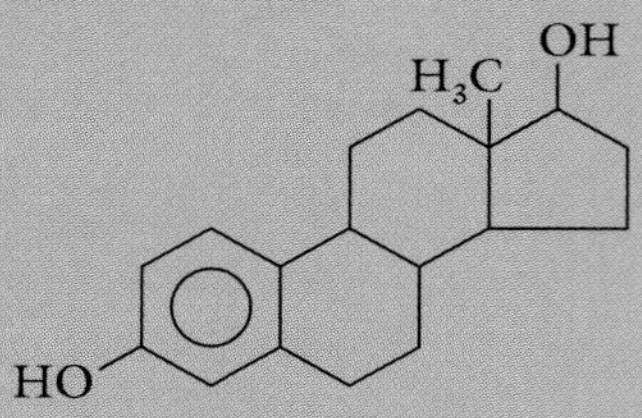

Estradiol

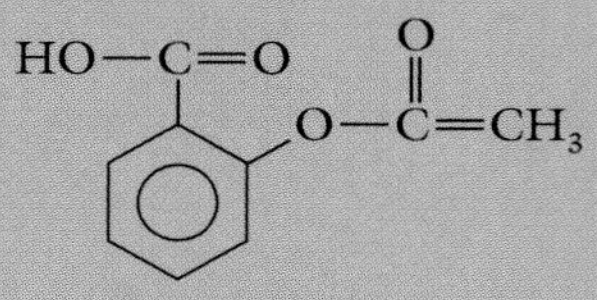

Aspirin

A number of chlorinated aromatics have been used as insecticides (DDT) and as heat-exchanging and electrical-insulating fluids (PCB). DDT and PCB are no longer used in the United States, however, or in many other highly industrialized countries, because they are harmful to humans and nonbiodegradable. Nevertheless, the use of DDT continues in Third World countries where people are more concerned about short-term starvation than about long-term health effects.

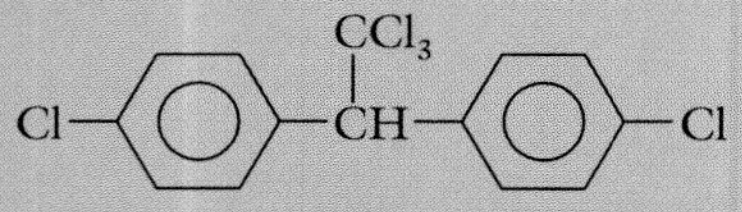

Dichlorodiphenyltrichloroethane (DDT)

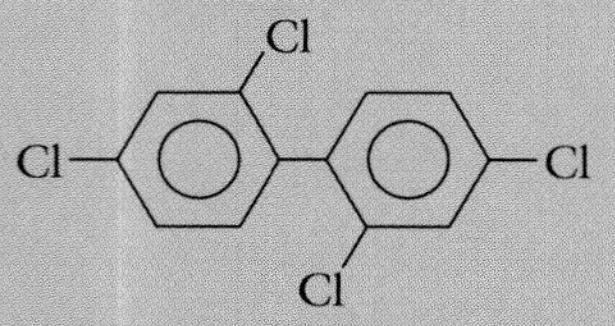

Polychlorinated biphenyl (PCB)

12.11 ISOMERS AND NAMES OF AROMATIC COMPOUNDS

In the IUPAC system, most monosubstituted benzenes are named by placing the name of the substituent in front of the parent name -**benzene:**

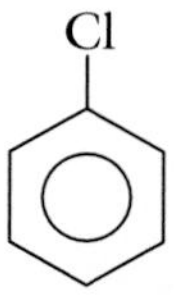

Chlorobenzene

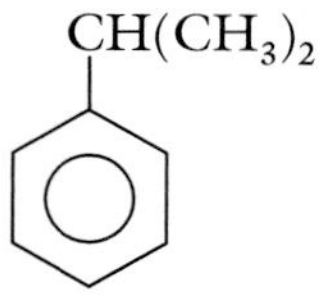

Isopropylbenzene

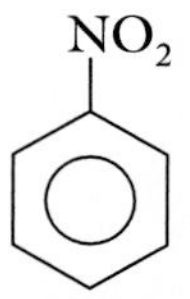

Nitrobenzene

Some monosubstituted benzenes have common names that have been adopted by IUPAC as the preferred names. Examples include toluene for methylbenzene, phenol for hydroxybenzene, benzoic acid for carboxybenzene, and aniline for aminobenzene:

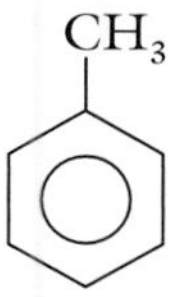

Toluene (Methylbenzene)

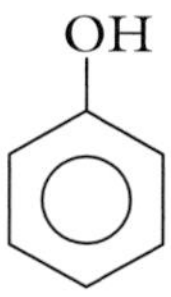

Phenol (Hydroxybenzene)

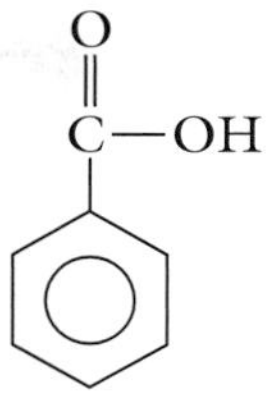

Benzoic acid (Carboxybenzene)

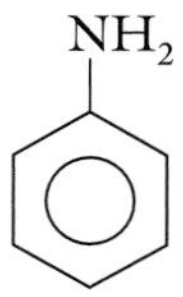

Aniline (Aminobenzene)

Disubstituted benzenes have three constitutional isomers, whose substituents take different positions around the benzene ring. There are two systems for indicating those positions: the numbering system (1,2-, 1,3-, 1,4-) or the *ortho*-, *meta*-, *para*-, system. Some disubstituted benzenes—for example, the dichlorobenzenes—are simply named as derivatives of benzene:

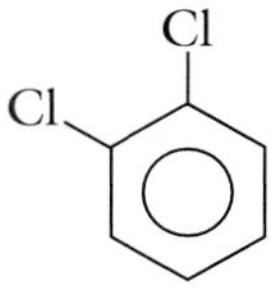

1,2-Dichlorobenzene (*o*-Dichlorobenzene)

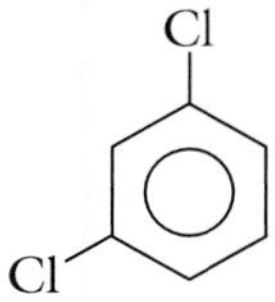

1,3-Dichlorobenzene (*m*-Dichlorobenzene)

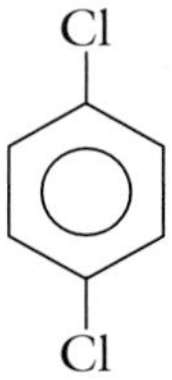

1,4-Dichlorobenzene (*p*-Dichlorobenzene)

Other disubstituted benzenes are named as derivatives of the monosubstituted benzene when the latter has a preferred common name:

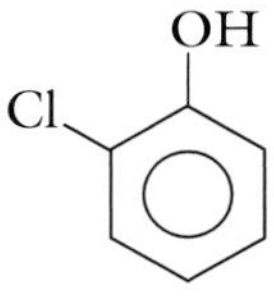

***o*-Clorophenol (2-Chlorophenol)**

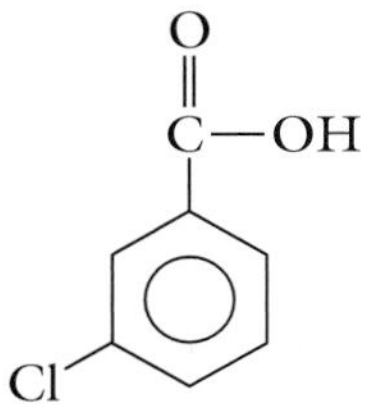

***m*-Chlorobenzoic acid (3-Chlorobenzoic acid)**

Trisubstituted benzenes are named in a similar fashion, except that numbers must be used (because the *ortho*-, *meta*-, and *para*- system is not applicable):

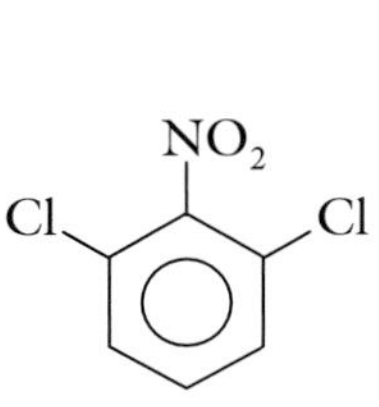

1,3-Dichloro-2-nitrobenzene

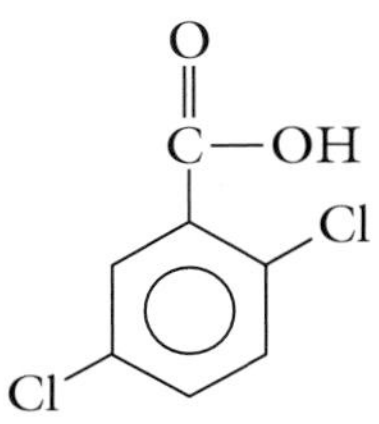

2,5-Dichlorobenzoic acid

Not all compounds containing a benzene ring are named by using "benzene" as the parent or base name. Some compounds contain an aliphatic part that is more complex than the aromatic component, and these compounds are usually given names derived from that of the aliphatic parent or base compound. In such cases, the prefix **phenyl** is added to indicate the presence of a benzene ring as a substituent:

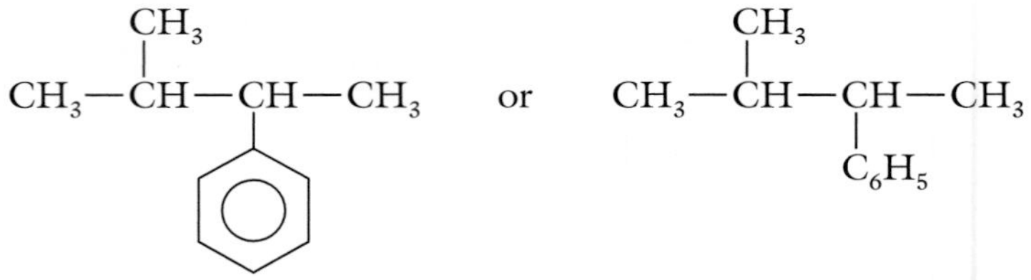

2-Methyl-3-phenylbutane

The phenyl group is often abbreviated by C_6H_5- or Ph- or the Greek letter ϕ (phi). We will use C_6H_5- in this book. A phenyl or substituted phenyl group is called an **aryl** group (symbol, Ar) to distinguish it from an alkyl group.

C_6H_5-CH_2- is known as a **benzyl** group:

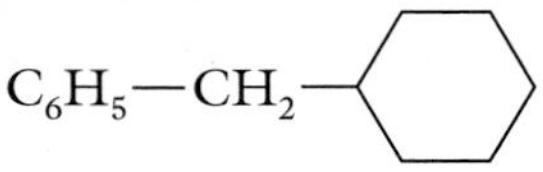

Benzylcyclohexane

Do not confuse phenyl and benzyl groups; they are different.

Unlike the nonaromatic cyclic compounds, substituted benzenes do not exist as cis and trans isomers, because the bonds extending from the carbons of the benzene ring are in the same plane as the ring (and only one bond extends from each of the carbons of the ring).

The basic approach to naming substituted benzenes is the same as that for substituted cycloalkanes (Section 11.7). The numbering of the ring carbons starts with a carbon holding a substituent and proceeds around the ring from there. When different sets of numbers can be obtained by starting at different carbons or by counting in different directions, the correct name is the one with the lowest set of numbers.

Example 12.11 Using the IUPAC nomenclature system to name aromatic compounds

Name each of the following compounds according to the IUPAC system:

(a) 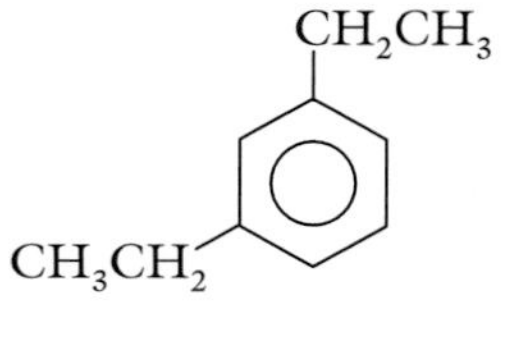

(b) 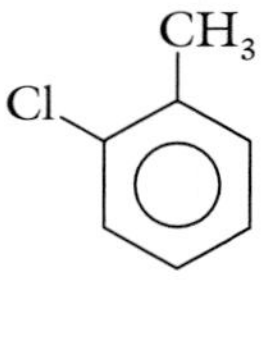

(c) 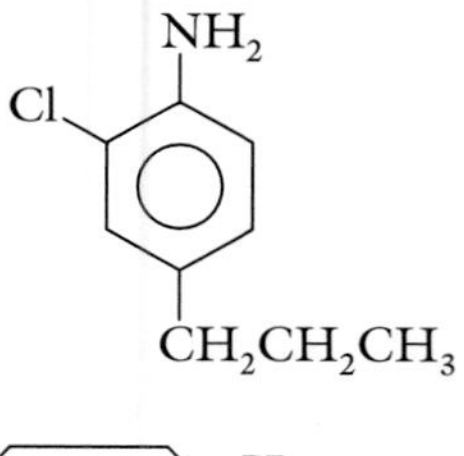

(d) 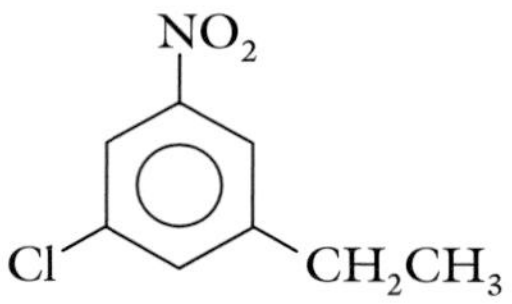

(e) 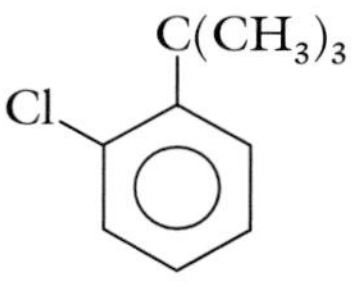

(f) 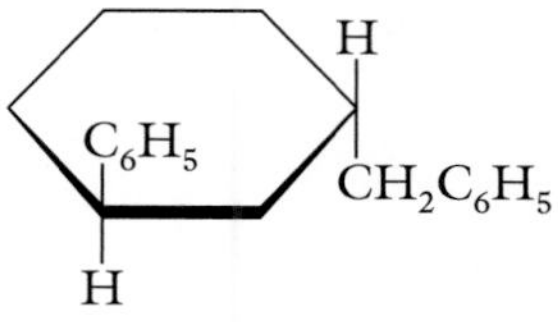

Solution

The numbering of the ring carbons starts with a carbon holding a substituent and proceeds around the ring from there. When different sets of numbers can be obtained by starting at different carbons or by counting in different directions, the correct name is the one with the set of lowest numbers.

(a) 1,3-Diethylbenzene or *m*-diethylbenzene.

(b) 2-Chlorotoluene or *o*-chlorotoluene.

(c) 2-Chloro-4-propylaniline.

(d) The set of numbers is 1,3,5, with different possibilities that depend on which carbon is chosen as C1 and whether numbering is clockwise or counterclockwise. The correct choices are those that give the lower numbers to substituents whose names start lower in the alphabet. The name is 1-chloro-3-ethyl-5-nitrobenzene.

(e) 1-*t*-Butyl-2-chlorobenzene or *o*-*t*-butylchlorobenzene.

(f) *trans*-1-Benzyl-3-phenylcyclohexane.

Problem 12.11 Draw structural formulas for the following compounds: (a) *p*-ethylbenzoic acid or 4-ethylbenzoic acid; (b) 2,4-dibromotoluene; (c) 4-phenyl-1-butene; (d) *m*-isopropylphenol or 3-isopropylphenol; (e) 1-chloro-3-ethyl-4-propylbenzene; (f) benzyl chloride.

12.12 REACTIONS OF AROMATIC COMPOUNDS

By now you might have the impression that benzene is chemically inert, but it is not. Although it is true that the carbon–carbon bonds making up the ring are highly resistant to addition, benzene undergoes substitution reactions in which some group substitutes for one of the hydrogen atoms extending from the ring. Halogenation, sulfonation, nitration, and alkylation are reactions in which substitution is by halogen (—F, —Cl, —Br), sulfonic acid ($—SO_3H$), nitro ($—NO_2$), and alkyl (—R) groups, respectively:

Halogenation:

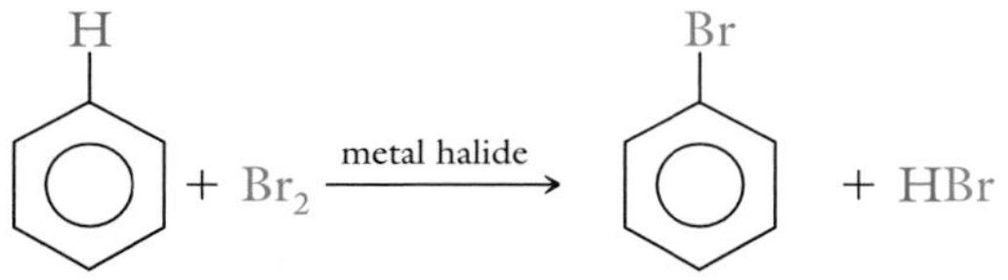

Bromobenzene

Nitration:

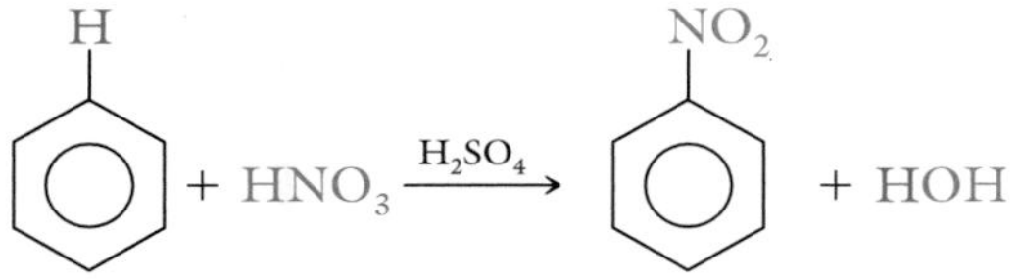

Nitrobenzene

Sulfonation:

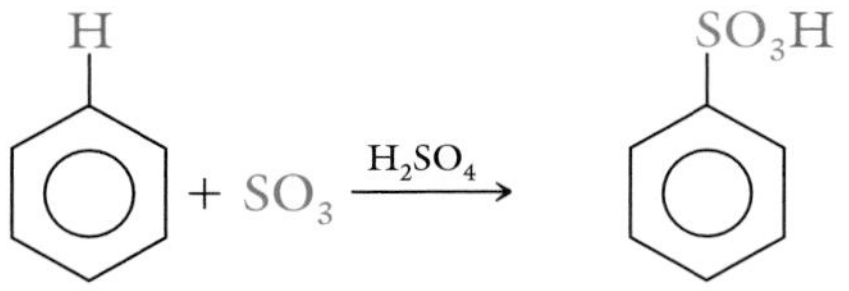

Benzenesulfonic acid

Alkylation:

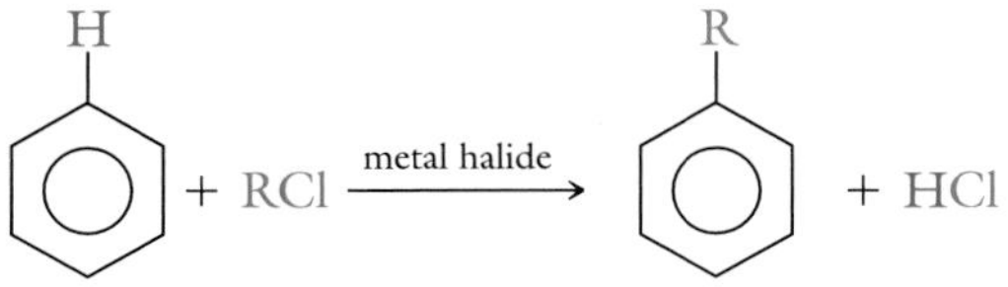

Alkylbenzene

Halogenation and alkylation of an aromatic compound require a metal halide such as $FeCl_3$ or $AlCl_3$ as a catalyst. Nitration and sulfonation require sulfuric acid as a catalyst. The reactions do not take place unless the appropriate catalyst is present.

The substitution reactions of benzene do not contradict our earlier discussion of the high stability of the benzene ring; instead they serve to emphasize it. These reactions leave the benzene structure intact. Note, in contrast, that, although the same reagents induce halogenation, nitration, sulfonation, and alkylation of alkenes, they do so by a process of addition to the carbon–carbon double bonds. (Nitration, sulfonation, and alkylation of alkenes were not covered in Chapter 11, but they do take place, as addition reactions.)

Aromatic compounds, like all organic compounds, undergo combustion to form carbon dioxide and water. Selective oxidations of alkylbenzenes under conditions less extreme than combustion show the high stability of aromatics relative to aliphatics. Alkylbenzenes undergo selective oxidation under moderately strong oxidizing conditions—for example, in the presence of hot acidic $KMnO_4$ or $K_2Cr_2O_7$. The benzene ring remains intact because of its stability, whereas the aliphatic part is oxidized:

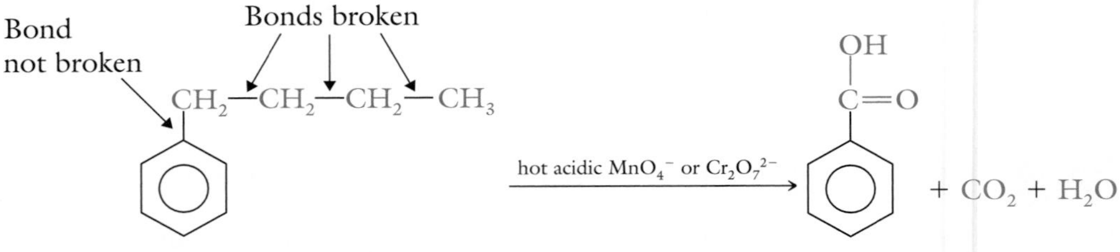

Whatever the size of the alkyl group, one by one all of its carbon–carbon bonds break, and the carbons and hydrogens are oxidized to CO_2 and H_2O, respectively. However, the carbon–carbon bond between the benzene ring and the alkyl group is not broken, because that would destroy the aromatic structure by subsequent oxidation of ring carbons. The carbon of the alkyl group attached to the benzene ring is oxidized to the carboxyl group (COOH), the highest oxidation state (Section 12.10) possible for carbon without breaking the bond to the aromatic ring.

Example 12.12 Writing equations for reactions of aromatic compounds

Write equations to show the product(s) formed in each of the following reactions. If no reaction takes place, write "no reaction." If more than one product is formed, indicate the major and minor products. (a) Benzene + Cl_2; (b) benzene + Cl_2 + $FeCl_3$; (c) benzene + SO_3 + H_2SO_4; (d) toluene + hot acidic $KMnO_4$.

Solution

(a) No reaction takes place, because halogenation requires a metal halide catalyst.

(b)

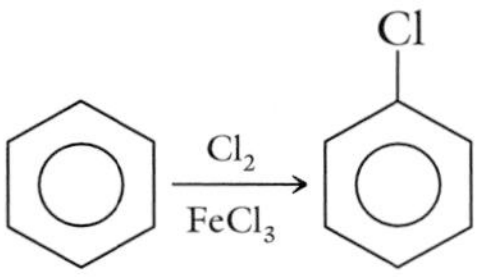

(c)

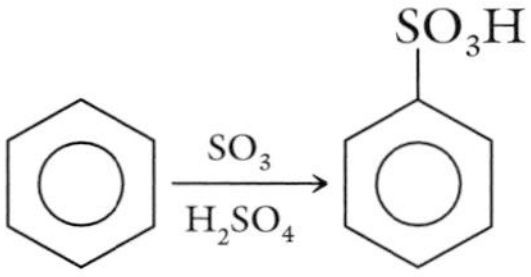

(d)

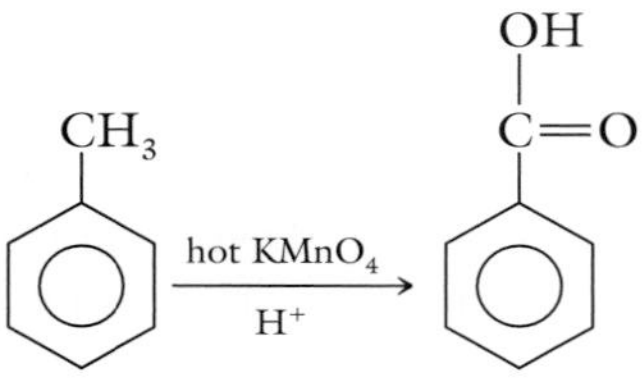

Problem 12.12 Write equations to show the product(s) formed in each of the following reactions. If no reaction takes place, write "no reaction." If more than one product is formed, indicate the major and minor products. (a) Benzene + CH_3CH_2Cl; (b) benzene + CH_3CH_2Cl + $AlCl_3$; (c) isopropylbenzene + hot acidic $K_2Cr_2O_7$.

Substitution and side-chain oxidation reactions are used industrially to produce a range of raw materials, which are converted into consumer products including detergents, dyes and pigments, and plastics.

To sum up what we have learned about reactions between organic molecules and halogens in this chapter and in Chapter 11, three different scenarios are possible:

Concept checklist

- ✔ Halogen substitution takes place at C–H bonds in alkanes in the presence of ultraviolet radiation or at high temperatures (Section 11.10).
- ✔ Halogen substitution takes place at C–H bonds of aromatic compounds in the presence of a metal halide catalyst.
- ✔ Halogen addition takes place at C–C double bonds in alkenes with or without the presence of high temperatures, UV radiation, or a metal catalyst (Section 12.6).

Example 12.13 Writing equations for the halogenation of various compounds

Bromine is added to a mixture of cyclohexane, cyclohexene, and benzene in the presence of $FeBr_3$. Which compound(s) undergo reaction? Write equations to show the product(s) formed in any reaction. (Note: Reactions are carried out at ambient temperature in the absence of ultraviolet radiation unless otherwise stated.)

Solution

Cyclohexane does not undergo reaction, because the temperature is not high and there is no ultraviolet radiation. Cyclohexene and benzene undergo addition and substitution, respectively:

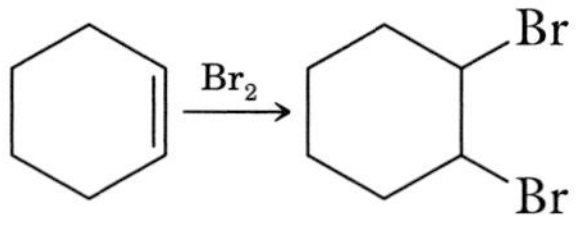

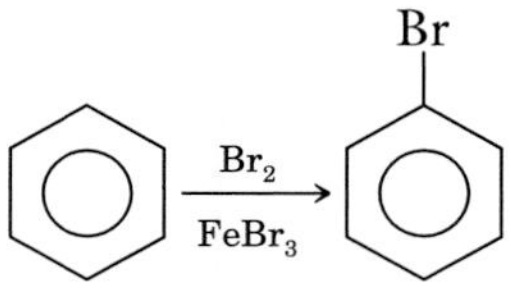

Problem 12.13 Bromine is added to a mixture of cyclohexane, cyclohexene, and benzene in the presence of ultraviolet radiation but in the absence of $FeCl_3$. Which compound(s) undergo reaction? Write equations to show the product(s) formed in any reaction.

Fused-ring aromatic compounds are described in Box 12.9 on the following page. Some of these compounds are used industrially; others are implicated as carcinogens.

12.9 Chemistry Around Us

Fused-Ring Aromatics

A **fused-ring compound** contains rings that share two or more ring atoms. The compound estradiol (Box 12.8), for example, contains four fused rings, one of which is aromatic. A number of **fused-ring aromatic compounds,** compounds in which all rings are aromatic, are commercially important: naphthalene is the active ingredient in moth balls; anthracene is used to manufacture a variety of dyes.

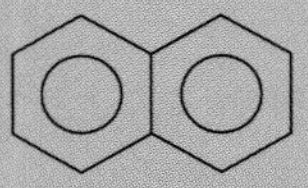

Naphthalene

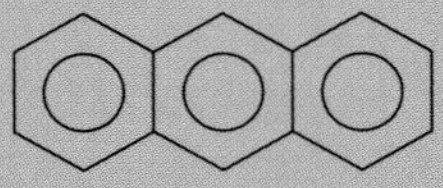

Anthracene

"Bent" fused-ring aromatics such as phenanthrene, benzanthracene, and benzopyrene are **carcinogens,** substances that cause cancer. These compounds are found in automobile exhaust, tobacco smoke, and burnt (especially barbecued) meats. The carcinogenic behavior of bent fused-ring aromatics is discussed in Section 21.8.

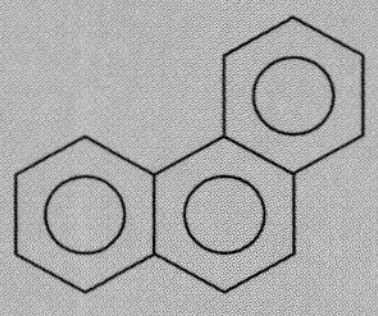

Phenanthrene

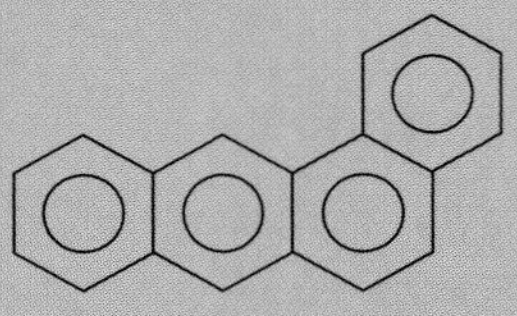

Benzanthracene

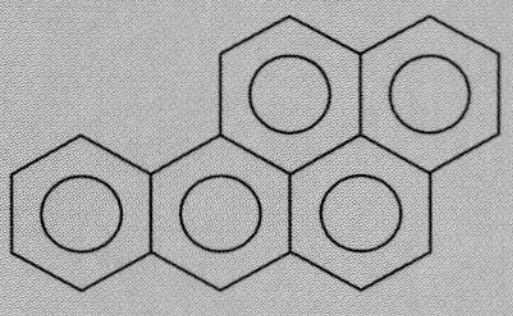

Benzopyrene

Tobacco smoke contains cancer-causing bent fused-ring aromatic compounds.

Summary

Unsaturated hydrocarbons, consisting of the alkene, alkyne, and aromatic families, contain carbon–carbon multiple bonds; that is, double or triple bonds between carbon atoms.

Alkenes Alkenes contain a double bond; that is, two adjacent carbons joined together with two bonds. The general formula for an alkene is C_nH_{2n}, which contains two fewer hydrogens than the general formula for an alkane. The carbon atoms of the double bond possess sp^2-hybridized orbitals with trigonal (120°) bond angles. The orbitals overlap to form one strong bond (a σ bond) and one weak bond (a π bond).

Constitutional Isomers of Alkenes Constitutional isomerism exists in alkenes not only because a given formula allows different carbon skeletons, as in the alkane family, but also because it allows different placements of the double bond within a carbon skeleton.

Naming Alkenes In the IUPAC system, alkenes are named on the basis of the longest continuous chain containing the double bond. The base name is preceded by a number indicating the position of the double bond. The family ending of alkene names is -ene. Prefixes with numbers identify any substituents attached to the longest chain.

Cis-Trans Stereoisomerism in Alkenes Cis-trans (geometrical) stereoisomers exist in alkenes when each carbon atom of the double bond has two different substituents. The cis isomer has substituents on the same side of the double bond. The substituents are on opposite sides of the double bond in the trans isomer.

Addition Reactions of Alkenes The presence of the weak π bond in the double bond greatly increases the reactivity of alkenes relative to the low reactivity of saturated hydrocarbons. Various addition reactions are possible: a reagent breaks into two fragments, each of which bonds (adds) to one of the carbons of the double bond. Hydrogen, halogens, hydrogen halides, and water are all able to add to the alkene double bond. The addition of asymmetric reagents (hydrogen halides, water) to asymmetric alkenes proceeds in a selective manner. The Markovnikov rule is followed, meaning the hydrogen fragment from the reagent adds to the carbon of the double bond that already holds more hydrogens prior to reaction.

Addition Polymerization Addition polymerization is a self-addition reaction in which large numbers of alkene molecules add to one another to form high-molecular-mass molecules called polymers.

Oxidation of Alkenes Alkenes, like other organic compounds, can also undergo combustion. Selective oxidation also takes place.

Alkynes Alkynes contain a triple bond, three bonds between two adjacent carbon atoms. Two of the three bonds are weak π bonds similar to those in the alkene double bond. The triple bond undergoes the same reactions as the double bond does but with twice as much of the inorganic reagent.

Aromatic Compounds Aromatic compounds contain the benzene ring, a six-membered ring with alternating double and single bonds. Substituted benzenes are named by IUPAC as derivatives of benzene, with prefixes indicating the substituents attached to the ring. Either a numbering system or the *ortho, meta, para* system is used to indicate the positions of substituents for disubstituted benzenes. Only the numbering system is used for trisubstituted benzenes. Some monosubsitituted benzenes have common names that are accepted by IUPAC and used as the basis for naming their di- and trisubstituted relatives.

Reactions of Aromatic Compounds The benzene ring has exceptional stability. Benzene is inert to the addition of halogens, hydrogen halides, sulfuric acid, and water—reagents that add to alkene double bonds.

Benzene and other aromatic compounds undergo substitution reactions with halogens, nitric acid, sulfur trioxide, and alkyl halides. Alkylbenzenes undergo oxidation of the alkyl group, but the benzene ring itself remains intact.

Summary of Key Reactions

ALKENES

Addition

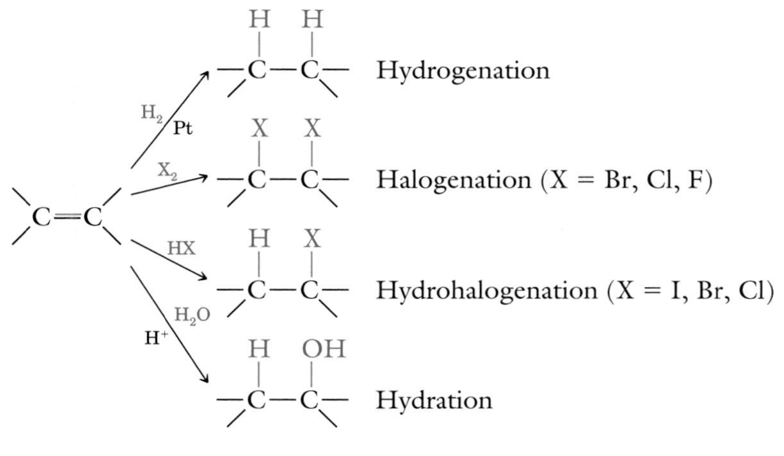

Addition polymerization

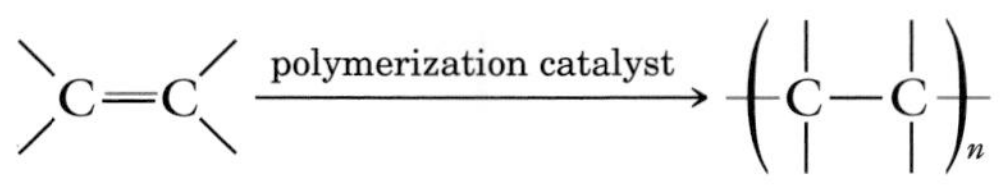

AROMATICS

Substitution

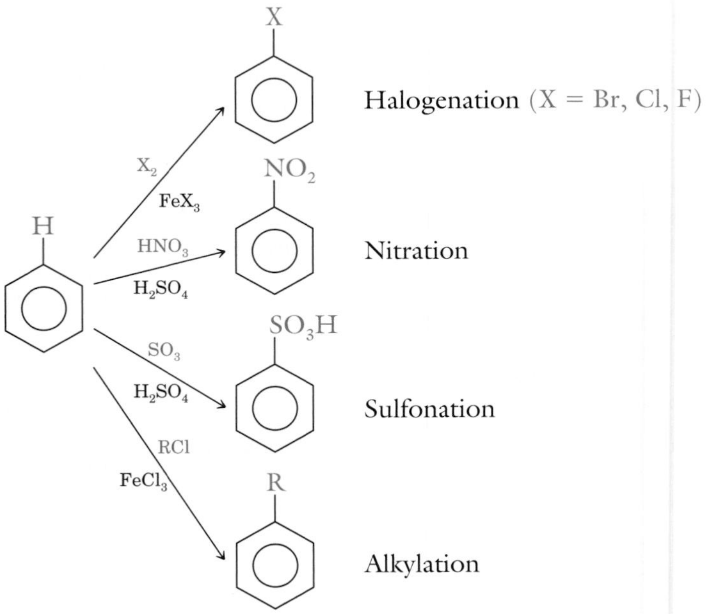

Selective oxidation

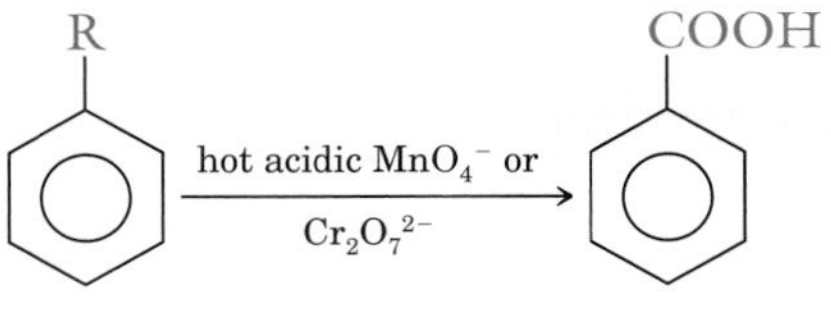

Key Words

addition polymerization, p. 346
addition reaction, p. 341
aliphatic, p. 352
alkene, p. 331
alkylation, p. 357
alkyne, p. 351
aromatic, p. 352
asymmetric reagent, p. 343
carbocation, p. 347
halogenation, pp. 342, 357
hydration, p. 344
hydrohalogenation, p. 343
hydrogenation, p. 342
Markovnikov rule, p. 345
mechanism, p. 346
monomer, p. 349
nitration, p. 357
pi (π) bond, p. 333
polymer, p. 346
repeat unit, p. 349
restricted rotation, p. 337
sigma (σ) bond, p. 333
sp^2, p. 333
sulfonation, p. 357
symmetric reagent, p. 341
trigonal bond angle, p. 333

Exercises

Alkenes

12.1 An unknown compound has the molecular formula C_5H_{12}. Is the compound an alkane, cycloalkane, or alkene?

12.2 An unknown compound has the molecular formula C_5H_{10}. Is the compound an alkane, cycloalkane, or alkene?

12.3 Which of the following are correct molecular formulas? (a) C_6H_{13}; (b) C_6H_{12}; (c) $C_6H_{13}Br$.

12.4 Which of the following are correct molecular formulas? (a) C_5H_{12}; (b) C_5H_{11}; (c) C_5H_{10}.

12.5 Give the molecular formula for each of the following compounds:

(a) $CH_3{-}CH{=}CH{-}CH_2CH_3$ (b) 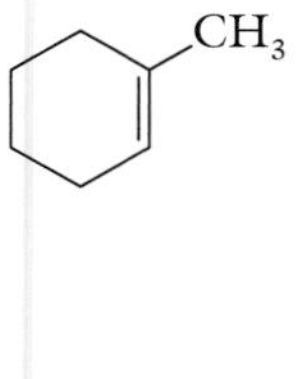

12.6 Give the molecular formula for each of the following compounds:

(a) $CH_2{=}C(CH_3){-}CH{=}C(CH_3)_2$

(b) 

Bonding in Alkenes

12.7 Which (if any) of carbons 1 through 12 in the following compound are sp^2 hybridized?

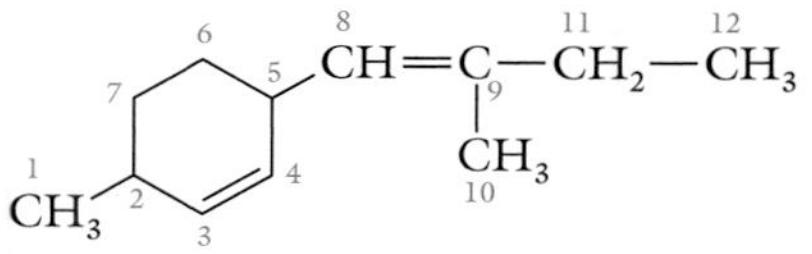

12.8 What are the values of bond angles A through E in the following compound?

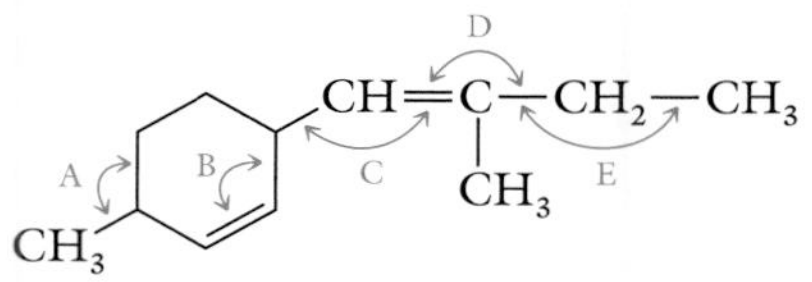

Constitutional Isomers of Alkenes

12.9 Draw structural formulas of all constitutional isomers of C_4H_8. Show only one structural formula for each isomer. Note that the wording of this exercise requires you to include cycloalkanes as well as alkenes of molecular formula C_4H_8.

12.10 Draw structural formulas of all constitutional isomers of C_5H_{10}. Show only one structural formula for each isomer. Note that the wording of this exercise requires you to include cycloalkanes as well as alkenes of molecular formula C_5H_{10}.

12.11 Draw structural formulas for each of the following compounds: (a) an alkene, C_5H_{10}, containing all five carbons in a continuous chain; (b) an alkene, C_5H_{10}, containing three methyl groups; (c) a cyclic compound, C_5H_{10}, containing one methyl group and three CH_2 groups; (d) an alkene, C_5H_{10}, containing an isopropyl group.

12.12 Draw structural formulas for each of the following compounds: (a) a cyclic compound, C_5H_{10}, containing one ethyl group; (b) an alkene, C_6H_{12}, containing two methyl groups and one ethyl group; (c) an alkene, C_6H_{12}, containing two ethyl groups; (d) an alkene, C_6H_{12}, containing a *t*-butyl group.

12.13 For each of the following pairs of structural formulas, indicate whether the pair represents (1) the same compound or (2) different compounds that are constitutional isomers or (3) different compounds that are not isomers.

(a) 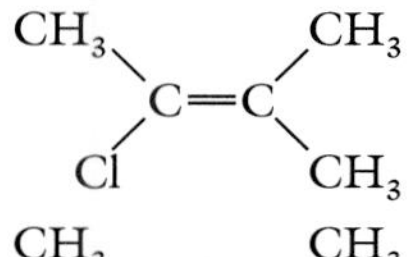 and 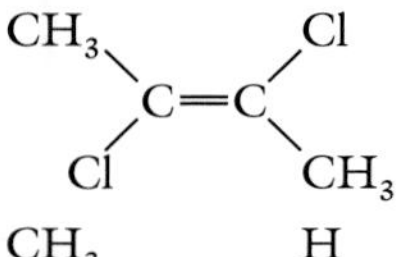

(b) 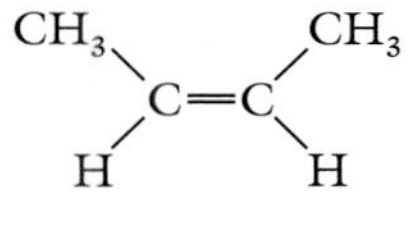 and 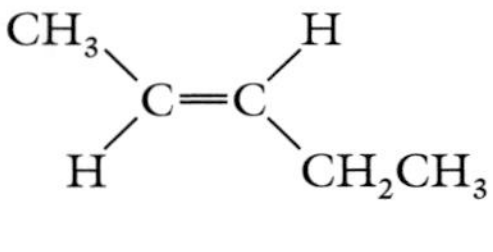

(c) 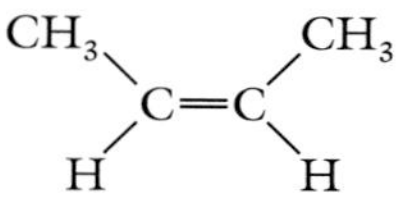 and 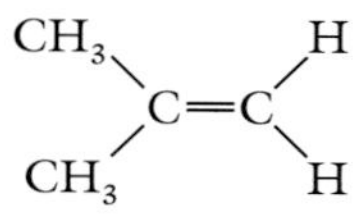

12.14 For each of the following pairs of structural formulas, indicate whether the pair represents (1) the same compound or (2) different compounds that are constitutional isomers or (3) different compounds that are not isomers.

(a) 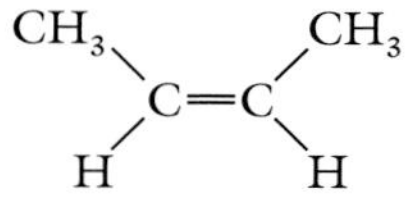 and 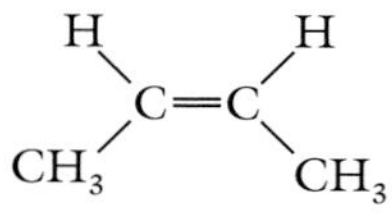

(b) 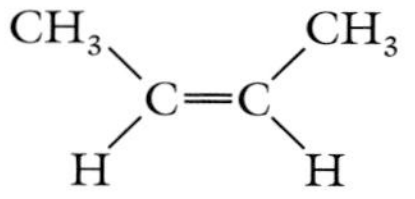 and 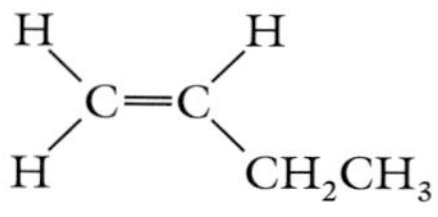

(c) 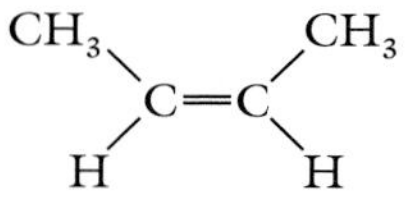 and 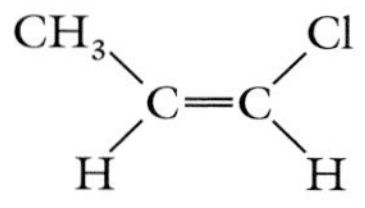

Naming Alkenes

12.15 Draw the structural formula for each of the following compounds: (a) 2-methyl-1-butene; (b) 3-ethyl-2-pentene; (c) 4-isopropyl-2,6-dimethyl-2-heptene; (d) 1,3-dimethylcyclohexene; (e) 3-*t*-butyl-2,4-dimethyl-1-pentene; (f) 5-methyl-1,4-hexadiene.

12.16 Name the following compounds by the IUPAC system:

(a) $CH_3{-}CH_2{-}CH_2{-}C({=}C(CH_3)_2){-}CH_2{-}CH_3$

(b) $CH_2{=}C(CH_3){-}CH{=}C(CH_3)_2$

(c) 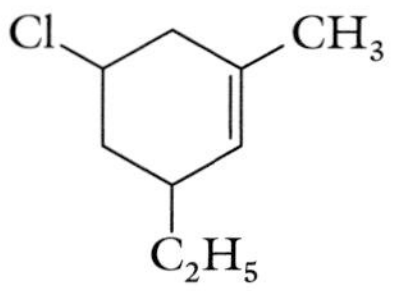

(d) $CH_3{-}CH(CH_2CH_3){-}CH(CH_3){-}CH{=}CH{-}CH(CH_3)_2$

Cis-Trans Isomerism in Alkenes

12.17 Which of the following compounds exist as separate cis and trans isomers? Draw structural formulas of the cis and trans isomers where applicable. (a) 2-Methyl-2-hexene; (b) 3-methyl-2-hexene; (c) 4-methyl-2-hexene; (d) 2-methyl-1-hexene; (e) 1,2-dimethylcyclopentene; (f) 1,2-dimethylcyclopentane.

12.18 Draw structural formulas of the following compounds: (a) *cis*-1-chloro-1-butene; (b) *trans*-3-hexene; (c) *trans*-4,4-dimethyl-2-pentene; (d) *cis*-1,3-dichlorocyclohexane.

12.19 For each of the following pairs of structural formulas, indicate whether the pair represents (1) the same compound or (2) different compounds that are constitutional isomers or (3) different compounds that are cis-trans isomers or (4) different compounds that are not isomers.

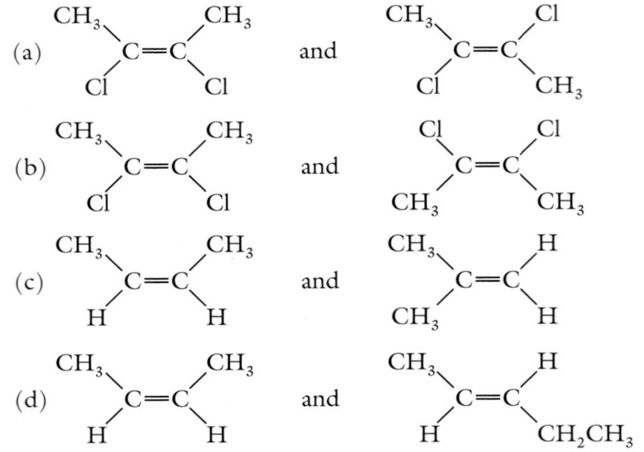

12.20 For each of the following pairs of structural formulas, indicate whether the pair represents (1) the same compound or (2) different compounds that are constitutional isomers or (3) different compounds that are cis-trans isomers or (4) different compounds that are not isomers.

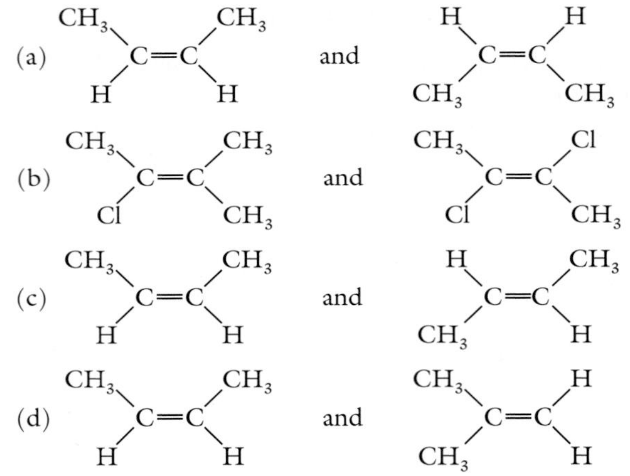

Addition Reactions of Alkenes

12.21 Give the organic product(s) formed in each of the following reactions. If no reaction takes place, write "no reaction." If there is more than one product, indicate only the major product(s).

(a) 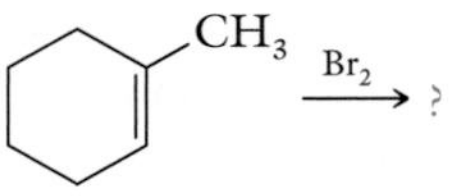

(b) 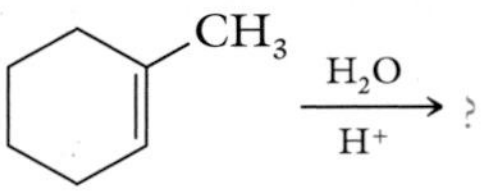

(c) $CH_2{=}C(CH_3)_2 \xrightarrow[Ni]{H_2}$?

(d) $CH_2{=}C(CH_3)_2 \xrightarrow{HCl}$?

(e) 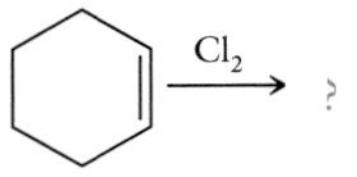

(f) $CH_3{-}CH{=}CH{-}CH_3 \xrightarrow{HCl}$?

(g) $CH_3{-}CH{=}CH{-}CH_2CH_3 \xrightarrow{HBr}$?

(h) $CH_3{-}CH{=}CH{-}CH_2CH_3 \xrightarrow{Br_2}$?

(i) $CH_3{-}CH{=}CH{-}CH_3 \xrightarrow{Cl_2}$?

(j) $CH_3{-}CH{=}CH{-}CH_2CH_3 \xrightarrow[H^+]{H_2O}$?

(k) $CH_3{-}CH{=}CH{-}CH_2CH_3 \xrightarrow[Ni]{H_2}$?

12.22 Indicate whether a reaction takes place when 1-hexene is treated with each of the following reagents: (a) NaOH; (b) H_2O, H^+; (c) H_2/Ni; (d) HCl; (e) Cl_2.

If a reaction takes place, give the organic product(s) formed. If more than one product is formed, indicate only the major product(s). If no reaction takes place, write "no reaction."

12.23 An unknown compound is hexane, methylcyclopentane, or 1-hexene. The red color of bromine is not decolorized when bromine is added to the unknown. Identify the unknown.

12.24 An unknown compound is hexane, methylcyclopentane, or 1-hexene. The red color of bromine is decolorized when bromine is added to the unknown. Identify the unknown.

Addition Polymerization

12.25 Give the equation for the polymerization of acrylonitrile, $CH_2{=}CH{-}CN$.

12.26 Give the equation for polymerization of styrene, $CH_2{=}CH{-}C_6H_5$.

Oxidation of Alkenes

12.27 Give the balanced equation for the combustion of 3-hexene.

12.28 Give the balanced equation for the combustion of 2-methyl-1-hexene.

12.29 An unknown compound is pentane, cyclopentane, or 2-pentene. The orange color of dichromate is not decolorized when dichromate is added to the unknown. Identify the unknown.

12.30 An unknown compound is pentane, cyclopentane, or 2-pentene. The purple color of permanganate is decolorized and a brown precipitate forms when permanganate is added to the unknown. Identify the unknown.

Alkynes

12.31 Draw the structural formula of 3,4-dimethyl-1-pentyne.

12.32 For each of the following systems, show the sequential reaction of the alkyne with the reagent. Show only the major product for each step in the reaction. (a) 1-Butyne + excess Cl_2; (b) propyne + excess HCl; (c) 2-butyne + excess Cl_2.

Aromatics

12.33 What structural features are possessed by aromatic compounds? What chemical property distinguishes aromatic compounds from aliphatic compounds?

12.34 Which of the following compounds are aromatic?

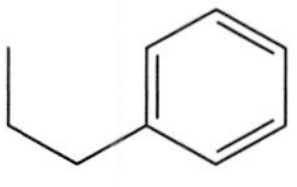
1

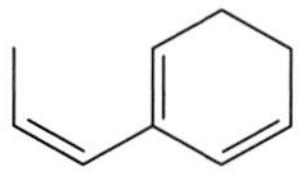
2

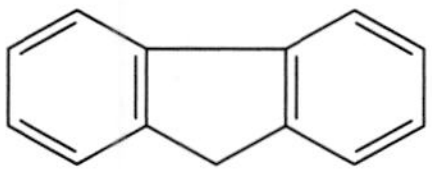
3

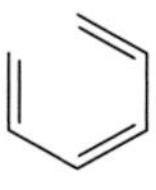
4

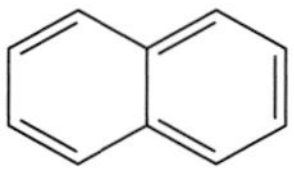
5

6

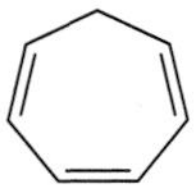
7

8

Isomers and Names of Aromatic Compounds

12.35 For each of the following pairs of structural formulas, indicate whether the pair represents (1) the same compound or (2) different compounds that are constitutional isomers or (3) different compounds that are cis-trans isomers or (4) different compounds that are not isomers.

(a) 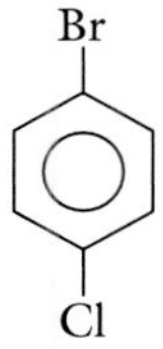 and 

(b) 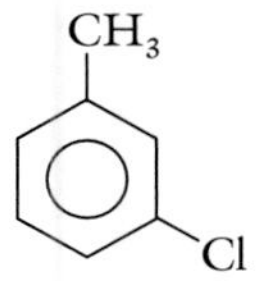 and 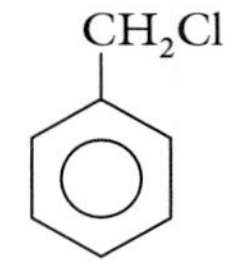

(c) 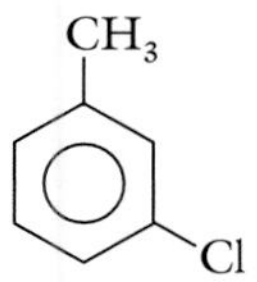 and 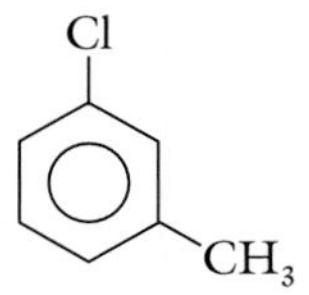

12.36 For each of the following pairs of structural formulas, indicate whether the pair represents (1) the same compound or (2) different compounds that are constitutional isomers or (3) different compounds that are cis-trans isomers or (4) different compounds that are not isomers.

(a) 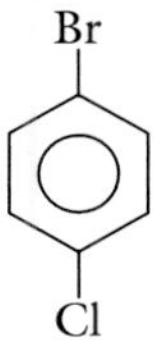 and 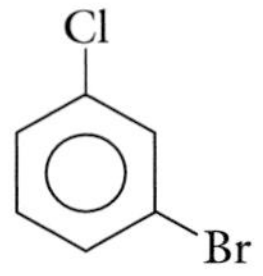

(b) 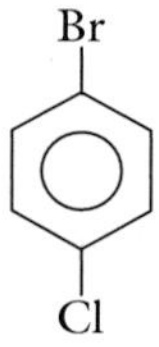 and 

(c)  and 

12.37 Name each of the following compounds by the IUPAC system:

(a) 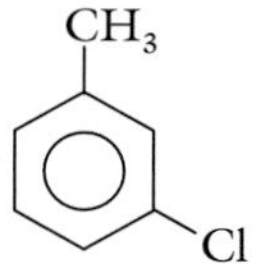

(b) 

(c) 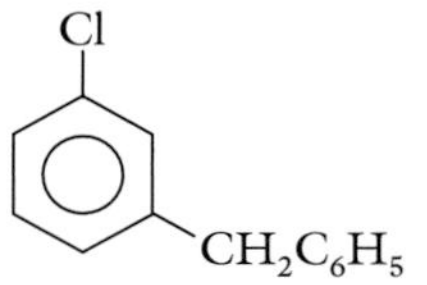

(d) 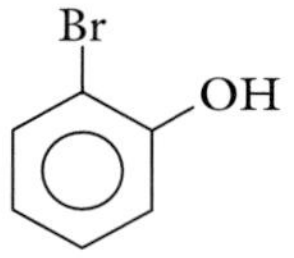

(e) 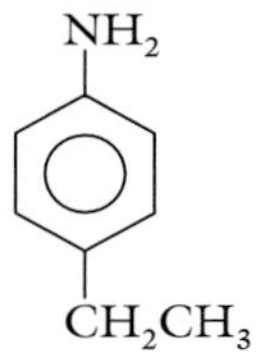

(f) 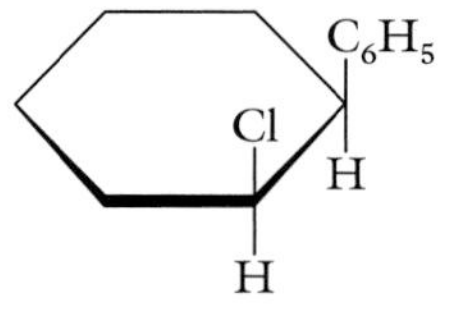

12.38 Name each of the following compounds by the IUPAC system:

(a) 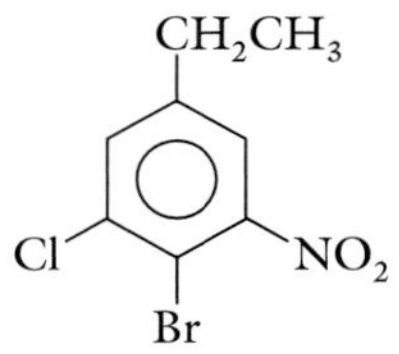

(b) 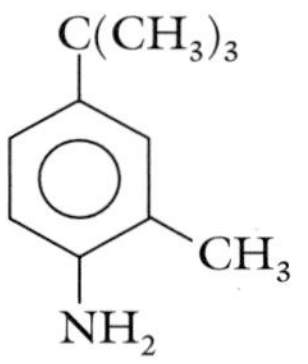

(c) 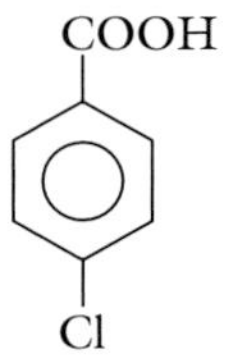

(d) 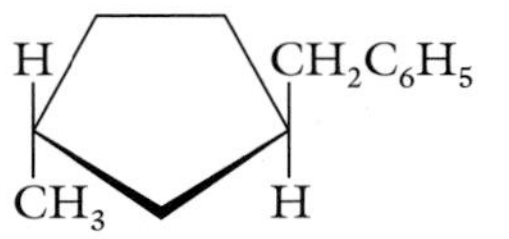

Reactions of Aromatics

12.39 Indicate whether a reaction takes place for each of the following reaction mixtures: (a) benzene + NaOH; (b) benzene + HBr; (c) benzene + SO_3 + H_2SO_4; (d) benzene + HNO_3 (with H_2SO_4 as catalyst); (e) combustion of benzene (give balanced equation).

If a reaction takes place, give the organic product(s) formed. If more than one product is formed, indicate only the major product(s). If no reaction occurs, write "no reaction."

12.40 Indicate whether a reaction takes place for each of the following reaction mixtures: (a) *t*-butylbenzene + hot acidic $K_2Cr_2O_7$; (b) benzene + H_2O; (c) benzene + $(CH_3)_2CHCl$ + $FeCl_3$; (d) combustion of ethylbenzene (give balanced equation); (e) benzene + Br_2 + $FeBr_3$.

If a reaction takes place, give the organic product(s) formed. If more than one product is formed, indicate only the major product(s). If no reaction takes place, write "no reaction."

Unclassified Exercises

12.41 Draw structural formulas of all possible isomers for compounds containing a benzene ring and having the molecular formula C_7H_7Br.

12.42 Draw structural formulas for each of the following compounds: (a) an alkene, C_6H_{12}, containing four methyl groups; (b) an alkyne, C_6H_{10}, containing a *t*-butyl group; (c) a cycloalkane, C_6H_{12}, containing one methyl group; (d) a cycloalkene, C_6H_{10}, containing no methyl groups; (e) a compound, C_8H_{10}, containing a benzene ring and two methyl groups; (f) a compound, C_8H_{10}, containing a benzene ring and one ethyl group.

12.43 Which of the following compounds exist as separate cis and trans isomers? Draw structural formulas of the cis and trans isomers where applicable. (a) 2-Hexene; (b) 1,2-dimethylbenzene; (c) 1-chloro-2-methylcyclohexane; (d) 1,2-dichlorocyclopentene; (e) 1-pentyne; (f) 2-pentyne.

12.44 Each of the following names is an incorrect IUPAC name. State why the name is incorrect. Indicate, if possible, what compound the namer of the compound had in mind and give the correct IUPAC name. (a) 4-Ethyl-1-pentene; (b) 5-*t*-butyl-1-isopropylcyclopentene; (c) 1,2,4-trimethyl-3-cyclohexene; (d) 4-butyl-1-methyl-2-hexene; (e) 1-propyl-5-methylbenzene; (f) *cis*-2-chloro-3-methyl-2-butene; (g) 2,4-pentadiene.

12.45 For each of the following pairs of structural formulas, indicate whether the pair represents (1) the same compound or (2) different compounds that are constitutional isomers or (3) different compounds that are cis-trans isomers or (4) different compounds that are not isomers.

(a)
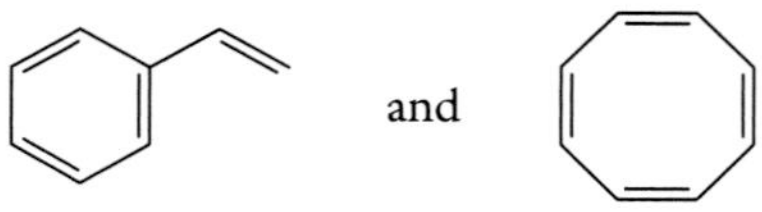

(b)
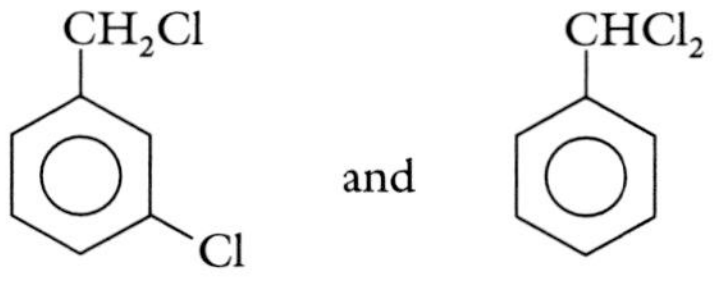

(c)
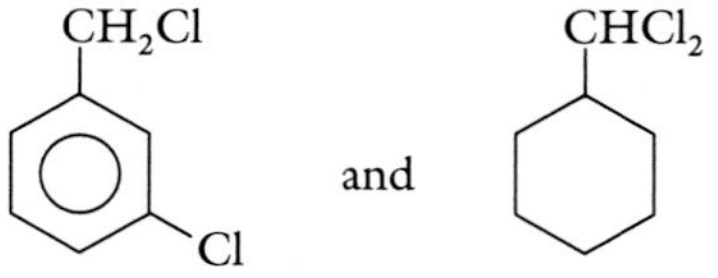

(d)
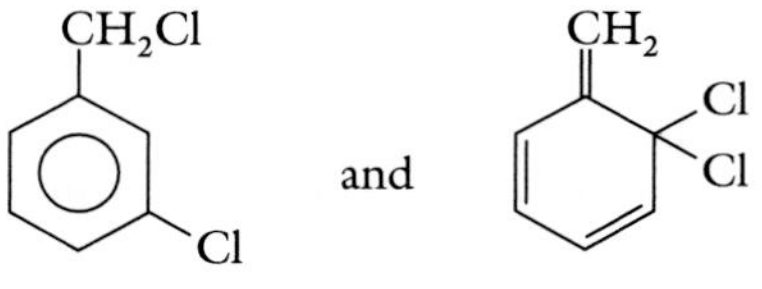

(e)
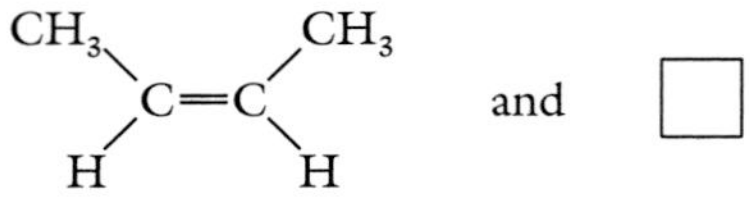

(f)
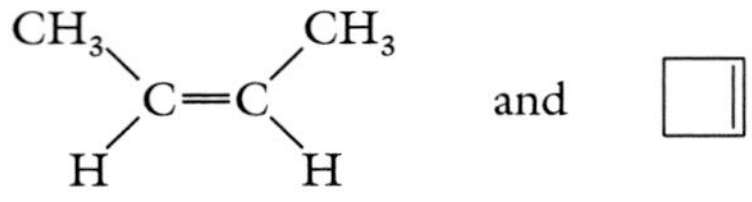

12.46 For each of the following pairs of structural formulas, indicate whether the pair represents (1) the same compound or (2) different compounds that are constitutional isomers or (3) different compounds that are cis-trans isomers or (4) different compounds that are not isomers.

(a)
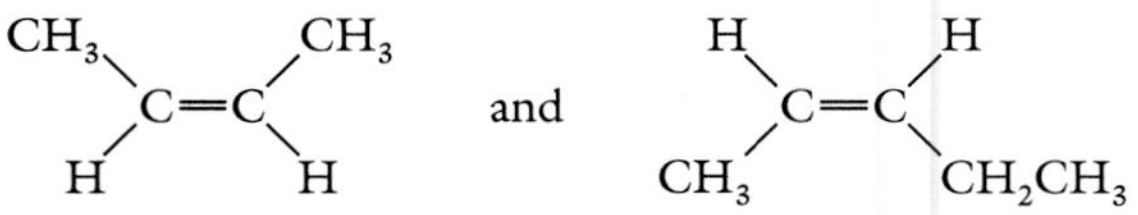

(b)
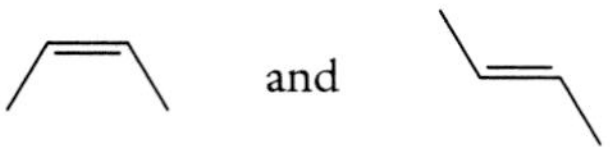

(c)
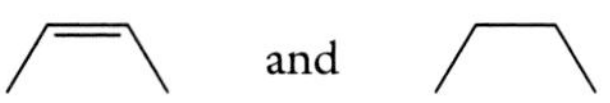

(d)
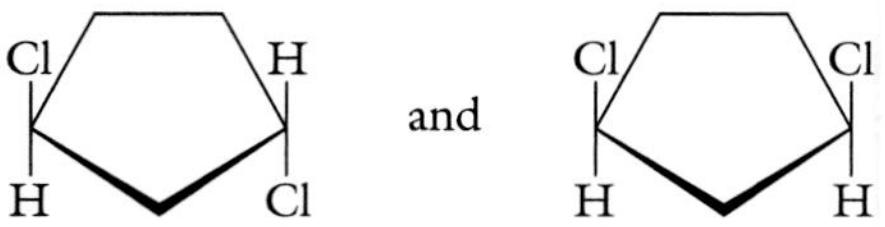

(e) $CH{\equiv}C{-}CH_2{-}CH_3$ and $CH_2{=}CH{-}CH{=}CH_2$

(f) $CH{\equiv}C{-}CH_2{-}CH_3$ and $CH_3{-}C{\equiv}C{-}CH_3$

12.47 Compare the boiling points and solubilities in water of cyclohexane, cyclohexene, and benzene.

12.48 Consider cyclohexane, cyclohexene, and benzene. Which will react with bromine under each of the following conditions? (a) Br_2; (b) Br_2 + high temperature or UV; (c) Br_2 + $FeBr_3$.

12.49 Consider cyclohexane, cyclohexene, and benzene. Which compound(s) will react with each of the following reagents? (a) Combustion (burning in air); (b) NaOH; (c) HCl; (d) H_2O/H^+; (e) hot acidic $KMnO_4$.

12.50 The addition of HCl to 1-pentene gives one major and one minor product, 2-chloropentane and 1-chloropentane, respectively. The addition of HCl to 2-pentene gives two products, 2-chloropentane and 3-chloropentane, in equal amounts. Explain the difference between the reactions of 1-pentene and 2-pentene.

12.51 What product is formed when 1,3-butadiene, $CH_2{=}CH{-}CH{=}CH_2$, reacts with excess Br_2?

12.52 Give the equation for the polymerization of each of the following monomers: (a) vinyl chloride (1-chloroethene); (b) isobutylene (2-methylpropene); (c) methyl methacrylate:

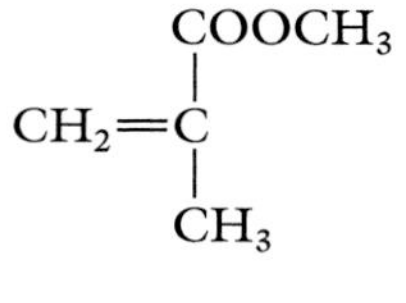

12.53 What is the molecular formula of each of the following compounds?

(a) 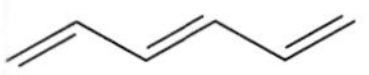

(b) 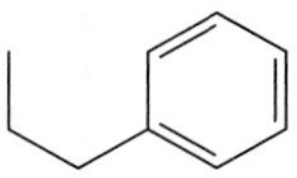

(c) 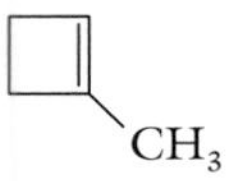

(d) 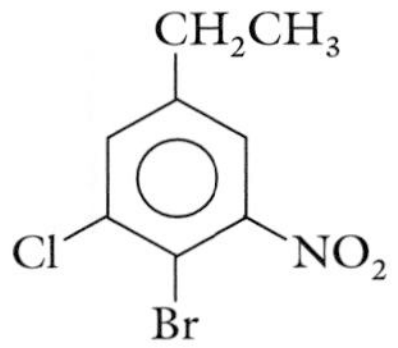

12.54 A chemist has two unknown samples, A and B. One of the samples is cyclohexene and the other is benzene, but the chemist does not know which sample is which. Both samples decolorize Br_2 when $FeBr_3$ is present, but only sample A decolorizes Br_2 in the absence of $FeBr_3$. Which sample is benzene and which sample is cyclohexene?

12.55 A chemist has two unknown samples, A and B. One of the samples is 1-hexene and the other is 1-hexyne, but the chemist does not know which sample is which. Both samples decolorize Br_2, but sample B decolorizes twice as much bromine as does sample A. Which sample is 1-hexene and which sample is 1-hexyne?

12.56 What is the hybridization of carbons 1 through 6 in the following compound? What are the bond angles about those carbons?

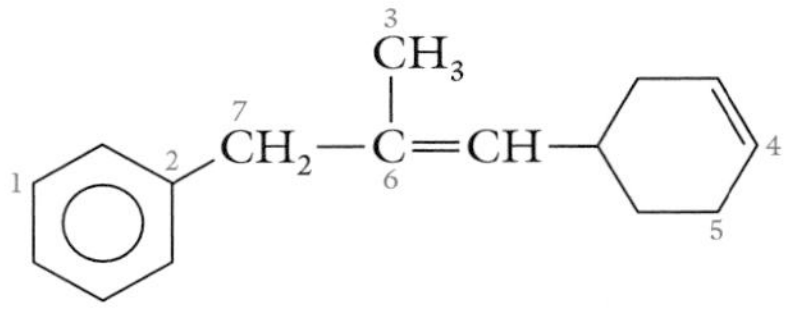

12.57 Give the organic product(s) formed in each of the following reactions. If no reaction takes place, write "no reaction." If more than one product is formed, indicate only the major product.

(a) 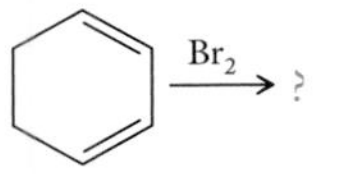

(b) 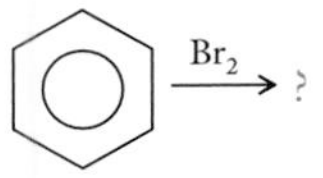

(c) 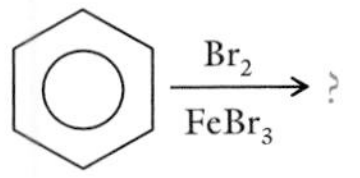

(d) 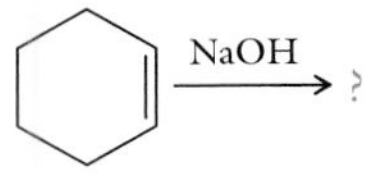

(e) 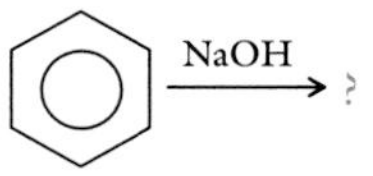

(f) 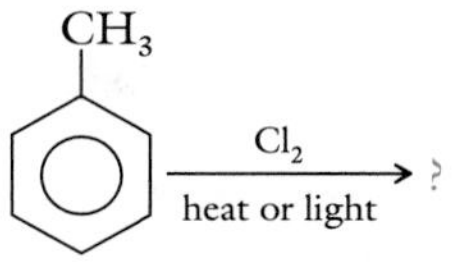

(g) 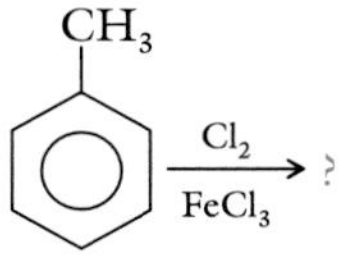

(h) 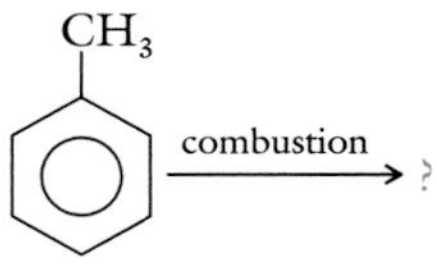

(i) 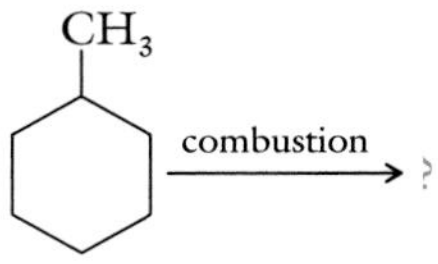

(j) 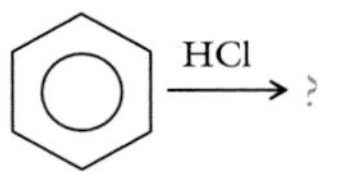

(k) 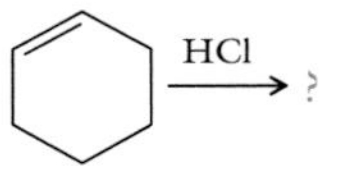

12.58 An unknown compound is hexane, methylcyclopentane, or 1-hexene. The purple color of permanganate is not decolorized when permanganate is added to the unknown. Identify the unknown.

12.59 An unknown compound is hexane, methylcyclopentane, or 1-hexene. The orange color of dichromate is replaced by a green-colored solution when dichromate is added to the unknown. Identify the unknown.

12.60 Which of the following compounds are aromatic?

(a) 

(b) 

(c) 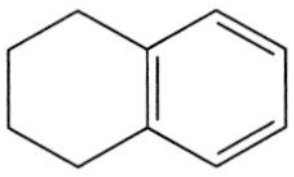

(d) 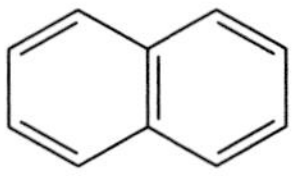

(e) 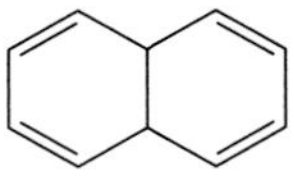

(f) 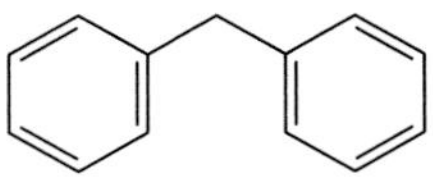

Chemical Connections

12.61 The president of the ABC company wants to produce rubbing alcohol (2-propanol) by the acid-catalyzed addition of water to propene. One chemist in his employ, Bill Smith, says that the proposed reaction will not work, because the major product will be 1-propanol instead of 2-propanol. Another chemist, Mary Jones, disagrees and supports her viewpoint by invoking Markovnikov's rule. Who is correct?

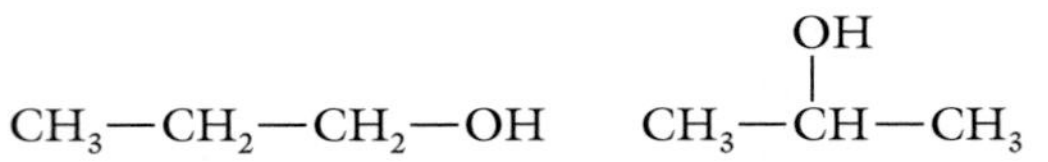

1-Propanol 2-Propanol

12.62 Compound A can exist as a pair of cis-trans isomers, but compound B cannot. Explain why.

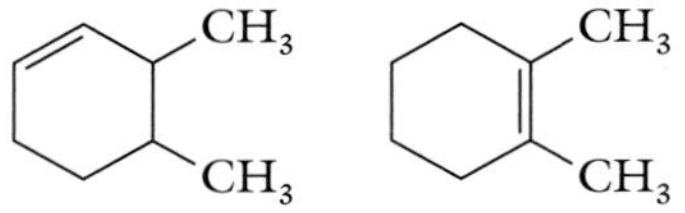

A B

12.63 Geraniol and myrcene are extracted from the oils of roses and bay leaves, respectively. The structures of geraniol and myrcene are shown in Box 12.1. A perfume manufacturer has ordered some geraniol for use in formulating a perfume. The shipment arrives, but there is a suspicion that myrcene was delivered instead of geraniol. Suggest a simple chemical test to distinguish between geraniol and myrcene.

12.64 Box 12.9 shows the structure of phenanthrene, a carcinogen found in tobacco smoke and barbecued meats. What is the molecular formula of phenanthrene?

12.65 The medical community advises that we lower the proportion of our dietary fat intake that contains saturated fats, thereby increasing the proportion that contains unsaturated fats. Although we have not yet studied the structures of fats, what do you think are the differences between saturated and unsaturated fats?

12.66 2,4.6-Trinitrotoluene (also called TNT) is used as an explosive. It is produced from toluene ($C_6H_5—CH_3$). What reagent(s) and reaction conditions are needed to convert toluene into 2,4,6-trinitrotoluene?

12.67 The rate of addition of HCl to a series of alkenes follows the order: 2-methylpropene > propene > ethene. Explain this order in relation to the mechanism for the addition reaction (Boxes 12.4 and 12.5).

12.68 The addition of HCl to 2-methylpropene yields 2-chloro-2-methylpropane as the only product, whereas 2-pentene (both the cis and trans isomers) yields about equal amounts of two products: 2-chloropentane and 3-chloropentane. Explain the difference in relation to the mechanism for the Markovnikov rule (Box 12.5).

CHAPTER 13

ALCOHOLS, PHENOLS, ETHERS, AND THEIR SULFUR ANALOGUES

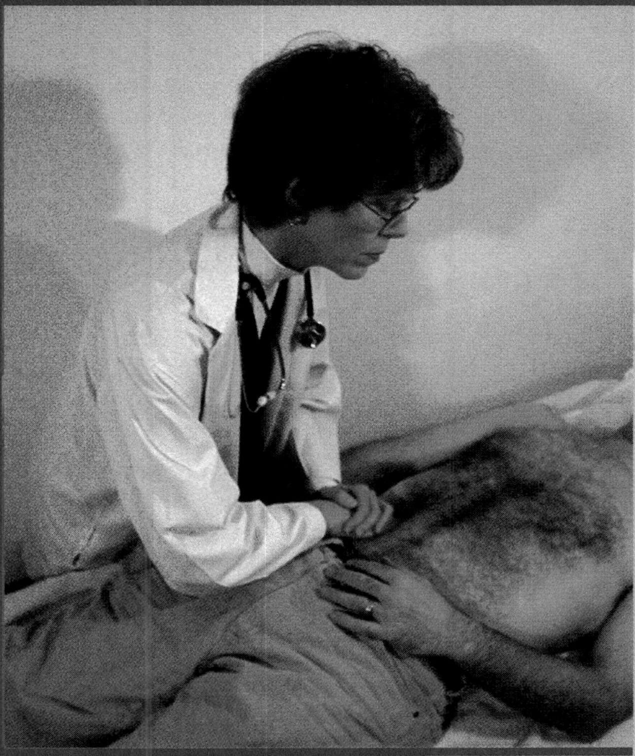

CHEMISTRY IN YOUR FUTURE

The patient in bed 4 of your ward, suffering from alcoholic liver disease and, as of this morning, pneumonia, has rapidly gone from bad to worse. The liver can lose 75% of its tissue without ceasing to function, but it is no match for the toxic effects of long-term chronic abuse of alcohol. Why does this particular organic chemical pose such a risk to the body's largest organ? The answer is complex, but some aspects are explained in Chapter 13.

LEARNING OBJECTIVES

- Draw and name alcohols.
- Describe the physical properties of alcohols.
- Describe and write equations for acid-base, dehydration, and oxidation reactions of alcohols.
- Draw and name phenols.
- Describe the properties of phenols.
- Draw and name ethers.
- Describe the properties of ethers.
- Draw and name thiols and disulfides.
- Describe the properties of thiols and disulfides.

—O—
Single bond

C═O
Double bond

Our study of organic compounds now turns to the many important compounds containing oxygen in addition to carbon and hydrogen. In this chapter, we will explore the chemistry of alcohols, phenols, and ethers—families in which oxygen is connected by single bonds to other atoms. We will also look briefly at some of the sulfur analogues of those three families. Then, in Chapters 14 through 16, we will cover the aldehydes, ketones, acids, esters, anhydrides, and amides—all families in which oxygen has a double bond to carbon.

We encounter alcohols in many aspects of our lives—in beer and other alcoholic beverages, in rubbing alcohol, as solvents for cosmetics and automotive antifreeze formulations, in cough medicines, and as starting materials for the synthesis of drugs and other widely used chemicals. Phenols find uses as antiseptics and disinfectants and as preservatives for food, gasoline, and other consumer goods. Ethers are employed as surgical anesthetics and in the manufacture of epoxy adhesives.

A number of physiologically important compounds, including carbohydrates, proteins, and lipids, possess the same functional groups as those found in alcohols, phenols, ethers, and their sulfur analogs. Biological molecules tend to possess the functional groups of two or more families. For example, carbohydrates are both alcohols and aldehydes or ketones. Carbohydrates cannot be understood without first studying the alcohols, aldehydes, and ketones, because the functional groups of those families interreact in a unique manner.

13.1 STRUCTURAL RELATIONS OF ALCOHOLS, PHENOLS, AND ETHERS

Table 13.1 shows the general formulas, along with specific examples, of alcohols, phenols, and ethers. **Alcohols** and **phenols** contain an —OH group—a **hydroxyl group**—attached to a carbon atom. The carbon atom bearing the —OH in alcohols is a saturated carbon; that is, it is sp^3 hybridized and connected by single bonds to four adjacent atoms. In phenols, the —OH group is attached to one of the sp^2-hybridized carbons in a benzene ring. Note that the hydroxyl group of alcohols and phenols is not the hydroxide anion, OH^-, of strong bases such as KOH and NaOH.

Ethers, like alcohols and phenols, contain an oxygen that has bonds to two different atoms, but, in contrast with alcohols and phenols, neither of the bonds is to hydrogen. The oxygen of an ether has single bonds to two different carbons, which may be either aliphatic or aromatic carbons.

TABLE 13.1 Alcohols, Phenols, and Ethers

	General Formula	Examples
alcohol	R—OH with R = aliphatic	CH_3—OH
phenol	R—OH with R = aromatic	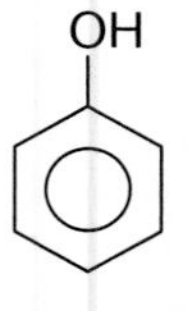
ether	R—O—R′ with R, R′ = same or different group, aliphatic or aromatic	CH_3—O—CH_3

Alcohols, phenols, and ethers can be thought of as organic derivatives of water. Replacement of one of the hydrogens in water with an organic group yields an alcohol or phenol, depending on whether the organic group is aliphatic or aromatic. Replacement of the H in the —OH group of an alcohol or phenol (that is, replacement of the second H of water) by an organic group yields an ether:

$$\text{H—O—H} \xrightarrow{\text{replacement of H by R}} \underset{\substack{\textbf{Alcohol (R = aliphatic)} \\ \textbf{Phenol (R = aromatic)}}}{\text{R—O—H}} \xrightarrow{\text{replacement of H by R}} \underset{\textbf{Ether}}{\text{R—O—R}}$$

Example 13.1 Distinguishing among alcohols, phenols, and ethers

Which of the following structures is an alcohol, a phenol, an ether, or something else?

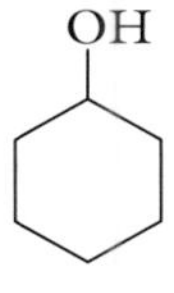
1

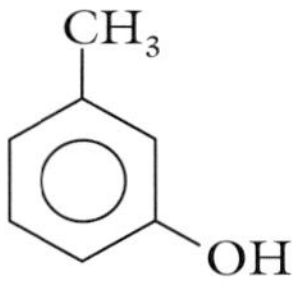
2

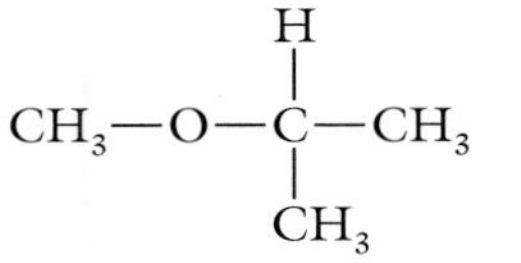
3

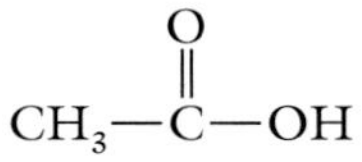
4

Solution

Structure 1 is an alcohol because the —OH is attached to a saturated carbon. Structure 2 is a phenol because the —OH is attached to a benzene ring. Structure 3 is an ether because the oxygen atom is bonded to two different carbons. Structure 4 has a hydroxyl group, but it is neither an alcohol nor a phenol, because the carbon holding the —OH is neither saturated nor part of a benzene ring. Structure 4 is a carboxylic acid (Chapter 15).

Problem 13.1 Which of the following structures is an alcohol, a phenol, an ether, or something else?

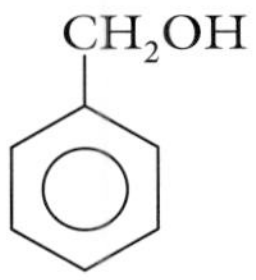
1

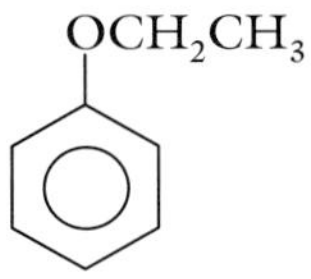
2

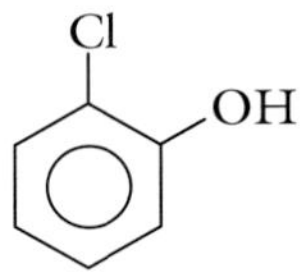
3

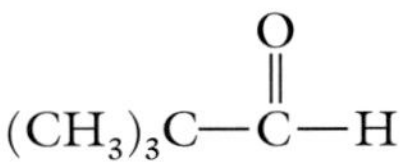
4

The oxygen atoms in alcohols, phenols, and ethers use sp^3-hybrid orbitals in forming bonds, analogous to sp^3-hybridized carbons but with a difference. The oxygen atom has six outer-shell electrons, two more than carbon. Whereas carbon has four half-filled sp^3 orbitals and is tetravalent, oxygen has two half-filled and two filled sp^3 orbitals and is divalent: the pairs of electrons in the two filled sp^3 orbitals do not participate in bonding to other atoms and are called **nonbonding electrons.** The two half-filled orbitals bond oxygen to other atoms.

The nonbonding electrons are shown (by pairs of electron dots) in Lewis structures but not in structural formulas (Section 3.7). Whichever representation is used, always keep in mind that the nonbonding electrons are present. The oxygen atom in water, alcohols, phenols, and ethers has a complete octet of electrons.

$$\underset{\textbf{Lewis structure}}{\text{R—}\ddot{\underset{\cdot\cdot}{\text{O}}}\text{H}} = \underset{\textbf{Structural formula}}{\text{R—OH}}$$

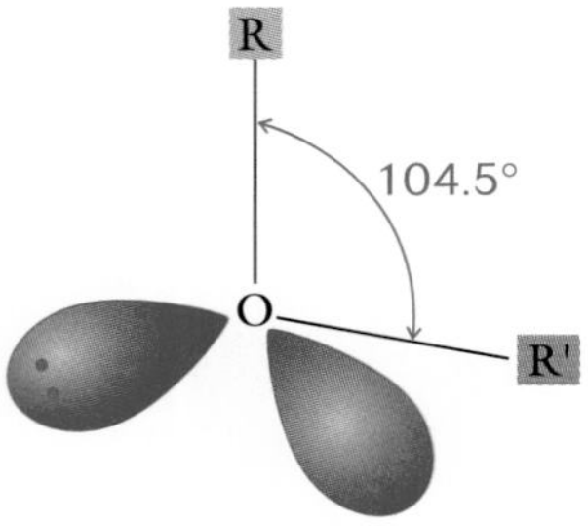

Figure 13.1 Bond angle for sp^3-hybridized oxgen in water (R = R′ = H), alcohols and phenols (R = aliphatic and aromatic, respectively, and R′ = H), and ethers (R, R′ = aliphatic or aromatic).

$CH_3—OH$
Methanol

$CH_3—CH_2—OH$
Ethanol

The bond angle about the oxygen atoms in water, alcohols, phenols, and ethers is close to 104.5° (Figure 13.1). This angle is slightly distorted from the tetrahedral bond angle of 109.5° because the two nonbonded electron pairs repulse the adjacent bonded electron pairs and compress their bond angle (Section 3.10).

13.2 CONSTITUTIONAL ISOMERISM IN ALCOHOLS

Constitutional isomerism in alcohols, as in alkenes, arises out of two kinds of connectivity differences:

- Different carbon skeletons
- Different placement of the functional group (—OH) on a carbon skeleton

No isomeric alcohols are possible for the two simplest alcohols—methanol and ethanol—which contain one carbon and two carbons, respectively. There are, however, two C_3H_8O alcohols, owing to two different placements of the —OH group on the three-carbon chain:

$CH_3—CH_2—CH_2—OH$
1-Propanol

$CH_3—CH(OH)—CH_3$
2-Propanol

Example 13.2 Drawing constitutional isomers of alcohols

Both types of connectivity differences (different carbon skeletons and different placements of functional group) exist for $C_4H_{10}O$ alcohols, resulting in four constitutional isomers. Give structural formulas of the isomers.

Solution

Two different C_4 carbon skeletons are possible:

C—C—C—C
1

C—C(C)—C
2

Two isomers are possible for skeleton 1 because the —OH group can bond at either of two locations—on an end carbon or on a next-to-end carbon of the carbon chain:

$CH_3—CH_2—CH_2—CH_2—OH$
1a

$CH_3—CH(OH)—CH_2—CH_3$
1b

Be careful not to draw duplicate structural formulas of the same isomer. Placement of the —OH on the far-left carbon gives a duplicate of skeleton 1a, whereas placement of the —OH on the second carbon from the right gives a duplicate of skeleton 1b.

Two additional C_4 alcohols are created by attaching the —OH group at different locations on skeleton 2:

$CH_3—C(CH_3)(OH)—CH_3$
2a

$CH_3—CH(CH_3)—CH_2—OH$
2b

Problem 13.2 How many alcohol structures are possible for the molecular formula $C_5H_{12}O$? Draw one structural formula for each isomer.

The presence of an —OH group in a compound does not change the C:H ratio compared with the corresponding alkane, cycloalkane, or alkene. Therefore, the molecular formula tells us whether an alcohol has a ring or double bond:

- The C:H ratio is the same as for an alkane, C_nH_{2n+2}, if the alcohol has no double bond or ring.
- The C:H ratio is the same as for an alkene or cycloalkane, C_nH_{2n}, if the alcohol has a double bond or ring.

Example 13.3 Determining the general formula for an alcohol

Show that alcohols 1 and 2 follow the general formulas $C_nH_{2n+2}O$ and $C_nH_{2n}O$, respectively.

$CH_3—CH_2—CH_2—OH$ (1) $CH_2=CH—CH_2—OH$ (2)

Solution

The formula of alcohol 1 is C_3H_8O with $n = 3$ and $2n + 2 = 8$. The formula of alcohol 2 is C_3H_6O with $n = 3$ and $2n = 6$.

Problem 13.3 Show that alcohols 1 and 2 follow the general formulas $C_nH_{2n+2}O$ and $C_nH_{2n}O$, respectively.

$(CH_3)_3C—CH_2—OH$

1

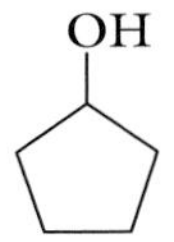

2

13.3 CLASSIFYING AND NAMING ALCOHOLS

Alcohols show differences in chemical reactivity, depending on their structural classification as **primary** (1°), **secondary** (2°), or **tertiary** (3°) **alcohols.** Alcohols are 1°, 2°, or 3° alcohols, depending on whether the carbon holding the —OH group is a 1°, 2°, or 3° carbon. Recall from Section 11.6 that a 1°, 2°, or 3° carbon is directly bonded to a total of one, two, or three other carbon atoms, respectively.

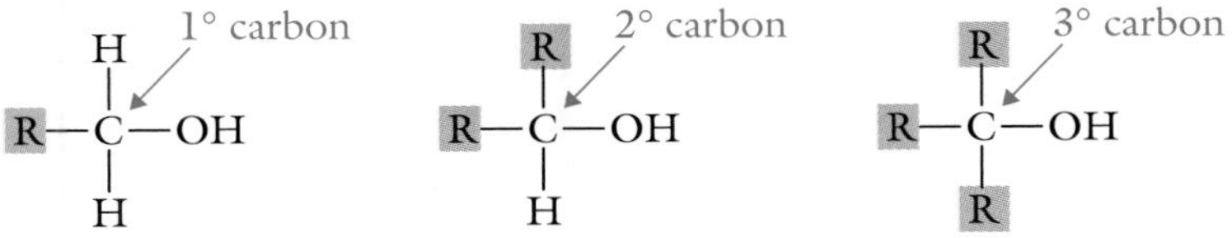

Primary alcohol **Secondary alcohol** **Tertiary alcohol**

Example 13.4 Classification of alcohols

Classify the following alcohols as primary, secondary, or tertiary.

(a) $CH_3—CH_2—CH_2—OH$ (b)

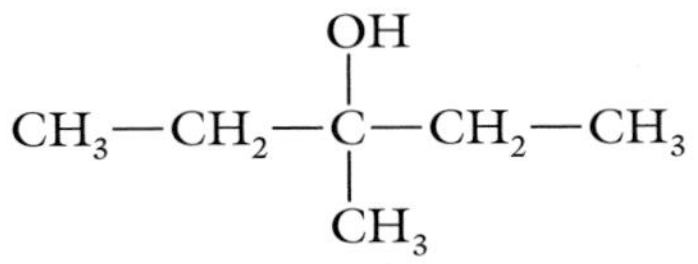

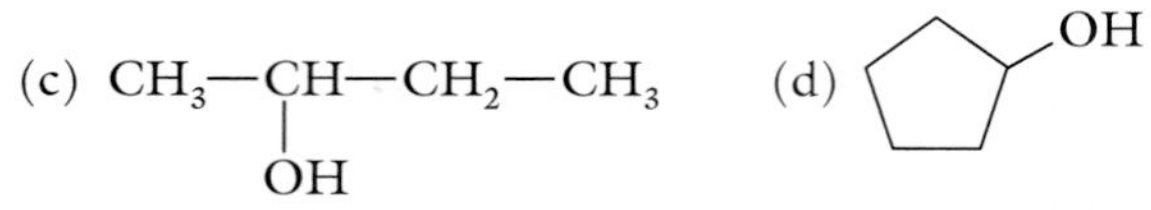

Solution

Locate the carbon bonded to the —OH group and then count the carbons directly attached to that carbon: (a) primary; (b) tertiary; (c) secondary; (d) secondary.

Problem 13.4 Classify the following alcohols as primary, secondary, or tertiary.

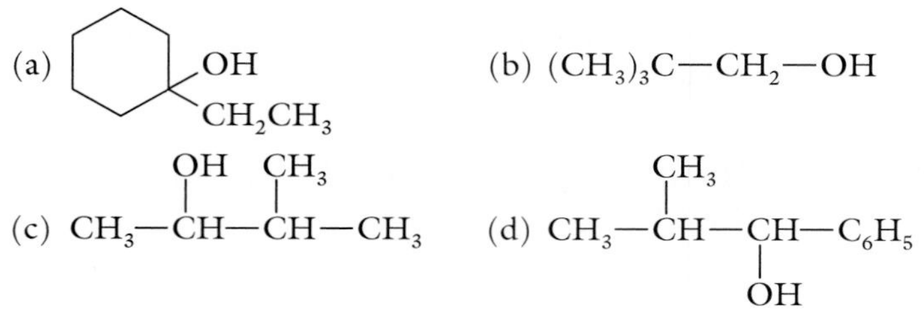

In the IUPAC nomenclature system, the rules for naming alcohols are a variation on the rules for naming alkanes:

Rules for naming alcohols

- Find the longest continuous chain that contains the carbon holding the —OH group.
- The ending **-e** of the alkane name is changed to **-ol.**
- The longest continuous chain is numbered from the end nearest the —OH group. The position of the —OH group is indicated by placing the number of the carbon holding the —OH in front of the base name.
- The names of substituents are added as prefixes, preceded by numbers indicating their positions on the longest chain.
- A cyclic compound with an —OH group attached to a ring carbon is named as a **cycloalkanol.** Numbering of the ring starts at the carbon holding the —OH and proceeds to other substituents on the ring in the direction that yields the lowest numbers for those substituents.
- A **polyfunctional alcohol** contains more than one —OH group in the molecule. Compounds with two and three —OH groups are called **diols** and **triols,** respectively. **Alkanediol** and **alkanetriol** form the bases of the IUPAC names of such compounds. Numbers are attached at the beginning of the names to specify the position of each —OH group (for example, 1,2-ethanediol, 1,2,3-propanetriol, and so forth).

Example 13.5 Naming alcohols by the IUPAC nomenclature system

Name the following compounds by the IUPAC system.

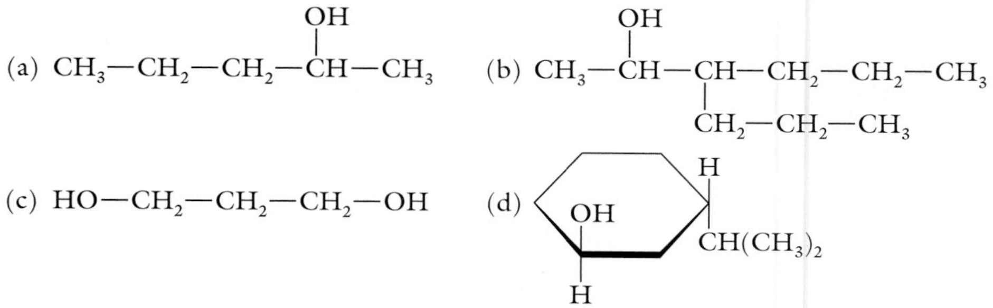

Solution

(a) The five-carbon chain holding the —OH group is numbered from the right side, the end nearest the —OH group, to yield the name 2-pentanol. Numbering from the other end is incorrect, because it yields 4-pentanol, containing a larger number for the position of the —OH group.

(b) The absolute longest chain contains seven carbons but does not contain the carbon holding the —OH. The longest chain holding the —OH group is the six-carbon chain shown here:

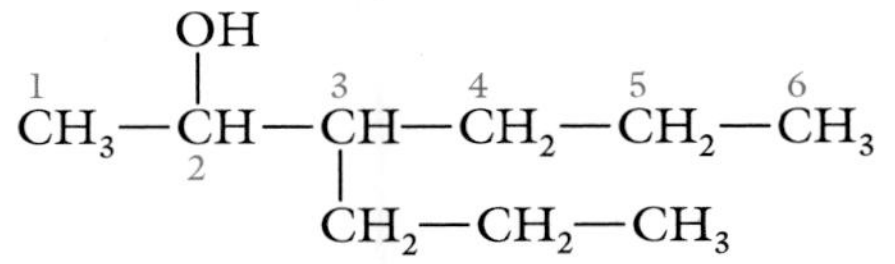

The chain is numbered from the left side, the side nearest the —OH. Attached to the six-carbon chain is a propyl group at C3. The name is 3-propyl-2-hexanol.

(c) The compound contains a three-carbon chain with —OH groups at C1 and C3 and is named 1,3-propanediol.

(d) The —OH group is at C1 of the ring. Proceeding counter clockwise to the isopropyl group yields the name 3-isopropylcyclohexanol. Proceeding, clockwise is incorrect, because it yields a larger number in the name (4-isopropylcyclohexanol) for the position of the isopropyl group. The complete name is *trans*-3-isopropylcyclohexanol because the structure shown has the —OH and isopropyl groups projecting on different sides of the plane of the ring.

Problem 13.5 Name the following compounds by the IUPAC system.

(a) 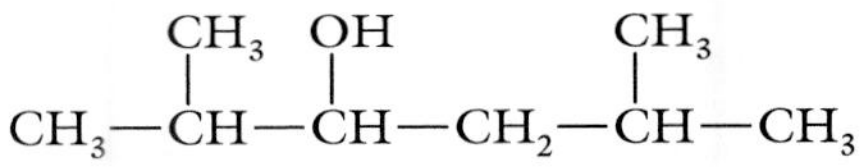

(b) 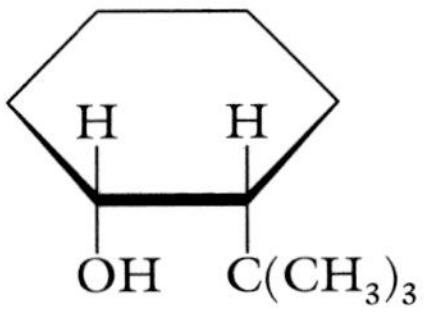

(c) 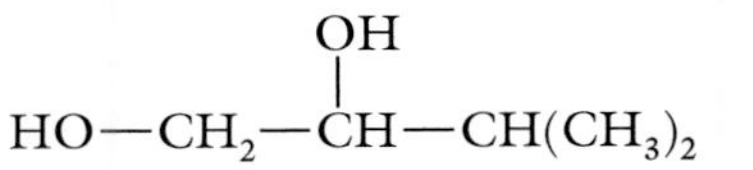

(d) 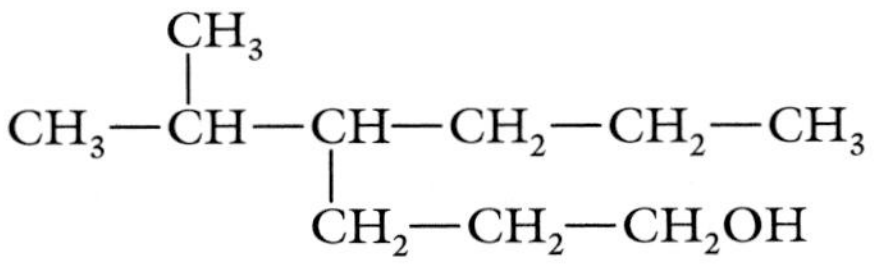

Simple alcohols are usually referred to by common names (Table 13.2 on the following page), which consist of the name of the alkyl group attached to the —OH group followed by a space and then the word alcohol. Simple alcohols are those containing one of the simple alkyl groups: unbranched alkyl groups from methyl, ethyl, propyl, butyl through decyl; branched three- and four-carbon alkyl groups (isopropyl, isobutyl, *s*-butyl, *t*-butyl).

Box 13.1 on page 378 describes many of the industrially important alcohols. Alcoholic beverages and their health aspects are described in Boxes 13.2 on page 379 and 13.3 on page 380.

13.4 PHYSICAL PROPERTIES OF ALCOHOLS

The physical properties of alcohols are very different from those of hydrocarbons (alkanes, alkenes, alkynes, and aromatics). Hydrocarbons, with their low secondary attractive forces (Section 11.9), have low boiling and melting points and are insoluble in water. Alcohols have high secondary attractive forces, and as a consequence are characterized by

- Much higher boiling and melting points than those of hydrocarbons
- Solubility in water

TABLE 13.2 IUPAC and Common Names of Alcohols

Alcohol	IUPAC Name	Common Name
$CH_3—OH$	methanol	methyl alcohol
$CH_3CH_2—OH$	ethanol	ethyl alcohol
$CH_3CH_2CH_2—OH$	1-propanol	propyl alcohol
$CH_3\overset{\overset{OH}{\vert}}{C}HCH_3$	2-propanol	isopropyl alcohol
$CH_3CH_2CH_2CH_2—OH$	1-butanol	butyl alcohol
$CH_3\overset{\overset{OH}{\vert}}{C}HCH_2CH_3$	2-butanol	*s*-butyl alcohol
$CH_3\overset{\overset{CH_3}{\vert}}{C}HCH_2—OH$	2-methyl-1-propanol	isobutyl alcohol
$CH_3\underset{\underset{CH_3}{\vert}}{\overset{\overset{CH_3}{\vert}}{C}}—OH$	2-methyl-2-propanol	*t*-butyl alcohol
$HO—CH_2CH_2—OH$	1,2-ethanediol	ethylene glycol
$HO—CH_2\overset{\overset{OH}{\vert}}{C}HCH_3$	1,2-propanediol	propylene glycol
$HO—CH_2\overset{\overset{OH}{\vert}}{C}HCH_2—OH$	1,2,3-propanetriol	glycerol; glycerin

The physical properties of alcohols arise from the hydrogen bonds (Section 6.3) that result from the presence of the highly polar $^{\delta-}O—H^{\delta+}$ bonds. There are strong attractive forces between the partly positive hydrogen end of the O—H bond of one alcohol molecule and the partly negative oxygen end of the O—H bond of a neighboring alcohol molecule. This attractive force, depicted by the dashed line in Figure 13.2, must be broken for an alcohol to boil. Figure 13.2 is an oversimplification because any one alcohol molecule is surrounded (three-dimensionally) by hydrogen bonds to several neighboring alcohol molecules. Higher temperatures are required to boil alcohols compared with hydrocarbons because the hydrogen-bonding attractions in alcohols are much stronger than the London forces in hydrocarbons.

Table 13.3 shows that the difference in boiling points between hydrocarbons and alcohols is very large. There is a 154 Celsius degree difference in the boiling points of ethane (CH_3CH_3) and methanol (CH_3OH), compounds of comparable molecular mass (30 amu and 32 amu, respectively). The difference in boiling points between hydrocarbons and alcohols of comparable molecular size decreases, however, as the molecules increase in size. For example, the difference between 1-butanol ($CH_3CH_2CH_2CH_2OH$) and pentane ($CH_3CH_2CH_2CH_2CH_3$), is 81 Celsius degrees. The reason for the decreasing difference in boiling points is that, as alcohols become larger, the part of the

INSIGHT INTO PROPERTIES

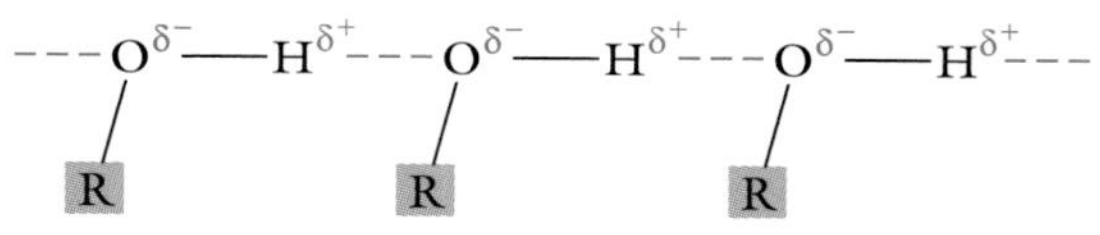

Figure 13.2 Hydrogen bonding in alcohols explains their high boiling points.

TABLE 13.3 Physical Properties of Alkanes, Alcohols, and Ethers

Compound	Structure	Molecular Mass (amu)	Boiling Point (°C)	Solubility in Water
ALCOHOLS				
methanol	CH_3OH	32	65	soluble
ethanol	CH_3CH_2OH	46	78	soluble
1-propanol	$CH_3CH_2CH_2OH$	60	97	soluble
1-butanol	$CH_3CH_2CH_2CH_2OH$	74	117	moderately soluble
1-pentanol	$CH_3CH_2CH_2CH_2CH_2OH$	88	138	slightly soluble
1-hexanol	$CH_3CH_2CH_2CH_2CH_2CH_2OH$	102	158	insoluble
1,2-ethanediol	$HOCH_2CH_2OH$	62	198	soluble
1,2,3-propanetriol	$HOCH_2CH(OH)CH_2OH$	92	290	soluble
ALKANES				
ethane	CH_3CH_3	30	−89	insoluble
propane	$CH_3CH_2CH_3$	44	−42	insoluble
butane	$CH_3CH_2CH_2CH_3$	58	−1	insoluble
pentane	$CH_3CH_2CH_2CH_2CH_3$	72	36	insoluble
ETHERS				
dimethyl ether	CH_3OCH_3	46	−23	soluble
diethyl ether	$CH_3CH_2OCH_2CH_3$	74	35	moderately soluble
WATER	H_2O	18	100	

molecule that participates in hydrogen bonding (the —OH group) becomes proportionately smaller, causing the alcohol molecule to increasingly resemble a nonpolar hydrocarbon.

Diols and triols, as might be expected, have especially high boiling points compared with alkanes of similar size. Alcohols containing more than one —OH group participate in more extensive hydrogen bonding, and this results in higher boiling points. For example, 1,2-ethanediol ($HOCH_2CH_2OH$) has a boiling point of 198°C compared with 97°C for 1-propanol ($CH_3CH_2CH_2OH$).

Hydrogen bonding also explains why alcohols of three or fewer carbons are completely miscible with water. Water alone and alcohol alone are extensively hydrogen bonded. The two compounds mix freely because alcohol molecules can hydrogen bond with water molecules. The O—H bonds attract one another irrespective of whether they reside in water or alcohol (Figure 13.3).

The amount of hydrogen bonding between alcohol and water decreases greatly as the number of carbons in the alcohol increases. The R part of an alcohol molecule is nonpolar, hydrophobic, and incapable of interaction (solvation) with the O—H bonds of water (Table 13.3); only the hydrophilic —OH

INSIGHT INTO PROPERTIES

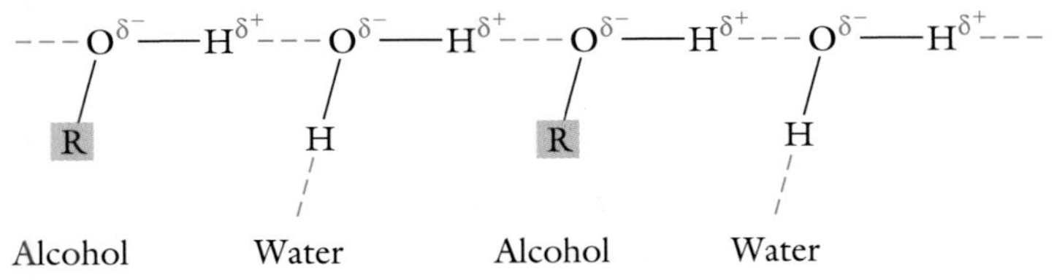

Figure 13.3 Hydrogen bonding between water and an alcohol explains why alcohols of three or fewer carbons are completely miscible with water.

13.1 Chemistry Around Us

Alcohols

Many alcohols, both naturally occurring and synthetic, have important industrial and everyday uses. Methanol (common name: methyl alcohol; see Table 13.2 for the structure of this and other alcohols) is used to manufacture formaldehyde (a raw material for certain plastics—see Box 14.2), as a solvent in which to carry out chemical reactions, and in Sterno heating pots.

Ethanol (common name: ethyl alcohol) comes from two sources: the fermentation of plant products and the hydration of ethene. Ethanol is the alcohol in alcoholic beverages. Its properties and effects as a beverage are described in Boxes 13.2 and 13.3. It is also used industrially and medically, in the production of plastics, drugs, and other organic ehcmicals; as a solvent for perfumes and varnishes; and as an antiseptic for sterilizing surgical instruments and the surface of a patient's body prior to surgery (as a 70% solution of ethanol in water).

2-Propanol (common name: isopropyl alcohol) is often called rubbing alcohol because it is applied to the skin to lower a fever. The isopropanol evaporates rapidly from the skin surface, removing heat and creating a cooling effect. 2-Propanol is also used as a solvent or carrier for cosmetics and skin creams and is used to manufacture acetone, an important industrial chemical (see Box 14.2).

Ethylene glycol (IUPAC name: 1,2-ethanediol) and propylene glycol (IUPAC name: 1,2-propanediol) are used in automotive cooling-system formulations. These mixtures of diol and water (typically combined in equal amounts) have higher boiling points and lower freezing points than those of water. They neither evaporate in summer nor freeze in winter and so prevent the damage that can happen to an engine when water alone is used to cool it. Another use of ethylene glycol is in the manufacture of poly(ethylene terephthalate), an important plastic (Mylar) and fiber (Dacron polyester) (Section 15.10).

1,2,3-Propanetriol (common names: glycerol, glycerin) is one of the products of our digestion of fats and oils. Unlike other alcohols, including the diols, glycerol is not toxic. Its lack of toxicity combined with its highly hydrophilic nature (a consequence of its three —OH groups) and sweet taste makes glycerol useful as a food additive to keep foods fresh by retention of moisture. It is also used in cosmetics, soaps, and skin creams. The reaction of glycerol with nitric acid produces nitroglycerin, a compound used both as a powerful explosive (employed in construction and, unfortunately, in warfare) and as a medicinal vasodilator. It is prescribed in the treatment of **angina pectoris,** a condition in which there is constriction of the blood vessels in heart muscle.

Menthol, obtained from peppermint oil, is used in cough drops and throat sprays to sooth the respiratory tract by increasing secretions from mucous membranes. Menthol is also an ingredient in shaving creams, toothpastes, and cigarettes.

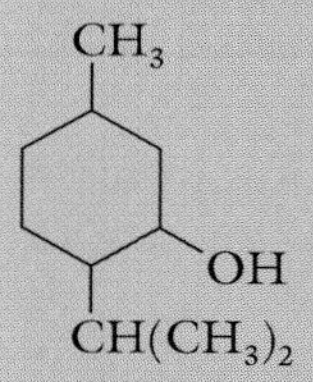

Menthol

group can interact with water. Whereas alcohols with three or fewer carbons are completely soluble, the four-carbon alcohol is only moderately soluble in water, the five-carbon alcohol is slightly soluble, and alcohols of more than five carbons show negligible water solubility. An alcohol of more than five carbons becomes water soluble only if it contains more than one —OH group.

Many biochemical processes take advantage of the ability of —OH groups to water solubilize otherwise insoluble molecules. For example, one of the mechanisms by which the liver detoxifies toxic substances is to attach sufficient —OH groups to such molecules to render them water soluble (Section 26.5). Their solubility allows them to be transported in the blood and excreted in the urine.

Example 13.6 Explaining the physical properties of alcohols

Explain each of the following properties:

(a) The boiling point of water (100°C) is considerably higher than that of methanol (65°C) even though the molecular mass of methanol is almost twice that of water.

13.2 Chemistry Around Us

Types of Alcoholic Beverages

The alcohol in alcoholic beverages is ethanol. Although ethanol is synthesized in large amounts by the hydration of ethylene (Section 12.6), federal law in the United States requires that ethanol for consumption be produced by the fermentation of grains, fruits, or vegetables. Ethanol used for consumption is heavily taxed. Ethanol used for other purposes is not taxed but is carefully controlled by government statutes.

Fermentation is the anaerobic (in the absence of oxygen) conversion of the carbohydrates in a grain, fruit, or vegetable into ethanol. The process requires yeast, which supplies a mixture of enzymes needed to catalyze the process. Starch and sucrose (table sugar) in the grain, fruit, or vegetable are first hydrolyzed to glucose and fructose, and they in turn are converted into ethanol:

$$\text{Starch, sucrose} \longrightarrow \text{glucose, fructose} \longrightarrow CH_3CH_2OH + CO_2$$

All alcoholic beverages consist almost entirely (>99%) of ethanol and water. These beverages differ from one another in the proportion of ethanol to water and in the chemicals that they contain other than ethanol and water. Beer contains from 3 to 8% v/v ethanol (lager beers contain less ethanol than do ale and stout beers). Wines contain from 10 to 14% v/v ethanol. Liquors have much higher ethanol contents, as much as 50% v/v ethanol. The ethanol content is usually expressed as **proof,** defined as twice the % v/v ethanol. Fermentation is limited to the production of beverages with contents no higher than 14% v/v ethanol, because yeast cannot survive at higher ethanol concentrations. Beers and wines are produced by fermentation followed by filteration, pasteurization, and carbonation. The higher ethanol content in liquors is produced by distillation of the fermentation mixture, taking advantage of the fact that ethanol has a lower boiling point than water.

Ethanol itself has no taste. The individual tastes of different beers, wines, and liquors result from the presence of small amounts (<1%) of other substances, known as **congeners,** compounds specific to the source of starch and sucrose or to the manufacturing process. Beers in the United States are made from barley; rice and sorghum seeds are used in Asia and Africa, respectively. Hops (flower clusters from the female vine *Humulus lupus*) are added to impart the bitter taste, whereas trace amounts of organic chemicals derived from the different grains are responsible for some of the more subtle flavors. Cask beers are aged in wooden casks and, in the process, acquire trace amounts of other congeners that originate in the wood. Light and dark beers are due to differences in the temperature (low and high, respectively) at which the grain is dried prior to fermentation. Most wines are produced from grapes, although rice is sometimes used, or other fruits. Wine color and flavor are affected by the variety of grape and the geographical location in which the grapes are grown. Sparkling wine (champagne) contains carbon dioxide. Some wines, such as vermouth, are flavored with herbs and spices.

Liquors include the various whiskeys, rum, gin, vodka, and brandy. Whiskey is produced from grains: scotch whiskey from barley, bourbon from corn, and rye from rye. Rum is produced from molasses (sugar cane). Brandy is made by adding additional ethanol to wine. Vodka, made from potatoes, has almost no congeners. Gin, made from grain alcohol by a second or even third distillation, is second only to vodka in low congener content and is flavored with juniper berries.

Rows of distillation stills produce whiskey at the Glenfiddich Distillery in Dufftown, Scotland.

(b) 1-Hexanol is not water soluble, whereas 1,3-hexanediol is water soluble.

(c) Butene is not water soluble, whereas 1-propanol is water soluble.

Solution

(a) The extent of hydrogen bonding in water is much greater than in methanol because each water molecule has two O—H bonds, whereas methanol has one.

13.3 Chemistry Within Us

Health Aspects of Alcohol Consumption

Most people believe, erroneously, that ethanol is a stimulant. Because one of its initial effects is to reduce social inhibitions, it seems to stimulate conversation, relieve tension, and bolster confidence. However, ethanol is a powerful central nervous system depressant.

A low level of alcohol consumption, a drink or two, does most people no harm, psychologically or physically. There is even some evidence that drinking small amounts has cardiovascular benefits. However, there is no doubt about the negative consequences of excessive alcohol consumption.

Continued acute consumption of alcohol (consumption of large amounts in a short time) depresses the central nervous system more and more, causing increasing loss of inhibitions and control, false feelings of euphoria and superiority, and loss of motor skills and coordination. There are more than 40,000 traffic fatalities annually in the United States, more than 40% of which are attributed to drivers whose judgments are impaired by alcohol consumption.

The point at which a person is legally considered in most localities to be drunk corresponds to a blood alcohol level of 0.10%. Acute consumption past this level can lead to blackouts (about 0.20% blood alcohol) and even death (0.30% and higher). Death occurs at those levels because ethanol shuts down the respiratory system by depressing the respiratory center in the brain—the person simply stops breathing.

There are more than 15,000 alcohol-related fatalities annually in the United States.

Chronic alcohol abuse (habitual consumption of excessive amounts over a long period of time) may not entail the acute consumption that results in immediate illness or death but, over time, also will result in major health problems. Slow liver destruction (cirrhosis) over many years has caused the deaths of many alcoholics.

The overall consequences of uncontrolled alcohol consumption include not only the aforementioned traffic fatalities, but also domestic and neighborhood violence and lost productivity at home and in the workplace. The human devastation caused by alcohol consumption exceeds that attributed to hard drugs such as crack cocaine.

The deleterious effects of alcohol consumption are caused not directly by ethanol itself but by the product acetaldehyde formed when the liver attempts to detoxify the body of ethanol. The liver is able to detoxify the acetaldehyde by further oxidation and excretion of the resulting products when the ethanol load on the system is low. However, the liver is not able to keep up with a high ethanol load, and this failure results in a buildup of acetaldehyde. Highly reactive toward proteins, acetaldehyde denatures (Section 20.9) them so that they are no longer capable of performing their physiological functions.

Other alcohols are much more toxic than ethanol, and consumption of even small amounts can result in death. Desperate people have been known to drink methanol, for example, and methanol is exceptionally dangerous because its detoxification by the liver produces formaldehyde, which is more reactive than acetaldehyde toward proteins. Ethylene glycol, often found in households because of its common use as an antifreeze, also is highly toxic. Intravenous feeding of ethanol is the medical treatment for methanol and ethylene glycol poisoning. In this treatment, the ethanol is preferentially detoxified in the liver, whereas the methanol or ethylene glycol is eliminated in the urine without being "detoxified" into deleterious substances.

Chemical-grade ethanol, produced for nonbeverage uses, is generally denatured (rendered unfit for consumption) by the addition of toxic materials such as methanol and 2-propanol. This addition prevents chemical-grade ethanol from being diverted to use in alcoholic beverages. Certain industrial processes, however, require pure ethanol, devoid of denaturing agents. The sale and distribution of pure ethanol are very tightly controlled, with heavy penalties (prison and fines) for diversion to consumption.

(b) The one —OH group in 1-hexanol cannot solubilize the six-carbon compound; the limit is three (possibly four) carbons per —OH group. The presence of two —OH groups in 1,3-hexanediol (that is, two —OH per six C, or a ratio of one —OH to three C) results in water solubilization.

(c) The "like dissolves like" rule applies. No attractive interactions are possible between butene and water, because butene is nonpolar. 1-Propanol, on the other hand, does hydrogen bond with water because, like water, it contains the —OH group.

Problem 13.6 Answer each of the following questions:

(a) Which has the higher boiling point, propanol or 1,2-ethanediol? Why?

(b) Why does butanol have a higher boiling point than *t*-butanol?

(c) Which is more soluble, propanol or butane, in a nonpolar solvent such as hexane?

13.5 ACIDITY AND BASICITY OF ALCOHOLS

Alcohols are usually considered neutral compounds. They neither turn blue litmus paper red nor turn red litmus paper blue. Aqueous solutions of alcohols have about the same pH value as that of water itself, pH = 7, a pH value often called neutral pH. However, alcohols are not neutral on an absolute basis. Whereas the hydrocarbon families of compounds show no acidic or basic behavior, alcohols resemble water in being very weakly amphoteric: they are both very weak acids and very weak bases (Section 9.6).

The very weak basicity of an alcohol is expressed by its ability to accept a proton from strong acids such as sulfuric acid (H_2SO_4) and phosphoric acid (H_3PO_4):

$$R{-}O{-}H + H_2SO_4 \rightleftharpoons R{-}\overset{+}{O}H_2 + HSO_4^-$$

Protonation takes place only in the presence of strong acids, and the equilibrium is far to the left. Less than 0.1% of the alcohol takes the protonated form, but this small concentration of protonated alcohol is important in dehydration reactions of alcohols (see Section 13.6, and Box 13.4, on page 383).

The acidity of an alcohol is far too weak to be observed by reaction with NaOH or many other strong bases. However, the very weak acidity of an alcohol is expressed in its reaction with active metals such as sodium (which are much stronger bases than NaOH) to yield gaseous hydrogen (Figure 13.4):

$$2\,R{-}O{-}H + 2\,Na \longrightarrow 2\,R{-}ONa + H_2$$

Figure 13.4 The weak acidity of an alcohol is expressed in its reaction with sodium metal to yield gaseous hydrogen (H_2).

13.6 DEHYDRATION OF ALCOHOLS TO ALKENES

Alcohols can be transformed into alkenes when heated in the presence of a catalytic amount of a strong acid such as sulfuric acid. An alcohol undergoes **intramolecular dehydration**—elimination of water—by losing an —OH group from a carbon atom and an H from an adjacent carbon atom:

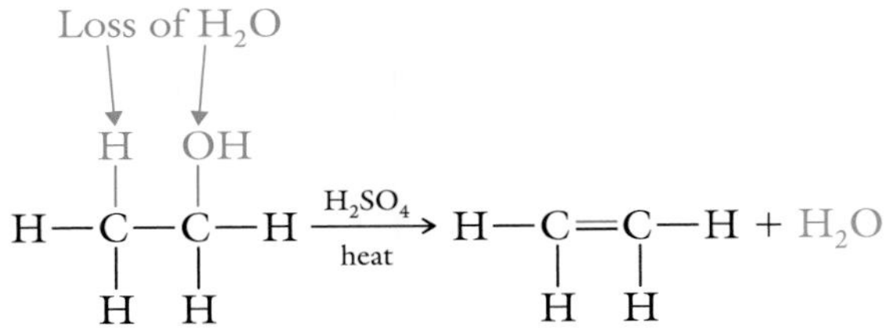

Note that dehydration cannot take place unless a strong acid catalyst is present.

All alcohols undergo intramolecular dehydration except alcohols without an H on a carbon atom adjacent to the carbon holding the —OH. Methanol is one such exception. Having only one carbon, it does not contain an adjacent carbon with an H.

Example 13.7 Writing equations for the dehydration of alcohols: I

Which of the following alcohols undergo dehydration in the presence of a strong acid catalyst? If dehydration takes place, show the product. (a) 1-Propanol; (b) phenylmethanol; (c) cyclohexanol.

Solution

Find the carbon attached to the —OH group, and then look to see if there is a carbon adjacent to it that is carrying an H.

(a) $CH_3CH_2{-}CH_2{-}OH \xrightarrow[\text{heat}]{H_2SO_4} CH_3CH{=}CH_2 + H_2O$

(b) 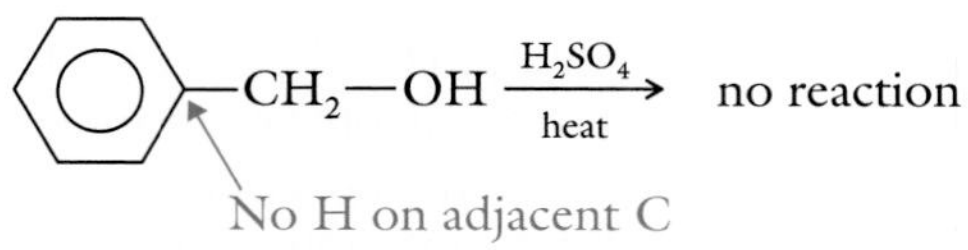

(c) 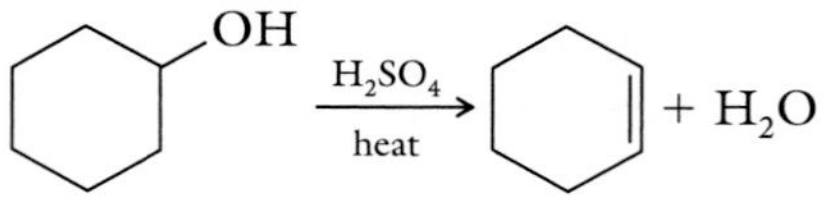

Problem 13.7 Which of the following alcohols undergo dehydration in the presence of a strong acid catalyst? If dehydration takes place, show the product. (a) 2-Propanol; (b) 1-butanol; (c) *t*-butyl alcohol.

◀◀ Hydration of an alkene is the addition of water to the double bond, as described in Section 12.6.

The dehydration of alcohols is the reverse of the hydration of alkenes. We can make the reaction proceed in either direction by putting Le Chatelier's principle to work (Section 8.8). When hydration is desired, we use a large excess of water relative to alkene to shift the equilibrium toward alcohol. When dehydration is desired, we shift the equilibrium toward alkene by taking advantage of the lower boiling point of the alkene compared with that of alcohol. We run the reaction at a temperature above the boiling point of alkene; alkene distills out of the reaction vessel, and this distillation shifts the equilibrium to produce more alkene. Box 13.4 describes the mechanism of dehydration.

The dehydration of many secondary and tertiary alcohols follows a more complicated course than that for primary alcohols. Such alcohols have hydrogens attached to carbons on both sides of the carbon holding the —OH and will yield two different alkenes on dehydration if the alcohol is asymmetric. For example, the dehydration of 2-butanol yields both 1-butene and 2-butene, by loss of hydrogens from C1 and C3, respectively:

Loss of H yields 2-butene

$$CH_3{-}\underset{|}{\overset{OH}{CH}}{-}CH_2{-}CH_3 \xrightarrow[-\,H_2O]{H_2SO_4} \underset{\textbf{1-Butene (Minor)}}{CH_2{=}CH{-}CH_2{-}CH_3} + \underset{\textbf{2-Butene (Major)}}{CH_3{-}CH{=}CH{-}CH_3}$$

Loss of H yields 1-butene

Dehydration is a selective process; the two alkenes are not formed in equal amounts. 2-Butene is the major product.

13.4 Chemistry in Depth

Mechanism of Dehydration of Alcohols

The dehydration of alcohols does not take place in the absence of a strong acid catalyst, because the bond connecting carbon to oxygen is too strong to allow the —OH group to break away. The presence of a strong acid catalyst is the key to dehydration.

The first step in the reaction mechanism for dehydration is protonation of the oxygen of the alcohol by the strong acid catalyst, as illustrated on the right. Protonation is the bonding of H^+ to one of the oxygen's free electron pairs, a consequence of the basicity of alcohols. The resulting electron deficiency of oxygen, symbolized by its plus charge, weakens the carbon-to-oxygen bond, which breaks to eliminate water and form a carbocation. The carbocation attains stability by loss of a proton and formation of an alkene double bond.

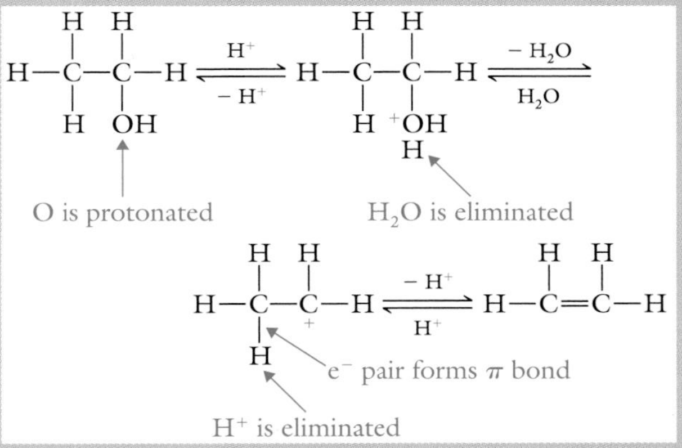

The reactivity of alcohols in dehydration follows the order 3° > 2° > 1°, which corresponds to the order of stability of the carbocations that formed as intermediates from the respective alcohols.

The relative stabilities of the different alkenes determine which alkene is the major product. The more stable the alkene, the higher its yield in the dehydration reaction. The stability of alkenes follows the order:

Most stable — **Least stable**

$$\underset{\text{Tetrasubstituted}}{R_2C{=}CR_2} > \overset{\text{Trisubstituted}}{R_2C{=}CHR} > \underset{\text{Disubstituted}}{R_2C{=}CH_2 = RCH{=}CHR} > \overset{\text{Monosubstituted}}{RCH{=}CH_2} > \underset{\text{Unsubstituted}}{H_2C{=}CH_2}$$

Most substituted — **Least substituted**

Stability increases with the degree of substitution of the double bond; that is, with increasing numbers of alkyl groups attached to the carbons of the double bond. The compound with the most substituted double bond is the major product. In the dehydration of 2-butanol, the major product (2-butene) has a disubstituted double bond, whereas the minor product (1-butene) has a monosubstituted double bond.

In actuality, the result of the dehydration of 2-butanol is more complicated than that just described. The 2-butene product is a mixture of *cis*-2-butene and *trans*-2-butene. The trans isomer is formed in greater abundance than the cis isomer because trans isomers are more stable than cis isomers.

Example 13.8 Writing equations for the dehydration of alcohols: II

Which of the following alcohols undergo dehydration in the presence of a strong acid catalyst? If dehydration takes place, show the product(s). If there is more than one product, indicate which is the major product. (a) 3-Hexanol; (b) 2-methyl-2-butanol.

Solution

(a) Look at the carbon atoms adjacent to the one bonded to the —OH.

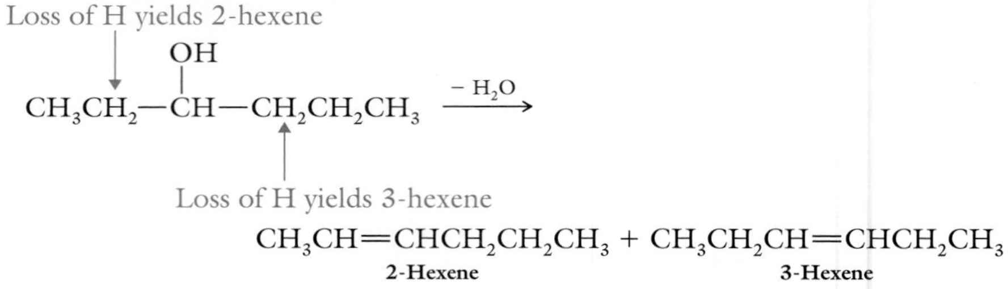

2-Hexene and 3-hexene are formed in equal amounts because each contains a disubstituted double bond. However, each is a mixture of cis and trans isomers with the trans isomer being present in larger amounts.

(b)

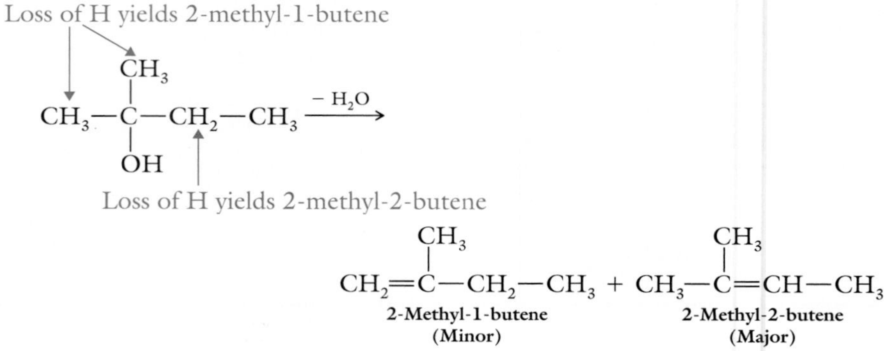

2-Methyl-2-butene is the major product because it contains a trisubstituted double bond, whereas 2-methyl-1-butene contains a disubstituted double bond.

Problem 13.8 Which of the following alcohols undergo dehydration in the presence of a strong acid catalyst? If dehydration takes place, show the product(s). If there is more than one product, indicate which is the major product. (a) 3-Pentanol; (b) 1-methylcyclopentanol.

▶▶ Glycolysis is a metabolic process that generates energy and is considered in Chapter 23.

The dehydration of alcohols takes place in a number of physiological processes, including glycolysis (Section 23.1).

Concept check

✔ Intramolecular dehydration of an alcohol in the presence of an acid catalyst produces an alkene. When more than one alkene is possible, the major product is the most substituted alkene.

13.7 OXIDATION OF ALCOHOLS

Alcohols undergo two types of oxidation, depending on reaction conditions.

- **Combustion:** All carbon atoms in the organic compound are oxidized to their highest oxidation state.
- **Selective (mild) oxidation:** Only the carbon holding the —OH group is oxidized.

Alcohols, like hydrocarbons, undergo combustion to carbon dioxide and water. The balanced combustion equation for ethanol is:

$$CH_3CH_2OH + 3\ O_2 \longrightarrow 2\ CO_2 + 3\ H_2O$$

Selective oxidations of organic compounds are oxidations at the weaker bonds in the molecules. Whereas combustion transforms an organic compound into carbon dioxide and water, selective oxidation transforms an organic compound into another organic compound, a member of a different family. Such transformations are important in biological systems as well as in industrial processes.

In an alcohol, the weaker bonds are the O—H bond and the adjacent C–H bond, and these weaker bonds are the bonds that undergo selective oxidation. Two of the most commonly used selective oxidizing agents are permanganate ion (MnO_4^-) and dichromate ion ($Cr_2O_7^{2-}$). Primary, secondary, and tertiary alcohols behave differently under selective oxidation conditions.

Primary alcohols are oxidized in two stages:

- Stage 1: Simultaneous loss of hydrogens (called **dehydrogenation**) from the —OH group and from the carbon holding the —OH produces a **carbonyl group** (C=O), as shown in the adjoining equation. The resulting compound is classified as an **aldehyde** because an H is attached to the carbonyl carbon (Section 11.2).
- Stage 2: Oxidation of the H attached to the carbonyl carbon to —OH converts the carbonyl group into a **carboxyl group** (COOH). The resulting compound is a **carboxylic acid** (Section 11.2).

▶▶ Aldehydes and carboxylic acids are considered in Chapters 14 and 15, respectively.

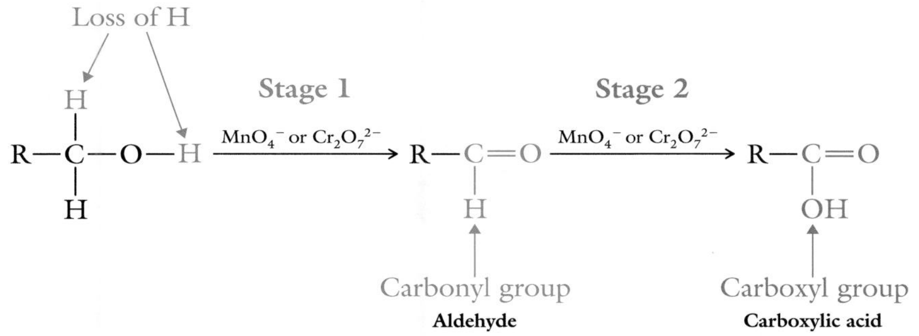

Oxidation in the laboratory proceeds from primary alcohol all the way to carboxylic acid without stopping at the aldehyde stage unless special procedures are used. One such procedure is to perform the reaction under conditions (temperature and pressure) that force the aldehyde to distill out of the reaction vessel as it forms. Removal of the aldehyde prevents its contact with the oxidizing agent, which in turn prevents it from being converted into the carboxylic acid.

Secondary alcohol oxidation follows a simpler course than the oxidation of primary alcohols. Oxidation cannot proceed past the carbonyl stage.

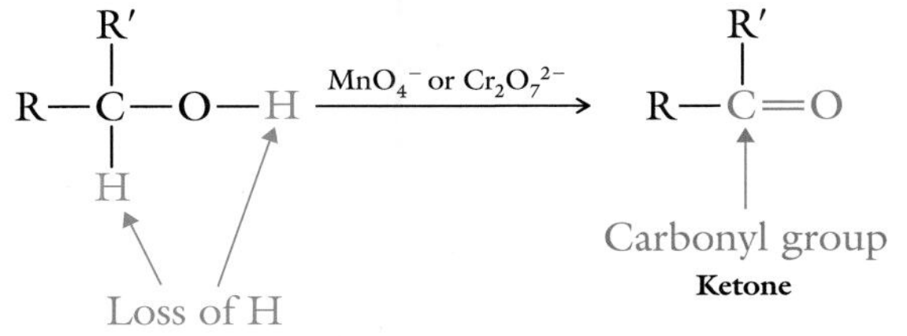

The product with the carbonyl group is a **ketone** (Section 11.2). (Note that ketones and aldehydes both have a carbonyl group. An aldehyde has one H and one R group attached to the carbonyl carbon, whereas a ketone has two R groups and no H attached to it. The R groups can be either aliphatic or aromatic.) Because a ketone lacks an H attached to the carbonyl carbon, there is no further oxidation under selective oxidation conditions. It is the H attached to the carbonyl carbon in an aldehyde that is oxidized in the second stage.

▶▶ Ketones are a subject of Chapter 14.

A PICTURE OF HEALTH

Examples of Alcohols, Phenols, Ethers, and Thiols

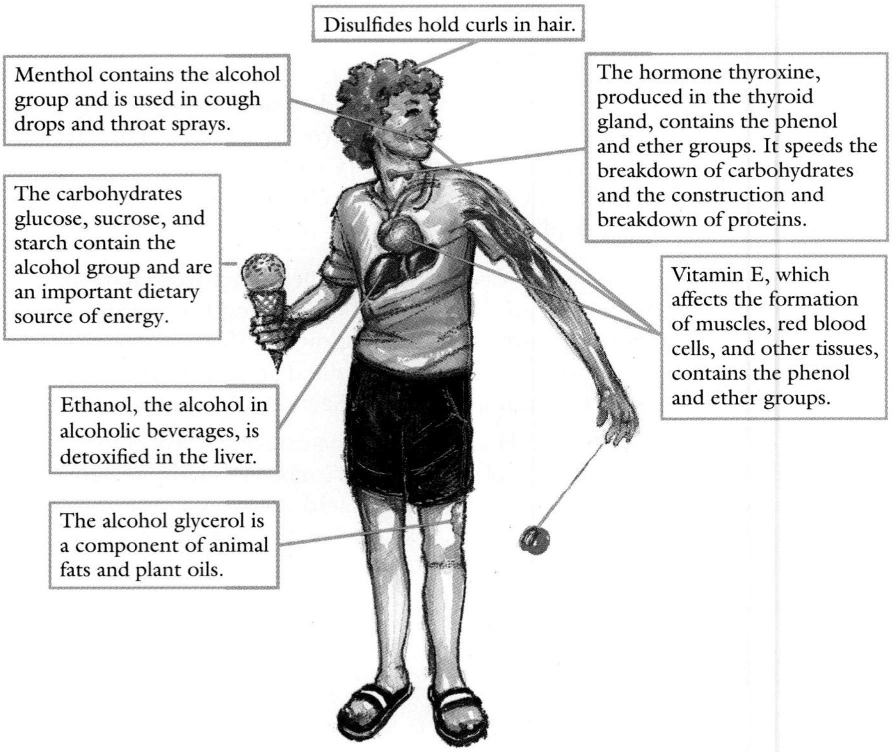
Disulfides hold curls in hair.
Menthol contains the alcohol group and is used in cough drops and throat sprays.
The hormone thyroxine, produced in the thyroid gland, contains the phenol and ether groups. It speeds the breakdown of carbohydrates and the construction and breakdown of proteins.
The carbohydrates glucose, sucrose, and starch contain the alcohol group and are an important dietary source of energy.
Vitamin E, which affects the formation of muscles, red blood cells, and other tissues, contains the phenol and ether groups.
Ethanol, the alcohol in alcoholic beverages, is detoxified in the liver.
The alcohol glycerol is a component of animal fats and plant oils.

Tertiary alcohols cannot undergo selective oxidation, because there is no H attached to the carbon to which the —OH is bonded. The OH hydrogen cannot be lost by itself.

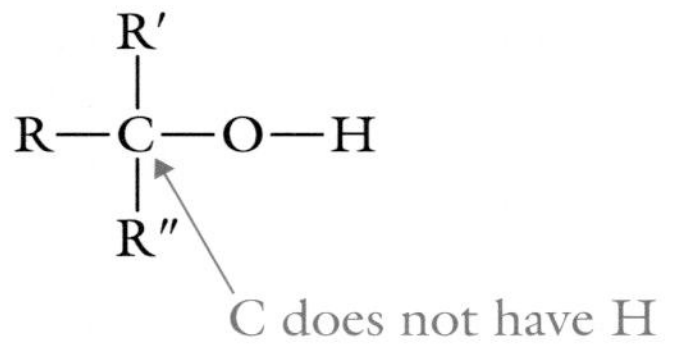
R′
R—C—O—H
R″
C does not have H

Oxidation of an alcohol requires the loss of a pair of hydrogens, one from the —OH and one from the carbon holding the —OH.

Oxidations of alcohols are important in living systems, which use specific enzymes to control the reaction. Malate dehydrogenase catalyzes the conversion of malic acid into oxaloacetic acid (by oxidizing the —OH group of malic acid) in a key step of the citric acid cycle, an important energy-generating process (Section 23.5).

Formaldehyde and acetone, important industrial chemicals (Box 14.2), are produced by the oxidations of the corresponding alcohols, methanol and 2-propanol, respectively.

Concept checklist

- ✔ Primary and secondary alcohols undergo selective oxidation by permanganate (MnO_4^-) and dichromate ($Cr_2O_7^{2-}$) ions.
- ✔ Primary alcohols are oxidized to aldehydes, which are subsequently oxidized to carboxylic acids. Secondary alcohols are oxidized to ketones with no further oxidation possible. Tertiary alcohols are not oxidized.

Example 13.9 Writing equations for the oxidation of alcohols

Which of the following alcohols undergo oxidation with dichromate or permanganate? If oxidation takes place, show the product. (a) 1-Propanol; (b) 1-methylcyclohexanol; (c) cyclohexanol.

Solution

(a) Primary alcohols undergo oxidation first to an aldehyde and then to a carboxylic acid:

$$CH_3CH_2CH_2{-}OH \xrightarrow[\text{or } MnO_4^-]{Cr_2O_7^{2-}} CH_3CH_2{-}\overset{\overset{\displaystyle O}{\|}}{C}{-}H \xrightarrow[\text{or } MnO_4^-]{Cr_2O_7^{2-}} CH_3CH_2{-}\overset{\overset{\displaystyle O}{\|}}{C}{-}OH$$

(b) Tertiary alcohols do not undergo selective oxidation:

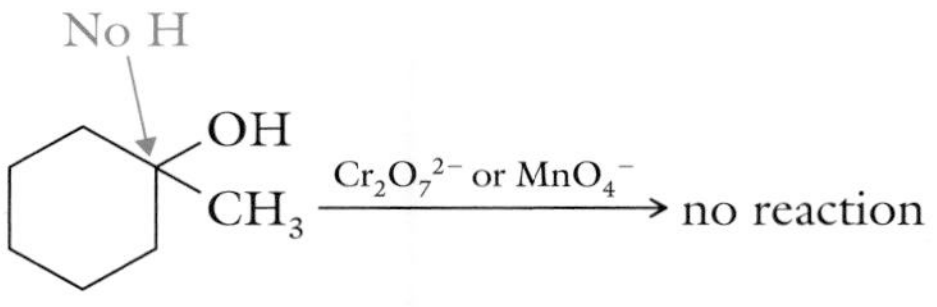

(c) Secondary alcohols undergo oxidation to ketones:

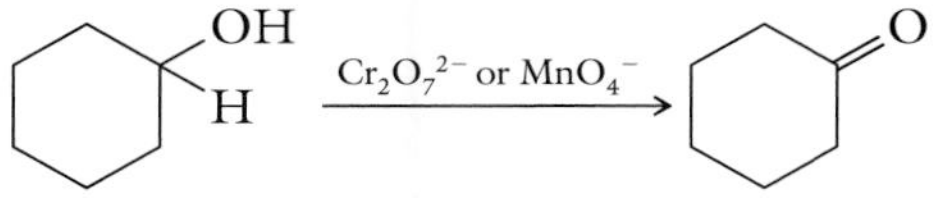

Problem 13.9 Which of the following alcohols undergo oxidation with dichromate or permanganate? If oxidation takes place, show the product. (a) 2-Propanol; (b) phenylmethanol; (c) *t*-butyl alcohol.

Permanganate and dichromate oxidations are useful **simple chemical diagnostic tests** for primary and secondary alcohols. Each oxidation is accompanied by a visible change:

- Permanganate oxidation: MnO_4^- (purple solution) is converted into MnO_2 (brown precipitate).
- Dichromate oxidation: $Cr_2O_7^{2-}$ (orange solution) is converted into Cr^{3+} (green solution).

These tests have the following uses and limitations:

- Primary and secondary alcohols are distinguished from tertiary alcohols because tertiary alcohols do not undergo permanganate and dichromate oxidations.
- Primary and secondary alcohols are distinguished from ethers because ethers do not undergo permanganate and dichromate oxidations (Section 13.9).
- Positive tests with permanganate and dichromate do not distinguish primary and secondary alcohols from alkenes, alkynes, and phenols, because those compounds also are oxidized (Sections 12.8, 12.9, 13.8).

Dichromate oxidation forms the basis of an instrument used to monitor for alcohol abuse in several settings—for example, to monitor inmates in prisons, probationers, or patients at alcohol-rehabilitation clinics. There is a direct relation between the concentration of ethanol in the blood and that in exhaled breath. The test subject fills a balloon with breath, the filled balloon is attached to the instrument, and the sampled breath is allowed to flow through the dichromate solution for a set time period. The extent to which the dichromate solution changes color from orange to green indicates the blood alcohol level. The balanced equation for the dichromate oxidation of ethanol is described in Box 4.2.

Example 13.10 Identifying an unknown compound

An unknown compound is either 2-methyl-2-propanol or 1-butanol. When a few drops of permanganate solution are added to it, the purple color of the permanganate fades and a brown precipitate forms. What is the identity of the unknown? Explain your answer.

Solution

The unknown is 1-butanol. 1-Butanol, a primary alcohol, gives a positive test with permanganate, whereas 2-methyl-2-propanol, a tertiary alcohol, does not react with permanganate.

Problem 13.10 An unknown compound is either 2-butanol or 1-butanol. When a few drops of dichromate solution (orange) are added to it, the mixture turns green. What is the identity of the unknown? Explain your answer.

13.8 PHENOLS

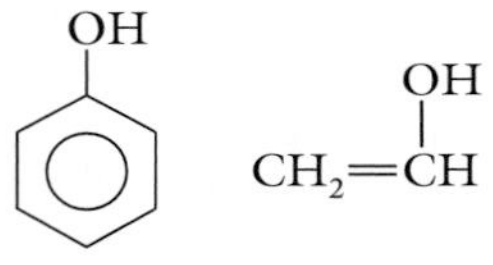

Phenol Enol

A phenol is an organic compound in which an —OH group is attached to one of the sp^2-hybridized carbons of a benzene ring (see Table 13.1). The phenol family includes the parent compound phenol as well as a variety of compounds with other substituents attached to the benzene ring in addition to the —OH group. Many naturally occurring compounds contain the phenol structure (Box 13.5).

Enols constitute another organic family in which an —OH is attached to an sp^2-hybridized carbon. They differ from phenols in that the —OH is attached not to a benzene ring but to a carbon–carbon double bond. Enols are generally unstable compounds that are encountered in very few systems. One exception is glycolysis, where an enol is an important intermediate in the process (Section 23.2).

In the IUPAC nomenclature system, substituted phenols are named as derivatives of the parent compound phenol.

Example 13.11 Naming phenols by the IUPAC nomenclature system

Name the following compounds by using the IUPAC system.

(a) 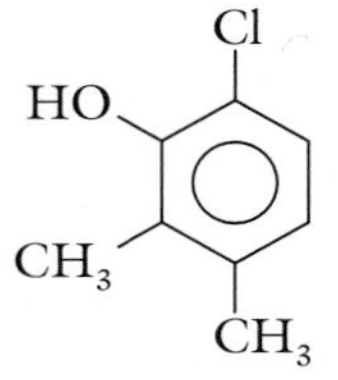

(b) 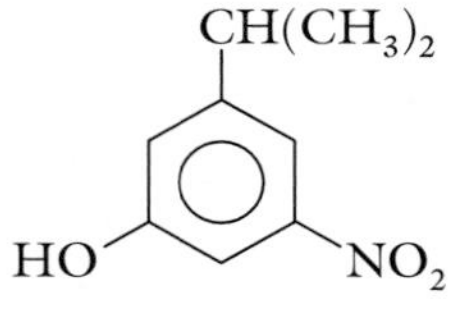

Solution

C1 is the carbon holding the —OH group. Number in the direction that assigns the set of lowest numbers to the other substituents on the benzene ring: (a) 6-chloro-2,3-dimethylphenol; (b) 3-isopropyl-5-nitrophenol.

Problem 13.11 Name the following compounds by using the IUPAC system.

(a) 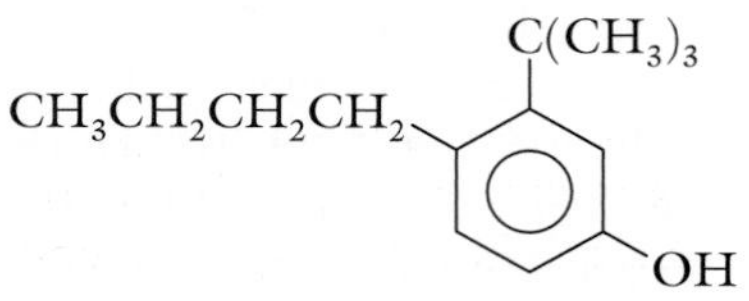

(b) 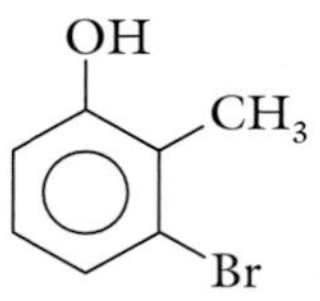

13.5 Chemistry Around Us

Phenols

Various phenols are used as antiseptics and disinfectants. Hexylresorcinol is commonly used in mouthwashes and throat lozenges, and *o*-phenylphenol serves as a disinfectant for walls, floors, furniture, and equipment both in hospitals and in homes.

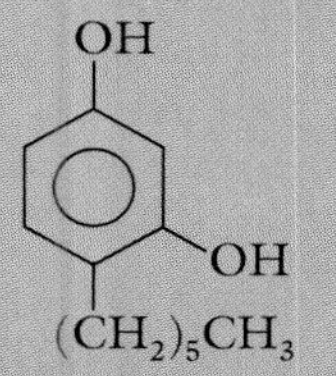

Hexylresorcinol

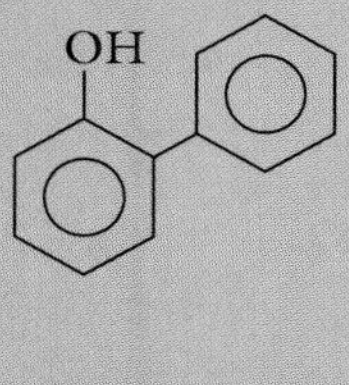

***o*-Phenylphenol**

The active irritants in poison oak, poison ivy, and poison sumac belong to a subclass of phenols known as the urushiols. All urushiols have a benzene ring with two —OH groups and a 15-carbon unbranched chain but differ in the number of double bonds (from 0 to 3) in that chain.

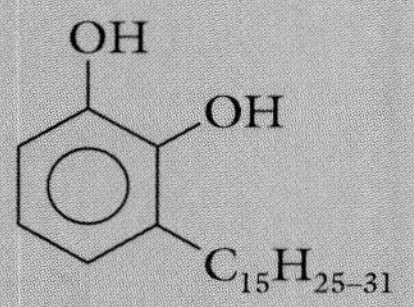

Urushiols

The drug marijuana, also called pot or grass, is obtained from the hemp plant *Cannabis sativa*. Leaves, flowers, and other parts of the plant are dried and smoked. The active ingredient tetrahydrocannibinol (THC), is a phenol. THC is a mild central-nervous-system depressant at low doses.

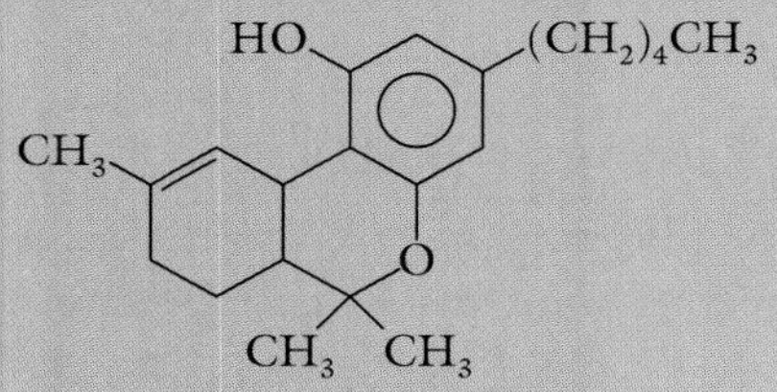

Tetrahydrocannabinol

BHT (butylated hydroxytoluene) and BHA (butylated hydroxyanisole) are used as **antioxidant** additives to protect foods, gasoline, lubricating oils, plastics, and rubbers from degradation due to oxidation. Vitamin E has the same function in the body. A phenol protects other materials in its vicinity against oxidation by undergoing sacrificial oxidation itself.

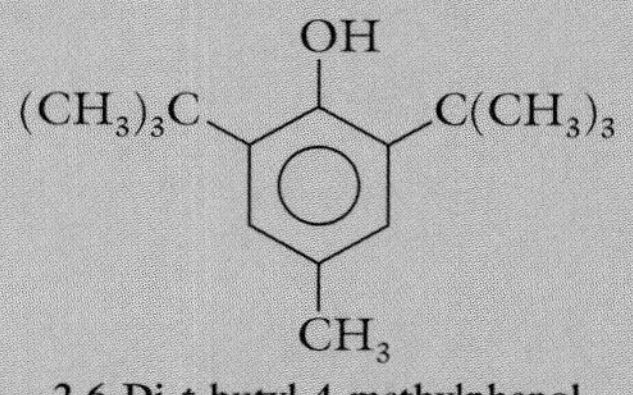

2,6-Di-*t*-butyl-4-methylphenol
(Butylated hydroxytoluene, BHT)

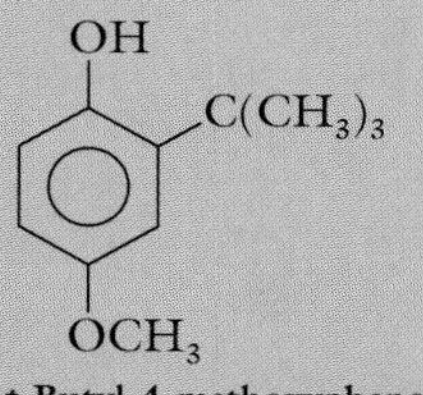

2-*t*-Butyl-4-methoxyphenol
(Butylated hydroxyanisole, BHA)

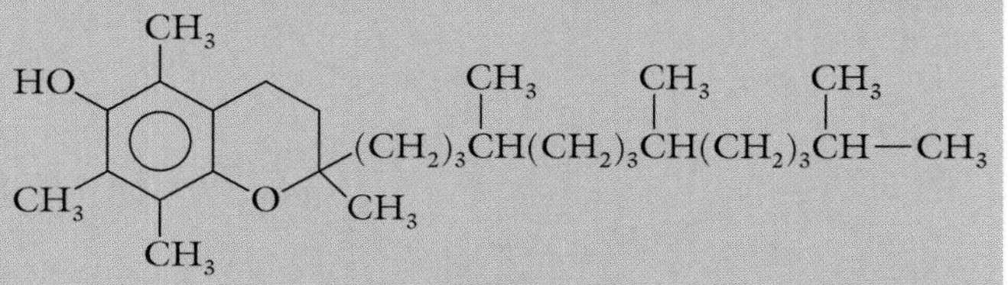

Vitamin E

Note that some of the phenols illustrated here are members of other families as well. BHA, vitamin E, and THC are ethers. THC is also an alkene.

The O—H bond in phenols is more polar than that in alcohols because, unlike an aliphatic group, the aromatic ring has an electron-withdrawing effect. All O—H bonds are polarized $^{\delta-}$O—H$^{\delta+}$ because oxygen is more electronegative than hydrogen. However, the electron-withdrawing effect of the aromatic ring (Ar) increases this polarization, causing the oxygen and hydrogen ends of the O—H bond to develop larger partial negative and partial positive charges, respectively, than are found on the O—H bond of an alcohol. The greater polarity of the O—H bond in phenols results in stronger hydrogen bonding and higher boiling and melting points as well as higher solubilities in water compared with alcohols. The boiling points of phenol and

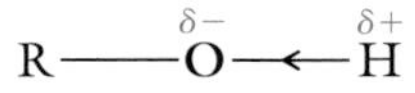

Alcohol
(Less polar)

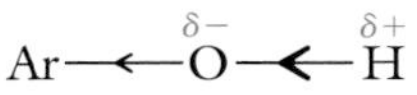

Phenol
(More polar)

cyclohexanol, for example, are 182°C and 162°C, respectively. Their water solubilities are 9.3 g/100 mL and 3.6 g/100 mL, respectively, at 20°C.

Example 13.12 Identifying the hydrogen bonding of phenols

Which of the following representations correctly describe the hydrogen bonding that is responsible for the high boiling point of phenol?

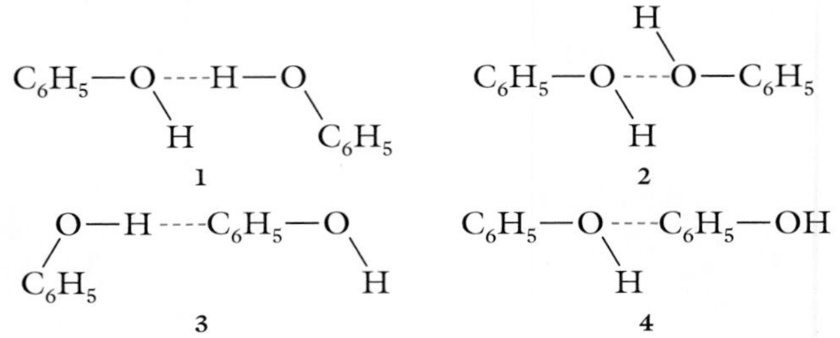

Solution

The correct representation of hydrogen bonding between phenol molecules is representation 1. There is an attraction between the oxygen of an O—H bond in one phenol molecule and the hydrogen of an O—H bond in another phenol molecule.

Problem 13.12 Which of the following representations correctly describe the hydrogen bonding that is responsible for the moderate solubility of phenol in water?

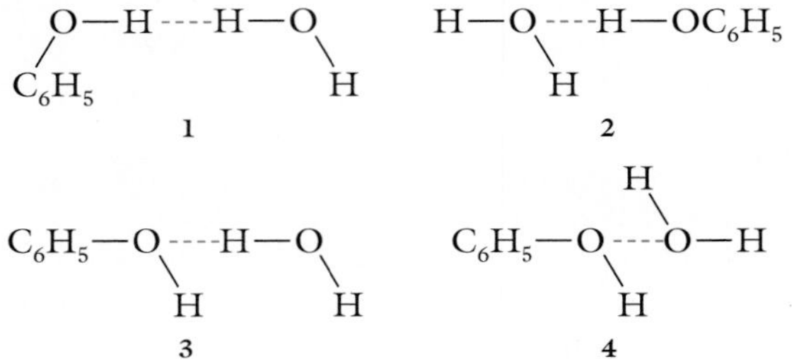

Phenols are weak acids but much more acidic than alcohols. The extent of ionization of phenol in water is much greater than that of alcohols.

$$C_6H_5—O—H + H_2O \rightleftharpoons C_6H_5—O^- + H_3O^+$$

For example, 0.1 molar aqueous solutions of cyclohexanol and phenol have pH values of about 7 and 5, respectively, corresponding to hydronium ion concentrations 100-fold higher in phenol than in cyclohexanol. The large difference in acidity between phenols and alcohols is easily tested by the use of an acid-base indicator such as litmus paper (Box 9.4). Phenols turn blue litmus paper red, whereas, like water, alcohols show neutral behavior: blue litmus paper does not change color.

The greater acidity of phenols compared with alcohols also shows itself by the reaction of phenol with a strong base, such as the hydroxide ion:

$$C_6H_5—O—H + NaOH \longrightarrow C_6H_5—O^- + Na^+ + H_2O$$

Phenol is completely converted into phenoxide ion ($C_6H_5—O^-$) by reaction with strong base: the equilibrium is completely to the right. The corresponding extent of reaction of an alcohol with NaOH is much less than 0.1%.

The higher acidity of phenols compared with that of alcohols is due to the electron-withdrawing effect of the phenyl ring compared with an R group. Electron withdrawal stabilizes the phenoxide anion by partly dispersing the negative charge around the phenyl ring, rather than leaving it entirely on the oxygen. This is called **charge dispersal:**

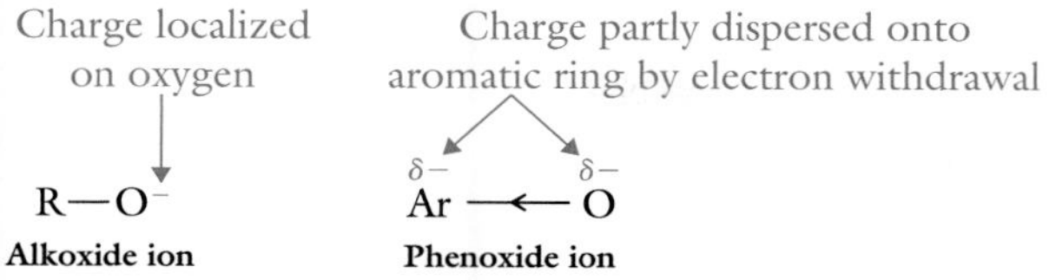

Its effect is to make the anion more stable and, therefore, it forms in higher concentration. The corresponding alkoxide anion is not stabilized by charge dispersal, because the R group is not electron withdrawing.

Phenols do not undergo dehydration, because dehydration would form a triple bond in the ring, resulting in destruction of the benzene structure. Phenols do undergo oxidation, however, even though it destroys the benzene ring. The oxidation reactions are quite complex and beyond the scope of this discussion. The oxidation of phenols forms the basis of their use as antioxidants, both in physiological systems and in commercial products (see Box 13.5).

13.9 ETHERS

Ethers, like alcohols and phenols, contain an oxygen that has bonds to two different atoms; but, in ethers, neither of those bonds is to hydrogen. Instead, the oxygen of an ether has single bonds to two different carbons, each of which may be either an aliphatic or an aromatic carbon (see Table 13.1). Review Example 13.1, Problem 13.1, and Exercise 13.1 if you are uncertain of the differences between ethers, alcohols, and phenols. Box 13.6 on the following page describes some important ethers.

A common nomenclature system is used for ethers containing simple groups. This system uses the names of the groups attached to the ether oxygen (in alphabetical order), followed by the word ether. There are spaces between the names of different groups and the word ether. The IUPAC nomenclature system is used for ethers containing other than the simple groups. In such cases, the ether is named as a member of some other family, with the more complex group determining the base name; the simpler group and the ether oxygen to which it is attached are treated as an alkoxy (RO) or aryloxy (ArO) group.

Example 13.13 Naming ethers

Name the following ethers:

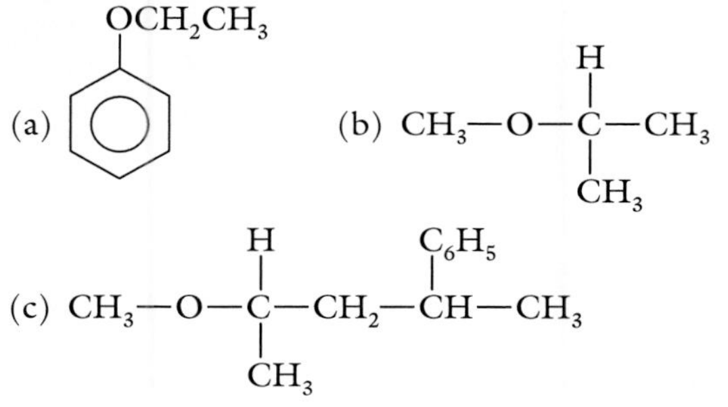

Solution

(a) Ethyl phenyl ether. (b) Isopropyl methyl ether. (c) The group on the right side of the oxygen does not have a simple name. The compound is named as a derivative of pentane. The pentane chain has a CH_3O (methoxy) group at C2 and a phenyl group at C4. Therefore, the name is 2-methoxy-4-phenylpentane.

13.6 Chemistry Around Us

Ethers

Before the 1850s, few patients ever survived surgery, and physicians were reluctant to operate except in dire circumstances. Surgical patients died either from the shock of undergoing surgery without anesthesia or from bacterial infections (which were not understood at the time).

An American dentist, William Morton, tackled the former problem in the 1850s by introducing the use of diethyl ether as an anesthetic for surgery. The English surgeon Joseph Lister addressed the other problem in 1865 by introducing the use of phenol as an antiseptic for sterilizing instruments and patients alike. Phenol itself is quite harsh in its effect and has long since been replaced by various substituted phenols (see Box 13.5). The use of anesthetics and antiseptics made modern surgery possible.

Diethyl ether and other ethers are general inhalation anesthetics: they produce both insensitivity to pain and unconsciousness, and they are introduced into the patient by inhalation. Most ethers have significant disadvantages, however: they are flammable and explosive, requiring great care in use, and they cause patients to suffer uncomfortable side effects, such as nausea, vomiting, and respiratory irritation. Fortunately, various halogenated ethers and halogenated alkanes with anesthetic properties have been found to be less flammable and to have milder side effects. Examples are enflurane and fluothane.

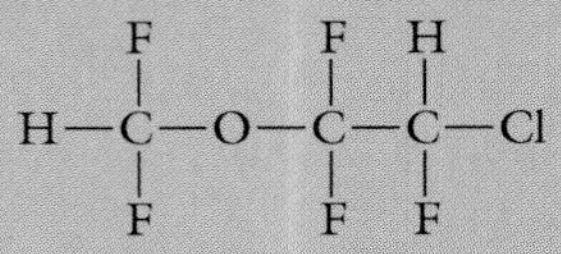

Enflurane

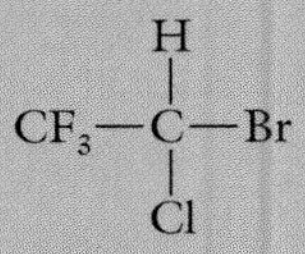

Fluothane

Ether functional groups are also found in many of the herbicides used to kill undesirable weeds, so as to improve the abundance of crops. Some of these herbicides were derived from polychlorinated phenols, a practice that has resulted in undesirable and unpredictable adverse side effects due to the presence of a certain class of impurities called **dioxins.** Dioxins are produced in minute amounts as by-products in the manufacture of herbicides. The most important dioxin is 2,3,7,8-tetrachlorodibenzo-*p*-dioxin (2,3,7,8-TCDD) because of its presence in the herbicide 2,4,5-T. 2,4,5-T was banned in the United States in 1985 because of concerns about its adverse effects on humans. 2,4,5-T was one component of the defoliant **Agent Orange** used in Vietnam by the United States military to remove enemy cover and destroy their crops. Many Vietnamese and Americans exposed to Agent Orange have suffered from a variety of health problems. Note that 2,3,7,8-TCDD is simultaneously a cyclic diether, an aromatic compound, and a chlorinated compound.

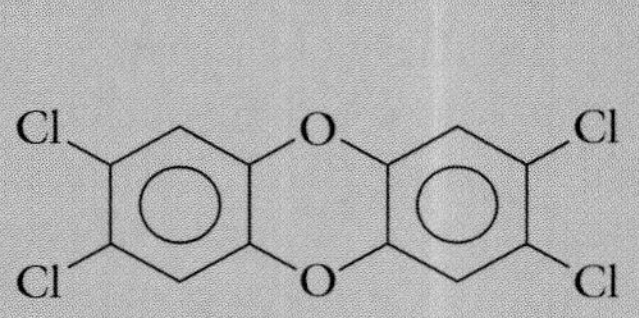

2,3,7,8-Tetrachlorodibenzo-*p*-dioxin (2,3,7,8-TCDD)

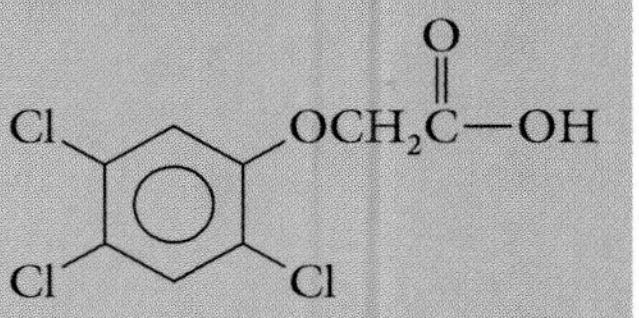

2,4,5-Trichlorophenoxyacetic acid (2,4,5-T)

Ethers in industrial use include tetrahydrofuran and dioxane, important solvents for carrying out chemical reactions. Unlike those two compounds, and ethers in general, ethylene oxide is a reactive compound and is used in the manufacture of various plastics, including epoxy adhesives. Another ether, *t*-butyl methyl ether, often incorrectly called methyl *t*-butyl ether or MTBE, is added to gasoline to reduce hydrocarbon emissions (a result of incomplete combustion) by improving the efficiency of the combustion process.

$$CH_3{-}O{-}C(CH_3)_3$$

t-Butyl methyl ether

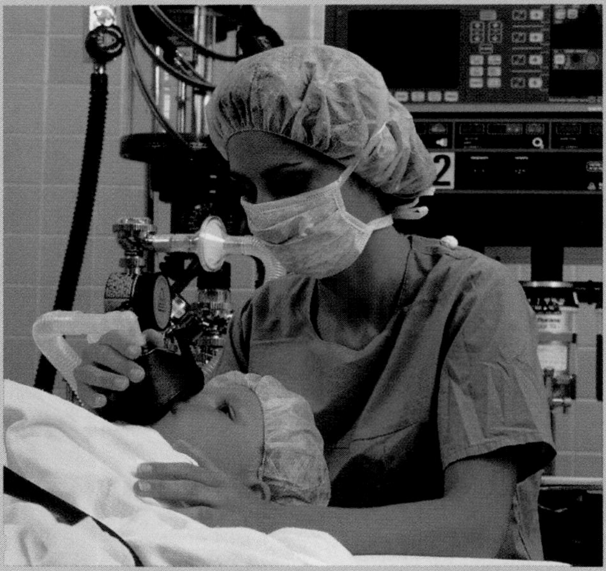

Administration of anesthesia to a child.

Problem 13.13 Name the following ethers:

(a) 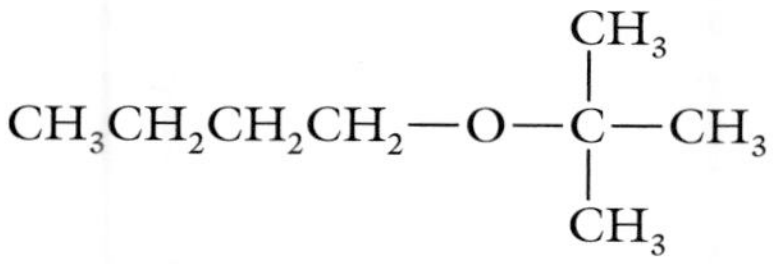

(b) 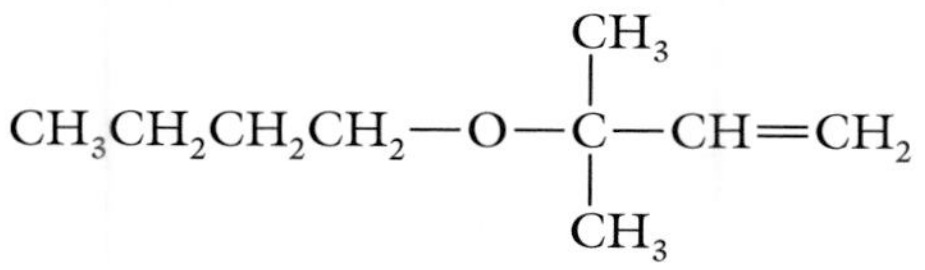

(c) 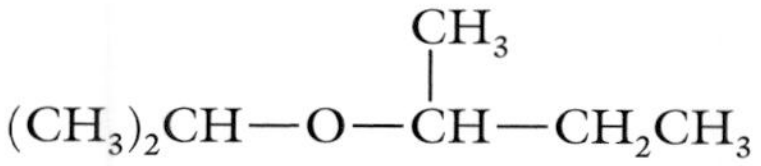

Common names are used for cyclic ethers:

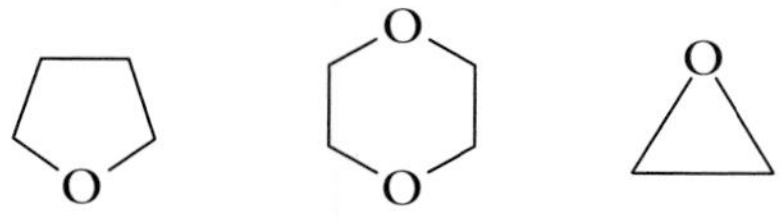

Tetrahydrofuran **Dioxane** **Ethylene oxide**

Any ether is a constitutional isomer of an alcohol containing the same number of carbons. For example, dimethyl ether and ethanol have the same molecular formula, C_2H_6O, but differ in their structural formulas and possess very different properties.

$CH_3—O—CH_3$
Dimethyl ether

$CH_3CH_2—O—H$
Ethanol

Example 13.14 Drawing constitutional isomers of ethers

Draw structural formulas and name the different ethers of molecular formula $C_4H_{10}O$.

Solution

$CH_3—O—CH_2CH_2CH_3$
Methyl propyl ether

$CH_3—O—CH(CH_3)_2$
Isopropyl methyl ether

$CH_3CH_2—O—CH_2CH_3$
Diethyl ether

Problem 13.14 Draw structural formulas of the compounds (alcohols) that are isomeric with the ethers in Example 13.14.

Refer again to Table 13.3, which compares the boiling points and water solubilities of several alkanes, alcohols, and ethers. The lower-molecular-mass ether, dimethyl ether, has a higher boiling point than the corresponding alkane because of the polar $^{\delta-}O—C^{\delta+}$ bond. The resulting dipole–dipole secondary forces are stronger than the London secondary forces in alkanes. However, the effect drops off rapidly with increasing molecular mass, and diethyl ether has the same boiling point as pentane. Ethers have considerably lower boiling points than alcohols because the hydrogen-bond secondary forces of alcohols are stronger than the dipole–dipole secondary forces of ethers.

Ethers are almost as soluble in water as are alcohols. Hydrogen bonding with water is more limited because there are no —OH groups in ethers. Ethers cannot hydrogen bond with themselves, but they can hydrogen bond with water because the oxygen in an ether is a hydrogen-bond acceptor. Alcohols are more water soluble than ethers because alcohols are hydrogen-bond donors as well as hydrogen-bond acceptors.

Ethers are unreactive toward acids, bases, and oxidizing agents and, for this reason, are often used as solvents in which to carry out chemical reactions. Ethers, like alkanes, participate in combustion and halogenation reactions only.

Example 13.15 Identifying the hydrogen bonding of ethers

Which of the following representations correctly describe the hydrogen bonding between dimethyl ether and water?

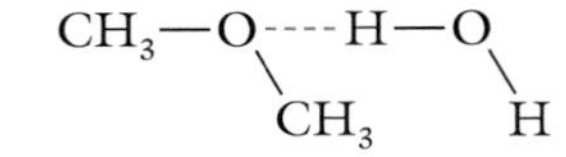

1

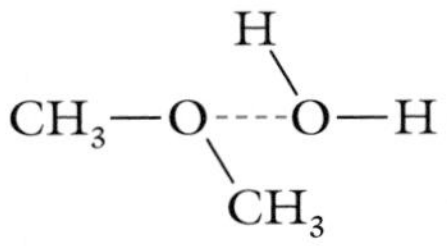

2

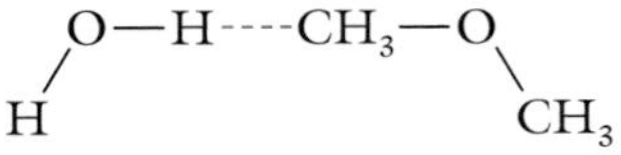

3

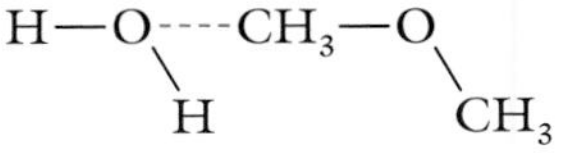

4

Solution

Hydrogen bonding is the attraction between the H of an O—H bond and the O of another bond (often but not limited to an O—H bond). Only representation 1 depicts hydrogen bonding.

Problem 13.15 Indicate which compound in each of the following pairs has the higher value of the specified property. If the two compounds in a pair have nearly the same value for the property, indicate that fact. Explain your conclusions.

(a) Methyl phenyl ether, *p*-methylphenol: boiling point.

(b) Dipropyl ether, 1-hexanol: solubility in heptane.

(c) Diethyl ether, 1-butanol: solubility in water.

13.10 FORMATION OF ETHERS BY DEHYDRATION OF ALCOHOLS

The discussion in Section 13.6 on the dehydration of alcohols to alkenes was an oversimplification for primary alcohols. Heating a primary alcohol in the presence of an acid catalyst produces not only an alkene, but also an ether:

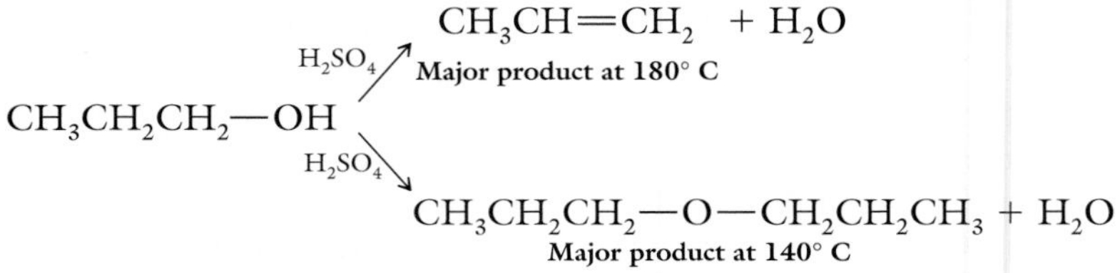

Intramolecular dehydration to form the alkene competes with intermolecular dehydration to form the ether. In **intramolecular dehydration,** an —OH group and a hydrogen atom are removed from adjacent carbons in the alcohol molecule. In **intermolecular dehydration,** the —OH group and H atom come from two different alcohol molecules. The reaction can be directed toward the production of alkene or the production of ether by controlling the reaction temperature.

- Alkene is the major product at 180°C.
- Ether is the major product at 140°C.

Unlike primary alcohols, secondary and tertiary alcohols do not form ethers when heated with an acid catalyst. Intramolecular dehydration to alkenes is the major product at all temperatures.

Example 13.16 Writing equations for the dehydration of alcohols

Write an equation to show the product(s) formed when ethanol is heated at 140°C in the presence of an acid catalyst. Indicate both the major and the minor products if more than one product is formed.

Solution

Intermolecular dehydration is the major reaction at 140°C:

$$CH_3CH_2{-}O{-}H \xrightarrow[-H_2O]{H_2SO_4} \underset{\textbf{Major}}{CH_3CH_2{-}O{-}CH_2CH_3} + \underset{\textbf{Minor}}{CH_2{=}CH_2}$$

Problem 13.16 Give the product(s) formed when ethanol is heated at 180°C in the presence of an acid catalyst. Indicate both the major and the minor products if more than one product is formed.

13.11 THIOLS

Thiols, also called **mercaptans,** are the sulfur analogues of alcohols. The —SH group is called the **mercapto** or **sulfhydryl** group.

In the IUPAC nomenclature system, the naming of thiols is similar to that of alcohols. The ending **-thiol** is added to the name of the alkane, but without dropping the -e from the alkane name.

$R{-}OH$
Alcohol

$R{-}SH$
Thiol

Thiols have distinct, often disagreeable, odors and flavors. The offensive liquid squirted by skunks in their defense is due to a mixture of thiols, mostly 3-methyl-1-butanethiol. Natural gas is odorless, which would make it difficult to detect gas leaks. However, the gas companies deliberately add small amounts of ethanethiol, which gives natural gas an odor and allows its safe use. Ethanethiol is useful as a warning for a gas leak because our odor receptors are remarkably sensitive to low-molecular-mass thiols. We can smell one part of ethanethiol in 10 billion parts of air.

$(CH_3)_2CHCH_2CH_2{-}SH$
3-Methyl-1-butanethiol

$CH_3CH_2{-}SH$
Ethanethiol

Thiols differ from alcohols in several ways:

- Thiols have considerably lower boiling points than alcohols, even though thiols have higher molecular masses, because thiols do not possess the strong hydrogen bonding of alcohols.
- Thiols are weak acids, much stronger than alcohols but weaker than phenols.
- Thiols are easily oxidized to disulfides by many oxidizing agents:

$$\underset{\textbf{Thiol}}{2\ R{-}SH} \xrightarrow{(O)} \underset{\textbf{Disulfide}}{R{-}S{-}S{-}R}$$

  Disulfides are named by naming the R groups attached to the sulfur atoms followed by the word **disulfide.**
- Disulfides are easily reduced back to thiols by many reducing agents:

$$R{-}S{-}S{-}R \xrightarrow{(H)} 2\ R{-}SH$$

  (In the preceding equations, the abbreviations (O) and (H) indicate general conditions for selective oxidation and reduction, respectively, without specification of the exact oxidizing or reducing agent.)

The sign warns that the fish in the lake are contaminated with mercury.

- Thiols react with heavy metal ions such as lead (Pb^{2+}) and mercury(II) (Hg^{2+}) to form insoluble salts:

$$2\ R{-}SH + M^{2+} \longrightarrow R{-}S{-}M{-}S{-}R + 2\ H^+$$

The reactions of thiols to form disulfides and insoluble heavy metal salts are of physiological importance (Sections 20.5 and 20.9). Many proteins contain thiol groups, and the formation of disulfides is often critical to their proper functioning. The formation of disulfides is also the basis for the permanent waving of hair (see Box 20.2). On the other hand, when such proteins form lead and mercury salts, the consequences to a living organism can be dire because the physiological functions of the proteins are altered. Brain development is retarded in infants who eat paint that has peeled or flaked off the walls in old houses, because, until very recently, house paint was manufactured with the use of lead salts. Adults who eat mercury-contaminated fish from certain lake regions can suffer similar effects.

The aromas and flavors of garlic and onions are due largely to disulfides—diallyl disulfide and allyl propyl disulfide, respectively.

$$CH_2{=}CHCH_2{-}S{-}S{-}CH_2CH{=}CH_2$$

Diallyl disulfide

$$CH_2{=}CHCH_2{-}S{-}S{-}CH_2CH_2CH_3$$

Allyl propyl disulfide

Example 13.17 Writing equations for the reactions of thiols

Write equations for each of the following reactions: (a) ionization of methanethiol in water; (b) oxidation of ethanethiol; (c) reaction of methanethiol with Pb^{2+}.

Solution

(a) $CH_3SH + H_2O \longrightarrow CH_3S^- + H_3O^+$

(b) $CH_3CH_2SH \xrightarrow{(O)} CH_3CH_2{-}S{-}S{-}CH_2CH_3$

(c) $2\ CH_3SH + Pb^{2+} \longrightarrow CH_3S{-}Pb{-}SCH_3 + 2\ H^+$

Problem 13.17 Write equations for each of the following reactions: (a) neutralization of 1-propanethiol with NaOH; (b) reduction of diethyl disulfide; (c) reaction of 1-propanethiol with Hg^{2+}.

Summary

Structural Relations of Alcohols, Phenols, and Ethers Alcohols contain an —OH (hydroxyl) group attached to a saturated (sp^3) carbon. Phenols contain an —OH attached to one of the sp^2-hybridized carbons of a benzene ring. Ethers, like alcohols and phenols, contain an oxygen that has bonds to two different atoms, but, in contrast with alcohols and phenols, neither of the bonds is to hydrogen. The oxygen of an ether has single bonds to two different carbons.

Constitutional Isomerism in Alcohols Constitutional isomerism in alcohols is due not only to different carbon skeletons, but also to different placements of the —OH group within a carbon skeleton. The C:H ratio for an alcohol is the same as that for the corresponding hydrocarbon with the same number of double bonds or rings.

Classifying and Naming Alcohols Alcohols are 1°, 2°, or 3° alcohols, depending on whether the carbon holding the —OH group is a 1°, 2°, or 3° carbon. In the IUPAC system, alcohols are named by using the longest continuous carbon chain containing the —OH group. The base name is preceded by a number indicating the position of the OH group, and the ending of the name is changed from -e to -ol. Prefixes with numbers are then added at the front to indicate substituents attached to the longest chain. Common names for simple alcohols are formed by using the name of the alkyl group attached to the —OH group followed by the word alcohol.

Physical Properties of Alcohols Alcohols participate in hydrogen bonding, which results in high boiling and melting points and solubility in water.

Acidity and Basicity of Alcohols Alcohols are very weak acids and bases.

Dehydration of Alcohols to Alkenes Intramolecular dehydration of an alcohol in the presence of an acid catalyst produces an alkene. When more than one alkene is possible, the major product is the most substituted alkene. A competing reaction for primary alcohols, but not secondary and tertiary alcohols, is intermolecular dehydration to an ether. At 180°C, the major product is alkene; at 140°C, the major product is ether.

Oxidation of Alcohols Primary and secondary alcohols undergo selective oxidation by permanganate (MnO_4^-) and dichromate ($Cr_2O_7^{2-}$) ions. Primary alcohols are oxidized to aldehydes, which are subsequently oxidized to carboxylic acids. Secondary alcohols are oxidized to ketones with no further oxidation possible. Tertiary alcohols are not oxidized. Permanganate and dichromate form the bases of simple chemical diagnostic tests for primary and secondary alcohols.

Phenols Substituted phenols are named by the IUPAC system as derivatives of the parent compound phenol. Phenols are more polar than alcohols and have higher boiling points and solubilities in water. Phenols are weak acids, but much stronger than alcohols. Phenols undergo oxidation but not dehydration.

Ethers Ethers are isomeric with alcohols. The names of ethers containing simple groups consist of the names of the groups attached to the ether oxygen, followed by the word ether. Ethers have considerably lower boiling points than alcohols but are almost as soluble as alcohols in water. Ethers are unreactive toward acids, bases, and oxidizing agents.

Thiols Thiols contain the —SH group. They have lower boiling points than alcohols. Thiols are stronger acids than alcohols (but weaker than phenols), form insoluble products with heavy metal ions, and are oxidized to disulfides. Disulfides can be reduced back to thiols.

Summary of Key Reactions

ALCOHOLS

Acidity

$$2\,R{-}O{-}H + 2\,Na \longrightarrow 2\,R{-}O^- + Na^+ + H_2$$

Basicity

$$R{-}O{-}H + H_2SO_4 \rightleftharpoons R{-}\underset{+}{\overset{H}{O}}{-}H + HSO_4^-$$

Dehydration to alkene

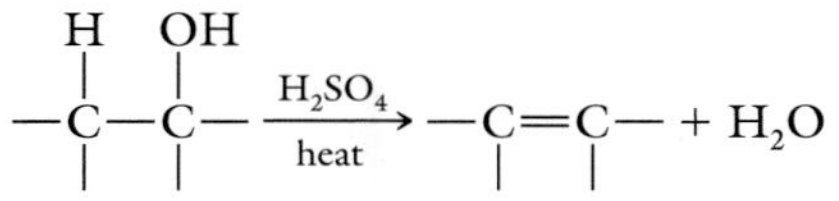

Most substituted double bond is major product

Dehydration to ether competes with alkene formation for 1° alcohols

$$2\,R{-}O{-}H \xrightarrow{H_2SO_4} R{-}O{-}R + H_2O$$

Ether is major product at 140°
Alkene is major product at 180°

Selective oxidation (3° alcohols do not oxidize)

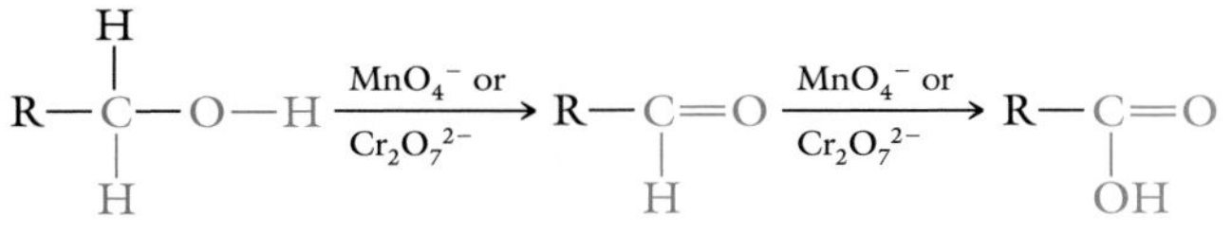

1° alcohol **Aldehyde** **Carboxylic acid**

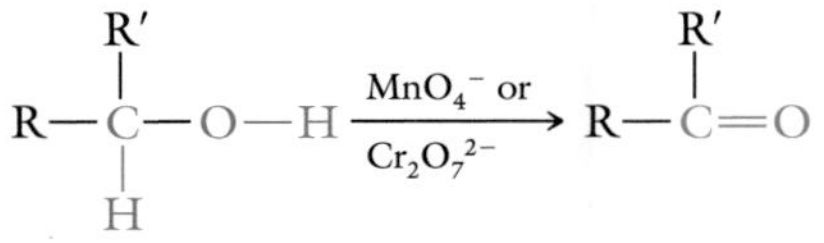

2° alcohol **Ketone**

PHENOLS

Stronger acids compared with alcohols

$$C_6H_5—O—H + NaOH \longrightarrow C_6H_5—O^- + Na^+ + H_2O$$

ETHERS (unreactive)

THIOLS

Oxidation to disulfide

$$2\ R—SH \xrightarrow{(O)} R—S—S—R$$

Disulfides are reduced back to thiols

$$R—S—S—R \xrightarrow{(H)} 2\ R—SH$$

React with heavy metal ions (M = Pb, Hg)

$$2\ R—SH + M^{2+} \longrightarrow \underset{\text{Precipitate}}{R—S—M—S—R} + 2\ H^+$$

Key Words

alcohol, p. 370
dehydration, p. 381, 394
dehydrogenation, p. 385
disulfide, p. 395
enol, p. 388
ether, p. 370, 391
phenol, p. 370, 388
primary alcohol, p. 373
secondary alcohol, p. 373
tertiary alcohol, p. 373
thiol, p. 395

Exercises

Structural Relation of Alcohols, Phenols, and Ethers

13.1 What structural features distinguish alcohols, phenols, and ethers?

13.2 Which of the following compounds is an alcohol, a phenol, an ether, or something else?

$(CH_3)_2CH—O—C_6H_5$

1

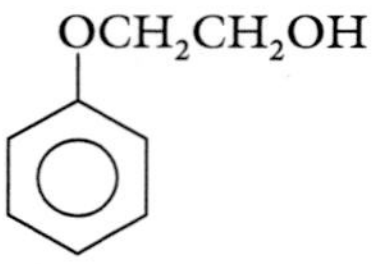

2

$(CH_3)_3C—OH$

3

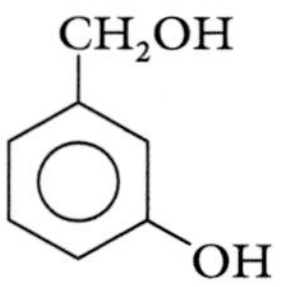

4

Constitutional Isomerism in Alcohols

13.3 Draw the structural formulas for (a) an alcohol, $C_5H_{12}O$, that contains one *t*-butyl group; (b) an alcohol, $C_4H_{10}O$, that contains one methyl group.

13.4 Draw the structural formulas for (a) a cyclic alcohol, $C_5H_{10}O$, that contains four CH_2 groups in the ring; (b) an alcohol, $C_6H_{14}O$, that contains four methyl groups but no *t*-butyl group.

13.5 For each of the following compounds, indicate whether the alcohol follows the general formula $C_nH_{2n+2}O$ or $C_nH_{2n}O$.

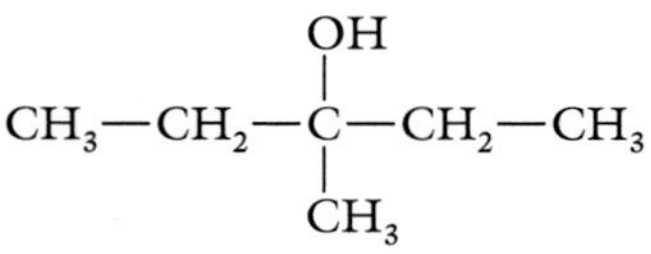

1

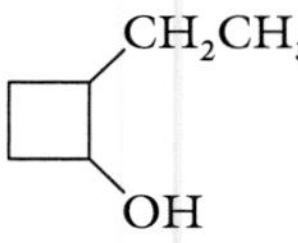

2

13.6 Each of the following compounds is an alcohol. Indicate whether each alcohol possesses one or more double bonds or rings or both. (a) $C_4H_{10}O$; (b) C_4H_8O; (c) C_4H_6O.

Classifying and Naming Alcohols

13.7 Give the IUPAC name and classify the following alcohols as primary, secondary, or tertiary:

(a) 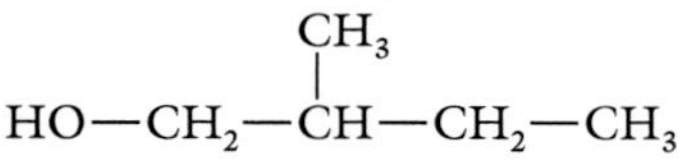

(b) 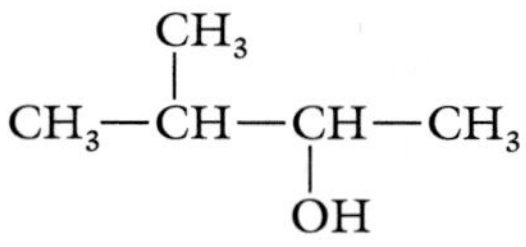

(c) 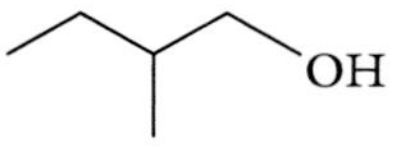

(d) $(CH_3)_3COH$

13.8 Draw the structural formula and classify the following alcohols as primary, secondary, or tertiary: (a) 5-chloro-4-methyl-2-hexanol; (b) 2,2-dimethylcyclobutanol; (c) 1,2,4-butanetriol; (d) *s*-butyl alcohol.

Physical Properties of Alcohols

13.9 Place the following compounds in order of boiling point: hexane, 1,4-butanediol, 1-pentanol, and 2-methyl-2-butanol. Explain the order.

13.10 Which of the following representations correctly describe the hydrogen bonding between methanol and water?

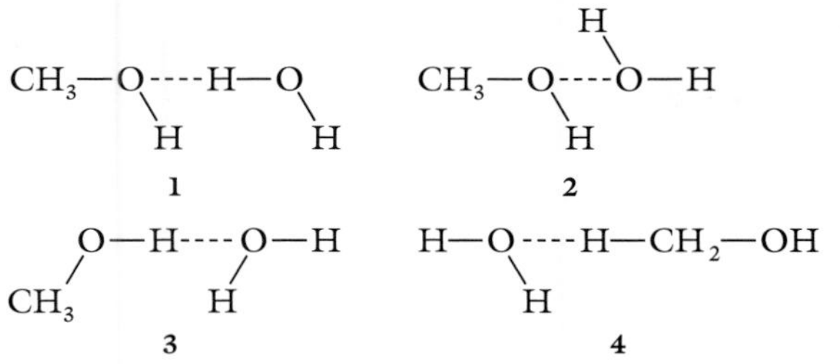

Acidity and Basicity of Alcohols

13.11 What is observed when a drop of ethanol is placed on blue litmus paper? Why?

13.12 What is observed when a drop of ethanol is placed on red litmus paper? Why?

13.13 Write the equation for the reaction of 1-butanol with HCl.

13.14 What occurs when 1-butanol is treated with each of the following reagents? If a reaction takes place, write an equation to show the product(s) formed. (a) NaOH; (b) Na.

Dehydration of Alcohols to Alkenes

13.15 Give the organic product(s) formed in each of the following reactions. If no reaction takes place, write "no reaction." If there is more than one product, indicate only the major product(s).

(a) $CH_3CH_2CH_2CH_2\text{—}OH \xrightarrow[\text{no heat}]{H_2SO_4} ?$

(b) $CH_3CH_2CH_2CH_2\text{—}OH \xrightarrow[\text{heat}]{} ?$

(c) $CH_3CH_2CH_2CH_2\text{—}OH \xrightarrow[\text{heat}]{H_2SO_4} ?$

13.16 Which of the following alcohols undergo dehydration in the presence of a strong acid catalyst? If dehydration takes place, show the product(s). If there is more than one product, indicate which is the major product. (a) 3-Methyl-2-butanol; (b) 2-methylcyclohexanol.

Oxidation of Alcohols

13.17 Which of the following alcohols undergo oxidation with dichromate or permanganate? If oxidation takes place, show the product. (a) 1-Pentanol; (b) 2-methyl-2-butanol.

13.18 Which of the following alcohols undergo oxidation with dichromate or permanganate? If oxidation takes place, show the product. (a) *t*-Butyl alcohol; (b) 3-methyl-2-butanol.

13.19 An unknown is either 1-butene or 1-butanol. Addition of a few drops of permanganate solution to the unknown results in decolorization of the purple color and a brown precipitate. What is the identity of the unknown? Explain your answer.

13.20 An unknown is either 2-methyl-2-propanol or 2-methyl-1-propanol. Addition of a few drops of dichromate solution (orange) to the unknown results in a green solution. What is the identity of the unknown? Explain your answer.

Phenols

13.21 Name the following compound by the IUPAC system.

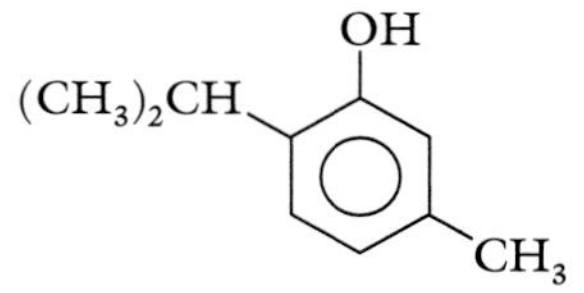

13.22 Draw the structure of 3-bromo-5-*t*-butylphenol.

13.23 Write an equation to show *p*-methylphenol acting as an acid when dissolved in water.

13.24 Place the following compounds in order of acidity: phenol, cyclohexanol, and HCl.

13.25 Indicate whether phenol reacts under each of the following conditions. If a reaction takes place, write the appropriate equation. (a) Combustion; (b) reaction with KOH.

13.26 Indicate whether phenol reacts under each of the following conditions. If a reaction takes place, write the appropriate equation. (a) Acid-catalyzed dehydration; (b) selective (mild) oxidation.

Ethers

13.27 Draw the structural formula for each of the following compounds: (a) 4-ethoxyphenol; (b) isobutyl propyl ether; (c) 3-isopropoxy-1-butanol.

13.28 Draw structural formulas of the ethers with molecular formula $C_5H_{12}O$. What other isomers exist for $C_5H_{12}O$?

13.29 Place the following compounds in order of boiling point: dimethyl ether, ethanol, and propane.

13.30 Which is more soluble in water, dimethyl ether or ethanol?

Formation of Ethers by Dehydration of Alcohols

13.31 What organic product(s) is formed when 1-methylcyclopentanol is heated at 180°C in the presence of an acid catalyst? What is the product at 140°C. Indicate only the major product(s) if more than one product is formed.

13.32 What alcohol(s) must be used to form each of the following ethers or alkenes on acid-catalyzed dehydration? Indicate the temperature that should be used to maximize the yield of the desired product. (a) Dibutyl ether; (b) 2-butene.

Thiols

13.33 Draw the structural formula for each of the following compounds: (a) 2-butanethiol; (b) dicyclopentyl disulfide.

13.34 Write equations for each of the following reactions: (a) neutralization of thiophenol (C_6H_5SH) with KOH; (b) oxidation of 2-propanethiol; (c) reaction of 1-propanethiol with Pb^{2+}; (d) reduction of diethyl disulfide.

Unclassified Exercises

13.35 Name the functional groups designated by A, B, C, D, and E.

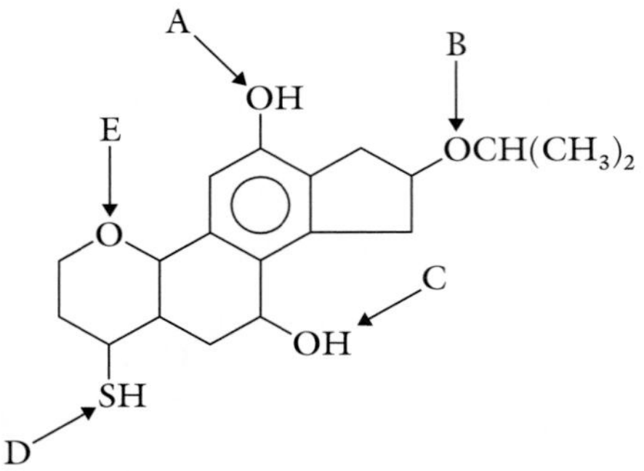

13.36 Identify each of the following compounds as an alcohol, an ether, a phenol, a thiol, a disulfide, or something else.

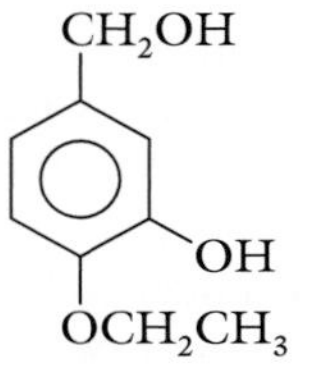

1

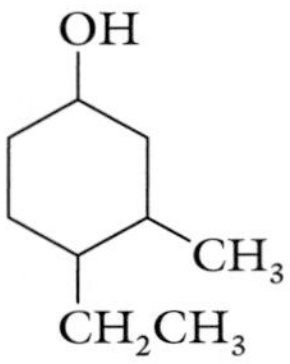

2

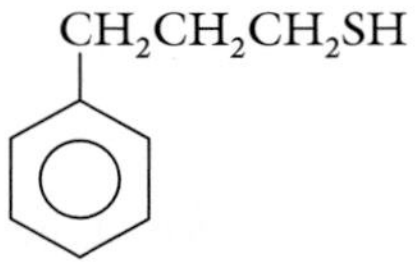

3

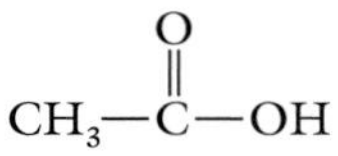

4

13.37 Identify each of the following compounds as an alcohol, an ether, a phenol, a thiol, a disulfide, or something else.

$CH_3CH_2—S—S—CH_2CH_3$

1

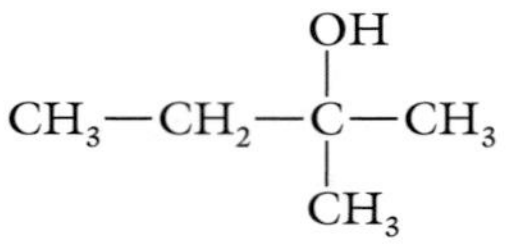

2

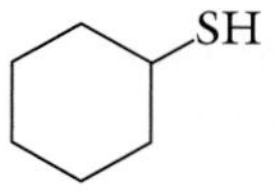

3

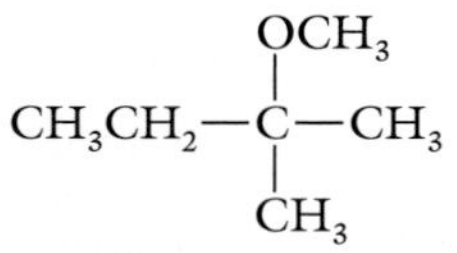

4

13.38 Draw the structural formulas for (a) a tertiary alcohol of molecular formula $C_5H_{12}O$; (b) a cyclic tertiary alcohol of five carbons and containing four carbons in the ring and one methyl group; (c) a phenol isomeric with *p*-chlorophenol; (d) an ether of molecular formula $C_6H_{14}O$ and containing a butyl group.

13.39 An unknown compound has the formula C_4H_8O. Is the unknown an ether or an alcohol? Does the unknown have a double bond or a ring?

13.40 For each of the following pairs of structural drawings, indicate whether the pair represents (1) the same compound or (2) different compounds that are constitutional isomers or (3) different compounds that are geometrical isomers or (4) different compounds that are not isomers.

(a) 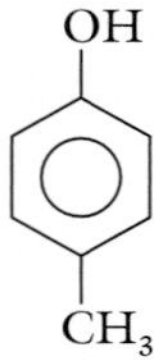

and 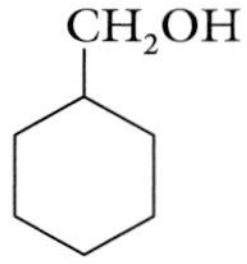

(b) $CH_3CH_2CH_2OCH(CH_3)_2$ and $CH_3CH_2CH_2CH_2CH_2OCH_3$

(c) $CH_3CH_2CH_2OCH(CH_3)_2$ and $CH_3CH_2CH_2CH_2CH_2CH_2OH$

(d) 

and 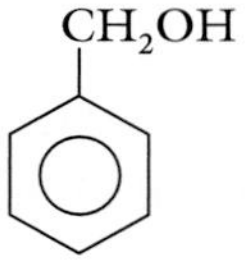

(e) $CH_3CH_2CH_2OCH(CH_3)_2$ and

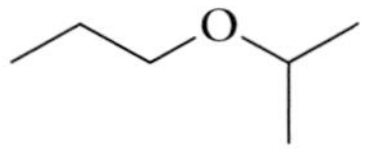

(f) 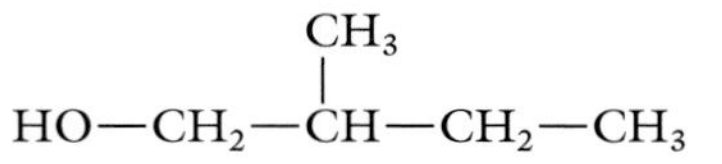

and

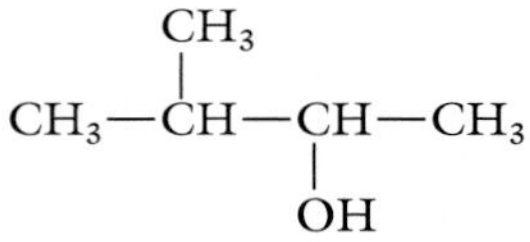

(g) 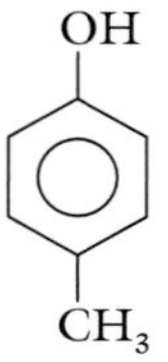

and 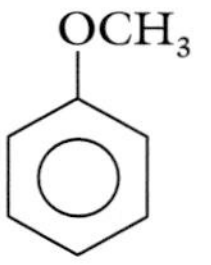

(h) 

and 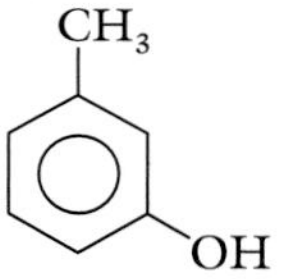

(i) 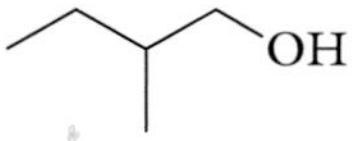

and 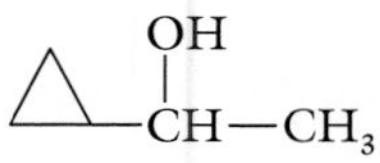

(j) 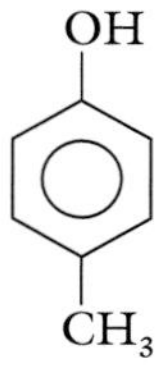

and 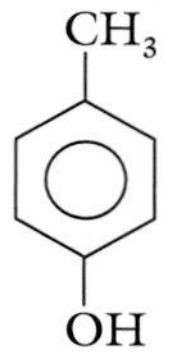

13.41 Draw the structural formula for each of the following: (a) *s*-butyl propyl ether; (b) *m*-propylphenol; (c) *cis*-1,2-cyclohexanediol; (d) 2-pentanethiol; (e) 1-pentene-3-ol.

13.42 Place the following compounds in order of solubility in water: pentane, 1,4-propanediol, 1-butanol, and diethyl ether. Explain the order.

13.43 Indicate whether hydrogen bonding is present in the following compounds or mixtures. If present, draw a representation of the hydrogen bonding. (a) Ethanol; (b) propanol and water mixture; (c) diethyl ether; (d) dimethyl ether and methanol mixture; (e) phenol and water mixture.

13.44 Indicate which compound in each of the following pairs has the higher value of the specified property. If the two compounds in a pair have nearly the same value for the property, indicate that fact. Explain your conclusions.

(a) Dimethyl ether, propane: solubility in water.

(b) Cyclopropyl methyl ether, cyclobutanol: boiling point.

(c) Ethanol, pentane: solubility in water.

(d) 1-Butanol, 1-butanethiol: boiling point.

13.45 Place the following compounds in order of acidity: phenol, HCl, cyclohexanol, and cyclohexanethiol.

13.46 Give the organic product(s) formed in each of the following reactions. If no reaction takes place, write "no reaction." If there is more than one product, indicate only the major product(s).

(a) $CH_3CH_2CH_2CH_2{-}OH \xrightarrow[180°C]{H_2SO_4} ?$

(b) $CH_3CH_2CH_2CH_2{-}OH \xrightarrow[140°C]{H_2SO_4} ?$

(c) $CH_3{-}C(C_6H_5)(OH){-}CH_3 \xrightarrow[180°C]{H_2SO_4} ?$

(d) $CH_3{-}C(C_6H_5)(OH){-}CH_3 \xrightarrow{Cr_2O_7^{2-}} ?$

(e) $CH_3CH_2CH_2CH_2{-}OH \xrightarrow{MnO_4^-} ?$

(f) $CH_3CH_2OCH_2CH_3 \xrightarrow{MnO_4^-} ?$

13.47 Give the organic product(s) formed in each of the following reactions. If no reaction takes place, write "no reaction." If there is more than one product, indicate only the major product(s).

(a) 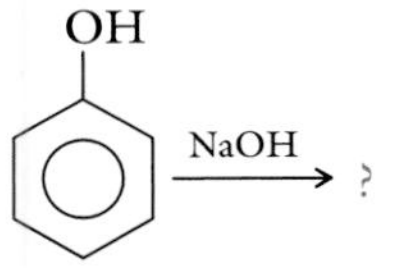

(b) 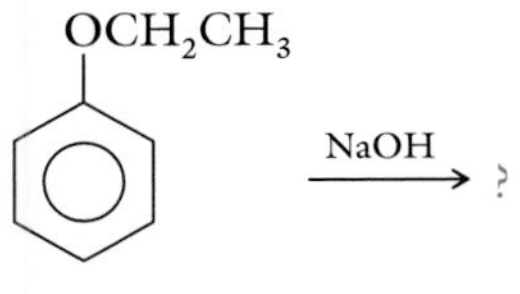

(c) 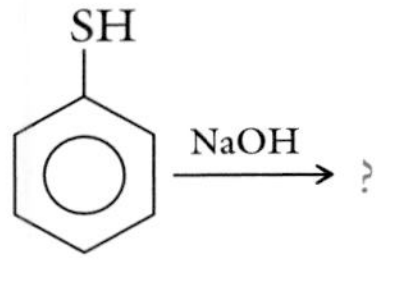

(d) 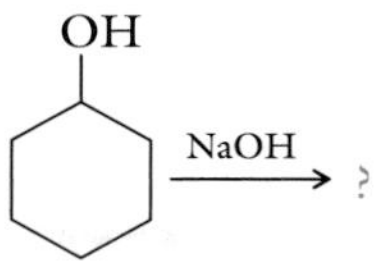

(e) 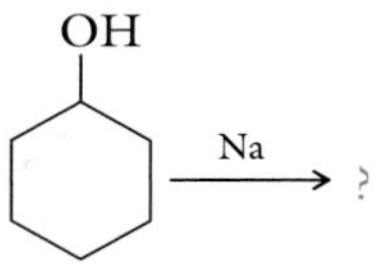

(f) $CH_3CH_2CH_2CH_2{-}SH \xrightarrow{Pb^{2+}} ?$

(g) $CH_3CH_2CH_2CH_2{-}SH \xrightarrow{(O)} ?$

(h) 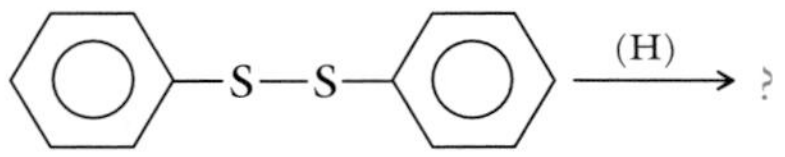

13.48 Write a balanced equation for the combustion of each of the following compounds: (a) 1-butanol; (b) diethyl ether.

13.49 For each of the following pairs of compounds, indicate which compound is more acidic: (a) 1-propanol and butane; (b) 1-propanol and HCl; (c) 1-propanol and water; (d) 1-propanol and phenol; (e) 1-propanol and 1-propanethiol.

13.50 An unknown is either *t*-butyl alcohol, ethyl methyl ether, pentane, or 1-butanol. Addition of a few drops of dichromate solution (orange) to the unknown results in a green solution. What is the identity of the unknown? Explain your answer.

13.51 An unknown is either *t*-butyl alcohol, ethyl methyl ether, pentane, or 1-butanol. The purple color of permanganate solution is not decolorized when permanganate is added to the unknown. What is the identity of the unknown? Explain your answer.

13.52 An unknown is either 1-butene or 1-butanol. Addition of a few drops of dichromate solution (orange) to the unknown results in a green solution. What is the identity of the unknown? Explain your answer.

13.53 An unknown is either cyclohexanol, toluene, or phenol. The unknown turns blue litmus paper red. What is the identity of the unknown? Explain your answer.

13.54 An unknown is either cyclohexanol, toluene, or phenol. The unknown is neutral to litmus paper: it does not turn blue litmus paper red or red litmus paper blue. What is the identity of the unknown? Explain your answer.

13.55 Give the bond angles for A, B, C, D, and E:

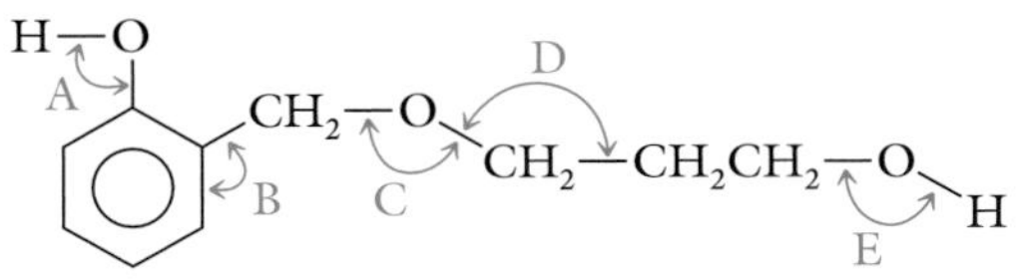

13.56 What is the hybridization (sp^3 or sp^2) of atoms 1 through 6?

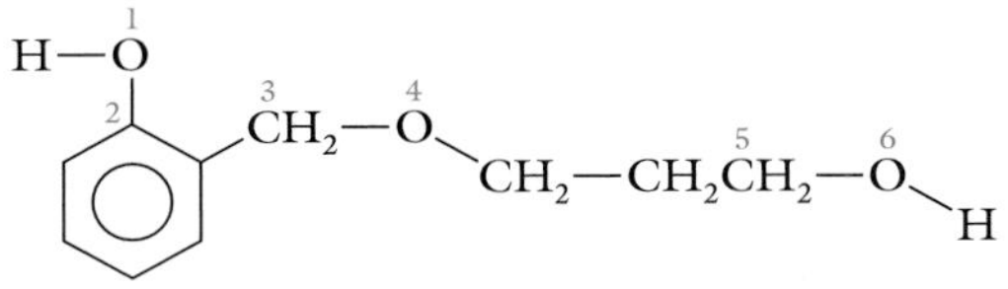

Chemical Connections

13.57 The XYZ Cough Drop Company uses hexylresorcinol as an ingredient in its cough drops. The plant manager suspects that a shipment from the supplier of hexylresorcinol contains 3-hexyl-1,2-cyclohexanediol instead of hexylresorcinol. What simple chemical test distinguishes between the two compounds?

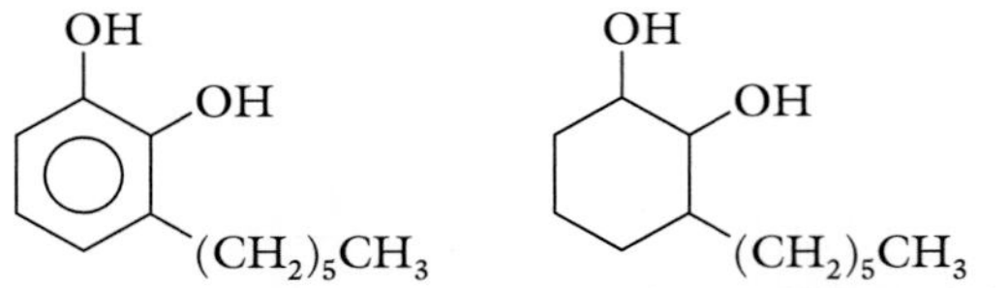

Hexylresorcinol **3-Hexyl-1,2-cyclohexanediol**

13.58 Ethanol has many important industrial uses in addition to its use in alcoholic beverages (see Box 13.1). Industrial ethanol is obtained from two sources: about 30% by fermentation of carbohydrates (see Box 13.2) and 70% by hydration of ethene. What reagent(s) and reaction conditions are required for producing ethanol by hydration of ethene?

13.59 2-Propanol (rubbing alcohol) is applied to the skin to lower a fever (see Box 13.1). Alcohols such as 1-decanol or 2-decanol are not useful for this purpose. How does 2-propanol lower a fever and why are 1-decanol and 2-decanol not useful?

13.60 Dehydration of alcohols requires a strong acid catalyst (see Box 13.4). Catalysts increase reaction rates by offering an alternate mechanism for reaction, one with a lower activation energy. What is the alternate mechanism for dehydration in the presence of an acid catalyst and why does it take place more rapidly?

13.61 1-Propanol and 2-propanol yield the same product, propene, on dehydration (by heating in the presence of acid), but 2-propanol reacts faster than 1-propanol. Explain the difference in reactivity in reference to the mechanism of dehydration (see Box 13.4).

13.62 The cyclic ether ethylene oxide (Section 13.9) is an important industrial chemical. It is produced by selective oxidation of ethene and subsequently reacted with water to produce 1,2-ethanediol, the compound used as an antifreeze in automobile cooling systems. Write the equation for the conversion of ethylene oxide into 1,2-ethanediol. The reaction is classified as a ring-opening addition reaction and is catalyzed by strong acids.

13.63 Whereas ethylene oxide readily undergoes ring-opening addition reactions such as that in Exercise 13.62, the cyclic ether dioxane is unreactive. Indicate the reason for the difference after referring to Box 11.2.

CHAPTER 14

ALDEHYDES AND KETONES

CHEMISTRY IN YOUR FUTURE

Your younger brother, a nursing student, enjoys visiting the kitchen that you manage in a long-term convalescent care facility. Today, as he samples the pound cake that you've been baking, he complains about the difficulty of the organic and biochemistry course that he is taking, especially this week's topic—aldehydes and ketones. You surprise him by saying, "Those aldehydes and ketones are great stuff; I use them all the time in my recipes." You then take him on a short tour of the kitchen, pointing to sugar, butter, cornstarch, and various flavorings and spices (almond, cinnamon, vanilla, mint), all of which are aldehydes or ketones. Chapter 14 describes the common features and properties of this family of organic compounds.

LEARNING OBJECTIVES

- Draw and name aldehydes and ketones.
- Describe the physical properties of aldehydes and ketones.
- Describe and write equations for oxidation and reduction reactions of aldehydes and ketones.
- Describe and write equations for hemiacetal and acetal formation by reaction of alcohols with aldehydes and ketones.
- Describe and write equations for hydrolysis of acetals.

We now begin the study of oxygen-containing organic compounds in which the carbon and oxygen are connected by a double bond. In this chapter, we examine two closely related families of compounds: the aldehydes and ketones.

The aldehyde and ketone families include many biological molecules: carbohydrates such as the sugar glucose, an essential source of energy throughout the plant and animal world; lipids such as the male and female sex hormones testosterone and progesterone; and a variety of chemical intermediates that take part in biochemical pathways, as we will see in Chapter 22 through 25. An assortment of naturally occurring compounds with strong flavors and odors—for example, vanilla, almonds, mint, and butter—also belong to these families.

Aldehydes and ketones have a number of commercial uses as solvents and as starting materials for manufacturing many industrial chemicals, including plastics and adhesives.

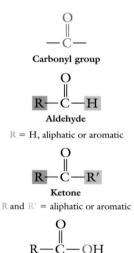

14.1 STRUCTURE OF ALDEHYDES AND KETONES

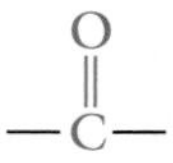
Carbonyl group

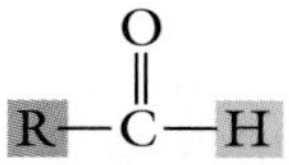
Aldehyde

R = H, aliphatic or aromatic

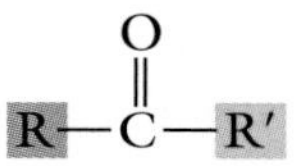
Ketone

R and R′ = aliphatic or aromatic

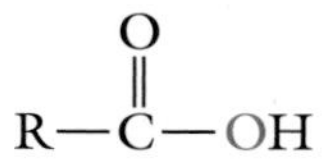
Carboxylic acid

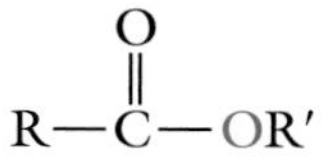
Ester

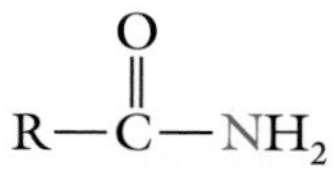
Amide

The common structural features of all **aldehydes** and **ketones** is the **carbonyl functional group,** a carbon–oxygen double bond (Section 11.2). Aldehydes are distinguished from ketones by the atoms directly attached to the carbon of the carbonyl group. This carbon is called the **carbonyl carbon.** In aldehydes, at least one hydrogen atom is directly attached to the carbonyl carbon, forming the **aldehyde group,** HC=O, often written as CHO. In ketones, only carbon atoms, no hydrogen atoms, are directly attached to the carbonyl carbon. The groups (represented by R and R′) attached to the carbonyl carbon can be either aliphatic or aromatic.

Other families also contain the carbonyl group (Table 14.1) but differ from aldehydes and ketones in having a **heteroatom**—any atom other than carbon or hydrogen—directly attached to the carbonyl carbon. In carboxylic acids and esters, the heteroatom is an oxygen; in amides, it is a nitrogen. Carboxylic acids, esters, and amides are the subject of Chapters 15 and 16.

Example 14.1 Recognizing aldehydes and ketones

Which of the following compounds are aldehydes or ketones and which are not? Classify each compound.

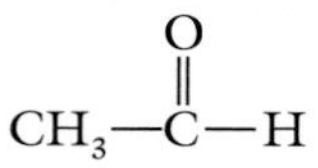
1

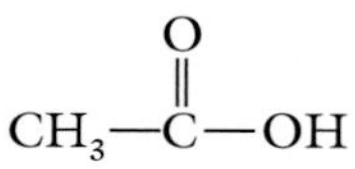
2

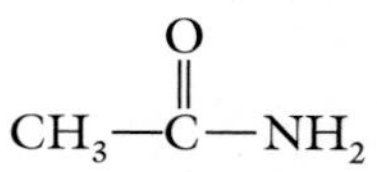
3

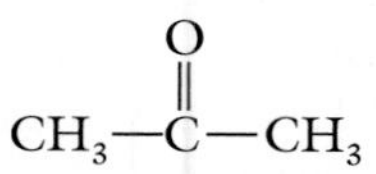
4

TABLE 14.1 Families Containing the Carbonyl Group (C=O)

	Family	Atoms Bonded to C=O: y	z
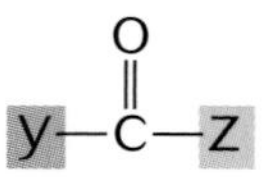	aldehyde	H	H or C
	ketone	C	C
	carboxylic acid, ester	H or C	O
	amide	H or C	N

Solution

Compound 1 is an aldehyde because it has at least one H and no heteroatom attached to the carbonyl carbon. Compound 2 is a carboxylic acid because of the heteroatom oxygen attached to the carbonyl carbon. Compound 3 is an amide because there is a nitrogen attached to the carbonyl carbon. Compound 4 is a ketone because it has only carbons, no H and no heteroatom, attached to the carbonyl carbon.

Problem 14.1 Classify each of the following compounds as an aldehyde, ketone, carboxylic acid, ester, or amide.

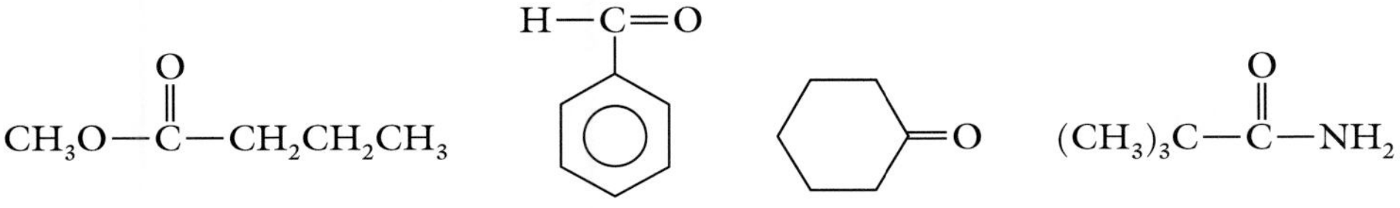

1 2 3 4

Aldehydes and ketones are related to alcohols in the same way that alkenes are related to alkanes. Loss of a hydrogen from each of two adjacent carbons of an alkane yields an alkene. Loss of the hydrogen from the —OH group of an alcohol, together with loss of a hydrogen from the carbon holding the OH, yields the carbonyl group:

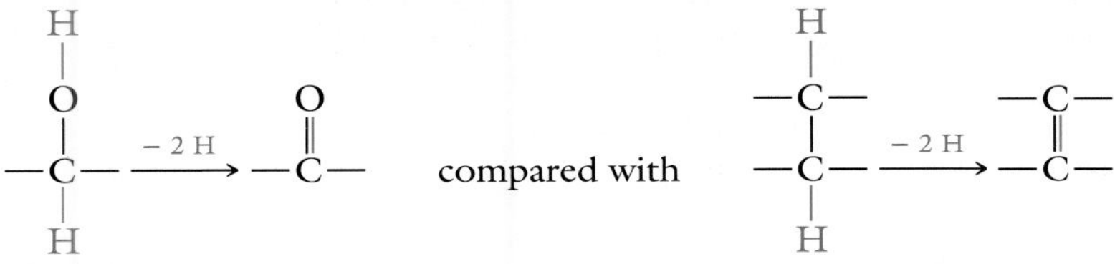

Alcohol Aldehyde or ketone Alkane Alkene

Recall that the presence of a double bond or ring lowers the C:H ratio in a compound by two hydrogens compared with alkanes. The presence of a C=O double bond has the same effect as that of a C=C double bond; aldehydes and ketones have the same C:H ratio as that of alkenes. The molecular formulas for the different families of organic compounds are shown in Table 14.2.

Another similarity between the aldehydes and ketones and the alkenes is in the bonding of the C=C and C=O double bonds. (Refer to Section 12.2 for a description of the carbon–carbon double bond.) Both the carbon and the oxygen atoms of the carbonyl group are sp^2 hybridized, and the carbonyl double bond contains one strong σ bond and one weak π bond. The bond angles about the carbonyl carbon are close to 120° (trigonal), as shown in Figure 14.1. Despite their structural similarities, the C=C and C=O double bonds do differ in important ways, as we shall soon see.

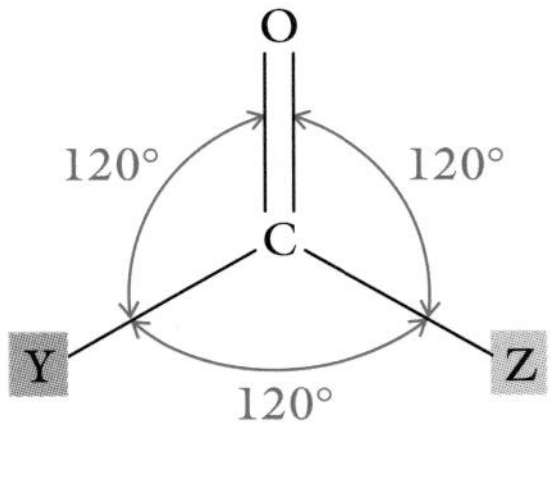

Figure 14.1 Bond angles in aldehydes (Y = H; Z = H, R) and ketones (Y = R; Z = R′).

TABLE 14.2 Molecular Formulas of Different Families

Family	Molecular Formula
alkane	C_nH_{2n+2}
cycloalkane	C_nH_{2n}
alkene	C_nH_{2n}
alcohol, ether	$C_nH_{2n+2}O$
aldehyde, ketone	$C_nH_{2n}O$

Like other families of organic compounds, aldehydes and ketones can have constitutional isomers based on different carbon skeletons and different placements of the carbonyl group. For example, both propanal and acetone have the molecular formula C_3H_6O:

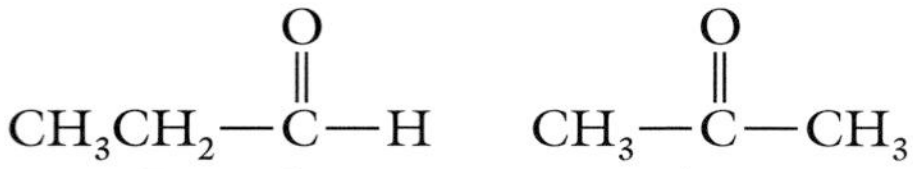

$$CH_3CH_2-\overset{\overset{\displaystyle O}{\|}}{C}-H \qquad CH_3-\overset{\overset{\displaystyle O}{\|}}{C}-CH_3$$

Propanal   Acetone

Example 14.2   Drawing constitutional isomers of aldehydes and ketones

How many aldehydes and ketones are possible for the molecular formula C_4H_8O? Draw the structural formula for each isomer.

Solution

Two different C_4 carbon skeletons are possible:

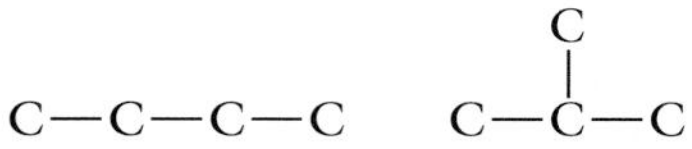

$$C-C-C-C \qquad \overset{\overset{\displaystyle C}{|}}{C-C-C}$$
   1   2

14.1   Chemistry Around Us

Aldehydes and Ketones in Nature

Many of the aldehydes and ketones in nature are polyfunctional, such as the carbohydrate (sugar) glucose. Note that glucose contains an aldehyde group and five hydroxyl groups. Closely related is the ketone sugar fructose. Glucose is found in combined form as starch in potatoes, corn, rice, and peas. Glucose and fructose are found combined together as sucrose, the table sugar used to sweeten our baked goods, coffee, and tea.

$$HOCH_2-\underset{\underset{\displaystyle OH}{|}}{CH}-\underset{\underset{\displaystyle OH}{|}}{CH}-\underset{\underset{\displaystyle OH}{|}}{CH}-\underset{\underset{\displaystyle OH}{|}}{CH}-\overset{\overset{\displaystyle O}{\|}}{CH}$$

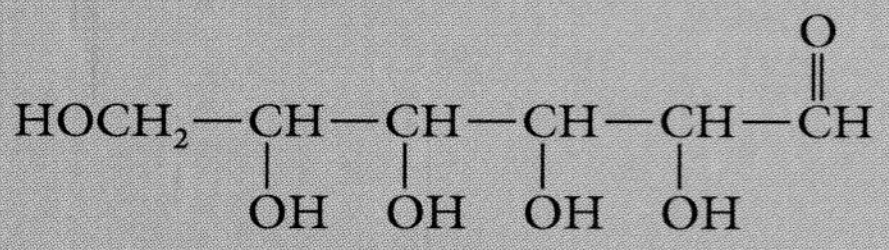

Glucose

$$HOCH_2-\underset{\underset{\displaystyle OH}{|}}{CH}-\underset{\underset{\displaystyle OH}{|}}{CH}-\underset{\underset{\displaystyle OH}{|}}{CH}-\overset{\overset{\displaystyle O}{\|}}{C}-CH_2OH$$

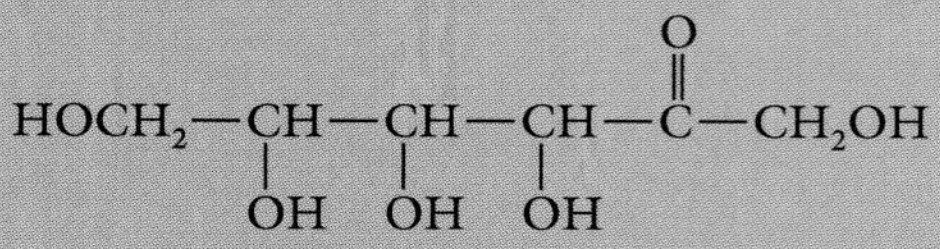

Fructose

The female and male sex hormones progesterone and testosterone, respectively, contain the ketone group. These hormones are members of the lipid family and belong to a class of lipids called steroids (Section 19.8).

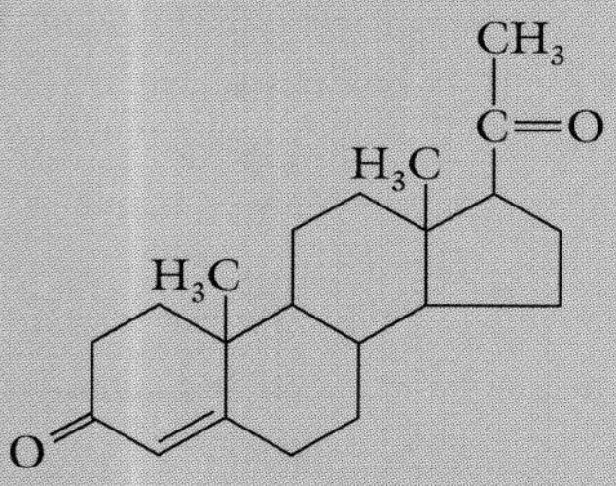

Progesterone

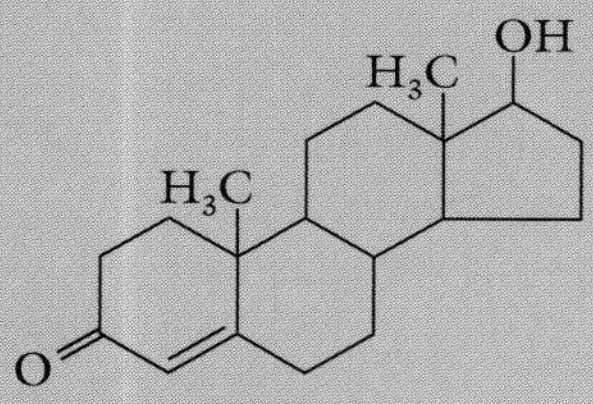

Testosterone

Aldehydes and ketones such as almond, cinnamon, and mint are used as spices and flavorings in cooking.

Both of these skeletons can yield an aldehyde, and skeleton 1 can also yield a ketone:

$$CH_3CH_2CH_2-\overset{\overset{\displaystyle O}{\|}}{C}-H \qquad CH_3\overset{\overset{\displaystyle CH_3}{|}}{CH}-\overset{\overset{\displaystyle O}{\|}}{C}-H \qquad CH_3-\overset{\overset{\displaystyle O}{\|}}{C}-CH_2CH_3$$

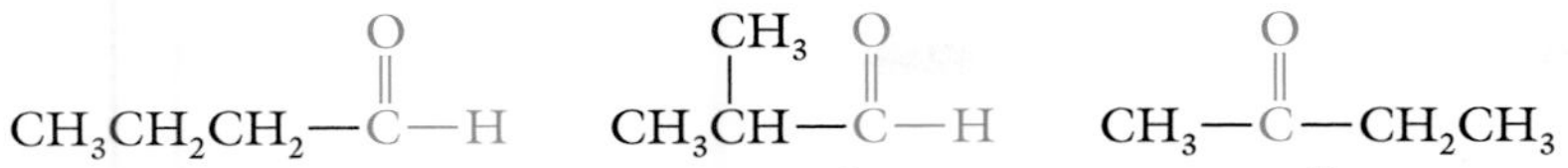

1a   2a   1b

No ketone is possible for skeleton 2, because the tertiary carbon already has three bonds to other carbon atoms and cannot also be a carbonyl carbon. We would have the impossibility of five bonds to carbon:

$$C-\overset{\overset{\displaystyle C}{|}}{\underset{\underset{\displaystyle O}{\|}}{C}}-C \qquad \leftarrow \text{C cannot have 5 bonds}$$

Problem 14.2   Draw the structural formula for each aldehyde and ketone with the molecular formula $C_5H_{10}O$.

Some of the naturally occurring aldehydes and ketones are described in Box 14.1. The commercial applications of formaldehyde and acetone, the simplest aldehyde and ketone, are described in Box 14.2 on the following page.

In Box 12.2, we learned that *cis*-11-retinal is essential to vision in most animals. *cis*-11-Retinal belongs to both the alkene and the aldehyde families, and both the C=C and the C=O functional groups are critical to its physiological function. The light-induced conversion of *cis*-11-retinal into *trans*-11-retinal, which triggers the electrical signal that the brain interprets as vision (Box 12.2), can take place only after *cis*-11-retinal reacts with the protein opsin to form rhodopsin. The presence of an aldehyde group in *cis*-11-retinal is critical to this event: rhodopsin is formed by a dehydration reaction between the carbonyl group of *cis*-11-retinal and an amine (NH₂) group of opsin.

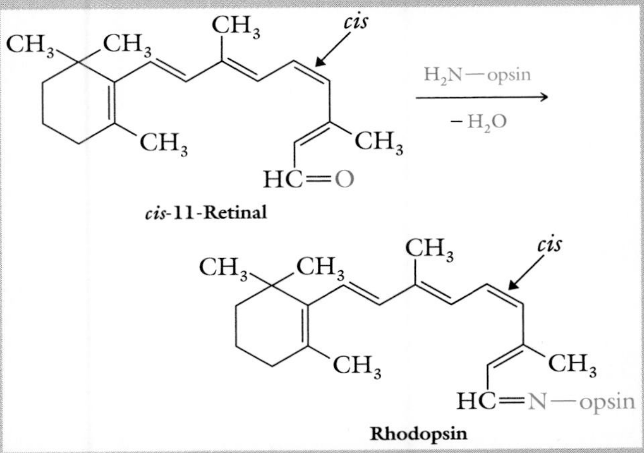

Plants produce a variety of aldehydes and ketones that have long been used as flavorings or spices—for example, benzaldehyde (almond), cinnamaldehyde (cinnamon), vanillin (vanilla), and menthone (mint). Biacetyl, a diketone, is the compound responsible for the taste of butter and is often added to margarine and other shortenings to impart a buttery flavor.

cis-11-Retinal

Rhodopsin

Benzaldehyde
(Almond)

Cinnamaldehyde
(Cinnamon)

Vanillin
(Vanilla)

Methone
(Mint)

$$CH_3-\overset{\overset{\displaystyle O}{\|}}{C}-\overset{\overset{\displaystyle O}{\|}}{C}-CH_3$$

Biacetyl
(Butter)

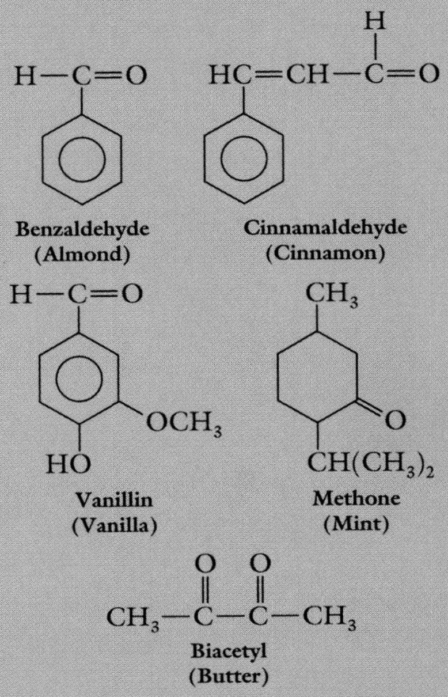

14.2 Chemistry Around Us

Important Aldehydes and Ketones

Formaldehyde (HCHO) is the simplest aldehyde, and acetone (CH_3COCH_3) is the simplest ketone. Both chemicals have a number of commercial applications.

Formaldehyde is used to preserve biological specimens and embalm cadavers, because it kills bacteria, viruses, and other microorganisms that would otherwise cause dead tissues to deteriorate. It kills the microorganisms by reacting with their proteins and destroying the physiological functions of these proteins. Similarly, it has been used to deactivate viruses for the preparation of vaccines. Formaldehyde has also been used for disinfecting buildings and ships, for tanning hides, and as a fungicide, but some of these uses have been curtailed because of the high toxicity of formaldehyde to humans. Formaldehyde is a gas at room temperature (boiling point = −21°C) and is generally used as an aqueous solution called **formalin.**

The extremely deleterious effects of drinking methanol noted in Box 13.3 are in fact due to the toxicity not of methanol itself but of the formaldehyde produced when the liver attempts to detoxify the body of methanol. Formaldehyde is toxic because it reacts with proteins and deactivates them—the same reaction that is useful for embalming purposes and also the same reaction that forms rhodopsin from *cis*-11-retinal and opsin (see Box 14.1). The consumption of ethanol is less a problem because acetaldehyde, produced by the detoxification of ethanol, is less reactive in deactivating proteins. However, long-term consumption of large amounts of ethanol eventually takes its toll on the body and the mind.

Formaldehyde preserves biological specimens such as this grasshopper.

Large amounts of formaldehyde are used in the manufacture of polymers. More than 5 billion pounds of polymers per year are produced in the United States by the reaction of formaldehyde with phenol, urea, and melamine. These polymers are used to manufacture plywood and particle board, printed circuit boards, and the protective and decorative surfaces for kitchen and bathroom countertops and tables.

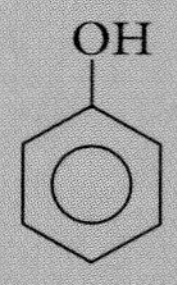

Phenol

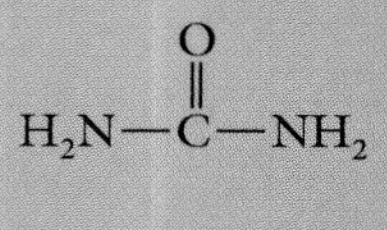

Urea

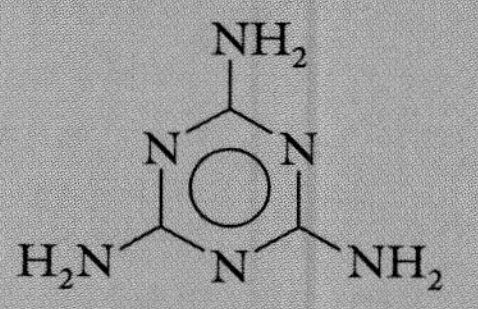

Melamine

Acetone is used in excess of 1 billion pounds per year in the United States. It and chemicals synthesized from it are useful as solvents for plastics, adhesives, pesticides, hydraulic fluids, and printing inks. Acetone is also used in nail polish remover formulations. Methyl methacrylate and bisphenol A are two important chemicals synthesized from acetone. Poly(methyl methacrylate) is manufactured from methyl methacrylate and is used for the manufacture of bathtubs, dentures, dental fillings, and windows for aircraft and factory buildings (Table 12.3). Bisphenol A is used to produce epoxy adhesives.

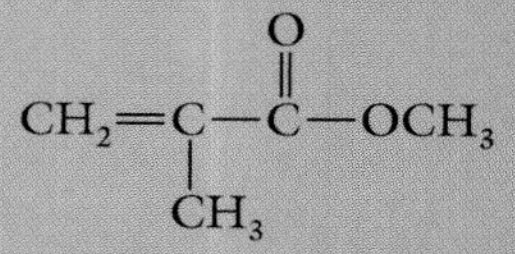

Methyl methacrylate

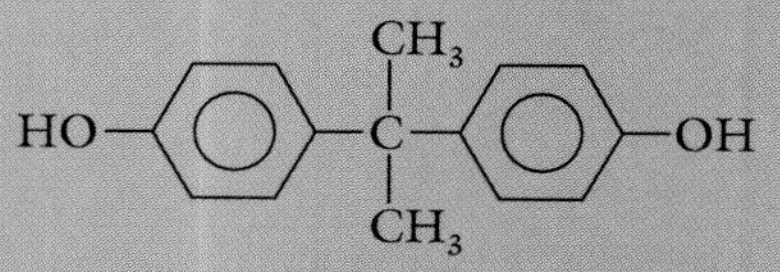

Bisphenol A

People with diabetes who are not receiving proper treatment may have a characteristic sweet odor of acetone on their breath. The diabetic does not adequately metabolize glucose, because of an insulin deficiency. This inadequacy causes a diversion of chemicals into other pathways, with the end result being an incomplete metabolism of fatty acids. Instead of conversion into carbon dioxide and water, the fatty acids are converted into acetone and other ketones called **ketone bodies** (Section 24.3) that are detected in the blood, urine, and breath. Section 26.11 describes the chemistry of diabetes in more detail.

14.2 NAMING ALDEHYDES AND KETONES

In the IUPAC nomenclature system, aldehydes and ketones are named by using a variation of the rules for naming alkanes:

Rules for naming aldehydes and ketones

- Find the longest continuous chain that contains the carbonyl carbon.
- The ending **-e** of the alkane name is changed to **-al** for aldehydes and **-one** for ketones.
- The longest continuous chain is numbered from the end nearest the C=O group. The position of the C=O group is indicated for ketones by placing the number of the carbonyl carbon in front of the base name. For aldehydes, the carbonyl carbon is always C1, and the number 1 is not used in naming an aldehyde.
- The names of substituents are used as prefixes, preceded by numbers indicating their positions on the longest chain.
- A cyclic compound containing a carbonyl carbon as part of the ring is named a **cycloalkanone.** Numbering of the ring starts at the carbonyl carbon and proceeds to other substituents on the ring in the direction that yields the lowest numbers for those substituents.
- C_6H_5–CHO is named benzaldehyde.

Example 14.3 Using the IUPAC nomenclature system

Name the following compounds by the IUPAC system:

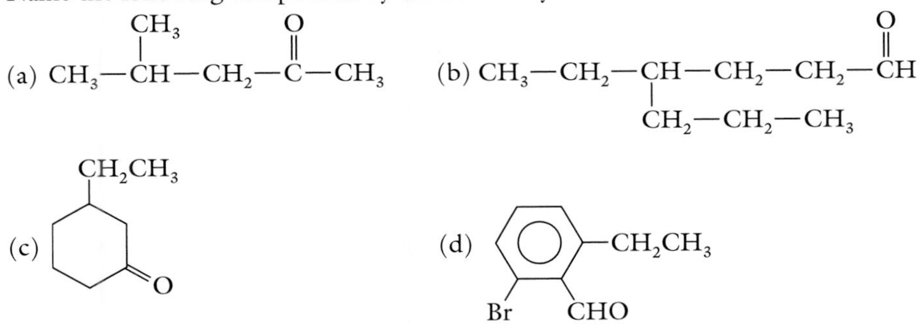

Solution

(a) This compound is a ketone. The longest continuous chain containing the carbonyl carbon is a five-carbon chain. The chain is numbered from the end nearest the carbonyl carbon to yield 2-pentanone as the base name. (Numbering the chain from the other end is incorrect, because it yields 4-pentanone, which has a higher number for the position of the carbonyl carbon.) The methyl group is at C4, which yields the name 4-methyl-2-pentanone.

(b) This compound is an aldehyde. The longest chain containing the carbonyl carbon is the seven-carbon chain:

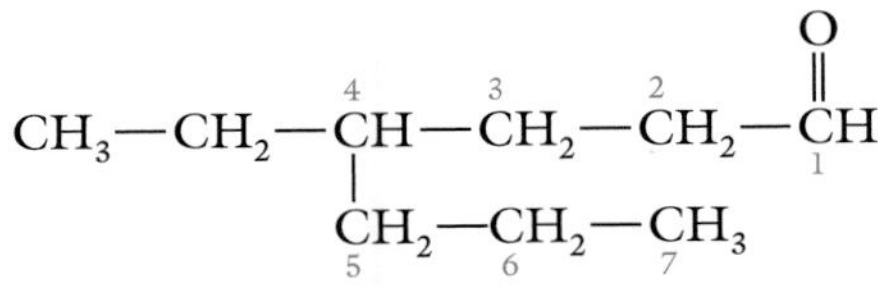

The carbonyl carbon is C1, and there is an ethyl group at C4. The name is 4-ethylheptanal.

(c) In this cyclic ketone, the carbonyl carbon is C1, which places the ethyl group at C3. The name is 3-ethylcyclohexanone.

(d) This compound is named as a derivative of benzaldehyde, C_6H_5–CHO. The aldehyde group is attached at C1 of the ring. The two possible names are 2-bromo-6-ethylbenzaldehyde and 6-bromo-2-ethylbenzaldehyde, numbering in the clockwise and counterclockwise directions, respectively. Both names have the same set of numbers. The correct name is 2-bromo-6-ethylbenzaldehyde because the lower number is assigned to substituents in alphabetical order.

Problem 14.3 Name the following compounds by the IUPAC system:

(a) $CH_3CH(CH_3)—C(=O)—CH_2CH(CH_3)CH_3$ (b) $(CH_3)_3CCH_2CH_2—C(=O)H$

(c) 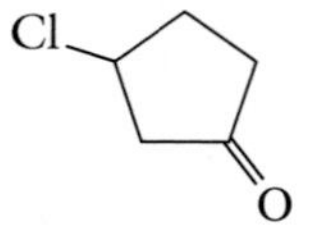

(d) 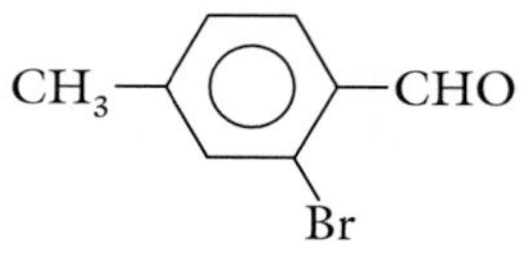

Unbranched acyclic aldehydes containing from one to four carbons are often referred to by common names. The common names are obtained by combining the prefixes **form-**, **acet-**, **propion-**, and **butyr-** with **-aldehyde** to give the names formaldehyde, acetaldehyde, propionaldehyde, and butyraldehyde (Table 14.3).

Common names for ketones are obtained by naming the groups attached to the carbonyl carbon followed by the word **ketone.** This method is used only when the groups attached to the carbonyl carbon are simple groups: the unbranched alkyl groups, from methyl, ethyl, propyl, butyl through decyl; the branched three- and four-carbon alkyl groups (isopropyl, isobutyl, *s*-butyl, *t*-butyl); and the unsubstituted cycloalkyl groups, such as cyclohexyl and cyclopentyl.

Note, however, that dimethyl ketone (IUPAC: 2-propanone), an important industrial compound, is usually called acetone.

TABLE 14.3 IUPAC and Common Names of Aldehydes and Ketones

Compound	IUPAC Name	Common Name
$H—C(=O)—H$	methanal	formaldehyde
$CH_3—C(=O)—H$	ethanal	acetaldehyde
$CH_3CH_2—C(=O)—H$	propanal	propionaldehyde
$CH_3CH_2CH_2—C(=O)—H$	butanal	butyraldehyde
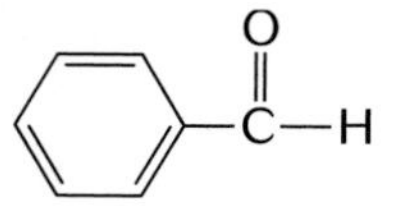	benzaldehyde	benzaldehyde
$CH_3—C(=O)—CH_3$	2-propanone	acetone
$CH_3—C(=O)—CH_2CH_3$	2-butanone	ethyl methyl ketone

Example 14.4 Using the common nomenclature system

Give the common name for each of the following ketones:

(a) 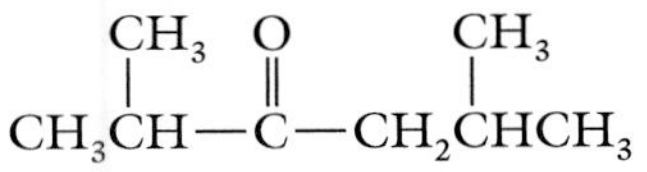 (b) 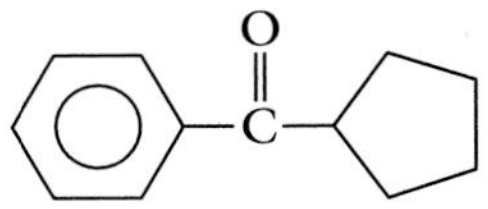

Solution

(a) The two groups attached to the carbonyl carbon are the isopropyl and isobutyl groups. The name is isobutyl isopropyl ketone because the groups are named in alphabetical order.

(b) The two groups attached to the carbonyl carbon are the phenyl and cyclopentyl groups. The name is cyclopentyl phenyl ketone.

Problem 14.4 Give the common name for each of the following ketones:

(a) 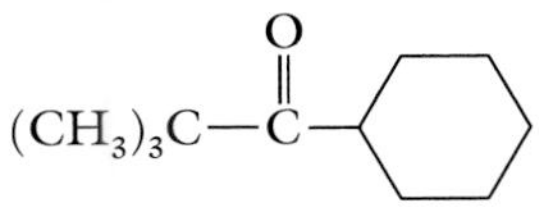 (b) 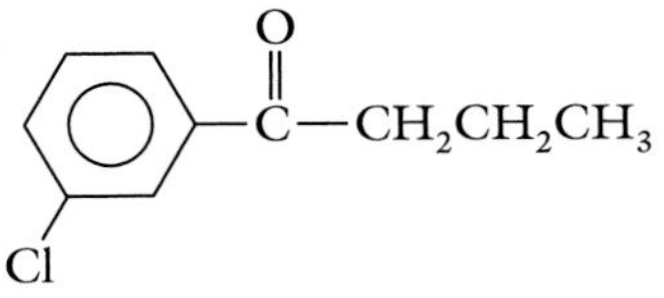

14.3 PHYSICAL PROPERTIES OF ALDEHYDES AND KETONES

Aldehydes and ketones are polar compounds because the carbonyl group is polar. The physical properties of aldehydes and ketones, the result of this polarity, are characterized by:

- Higher boiling and melting points than those of the corresponding hydrocarbons but lower than those of the corresponding alcohols
- Almost the same water solubility as that of alcohols

Unlike the carbon–carbon double bond of alkenes, the C=O bond is highly polarized. The weakly held electrons of the π bond are pulled closer to the oxygen of the carbonyl group than to the carbon because oxygen is much more electronegative than carbon. This polarization of the π bond electrons is often expressed by using the ionic structure shown on the right-hand side below. However, the π bond electrons are not completely on oxygen; there are not full 1+ and 1− charges on carbon and oxygen, respectively. Neither the covalent structure nor the ionic structure accurately represents a carbonyl group. The π electrons are neither equally shared by carbon and oxygen nor completely on oxygen. The situation is in between those two extremes. The carbonyl group is a hybrid (blend) of the covalent and ionic structures, and we represent the situation by drawing the two structures separated by a two-headed arrow, as shown in the margin.

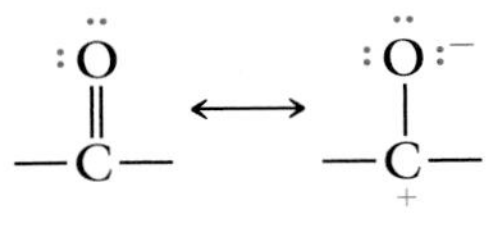

Covalent Ionic

The carbonyl group is said to be a **resonance hybrid** of the two structures, but most of the time only one or the other of the two structural drawings is shown. Irrespective of which structure is drawn, the carbonyl group has partial charges, δ^+ on carbon and δ^- on oxygen, that account for the polar nature of the carbonyl group. An alternate representation of the carbonyl group is often used instead of having the covalent and ionic structures separated by the double-headed arrow:

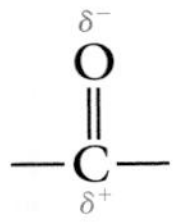

INSIGHT INTO PROPERTIES

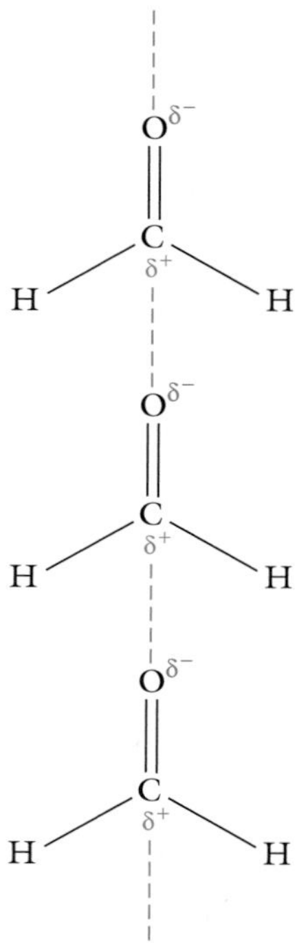

Figure 14.2 Dipole–dipole attractive forces in formaldehyde, which has a higher boiling point than that of ethane.

The polar carbonyl group is responsible for the strong secondary attractive forces in aldehydes and ketones. There are strong attractive forces between the partly positive carbon end of the C=O dipole of one molecule and the partly negative oxygen end of the C=O dipole of a neighboring molecule. These attractive forces are depicted for formaldehyde by the dashed lines in Figure 14.2.

Aldehydes and ketones have considerably higher boiling and melting points than those of the corresponding hydrocarbons because the dipole–dipole secondary forces in aldehydes and ketones are much stronger than the London forces in hydrocarbons. Higher temperatures are needed to overcome the attractive forces in aldehydes and ketones compared with those in hydrocarbons. For example, Table 14.4 shows the boiling points of butanal, 2-butanone, and pentane—all compounds having the same molecular mass. The boiling points of the aldehyde and ketone are considerably higher than the boiling point of the alkane (76°C and 80°C versus 36°C). However, aldehydes and ketones have boiling points considerably lower than those of alcohols of similar molecular mass. For example, the boiling point of 1-butanol is 117°C versus 76°C for butanal. These differences are also evident in Figure 14.3 and result from the differences in secondary attractive forces. Hydrogen-bonding attractive forces (alcohols) are stronger than dipole–dipole attractive forces (aldehydes, ketones), which are stronger than London forces (hydrocarbons), as illustrated in Figure 14.3. The differences in boiling points between alkanes, aldehydes and ketones, and alcohols narrow with increasing molecular mass (Figure 14.4). The effect of a C=O group in an aldehyde or ketone or an —OH group in an alcohol on the strength of the secondary attractive forces decreases as the fraction of the molecule that the C=O or —OH group constitutes becomes smaller.

Although aldehydes and ketones have boiling points considerably lower than those of alcohols, they have almost the same water solubility as that of alcohols. The reason for this similarity is that aldehydes and ketones are

TABLE 14.4 Boiling Points of Aldehydes and Ketones

Compound	Structure	Molecular Mass (amu)	Boiling Point (°C)
ALDEHYDES AND KETONES			
formaldehyde	$HCHO$	30	−21
acetaldehyde	CH_3CHO	44	21
propanal	CH_3CH_2CHO	58	49
acetone	CH_3COCH_3	58	56
butanal	$CH_3CH_2CH_2CHO$	72	76
2-butanone	$CH_3COCH_2CH_3$	72	80
pentanal	$CH_3CH_2CH_2CH_2CHO$	86	103
2-pentanone	$CH_3COCH_2CH_2CH_3$	86	102
3-pentanone	$CH_3CH_2COCH_2CH_3$	86	102
ALKANE			
pentane	$CH_3CH_2CH_2CH_2CH_3$	72	36
ETHER			
diethyl ether	$CH_3CH_2OCH_2CH_3$	74	35
ALCOHOL			
1-butanol	$CH_3CH_2CH_2CH_2OH$	74	117

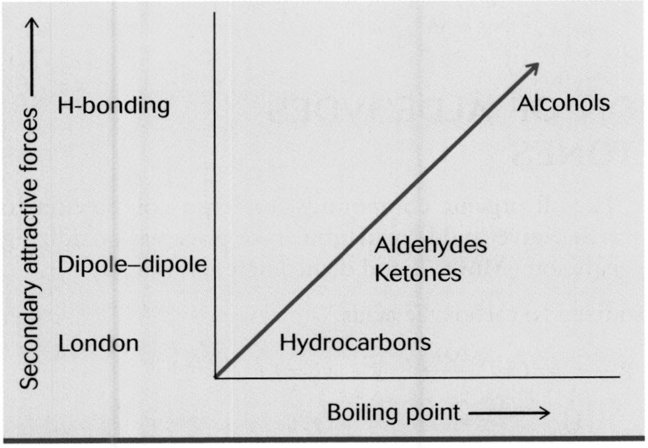

Figure 14.3 Relation between boiling point and strength of secondary attractive forces in hydrocarbons, aldehydes and ketones, and alcohols.

INSIGHT INTO PROPERTIES

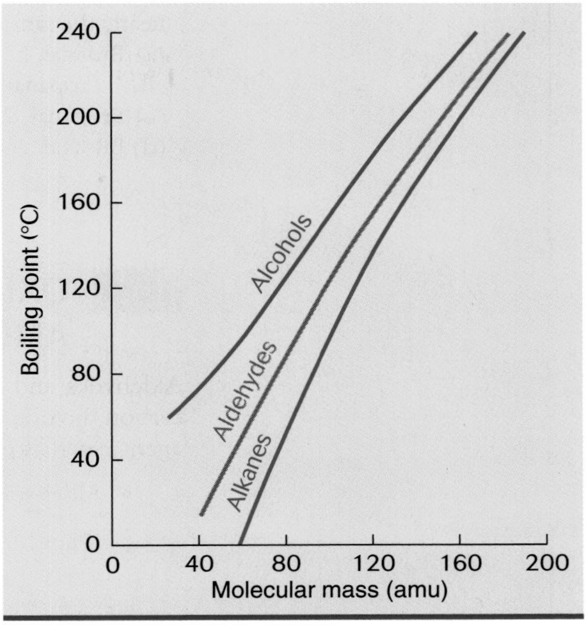

Figure 14.4 Boiling-point dependence on molecular mass for linear alkanes, aldehydes, and alcohols.

hydrogen-bond acceptors and, therefore, they hydrogen bond with water even though they do not hydrogen bond with themselves.

There is very little difference in physical properties between aldehydes and ketones. They have very nearly the same boiling and melting points and water solubilities.

Example 14.5 Recognizing hydrogen bonding in aldehydes and ketones

Which of the following representations correctly describes the hydrogen bonding that takes place between acetaldehyde (ethanal) and water?

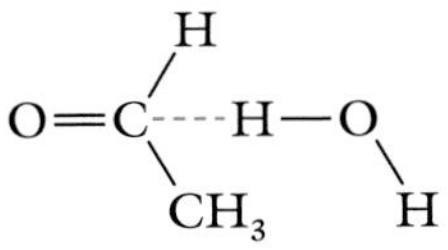

1

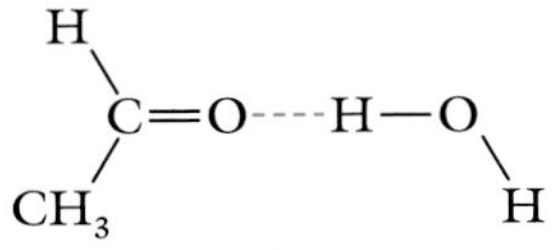

2

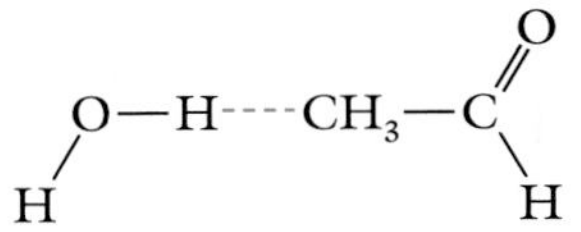

3

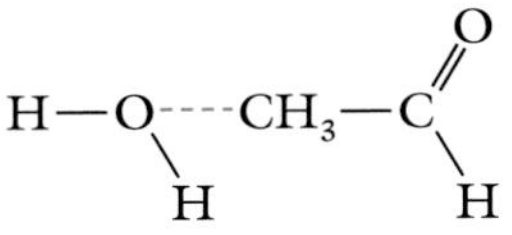

4

Solution

Only representation 2 describes hydrogen bonding. Hydrogen bonding here is the attraction between the H of an O—H bond of water and the O of the carbonyl group of the ketone. Hydrogen bonding does not take place between H and C, as described in representations 1 and 3, or between O and C, as described in representation 4.

Problem 14.5 Indicate which compound in each of the following pairs has the higher value of the specified property. If the two compounds in a pair have nearly the same value for the property, indicate that fact. Explain your conclusions.
(a) Butanal, hexanal: boiling water.
(b) 1-Propanal, pentanal: solubility in water.
(c) Propanal, 2-propanol: boiling point.
(d) Butanal, 1-butanol: solubility in water.

14.4 OXIDATION OF ALDEHYDES AND KETONES

Aldehydes and ketones, like all organic compounds, undergo combustion to carbon dioxide and water. Selective (mild) oxidations take place with oxidizing agents such as permanganate ion (MnO_4^-) and dichromate ion ($Cr_2O_7^{2-}$):

- Aldehydes are oxidized to carboxylic acids.

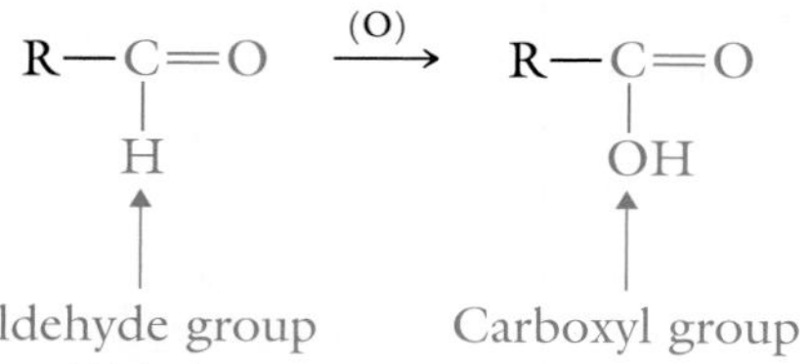

Aldehyde **Carboxylic acid**

▶▶ Carboxylic acids are a topic of Chapter 15.

- Ketones are not oxidized

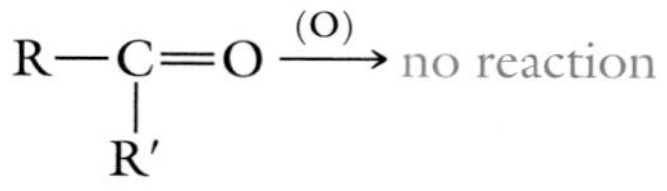

Ketone

We encountered these same selective oxidations in our study of alcohols (Section 13.7). Aldehydes and ketones are the oxidation products of primary and secondary alcohols, respectively. Whereas aldehydes are oxidized further to carboxylic acids, ketones are inert to further oxidation under mild conditions.

Selective oxidation of aldehydes is useful for the industrial synthesis of carboxylic acids.

Example 14.6 Recognizing the products of selective oxidation

Which of the following compounds undergo mild oxidation with dichromate or permanganate? If oxidation takes place, show the product. (a) Propanal; (b) acetone (2-propanone).

Solution

(a) Propanal, an aldehyde, is oxidized to a carboxylic acid.

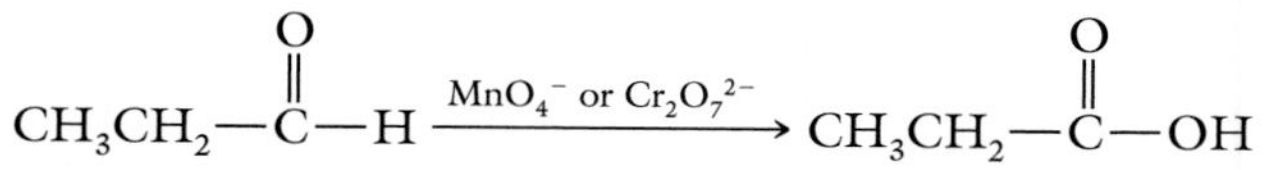

(b) Oxidation does not take place, because the compound is a ketone.

Problem 14.6 Which of the following compounds undergo mild oxidation with dichromate or permanganate? If oxidation takes place, show the product. (a) Cyclopentanone; (b) pentanal.

Dichromate and permanganate oxidations are useful simple chemical diagnostic tests for distinguishing aldehydes from ketones. Positive tests for an aldehyde are easily observed as color changes (Section 13.7). However, permanganate and dichromate oxidations cannot distinguish aldehydes from primary and secondary alcohols and alkenes, which also undergo oxidation under these conditions (Sections 12.8 and 13.7). Fortunately, there are other selective oxidizing reagents that are more selective and distinguish aldehydes not only from ketones, but also from primary and secondary alcohols and alkenes. Tollens's reagent is one such reagent. It oxidizes aldehydes but not the other compounds. **Tollens's reagent** is a colorless, alkaline solution of silver ion in ammonia, in which the silver ion is stabilized as its polyatomic ion, $[Ag(NH_3)_2]^+$. A positive test for an aldehyde is the formation of a shiny gray precipitate of silver metal, formed by reduction of silver ion. The silver usually coats the inside of the test tube and gives the appearance of a silver mirror.

Figure 14.5 Benedict's reagent (blue Cu^{2+}) is used in a simple diagnostic chemical test for distinguishing α-hydroxy aldehydes and α-hydroxy ketones from simple aldehydes and ketones. In the presence of an α-hydroxy aldehyde or ketone (right test tube), the blue color disappears and a red precipitate (Cu_2O) forms. In the presence of a simple aldehyde or ketone (left test tube), no reaction takes place and the blue color of the reagent is unchanged.

Benedict's reagent is another useful oxidizing agent, but its oxidizing action is different from Tollens's reagent. Benedict's reagent is a blue, alkaline solution of cupric ion (Cu^{2+}) stabilized as its polyatomic ion with citrate anion (citrate is the anion of citric acid). The blue color of cupric ion disappears and a red precipitate of Cu_2O forms in a positive test as cupric ion is reduced to cuprous ion (Cu^+). These color changes can be seen in Figure 14.5. A positive test is one in which a substantial amount of red Cu_2O precipitate is rapidly formed. Simple aldehydes such as acetaldehyde and benzaldehyde, although oxidized by Tollens's reagent, do not give positive tests with Benedict's reagent. Some simple aldehydes give weak positive tests, with the red Cu_2O precipitate forming slowly or in only small amounts. Such results are considered to be negative tests, not positive tests. Only α-hydroxy aldehydes give positive tests with Benedict's reagent. The α notation refers to the carbon adjacent to the carbonyl carbon, so an α-hydroxyl group is an —OH group on an α-carbon:

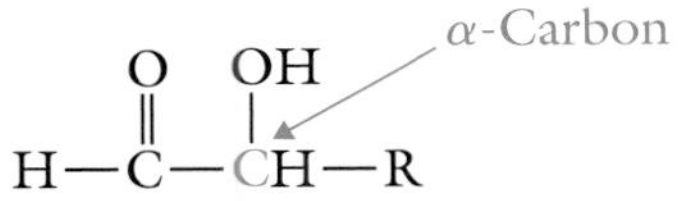

α-Hydroxy aldehyde
Positive Benedict's test

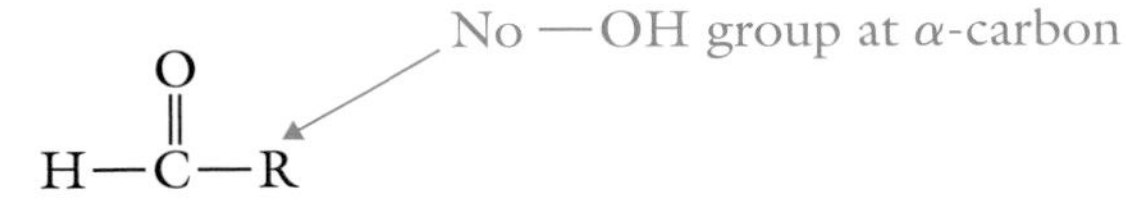

Simple aldehyde
Weak or negative Benedict's test

One group of ketones, α-hydroxy ketones, also give a positive Benedict's test. α-Hydroxy ketones are ketones with an —OH group on a carbon atom adjacent to the carbonyl carbon:

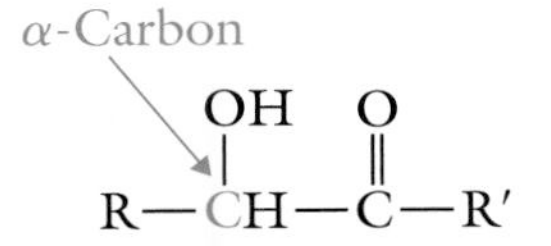

α-Hydroxy ketone
Positive Benedict's test

Although the use of the Benedict's reagent is limited to α-hydroxy aldehydes and α-hydroxy ketones, it is an important reagent in both chemical and clinical laboratory settings because it detects glucose and certain other carbohydrates (sugars). Glucose is not a simple aldehyde but a multifunctional molecule containing five hydroxyl groups as well as an α-hydroxy aldehyde group (see Box 14.1). Fructose (see Box 14.1), also an important carbohydrate, contains an α-hydroxy ketone group. Thus, glucose, fructose, and a number of other carbohydrates can be distinguished from simple aldehydes and ketones, alcohols, and other compounds by the Benedict's test.

▶▶ **Carbohydrates are important biological compounds and the subject of Chapter 18.**

Example 14.7 Recognizing compounds that react with Tollens's and Benedict's reagents

For each of the following compounds, indicate whether the compound gives a positive or negative test with (a) Tollens's reagent and (b) Benedict's reagent.

1

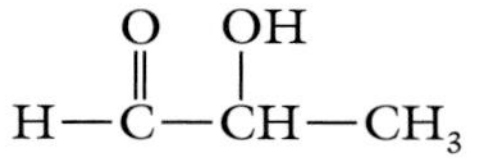

2

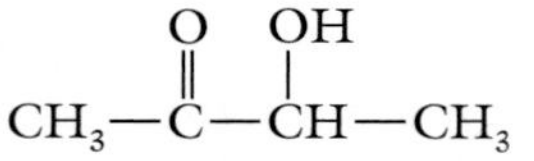

3

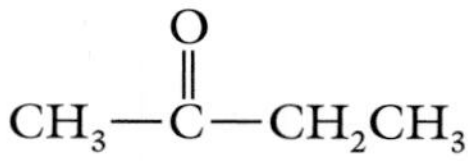

4

Solution

(a) Tollens's reagent gives positive tests with all aldehydes, but not with any ketones. Compounds 1 and 2 are aldehydes and give positive tests with Tollens's reagent. Compounds 3 and 4 are ketones and give negative tests.

(b) Only α-hydroxy aldehydes and α-hydroxy ketones give positive tests with Benedict's reagent. Compounds 2, an α-hydroxy aldehyde, and 3, an α-hydroxy ketone, give positive tests with Benedict's reagent. Compound 1, a simple aldehyde, gives a negative test. Compound 4, a simple ketone, gives a negative test.

Problem 14.7 For each of the following compounds, indicate whether the compound gives a positive or negative test with (a) Tollens's reagent and (b) Benedict's reagent.

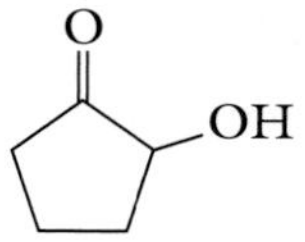

1

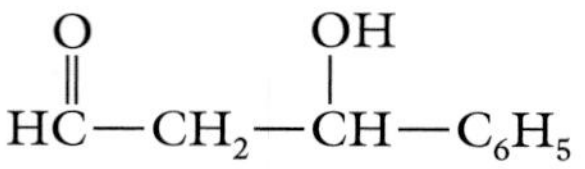

2

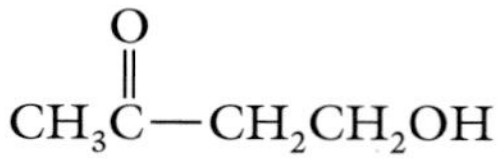

3

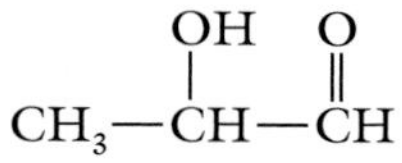

4

We have now considered the use of several simple chemical tests for analytical purposes—the bromine test for alkenes (Section 12.6) and various oxidation tests for alkenes, alcohols, and aldehydes (Sections 12.8, 13.7, and 14.4). Other much more powerful analytical tools are available for elucidating the structures of organic and biochemical molecules (Figures 14.6 and 14.7 and Box 14.3 on page 418).

14.5 REDUCTION OF ALDEHYDES AND KETONES

Aldehydes and ketones can be reduced to alcohols—aldehydes to primary alcohols, and ketones to secondary alcohols. The overall process consists of the addition of one hydrogen each to the carbon and oxygen of the carbonyl group. Reduction of aldehydes and ketones to alcohols is the reverse of the oxidation of alcohols to aldehydes and ketones described in Section 13.7.

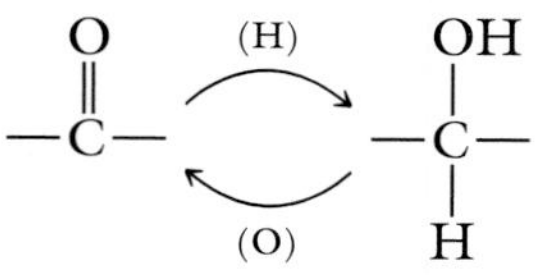

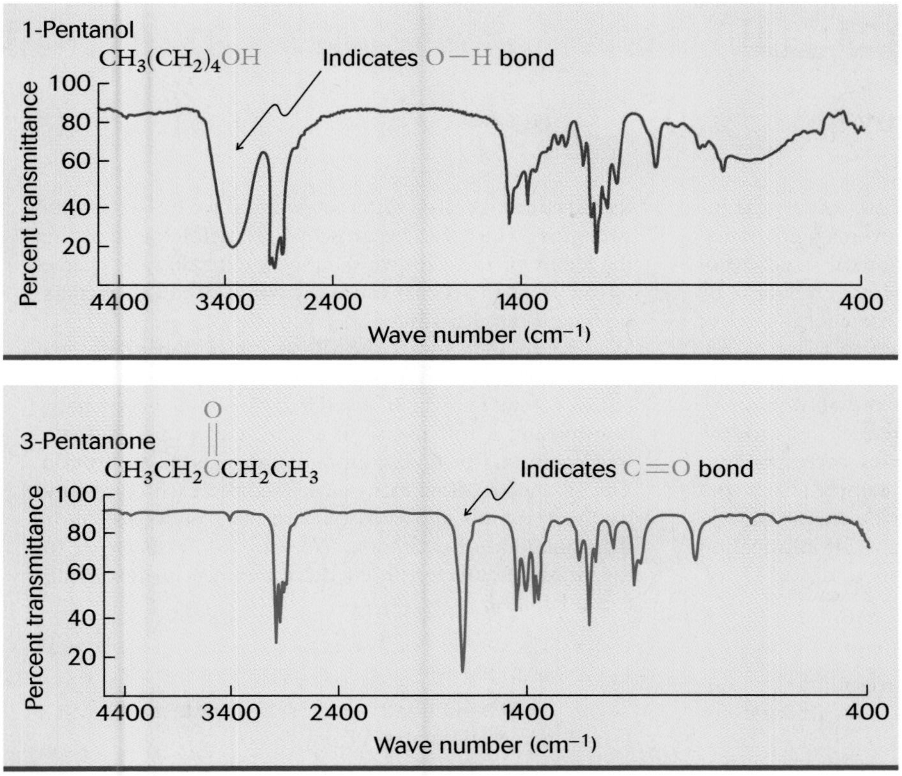

Figure 14.6 IR spectra of 1-pentanol and 3-pentanone.

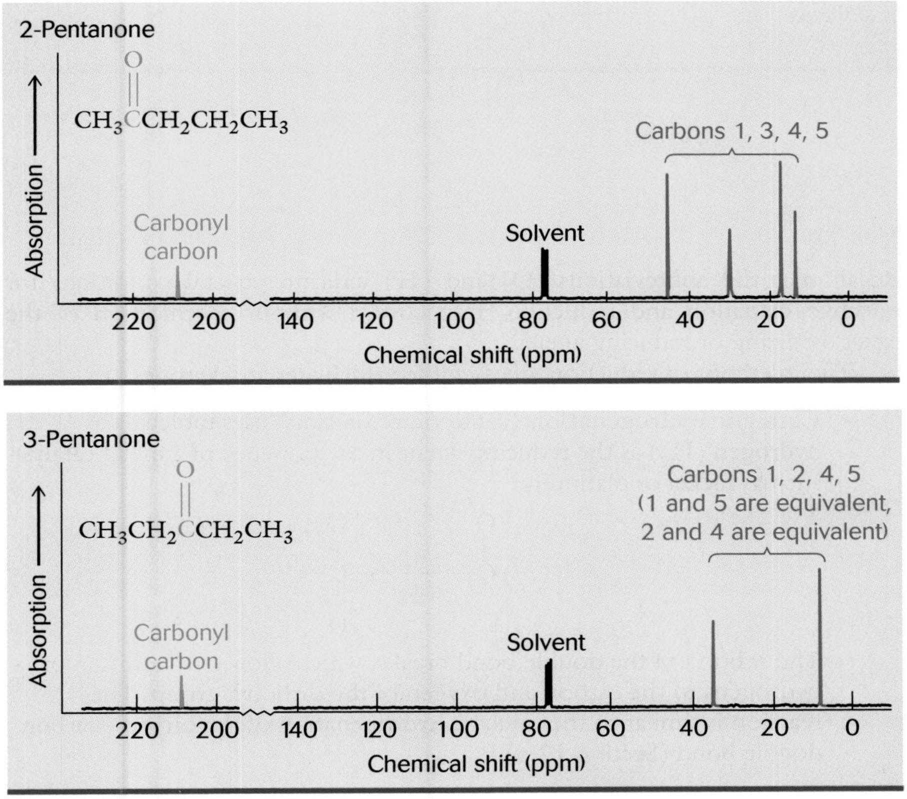

Figure 14.7 NMR spectra of 2-pentanone and 3-pentanone

14.3 Chemistry in Depth

Absorption Spectroscopy (IR, NMR)

Absorption spectroscopy (Section 3.10) entails the interaction of a compound with electromagnetic radiation and is useful for analyzing the chemical structure of unknown compounds. Molecules undergo excitation by the absorption of energy, which raises the energy level of the molecule from a lower, ground state to a higher, excited state. A sample is subjected to incident radiation of different frequencies, and one detects the absorption of energy by comparing the transmitted frequencies with the incident frequencies. Missing frequencies correspond to frequencies that were absorbed by the sample. Each specific frequency (corresponding to a specific energy of radiation) that causes a molecular excitation yields information about the chemical structure of the compound.

Incident frequencies ⟶ Sample ⟶ Transmitted frequencies

Different types of spectroscopy, corresponding to different types of molecular excitations, are available for studying chemical structure.

Infrared (IR) spectroscopy

Energies in the infrared region cause a variety of bond excitations, such as the stretching of bonds. The presence of specific bonds in a compound is detected by observing which frequencies of infrared radiation are absorbed by the compound. Different bonds require different energies for excitation because different bonds have different bond strengths. Thus, IR spectroscopy is useful for detecting the family of an unknown sample by detecting the characteristic frequencies that cause excitation of the characteristic bond(s) of the family.

Figure 14.6 shows the IR spectra of 1-pentanol and 2-pentanone. Energy transmission is plotted against wave number (cm^{-1}), which is related to frequency. Energy transmission is the inverse of energy absorption. A downward-pointing peak corresponds to energy absorption. The IR spectra show many peaks because each compound has many bonds and each bond usually undergoes more than one kind of excitation. We search each spectrum for the peak(s) characteristic of the functional groups of different families:

- The O—H bond of an alcohol shows a broad peak at 3300 cm^{-1}.
- The C═O bond of an aldehyde or ketone shows a narrow peak at 1700 cm^{-1}.

Alcohols are easily distinguished from aldehydes and ketones by looking for these characteristic peaks in the IR spectrum. Thus, the IR spectrum of 1-pentanol shows the 3300 cm^{-1} peak (O—H), but no peak at 1700 cm^{-1} (C═O). The IR spectrum of 2-pentanone shows the 1700 cm^{-1} peak (C═O), but no peak at 3300 cm^{-1} (O—H).

Recall that the abbreviations (O) and (H) indicate general conditions for selective oxidation and reduction, respectively, without specification of the exact oxidizing or reducing agent.

Two methods of reduction are useful for aldehydes and ketones:

- **Catalytic hydrogenation (catalytic reduction)** uses molecular hydrogen (H_2) as the reducing agent in the presence of a metal catalyst such as nickel or platinum:

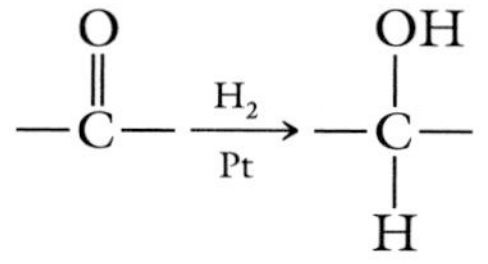

The π bond of the double bond breaks, which allows the addition of hydrogens to the carbon and oxygen of the carbonyl group. The reaction is similar to the catalytic hydrogenation of the carbon–carbon double bond (Section 12.6).

IR spectroscopy is generally limited to detecting the family of an unknown sample. It cannot easily differentiate between different members of the same family.

Nuclear magnetic resonance (NMR) spectroscopy

NMR spectroscopy is the most powerful of the different spectroscopic methods. Not only can it detect the family of an unknown sample, but it usually differentiates between different members of the same family. In NMR spectroscopy, a compound is placed in the magnetic field of a NMR spectrometer and then subjected to different frequencies of the radiofrequency region of electromagnetic radiation. The nuclei of certain atoms of a compound behave like tiny magnets—a property called **nuclear spin.** In the NMR spectrometer, there are two possible nuclear spin states—alignment of nuclear spin with or against the magnetic field of the spectrometer. Alignment with the magnetic field is lower in energy than alignment against the magnetic field. Radiofrequency energy causes the excitation of nuclear spins from alignment with the magnetic field to alignment against the magnetic field.

Proton NMR entails excitation of the nuclear spins of a compound's hydrogen atoms. Carbon-13 NMR entails excitation of the nuclear spins of a compound's C-13 carbons. (The carbon atoms in compounds consist of a mixture of C-12 and C-13 isotopes; only C-13 nuclei undergo the NMR excitation.) We will illustrate the great utility of NMR spectroscopy by showing how easily 2-pentanone and 3-pentanone can be distinguished from each other by C-13 NMR. C-13 NMR spectroscopy has two important features:

- Each nonequivalent carbon in a compound produces a different peak.
- The frequency of a peak depends on both the type of carbon (for example, CH_2, CH_3, C=O) and the atoms or groups attached to the carbon.

Figure 14.7 shows the C-13 NMR spectra of the two ketones. Energy absorption is plotted against chemical shift (ppm), which is related to frequency. Absorption peaks point upward in NMR spectra instead of downward as in IR spectra. The C=O carbon peak is seen near 210 ppm. The trio of peaks near 80 ppm is due to the solvent used in the NMR experiment. 2-Pentanone has four additional peaks corresponding to carbons 1, 3, 4, and 5. 3-Pentanone has only two additional peaks because carbons 1 and 5 are equivalent and carbons 2 and 4 are equivalent. 2-Pentanone has a total of five peaks, whereas 3-pentanone has a total of three peaks. The two compounds are easily distinguished from each other.

The NMR technique has been developed as magnetic resonance imaging (MRI) for use in medical diagnosis (Section 10.5).

- **Hydride reduction** uses a metal hydride as the reducing agent—for example, lithium aluminum hydride ($LiAlH_4$). The aldehyde or ketone is first reacted with the metal hydride followed by reaction with water:

$$-\overset{\overset{\large O}{\|}}{C}- \xrightarrow[2.\ H_2O]{1.\ LiAlH_4} -\overset{\overset{\large OH}{|}}{\underset{\underset{\large H}{|}}{C}}- \qquad (1)$$

To understand this reaction, we need to consider the nature of metal hydrides. Compounds containing H bonded to a more electronegative element (for example, Cl in HCl and O in H_2O or ROH) are sources of H^+ because H gives up the shared pair of electrons to the more electronegative element (Cl or O). Metal hydrides are different, being sources of hydride ion ($H:^-$, or simply H^-) because H is more electronegative than a metal.

$$H \!\rightarrow\! Cl \longrightarrow H^+ + Cl^- \qquad H \!\leftarrow\! \text{metal} \longrightarrow H^- + \text{metal}^+$$

Cl more electronegative than H | **H more electronegative than metal**

Reduction of the carbonyl group by metal hydride proceeds because of the partly ionic character of the carbonyl group. In the first step, the $H\!:^-$ from the metal hydride bonds to the C^+ of the carbonyl group. In the second step, H^+ from H_2O bonds to the O^- of the carbonyl group:

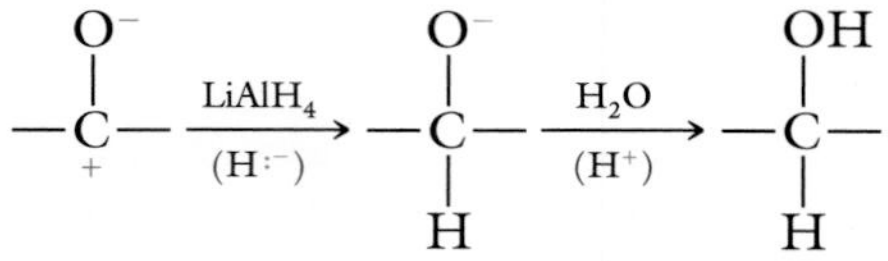

The reagents must be added sequentially, metal hydride followed by water, rather than simultaneously to achieve reduction of the carbonyl group. In chemical equations, sequential addition is indicated by numbers in front of the reagents, as in Equation 14.1. Water must be added only after the aldehyde or ketone and metal hydride have been allowed an appropriate period of time to react with each other. Simultaneous addition of metal hydride and water results in reaction between hydride ion and water:

$$H\!:^- + H_2O \longrightarrow H_2 + HO^-$$

Reduction of the carbonyl group does not take place, because water is more reactive toward hydride than is the carbonyl group.

Hydride reductions of carbonyl groups are important in living cells. In the cell, however, these reactions do not use metal hydrides, because metal hy-

A PICTURE OF HEALTH

Examples of Aldehydes and Ketones

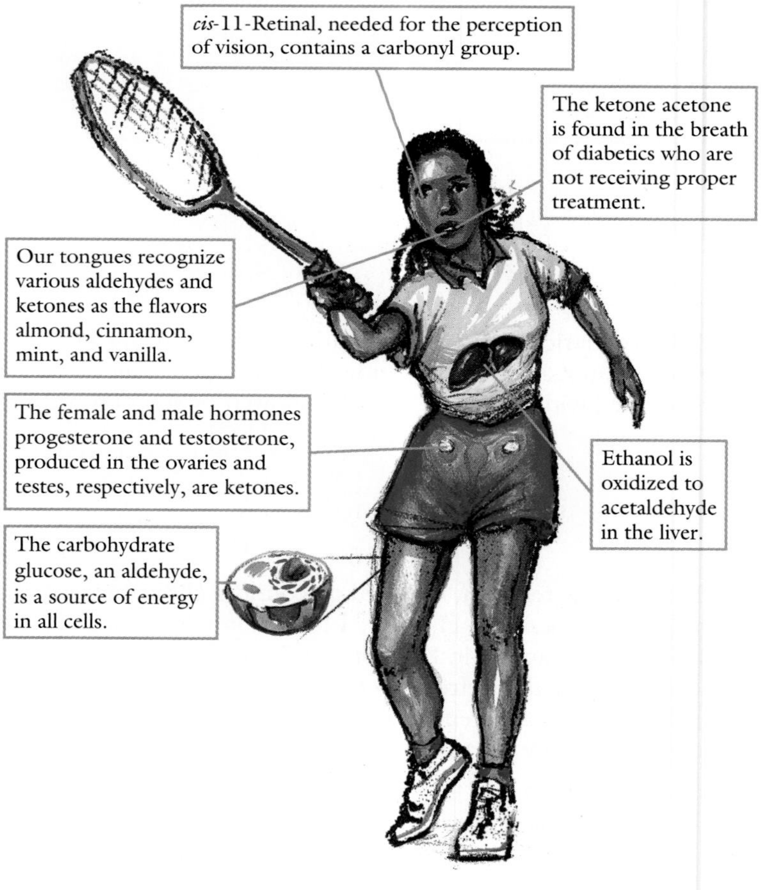

cis-11-Retinal, needed for the perception of vision, contains a carbonyl group.
The ketone acetone is found in the breath of diabetics who are not receiving proper treatment.
Our tongues recognize various aldehydes and ketones as the flavors almond, cinnamon, mint, and vanilla.
The female and male hormones progesterone and testosterone, produced in the ovaries and testes, respectively, are ketones.
Ethanol is oxidized to acetaldehyde in the liver.
The carbohydrate glucose, an aldehyde, is a source of energy in all cells.

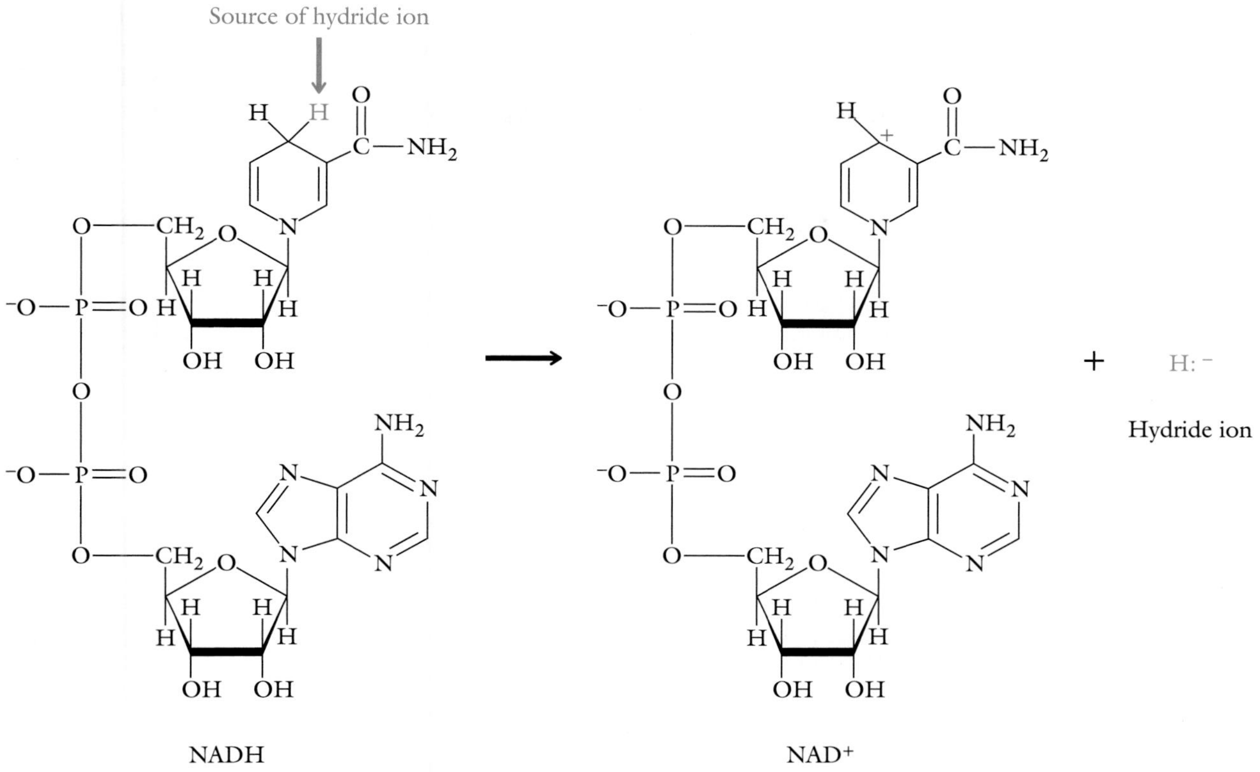

Figure 14.8 NADH is a source of hydride ion (H:$^-$) for the reduction of carbonyl groups in biological systems.

drides are not stable in aqueous environments. Metal hydrides react directly with water; hydride ion would not survive to reduce a carbonyl group. Instead, cells use a complex organic compound called **nicotinamide adenine dinucleotide (NADH),** to carry out hydride reductions. NADH is stable in aqueous environments. Figure 14.8 shows the structure of NADH and the location of its hydride ion. The structure of NADH is quite complicated. It is drawn in detail here to give you some idea of the complexity of biological molecules. We will refer to it simply as NADH.

▶▶ NADH is an important biological reducing agent that will be encountered again in Chapters 22 through 26.

An example of a hydride reduction in living cells is the reduction of pyruvic acid to lactic acid during strenuous exercise, such as running or bicycling. The reduction is a two-step process and is catalyzed by the enzyme alcohol dehydrogenase. It consists of the additions of H:$^-$ and H$^+$ to the carbonyl group. H:$^-$ is supplied by NADH and adds to the carbonyl carbon; this step is followed by the addition of H$^+$ from water to the carbonyl oxygen:

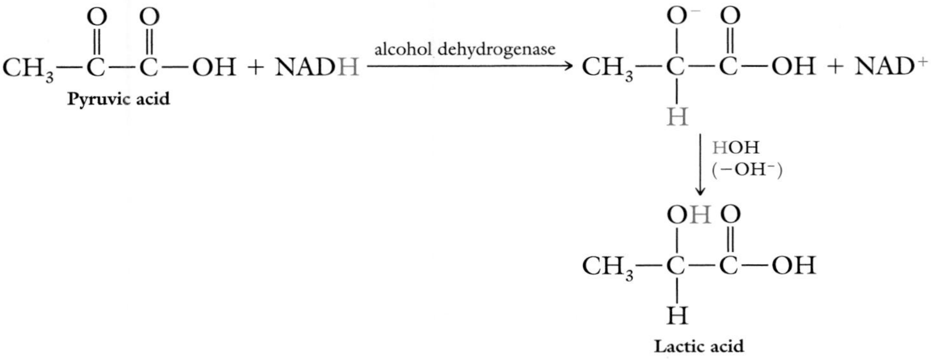

The oxidized form of NADH, after it has donated its hydride ion to another compound, is called NAD^+ and is used elsewhere in the organism as an oxidizing agent—for example, to oxidize ethanol to acetaldehyde. The role of NADH in this and other biological reductions is discussed further in Chapters 22 through 26.

Example 14.8 Writing equations for the reduction of aldehydes and ketones

Complete each of the following reactions by writing the structure of the organic product. If no reaction takes place, write "no reaction." If there is more than one product, show only the major product(s).

(a) $CH_3CH_2CH_2\overset{\overset{\displaystyle O}{\|}}{C}H \xrightarrow[2.\ H_2O]{1.\ LiAlH_4} ?$

(b) $CH_3CH_2CH_2\overset{\overset{\displaystyle O}{\|}}{C}H \xrightarrow[H_2O]{LiAlH_4} ?$

(c) $CH_3\overset{\overset{\displaystyle O}{\|}}{C}CH_2CH_3 \xrightarrow[Ni]{H_2} ?$

(d) $CH_3\overset{\overset{\displaystyle O}{\|}}{C}CH_2CH_3 \xrightarrow{H_2} ?$

(e) $CH_3CH_2CH_2\overset{\overset{\displaystyle O}{\|}}{C}H \xrightarrow[2.\ H_2O]{1.\ NADH,\ \text{alcohol dehydrogenase}} ?$

Solution

(a) Sequential addition of metal hydride and water reduces the carbonyl group:

$$CH_3CH_2CH_2\overset{\overset{\displaystyle O}{\|}}{C}H \xrightarrow[2.\ H_2O]{1.\ LiAlH_4} CH_3CH_2CH_2\overset{\overset{\displaystyle OH}{|}}{\underset{\underset{\displaystyle H}{|}}{C}}H$$

(b) The aldehyde is not reduced, because $LiAlH_4$ and H_2O are not added sequentially. $LiAlH_4$ reacts with H_2O instead of with the carbonyl group.

(c) Catalytic hydrogenation reduces the ketone to a secondary alcohol:

$$CH_3\overset{\overset{\displaystyle O}{\|}}{C}CH_2CH_3 \xrightarrow[Ni]{H_2} CH_3\overset{\overset{\displaystyle OH}{|}}{\underset{\underset{\displaystyle H}{|}}{C}}CH_2CH_3$$

(d) Reduction does not take place, because the metal catalyst (Ni or Pt) is absent.

(e) Sequential reaction with NADH and water reduces the carbonyl group, exactly as in the metal hydride reduction with $LiAlH_4$:

$$CH_3CH_2CH_2\overset{\overset{\displaystyle O}{\|}}{C}H \xrightarrow[2.\ H_2O]{1.\ NADH,\ \text{alcohol dehydrogenase}} CH_3CH_2CH_2\overset{\overset{\displaystyle OH}{|}}{\underset{\underset{\displaystyle H}{|}}{C}}H$$

Problem 14.8 Write an equation for each of the following reactions: (a) reduction of cyclohexanone by using $LiAlH_4$; (b) catalytic reduction of benzaldehyde; (c) reduction of 2-pentanone with NADH.

14.6 HEMIACETAL AND ACETAL FORMATION BY REACTION WITH ALCOHOL

Aldehydes and ketones react with alcohols to form **hemiacetals** and **acetals.** Hemiacetals and acetals are of interest because they are important structures in carbohydrates.

The reaction of aldehydes and ketones with alcohols requires an acid catalyst and takes place in a two-step process. First, the alcohol adds to the carbonyl group to form a hemiacetal. Second, if excess alcohol is present, the hemiacetal reacts with a second molecule of alcohol to form an acetal. This second step is a dehydration reaction similar to the intermolecular dehydration of alcohols to form ethers (Section 13.10). The reaction is illustrated for propanal and methanol:

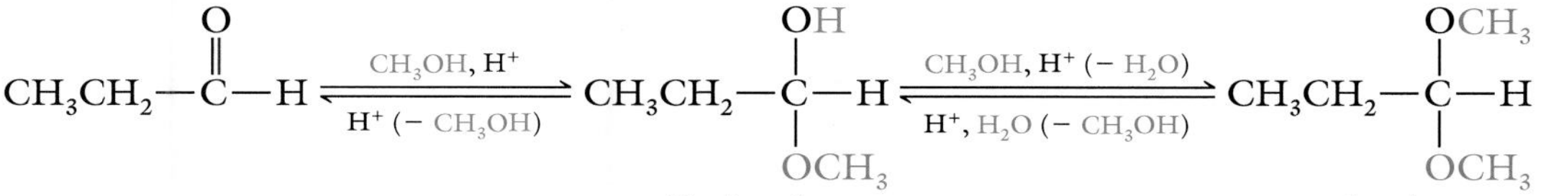

Hemiacetal Acetal

The reaction proceeds in a similar manner for ketones. The products formed from ketones were formerly called **hemiketals** and **ketals** to distinguish them from the products formed from aldehydes. However, the most recent convention uses the names hemiacetals and acetals for the products derived from both aldehydes and ketones.

Most hemiacetals are unstable and cannot be isolated; that is, the equilibrium in the first reaction is far to the left. However, the equilibrium in the second reaction is far to the right, and the acetal product is isolated if excess alcohol is present. This is an example of Le Chatelier's principle (Section 8.8): the favorable equilibrium in the second reaction shifts the first reaction, which has an unfavorable quilibrium, to the right.

Example 14.9 Drawing hemiacetal and acetal structures

Show the structures of the hemiacetal and acetal formed from the reaction of acetone (2-propanone) with excess ethanol.

Solution

To write the structure of the hemiacetal, first visualize the cleavage of the alcohol at the O–H bond to form RO$^-$ and H$^+$ fragments:

$$CH_3CH_2OH \longrightarrow CH_3CH_2O^- + H^+$$

Next, visualize the carbonyl group of acetone as an ionic structure:

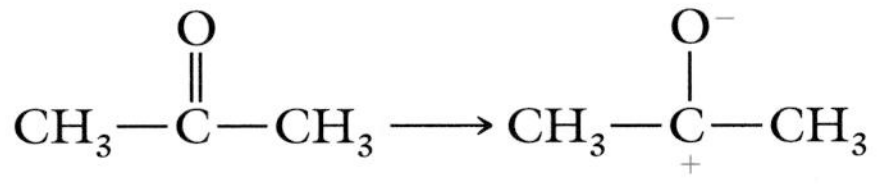

Covalent Ionic

The hemiacetal is formed by attachment of the RO$^-$ and H$^+$ fragments to the oppositely charged ends of the carbonyl group:

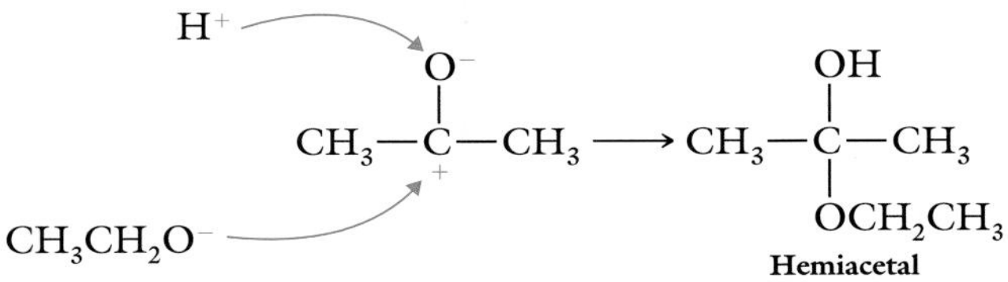

Hemiacetal

The acetal is formed by dehydration between the —OH group of the hemiacetal and the —OH group of the second molecule of ethanol:

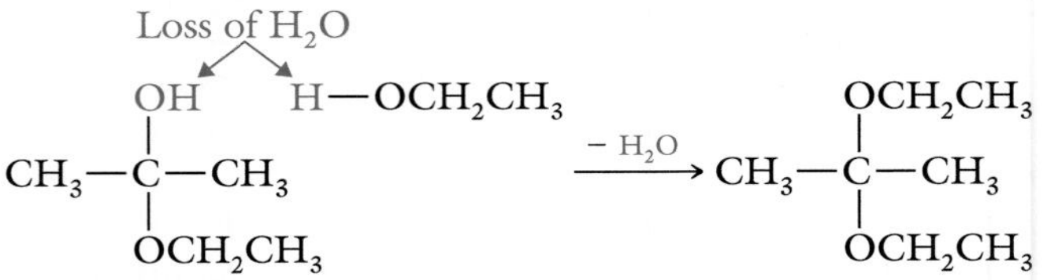

Hemiacetal **Acetal**

Problem 14.9 Show the structures of the hemiacetal and acetal formed from the reaction of butanal with excess methanol.

Five- and six-membered cyclic hemiacetals are an important exception to the generalization that hemiacetals are unstable and cannot be isolated. Five- and six-membered cyclic hemiacetals are stable and can be isolated, although the reasons are not completely understood. Cyclic hemiacetals are formed when a compound possesses both alcohol and carbonyl groups. Cyclic hemiacetal structures are found extensively in carbohydrates (Section 18.4).

Example 14.10 Drawing the structure of a cyclic hemiacetal

Show the structure of the cyclic hemiacetal formed by intramolecular reaction between the alcohol and carbonyl groups of 5-hydroxypentanal.

Solution

To write the structure of the hemiacetal, first visualize cleavage of the alcohol O–H bond to yield RO^- and H^+ fragments. Next, visualize the carbonyl group as an ionic structure. The hemiacetal is formed by attachment of the RO^- and H^+ fragments to the oppositely charged ends of the carbonyl group:

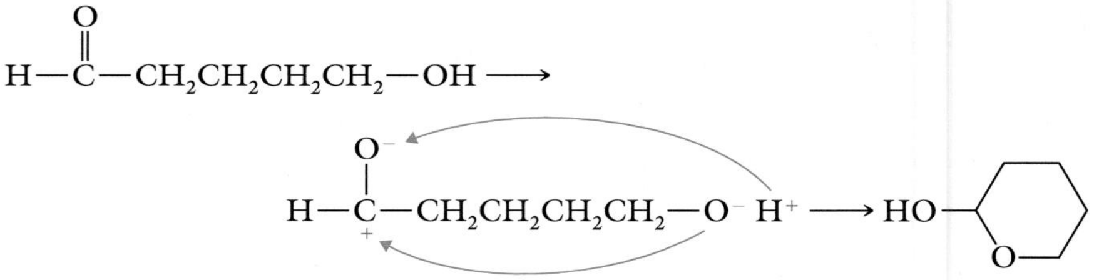

Problem 14.10 Show the structure of the hemiacetal formed by intramolecular reaction between the —OH and carbonyl groups of 6-hydroxyl-2-hexanone.

Recognition of hemiacetal and acetal structures is important because of their presence in carbohydrates. Both hemiacetals and acetals have two different oxygens bonded to the same carbon. In acetals, both oxygens are in —OR groups; in hemiacetals, one of the oxygens is in an —OH group and the other in an —OR group:

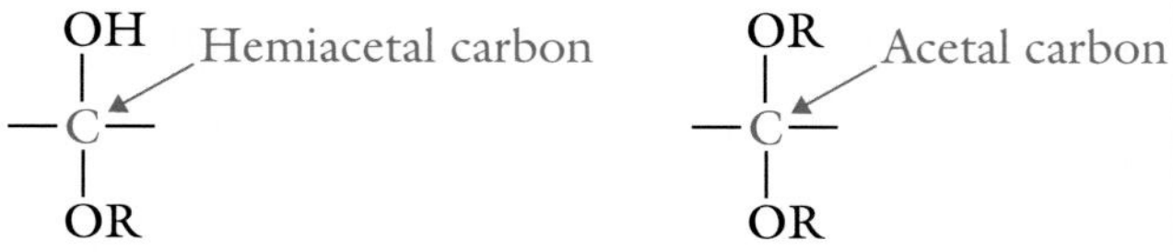

Hemiacetal **Acetal**

Example 14.11 Recognizing hemiacetals and acetals

For each of the following compounds, indicate whether it is a hemiacetal, an acetal, or something different.

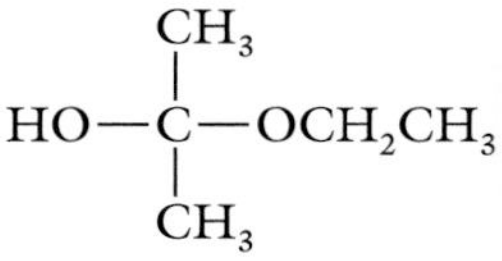

$$
\begin{array}{cc}
\underset{\displaystyle 1}{HO-\overset{\displaystyle CH_3}{\underset{\displaystyle CH_3}{\overset{|}{\underset{|}{C}}}}-OCH_2CH_3} &
\underset{\displaystyle 2}{HO-\overset{\displaystyle CH_3}{\underset{\displaystyle CH_3}{\overset{|}{\underset{|}{C}}}}-CH_2OCH_3}
\end{array}
$$

$$
\underset{\displaystyle 3}{CH_3O-\overset{\displaystyle CH_3}{\underset{\displaystyle CH_3}{\overset{|}{\underset{|}{C}}}}-OCH_3}
$$

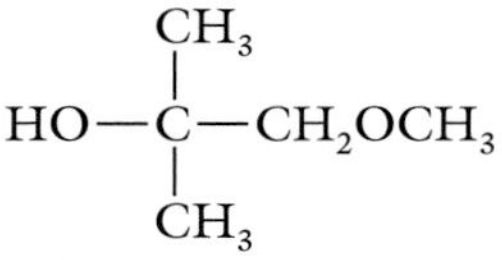

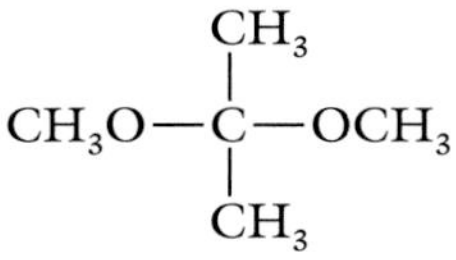

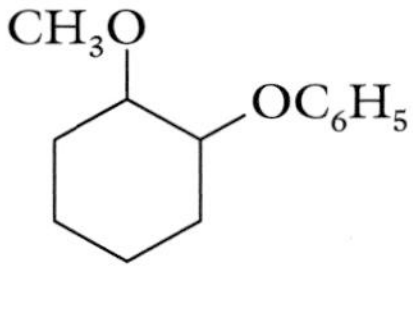

Solution

Compound 1 is a hemiacetal because it has a carbon with both an —OH and an —OR group attached:

Hemiacetal carbon

$$
HO-\overset{\displaystyle CH_3}{\underset{\displaystyle CH_3}{\overset{|}{\underset{|}{C}}}}-OCH_2CH_3
$$

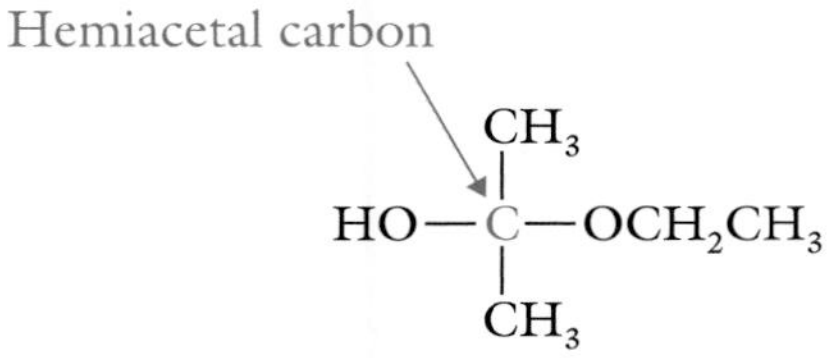

Compound 3 is an acetal because it has a carbon with two —OR groups attached:

Acetal carbon

$$
CH_3O-\overset{\displaystyle CH_3}{\underset{\displaystyle CH_3}{\overset{|}{\underset{|}{C}}}}-OCH_3
$$

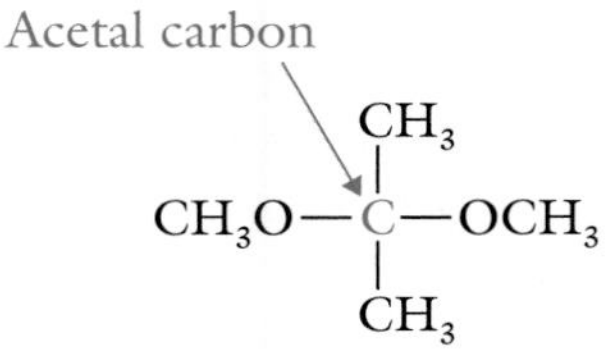

Compounds 2 and 4 are neither hemiacetals or acetals, because no carbon is bonded to two oxygens.

Problem 14.11 For each of the following compounds, indicate whether it is a hemiacetal, an acetal, or something different.

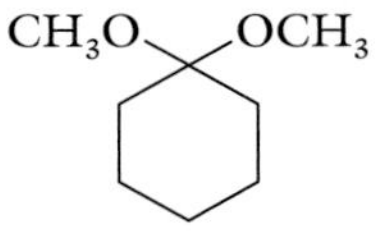

$$
\begin{array}{cc}
1 & 2
\end{array}
$$

$$
\begin{array}{cc}
3 & 4
\end{array}
$$

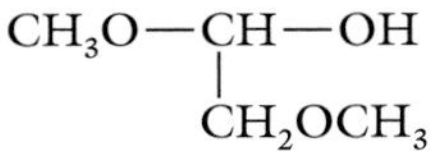

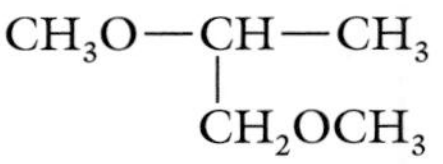

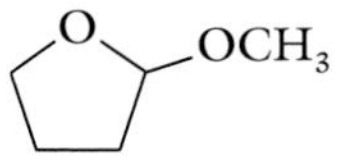

Hemiacetal and acetal formations are reversible reactions. In the presence of excess water and an acid catalyst, an acetal is converted into the aldehyde or ketone and alcohol. This is an example of a **hydrolysis** reaction—the reaction of an organic compound with water resulting in cleavage of the compound into two organic fragments each of which combines with a fragment (H^+ or OH^-) from water. Each of the organic products contains fewer carbons than did the original organic compound. The hydrolysis of an acetal proceeds through the intermediate formation of the hemiacetal, but the hemiacetal is not isolated, because of its instability. As mentioned earlier, Le Chatelier's principle determines the direction of the reaction. Acetal formation from an aldehyde or ketone requires an excess of alcohol, whereas hydrolysis of an acetal requires an excess of water.

In animals, hydrolysis of acetal groups is the first step in the digestive breakdown of starch and other carbohydrates in the digestive tract (Section 18.7).

Example 14.12 | Writing the products of acetal or hemiacetal hydrolysis

Complete each of the following reactions by writing the structure of the product. If no reaction takes place, write "no reaction." If there is more than one product, show only the major product(s).

(a) $CH_3OCH_2 - \overset{\overset{\displaystyle CH_3}{|}}{\underset{\underset{\displaystyle H}{|}}{C}} - OCH_3 \xrightarrow[\text{hydrolysis}]{H_2O, \ H^+}$? 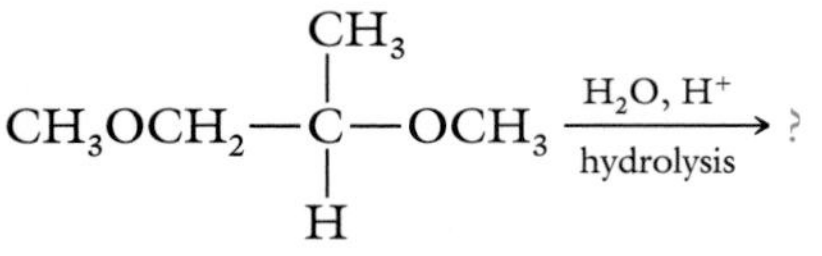

(b) $CH_3CH_2 - \overset{\overset{\displaystyle OCH_3}{|}}{\underset{\underset{\displaystyle H}{|}}{C}} - OCH_3 \xrightarrow[\text{hydrolysis}]{H_2O, \ H^+}$? 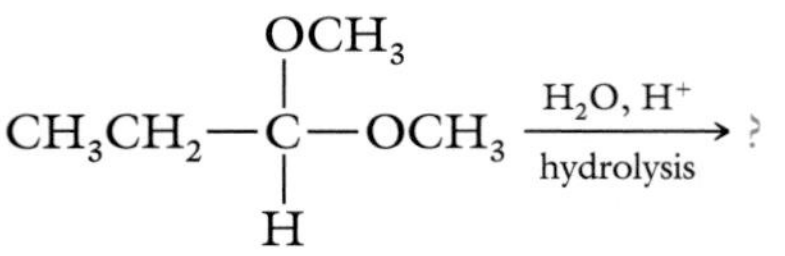

Solution

(a) No reaction. The compound is neither a hemiacetal nor an acetal. It is an ether (in fact, it has two ether groups), and ethers do not undergo hydrolysis.

(b) The compound is an acetal that undergoes hydrolysis. Hydrolysis of the acetal to the hemiacetal is the reverse of what occurs when a hemiacetal is converted into an acetal. One of the $-OCH_3$ groups leaves the acetal and combines with an H of water to form CH_3OH. The OH from water attaches to the carbon from which the $-OCH_3$ group left:

$$CH_3CH_2 - \overset{\overset{\displaystyle OCH_3}{|}}{\underset{\underset{\displaystyle H}{|}}{C}} - OCH_3 \xrightarrow{- \ CH_3OH} CH_3CH_2 - \overset{\overset{\displaystyle O-H}{|}}{\underset{\underset{\displaystyle H}{|}}{C}} - OCH_3$$

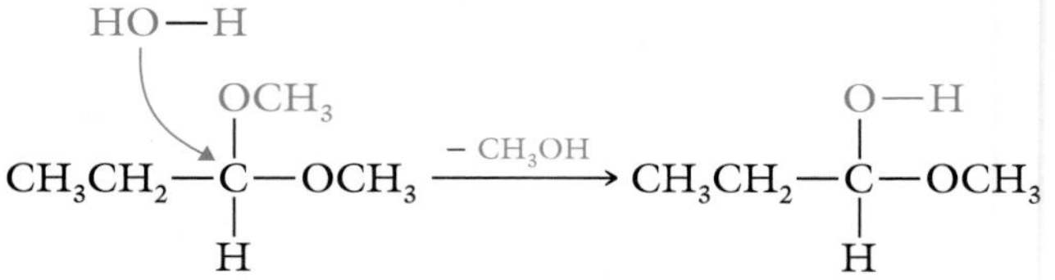

The hemiacetal is not isolated. Hemiacetals other than cyclic hemiacetals are unstable and are quickly converted into the aldehyde or ketone by loss of CH_3OH:

$$CH_3CH_2 - \overset{\overset{\displaystyle O-H}{|}}{\underset{\underset{\displaystyle H}{|}}{C}} - OCH_3 \xrightarrow{- \ CH_3OH} CH_3CH_2 - \overset{\overset{\displaystyle O}{\|}}{\underset{\underset{\displaystyle H}{|}}{C}}$$

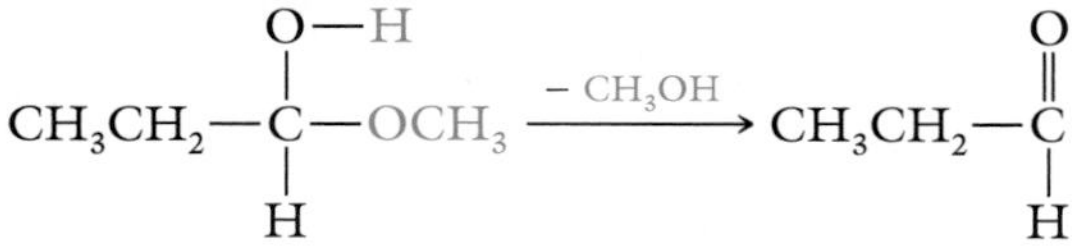

The overall reaction consists of one molecule each of acetal and water reacting to yield one aldehyde and two alcohol molecules:

$$CH_3CH_2 - \overset{\overset{\displaystyle OCH_3}{|}}{\underset{\underset{\displaystyle H}{|}}{C}} - OCH_3 + H_2O \xrightarrow{H^+} CH_3CH_2 - \overset{\overset{\displaystyle O}{\|}}{\underset{\underset{\displaystyle H}{|}}{C}} + 2 \ CH_3OH$$

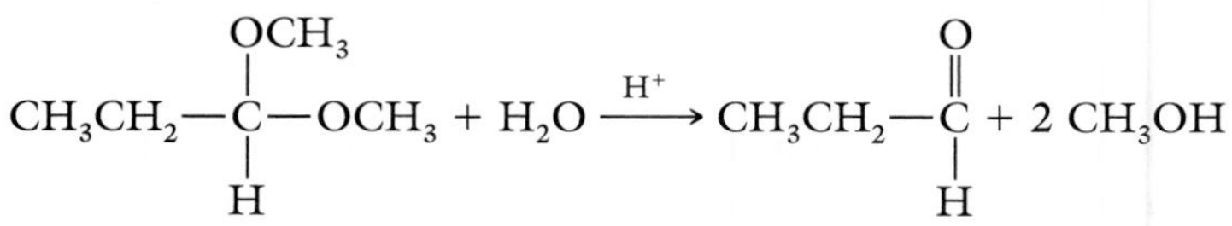

Problem 14.12 Show the product(s) formed when each of the following compounds undergoes acid-catalyzed hydrolysis:

(a) $CH_3CH_2O - \overset{}{\underset{\underset{\displaystyle CH_2C_6H_5}{|}}{CH}} - OCH_2CH_3$ 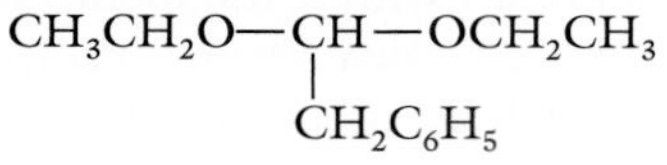(b) 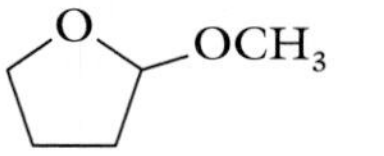

Summary

Aldehydes and ketones contain the carbonyl group (C=O). An aldehyde has at least one hydrogen atom directly attached to the carbon of the carbonyl group. A ketone has only carbon atoms, no hydrogen atoms, directly attached to the carbon of the carbonyl group. The C:H ratio for an aldehyde or ketone is two hydrogens fewer than the corresponding alcohol or alkane.

Naming Aldehydes and Ketones In the IUPAC system, aldehydes and ketones are named by identifying the longest continuous carbon chain containing the carbonyl group. Aldehydes and ketones are named as alkanals and alkanones, respectively. For ketones, the position of the carbonyl carbon is indicated by a number in front of the name. For aldehydes, no number is needed, because the carbonyl carbon is always C1. Prefixes preceded by numbers indicate the substituents attached to the longest chain. C_6H_5—CHO is named benzaldehyde. Common names are used for some aldehydes and ketones. Common ketone names are obtained by naming the groups attached to the carbonyl carbon, followed by the word ketone.

Physical Properties of Aldehydes and Ketones Aldehydes and ketones have boiling and melting points intermediate between those of hydrocarbons and alcohols, a consequence of their dipole–dipole secondary forces being intermediate in strength between the London forces in hydrocarbons and the hydrogen-bond forces in alcohols. Aldehydes and ketones are only slightly less soluble than alcohols in water. Although aldehydes and ketones cannot hydrogen bond with themselves, they can hydrogen bond with water.

Chemical Reactions of Aldehydes and Ketones Aldehydes, but not ketones, undergo oxidation with several selective oxidizing agents—dichromate, permanganate, and Benedict's and Tollens's reagents. Dichromate and permanganate distinguish aldehydes from ketones but not from primary and secondary alcohols and alkenes, which also give positive tests. Tollens's reagent is more selective, distinguishing aldehydes from ketones, alcohols, and alkenes. Benedict's reagent is useful for distinguishing α-hydroxy aldehydes and α-hydroxy ketones from other compounds.

Aldehydes and ketones undergo addition reactions with molecular hydrogen, hydrides, and alcohols. Aldehydes and ketones are reduced to primary and secondary alcohols, respectively. Both catalytic (H_2 with metal catalyst) and hydride (metal hydrides, NADH + enzyme) reductions are important.

Alcohol adds to the carbonyl group of aldehydes and ketones to form hemiacetals. The hemiacetals undergo dehydration with excess alcohol to form acetals. An acid catalyst is required for both reactions. Acid-catalyzed hydrolysis reverses these reactions. Most hemiacetals are unstable and cannot be isolated, but cyclic hemiacetals are stable and can be isolated.

Summary of Key Reactions

SELECTIVE OXIDATION

MnO_4^- and $Cr_2O_7^{2-}$

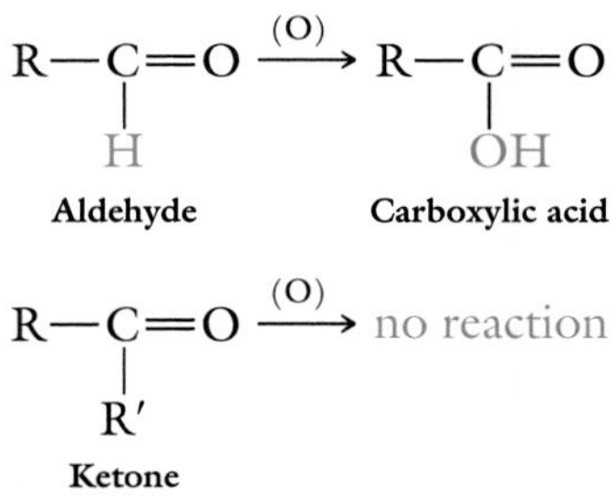

- Primary and secondary alcohols, as well as alkenes, are oxidized by MnO_4^- and $Cr_2O_7^{2-}$.
- Tollens's reagent (Ag^+) oxidizes aldehydes but not ketones, alcohols, or alkenes.
- Benedict's reagent (Cu^{2+}) oxidizes α-hydroxy aldehydes and α-hydroxy ketones. Simple aldehydes give weak or negative results.

REDUCTION

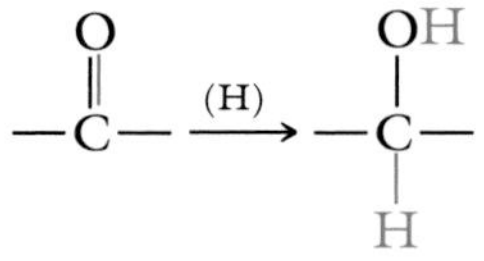

- Reduction is carried out with H_2/Pt or sequential addition of hydride followed by water or, in biological systems, with NADH.

HEMIACETAL AND ACETAL FORMATION BY REACTION WITH ALCOHOL

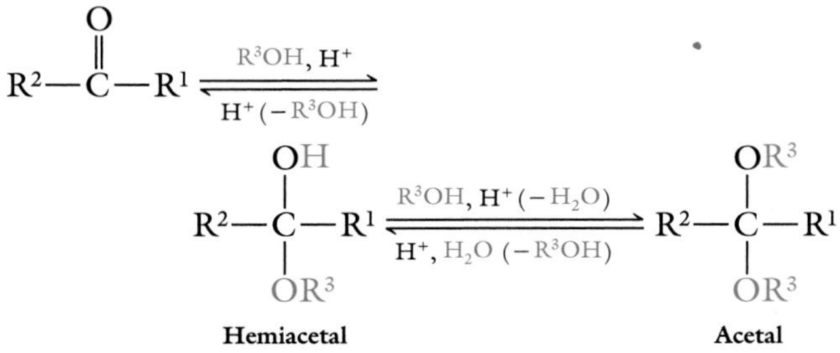

- Reaction requires an acid catalyst.
- Reaction is reversible. Hydrolysis yields the aldehyde or ketone and alcohol.

Key Words

acetal, p. 424
aldehyde, p. 404
Benedict's reagent, p. 415
carbonyl group, p. 404
catalytic reduction, p. 418
hemiacetal, p. 424
hydride reduction, p. 419
hydrolysis, p. 425
α-hydroxy aldehyde, p. 415
ketone, p. 404
oxidation, p. 414
Tollens's reagent, p. 415

Exercises

Structure of Aldehydes and Ketones

14.1 Does the molecular formula $C_5H_{12}O$ fit an alcohol, ether, aldehyde, or ketone?

14.2 Does the molecular formula $C_5H_{10}O$ fit an alcohol, ether, aldehyde, or ketone?

14.3 Draw structural formulas of all aldehydes having the molecular formula $C_6H_{12}O$.

14.4 Draw structural formulas of all ketones having the molecular formula $C_6H_{12}O$.

14.5 Which of the following compounds are aldehydes or ketones and which are not? Classify each compound. (Refer to Table 14.1 as needed.)

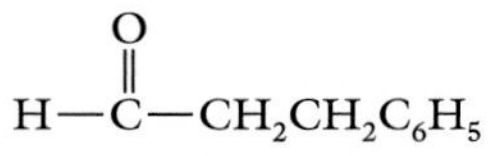
1

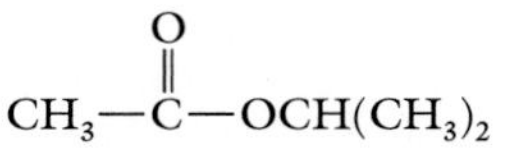
2

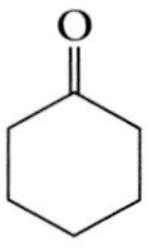
3

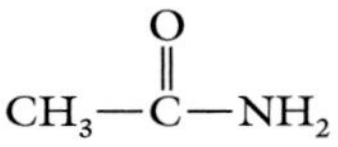
4

14.6 For each of the following pairs of structural formulas, indicate whether the pair represents (1) the same compound or (2) different compounds that are constitutional isomers or (3) different compounds that are geometrical isomers or (4) different compounds that are not isomers.

(a) 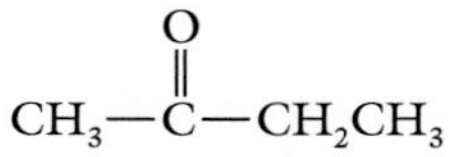

and 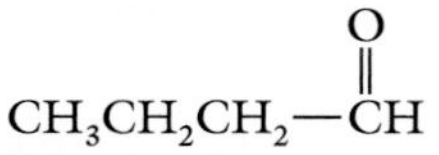

(b) 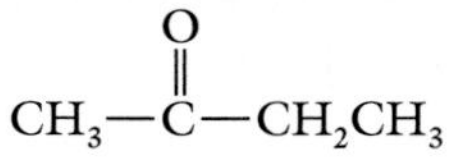

and 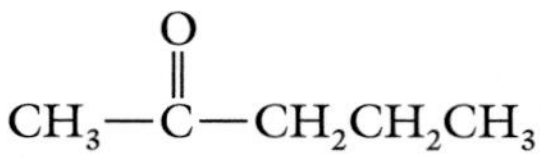

(c) 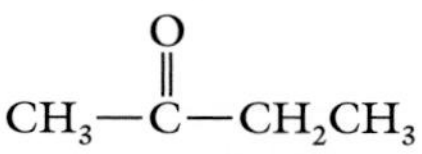

and $CH_2{=}CHCH_2CH_2OH$

(d) 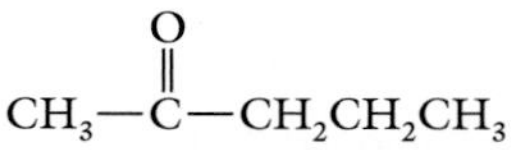

and 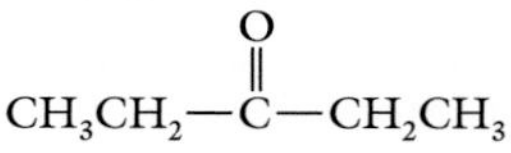

(e) 

and 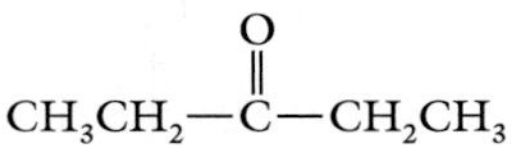

(f) 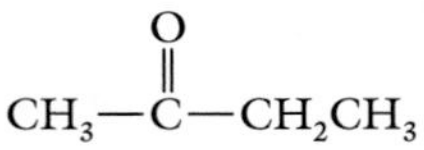

and 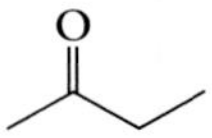

14.7 What are the hybridizations of atoms a, b, c, d, and e in the following compound?

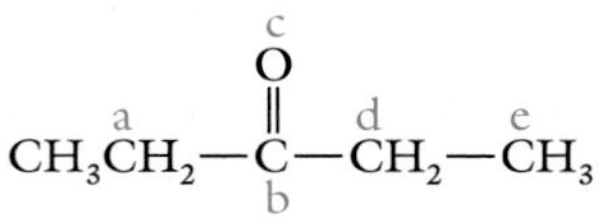

14.8 What are bond angles A, B, and C in the following compound?

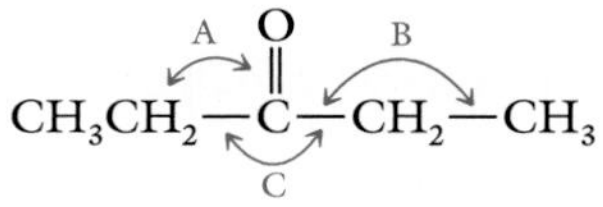

Naming Aldehydes and Ketones

14.9 Draw the structural formula for each of the following compounds: (a) 2,3-dimethylhexanal; (b) ethyl isopropyl ketone; (c) 3-*s*-butyl-4-ethylbenzaldehyde.

14.10 Draw the structural formula for each of the following compounds: (a) 4,5-dimethyl-2-hexanone; (b) 2-phenyl-3-heptanone; (c) 3-methylcyclohexanone.

14.11 Name each of the following compounds:

(a) 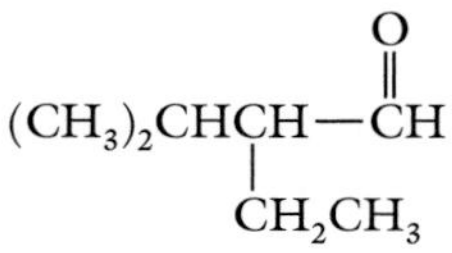

(b) 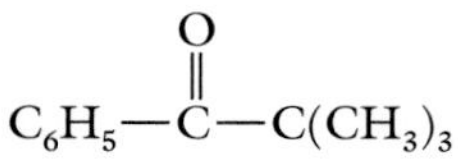

(c) 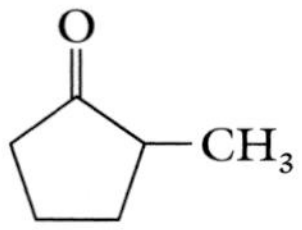

(d) 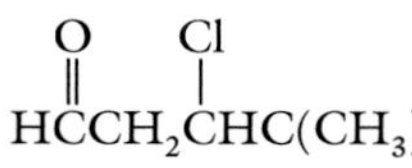

14.12 Name each of the following compounds:

(a) 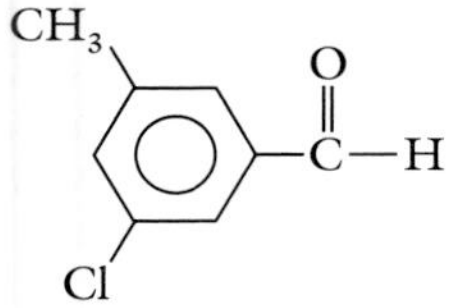

(b) $CH_3\overset{CH_3}{\overset{|}{C}}HCH_2\overset{O}{\overset{\|}{C}}CH_2\overset{CH_2CH_2CH_3}{\overset{|}{C}}HC(CH_3)_3$

(c) 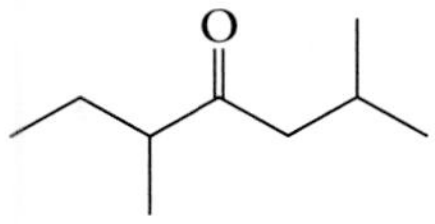

Physical Properties of Aldehydes and Ketones

14.13 Indicate which compound in each of the following pairs has the higher value of the specified property. If the two compounds in a pair have nearly the same value for the property, indicate that fact. Explain your conclusions.

(a) 2-Butanone, 3-pentanone: boiling point.
(b) Butanal, 2-butanone: boiling point.
(c) Butanal, 2-butanone: solubility in water.
(d) Butanal, pentane: boiling point.
(e) Butanal, pentane: solubility in water.
(f) 2-Butanone, 3-pentanone: solubility in water.
(g) 2-Butanone, 2-butanol: boiling point.
(h) 2-Butanone, 2-butanol: solubility in water.

14.14 Indicate whether hydrogen bonding is present in each of the following compounds or mixtures. Draw a representation of the hydrogen bonding if present.
(a) Ethanal; (b) water; (c) ethanal and water; (d) acetone (2-propanone) and ethanol.

Oxidation of Aldehydes and Ketones

14.15 Which of the following compounds undergo oxidation with dichromate and permanganate reagents? If oxidation takes place, write the structure(s) of the product(s).

$$CH_3CH_2-\overset{O}{\overset{\|}{C}}-CH_3 \qquad CH_3CH_2-\overset{O}{\overset{\|}{C}}-H$$

1 2

$$CH_3CH_2CH_2CH_2OH$$

3

14.16 Which of the compounds in Exercise 14.15 undergo oxidation with (a) Tollens's reagent and (b) Benedict's reagent? Briefly explain what is observed visually when there is a positive test.

14.17 For each of the following compounds, indicate whether the compound gives a positive test with (a) Tollens's reagent and (b) Benedict's reagent?

$$CH_3CH_2\overset{O}{\overset{\|}{C}}-\overset{OH}{\overset{|}{C}}H-CH_3$$

1

$$HO-\overset{O}{\overset{\|}{C}}CH_2\overset{OH}{\overset{|}{C}}H-CH_3$$

2

$$CH_3CH_2-\overset{O}{\overset{\|}{C}}-CH_2-\overset{O}{\overset{\|}{C}}H$$

3

$$H-\overset{O}{\overset{\|}{C}}-\overset{OH}{\overset{|}{C}}H-CH_2CH_3$$

4

For both Tollens's and Benedict's reagents, briefly explain what is observed visually when there is a positive test.

14.18 An unknown compound is either 2-pentanone or pentanal. Addition of a few drops of Tollens's reagent to the unknown results in the formation of a silver mirror. Identify the unknown.

14.19 An unknown compound is either 1-pentanol or pentanal. When a few drops of basic permanganate solution is added to the unknown, the purple color of the permanganate disappears and a brown precipitate forms. Identify the unknown.

14.20 An unknown compound is either 1-pentanol or pentanal. A shiny mirror forms when Tollens's reagent is added to the unknown. Identify the unknown.

Reduction of Aldehydes and Ketones

14.21 Which of the following compounds undergo reduction with metal hydrides? If reduction takes place, give the product(s).

(a) $CH_3CH_2-\overset{O}{\overset{\|}{C}}-CH_3$

(b) $CH_3CH_2-\overset{O}{\overset{\|}{C}}-H$

(c) $CH_3CH_2CH_2CH_2OH$

14.22 Which of the compounds in Exercise 14.21 undergo reduction with H_2/Pt? If reduction takes place, give the structure(s) of the product(s).

14.23 Write the equation showing hydride ion, $H:^-$, reacting with water. Is the hydride ion an acid or a base in this reaction?

14.24 For each of the following compounds, indicate whether it can be synthesized from an aldehyde or ketone by reduction. If yes, give the structure of the aldehyde or ketone.

(a) $CH_3-\overset{CH_3}{\overset{|}{\underset{OH}{\underset{|}{C}}}}-CH_3$ (b) $CH_3-\overset{CH_3}{\overset{|}{C}}H-CH_2-OH$

(c) 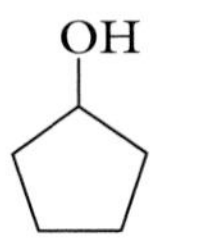 (d) 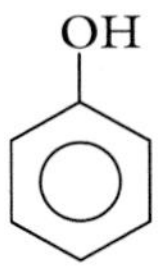

Hemiacetal and Acetal Formation by Reaction with Alcohol

14.25 Write equations to show the reaction of benzaldehyde with excess methanol in the presence of an acid catalyst. Show the formation of the hemiacetal and acetal.

14.26 For each of the following compounds, indicate whether it is a hemiacetal, an acetal, or something different.

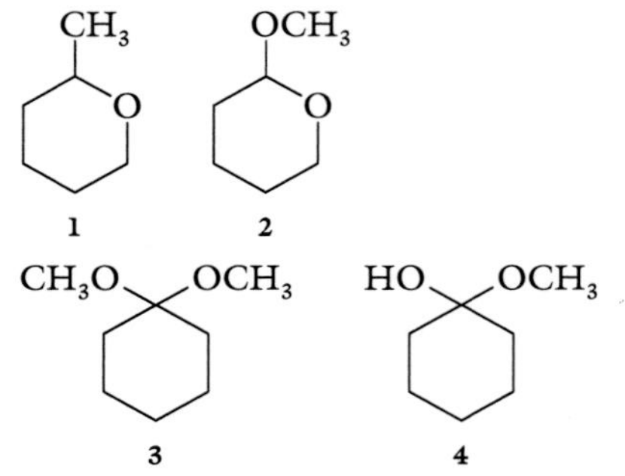

14.27 Give the structures of the aldehyde or ketone and the alcohol that are needed to synthesize each of the following hemiacetals or acetals:

(a) $C_6H_5CH(OCH_2CH_3)_2$

(b) 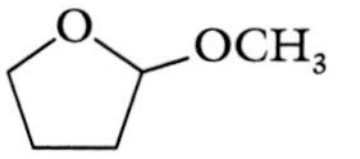

(c) 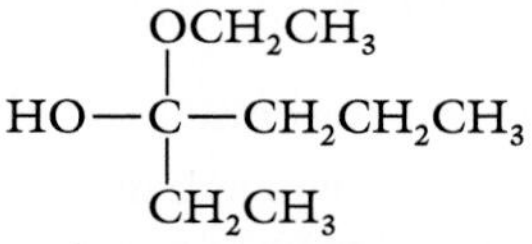

(d) 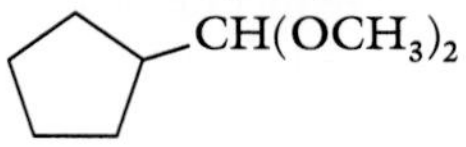

14.28 Which of the following compounds undergo hydrolysis? Give the product(s) of hydrolysis for any compound that undergoes hydrolysis. Indicate "no reaction" if hydrolysis does not take place.

(a) $C_6H_5CH(OCH_2CH_3)_2$
(b) $HOCH_2CH_2OCH_3$
(c) 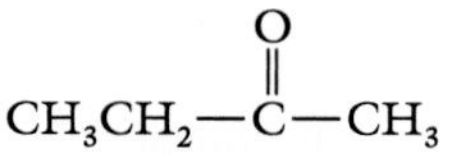
(d) 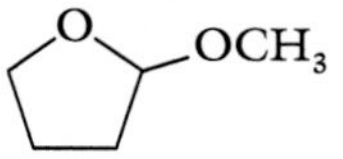

Unclassified Exercises

14.29 Compare the C=C and C=O double bonds with respect to: (a) hybridization of the atoms; (b) bond angles about the atoms; (c) polarity; (d) chemical reactions.

14.30 Identify the families of functional groups 1–9 in the following compound:

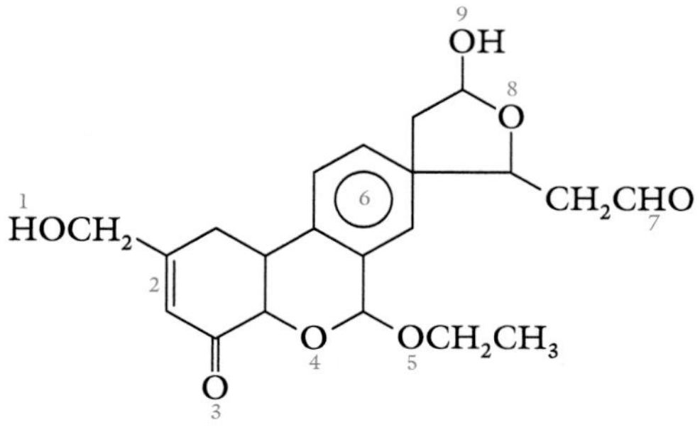

14.31 Draw structural formulas for each of the following compounds: (a) $C_5H_{10}O$ aldehyde containing one methyl group; (b) cyclic ketone of five carbons with all carbons in the ring; (c) C_8 ketone containing a benzene ring; (d) $C_6H_{12}O$ aldehyde containing a *t*-butyl group; (d) $C_6H_{12}O$ ketone containing a *t*-butyl group.

14.32 For each of the following pairs of structural formulas, indicate whether the pair represents (1) the same compound or (2) different compounds that are constitutional isomers or (3) different compounds that are geometrical isomers or (4) different compounds that are not isomers.

(a) 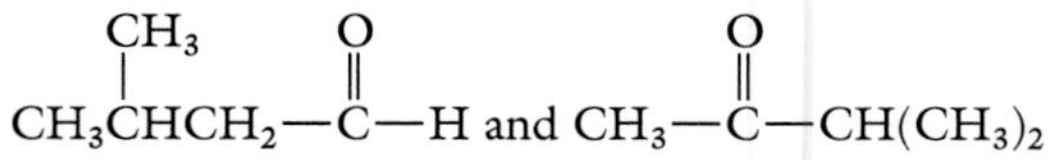

(b) 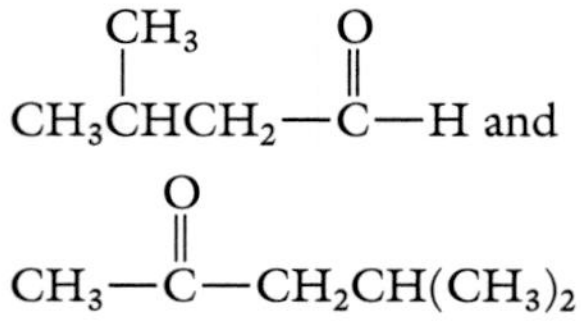

(c) 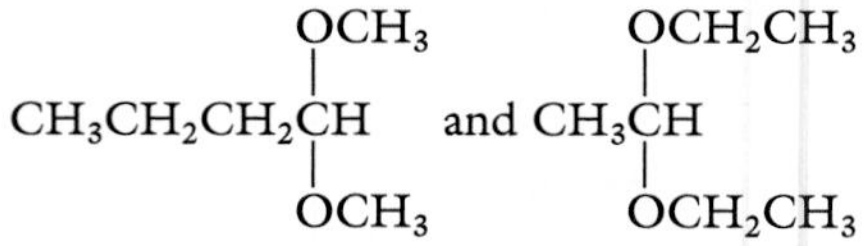

(d) 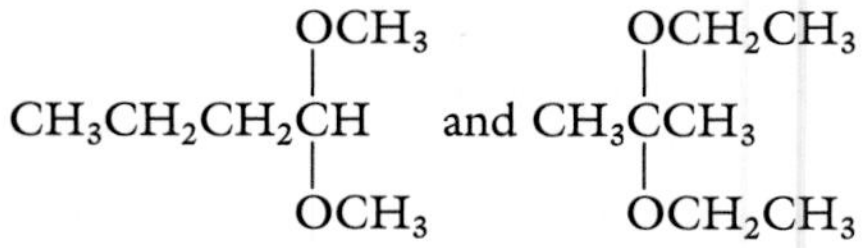

(e) 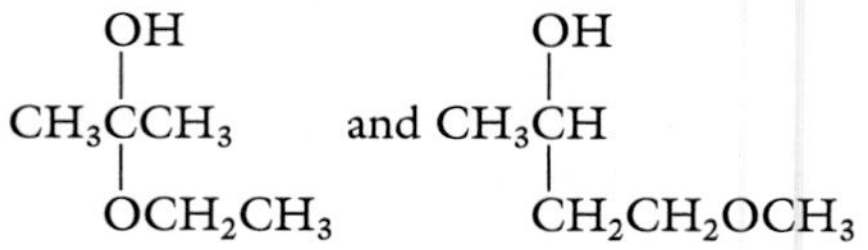

(f) 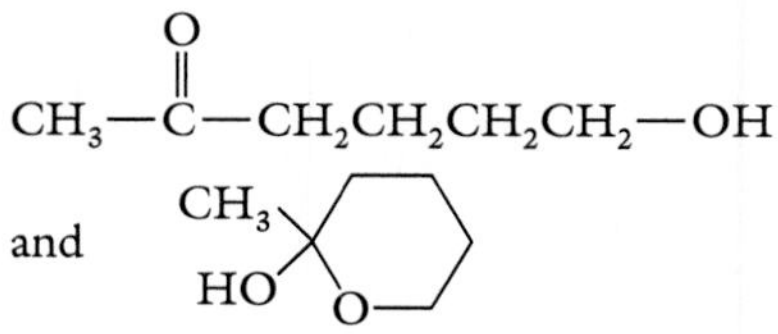

(g) 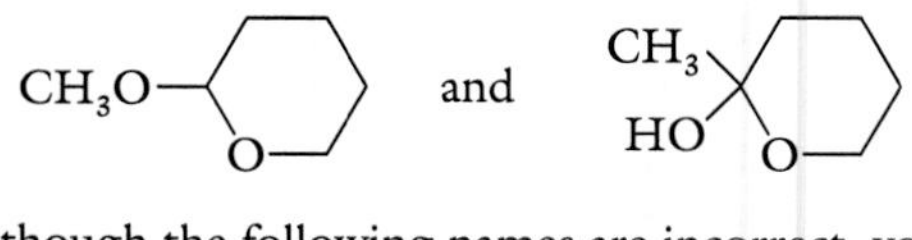

14.33 Although the following names are incorrect, you can deduce the structure for each compound. Give the correct names. (a) 4-Methylcyclopentanone; (b) 2-propylbutanal; (c) 1-pentanone; (d) 3-chloro-4-pentanone.

14.34 A cyclic, nonaromatic compound containing an attached aldehyde group is named by assigning a name to the cyclic compound exclusive of the aldehyde group and then adding the suffix **-carbaldehyde.** Name each of the following compounds:

(a) 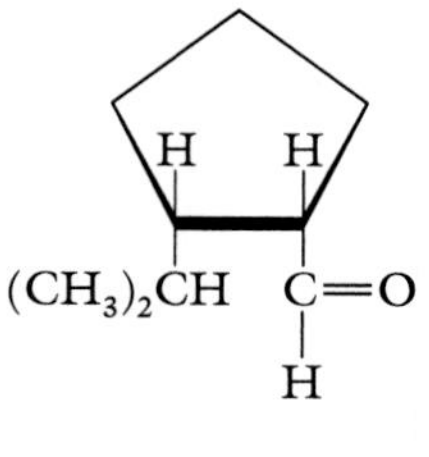

(b) 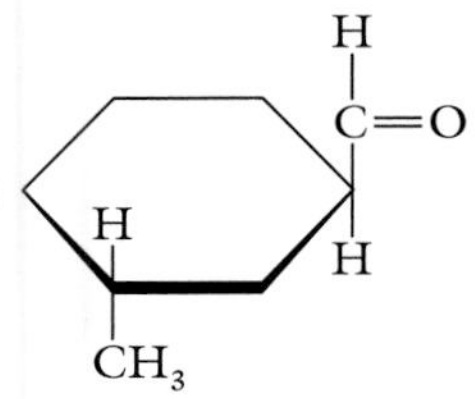

14.35 Indicate which compound in each of the following pairs has the higher value of the specified property. If the two compounds in a pair have nearly the same value for the property, indicate that fact. Explain your conclusions.

(a) Butanal, 1-butanol: boiling point.
(b) 1-Butanol, 2-butanone: solubility in water.
(c) Butanal, diethyl ether: boiling point.
(d) Butanal, diethyl ether: solubility in water.

14.36 Place the following compounds in order of boiling point: pentane, butanal, 1-butanol, diethyl ether. Explain the order.

14.37 Place the following compounds in order of solubility in water: pentane, butanal, 1-butanol, diethyl ether. Explain the order.

14.38 Indicate whether hydrogen bonding is present in each of the following compounds or mixtures. Draw a representation of the hydrogen bonding if present. (a) Propanal; (b) propanal and propanol; (c) propanal and diethyl ether.

14.39 An unknown is either 2-pentanone or 2-methyl-2-butanol. Identify the unknown if it gives a negative test with Tollens's reagent.

14.40 The structure of glucose is shown here. Does glucose give a positive test with Benedict's and Tollens's reagents?

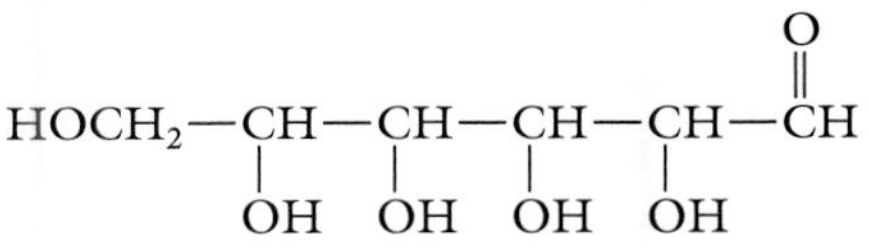

Glucose

14.41 Does glucose undergo reaction with H_2/Pt? If reduction takes place, write the structure(s) of the product(s).

14.42 Show the structure of the cyclic hemiacetal formed by intramolecular reaction between the alcohol and carbonyl groups of 5-hydroxyl-2-hexanone.

14.43 Write equations to show the reaction of 2-butanone with excess methanol. Show the formation of the hemiacetal and acetal.

14.44 Give the organic product(s) formed in each of the following reactions. If no reaction takes place, write "no reaction." If there is more than one product, indicate only the major product(s). If one molecule of the compound is capable of reacting with more than one molecule of reagent, assume that the reaction is carried out with excess reagent.

(a) $CH_3\overset{\overset{O}{\|}}{C}CH_2CH_3 \xrightarrow[\text{2. } H_2O]{\text{1. } LiAlH_4}$?

(b) $CH_3\overset{\overset{O}{\|}}{C}CH_2CH_3 \xrightarrow[H_2O]{LiAlH_4}$?

(c) 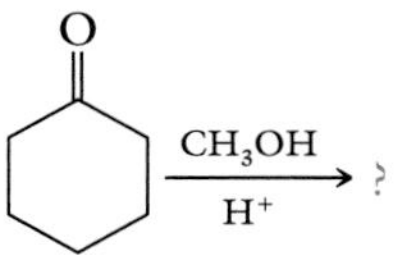

(d) $CH_3-C(OCH_3)_2-C_6H_5 \xrightarrow{H^+,\ H_2O}$?

(e) $CH_3-C(CH_2OCH_3)_2-C_6H_5 \xrightarrow{H^+,\ H_2O}$?

(f) $CH_3CH_2CH_2\overset{\overset{O}{\|}}{C}H \xrightarrow{MnO_4^-}$?

(g) $CH_3CH_2CH_2CH_2OH \xrightarrow{MnO_4^-}$?

(h) 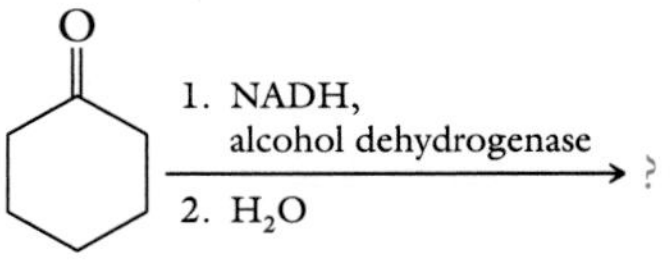

14.45 Give the organic product(s) formed in each of the following reactions. If no reaction takes place, write "no reaction." If there is more than one product, indicate only the major product(s). If one molecule of the compound is capable of reacting with more than one molecule of reagent, assume that the reaction is carried out with excess reagent.

(a) 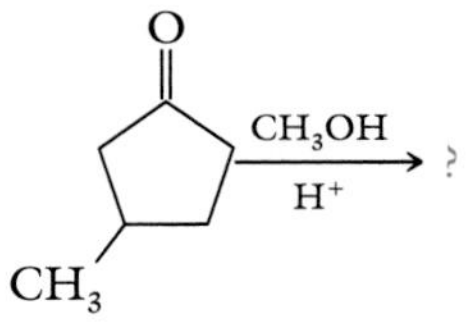

(b) $CH_3-\overset{\overset{O}{\|}}{C}-C_6H_5 \xrightarrow[Ni]{H_2}$?

(c) $CH_3-C(OCH_3)_2-CH_3 \xrightarrow[Ni]{H_2}$?

(d) $CH_3CH_2CH_2\overset{\overset{O}{\|}}{C}H \xrightarrow{\text{Tollens's reagent}}$?

(e) $CH_3-C(OCH_3)_2-C_6H_5 \xrightarrow{\text{Benedict's reagent}}$?

(f) $H-\overset{\overset{O}{\|}}{C}-CH_2C_6H_5 \xrightarrow[\text{2. } H_2O]{\text{1. NADH, alcohol dehydrogenase}}$?

(g) 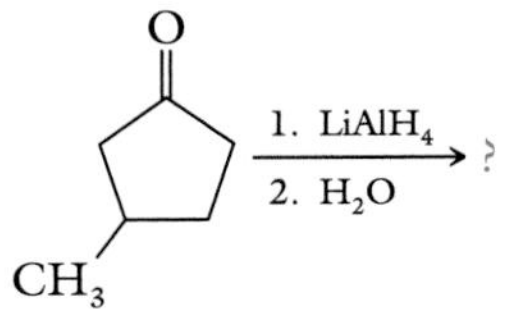

(h) 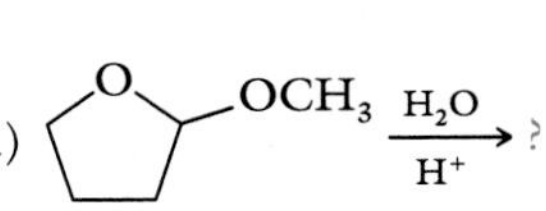

14.46 An unknown compound C_3H_6O is converted into $C_3H_6O_2$ on reaction with dichromate. Identify the compound.

14.47 Identify the compound that yields 2-butanol on reduction with a metal hydride.

14.48 An unknown compound C_4H_8O, either an aldehyde or ketone, gives a negative test with Tollens's reagent but undergoes hydride reduction to form an alcohol. Identify the compound.

14.49 What aldehyde or ketone, if any, can be synthesized by oxidation of each of the following compounds?

(a) 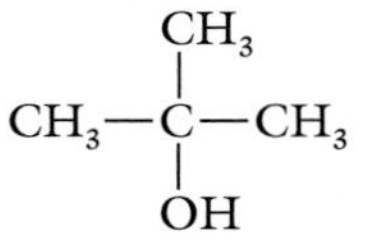

(b) 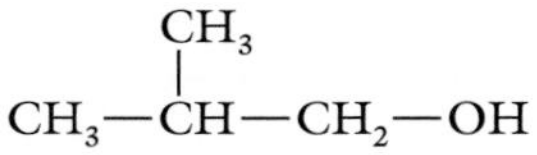

(c) 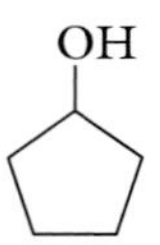

(d) 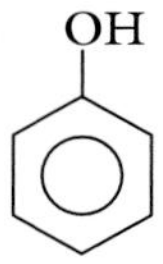

14.50 What aldehyde yields each of the following carboxylic acids by selective oxidation?

(a) 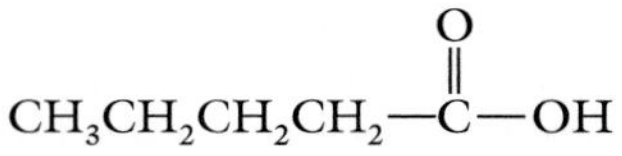

(b) 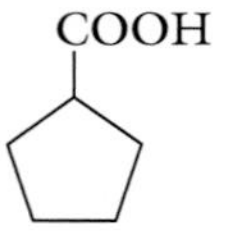

(c) 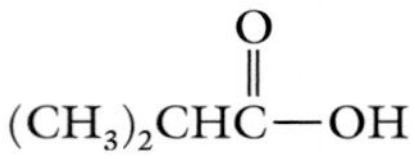

Chemical Connections

14.51 The structures of menthone (mint) and vanillin (vanilla) are shown in Box 14.1. What simple chemical tests distinguish the two compounds.

14.52 One of the industrial processes for producing acetone (2-propanone) uses 2-propanol as the starting material. What reaction is used to convert 2-propanol into acetone?

14.53 The structure of cinnamaldehyde (cinnamon) is shown in Box 14.1. Cinnamaldehyde, C_9H_8O, can be reduced by both the catalytic hydrogenation and the hydride reduction methods. However, the products are different, having the molecular formulas $C_9H_{12}O$ and $C_9H_{10}O$, respectively. Explain the difference.

14.54 Formaldehyde is useful for preserving biological specimens and embalming cadavers. It kills microorganisms (see Box 14.2) by inactivating their proteins by reacting with their amine groups ($-NH_2$). The reaction is a dehydration between the amine and carbonyl groups with the formation of a C=N double bond. Write the equation for this reaction with the simple amine CH_3-NH_2.

14.55 Although Exercise 14.40 showed the structure of glucose as 2,3,4,5,6-pentahydroxyhexanal, there is less than 0.2% of that acyclic structure in aqueous solutions of glucose. Glucose contains greater than 99.8% of a mixture of the following two cyclic structures:

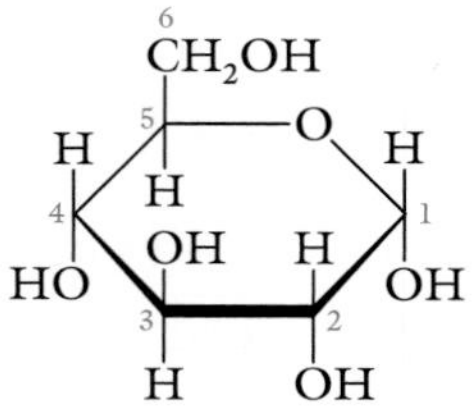

α-D-Glucose

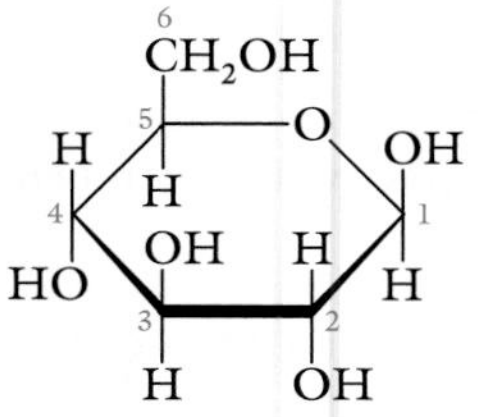

β-D-Glucose

What is the relation between these two cyclic structures and the acyclic structure shown in Exercise 14.40?

CHAPTER 15

CARBOXYLIC ACIDS, ESTERS, AND OTHER ACID DERIVATIVES

CHEMISTRY IN YOUR FUTURE

Because of his circulatory system problems, your father has been advised by his physician to take an aspirin (acetylsalicylic acid) a day to reduce his risk of heart attack. Coincidentally, you have been reading about this risk reduction in one of your professional journals. As a quality-control manager in a food processing plant, you are aware that some commonly used flavor additives for foods are salicylates and thus related to aspirin. One recent article suggested that the presence of these salicylates in foods may be responsible for the general decline in heart disease in the United States since World War II. However, the situation may not be so simple. Aspirin is both a carboxylic acid and an ester, but the salicylate flavor additives are only esters. Chapter 15 describes and compares these groups.

LEARNING OBJECTIVES

- Draw and name carboxylic acids and their derivatives (esters, acid anhydrides, and acid halides).
- Describe the physical properties of carboxylic acids and their esters.
- Describe and write equations for the synthesis of carboxylic acids and their esters.
- Describe and write equations to demonstrate the acidity of carboxylic acids.
- Draw and name carboxylate salts and describe the cleaning action of soaps.
- Describe and write equations for the hydrolysis and saponification of carboxylic acid esters.
- Draw and name phosphate esters and anhydrides.

Carboxylic acids and their derivatives are all around us. The tart taste of citrus fruits, vinegar, and rhubarb; the sharp sting of red ants; the unsavory smell and taste of rancid butter—all are due to carboxylic acids. Vitamin C, which is an essential part of our diet, is a carboxylic acid.

The pleasant odors and tastes of many fruits are due to carboxylic acid derivatives called esters, which are therefore commonly used in the manufacture of flavors. Esters are also used in the manufacture of plastics and textile fibers—not only for clothing and household goods, but also for medical and surgical applications—and in medicines such as aspirin, ibuprofen, and acetaminophen. Animal fats and plant oils are ester derivatives of carboxylic acids called fatty acids, and the sodium and potassium salts of these acids are used as soaps. The salts of other carboxylic acids are used as food preservatives. Proteins are amides, which are nitrogen-containing derivatives of carboxylic acids.

This chapter focuses mainly on carboxylic acids and their ester derivatives. Amides are so important to biological systems that a separate chapter (Chapter 16) deals with them. Also very important to living organisms are the phosphoric acid anhydrides and esters, which we consider at the end of this chapter and will meet again in our study of biochemistry. In contrast, the carboxylic acid halides and anhydrides are derivatives that are not present as such in living organisms and are discussed only briefly later in the chapter.

15.1 COMPARISON OF CARBOXYLIC ACIDS AND THEIR DERIVATIVES

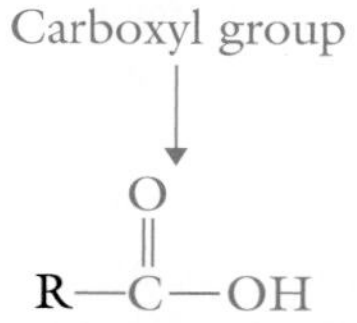

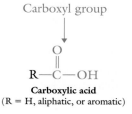

Carboxylic acid
(R = H, aliphatic, or aromatic)

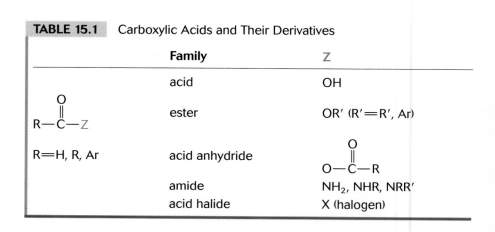

Carboxylic acids are another family of organic compounds that, like aldehydes and ketones, contain the carbonyl group, C═O. The identifying characteristic of carboxylic acids (often simply called organic acids) is that the C═O is present in a **carboxyl group,** a carbonyl group with an —OH attached. The carboxyl group is often abbreviated as —COOH or —CO_2H and carboxylic acids as RCOOH or RCO_2H.

The **carboxylic acid derivatives** are four additional families derived from the carboxylic acids—esters, acid anhydrides, amides, and acid halides:

Ester **Acid anhydride** **Amide** **Acid halide** (X = halogen)

These derivatives differ from one another and from carboxylic acids in the group attached to the carbonyl carbon (Table 15.1).

TABLE 15.1 Carboxylic Acids and Their Derivatives

	Family	Z
	acid	OH
R—C(═O)—Z	ester	OR′ (R′═R′, Ar)
R═H, R, Ar	acid anhydride	O—C(═O)—R
	amide	NH_2, NHR, NRR′
	acid halide	X (halogen)

Esters and **anhydrides,** like carboxylic acids, have a second oxygen atom bonded to the carbonyl carbon, but the three families differ in the group that contains that oxygen: in esters, an —OR′ group; in anhydrides, an —O—CO—R group; and, in carboxylic acids, an —OH group. Anhydrides can also be defined as compounds with two different carbonyl carbons attached to the same oxygen.

Amides have a nitrogen attached in the form of an amino group—often the unsubstituted amino group $—NH_2$, as just shown, but also the substituted amino groups —NHR and —NRR′. **Acid halides** have a halogen connected to the carbonyl carbon.

All four types of carboxylic acid derivatives can be synthesized from and hydrolyzed back to carboxylic acids.

Example 15.1 Identifying carboxylic acids and their derivatives

Classify each of the following compounds as a carboxylic acid, ester, amide, anhydride, acid halide, or something else.

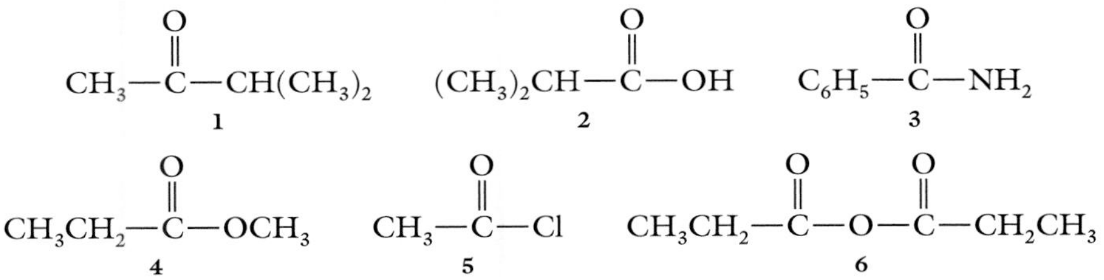

Solution

Compound 1 is a ketone. It is neither a carboxylic acid nor a carboxylic acid derivative, because there are only carbons, no H and no heteroatom, attached to the carbonyl carbon. All of the other compounds are recognizable as acids or acid derivatives by the atom or group attached to the carbonyl carbon. Compound 2 is a carboxylic acid (—OH attached to the carbonyl carbon). Compound 3 is an amide (N attached to the carbonyl carbon). Compound 4 is an ester (—OR attached to the carbonyl carbon). Compound 5 is an acid halide (—Cl attached to the carbonyl carbon). Compound 6 is an acid anhydride (two carbonyl carbons attached to the same oxygen).

Problem 15.1 Classify each of the following compounds as a carboxylic acid, ester, amide, anhydride, or something else.

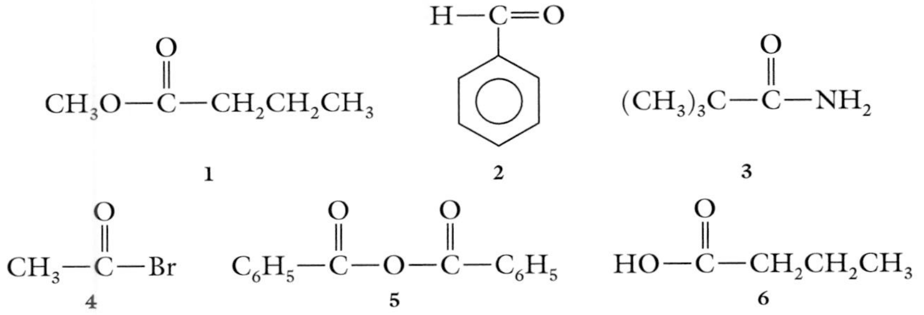

A compound is not a carboxylic acid, ester, amide, anhydride, or halide unless the —OH, —OR′, $—NH_2$, —O—CO—OR, or halogen group is directly attached to the carbonyl carbon. For example, consider isomeric compounds I and II shown in the margin. Compound I is a carboxylic acid. Compound II is not a carboxylic acid, and its properties are not those of a carboxylic acid, even though it contains both carbonyl and —OH groups. Instead, the properties of compound II are the sum of the properties of an

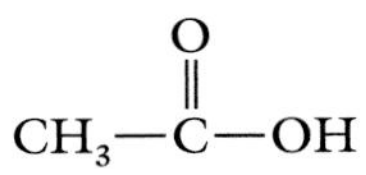

I

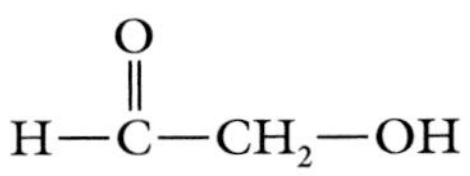

II

aldehyde and an alcohol. It is the specific combination of the carbonyl and hydroxyl groups in the carboxyl group that gives a carboxylic acid its unique properties.

As in other families of organic compounds, constitutional isomers (based on different carbon skeletons and different placements of the carboxyl group) are possible for carboxylic acids and acid derivatives. The orbital hybridization of the atoms in the C═O group and the bond angles about the C═O group in carboxylic acids and their derivatives are the same as those in aldehydes and ketones (*sp*^{2} and 120°). The atoms (C, N, O) connected to the carbonyl carbon are all *sp*^{3} hybridized with tetrahedral bond angles.

15.2 SYNTHESIS OF CARBOXYLIC ACIDS

Methods of synthesizing carboxylic acids were encountered in preceding chapters, in the course of considering the reactions of other organic families. Benzoic acid, the simplest aromatic carboxylic acid, is synthesized from alkyl benzenes by selective oxidation of the alkyl group to a carboxyl group (Section 12.12):

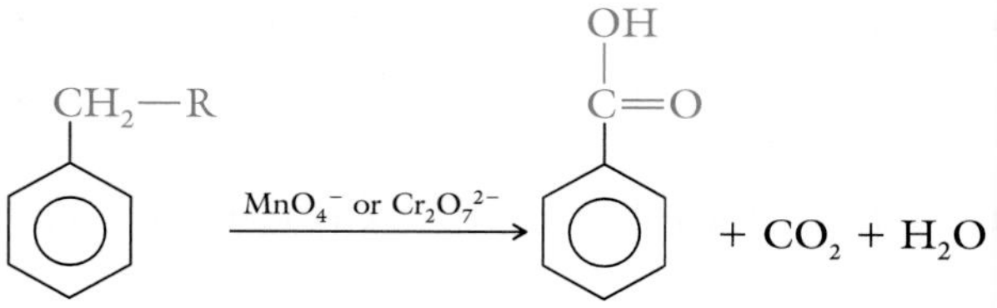

An aromatic ring is more stable than an aliphatic group, as described in Section 12.12.

Carboxylic acids are the end products of selective oxidation of primary alcohols and aldehydes (Sections 13.7 and 14.4):

$$\text{R—CH}_2\text{—OH} \xrightarrow{\text{MnO}_4^- \text{ or Cr}_2\text{O}_7^{2-}} \text{R—}\overset{\text{H}}{\overset{|}{\text{C}}}\text{═O} \xrightarrow{\text{MnO}_4^- \text{ or Cr}_2\text{O}_7^{2-}} \text{R—}\overset{\text{OH}}{\overset{|}{\text{C}}}\text{═O}$$

Example 15.2 Synthesizing carboxylic acids by selective oxidation

Write the equation for the selective oxidation of each of the following compounds to form a carboxylic acid: (a) toluene (methylbenzene); (b) 1-propanol; (c) propanal.

Solution

(a) 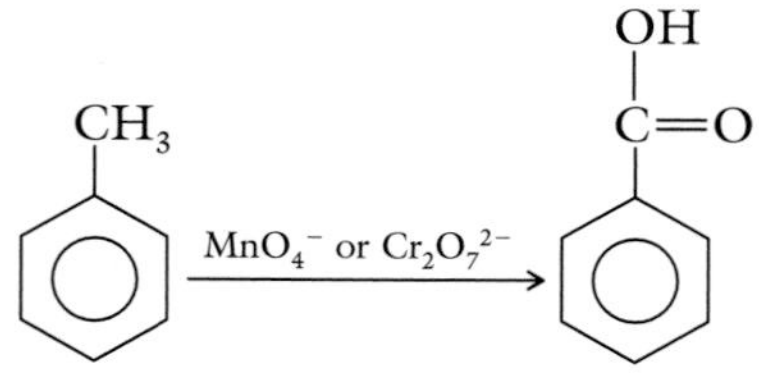

(b) $$\text{CH}_3\text{CH}_2\text{CH}_2\text{—OH} \xrightarrow{\text{MnO}_4^- \text{ or Cr}_2\text{O}_7^{2-}} \text{CH}_3\text{CH}_2\overset{\text{OH}}{\overset{|}{\text{C}}}\text{═O}$$

(c) $$\text{CH}_3\text{CH}_2\overset{\text{H}}{\overset{|}{\text{C}}}\text{═O} \xrightarrow{\text{MnO}_4^- \text{ or Cr}_2\text{O}_7^{2-}} \text{CH}_3\text{CH}_2\overset{\text{OH}}{\overset{|}{\text{C}}}\text{═O}$$

Problem 15.2 Write the equation for the selective oxidation of each of the following compounds: (a) ethylbenzene; (b) ethanol; (c) acetaldehyde (ethanal); (d) acetone (2-propanone).

15.3 NAMING CARBOXYLIC ACIDS

In the IUPAC nomenclature system, carboxylic acids are named according to the same rules as those for naming alkanes, with some modifications:

Rules for naming carboxylic acids

- Find the longest continuous chain that contains the carboxyl carbon. The carboxyl carbon is C1.
- The ending **-e** of the alkane name is changed to **-oic acid.**
- The names of substituents are used as prefixes, preceded by numbers to indicate their positions on the longest chain.
- A compound with two —COOH groups is named an **alkanedioic acid.** Separate numbers are used to specify the position of each —COOH group.
- The aromatic compound C_6H_5COOH is called **benzoic acid.** The ring carbon holding the —COOH is C1.

Many carboxylic acids are referred to by common names (Table 15.2). The prefixes used for the common names of carboxylic acids are the same as those used for the corresponding aldehydes—for example, formaldehyde and acetaldehyde (see Table 14.4) correspond to formic and acetic acids, respectively (Table 15.2).

Example 15.3 Naming carboxylic acids by the IUPAC system

Name each of the following compounds by the IUPAC system:

(a) 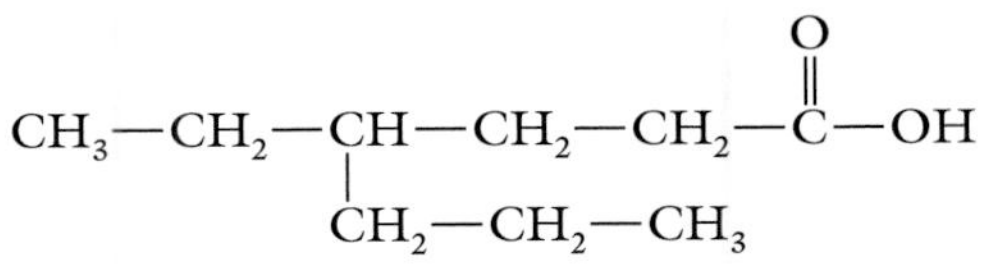

(b) 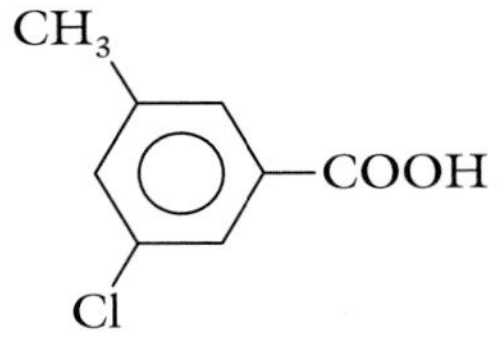

(c) 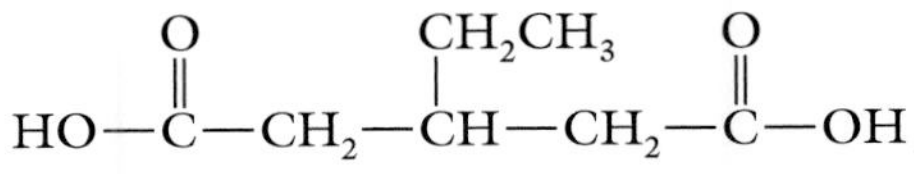

TABLE 15.2 IUPAC and Common Names of Carboxylic Acids

Compound	IUPAC Name	Common Name	Melting Point (°C)	Boiling Point (°C)
H—COOH	methanoic acid	formic acid		101
CH_3—COOH	ethanoic acid	acetic acid		118
CH_3CH_2—COOH	propanoic acid	propionic acid		141
$CH_3CH_2CH_2$—COOH	butanoic acid	butyric acid		164
$CH_3(CH_2)_{10}$—COOH	dodecanoic acid	lauric acid	44	
$CH_3(CH_2)_{16}$—COOH	octadecanoic acid	stearic acid	71	
HOOC—COOH	1,2-ethanedioic acid	oxalic acid	190	
HOOC—CH_2—COOH	1,3-propanedioic acid	malonic acid	135	
HOOC—CH_2CH_2—COOH	1,4-butanedioic acid	succinic acid	188	
C_6H_5—COOH	benzoic acid	benzoic acid	122	249

Solution

(a) The longest chain containing the COOH carbon is the seven-carbon chain. The COOH carbon is C1, and there is an ethyl group at C4. The name is 4-ethylheptanoic acid.

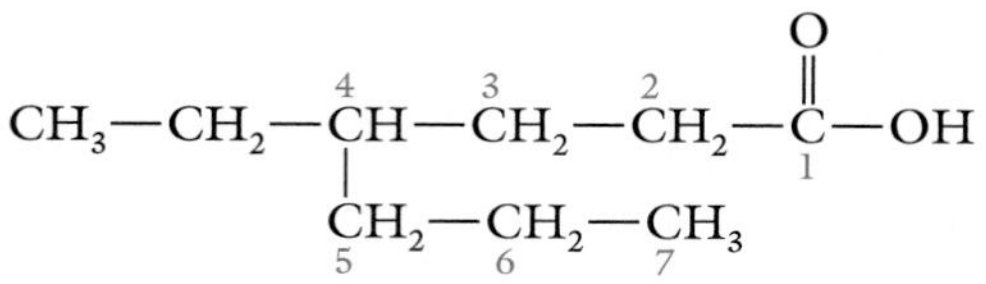

(b) The —COOH group is at C1 of the benzene ring. The name is 3-chloro-5-methylbenzoic acid.

(c) This compound is an alkanedioic acid. The longest chain containing both —COOH groups has five carbons, with the —COOH groups at C1 and C5 and an ethyl group at C3. The name is 3-ethyl-1,5-pentanedioic acid.

Problem 15.3 Name each of the following compounds by the IUPAC system:

(a) 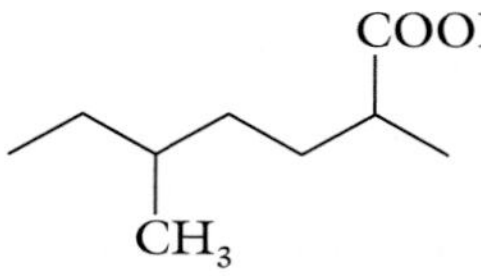

(b) 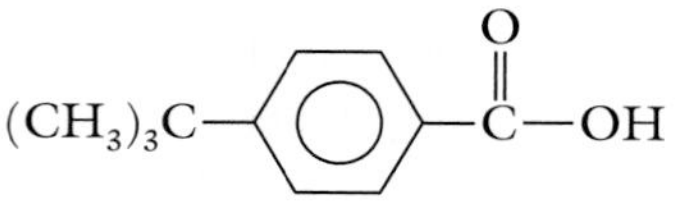

(c) 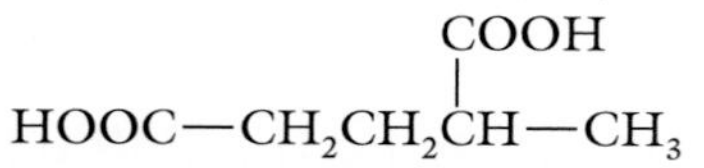

Box 15.1 on page 441 describes some of the carboxylic acids that are commonly encountered in nature. Many of these carboxylic acids are polyfunctional compounds: either they contain two or more carboxyl groups or they contain other functional groups besides a carboxyl group. In the latter category are **hydroxy-, keto-,** and **aminocarboxylic acids**—carboxylic acids that contain hydroxyl, carbonyl, and amino groups, respectively.

15.4 PHYSICAL PROPERTIES OF CARBOXYLIC ACIDS

So far in our study of organic compounds, alcohols have been the family with the strongest secondary attractive forces and highest melting and boiling points, but carboxylic acids have even higher secondary attractive forces and even higher boiling and melting points than those of alcohols (Table 15.3). The —OH and carbonyl groups of the carboxyl group are both polar:

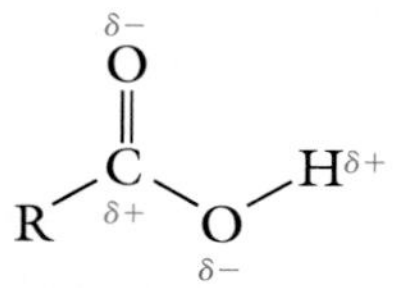

The presence of the two polar groups results in very strong hydrogen bonding between pairs of carboxylic acid molecules, with strong attraction between the carbonyl oxygen and hydroxyl hydrogen:

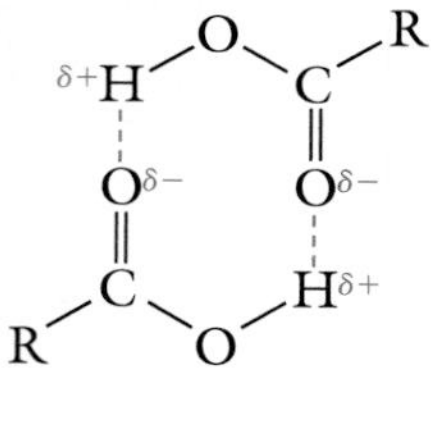

TABLE 15.3 Comparison of Boiling Points of Compounds from Different Families

Compound	Structure	Molecular Mass (amu)	Boiling Point (°C)
pentane	$CH_3CH_2CH_2CH_2CH_3$	72	36
diethyl ether	$CH_3CH_2OCH_2CH_3$	74	35
methyl acetate	CH_3COOCH_3	74	57
butanal	$CH_3CH_2CH_2CHO$	72	76
2-butanone	$CH_3COCH_2CH_3$	72	80
1-butanol	$CH_3CH_2CH_2CH_2OH$	74	117
propanoic acid	CH_3CH_2COOH	74	141

Thus, carboxylic acids exist as **dimers** (pairs of associated molecules) under most conditions. Because of the high level of attraction between the molecules in the dimer, high temperatures are needed to achieve melting and boiling.

Carboxylic acids show slightly higher solubility in water than do alcohols because of the three hydrogen-bond interactions with water shown in the margin: at the carbonyl oxygen (I), hydroxyl oxygen (II), and hydroxyl hydrogen (III).

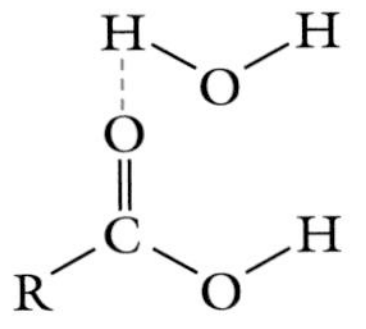

I

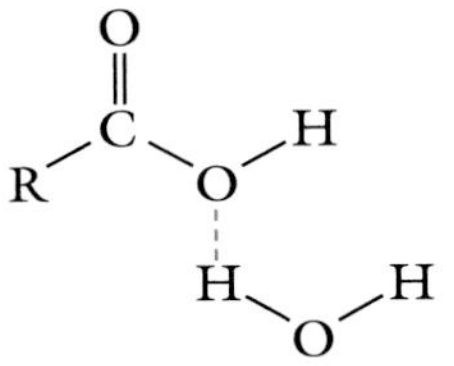

II

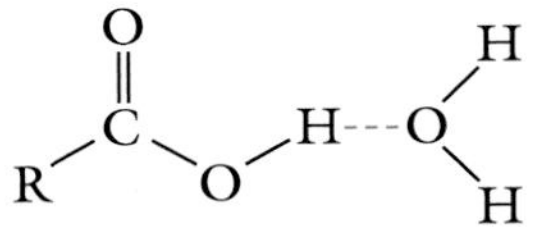

III

The slightly greater water solubility of carboxylic acids compared with alcohols is observed as a more gradual decrease in solubility with increasing number of carbons. The water solubilities of 1-pentanol and pentanoic acid are 2.4 g/100 mL and 5.0 g/100 mL, respectively, at 20°C.

15.5 ACIDITY OF CARBOXYLIC ACIDS

Carboxylic acids are weak acids, undergoing ionization in water to form **carboxylate ion** and hydronium ion:

$$\underset{\text{Carboxylic acid}}{\overset{\overset{\large O}{\|}}{RC}{-}O{-}H} + \underset{\text{Water}}{H_2O} \rightleftharpoons \underset{\text{Carboxylate ion}}{\overset{\overset{\large O}{\|}}{RC}{-}O^-} + \underset{\text{Hydronium ion}}{H_3O^+}$$

The extent of ionization is less than about 1 to 2%; that is, the equilibrium is far to the left, as indicated by the unequal arrows. Carboxylic acids are much weaker acids than the strong inorganic acids such as HCl, H_2SO_4, HNO_3, and $HClO_4$, which are almost completely ionized in water. However, carboxylic acids are stronger acids than phenols and much stronger acids than alcohols. Table 15.4 on the following page compares the $[H_3O^+]$, pH, and percent ionization of 0.1 *M* solutions of different compounds.

The carboxylic acids are more acidic than phenols because of the electron-withdrawing effect of the carbonyl group, a consequence of the carbonyl carbon's partial positive charge. Electron withdrawal stabilizes the carboxylate ion by charge dispersal: the negative charge does not reside entirely on the oxygen but is partly dispersed onto the carbonyl group. The extent of charge dispersal is even greater for the carboxylate ion than for the phenolate ion because the carbonyl group is more strongly electron withdrawing than is a benzene ring.

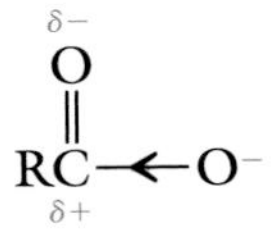

Carboxylate ion

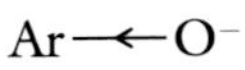

Phenolate ion

Carboxylic acids are sufficiently acidic to turn blue litmus paper red—a simple chemical test that distinguishes carboxylic acids from alcohols (which do not react with the indicator dye in litmus paper) but not from phenols (which also turn blue litmus paper red).

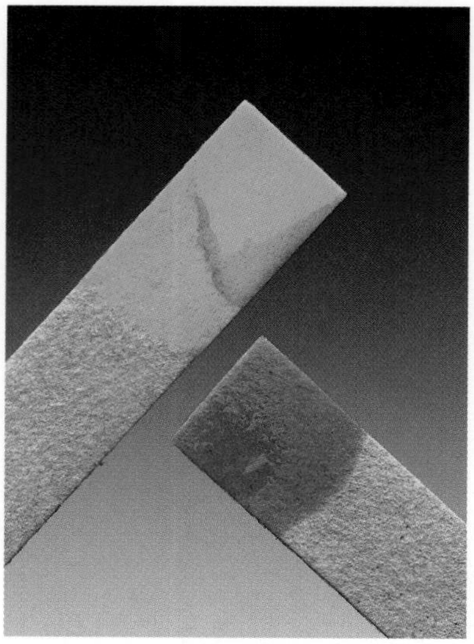

Blue litmus paper turns red when treated with a carboxylic acid (left) but is unaffected by an alcohol (right).

TABLE 15.4 Relative Acidities of Organic Compounds

		Values for 0.10 *M* Aqueous Solution		
Compound	**Structure**	**$[H_3O^+]$**	**pH**	**Ionization (%)**
hydrochloric acid	HCl	0.10	1.0	100
benzoic acid	C_6H_5COOH	2.5×10^{-3}	2.6	2.5
acetic acid	CH_3COOH	1.3×10^{-3}	2.9	1.3
phenol	C_6H_5OH	3.3×10^{-6}	5.4	0.0032
ethanol	CH_3CH_2OH	1.0×10^{-7}	7.0	0.0001
water	H_2O	1.0×10^{-7}	7.0	0.0001

Sodium hydroxide and sodium bicarbonate are sufficiently strong bases that their reactions with a carboxylic acid go to completion, forming the carboxylate salt:

$$\underset{\text{Carboxylic acid}}{R\overset{O}{\overset{\|}{C}}{-}O{-}H} + \underset{\text{Sodium hydroxide}}{NaOH} \longrightarrow \underset{\text{Carboxylate salt}}{R\overset{O}{\overset{\|}{C}}{-}O^-\ Na^+} + H_2O$$

$$R\overset{O}{\overset{\|}{C}}{-}O{-}H + \underset{\text{Sodium bicarbonate}}{NaHCO_3} \longrightarrow \underset{\text{Carboxylate salt}}{R\overset{O}{\overset{\|}{C}}{-}O^-\ Na^+} + H_2CO_3$$

$$H_2CO_3 \longrightarrow H_2O + CO_2$$

◀◀ $pH = -\log[H_3O^+]$ (Section 9.3)

A solution of a carboxylic acid in pure water has a pH near 3, and only a small percentage of the carboxylic acid exists as ionized molecules (see Table 15.4). In more alkaline media (higher pH), the fraction of ionized, carboxylate ions increases. Almost all physiological fluids are maintained at near-neutral pH by the bicarbonate/carbonic acid buffer (Sections 9.6 and 9.8). Blood pH is 7.35 and the pH in most cells ranges from 6.8 to 7.1. This physiological pH is 10,000 times as basic as pH 3. The greater basicity of physiological fluids relative to carboxylic acids results in neutralization of the carboxylic acids and conversion into carboxylate salts. Therefore, carboxylic acids, such as pyruvic, lactic and citric acids (Box 15.1), in living organisms are present almost entirely (>99%) in their carboxylate (ionized) form.

Example 15.4 Ionization of carboxylic acids

Indicate whether a reaction takes place under each of the following conditions. If a reaction takes place, write the appropriate equation. (a) Benzoic acid + water; (b) benzoic acid + aqueous NaOH.

Solution

(a) Benzoic acid is only weakly ionized in water. The equilibrium is far to the left, as indicated by the unequal arrows:

$$C_6H_5\overset{O}{\overset{\|}{C}}{-}O{-}H + H_2O \rightleftharpoons C_6H_5\overset{O}{\overset{\|}{C}}{-}O^- + H_3O^+$$

(b) In the presence of a strong base such as NaOH, benzoic acid is almost completely ionized (indicated by the single arrow):

$$C_6H_5\overset{O}{\overset{\|}{C}}{-}O{-}H + NaOH \longrightarrow C_6H_5\overset{O}{\overset{\|}{C}}{-}O^-\ Na^+ + H_2O$$

Problem 15.4 Indicate whether a reaction takes place under the following conditions. If a reaction takes place, write the appropriate equation. (a) Propanoic acid + water; (b) propanoic acid + aqueous NaOH; (c) propanoic acid + aqueous $NaHCO_3$; (d) phenol + aqueous $NaHCO_3$.

Aromatic carboxylic acids are somewhat more acidic than aliphatic ones. Table 15.4 shows that $[H^+]$ for a benzoic acid solution is about twice that for

15.1 Chemistry Around Us

Carboxylic Acids in Nature

Carboxylic acids are found in considerable abundance in nature. Most have sour (tart) tastes, and some have distinctive, often disagreeable, odors. Formic acid (IUPAC name: methanoic acid), the simplest of all carboxylic acids, is responsible for the irritation from the sting of red ants. Vinegar is a dilute (about 3–5%) aqueous solution of acetic acid (IUPAC name: ethanoic acid) formed by the oxidation of ethanol in wine. Acetic acid is also an important industrial chemical, with large amounts used to manufacture cellulose acetate, which is used in fabrics and textiles, cigarette filters, and plastics applications. Propionic acid (IUPAC name: propanoic acid) is responsible for the odor and flavor of Swiss cheese, whereas the off-flavor and odor of rancid (spoiled) butter is due to butyric acid (IUPAC name: butanoic acid).

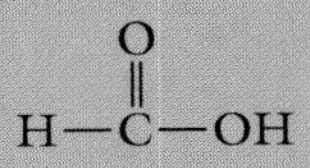
Formic acid

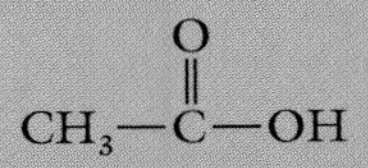
Acetic acid

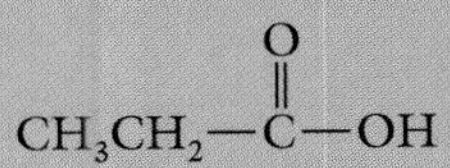
Propionic acid

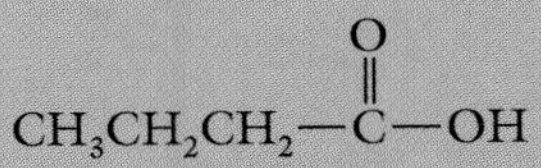
Butyric acid

Many of the carboxylic acids in nature are polyfunctional—for example, oxalic, citric, lactic, and pyruvic acids. Oxalic acid (IUPAC name: 1,2-ethanedioic acid) is a toxic compound found in spinach and rhubarb. The concentration of oxalic acid in spinach leaves and rhubarb stalks (stems) is very low, well below levels that are toxic to humans. However, rhubarb leaves contain much higher concentrations, and ingestion of even a small amount of the rhubarb leaves can be dangerous.

Citric acid (IUPAC name: 3-hydroxy-1,3,5-propanetricarboxylic acid) is responsible for the tart taste of citrus

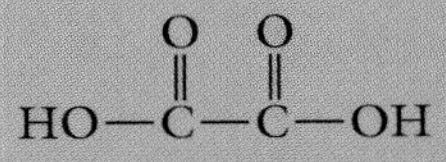
Oxalic acid

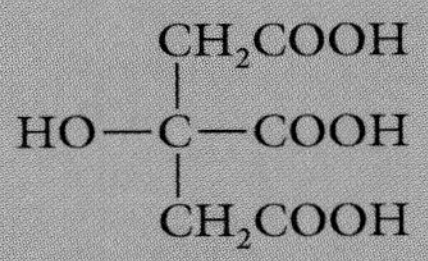
Citric acid

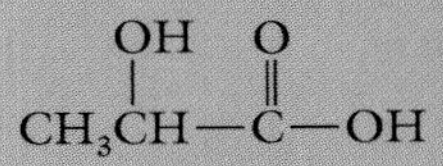
Lactic acid

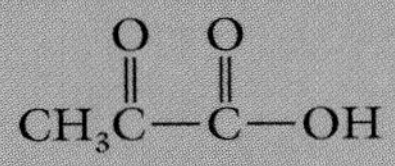
Pyruvic acid

fruits such as lemons, limes, oranges, and grapefruits. Lemon juice is more than 5% citric acid. Lactic acid, produced by the bacterial fermentation of lactose (milk sugar), gives sour milk its disagreeable taste.

Lactic acid is used in cosmetic lotions and creams for treating dry, flaking, itchy skin and has been promoted as a beauty aid for removing wrinkles and moisturizing the skin. It is thought to hasten the sloughing off of dead cells from the skin. Such cosmetic preparations are often referred to as α-hydroxys, because there is a hydroxy group on the α-carbon (Section 14.4).

Our bodies produce certain carboxylic acids in the course of cellular metabolism. Lactic acid and pyruvic acid are produced in muscle tissue during vigorous exercise (Section 23.2). Citric and other acids are intermediates in the citric acid cycle (Section 23.5).

In Chapter 20, we will see how carboxylic acids containing α-amino (NH_2) groups, such as alanine, are used in nature as the building blocks for synthesizing proteins.

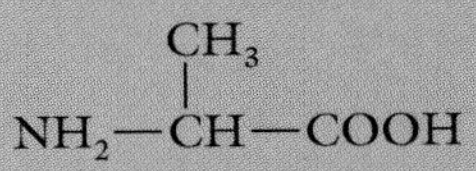
Alanine

Grapefruit, orange, and lemon contain citric acid, spinach contains oxalic acid, vinegar contains acetic acid, and Swiss cheese contains propionic acid.

an acetic acid solution. This difference is due to the electron-withdrawing characteristic of the benzene ring compared with an alkyl group. Greater electron withdrawal helps stabilize the aromatic carboxylate anion relative to the alkyl carboxylate anion.

15.6 CARBOXYLATE SALTS

A **carboxylate salt,** the product of reaction between a carboxylic acid and a strong base, is named as follows:

Rules for naming carboxylate salts

- The positive ion is named first.
- The carboxylate ion is named by changing the ending of the acid's name (either IUPAC or common) from **-ic acid** to **-ate.**

For example, the sodium salt of butanoic acid is named sodium butanoate.

Example 15.5 Naming carboxylate salts

Name the following carboxylate salts:

(a) $H-\overset{\overset{\large O}{\|}}{C}-O^-\ NH_4^+$ (b) $(C_6H_5-COO)_2Ca$

Solution

(a) This compound is the ammonium salt of methanoic (common name: formic) acid. The name is ammonium methanoate (common name: ammonium formate).

(b) This compound is the calcium salt of benzoic acid and is named calcium benzoate.

Problem 15.5 Name the following carboxylate salts:

(a) $(CH_3)_2CHCH_2\overset{\overset{\large O}{\|}}{C}ONa$ (b) $(CH_3COO^-)_3\ Al^{3+}$

Carboxylate salts are ionic compounds and, therefore, possess much higher melting and boiling points than those of the corresponding carboxylic acids, which are covalent compounds. The attractive forces between ions in an ionic compound are stronger than any of the secondary attractive forces in covalent compounds (Sections 6.2 through 6.4). As a consequence, all carboxylate salts are solids at room temperature. For example, sodium formate is a solid with a melting point of 253°C, whereas formic acid is a liquid with a melting point of 8°C.

Carboxylate salts are much more soluble in water than the corresponding carboxylic acids. Carboxylic acids with more than 6 carbons are weakly soluble or insoluble in water, but the salts are soluble up to a much larger number of carbons. For example, sodium stearate, with 18 carbons is water soluble, whereas stearic acid is water insoluble. This increased solubility of carboxylate salts is due to the strong attractive forces between ions and water, much stronger than the hydrogen-bond attractions between carboxylic acids and water.

$CH_3(CH_2)_{16}\overset{\overset{\large O}{\|}}{C}-OH$

Stearic acid
Water insoluble

$CH_3(CH_2)_{16}\overset{\overset{\large O}{\|}}{C}-O^-Na^+$

Sodium stearate
Water soluble

The solubility of a carboxylic acid is greatly increased in an environment of neutral or higher pH, because the carboxylic acid is converted into the carboxylate salt. This is typical of physiological conditions, where carboxylic acids are solubilized by conversion into the carboxylate salt. Medicines that contain carboxyl groups and need to be administered by injection are usually synthesized as their carboxylate salts. The carboxylate salt allows the preparation of higher concentrations of the medicine in aqueous solution, which results in faster absorption into the body on injection.

Example 15.6 Predicting the melting points of compounds

Place the following compounds in order of increasing melting point: sodium propanoate, 1-hexanol, pentanoic acid.

Solution

1-Hexanol < pentanoic acid ≪ sodium propanoate. The ionic compound sodium propanoate has a much higher melting point than those of the two covalent compounds. Between the two covalent compounds of the same molecular mass, the carboxylic acid has the higher melting point because it has stronger hydrogen bonding (because of dimer formation) than the alcohol does.

Problem 15.6 Place the following compounds in order of increasing water solubility: $CH_3(CH_2)_{14}COOK$, $CH_3(CH_2)_{14}COOH$, $CH_3(CH_2)_{14}OH$.

A carboxylate salt, like the salt of any weak acid, is a weak base (proton acceptor) and undergoes acid-base reaction with water and strong acids (Section 9.5):

$$R\overset{\overset{\displaystyle O}{\|}}{C}{-}O^-\,K^+ + H_2O \rightleftharpoons R\overset{\overset{\displaystyle O}{\|}}{C}{-}OH + KOH$$

$$R\overset{\overset{\displaystyle O}{\|}}{C}{-}O^-\,K^+ + HCl \longrightarrow R\overset{\overset{\displaystyle O}{\|}}{C}{-}OH + KCl$$

Carboxylate salts are sufficiently basic to turn red litmus paper blue.

Some carboxylate salts have antibacterial and antifungal properties—they stop the growth of bacteria and fungi—and are used as food preservatives and in medications (Box 15.2 on the following page).

15.7 SOAPS AND THEIR CLEANING ACTION

The soaps that we use to clean our bodies are the sodium or potassium carboxylate salts of long-chain carboxylic acids, called **fatty acids** and typically mixtures of unbranched aliphatic chains of 12 to 20 carbon atoms. The fatty acids are called fatty acids because they are obtained from animal fats and plant oils (for example, beef and pig fats, coconut oil). The carboxylate salts are derived from fats and oils by a process called saponification (Sections 15.11 and 19.4). An example of a soap molecule is sodium stearate, the sodium carboxylate salt of stearic acid, $CH_3(CH_2)_{16}COO^-Na^+$.

▶▶ Fats and oils are lipids and the subject of Chapter 19.

Dirt adheres to surfaces rather than dissolving in water, either because the dirt is hydrophobic (water hating) or because it is physically combined with body oils, cooking fats, greases, and other substances that are hydrophobic. Soap molecules in water allow greasy dirt to be solubilized because soap molecules are amphipathic: that is, they contain both hydrophobic and hydrophilic (water-loving) parts. The long hydrocarbon chain of the carboxylate salt is hydrophobic, and the ionic COO^-Na^+ end is hydrophilic. The hydrophobic and

15.2 Chemistry Around Us

Carboxylate Salts

Carboxylate salts are used in foods and other consumer products to stop or inhibit the growth of fungi (mold) and bacteria. Propionate salts, especially calcium and to a lesser extent sodium, are used as preservatives for baked goods, such as bread and cakes, and cheeses.

Two slices of moldy homemade wheat bread are at the left of commercial bread made with preservatives such as propionate salts.

Sodium benzoate exists naturally in cranberries, prunes, and other foods and is added as a food preservative to acidic food products such as carbonated beverages, ketchup, jams, and jellies, and canned fruit. Sorbic acid (IUPAC name: 2,4-hexadienoic acid) and its sodium and potassium salts are used as preservatives for wines, carbonated beverages, pickled products, and fruit juices.

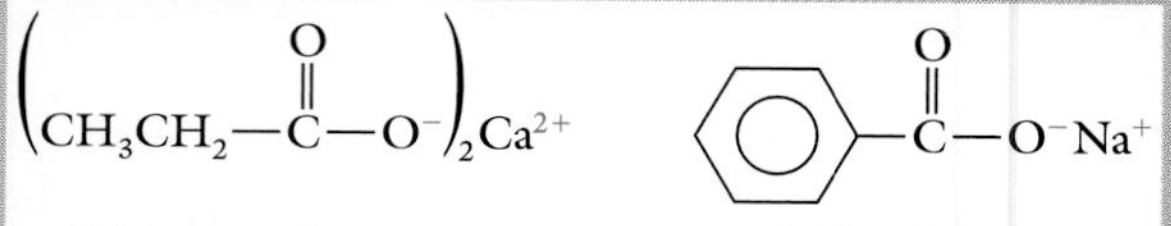

Calcium propionate   Sodium benzoate

$CH_3CH{=}CHCH{=}CHCOOH$

Sorbic acid

Zinc 10-undecenoate is the active antifungal ingredient in athlete's-foot preparations. Aluminum acetate is used in creams and lotions for the treatment of diaper rash and acne.

$[CH_2{=}CH(CH_2)_8COO^-]_2Zn^{2+}$ $[CH_3COO^-]_3Al^{3+}$

Zinc 1-undecenoate   Aluminum acetate

hydrophilic parts, called the hydrophobic tail and hydrophilic head, are represented by the wavy line and circle, respectively, in Figure 15.1.

The hydrophobic tail of a soap molecule has an affinity for the greasy dirt, whereas the hydrophilic head has an affinity for water, both consequences of the like dissolves like principle. Soap molecules, being attracted strongly to both dirt and water, break up the dirt particles into smaller particles. The surfaces of these particles are surrounded by soap molecules, which allows them to become solubilized in water.

The requirements for a good soap are an ionic group as the hydrophilic head and a long hydrocarbon group as the hydrophobic tail. A long-chain carboxylic acid, such as stearic acid, has only one of the requirements; it does not function well as a soap, because the carboxyl group is not nearly as hydrophilic as the carboxylate salt. Likewise, the carboxylate salt of a short-chain car-

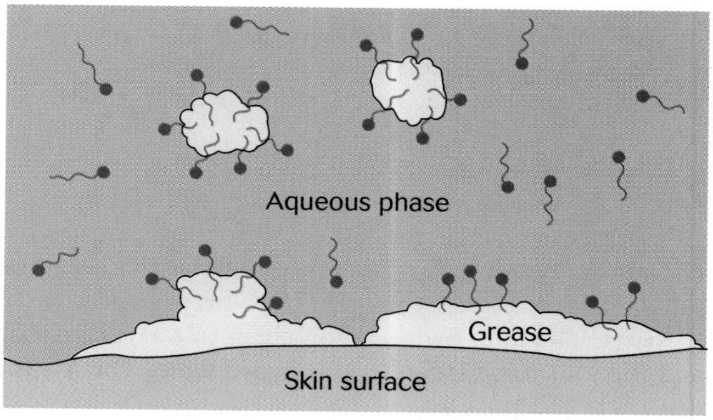

Figure 15.1 The cleaning action of soap.

15.3 Chemistry Around Us

Hard Water and Detergents

Soaps—sodium or potassium carboxylates—are one type of cleaning agent, or **detergent.** Soaps work best in **soft water**—water that does not contain significant amounts of Ca^{2+}, Mg^{2+}, Fe^{2+}, and Fe^{3+}. When a soap is used in **hard water**—water containing these ions, the ions exchange with the Na^+ and K^+ of the soap to form insoluble calcium, magnesium, and iron carboxylate salts. For example:

$$\underset{\text{Water-soluble soap}}{2\ CH_3(CH_2)_{16}COO^-Na^+} + Ca^{2+} \longrightarrow \underset{\text{Water-insoluble curd}}{[CH_3(CH_2)_{16}COO^-]_2\ Ca^{2+}} + 2\ Na^+$$

These salts precipitate from solution as **soap curd,** the scum responsible for the off-white or light gray rings in bathtubs and sinks and the dull finish that sometimes appears on washed clothes. Only when the Ca^{2+}, Mg^{2+}, Fe^{2+}, and Fe^{3+} ions have been removed from hard water by the formation of curd can the remaining soap begin its cleaning action. Thus, in regions of hard water, much larger amounts of soap are required for cleaning.

One solution to the problem of hard water is to install a **water softener** in the water line in the home. This device converts hard water into soft water by replacing Ca^{2+}, Mg^{2+}, Fe^{2+}, and Fe^{3+} with Na^+ ions, a process called **ion exchange.** Another solution is to use a different kind of detergent, such as sodium or potassium alkylbenzene sulfonates.

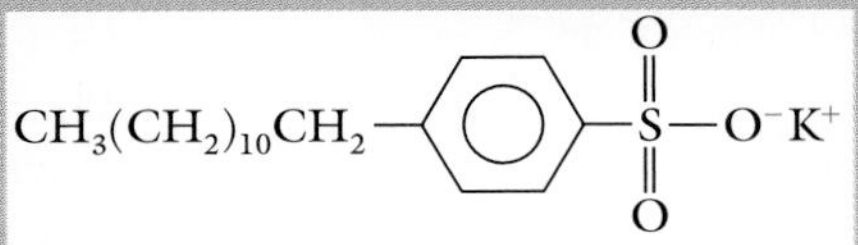

Alkylbenzene sulfonate

Like the carboxylate soaps, these compounds exchange their Na^+ and K^+ ions for the Ca^{2+}, Mg^{2+}, Fe^{2+}, and Fe^{3+} ions in the hard water, but the resulting calcium, magnesium, and iron alkylbenzene sulfonates are water soluble. Thus, they stay in solution and exert their cleaning action.

boxylic acid, such as $CH_3(CH_2)_2COONa$, has only one of the requirements. It does not function well as a soap, because the small hydrocarbon tail is not sufficiently hydrophobic.

The soaps that we use contain small amounts of other materials such as perfumes, deodorants, skin moisturizers, and colorants. Some soaps have air blown into them so that they float on water. Antiseptics are added to make medicated soaps, and pumice is added to make scouring soaps.

Box 15.3 describes the special requirements of soaps used in regions with hard water.

15.8 ESTERS FROM CARBOXYLIC ACIDS AND ALCOHOLS

Esters (RCOOR′) are formed from carboxylic acids by **esterification,** the reaction of an acid with an alcohol (R′ = aliphatic) or a phenol (R′ = aromatic). The two reactants are heated in the presence of a strong acid, such as sulfuric acid, as catalyst:

Bonds broken — Bond formed

$$\underset{\text{Acid}}{R{-}\overset{\displaystyle O}{\overset{\|}{C}}{-}OH} + \underset{\substack{\text{Alcohol}\\ \text{or phenol}}}{H{-}O{-}R'} \overset{H^+}{\rightleftharpoons} \underset{\text{Ester}}{R{-}\overset{\displaystyle O}{\overset{\|}{C}}{-}O{-}R'} + H_2O$$

Esterification entails the breaking of two bonds—the O—H bond in the alcohol (or phenol) and the C—O bond between the hydroxyl oxygen and the carbonyl carbon of the carboxylic acid. Bond breakage is followed by an

Esters are responsible for the pleasant odors and tastes of many fruits.

exchange of fragments and new bond formation. The OH from the carboxylic acid bonds to the H from the alcohol to form water. The alcohol oxygen bonds to the carbonyl carbon to form the ester.

Esterification is a reversible reaction. The yield of ester is increased by removal of water to shift the equilibrium to the right side (Le Chatelier's principle). This removal is easily achieved in esterification because the reaction is usually run at a temperature considerably above the boiling point of water, causing water to distill out of the reaction vessel as it is formed.

Living organisms synthesize esters by reactions catalyzed by enzymes. These esters include a variety of plant and animal waxes, fats, and oils (Chapter 19). Esters are also synthesized from the reaction of an alcohol (or phenol) with a carboxylic acid anhydride or halide, rather than the carboxylic acid (Section 15.12).

Whereas most carboxylic acids have disagreeable odors, esters of carboxylic acids have pleasant odors and flavors. The lower-molecular-mass esters are responsible for the odors and tastes of many fruits. Synthetic esters are often added as flavoring agents to soft drinks, ice creams, yogurts, and other food products. For example, ethyl butanoate is used as the flavoring agent to add pineapple taste and odor to a food product. Esters are also used as perfume fragrances.

One of the most widely used medicines in the world, aspirin, is an ester derived from *p*-hydroxybenzoic acid (common name: salicylic acid). Box 15.4 on page 448 discusses aspirin and related medicines.

Example 15.7 Writing the equation for the synthesis of an ester

Give the equation for ester formation between propanoic acid and 2-propanol.

Solution

$$CH_3CH_2{-}\overset{\overset{\displaystyle O}{\|}}{C}{-}OH + H{-}OCH(CH_3)_2 \xrightarrow[-H_2O]{H^+} CH_3CH_2{-}\overset{\overset{\displaystyle O}{\|}}{C}{-}OCH(CH_3)_2$$

Problem 15.7 Give the equation for ester formation between butanoic acid and phenol.

Thioesters are compounds formed by esterification of carboxylic acids with thiols (Section 13.11):

$$\underset{\text{Acid}}{R{-}\overset{\overset{\displaystyle O}{\|}}{C}{-}OH} + \underset{\text{Thiol}}{H{-}S{-}R'} \overset{H^+}{\rightleftharpoons} \underset{\text{Thioester}}{R{-}\overset{\overset{\displaystyle O}{\|}}{C}{-}S{-}R'} + H_2O$$

Thioesters derived from the thiol coenzyme A play important roles in several metabolic pathways, including the citric acid cycle (Section 23.5).

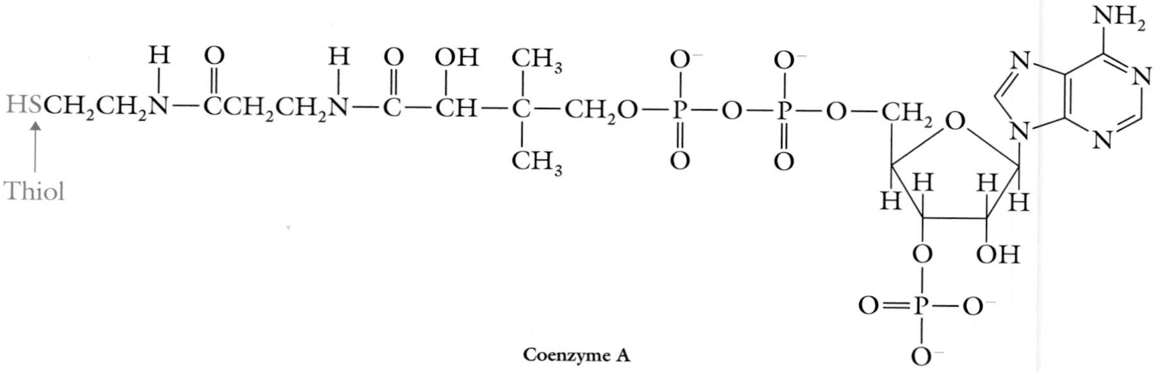

Coenzyme A

15.9 NAMES AND PHYSICAL PROPERTIES OF ESTERS

The names of esters combine the names of the **alkoxy** or **aryloxy** (that is, the alcohol or phenol) part and the **acyl** (that is, the carboxylic acid) part of the compound.

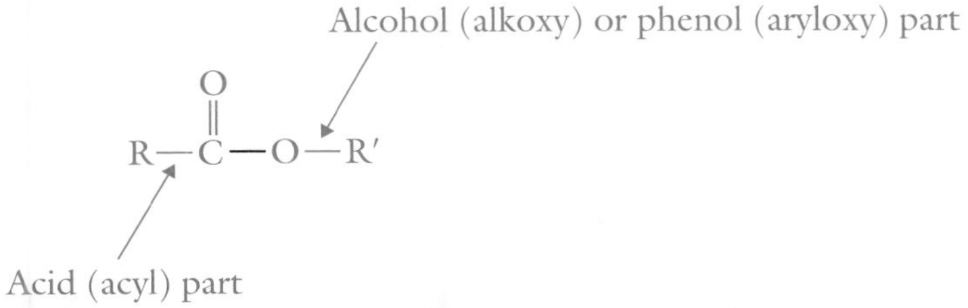

Rules for naming esters

- The alkyl (or aryl) group of the alkoxy (or aryloxy) part of an ester is named first as a separate word.
- The acyl part is named by changing the ending of the name of the carboxylic acid (either IUPAC or common) from **-ic acid** to **-ate.**

Example 15.8 Naming esters by the IUPAC nomenclature system

Give the IUPAC names for the following compounds:

(a) $CH_3CH_2-\overset{\displaystyle O}{\overset{\|}{C}}-OCH(CH_3)_2$ (b) $CH_3CH_2O-\overset{\displaystyle O}{\overset{\|}{C}}-CH(CH_3)_2$

(c) 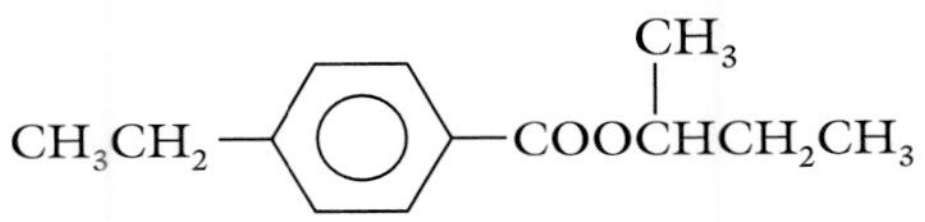

Solution

(a) The alkoxy part contains an isopropyl group. The acyl part comes from propanoic acid (common name: propionic acid). The IUPAC name of the ester is isopropyl propanoate (common name: isopropyl propionate).

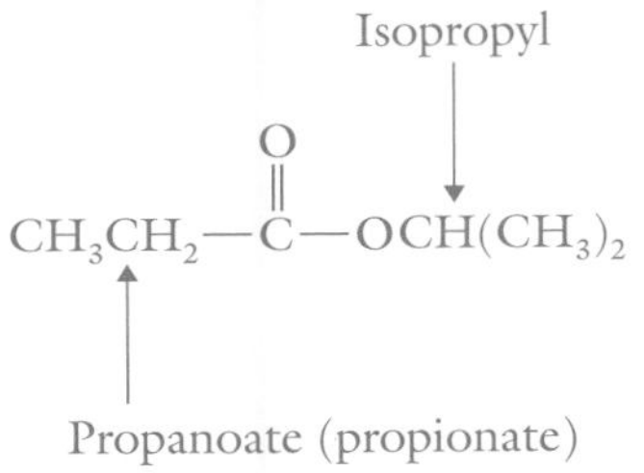

(b) Do not assume that the structure of an ester is always written with the acyl part on the left and the alkoxy part on the right. This ester is drawn in the opposite manner, acyl part on the right and alkoxy part on the left. Always remember that the acyl part does not contain an oxygen connected to the carbonyl carbon. The IUPAC name is ethyl 2-methylpropanoate.

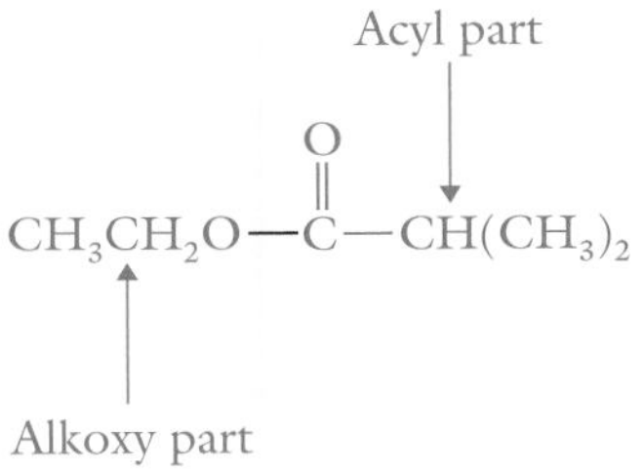

(c) The alkoxy part contains the *s*-butyl group, and the acyl part comes from *p*-ethylbenzoic acid. The ester is named *s*-butyl *p*-ethylbenzoate.

Problem 15.8 Give the IUPAC names of the following compounds:

(a) 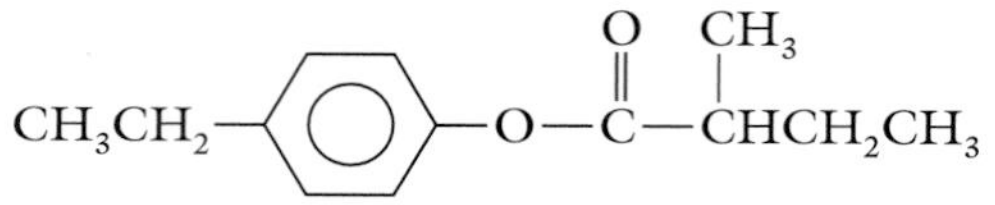

(b) $HCOOCH_2CH_2CH_2CH_2CH_3$ (c) $CH_3OOCCH_2CH_2CH_2CH_2CH_3$

15.4 Chemistry Around Us

Aspirin and Aspirin Substitutes

Aspirin (acetylsalicylic acid) is an **analgesic, antipyretic,** and **anti-inflammatory** medicine. That is, it relieves pain, such as the pain of headache, toothache, and sprains; it reduces fever but does not affect normal body temperature; and it reduces inflammation, such as the inflammation accompanying arthritis. Aspirin works by inhibiting the formation of prostaglandins (Section 19.8), a group of lipids responsible for the body's inflammatory and febrile responses and for the sensation and transmission of pain in the nervous system.

Acetylsalicylic acid, both a carboxylic acid and an ester, is synthesized by the acyl transfer reaction (Section 15.12) from acetic anhydride to the —OH group of salicylic acid:

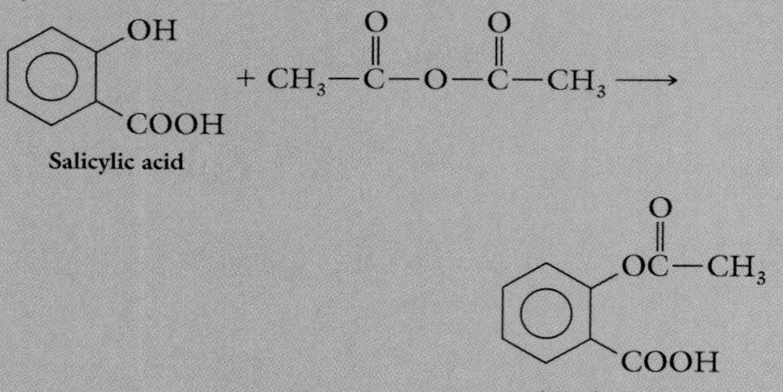

Salicylic acid

Acetylsalicylic acid

Aspirin can also have adverse effects in some people. It decreases the ability of blood to clot, which can result in gastrointestinal bleeding, general susceptibility to tissue bruising, and hemorrhagic stroke—that is, stroke resulting from excessive bleeding from ruptured blood vessels in the brain. The risk of gastrointestinal bleeding can be reduced by using enteric-coated aspirin—aspirin coated with a material that dissolves only in alkaline pH. Enteric-coated aspirin passes intact through the interior of the highly acidic stomach, and the coating does not dissolve until it reaches the alkaline contents of the intestine. Another potential adverse effect of aspirin is Reye's syndrome, a devastating and sometimes fatal illness including brain and liver dysfunction. It affects a small percentage of children who take aspirin in bouts of chickenpox or flu.

The anticlotting action of aspirin is used to protect people at risk for heart disease. Daily or every-other-day doses of aspirin prevent heart attacks by suppressing the formation of blood clots in clogged arteries feeding the heart.

Acetaminophen (an amide), ibuprofen (a carboxylic acid), and naproxen sodium are alternative medicines for people who cannot take aspirin. Acetaminophen (Tylenol, Panadol, Datril) is a useful alternative to aspirin as an analgesic and antipyretic, but it is not an anti-inflammatory. Ibuprofen (Advil, Motrin, Nuprin) is an anti-inflammatory, perhaps superior to aspirin, as well as an analgesic and antipyretic. Naproxen sodium (Aleve) is an anti-inflammatory as well as an analgesic and antipyretic having a less-adverse effect on the gastrointestinal system than aspirin.

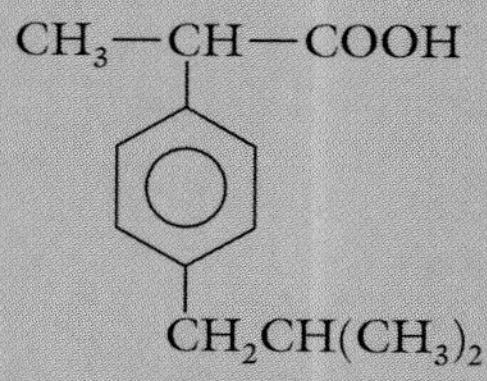

Ibuprofen

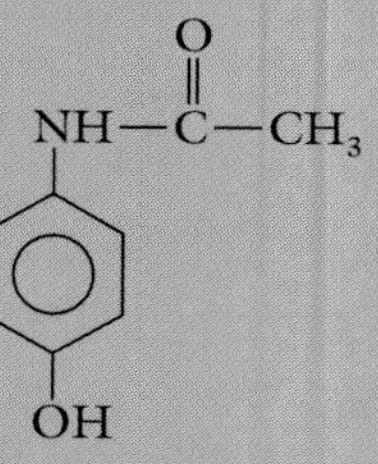

Acetaminophen

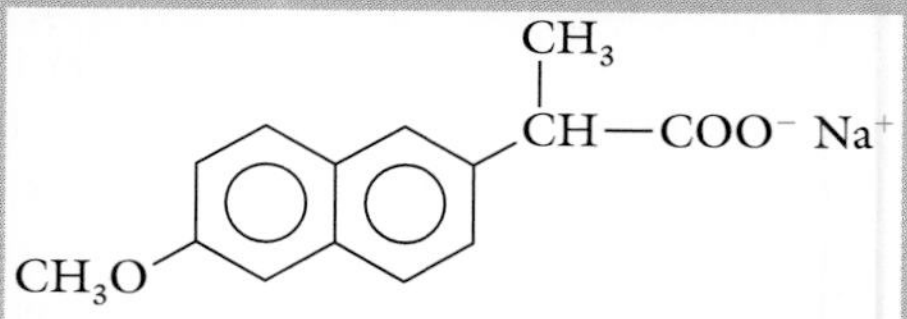

Naproxen sodium

These medicines need to be used with caution and moderation. Indiscriminate use of acetaminophen has been linked to liver (especially when used with alcohol) and kidney damage. Long-term use of naproxen sodium also is implicated in kidney damage.

The polarity and secondary attractive forces in esters are lower than those in aldehydes and ketones, which results in lower melting and boiling points for esters relative to aldehydes and ketones. For example, the boiling points of methyl acetate and butanal are 57°C and 76°C, respectively (see Table 15.3). Esters have about the same water solubilities as those of aldehydes and ketones because they hydrogen bond equally well with water.

15.10 POLYESTER SYNTHESIS

The esterification reaction becomes a polymerization reaction if one uses **bifunctional reactants;** that is, reactants having two functional groups per molecule. They behave as monomers, joining together in large numbers (>50) to produce the much larger sized polymer molecules.

For example, the acid-catalyzed reaction of a dicarboxylic acid (which contains two —COOH groups) and a diol (which contains two —OH groups) produces a **polyester.** The reactants initially yield dimer molecules with —COOH at one end and —OH at the other end. The dimer molecules react to form a tetramer, which also contains a carboxyl group at one end and a hydroxyl at the other end. The tetramers react, and the process repeats itself over and over again to produce larger and larger molecules. When the molecular mass of the polyester is high enough to produce a solid material with the desired mechanical strength, the reaction may be stopped by cooling.

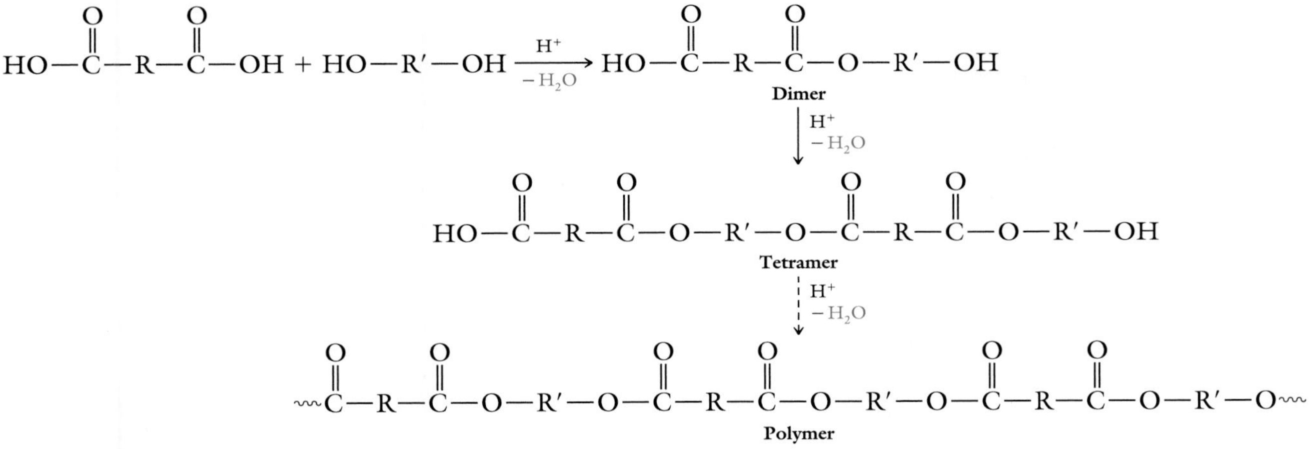

The wavy lines in the polymer structure indicate that the structure goes on and on with large numbers of monomer units linked together.

The polymer structure is usually drawn in much more abbreviated form, with brackets enclosing the **repeat unit:**

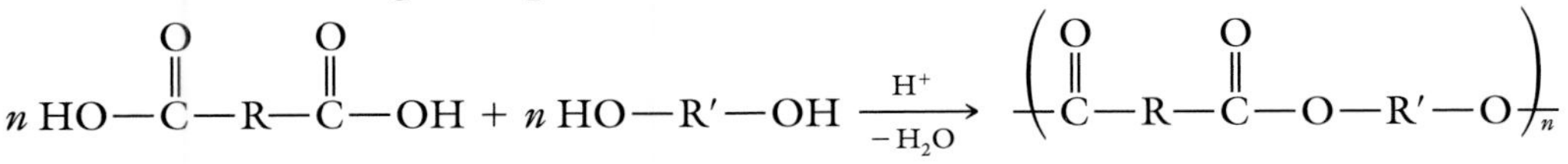

The subscript n indicates that the repeat unit repeats over and over many times (that is, n is a large number).

This type of polymerization reaction is called a **condensation polymerization** because a small molecule (water in this case) is a by-product of the process. Condensation polymerization is one of two types of polymerization reactions. The other type, in which no small by-product results, is called addition polymerization, and is the type of polymerization that occurs with alkenes (Section 12.7 and Box 12.6).

Example 15.9 Writing the equation for a polyesterification

Write the equation to describe the polyesterification reaction between ethylene glycol ($HOCH_2CH_2OH$) and terephthalic acid (p-$HOOC—C_6H_4—COOH$).

Solution

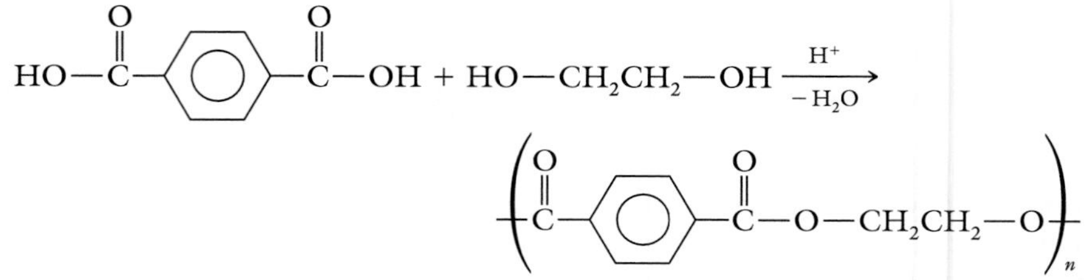

Problem 15.9 Each of the following pairs of reactants undergoes esterification. Which of the pairs produces a polymer? Write the equation for any reaction(s) that produce(s) a polymer.

(a) $HO—\overset{O}{\overset{\|}{C}}CH_2CH_2CH_2CH_2\overset{O}{\overset{\|}{C}}—OH + HO—CH_2CH_3$

(b) $HOOCCH_2CH_2CH_2CH_3 + HOCH_2CH_2OH$

(c) $HO—\overset{O}{\overset{\|}{C}}CH_2CH_2CH_2CH_2\overset{O}{\overset{\|}{C}}—OH + HO—CH_2CH_2—OH$

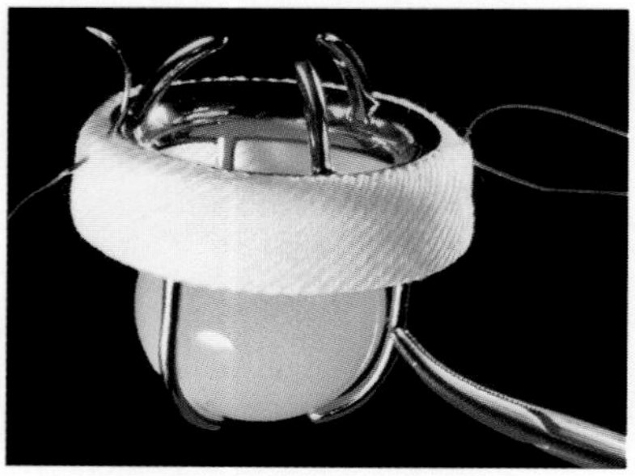

Polyester fabric is used in constructing artificial heart valves.

The polymer formed by the reaction in Example 15.9 is poly(ethylene terephthalate), or PET (also called Dacron, Mylar, polyester). This commercially important material has an annual production of 4 billion pounds in the United States. About 80% of the production goes into textile applications—including clothing, curtains, upholstery, tire cord—and medical applications, such as the fabric used to repair blood vessels and the knitted tubing used in artificial heart valves. The remaining 20% of PET production goes into plastics applications. They include photographic, magnetic, and X-ray films and tapes and bottles for soft drinks, beer, and other food products.

15.11 HYDROLYSIS OF ESTERS

Hydrolysis reactions were last encountered in Section 14.6, when we studied the hydrolysis of hemiacetals and acetals. Recall that hydrolysis is a reaction in which a molecule reacting with water is cleaved into two parts. Esters hydrolyze to the corresponding carboxylic acid and alcohol (or phenol) under either acidic or basic reaction conditions. Acidic hydrolysis requires the presence of a strong acid catalyst such as sulfuric acid:

$$\underset{\text{Ester}}{R—\overset{O}{\overset{\|}{C}}—O—R'} + H_2O \overset{H+}{\rightleftharpoons} \underset{\text{Acid}}{R—\overset{O}{\overset{\|}{C}}—OH} + \underset{\text{Alcohol or phenol}}{H—O—R'}$$

Acidic hydrolysis of an ester is the reverse of acid-catalyzed ester formation from a carboxylic acid and alcohol (or phenol). Carrying out the reaction with a large excess of water shifts the equilibrium toward the hydrolysis products (Le Chatelier's principle). This is the reverse of the conditions used to achieve esterification.

The hydrolysis and formation of esters take place continuously in living organisms. Triacylglycerols and many other lipids in our bodies and in our diets are esters. Digestion of these compounds is a simple matter of ester hydrolysis (Section 19.4). Our cells then synthesize new esters (to construct cell membranes, for example). These syntheses are catalyzed not by strong acid, such as sulfuric acid, but by special proteins called enzymes.

Basic hydrolysis requires a strong base such as sodium or potassium hydroxide:

$$\underset{\text{Ester}}{R{-}\overset{\overset{\displaystyle O}{\|}}{C}{-}O{-}R'} + NaOH \xrightarrow{H_2O} \underset{\text{Carboxylate salt}}{R{-}\overset{\overset{\displaystyle O}{\|}}{C}{-}ONa} + \underset{\text{Alcohol or phenol}}{H{-}O{-}R'}$$

Basic hydrolysis of an ester produces the metal carboxylate salt instead of the carboxylic acid because the reaction is performed in the presence of the base. This reaction, also called **saponification,** is used to manufacture soaps from certain animal fats and vegetable oils (Section 19.4).

Basic hydrolysis and acidic hydrolysis of esters differ in two important ways:

- Basic hydrolysis, unlike acidic hydrolysis, is not an equilibrium reaction. The formation of an ionic compound (the carboxylate salt) drives the equilibrium to the right, making the reaction irreversible. Basic hydrolysis is often preferred over acidic hydrolysis for this reason.
- The acid in acidic hydrolysis is a catalyst—it is not used up in the reaction and only small amounts are needed relative to the ester. The base in basic hydrolysis (saponification) is a reactant, needed in equimolar amounts relative to the ester, and is consumed in the reaction.

Example 15.10 Recognizing and writing esterification reactions

Complete the following equations. If no reaction takes place, write "no reaction."

(a) $CH_3\overset{\overset{\displaystyle O}{\|}}{C}{-}OCH_2CH_3 \xrightarrow{H_2O} ?$

(b) $CH_3\overset{\overset{\displaystyle O}{\|}}{C}{-}OCH_2CH_3 \xrightarrow[H^+]{H_2O} ?$

(c) $CH_3\overset{\overset{\displaystyle O}{\|}}{C}{-}OCH_2CH_3 \xrightarrow[NaOH]{H_2O} ?$

Solution

(a) No reaction, because there is no acid catalyst present.

(b) Hydrolysis takes place because an acid catalyst is present. Two bonds break, an O—H of water and the C—O of the ester (as indicated below by the dashed line). Fragments from the ester and water bond together to form the products. The —OH from water attaches to the acyl part of the ester to form the carboxylic acid. The H from water attaches to the alkoxy part of the ester to form the alcohol. This reaction is the reverse of the esterification reaction described in Section 15.9.

$$\begin{array}{l} CH_3\overset{\overset{\displaystyle O}{\|}}{C}\,\vdots\,OCH_2CH_3 \\ HO\,\vdots\,H \end{array} \xrightarrow[H^+]{H_2O} CH_3\overset{\overset{\displaystyle O}{\|}}{C}{-}OH + H{-}OCH_2CH_3$$

(c) Saponification takes place when an ester is treated with aqueous strong base. The simplest way to determine the products is to write the equation for acidic hydrolysis and then react the carboxylic acid product with base to convert it into the carboxylate salt:

$$CH_3\overset{\overset{\displaystyle O}{\|}}{C} {-}OCH_2CH_3 \xrightarrow[\text{NaOH}]{H_2O} CH_3\overset{\overset{\displaystyle O}{\|}}{C}{-}OH + H{-}OCH_2CH_3$$
$$HO{-}H \qquad\qquad\qquad\qquad\qquad \downarrow \text{NaOH}$$
$$CH_3\overset{\overset{\displaystyle O}{\|}}{C}{-}ONa$$

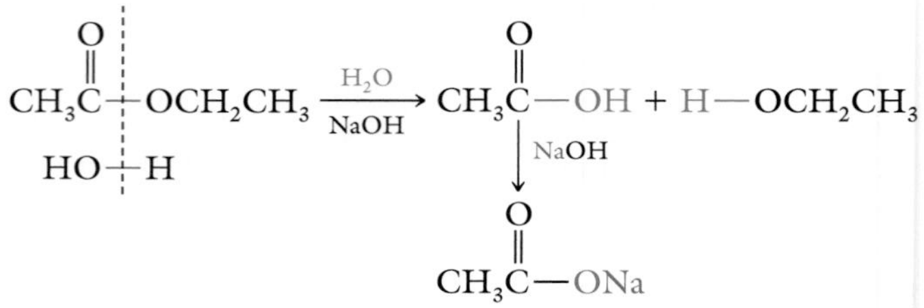

The overall reaction is written in abbreviated form as:

$$CH_3\overset{\overset{\displaystyle O}{\|}}{C}{-}OCH_2CH_3 \xrightarrow[\text{NaOH}]{H_2O} CH_3\overset{\overset{\displaystyle O}{\|}}{C}{-}ONa + H{-}OCH_2CH_3$$

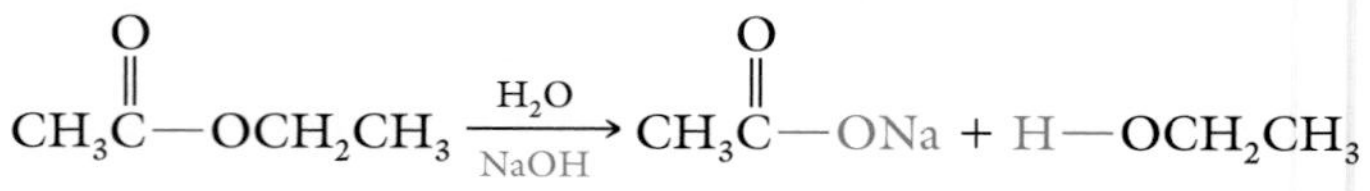

Problem 15.10   Write the equation for each of the following reactions: (a) acidic hydrolysis of phenyl butanoate; (b) saponification of propyl propanoate with KOH.

15.12   CARBOXYLIC ACID ANHYDRIDES AND HALIDES

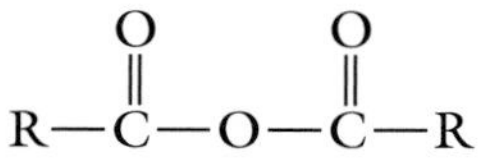

Acid anhydride

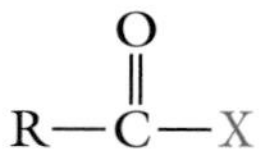

Acid halide
(X = halogen)

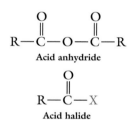

Carboxylic acid anhydrides are named by changing the end of the name of the corresponding carboxylic acid from **-acid** to **-anhydride.** Acid halides (also called acyl halides) are named by changing the ending from **-ic acid** to **-yl halide.**

Example 15.11	**Naming carboxylic acid anhydrides and halides**

Name the following compounds:

(a) $CH_3CH_2{-}\overset{\overset{\displaystyle O}{\|}}{C}{-}Cl$ 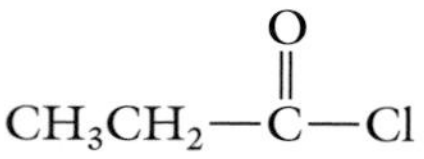

(b) $C_6H_5{-}\overset{\overset{\displaystyle O}{\|}}{C}{-}O{-}\overset{\overset{\displaystyle O}{\|}}{C}{-}C_6H_5$ 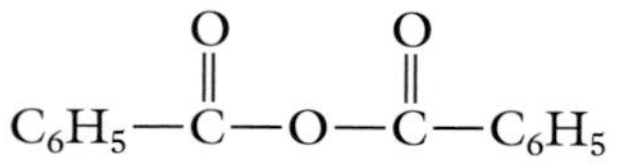

Solution

(a) This compound is the acid chloride of propanoic acid (common name: propionic acid). The name is propanoyl chloride (common name: propionyl chloride).

(b) This compound is the anhydride of benzoic acid. The name is benzoic anhydride.

Problem 15.11   Give the structure of each of the following compounds: (a) butanoyl bromide; (b) acetic anhydride.

     The term anhydride indicates a compound that is formed by dehydration. The anhydride group forms when two acid molecules are bonded together by dehydration. The next section will show that carboxylic acids are not the only acids to form anhydrides.

Acid anhydrides and halides are highly reactive with water and cannot exist in biological systems. They react with water to form the corresponding carboxylic acid:

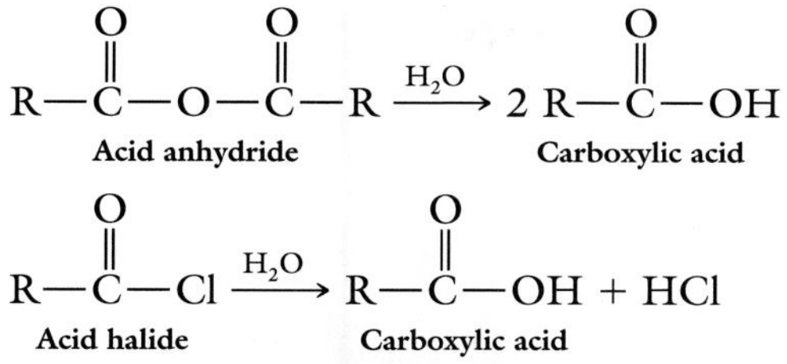

$$\underset{\text{Acid anhydride}}{R-\overset{\overset{\displaystyle O}{\|}}{C}-O-\overset{\overset{\displaystyle O}{\|}}{C}-R} \xrightarrow{\ H_2O\ } 2\ \underset{\text{Carboxylic acid}}{R-\overset{\overset{\displaystyle O}{\|}}{C}-OH}$$

$$\underset{\text{Acid halide}}{R-\overset{\overset{\displaystyle O}{\|}}{C}-Cl} \xrightarrow{\ H_2O\ } \underset{\text{Carboxylic acid}}{R-\overset{\overset{\displaystyle O}{\|}}{C}-OH} + HCl$$

Acid anhydrides are useful in the laboratory for synthesizing esters by reaction with alcohols (or phenols):

$$R-\overset{\overset{\displaystyle O}{\|}}{\underset{\underset{\text{Acyl group}}{\uparrow}}{C}}-O-\overset{\overset{\displaystyle O}{\|}}{C}-R + R'O-H \longrightarrow R-\overset{\overset{\displaystyle O}{\|}}{C}-OR' + R-\overset{\overset{\displaystyle O}{\|}}{C}-O-H$$

$$R-\overset{\overset{\displaystyle O}{\|}}{\underset{\underset{\text{Acyl group}}{\uparrow}}{C}}-Cl + R'O-H \longrightarrow R-\overset{\overset{\displaystyle O}{\|}}{C}-OR' + H-Cl$$

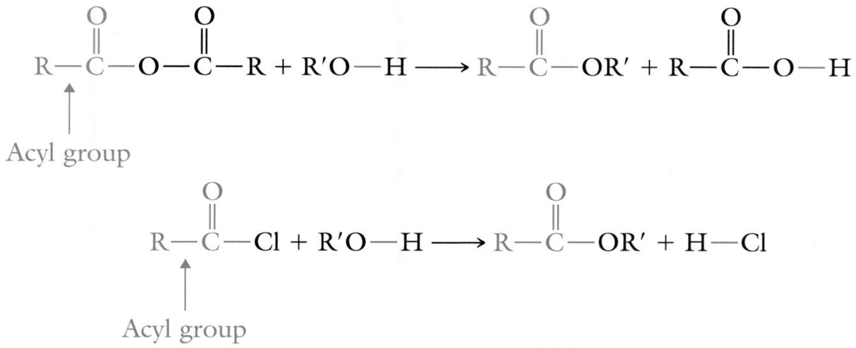

The synthesis of an ester from acid anhydrides and halides is more efficient than ester synthesis from the carboxylic acid. The industrial production of aspirin is carried out by using the reaction of acetic anhydride and salicylic acid (see Box 15.4).

Acid anhydrides and halides, as well as carboxylic acids, are often called **acyl transfer agents** because they transfer the **acyl group,** RCO, to the oxygen of an alcohol (or phenol). Reactions that include acyl transfer, called **acyl transfer reactions,** are important in biological systems. Protein synthesis takes place through acyl transfer reactions (Section 21.7). Thioesters derived from coenzyme A participate in metabolic processes through acyl transfer reactions (Chapters 23 and 24).

Example 15.12 | **Writing the equation for an acyl transfer reaction**

Write the equation for the acyl transfer reaction between acetic anhydride and ethanol.

Solution

$$CH_3-\overset{\overset{\displaystyle O}{\|}}{C}-O-\overset{\overset{\displaystyle O}{\|}}{C}-CH_3 + CH_3CH_2O-H \longrightarrow CH_3-\overset{\overset{\displaystyle O}{\|}}{C}-OCH_2CH_3 + CH_3-\overset{\overset{\displaystyle O}{\|}}{C}-O-H$$

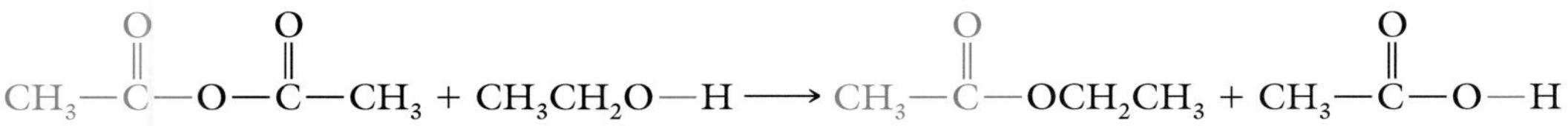

Problem 15.12 Write the equation for the acyl transfer reaction between benzoyl chloride and methanol.

15.13 PHOSPHORIC ACIDS AND THEIR DERIVATIVES

Although carboxylic acid anhydrides are not found in biological systems, other types of acid anhydrides—those derived from phosphoric acids—do play

A PICTURE OF HEALTH

Examples of Carboxylic Acids, Esters, and Other Acid Derivatives

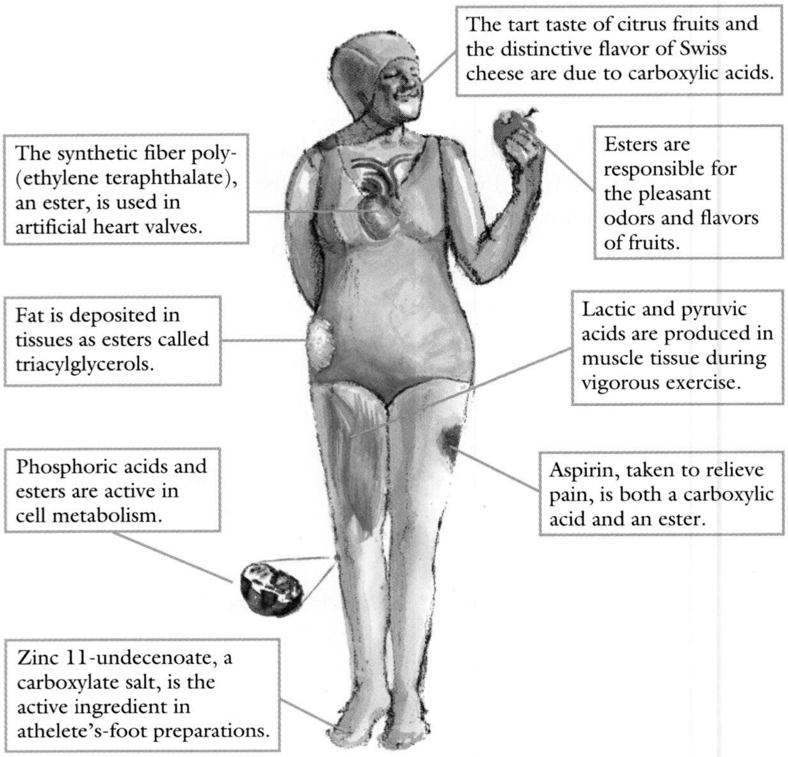

The tart taste of citrus fruits and the distinctive flavor of Swiss cheese are due to carboxylic acids.
The synthetic fiber poly-(ethylene teraphthalate), an ester, is used in artificial heart valves.
Esters are responsible for the pleasant odors and flavors of fruits.
Fat is deposited in tissues as esters called triacylglycerols.
Lactic and pyruvic acids are produced in muscle tissue during vigorous exercise.
Phosphoric acids and esters are active in cell metabolism.
Aspirin, taken to relieve pain, is both a carboxylic acid and an ester.
Zinc 11-undecenoate, a carboxylate salt, is the active ingredient in athelete's-foot preparations.

▶▶ Phospholipids are considered in Section 19.6.

important roles in cell processes. Esters derived from phosphoric acids also are important in living systems—for example, phospholipids are used in constructing cell membranes.

The **phosphoric acids** are a group of three acids: phosphoric acid, diphosphoric acid, and triphosphoric acid. The simplest, **phosphoric acid,** undergoes intermolecular dehydration to form **diphosphoric acid:**

$$\underset{\text{Phosphoric acid}}{\text{HO}-\overset{\overset{\displaystyle\text{O}}{\|}}{\underset{\underset{\displaystyle\text{OH}}{|}}{\text{P}}}-\text{OH}} + \underset{\text{Phosphoric acid}}{\text{HO}-\overset{\overset{\displaystyle\text{O}}{\|}}{\underset{\underset{\displaystyle\text{OH}}{|}}{\text{P}}}-\text{OH}} \xrightarrow{-\text{H}_2\text{O}} \underset{\text{Diphosphoric acid}}{\text{HO}-\overset{\overset{\displaystyle\text{O}}{\|}}{\underset{\underset{\displaystyle\text{OH}}{|}}{\text{P}}}-\text{O}-\overset{\overset{\displaystyle\text{O}}{\|}}{\underset{\underset{\displaystyle\text{OH}}{|}}{\text{P}}}-\text{OH}}$$

Triphosphoric acid is formed by dehydration between diphosphoric and phosphoric acids:

$$\text{HO}-\overset{\overset{\displaystyle\text{O}}{\|}}{\underset{\underset{\displaystyle\text{OH}}{|}}{\text{P}}}-\text{OH} + \text{HO}-\overset{\overset{\displaystyle\text{O}}{\|}}{\underset{\underset{\displaystyle\text{OH}}{|}}{\text{P}}}-\text{O}-\overset{\overset{\displaystyle\text{O}}{\|}}{\underset{\underset{\displaystyle\text{OH}}{|}}{\text{P}}}-\text{OH} \xrightarrow{-\text{H}_2\text{O}} \underset{\text{Triphosphoric acid}}{\text{HO}-\overset{\overset{\displaystyle\text{O}}{\|}}{\underset{\underset{\displaystyle\text{OH}}{|}}{\text{P}}}-\text{O}-\overset{\overset{\displaystyle\text{O}}{\|}}{\underset{\underset{\displaystyle\text{OH}}{|}}{\text{P}}}-\text{O}-\overset{\overset{\displaystyle\text{O}}{\|}}{\underset{\underset{\displaystyle\text{OH}}{|}}{\text{P}}}-\text{OH}}$$

The HO—P═O groups in phosphoric, diphosphoric, and triphosphoric acids have acidic properties similar to those of the HO—C═O group in a carboxylic acid. The phosphoric acids are somewhat stronger acids than carboxylic acids but are still weak acids compared with the strong acids such as HCl, H_2SO_4, HNO_3, and $HClO_4$. Phosphoric, diphosphoric, and triphosphoric acids are polyprotic acids because each —OH group is acidic.

Example 15.13 Writing an equation for the reaction of phosphoric acid with strong base

Write equations to show successive reactions of each of the —OH groups in phosphoric acid with NaOH.

Solution

Each —OH group undergoes acid-base reaction in succession:

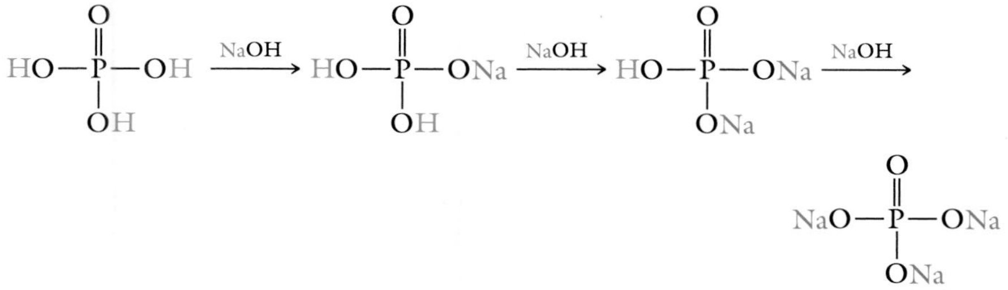

Problem 15.13 Show the final product after diphosphoric acid reacts with excess KOH.

Diphosphoric and triphosphoric acids are anhydrides as well as acids. Notice also that diphosphoric and triphosphoric acids have a close relation to carboxylic acid anhydrides:

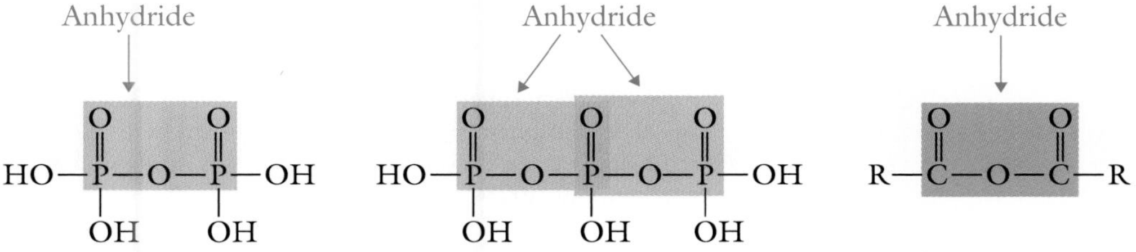

Diphosphoric acid **Triphosphoric acid** **Carboxylic acid anhydride**

The HO—P=O groups in phosphoric, diphosphoric, and triphosphoric acids can undergo esterification with alcohols or phenols analogous to the esterification of the HO—C=O group of a carboxylic acid. The products are called **phosphoric acid esters** or **phosphate esters.** An example is the esterification of diphosphoric acid with ethanol:

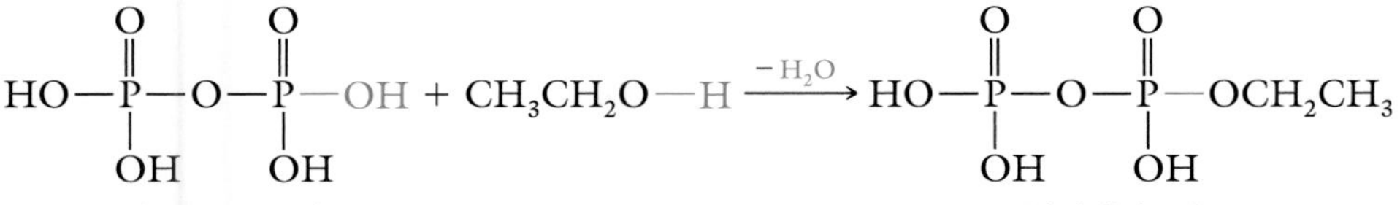

Diphosphoric acid **Ethyl diphosphate**

Phosphate esters are named in the same way as carboxylic esters. The group derived from the alcohol (or phenol) is named followed by the name of the acid with the ending changed from **-ic acid** to **-ate.**

Example 15.14 Writing the equation for phosphate ester formation

Write the equation for esterification of triphosphoric acid with 1-butanol. Name the ester.

Solution

$$HO-\overset{\overset{\displaystyle O}{\|}}{\underset{\underset{\displaystyle OH}{|}}{P}}-O-\overset{\overset{\displaystyle O}{\|}}{\underset{\underset{\displaystyle OH}{|}}{P}}-O-\overset{\overset{\displaystyle O}{\|}}{\underset{\underset{\displaystyle OH}{|}}{P}}-OH \;+\; H-OCH_2CH_2CH_2CH_3 \xrightarrow{-H_2O}$$

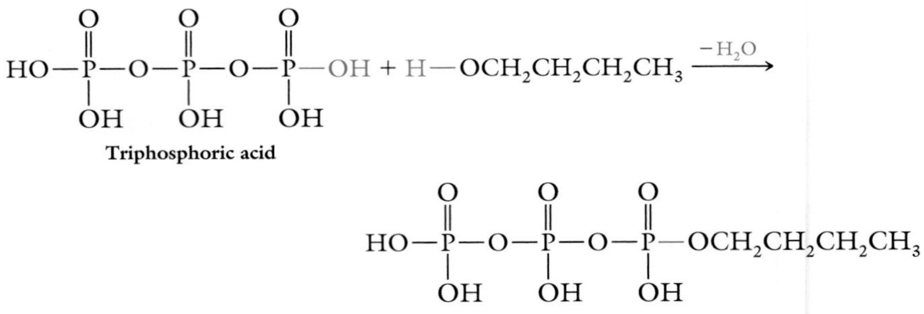

Triphosphoric acid

$$HO-\overset{\overset{\displaystyle O}{\|}}{\underset{\underset{\displaystyle OH}{|}}{P}}-O-\overset{\overset{\displaystyle O}{\|}}{\underset{\underset{\displaystyle OH}{|}}{P}}-O-\overset{\overset{\displaystyle O}{\|}}{\underset{\underset{\displaystyle OH}{|}}{P}}-OCH_2CH_2CH_2CH_3$$

Butyl triphosphate

Problem 15.14 Write the equation for esterification of two of the —OH groups in phosphoric acid with methanol. Name the ester.

Phosphate esters are important in living systems. Many biological molecules that we will encounter in our study of biochemistry are phosphate esters: ATP, NADH/NAD$^+$, phospholipids, nucleic acids (DNA and RNA), and phosphate esters of carbohydrates, to name a few. Some of these compounds, such as ATP (Box 15.5) and NADH (see Figure 14.6), contain both phosphate ester and phosphate anhydride groups.

15.5 Chemistry Within Us

Phosphate Esters in Biological Systems

Adenosine triphosphate (ATP) is the principal molecule in the process by which cells use energy. Acting as an energy-transferring intermediate, it makes the energy obtained from food available for all cell activities (see Section 22.9). Its role in energy transfer requires its hydrolysis to adenosine diphosphate (ADP), a process that releases energy. The phosphate —OH groups are shown here in ionized form because they are ionized under most physiological conditions (pH 7).

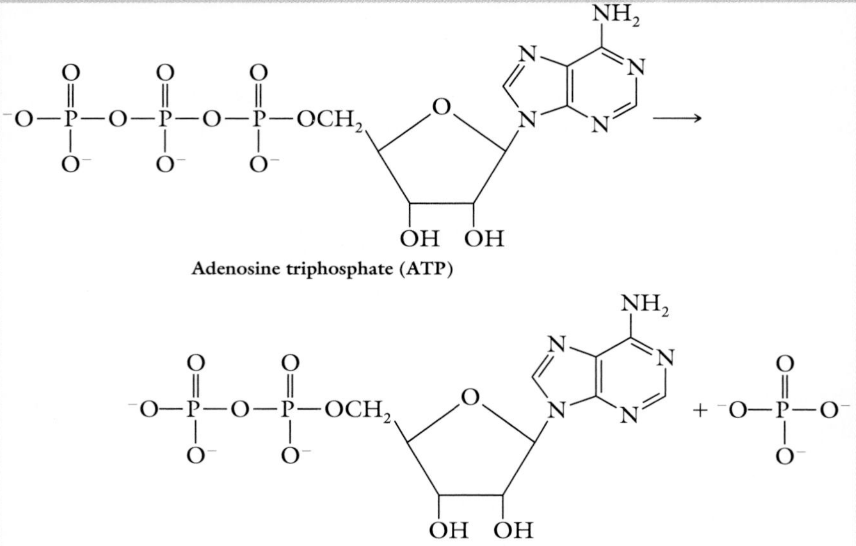

Adenosine triphosphate (ATP)

Adenosine diphosphate (ADP)

Summary

Comparison of Carboxylic Acids and Their Derivatives Carboxylic acids contain the carboxyl group, —COOH, a carbonyl group with an —OH group attached to it. Carboxylic acids and acid derivatives differ in the group attached to the carbonyl carbon.

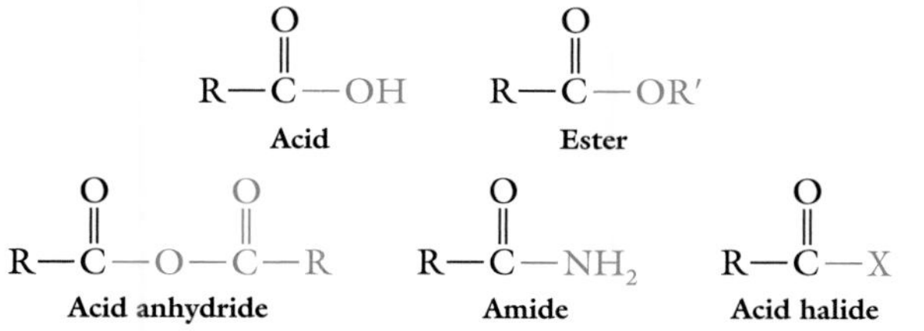

Esters and anhydrides, like carboxylic acids, have an oxygen group bonded to the carbonyl carbon but differ in the type of oxygen group. Esters and anhydrides have —OR′ and —O—CO—R groups, respectively, whereas carboxylic acids have —OH. Amides have nitrogen (in an $—NH_2$, —NHR, or $—NR_2$ group), and acid halides have a halogen connected to the carbonyl carbon.

Carboxylic Acids Carboxylic acids are synthesized by selective oxidation of alkylbenzenes, primary alcohols, and aldehydes.

Carboxylic acids are named by the IUPAC system by naming the longest continuous carbon chain containing the carboxyl group. The ending of the corresponding alkane is changed from -e to -oic acid. The carboxyl carbon is C1. Prefixes preceded by numbers indicate substituents attached to the longest chain. C_6H_5COOH is called benzoic acid.

Carboxylic acids have considerably higher boiling and melting points than those of alcohols because of dimer formation by hydrogen bonding between pairs of acid molecules. Carboxylic acids have good water solubility, slightly higher than that of alcohols, owing to strong hydrogen-bond attractions with water.

Carboxylic acids are weak acids but are stronger acids than phenols and much stronger acids than alcohols. Carboxylic acids and phenols can be distinguished from alcohols by litmus paper.

Carboxylate Salts Carboxylate salts, formed by reaction of carboxylic acids with strong base, are named by naming the positive ion followed by the name of the carboxylic acid with the ending changed from -ic acid to -ate. Carboxylate salts are ionic compounds and therefore have much higher melting and boiling points and water solubility than those of carboxylic acids. Sodium and potassium carboxylate salts derived from unbranched long-chain acids with 12 to 20 carbons find use as soaps.

Carboxylic Acid Esters An ester is formed by acid-catalyzed dehydration between a carboxylic acid and alcohol (or phenol). Esters are named by naming the alkyl (or aryl) substituent of the alcohol (or phenol) followed by the name of the carboxylic acid with the ending changed from -ic acid to -ate. Esters have slightly lower melting and boiling points than those of aldehydes and ketones and about the same water solubilities.

Polyesters are formed by reaction of bifunctional carboxylic acids and bifunctional alcohols (or phenols).

An ester undergoes acidic hydrolysis to the corresponding carboxylic acid and alcohol (or phenol). Basic hydrolysis (saponification) produces the carboxylate salt instead of the carboxylic acid.

Carboxylic Acid Anhydrides and Halides Carboxylic acid anhydrides and halides undergo hydrolysis to form esters. They react with alcohols and phenols to form esters.

Phosphoric Acids and Their Derivatives Phosphoric, diphosphoric, and triphosphoric acids are stronger acids than carboxylic acids but are weaker than the strong acids such as HCl, H_2SO_4, $HClO_4$, and HNO_3. Like carboxylic acids, they undergo esterification with alcohols or phenols.

Summary of Key Reactions

CARBOXYLIC ACIDS

Acidity

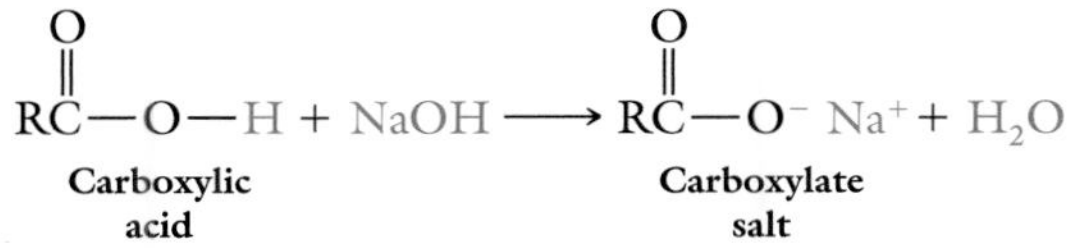

Esterification

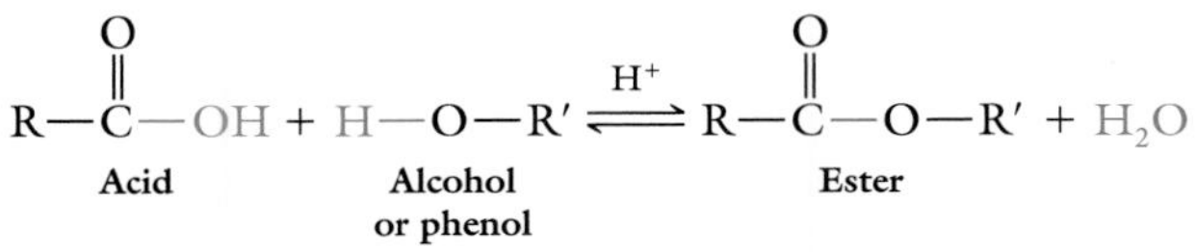

ESTERS

Acidic hydrolysis is the reverse of esterification:

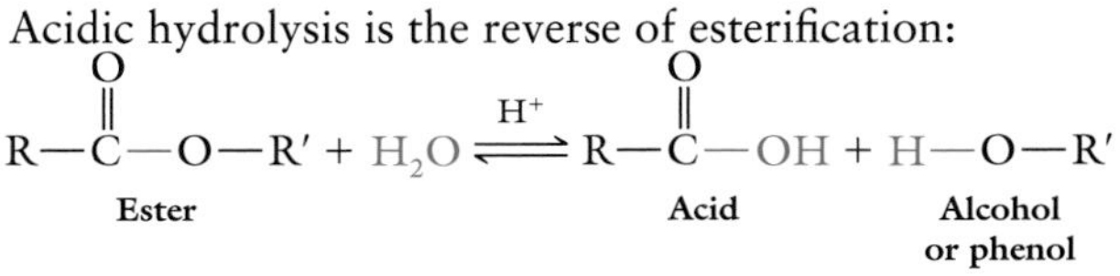

Saponification is hydrolysis with a base:

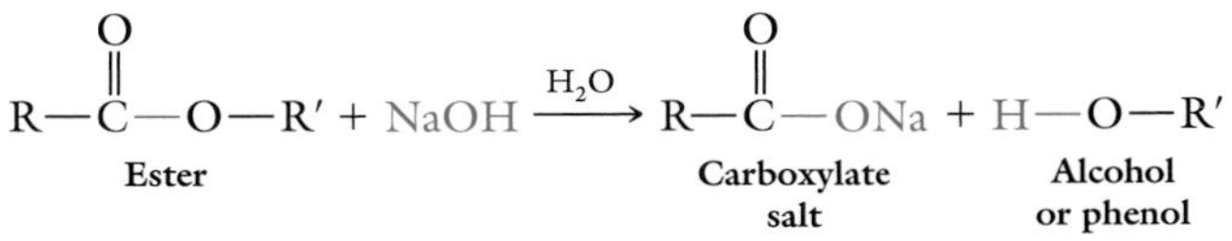

CONDENSATION POLYMERIZATION

$$n\,HO-\overset{O}{\overset{\|}{C}}-R-\overset{O}{\overset{\|}{C}}-OH + n\,HO-R'-OH \xrightarrow[-H_2O]{H^+} \left(\overset{O}{\overset{\|}{C}}-R-\overset{O}{\overset{\|}{C}}-O-R'-O\right)_n$$

CARBOXYLIC ACID ANHYDRIDES (Z = —O—CO—R) AND HALIDES (Z = Cl, Br)

Hydrolysis

$$R-\overset{O}{\overset{\|}{C}}-Z \xrightarrow{H_2O} \underset{\textbf{Carboxylic acid}}{R-\overset{O}{\overset{\|}{C}}-OH} + HZ$$

Esterification

$$R-\overset{O}{\overset{\|}{C}}-Z \xrightarrow{R'OH} \underset{\textbf{Ester}}{R-\overset{O}{\overset{\|}{C}}-OR'} + HZ$$

ESTERIFICATION OF PHOSPHORIC ACIDS

$$-O-\underset{OH}{\underset{|}{\overset{O}{\overset{\|}{P}}}}-OH + ROH \xrightarrow{-H_2O} -O-\underset{OH}{\underset{|}{\overset{O}{\overset{\|}{P}}}}-OR$$

Key Words

acid halide, pp. 434, 452
acyl group, p. 453
acyl transfer, p. 453
amide, p. 434
anhydride, pp. 434, 452
bifunctional reactant, p. 449
carboxylate ion, p. 439
carboxyl group, p. 434
carboxylic acid, p. 434
condensation polymerization, p. 449
diphosphoric acid, p. 454
ester, pp. 434, 445
esterification, p. 445
monomer, p. 449
phosphate ester, p. 455
phosphoric acid, p. 454
phosphoric acid anhydride, p.455
saponification, p. 451
soap, p. 443
thioester, p. 446
triphosphoric acid, p. 454

Exercises

Comparison of Carboxylic Acids and Derivatives

15.1 Classify each of the following compounds as a carboxylic acid, ester, amide, anhydride, acid halide, or something else.

$$\underset{\mathbf{1}}{CH_3-\overset{O}{\overset{\|}{C}}-CH_2-OCH_3} \qquad \underset{\mathbf{2}}{CH_3CH_2-\overset{O}{\overset{\|}{C}}-OCH_3}$$

$$\underset{\mathbf{3}}{C_6H_5-\overset{O}{\overset{\|}{C}}-Cl} \qquad \underset{\mathbf{4}}{CH_3-\overset{O}{\overset{\|}{C}}-CH_2-OH}$$

$$\underset{\mathbf{5}}{CH_3CH_2-\overset{O}{\overset{\|}{C}}-OH} \qquad \underset{\mathbf{6}}{CH_3CH_2-\overset{O}{\overset{\|}{C}}-NH_2}$$

$$\underset{\mathbf{7}}{HO-\overset{O}{\overset{\|}{C}}-CH_2-OCH_3} \qquad \underset{\mathbf{8}}{H-\overset{O}{\overset{\|}{C}}-O-\overset{O}{\overset{\|}{C}}-H}$$

15.2 For each of the following pairs of structural formulas, indicate whether the pair represents (1) the same compound or (2) different compounds that are constitutional isomers or (3) different compounds that are geometrical isomers or (4) different compounds that are not isomers.

(a) $CH_3-\overset{O}{\overset{\|}{C}}-OCH_2CH_3$ and $CH_3CH_2CH_2-\overset{O}{\overset{\|}{C}}-OH$

(b) $CH_3CH_2CH_2CH_2-\overset{O}{\overset{\|}{C}}-OH$ and $CH_3CH_2\overset{CH_3}{\overset{|}{C}H}-\overset{O}{\overset{\|}{C}}-OH$

(c) $CH_3CH_2CH_2CH_2-\overset{O}{\overset{\|}{C}}-OH$ and $CH_3CH_2\overset{CH_3}{\overset{|}{C}H}CH_2-\overset{O}{\overset{\|}{C}}-OH$

(d) $CH_3-\overset{O}{\overset{\|}{C}}-CH_2CH_2OH$ and $CH_3CH_2CH_2-\overset{O}{\overset{\|}{C}}-OH$

15.3 Draw a structural formula for each carboxylic acid with the molecular formula $C_5H_{10}O_2$.

15.4 Draw a structural formula for each ester with the molecular formula $C_4H_8O_2$.

Synthesis of Carboxylic Acids

15.5 Indicate which of the following compounds yield carboxylic acids on selective oxidation. For those that do, write an equation showing the conversion. Use permanganate or dichromate as the oxidizing agent. (a) Isopropylbenzene; (b) 1-pentanol; (c) 2-pentanol.

15.6 Indicate which of the following compounds yield carboxylic acids on selective oxidation. For those that do, write an equation showing the conversion. Use permanganate or dichromate as the oxidizing agent. (a) Butanal; (b) 2-butanone.

Naming Carboxylic Acids

15.7 Name each of the following carboxylic acids:

(a) $CH_3-C(CH_2CH_2CH(CH_3)_2)(COOH)-CH(CH_3)_2$

(b) $(CH_3)_3CCH(Cl)CH_2COOH$

(c)
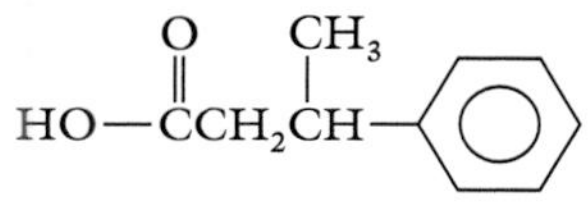

(d) $HOOC-CH(COOH)CH_2CH_3$

15.8 Draw the structural formula for each of the following compounds: (a) 2-ethyl-4-methylbenzoic acid; (b) 2-isopropyl-1, 4-butanedioic acid.

Physical Properties of Carboxylic Acids

15.9 Draw a representation of the hydrogen bonding responsible for the high boiling point of propanoic acid.

15.10 Draw a representation of the hydrogen bonding present when propanoic acid dissolves in water.

15.11 Indicate which compound in each of the following pairs has the higher value of the specified property. If the two compounds in a pair have nearly the same value for the property, indicate that fact. Explain your conclusions.
(a) Ethanoic acid, 1-propanol: boiling point.
(b) Ethanoic acid, 1-propanol: solubility in water.

15.12 Indicate which compound in each of the following pairs has the higher value of the specified property. If the two compounds in a pair have nearly the same value for the property, indicate that fact. Explain your conclusions.
(a) Heptanoic acid, 1-octanol: solubility in water.
(b) Propanoic acid, butanal: boiling point.

Acidity of Carboxylic Acids

15.13 Isomers 1 and 2 both have —OH and C=O groups, but compound 1 is 10,000-fold more acidic than compound 2. Why?

$CH_3CH_2-C(=O)-OH$ (1) $\quad CH_3-C(=O)-CH_2OH$ (2)

15.14 Place the following compounds in order of increasing acidity: phenol, propanol, propanoic acid. Explain the order of acidity in terms of the relative stability of the anion derived from each compound.

15.15 Indicate whether an acid-base reaction takes place under each of the conditions given. If reaction takes place, write the appropriate equation. (a) Ethanoic acid + water; (b) ethanoic acid + NaOH; (c) ethanoic acid + $NaHCO_3$; (d) phenol + water; (e) phenol + NaOH.

15.16 Show the reaction of 1,6-hexanedioic acid, $HOOC(CH_2)_4COOH$, with excess NaOH.

15.17 What is the structure of the predominant species present when propanoic acid is dissolved in water? What structure predominates when propanoic acid is added to a buffered solution at pH = 7 (similar to most physiological pH conditions)?

15.18 Compare the acidity of benzoic acid and hexanoic acid. What is the reason for the difference?

15.19 An unknown compound is either 1-propanol or propanoic acid. The unknown turns blue litmus paper red. Identify the unknown.

15.20 An unknown compound is either 1-propanol or propanoic acid. The unknown does not turn blue litmus paper red. Identify the unknown.

Carboxylate Salts

15.21 Draw the structural formula for each of the following compounds: (a) sodium hexanoate; (b) calcium benzoate.

15.22 Name the following carboxylate salts:

(a) $KO-C(=O)-CH(CH_3)CH_2CH_3$

(b) $NaO-C(=O)CH_2CH_2CH_2CH_2C(=O)-ONa$

15.23 Indicate which compound in each of the following pairs has the higher value of the specified property. If the two compounds in a pair have nearly the same value for the property, indicate that fact. Explain your conclusions.
(a) Sodium propanoate, propanoic acid: melting point.
(b) Sodium hexanoate, octanoic acid: solubility in water.
(c) Sodium hexanoate, octanoic acid: solubility in hexane.

15.24 Give the equations showing sodium propanoate acting as a base toward: (a) water; (b) HCl.

15.25 An unknown is either butanoic acid or sodium butanoate. The unknown turns red litmus paper blue. Identify the unknown.

15.26 An unknown is either butanoic acid or sodium butanoate. The unknown turns blue litmus paper red. Identify the unknown.

Soaps and Their Cleaning Action

15.27 Which of the following compounds is a soap? Explain.

$$\underset{\mathbf{1}}{CH_3(CH_2)_2\overset{O}{\overset{\|}{C}}ONa} \qquad \underset{\mathbf{2}}{CH_3(CH_2)_2\overset{O}{\overset{\|}{C}}OH}$$

$$\underset{\mathbf{3}}{CH_3(CH_2)_{14}\overset{O}{\overset{\|}{C}}ONa} \qquad \underset{\mathbf{4}}{CH_3(CH_2)_{14}\overset{O}{\overset{\|}{C}}OH}$$

15.28 Explain how a soap removes greasy dirt from our bodies. Describe the process in relation to the secondary attractive forces that are operative.

Synthesis of Esters from Carboxylic Acids and Alcohols

15.29 Give the equation for ester formation between each of the following pairs of compounds: (a) pentanoic acid and ethanol; (b) benzoic acid and butanol.

15.30 Give the equation for ester formation between each of the following pairs of compounds: (a) 2-methylpropanoic acid and phenol; (b) pentanoic acid and ethanethiol.

15.31 What carboxylic acid and alcohol or phenol are needed to synthesize each of the following esters?

(a) $(CH_3)_3CCH_2CH_2-\overset{O}{\overset{\|}{C}}-OCH(CH_3)_2$

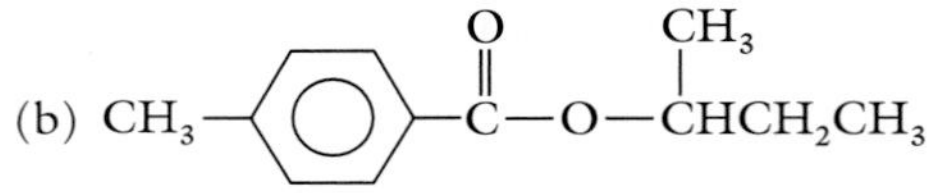

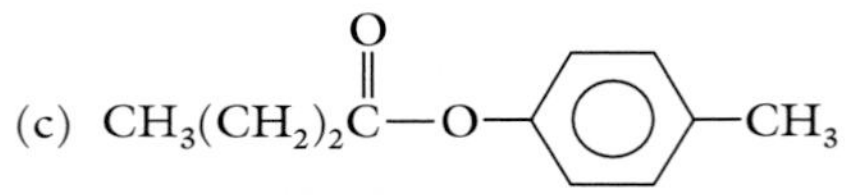

15.32 What carboxylic acid and alcohol or phenol are needed to synthesize each of the following esters?

(a) $CH_3CH_2O-\overset{O}{\overset{\|}{C}}-CH_2CH_2CH_2-\overset{O}{\overset{\|}{C}}-OCH_2CH_3$

(b) $CH_3CH_2\overset{O}{\overset{\|}{C}}-O-CH_2CH_2CH_2-O-\overset{O}{\overset{\|}{C}}CH_2CH_3$

Names and Physical Properties of Esters

15.33 Draw the structural formula for each of the following esters: (a) *t*-butyl butanoate; (b) cyclohexyl benzoate.

15.34 Name the following compounds:

(a) $CH_3CH_2CH_2CH_2CH_2\overset{O}{\overset{\|}{C}}-O-CH_2\overset{CH_3}{\overset{|}{C}}HCH_3$

(b) $(CH_3)_2CHO-\overset{O}{\overset{\|}{C}}-\overset{O}{\overset{\|}{C}}-OCH(CH_3)_2$

15.35 Indicate which compound in each of the following pairs has the higher value of the specified property. If the two compounds in a pair have nearly the same value for the property, indicate that fact. Explain your conclusions.
(a) Methyl propanoate, pentanal: boiling point.
(b) Methyl butanoate, hexanal: solubility in water.
(c) Methyl propanoate, 1-pentanol: boiling point.
(d) Methyl pentanoate, hexanoic acid: solubility in water.
(e) Methyl propanoate, butanoic acid: boiling point.

15.36 Which of the following representations describe the hydrogen bonding present when methyl formate dissolves in water?

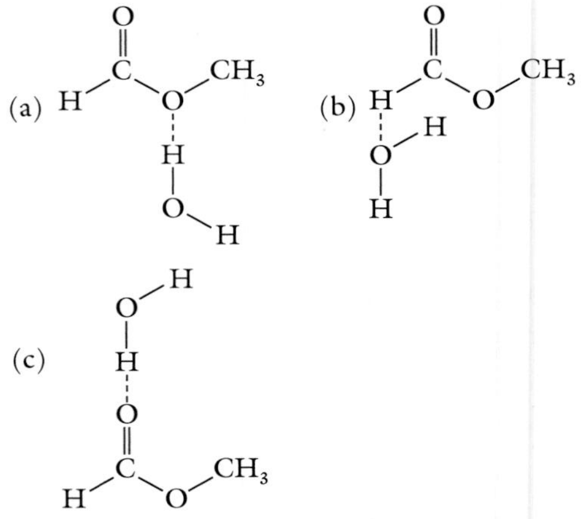

Polyester Synthesis

15.37 What reactants are needed to synthesize the following polyester?

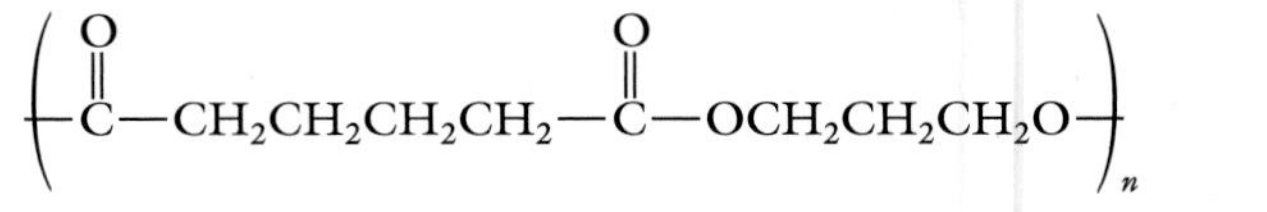

15.38 Draw the structure of the polyester formed from the reaction of 1,4-butanedioic acid and 1,4-dihydroxybenzene.

Hydrolysis of Esters

15.39 Write the equation for each of the following reactions: (a) acidic hydrolysis of propyl propanoate; (b) saponification of phenyl butanoate with NaOH.

15.40 Give the product(s) formed in each of the following reactions. If no reaction takes place, write "no reaction." If more than one product is formed, indicate only the major product(s).

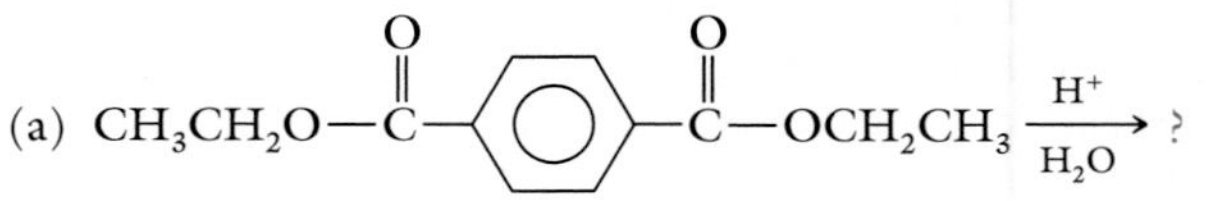

(b) $CH_3CH_2\overset{O}{\overset{\|}{C}}-O-CH_2CH_2CH_2-O-\overset{O}{\overset{\|}{C}}CH_2CH_3 \xrightarrow[H_2O]{NaOH} ?$

Carboxylic Acid Anhydrides and Halides

15.41 Draw the structural formula for each of the following compounds: (a) 2-methylpropanoyl chloride; (b) ethanoic anhydride.

15.42 Name the following compounds:

(a) $C_6H_5-\overset{O}{\overset{\|}{C}}-Br$

(b) $CH_3CH_2-C(=O)-O-C(=O)-CH_2CH_3$

15.43 Write an equation to show the reaction of benzoyl chloride with water.

15.44 Write an equation to show the reaction of benzoic anhydride with water.

Phosphoric Acids and Their Derivatives

15.45 Write the equation for esterification of diphosphoric acid with an equimolar amount of 1-butanol. Name the ester.

15.46 Give the structural formula of isopropyl triphosphate.

15.47 Give the structural formula for the form of diphosphoric acid present at pH = 7.

15.48 Give the structural formula for the form of isopropyl triphosphate present at pH = 7.

Unclassified Exercises

15.49 A cyclic, nonaromatic compound containing an attached carboxyl group is named by naming the cyclic compound exclusive of the —COOH group and then adding the suffix **-carboxylic acid.** Draw the structure of *cis*-2-methylcyclopentanecarboxylic acid.

15.50 Carboxylic acids with a C=C in the longest chain containing the carboxyl group are named as alkenoic acids. Draw the structure of *trans*-2-pentenoic acid.

15.51 A compound that contains both hydroxyl and carboxyl groups undergoes intramolecular ester formation. The product is called a **lactone.** Show the lactone formed from 5-hydroxypentanoic acid.

15.52 Draw structural formulas for: (a) an unbranched carboxylic acid of molecular formula $C_5H_{10}O_2$; (b) a branched carboxylic acid of molecular formula $C_5H_{10}O_2$ and containing an isopropyl group; (c) a lactone (cyclic ester) of four carbons with all carbons in the ring; (d) an ester of molecular formula $C_5H_{10}O_2$ and containing two ethyl groups; (e) a carboxylic acid of molecular formula $C_6H_{12}O_2$ and containing a *t*-butyl group; (f) an anhydride derived from a carboxylic acid of three carbons.

15.53 Draw the structural formula for each of the following compounds: (a) *trans*-3-chlorocyclohexanecarboxylic acid; (b) calcium acetate; (c) isopropyl benzoate; (d) pentanoic anhydride; (e) propanoyl chloride; (f) methyl diphosphate; (g) dimethyl phosphate.

15.54 Name the following compounds:

(a) $CH_3CH_2CH_2C(=O)-OCH_2CH_2CH_3$

(b) $CH_3CH_2CH(Cl)CH_2CH_2C(=O)-ONa$

(c) $HO-P(=O)(OH)-O-P(=O)(OCH_2CH_3)-OCH_2CH_3$

(d) 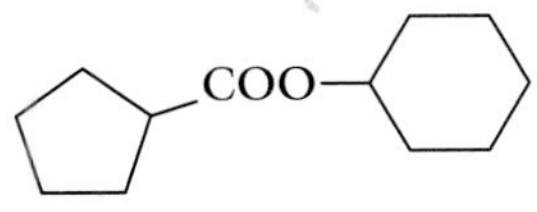

15.55 What are the hybridizations of atoms a, b, c, and d in the following compound? What are bond angles A, B, and C?

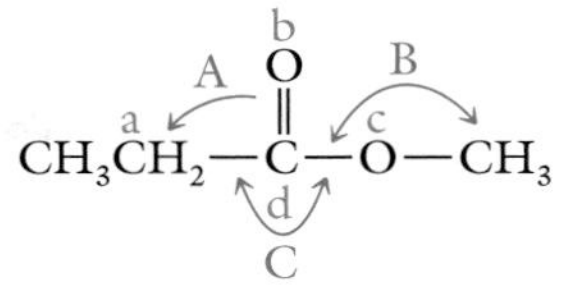

15.56 Indicate which compound in each of the following pairs has the higher value of the specified property. If the two compounds in a pair have nearly the same value for the property, indicate that fact. Explain your conclusions.
(a) Methyl butanoate, hexanal: boiling point.
(b) Methyl butanoate, 2-hexanone: solubility in water.
(c) Methyl propanoate, 1-pentanol: boiling point.
(d) Methyl propanoate, butanoic acid: boiling point.

15.57 An unknown compound is either phenol, propanol, or propanoic acid. The unknown turns blue litmus paper red. Identify the unknown.

15.58 An unknown compound is either phenol, propanol, or propanoic acid. The unknown does not turn blue litmus paper red. Identify the unknown.

15.59 Give the equation for esterification of acetic acid with 1-butanethiol.

15.60 Give the product formed when the ester group in the following lactone (cyclic ester) undergoes acid hydrolysis:

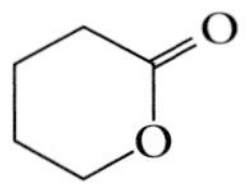

15.61 Give the product(s) formed in each of the following reactions. If no reaction takes place, write "no reaction." If more than one product is formed, indicate only the major product.

(a) $CH_3CH_2C(=O)-OCH_2C_6H_5 \xrightarrow{H_2O} ?$

(b) $CH_3CH_2C(=O)-OCH_2C_6H_5 \xrightarrow[H^+]{H_2O} ?$

(c) $CH_3CH_2C(=O)-OCH_2C_6H_5 \xrightarrow[NaOH]{H_2O} ?$

(d) 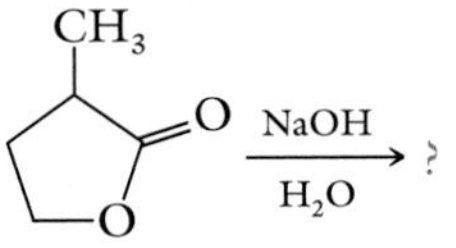

(e) $HO-P(=O)(OH)-O-P(=O)(OH)-O-P(=O)(OH)-OH \xrightarrow{CH_3CH_2CH_2OH} ?$

(f) $CH_3CH_2CH_2C(=O)-OH \xrightarrow{NaHCO_3} ?$

(g) $(CH_3)_2CHCH_2CH_2C(=O)-OH \xrightarrow[H^+]{C_6H_5OH} ?$

$$\text{(h)}\quad CH_3CH_2-\overset{\overset{\displaystyle O}{\|}}{C}-O-\overset{\overset{\displaystyle O}{\|}}{C}-CH_2CH_3 \xrightarrow{H_2O} ?$$

$$\text{(i)}\quad CH_3CH_2-\overset{\overset{\displaystyle O}{\|}}{C}-Cl \xrightarrow{H_2O} ?$$

$$\text{(j)}\quad CH_3CH_2\overset{\overset{\displaystyle O}{\|}}{C}-SCH_2C_6H_5 \xrightarrow[H^+]{H_2O} ?$$

15.62 The ester $C_5H_{10}O_2$ yields ethanol on acid-catalyzed hydrolysis. Identify the other product(s) of hydrolysis.

15.63 An unknown compound is either an ester or a carboxylic acid. What simple chemical diagnostic test can be used to distinguish between the two possibilities?

Chemical Connections

15.64 Low-molecular-mass carboxylic acids have distinct and often disagreeable odors. Neutralization of such acids with NaOH or other base results in a much less intense odor. Why?

15.65 Nitroglycerine is used as an explosive and to relieve chest pain due to angina pectoris (see Box 13.1). Nitroglycerine is the trinitroester of glycerol (also called glycerine), formed by dehydration between each of the —OH groups of glycerol with a molecule of nitric acid. Write an equation to show the synthesis of nitroglycerine from glycerol and nitric acid.

$$\begin{array}{lcl} CH_2-OH & & \\ | & & \\ CH-OH & \qquad & HO-NO_2 \\ | & & \\ CH_2-OH & & \\ \textbf{Glycerol} & & \textbf{Nitric acid} \end{array}$$

15.66 Example 15.9 shows the conversion of terephthalic acid into the important plastic and fiber poly(ethylene terephthalate). Terephthalic acid is produced from *p*-xylene (*p*-dimethylbenzene). What reaction converts *p*-xylene into terephthalic acid?

15.67 By reference to Box 9.1, calculate the $[H_3O^+]$ and pH of a 1.00 M aqueous solution of acetic acid. K_a for acetic acid is 1.74×10^{-5}.

15.68 Two chemists at the XYZ Pharmaceutical Company are asked to devise synthetic strategies for the production of the active ingredient in aspirin formulations, acetylsalicylic acid (see Box 15.4). Betty Brown proposes esterification of salicylic acid by acetic anhydride, whereas John Smith proposes esterification of salicylic acid with methanol. John Smith's proposal is rejected because it produces the wrong ester. Betty Brown is placed in charge of the project to produce aspirin with the use of her proposed strategy. What is wrong with John Smith's strategy? Why is Betty Brown's strategy correct?

CHAPTER 16

AMINES AND AMIDES

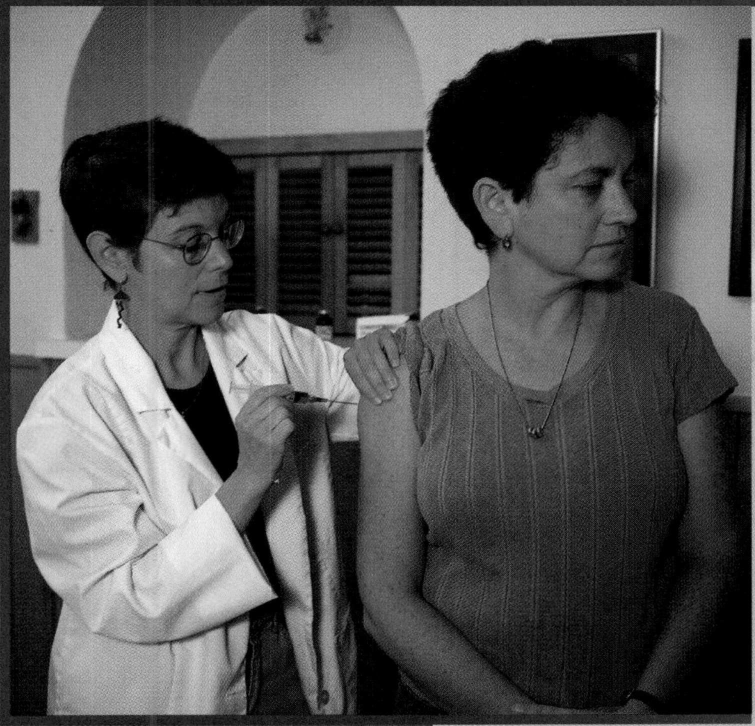

CHEMISTRY IN YOUR FUTURE

Mrs. Carter is suffering badly from hay fever this season, and her doctor has prescribed an injection of diphenhydramine hydrochloride (Benedryl). As you give her the injection, she asks if this drug is the same as the diphenhydramine ointment (Caladryl) that the doctor prescribed for her son's poison ivy last year. You tell her that, yes, the active ingredients in the injection and ointment are essentially the same and that you think that, when "hydrochloride" is attached to the name of a drug, it has something to do with making that particular compound more soluble in body fluids. Is this assumption of yours true, and, if so, to what kinds of compounds does it apply and why? The answer is contained in this chapter on amines and amides.

LEARNING OBJECTIVES

- Draw and name amines.
- Describe the physical properties of amines.
- Explain and write equations to demonstrate the basicity of amines.
- Draw and name amine salts and describe their properties.
- Draw and name amides.
- Describe the physical properties of amides.
- Describe and write equations for the synthesis and hydrolysis of amides.

We now turn our attention from organic compounds containing oxygen to those containing nitrogen. These compounds are the amine and amide families—two families that play major roles in biological systems. Amino acids, the building blocks of proteins, are both amines and carboxylic acids; the proteins themselves are amides. The nucleic acids, deoxyribonucleic (DNA) and ribonucleic (RNA) acids, which are the molecules responsible for transmitting and expressing hereditary information, contain amine and amide groups (along with carbohydrate and phosphate ester groups). Some of the lipids that make up cell membranes contain amine groups. The polysaccharide chitin—the main structural component of the exoskeletons of crustaceans, insects, and spiders—contains amide groups.

A wide range of naturally occurring and synthetic amines and amides are physiologically active; that is, they affect the control and regulation of organs, tissues, or cells. Among them are the hormones adrenaline and melatonin; antibiotics, such as penicillin and tetracyclin; and a vast number of substances used for medicinal and other purposes: caffeine, nicotine, narcotics, anesthetics, antidepressants, high blood pressure medications, and tranquilizers. An amide of enormous commercial importance is the textile fiber and plastic known as nylon.

16.1 COMPARISON OF AMINES AND AMIDES

Amines and **amides** are organic derivatives of ammonia in which one or more of the hydrogens are replaced by organic groups (see below). The bonding of nitrogen directly to a carbonyl carbon forms the **amide group,** which distinguishes an amide from an amine and makes their properties quite different. In the amine structure, two of the R groups can be H; all three R groups can be H in the amide. The R groups can be aliphatic or aromatic in both amines and amides.

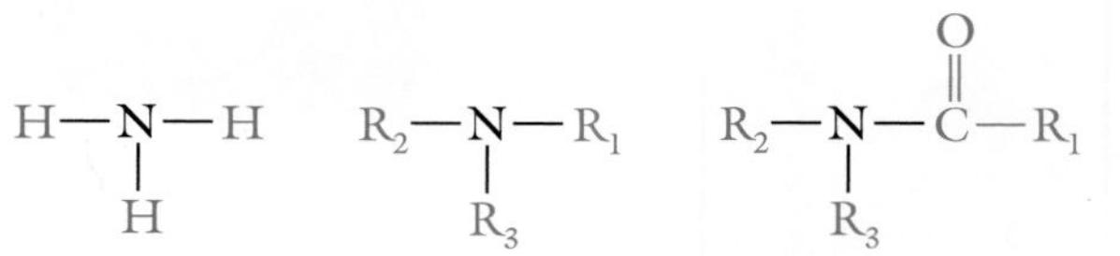

Ammonia **Amine** **Amide**

◀◀ *sp*³ orbitals were described in Section 11.3.

In forming the single bonds in ammonia, amines, and amides, nitrogen is sp^3 hybridized—like the oxygen in ethers and alcohols and the carbons in alkanes. However, the three elements differ in an important way. Nitrogen atoms contain one more electron than carbon and one less electron than oxygen. Carbon is tetravalent with no nonbonding electrons, nitrogen is trivalent with one pair of nonbonding electrons, and oxygen is divalent with two pairs of nonbonding electrons. The nonbonded electron pairs of nitrogen and oxygen, shown below in red, are omitted from most structural drawings but are understood to be present. These nonbonded electrons contribute to the properties of amines and amides.

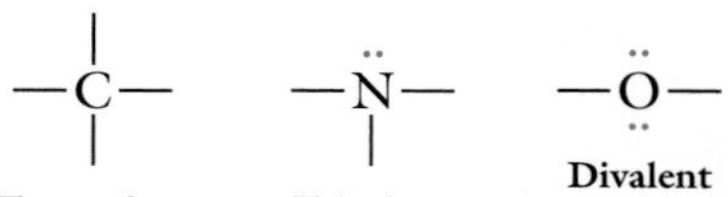

Tetravalent **Trivalent** **Divalent**

The bond angles about the nitrogen atom in an amine are close to 107°. This is slightly distorted from the tetrahedral bond angle of 109.5° because the nonbonded pair of electrons repulse the adjacent electron pairs and compress their bond angles (Section 3.10). The bond angles in amides differ from those in amines (Section 16.11).

16.2 CLASSIFICATION OF AMINES

Amines are classified according to the number of carbons directly bonded to the nitrogen. **Primary, secondary,** and **tertiary amines** have one, two, and three carbons directly bonded to the nitrogen. Put another way, 1°, 2°, and 3° amines are derived from ammonia by replacement of one, two, and three of the hydrogens by aliphatic or aryl groups.

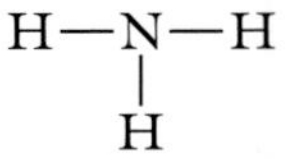

Ammonia

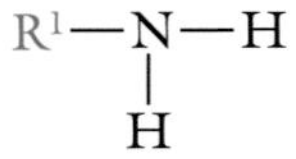

1° Amine

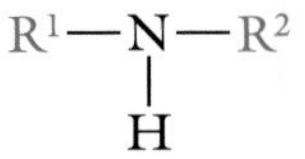

2° Amine

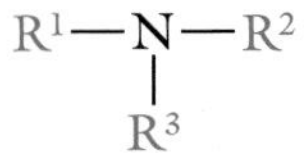

3° Amine

Amines are also classified as aromatic (aryl) or aliphatic amines. An **aromatic (aryl) amine** is an amine in which at least one benzene or other aromatic group is directly bonded to nitrogen. An **aliphatic amine** is an amine in which no benzene or other aromatic groups are directly bonded to nitrogen.

Example 16.1 Classifying amines

Classify each of the following compounds as (a) amines, amides, or something else; (b) primary, secondary, or tertiary, if amines; (c) aliphatic or aromatic, if amines.

$CH_3CH_2CH_2CH_2-NH_2$

1

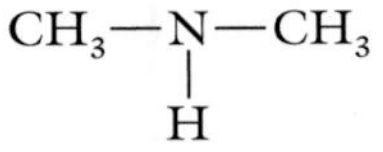

2

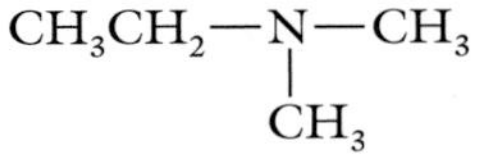

3

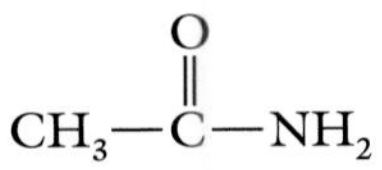

4

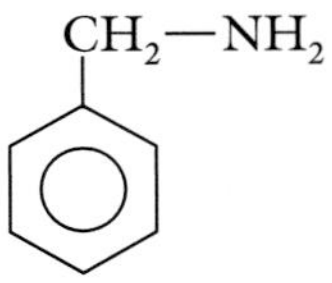

5

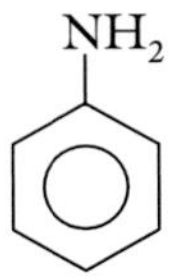

6

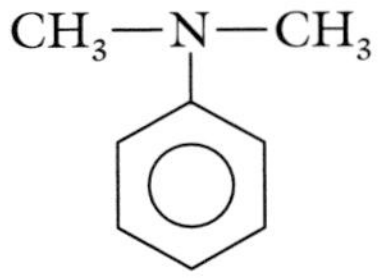

7

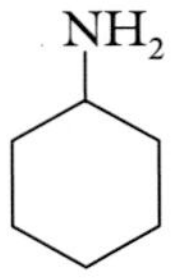

8

Solution

(a) All compounds except 4 are amines because at least one organic group, but no carbonyl carbon, is attached to nitrogen. Compound 4 is an amide because a carbonyl carbon is attached to nitrogen.

(b) Compounds 1, 5, 6, and 8 are primary amines because only one carbon is directly attached to nitrogen. Compound 2 is a secondary amine because two carbons are directly attached to nitrogen. Compounds 3 and 7 are tertiary amines because three carbons are directly attached to nitrogen.

(c) Compounds 1–3, 5, and 8 are aliphatic amines because no aromatic group is bonded directly to nitrogen. Compounds 6 and 7 are aromatic amines because an aromatic group is directly bonded to nitrogen. Note that, in compound 7, both aliphatic and aromatic groups are bonded to nitrogen. Such amines are classified as aromatic amines. Compound 5 is an aliphatic amine because the aromatic group is not directly bonded to nitrogen. Compound 8 is not an aromatic amine, because the ring is an aliphatic ring, not an aromatic one.

Problem 16.1 Classify each of the following compounds as (a) amines, amides, or something else; (b) primary, secondary, or tertiary, if amines; (c) aliphatic or aromatic, if amines.

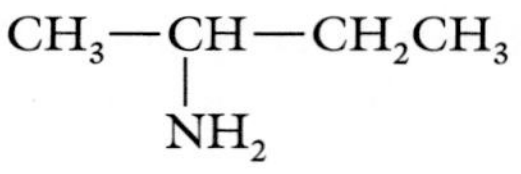
1

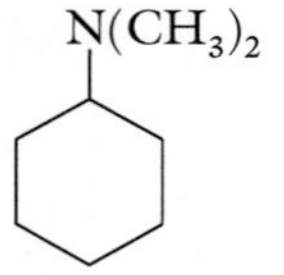
2

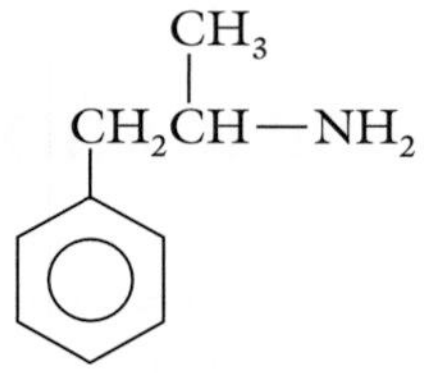
3

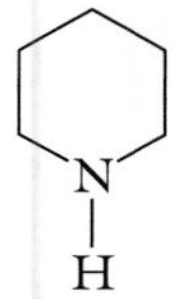
4

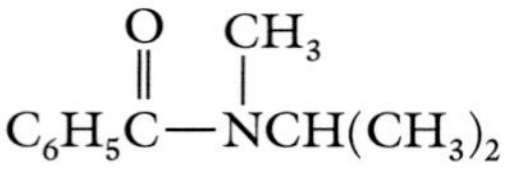
5

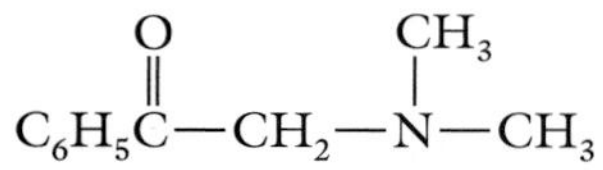
6

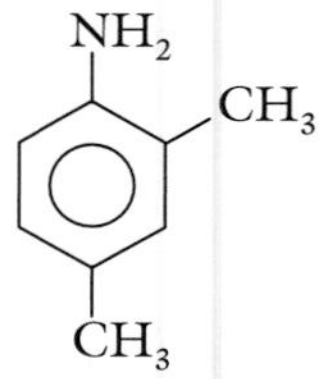
7

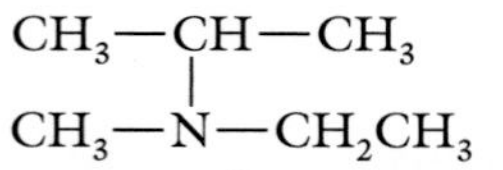
8

Constitutional isomerism in amines, which is analogous to that in other families (Sections 12.3 and 13.2), is possible from two kinds of connectivity differences:

- Different carbon skeletons
- Different placements of the nitrogen on a carbon skeleton

Example 16.2 Drawing constitutional isomers

Draw the constitutional isomers of C_3H_9N.

Solution

The three carbons can make up one group, either a propyl or an isopropyl group, attached to nitrogen:

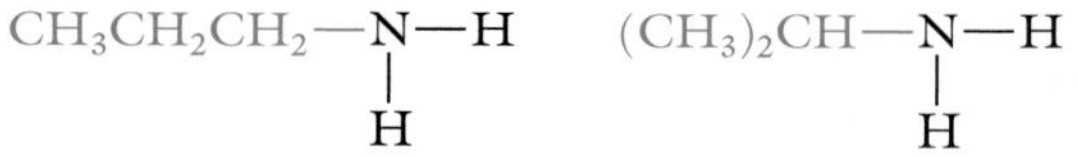

Alternately, the three carbons need not be part of the same group. We can have either three methyl groups or one methyl and one ethyl group attached to nitrogen:

Problem 16.2 Draw the constitutional isomers of $C_4H_{11}N$.

16.3 NAMING AMINES

The systems for naming amines are more varied than for any other family. Amines containing only the simple groups are best named by the common nomenclature system. The simple groups consist of: the unbranched alkyl groups from methyl, ethyl, propyl, and butyl through decyl; the branched three- and four-carbon alkyl groups (isopropyl, isobutyl, *s*-butyl, *t*-butyl); and the unsubstituted cycloalkyl groups, such as cyclohexyl and cyclopentyl. In this common system, the groups (other than H) attached to nitrogen are named in alphabetical order followed without space by the suffix **-amine;** for example:

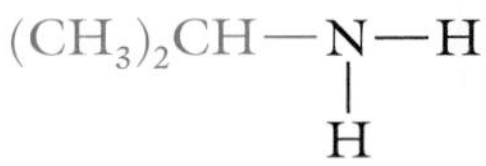

Isopropyl**amine**

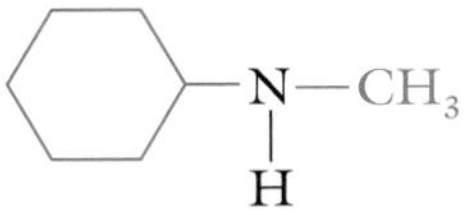

Cyclohexylmethyl**amine**

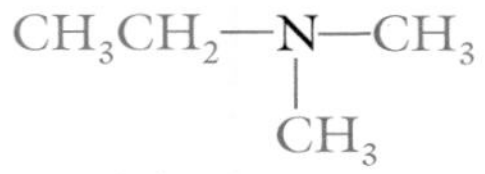

Ethyldimethyl**amine**

$C_6H_5NH_2$ is called **aniline,** and substituted anilines are named as derivatives of aniline. Substituents other than H on nitrogen are distinguished from substituents on the aromatic ring by using an *N*- prefix instead of a number or an *o*-, *m*-, or *p*- prefix; for example:

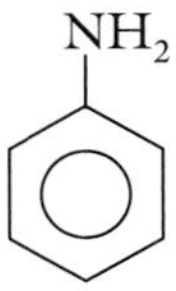

Aniline

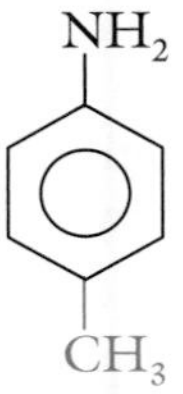

p-Methyl**aniline**
(4-Methylaniline)

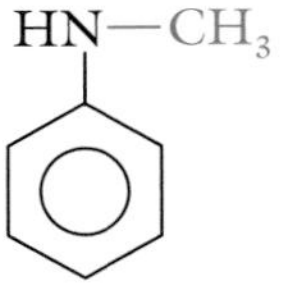

N-Methyl**aniline**

Because the IUPAC system for naming amines is not consistent with IUPAC nomenclature for other organic families, the most widely used system for naming amines is that recommended by ***Chemical Abstracts,*** *CA* for short. Published by the American Chemical Society, *CA* is a journal that abstracts the world's chemical literature. The *CA* rules for naming amines are similar to the IUPAC rules for naming alcohols:

Rules for naming amines

- Name primary amines on the basis of the longest continuous chain that has the attached nitrogen.
- The amine is named as an **alkanamine** by changing the ending of the name of the alkane from **-e** to **-amine.** The position of attachment of nitrogen to the carbon chain is indicated by a number in front of the name. The carbon chain is numbered from the end nearest the nitrogen.
- Prefixes with numbers are used to indicate substituents attached to the longest chain.
- Compounds with two NH_2 groups are named as alkanediamines.
- A cyclic, nonaromatic compound containing an attached NH_2 group is named as a **cycloalkanamine.** The ring carbon holding the nitrogen is C1.
- Name secondary and tertiary amines on the basis of the group with the longest carbon chain attached to nitrogen. The names of the other groups attached to nitrogen are placed in front of the alkanamine or cycloalkanamine name. *N*- and *N,N*- prefixes are used to indicate that the other groups are attached to nitrogen.

Although the *CA* nomenclature system uses the name benzenamine for $C_6H_5NH_2$, we will use the more widely accepted name aniline from the common nomenclature system.

Example 16.3 Naming amines

Give the common name or *CA* name, whichever is appropriate, for each of the following compounds:

(a) 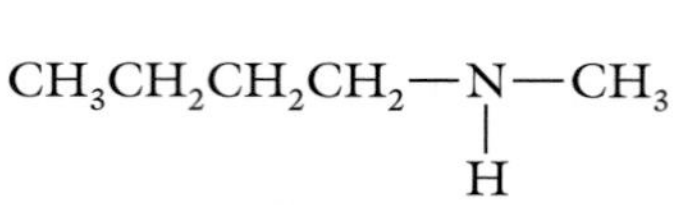

(b) 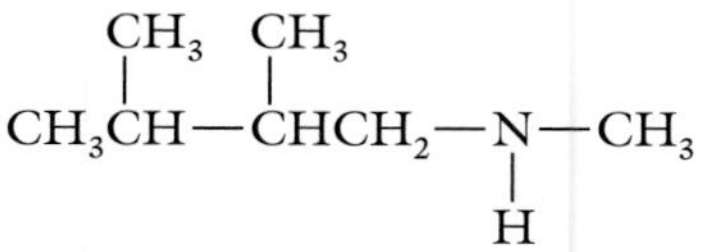

(c) 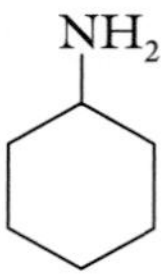

(d) 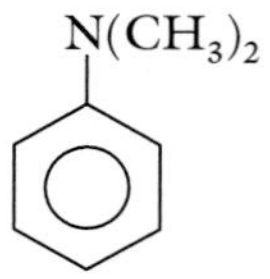

(e) 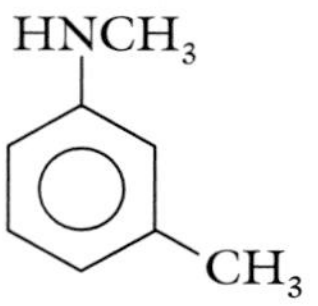

Solution

(a) Because only simple groups, butyl and methyl, are attached to nitrogen, the common name is used: butylmethylamine.

(b) The common nomenclature system is not applicable, because the larger group is not one of the simple groups. The *CA* nomenclature system must be used. The longest chain attached to the nitrogen is four carbons, with nitrogen attached at C1. The name is 2,3,*N*-trimethyl-1-butanamine.

(c) The common name, cyclohexylamine, is used because the cyclohexyl group is a simple group. (The *CA* name is cyclohexanamine.)

(d) The common name is used: *N,N*-dimethylaniline.

(e) The compound is a substituted aniline with one methyl group on nitrogen and one methyl substituted on the ring. The name is *m,N*-dimethylaniline (or 3,*N*-dimethylaniline).

Problem 16.3 Give the common name or *CA* name, whichever is appropriate, for each of the following compounds:

(a) 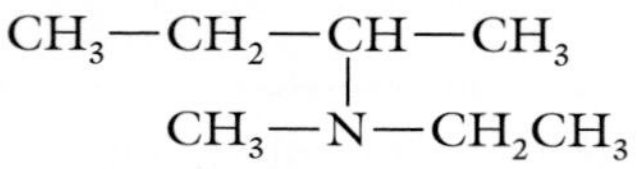

(b) 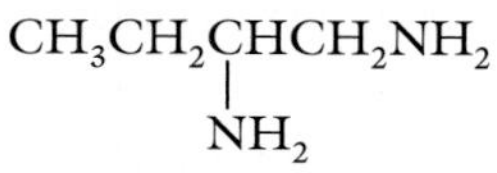

(c) 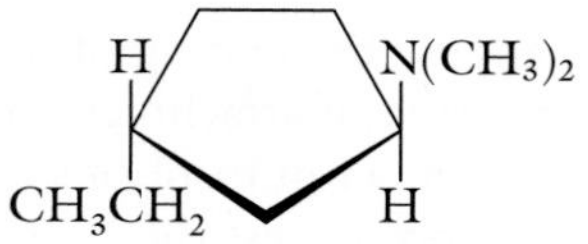

(d) 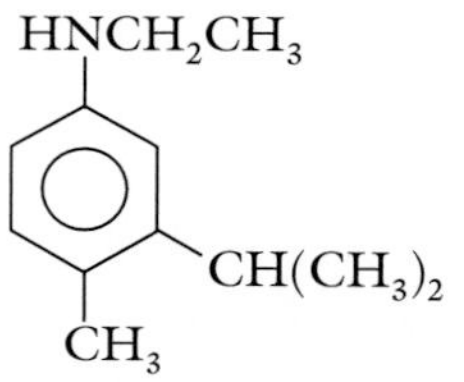

(e) 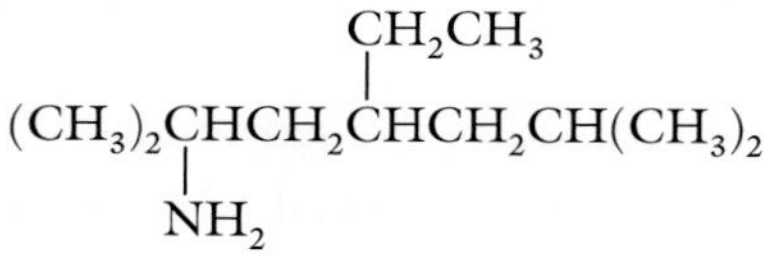

4-Aminobutanoic acid

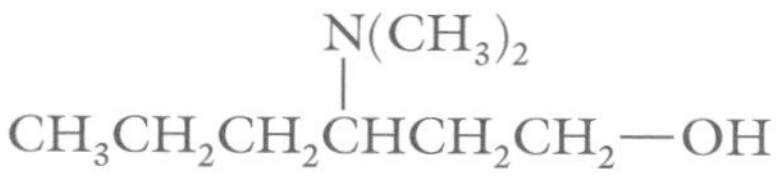

3-(*N,N*-Dimethylamino)-1-hexanol

The groups $—NH_2$, —NHR, and $—NR_2$ are called amino, *N*-alkylamino, and *N,N*-dialkylamino groups, respectively. These names are used as prefixes in the names of compounds that contain both an amine group and an oxygen-containing functional group such as —COOH or —OH. Such compounds are named not as amines but as members of the family represented by the oxygen-containing group. In other words, the oxygen-containing functional groups have preference over nitrogen for nomenclature purposes. Examples are 4-aminobutanoic acid and 3-(*N,N*-dimethylamino)-1-hexanol.

Cyclic compounds containing one or more nitrogen atoms in the ring (rather than attached to the ring) are called **nitrogen heterocyclics** or **hetero-**

TABLE 16.1 Common Names of Heterocyclic Structures

Name	Structure	Present in
imidazole	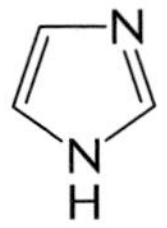	histidine (amino acid)
piperidine	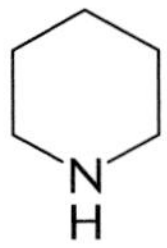	quinine and other drugs
purine	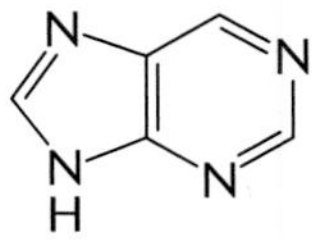	nucleic acids (DNA, RNA)
pyridine	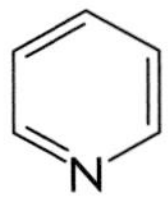	nicotine, vitamin B
pyrimidine	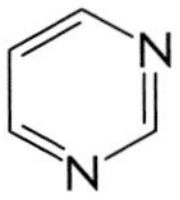	nucleic acids (DNA, RNA)
pyrrole	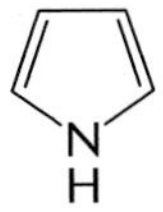	hemoglobin
pyrrolidine	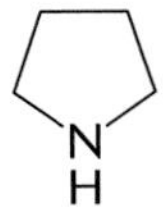	nicotine, proline (amino acid)

cyclic amines. Table 16.1 lists the most important types of nitrogen heterocyclics, many of which play important roles in biological systems.

Alkaloids are amines produced by plants. They are usually heterocyclic amines, taste bitter, and have significant physiological effects. Alkaloids are used, either as the plant part itself or as the extracted compounds, for medicinal and other physiological purposes. Their effects can be put to good use or they can be disastrous and life threatening when used unwisely (Box 16.1 on page 471).

Alkaloids are but one group of physiologically active amines. More physiologically active compounds, including synthetic and natural medications, belong to the amine family than to any other (Boxes 16.2 on page 473 and 16.3 on page 474). Some physiologically active compounds are amides, often in addition to being amines, and many also possess other functional groups such as ether, ester, or alcohol.

16.4 PHYSICAL PROPERTIES OF AMINES

The physical properties of an amine depend on whether it is a primary, secondary, or tertiary amine:

- Primary and secondary amines have melting and boiling points comparable to those of aldehydes and ketones (Table 16.2 on the next page).

TABLE 16.2 Comparison of Boiling Points of Amines and Compounds from Different Families

Compound	Structure	Molecular Mass (amu)	Boiling Point (°C)
butane	$CH_3CH_2CH_2CH_3$	58	− 1
2-methylpropane	$(CH_3)_3CH$	58	− 12
trimethylamine (3°)	$(CH_3)_3N$	59	3
N-methylethylamine (2°)	$CH_3CH_2NHCH_3$	59	36
propylamine (1°)	$CH_3CH_2CH_2NH_2$	59	47
propanal	CH_3CH_2CHO	58	49
methyl formate	$HCOOCH_3$	60	32
2-propanone	CH_3COCH_3	58	56
1-propanol	$CH_3CH_2CH_2OH$	60	97
ethanoic acid	CH_3COOH	60	118
triethylamine (3°)	$(CH_3CH_2)_3N$	101	89
dipropyl ether	$(CH_3CH_2CH_2)_2O$	102	91
heptane	$CH_3(CH_2)_5\,CH_3$	100	98

- Tertiary amines have melting and boiling points considerably lower than those of primary and secondary amines and comparable to those of ethers and hydrocarbons.
- All amines have water solubility comparable to those of aldehydes and ketones and slightly lower than that of alcohols.

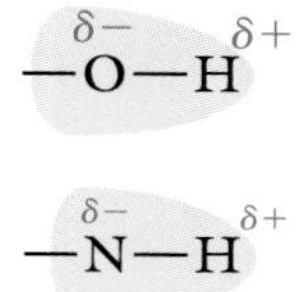

O–H bond is more polar than N–H bond

Primary and secondary amines, unlike tertiary amines, possess N–H bonds and participate in hydrogen bonding (Figure 16.1). This hydrogen bonding results in strong secondary forces and high melting and boiling points, slightly higher for primary amines than for secondary amines because primary amines have two N–H bonds, whereas secondary amines have only one. The N–H bond is less polar than the O–H bond, because nitrogen is less electronegative than oxygen and, therefore, the hydrogen bonding in primary and secondary amines is not as strong as that in alcohols. For this reason, alcohols have higher melting and boiling points than those of primary and secondary amines. The strength of secondary forces due to N–H hydrogen bonding is comparable to that of the dipole–dipole forces in aldehydes and ketones (see Table 16.2).

Tertiary amines have much weaker secondary forces than those of primary and secondary amines because tertiary amines possess only the weakly polar C–N bonds, no N–H bonds. The strength of these attractive forces is comparable to those in ethers and hydrocarbons.

Many amines have distinct and disagreeable fishy odors. The common names of some amines are based on their origin and odor—for example, putrescine (1,4-butanediamine) and cadaverine (1,5-pentanediamine) are found in rotting meat.

Amines of as many as three or four carbons are soluble in water in all proportions, with solubility dropping off as the number of carbons increases. Solubility is very low or negligible for amines with more than six or seven carbons. Tertiary amines are slightly less water soluble than primary and secondary amines because of differences in hydrogen bonding. The N–H bonds of primary and secondary amines participate in hydrogen-bonding attractions with

INSIGHT INTO PROPERTIES

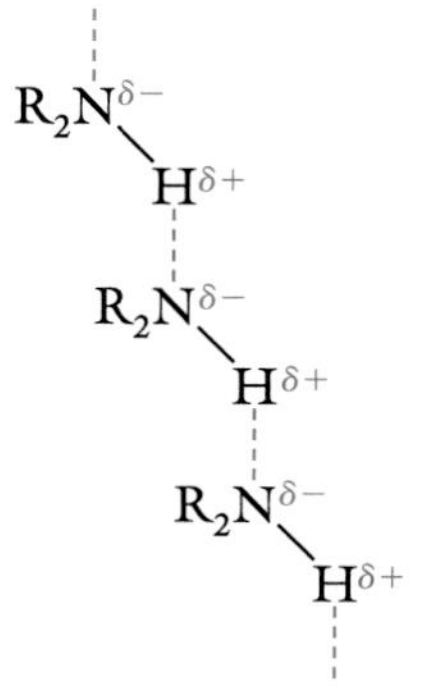

Figure 16.1 Hydrogen bonding in secondary amines.

16.1 Chemistry Within Us

Opium Alkaloids

The **opium alkaloids,** also called **opiates,** include compounds obtained from the unripened seeds of the opium poppy *Papaver somniferum* (which is not the common poppy) as well as synthetic compounds of similar structure, all possessing similar physiological activity. Opiates are used medically as analgesics (pain relievers) and hypnotics (sleep inducers). They also induce a sense of euphoria (a "high"), which accounts for their illegal, nonmedical use. The downside of opium alkaloids is that they are addictive and induce confusion and lethargy. Morever, they depress the respiratory system; that is, reduce its activity by affecting the brain centers that control respiration. Extreme depression of the brain's respiratory centers can result in death from respiratory arrest.

The raw **opium** extracted from the opium poppy contains almost two dozen different compounds. In the United States, raw opium was used in many patented medicines, such as Mrs. Winslow's Soothing Syrup, until the beginning of the twentieth century.

Morphine, the principal alkaloid in opium, was isolated from raw opium early in the nineteenth century. The invention of the hypodermic syringe in the middle of the nineteenth century led to large-scale use of morphine for pain relief in the American Civil War (1861–1865). Direct injection of an aqueous solution of morphine (in the form of its amine salt) into the bloodstream allowed quick relief of pain from battle injuries but resulted in large numbers of soldiers addicted to morphine. Morphine and other narcotics became controlled substances under federal law in 1914. Physicians can still prescribe morphine for pain relief, but great care is taken to avoid prolonged use or high doses, either of which may lead to addiction.

Codeine, the monomethyl ether of morphine, is present in raw opium in much smaller amounts than is morphine. Codeine is usually produced commercially from morphine by converting one of the —OH groups into a methyl ether group. Codeine is similar to morphine in its physiological effects but is less potent and less addictive. It is used in prescription cough suppressant syrups.

Heroin was first synthesized in the 1890s by reacting morphine with acetic anhydride to convert the —OH groups into acetate (ester) groups (an acyl transfer reaction; see Section 15.12). At that time, heroin was considered an antidote for morphine addiction, but nothing could have been farther from reality. Heroin produces a stronger high than morphine and is much more addictive, and the addiction is harder to overcome. Heroin use is illegal, even by prescription, in the United States.

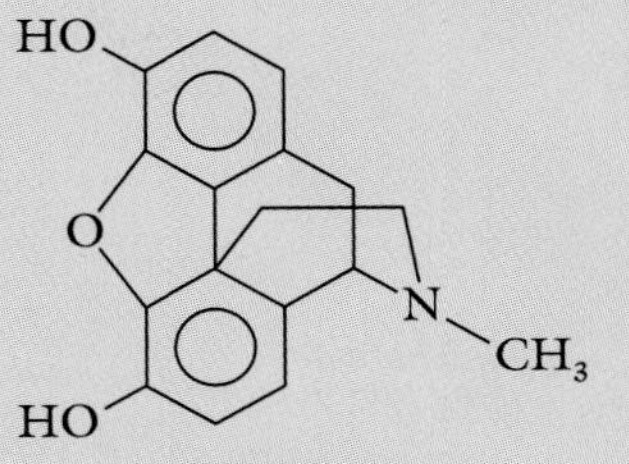

Morphine

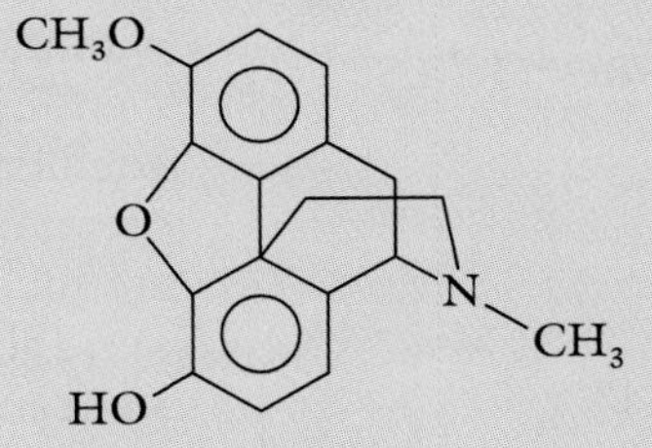

Codeine

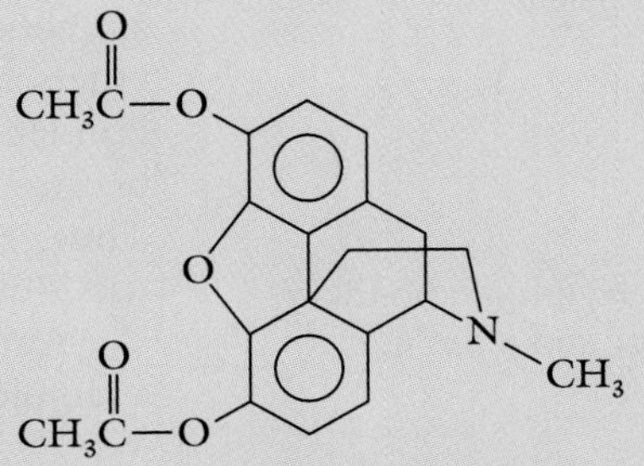

Heroin

Opium poppies and seed pod. The seed pod is cut to collect the raw opium.

If you compare the structures of morphine, codeine, and heroin, you see what appear to be relatively modest differences, yet these differences result in significant differences in physiological activity. Opiates produce their physiological effects such as pain relief or euphoria by binding to receptor sites in the brain. Small differences in chemical structures result in different extents of binding at receptor sites and thus different physiological effects.

water (Figure 16.2). In tertiary amines, hydrogen bonding with water is more limited because there are no N–H bonds. Hydrogen bonding takes place between the nitrogen of a tertiary amine and a hydrogen of water (Figure 16.3), analogous to the hydrogen bonding between an ether and water.

The slightly lower water solubility of amines compared with that of alcohol results from the slightly weaker attractive forces between water and the less polar N–H bonds of primary and secondary amines or the nitrogen of tertiary amines. The attractive forces between water and the O–H of an alcohol are stronger. The water solubilities of amines are about the same as those of aldehydes and ketones, because the hydrogen-bond attractive forces between amines and water are comparable to those between water and the carbonyl oxygen of aldehydes or ketones.

INSIGHT INTO PROPERTIES

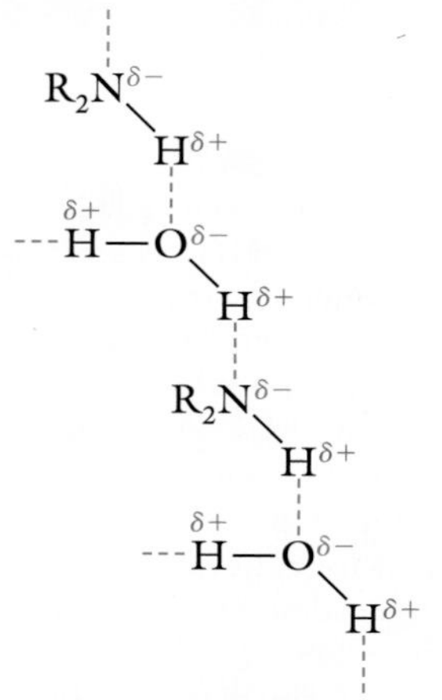

Figure 16.2 Hydrogen bonding between water and a secondary amine.

16.5 BASICITY OF AMINES

Like ammonia (Section 9.4), amines are basic because of the ability of the nonbonded pair of electrons on nitrogen to accept (bond to) a proton from an acid such as water or HCl. The reaction is illustrated for methylamine reacting with water:

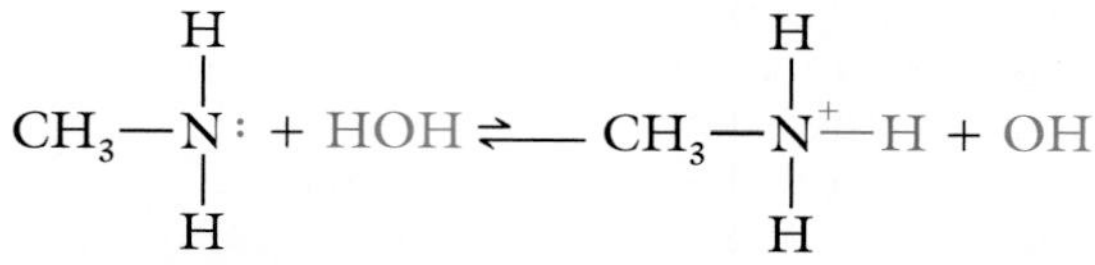

Methylamine

Protonation of the nitrogen converts the amine into a positive ion because the nitrogen becomes electron deficient by sharing its pair of nonbonded electrons with the proton from water.

The reaction of an amine with water is an equilibrium reaction with the equilibrium far to the left. Methylamine is a typical aliphatic amine in this respect: only 6.6% of it is converted into the amine salt in water. Thus, amines are much weaker bases than strong inorganic bases such as NaOH and KOH.

Table 16.3 compares the $[OH^-]$, $[H_3O^+]$, pH, and percent ionization of some inorganic and organic bases. Amines are considerably stronger bases than water. Amines are also stronger bases than alcohols and ethers, whose basicities are about the same as that of water. The oxygen atoms in water, alcohols, and ethers have the ability, like nitrogen, to bond to H^+ by using a pair of nonbonded electrons. However, oxygen is a poorer electron donor than nitrogen because oxygen, being more electronegative, holds its electrons more tightly. Thus, the oxygen-containing organic compounds are weaker proton acceptors—weaker bases—than amines.

INSIGHT INTO PROPERTIES

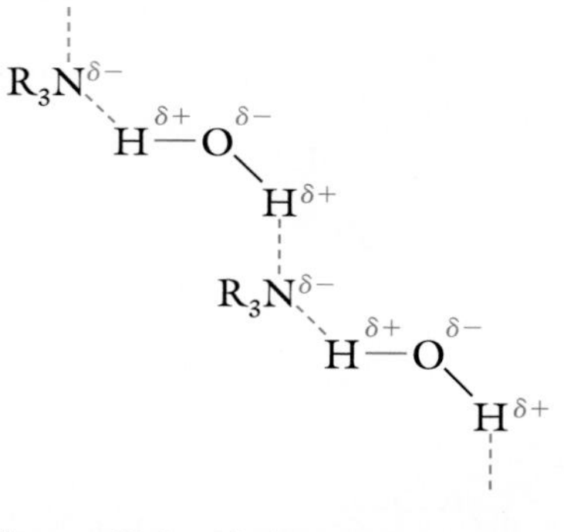

Figure 16.3 Hydrogen bonding between water and a tertiary amine.

Amines are sufficiently basic to turn red litmus paper blue, which serves as a simple chemical diagnostic test for amines. The only other type of organic

TABLE 16.3 Relative Basicities of Amines

	Values for 0.10 *M* Aqueous Solution			
Compound	**$[OH^-]$**	**$[H_3O^+]$**	**pH**	**Ionization (%)**
NaOH	0.10	1.0×10^{-13}	13.0	100
CH_3NH_2	6.8×10^{-3}	1.5×10^{-12}	11.8	6.3
$C_6H_5NH_2$	6.5×10^{-6}	2.4×10^{-9}	8.3	0.0062
H_2O	1.0×10^{-7}	1.0×10^{-7}	7.0	0.0001

16.2 Chemistry Within Us

Drugs for Controlling Blood Pressure

High blood pressure, often called the "silent killer" because it has no or very few symptoms in its early stages, has an unrelenting long-term adverse effect on the cardiovascular system. If undetected and untreated, the end result can be stroke or heart attack. Blood pressure is the pressure developed in blood vessels as the heart pumps blood through the blood vessels (see Box 5.2). A variety of drugs, mostly synthetic, are available to control high blood pressure, with an excellent prognosis for long-term health and quality of life.

Different types of drugs control blood pressure through different mechanisms. **Adrenergic inhibitors** lower blood pressure by interfering with the sympathetic nervous system, which controls the body's automatic response to stress, changes in blood pressure, heartbeat, and perspiration, for example. Adrenergic inhibitors include α-blockers such as terazosin (Hytrin) and β-blockers such as atenolol (Tenormin). **α-Blockers** relax blood vessels by blocking α-receptors that cause blood vessel constriction. **β-Blockers** have a more general effect, acting on the β-receptors in heart, lung, and blood vessels that cause the heart to beat harder and faster and the blood vessels to constrict.

Diuretics such as furosemide (Lasix) cause the kidneys to produce more urine. The increased production of urine reduces blood pressure by lowering the volume of blood that the heart pumps to all parts of the body. **Calcium channel blockers** such as diltiazem (Cardizem) inhibit the entry of Ca^{2+}—an ion essential for muscle contraction—into the muscle cells of blood vessels. This results in the relaxation of blood vessel walls, which decreases blood pressure and heart rate. **Vasodilators** such as hydralazine (Apresoline) act on the walls and tissues of small arteries to decrease blood vessel tension and resistance. **Angiotensin-converting enzyme (ACE)** inhibitors such as lisinopril (Prinivil) inhibit the conversion of angiotensin I into angiotensin II. Angiotensin II causes the constriction of blood vessels, and the inhibition of its formation results in a lowering of blood pressure.

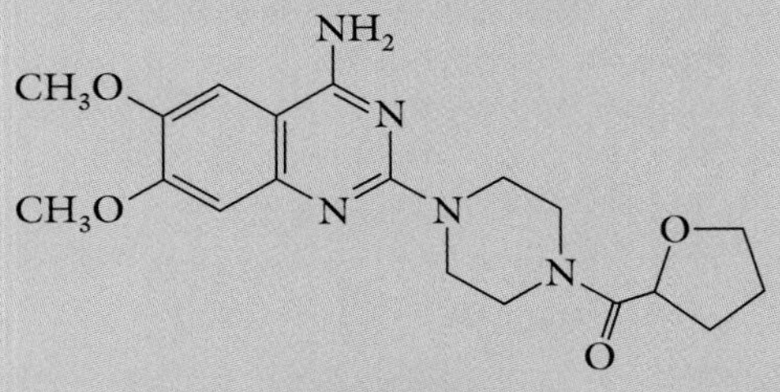

Terazosin (Hytrin)

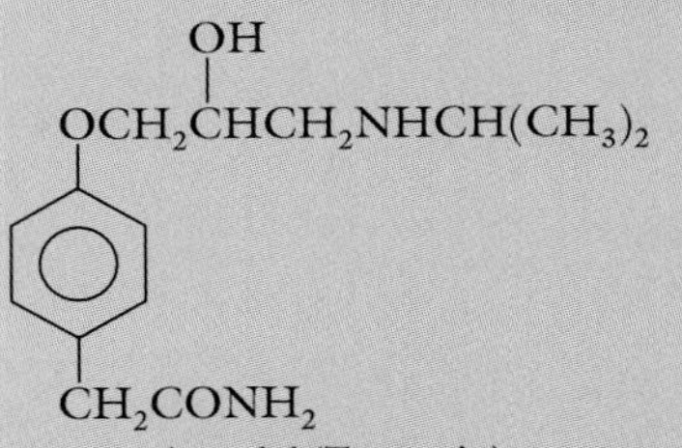

Atenolol (Tenormin)

compound sufficiently basic to turn red litmus paper blue is a carboxylate salt ($RCOO^-$) (Section 15.6).

Even though an amine is a weak base, its reaction with an acid goes to completion with a strong acid such as HCl and the amine is converted into an **amine salt:**

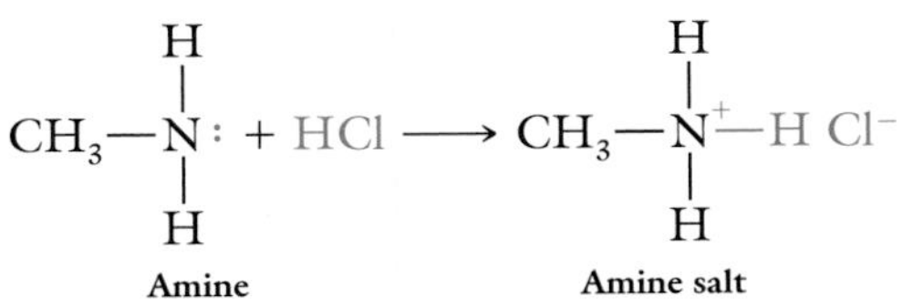

The amine is often called **free amine** or **free base** to distinguish it from the amine salt.

Amines present in the body are almost entirely (>99%) in the amine salt (ionized) form. The pH of a solution of an amine in pure water is near 12 (see Table 16.3), and amines exist mostly in the un-ionized form at this high pH. However, most physiological fluids are buffered (by the bicarbonate/carbonic acid buffer) to a pH very near 7—five orders of magnitude more acidic than pH 12. At this higher acidity, amines exist as their amine salts. Compare this

16.3 Chemistry Within Us

Other Amines and Amides with Physiological Activity

Caffeine is probably the most widely and often used mild central nervous system (CNS) stimulant. It increases the basal metabolism rate, the heart rate, the secretion of stomach acid, and urine production. Caffeine exists naturally in coffee and tea and is added to soft drinks. Most people claim to need a cup or two of coffee to get started in the morning or to stay awake if they are working late. Not surprisingly, caffeine is the main ingredient in stay-awake medications such as No Doz. There is evidence that caffeine is moderately addictive, so caffeine-free soft drinks, coffee, and tea are now readily available. Caffeine is also a component of many nonprescription drugs, including pain relievers, antihistamines, and cold medications. Its purpose is to counteract the unwanted sedative effects of the main ingredients of these medications.

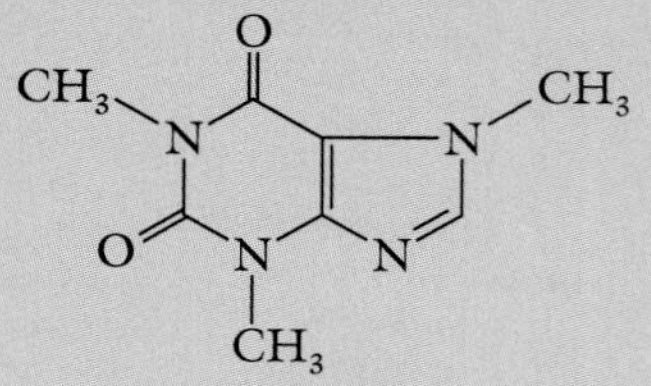

Caffeine

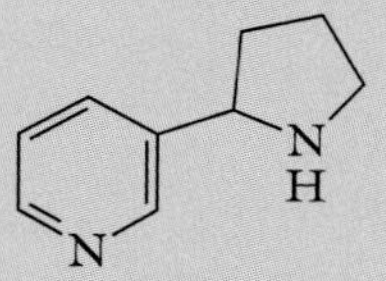

Nicotine

Nicotine, a mild CNS stimulant found in tobacco, has been categorized by the U.S. surgeon general as an addictive drug. It is inhaled in tobacco smoke and absorbed orally when tobacco is chewed. Nicotine is highly toxic if injected directly into the bloodstream, but the amounts delivered into the bloodstream by smoking or chewing tobacco are far below the acute toxic dose. Nevertheless, regular tobacco use of any kind has long-term effects on the cardiovascular system, and smoking tobacco creates respiratory problems. Cigarette smoke contains significant amounts of carbon monoxide and fused-ring aromatic compounds. Carbon monoxide is absorbed by hemoglobin, interfering with the ability of blood to transport oxygen from the lungs to the rest of the body. Fused-ring aromatic compounds (see Box 12.9) are cancer-causing agents.

A woman is absorbing the central-nervous-system stimulants nicotine, by smoking, and caffeine, by drinking coffee.

with what we learned about carboxylic acids, which exist only in their ionized (carboxylate) form at physiological pH (Section 15.5).

Example 16.4 Writing equations for the basicity of amines

Write the equation for each of the following reactions: (a) aniline is dissolved in water; (b) ethyldimethylamine is neutralized with HCl.

Solution

(a) Aniline ionizes weakly in water; the arrows show the equilibrium is far to the left.

$$C_6H_5NH_2 + H_2O \rightleftharpoons C_6H_5NH_3^+ + OH^-$$

Epinephrine (adrenaline), a hormone secreted by the adrenal glands when we are stressed or frightened, prepares the body to respond to a perceived threat. Several synthetic drugs called **amphetamines** are structurally related to adrenaline and mimic its stimulant effect. These drugs include isoproterenol, prescribed for emphysema and asthma, and methedrine, prescribed as an antidepressant and sometimes for weight control. Illegal use of amphetamines, called "uppers" or "pep pills" on the street, is a serious drug-abuse problem. Overuse can result in sleeplessness, paranoia, and hallucinations.

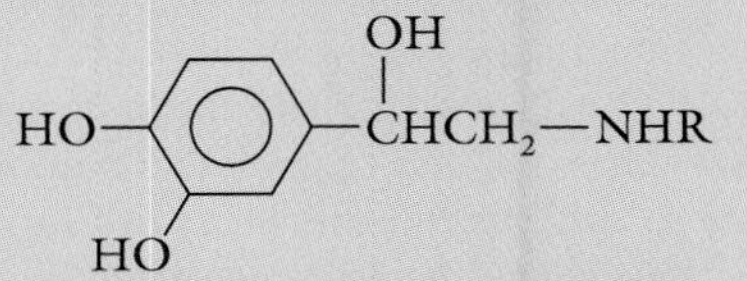

Epinephrine (Adrenalin), R = CH_3
Isoproterenol, R = $CH(CH_3)_2$

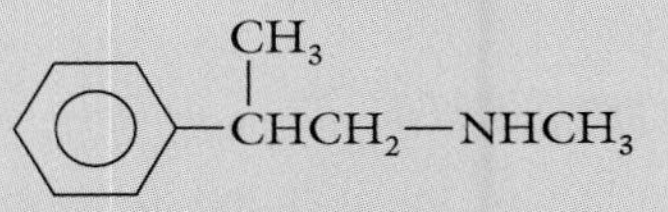

Methamphetamine (Methedrine)

Barbiturates, a class of amides that induce sedation and sleep, are called "downers" because of their mood-depressing effect. They are particularly dangerous when used in combination with alcohol because of a synergistic effect between the two in which their combined effect is greater than the effect of either drug alone. This combination is the agent of some suicides, suicide attempts, and accidental suicides. The barbiturate phenobarbital (Luminal) is used as an anticonvulsant agent for treating epilepsy and brain trauma.

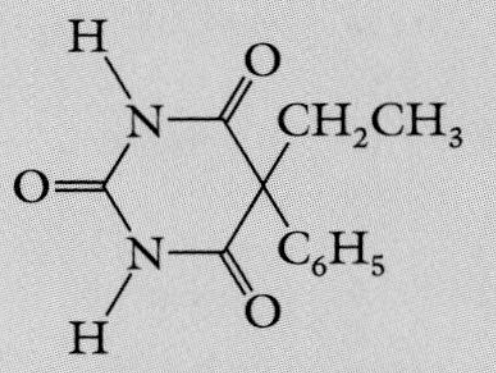

Phenobarbital (Luminal)

Other synthetic drugs that contain amine or amide functional groups are antibiotics such as ampicillin, tranquilizers (antianxiety drugs) such as diazepam (Valium), antidepressants such as fluoxetine (Prozac), ulcer-healing medications such as cimetidine (Tagament), and local anesthetics such as procaine (Novocain).

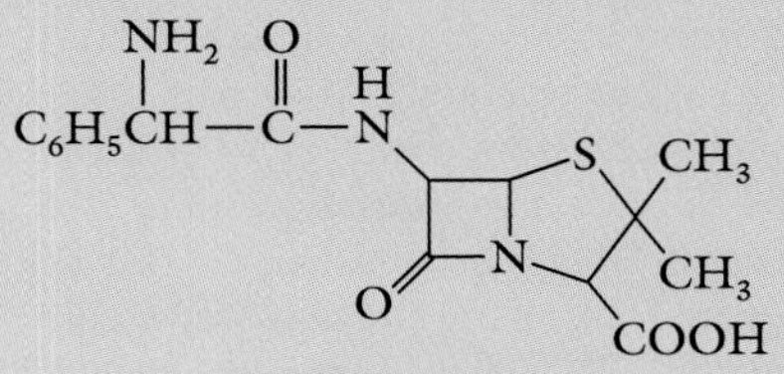

Ampicillin

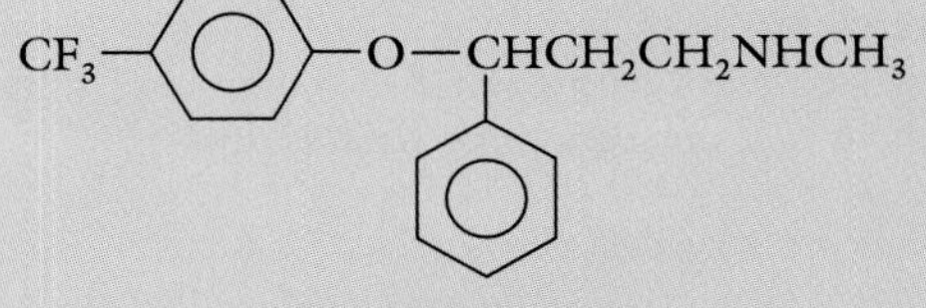

Fluoxetine (Prozac)

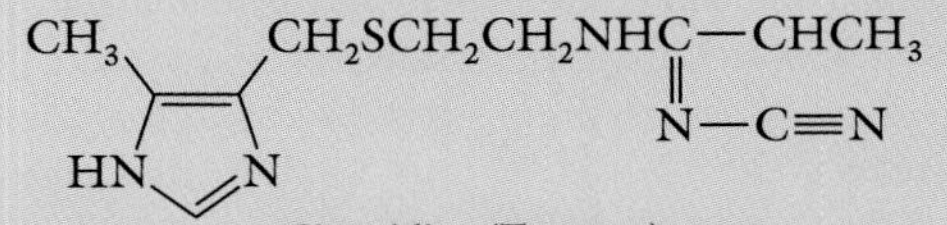

Cimetidine (Tagamet)

(b) This tertiary amine reacts with strong acid to form the amine salt; the equilibrium is completely to the right.

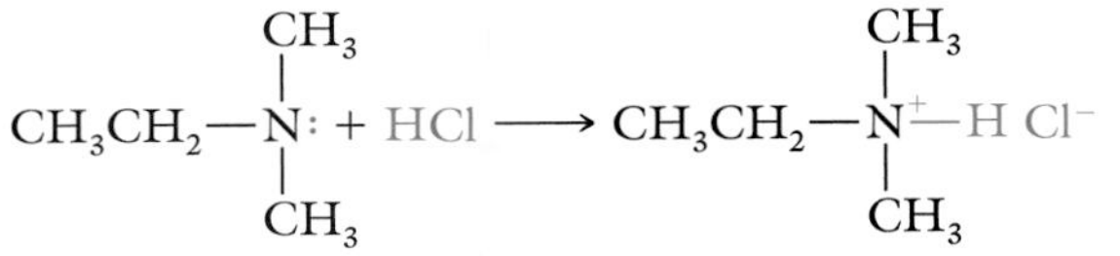

Problem 16.4 Write the equation for each of the following reactions: (a) aniline is neutralized with H_2SO_4; (b) diethylamine is dissolved in water.

$R—\ddot{N}H_2$
Alkylamine
Nonbonded electrons are more available

$Ar \leftarrow \ddot{N}H_2$
Arylamine
Nonbonded electrons are less available

Arylamines are much less basic than aliphatic amines. Table 16.3 shows that $[OH^-]$ is about 1000-fold as great for a methylamine solution as for an aniline solution. This difference is due to the electron-withdrawing effect of

PROFESSIONAL CONNECTIONS

Rose Eva Constantino, Ph.D., J.D. Associate Professor, Psychiatric Mental Health Nursing

Rose Eva Constantino studied chemistry as an undergraduate in Manila, at Philippine Union College.

Why did you choose the position that you are in today?

Because of the people orientation, the opportunity to help people who are ill. I already had my B.S.N. when I moved to the United States from the Philippines in 1964 and wanted to take advantage of the American system of education. In 1969, I enrolled at the University of Pittsburgh School of Nursing's newly opened psychiatric masters program and received my M.S.N. in 1971. I was employed at the School of Nursing to teach undergraduate students in psychiatric nursing. While teaching, I took part-time courses toward a Ph.D, and received my Ph.D. in 1979 with a major in nursing. After I finished my dissertation, I enrolled in the J.D. (law) program at Duquesne University. I did this while working full time and took all my law classes at night.

What do you enjoy most?

I enjoy teaching psychiatric mental health nursing, but I also enjoy my work with abused women and children, helping them sort out their topsy-turvy lives.

Please describe a typical day.

If I'm in court, I get there at 8:30 a.m. and typically handle four cases. I file the court papers and give the clients some emotional support after the resolution of the case. My law work is done on a pro bono basis. When this is done, I ask the women if they would be interested in participating in a research study related to my interest in providing the best care for the abused. This is the same nursing "postvention" care I developed for widows whose spouses committed suicide. At school, I teach undergraduate and doctoral students. I lecture once a week, for 2 to 3 hours, and take ten students to a psychiatric hospital for their clinical experience.

What advances have been made in your field?

Several advances have shifted our work to the biomedical perspective. Research on antidepressants and the neuroleptics, cytochromes, enzymes, and so on, have changed nursing intervention from a caregiver/custodial role into an educator/facilitator role. We used to primarily see the patient's psychiatric illness as related to the environment and the family. We know now that a great portion of the illness may be genetic and biochemical.

Have these advances changed your everyday life?

The nursing care that we provide to psych patients goes beyond emotional support. Medication education is very important. We teach patients the interactions of medications—what changes the chemical components when combined with other medications or with food and drink. The interactions between different drugs are very important to patients. It's basic chemistry but, at the same time, it's very complex and it's fairly new to nursing.

Do you use chemistry in your job?

Yes, I need it for my research and in my class. In my research, I not only study the psychological responses of women to abuse, but also assess their immune systems by drawing blood for the T-cell mitogenesis-to-phytohemaglutinin assay. It measures the responsiveness of their T-cells to possible infections and illnesses such as colds, and so forth. In class, I need chemistry to help students understand the interactions of psychotropic medications.

What advice do you have for students who are studying chemistry today?

They should try to enjoy it first and not give it up. It's very important. They will find it valuable later on, even if it is not their favorite subject; don't turn your back on it! Technology helps make it interesting, so they and their instructors should make use of technology in learning chemistry. There are so many three-dimensional audiovisual and interactive programs now available on chemical structures and interactions. Instructors and students should make use of them to make learning chemistry fun and memorable.

the benzene ring compared with an alkyl group. Greater electron withdrawal from nitrogen makes the nonbonded electron pair of nitrogen less available for bonding to a proton.

16.6 AMINE SALTS

An amine salt, the product of a reaction between an amine and a strong acid, is named by naming the positive ion first and then naming the negative ion. The positive ion is named differently, depending on whether the amine is aliphatic, aromatic, or heterocyclic:

Rules for naming amine salts

- For aromatic and heterocyclic amines, the final **-e** of the name of the amine is replaced by **-ium.**
- For aliphatic amines, the ending of the name of the amine is changed from **-amine** to **-ammonium.**

Pharmaceutical companies often use an alternate approach for naming amine salts used as medicines. The name of the amine is followed by the name of the acid used to synthesize the amine salt. Amine salts from HCl are called amine hydrochlorides; those from H_2SO_4 are called amine hydrogen sulfates.

Example 16.5 Naming amine salts

Name the following amine salts:

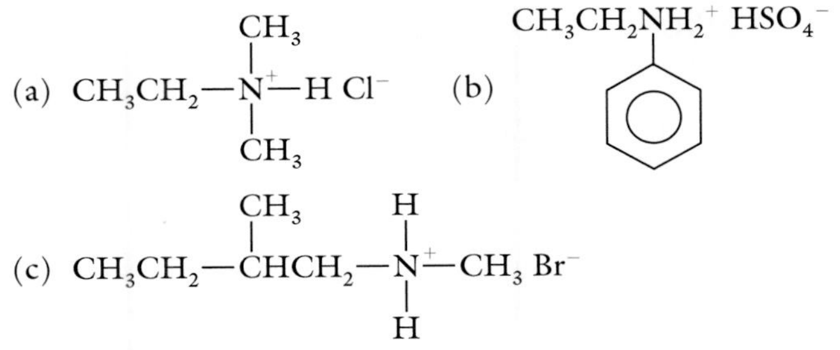

Solution

(a) The amine salt is derived from ethyldimethylamine and HCl. The name is ethyldimethylammonium chloride.

(b) The amine salt is derived from *N*-ethylaniline and H_2SO_4. The name is *N*-ethylanilinium hydrogen sulfate.

(c) The amine salt is derived from 2,*N*-dimethyl-1-butanamine and HBr. The name is 2,*N*-dimethyl-1-butanammonium bromide.

Problem 16.5 Draw the structures of the following amine salts: (a) cyclohexylammonium chloride; (b) *N,N*-dimethyl-1-butanammonium hydrogen sulfate; (c) pyridinium bromide (refer to Table 16.1).

Properties of Amine Salts

Amine salts are ionic and possess much higher secondary attractive forces than any covalent organic compounds do, even higher than carboxylic acids, which possess strong hydrogen-bond secondary attractions. The melting and boiling points of amine salts are much higher than those of the corresponding amines. All amine salts therefore have high melting points and are solids at room temperature. For example, methylammonium chloride is a solid with a melting point of 232°C and decomposes without boiling. Compare it with methylamine, which is a gas with a melting point of −94°C and a boiling point of −6°C. The difference in the melting and boiling points of amines and amine salts is having a significant and disturbing effect on cocaine use, increasing the number of people addicted to cocaine (Box 16.4).

Amine salts are also much more water soluble than are amines. Amines of more than six carbons have very low or no solubility in water, whereas the solubility of amine salts extends to much larger numbers of carbons. This difference results from the much greater strength of the attractive forces between ions and water than that of the hydrogen-bond attractions between amines and water. Thus, the solubility of an amine is greatly increased in an environment of neutral or lower pH because the amine is converted into the amine salt. Amines such as hormones and neurotransmitters are solubilized in the body because they are in their salt form at physiological pH. The salt form allows them to be transported by the blood at the concentrations required for their specialized functions.

Amines that are used as medicines to be administered intravenously are almost always converted into amine salts to increase their water solubility so that they may be administered as aqueous solutions. There are relatively few drugs with enough water-solubilizing groups (such as hydroxyl and amino groups) relative to the number of carbons to allow their use as free amines. For medicines taken orally, this is not a problem, because free amines are readily converted into the amine salts in the highly acidic environment in the stomach (pH ≈ 1.5–2.0). The amine salts are water soluble and readily pass into the bloodstream through the cells of the stomach and intestinal linings.

Example 16.6 Predicting the physical properties of amines and amine salts

Place the following compounds in order of increasing melting point: *N*-ethyl-*N*-methylaniline, anilinium chloride, *N*-propylaniline.

Solution

N-Ethyl-*N*-methylaniline < *N*-propylaniline ≪ anilinium chloride. The ionic compound anilinium chloride has a much higher melting point than those of the covalent compounds. Of the two covalent compounds of the same molecular mass, the secondary amine *N*-propylaniline has the higher melting point because it possesses hydrogen bonding. The tertiary amine *N*-ethyl-*N*-methylaniline has weaker secondary forces—dipole–dipole instead of hydrogen bonding—because it does not have N–H bonds.

Problem 16.6 Place the following compounds in order of increasing water solubility: $CH_3(CH_2)_6NH_2$, $CH_3(CH_2)_{14}NH_3{}^+Cl^-$, $CH_3(CH_2)_{14}NH_2$.

An amine salt, like any salt of a weak base, is a weak acid and undergoes acid-base reaction with water and strong bases:

$$CH_3NH_3{}^+\ Cl^- + H_2O \rightleftharpoons CH_3NH_2 + H_3O^+ + Cl^-$$

$$CH_3NH_3{}^+\ Cl^- + NaOH \longrightarrow CH_3NH_2 + H_2O + NaCl$$

Amine salts are sufficiently acidic to turn blue litmus paper red.

Quaternary Ammonium Salts

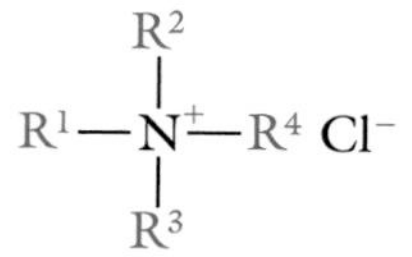
Quaternary ammonium salt

The ammonium salts just described have one, two, or three alkyl groups, but it is also possible to synthesize amine salts having four alkyl groups. Such salts are called **quaternary ammonium salts** and are named in the same way as amine salts, taking into account the need to name four instead of three groups attached to nitrogen. Without a hydrogen attached to nitrogen that can be

16.4 Chemistry Within Us

Cocaine: Free Base Versus Amine Salt

Cocaine is extracted from the leaves of the coca plant, which is found almost exclusively on the slopes of the Andes of South America. It has been used medically as a local anesthetic to relieve pain in severe injuries and in nasal surgery, but such use is restricted because of its highly addictive and destructive nature. Cocaine has become one of the most abused narcotics by so-called recreational users.

Cocaine is a powerful central nervous system stimulant, speeding up the heart rate, raising the blood pressure, and stimulating the pleasure centers in the brain. Cocaine gives the user a great sense of euphoria, power, confidence, and increased stamina. The effect is short lived (30–60 min), however, and is followed by severe depression and a craving for more cocaine. This leads to repeated use, setting off a cycle of euphoria and depression that quickly results in psychological addiction. Long-term users often develop a paranoid psychosis and run a high risk of death from respiratory or heart failure.

Treating crushed coca leaves with aqueous HCl converts the cocaine into its amine salt **cocaine hydrochloride,** which dissolves in the aqueous solution. The crushed leaves are discarded, and the water is evaporated from the aqueous solution to yield solid cocaine hydrochloride, the original form of cocaine illegally sold on the streets. Cocaine hydrochloride is used in two ways: "snorting" and injection. Cocaine hydrochloride inhaled into the nasal passages dissolves in the watery mucous membranes of the nose and throat and enters the bloodstream. Long-term use results in the deterioration of the mucous membranes and nose cartilage and a chronic runny nose. Injection of an aqueous solution of cocaine hydrochloride directly into the blood stream gives a much faster and higher "high" than snorting does because the cocaine concentration in the bloodstream is much higher. The risks associated with injection, as for all injectable drugs, include hepatitis, AIDS, and other infections spread through the sharing of needles.

Today, much of the cocaine used illegally is free base cocaine, called **crack cocaine** or just **crack.** Free base cocaine is made by neutralizing the amine salt with aqueous base (NaOH) and then adding diethyl ether to extract the cocaine from the aqueous solution. Evaporation of the ether solvent yields free base cocaine. Crack can be smoked, which makes it a much more dangerous drug than cocaine hydrochloride. People are much less hesitant about smoking a drug than about injecting one.

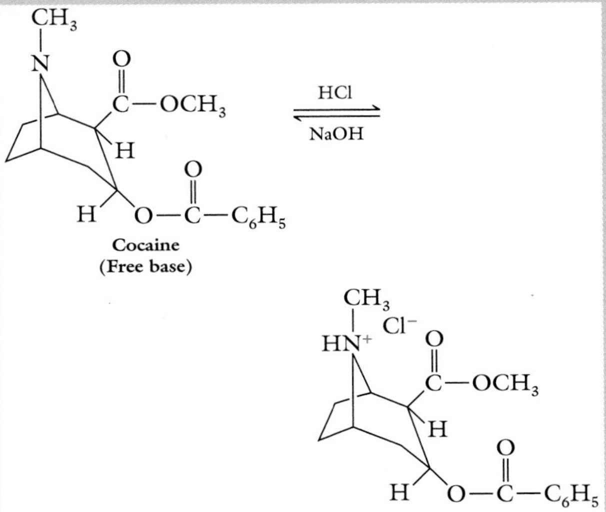

Free-base cocaine can be smoked because it is a covalent compound; it has a lower melting point (98°C) and is much more volatile than the ionic compound cocaine hydrochloride. The latter cannot be smoked, because it does not melt and is not volatile. When smoked, free base cocaine is rapidly absorbed through the membranes of the respiratory system, giving an immediate high. The effect is the equivalent of injecting cocaine hydrochloride directly into the bloodstream, producing much faster and stronger euphoric effects than are achieved by snorting cocaine hydrochloride. The result has been a sharp rise in cocaine use and in the number of people addicted.

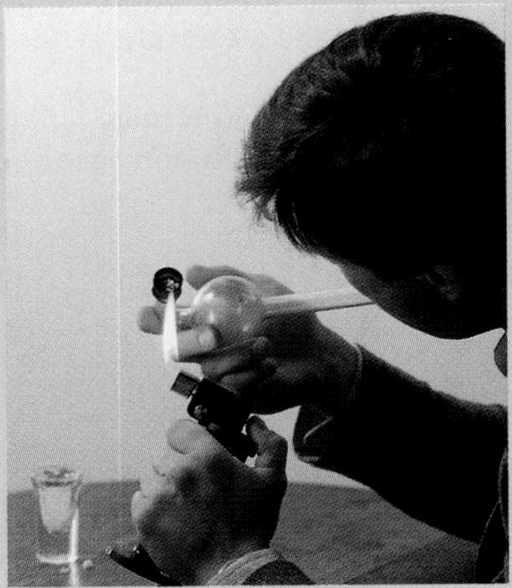

Crack (free-base) cocaine is more dangerous than the cocaine amine salt because crack can be smoked, which is easier than snorting or injecting cocaine.

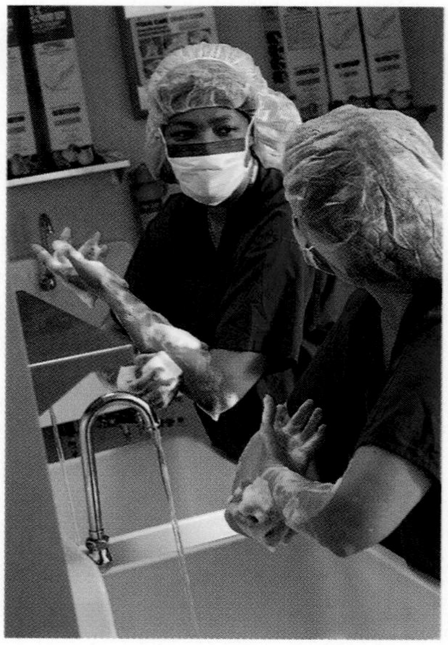

Surgical nurses scrub their hands with disinfectant before surgery. Many hospital disinfectants contain quaternary ammonium salts.

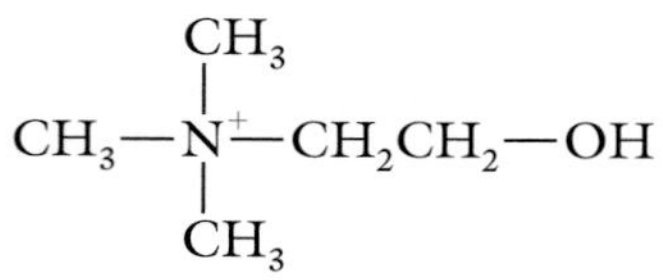

Choline

▶▶ Neurotransmitters are chemicals that carry information from one nerve cell to another, as shown in Section 26.4

removed by base and without a nonbonded electron pair on nitrogen that can react with a proton, the structures of quaternary ammonium salts are not affected by acid or base as are those of amines.

Quaternary ammonium salts in which one of the alkyl groups is a long hydrocarbon chain (such as the chain $CH_3(CH_2)_{17}$—) are used as soaps (Section 15.7), germicides, antiseptic soaps, and mouthwashes; they are used for disinfecting skin and hands preparatory to surgery and for sterilizing surgical instruments. The quaternary ammonium salt choline is a constituent of phospholipids used to construct cell membranes (Sections 19.6 and 19.10). It also functions as a neurotransmitter.

16.7 CLASSIFICATION OF AMIDES

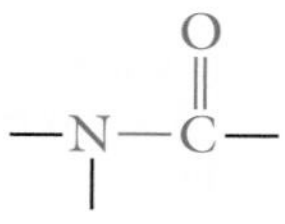

Amide

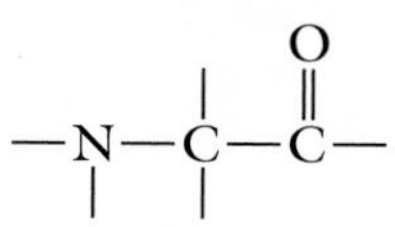

Not an amide

Amides contain the amide group—a carbonyl group attached directly to a nitrogen. A compound containing a nitrogen and a carbonyl is not an amide unless the two are bonded directly together.

Amides are classified as primary, secondary, and tertiary amides when two, one, and zero hydrogens, respectively, are attached to nitrogen:

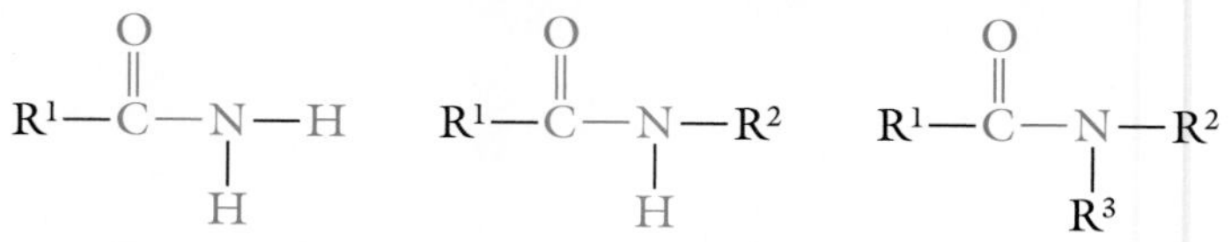

Primary Secondary Tertiary

R^1, R^2, and R^3 can be aliphatic or aromatic.

Proteins are amides of central biological importance and will be discussed in Chapter 20. Proteins have a variety of physiological functions in all living organisms. They are used in construction, protection, and movement (skin, bone, muscle); as catalysts in biochemical reactions (enzymes); in the regulation and control of physiological processes (hormones, neurotransmitters); and in the transport of oxygen throughout the body (hemoglobin).

Many physiologically active compounds are amides (see Boxes 16.2 and 16.3), including the sweetener aspartame (NutraSweet) (Box 18.1).

Example 16.7 Classifying amides

All of the following compounds are constitutional isomers of C_4H_9NO. Which of the compounds are amides? Classify the amides as primary, secondary, or tertiary.

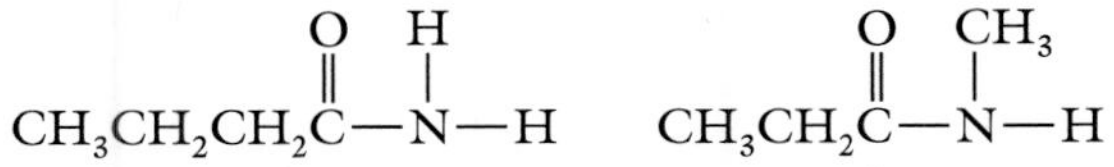

1 2

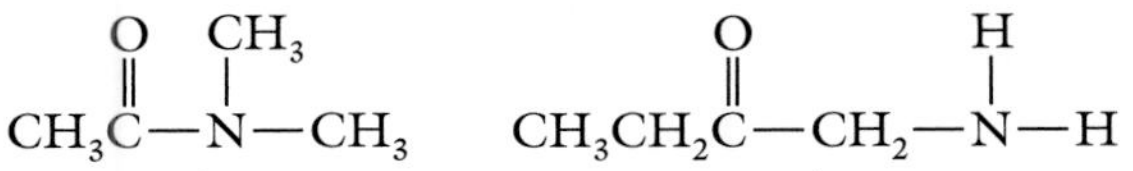

3 4

Solution

Compounds 1, 2, and 3 are amides because the nitrogen is directly bonded to the carbonyl carbon. Compound 4 is not an amide, because the nitrogen is not directly bonded to the carbonyl carbon. Compound 4 is an amine and a ketone. Compounds 1, 2, and 3 are primary, secondary, and tertiary amides, respectively, because there are two, one, and zero hydrogens bonded to nitrogen.

Problem 16.7 Which of the following compounds are amides? Classify the amides as primary, secondary, or tertiary.

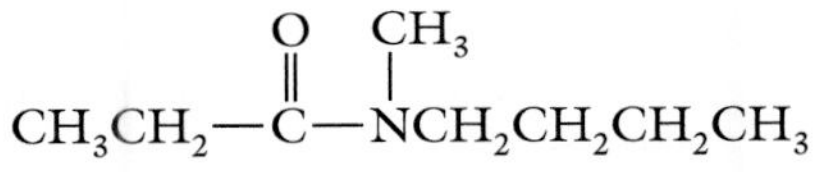

1

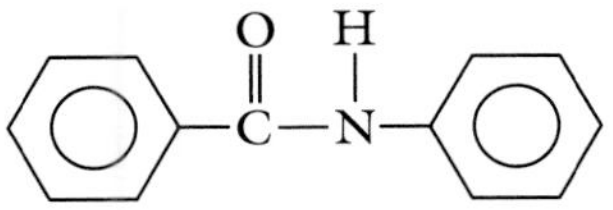

2

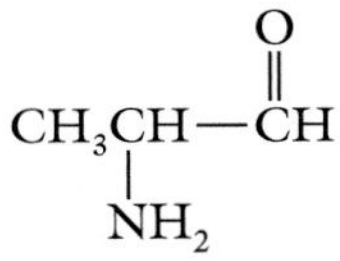

3

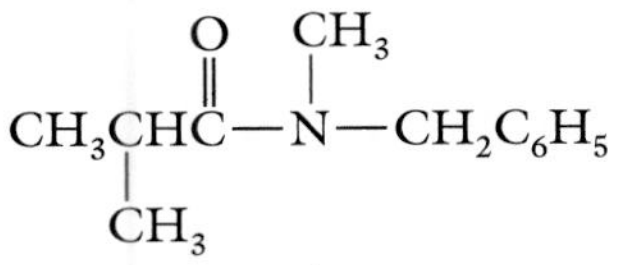

4

A PICTURE OF HEALTH

Examples of Amines and Amides

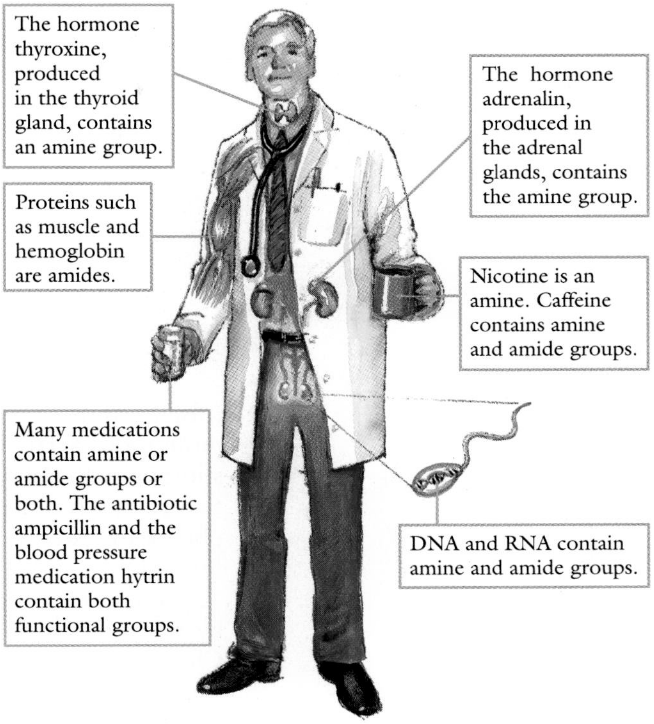

The hormone thyroxine, produced in the thyroid gland, contains an amine group.
Proteins such as muscle and hemoglobin are amides.
Many medications contain amine or amide groups or both. The antibiotic ampicillin and the blood pressure medication hytrin contain both functional groups.
The hormone adrenalin, produced in the adrenal glands, contains the amine group.
Nicotine is an amine. Caffeine contains amine and amide groups.
DNA and RNA contain amine and amide groups.

16.8 SYNTHESIS OF AMIDES

An amide is synthesized by the reaction of a carboxylic acid with ammonia or with a primary or secondary amine. However, a successful synthesis requires control of the reaction temperature. At ambient-to-moderate temperatures (about 20–50°C), there is no amide formation; there is only an acid-base reaction to form an ammonium carboxylate salt:

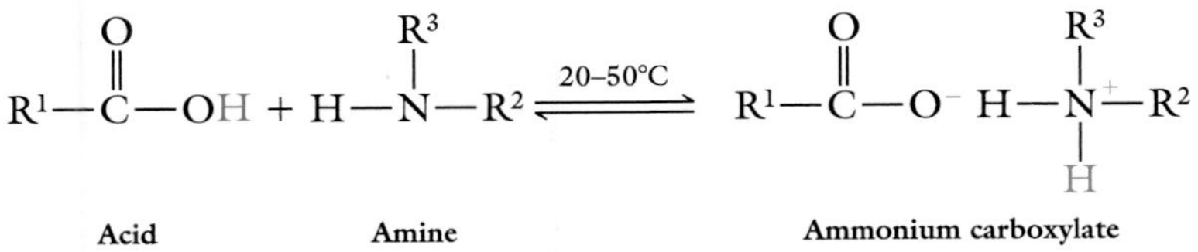

R^1, R^2, and R^3 are hydrogen, aliphatic, or aryl.

When the reactions are carried out at elevated temperatures ($>100°C$), amides are formed in high yield by dehydration between the two reactants:

$$R^1-\overset{\overset{\displaystyle O}{\|}}{C}-OH + H-\overset{\overset{\displaystyle H}{|}}{N}-H \xrightarrow{>100°C} R^1-\overset{\overset{\displaystyle O}{\|}}{C}-\overset{\overset{\displaystyle H}{|}}{N}-H + H_2O$$

Carboxylic acid   Ammonia   Primary amide

$$R^1-\overset{\overset{\displaystyle O}{\|}}{C}-OH + H-\overset{\overset{\displaystyle H}{|}}{N}-R^2 \xrightarrow{>100°C} R^1-\overset{\overset{\displaystyle O}{\|}}{C}-\overset{\overset{\displaystyle H}{|}}{N}-R^2 + H_2O$$

Carboxylic acid   Primary amine   Secondary amide

$$R^1-\overset{\overset{\displaystyle O}{\|}}{C}-OH + H-\overset{\overset{\displaystyle R^3}{|}}{N}-R^2 \xrightarrow{>100°C} R^1-\overset{\overset{\displaystyle O}{\|}}{C}-\overset{\overset{\displaystyle R^3}{|}}{N}-R^2 + H_2O$$

Carboxylic acid   Secondary amine   Tertiary amide

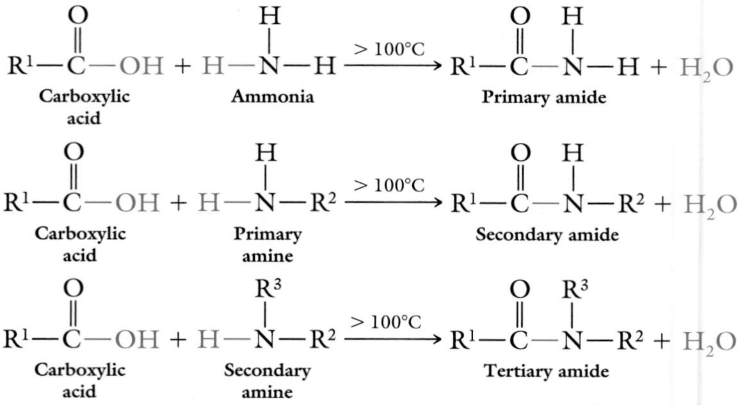

In amide formation, ammonia or an amine (minus one H) substitutes for (displaces) the —OH of the carboxylic acid. The reaction is very similar to esterification (Section 15.8). In both reactions, a carboxylic acid undergoes substitution with dehydration between the carboxylic acid and other reagent. The two reactions differ regarding the identity of the other reagent: an alcohol or phenol for esterification; ammonia or a primary or secondary amine for amide formation.

Amide formation is a reversible reaction. The yield of amide is increased by the removal of water, which shifts the equilibrium to the right side (Le Chatelier's principle). As in ester formation, this rightward shift is achieved by running the reaction at a temperature higher than the boiling point of water. Water distills out of the reaction vessel as it is formed.

Note that tertiary amines cannot form amides, because they do not have the necessary hydrogen on the nitrogen to participate in the dehydration. At any temperature, a tertiary amine and a carboxylic acid undergo an acid-base reaction but not amide formation.

Like ester formation, amide formation is an acyl transfer reaction with the carboxylic acid acting as an acyl transfer agent. We will consider acyl transfer reactions in detail when we study the biosynthesis of proteins in Chapter 21.

Example 16.8   **Recognizing the conditions for amide formation**

Complete each of the following reactions. If no reaction takes place, write "no reaction."

(a) $CH_3CH_2-\overset{\overset{\displaystyle O}{\|}}{C}-OH + HN(CH_3)_2 \xrightarrow{20-50°C}$?

(b) $CH_3CH_2-\overset{\overset{\displaystyle O}{\|}}{C}-OH + HN(CH_3)_2 \xrightarrow{>100°C}$?

(c) $CH_3CH_2-\overset{\overset{\displaystyle O}{\|}}{C}-OH + N(CH_3)_3 \xrightarrow{>100°C}$?

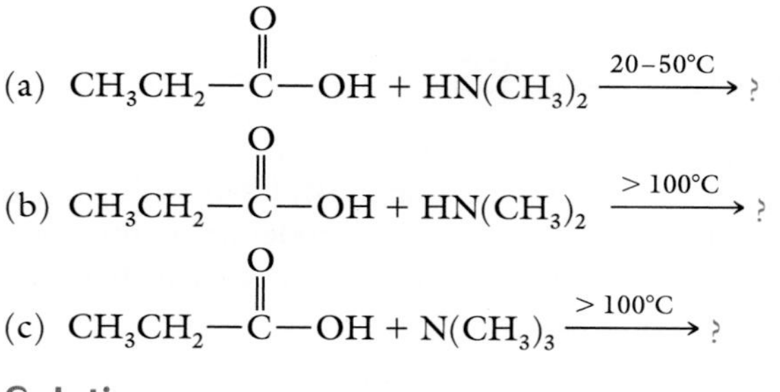

Solution

Acid-base reactions take place at the lower temperatures. Amide formation requires temperatures above 100°C.

(a) $CH_3CH_2-\overset{\overset{\displaystyle O}{\|}}{C}-OH + HN(CH_3)_2 \xrightarrow{20-50°C} CH_3CH_2-\overset{\overset{\displaystyle O}{\|}}{C}-O^- \; H_2\overset{+}{N}(CH_3)_2$

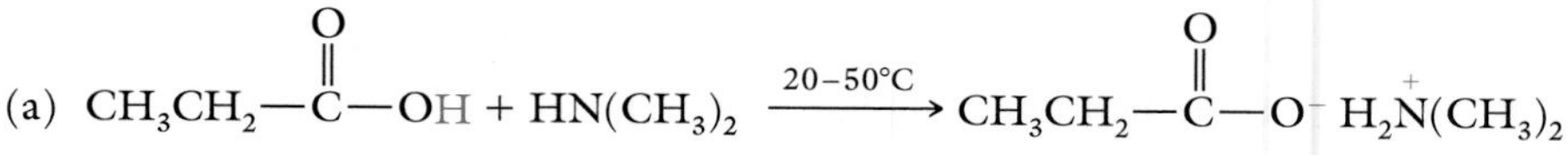

(b) $$CH_3CH_2-\overset{\overset{\displaystyle O}{\|}}{C}-OH + H-N(CH_3)_2 \xrightarrow[-H_2O]{>100^\circ C} CH_3CH_2-\overset{\overset{\displaystyle O}{\|}}{C}-N(CH_3)$$

(c) Amide formation does not take place with tertiary amines at any temperature. The only reaction is an acid-base reaction:

$$CH_3CH_2-\overset{\overset{\displaystyle O}{\|}}{C}-OH + N(CH_3)_3 \xrightarrow{>100^\circ C} CH_3CH_2-\overset{\overset{\displaystyle O}{\|}}{C}-O^- \ H\overset{+}{N}(CH_3)_3$$

Problem 16.8 Complete each of the following reactions. If no reaction takes place, write "no reaction." (a) Benzoic acid + methylamine at 20–50°C; (b) acetic acid + pyridine at >100°C; (c) benzoic acid + dimethylamine at >100°C.

Acid halides and anhydrides are often used instead of carboxylic acids to form amides because the reactions proceed considerably faster, even at ambient temperatures. Moreover, the reactions easily proceed to completion because they are not reversible.

$$R^1-\overset{\overset{\displaystyle O}{\|}}{C}-Cl + H-\overset{\overset{\displaystyle R^3}{|}}{N}-R^2 \longrightarrow R^1-\overset{\overset{\displaystyle O}{\|}}{C}-\overset{\overset{\displaystyle R^3}{|}}{N}-R^2 + HCl$$

$$R^1-\overset{\overset{\displaystyle O}{\|}}{C}-O-\overset{\overset{\displaystyle O}{\|}}{C}-R^1 + H-\overset{\overset{\displaystyle R^3}{|}}{N}-R^2 \longrightarrow R^1-\overset{\overset{\displaystyle O}{\|}}{C}-\overset{\overset{\displaystyle R^3}{|}}{N}-R^2 + R^1-\overset{\overset{\displaystyle O}{\|}}{C}-O-H$$

Example 16.9 Writing the equation for amide synthesis from an acid halide

Give the equation for amide formation between propanoyl chloride and dimethylamine.

Solution

$$CH_3CH_2-\overset{\overset{\displaystyle O}{\|}}{C}-Cl + H-N(CH_3)_2 \xrightarrow{-HCl} CH_3CH_2-\overset{\overset{\displaystyle O}{\|}}{C}-N(CH_3)_2$$

Problem 16.9 Give the equation for amide formation between propanoic anhydride and dimethylamine.

16.9 POLYAMIDE SYNTHESIS

Polyamides, like polyesters (Section 15.10), are produced by polymerization of bifunctional reagents (also called monomers). Whereas polyesters are produced from the reaction of dicarboxylic acids and diols, polyamide formation depends on the reaction of dicarboxylic acids and diamines. Large numbers (>50) of monomer molecules react together to produce the much larger sized polymer:

$$n\ HO-\overset{\overset{\displaystyle O}{\|}}{C}-R-\overset{\overset{\displaystyle O}{\|}}{C}-OH + n\ H_2N-R'-NH_2 \xrightarrow{-H_2O} \left(\overset{\overset{\displaystyle O}{\|}}{C}-R-\overset{\overset{\displaystyle O}{\|}}{C}-\overset{\overset{\displaystyle H}{|}}{N}-R'-\overset{\overset{\displaystyle H}{|}}{N}\right)_n$$

This polymerization, like polyesterification, is a condensation polymerization because water is a by-product of the reaction.

R and R′ in the preceding structures represent divalent units such as:

$-CH_2CH_2-$ $\quad -CH_2CH_2CH_2CH_2-$ 

As stated in Section 15.10, the repeat unit is the structure inside the brackets and the subscript n is used to indicate that the repeat unit repeats over and over many times; that is, n is a large number.

The polymerization reaction between 1,6-hexanediamine, $H_2N(CH_2)_6NH_2$, and 1,6-hexanedioic acid, $HOOC(CH_2)_4COOH$, produces the important commercial material called poly(hexamethylene adipamide), or nylon-66:

$$HO-\overset{O}{\overset{\|}{C}}-(CH_2)_4-\overset{O}{\overset{\|}{C}}-OH + H_2N-(CH_2)_6-NH_2 \xrightarrow{-H_2O} \left(\overset{O}{\overset{\|}{C}}-(CH_2)_4-\overset{O}{\overset{\|}{C}}-\overset{H}{\overset{|}{N}}-(CH_2)_6-\overset{H}{\overset{|}{N}}\right)_n$$

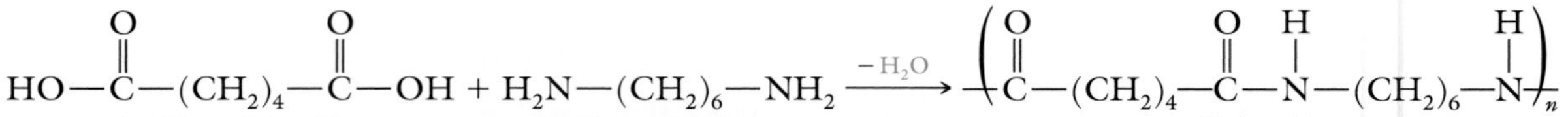

1,6-Hexandioic acid   1,6-Hexanediamine   Nylon-66

The annual production of nylon-66 in the United States is more than 4 billion pounds. About 80% of it goes into textile applications such as wearing apparel, carpeting, automobile seat belts, tire cord, and parachutes. The remainder goes into plastics applications such as kitchen utensils, power tool housings, automobile components (engine fans, brake and power steering reservoirs, lamp housings), motor gears and bearings, ski boots, and tennis racquet frames. Other polyamides are used in fire-resistant clothing for fire fighters and bullet-proof vests for law-enforcement personnel. Nylon polymers are also used in surgical applications such as suture material.

16.10 NAMING AMIDES

Like all derivatives of carboxylic acids, amide names are based on the name of the corresponding carboxylic acid:

Rules for naming amides

- The ending of the name of the corresponding carboxylic acid is changed from **-ic acid** (common) or **-oic acid** (IUPAC) to **-amide.**
- The names of groups attached to nitrogen are placed first, using an *N*- prefix for each group.

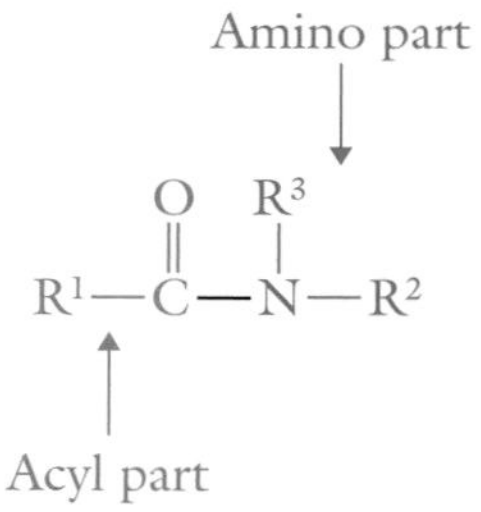

To name an amide, we need to distinguish the acyl and amino parts of the amide. The acyl and amino parts are those derived from the carboxylic acid and amine, respectively.

Example 16.10   Naming amides

Name the following compounds:

(a) $CH_3CH_2CH_2\overset{O}{\overset{\|}{C}}-\overset{CH_3}{\overset{|}{N}}-CH_2CH_3$ (b) $CH_3CH_2\overset{H}{\overset{|}{N}}-\overset{O}{\overset{\|}{C}}-CH_3$

(c) $CH_3CH_2\overset{O}{\overset{\|}{C}}-CH_2-\overset{H}{\overset{|}{N}}-H$

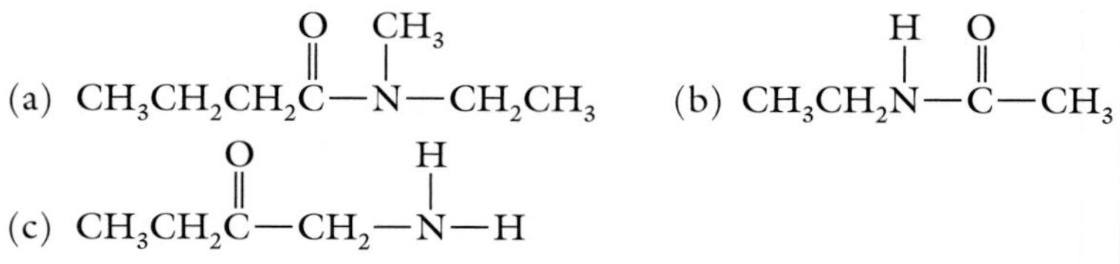

Solution

(a) The acyl and amino parts of the amide are:

Amino part

$$CH_3CH_2CH_2\overset{O}{\overset{\|}{C}}-\overset{CH_3}{\overset{|}{N}}-CH_2CH_3$$

Acyl part

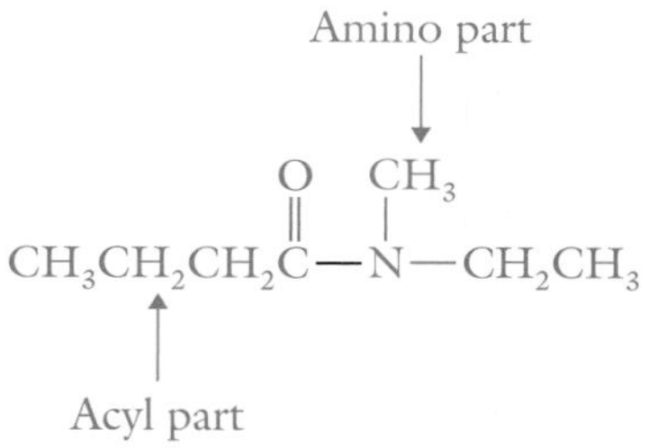

The carboxylic acid is butanoic acid (IUPAC name) or butyric acid (common name). One ethyl and one methyl group are attached to the nitrogen. The IUPAC name is *N*-ethyl-*N*-methylbutanamide; the common name is *N*-ethyl-*N*-methylbutyramide.

(b) This amide is not drawn in the usual manner with the acyl part on the left and the amino part on the right. It is drawn with the acyl part on the right and the amino part on the left. The carboxylic acid is ethanoic acid (IUPAC name) or acetic acid (common name). An ethyl group is attached to the nitrogen. The IUPAC name is *N*- ethylethanamide; the common name is *N*-ethylacetamide.

(c) The compound is not an amide. It is a ketone and an amine. It is named as a ketone because oxygen-containing functional groups have preference over amines (Section 16.3). The name is 1-amino-2-butanone.

Problem 16.10 Name the following compounds:

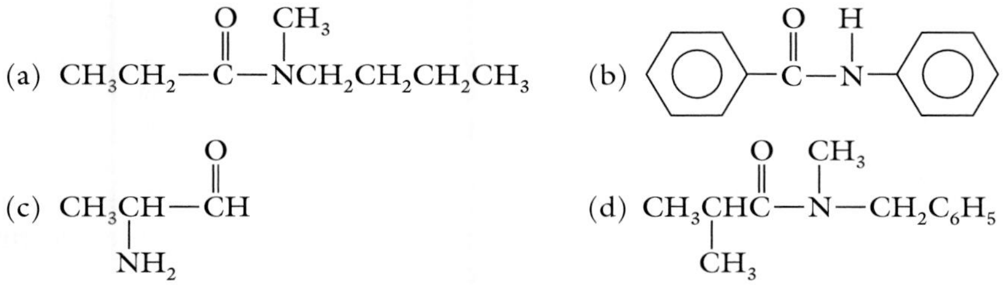

16.11 PHYSICAL AND BASICITY PROPERTIES OF AMIDES

The physical properties of amides are summarized by three observations:

- Amides have higher melting and boiling points than those of carboxylic acids (Table 16.4).
- The order of melting and boiling points of amides is primary ≅ secondary > tertiary.
- Amides have slightly higher water solubility than that of carboxylic acids.

Amides have the strongest secondary attractive forces and the highest melting and boiling points of any covalent organic compounds. Recall that the melting and boiling points of carboxylic acids are much higher than those of other families (see Table 16.2), yet the amides have even higher melting and boiling points (Table 16.4). For example, the melting and boiling points of

TABLE 16.4 Comparison of Carboxylic Acids and Amides

Compound	Structure	Molecular Mass (amu)	Melting Point (°C)	Boiling Point (°C)
acetic acid	CH_3COOH	60	17	118
acetamide	CH_3CONH_2	59	82	221
N-methylacetamide	$CH_3CONHCH_3$	73	28	204
N,N-dimethylacetamide	$CH_3CON(CH_3)_2$	87	−20	165

acetamide are 82°C and 221°C, respectively, compared with 17°C and 118°C for acetic acid.

Amides possess very strong dipole–dipole attractive forces, stronger even than the very strong hydrogen-bond attractions in carboxylic acids. In an amide group, the polarized $^{\delta+}C{-}O^{\delta-}$ bond of the carbonyl group interacts strongly with the nonbonded electron pair of nitrogen. The electron pair of nitrogen is pulled toward the carbonyl carbon, and the amide group is best described as a resonance hybrid of the following structures:

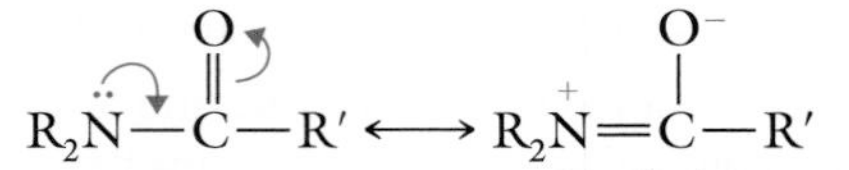

Dipolar ion

The structure on the right is called a **dipolar ion** because atoms bearing opposite charges are contained in the same molecule. This dipolar ion contains positive and negative charges on nitrogen and oxygen, respectively, separated by three atoms (instead of the usual two atoms as in the C=O group). The large charge separation results in greatly increased polarity and secondary attractive forces. In line with the N=C double bond structure of the amide group, the bond angles about the nitrogen are close to the 120° value about the carbon–carbon double bonds in alkenes (Section 12.2). We will see that the dipolar ion structure of an amide is an important contribution to the structure and physiological functions of proteins (Section 20.5).

In primary and secondary amides, but not in tertiary amides, hydrogen-bond attractions are superimposed on the very strong dipole–dipole attractions. Thus, primary and secondary amides have higher secondary forces than those of tertiary amides, and this difference is apparent in the differences in their melting and boiling points (see Table 16.4). The melting and boiling points of primary and secondary amides are not only significantly higher than those of tertiary amides, but also much higher than those of carboxylic acids.

Amides are slightly more water solubile than carboxylic acids. Amides undergo very strong attractive interactions (both dipole–dipole and hydrogen bond) with water because of the high polarity of the $^{+}N{=}C{-}O^{-}$ structure.

Whereas amines are weak bases, amides are not more basic than water, ethers, and alcohols. The structural feature that results in the very high boiling and melting points of amides—the electron-withdrawing effect of the carbonyl group on the amide nitrogen—is also responsible for the lack of basicity in amides. This electron-withdrawing effect makes the nonbonded electron pair of nitrogen much less available for accepting a proton.

16.12 HYDROLYSIS OF AMIDES

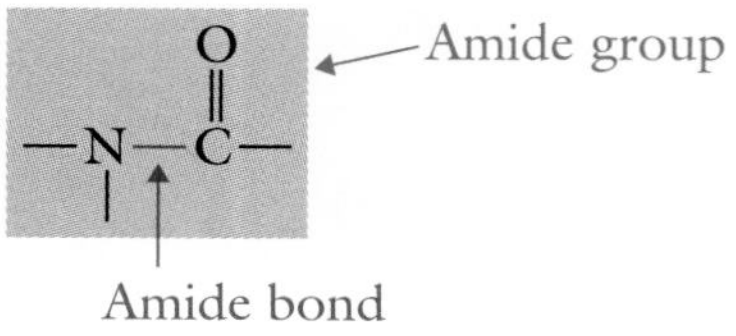

Amides undergo hydrolysis (a cleavage reaction with water) less readily than do esters. In fact, the **amide bond** (the bond between the carbonyl carbon and the nitrogen) is very resistant to hydrolysis at neutral pH, a property of great importance in biological systems, given that the stability of proteins depends on the stability of the amide bond. However, there are two ways in which amides are made to hydrolyze to the corresponding carboxylic acid and ammonia or amine: by the use of a strong acid (acidic hydrolysis) or a strong base (basic hydrolysis). The reaction is similar to the hydrolysis of esters (Section 15.11).

In the absence of acid or base, the hydrolysis equilibrium is so unfavorable that we can say that hydrolysis does not occur. In the presence of strong acid or base, one of the products of hydrolysis is converted into a salt, and this conversion drives the equilibrium to the right. Acidic hydrolysis forms the amine salt; basic hydrolysis forms the carboxylate salt.

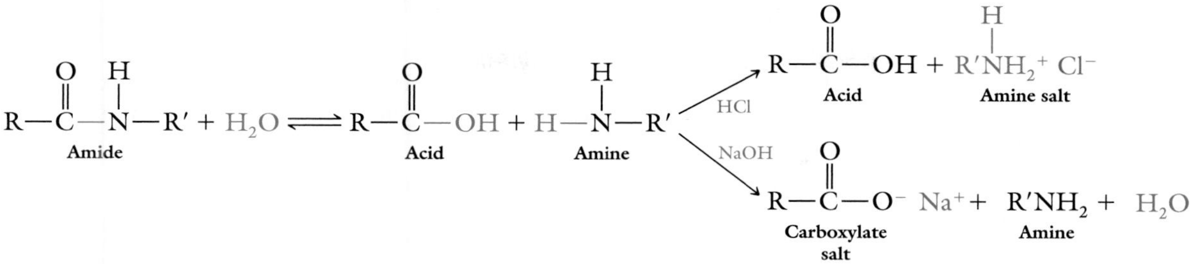

After acidic or basic hydrolysis has taken place, the product can be neutralized by adding strong base or acid, respectively, to convert the amine salt or carboxylate salt into the corresponding amine or carboxylic acid. Note that the acid and base are not catalysts in the hydrolysis reactions. The acid or base is a reactant and is required in equimolar amounts relative to the amide.

Amide synthesis and acidic hydrolysis take place continuously in living organisms. Proteins are amides, and their digestion is simply the acidic hydrolysis of amide bonds (Section 20.4). Living organisms also need to synthesize various proteins (for example, bone, muscle, enzymes). Both the protein synthesis and digestion reactions are catalyzed by enzymes.

▶▶ Enzymes are proteins that catalyze biological reactions, as described in Section 22.3.

Example 16.11 Writing equations for the hydrolysis of amides

Complete the following equations. If no reaction takes place, write "no reaction."

(a) $CH_3C(=O)-N(H)CH_2CH_3 \xrightarrow{H_2O} ?$

(b) $CH_3C(=O)-N(H)CH_2CH_3 \xrightarrow[HCl]{H_2O} ?$

(c) $CH_3C(=O)-N(H)CH_2CH_3 \xrightarrow[NaOH]{H_2O} ?$

Solution

(a) No reaction takes place, because of the absence of either a strong acid or a strong base.

(b) Hydrolysis proceeds to completion because a strong acid is present. Amide hydrolysis requires the breakage of one bond in each reactant, as indicated in the following reaction by the dotted line. The OH from water attaches to the acyl part of the amide to form the carboxylic acid. The H from water attaches to the nitrogen fragment from the amide to form the amine. The amine produced by hydrolysis is immediately converted into the amine salt by neutralization with HCl.

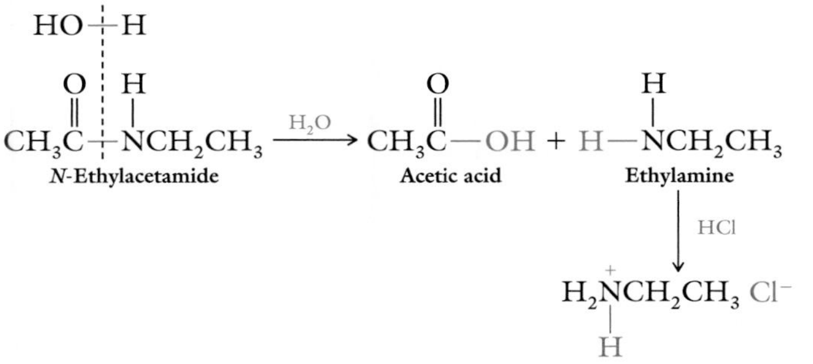

The overall reaction may be written in the abbreviated form:

$$CH_3\overset{O}{\underset{\|}{C}}-\overset{H}{\underset{|}{N}}CH_2CH_3 \xrightarrow[HCl]{H_2O} CH_3\overset{O}{\underset{\|}{C}}-OH + H_3\overset{+}{N}CH_2CH_3 \; Cl^-$$

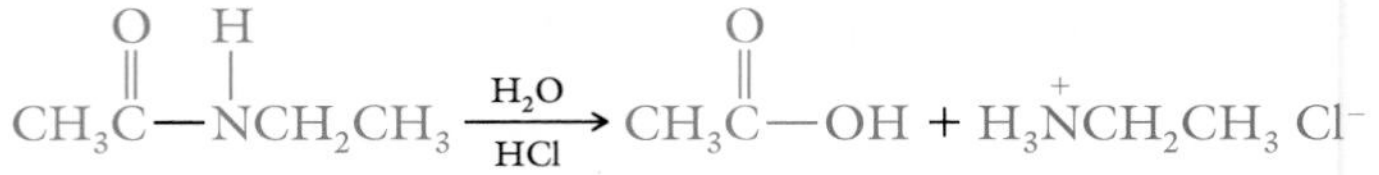

(c) Basic hydrolysis proceeds in a manner similar to acidic hydrolysis except that the presence of strong base converts the carboxylic acid product into the carboxylate salt:

$$CH_3\overset{O}{\underset{\|}{C}}\!\!\not|\,\overset{H}{\underset{|}{N}}CH_2CH_3 \xrightarrow{H_2O} CH_3\overset{O}{\underset{\|}{C}}-OH + H-\overset{H}{\underset{|}{N}}CH_2CH_3$$
$$HO\!\!\not|\,H \qquad\qquad\qquad \downarrow{\scriptstyle NaOH \atop -H_2O}$$
$$CH_3\overset{O}{\underset{\|}{C}}-O^-\,Na^+$$

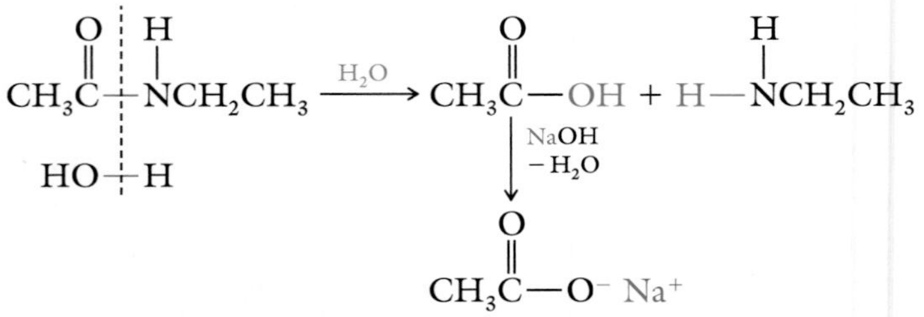

The overall reaction is written in the abbreviated form:

$$CH_3\overset{O}{\underset{\|}{C}}-\overset{H}{\underset{|}{N}}CH_2CH_3 \xrightarrow{NaOH} CH_3\overset{O}{\underset{\|}{C}}-O^-\,Na^+ + H_2NCH_2CH_3$$

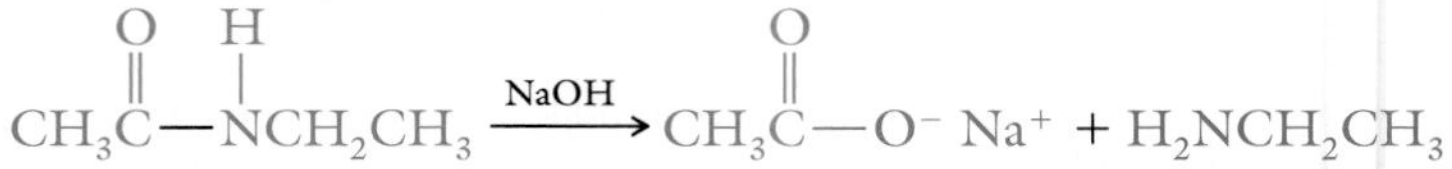

Problem 16.11   Complete the following equations. If no reaction takes place, write "no reaction."

(a) $CH_3CH_2CH_2\overset{O}{\underset{\|}{C}}-\overset{H}{\underset{|}{N}}-C_6H_5 \xrightarrow[H_2O]{HCl} ?$ 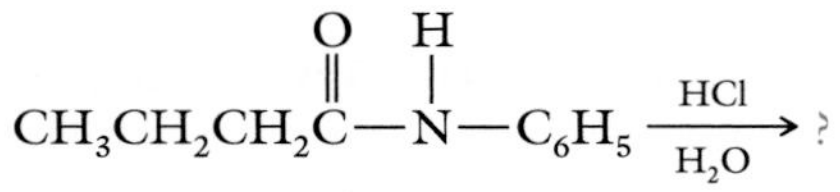

(b) $CH_3CH_2\overset{O}{\underset{\|}{C}}-\overset{H}{\underset{|}{N}}-CH_2CH_2CH_3 \xrightarrow[H_2O]{NaOH} ?$ 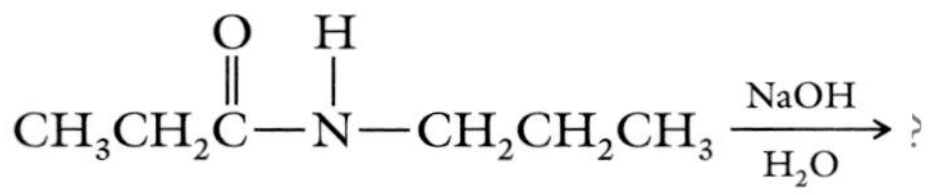

Summary

Comparison of Amines and Amides   Amines and amides are organic derivatives of ammonia in which one or more of the hydrogens are replaced by organic groups:

$$\underset{\text{Ammonia}}{H-\overset{}{\underset{|}{\underset{H}{N}}}-H} \qquad \underset{\text{Amine}}{-\overset{}{\underset{|}{N}}-} \qquad \underset{\text{Amide}}{-\overset{}{\underset{|}{N}}-\overset{O}{\overset{\|}{C}}-}$$

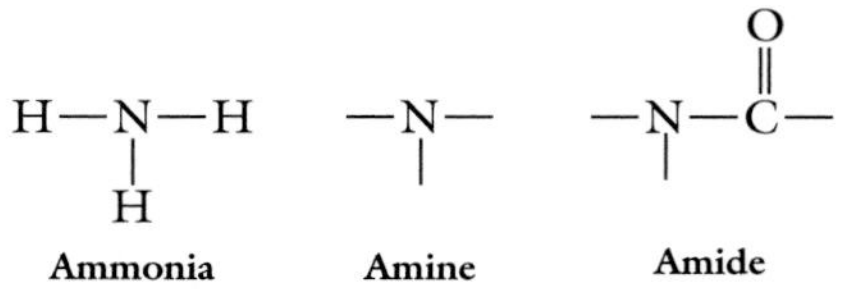

A compound is an amide instead of an amine when it has a carbonyl carbon attached to nitrogen. Amines and amides are classified as primary, secondary, or tertiary, respectively, when one, two, or three of the hydrogens of ammonia are replaced by organic groups.

Naming Amines   Amines containing simple groups are named by a common nomenclature system in which the names of the groups attached to nitrogen are followed without space by the suffix -amine.

In the *CA* system, amine names are based on the longest continuous carbon chain attached to nitrogen. The ending of the corresponding alkane's name is changed from -e to -amine. The position of attachment of nitrogen to the carbon chain is indicated by a number in front of the name. The carbon chain is numbered from the end nearest the nitrogen. Prefixes preceded by numbers indicate groups attached to the longest chain. Prefixes preceded by *N*- and *N,N*- indicate other groups attached to nitrogen.

A cyclic, nonaromatic compound containing an attached NH_2 group is named as a cycloalkanamine. The ring carbon holding the nitrogen is C1.

$C_6H_5NH_2$ is called aniline, and substituted anilines are named as derivatives of aniline. Substituents on nitrogen are distinguished from substituents on the aromatic ring by using an *N*- prefix instead of a number (or the *o*-, *m*-, *p*- prefixes).

Physical Properties of Amines   Primary and secondary amines, unlike tertiary amines, possess hydrogen-bonding capabilities and melting and boiling points comparable to those of aldehydes and ketones. Tertiary amines have melting and boiling points comparable to those of hydrocarbons and ethers. Amines of as many as three or four carbons are soluble in water in all proportions.

Basicity of Amines   Amines are weak bases because the nonbonded pair of electrons on nitrogen can accept (bond to) a proton from an acid. Aromatic amines are weaker bases than aliphatic amines.

Amine Salts   Amine salts, formed by reaction of amines with acids, are named by changing the ending of the name of the amine from -amine to -ammonium for aliphatic amines and from -e to -ium for aromatic and heterocyclic amines, followed by the name of the negative ion. Amine salts are ionic compounds that have higher melting and boiling points and greater water solubilities than covalent organic compounds have.

Synthesis and Naming of Amides Amides are formed by reactions of ammonia or primary or secondary amines with carboxylic acids, acid halides, or acid anhydrides. Amides are named by changing the ending of the name of the corresponding carboxylic acid from -ic acid or -oic acid to -amide. The names of groups attached to nitrogen are given first, using an *N*-prefix for each group.

Polyamides are formed by reactions of bifunctional carboxylic acids and amines.

Physical Properties of Amides Amides have the highest melting and boiling points of all covalent organic compounds because of very strong dipole–dipole secondary attractions, a result of interaction between the carbonyl group and the nonbonded electron pair of nitrogen. Primary and secondary amides, which, unlike tertiary amides, have N–H bonds, also have hydrogen-bond secondary attractions and thus higher melting and boiling points than tertiary amides. Amides have slightly greater water solubility than that of carboxylic acids.

Chemical Properties of Amides Amides are far less basic than amines because the nonbonded electron pair of nitrogen is much less available, a consequence of electron withdrawal by the carbonyl group. Amides are no more basic than water or alcohols.

Amides undergo acidic or basic hydrolysis to the corresponding carboxylic acid and amine or ammonia. The basic reaction produces the carboxylate salt, whereas the acidic reaction produces the amine salt.

Summary of Key Reactions

AMINES

Basicity

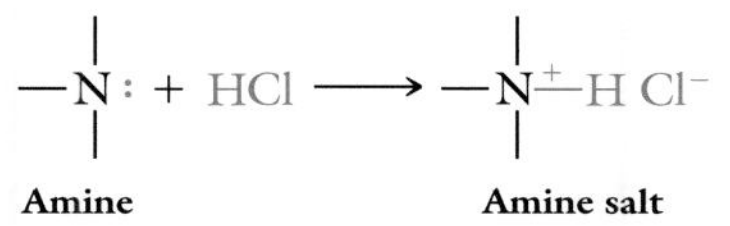

AMIDES

Formation

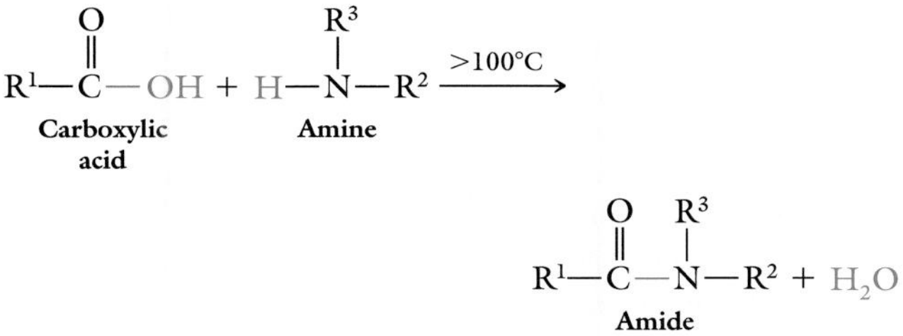

Amide is primary if both R^2 and R^3 are H, secondary if only R^2 or R^3 is H, and tertiary if neither R^2 nor R^3 is H.

Polymerization

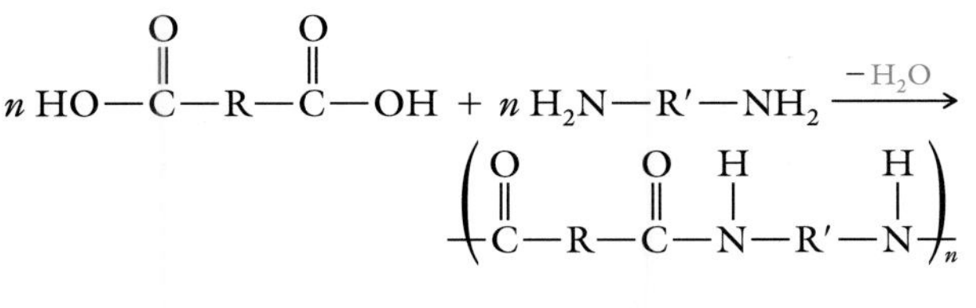

Hydrolysis

ACIDIC

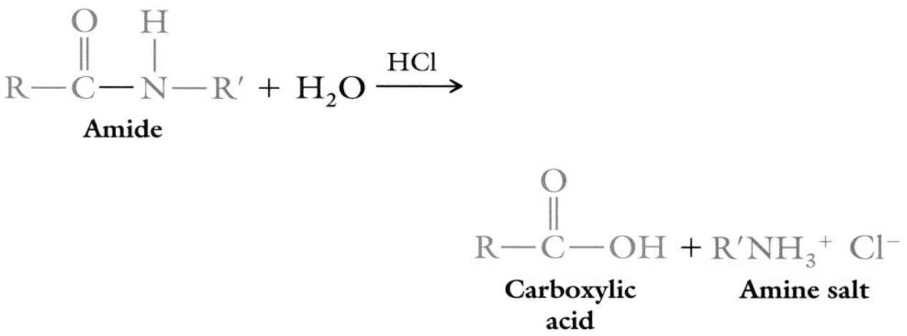

BASIC

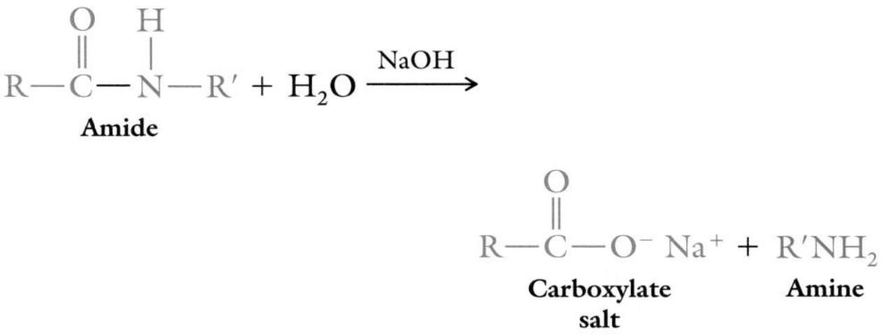

Key Words

aliphatic amine, p. 465
alkaloid, p. 469
amide, pp. 464, 480
amide bond, p. 486
amide group, p. 464
amine, p. 464
amine salt, p. 473
aromatic amine, p. 465
aryl amine, p. 465
dipolar ion, p. 486
heterocyclic amine, p. 468
nitrogen heterocyclic, p. 468
polyamide, p. 483
quaternary ammonium salt, p. 478

Exercises

Classification of Amines

16.1 Draw structural formulas for tertiary amines with the molecular formula $C_5H_{13}N$.

16.2 Draw structural formulas for secondary amines with the molecular formula $C_5H_{13}N$.

16.3 Classify each of the following compounds as amines, amides, or something else; as primary, secondary, or tertiary, if amines; as aliphatic or aromatic, if amines.

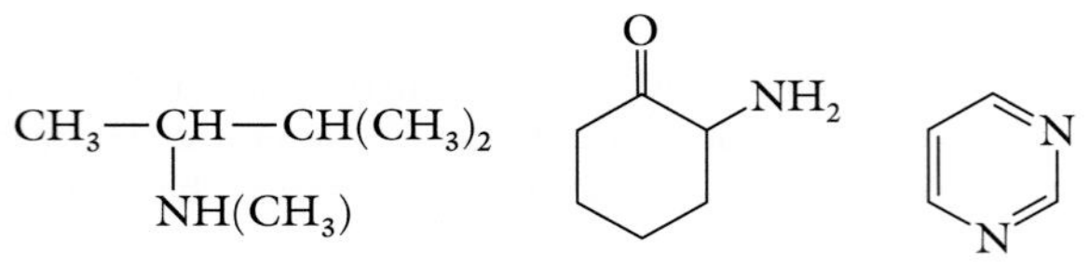

1 2 3

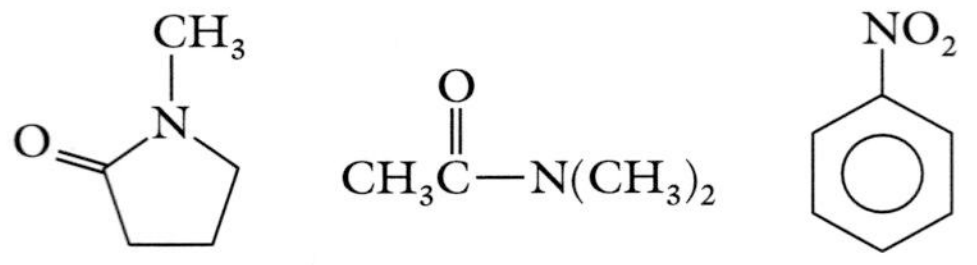

4 5 6

16.4 Classify each of the following compounds as amines, amides, or something else; as primary, secondary, or tertiary, if amines; as aliphatic or aromatic, if amines.

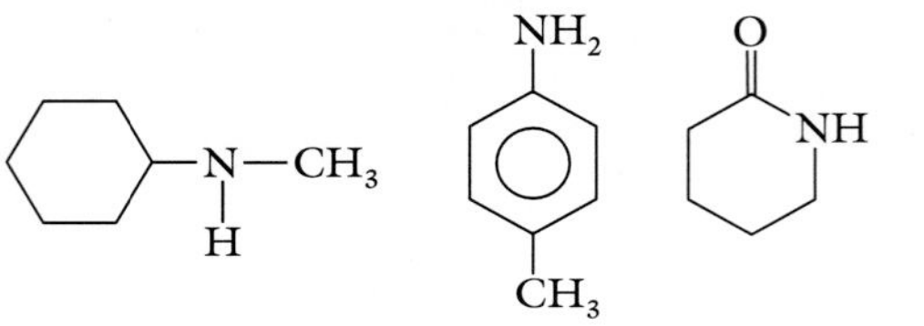

1 2 3

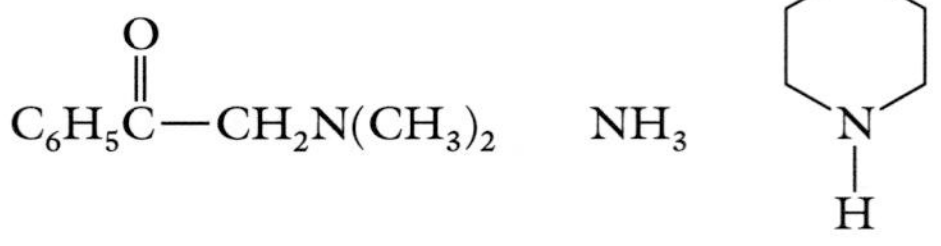

4 5 6

Naming Amines

16.5 Draw the structure of each of the following compounds: (a) *N*-methyl-1-pentanamine; (b) isopropylmethylamine; (c) *cis*-4-ethyl-*N*-propylcyclohexanamine; (d) *N*-ethylaniline.

16.6 Draw the structure of each of the following compounds: (a) *t*-butylamine; (b) 1,5-pentanediamine; (c) *trans*-3-ethyl-*N,N*-dimethylcyclopentanamine; (d) *N*-isopropyl-3,4-dimethylaniline.

16.7 Give the common name or *CA* name, whichever is appropriate, for each of the following compounds:

(a) 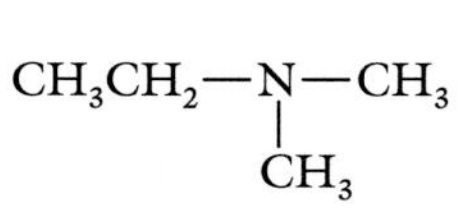 (b) 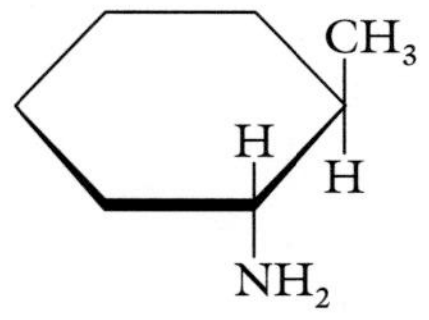

(c) 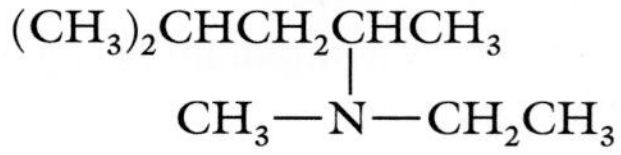

(d) 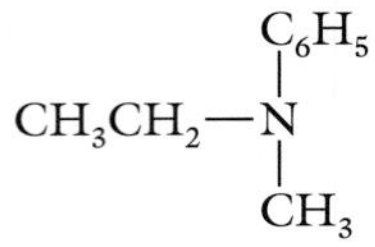

16.8 Give the common name or *CA* name, whichever is appropriate, for each of the following compounds:

(a) $CH_3CH_2CH_2CH_2NHCH(CH_3)_2$

(b) 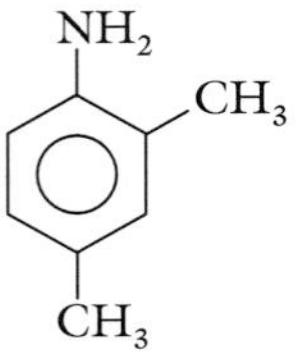 (c) 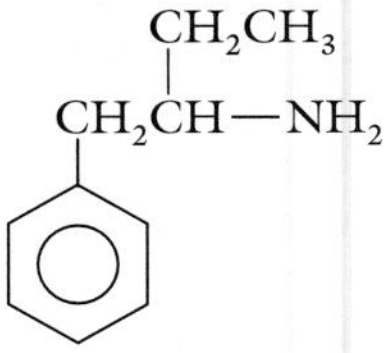

(d) 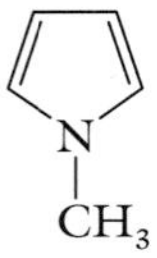

Physical Properties of Amines

16.9 Draw structural representation(s) to describe the hydrogen bonding responsible for the high boiling point of ethylamine.

16.10 Draw structural representation(s) to describe the hydrogen bonding present when ethylamine dissolves in water.

16.11 Indicate which compound in each of the following pairs has the higher value of the specified property. If the two compounds in a pair have nearly the same value for the property, indicate that fact. Explain your conclusions.
(a) Butylamine, 1-butanol: boiling point;.
(b) Butylamine, butanal: solubility in water.
(c) Butylamine, ethyldimethylamine: boiling point.
(d) Butylamine, diethylamine: solubility in water.

16.12 Indicate which compound in each of the following pairs has the higher value of the specified property. If the two compounds in a pair have nearly the same value for the property, indicate that fact. Explain your conclusions.
(a) Butylamine, butanal: boiling point.
(b) Butylamine, diethyl ether: boiling point.
(c) Butylamine, ethyldimethylamine: solubility in water.
(d) Butylamine, diethylamine: boiling point.

Basicity of Amines

16.13 Give the equation to show propylamine acting as a base toward HCl.

16.14 Indicate whether any acid-base reaction takes place under each of the following conditions. If a reaction takes place, write the appropriate equation. Make sure to indicate whether the reaction goes to completion or is an equilibrium reaction. (a) Butylamine + water; (b) butylamine + H_2SO_4.

16.15 Place the following compounds in order of increasing basicity: propylamine, KOH, aniline, water, 1-propanol. Explain the order of basicity.

16.16 Which is the stronger base, aniline or cyclohexylamine? Why?

16.17 An unknown compound is either 1-propanol or propylamine. The unknown turns red litmus paper blue. Identify the unknown.

16.18 An unknown compound is either 1-propanol or propylamine. The unknown does not turn red litmus paper blue. Identify the unknown.

16.19 Show the reaction of 1,6-hexanediamine, $H_2N(CH_2)_6NH_2$, with excess HCl.

16.20 What is the structure of the predominant species present when ethylamine is dissolved in water? What structure predominates when ethylamine is added to a buffered solution at pH = 7 (similar to body pH)?

Amine Salts

16.21 Give the structural formula of each of the following compounds: (a) *N,N*-dimethylcyclopentanammonium chloride; (b) *N*-ethyl-*N*-methylpiperidinium sulfate; (c) ethylmethylpropylammonium bromide.

16.22 Name each of the following compounds:

(a) 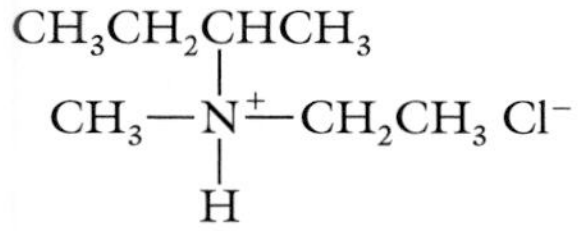

(b) 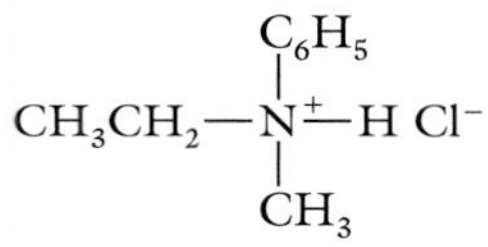

(c) 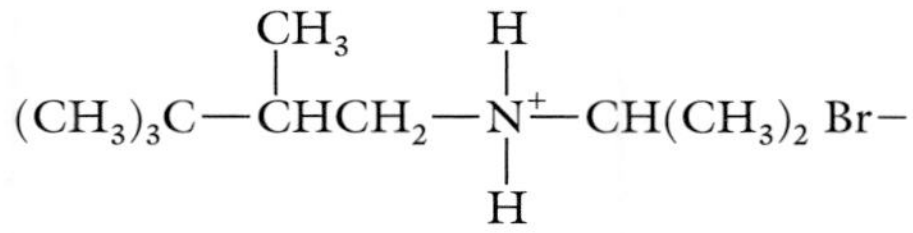

(d) 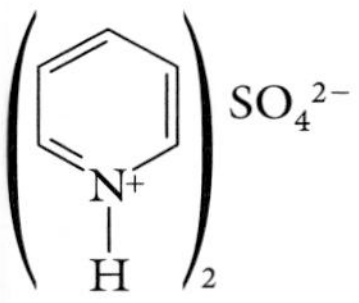

16.23 Indicate which compound in each of the following pairs has the higher value of the specified property. If the two compounds in a pair have nearly the same value for the property, indicate that fact. Explain your conclusions.
(a) Trimethylammonium chloride, triethylamine: boiling point.
(b) Hexylamine, hexylammonium chloride: solubility in water.

16.24 Give the equations showing trimethylammonium chloride acting as an acid toward the following bases. Make sure to indicate whether the reaction goes to completion or is an equilibrium reaction. (a) Water; (b) NaOH.

16.25 An unknown is either butylamine or butylammonium chloride. The unknown turns red litmus paper blue. Identify the unknown.

16.26 An unknown is either butylamine or butylammonium chloride. The unknown turns blue litmus paper red. Identify the unknown.

Classification of Amides

16.27 Which of the following compounds are amides? Classify amides as primary, secondary, or tertiary.

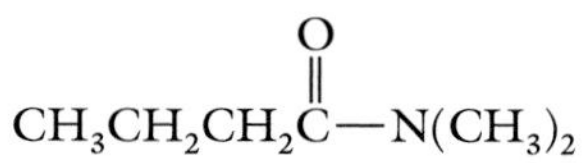

1

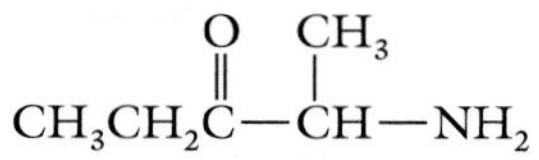

2

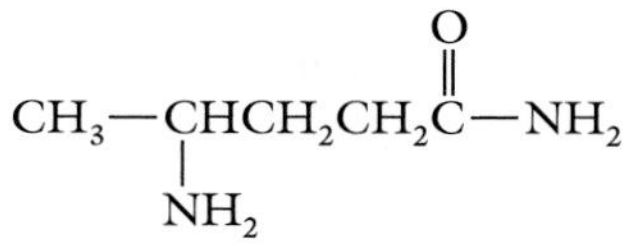

3

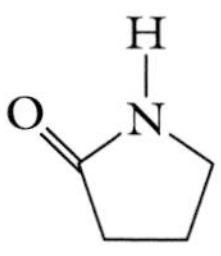

4

16.28 Draw the structures of all amides of formula C_3H_7NO.

Synthesis of Amides

16.29 Write the equation for the reaction that takes place between each of the following pairs of reactants: (a) benzoic acid and butylamine at 25°C; (b) benzoic acid and butylamine at > 100°C; (c) benzoyl chloride and ethylamine; (d) acetic anhydride and *N*-methylaniline.

16.30 Write the equation for the reaction that takes place between each of the following pairs of reactants: (a) acetic acid and trimethylamine at > 100°C; (b) butanoic acid and aniline at > 100°C; (c) acetyl bromide and aniline; (d) benzoic anhydride and ethylmethylamine.

16.31 What carboxylic acid and amine are needed to synthesize each of the following amides?

(a) 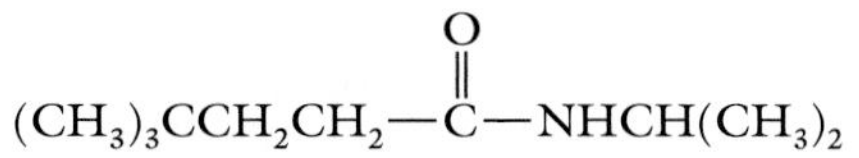

(b) 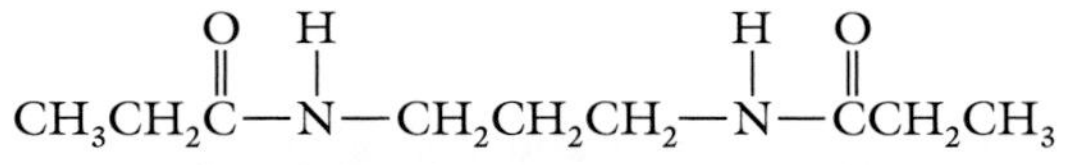

16.32 What carboxylic acid and amine are needed to synthesize each of the following amides?

(a) 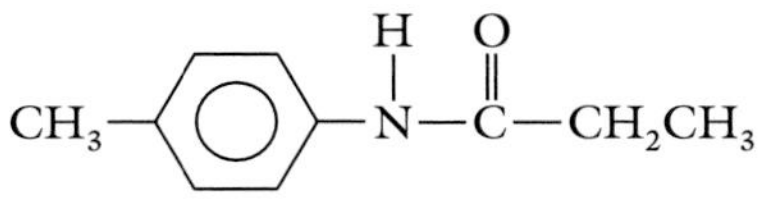

(b) 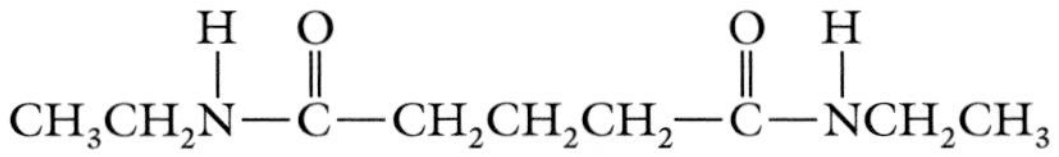

Polyamide Synthesis

16.33 What diacid and diamine reactants are needed to synthesize the following polyamide?

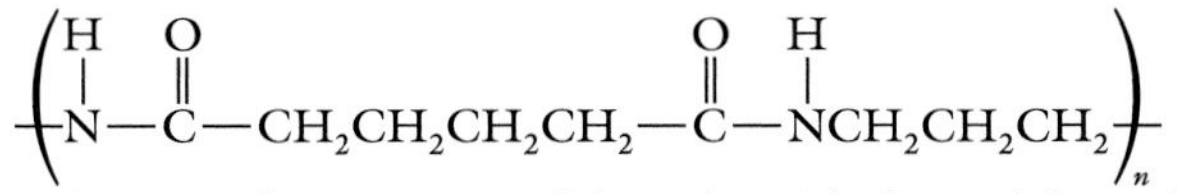

16.34 Draw the structure of the polyamide formed from the reaction of 1,6-hexanedioic acid and 1,4-benzenediamine.

Naming Amides

16.35 Give the structure of each of the following amides:
(a) *N-t*-butylbutanamide.
(b) *N*-phenyl-3-methylhexanamide.

16.36 Give the structure of each of the following amides:
(a) N-ethyl-N-propylbenzamide;
(b) 2,N-dimethyl-N-phenylpropanamide.

16.37 Name each of the following compounds:

(a) $CH_3CH_2CH_2CH_2$—$\overset{\overset{\displaystyle O}{\|}}{C}$—$NHCH(CH_3)_2$

(b) $\langle O \rangle$—$\overset{\overset{\displaystyle H}{|}}{N}$—$\overset{\overset{\displaystyle O}{\|}}{C}$—$CH_2CH_3$

(c) $(CH_3CH_2)_2N$—$\overset{\overset{\displaystyle O}{\|}}{C}$—$C_6H_5$

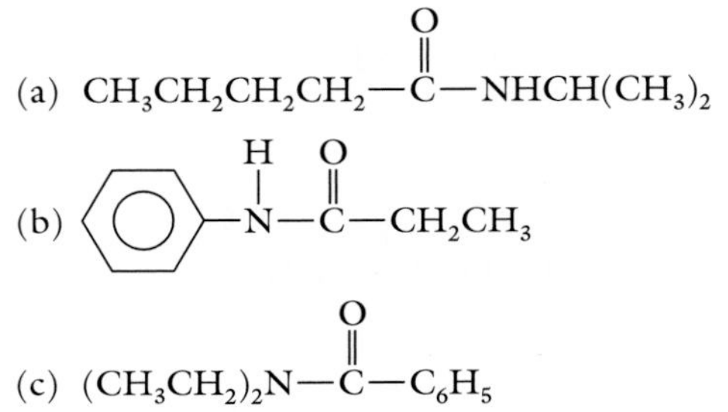

16.38 Name each of the following compounds:

(a) $CH_3CH_2CH_2\overset{\overset{\displaystyle O}{\|}}{C}$—$\overset{\overset{\displaystyle CH_3}{|}}{N}$—$CH_2CH_3$

(b) $CH_3CH_2\overset{\overset{\displaystyle H}{|}}{N}$—$\overset{\overset{\displaystyle O}{\|}}{C}$—$CH_3$

(c) CH_3—$\overset{\overset{\displaystyle}{\underset{\underset{\displaystyle NH_2}{|}}{CH}}}CH_2CH_2\overset{\overset{\displaystyle O}{\|}}{C}$—$NH_2$

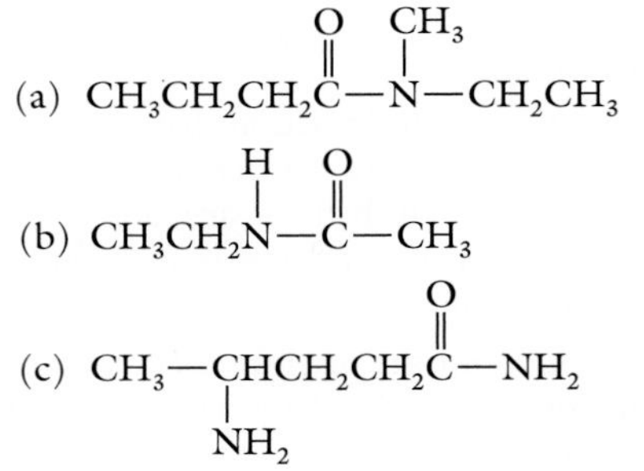

Physical and Basicity Properties of Amides

16.39 Indicate which compound in each of the following pairs has the higher value of the specified property. If the two compounds in a pair have nearly the same value for the property, indicate that fact. Explain your conclusions.
(a) Pentanamide, N,N-dimethylpropanamide: boiling point.
(b) Pentanamide, N,N-dimethylpropanamide: solubility in water.
(c) Butanamide, butanoic acid: boiling point.
(d) Hexanamide, hexanoic acid: solubility in water.
(e) Butanamide, 1-pentanol: boiling point.

16.40 Which of the following representations correctly describe hydrogen bonding?

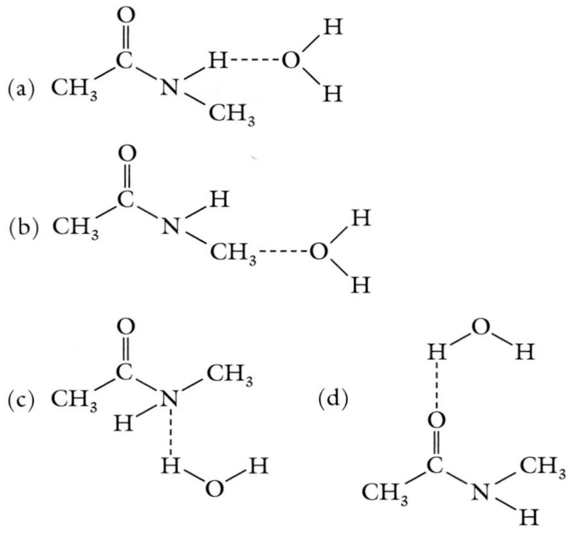

16.41 Place the following compounds in order of increasing basicity: ethanamide, propanamine, 1-propanol, NaOH, water. Explain the reasons for the order.

16.42 Isomers 1 and 2 both have NH_2 and $C{=}O$ groups but isomer 1 is 10,000-fold less basic than isomer 2. What is the reason for the difference?

CH_3CH_2—$\overset{\overset{\displaystyle O}{\|}}{C}$—$NH_2$ CH_3—$\overset{\overset{\displaystyle O}{\|}}{C}$—$CH_2NH_2$
       1   **2**

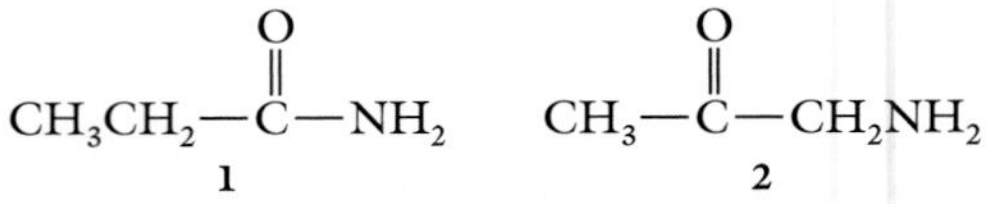

Hydrolysis of Amides

16.43 Write the equation for: (a) acidic hydrolysis of N-ethyl-N-methylpropanamide with HCl; (b) basic hydrolysis of N-phenylbenzamide with KOH.

16.44 Give the structure of the organic product in each of the following reactions. If no reaction takes place, write "no reaction." If there is more than one product, show only the major product(s).

(a) $CH_3CH_2\overset{\overset{\displaystyle H}{|}}{N}$—$\overset{\overset{\displaystyle O}{\|}}{C}$—$\langle O \rangle$ $\xrightarrow[\text{HCl}]{H_2O}$?

(b) $CH_3CH_2\overset{\overset{\displaystyle O}{\|}}{C}$—$\overset{\overset{\displaystyle H}{|}}{N}$—$CH_2CH_2CH_3$ $\xrightarrow[\text{H}_2\text{O}]{\text{NaOH}}$?

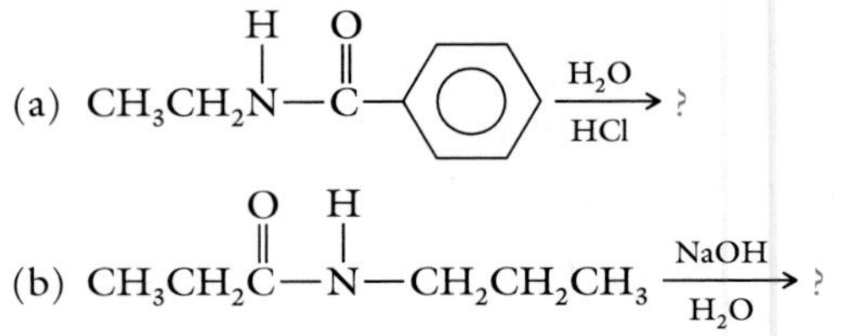

Unclassified Exercises

16.45 Draw structural formulas for each of the following compounds: (a) primary amide containing three carbons; (b) tertiary amide of four carbons and containing three methyl groups; (c) lactam (cyclic amide) of five carbons with all carbons in the ring; (d) secondary amine of molecular formula C_7H_9N and containing a benzene ring; (e) tertiary amine of molecular formula $C_6H_{15}N$ and containing a t-butyl group.

16.46 Give the structural formula of each of the following compounds: (a) N-isobutylaniline; (b) p-isobutylaniline; (c) N,N-dimethylpiperidinium chloride; (d) N-methyl-3-chlorobutanamide; (e) 3-aminopentanamide; (f) butylisopropylmethylamine.

16.47 Give the structural formula of each of the following compounds: (a) 5-amino-3-methyl-2-pentanone; (b) 4-pentene-1-amine.

16.48 What are the hybridizations of atoms a, b, c, and d in the following compound? What are bond angles A, B, and C?

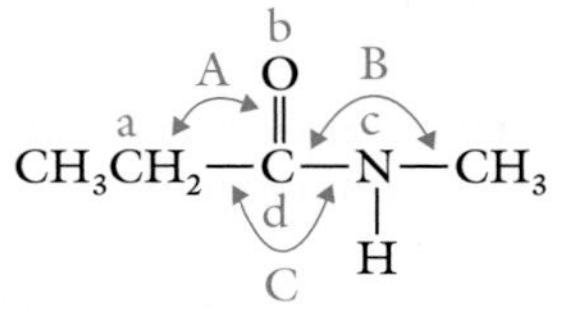

16.49 What are the hybridizations of atoms a, b, and c in the following compounds? What are bond angles A and B?

(a) CH_3CH_2—$\overset{\overset{\displaystyle A}{\frown}}{\underset{\underset{\displaystyle H}{|}}{N}}$—$CH_3$ (b) CH_3CH_2—$\overset{\overset{\displaystyle H \;\; A}{|}}{\underset{\underset{\displaystyle H}{|}}{N^{\pm}}}$—$CH_3$ $\;Cl^-$

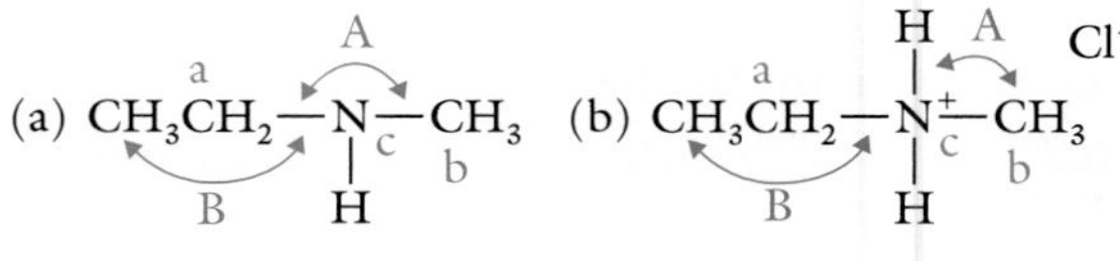

16.50 Identify amine and amide groups in the following compounds. Classify amines and amides as primary, secondary, or tertiary. Which amines are heterocyclic amines?

(a) 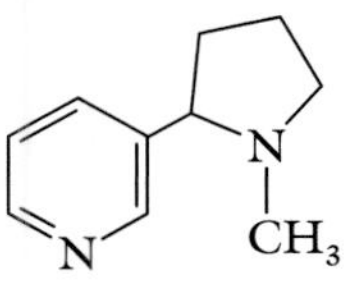

Nicotine

(b) 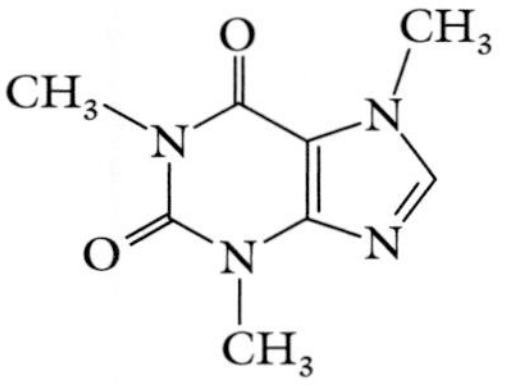

Caffeine

(c) 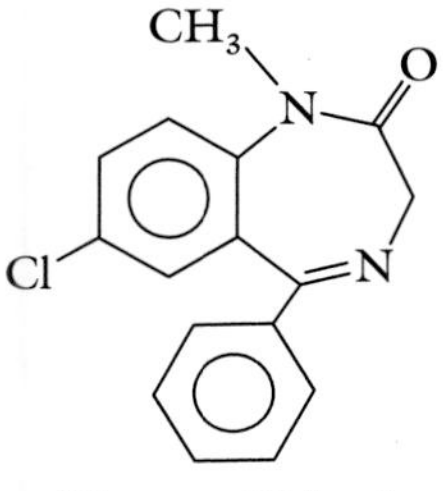

Diazepam (Valium)

16.51 Indicate which compound in each of the following pairs has the higher value of the specified property. If the two compounds in a pair have nearly the same value for the property, indicate that fact. Explain your conclusions.
(a) Butanamide, hexanamine: boiling point.
(b) Butanamide, hexanamine: solubility in water.
(c) Pentanamide, pentanoic acid: boiling point.
(d) Pentanamide, pentanoic acid: solubility in water.
(e) Trimethylammonium chloride, hexanamide: boiling point.
(f) Pentanamide, sodium butanoate: boiling point.
(g) Butanamine, 1,8-octanediamine: solubility in water.

16.52 Place the following compounds in order of boiling point: methyl propanoate, 1-pentanol, pentylamine, butanoic acid, pentanal, trimethylammonium chloride, butanamide. Explain the order.

16.53 An unknown compound is either an amine or an amide. What simple chemical diagnostic test can be used to distinguish between the two possibilities?

16.54 An unknown compound is either propylamine, propanol, or ethanamide. The unknown turns red litmus paper blue. Identify the unknown.

16.55 An unknown compound is either propylamine, propanol, or ethanamide. The unknown does not turn blue litmus paper red. Identify the unknown.

16.56 Give the structure of the cyclic amide (lactam) formed from 5-aminopentanoic acid.

16.57 Each of the following pairs of reactants undergoes amide formation. Which of the pairs produces a polymer? Write the equation for the reaction(s) that produce(s) polymer.

(a) $HO-\overset{O}{\overset{\|}{C}}CH_2CH_2CH_2CH_2\overset{O}{\overset{\|}{C}}-OH + H_2NCH_2CH_3$

(b) $HOOCCH_2CH_2CH_2CH_3 + H_2NCH_2CH_2NH_2$

(c) $HO-\overset{O}{\overset{\|}{C}}CH_2CH_2CH_2CH_2\overset{O}{\overset{\|}{C}}-OH + H_2NCH_2CH_2NH_2$

16.58 Give the structure of the organic product in each of the following reactions. If no reaction takes place, write "no reaction." If there is more than one product, show only the major product(s).

(a) $CH_3CH_2\overset{O}{\overset{\|}{C}}-NHCH_2C_6H_5 \xrightarrow[HCl]{H_2O} ?$

(b) $CH_3CH_2\overset{O}{\overset{\|}{C}}-NHCH_2C_6H_5 \xrightarrow[NaOH]{H_2O} ?$

(c) 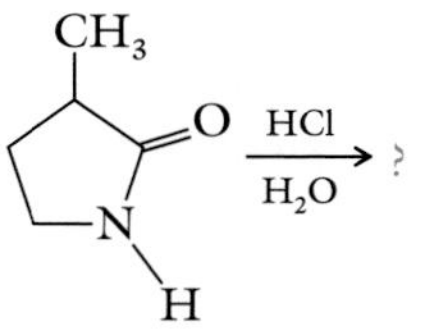

(d) $\left(\overset{H}{\overset{|}{N}}-\overset{O}{\overset{\|}{C}}(CH_2)_4\overset{O}{\overset{\|}{C}}-\overset{H}{\overset{|}{N}}(CH_2)_3 \right)_n \xrightarrow{\text{hydrolysis}} ?$

(e) $CH_3CH_2CH_2CH_2\overset{O}{\overset{\|}{C}}-OH \xrightarrow[25°C]{C_6H_5NH_2} ?$

(f) $CH_3CH_2CH_2CH_2\overset{O}{\overset{\|}{C}}-OH \xrightarrow[>100°C]{C_6H_5NH_2} ?$

(g) $CH_3CH_2-\underset{H}{\underset{|}{\overset{C_6H_5}{\overset{|}{N}}}} \xrightarrow{H_2SO_4} ?$

(h) $CH_3CH_2CH_2CH_2\overset{O}{\overset{\|}{C}}-Cl \xrightarrow{CH_3NH_2} ?$

(i) 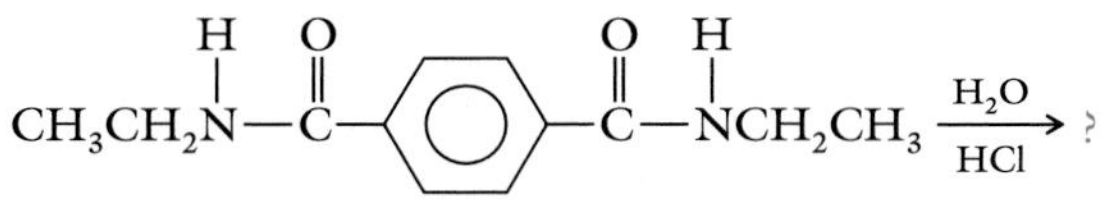

16.59 For each of the following pairs of structural formulas, indicate whether the pair represents (1) the same compound or (2) different compounds that are constitutional isomers or (3) different compounds that are geometrical isomers or (4) different compounds that are not isomers.

(a) $CH_3CH_2-\overset{O}{\overset{\|}{C}}-N(CH_3)_2$ or $CH_3CH_2-\overset{O}{\overset{\|}{C}}-CH_2N(CH_3)_2$

(b) $CH_3CH_2-\overset{O}{\overset{\|}{C}}-N(CH_3)_2$ or $H-\overset{O}{\overset{\|}{C}}-CH_2CH_2N(CH_3)_2$

(c) $(CH_3)_3N$ or $CH_3CH_2CH_2NH_2$

(d) 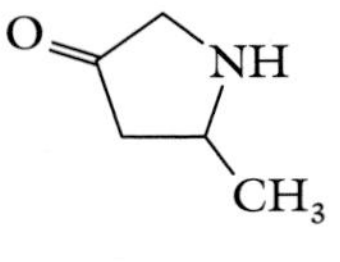 and 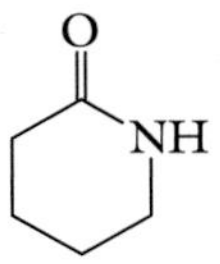

Chemical Connections

16.60 Low-molecular-mass amines have distinct and disagreeable fishy odors. Cooks often squeeze lemon juice, which contains citric acid (Box 15.1), on fish to cut down on the fishy odors resulting from such amines. How does the addition of citric acid result in a less intense fishy odor?

16.61 Using the method described in Box 9.1, calculate the pH of a 1.00 *M* aqueous solution of methylamine. The value of K_b for methylamine is 4.59×10^{-4}.

16.62 Ephedrine is used as a bronchodilator and decongestant. Phenobarbital is used as an anticonvulsant and sedative. What simple chemical test distinguishes these two medicines?

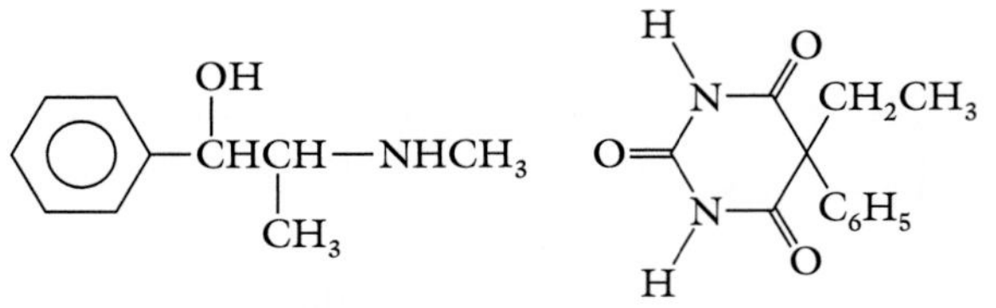

Ephedrine **Phenobarbital**

16.63 The amide $C_5H_{11}NO$ yields ethylamine on hydrolysis. Identify the other product(s) of hydrolysis.

16.64 1-Decanol and 1-decanamine are products marketed by the XYZ Chemical Company. A mistake in the plant results in the mixing of these two products. There is a need to quickly separate the mixture into the individual components because customers are awaiting delivery of the two chemicals. Distillation would be a useful method for the needed separation, but the distillation equipment is broken and will not be repaired for several days. Describe an alternate method of separating 1-decanol from 1-decanamine. Note that both 1-decanol and 1-decanamine are insoluble in water.

16.65 A 10% aqueous ammonia (NH_3) solution is useful as a reflex respiratory stimulant ("smelling salt"). Explain why a solution of ammonium chloride (NH_4Cl) is not useful for the same purpose.

CHAPTER 17

STEREOISOMERISM

CHEMISTRY IN YOUR FUTURE

On his way back to see the doctor for an examination, a patient pauses to discuss his high-blood-pressure drug with you. He says that he's been meaning to ask why his drug suddenly underwent an increase in price a few months ago. You explain that the FDA recently approved a "single enantiomer" version of the drug. The new version is purer and more effective and causes fewer side effects, so the clinic feels that it is worth the cost of the more expensive manufacturing process. "Purer in what way?" he asks. This chapter helps you to answer his question.

LEARNING OBJECTIVES

- Distinguish between constitutional isomers and stereoisomers.
- Distinguish between chiral and achiral molecules.
- Identify tetrahedral stereocenters.
- Draw and name enantiomers of compounds with one tetrahedral stereocenter.
- Describe the optical activity of chiral compounds.
- Describe the process of chiral recognition.
- Draw enantiomers and diastereomers of compounds with two tetrahedral stereocenters.
- Draw enantiomers and diastereomers of cyclic compounds with tetrahedral stereocenters.

In Chapter 11, where we began our study of organic chemistry, two kinds of isomers were defined: constitutional isomers and stereoisomers. In reality, there are two classes of stereoisomers—enantiomers and diastereomers—and there are several subclasses of diastereomers. In Chapters 11 and 12, we studied one subclass of diastereomers: geometric (also called cis-trans) isomers of alkenes and cycloalkanes. Before beginning our study of biochemistry, we need to introduce enantiomers and the remaining subclasses of diastereomers, because they are found in biological molecules (carbohydrates, lipids, proteins, and nucleic acids). Evolution has occurred in such a way that, for any molecule that can exist as two or more stereoisomers, biological systems use only one of the stereoisomers. Thus, stereoisomerism is crucial to the specificity of biochemical processes. We observed such specificity in the effect of cis-trans isomerism on our perception of light and in the action of pheromones in insects (Boxes 12.2 and 12.3). In biochemical reactions, differences in stereoisomers can be as important as or more important than differences in functional groups.

Before considering the stereoisomers that we have not previously encountered, let's review what we have learned about isomers in earlier chapters.

17.1 REVIEW OF ISOMERISM

Isomers are different compounds that have the same molecular formula. Figure 17.1 shows that there are two different types of isomers: constitutional isomers and stereoisomers. **Constitutional isomers,** also called **structural isomers,** are different compounds with the same molecular formula but different **connectivity,** the order in which atoms are attached to each other (Sections 11.5, 12.3, 13.2). There are three types of connectivity differences, any one of which gives rise to constitutional isomers:

- Different carbon skeletons, as in butane and 2-methylpropane:

$$CH_3-CH_2-CH_2-CH_3 \qquad CH_3-\underset{\underset{CH_3}{|}}{CH}-CH_3$$

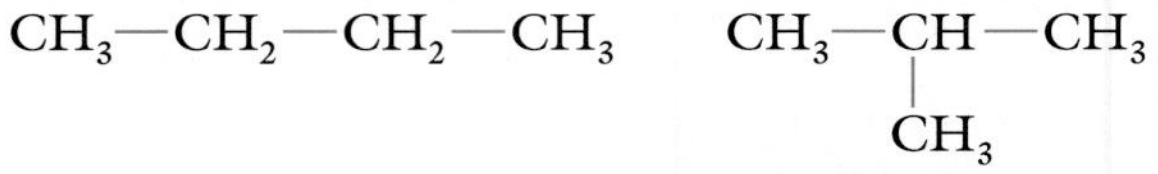

Butane   **2-Methylpropane**

- Different placements of the functional group(s) on the same carbon skeleton, as in 1-propanol and 2-propanol:

$$CH_3-CH_2-CH_2-OH \qquad CH_3-\underset{\underset{OH}{|}}{CH}-CH_3$$

1-Propanol   **2-Propanol**

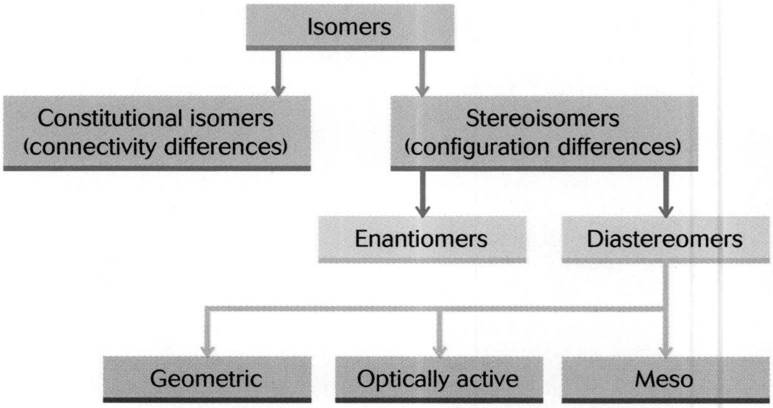

Figure 17.1 Different types of isomers.

- Different functional groups together with different carbon skeletons, as in ethanol and dimethyl ether:

$$\underset{\textbf{Ethanol}}{CH_3CH_2{-}OH} \qquad \underset{\textbf{Dimethyl ether}}{CH_3{-}O{-}CH_3}$$

Stereoisomers have the same connectivity but differ in their configurations. **Configuration** describes the relative orientations in space of the atoms of a stereoisomer, independent of changes that occur by rotation about single bonds. (Configuration must not be confused with conformation. **Conformation** describes the relative orientations of the atoms of a compound caused by rotation about single bonds—Section 11.4.)

There are two classes of stereoisomers (see Figure 17.1):

- **Enantiomers** are stereoisomers that are nonsuperimposable mirror images of each other.
- **Diastereomers** are stereoisomers that are not enantiomers; that is, they are not mirror images of each other.

One type of diastereomer, **geometric stereoisomers** (also called **cis-trans stereoisomers**), was discussed in Sections 11.9 and 12.5. Geometric stereoisomers are found in alkenes and cyclic compounds when the two carbons of the double bond or any two carbons of a ring each contain two different substituents. Such carbon atoms are called **stereocenters** because a hypothetical exchange of the positions of any two substituents would convert one stereoisomer into the other. Examples of geometric isomers are *cis*- and *trans*-2-butene and *cis*- and *trans*-1,2-dimethylcyclopropane. The cis isomer differs from the trans isomer in the configuration at a stereocenter.

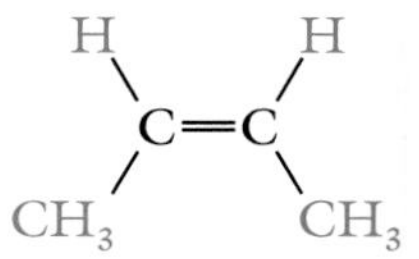

***cis*-2-Butene**

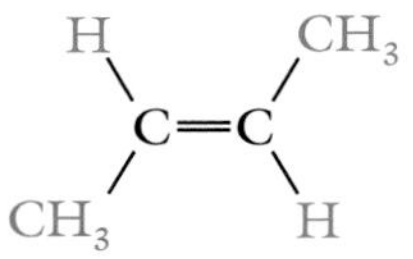

***trans*-2-Butene**

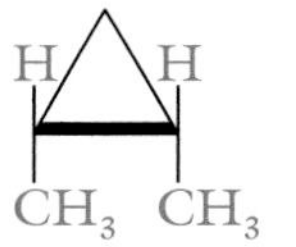

***cis*-1,2-Dimethylcyclopropane**

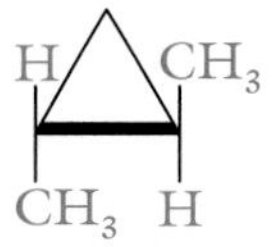

***trans*-1,2-Dimethylcyclopropane**

This chapter deals with enantiomers as well as those diastereomers not previously encountered.

17.2 ENANTIOMERS

Many of the organic compounds that we have come across in preceding chapters exist not as single compounds but as pairs of compounds called **enantiomers.** Two compounds constitute a pair of enantiomers if they meet both of the following conditions:

- The molecules of the two compounds are mirror images of each other.
- The molecules of the two compounds are nonsuperimposable on each other.

How do we test two molecules to decide if they meet these two conditions?

- **Mirror-image condition:** Imagine each molecule looking into a mirror. The two molecules are mirror images if each sees the other as its mirror image.

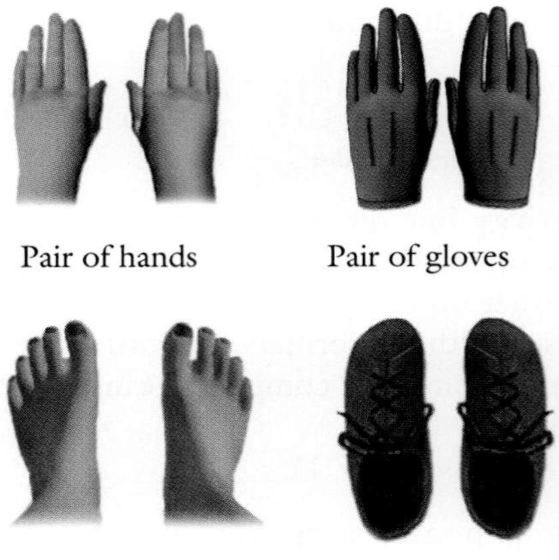

Figure 17.2 Chiral objects.

Figure 17.3 Our hands are chiral—nonsuperimposable, mirror-image objects. This figure shows that the left hand is identical to the mirror image of the right hand.

- **Nonsuperimposable condition:** Imagine the process of moving one molecule so that it merges into the other. The two molecules are superimposable if they become identical; that is, each and every part of one molecule coincides exactly with its corresponding part in the other molecule. The two molecules are nonsuperimposable if they do not become identical.

A molecule that is nonsuperimposable on its mirror-image molecule is said to be **chiral** or to have **chirality.** Molecules that do not possess chirality are **achiral.**

It is easier to understand chirality in molecules if we look at some chiral and achiral macroscopic objects in everyday life. Our hands and feet are chiral, as are a pair of shoes and a pair of gloves (Figure 17.2). Consider the relation between a pair of hands (Figure 17.3). A left hand is the nonsuperimposable mirror image of a right hand. By analogy, objects and molecules that are chiral are said to possess **handedness.** Our feet, too, are chiral; but most kinds of socks for our feet are not—they are achiral. Other familiar achiral objects are coffee cups and forks (Figure 17.4). These achiral objects are superimposable on their mirror image objects. Some achiral objects become chiral objects when they are decorated. For example, the coffee cup without any design is achiral, but the same coffee cup with the name MARY on it is chiral.

An example of a chiral molecule is 2-butanol. Two 2-butanols exist, whose molecules are nonsuperimposable mirror images of each other (Figure 17.5). Sliding one molecule over and into the other superimposes the $—OH$ and $—C_2H_5$ groups but not the $—CH_3$ and $—H$ groups. Each of the molecules is chiral.

The spiraled seashell is chiral.

Figure 17.4 Achiral objects.

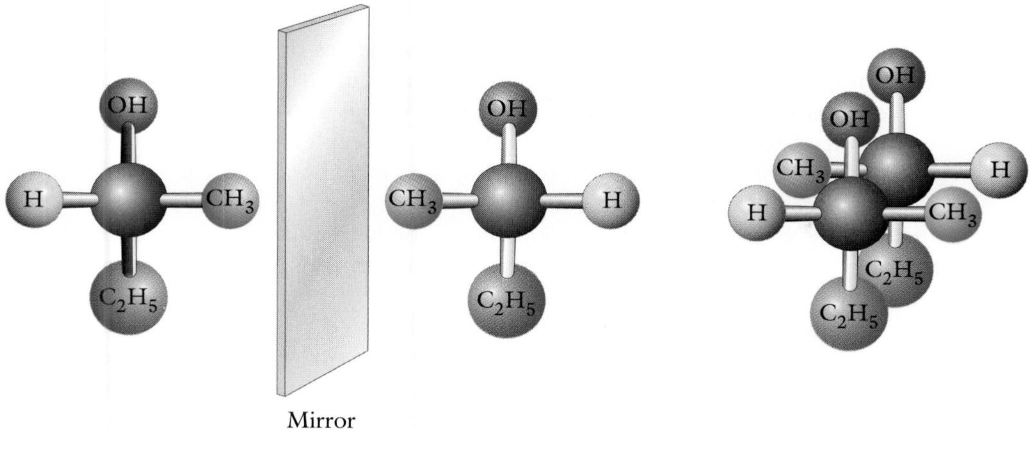

(a) (b)

Figure 17.5 Enantiomers of 2-butanol. (a) The two compounds are nonsuperimposable mirror images. (b) The two compounds cannot be superimposed by sliding one over and merging it into the other, because the —H and —CH_3 substituents do not superimpose.

Figure 17.6 shows two alternate structural representations of the 2-butanol enantiomers, **wedge-bond** and **Fischer-projection** representations. The wedge-bond representations retain some three-dimensional perspective but not as much as the molecular models in Figure 17.5. These drawings use the convention that the solid, horizontal bonds (▬) extend in front of the plane of the paper, whereas the dashed, vertical bonds (┅┅) extend behind the plane of the paper. Both solid and dashed wedge bonds are oriented with their wider ends nearer the viewer. Most chemists prefer to use Fischer projections because they can be drawn much more quickly.

The Fischer projections' lack of three-dimensional perspective would be a major drawback except that we rigidly adhere to the same convention used in the wedge-bond representations: horizontal bonds extend in front of the plane of the paper and vertical bonds extend behind the plane of the paper. This

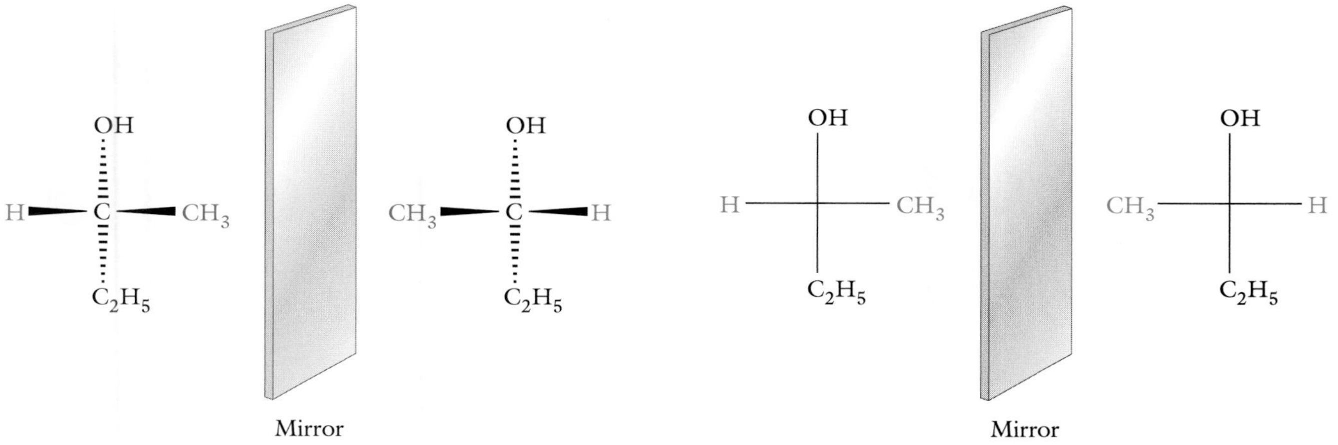

(a) (b)

Figure 17.6 Alternate representations of the 2-butanol enantiomers: (a) wedge-bond structural drawing; (b) Fisher-projection structural drawing.

convention is an absolute requirement for representing chiral molecules when we need to distinguish between the two molecules of an enantiomeric pair. It is not as critical for representing achiral molecules, because there is only one compound with that particular connectivity. It is also not critical when we do not need to distinguish between the two molecules of an enantiomeric pair, which is why the convention regarding horizontal and vertical bonds was not considered in earlier chapters.

The use of a molecular model kit can be helpful if you have trouble visualizing three-dimensional structures and understanding the wedge-bond and Fischer-projection representations.

Figure 17.7 shows mirror-image molecules of 2-methyl-2-butanol. Unlike those of 2-butanol, the mirror-image molecules of 2-methyl-2-butanol are the same compound because they are superimposable on each other. 2-Methyl-2-butanol is not a chiral molecule and exists as one compound, not as a pair of enantiomers.

The ultimate test of whether some compound is one stereoisomer of a pair of enantiomers is to determine whether a molecule of the compound is superimposable on its mirror image. However, a simpler test can be applied:

- A pair of enantiomers is possible only when a compound contains a **tetrahedral stereocenter;** that is, a carbon atom with four different substituents. (In older terminology, tetrahedral stereocenters were called chiral centers.)

The C2 of 2-butanol is a tetrahedral stereocenter (see Figure 17.5). Four different substituents are attached to it: —H, —OH, $—CH_3$, and $—C_2H_5$. As mentioned earlier, a stereocenter is an atom situated in a compound in such a way that a hypothetical exchange of the positions of any two substituents would convert one stereoisomer into the other (Section 17.1). This is exactly the case for a tetrahedral stereocenter. Referring to Figure 17.5, we can see that exchanging the positions of the —H and $—CH_3$ groups converts the left stereoisomer into the right one, and vice versa. The two enantiomers of 2-butanol differ in their configurations at the tetrahedral stereocenter.

None of the carbons in 2-methyl-2-butanol is a tetrahedral stereocenter. For example, C2 does not have four different substituents; two of the substituents attached to C2 are the same (two CH_3 groups). 2-Methyl-2-butanol is achiral, as indicated earlier.

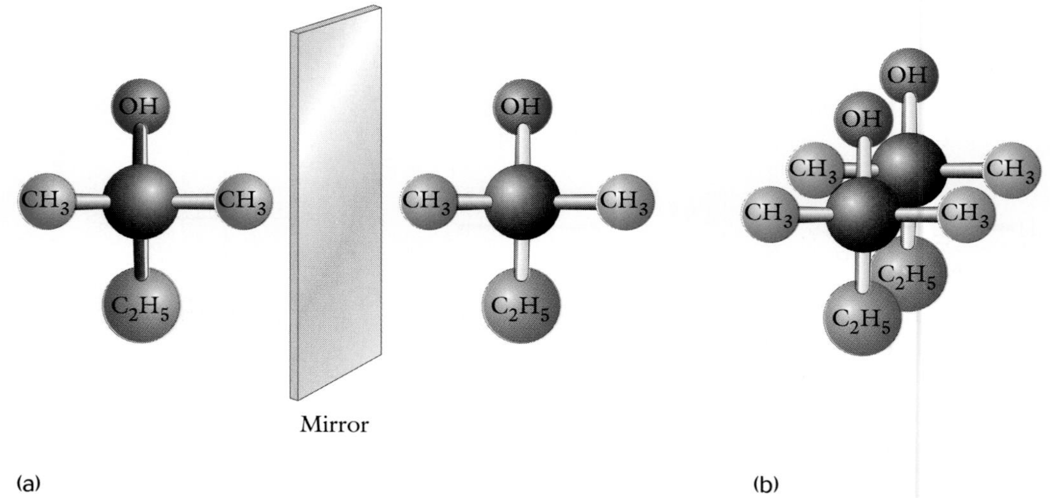

Figure 17.7 (a) Mirror-image molecules of 2-methyl-2-butanol. (b) The two molecules can be superimposed by sliding one over and merging it into the other.

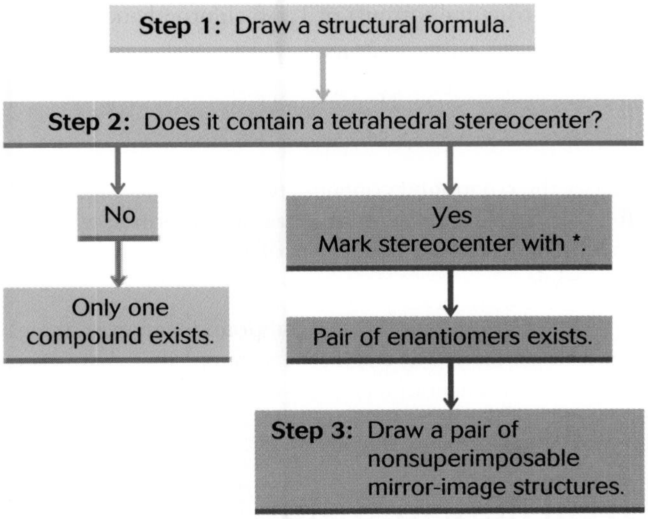

Figure 17.8 Procedure for determining whether a structural formula represents a single compound or a pair of enantiomers.

Figure 17.8 outlines a procedure for determining whether a compound is chiral. The procedure is illustrated in Example 17.1.

Example 17.1 Determining whether a compound is chiral

Which of the following compounds can exist as a pair of enantiomers and which cannot? If enantiomers are possible, draw structural formulas to represent the enantiomers. (a) 1-Bromo-1-chloroethane; (b) 1,1-dichloroethane.

Solution

Follow the procedure described in Figure 17.8.

(a) **Step 1.** Draw a structural formula of 1-bromo-1-chloroethane.

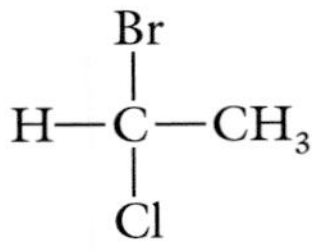

Step 2. Does the compound contain a tetrahedral stereocenter? Yes, C1 is a tetrahedral stereocenter because it has four different substituents (—H, —Br, —Cl, $—CH_3$). Marking the stereocenter with an asterisk is a useful way of keeping track of which carbon is the stereocenter.

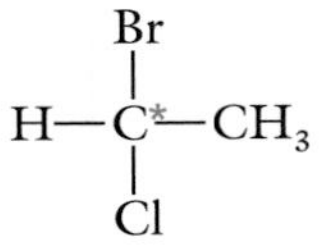

Step 3. Draw a pair of nonsuperimposable mirror-image structures. Show the tetrahedral stereocenter as a central carbon to which the four different substituents are attached.

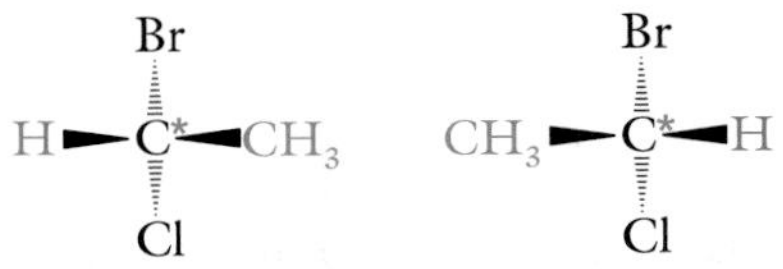

1 2

(b) **Step 1.** Draw a structural formula of 1,1-dichloroethane.

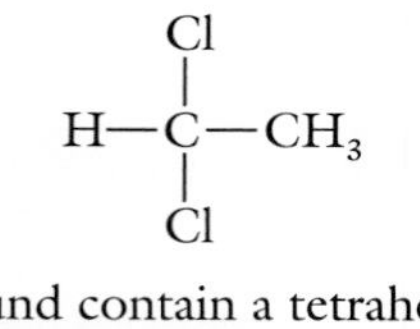

Step 2. Does the compound contain a tetrahedral stereocenter? No, neither carbon has four different substituents. C1 has two chlorines, whereas C2 has three hydrogens. There are no enantiomers of 1,1-dichloroethane; only one compound is possible.

Problem 17.1 Which of the following compounds can exist as a pair of enantiomers and which cannot? If enantiomers are possible, draw structural formulas to represent the enantiomers. (a) 3-Hydroxypropanal; (b) 2-hydroxypropanal.

17.3 INTERPRETING STRUCTURAL FORMULAS OF ENANTIOMERS

In this section, we will develop some guidelines for interpreting structural drawings of enantiomers. Consider structures 1 and 2 in the Solution to Example 17.1a. These structures are one way of representing the pair of enantiomers of 1-bromo-1-chloroethane, but several other structural representations are possible (Figure 17.9). This single pair of enantiomers has several different representations because molecules, being three-dimensional, can be viewed from different directions: from the left or the right, from below or above, from the front or the back, and so forth. Any one pair of structural drawings, 1 and 2 or 3 and 4 or 5 and 6 or 7 and 8, is a valid representation of the two enantiomers, but only one pair of drawings should be used to represent them because there is only one pair of enantiomers.

There is no such ambiguity when we work with molecular models; such models show the molecules as the three-dimensional structures that they are. The ambiguity arises only when we draw and observe three-dimensional structures on a two-dimensional (flat) surface.

Even experienced chemists can find it difficult to compare a pair of structural drawings and decide whether they represent a pair of enantiomers or the same compound (which may be an achiral compound or the same enantiomer of a pair of enantiomers). It is not easy, for example, to compare structures 1 and 4 or 2 and 7, and so forth, of Figure 17.9 and determine that they are the same molecule seen from different directions. We will limit ourselves in this

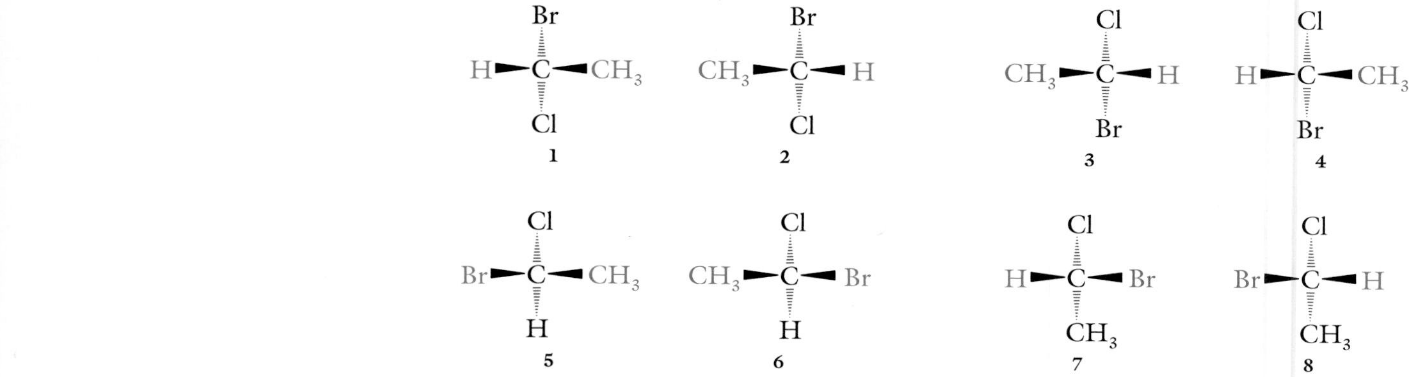

Figure 17.9 Different representations of the one pair of enantiomers of 1-bromo-1-chloroethane: 1 = 3 = 5 = 7 and 2 = 4 = 6 = 8.

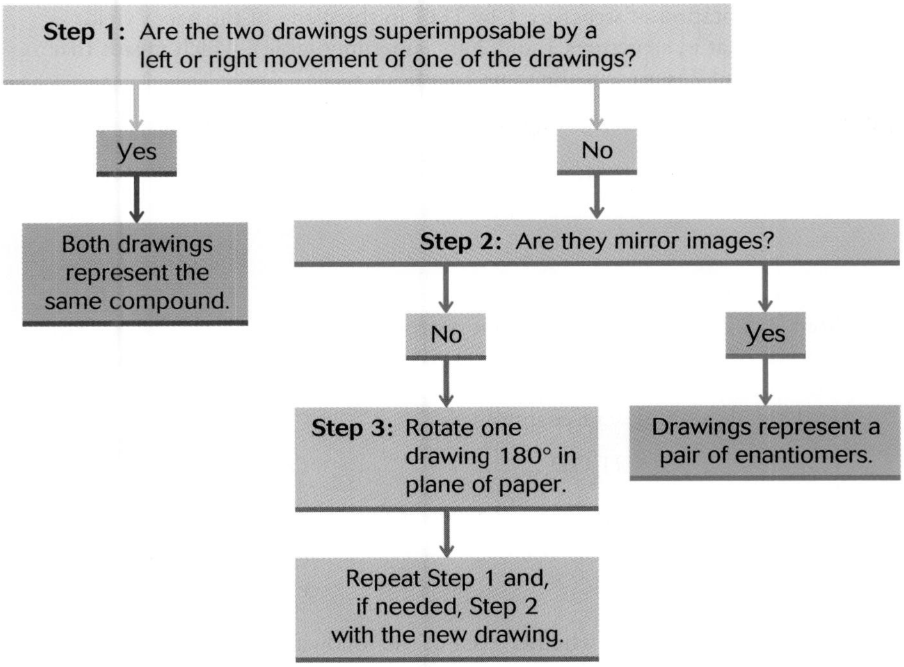

Figure 17.10 Procedure for determining whether two structural drawings represent the same compound or a pair of enantiomers.

book to comparing structural drawings originating from the same observation site. This limitation is important because it means that any two structural drawings can be compared either as shown or after a simple rotation of one or the other drawing (but not both) 180° in the plane of the paper. No other rotations of the drawings will be allowed. Rotations of any degrees other than 180° in the plane of the paper and rotations of any kind out of the plane of the paper will not be allowed, because they would give structures that violate the convention of horizontal bonds extending forward and vertical bonds extending backward. The procedure for comparing two structural drawings is detailed in Figure 17.10 and applied in Example 17.2.

Example 17.2 Determining whether two structural drawings represent a pair of enantiomers or the same compound

Indicate whether each of the following pairs of structural drawings represents the same enantiomer or a pair of enantiomers.

(a)
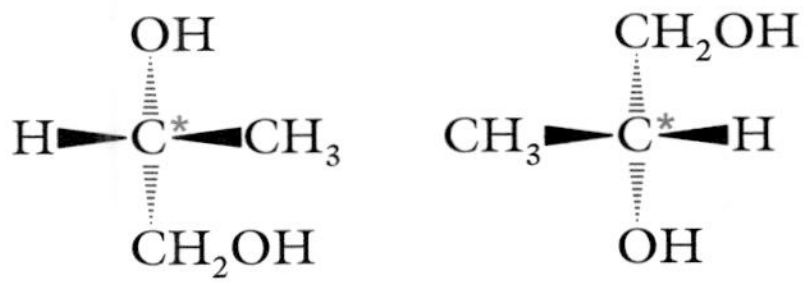

1 2

(b)
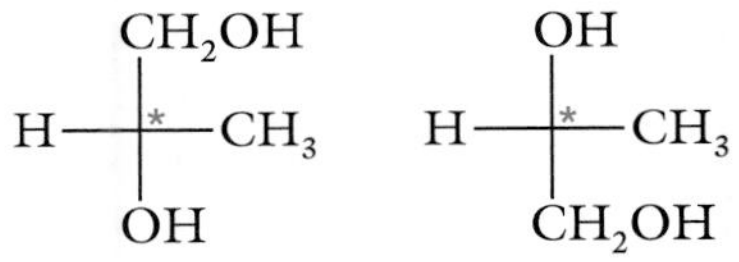

3 4

Solution

(a) **Step 1.** Structures 1 and 2 are not superimposable.
Step 2. Structures 1 and 2 are not mirror images.

Step 3. Rotation of structure 1 by 180° in the plane of the paper yields structure 2 (that is, structures 1 and 2 are superimposable), which means that structures 1 and 2 represent the same enantiomer.

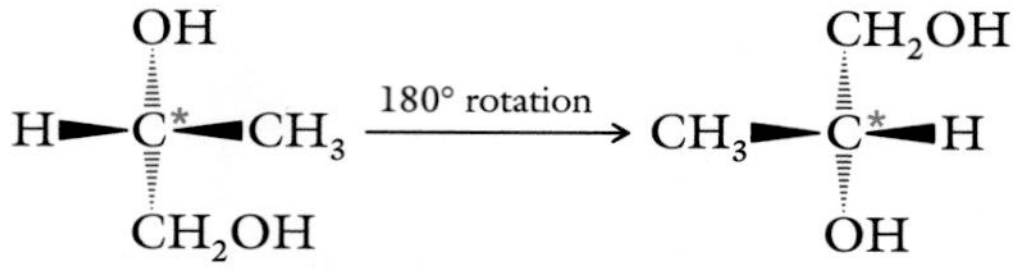

1 2

(b) **Step 1.** Structures 3 and 4 are not superimposable.

Step 2. Structures 3 and 4 are not mirror images.

Step 3. Rotation of structure 3 by 180° in the plane of the paper yields structure 3′, which is the nonsuperimposable mirror image of structure 4. Structures 3 and 4 represent enantiomers.

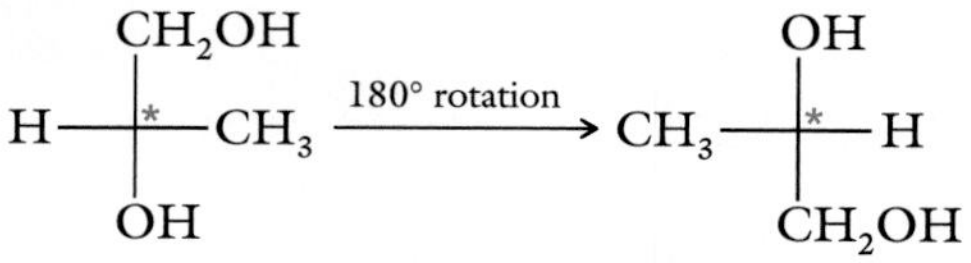

3 3′

Problem 17.2 Indicate whether each of the following pairs of structural drawings represents the same enantiomer or a pair of enantiomers.

(a)

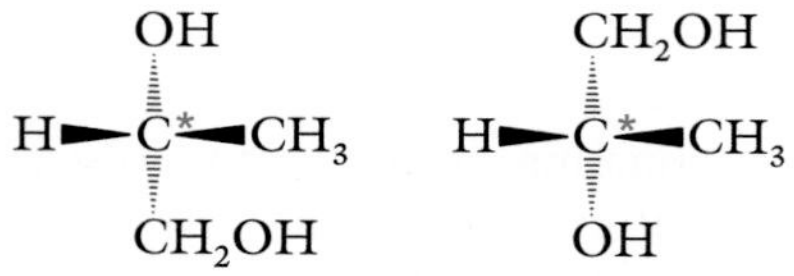

1 2

(b)

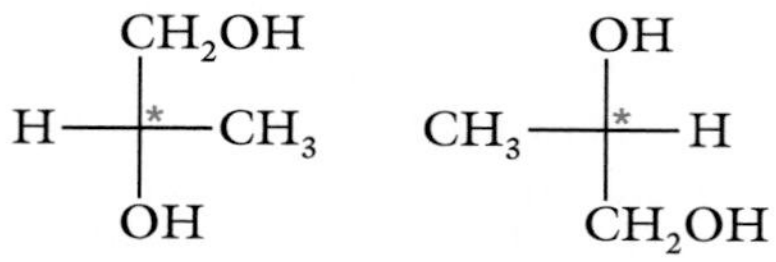

3 4

17.4 NOMENCLATURE OF ENANTIOMERS

One enantiomer of a pair of enantiomers is distinguished from the other enantiomer by adding a prefix before the name of the compound. The prefix indicates the configuration—the order of attachment of the four substituents—at the tetrahedral stereocenter. Before adding this prefix, chemists must experimentally determine the configuration at the tetrahedral stereocenter of each enantiomer by a technique called X-ray crystallography. In X-ray crystallography, the scattering of X-ray energy by the atomic nuclei of a molecule is used to determine the three-dimensional arrangement of the atoms of the compound (Section 3.10).

◀◀ The properties of X-rays are discussed in Sections 2.7 and 10.1.

There are two systems of nomenclature for enantiomers: the original nomenclature system uses the prefixes D- and L-; a more recent system uses the prefixes (*R*)- and (*S*)-. The D/L system is simpler and, though more limited in its application than the (*R*)/(*S*) system, is used extensively by biochemists and biologists in naming carbohydrates, amino acids, and other important biochemicals. We concentrate here on the D/L system.

The D/L system applies only when the tetrahedral stereocenter has the following substituents:

- Hydrogen

- A heteroatom substituent X (such as OH or NH_2) with the heteroatom bonded to the tetrahedral stereocenter
- Two different R substituents, R^1 and R^2, each of which has a carbon bonded to the tetrahedral stereocenter

Naming enantiomers by the D/L system requires that Fischer projections (or wedge-bond representations) of the enantiomers be drawn from an observation site such that the placement of the R^1 and R^2 substituents is vertical (they extend backward), whereas the placement of the H and hetero atom substituents is horizontal (they extend forward). The Fischer projections are then oriented so that R^1, the substituent having the most-substituted carbon atom (the carbon with the fewest hydrogens) attached to the tetrahedral stereocenter, is located upward and that R^2, the substituent having the least-substituted carbon atom (the carbon with the most hydrogens) attached to the tetrahedral stereocenter, is located downward. If necessary, the Fischer projection is rotated 180° in the plane of the paper to achieve this orientation. The enantiomer with the heteroatom bonded on the right side is the D-enantiomer, whereas the enantiomer with the heteroatom bonded on the left side is the L-enantiomer:

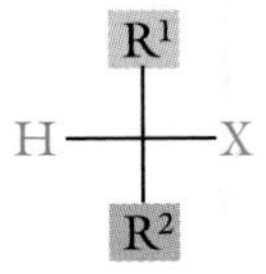

D

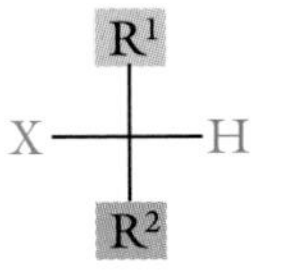

L

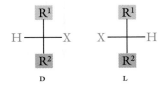

Example 17.3 Identifying the D- and L-enantiomers

The chemist Emil Fischer originated and first applied the D/L nomenclature system in 1891 for the two enantiomers of glyceraldehyde. Which of the following compounds is D-glyceraldehyde and which is L-glyceraldehyde?

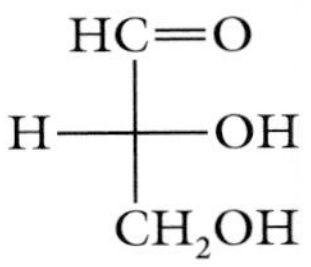

1

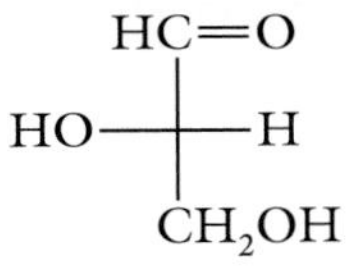

2

Solution

Structures 1 and 2 are correctly oriented in space for application of the D/L nomenclature system: the carbon substituents are oriented vertically with the most (HC=O) and least (CH_2OH) substituted carbons placed up and down, respectively. The enantiomer with the heteroatom (OH) bonded on the right side of the tetrahedral stereocenter, structure 1, is D-glyceraldehyde; the enantiomer with the OH on the left, structure 2, is L-glyceraldehyde.

Problem 17.3 Name the following enantiomer of alanine, an amino acid used by nature in synthesizing proteins, as D- and L-.

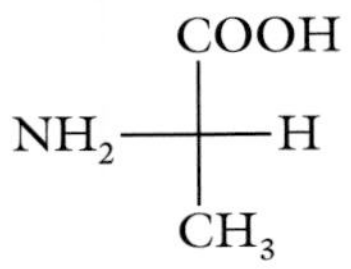

17.5 PROPERTIES OF ENANTIOMERS

Chiral compounds differ from achiral compounds in one physical property (optical activity) and one chemical property (chiral recognition).

Optical Activity

Chiral compounds exhibit **optical activity,** and are said to be **optically active.** Achiral compounds are **optically inactive.** The enantiomers of an enantiomeric pair are identical in all their physical properties (such as boiling and melting points and solubility) except optical activity. Optical activity is the ability of a compound to rotate the plane of plane-polarized light. To understand what this means, we need to look briefly at the nature of light and the measurement of optical activity.

Optical activity is measured by an instrument called a **polarimeter** (Figure 17.11), composed of a light source, polarizer, sample cell (where a solution of the chemical compound to be studied is placed), and analyzer, together with the observer (human eye). A beam of ordinary light such as that from a light bulb contains electromagnetic energy vibrating in all angular orientations around the axis of the light beam. Light vibrating in only one angular orientation, called **plane-polarized light,** is isolated by passing ordinary light through a **polarizer.** Materials such as calcite and polaroid filters (the same polaroid filters used in sunglasses to reduce glare) are used as polarizers. The polarizer is set up so that the plane-polarized light emerges in a specified plane—for instance, the vertical plane as in Figure 17.11. The **analyzer,** made of the same material as the polarizer, allows light vibrating in only one direction to

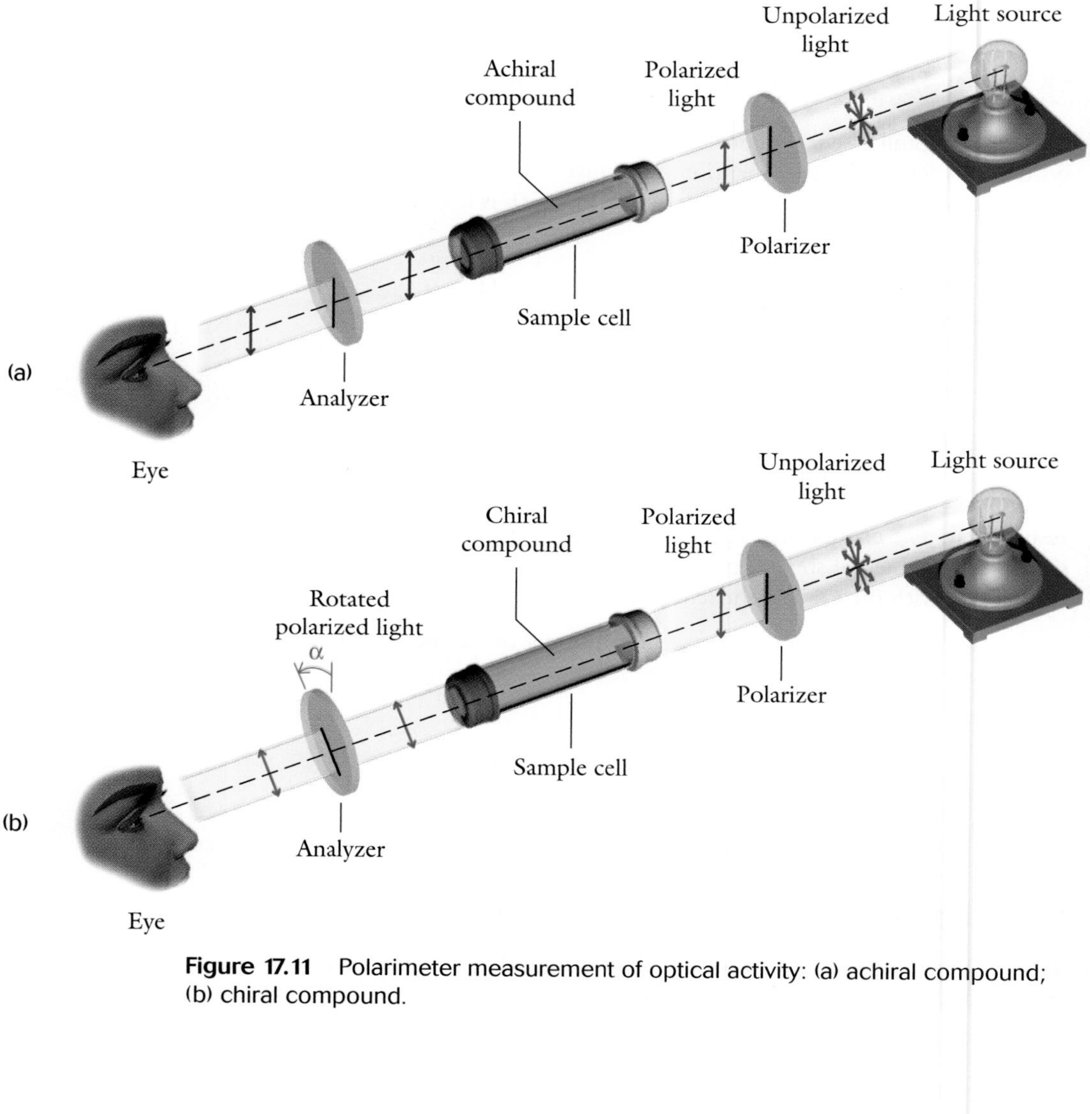

Figure 17.11 Polarimeter measurement of optical activity: (a) achiral compound; (b) chiral compound.

pass through it. The analyzer is initially set up to match the polarizer, allowing only light in the same plane—in this case, the vertical plane—to pass through.

When the sample cell contains a solution of an achiral compound (Figure 17.11a), there is no observed interaction of the compound with the plane-polarized light. The plane-polarized light passes through unchanged; that is, the light emerging from the sample cell is still polarized in the same plane. The light passes through the analyzer and can be observed.

When the sample cell contains a solution of a chiral compound (Figure 17.11b), the plane-polarized light interacts with the compound as it travels through the sample cell. The plane of the plane-polarized light that emerges is rotated at an angle α to the vertical plane. The rotated light cannot pass through the analyzer and cannot be observed. The experimenter then rotates the analyzer until light passes through and can be observed. This angle of rotation also is α, the amount to which the plane of the plane-polarized light has been rotated.

The two enantiomers of a pair of enantiomers rotate the plane of plane-polarized light by the same number of degrees but in opposite directions. One enantiomer rotates light in the clockwise direction and is called **dextrorotatory;** the other enantiomer rotates light in a counterclockwise direction and is called **levorotatory.** The direction of rotation of plane-polarized light is indicated in the names of enantiomers by placing (+)- and (−)- as prefixes to their names; for example, (+)-glyceraldehyde and (−)-glyceraldehyde.

It is important to keep in mind that the prefixes (+)- and (−)- (corresponding to dextrorotatory and levorotatory) are not related to the prefixes D- and L-. The direction of optical rotation—whether (+) or (−)—is determined by experiments with the polarimeter. The configuration at the tetrahedral stereocenter—whether D or L—is determined by X-ray crystallography. There is no general relation between configuration and direction of optical rotation. Some D-enantiomers rotate in a clockwise (dextrorotatory) direction; other D-enantiomers rotate in a counterclockwise (levorotatory) direction.

If both the configuration and the optical rotation of a chiral compound are known, both are indicated in the name. For example, the dextrorotatory enantiomer of glyceraldehyde is the D-enantiomer and the name is D-(+)-glyceraldehyde. The levorotatory enantiomer of fructose, an important carbohydrate, is the D-enantiomer and the name is D-(−)-fructose.

Specific Rotation

The degree of rotation α observed for a chiral compound in a polarimeter experiment depends on the number of chiral molecules encountered by the light beam as it passes through the sample tube. The value of α increases if the experimenter uses a more concentrated solution or a longer sample cell or both. To place the observed rotation on a standard basis, chemists calculate a quantity called the **specific rotation,** $[\alpha]$, defined by

$$[\alpha] = \frac{\alpha}{CL}$$

where α is the observed rotation in degrees for a sample concentration of C grams per milliliter (g/mL) and sample-cell length of L decimeters (1 dm = 10 cm). (For pure liquids, C is the density of the liquid.)

Whereas α values for the same chiral compound will be different for experiments performed by different workers (using different concentrations and different sample-cell lengths), $[\alpha]$ is independent of the details of the particular polarimeter experiment. All polarimeter experiments with the same compound yield the same value of $[\alpha]$. Specific rotation is a physical property of the enantiomer and has a definite and constant value, just as boiling and melting points, density, color, and solubility have. The specific rotations of the enantiomers in an enantiomeric pair are the same in degrees but in the opposite direction; for example, $[\alpha] = +13.5°$ and $-13.5°$ for the enantiomers of glyceraldehyde. Like other physical properties, $[\alpha]$ is useful for identifying a compound.

The details of the measurement of specific rotation are more complicated than indicated because $[\alpha]$ varies somewhat, depending on the frequency of light, the solvent used, and the temperature of the experiment. To compare $[\alpha]$ values obtained by different workers, the values must have been obtained under exactly the same conditions: the same temperature, solvent, and (frequency) source of light.

An equimolar (1 : 1) mixture of two enantiomers (of a pair of enantiomers) does not exhibit optical activity, because, in such a mixture, known as a **racemic mixture** or **racemate,** the rotation in one direction by one enantiomer is canceled by rotation in the opposite direction by the other.

Example 17.4 Calculation of specific rotation, $[\alpha]$

A solution of L-valine at a concentration of 1.50 g/mL gives an optical rotation of $-20.3°$ when observed in the 5.00-cm-long sample cell of a polarimeter. Calculate the specific rotation of L-valine.

Solution

Substitute the values of α, C, and L into the equation $[\alpha] = \alpha/CL$, making sure that the correct units are used for C (g/mL) and L (dm). The concentration is given in the correct units, but the sample-cell length is not; 5.00 cm must be converted into 0.500 dm before substitution into the equation for $[\alpha]$.

$$[\alpha] = \frac{\alpha}{CL} = \frac{-20.3°}{1.50\ \text{g/mL} \times 0.50\ \text{dm}} = \frac{-27.1°}{\text{g/mL} \times \text{dm}} = -27.1°$$

Although the units of specific rotation are usually reported simply as degrees, it should always be understood that the actual units are degrees per gram per milliliter per decimeter.

Problem 17.4 D-Glucose has a specific rotation of $+53°$. Calculate the rotation of a solution of D-glucose at a concentration of 30 g/L when measured in a 15-cm-long sample cell.

Chemical Properties: Chiral Recognition

Whether the enantiomers of a chiral compound have the same or different chemical reactivity depends on the nature of the other reactants or the enzyme (biological catalyst) or both.

- When reacting with achiral compounds, enantiomers usually have the same reactivity.
- When reacting with chiral compounds or in reactions catalyzed by chiral enzymes, enantiomers usually have very large differences in reactivity.

These differences in the interaction of chiral compounds with achiral compounds and with other chiral compounds or enzymes can be understood, again, by considering feet, socks, and shoes (see Figures 17.2 and 17.4). Most socks are not chiral, though feet are. Either one of a pair of socks can be worn on either foot, but there are very strict requirements for wearing shoes. Only the left shoe fits well on the left foot, and only the right shoe fits well on the right foot.

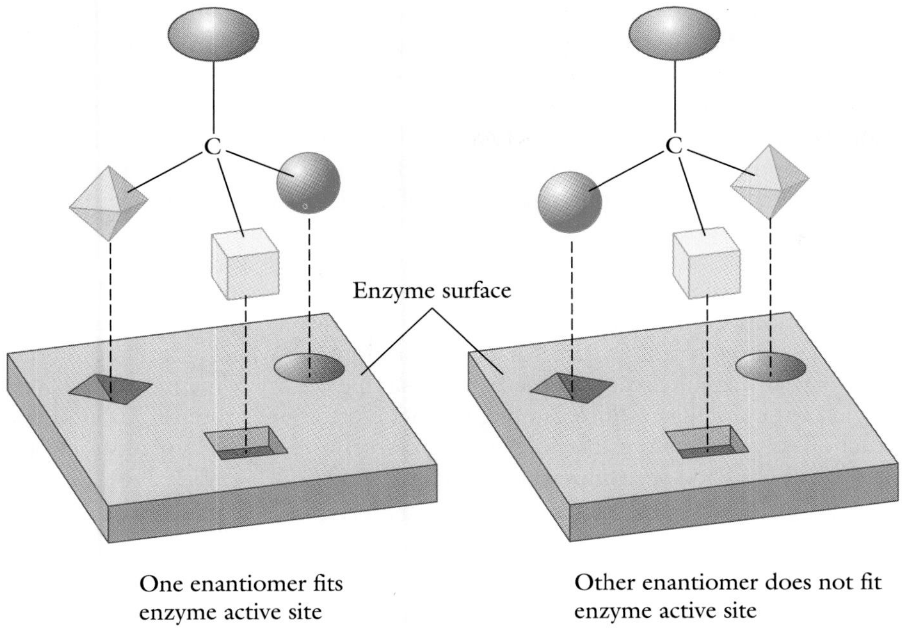
C
C
Enzyme surface
One enantiomer fits
enzyme active site
Other enantiomer does not fit
enzyme active site

Figure 17.12 Chiral recognition of enantiomers by chiral enzyme.

Many reactions in living cells include one or more chiral molecules and are catalyzed by enzymes, which are chiral. Such reactions require a unique fit between two chiral species for reaction to take place. In enzyme-catalyzed reactions, one or more of the reactants bind to a site on the enzyme known as the **active site** (Section 8.4). For many enzymes, the active site is chiral. Figure 17.12 illustrates the difference in fit between two enantiomers at the active site of a chiral enzyme. Only one enantiomer fits into (binds to) the active site, and reaction proceeds only with this enantiomer. This phenomenon, called **chiral recognition** or **chiral discrimination,** is one mechanism by which enzymes show discrimination (specificity) in biological systems.

Not all biological discriminations are based on chirality; some are based on geometric isomerism and others on molecular (or ionic) size, shape, and polarity (or charge). These bases constitute the complementarity principle (Sections 8.4 and 22.3, Box 12.2). Boxes 12.2 and 12.3 describe the effects of geometric isomerism on vision and pheromone action, respectively.

There are examples of chiral recognition in many physiological processes. For example, we ingest starch (a carbohydrate) and a variety of proteins and digest—that is, hydrolyze—them into simpler substances (glucose and amino acids, respectively). We then use these smaller compounds for a variety of purposes: to produce energy; to synthesize energy-storing materials such as glycogen;

A PICTURE OF HEALTH

Examples of Stereospecific Molecules

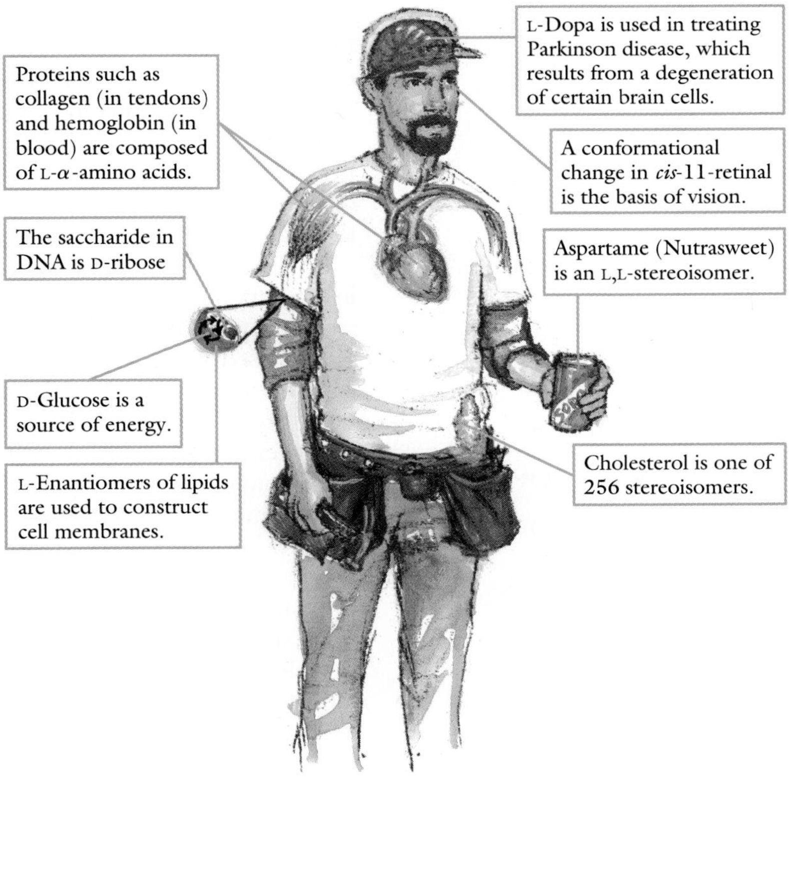
L-Dopa is used in treating Parkinson disease, which results from a degeneration of certain brain cells.
Proteins such as collagen (in tendons) and hemoglobin (in blood) are composed of L-α-amino acids.
A conformational change in *cis*-11-retinal is the basis of vision.
The saccharide in DNA is D-ribose
Aspartame (Nutrasweet) is an L,L-stereoisomer.
D-Glucose is a source of energy.
Cholesterol is one of 256 stereoisomers.
L-Enantiomers of lipids are used to construct cell membranes.

and to synthesize structural materials such as the components of bone, skin, and muscle. The enzymes used in these processes have evolved as highly specific chiral catalysts: only D-carbohydrates and L-proteins are used or synthesized in humans and most other organisms. In fact, evolution has produced a biological world in which, with very few exceptions, all carbohydrates are D-carbohydrates and all proteins are L-proteins (Sections 18.3 and 20.1). Whereas reactions in living cells produce only one enantiomer of a pair of enantiomers, synthetic reactions in the laboratory usually produce racemic mixtures.

▶▶ Carbohydrates and proteins are the subjects of Chapters 18 and 20, respectively.

Chiral recognition in biological systems does not always mean that one enantiomer participates in a reaction and the other does not react at all. Different enantiomers may simply produce different effects, as in our sense of taste and smell (Box 17.1). Or the "inactive enantiomer," when it enters the body, may participate in a reaction that is undesirable and detrimental to the well-being of the organism. For example, synthetic drugs are produced in industrial reactions that often result in racemic mixtures, not a single enantiomer, and this result has had some important consequences (Box 17.2 on page 512).

Concept checklist

- ✔ Enantiomers are a pair of stereoisomers whose molecules are nonsuperimposable mirror images.
- ✔ The only difference in physical properties of a pair of enantiomers is the direction of rotation of plane-polarized light.
- ✔ Enantiomers show no difference in chemical reactivity toward achiral compounds.
- ✔ Enantiomers show large differences in chemical reactivity toward chiral compounds or in reactions catalyzed by chiral enzymes.

17.6 COMPOUNDS CONTAINING TWO OR MORE TETRAHEDRAL STEREOCENTERS

Our discussion so far has focused on compounds with one tetrahedral stereocenter, but many molecules (especially biologically important ones) contain more than one tetrahedral stereocenter. The number of possible stereoisomers increases as the number of tetrahedral stereocenters increases. The maximum number of stereoisomers possible for a compound with n tetrahedral stereocenters is 2^n. Thus, as many as 4, 8, and 16 stereoisomers are possible when there are two, three, and four tetrahedral stereocenters, respectively.

It is important to recognize the presence of more than two stereocenters in a compound because many biologically important compounds contain many stereocenters. D-Glucose, the most plentiful carbohydrate and an energy source used by most plant and animal cells, contains four tetrahedral stereocenters (Section 18.2). Starch and cellulose, consisting of hundreds of glucose units bonded together (Section 18.6), contain many hundreds of tetrahedral stereocenters. Proteins, consisting of large numbers of combined amino acids, also contain a great many tetrahedral stereocenters. A biologically active molecule containing stereocenters is always one specific stereoisomer out of a large number of possibilities. D-Glucose is one stereoisomer out of a total of 16. Starch, cellulose, and proteins are each found in nature as one of an astronomical number of possible stereoisomers.

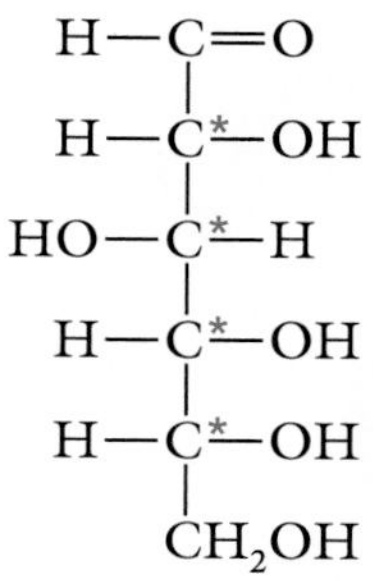

D-Glucose

Compounds containing more than one tetrahedral stereocenter are too complex for us to consider in great detail. As we shall see, not all the stereoisomers of compounds with multiple tetrahedral stereocenters are enantiomers, and not all the stereoisomers are optically active. We will limit our detailed exploration to compounds with two tetrahedral stereocenters.

17.1 Chemistry Within Us

Senses of Smell and Taste

The senses of taste (**gustation**) and smell (**olfaction**) depend on **chemoreceptors,** sensory cells that detect certain molecules or ions by binding to them through secondary attractive forces. Taste chemoreceptors are organized into **taste buds** on the upper surface of the tongue and, to a lesser extent, on the roof of the mouth. There are four primary taste perceptions—sweet, sour, salt, bitter. Sour and salt tastes are associated with chemoreceptors sensitive to H_3O^+ and Na^+ (or other metal cations), respectively. Sweet and bitter tastes are associated with a variety of organic compounds; a number of different compounds bind to these chemoreceptors.

Binding takes place at specific proteins located on the cell membranes of chemoreceptor cells. The structural requirements for binding of a molecule to a chemoreceptor include overall molecular size and shape, stereoisomerism, and secondary attractive forces in different parts of the molecule and the chemoreceptor protein. These requirements constitute the complementarity principle (Sections 8.4 and 22.3, Box 12.2). Binding is generally possible only between specific parts of the chemoreceptor protein and the molecule. Binding to the chemoreceptor cell generates an electrical impulse, which is transmitted to the end of the sensory cell and then from neuron (nerve cell) to neuron until it reaches the sensory centers in the brain. The brain interprets the electrical impulse as a specific taste.

A central taste bud is surrounded by many papillae in this electron micrograph of a part of the tongue surface.

Smell is perceived in a similar manner, through olfactory chemoreceptors lining the upper part of the nasal cavity.

There is considerable interaction between the sensations of taste and smell; so each reinforces the other. In actuality, a good deal of what we call taste is smell. For example, taste is sharply reduced when the nasal passages are blocked by a head cold. And foods often have more taste when hot because higher temperatures volatilize chemicals, which reach the olfactory chemoreceptors in the nasal passages.

The effect of stereoisomerism on taste and smell is often very striking, indicating that some chemoreceptor sites are themselves chiral and able to bind only one enantiomer of a pair of enantiomers. Consider the enantiomers of carvone. One enantiomer is the principal component of spearmint oil and the other the principal component of caraway seed oil, each with its characteristic taste and odor.

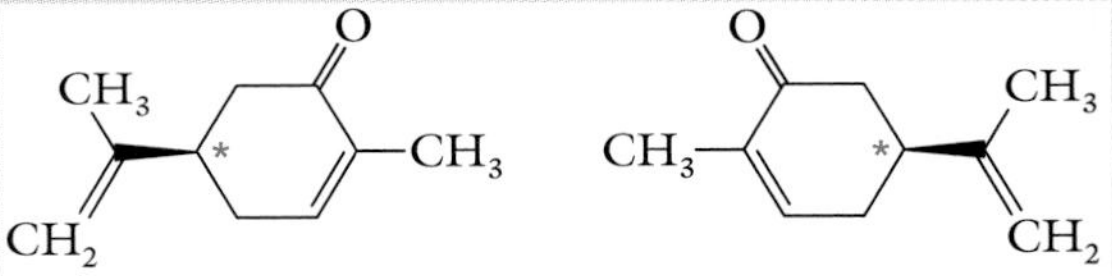

(−)-Carvone (Spearmint oil) (+)-Carvone (Caraway seed oil)

The amino acid asparagine has a pair of enantiomers. D- and L-asparagines are found in the plants vetch and asparagus, respectively. D-Asparagine has a sweet taste. L-Asparagine is tasteless to some people but bitter-tasting to others. The contradictory reports on the taste of L-asparagine illustrate the variations in the perceptions of taste. These variations arise from individual differences in the numbers and types of chemoreceptor cells as well as differences in the response of individual nervous systems (including the sensory centers in the brain).

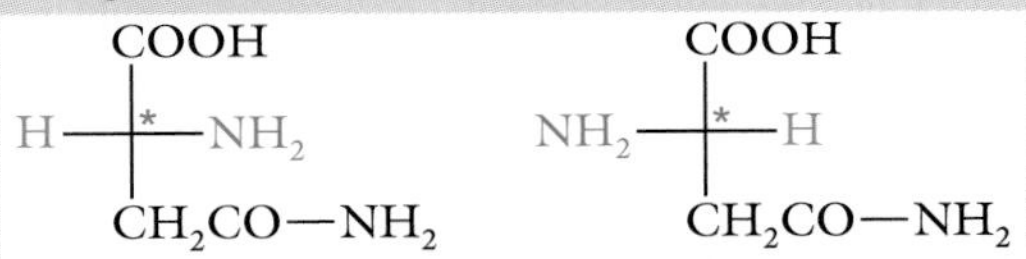

D-Asparagine L-Asparagine

17.2 Chemistry Within Us

Synthetic Chiral Drugs

About 40% to 50% of synthetic drugs consist of compounds containing one or more tetrahedral stereocenters, and the typical manufacturing process, unlike biological reactions, yields racemic mixtures (racemates) of enantiomers.

The body responds to a racemate in one of two ways:

- Both enantiomers react but in different processes, one of which may produce an undesired effect on the organism.
- One enantiomer reacts in a biological process and the other passes through the organism with no effect.

An example of the first response is in the treatment of people with Parkinson disease. In this disease, the production of a neurotransmitter called dopamine is reduced. The consequence is an impaired ability to transmit information to the brain centers that control muscle movement, resulting in uncontrolled tremors, especially in the limbs, and difficulty in walking. The synthetic drug L-dopa (L-3,4-dihydroxyphenylalanine) is used to treat this condition. L-Dopa is actually L-(−)dopa and is called levodopa because it is levorotatory. The original medication used was racemic dopa, which showed some adverse side effects, notably granulocytopenia—a decreased leukocyte count in the blood. It was subsequently found that the D-dopa enantiomer was both responsible for the adverse side effects and ineffective in treating Parkinson disease.

L-Dopa is taken orally and subsequently absorbed into the bloodstream and transported to the brain, where it enters brain cells and is converted into dopamine. Dopamine itself cannot be used to treat the disease, because it cannot cross from the bloodstream into brain cells—it cannot cross the so-called blood–brain barrier. Unfortunately, L-dopa becomes progressively less effective with time because Parkinson disease also causes degeneration of the dopamine receptors.

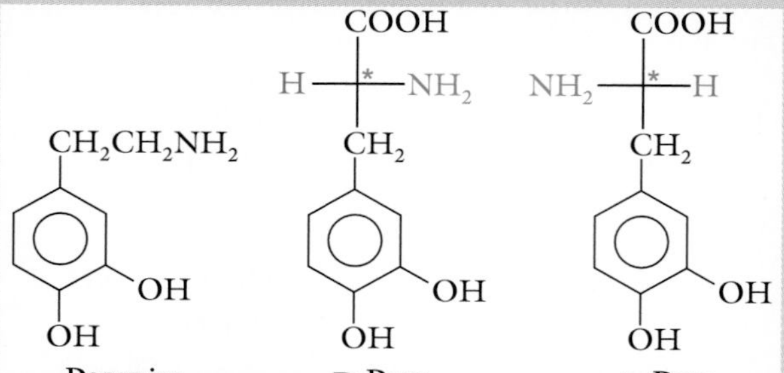

Dopamine D-Dopa L-Dopa

An example of the second type of response to taking a drug as a racemate is provided by propranolol (Inderal), a β-blocker used as an antihypertensive, antiarrhythmic, and antianginal drug. The drug that is currently available is racemic propanolol, but only (−)-propranolol is effective. (+)-Propranolol is not effective but has no known adverse effects.

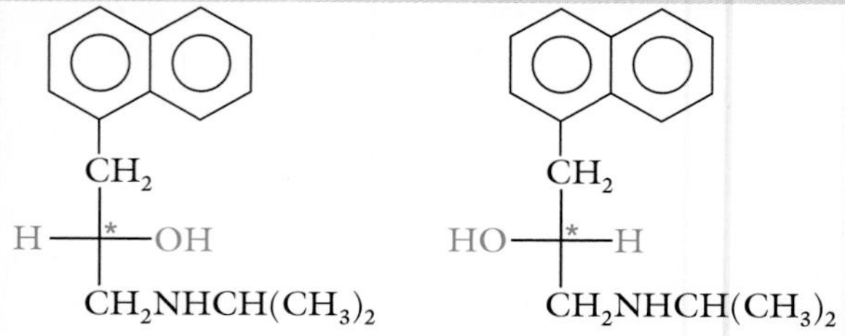

Enantiomers of propranolol

Drug companies have begun to synthesize and market **single-enantiomer drugs** in place of racemates to optimize the effectiveness of drugs even when the second enantiomer has no effect, whether desirable or adverse. The body is taxed by having to metabolize and excrete the unnecessary enantiomer, which may limit the dosage of the desirable enantiomer that can be prescribed. Among the single-enantiomer drugs now used are L-dopa, lisinopril (an ACE inhibitor for high blood pressure; Box 16.1), sertraline (an antidepressant), and ibuprofen (an anti-inflammatory drug; Box 15.4).

The synthesis of a single-enantiomer drug requires special techniques. Biological reactions produce single-enantiomer compounds under the direction of chiral enzymes, but most syntheses carried out in the laboratory result in racemates because they are not controlled by chiral catalysts. Most successful syntheses of single-enantiomer drugs have used one of two approaches:

- **Resolution:** The drug is produced as a racemate; then special separation techniques are used to separate the enantiomers.
- **Stereospecific synthesis:** Researchers devise a synthetic route that has a reaction step that can be controlled by an available chiral catalyst to produce only the desired enantiomer.

Two Tetrahedral Stereocenters with Different Sets of Substituents

Consider a compound with two tetrahedral stereocenters. The maximum number of four stereoisomers exists when the two tetrahedral stereocenters do not have the same set of four different substituents. In 2,3-pentanediol, for example, one tetrahedral stereocenter has $-H$, $-CH_3$, $-OH$, and $-CH(OH)C_2H_5$ and the other has $-H$, $-C_2H_5$, $-OH$, and $-CH(OH)CH_3$. There are $2^2 = 4$ stereoisomers of 2,3-pentanediol, differing from each other in the configurations at C2 and C3:

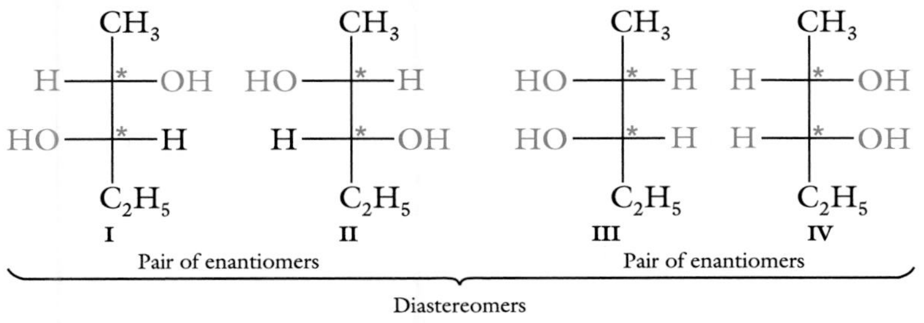

Note that one of the vertical bonds in each of these structural drawings does not conform to the convention described in Section 17.2. The bonds from C1 to C2 and from C3 to C4 extend behind the plane of the paper, but the bond between C2 and C3 is in the plane of the paper.

The four stereoisomers consist of two pairs of enantiomers: I and II are one pair, and III and IV are the other pair. Neither I nor II is an enantiomer of III or IV, and none of these compounds are superimposable on each other—all four are different stereoisomers. The term **diastereomer** (Section 17.1) describes the relation of either compound from one enantomeric pair to either compound from the other enantiomeric pair. Enantiomers I and II are diastereomers of III and IV, and vice versa. Diastereomers are stereoisomers that are not enantiomers; that is, they are neither superimposable nor mirror images. (Recall that the term diastereomer also describes the relation between a pair of cis and trans stereoisomers; Sections 11.9 and 12.5.).

All diastereomers (except meso compounds, which are described in the next subsection, and cis and trans compounds) are optically active. However, unlike enantiomers, diastereomers have specific rotations that usually differ in their numerical value and may have the same or opposite direction of rotation. If the specific rotations are the same, it is entirely by chance. Even the direction of optical rotation can be different for two diastereomers. Another difference from enantiomers is that diastereomers differ in other physical properties, such as melting and boiling points and solubility. Usually, diastereomers also differ in their chemical properties.

The artificial sweetener aspartame (NutraSweet), *N*-L-aspartyl-L-phenylalanine methyl ester, is an example of chiral recognition and of the differences in properties seen in enantiomers and diastereomers. This compound contains two tetrahedral stereocenters that do not have the same set of four different substituents. Of the four possible stereoisomers, only the L,L-stereoisomer (the one in which both tetrahedral stereocenters have the L configuration) is sweet—almost 200 times as sweet as sucrose (ordinary table sugar). The L,D-stereoisomer not only is not sweet, but is bitter tasting.

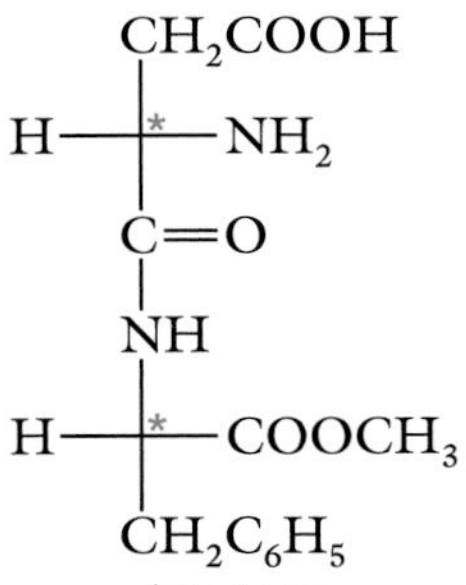

Aspartame

Two Tetrahedral Stereocenters with the Same Set of Substituents

Let's consider a compound with two tetrahedral stereocenters that have the same set of four different substituents. An example is 2,3-butanediol. Each stereocenter has $-H$, $-CH_3$, $-OH$, and $-CH(OH)CH_3$. We can refer to the structural drawings of the four stereoisomers of 2,3-pentanediol on page 513 to draw the 2,3-butanediols. All that is needed is to replace the ethyl groups with methyls:

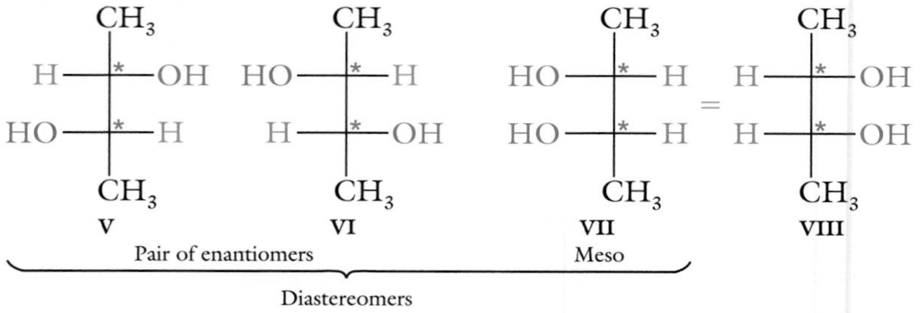

Note the equal sign between drawings VII and VIII. These drawings represent the same compound, not different compounds. Drawings VII and VIII represent the same compound because a 180° rotation of either structure makes it superimposable with the other structure.

When the two tetrahedral stereocenters of a compound have the same set of four different substituents, there are only three, not four, stereoisomers. There is a pair of enantiomers (V and VI) and a lone, third stereoisomer (VII = VIII) called a **meso** compound. The meso compound is not an enantiomer of V or VI; it is neither a mirror image nor superimposable on V or VI. Therefore, V and VI are diastereomers of the meso compound, and vice versa. The meso stereoisomer can be shown as either VII or VIII but not both. If you draw both VII and VIII, you must place an equal sign between them. Meso compounds are achiral and optically inactive even though they contain two tetrahedral stereocenters.

Example 17.5 Determining the stereoisomers of compounds with two tetrahedral stereocenters

Draw structural formulas of all stereoisomers of 2-methyl-1,3-butanediol:

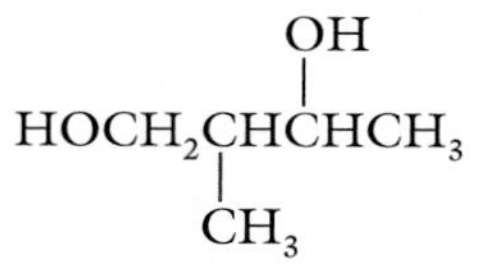

Which stereoisomers are optically active? Which are enantiomers and which are diastereomers?

Solution

Because this compound has two tetrahedral stereocenters with different sets of four different substituents, four stereoisomers are possible:

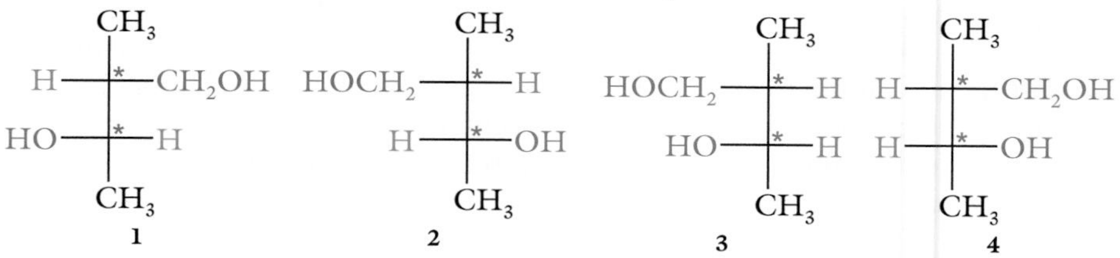

All four compounds are optically active. Each stereoisomer is simultaneously both an enantiomer and a diastereomer. Stereoisomers 1 and 2 are enantiomers of each other, and each is a diastereomer of both 3 and 4. Stereoisomers 3 and 4 are enantiomers of each other, and each is a diastereomer of both 1 and 2.

Problem 17.5 Draw structural formulas of all stereoisomers of tartaric acid (2,3-dihydroxy-1,4-butanedioic acid):

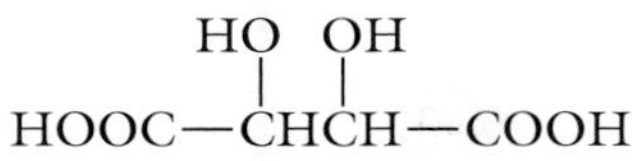

Which stereoisomers are optically active? Which are enantiomers and which are diastereomers?

Concept checklist

- ✔ There are two pairs of enantiomers when the two tetrahedral stereocenters do not have the same set of four different substituents.
- ✔ There is one pair of enantiomers and a meso compound when the two tetrahedral stereocenters have the same set of four different substituents.

17.7 CYCLIC COMPOUNDS CONTAINING TETRAHEDRAL STEREOCENTERS

Cyclic compounds that have tetrahedral stereocenters also can exist as enantiomers. The basic criterion remains the same. Enantiomers are possible when a compound is nonsuperimposable on its mirror image.

A ring carbon is a tetrahedral stereocenter if:

- the two nonring substituents are different and
- the ring is not symmetrical with respect to that carbon.

Consider methylcyclohexane (Figure 17.13a). It fulfills the first requirement—it has two different nonring substituents attached to C1. However, the second requirement is not fulfilled—the two halves of the ring on each side of C1 are the same ($CH_2CH_2CH_2$). There is no tetrahedral stereocenter, and the

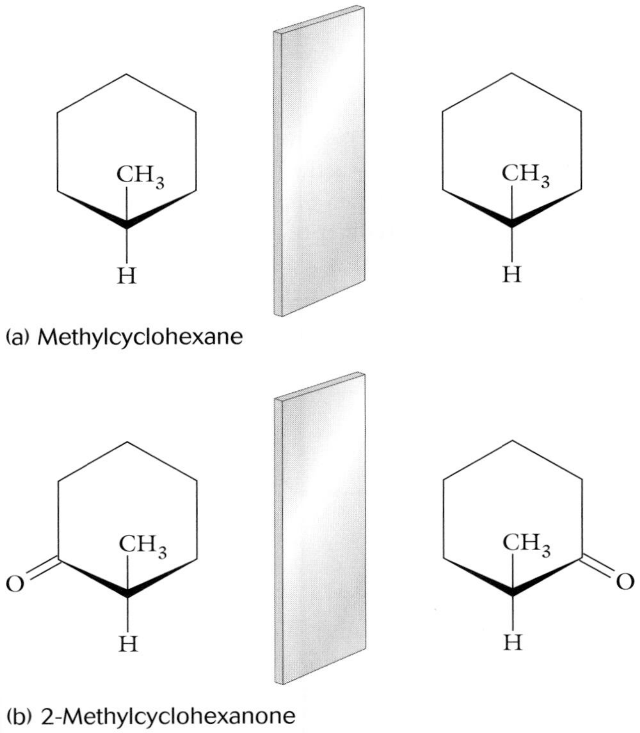

Figure 17.13 Mirror-image molecules of (a) methylcyclohexane (superimposable) and (b) 2-methylcyclohexanone (nonsuperimposable).

mirror-image molecules are superimposable. Thus, methylcyclohexane exists as a single compound.

2-Methylcyclohexanone, however, exists as a pair of enantiomers (Figure 17.13b). Carbon 2 is a tetrahedral stereocenter: the two nonring substituents are different, and the two halves of the ring on each side of C2 are different ($COCH_2CH_2$ versus $CH_2CH_2CH_2$). The mirror-image molecules are nonsuperimposable, and 2-methylcyclohexanone exists as a pair of enantiomers.

Example 17.6 Determining the stereoisomers of cyclic compounds

Draw the stereoisomers of: (a) 1-chloro-2-methylcyclohexane; (b) 1,2-dimethylcyclohexane. Which stereoisomers are optically active? Which are enantiomers and which are diastereomers?

Solution

Both 1-chloro-2-methylcyclohexane and 1,2-dimethylcyclohexane possess two tetrahedral stereocenters. Each stereocenter has two different nonring substituents, and the two halves of the ring are different at both C1 and C2. The situation is similar to that for acyclic compounds with two tetrahedral stereocenters.

(a) Because the two tetrahedral stereocenters of 1-chloro-2-methylcyclohexane do not have the same set of four different substituents, the maximum number of four stereoisomers is possible. There are a trans pair of enantiomers and a cis pair of enantiomers:

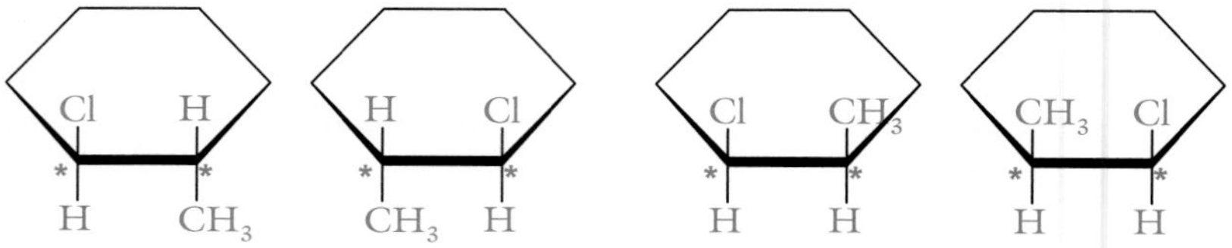

1 2
Trans pair of enantiomers

3 4
Cis pair of enantiomers

The trans and cis designations refer to the placements of substituents on the opposite or same sides of the ring, respectively (Section 11.9).

All four compounds are optically active. Each stereoisomer is simultaneously both an enantiomer and a diastereomer. Stereoisomers 1 and 2 are enantiomers of each other, and each is a diastereomer of both 3 and 4. Stereoisomers 3 and 4 are enantiomers of each other, and each is a diastereomer of both 1 and 2.

(b) 1,2-Dimethylcyclohexane has two tetrahedral stereocenters with the same set of four different substituents, so only three stereoisomers are possible. There are a trans pair of enantiomers and a cis meso compound:

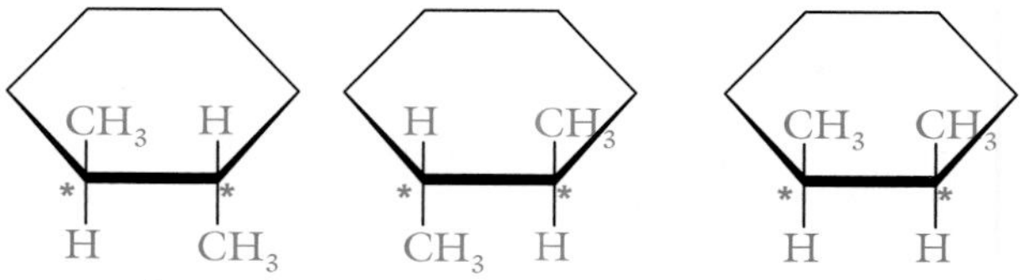

5 6
Trans pair of enantiomers

7
Cis meso

The trans compounds are optically active, but the meso compound is optically inactive. Stereoisomers 5 and 6 are enantiomers of each other, and each is a diastereomer of 7. Stereoisomer 7 is a diastereomer of both 1 and 2. The cis meso compound is easy to spot because it is superimposable on its mirror image. A pair of cis enantiomers exist for 1-chloro-2-methylcyclohexane, but only one cis compound exists for 1,2-dimethylcyclohexane because the nonring substituents at the tetrahedral stereocenters are not the same for 1-chloro-2-methylcyclohexane, whereas they are the same for 1,2-dimethylcyclohexane.

Problem 17.6 Draw the stereoisomers of each of (a) chlorocyclohexane; (b) 1-chloro-3-methylcyclohexane. Which stereoisomers are optically active? Which are enantiomers and which are diastereomers?

Cyclic compounds with tetrahedral stereoisomers are common in nature. Cholesterol is such a compound. It has eight tetrahedral stereocenters, and there are 2^8, or 256, possible stereoisomers. Only one stereoisomer (the one called cholesterol) exists in nature. Cholesterol is an important component of cell membranes and the starting material for the biosynthesis of many important compounds, including vitamin D and the sex hormones progesterone and testosterone (Sections 19.7 and 19.10).

▶▶ Cholesterol, progesterone, and testosterone are lipids and the subjects of Chapter 19.

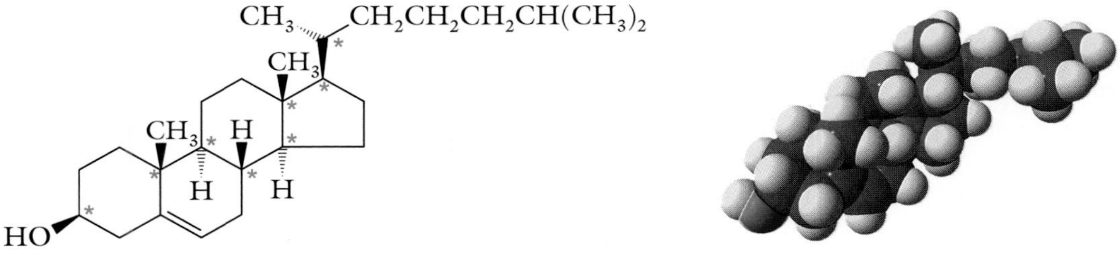

Cholesterol

Summary

Isomers Isomers are different compounds that have the same molecular formula. There are two types of isomers: constitutional isomers and stereoisomers. Constitutional (structural) isomers differ in connectivity; that is, the order of attachment of atoms to each other. Stereoisomers have the same connectivity but differ in configuration; that is, the relative orientations in space of the atoms of a compound, independent of changes that occur by rotation about single bonds. Stereoisomers that are mirror images of each other are called enantiomers. Diastereomers are stereoisomers (including geometric isomers) that are not enantiomers; that is, they are not mirror images of each other.

Enantiomers A pair of enantiomers—nonsuperimposable mirror-image compounds—exists when there is a tetrahedral stereocenter, a carbon with four different substituents. Such compounds are also called chiral compounds. Enantiomers can be differentiated by a naming system that uses the prefixes D- and L-. Chiral compounds are optically active: they rotate plane-polarized light, as measured by a polarimeter. Each enantiomer in a pair of enantiomers rotates plane-polarized light the same number of degrees but in opposite directions. A racemic mixture—an equimolar mixture of enantiomers—is optically inactive.

Enantiomers have the same chemical reactivity when reacting with achiral reactants, but often show very large differences in reactivity when reacting with chiral reactants or in reactions catalyzed by chiral enzymes, a phenomenon called chiral recognition or chiral discrimination.

Compounds Containing Two or More Tetrahedral Stereocenters The maximum number of stereoisomers possible for a compound with n tetrahedral stereocenters is 2^n. For a compound with two tetrahedral stereocenters that do not have the same four different substituents, two pairs of enantiomers are possible. Either compound from one pair of enantiomers is a diastereomer of either compound of the other pair of enantiomers. The maximum number of stereoisomers is not observed when the two tetrahedral stereocenters have the same set of four different substituents. In this case, there are three stereoisomers: a pair of enantiomers and a meso compound. Either compound of the pair of enantiomers is a diastereomer of the meso compound and vice versa. Meso compounds are achiral and not optically active.

Cyclic Compounds Containing Tetrahedral Stereocenters Enantiomers are also possible for cyclic compounds having one or more tetrahedral stereocenters. A ring carbon is a tetrahedral stereocenter if the two nonring substituents are different and the ring is not symmetrical with respect to that carbon.

Key Words

achiral, p. 498
chiral, p. 498
chiral discrimination (chiral recognition), p. 508
diastereomer, p. 497
enantiomer, p. 497

Fischer projection, p. 499
meso compound, p. 514
nonsuperimposable, p. 498
optical activity, p. 506
optical rotation, p. 507
racemic mixture, p. 508
specific rotation, p. 507
stereocenter, p. 497
stereoisomer, p. 497
superimposable, p. 498
tetrahedral stereocenter, p. 501
wedge bond, p. 499

Exercises

Review of Isomerism

17.1 Draw the structural formulas of constitutional isomers of C_4H_{10}.

17.2 Draw the structural formulas of constitutional isomers of C_5H_{12}.

17.3 Draw the structural formulas of ethers that are constitutional isomers of butanol.

17.4 Draw the structural formulas of ketones that are constitutional isomers of pentanal.

17.5 Draw the structural formula of the diastereomer of *cis*-2-butene.

17.6 Draw the structural formula of the diastereomer of *cis*-1,2-dimethylcyclobutane.

Enantiomers

17.7 Which of the following objects are chiral and which achiral? (a) A person; (b) an automobile; (c) a basketball without any design or name imprinted on it; (d) a basketball with the name MICHAEL JORDAN imprinted on it.

17.8 Which of the following objects are chiral and which achiral? (a) A clear glass coffee mug; (b) a coffee mug with your name imprinted on it; (c) a dog; (d) a television set.

17.9 Which of the following structures can exist as a pair of enantiomers and which cannot? Draw structural formulas of enantiomers, placing an asterisk next to each tetrahedral stereocenter.

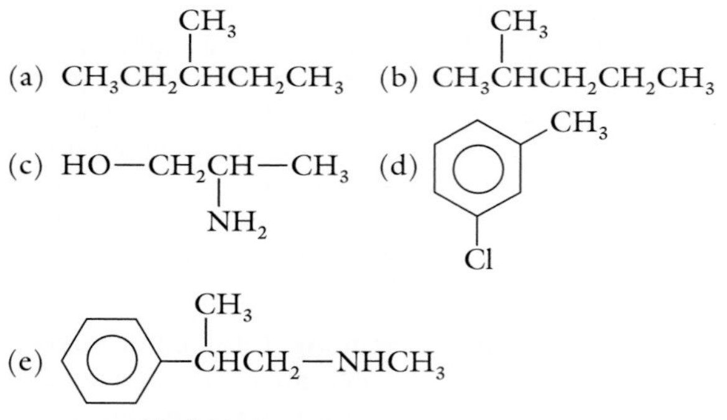

Methamphetamine

17.10 Which of the following structures can exist as a pair of enantiomers and which cannot? Draw structural formulas of enantiomers, placing an asterisk next to each tetrahedral stereocenter.

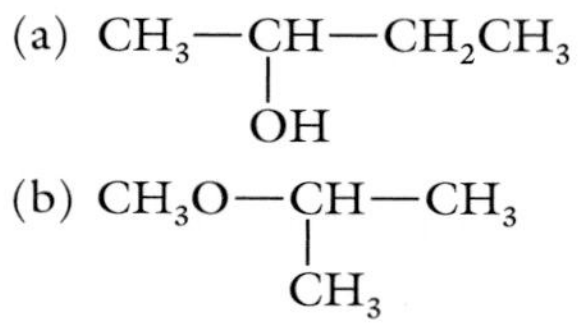

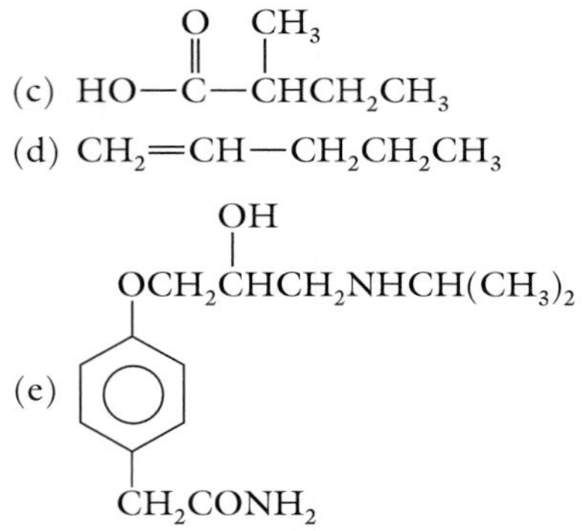

Metoprolol (Lopressor)

Interpreting Structural Formulas of Enantiomers

17.11 For each of the following pairs of structural formulas, indicate whether the pair represents (1) the same compound or (2) different compounds that are constitutional isomers or (3) different compounds that are enantiomers or (4) different compounds that are diastereomers or (5) different compounds that are not isomers.

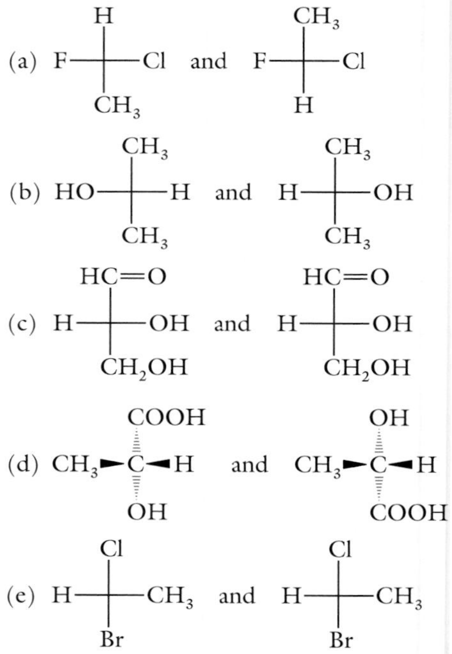

17.12 For each of the following pairs of structural formulas, indicate whether the pair represents (1) the same compound or (2) different compounds that are constitutional isomers or (3) different compounds that are enantiomers or (4) different compounds that are diasteromers or (5) different compounds that are not isomers.

(a) 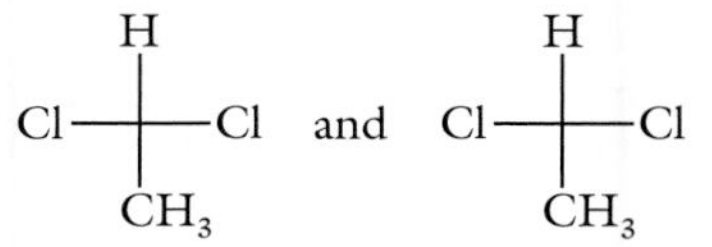

(b) 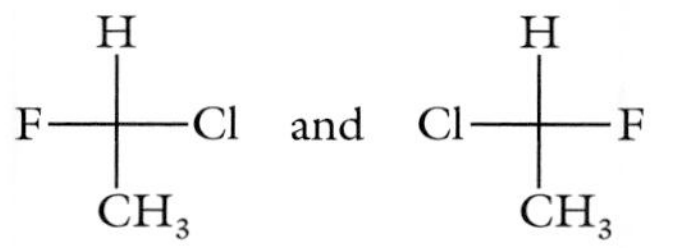

(c) 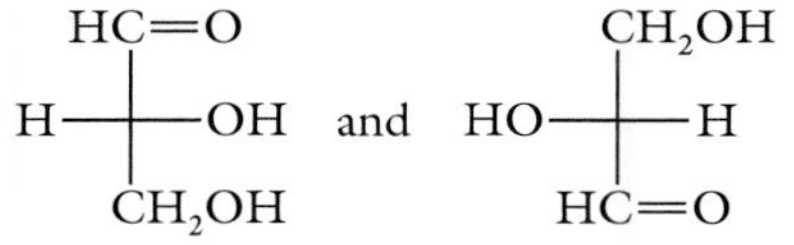

(d) 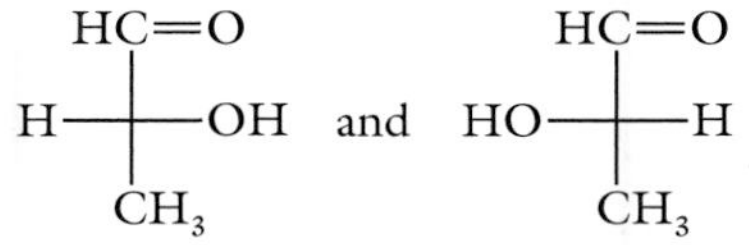

(e) 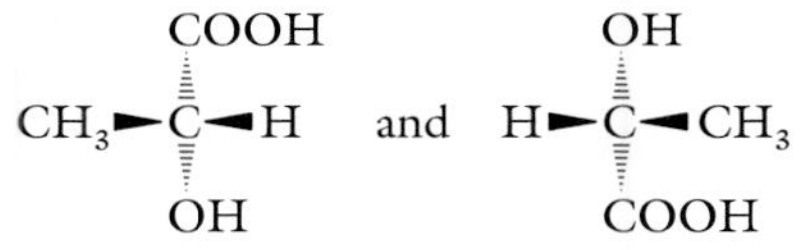

Nomenclature of Enantiomers

17.13 Identify each of the following stereoisomers as the D- or L-enantiomer. Check to make sure that the drawing is correctly oriented for making its assignment as D or L. If necessary, rotate the drawing by 180°.

(a) 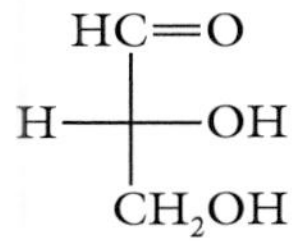 (b) 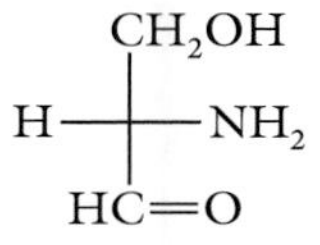

(c) 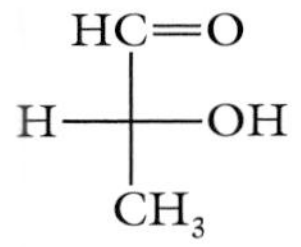 (d) 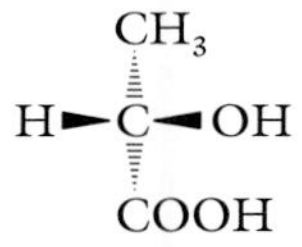

17.14 Identify each of the following stereoisomers as the D- or L-enantiomer. Check to make sure that the drawing is correctly oriented for making its assignment as D or L. If necessary, rotate the drawing by 180°.

(a) 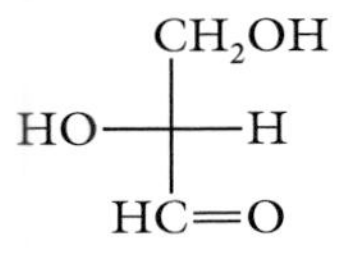 (b) 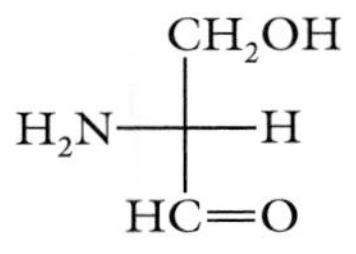

(c) 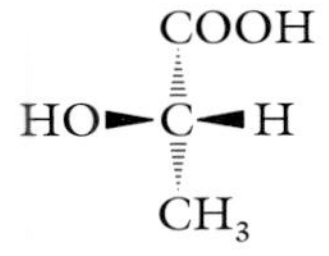 (d) 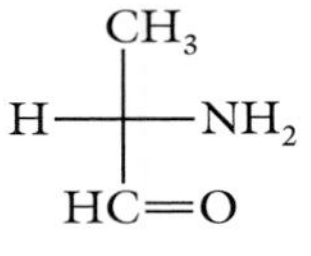

Properties of Enantiomers

17.15 Which of the following compounds rotate plane-polarized light? If so, in which direction? (a) Ethanol; (b) D-glucose; (c) (+)-phenylalanine; (d) racemic glutamic acid.

17.16 Which of the following compounds rotate plane-polarized light? If so, in which direction? (a) Pentane; (b) (−)-glucose; (c) L-phenylalanine; (d) racemic lactic acid.

17.17 D-Glutamic acid has a specific rotation of − 31.5°. What is the specific rotation of L-glutamic acid?

17.18 Is D or L the correct designation for the enantiomer of lactic acid with $[\alpha] = -13.5°$?

17.19 The following compound is (−)-alanine. Draw (+)-alanine.

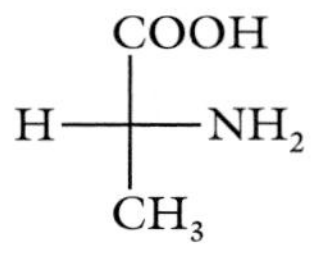

17.20 The following compound is L-valine. Draw D-valine.

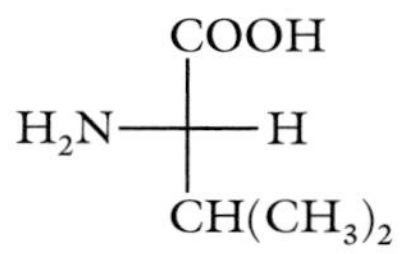

17.21 The water solubility of L-alanine is 127 g/L at 25°C. What is the water solubility of D-alanine?

17.22 The density of L-valine is 1.316 g/mL at 25°C. What is the density of D-valine at 25°C?

17.23 A solution of D-fructose (4.0 g/100 mL) gives an optical rotation of − 3.6° when measured in a polarimeter sample cell 10.0 cm long. Calculate the specific rotation of D-fructose.

17.24 A solution of (+)-menthol (10.0 g/100 mL) gives an optical rotation of + 2.6° when measured in a polarimeter sample cell 5.0 cm long. Calculate the specific rotation of (−)-menthol.

17.25 The specific rotation of (+)-penicillin V is + 223°. What will be the observed rotation when using a solution concentration of 60.0 g per 1000 mL of solution and a polarimeter sample cell of 15.0-cm length?

17.26 The specific rotation of (−)-morphine is − 132°. What will be the observed rotation when using a solution concentration of 10.0 g per 250 mL of solution and a polarimeter sample cell of 20.0-cm length?

Compounds Containing Two or More Tetrahedral Stereocenters

17.27 Use an asterisk to indicate each tetrahedral stereocenter (if any) in each of the following structures. Draw all possible stereoisomers for each structure. Which stereoisomers are optically active? Which stereoisomers are enantiomers and which are diastereomers?

(a) 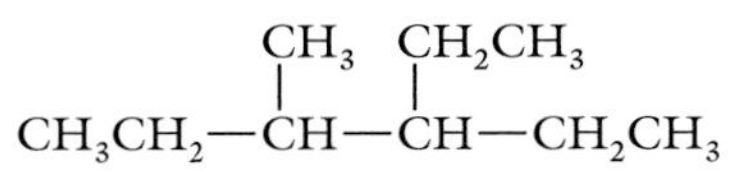

(b) 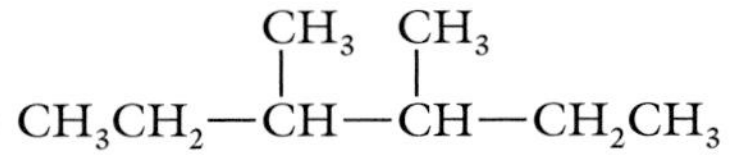

(c) 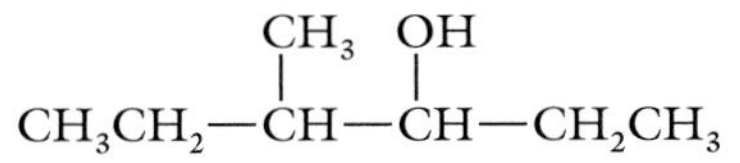

17.28 Use an asterisk to indicate each tetrahedral stereocenter (if any) in the following structures. Draw all possible stereoisomers for each structure. Which stereoisomers are optically active? Which stereoisomers are enantiomers and which are diastereomers?

(a) $HOOC-\overset{\overset{\displaystyle CH_3}{|}}{CH}-\overset{\overset{\displaystyle COOH}{|}}{CH}-COOH$

(b) $HOOC-\overset{\overset{\displaystyle OH}{|}}{CH}-\overset{\overset{\displaystyle OH}{|}}{CH}-COOH$

(c) $HOOC-\overset{\overset{\displaystyle CH_3}{|}}{CH}-\overset{\overset{\displaystyle OH}{|}}{CH}-COOH$

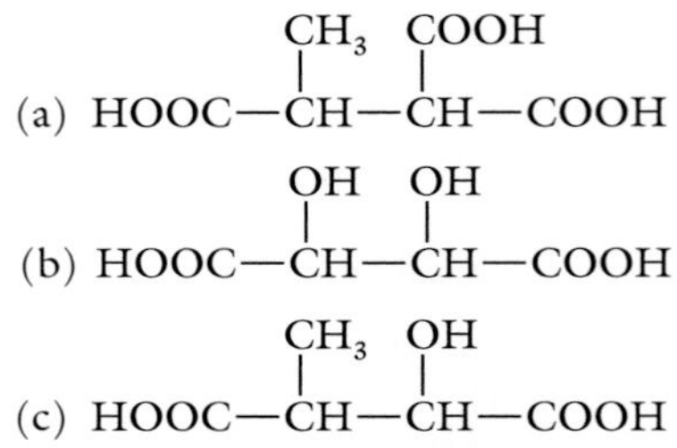

17.29 There are four stereoisomers (1, 2, 3, and 4) of 2-methyl-1,3-butanediol. The structures of two of the stereoisomers are:

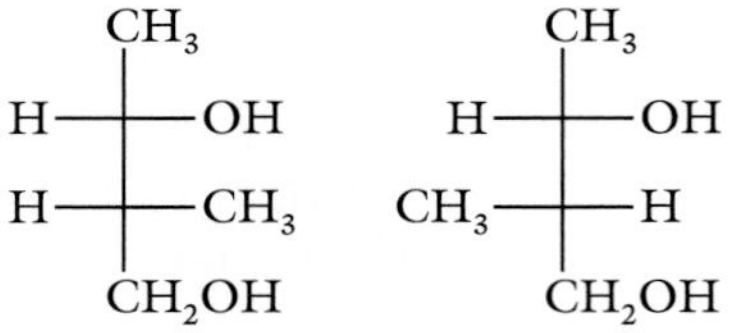

Stereoisomer 1 has a specific rotation of + 15° and a boiling point of 180–182°C. Stereoisomer 2 has a specific rotation of + 26° and a boiling point of 163–165°C. Draw the structures, and give the boiling points and specific rotations of stereoisomers 3 and 4.

17.30 There are three stereoisomers (1, 2, and 3) of tartaric acid. The structures of two of the stereoisomers are:

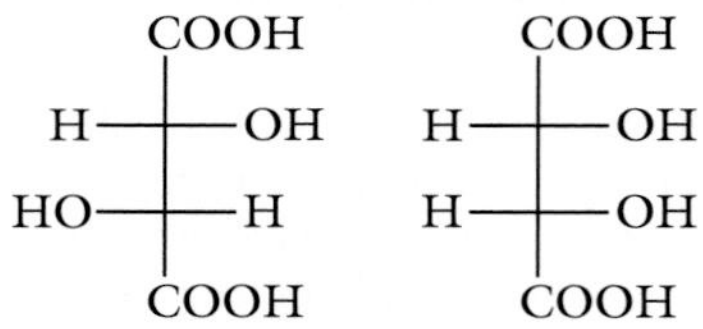

Stereoisomer 1 has a specific rotation of + 12.7° and a melting point of 172–174°C. Stereoisomer 2 is not optically active and has a melting point of 146–148°C. Draw the structure, and give the melting point and specific rotation of stereoisomer 3. Why is stereoisomer 2 not optically active?

17.31 Use an asterisk to indicate each tetrahedral stereocenter (if any) in each of the following structures. What is the maximum number of stereoisomers possible for each of these structures?

(a) $\begin{array}{l} HC{=}O \\ | \\ HC{-}OH \\ | \\ HC{-}OH \\ | \\ HC{-}OH \\ | \\ HC{-}OH \\ | \\ H_2C{-}OH \end{array}$

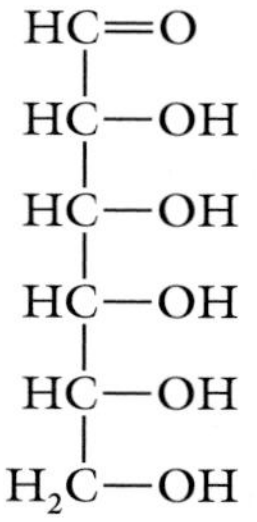

(b) $H_2N-\overset{\overset{\displaystyle}{|}}{\underset{\underset{\displaystyle CH_3}{|}}{CH}}-\overset{\overset{\displaystyle O}{\|}}{C}-\overset{\overset{\displaystyle H}{|}}{N}-\overset{\overset{\displaystyle}{|}}{\underset{\underset{\displaystyle CH_2OH}{|}}{CH}}-\overset{\overset{\displaystyle O}{\|}}{C}-\overset{\overset{\displaystyle H}{|}}{N}-\overset{\overset{\displaystyle}{|}}{\underset{\underset{\displaystyle CH_2COOH}{|}}{CH}}-\overset{\overset{\displaystyle O}{\|}}{C}-OH$

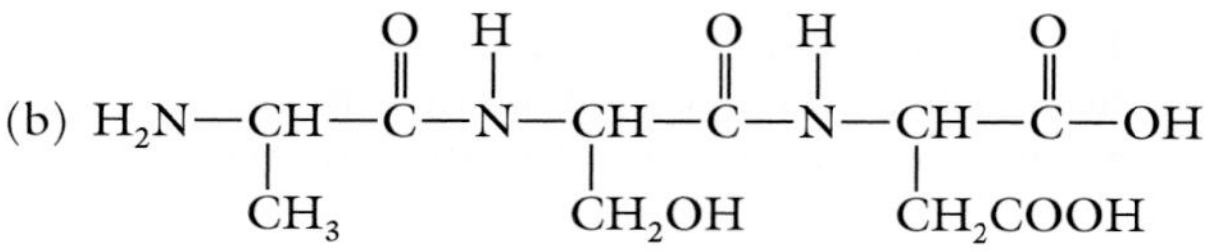

17.32 Use an asterisk to indicate each tetrahedral stereocenter (if any) in each of the following structures. What is the maximum number of stereoisomers possible for each of these structures?

(a) $\begin{array}{l} H_2C{-}OH \\ | \\ C{=}O \\ | \\ HC{-}OH \\ | \\ HC{-}OH \\ | \\ HC{-}OH \\ | \\ H_2C{-}OH \end{array}$

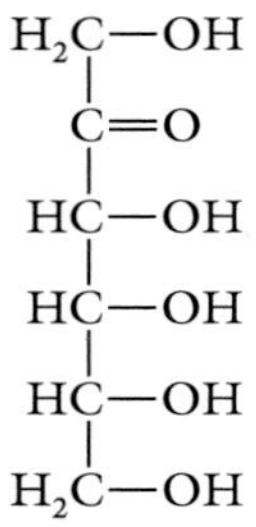

(b) $H_2N-\overset{\overset{\displaystyle}{|}}{\underset{\underset{\displaystyle CH(CH_3)_2}{|}}{CH}}-\overset{\overset{\displaystyle O}{\|}}{C}-\overset{\overset{\displaystyle H}{|}}{N}-\overset{\overset{\displaystyle}{|}}{\underset{\underset{\displaystyle CH_2COOH}{|}}{CH}}-\overset{\overset{\displaystyle O}{\|}}{C}-\overset{\overset{\displaystyle H}{|}}{N}-CH_2-\overset{\overset{\displaystyle O}{\|}}{C}-OH$

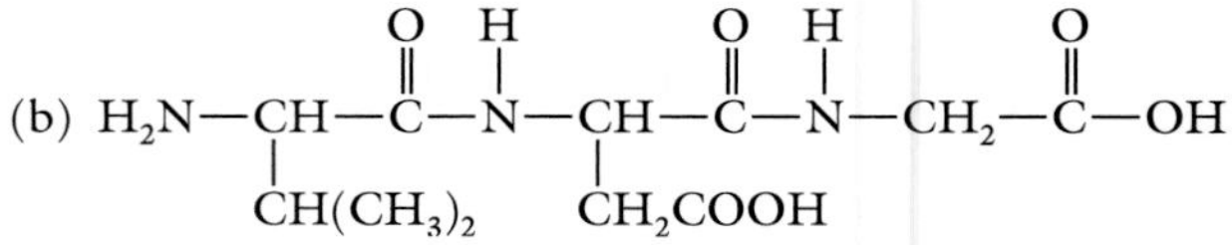

Cyclic Compounds Containing Tetrahedral Stereocenters

17.33 Draw the stereoisomers (if any) of each of the following compounds: (a) 3-methylcyclohexanone; (b) 1,1-dimethylcyclopentane; (c) 2-methyl-1,3-cyclohexanedione; (d) 1-hydroxy-2-methylcyclohexane; (e) 1,2-dihydroxycyclohexane. Which stereoisomers are optically active? Which stereoisomers are enantiomers and which are diastereomers?

17.34 Draw the stereoisomers (if any) of each of the following compounds: (a) chlorocyclopentane; (b) 1,1-dichloro-2-chlorocyclopentane; (c) 1,1,3,3-tetrachloro-2-methylcyclopentane; (d) 1-chloro-2-methylcyclopentane; (e) 1,2-dichlorocyclopentane. Which stereoisomers are optically active? Which stereoisomers are enantiomers and which are diastereomers?

17.35 Use an asterisk to indicate each tetrahedral stereocenter (if any) in each of the following structures. How many stereoisomers are possible for each of these structures?

(a) Cocaine

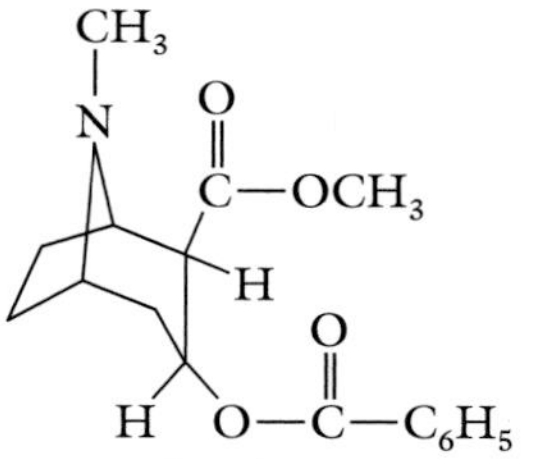

(b) Nicotine

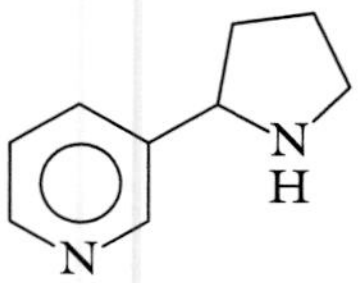

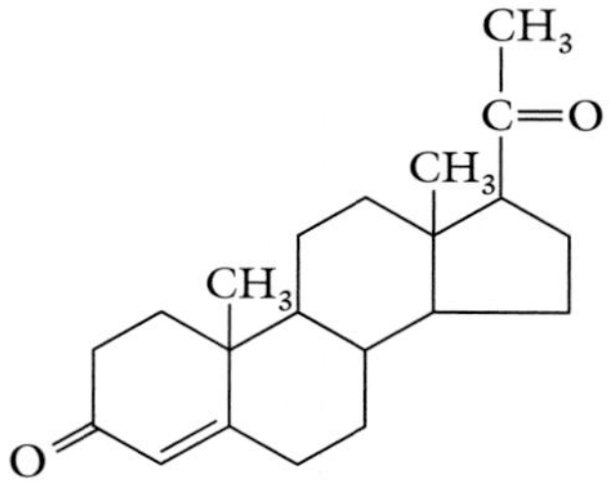

(c) Progesterone

17.36 Use an asterisk to indicate each tetrahedral stereocenter (if any) in each of the following structures. How many stereoisomers are possible for each of these structures?

(a) 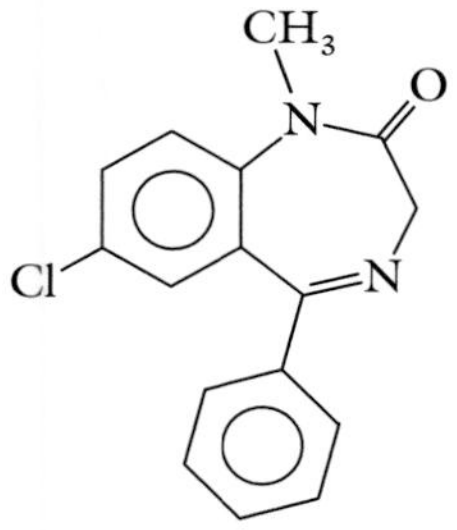

Diazepam (Valium)

(b) 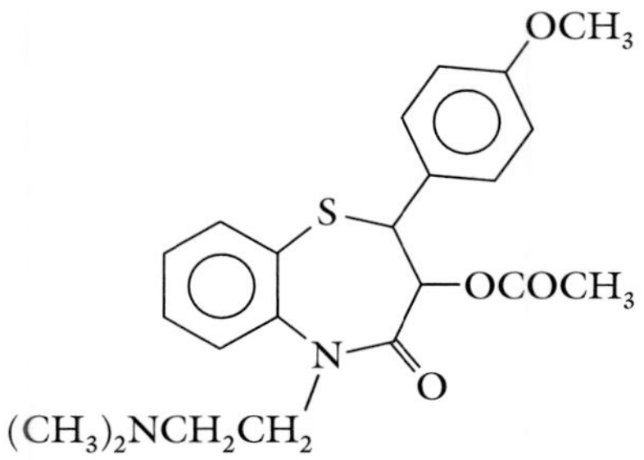

Diltiazem (Cardizem)

(c) 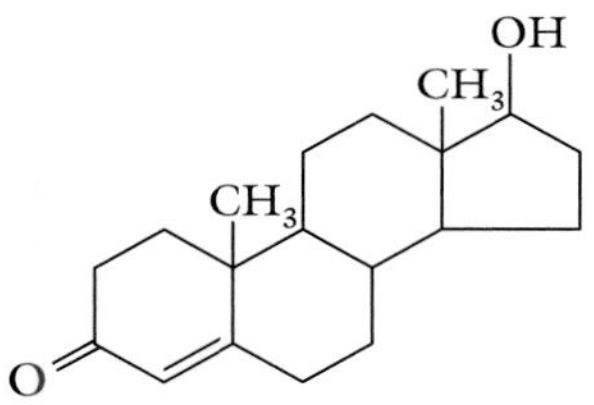

Testosterone

Unclassified Exercises

17.37 For each of the following pairs of structural formulas, indicate whether the pair represents (1) the same compound or (2) different compounds that are constitutional isomers or (3) different compounds that are enantiomers or (4) different compounds that are diasteromers or (5) different compounds that are not isomers.

(a) $CH_3CH_2CH_2OCH(CH_3)_2$ and $CH_3CH_2CH_2CH_2CH_2OCH_3$

(b) 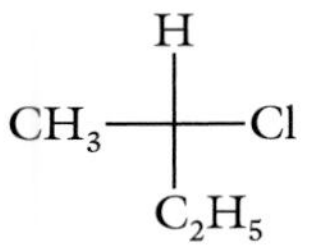 and 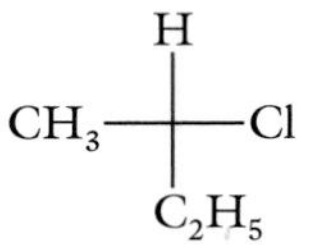

(c) 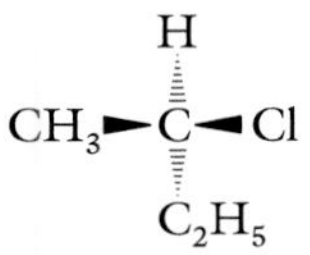 and 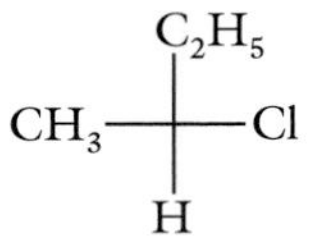

(d) $CH_3CH_2CH_2OCH(CH_3)_2$ and $CH_3CH_2CH_2CH_2CH_2CH_2OH$

(e) $CH_3CH_2CH_2OCH(CH_3)_2$ and 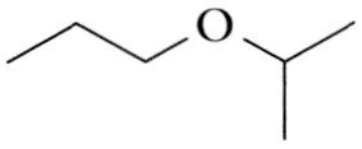

(f) 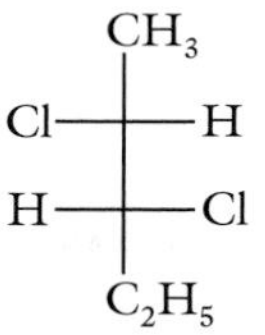 and 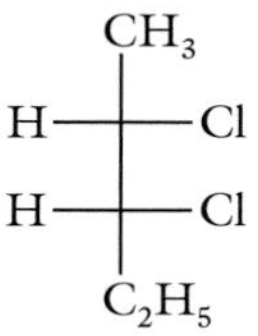

(g) 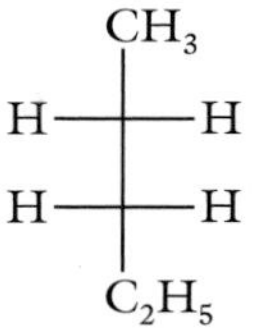 and 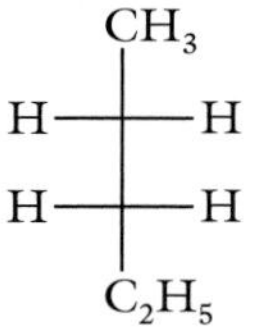

(h) 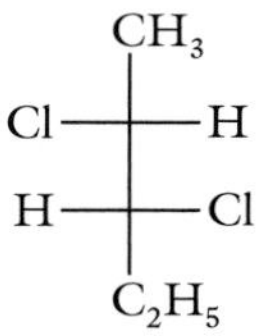 and 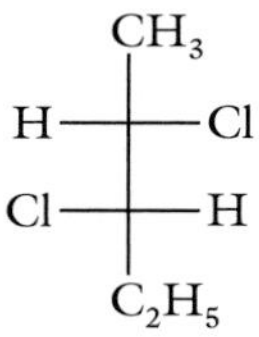

(i) 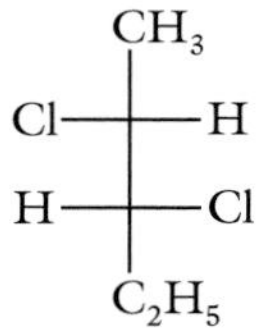 and 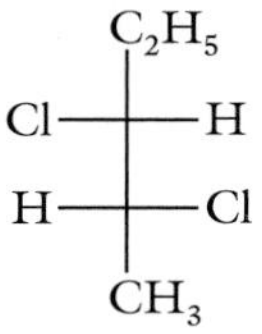

(j) 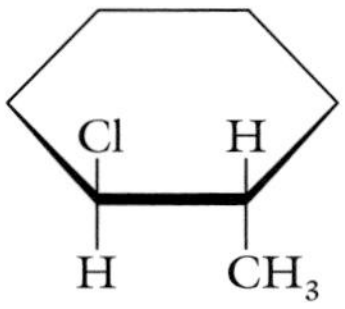 and 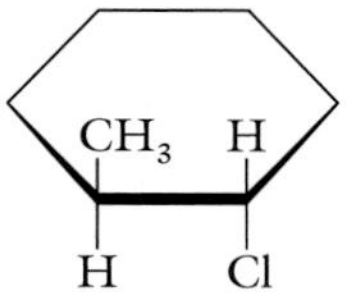

(k) 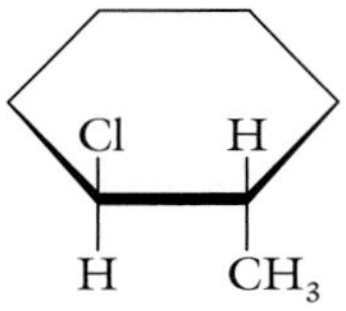 and 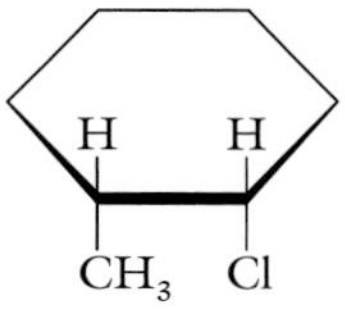

(l) 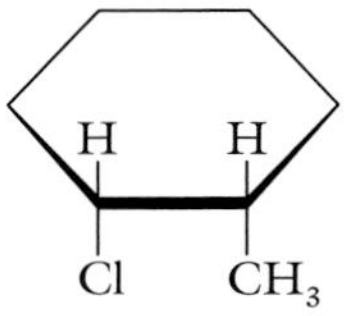 and 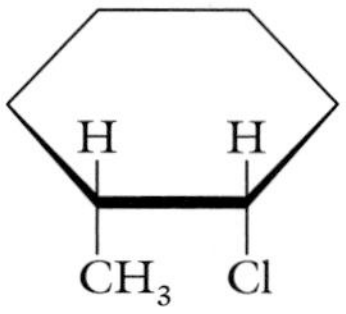

(m) 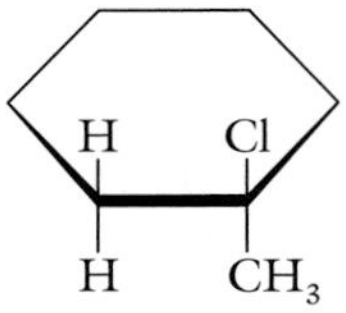 and 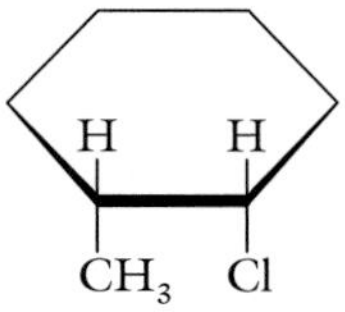

(n) 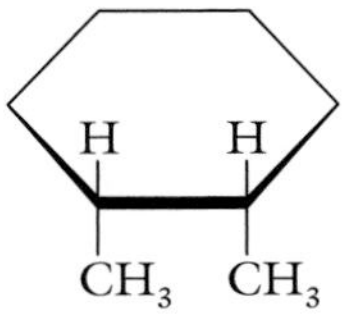 and 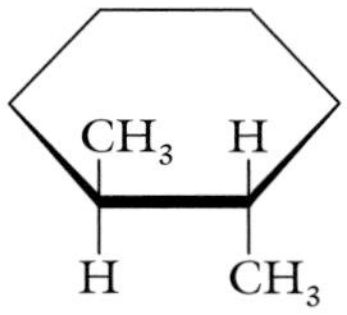

(o) 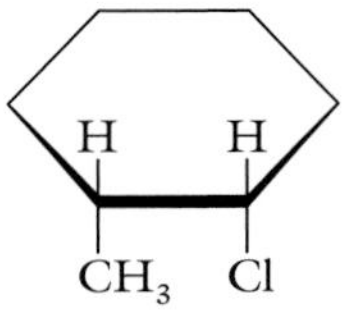 and 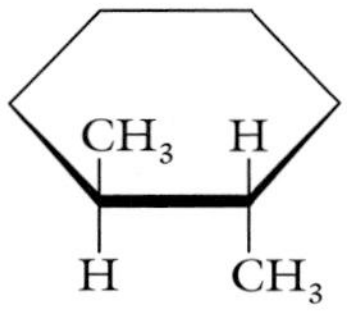

17.38 What is chiral recognition in biological systems?

17.39 Draw the four stereoisomers of 4-chloro-2-pentene. Use an asterisk to indicate each tetrahedral stereocenter.

Chemical Connections

17.40 (+)-2-Methylbutanoic acid and (−)-2-methylbutanoic acid react at the same rate of esterification with ethanol. However, in esterification with (+)-*s*-butyl alcohol, (−)-2-methylbutanoic acid reacts much faster than does (+)-2-methylbutanoic acid. Describe the mechanism responsible for this behavior.

17.41 Addition of HCl to 1-butene produces optically inactive 2-chlorobutane. Why is the product optically inactive even though C2 of 2-chlorobutane is a tetrahedral stereocenter?

17.42 Racemic thalidomide was used as a sedative and antinausea agent by pregnant women in Germany and England in the late 1950s and early 1960s. Many of those women gave birth to babies with severely deformed or missing arms and legs and often with abnormalities of the eyes, ears, and digestive system. For many years, these birth defects were thought to have been caused solely by L-thalidomide. Subsequently, it was found that, although the D-enantiomer does not directly cause the birth defects, it does do so indirectly because it is converted under physiological conditions into the racemic mixture. The structure of L-thalidomide is shown belown. Draw D-thalidomide.

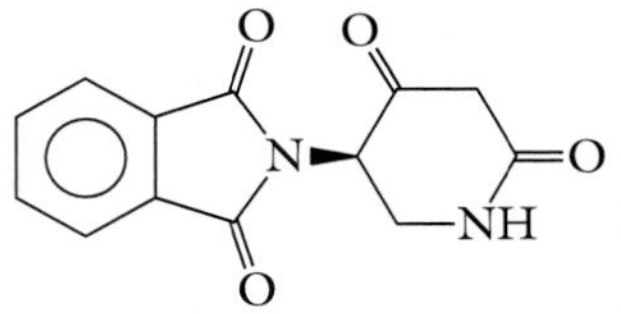

L-Thalidomide

17.43 The sex attractant for the common housefly is muscalure (*cis*-tricosene; Box 12.3), the trans isomer being totally inactive. Is chiral recognition responsible for the large difference in physiological effect of the two isomers?

PART 3

BIOCHEMISTRY

We now apply organic chemistry to biochemistry, a study of the structures and physiological functions of biochemicals—such as the physical processes and chemical reactions by which organisms extract, transform, and use energy and materials from their environment. Most biochemicals have complex structures that are uniquely designed for their biological functions. An example is hemoglobin—the protein that enables oxygen to be transported throughout the body. Hemoglobin is composed of four polypeptide molecules, each of which has a heme molecule in its interior. Each heme has iron at its center, and it is the iron that holds and stores oxygen.

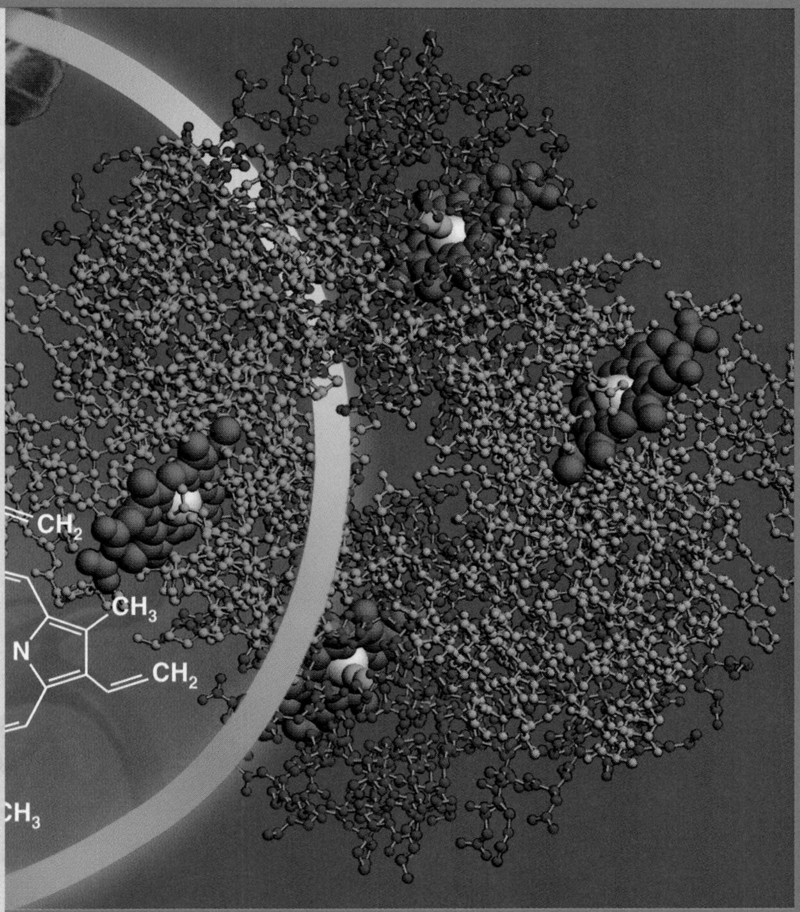

CHAPTER 18

CARBOHYDRATES

CHEMISTRY IN YOUR FUTURE

Two children who are raising a heifer as a 4H project have brought her to the animal hospital where you work as a technician; they are worried that she is not gaining weight. As you weigh the animal, take her temperature, and ask questions about her eating habits, they ask you to explain why grass, which can't be digested by humans, is good food for a cow. You tell her that both the cellulose in grass and, say, the starch in the potatoes that humans eat are made from glucose. The difference is in the way in which the units of glucose are joined together, so grass and potatoes require different enzymes for digestion. Humans don't have the enzyme necessary for digesting cellulose. Chapter 18 describes the structures of these carbohydrates and the differences in their digestibility.

LEARNING OBJECTIVES

- Define the scope of biochemistry.
- Describe the biological roles of carbohydrates and distinguish between monosaccharides, oligosaccharides, and polysaccharides.
- Describe how monosaccharides are classified and named.
- Describe the characteristics of the D families of aldoses and ketoses.
- Draw the cyclic hemiacetal structures of saccharides.
- Describe and write equations for mutarotation, oxidation, and acetal-formation reactions of monosaccharides.
- Describe the structures and functions of disaccharides.
- Describe the structures and functions of polysaccharides.
- Describe photosynthesis and the interdependence of plants and animals.

With this chapter we begin our study of **biochemistry,** the study of the chemical activities within individual cells and among communities of cells. The relation between anatomical structure and biological function is a powerful theme that is traditionally used to organize data and provide a context for biological information. For example, a description of the human kidney or heart would be incomplete if it did not couple the anatomical details with their physiological functions. Biochemistry extends the theme of structure and function downward from the cellular level to the subcellular and molecular levels of organization.

Life is not simple to define, and few of us would attempt to do it in one or two sentences. However, the functions of life are well recognized: to extract materials and energy from the environment; to use the materials and energy to maintain the structure of the organism and carry out essential processes; to compete with other life forms and survive to replicate and produce succeeding generations. Despite the enormous diversity among species, both the molecules of which they are composed **(biomolecules)** and the biochemical processes that they use to maintain and propagate life are remarkably similar. We begin our consideration of biochemistry with the study of biomolecules, and our presentation emphasizes three principal aspects:

- **Structure:** The chemical structure of biomolecules; their organization into cellular components, cells, tissues, and macroscopic structures such as muscle and bone; and the relation of molecular structure to physiological function. Structure will be explored in Chapters 18 through 21.
- **Transmission of information:** The molecular basis of heredity, the mechanism by which genetic information is stored and transmitted. Genetic transmission is the focus of Chapter 21.
- **Metabolism:** Collectively, the physical processes and chemical reactions by which organisms extract, transform, and use energy and materials from their environment. These subjects are treated in Chapters 22 through 26.

Organisms require both inorganic and organic materials to sustain life. The necessary inorganic substances are water and a host of inorganic ions such as Na^+, K^+, Fe^{2+}, Ca^{2+}, Cl^-, $HPO_4{}^{2-}$, and $HCO_3{}^-$. Many organic biomolecules, particularly proteins, function only in the presence of a specific inorganic ion; sometimes the ion is an integral part of the molecular structure. For example, hemoglobin, the protein that transports oxygen from the lungs to other tissues, contains Fe^{2+} (Chapter 20). Many physiological functions depend on the difference in concentrations of certain ions inside and outside cells, a difference maintained by the cell membrane (Chapter 19).

The four principal families of organic biomolecules are the carbohydrates, lipids, proteins, and nucleic acids. Proteins have many functions: they are the structural materials of muscle and bone; they are the enzymes that catalyze biochemical reactions; they transport oxygen through the blood to tissues; they protect the organism against viruses and bacteria; and they function as hormones to regulate physiological functions (Chapter 20). Lipids are the main components of the membranes that enclose all living cells; they are sources of energy; and some function as hormones to regulate physiological functions (Chapter 19). Nucleic acids direct and control the transmission of hereditary information and the synthesis of proteins and, ultimately, all cellular materials (Chapter 21).

Carbohydrates and their various functions are the subject of the remainder of this chapter.

18.1 INTRODUCTION TO CARBOHYDRATES

Carbohydrates, which are also called **saccharides,** are the single most abundant family of organic compounds to be found in nature. They have a variety of functions:

- The metabolism (breakdown) of the carbohydrate glucose generates the energy required for all life processes. Starch in plants and glycogen in animals are carbohydrates serving as storage forms of glucose.
- Carbohydrates serve as structural and protective materials. One example is cellulose, used in cell walls and in the extracellular structures of plants; another is chitin, which forms the exoskeletons of crustaceans and insects.
- Carbohydrates are precursors for the biosynthesis of proteins, lipids, and nucleic acids.
- Carbohydrates attached to proteins or lipids in cell membranes help cells recognize each other as well as specific molecules, triggering physiological processes such as fertilization, cell growth, and immune responses.
- The carbohydrates ribose and deoxyribose are components of ribonucleic (RNA) and deoxyribonucleic (DNA) acids, respectively.

Carbohydrates are familiar in our lives. Starch is a major component of our diet—in potatoes, cereals, beans, peas, corn, rice, pasta, and bread. We use table sugar, or sucrose, to sweeten our baked goods, coffee, tea, and soft drinks. The wood used to build our houses, the paper on which this book is printed, and the cotton in our clothes are largely composed of cellulose.

Every carbohydrate is either a **polyhydroxyaldehyde** or a **polyhydroxyketone,** polyfunctional molecules containing an aldehyde or ketone group and two or more hydroxyl groups. Derivatives and polymers of these molecules are also considered as carbohydrates. Carbohydrates are further classified as monosaccharides, oligosaccharides, and polysaccharides. **Monosaccharides,** the simplest saccharides, have important biological roles themselves and also serve as the building blocks (called **monomers, residues,** or **repeat units**) for synthesizing larger saccharides. Oligosaccharides and polysaccharides can be hydrolyzed to monosaccharides, but monosaccharides cannot be further hydrolyzed to still smaller saccharides. Glucose, fructose, and galactose are important monosaccharides.

Polysaccharides contain large numbers, usually hundreds or even thousands, of monosaccharide units bonded together. Important examples are starch, cellulose, glycogen, and chitin. The name **oligosaccharide** is loosely used for any saccharide larger than a monosaccharide but smaller than a polysaccharide. There is no well-defined demarcation between oligosaccharides and polysaccharides. Most biochemists consider oligosaccharides to contain as many as 10 to 20 monosaccharide units. More specific terms such as **disaccharide, trisaccharide,** and **tetrasaccharide** are more useful. Sucrose and lactose (milk sugar) are important disaccharides. Monosaccharides and disaccharides are also called **sugars** because almost all of them taste sweet.

IUPAC names are not generally used for carbohydrates, because they are excessively long and cumbersome, and common names have become strongly entrenched in the chemical and biological literature. The common names of most, but not all, saccharides have the ending **-ose,** as in glucose, fructose, lactose, sucrose, and cellulose. Exceptions include most polysaccharides, such as starch, chitin, and glycogen.

18.2 MONOSACCHARIDES

Monosaccharides have one of the following structures:

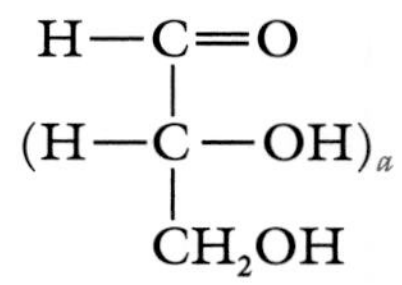

Aldose
(Polyhydroxyaldehyde)

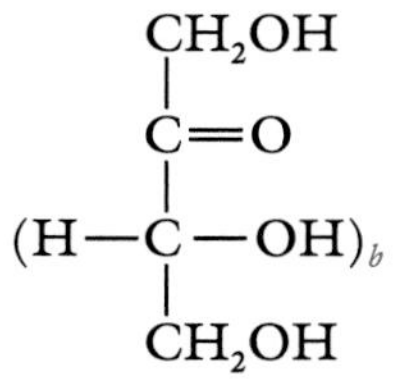

Ketose
(Polyhydroxyketone)

For monosaccharides of biological importance, a = 1 to 4 and b = 0 to 3; that is, they are monosaccharides containing from three to six carbons. The carbon chains are numbered from the end nearest the carbonyl carbon. In polyhydroxyaldehydes, the carbonyl carbon is C1; in polyhydroxyketones, the carbonyl carbon is C2.

Classification and Nomenclature

Monosaccharide names classify the compounds in two ways simultaneously, by combining two kinds of prefixes before **-ose:**

Rules for naming monosaccharides

- Monosaccharides with an aldehyde group are **aldoses;** those with a ketone group are **ketoses.**
- Monosaccharides with three, four, five, and six carbons are **trioses, tetroses, pentoses,** and **hexoses,** respectively.

For example, a five-carbon monosaccharide with a ketone group is called a ketopentose.

Example 18.1 Classifying monosaccharides

Classify each of the following monosaccharides to indicate both the type of carbonyl group and the number of carbons present:

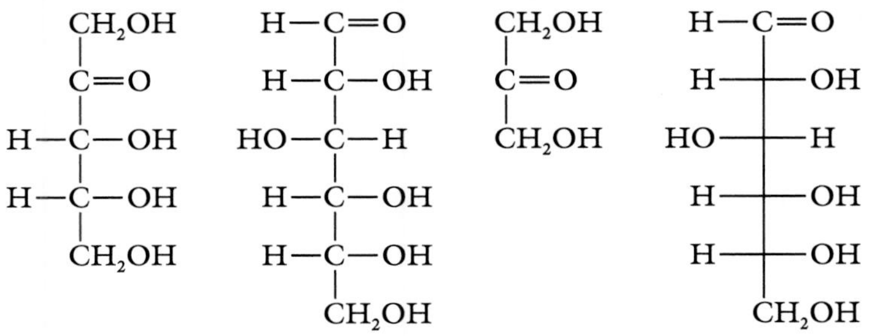

1 2 3 4

Solution

First determine whether the molecule is an aldehyde or ketone, and then count the number of carbons. Monosaccharide 1 is a ketopentose, 2 is an aldohexose, 3 is a ketotriose, 4 is an aldohexose. Note that monosaccharides 2 and 4 are identical. Monosaccharide 4 is the Fischer projection of 2. Fischer projections are used more often than not.

Problem 18.1 Classify each of the following monosaccharides to indicate both the type of carbonyl group and the number of carbons present:

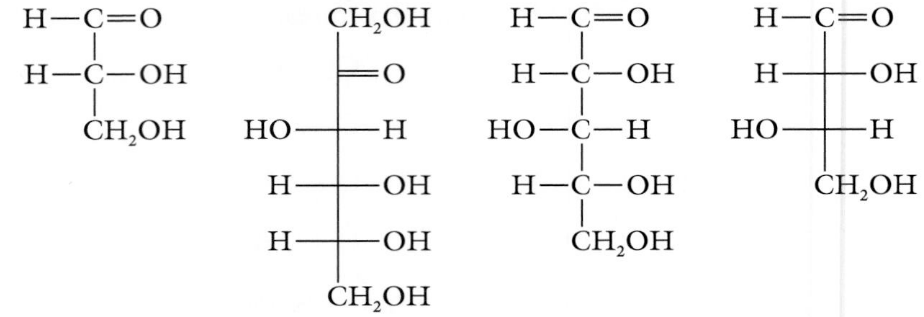

1 2 3 4

Stereoisomerism

Figures 18.1 and 18.2 show the D-aldoses and D-ketoses. Each figure starts with the triose at the top and proceeds downward through the tetroses, pentoses, and hexoses. The aldotriose D-glyceraldehyde (Figure 18.1) has one tetrahedral stereocenter, and each additional carbon in tetroses, pentoses, and

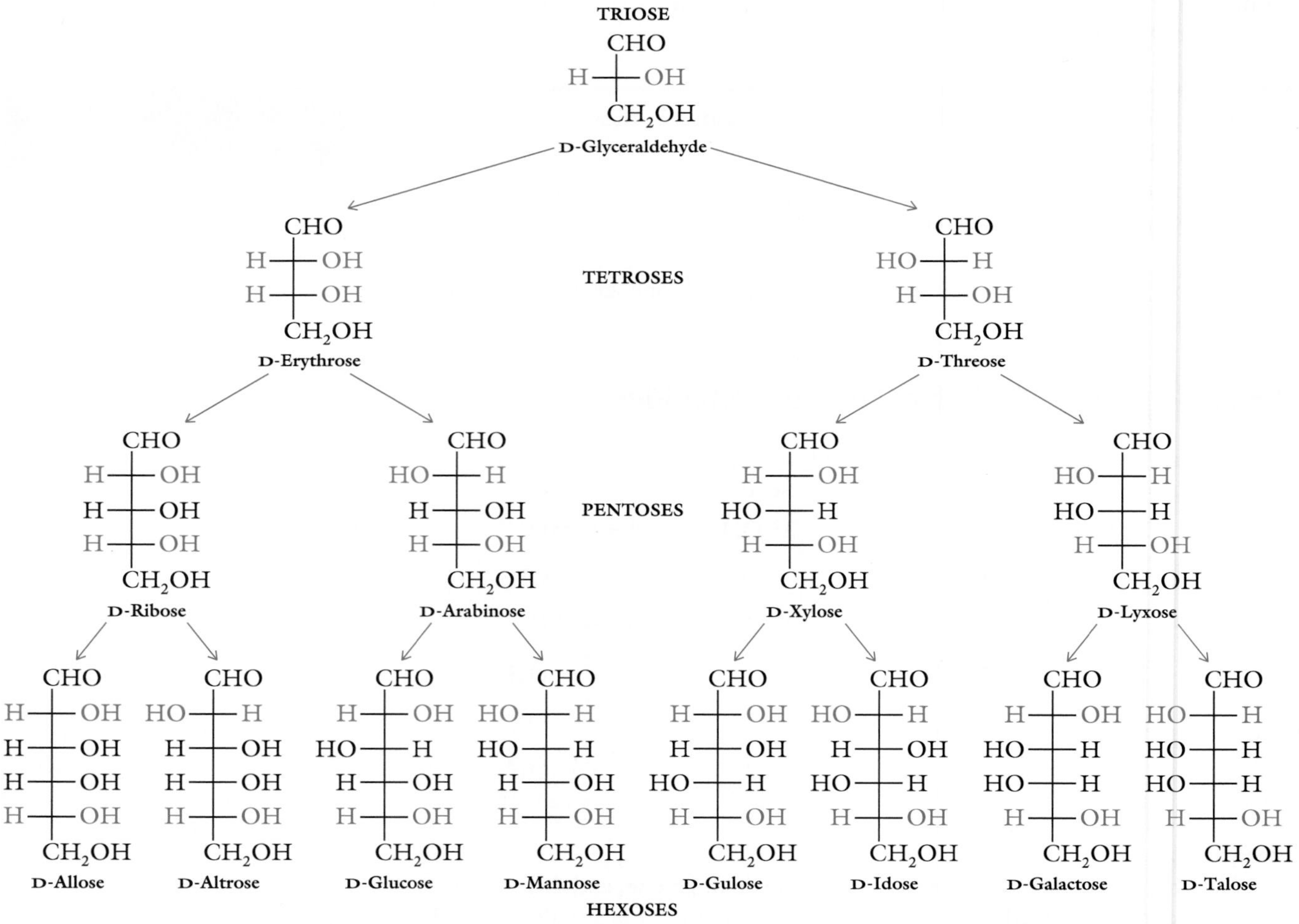

Figure 18.1 D-Aldoses. Each pair of arrows denotes a pair of diastereomers whose configurations are identical at all tetrahedral stereocenters except C2 (red). All D-aldoses have the same configuration at the stereocenter farthest from the carbonyl group (blue).

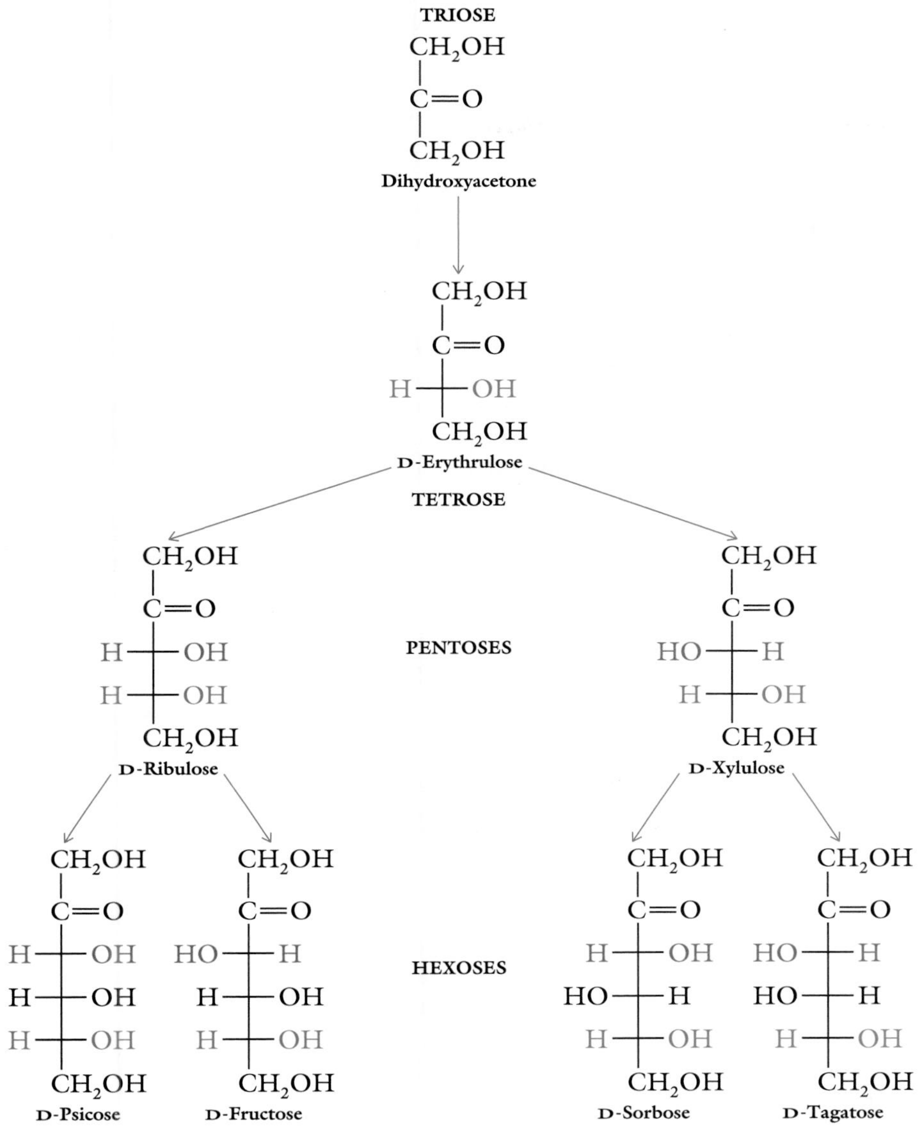

Figure 18.2 D-Ketoses. Each pair of arrows denotes a pair of diastereomers whose configurations are identical at all tetrahedral stereocenters except C3 (red). All D-ketoses have the same configuration at the stereocenter farthest from the carbonyl group (blue).

hexoses adds another tetrahedral stereocenter. The ketotriose dihydroxyacetone (Figure 18.2) does not have a tetrahedral stereocenter. The ketotetrose D-erythrulose is the smallest ketose to possess a tetrahedral stereocenter; each additional carbon adds a tetrahedral stereocenter.

A complete description of the stereoisomerism of an aldose or ketose requires a separate D or L designation (Section 17.4) for each tetrahedral stereocenter, but this designation is not necessary for our purposes. Our primary interest is the relations between the various aldoses and those between the various ketoses. All of the monosaccharides in Figures 18.1 and 18.2 have the same configuration at the tetrahedral stereocenter (shown in blue) farthest from the carbonyl carbon. That configuration is the D configuration—the —OH group is on the right side of the Fischer projection—and all of the monosaccharides are called D-saccharides to indicate this common structural feature.

In theory, each of the D-saccharides in Figures 18.1 and 18.2 has a corresponding mirror-image enantiomer, but these L-saccharides are rarely found in nature. The D-sugars predominate in nature because they are synthesized and

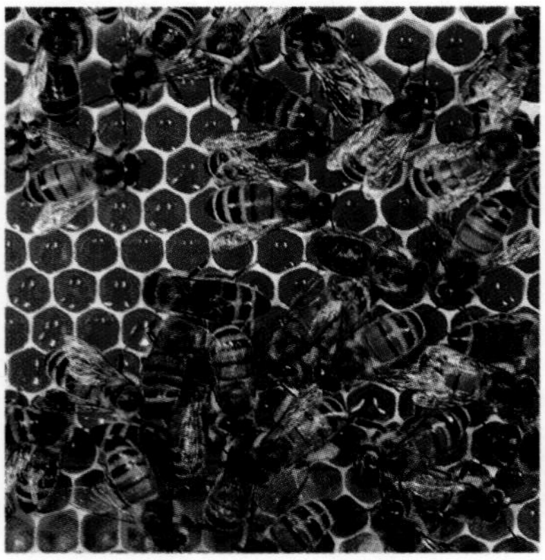

The honey produced by bees is a concentrated aqueous solution that is more than 80% carbohydrates, most of which are D-fructose and D-glucose.

utilized by stereospecific enzymes that function through the principle of chiral recognition (Section 17.5).

In discussions of saccharides, the names of saccharides are often used without the prefix D-. Whenever the stereochemistry of a saccharide is not indicated, such as in "glucose" or "fructose," you should assume that the D-enantiomer is meant.

The saccharides in any horizontal grouping in Figure 18.1 or in Figure 18.2 are diastereomers of each other; that is, they are stereoisomers that are not enantiomers.

Example 18.2 Recognizing structural relationships

For each of the following pairs of compounds, indicate whether the pair represents different compounds that are (1) constitutional isomers or (2) stereoisomers that are enantiomers or (3) stereoisomers that are diasteromers or (4) not isomers. Refer to Figures 18.1 and 18.2 for the structures of the compounds. (a) D-Glucose and D-mannose; (b) D-ribose and D-xylulose; (c) D-fructose and D-arabinose; (d) D-sorbose and L-sorbose; (e) D-sorbose and D-fructose.

Solution

(a) Diastereomers; they are stereoisomers that are not enantiomers.

(b) Constitutional isomers; D-ribose is an aldopentose and D-xylulose is a ketopentose.

(c) Not isomers; D-fructose is a hexose and D-arabinose is a pentose.

(d) Enantiomers; they are nonsuperimposable mirror-image stereoisomers.

(e) Diastereomers; they are stereoisomers that are not enantiomers.

Problem 18.2 For each of the following pairs of compounds, indicate whether the pair represents different compounds that are (1) constitutional isomers or (2) stereoisomers that are enantiomers or (3) stereoisomers that are diasteromers or (4) not isomers. Refer to Figures 18.1 and 18.2 for the structures of the compounds. (a) D-Glucose and D-talose; (b) D-ribose and D-sorbose; (c) D-fructose and L-fructose; (d) D-erythrose and L-erythrulose; (e) D-sorbose and D-psicose.

Some Important Monosaccharides

D-Glucose is overwhelmingly the most abundant monosaccharide in nature, occurring mostly in combined form in starch, cellulose, glycogen, chitin, lactose, and sucrose. Dietary starch and stored glycogen (in the liver) are broken down to glucose, which is transported to and taken into cells for energy production. A 5% aqueous solution of D-glucose is used for the intravenous feeding of patients unable to take nourishment by mouth. Glucose is also known as **blood sugar** or **dextrose.**

Two other hexoses are very important. D-Fructose, also called **levulose** (because it is levorotatory; Section 17.5), is bonded to D-glucose in the disaccharide sucrose, the sugar in fruits and table sugar. D-Galactose is bonded to D-glucose in the disaccharide lactose, the sugar in mammalian milk. Although D-glucose is the most common component of polysaccharides, D-fructose and D-mannose also are present in some polysaccharides.

The high stereospecificity in nature is evident in the disproportionately high utilization of D-glucose relative to the other aldohexoses. There are four tetrahedral stereocenters and 2^4 or 16 possible stereoisomers for D-glucose, the 8 D-aldohexoses of Figure 18.1 plus the corresponding mirror-image

L-aldohexoses. But the amount of D-glucose found in nature far exceeds the total of all other aldohexoses, as well as all other monosaccharides.

D-Ribose and D-xylose are important pentoses. D-Xylose is a component of some plant polysaccharides. D-Ribose and 2-deoxy-D-ribose are components of ribonucleic (RNA) and deoxyribonucleic (DNA) acids, respectively. 2-Deoxy-D-ribose is a **deoxysaccharide,** or **deoxysugar,** having one less oxygen than its corresponding saccharide at C2: an —OH group of the saccharide is replaced by —H in the deoxysaccharide.

The trioses D-glyceraldehyde and dihydroxyacetone are important intermediates in metabolic processes (Sections 23.1, 23.7, and 26.3).

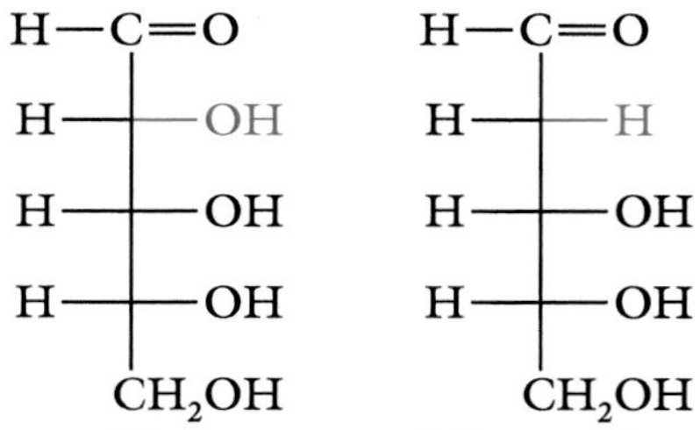

D-Ribose **2-Deoxy-D-Ribose**

18.3 CYCLIC HEMIACETAL STRUCTURES

So far, the structural representations that we have used for pentoses and hexoses have not been entirely correct. Recall from Section 14.6 that the —OH group of an alcohol and the carbonyl group of an aldehyde or ketone undergo an acid-catalyzed addition reaction to form a hemiacetal:

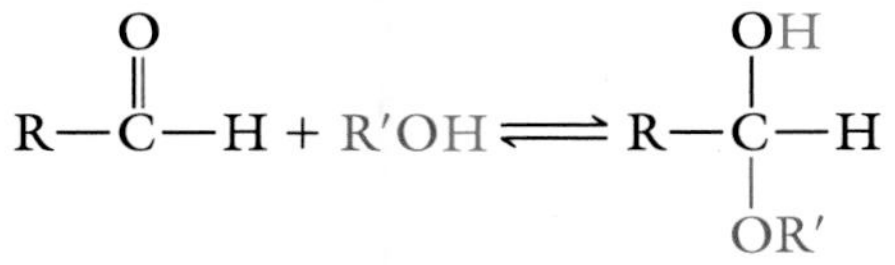

Hemiacetal

Furthermore, we saw that, in a molecule with both an alcohol —OH group and an aldehyde or ketone carbonyl group, intramolecular hemiacetal formation produces a **cyclic hemiacetal.** For example, addition of the —OH group at C5 of D-glucose to its own carbonyl group at C1 forms a six-membered cyclic hemiacetal, called a **pyranose ring** (Figure 18.3). Intramolecular hemiacetal formation is the key to understanding the structures and reactions of saccharides.

We can visualize the conversion of the acyclic D-glucose structure into the cyclic hemiacetal by the following process:

- Draw the Fischer projection of the acyclic structure and then turn it sideways.

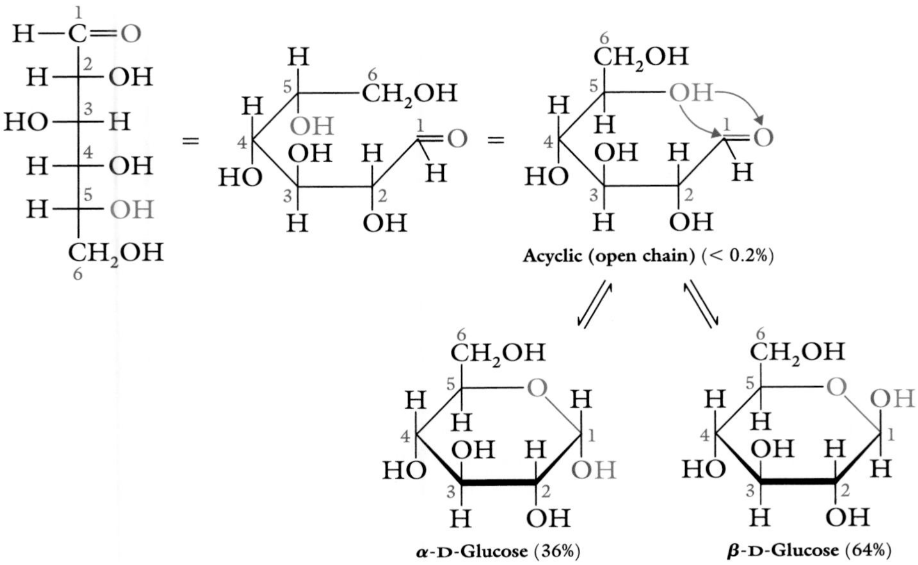

Figure 18.3 Hemiacetal formation in D-glucose. The oxygen of the OH at C5 of the acyclic structure becomes the oxygen in the hemiacetal ring, whereas the carbonyl oxygen becomes the OH at C1.

- Rotate the bond between C4 and C5 to bring the C5 —OH group close to the carbonyl group.
- Add the alcohol —OH to the carbonyl group: break the O–H bond, and then bond the H of the —OH to the carbonyl O and bond the O of the —OH to the carbonyl C.

Hemiacetal formation converts the carbonyl carbon into a new tetrahedral stereocenter. This new stereocenter is C1 of the cyclic hemiacetal in Figure 18.3 and is called the **hemiacetal carbon.** You can recognize the hemiacetal carbon as the carbon with two different oxygen groups attached: an alcohol —OH and an ether —OR.

Two stereoisomers of the cyclic hemiacetal, called the α and β configurations, are possible; these stereoisomers are diastereomers. Ring closure produces a mixture of the two diastereomers because neither configuration is excluded from forming during the reaction. Thus, D-glucose is a mixture of the two hemiacetals α-D-glucose and β-D-glucose. These diastereomers, differing only in the configuration at the hemiacetal carbon, are called **anomers.**

- The α-anomer of D-glucose has a trans relation between the —OH at the hemiacetal carbon (C1) and the $—CH_2OH$ at C5.
- The β-anomer of D-glucose has a cis relation between the —OH at the hemiacetal carbon (C1) and the $—CH_2OH$ at C5.

As we shall see later in the chapter, the difference between the α and β configurations at the hemiacetal carbon is highly significant to saccharide structure and function.

As noted in Section 14.6, hemiacetal formation between simple alcohols and aldehydes or ketones is not important, because the equilibrium is overwhelmingly to the left; the hemiacetal is very unstable. This is not the case for cyclic hemiacetal formation in saccharides: the equilibrium is overwhelmingly in favor of the cyclic hemiacetal, which is a stable product. The acyclic structure, often called the **open-chain structure,** is not stable relative to the cyclic hemiacetal. In solutions of D-glucose, more than 99.8% of the D-glucose is present as the cyclic hemiacetal structures with less than 0.2% present as the acyclic structure. The β-anomer is about twice as abundant as the α-anomer.

Concept checklist

✔ Hexoses and pentoses exist predominantely as cyclic hemiacetals, not as acyclic compounds containing alcohol —OH and aldehyde or ketone C=O groups.

✔ Even when saccharides are represented by open-chain structures, the actual structure of the saccharide is predominately the cyclic hemiacetal.

D-Fructose and other ketohexoses also exist as cyclic hemiacetals. Addition of the —OH at C5 to the carbonyl group forms a five-membered cyclic hemiacetal called a **furanose ring** (Figure 18.4). The hemiacetal carbon is C2. Again, a pair of anomers is formed. In α-D-fructose, the —OH group at C2 and the $—CH_2OH$ group at C5 have a trans relation; in β-D-fructose the two groups have a cis relation.

The cyclic hemiacetal structures of saccharides shown in Figures 18.3 and 18.4 are called **Haworth structures** or **Haworth projections;** such projections are drawn with bold bonds in the lower half of the ring to convey the three-dimensional aspects of the cyclic hemiacetals. We used the same convention previously described for cyclic compounds (Section 11.8): the ring is perpendicular to the plane of the paper, and the groups attached to the carbons of the ring are parallel to the plane of the paper. Haworth projections are not always drawn with bold bonds.

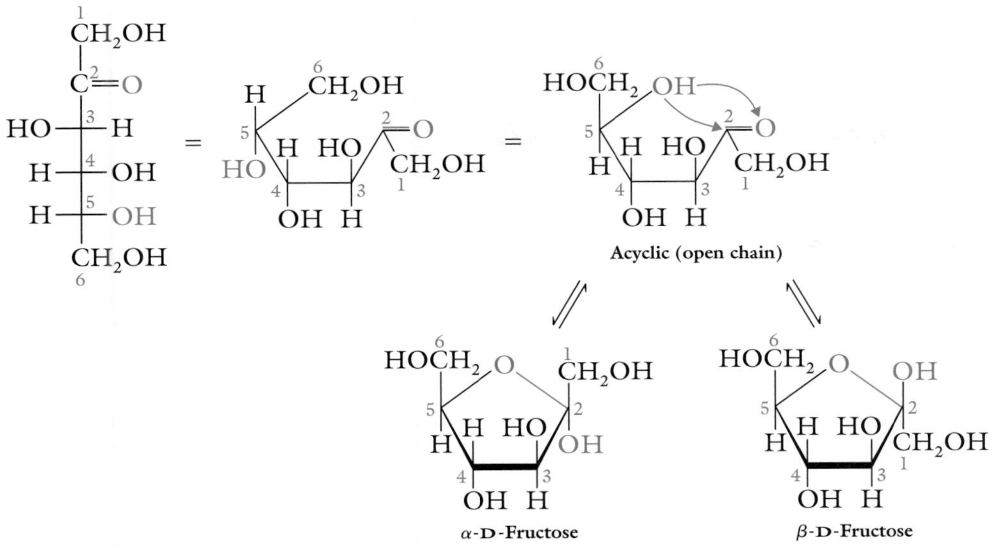

Figure 18.4 D-Fructose is an equilibrium mixture of α- and β-hemiacetal structures and an acyclic structure. The oxygen of the OH at C5 of the acyclic structure becomes the oxygen in the hemiacetal ring, whereas the carbonyl oxygen becomes the OH at C2.

The horizontal bonds extending forward and vertical bonds extending backward convention of Fischer projections should not be confused with the spatial orientations in Haworth projections. There is a simple translation between the two structures: groups appearing to the right in Fischer projections extend downward from the ring in Haworth projections; those to the left in the Fischer projection extend upward from the ring in Haworth projections.

Example 18.3 Drawing the cyclic hemiacetal structure of a monosaccharide

Draw the cyclic structure of α-D-galactose, referring only to Figure 18.1. Indicate how this task is simplified if you are allowed to directly refer to both Figures 18.1 and 18.3.

Solution

Use the three-step process described near the beginning of Section 18.3 and illustrated in Figure 18.3 for D-glucose: draw the Fischer projection, turn it sideways, rotate the bond between C4 and C5, and add the alcohol O–H to the carbonyl group.

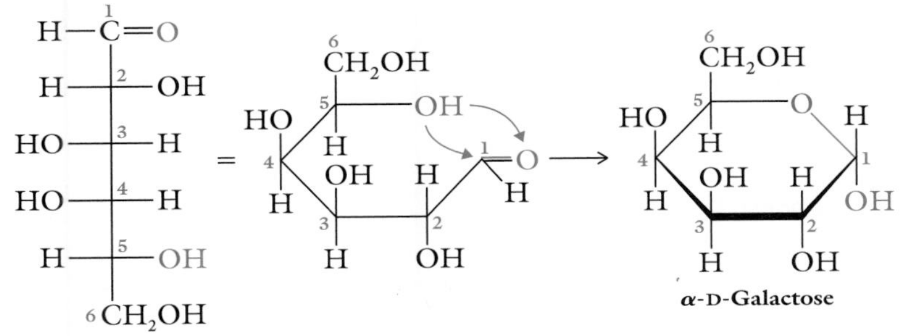

Figure 18.1 shows that D-galactose differs from D-glucose only at the configuration at C4. If Figure 18.3 is available for reference, all that is needed is to take the structure of α-D-glucose and reverse the positions of the —H and —OH groups at C4:

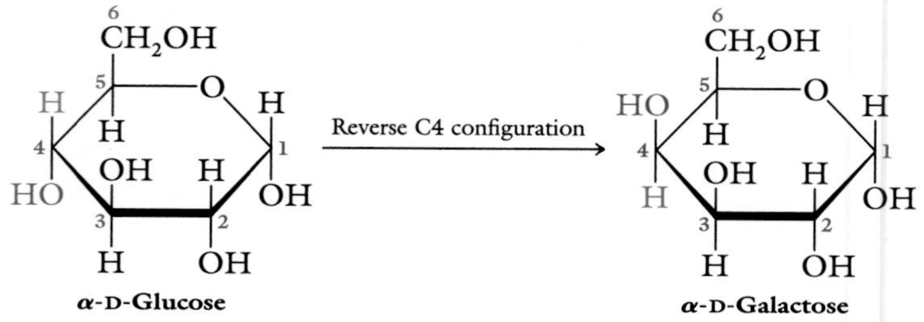

Problem 18.3 Draw the cyclic structure of β-D-sorbose by reference to Figures 18.2 and 18.4.

18.4 CHEMICAL AND PHYSICAL PROPERTIES OF MONOSACCHARIDES

As you might expect of polyhydroxyaldehydes and polyhydroxyketones, monosaccharides undergo considerable hydrogen bonding with themselves and with water. Thus, monosaccharides are solids at room temperature and are very soluble in water. Highly concentrated solutions are very viscous liquids; think of the consistency of honey, maple syrup, and molasses. Monosaccharides are only slightly soluble in alcohols such as methanol and ethanol and insoluble in less-polar solvents such as ethers and hydrocarbons. Almost all monosaccharides (and disaccharides) taste sweet (Box 18.1 and Table 18.1).

A solution of a monosaccharide such as D-glucose is an equilibrium mixture of α-D-glucose, β-D-glucose, and the acyclic, or open-chain, structure. Because the molecules interconvert rapidly from one structure into another, the properties of D-glucose are simultaneously those of an aldehyde with four —OH groups and those of a hemiacetal with five —OH groups. The presence of this equilibrium has been proved by the mutarotation and oxidation experiments described next.

TABLE 18.1 Sweetness of Various Compounds Relative to Sucrose

Compound	Sweetness*
lactose	0.16
galactose	0.32
maltose	0.33
glucose	0.74
sucrose	1.00
fructose	1.73
calcium cyclamate	30
aspartame	160
acesulfame K	200
saccharin	500

* Relative to sucrose = 1.00

18.1 Chemistry Within Us

How Sweet Is It?

Sweet taste is one of the four primary taste perceptions, the others being sour, salty, and bitter (see Box 17.1). Almost all monosaccharides and disaccharides taste sweet, as do many other compounds with —OH groups on adjacent carbon atoms. For saccharides, sweetness depends on molecular structure, size, and stereoisomerism. The relative sweetness of various monosaccharides, disaccharides, and artificial sweeteners are listed in Table 18.1. Sweetness is measured relative to sucrose, set at 1.00.

Excessive sugar intake resulting in obesity is a major health problem in the developed countries, and this problem has spurred the growth of the artificial sweetener industry. A number of synthetic organic compounds are far sweeter than sucrose, which allows their use as artificial sweeteners without significant caloric value because they are used in very small amounts. The greater sweetness of artificial sweeteners surprises many people; in fact, the artificial sweeteners are structurally very different from natural sugars. It clearly indicates a complex relation between the taste chemoreceptors and the structures that they respond to most strongly. Artificial sweeteners are also useful for diabetics who must restrict their intake of sugars (Box 20.3).

Saccharin and cyclamates were among the first artificial sweeteners. Cyclamates were banned in the United States in 1970 because some studies showed that they cause cancer in laboratory animals. Saccharin was implicated in bladder cancer in laboratory animals in 1978 but was not banned, because legislators were reluctant to ban the only artificial sweetener available. Saccharin does, however, carry a warning that it may be a health hazard.

Newer artificial sweeteners include aspartame (NutraSweet; Section 17.6) and acesulfame K. Extensive testing suggests no major ill effects on health. Aspartame has largely replaced saccharin as an artificial sweetener. Acesulfame K is useful in food products that require cooking, because higher temperatures degrade aspartame but not acesulfame K.

Aspartame should not be used by people with phenylketonuria (PKU), a disorder of phenylalanine metabolism that can lead to mental retardation. After ingestion, aspartame is hydrolyzed to phenylalanine, aspartic acid, and methanol. Although methanol is toxic (Box 13.3), it is not dangerous at the minute concentrations formed from aspartame intake. However, even the small amounts of phenylalanine produced are a serious problem for phenylketonurics. Instead of transforming phenylalanine into tyrosine, people with PKU transform it into phenylpyruvate, which causes severe mental retardation.

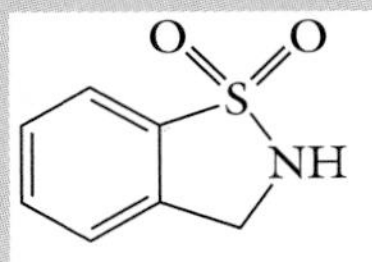

Saccharin

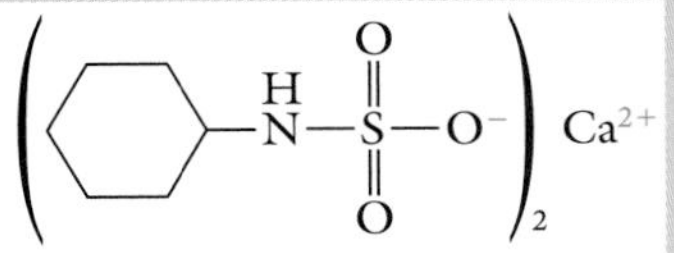

Calcium cyclamate

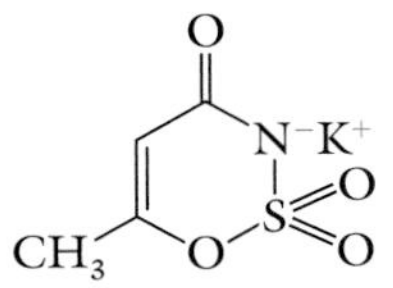

Acesulfame K

Mutarotation

Pure crystalline samples of α-D-glucose or β-D-glucose can be obtained by crystallization from different solvents. But all aqueous solutions of D-glucose consist of the equilibrium mixture of about 36% α-D-glucose, 64% β-D-glucose, and less than 0.2% open-chain structure. This equilibrium is responsible for an optical rotation phenomenon called **mutarotation.** The specific rotations of α-D-glucose and β-D-glucose are $+112.2°$ and $+18.7°$, respectively; these rotations can be measured immediately after dissolving each pure anomer separately in water. On standing, however, each solution undergoes a change in specific rotation until the equilibrium value of $+52.7°$ is obtained. Each of the pure anomers undergoes equilibration to the same mixture of α-D-glucose, β-D-glucose, and open-chain compound.

Concept check

✔ Each D-glucose anomer undergoes ring opening to the open-chain structure followed by ring closure to the mixture of α- and β-anomers.

Monosaccharides—and dissacharides, as we shall see—are often drawn as either the α- or β-anomer to conserve space. This convention is not meant to denote the presence of only one anomer. Every solution of a saccharide with a hemiacetal group consists of an equilibrium mixture of the two anomers and the open-chain structure, whether the three structures are drawn or not.

Oxidation of the Aldehyde Group

Mild oxidizing reagents such as Benedict's reagent (Cu^{2+} complexed with citrate ion in alkaline solution) oxidize the aldehyde group in an aldose to a carboxyl group (Section 14.4). Note that this oxidation is a reaction of an aldehyde (the open-chain structure of D-glucose); it is not a reaction of a hemiacetal (the ring structure of D-glucose):

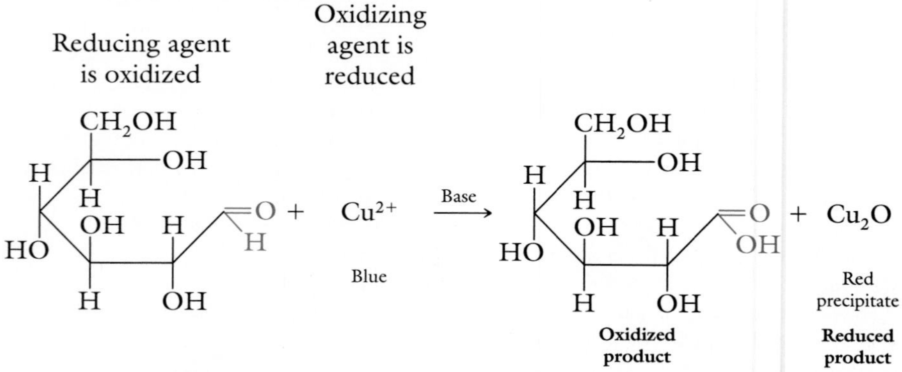

(The oxidized product is shown with its carboxyl group as COOH but will actually exist in its ionized form COO^- under the basic conditions of the reaction.)

The oxidation reaction demonstates that α-D-glucose and β-D-glucose are in equilibrium with the open-chain aldehyde. The oxidation is a quantitative reaction: one mole of aldose is oxidized, even though, at equilibrium, less than 0.2% of the aldose is present as the aldehyde structure. As the small amount of aldehyde is oxidized, some α-D-glucose and β-D-glucose convert into the aldehyde, which is then oxidized; more α-D-glucose and β-D-glucose convert into the aldehyde, which is then oxidized; and so forth. Eventually, all of the α-D-glucose and β-D-glucose are converted into the aldehyde and oxidized, according to Le Chatelier's principle.

The oxidation of an aldose by Benedict's reagent is used in a clinical test for monosaccharides in urine. The blue color of the reagent is replaced by the red color of the Cu_2O precipitate. The intensity of the red color observed is directly proportional to the concentration of monosaccharide in the urine. The clinical test uses a paper strip impregnated with Benedict's reagent. The paper strip is dipped in the urine sample, and the resulting red color is compared with a standard color chart to determine the monosaccharide concentration.

As noted in Section 14.4, most ketones are not oxidized by Benedict's reagent, but α-hydroxy ketones such as D-fructose and other ketoses give positive tests because they are converted into aldoses by the alkaline conditions of Benedict's reagent. Aldoses and ketoses are called **reducing sugars** because the sugar is the reducing agent in the oxidation reaction with Benedict's reagent.

Acetal Formation: Production of Glycosides

Monosaccharides, like all hemiacetals, are converted into acetals by acid-catalyzed dehydration of the hemiacetal —OH group with an alcohol (Section 14.6). For example, reaction of D-glucose with methanol yields a mixture of methyl α- and β-acetals. Dehydration takes place preferentially at the hemiacetal —OH group because it is the most reactive —OH group in the molecule.

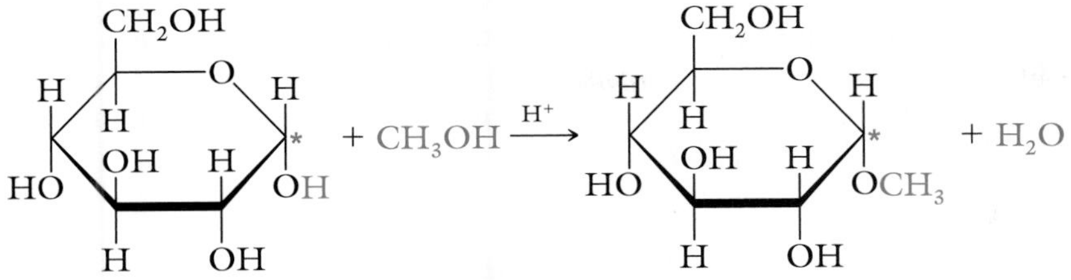

α-D-Glucose Methyl-α-D-glycoside

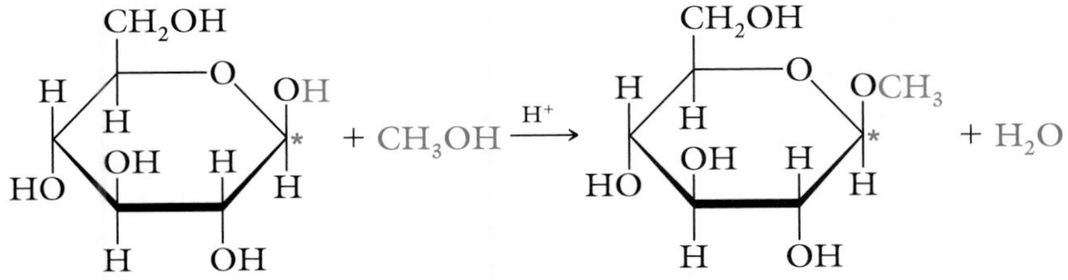

β-D-Glucose Methyl-β-D-glycoside

The hemiacetal and acetal carbons are marked with asterisks. You can recognize the acetal carbon by its attachment to two ether, —OR, groups.

Acetals of carbohydrates are called **glycosides,** and the acetal carbon and its two —OR groups form a **glycosidic linkage.** A glycosidic linkage forms as either an α- or a β-glycosidic linkage, depending on whether the α- or β-acetal takes part in the reaction. In laboratory reactions catalyzed by acids such as sulfuric acid, a mixture of the two acetals is formed. In living systems, the formation of glycosidic linkages is catalyzed by enzymes stereospecifically—that is, one or the other glycosidic linkage is formed exclusively—which has important consequences for the physiological functioning of the acetal.

Example 18.4 Writing equations for reactions of monosaccharides

Show the formation of methyl α- and β-D-glycosides from the reaction of D-fructose with methanol. Place asterisks next to the hemiacetal and acetal carbons.

Solution

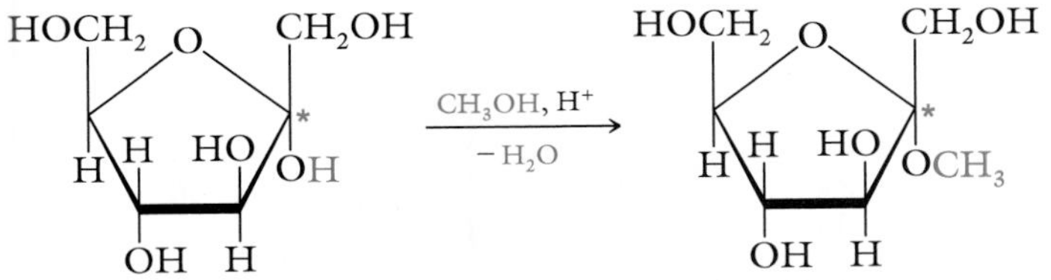

α-D-Fructose Methyl α-D-glycoside

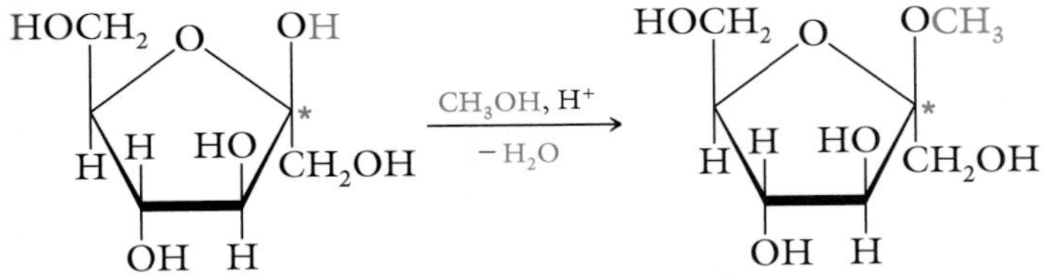

β-D-Fructose Methyl β-D-glycoside

Problem 18.4 Show the product(s) formed when D-fructose is oxidized by Benedict's reagent.

Glycosides are not reducing sugars, because they are not in equilibrium with an open-chain compound, whether α-hydroxy aldehyde or ketone. For the same reason, they do not undergo mutarotation. Glycosides can be hydrolyzed to the saccharide and alcohol by reaction with water in the presence of an acid catalyst or appropriate enzyme.

The glycosidic linkage is the type of bond through which monosaccharide residues are linked to form disaccharides and polysaccharides. Whether the bond is an α- or β-glycosidic linkage is important to the structure and function of disaccharides and polysaccharides.

Other Derivatives of Monosaccharides

Phosphate esters are produced by dehydration between an —OH group of phosphoric acid, H_3PO_4, and an —OH of a saccharide (Section 15.13). D-Ribose-5-phosphate is a building block in the synthesis of nucleic acids (Section 21.1). A number of other phosphate esters, such as D-glucose-6-phosphate and D-fructose-6-phosphate, are intermediates in the metabolism of saccharides (Section 23.1). D-Glucuronic acid, in which the $—CH_2OH$ group of glucose has been oxidized to a carboxyl group, is an **acidic sugar.** D-Glucosamine, in which the —OH group at C2 is replaced by an amino (NH_2) group, is an **aminosugar.** (Note that the phosphate, carboxyl, and amino groups are shown in the ionized form in which they exist at physiological pH.)

N-Acetyl-D-glucosamine, derived from D-glucosamine, is the building block for chitin—the hard external covering (exoskeleton) of crustaceans. *N*-Acetyl-D-glucosamine and D-glucuronic acid are building blocks for hyaluronic acid, which is found in the synovial fluid of joints, where it acts as a shock absorber, and in the vitreous humor of the eye ball, where it serves to hold the retina in place.

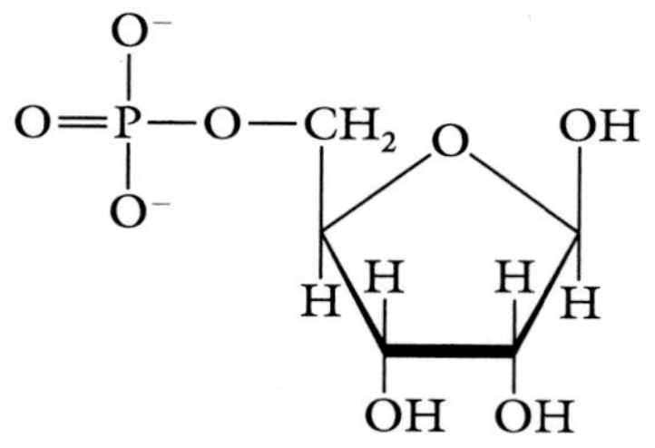

β-D-Ribose-5-phosphate

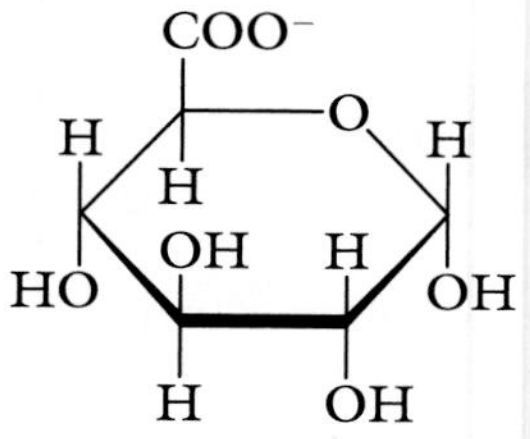

α-D-Glucuronic acid

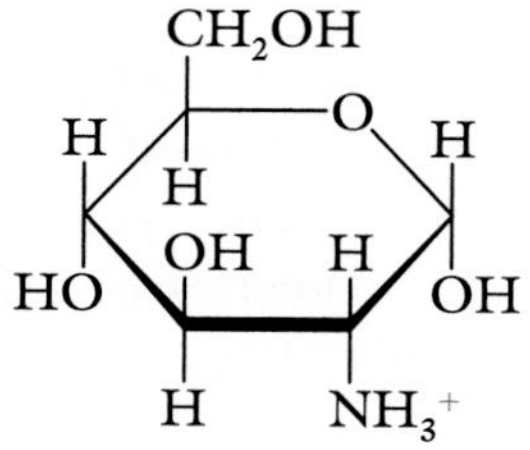

α-D-Glucosamine

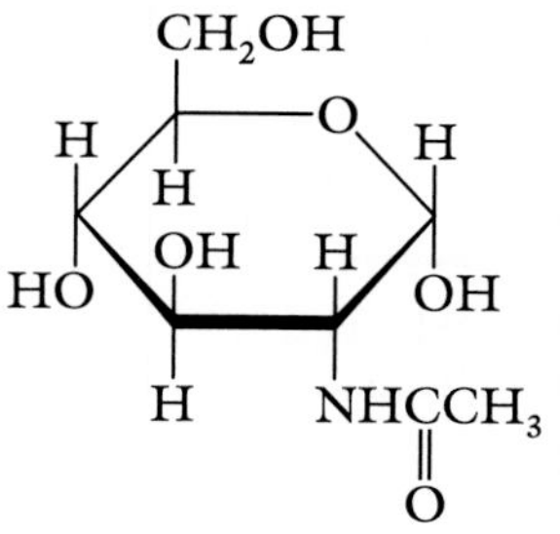

***N*-Acetyl-α-D-glucosamine**

18.5 DISACCHARIDES

Disaccharides consist of two monosaccharide residues, either the same or different monosaccharides, joined by a glycosidic linkage. We can visualize the formation of the glycosidic linkage as dehydration between the hemiacetal —OH of one monosaccharide and an —OH group (either the hemiacetal

—OH or an alcohol —OH) of the other monosaccharide. This linkage is the same as that described in Section 18.4 for the reaction of a monosaccharide with an alcohol.

To characterize a disaccharide, we need to determine several structural features:

- What are the monosaccharides?
- Is the glycosidic linkage an α or a β linkage? Laboratory reactions produce mixtures of α- and β-glycosides. Nature, however, is stereospecific, forming exclusively one or the other.
- Do both monosaccharides link through their hemiacetal —OH groups? If not, for the monosaccharide that does not link through its hemiacetal —OH, which —OH group participates in the linkage? For aldohexoses such as D-glucose, it is usually (but not always) the —OH at C4.

We will consider four disaccharides: lactose and sucrose, the two most important naturally occurring disaccharides; and maltose and cellobiose, which are not present as such in nature but are produced in the breakdown of the naturally occurring polysaccharides starch and cellulose.

Maltose

Maltose, also called **malt sugar** or **corn sugar,** is produced by the partial digestion (hydrolysis) of starch by the enzyme **amylase,** which is secreted into the mouth by the salivary glands and into the small intestine by the pancreas. Plants also produce amylase, which is responsible for the formation of maltose when grains such as barley are allowed to soften in water and germinate. Subsequent fermentation by yeast produces alcoholic beverages (Box 13.2). Maltose is also produced in processed corn syrup, which is used in beverages such as malted milk. Hydrolysis of maltose, using either the enzyme **maltase** or an acid catalyst, shows that maltose is diglucose. The two D-glucose residues are linked together by an $\alpha(1 \rightarrow 4)$-glycosidic linkage. Glycosidic linkages are denoted by the following convention:

- The α in $\alpha(1 \rightarrow 4)$- indicates the anomeric configuration of the residue that is linked through its hemiacetal —OH.
- The first number in $\alpha(1 \rightarrow 4)$- indicates the position of the hemiacetal carbon in the same residue: C1 in aldoses; C2 in ketoses.
- The second number in $\alpha(1 \rightarrow 4)$- indicates the position of the carbon on the other residue to which it is attached.

The linkage between monosaccharide residues is most easily understood by visualizing the formation of D-maltose from two molecules of D-glucose by a dehydration reaction, as shown in Figure 18.5 on the following page. (The linking together of monosaccharides in living cells is more complicated than a simple dehydration—Section 23.9.)

Because maltose has a free hemiacetal —OH on one of the two D-glucose residues (the one on the right side in Figure 18.5), it consists of an equilibrium mixture of α-maltose, β-maltose, and open-chain maltose. Thus, maltose is a reducing sugar, it gives a positive test with Benedict's reagent, and it undergoes mutarotation.

Do not confuse the α- in α-maltose with that in $\alpha(1 \rightarrow 4)$-. The α- in α-maltose refers to the α-anomer of maltose. The α- in $\alpha(1 \rightarrow 4)$- refers to the configuration of the glycosidic linkage between the two monosaccharide residues.

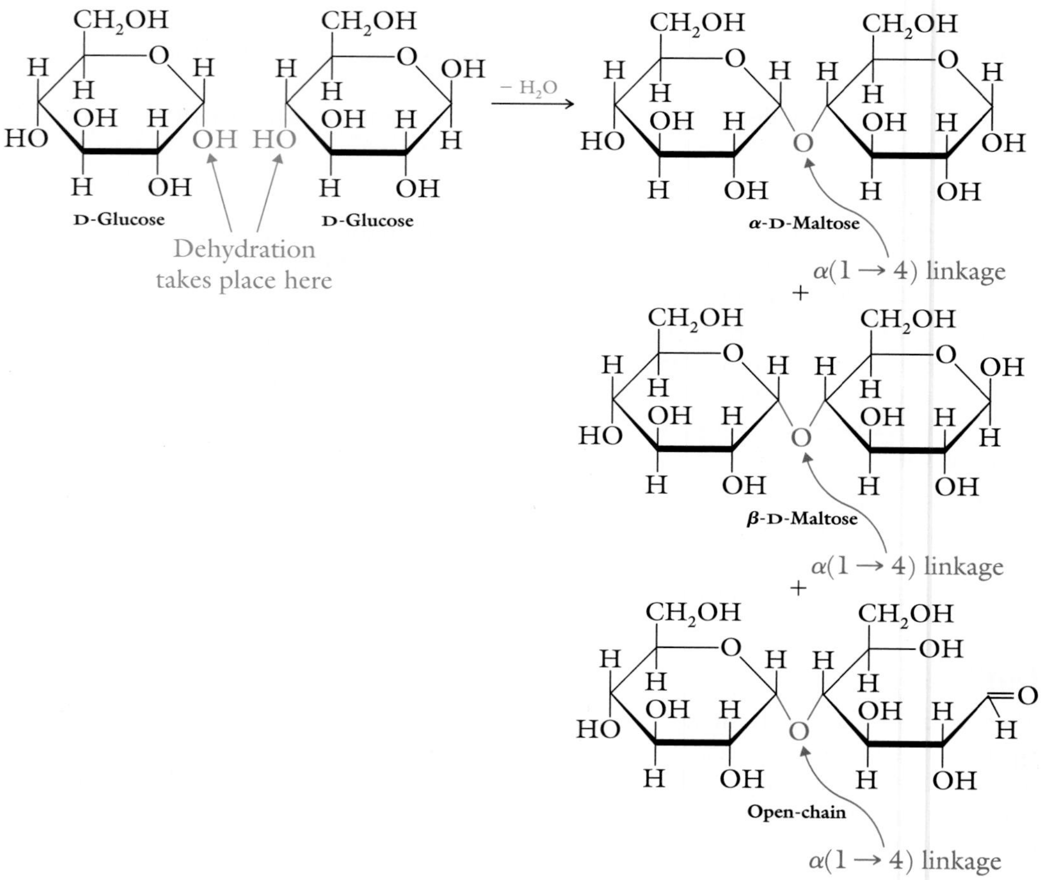

Figure 18.5 Maltose has two D-glucose residues bonded by an $\alpha(1 \rightarrow 4)$-glycosidic linkage.

Note that we often omit the "D-" from the names of maltose and other saccharides, especially those larger than monosaccharides. Always remember that all saccharides are D-saccharides, unless otherwise noted.

Milk contains the disaccharide lactose; the cookies contain the disaccharide sucrose and the polysaccharide starch.

Cellobiose

Cellobiose is produced by partial hydrolysis of cellulose, brought about by the enzyme **cellulase** or an acid catalyst. Hydrolysis of cellobiose by the enzyme **cellobiase** or an acid catalyst shows that cellobiose is diglucose. The only difference between maltose and cellobiose is the linkage between D-glucose residues: cellobiose has a $\beta(1 \rightarrow 4)$-glycosidic linkage rather than an $\alpha(1 \rightarrow 4)$-glycosidic linkage (Figure 18.6). Like maltose, cellobiose is a reducing sugar and undergoes mutarotation.

Lactose

Lactose, or **milk sugar,** constitutes from about 4 to 8% of mammalian milk and is a major energy source for nursing young. Lactose contains D-galactose and D-glucose joined by a $\beta(1 \rightarrow 4)$-glycosidic linkage, from the hemiacetal —OH (at C1) of D-galactose to the C4 position of D-glucose (Figure 18.7). Because the hemiacetal —OH of D-glucose is retained in lactose, lactose is a reducing sugar and undergoes mutarotation.

Lactose intolerance and galactosemia are significant hereditary diseases (Box 18.2 on page 543 and Section 26.3).

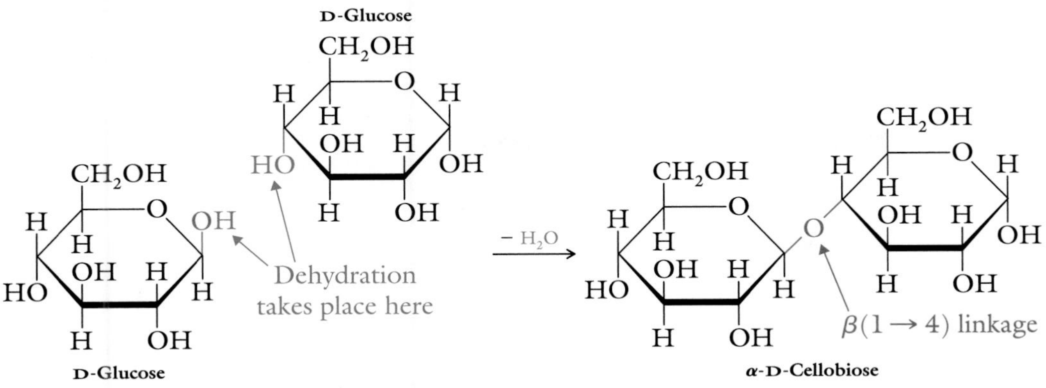

Figure 18.6 Cellobiose has two D-glucose residues bonded by a $\beta(1 \rightarrow 4)$-glycosidic linkage. Cellobiose is an equilibrium mixture of α- and β-anomers and the open-chain (acyclic) structure. (Only the α-anomer is shown.)

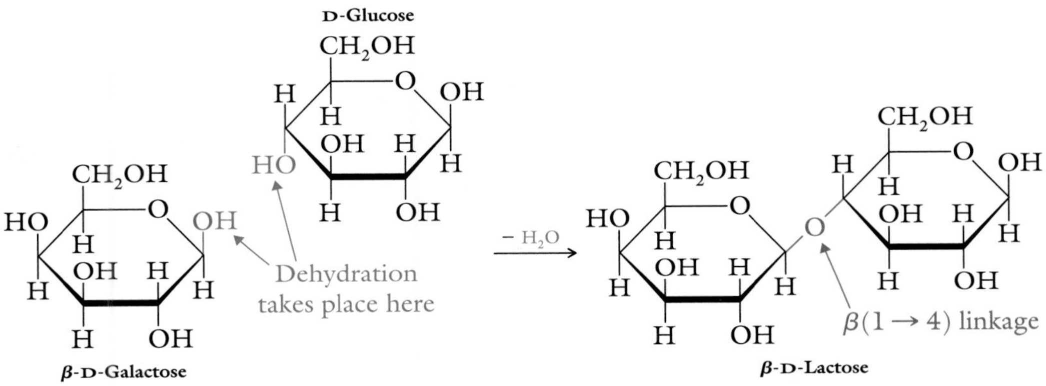

Figure 18.7 Lactose has D-galactose bonded to D-glucose by a $\beta(1 \rightarrow 4)$-glycosidic linkage. Lactose is an equilibrium mixture of α- and β-anomers and the open-chain (acyclic) structure. (Only the β-anomer is shown.)

Sucrose

Sucrose is the most abundant disaccharide in nature. It is found in the fruits and vegetables of our diet. Sugar cane and sugar beets are the commercial sources of sucrose used as table sugar to sweeten coffee, tea, soft drinks, ice cream, and baked goods. Sucrose is formed by acetal formation between the hemiacetal —OH groups of α-D-glucose and β-D-fructose (Figure 18.8 on the following page). The linkage between the monosaccharide residues is an $\alpha,\beta(1 \rightarrow 2)$-glycosidic linkage; that is, the linkage is an α-glycosidic linkage with respect to D-glucose but a β-glycosidic linkage with respect to D-fructose. Unlike the other disaccharides, sucrose does not contain a hemiacetal linkage and is therefore not a reducing sugar; it does not give a positive Benedict's test and does not undergo mutarotation.

Example 18.5 Drawing the structure of a disaccharide

Melibiose, a disaccharide found in some plants, contains D-galactose linked to D-glucose by an $\alpha(1 \rightarrow 6)$-glycosidic linkage—an α-glycosidic linkage from C1 of D-galactose to C6 of D-glucose. Draw the structure of melibiose and indicate whether it is a reducing sugar. Does it undergo mutarotation?

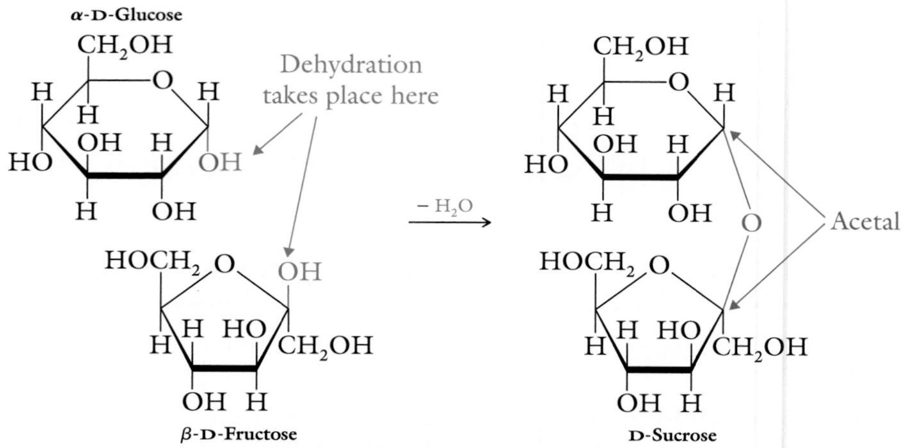

Figure 18.8 Sucrose has D-glucose and D-fructose bonded by an $\alpha,\beta(1 \rightarrow 2)$-glycosidic linkage. There are no anomers and no open-chain structure for sucrose.

Solution

Draw α-D-galactose above and to the left of D-glucose so that C1 of α-D-galactose is near C6 of D-glucose. Removing a molecule of water from the hemiacetal —OH of α-D-galactose and the —OH at C6 of D-glucose gives the structure of melibiose.

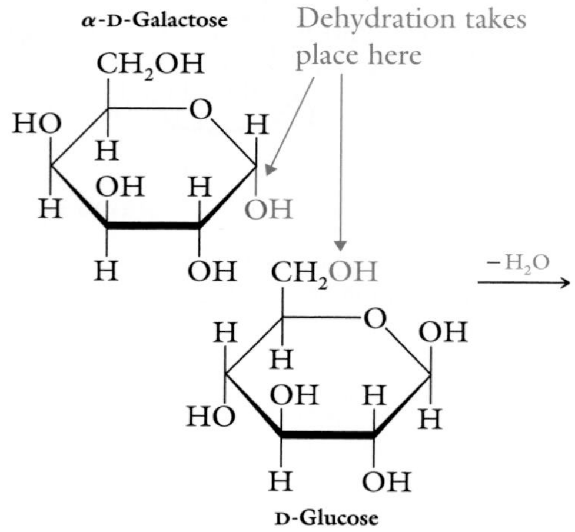

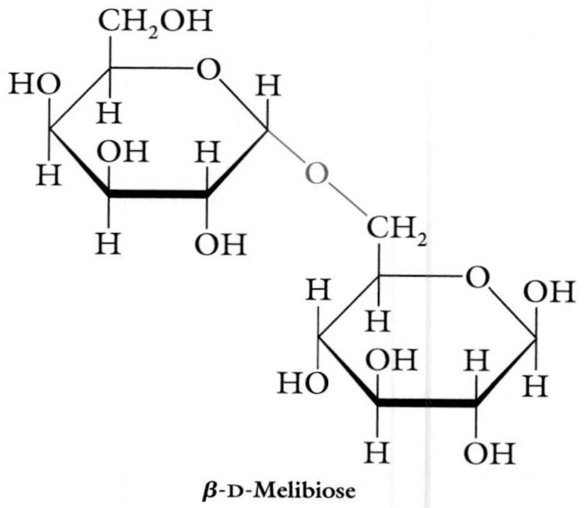

Melibiose has a hemiacetal group on the D-glucose residue and therefore is a reducing sugar and undergoes mutarotation.

Problem 18.5 What monosaccharides are obtained when the following disaccharide is hydrolyzed? (Refer to Figures 18.1 through 18.4 as needed.) Is the

18.2 Chemistry Within Us

Hereditary Problems of Lactose Utilization

Milk and milk products, such as cheese and ice cream, are not suitable foods for everyone. Some people are unable to utilize—in fact, unable to tolerate—lactose, the disaccharide present in milk. Digestion of lactose to D-galactose and D-glucose requires the enzyme lactase. Most infants and children produce sufficient lactase to digest lactose, but, in adulthood, many people develop a significant lactase deficiency that severely limits their ability to digest milk and milk products, a hereditary condition known as **lactose intolerance.** Unhydrolyzed lactose accumulates in the small intestine after ingestion because disaccharides are too large to pass through the intestinal membrane and enter the metabolic pathways. Two problems result: (1) osmotic pressure in the intestine increases, causing an influx of fluid into the intestine to dilute the lactose solution, and (2) bacterial fermentation of lactose in the colon produces large amounts of carbon dioxide and irritating acids. The consequences are abdominal distention and cramping, nausea, pain, and diarrhea.

People can manage their lactose intolerance by limiting the intake of milk and milk products, taking commercially prepared lactase orally with food, and using milk and milk products in which the lactose has been hydrolyzed enzymatically.

The incidence of lactose intolerance in population groups varies with geographical origins. Most Asians, almost 80% of Africans, and 20% of Caucasians (but less than 5% of Scandinavians) are lactose intolerant.

In another hereditary condition, **galactosemia,** lactose is hydrolyzed to D-galactose and D-glucose, but the enzyme that catalyzes the conversion of D-galactose into D-glucose, a step necessary for normal metabolism, is defective. Infants with galactosemia fail to thrive on milk products. D-Galactose accumulates in the bloodstream and causes mental retardation, impaired liver function, cataracts, and even death. Galactosemia is treated only by a galactose-free diet. If the disorder is recognized in infancy, its effects can be eliminated by removing milk and milk products and other sources of D-galactose from the diet. Except for mental retardation, which can develop in a child before the disorder is diagnosed, most of the symptoms of galactosemia are minimized by controlling the diet. Galactosemia is generally not a problem in people other than infants, because the body eventually produces the required enzyme for galactose metabolism.

disaccharide a reducing sugar? Does it undergo mutarotation? Characterize the linkage between monosaccharide residues as α or β.

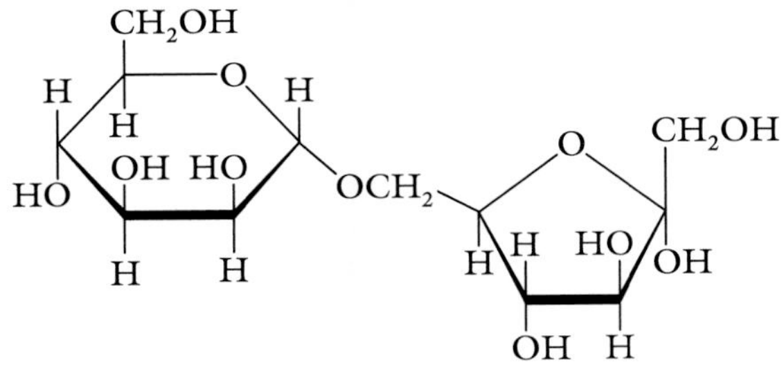

Digestion and Absorption of Carbohydrates

Maltose (from partial digestion of starch), lactose, and sucrose are part of the diet of humans and other mammals. However, these disaccharides cannot be absorbed as such from the intestinal tract (Section 26.3). Only monosaccharides are small enough to pass through the membranes of the intestinal cells and into the bloodstream (Box 7.4). Larger saccharides must first be digested to monosaccharides, which is accomplished by the enzymes maltase, lactase, and sucrase, respectively.

Note that enzymes are generally named by using the prefix of the name of the compound undergoing reaction (digestion in this case) and adding **-ase** as a suffix.

18.6 POLYSACCHARIDES

Polysaccharides are polymers containing large numbers (usually hundreds or even thousands) of monosaccharide residues (repeat units) bonded together. They have a variety of structures and functions in living organisms. Some, such as starch and glycogen, serve as storage forms of D-glucose. Other polysaccharides, such as cellulose and chitin, serve as structural materials. Some contain only one type of monosaccharide residue; others contain more than one. Some oligosaccharides combine with proteins or lipids to form complex compounds that permit cells to recognize other cells or molecules in their surroundings.

Polysaccharides differ structurally in several ways:

- In the monosaccharide(s) that constitute the residues.
- In the —OH group(s) that participate in linking the monosaccharide residues.
- In the glycosidic linkage (α or β), when one of the —OH groups is the hemiacetal —OH (as is usual).
- In the presence or absence of branching in the polysaccharides (as described next).

Starch and Glycogen

Starch and glycogen are **storage,** or **nutritional, polysaccharides:** D-glucose is stored as starch in plants and as glycogen in animals. **Starch** is deposited in plant cells as insoluble granules composed of two different types of polyglucose molecules, **amylose** and **amylopectin.** Most starches are from 10 to 30% amylose and from 70 to 90% amylopectin.

Amylose is a linear (unbranched) polymer of D-glucose residues linked through $\alpha(1 \rightarrow 4)$-glycosidic linkages. Starch contains a mixture of amylose molecules, ranging from a few thousand to one-half million amu in molecular mass. Figure 18.9 shows the structure of amylose.

Amylopectin, the major component of starch, is a branched polymer. Like amylose, it contains D-glucose residues linked through $\alpha(1 \rightarrow 4)$-glycosidic linkages, but, in addition, it branches repeatedly through $\alpha(1 \rightarrow 6)$-glycosidic linkages (Figure 18.10). Not only are there branches, but there are branches on branches (Figure 18.11). Except at the branch points, the linkages between D-glucose residues are $\alpha(1 \rightarrow 4)$-glycosidic linkages. Amylopectins have molecular masses as high as 1 million amu or more.

Glycogen is similar to amylopectin but more extensively branched.

Recall that reducing sugars are those that have a hemiacetal group and thus an aldehyde group. One of the polymer chain ends of amylose—the chain end on the right side of the molecule described by Figure 18.9—is a hemiacetal and has reducing power, but the other end does not. Only one chain end of amylopectin is a hemiacetal; all other chain ends are acetals and have no reducing power. Even though both amylose and amylopectin have a

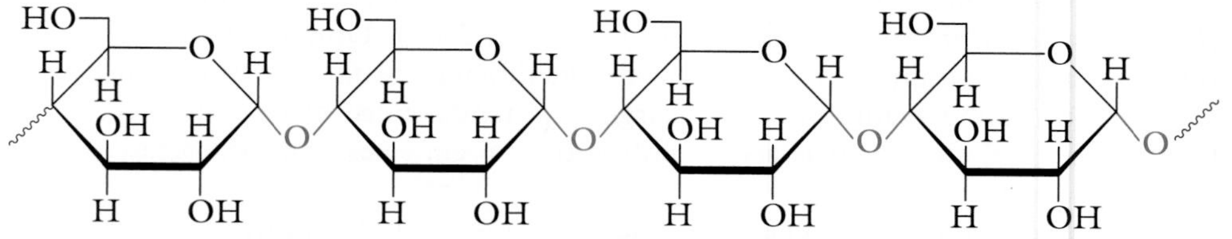

Figure 18.9 Amylose is a linear polymer of D-glucose residues bonded together by $\alpha(1 \rightarrow 4)$-glycosidic linkages.

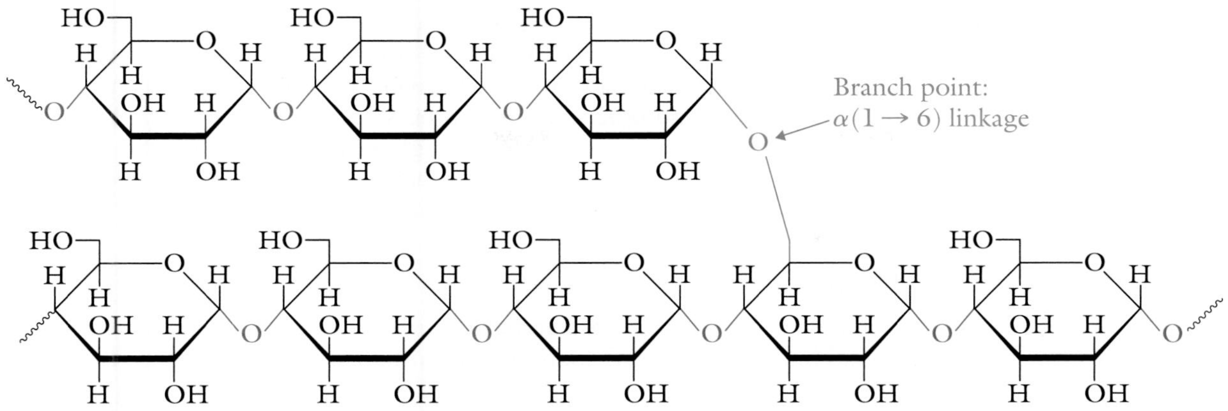

Figure 18.10 Amylopectin contains D-glucose residues bonded together by $\alpha(1 \rightarrow 4)$-glycosidic linkages (red) with branching through $\alpha(1 \rightarrow 6)$-glycosidic linkages (blue).

reducing end, starch does not give a positive test with Benedict's reagent and does not undergo mutarotation. The reducing end is a minute proportion of the overall amylose and amylopectin structures, and these tests are not sensitive enough to detect this small amount of reducing material.

Concept check

✔ All polysaccharides are nonreducing sugars; they do not give a positive Benedict's test, and they do not undergo mutarotation.

Starch and Glycogen in Digestion and Metabolism

Starch is the principal carbohydrate of our diet and is digested to D-glucose. Amylase in the digestive tract hydrolyzes the amylose and amylopectin to maltose; maltase cleaves maltose to D-glucose (Section 18.5). Amylase and maltase are unable to hydrolyze $\alpha(1 \rightarrow 6)$-glycosidic linkages, however, and thus leave the parts of amylopectin at the branch points unhydrolyzed. This product, **dextrin,** is hydrolyzed by the enzyme **dextrinase** to D-glucose.

Some of the D-glucose produced by the digestion of starch is used immediately for energy production. Some of the excess D-glucose is polymerized into

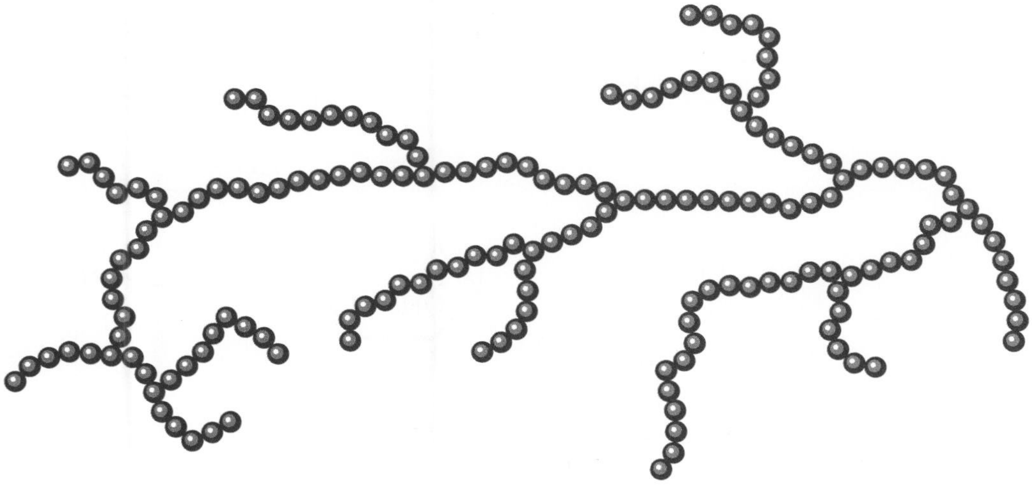

Figure 18.11 Schematic representation of the branching in amylopectin and glycogen. Each sphere represents a D-glucose residue.

▶▶ The roles of glycogenesis and glycogenolysis in utilizing D-glucose are considered in Chapter 23.

glycogen by a process called **glycogenesis** and stored in the liver and skeletal muscle. D-Glucose in excess of the amount needed to maintain the glycogen reserve is converted into lipids, which are deposited in fat tissue. When required for muscle action and other metabolic processes, D-glucose is cleaved from the chain ends of glycogen by a process called **glycogenolysis**. The highly branched glycogen structure is very efficient for fast generation of D-glucose because cleavage takes place at the multitude of chain ends simultaneously.

Cellulose

Cellulose, a **structural polysaccharide,** is the most abundant organic compound in the biosphere, accounting for more than half of all organic carbon. It is a component of the construction material of plant-cell walls that gives plants their overall shape and physical strength. Cotton is almost pure cellulose. Cellulose is linear polyglucose with $\beta(1 \rightarrow 4)$-glycosidic linkages between D-glucose residues (Figure 18.12), as described for cellobiose in Section 18.5.

Cellulose is a component of the construction materials that give this sunflower its overall shape and physical strength.

Cellulose and starch—both polyglucose molecules produced by plants—are very different materials. Consider what occurs when you add water to starch—say, corn starch. The starch forms a paste, or colloidal dispersion. We use this property of starch to thicken gravies and sauces. Now consider adding water to cellulose—say, a piece of wood. The wood absorbs water but generally retains its shape and much of its physical strength. The difference in properties is a consequence of differences between the three-dimensional shapes, or conformations (Section 11.3), of the cellulose and starch molecules, which in turn depend on the difference in glycosidic linkages.

Many different conformations are possible for polymers because of rotations about the many single bonds in the polymer chain. We generally find that some feature of the polymer's chemical structure (the glycosidic linkage in this case) results in the polymer's existing in a specific conformation because that conformation is more stable than all others. Most importantly, the physiological functioning of biological polymers such as polysaccharides is critically dependent on conformation. (Chapter 20 contains a more detailed discussion of polymer conformation and its effect on physiological function.)

Cellulose molecules exist in extended chain conformations because of the $\beta(1 \rightarrow 4)$-glycosidic linkages between glucose units. In the extended conformations, the chains form ribbons that pack side by side and one on top of the other and are stabilized by intramolecular and intermolecular hydrogen bonding (Figure 18.13). This property of cellulose makes it useful as a structural material in nature, as well as in the industrial manufacture of paper, cardboard, structural materials, textiles, and plastics (Box 18.3 on page 549). The $\alpha(1 \rightarrow 4)$-glycosidic linkages of amylose and amylopectin, on the other hand,

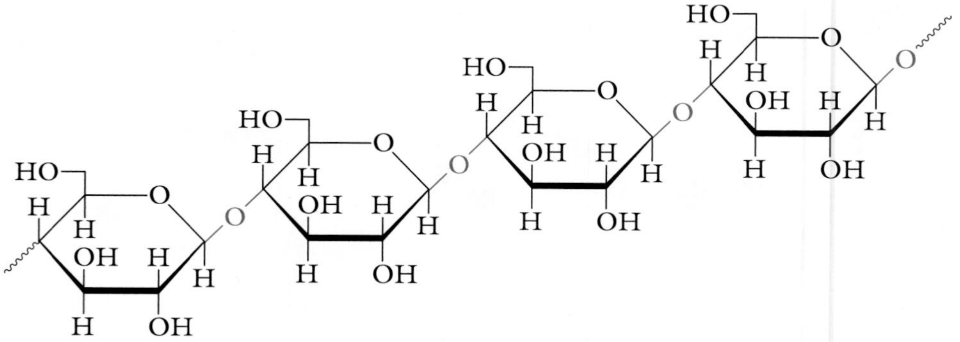

Figure 18.12 Cellulose is a linear polymer of D-glucose residues bonded by $\beta(1 \rightarrow 4)$-glycosidic linkages.

INSIGHT INTO FUNCTION

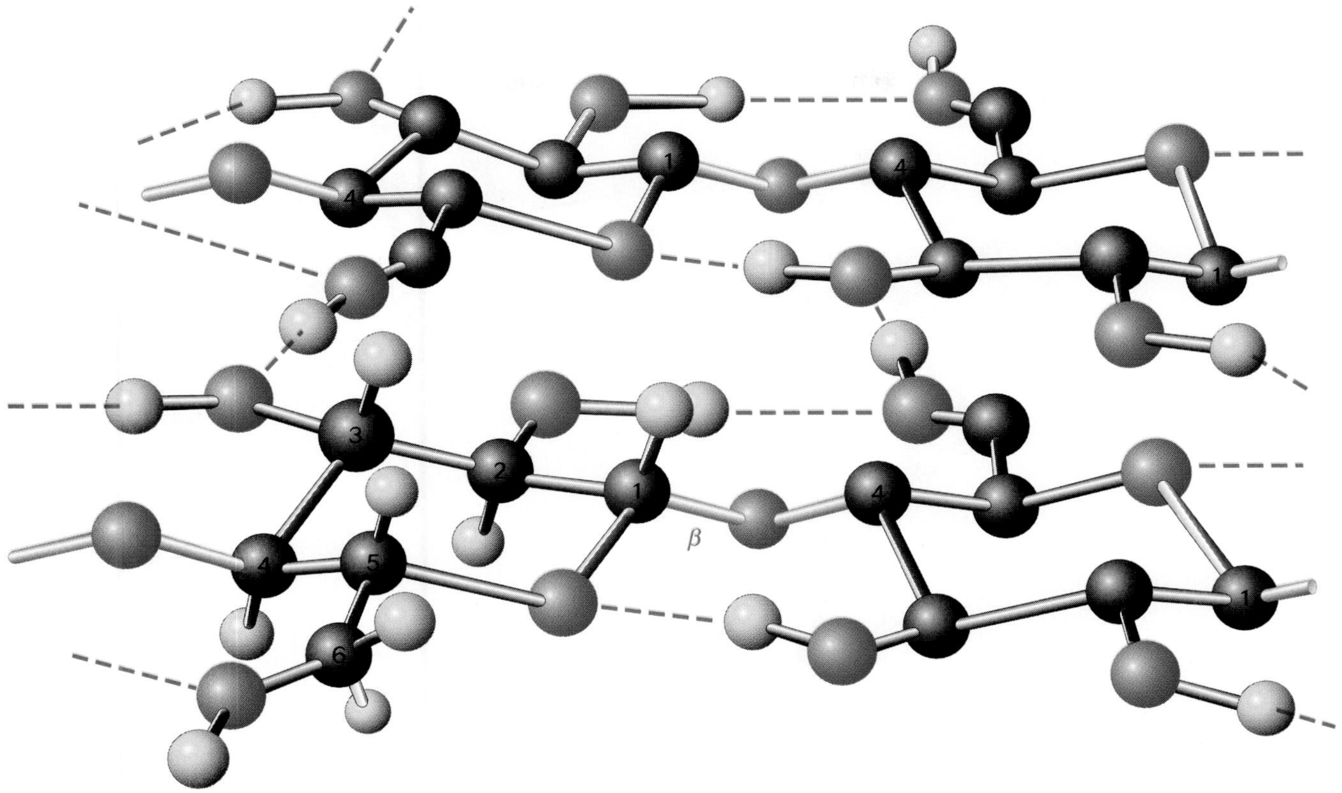

Figure 18.13 Extended chain conformations of parts of two different cellulose chains are shown held together by intermolecular hydrogen bonding. Hydrogen bonding continues in the same plane with other chains as well as in planes above and below this plane to form strong, fibrous bundles. For simplicity, only the glucose residue in the left forefront is shown with all hydrogens attached to carbons. Adapted from C. K. Mathews and K. E. Van Holde, *Biochemistry,* 2nd ed. Benjamin Cummins, Menlo Park, CA, 1996. ©Irving Geis.

prevent the formation of extended chain conformations. Instead, amylose and amylopectin form helices stabilized by intramolecular hydrogen bonding (Figure 18.14 on the following page).

Concept checklist

- ✔ Extensive intermolecular hydrogen bonding between adjacent molecules gives cellulose its great physical strength. Cellulose molecules are bound together in fiberous bundles that exclude large attractive interactions with water.
- ✔ The lack of intermolecular hydrogen bonding imparts only modest physical strength to starch. The amylose and amylopectin helices are easily solvated by water, which allows for rapid hydrolysis to glucose.

The difference in the glycosidic linkages of cellulose and starch have another important consequence: the enzymes that cleave $\alpha(1 \rightarrow 4)$ linkages do not cleave $\beta(1 \rightarrow 4)$ linkages and vice versa. Humans cannot digest cellulose, because we do not possess the enzyme cellulase, which cleaves $\beta(1 \rightarrow 4)$ linkages. However, cellulose is the main source of nutritional carbohydrate for grazing animals such as cows, sheep, horses, and deer and for insects such as termites. These species do not themselves possess cellulase, but their digestive tracts contain symbiotic microorganisms that secrete it. Although cellulose

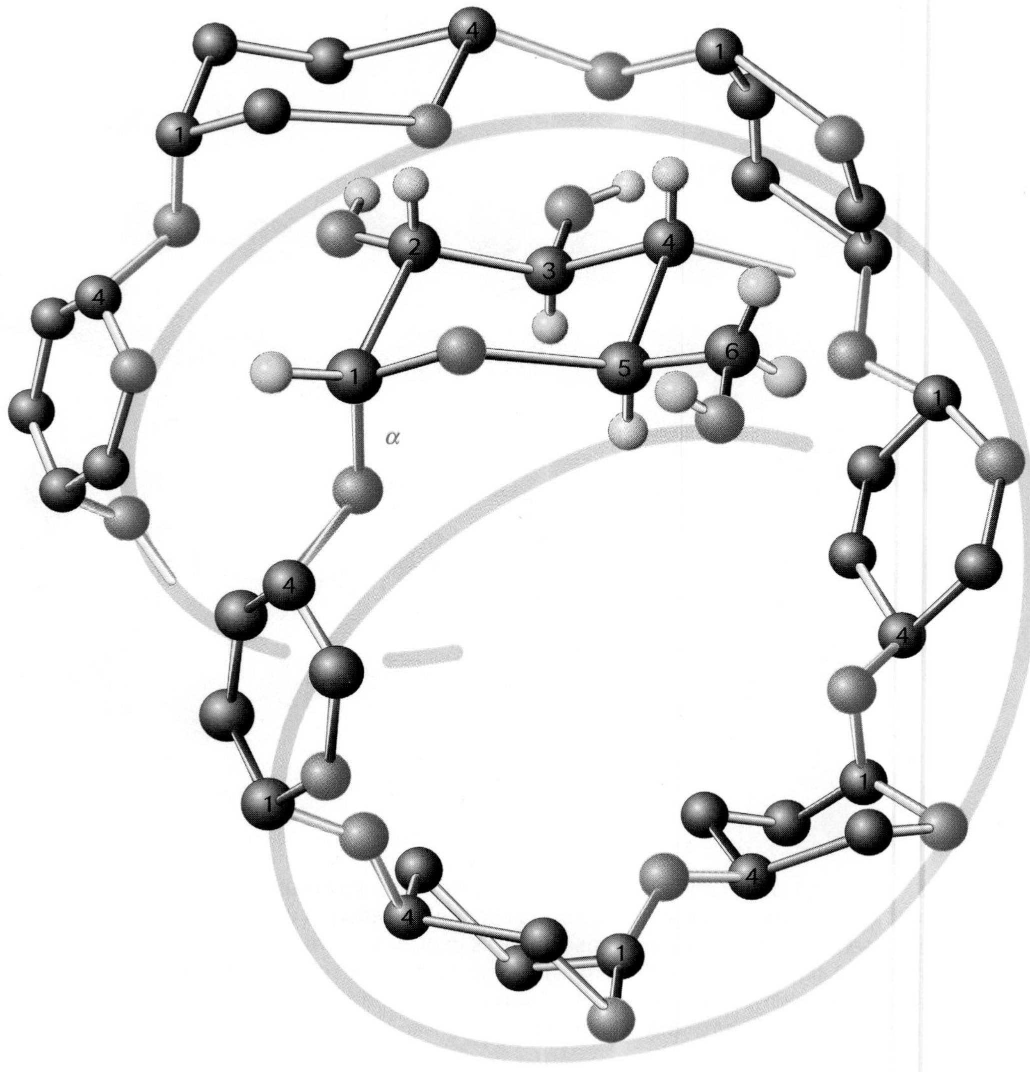

Figure 18.14 Helical (coiled) conformation of amylose. The helix is stabilized by intramolecular hydrogen bonding between —OH groups. There is very little intermolecular hydrogen bonding between different amylose helices. Adapted from C. K. Mathews and K. E. Van Holde, *Biochemistry,* 2nd ed. Benjamin Cummins, Menlo Park, CA, 1996. ©Irving Geis.

does not play a nutritional role for humans, dietary cellulose has some beneficial effects (Box 18.4 on page 550).

Note that the six-membered rings are shown in their chair conformations in Figures 18.13 and 18.14. Because of the tetrahedral bond angles for carbon atoms, the chair is the actual structure assumed by the six-membered ring (see Box 11.1). The chair structure and the type of glycosidic linkage are responsible for the formation of extended-chain or helical conformations of the different polysaccharides. For simplicity, however, we usually show the six-membered ring as a flat planar structure.

Cell Recognition: Glycolipids and Glycoproteins

Every multicellular organism has a variety of cell types, each performing specialized tasks. It is essential that different kinds of cells recognize and interact

18.3 Chemistry Around Us

Plastics and Textile Fibers from Cellulose

The high cellulose content of wood and cotton is responsible for the wide-ranging utility of both materials. The hairs on the seed of the cotton plant, for example, are more than 95% cellulose. Fabrics suitable for clothing are produced when the seed hairs are spun into thread and then woven. Not only do cotton fabrics hold up well to laundering, but they are comfortable on the body, the latter because of cotton's high moisture absorption and heat conduction. The annual production of cotton in the United States is more than 2 billion pounds.

The overall composition of wood is about 50% cellulose, with most of the remainder made up of **lignin,** an organic polymer containing aromatic groups. The cellulose fibers are imbedded in lignin, thus creating a composite material of high physical strength. (The same approach is used to strengthen concrete by reinforcing it with steel rods.) The composite structure gives large trees their enormous strength and ability to withstand their own weight as well as very strong winds.

Wood is processed into a variety of products. It is cut and used to construct furniture, housing, and other structures. It is converted into paper for books, newspapers, magazines, and cardboard boxes. Cellulose separated from wood is the starting material for various fiber (rayon) and plastic (cellophane, cellulose acetate) products. **Cellophane** is used in packaging wrap for food and tobacco products. **Rayon** is a widely used textile fiber, although not as strong as cotton fiber.

Cellulose acetate, produced by reaction of cellulose with acetic anhydride, $(CH_3CO)_2O$, is used in cigarette filters, photographic film, lacquers and protective coatings for automobiles and furniture, and eyeglass frames.

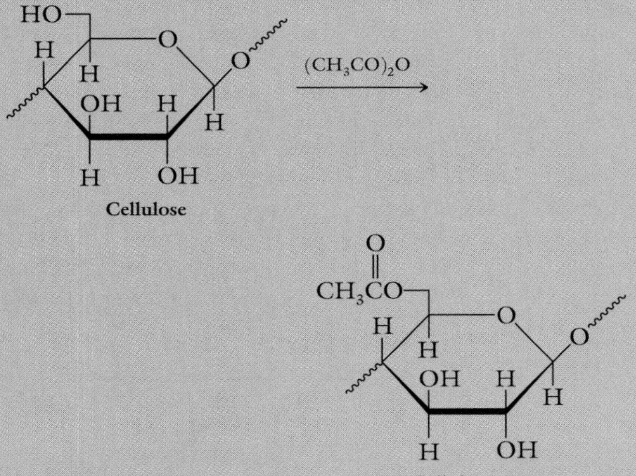

Cellulose → Cellulose acetate

properly with other cells and molecules and that an organism must recognize its own cells as distinct from foreign cells. In many cases, these tasks are accomplished by a process called **cell recognition,** whereby cells recognize one another because of saccharides attached to cell surfaces. The saccharides, usually oligosaccharides, are present as **glycolipids** or **glycoproteins,** molecules having saccharide parts bonded to lipid and protein molecules, respectively. The lipid or protein part of the glycolipid or glycoprotein is integrated into the cell membrane structure, with the saccharide part located on the external membrane surface; so the glycolipids and glycoproteins are actually part of the cell membrane.

▶▶ Every cell is separated from its extracellular environment by a cell membrane, as considered in Section 19.10.

Cell recognition is usually through secondary attractive forces between the saccharide on the cell surface and a protein on another cell surface or in the extracellular matrix. An example of this phenomenon is fertilization, the union of an ovulated egg and a sperm. The cell membrane of an ovulated egg has oligosaccharide units projecting from its glycoproteins. Protein receptor sites on a sperm cell of the same species recognize and bind to these oligosaccharides; binding is followed by the release of enzymes by the sperm cell. The enzymes dissolve the coat of the egg and allow entry of the sperm.

Scanning electron micrograph of human sperm and egg at the moment of penetration.

Cell recognition is critical to success in performing blood transfusions. The four types of human red blood cells (A, B, AB, and O) have different oligosaccharides (called **antigens**) projecting from their cell membranes. The body produces specific **antibodies** (immunoglobulin proteins) that recognize

18.4 Chemistry Within Us

Dietary Fiber

Not all components of the foods that we ingest are digestible. Nondigestible polysaccharides called **dietary fiber, bulk,** or **roughage** are part of any diet that includes fruits, vegetables, and grains. These polysaccharides have beneficial effects even though they have no nutritional value. Dietary fiber softens stools by absorbing water and helps stimulate the peristaltic movements of the bowel—contractions of the muscles of the intestinal wall—thus facilitating the more rapid passage of material through the intestinal tract and improving the regularity of bowel movements. When stools move more quickly through the colon, intestinal bacteria have less time in which to produce harmful chemicals. An additional benefit of dietary fiber is control of appetite and weight by giving a feeling of fullness.

Dietary fiber can be insoluble, as is the cellulose present in vegetable stalks and leaves, wheat, bran, and brown rice, or it can be soluble, as is the pectin present in fruits, barley, and oats. **Pectin** is a polysaccharide containing mostly residues of D-galacturonic acid with $\alpha(1 \rightarrow 4)$-glycosidic linkages. Legumes such as beans and peas contain both insoluble and soluble fiber.

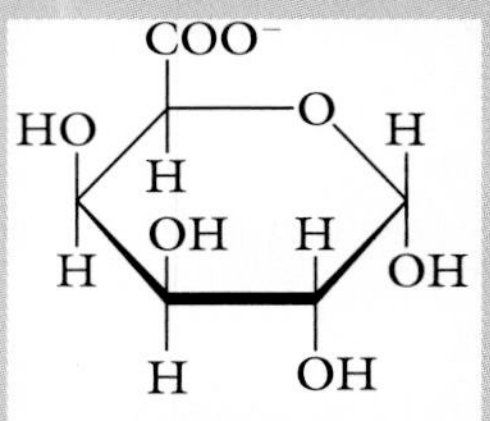

α-D-Galacturonic acid

Although the separate roles of insoluble and soluble fiber are not understood, there is speculation that soluble fiber is effective in absorbing harmful materials, including carcinogens and cholesterol.

Epidemiological evidence suggests that high-fiber diets reduce the incidence of colon and rectal cancer, hemorrhoids, diverticulitis, diabetes, and cardiovascular disease. On the other hand, the principal reason for the reduction may be that diets high in fiber are generally low in meats and fats.

and distinguish the oligosaccharide antigen of its own blood type from those of other blood types. The term antigen is often used more broadly to indicate the complete cell, including its oligosaccharide recognition site, or any other type of cell or particle (for example, an invading virus or bacterium) with its oligosaccharide site for antibody-antigen recognition.

Certain blood types are compatible for transfusion and others are not, depending on the antigen-antibody combination, as described in Table 18.2. For example, people with type O blood are universal donors (can donate blood to people of any type), whereas those with type AB blood are universal recipients (can receive blood from people of any type). Incompatible antigen-antibody combinations can result in death on transfusion. An example is the transfusion

TABLE 18.2 ABO Blood-Typing System

	Blood Type			
Characteristic	**A**	**B**	**AB**	**O**
antigen on red blood cell surface	A	B	A, B	none
makes antibodies against antigen	B	A	none	A, B
can receive from blood type	O, A	O, B	O, A, B, AB	O
can donate to blood type	A, AB	B, AB	AB	O, A, B, AB

of type A blood into a type B recipient. The antibodies in the recipient's blood recognize the transfused blood cells as foreign and bind to them. This antibody-antigen binding results in clumping together of the transfused blood cells and possible death due to clogging of arteries.

Other processes that require cell recognition include:

- Inhibition of cell growth. Normal cells cease growing at the outer boundaries of organs. Cancerous cells do not possess this inhibition of growth.
- Infection of cells by bacteria and viruses.

18.7 PHOTOSYNTHESIS

Life as we know it, both plant and animal, is possible only because of photosynthesis, the process that plants use to trap solar energy and use it to incorporate carbon into organic compounds, in the form of D-glucose and other saccharides. Before the evolution of photosynthetic organisms 3 billion years ago, Earth's atmosphere was lacking in O_2 but rich in CO_2, and primitive organisms were generating energy from molecules produced by nonbiological, nonrenewable means. Without photosynthesis, these energy sources would have eventually run out, and life would have ended. The new, photosynthetic organisms synthesized carbohydrates from carbon dioxide, water, and sunlight. In the process, they added oxygen to Earth's atmosphere and paved the way for oxygen-using, or aerobic, metabolism and the evolution of higher plants and animals. Photosynthesis is the ultimate source of energy and of all organic molecules used by all plant and animal life forms.

Photosynthesis, then, is the process of **carbon fixation**—the conversion of inorganic carbon (CO_2) into organic carbon in the form of carbohydrates, with release of oxygen into the environment. It takes place in algae and higher plants. About 20% of Earth's photosynthesis takes place in land plants, and about 80% takes place in aquatic plants. The overall equation describing photosynthesis is deceptively simple:

$$x\,CO_2 + x\,H_2O + \text{sunlight} \xrightarrow{\text{chlorophyll}} (CH_2O)_x + x\,O_2$$

where $(CH_2O)_x$ represents monosaccharides with $x = 3–6$. The details of the process are quite complicated. The green pigment chlorophyll absorbs the energy of sunlight, which drives a series of reactions to produce the high-energy saccharides from low-energy carbon dioxide and water. The energy stored in saccharides is subsequently retrieved when the saccharides are metabolized by the plants or by animals, which are dependent on plants. The energy is retrieved in a form that is useful to the organisms through a complicated series of reactions that consume oxygen and release carbon dioxide and water.

Plants convert the D-glucose and other monosaccharides formed by photosynthesis into starch, cellulose, and other saccharides, which serve various nutritional and structural functions, as described in Section 18.6. Plants supply not only carbohydrates, but also many other organic biochemicals needed by animals. For example, plants convert carbohydrates into lipids, some of which are essential for animals (that is, lipids that the animals require but cannot themselves synthesize), and plants also convert carbohydrates into amino acids, proteins, nucleic acids, and other nitrogen-containing compounds, some of which are also essential in animal diets. The nitrogen required for these conversions is made available to plants by microorganisms living in and around their roots. Thus, plants are also the source of nitrogen for animals.

In summary, the ultimate source of all organic biomolecules for animals is plants. Herbivores obtain all of their carbon and nitrogen directly from plants;

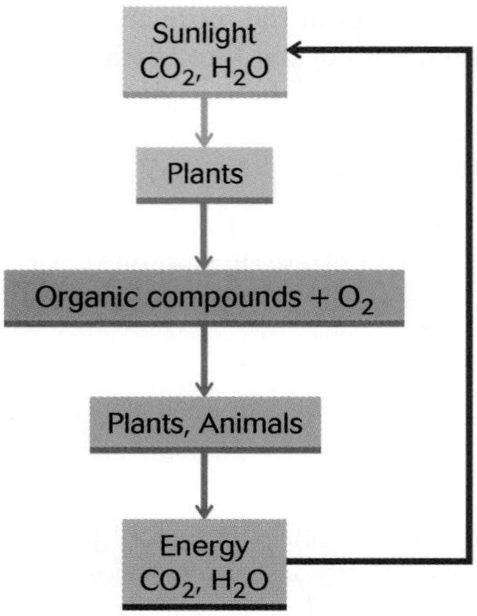

Figure 18.15 Photosynthesis.

carnivores obtain them from plants indirectly, from the flesh and viscera of the animals in their diets; omnivores such as humans obtain them directly and indirectly from plants and animals.

Plants and animals are interdependent (Figure 18.15). Plants transform the inorganic carbon of carbon dioxide into organic carbon compounds—lipids, proteins, nucleic acids, and other biomolecules—using the energy of the sun. In the process, they produce oxygen, which both plants and animals need to metabolize the organic carbon compounds so as to generate energy, synthesize other biomolecules, and carry out numerous life processes. In doing so, they return carbon dioxide along with water to the biosphere. Thus, there is a **carbon cycle,** in which carbons are shuffled back and forth between inorganic and organic forms, and an **oxygen cycle,** with both plants and animals using oxygen for energy-producing metabolism and plants regenerating oxygen by photosynthesis.

Summary

Biochemistry is the study of the structure of biomolecules and of how they transmit genetic information and store and use energy. Carbohydrates, lipids, proteins, and nucleic acids are the principal families of organic biomolecules.

Monosaccharides Carbohydrates (saccharides), the most abundant of the organic biomolecules, are aldoses (polyhydroxyaldehydes) or ketoses (polyhydroxyketones) or compounds that can be hydrolyzed to aldoses or ketoses or are derived from them. Monosaccharides, the simplest saccharides, are the building blocks for producing larger saccharides, including disaccharides and polysaccharides.

Monosaccharides of biological interest are those containing from three to six carbons. They are classified as aldoses or ketoses and by the number of carbon atoms in the chain containing the carbonyl group. Aldoses of three or more carbons and ketoses of four or more carbons contain tetrahedral stereocenters. Almost all monosaccharides of biological interest belong to the D family, those having the D configuration (the —OH group on the right side in the Fischer projection) at the tetrahedral stereocenter farthest from the carbonyl group.

D-Glucose, D-fructose, D-glyceraldehyde, dihydroxyacetone, and D-galactose are important monosaccharides. D-Glucose is combined with D-fructose and D-galactose in sucrose and lactose, respectively. Starch, glycogen, and cellulose are polymers of D-glucose.

A monosaccharide exists as cyclic hemiacetal α- and β-anomers in equilibrium with very low concentrations of the acyclic (open-chain) compound.

Diastereomers exist for monosaccharides containing more than one tetrahedral stereocenter. Anomers are diastereomers that differ in the configuration only at the cyclic hemiacetal carbon.

Monosaccharides undergo mutarotation, oxidation of the aldehyde group with Benedict's reagent, and acetal formation.

Disaccharides and Polysaccharides Disaccharides and polysaccharides are formed by glycosidic (acetal) linkages between monosaccharide residues. Disaccharides and polysaccharides are characterized by the monosaccharides from which they are produced, the positions of linkage between the monosaccharide rings, and the type of glycosidic linkage, α or β. Maltose, cellobiose, lactose, and sucrose are important disaccharides. Starch (composed of amylose and amylopectin), glycogen, cellulose, and chitin are important polysaccharides. Glycolipids and glycoproteins take part in cell recognition processes.

Photosynthesis Photosynthesis, the process for carbon fixation, is the source of all carbohydrates. Plants use carbon dioxide, water, and sunlight to form monosaccharides, such as D-glucose, from which they synthesize starch, cellulose, and other saccharides, as well as lipids, proteins, and nucleic acids. Animals depend on plants for organic compounds and energy.

Key Words

aldose, p. 527
anomer, p. 532
carbohydrate, p. 526
carbon fixation, p. 551
cell recognition, p. 548
glycoside, p. 537
glycosidic linkage, p. 537
Haworth projection, p. 532
hemiacetal carbon, p. 532
ketose, p. 527
mutarotation, p. 535
photosynthesis, p. 551
reducing sugar, p. 536
saccharide, p. 526
storage polysaccharide, p. 544
structural polysaccharide, p. 546
sugar, p. 526

Exercises

Monosaccharides

18.1 Define each of the following terms: (a) Aldose; (b) hexose; (c) ketopentose; (d) aldotetrose; (e) D-family.

18.2 Define each of the following terms: (a) Ketose; (b) pentose; (c) aldohexose; (d) ketotetrose; (e) L-family.

18.3 How many aldopentoses are possible? How many L-aldopentoses are possible?

18.4 How many ketohexoses are possible? How many D-ketohexoses are possible?

18.5 Why are there eight D-aldohexoses but only four D-ketohexoses?

18.6 Why are there four D-aldopentoses but only two D-ketopentoses?

18.7 For each of the following pairs of compounds, indicate whether the pair represents different compounds that are (1) constitutional isomers or (2) stereoisomers that are enantiomers or (3) stereoisomers that are diasteromers or (4) not isomers. Refer to Figures 18.1 through 18.4 as needed for the structures of the compounds. (a) D-Glucose and D-sorbose; (b) D-fructose and D-sorbose; (c) (+)-tagatose and (−)-tagatose; (d) D-sorbose and L-sorbose; (e) D-ribose and 2-deoxy-D-ribose.

18.8 For each of the following pairs of compounds, indicate whether the pair represents different compounds that are (1) constitutional isomers or (2) stereoisomers that are enantiomers or (3) stereoisomers that are diasteromers or (4) not isomers. Refer to Figures 18.1 through 18.4 as needed for the structures of the compounds. (a) D-Glucose and 3-deoxy-D-glucose; (b) D-ribose and D-glucose; (c) D-mannose and L-mannose; (d) D-galactose and D-mannose; (e) D-galactose and D-fructose.

Cyclic Hemiacetal Structures

18.9 When are Haworth projections used for monosaccharides?

18.10 When are Fischer projections used for monosaccharides?

18.11 What is a pyranose?

18.12 What is a furanose?

18.13 What structural feature distinguishes between the α- and the β-anomers of D-glucose?

18.14 What structural feature distinguishes between the α- and the β-anomers of D-fructose?

18.15 Draw the cyclic structure of α-D-allose. Refer to Figures 18.1 and 18.3.

18.16 Draw the cyclic structure of β-D-mannose. Refer to Figures 18.1 and 18.3.

18.17 Draw the cyclic structure of β-D-psicose. Refer to Figures 18.2 and 18.4.

18.18 Draw the cyclic structure of α-D-tagatose. Refer to Figures 18.2 and 18.4.

Chemical and Physical Properties of Monosaccharides

18.19 A solid sample of β-D-galactose is 100% β-D-galactose. What accounts for a solution of β-D-galactose giving a positive Benedict's test? What is experimentally observed in performing the Benedict's test?

18.20 A solid sample of α-D-fructose is 100% α-D-fructose. What accounts for a solution of α-D-fructose undergoing mutarotation? What is experimentally observed during mutarotation?

18.21 How many tetrahedral stereocenters are present in each of the following monosaccharides? (a) D-Glyceraldehyde; (b) D-sorbose; (c) L-mannose; (d) 2-deoxy-D-ribose.

18.22 How many tetrahedral stereocenters are present in each of the following monosaccharides? (a) Dihydroxyacetone; (b) D-ribose; (c) 3-deoxy-D-sorbose; (d) D-galactose.

18.23 Write an equation for the reaction of D-glucose with ethanol to form the ethyl β-D-glycoside.

18.24 Write an equation for the reaction of D-fructose with ethanol to form the ethyl α-D-glycoside.

18.25 Give the structure of the organic product produced when D-galactose reacts with Benedict's reagent.

18.26 Draw the structure of α-D-glucose-1-phosphate, the phosphate ester formed from α-D-glucose at C1.

18.27 Which of the following compounds are reducing sugars? Which undergo mutarotation? (a) D-Glucose; (b) the ethyl β-glycoside of D-glucose; (c) D-mannose; (d) L-glucose.

18.28 Which of the following compounds are reducing sugars? Which undergo mutarotation? (a) D-Sorbose; (b) D-fructose; (c) the methyl α-glycoside of D-fructose; (d) L-fructose.

Diasaccharides

18.29 What is an $\alpha(1 \rightarrow 4)$-glycosidic linkage between two aldohexoses?

18.30 What is a $\beta(1 \rightarrow 4)$-glycosidic linkage between two aldohexoses?

18.31 What are the products of human digestion of each of the following disaccharides? (a) Maltose; (b) lactose; (c) sucrose; (d) cellobiose.

18.32 What is obtained when each of the following disaccharides undergoes acid-catalyzed hydrolysis? (a) Maltose; (b) lactose; (c) sucrose; (d) cellobiose.

18.33 The disaccharide sucrose contains D-glucose linked to D-fructose by an $\alpha,\beta(1 \rightarrow 2)$-glycosidic linkage; that is, the linkage is an α-glycosidic linkage with respect to D-glucose but a β-glycosidic linkage with respect to D-fructose. Draw the structure of sucrose. Explain why sucrose is not a reducing sugar and does not undergo mutarotation.

18.34 The disaccharide lactose contains D-galactose linked by a $\beta(1 \rightarrow 4)$-glycosidic linkage to D-glucose. Draw the structure of lactose. Explain why lactose is a reducing sugar and undergoes mutarotation.

18.35 Gentiobiose, a disaccharide found in some plants, consists of two D-glucose units linked together by a $\beta(1 \rightarrow 6)$-glycosidic linkage. Draw the structure of gentiobiose. Is gentiobiose a reducing sugar? Does it undergo mutarotation?

18.36 Melibiose consists of D-galactose and G-glucose residues. There is an $\alpha(1 \rightarrow 6)$-glycosidic linkage from D-galactose to D-glucose; that is, D-glucose is linked through its C6, and its hemiacetal group is intact. Draw the structure of melibiose. Is melibiose a reducing sugar? Does it undergo mutarotation?

18.37 An unknown is either sucrose or lactose. The blue color of Benedict's reagent is not discharged when the reagent is added to the unknown. Identify the unknown and indicate how you arrive at your answer.

18.38 An unknown is either sucrose or lactose. The blue color of Benedict's reagent is discharged and a red precipitate forms when the reagent is added to the unknown. Identify the unknown and indicate how you arrive at your answer.

Polysaccharides

18.39 Compare the structures and functions of amylose and amylopectin.

18.40 Compare the structures and functions of glycogen and cellulose.

18.41 What are the products of human digestion of each of the following polysaccharides? What are the products of digestion of these polysaccharides by grazing animals such as cows? (a) Amylose; (b) amylopectin; (c) glycogen; (d) cellulose.

18.42 What is obtained when each of the following polysaccharides undergoes acid-catalyzed hydrolysis? (a) Amylose; (b) amylopectin; (c) glycogen; (d) cellulose.

18.43 Chitin, the second most abundant polysaccharide after cellulose, contains *N*-acetyl-D-glucosamine residues (Section 18.4) bonded through $\beta(1 \rightarrow 4)$-glycosidic linkages. Draw the structure of chitin.

18.44 Dextran, the storage polysaccharide in yeasts and bacteria, contains D-glucose residues linked together by $\alpha(1 \rightarrow 6)$-glycosidic linkages. Draw the structure of dextran.

Photosynthesis

18.45 What are the immediate sources of organic compounds for animals? What is the ultimate source of organic compounds for animals?

18.46 What is the source of organic compounds for plants?

Unclassified Exercises

18.47 What structural feature(s) define a carbohydrate?

18.48 What is the difference between monosaccharides, disaccharides, and polysaccharides?

18.49 Which carbohydrates are called sugars?

18.50 What structural characteristic in the Fischer projection defines a D-aldohexose rather than an L-aldohexose? in the Haworth representation? Refer to Figure 18.3 if needed.

18.51 What structural characteristic in the Fischer projection defines a D-ketohexose rather than an L-ketohexose? in the Haworth representation? Refer to Figure 18.4 if needed.

18.52 Why are there eight D-aldohexoses but only two D-ketopentoses?

18.53 Use the terms monosaccharide, disaccharide, and polysaccharide to categorize each of the following compounds: (a) Sucrose; (b) galactose; (c) lactose; (d) amylose; (e) amylopectin; (f) fructose; (g) cellulose.

18.54 Which of the following compounds give a positive Benedict's test? Which undergo mutarotation? (a) Sorbose; (b) glycogen; (c) sucrose; (d) lactose; (e) starch; (f) talose; (g) cellulose.

18.55 For each of the following pairs of compounds, indicate whether the pair represents different compounds that are (1) constitutional isomers or (2) stereoisomers that are enantiomers or (3) stereoisomers that are diasteromers or (4) not isomers. Refer to Figures 18.1 through 18.4 as needed for the structures of the compounds. (a) α-D-Glucose and β-D-glucose; (b) (+)-galactose and (−)-fructose; (c) (+)-galactose and (−)-galactose; (d) maltose and cellobiose; (e) glucose and amylose; (f) glucose and maltose; (g) cellulose and amylose, both having the same molecular mass.

18.56 Draw the cyclic structure of β-D-idose. Refer to Figures 18.1 and 18.3.

18.57 What is the difference in function between structural and storage polysaccharides?

18.58 Why are humans unable to digest cellulose?

18.59 Why are grazing animals such as cows and sheep able to digest cellulose? Why are they also able to digest starch?

18.60 The structure of raffinose, a trisaccharide found in some plants, is shown here. Identify the monosaccharide residues in raffinose. Identify the linkages between monosaccharide residues as acetal or hemiacetal linkages. Is raffinose a reducing sugar? Does it undergo mutarotation?

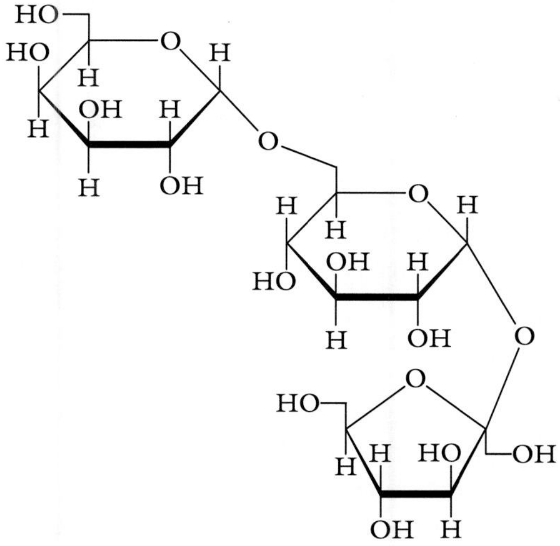

18.61 Hydrolysis of 1 mol of a carbohydrate yields 4 mol of monosaccharide. What is the structure of the carbohydrate?

18.62 Draw the cyclic structure of α-L-glucose. Refer to Figures 18.1 and 18.3 as needed.

18.63 Isomaltose is the disaccharide formed by partial hydrolysis of dextran, a storage polysaccharide in yeasts and bacteria. Isomaltose is diglucose with an $\alpha(1 \rightarrow 6)$-glycosidic linkage. Draw the structure of isomaltose. Is isomaltose a reducing sugar? Does it undergo mutarotation?

18.64 Draw the structure of the polysaccharide that consists of D-fructose residues linked together by $\alpha(2 \rightarrow 6)$-glycosidic linkages.

18.65 Write the balanced overall equation for the photosynthesis of D-glucose.

18.66 Where is most of Earth's photosynthesis carried out?

18.67 Describe the oxygen cycle.

Chemical Connections

18.68 D-Sorbitol, $C_6H_{14}O_6$, is found in apples, cherries, and other fruits. It can be synthesized by the reduction of D-glucose by the addition of H_2 in the presence of a catalyst such as Pt. Write an equation that shows this synthesis.

18.69 How many glucose residues are there in each molecule of amylose of molecular mass 162,000 amu? in cellulose of molecular mass 162,000 amu?

18.70 The polysaccharide hyaluronic acid is found in the synovial fluid of joints, where it acts as a shock absorber, and in the vitreous humor of the eye ball, where it serves to hold the retina in place. Hyaluronic acid contains alternating D-glucuronic acid and *N*-acetyl-D-glucosamine residues (Section 18.4). D-Glucuronic acid is connected to *N*-acetyl-D-glucosamine by a $\beta(1 \rightarrow 3)$-glycosidic linkage; *N*-acetyl-D-glucosamine is connected to D-glucuronic acid by a $\beta(1 \rightarrow 4)$-glycosidic linkage. Draw the structure of hyaluronic acid.

18.71 Both D-glucose and D-fructose are mixtures of the α- and β-anomers and the acyclic compound in aqueous solution. Sucrose is formed by the exclusive linking of α-D-glucose through C1 and β-D-fructose through C2. Speculate on the mechanism by which this exclusive link is formed.

18.72 The cell-surface oligosaccharides that determine blood typing contain L-fucose, one of the rare L-saccharides found in nature. L-Fucose is 6-deoxy-L-galactose. Draw the structure of L-fucose.

18.73 Explain why galactosemia is a much more serious genetic disease than lactose intolerance (Box 18.2).

CHAPTER 19

LIPIDS

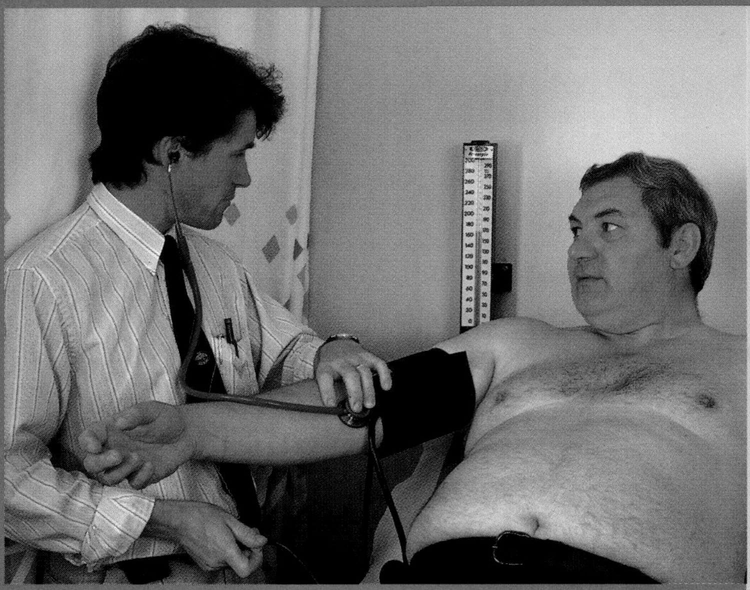

CHEMISTRY IN YOUR FUTURE

A month after coronary angioplasty, a patient visits the HMO for a follow-up exam. As you take his blood pressure, you ask if he has made any of the dietary changes that you recommended at the time of his discharge. Proudly, he announces that, instead of eggs, bacon, and potatoes every morning (20 grams of fat, at least), he is now enjoying a big bowl of granola. Unfortunately, you discover, he has been buying a brand loaded with palm and coconut oils, probably containing as much saturated fat as his former breakfast did. How can the fat from a plant oil, he asks, be as unhealthy as the fat from a pig? Chapter 19 has some of the answers.

LEARNING OBJECTIVES

- Draw the structures and describe the physical properties of saturated and unsaturated fatty acids.
- Draw the structures and describe the physical properties of triacylglycerols.
- Describe and write equations for hydrolysis, catalytic hydrogenation, and rancidity reactions of triacylglycerols.
- Describe and draw the structures of waxes.
- Describe and draw the structures of glycerophospholipids and sphingolipids, and write equations for their hydrolysis.
- Describe and draw the structures of steroids, eicosanoids, and fat-soluble vitamins, and outline their functions.
- Describe the structure of biological membranes.
- Describe the processes by which molecules and ions are transported across biological membranes.

Lipids are the predominant and most efficient form of stored energy in animals. Together with carbohydrates and proteins, they are the major components of our diet. We obtain lipids from animal fats and plant oils and use them as a source of energy. Any excess lipids taken in are stored as fat. Among our dietary lipids are beef, pork, and poultry fats; fats in milk and milk products such as butter and cheese; fish oils; plant oils such as peanut, corn, and olive oils, and margarine manufactured from plant oils.

Lipids come in a variety of molecular structures with a broad array of functions:

- Triacylglycerols—animal fats and plant oils—are sources of energy and a storage form of energy not required for immediate use.
- Glycerophospholipids, sphingolipids, and cholesterol (together with proteins) are the primary structural components of the membranes that surround all cells and organelles.
- Steroid hormones, including the sex hormones, and other hormonelike lipids act as chemical messengers, initiating or altering activity in specific target cells.
- The lipid-soluble (fat-soluble) vitamins A, D, E, and K are essential to a variety of physiological functions.
- Bile salts are needed for the digestion of lipids in the intestinal tract.

19.1 CLASSIFICATION OF LIPIDS

Other families of organic compounds, including carbohydrates, proteins, and nucleic acids, are defined by the presence of a common functional group. This is not true of lipids, a family that includes compounds with diverse functional groups: ester, amide, alcohol, phosphate, acetal, and others. The classification of compounds as lipids is based on their solubility behavior. To test for solubility, animal or plant matter is crushed into small pieces or powder and then mixed with a nonpolar or low-polarity solvent, such as toluene, dioxane, or carbon tetrachloride. Those compounds that dissolve in the solvent are classified as lipids.

Concept checklist

✔ Lipids are compounds, present in plants and animals, that are soluble in nonpolar or low-polarity solvents.

Lipids are nonpolar compounds or compounds of low polarity; they are insoluble (or barely soluble, at most) in water. Most lipids other than triacylglycerols are **amphipathic:** one part of the molecule is hydrophobic and another part is hydrophilic (Sections 6.4 and 15.7). The reason that lipids have little or no solubility in water is that the hydrophobic part is much larger than the hydrophilic part.

Lipids are classified as hydrolyzable or nonhydrolyzable. **Hydrolyzable lipids** undergo hydrolytic cleavage into two or more considerably smaller compounds in the presence of acid, base, or digestive enzyme. All hydrolyzable lipids contain at least one ester group; some also contain an amide, phosphate, or acetal group. **Nonhydrolyzable lipids** do not undergo hydrolytic cleavage into considerably smaller compounds, because they do not contain ester, amide, phosphate, or acetal groups. The terms **saponifiable** and **nonsaponifiable lipids** are also used; saponification is hydrolysis with a base (Section 15.11). Hydrolyzable lipids include the triacylglycerols, waxes, glycerophospholipids, and sphingolipids (Figure 19.1 on the following page). Nonhydrolyzable lipids include steroids, eicosanoids, and fat-soluble vitamins.

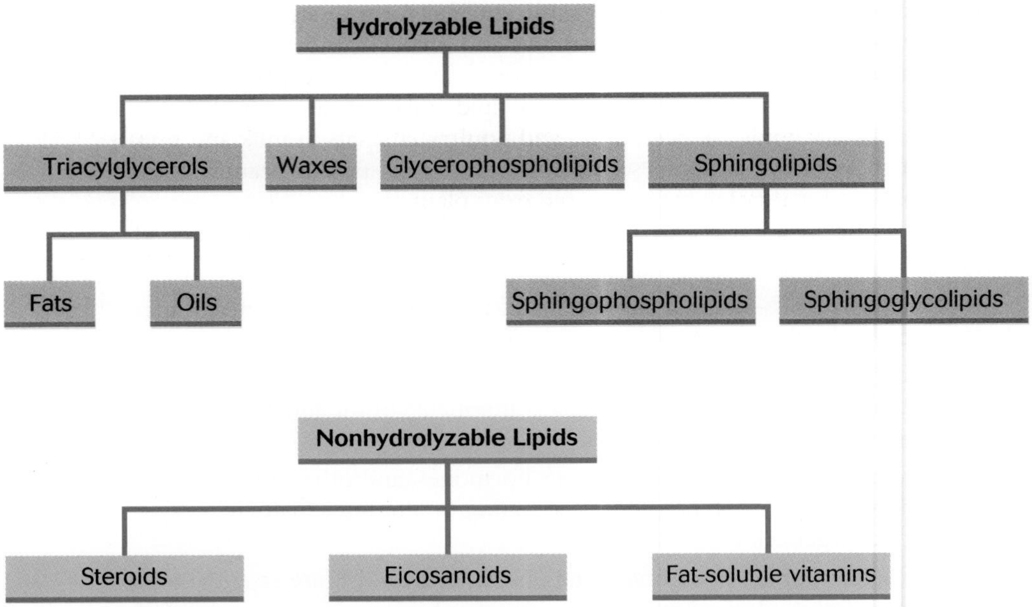

Figure 19.1 Classification of lipids.

In this chapter we describe the different lipids and their functions, as well as the structure of biological membranes and how they serve as selective membranes.

19.2 FATTY ACIDS

All hydrolyzable lipids contain one or more **fatty acids**—carboxylic acids with long hydrocarbon chains (Section 15.7). Almost all fatty acids are found in nature in combined form; only very small amounts are found as uncombined, free acids. The fatty acids used in nature to construct hydrolyzable lipids are almost exclusively the linear (unbranched) acids with an even number of carbons listed in Table 19.1.

Fatty acids may be either saturated or unsaturated. **Saturated fatty acids** contain no C=C double bonds. **Unsaturated fatty acids** contain one or more C=C double bonds; they are the **monounsaturated** and **polyunsaturated fatty acids,** respectively. In naturally occurring fatty acids, almost all double bonds have the cis configuration. Palmitic and stearic acids are the most abundant saturated fatty acids. Oleic acid is the most abundant unsaturated fatty acid. Two unsaturated fatty acids, linoleic and linolenic, are **essential fatty acids**—fatty acids that are synthesized only by plants but are required by animals and thus must be included in their diets. All other fatty acids are **nonessential fatty acids** that animals synthesize either from other fatty acids or from other precursors (derived from other foodstuffs).

The position of the double bond nearest the methyl group in unsaturated fatty acids is often indicated by an **omega number** (ω-). The ω-1 carbon is the methyl carbon in the fatty acid chain; that is, the carbon farthest from the carboxyl group. The omega number of the fatty acid is the number of the first carbon of the first double bond, counting from the ω-1 carbon. (This numbering method is the opposite of the IUPAC nomenclature system in which the counting begins at the carboxyl carbon, as C1.) Linoleic and linolenic acids are ω-6 and ω-3 fatty acids, respectively.

TABLE 19.1 Common Saturated and Unsaturated Fatty Acids

Name of Acid	Number of Double Bonds	Number of Carbons	Structure	Melting Point (°C)
SATURATED FATTY ACIDS				
lauric	0	12	$CH_3(CH_2)_{10}COOH$	44
myristic	0	14	$CH_3(CH_2)_{12}COOH$	54
palmitic	0	16	$CH_3(CH_2)_{14}COOH$	63
stearic	0	18	$CH_3(CH_2)_{16}COOH$	69
arachidic	0	20	$CH_3(CH_2)_{18}COOH$	77
behenic	0	22	$CH_3(CH_2)_{20}COOH$	81
lignoceric	0	24	$CH_3(CH_2)_{22}COOH$	84
MONOUNSATURATED FATTY ACIDS				
palmitoleic	1	16	$CH_3(CH_2)_5CH{=}CH(CH_2)_7COOH$	1
oleic	1	18	$CH_3(CH_2)_7CH{=}CH(CH_2)_7COOH$	13
nervonic	1	24	$CH_3(CH_2)_7CH{=}CH(CH_2)_{13}COOH$	42
POLYUNSATURATED FATTY ACIDS				
linoleic	2	18	$CH_3(CH_2)_4(CH{=}CHCH_2)_2(CH_2)_6COOH$	−5
linolenic	3	18	$CH_3CH_2(CH{=}CHCH_2)_3(CH_2)_6COOH$	−11
arachidonic	4	20	$CH_3(CH_2)_4(CH{=}CHCH_2)_4(CH_2)_2COOH$	−49
eicosapentaenoic	5	20	$CH_3CH_2(CH{=}CHCH_2)_5(CH_2)_2COOH$	−54

Example 19.1 Recognizing naturally occurring fatty acids

Which of the following carboxylic acids are present in animal and plant lipids?

$CH_3(CH_2)_{13}COOH$

1

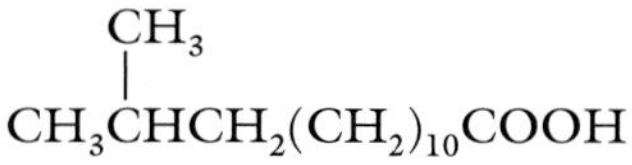

2

trans-$CH_3(CH_2)_5CH{=}CH(CH_2)_7COOH$

3

Solution

None of these carboxylic acids are components of animal and plant lipids. Acid 1 has an odd number of carbons. Acid 2 both has an odd number of carbons and is branched. Acid 3 has a trans double bond. The fatty acids of animal and plant lipids are unbranched, with an even number of carbons; double bonds, when present, have a cis configuration.

Problem 19.1 Which of the following carboxylic acids are present in animal and plant lipids?

$CH_3(CH_2)_{14}COOH$

1

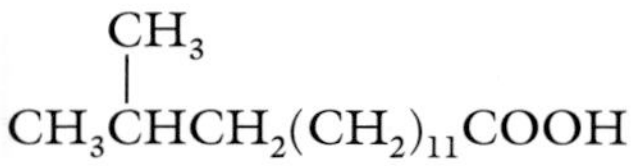

2

cis-$CH_3(CH_2)_5CH{=}CH(CH_2)_7COOH$

3

The physical properties of fatty acids depend on the number of carbons and double bonds. Fatty acids have very low water solubility, and solubility decreases as the number of carbons increases. The water solubilities of lauric and stearic acids at 30°C are 0.063 g/L and 0.0034 g/L, respectively. Although the carboxyl group of a fatty acid can hydrogen bond with water, the much larger, nonpolar, hydrocarbon part cannot. The overall nonpolar nature of fatty acids is apparent when we compare the water solubilities of D-glucose and lauric acid, which have similar molecular masses. D-Glucose is much more soluble (1100 g/L) than lauric acid (0.063 g/L). The large hydrocarbon part of a fatty acid is responsible for its solubility in nonpolar solvents such as benzene. The solubilities of lauric and stearic acids in benzene are 124 and 2600 g/L, respectively.

The melting points of fatty acids (and lipids that contain them) increase as the number of carbons increases and decrease as the number of double bonds increases (see Table 19.1). The effect of double bonds on melting point is large. Stearic acid, a saturated fatty acid, has a melting point of 70°C. The melting point is lowered to 13, −5, and −11°C with the presence of one, two, and three double bonds, respectively, in oleic, linoleic, and linolenic acids.

Cis double bonds decrease the melting points by altering molecular shape, thus affecting the ability of the molecules to pack tightly together. Figure 19.2 shows the shapes of stearic, oleic, and linoleic acids. Saturated fatty acid molecules, with elongated conformations, pack closely together, with strong secondary attractive forces and thus high melting points. The cis double bonds in unsaturated fatty acids create kinks, or bends, in the hydrocarbon chain, and these kinks prevent tight molecular packing. With weaker secondary attractive forces, lower temperatures are needed to melt the compound.

Trans double bonds also lower the melting points of fatty acids, but much less than do cis double bonds. For example, the melting point of elaidic acid, the *trans*-isomer of oleic acid, is 45°C, lower than that of stearic acid (69°C) but higher than that of oleic acid (13°C). The molecular kinks placed in molecules by trans double bonds do not distort the elongated conformation of the hydrocarbon chain nearly as much as cis double bonds do.

INSIGHT INTO PROPERTIES

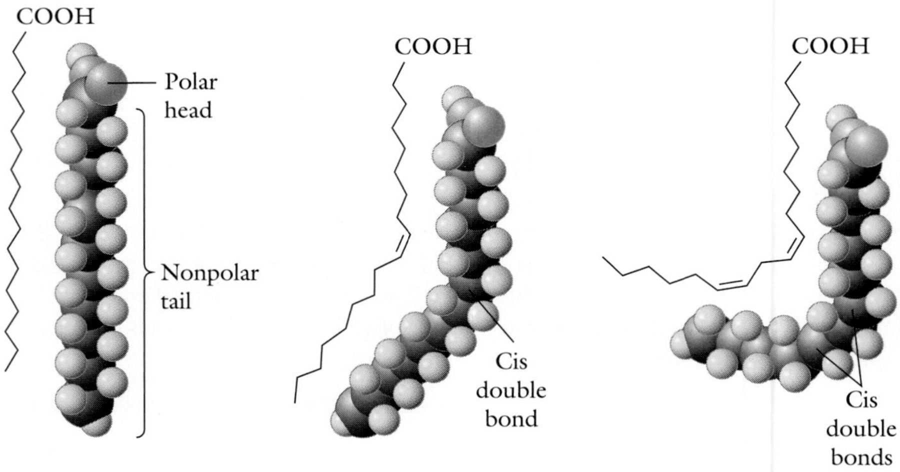

Stearic acid Melting point 69°C

Oleic acid Melting point 13°C

Linoleic acid Melting point −5°C

Figure 19.2 Molecular shapes of saturated and unsaturated fatty acids. These differences in shape are reflected in melting-point differences.

19.3 STRUCTURE AND PHYSICAL PROPERTIES OF TRIACYLGLYCEROLS

Triacylglycerols, also called **triglycerides,** make up about 90% of our dietary lipid intake. All **animal fats** such as beef, butter, pork, and poultry fats and all **plant oils** such as corn, olive, peanut, and soybean oils are triacylglycerols. They are esters (Section 15.8)—in fact, triesters—of glycerol. We can visualize their formation by the esterification of glycerol (1,2,3-propanetriol) with fatty acids:

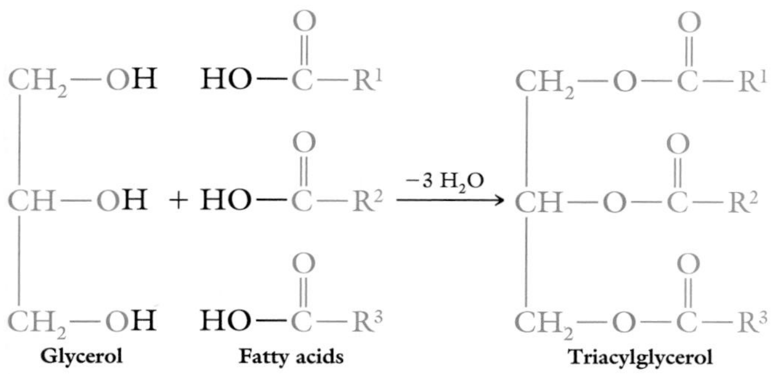

The biosynthesis of triacylglycerols is a more complex process than indicated (Section 24.5).

In **simple triacylglycerols,** $R^1 = R^2 = R^3$. In **complex,** or **mixed, triacylglycerols,** R^1, R^2, and R^3 are different.

Example 19.2 Drawing the structures of triacylglycerols

Draw the structure of the triacylglycerol containing equimolar amounts of myristic, palmitic, and oleic acids.

Solution

Draw glycerol. Draw the three fatty acids, each lined up with its carboxyl group alongside an —OH group of glycerol. Form the ester group at each of the three carbons of glycerol by carrying out the dehydrations between glycerol and fatty acid —OH groups indicated in blue.

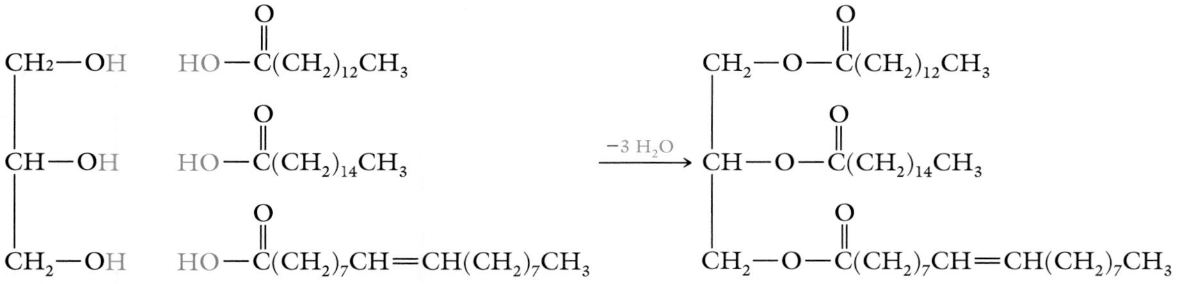

Problem 19.2 Draw the structure of the triacylglycerol containing equimolar amounts of lauric, palmitoleic, and linoleic acids.

Naturally occurring triacylglycerols contain a variety of fatty acid components. Moreover, any fat or oil is a mixture of many different triacylglycerols, very few of which are simple triacylglycerols. Table 19.2 on the following page gives the approximate fatty acid composition of various animal and plant triacylglycerols. Triacylglycerols from animal sources other than fish contain a lower percentage of unsaturated fatty acids than do most plant triacylglycerols,

TABLE 19.2 Fatty Acid Composition of Triacylglycerols

Source of Fat or Oil	Percent Fatty Acid								
	Saturated				Monounsaturated			Polyunsaturated	
	$<C_{14}$	C_{14}	C_{16}	C_{18}	C_{14}	C_{16}	C_{18}	C_{18}	$>C_{18}$
ANIMAL									
beef (tallow)		6	27	14			50	3	
butter	14	11	29	9		5	27	4	1
human		3	24	8		5	47	10	3
pork (lard)		1	28	12		3	48	6	2
herring		7	13			5		21	53
sardine		5	15	3	15	12	←18→		32
PLANT									
corn		1	10	3		2	34	50	
olive			7	2			85	3	
peanut			8	3			56	26	7
soybean			10	2			29	58	1
coconut	60	19	11	2			8		
palm		1	40	6			43	10	

from about 40 to 60% versus 85 to 90%. The terms **saturated** and **unsaturated** are used to indicate the different types of fatty acid content.

Fish triacylglycerols differ considerably in composition from those of other animals: from 75 to 80% of the fatty acids are unsaturated, compared with 40 to 60% in other animals, and a much greater proportion of the unsaturated component comprises polyunsaturated fatty acids. A few plants have triacylglycerol compositions much like those of animals other than fish. Examples are palm and coconut oils. Coconut triacylglycerol is extremely atypical of plant triacylglycerols; it has a saturated fatty acid content of 92%, far exceeding that of any animal triacylglycerol.

The melting points of plant triacylglycerols are lower than those of the animal triacylglycerols because of the higher content of unsaturated fatty acids in the plant lipids. Cis double bonds lower secondary attractive forces in triacylglycerols, similar to their effect in fatty acids (Section 19.2). Plant triacylglycerols are liquid, whereas animal triacylglycerols are solid at ambient temperatures (about 20°C). The terms **fats** and **oils** are used to indicate solid and liquid triacylglycerols, respectively.

All triacylglycerols, irrespective of the source, are water insoluble; they are even less soluble than the corresponding fatty acids, because of the low polarity of the ester groups.

19.4 CHEMICAL REACTIONS OF TRIACYLGLYCEROLS

Triacylglycerols undergo hydrolysis, catalytic hydrogenation, and rancidity reactions.

Hydrolysis

Triacylglycerols undergo hydrolysis at the ester groups (Section 15.11) to produce glycerol and fatty acids. Hydrolysis in the laboratory requires an acid or base: acidic hydrolysis yields the fatty acids; basic hydrolysis yields the fatty acid carboxylate salts:

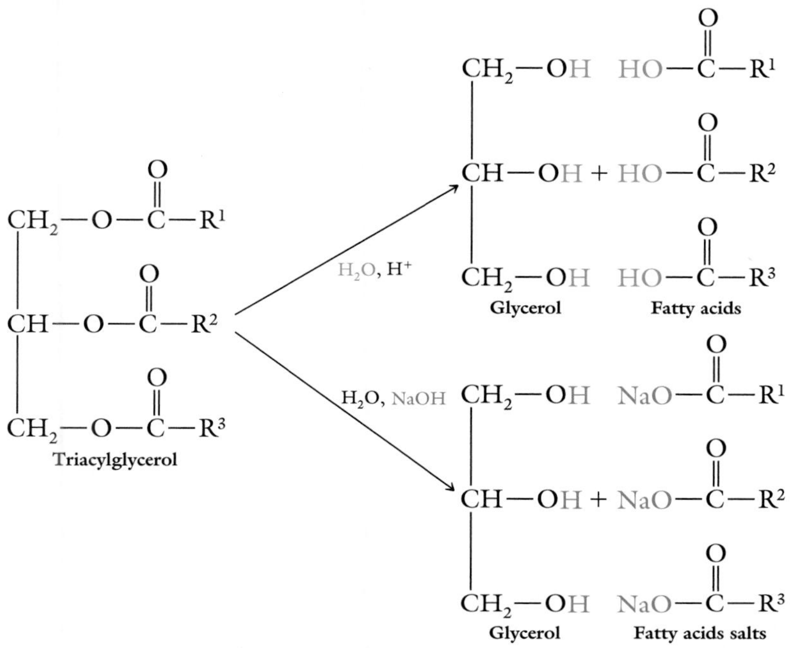

Basic hydrolysis, or **saponification,** of animal fats and some plant oils produces soaps, which are the mixtures of fatty acid sodium (or potassium) salts that remain after removal of glycerol. Various additives are used to enhance the cleaning power of commercial soaps or to give them other properties desirable to the consumer (Section 15.7).

Example 19.3 Writing equations for the hydrolysis of triacylglycerols

Write the equation for the acid-catalyzed hydrolysis of the triacylglycerol containing equimolar amounts of myristic, stearic, and palmitoleic acids.

Solution

Hydrolyze each ester group to an alcohol and carboxylic acid:

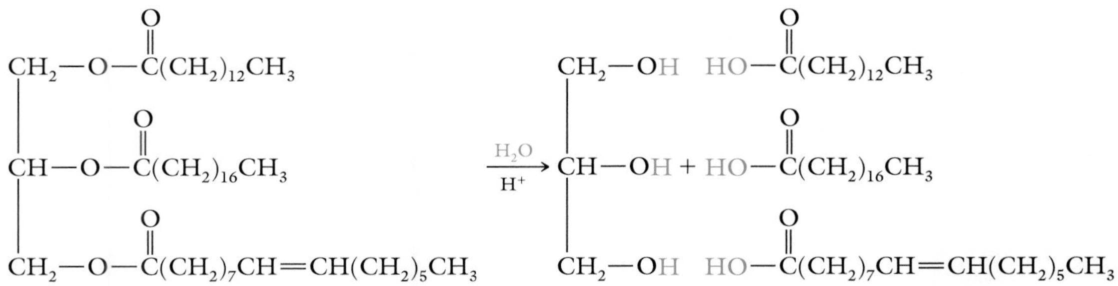

Problem 19.3 Write the equation for the saponification with NaOH of the triacylglycerol containing equimolar amounts of myristic, stearic, and palmitoleic acids.

Digestion and Storage of Triacylglycerols

Digestion (hydrolysis) of dietary triacylglycerols to smaller molecules is necessary for their utilization because triacylglycerols are too large to diffuse through the intestinal membranes. This digestion is carried out with enzymes called **lipases,** with the help of bile salts (Section 19.7), and takes place in the small intestine, which has a basic pH. Digestion is more complicated than hydrolysis in the laboratory (Section 26.3). Digestion of triacylglycerols produces a mixture of products, most of which are monoacylglycerols and fatty acids and some of which are diacylglycerols and glycerol. The cells of the intestinal lining then rebuild triacylglycerols from their smaller components and package them, with proteins, into lipoprotein particles called **chylomicrons,** which are transported through the lymphatic system to the bloodstream. The blood carries the chylomicrons to various tissues where the triacylglycerol is separated from the protein and hydrolyzed to fatty acids and glycerol, which are then metabolized to generate energy. Fatty acids not needed for immediate energy generation are reconverted into triacylglycerols and stored as fat droplets in **adipose cells,** also called **adipocytes** or **fat cells.** Such cells make up most of the fatty, or adipose, tissue in animals.

▶▶ The structure and properties of chylomicrons are considered in Section 26.15.

Triacylglycerols are the main storage form of energy in animals. Recall that animals also store energy as glycogen (Section 18.7), but triacylglycerols are a far more efficient form of energy storage: the energy produced by metabolic oxidation of triacylglycerols is 9 kcal/g (dry weight) versus 4 kcal/g for glycogen (dry weight)(1 cal = 4.184 J). The reason for this difference in the amount of energy produced is that the carbon atoms of triacylglycerols are more reduced than those of glycogen; that is, they contain more hydrogen and less oxygen. Furthermore, triacylglycerols are stored in highly anhydrous form, whereas glycogen is stored in hydrated form. One gram of stored triacylglycerol is essentially pure triacylglycerol, but one gram of stored glycogen is two-thirds water. If we take the difference in hydration into account, 1 g of stored triacylglycerol yields about seven times the energy of 1 g of stored glycogen.

A dormouse hibernating in the winter. Fat reserves are used for energy, and they insulate the dormouse from cold temperatures.

The glycogen stored in your liver and muscle is enough to fill your energy needs for about 12 h. The triacylglycerol stored in adipose cells can fill your energy needs for several weeks to a couple of months. Hibernating animals accumulate sufficient triacylglycerol reserves before hibernation to last through a winter season. Plants store triacylglycerols in their seeds to provide the energy required for germination. Layers of adipose tissue under the skin also serve as thermal insulation against very low temperatures for penguins, seals, walruses, and other warm-blooded polar animals.

As described in Box 18.1, the increase in obesity in Western developed countries led to the development of artificial sweeteners. More recently, the use of artificial, or noncaloric, fats was approved in the United States (Box 19.1).

Catalytic Hydrogenation

Catalytic hydrogenation is the addition of hydrogen to alkene double bonds in the presence of a metal catalyst such as nickel (Ni) or platinum (Pt) (Section 12.6). This process is the basis for the conversion of plant oils into margarine and other products.

Shortening, peanut butter, and margarine are produced by the hydrogenation of vegetable oils.

Butter (and other animal triacylglycerols) are implicated in the development of atherosclerosis because of their high content of saturated triacylglycerols and cholesterol (Box 24.1). Vegetable oils such as corn oil are preferable dietary lipids because of their higher content of unsaturated triacylglycerols and almost total absence of cholesterol. However, most people would find it quite unpalatable to spread corn oil on their toast or baked potatoes. An alter-

19.1 Chemistry Within Us

Noncaloric Fat

Artificial sweeteners (see Box 18.1) have been used for many years to limit caloric intake without eliminating pastries and other sweet foods from the diet. Now, food products with artificial fats also are on the market. In 1996, the U.S. Food and Drug Administration approved the use of Olestra, a sucrose octaester, in snack foods such as corn and potato chips. A sucrose octaester is produced by esterification of the eight hydroxyl groups of sucrose with fatty acids derived from edible fats and oils (soybean, corn, cottonseed).

Frying foods in sucrose octaester gives them the flavor of foods fried in fat. Unlike other fats used for frying, however, sucrose octaester is noncaloric, because it cannot be digested. The lipase enzymes that hydrolyze the ester groups in triacylglycerols and other hydrolyzable lipids cannot hydrolyze the ester groups of sucrose octaester, because the octaester is too bulky to bind to the active sites of the enzymes. It is also too large to enter the cells lining the intestine. Sucrose octaester therefore passes through the intestinal tract and is excreted in the feces. The only caloric intake from a bag of potato or corn chips fried in sucrose octaester is from the potato or corn.

There are possible drawbacks to the use of sucrose octaester, so it should not be consumed in large amounts. It can cause gastrointestinal cramping, flatulence, anal leakage, and loose bowels in some people who consume too much of it, and it may also interfere slightly with the absorption of fat-soluble vitamins and other nutrients.

Potato chips made from Olestra are free of calories from fat.

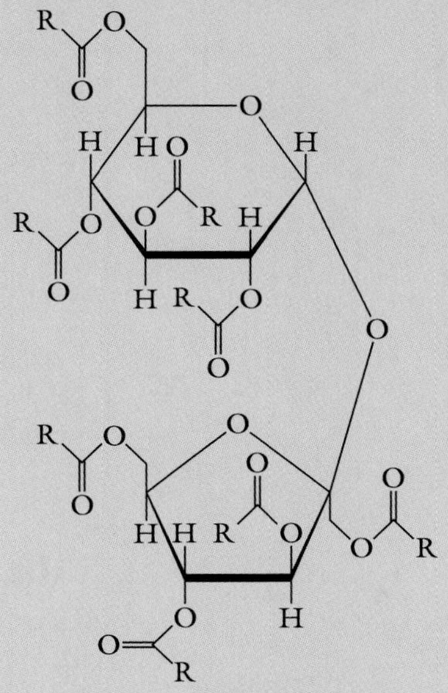
Sucrose octaester

native is margarine, produced by **hardening** of corn or other vegetable oils by partial hydrogenation that converts the oil into a semisolid or solid product acceptable to our palates. Milk and artificial color added to the product give it an attractive flavor, color, and consistency. Partial hydrogenation is also used to produce cooking oils such as Crisco and Spry from vegetable oils, as well as peanut butter from peanut oil.

For some time, margarine has been considered a more healthy dietary fat than butter, but recent evidence shows that the situation is more complicated. In the partial hydrogenation process, some of the cis double bonds undergo conversion into trans double bonds, a process called **isomerization.** Fatty acids with trans double bonds have physical properties much more like those of saturated fatty acids than like cis unsaturated fatty acids (Section 19.1). Furthermore, there is evidence that trans fatty acids raise blood cholesterol levels more than cis fatty acids do, possibly even more than saturated fatty acids do.

Example 19.4   Writing equations for hydrogenation of triacylglycerols

Write the equation for the Pt-catalyzed hydrogenation of the triacylglycerol containing equimolar amounts of lauric, palmitoleic, and linoleic acids. (Assume complete hydrogenation of all double bonds unless otherwise indicated.)

Solution

Hydrogen is added to each C–C double bond.

$$
\begin{array}{l}
CH_2-O-\overset{\displaystyle O}{\overset{\|}{C}}(CH_2)_{10}CH_3 \\[2mm]
CH-O-\overset{\displaystyle O}{\overset{\|}{C}}(CH_2)_7CH{=}CH(CH_2)_5CH_3 \qquad \xrightarrow[\text{Pt}]{\text{H}_2} \\[2mm]
CH_2-O-\overset{\displaystyle O}{\overset{\|}{C}}(CH_2)_6(CH_2CH{=}CH)_2(CH_2)_4CH_3
\end{array}
$$

$$
\begin{array}{l}
CH_2-O-\overset{\displaystyle O}{\overset{\|}{C}}(CH_2)_{10}CH_3 \\[2mm]
CH-O-\overset{\displaystyle O}{\overset{\|}{C}}(CH_2)_{14}CH_3 \\[2mm]
CH_2-O-\overset{\displaystyle O}{\overset{\|}{C}}(CH_2)_{16}CH_3
\end{array}
$$

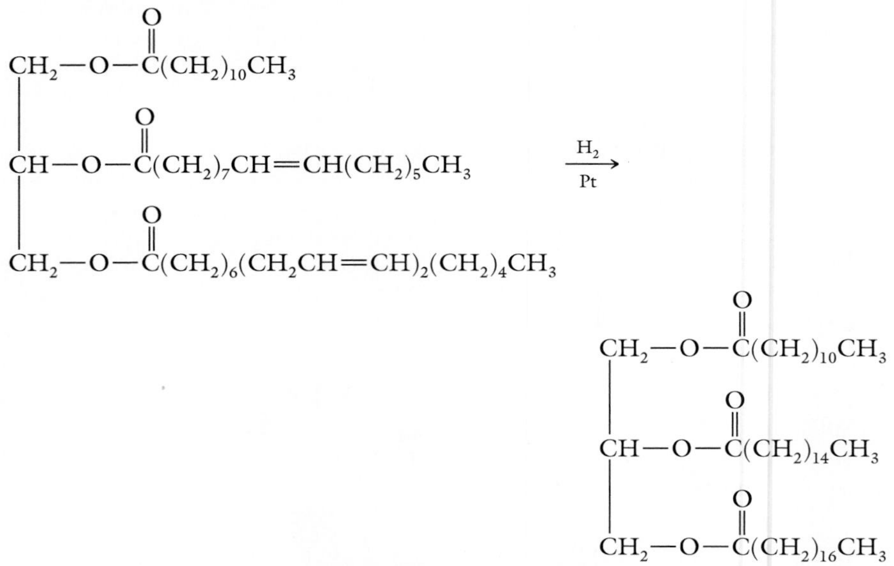

Problem 19.4   Write the equation for Pt-catalyzed hydrogenation of the triacylglycerol that contains equimolar amounts of palmitic, oleic, and linoleic acid units.

Rancidity

Butter, salad and cooking oils, mayonnaise, and fatty meats can become **rancid,** developing unpleasant odors and flavors. Rancidity is due to a combination of two reactions:

- Bacterial hydrolysis of ester groups
- Air oxidation of alkene double bonds

In fats containing triacylglycerols with some low-molecular-mass carboxylic acids, hydrolysis by airborne bacteria under moist, warm conditions is directly responsible for rancid odors and flavors. For example, butter contains triacylglycerols with 3 to 4% butanoic acid and 1 to 2% hexanoic acid. Bacterial hydrolysis (digestion) of butter releases these volatile, malodorous, and off-flavor acids.

Foul-tasting and malodorous carboxylic acids are also generated by air oxidation of double bonds in fats and oils or in the unsaturated acids released by bacterial hydrolysis of fats and oils. Oxidation cleaves double bonds, with each carbon of the double bond being converted into a —COOH group. The end result is the creation of low-molecular-mass, volatile, offensive-smelling carboxylic and dicarboxylic acids.

The rancidity reactions are slowed down by refrigeration. Fats and oils stored at room temperature, such as salad oil, should be tightly capped to impede the entry of moisture, bacteria, and oxygen—all of which are required for the hydrolysis and oxidation reactions. Some manufacturers of vegetable oils and other products add substituted phenols such as BHA (butylated hydroxyanisole) and BHT (butylated hydroxytoluene) as antioxidants to inhibit the oxidation reaction (see Box 13.5).

Example 19.5 Writing equations for rancidity reactions

A triacylglycerol undergoes hydrolysis by bacteria to yield a mixture of lauric, palmitic, and palmitoleic acids. Write the equation for the subsequent air oxidation of palmitoleic acid. Why don't lauric and palmitic acids undergo air oxidation?

Solution

$$CH_3(CH_2)_5CH{=}CH(CH_2)_7\overset{\overset{\displaystyle O}{\|}}{C}{-}OH \xrightarrow{(O)} CH_3(CH_2)_5\overset{\overset{\displaystyle O}{\|}}{C}{-}OH + HO{-}\overset{\overset{\displaystyle O}{\|}}{C}(CH_2)_7\overset{\overset{\displaystyle O}{\|}}{C}{-}OH$$

Lauric and palmitic acids do not contain double bonds; oxidation takes place only at double bonds.

Problem 19.5 Write the equation for the air oxidation of oleic acid.

19.5 WAXES

Waxes have a variety of protective functions in plants and animals. Many fruits, vegetables, and plant leaves are coated with waxes that protect against parasites and mechanical damage and prevent excessive water loss. The skin glands of many vertebrates secrete waxes that protect hair, furs, feathers, and skin, keeping them lubricated, pliable, and waterproof. Waterfowl depend on waxes to make their feathers water repellent and to prevent them from becoming water-soaked and less buoyant. Bees secrete wax to construct the honeycomb in which eggs are laid and new bees develop. In some marine organisms, notably the plankton, waxes are used instead of triacylglycerols for energy storage.

Waxes coat the duck's body and feathers and enable it to float effortlessly in the water.

Most waxes are mixtures of esters of fatty acids with long-chain alcohols having one —OH group. The acids and alcohols usually contain an even number of carbons, in the range of 14 to 36 carbons, and are unbranched. For example, the major component of beeswax is melissyl palmitate, the ester derived from palmitic acid and melissyl alcohol:

$$\underbrace{CH_3(CH_2)_{14}{-}\overset{\overset{\displaystyle O}{\|}}{C}}_{\textbf{Palmitate}}{-}\underbrace{O{-}(CH_2)_{29}CH_3}_{\textbf{Melissyl}}$$

Waxes have many commercial uses. Carnauba wax, from the leaves of Brazilian palm trees, is used for high gloss, hard finishes for automobiles, boats, and floors. Lanolin, a component of wool wax, is used in cosmetic and pharmaceutical creams, lotions, and other products. Spermaceti, a component of the oil obtained from the head of a sperm whale, was also used in these products until such use was banned in the late 1970s.

19.6 AMPHIPATHIC HYDROLYZABLE LIPIDS

The predominant lipids of cell membranes are the amphipathic hydrolyzable lipids. There are two broad groups: those based on glycerol, the **glycerophospholipids (phosphoglycerides),** and those based on sphingosine, the **sphingolipids.** Unlike triacylglycerols, which have only nonpolar groups, the glycerophospholipids and sphingolipids have one highly hydrophilic group

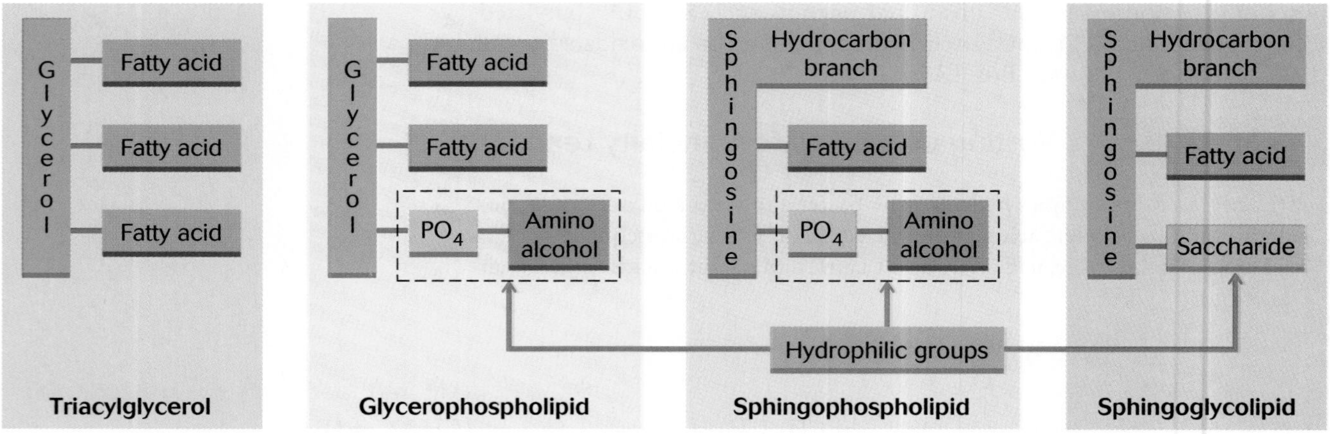

Figure 19.3 Structural comparison of different hydrolyzable lipids.

(Figure 19.3). The hydrophilic group is responsible for the amphipathic nature of these lipids—which allows their assembly into cell membranes (Section 19.10).

Glycerophospholipids

Like triacylglycerols, glycerophospholipids have fatty acid ester groups at two of the carbons of glycerol, but, at the third carbon, they have a phosphodiester group. Esterification of glycerol with two fatty acids and one phosphoric acid forms a **phosphatidic acid;** one of the ester groups is a phosphate ester (Section 15.13). Further esterification of the phosphate ester with an alcohol produces a glycerophospholipid by formation of a phosphodiester group:

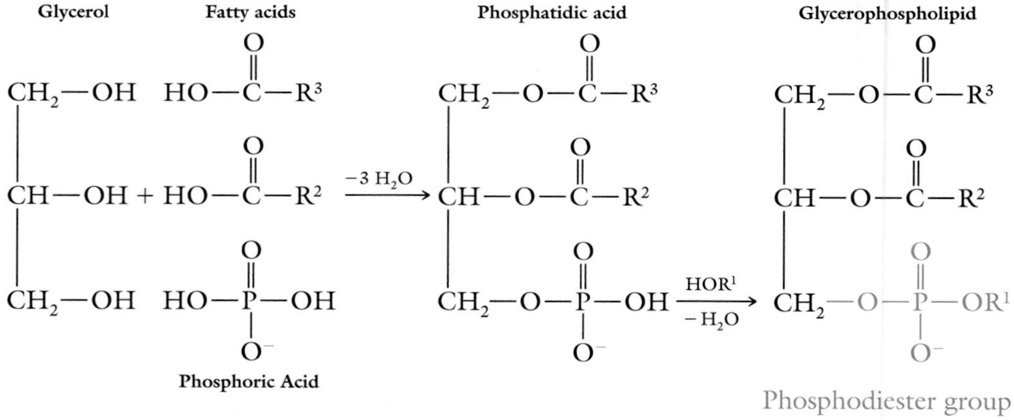

Although all P—OH groups actually exist as P—O$^-$ groups at physiological pH (pH ~ 7), the phosphoric and phosphatidic acids, respectively, are shown with one and two P—OH groups in unionized form. This allows us to more easily visualize the formations of the phosphatidic acid and glycerophospholipid by dehydration reactions.

Table 19.3 lists the alcohols (R^1OH in the preceding reaction) found in the phosphodiester group of various glycerophospholipids. Glycerophospholipids are often named by placing **phosphatidyl**- before the name of the alcohol—for example, phosphatidylcholine when the alcohol is choline. However, this nomenclature does not specify the identities of the two fatty acid units. Some glycerophospholipids have other common names; those containing choline and ethanolamine are called **lecithins** and **cephalins,** respectively.

TABLE 19.3   Alcohols in Glycerophospholipids

Alcohol (R¹OH)	Structure of R¹
ethanolamine	$-CH_2CH_2\overset{+}{N}H_3$
choline	$-CH_2CH_2\overset{+}{N}(CH_3)_3$
serine	$\underset{-CH_2\overset{\underset{\mid}{}}{C}H\overset{+}{N}H_3}{\overset{COO^-}{}}$
inositol	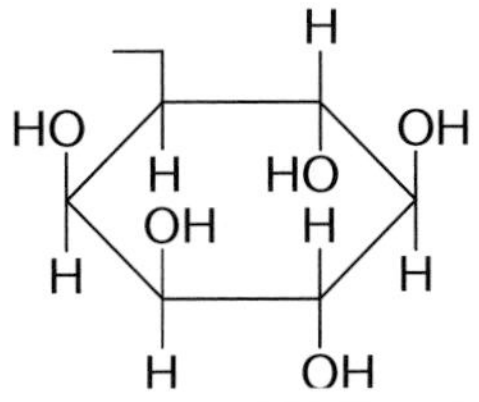

The laboratory hydrolysis of a glycerophospholipid yields glycerol, fatty acids, phosphoric acid, and alcohol. Like triacylglycerols, glycerophospholipids yield a more complex mixture of products on digestive hydrolysis.

Example 19.6   **Drawing the structures of glycerophospholipids**

Draw the glycerophospholipid whose hydrolysis yields equimolar amounts of glycerol, palmitic acid, oleic acid, phosphoric acid, and ethanolamine.

Solution

Draw glycerol. Draw palmitic, oleic, and phosphoric acids alongside the —OH groups of glycerol. Draw ethanolamine alongside phosphoric acid. Carry out the dehydrations indicated in blue.

$$CH_2-OH \qquad HO-\overset{O}{\overset{\|}{C}}(CH_2)_{14}CH_3$$

$$CH-OH \qquad HO-\overset{O}{\overset{\|}{C}}(CH_2)_7CH=CH(CH_2)_7CH_3 \quad \xrightarrow{-4\,H_2O}$$

$$CH_2-OH \qquad HO-\overset{O}{\overset{\|}{\underset{\underset{O^-}{\mid}}{P}}}-OH \qquad HOCH_2CH_2\overset{+}{N}H_3$$

$$CH_2-O-\overset{O}{\overset{\|}{C}}(CH_2)_{14}CH_3$$

$$CH-O-\overset{O}{\overset{\|}{C}}(CH_2)_7CH=CH(CH_2)_7CH_3$$

$$CH_2-O-\overset{O}{\overset{\|}{\underset{\underset{O^-}{\mid}}{P}}}-OCH_2CH_2\overset{+}{N}H_3$$

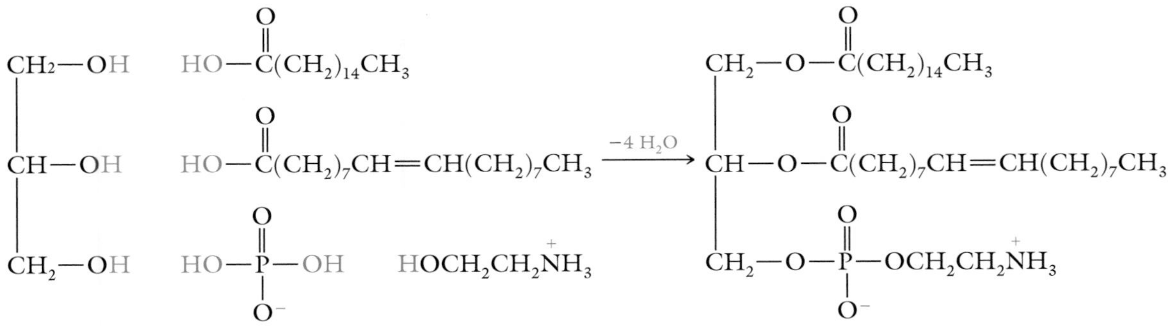

Problem 19.6   Draw the glycerophospholipid whose hydrolysis yields equimolar amounts of glycerol, stearic acid, linoleic acid, phosphoric acid, and choline.

Sphingolipids

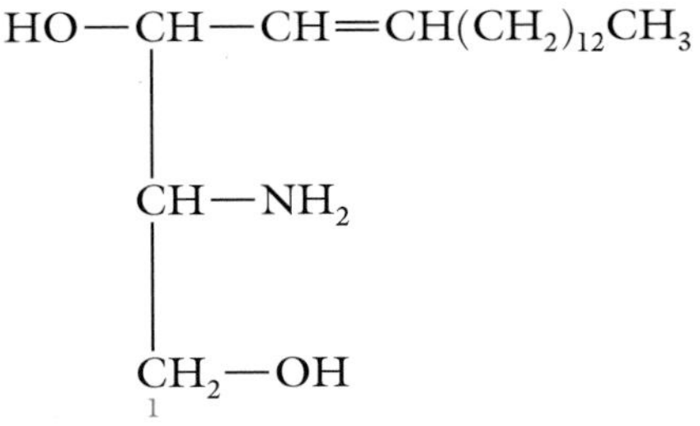

Sphingolipids are amphipathic hydrolyzable lipids based on sphingosine instead of glycerol. Sphingosine is a long-chain, unsaturated amino alcohol. Sphingolipids have an amide group at C2, formed by reaction of a fatty acid with the amine group. There are two types of sphingolipids—sphingophospholipids and sphingoglycolipids—differing in the structural unit at C1, which is the carbon at the bottom of the sphingosine structure shown in the margin.

Like glycerophospholipids, **sphingophospholipids** have a phosphodiester group at C1. That is why glycerophospholipids and sphingophospholipids are

$$HO-CH-CH=CH(CH_2)_{12}CH_3$$
$$\mid$$
$$CH-NH_2$$
$$\mid$$
$$\underset{1}{CH_2}-OH$$

Sphingosine

often classified together as **phospholipids.** In the following structure, $—R^1$ represents one of the alcohol units in Table 19.3; that is, the same alcohols are found in both glycerophospholipids and sphingophospholipids. **Sphingoglycolipids,** or **glycolipids** for short, contain a saccharide unit at C1, joined through an acetal linkage. Some sphingoglycolipids contain monosaccharide units; others contain oligosaccharide units.

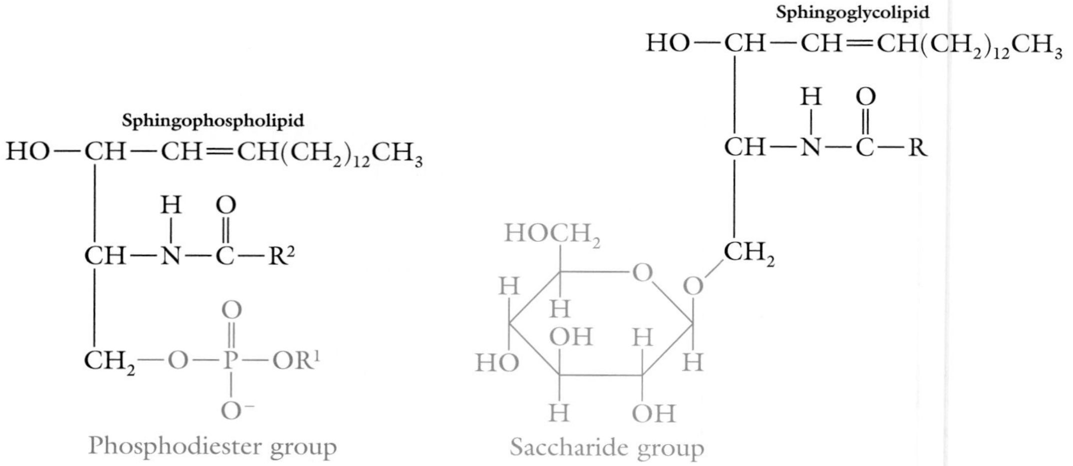

Example 19.7 Drawing the structures of sphingolipids

Draw the sphingolipid whose hydrolysis yields equimolar amounts of sphingosine, palmitoleic and phosphoric acids, and choline.

Solution

Draw sphingosine. Draw palmitoleic acid alongside the $—NH_2$ group and phosphoric acid alongside the $—OH$ group. Draw choline alongside phosphoric acid. Carry out the dehydrations indicated in blue.

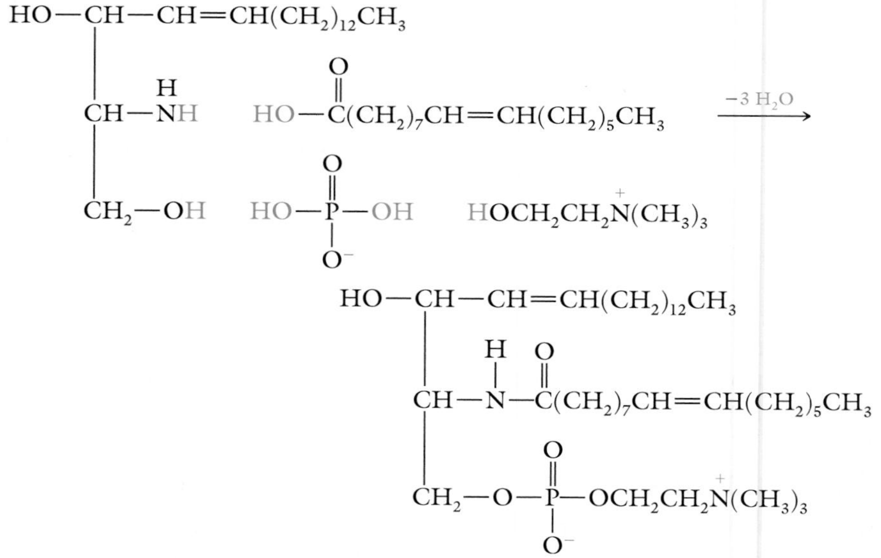

Problem 19.7 Draw the sphingolipid whose hydrolysis yields equimolar amounts of sphingosine, oleic acid, and β-D-glucose.

Cell membranes contain mixtures of different glycerophospholipids and sphingolipids (as well as cholesterol). Because different cell membranes have different functions, the relative amounts of the different amphipathic hydrolyzable lipids vary from one type of cell to another. For example, the membranes of the myelin sheath that surrounds and electrically insulates many nerve cell axons are rich in sphingophospholipids. Sphingoglycolipids containing oligosaccharides are usually found at cell surfaces where they perform cell recognition functions (Section 18.6). Many illnesses arise from inadequate metabolism and utilization of glycerophospholipids and sphingolipids—for example, Niemann-Pick, Tay-Sachs, Gaucher, and Fabry diseases (Section 24.7).

19.7 STEROIDS: CHOLESTEROL, STEROID HORMONES, AND BILE SALTS

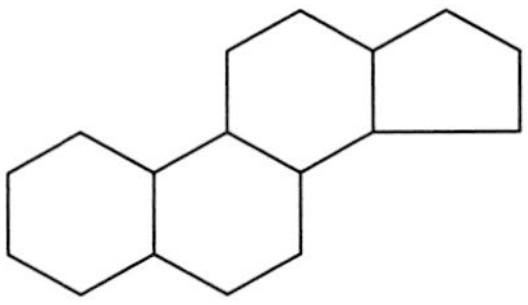

Steroid ring structure

Steroids are nonhydrolyzable lipids that contain the **steroid ring structure,** which consists of four fused rings, three of them six-membered rings and one of them a five-membered ring. This group includes cholesterol, adrenocortical and sex hormones, and bile salts.

Cholesterol is the major steroid in animals. Plants contain very little cholesterol but do contain related compounds in their membranes. The cholesterol molecule contains eight tetrahedral stereocenters (indicated by asterisks) and thus $2^8 = 256$ stereoisomers (128 pairs of enantiomers), but it exists in nature as only one stereoisomer—the one shown here:

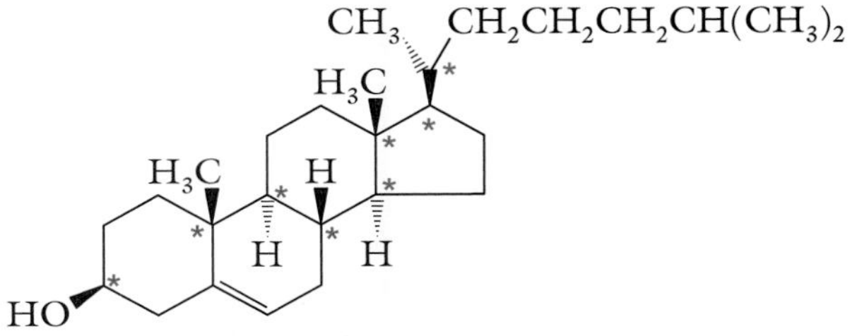

Cholesterol

This stereoisomer is another example of the remarkable stereospecificity of biological systems.

We now know that cholesterol plays an important role in cardiovascular disease and are advised to lower our intake of this lipid (Box 24.1). Nevertheless, we must keep in mind that cholesterol is critical to many physiological functions. In particular, it serves as a component of membranes (Section 19.10) and as a precursor for all other animal steroids, including the adrenocortical and sex hormones and the bile salts (Figure 19.4 on the following page). The body synthesizes all the cholesterol that it needs if cholesterol is excluded from the diet.

Steroid Hormones

The steroid hormones are just one group of **hormones,** substances synthesized in and secreted by endocrine glands and then transported in the bloodstream to target tissues, where they regulate a variety of cell functions. Hormones are often called chemical messengers because they relay messages between different parts of the body. They are effective at very low concentrations, often in the picomolar to nanomolar range (10^{-12}–10^{-9} M).

The **adrenocortical hormones,** produced in the adrenal glands located at the top of the kidneys, include cortisol and aldosterone. Cortisol helps the body respond to short-term stress. Because a high blood-glucose level is needed for your brain to function, cortisol stimulates other cells to decrease their use of glucose and, instead, metabolize fats and proteins for energy. Cortisol also blocks the immune-system, because a stress situation is not the time

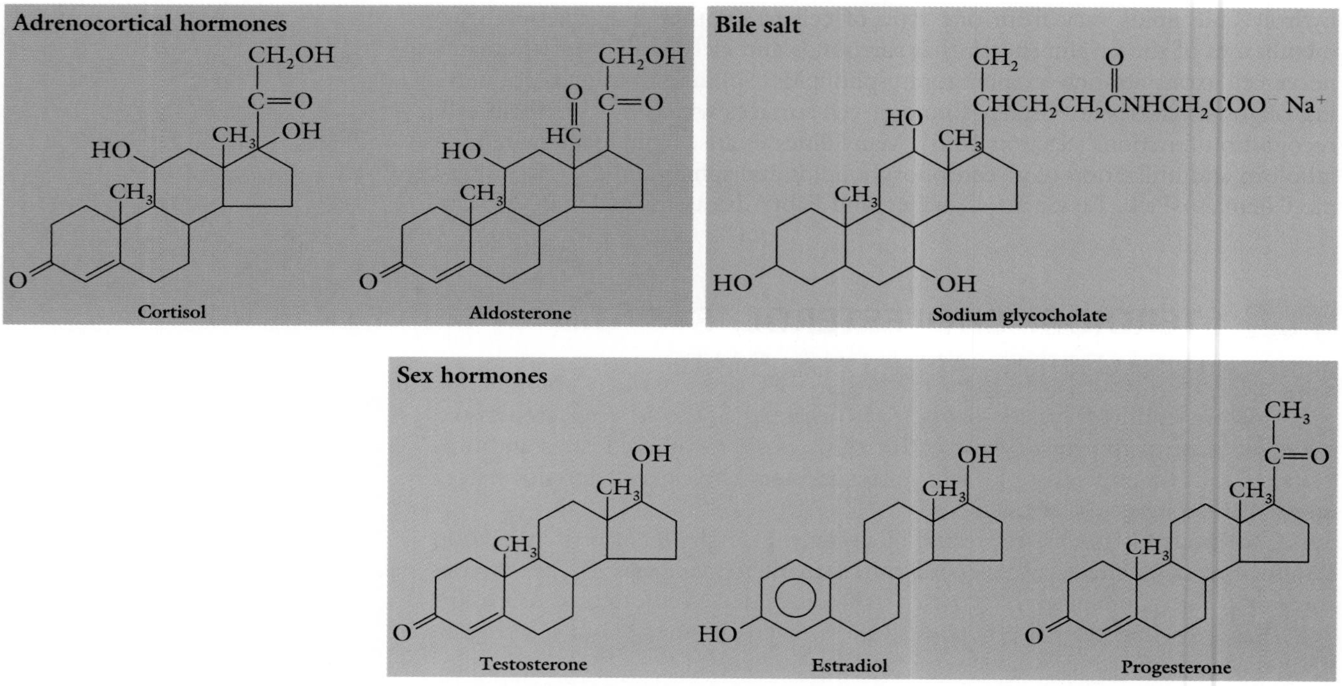

Figure 19.4 Steroids.

to feel sick or have an allergic reaction. For this reason, cortisol is used as an anti-inflammatory and antiallergy medication. However, long-term use of cortisol can be damaging to the body because of its effect on the immune system.

Aldosterone takes part in the regulation of electrolyte concentrations in tissues. It stimulates Na^+, K^+, and water reabsorption by the kidney, helping to maintain blood volume and blood pressure.

The **sex hormones** are produced in the gonads: ovaries in females and testes in males. There are two kinds of female sex hormones—the **estrogens** and **progestins**—and one kind of male sex hormone—the **androgens.** The ovaries produce mostly estrogens, such as estradiol; they also produce small quantities of androgens. The testes produce mostly androgens, such as testosterone; they also produce small quantities of estrogens. Progestins are female hormones produced only in the ovaries. The most abundant progestin is progesterone, which together with estradiol regulates the menstrual cycle. Box 19.2 describes the hormonal changes in the menstrual cycle and the use of hormones in contraceptive drugs.

The androgens and estrogens regulate the development of the sex organs, the production of sperm and ova, and the development of secondary sex characteristics: lack of facial hair, increased breast size, and high voice in women; facial hair, increased musculature, and deep voice in men. Synthetic androgens are used by some athletes to promote muscle growth (Box 19.3 on page 574).

Bile Salts

Bile salts, also called **bile acids,** have a carboxylate salt group attached to the steroid ring structure. They play a crucial role in lipid digestion. Bile salts are synthesized from cholesterol in the liver and stored in the gall bladder as a solution called **bile,** which is a mixture of bile salts, cholesterol, and pigments from the breakdown of red blood cells. Sodium glycocholate (see Figure 19.4) is a bile salt.

19.2 Chemistry Within Us

The Menstrual Cycle and Contraceptive Drugs

After puberty, a woman secretes sex hormones in a repeating pattern, called the **menstrual cycle,** that causes periodic changes in her reproductive organs. The most important changes are the development and release of a single egg cell (ovum) and the growth of a uterine lining (endometrium), which prepares the uterus to receive a fertilized ovum.

A cycle, lasting 28 days on average, begins on the first day of menstruation (day 1). The levels of estrogens and progestins, mainly estradiol and progesterone, are low at this time, which triggers the hypothalamus to release gonadotropin-releasing hormone (GnRH). GnRH travels in the bloodstream to the pituitary gland, where it stimulates the release of follicle-stimulating hormone (FSH) and luteinizing hormone (LH) into the bloodstream.

FSH and LH stimulate ovum development in an ovarian structure called a **follicle.** Ovulation—the release of an ovum into one of the fallopian tubes where it can be fertilized by a sperm—occurs on or close to day 14. The ovum travels through the fallopian tube to the uterus. The follicle that released the ovum becomes a corpus luteum—an endocrine gland that secretes estrogen and progestin; these hormones control the development and maintenance of the uterine endometrium for receiving a fertilized ovum. Negative feedback by estradiol and progesterone to the hypothalamus stops the release of GnRH, which in turn halts the secretion of FSH and LH. In a menstrual cycle, when the ovum is unfertilized, the corpus luteum shrinks, secretion of estradiol and progesterone decreases, and the endometrium begins to slough off. The negative feedback on the hypothalamus is relieved, and the pituitary increases its secretion of FSH and LH. Thus the cycle begins again.

If the ovum is fertilized as it travels through the fallopian tube, the corpus luteum continues to secrete estradiol and progesterone, promoting implantation of the fertilized egg in the uterine wall and subsequent development of an embryo.

Oral contraceptives (birth-control pills) are combinations of a synthetic estrogen, such as ethynyl estradiol, and a synthetic progestin, such as norethindrone, that act by preventing ovulation. Natural estrogens and progestins cannot be used, because they are broken down after ingestion. This combination of hormones, taken for days 5 through 24 of the menstrual cycle, mimics the hormone levels present in early pregnancy and tricks the body into thinking that it is pregnant. Negative feedback inhibits GnRH, FSH, and LH secretion, and thus ovulation does not occur.

The negative side effects of oral contraceptives—chiefly an increased risk of cardiovascular complications such as blood clots, hypertension, stroke, and heart attack—were mostly associated with the high hormone levels used in early contraceptive pills. Modern contraceptive pills carry very low risks, except for women older than 35 who smoke or have hypertension.

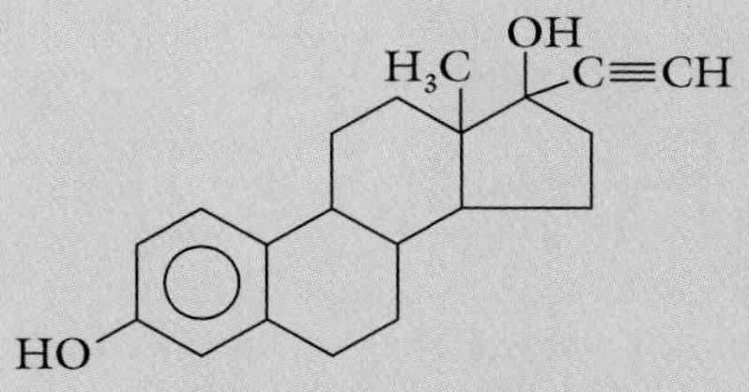

Ethynyl estradiol

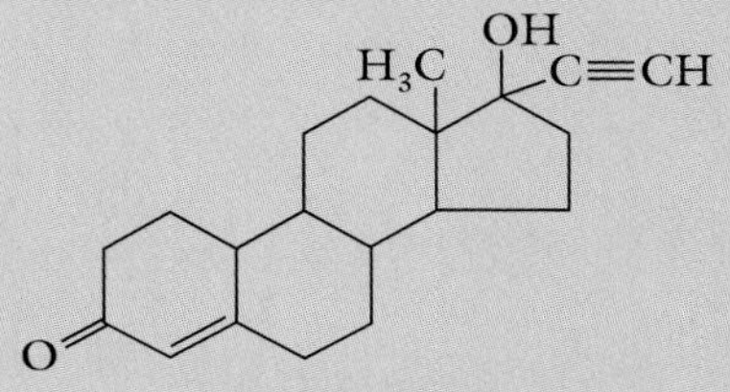

Norethindrone

Mifepristone (RU-486) is a postcoital contraceptive pill. Often called the **morning-after pill,** it is almost completely effective when taken within 72 h after intercourse. Mifepristone works by blocking the progesterone receptors in the uterine lining, thus preventing progesterone from exerting its endometrium maintenance effects. The endometrium sloughs off, as in menstruation. The effectiveness of mifepristone is enhanced by a dose of a prostaglandin taken 2 days later to induce uterine contractions. Mifepristone with a prostaglandin causes a spontaneous abortion and is used for that purpose, as an alternative to aspiration abortion, as late as 1 to 2 months after the last menstrual period.

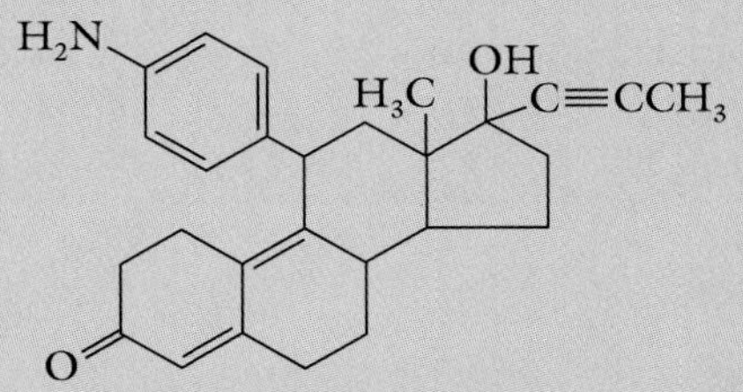

Mifepristone (RU-486)

Mifepristone is used widely in Europe, especially France, and in the Far East. It is not yet available in the United States, but clinical trials are in progress. Antiabortion advocates have strongly opposed the legalization of mifepristone.

19.3 Chemistry Within Us

Anabolic Steroids

Anabolic steroids are steroids that stimulate the biosynthesis of proteins, the primary constituent of muscle. Some athletes use them to enhance their performance, especially in sports where increased size, strength, and stamina are advantageous: football, weight lifting, bodybuilding, and track and field events such as the discus, shotput, and hammer throw. The enhanced performance is achieved through both increased muscle development and increased aggressiveness.

Testosterone is a natural anabolic steroid. When taken orally or by intramuscular injection, testosterone is not very effective in enhancing atheletic performance, because most of it undergoes metabolic breakdown. Laboratory modification of testosterone has produced synthetic anabolic steroids, such as nandrolone decanoate and stanozolol, that are more resistant to metabolic breakdown and more effective than testosterone.

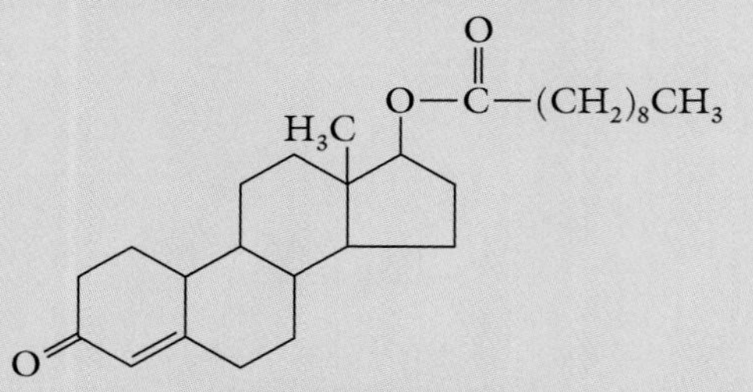

Nandrolone decanoate

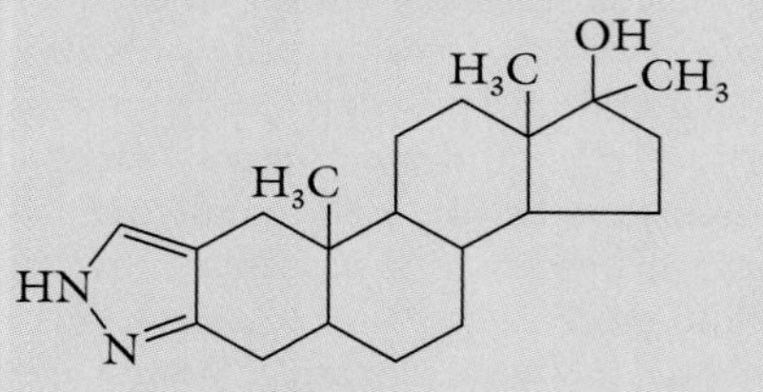

Stanozolol

The use of anabolic steroids in preparation for athletic competitions is banned by all sports organizations. Blood and urine testing of athletes has become routine before and after major athletic events, and atheletes with positive test results have been disqualified. However, new synthetic anabolic steroids are always being invented, and testing methods may not be able to keep ahead of these developments.

Participants in a drug-free body-building championship in Austin, Texas.

Anabolic steroids are controlled substances in the United States, under the Anabolic Steroids Act of 1990, although some are approved for therapeutic uses: testosterone replacement in hypogonadal men, treatment of endometriosis and fibrocystic disease of the breast in women, and rehabilitation of patients with muscle atrophy resulting from severe injury.

Anabolic steroids can have severe side effects, ranging from temporary acne to increased liver and cardiovascular disease and including psychological effects such as mood swings and antisocial and violent behavior. Men may experience testicular atrophy, transient infertility, and accelerated male pattern baldness. Women may experience altered menstruation, clitoral enlargement, deepened voice, increased facial and body hair, and male-pattern baldness.

Bile is secreted into the small intestine after a fatty meal and participates in the digestion of hydrolyzable lipids. Dietary lipid arrives in the intestine in the form of insoluble lipid globules. Hydrolysis by lipases can take place only at the surface of the globules; and, the larger the globules, the smaller the overall surface area of lipid available for digestion. Bile salts, like soap (Section 15.7 and Box 6.1), break up large lipid globules into much smaller ones, greatly increasing the surface area available to the lipases.

The conversion of cholesterol into bile salts, as well as the solubilization of cholesterol by bile salts in the bile, is the way in which excess cholesterol is eliminated through the intestinal tract. When the cholesterol concentration in bile is too high, however, cholesterol precipitates in the form of gallstones. Some gallstones are small enough to pass through the bile duct to the intestine, causing considerable pain at times but no long-term consequences. Larger gallstones may become stuck in the bile duct, however, and the consequences are severe. The duct blockage prevents the person from digesting fats, because bile cannot enter the intestine. The resulting abdominal pain and nausea do not subside; the skin becomes yellow—a condition called jaundice—as bile pigments enter the blood; and the stools lose their brown color. Ultrasound or laser surgery are used to break up gallstones into pieces that can pass through the bile duct.

Bile salts are also needed for the efficient intestinal absorption of the fat-soluble vitamins (A, D, E, and K).

19.8 EICOSANOIDS

The **eicosanoids** are nonhydrolyzable lipids derived from the polyunsaturated C_{20} fatty acid called arachidonic acid (Figure 19.5). There are three groups of eicosanoids: the leukotrienes, prostaglandins, and thromboxanes. In a **leukotriene,** the 20-carbon chain of arachidonic acid and its carboxyl group is unchanged. A **prostaglandin** is similar to a leukotriene but has a cyclopentane ring formed by bond formation between C8 and C12 of the 20-carbon chain. A **thromboxane** is similar to a prostaglandin but, instead of the cyclopentane ring, C8 through C12 form a six-membered cyclic ether structure. As shown in Figure 19.5, individual leukotrienes, prostaglandins, and thromboxanes are distinguished by a letter and subscript number.

Eicosanoids have hormonelike physiological functions. Like hormones, eicosanoids produce their regulatory effects at low concentrations. Unlike hormones, eicosanoids are not transported in the bloodstream from their sites of synthesis to their sites of action. Eicosanoids are **local hormones** or **local mediators,** acting in the same tissues in which they are synthesized. They are produced in most tissues. Eicosanoids play roles in:

- the inflammatory response in joints (rheumatoid arthritis), skin (psoriasis), muscle (overexertion), and eyes;

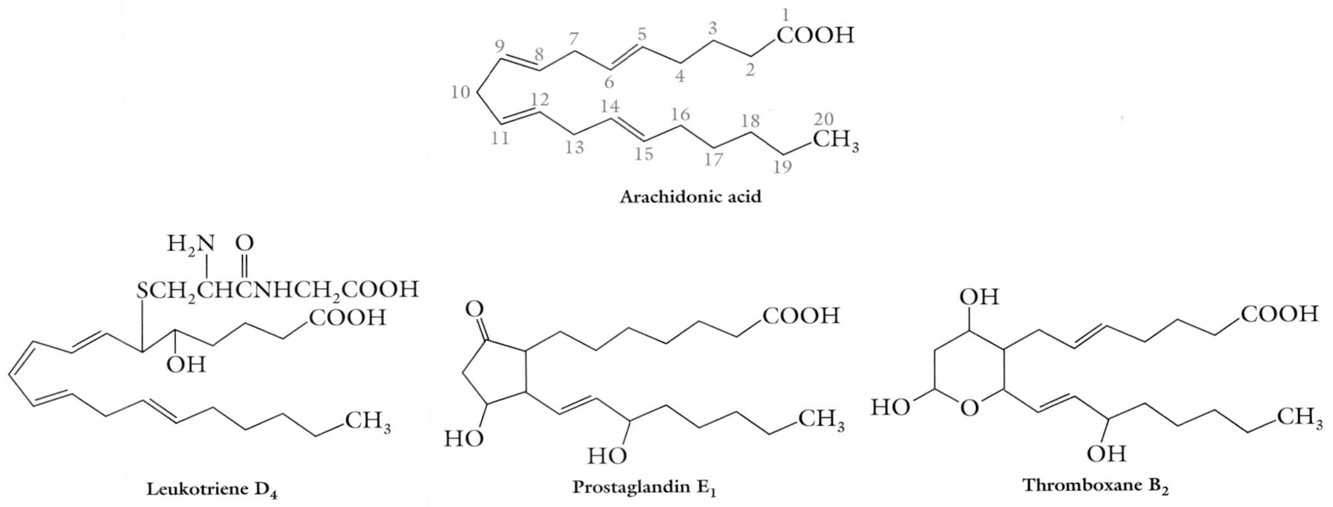

Figure 19.5 Examples of the three categories of eicosanoids, derived from arachidonic acid.

- the production of pain and fever in disease and injury;
- the regulation of blood pressure;
- blood clotting;
- the induction of labor;
- the regulation of the wake–sleep cycle; and
- allergic and asthmatic reactions.

19.9 FAT-SOLUBLE VITAMINS

Vitamins are organic compounds that are required in trace amounts for normal metabolism but are not synthesized by the organism that requires them. They must be included in the diet. Vitamins are classified as water soluble or fat soluble, on the basis of their solubility characteristics. The **water-soluble vitamins,** which are the B and C vitamins, contain sufficient polar functional groups such as hydroxyl and amine to make them water soluble. The **fat-soluble vitamins,** which are the A, D, E, and K vitamins, are a group of nonhydrolyzable lipids (Table 19.4). Most vitamins consist of a closely related group of compounds.

Strictly speaking, vitamin D is a steroid hormone and not a vitamin, because it can be synthesized from cholesterol. Recall that we are not dependent on our diet for cholesterol. Vitamin D has traditionally been classified as a vitamin because, for many years, scientists did not know that it is synthesized from cholesterol.

Fat-soluble vitamins are obtained mostly from vegetables and fruits, fish, liver, and dairy products. Deficiences of these vitamins cause serious health problems, as indicated in Table 19.4; however, excess fat-soluble lipids are stored in fat tissues, so they need not be ingested daily. In fact, consumption of megadoses of fat-soluble vitamins can produce problems because of their accumulation in the body. For example, excess vitamin D causes deposition of abnormal amounts of calcium in bone, resulting in bone pain. It also causes calcification of soft tissue such as the kidneys and brain, resulting in impaired kidney function and mental retardation. High consumption of water-soluble vitamins presents much less risk, because the excess is excreted in the urine.

A PICTURE OF HEALTH

Examples of Lipids in the Body

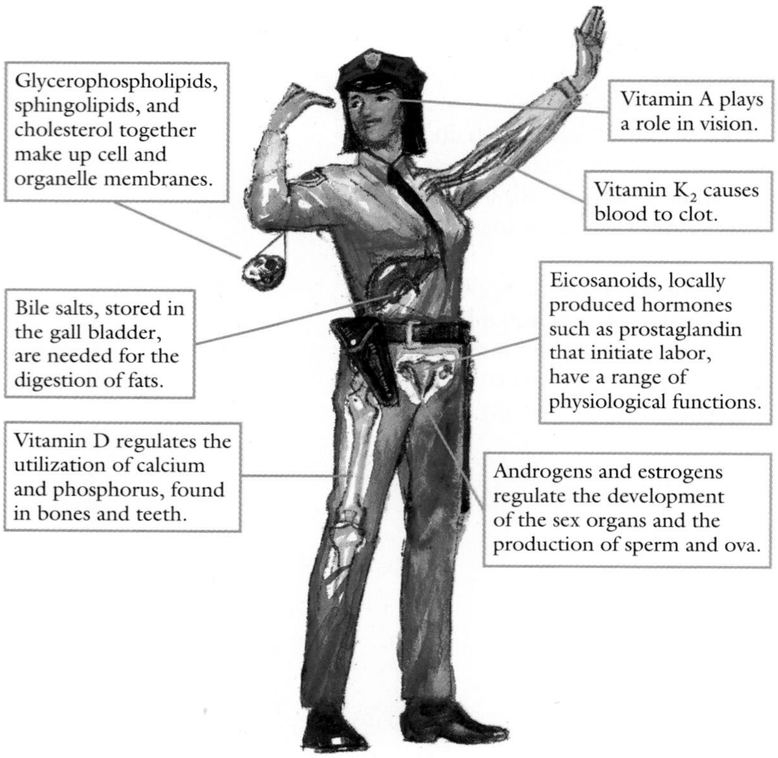

Glycerophospholipids, sphingolipids, and cholesterol together make up cell and organelle membranes.
Vitamin A plays a role in vision.
Vitamin K_2 causes blood to clot.
Bile salts, stored in the gall bladder, are needed for the digestion of fats.
Eicosanoids, locally produced hormones such as prostaglandin that initiate labor, have a range of physiological functions.
Vitamin D regulates the utilization of calcium and phosphorus, found in bones and teeth.
Androgens and estrogens regulate the development of the sex organs and the production of sperm and ova.

▶▶ Water-soluble vitamins are considered in Chapter 23.

19.10 BIOLOGICAL MEMBRANES

Every cell, whether prokaryotic (bacteria) or eukaryotic (higher organisms), is separated from its extracellular environment by a cell membrane (Section 22.1). Eukaryotic cells are more highly organized into separate membrane-enclosed **organelles** (see Figure 22.1). Each of the organelles performs a spe-

TABLE 19.4 Fat-Soluble Vitamins

Name and Structure	Function
vitamin A (*trans*-retinol) 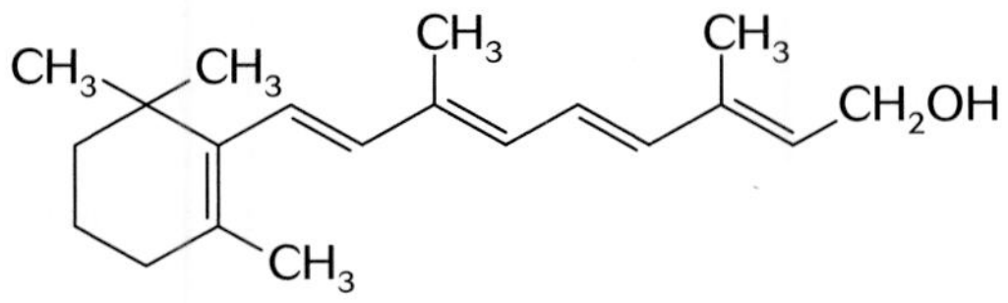	Plays key role in vision by its conversion into *cis*-11-retinal and subsequently into rhodopsin (see Box 12.2). Aids proper functioning of mucous membranes and epithelial tissues. Deficiency: dry eyes, skin, sterility in males, night blindness
1,25-dihydroxyvitamin D_3 (active form of Vitamin D) 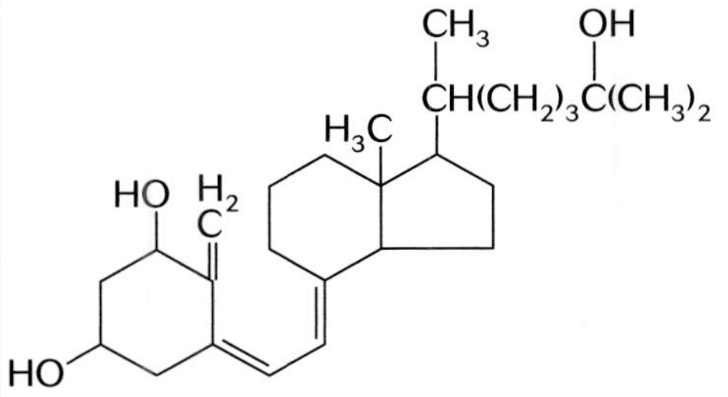	Regulates calcium and phosphate utilization and deposition in bone and cartilage. Deficiency: rickets in children (bowlegs, spinal curvature, knock-knees, pelvic and thoracic deformaties); osteomalacia in adults (weakened bones susceptible to fracture)
vitamin E (α-tocopherol) 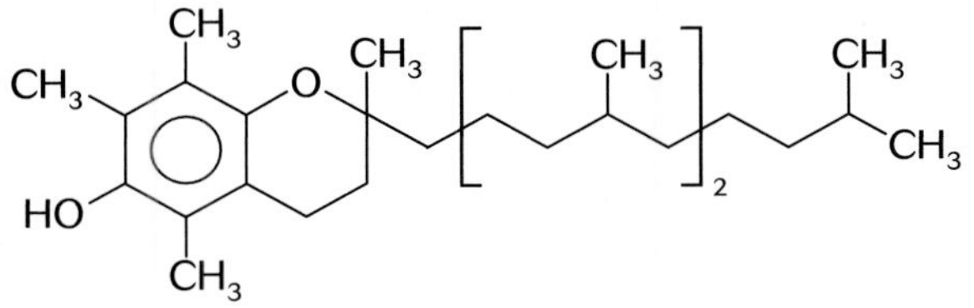	Acts as antioxidant to protect unsaturated fatty acid components of cell-membrane lipids against oxidative cleavage by O_2 and free radicals (see Section 10.2). Deficiency: scaly skin, muscular weakness and atrophy, sterility
vitamin K_2 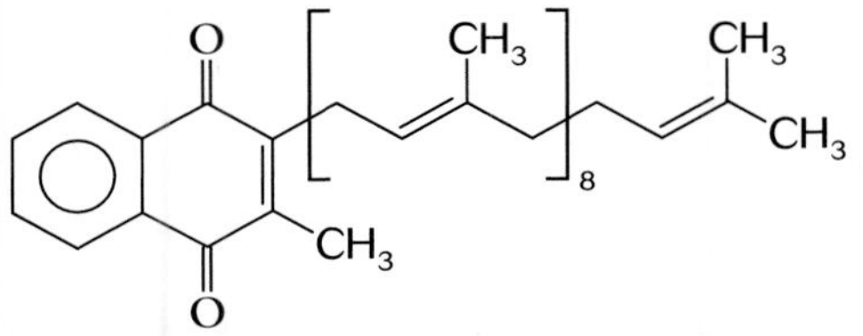	Regulates formation of prothrombin, needed for blood clotting. Deficiency: increased time for blood clotting, a serious problem when a person is bruised, wounded, or undergoing surgery

cialized function: the nucleus synthesizes nucleic acids and stores the cell's genetic information; lysosomes digest macromolecules into their smaller components, which are then recycled by cells; mitochondria metabolize carbohydrates and lipids to generate energy; the endoplasmic reticulum synthesizes proteins and lipids; and the Golgi apparatus synthesizes oligosaccharides and adds to them lipids and proteins to form glycolipids and glycoproteins.

These membranes are not inert barriers. They are highly selective permeability barriers that regulate the molecular and ionic composition within cells and organelles:

- Membranes control the entry into cells and organelles of the materials required for synthesizing biomolecules, the metabolic fuels to generate energy, and other materials to be modified or broken down. Membranes also control the exit of products from organelles for use elsewhere in the cell and from cells for use elsewhere in the organism and the exit of waste products of cellular processes.

- Membranes play the central role in cell recognition (Section 18.6).
- Membranes take part in cell communication. Receptor molecules in membranes receive information from other cells—in the form of hormones, for example—and translate it into molecular changes within the cell.
- Membranes maintain different concentrations of ions on each side of the membrane barrier. These concentration gradients are used to generate chemical or electrical signals for the transmission of nerve impulses and muscle action and to transport other ions across the membrane.

Membrane Structure

Biological membranes consist mostly of lipids and proteins, the relative amounts varying considerably from one type of membrane to another. At one extreme, the myelin sheath—multiple layers of membranes that insulate nerve cells in some parts of the nervous system—is about 80% lipid and 20% protein by weight. At the other extreme, the inner membrane of the mitochondrion is about 20% lipid and 80% protein. Most membranes, such as the plasma membrane of human erythrocytes (red blood cells), contain about equal amounts of lipid and protein. The typical membrane is also from about 2 to 10% carbohydrate.

The lipids of membranes are the amphipathic lipids—glycerophospholipids, sphingolipids, and cholesterol—all of which have:

- a long hydrophobic (nonpolar) part, or **tail**—the two hydrocarbon chains in glycerophospholipids and sphingolipids or the steroid ring structure with its attached alkyl group in cholesterol
- a short hydrophilic (polar or ionic) part, or **head**—the phosphodiester group in glycerophospholipids and sphingophospholipids, the saccharide in sphingoglycolipids, or the hydroxyl group in cholesterol

Figure 19.6 shows the **fluid-mosaic model** that biochemists use to conceptualize biological membranes. The membrane lipids are organized into a **lipid bilayer.** That is, two lipid layers, each having a hydrophobic side and a hydrophilic side, are arranged with the hydrophobic sides in contact and the hydrophilic sides forming the inner and outer surfaces of the membrane, in contact with the internal and external aqueous environments. The driving force for formation of this lipid bilayer comprises secondary attractive and repulsive forces. The orientation of the tails toward the membrane interior and of the heads toward the inner and outer membrane surfaces:

- prevents the repulsive interactions between the hydrophobic tails and the intracellular and extracellular aqueous environments;
- maximizes the attractive forces between the hydrophobic tails, both within each layer and between the two layers of the bilayer;
- maximizes the attractive forces of the hydrophilic heads with each other, as well as between the hydrophilic heads and the intracellular and extracellular aqueous environments.

Stereospecificity of the amphipathic lipids is important in their ability to aggregate together into the lipid bilayer. All of the glycerophospholipids and sphingolipids are L-enantiomers (C2 in the glycerol and sphingosine compo-

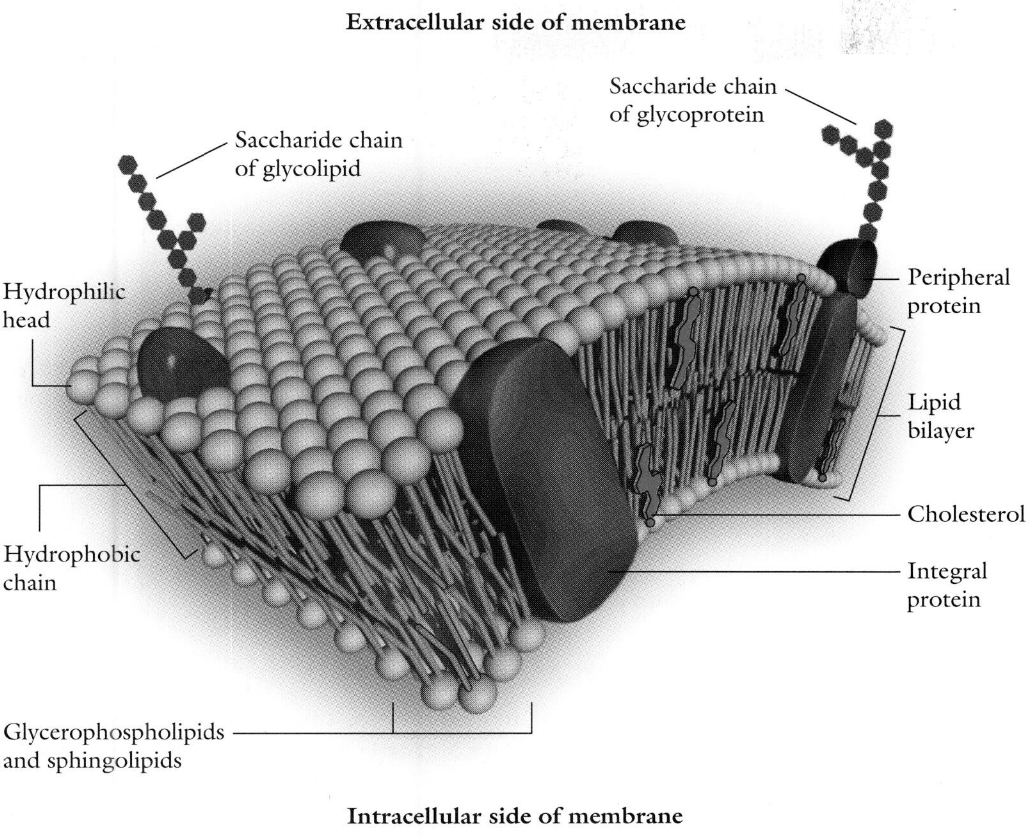

Figure 19.6 Fluid-mosaic model of the cell membrane.

nents is a tetrahedral stereocenter). As stated earlier, cholesterol is one specific stereoisomer of a possible 256 stereoisomers (Section 17.7).

Membrane lipids establish the compartments in which the cell's metabolic processes take place, but membrane proteins play the crucial role in mediating those processes, by functioning as enzymes and as transporters of molecules and ions through the membrane. There are two general types of membrane proteins: integral and peripheral proteins. **Integral,** or **intrinsic, proteins** are embedded in the membrane. Some penetrate only part way through the lipid bilayer, on one side or the other of the membrane; others penetrate completely from one side to the other. **Peripheral,** or **extrinsic, proteins** are located on the membrane surfaces and do not penetrate into the membrane. Membrane proteins associate with the lipid bilayer through secondary attractive forces. Peripheral proteins undergo ionic and hydrogen-bond interactions with the lipid heads at the membrane surface. Integral proteins have both hydrophobic parts, which allow them to penetrate the hydrophobic interior of the lipid bilayer, and hydrophilic parts, which prefer contact with the aqueous medium.

The carbohydrates of membranes are on the extracellular hydrophilic surfaces, covalently bound to membrane lipids as glycolipids and to proteins as glycoproteins. The carbohydrate parts of glycolipids and glycoproteins serve as the receptor sites in cell recognition and cell communication processes.

The fluid-mosaic model of membrane structure seen in Figure 19.6 is aptly named. The lipid bilayer is a mosaic—a complex pattern of different lipids and proteins. Membranes are structurally and functionally asymmetric. The outer and inner surfaces have different components and functions, and each surface is itself asymmetric, with different components and functions at different locations. The lipid bilayer is fluid because the individual lipid and protein molecules have lateral mobility: they can diffuse sideways in the plane of each layer of the bilayer. This mobility is possible because the lipids and proteins are held together by secondary attractive forces, not covalent bonds.

The fluidity of membranes varies with membrane composition and temperature:

- Shorter and unsaturated fatty acid components in the glycerophospholipids and sphingolipids reduce secondary attractive forces and increase fluidity.
- Lower temperatures decrease fluidity through decreased motion of the lipid tails.
- Cholesterol modulates the fluidity of the membrane. Low cholesterol levels strengthen the membrane, but higher levels make the membrane too rigid.

The precise fluidity needed for a membrane depends on its function. The relative amounts of cholesterol, saturated chains, and unsaturated chains vary between species of organisms, among cell membranes and organelles in the same organism, and even in the same type of membrane over time with changes in temperature and dietary lipid composition.

Transport Through Membranes

Small molecules such as water, inorganic ions, simple carbohydrates, amino acids, and lipids pass through cell and organelle membranes by three processes: simple transport, facilitated transport, and active transport (Figure 19.7).

Simple transport (diffusion) is the passage of a solute across a membrane along a concentration gradient from the high-concentration side to the low-concentration side. Simple transport accommodates only a few small mol-

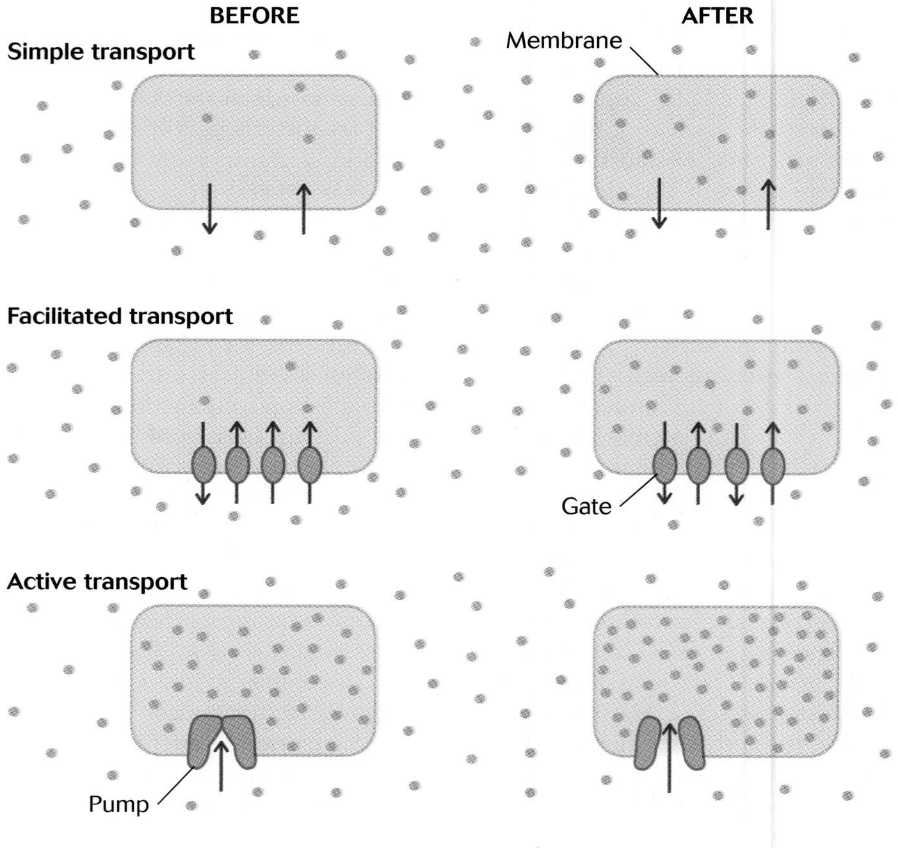

Figure 19.7 Transport across a biological membrane. Both simple and facilitated transport equalize solute concentrations across a membrane. Active transport proceeds against a concentration gradient and either maintains or increases the concentration gradient.

ecules such as O_2, N_2, H_2O, urea (NH_2CONH_2), and ethanol. For ions and larger polar molecules, simple transport is prevented by the large secondary repulsive forces encountered in the hydrophobic regions of the lipid bilayer. Transport of these solutes requires facilitated or active transport.

Facilitated transport also is along a concentration gradient, but, in contrast with simple transport, specific integral proteins—called **transporters** or **permeases**—facilitate and speed up the transport of solute. Permeases act as selective **gates,** or **channels,** for the binding and transport of specific solutes. Glucose, chloride ion (Cl^-), and bicarbonate ion (HCO_3^-), for example, cross membranes by facilitated transport.

Active transport is transport across a membrane against a concentration gradient, from the low-concentration side to the high-concentration side. Like facilitated transport, it requires specific permeases. In active transport, the permeases act as **pumps** to drive solute in the energetically unfavorable direction against the concentration gradient. Active transport requires energy, which is supplied by coupling the transport process to a metabolic process that is energetically favorable. For example, nerve-impulse transmission and muscle contraction require nerve and muscle cells to maintain a lower concentration of Na^+ and a higher concentration of K^+ in the cell than in the extracellular fluid (see Figure 19.7). The energetically unfavorable transport of Na^+ out and K^+ in (called the **Na^+/K^+ pump**) is coupled with and balanced by the energetically favorable hydrolysis of adenosine triphosphate (ATP) to adenosine diphosphate (ADP). Many other ions and molecules, including H^+, Ca^{2+}, amino acids, and monosaccharides pass through cell and organelle membranes by active transport.

▶▶ The function of the Na^+/K^+ pump in nerve membranes is described in Section 26.5.

Macromolecules, such as polysaccharides and proteins, and large particles, such as lipoproteins (carriers of lipids; Section 26.15), are transported into and out of cells and organelles by a different mechanism. The process is called **exocytosis** for the export of material and **endocytosis** for the import of material (Figure 19.8). Macromolecules and other materials for export are packaged into membrane-enclosed **vesicles** that move to the cell membrane. The lipid

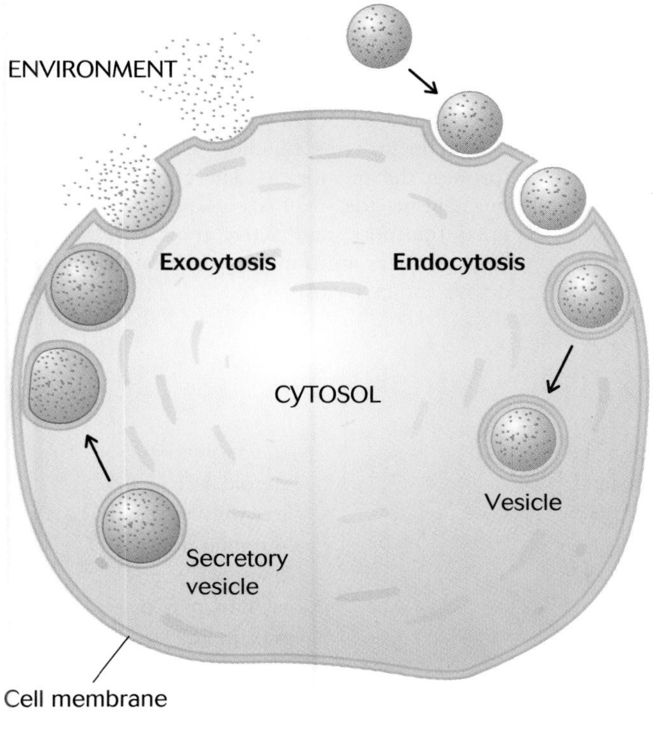

Figure 19.8 In exocytosis (left side of cell), a secretory vesicle merges with the cell wall and releases its contents into the environment. In endocytosis (right side of cell), the membrane forms a pocket to receive material from the environment. The pocket closes to become a vesicle that moves into the cell interior and subsequently releases its contents.

bilayers of the vesicle and cell membranes merge together, an opening develops to the outside of the cell, and the vesicle contents are released. Endocytosis proceeds in essentially the reverse manner. The cell or organelle membrane forms a pocket that receives material from the environment. The pocket develops into a vesicle that moves into the cell or organelle and subsequently releases its contents.

Summary

Classification of Lipids The lipids are biomolecules soluble in nonpolar or low-polarity solvents. They are classified as hydrolyzable or nonhydrolyzable, depending on the presence or absence of hydrolyzable groups. Triacylglycerols, waxes, glycerophospholipids, and sphingolipids are the hydrolyzable lipids; steroids, eicosanoids, and fat-soluble vitamins are the nonhydrolyzable lipids.

Fatty Acids Fatty acids are the carboxylic acid components of hydrolyzable lipids that usually contain from 12 to 24 carbons. They are almost exclusively linear (unbranched) acids with an even number of carbons. Saturated fatty acids contain no C=C double bonds. Unsaturated fatty acids contain one or more cis C=C double bonds: the monounsaturated and polyunsaturated fatty acids.

Triacylglycerols Animal fats and plant oils are triacylglycerols, the triesters formed by esterification of glycerol with fatty acids. Animal fats are generally richer in saturated fatty acids, and plant oils are richer in unsaturated fatty acids. These lipids undergo hydrolysis (including saponification and digestion), catalytic hydrogenation, and rancidity reactions. Triacylglycerols are the main storage form of energy in animals. Energy is produced more efficiently by triacylglycerol metabolism than by glycogen metabolism.

Waxes Waxes, the esters of long-chain monohydric alcohols and fatty acids, perform a variety of protective functions in plants and animals.

Amphipathic Hydrolyzable Lipids Glycerophospholipids, amphipathic lipids containing a highly hydrophilic ionic group, are the major component of biological membranes. Sphingolipids, amphipathic lipids based on sphingosine instead of glycerol, are another component.

Steroids Steroids are nonhydrolyzable lipids that contain the steroid ring structure, consisting of four fused rings, of which three are six-membered rings and one is a five-membered ring. Cholesterol, steroid hormones, and bile salts are steroids. Cholesterol is a component of biological membranes and is the precursor to all other steroids. The adrenocortical and sex hormones are synthesized in endocrine glands and transported in the bloodstream to target tissues, where they regulate various functions ranging from the control of metabolism to the development of the sex organs and secondary sexual characteristics. Hormones are effective at very low concentrations. Bile salts are needed for the efficient digestion of hydrolyzable lipids and absorption of fat-soluble vitamins.

Eicosanoids The eicosanoids—consisting of prostaglandins, leukotrienes, and thromboxanes—act as local hormones. Unlike hormones, eicosanoids regulate functions in the tissues in which they are produced. They mediate such functions as the inflammatory response, muscle contraction, and blood clotting.

Fat-Soluble Vitamins Vitamins are organic compounds required in trace amounts for an organism's normal metabolism but cannot be synthesized by the organism and must be included in the diet. The fat-soluble vitamins are A, D, E, and K; the water-soluble vitamins are B and C.

Biological Membranes Biological membranes consist mostly of lipids and proteins. The lipids utilized for membranes are the amphipathic lipids—glycerophospholipids, sphingolipids, and cholesterol. Membrane lipids are organized into a bilayer in which membrane proteins are embedded. The proteins function as permeases, carrying solutes through the membrane. There are three types of transport mechanisms for small-sized solutes: simple transport, facilitated transport, and active transport. Macromolecules and large particles are transported through membranes by endocytosis and exocytosis.

Key Words

bile salt, p. 572
biological membrane, p. 576
cholesterol, p. 571
eicosanoid, p. 575
fat-soluble vitamin, p. 576
fatty acid, p. 558
fluid-mosaic model, p. 578
glycerophospholipid, p. 568
hormone, p. 571
hydrolyzable lipid, p. 557
leukotriene, p. 575
lipid, p. 557
lipid bilayer, p. 578
local hormone, p. 575
permease, p. 581
prostaglandin, p. 575
sphingolipid, p. 569
steroid, p. 571
triacylglycerol, p. 561
thromboxane, p. 575
transport, p. 581
vitamin, p. 576
wax, p. 567

Exercises

Fatty Acids

19.1 Which of the following carboxylic acids are present in animal and plant lipids?

$$CH_3(CH_2)_{16}COOH$$
1

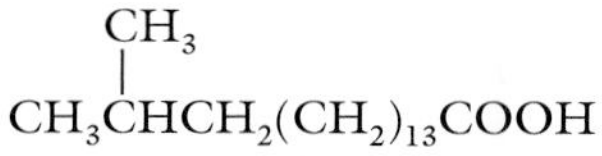

2

$$cis\text{-}CH_3(CH_2)_5CH{=}CH(CH_2)_8COOH$$
3

19.2 Which of the following carboxylic acids are present in animal and plant lipids?

$$CH_3(CH_2)_{13}COOH$$
1

$$CH_3(CH_2)_{14}COOH$$
2

$$trans\text{-}CH_3(CH_2)_5CH{=}CH(CH_2)_7COOH$$
3

19.3 What structural features distinguish between saturated, monounsaturated, and polyunsaturated fatty acids?

19.4 Categorize the following fatty acids as saturated, monounsaturated, polyunsaturated, ω-3, ω-6, and ω-9:
(a) $CH_3(CH_2)_{16}COOH$;
(b) $CH_3(CH_2)_4(CH{=}CHCH_2)_2(CH_2)_6COOH$;
(c) $CH_3CH_2(CH{=}CHCH_2)_3(CH_2)_6COOH$;
(d) $CH_3(CH_2)_7CH{=}CH(CH_2)_7COOH$.

19.5 Which of the following compounds has the higher melting point and why?

$$cis\text{-}CH_3(CH_2)_5CH{=}CH(CH_2)_7COOH$$
1

$$CH_3(CH_2)_{14}COOH$$
2

19.6 Which of the following compounds has the higher melting point and why?

$$trans\text{-}CH_3(CH_2)_5CH{=}CH(CH_2)_7COOH$$
1

$$cis\text{-}CH_3(CH_2)_5CH{=}CH(CH_2)_7COOH$$
2

Structure and Physical Properties of Triacylglycerols

19.7 Draw the structure of the triacylglycerol containing equimolar amounts of lauric, myristic, and oleic acids.

19.8 Draw the structure of the triacylglycerol containing equimolar amounts of palmitic, stearic, and palmitoleic acids.

19.9 Consider triacylglycerols with the following fatty acid compositions: (a) palmitic, stearic, stearic; (b) palmitic, oleic, linoleic; (c) palmitic, stearic, oleic. Place the triacylglycerols in order of melting points. Explain the order.

19.10 Explain why fish triacylglycerols are liquid, whereas beef triacylglycerols are solid.

Chemical Reactions of Triacylglycerols

19.11 Consider the following triacylglycerol:

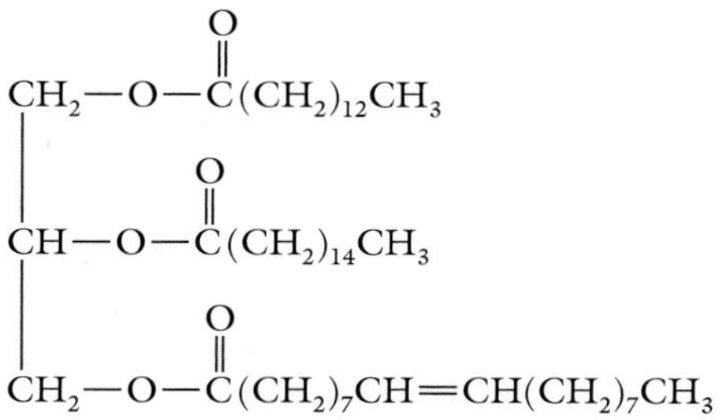

Write equations to describe the following reactions of this compound: (a) saponification (hydrolysis) with NaOH; (b) catalytic hydrogenation (H_2/Pt).

19.12 Write the equation for the acid-catalyzed hydrolysis of the triacylglycerol in Exercise 19.11. How does digestion with lipase differ from the acid-catalyzed hydrolysis?

19.13 Explain why bacterial hydrolysis of the following triacylglycerol produces a malodorous mixture. Write the equation for the reaction.

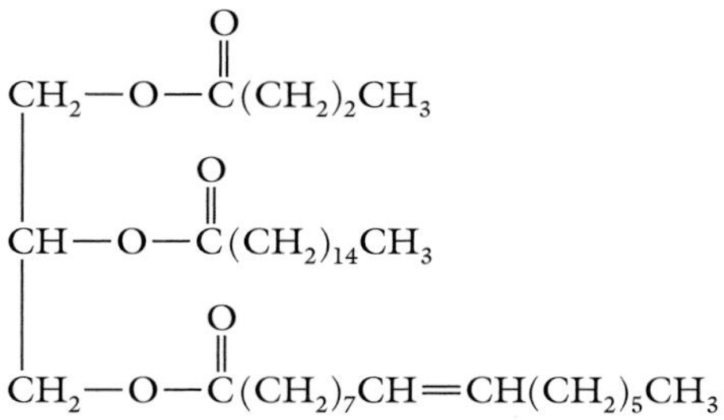

19.14 Explain why air oxidation of the triacylglycerol in Exercise 19.13 produces a malodorous mixture. Write the equation for the reaction.

19.15 What reaction is used to harden corn oil in the manufacture of margarine?

19.16 What reaction is used to produce a soap from beef fat?

Waxes

19.17 One component of the wax obtained from the leaves of a South American tree has the formula $C_{40}H_{80}O_2$. What is the structure of this wax component, 1, 2, or 3?

$$CH_3CH_2CH_2COO(CH_2)_{35}CH_3$$
1

$$CH_3(CH_2)_{16}COO(CH_2)_{21}CH_3$$
2

$$CH_3(CH_2)_{15}COO(CH_2)_{22}CH_3$$
3

19.18 Spermaceti, a fragrant mixture of lipids isolated from sperm whales, was used in formulating cosmetics until banned in the late 1970s. A major component of spermacetti is the wax cetyl palmitate formed from palmitic acid and cetyl alcohol, $CH_3(CH_2)_{15}OH$. Draw the structure of cetyl palmitate.

Amphipathic Hydrolyzable Lipids

19.19 What structural feature distinguishes glycerophospholipids from triacylglycerols?

19.20 What structural feature distinguishes sphingolipids from triacylglycerols?

19.21 Draw the structure of the glycerophospholipid whose hydrolysis yields equimolar amounts of glycerol, myristic acid, oleic acid, phosphoric acid, and serine.

19.22 Draw the structure of the glycerophospholipid whose hydrolysis yields glycerol, lauric acid, linolenic acid, phosphoric acid, and choline.

19.23 What are the products when the following lipid is hydrolyzed in the presence of an acid?

$$CH_2-O-\overset{O}{\overset{\|}{C}}(CH_2)_{14}CH_3$$
$$CH-O-\overset{O}{\overset{\|}{C}}(CH_2)_{16}CH_3$$
$$CH_2-O-\overset{O}{\underset{O^-}{\overset{\|}{P}}}-OCH_2CH_2\overset{+}{N}(CH_3)_3$$

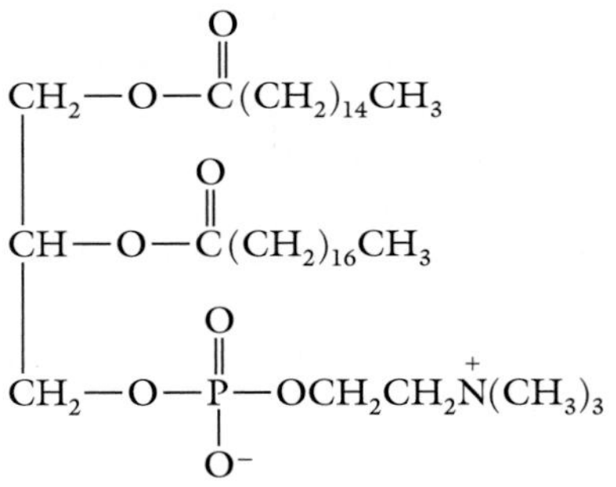

19.24 What are the products when the lipid in Exercise 19.23 undergoes saponification with KOH?

19.25 Draw the structure of the sphingolipid containing linoleic acid, phosphoric acid, and choline.

19.26 Draw the structure of the sphingolipid containing palmitic acid and α-D-glucose.

19.27 What are the products when the following lipid is hydrolyzed in the presence of an acid?

$$HO-CH-CH=CH(CH_2)_{12}CH_3$$
$$CH-\overset{H}{\underset{}{N}}-\overset{O}{\overset{\|}{C}}(CH_2)_7CH=CH(CH_2)_7CH_3$$
$$CH_2-O-\overset{O}{\underset{O^-}{\overset{\|}{P}}}-OCH_2CH_2\overset{+}{N}(CH_3)_3$$

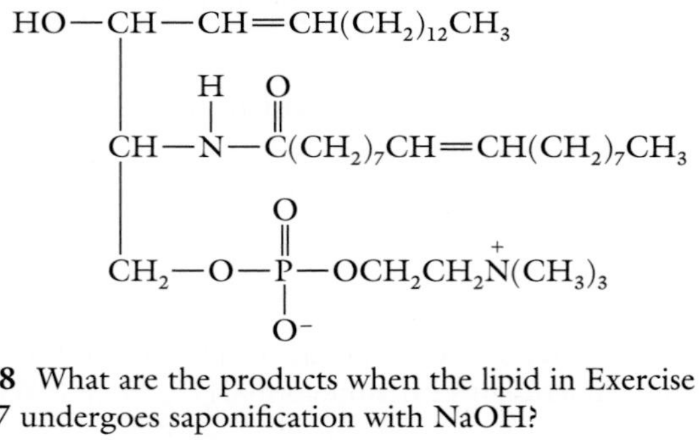

19.28 What are the products when the lipid in Exercise 19.27 undergoes saponification with NaOH?

Steroids

19.29 Why are steroids not classified as hydrolyzable lipids?

19.30 What structural feature is characteristic of all compounds classified as steroids?

19.31 Testosterone is one of many stereoisomers. Draw the structure of testosterone (Figure 19.4) and place an asterisk next to each tetrahedral stereocenter. How many stereoisomers are possible for testosterone?

19.32 Estradiol is one of many stereoisomers. Draw the structure of estradiol (Figure 19.4) and place an asterisk next to each tetrahedral stereocenter. How many stereoisomers are possible for estradiol?

19.33 What are the physiological functions of the female and male sex hormones?

19.34 What are the physiological functions of the adrenocortical hormones?

19.35 What compound is the precursor to the steroid hormones and bile salts?

19.36 What structural differences are there between testosterone and estradiol (Figure 19.4)?

19.37 What structural feature of bile salts distinguishes them from other steroids?

19.38 How do bile salts aid in the digestion of lipids?

Eicosanoids

19.39 Even though leukotriene D_4 (Figure 19.5) has an amide group that undergoes hydrolysis, it is not classified as a hydroyzable lipid. Why?

19.40 What feature is characteristic of all compounds classified as eicosanoids?

19.41 What structural feature distinguishes prostaglandins from leukotrienes?

19.42 What structural feature distinguishes leukotrienes from thromboxanes?

19.43 What is a hormone? Why are eicosanoids called local hormones, whereas steroid hormones are called hormones?

19.44 What are the physiological functions of eicosanoids?

Fat-Soluble Vitamins

19.45 Define vitamin.

19.46 What distinguishes vitamins from hormones?

19.47 What property distinguishes fat-soluble from water-soluble vitamins?

19.48 Why is there a much greater danger of overdosing on fat-soluble vitamins than on water-soluble vitamins?

19.49 Describe the physiological functions of vitamins A, D, E, and K.

19.50 Some textbooks classify vitamin D as a vitamin; others classify it as a steroid hormone. Why?

Biological Membranes

19.51 Where are membranes found in the body?

19.52 What are the functions of biological membranes?

19.53 What lipids are present in biological membranes? What is their function?

19.54 What nonlipids are present in biological membranes? What is their function?

19.55 Why does the flexibility of cell membranes increase with increasing content of unsaturated fatty acids in the glycerophospholipids and sphingolipids?

19.56 Why does the flexibility of cell membranes decrease with increasing content of saturated fatty acids in the glycerophospholipids and sphingolipids?

19.57 Why do ions such as Na^+ and HCO_3^- and molecules such as D-glucose not pass through membranes by simple transport? Such ions and molecules pass through membranes only by facilitated transport or active transport.

19.58 What is the difference between simple transport and facilitated transport?

19.59 Why does active transport require energy?

19.60 Distinguish between integral and peripheral proteins.

19.61 Lactose is transported into cells through integral proteins under conditions in which the intracellular concentration of lactose exceeds the extracellular concentration. Does this constitute simple transport, facilitated transport, or active transport?

19.62 Glucose is transported into cells through integral proteins under conditions in which the extracellular concentration of glucose exceeds the intracellular concentration. Does this constitute simple transport, facilitated transport, or active transport?

Unclassified Exercises

19.63 Which solvent would be least effective in dissolving lipids? $CH_3(CH_2)_4CH_3$; CH_3OH; CCl_4; $C_2H_5OC_2H_5$.

19.64 What is the difference between essential and nonessential fatty acids?

19.65 Give the ω number for each of the following fatty acids: (a) $CH_3(CH_2)_{14}COOH$;
(b) $CH_3(CH_2)_5CH{=}CH(CH_2)_7COOH$;
(c) $CH_3(CH_2)_4(CH{=}CHCH_2)_2(CH_2)_6COOH$.

19.66 Consider the following triacylglycerol:

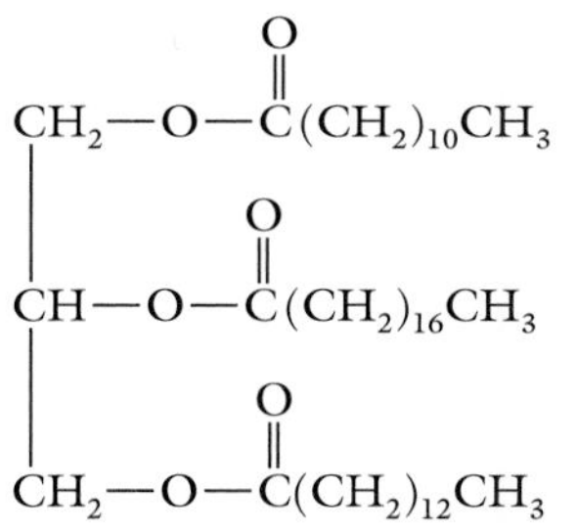

What are the products in each of the following reactions of this compound? (a) Acid-catalyzed hydrolysis; (b) saponification with NaOH; (c) catalytic hydrogenation (H_2/Pt).

19.67 What are the biological functions of waxes?

19.68 Consider the following glycerophospholipid:

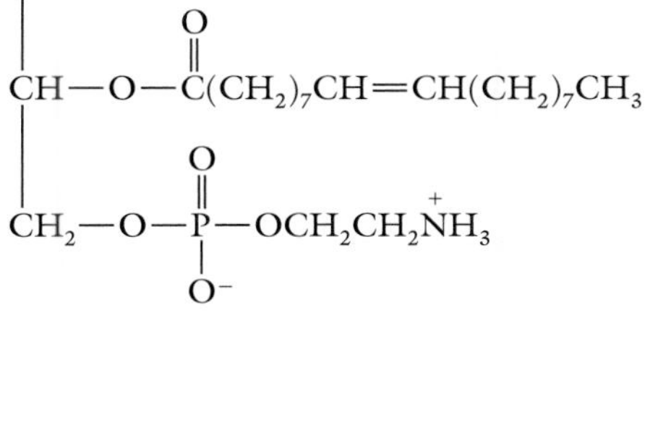

What are the products in each of the following reactions of this compound? (a) Acid-catalyzed hydrolysis; (b) saponification with KOH; (c) catalytic hydrogenation (H_2/Pt).

19.69 How does digestion of triacylglycerols and other hydrolyzable lipids differ from hydrolysis in the laboratory with the use of an acid or base?

19.70 Which of the following lipids are based on glycerol? Which are based on sphingosine? (a) Sphingolipids; (b) leukotrienes; (c) glycerophospholipids; (d) prostaglandins; (e) steroids; (f) triacylglycerols; (g) waxes; (h) sex hormones.

19.71 Which of the following lipids are hydrolyzable? (a) Sphingolipids; (b) leukotrienes; (c) glycerophospholipids; (d) prostaglandins; (e) steroids; (f) triacylglycerols; (g) waxes; (h) sex hormones.

19.72 Which of the following lipids are components of biological membranes? (a) Sphingolipids; (b) leukotrienes; (c) glycerophospholipids; (d) prostaglandins; (e) cholesterol; (f) triacylglycerols; (g) waxes; (h) sex hormones.

19.73 Which of the following lipids undergo saponification? (a) Sphingolipids; (b) leukotrienes; (c) glycerophospholipids; (d) prostaglandins; (e) steroids; (f) triacylglycerols; (g) waxes; (h) sex hormones.

19.74 Write the equation for the reaction of the lipid in Exercise 19.68 with Br_2.

19.75 A triacylglycerol contains equimolar amounts of myristic, palmitic, and stearic acids. Three constitutional isomers exist for this composition. Draw the isomers.

19.76 What reactions are responsible for rancidity in butter and other fats and oils?

19.77 Triacylglycerol A contains equimolar amounts of lauric, stearic, and myristic acids, whereas triacylglycerol B contains equimolar amounts of oleic, stearic, and myristic acids. Air oxidation produces a malodorous mixture from triacylglycerol B but not from triacylglycerol A. Explain.

19.78 What important compounds are synthesized in the body from cholesterol? What other purpose does cholesterol serve?

19.79 Draw the structure of progesterone (Figure 19.4) and place an asterisk next to each tetrahedral stereocenter. How many stereoisomers are possible for progesterone?

19.80 What is the common feature of all lipids in cell membranes?

19.81 Why does the lipid bilayer not include triacylglycerols?

19.82 Oxygen is transported into cells without the aid of proteins under conditions in which the extracellular concentration of oxygen exceeds the intracellular concentration. Does this constitute simple transport, facilitated transport, or active transport?

19.83 Certain cells that line the stomach pump out protons against a concentration gradient of more than 1 million-to-1. Does this constitute simple transport, facilitated transport, or active transport?

Chemical Connections

19.84 (a) Compare the energies generated by the oxidation (metabolism) of 100 g each of dry triacylglycerol and dry glycogen. (Refer to Section 19.4.)

(b) In the body, triacylglycerol is stored in anhydrous form but glycogen is hydrated (67% water). Compare the energies generated by the oxidation of 100 g each of dry triacylglycerol and hydrated glycogen.

19.85 Coconut oil is an atypical plant triacylglycerol. Explain why coconut oil is a liquid instead of a solid in spite of its very high saturated fatty acid composition (Table 19.2).

19.86 Triacylglycerol A contains equimolar amounts of lauric, stearic, and myristic acids, whereas triacylglycerol B contains equimolar amounts of butanoic, stearic, and myristic acids. Bacterial hydrolysis produces a malodorous mixture from triacylglycerol B but not from triacylglycerol A. Explain.

19.87 One of the cholesteryl esters used to transport cholesterol in the blood is cholesteryl linoleate, the ester of cholesterol and linoleic acid. Draw the structure of cholesteryl linoleate.

19.88 Membrane compositions of fish and other cold-blooded animals change when their environmental temperature is lowered. The unsaturated fatty acid content of the lipids in the cell membranes increases when the organism becomes adapted to the lower temperature. What is the purpose of this increase?

19.89 A clogged kitchen sink is often due to the buildup of fats in the trap. Explain how the addition of an aqueous NaOH solution eliminates these clogs.

19.90 Punctures and other mechanical disruptions in cell membranes are quickly resealed. What mechanism is responsible for the self-sealing property of cell membranes?

19.91 Imagine the planet Gibo in a far-off galaxy where the life forms have a similar appearance to those on Earth. However, there is one huge difference: heptane on Gibo has the role played by water on Earth. If the molecules that form cell membranes are the same as those on Earth, what is the major difference in the construction of cell membranes on Gibo compared with Earth?

CHAPTER 20

PROTEINS

CHEMISTRY IN YOUR FUTURE

As a USDA meat inspector, you are often asked about the outbreaks of bovine spongiform encephalopathy ("mad cow disease") that occurred in Great Britain. The pathogen is neither a virus nor a bacterium. One day, you read about the 1997 Nobel Prize in Medicine awarded to Stanley Prusiner, who believes that the disease is caused by a prion, a rogue form of a protein that is normally present in mammals. The protein can exist in two conformations: the normal, harmless conformation and an improperly folded one that causes mad cow disease. Evidence that humans might contract a version of the disease by ingesting beef from diseased cattle created something of a panic in Great Britain and concern in other countries. This chapter describes the importance of protein conformation and folding and their effects on biological functions.

LEARNING OBJECTIVES

- Draw and categorize the α-amino acids.
- Describe the effect of pH changes on α-amino acid structure.
- Describe the structures of peptides, and write equations for their synthesis from α-amino acids.
- Describe and write equations for reactions of peptides.
- Distinguish between the primary, secondary, tertiary, and quaternary structures of proteins.
- Describe the forces that stabilize proteins.
- Describe the four types of protein conformation.
- Describe the differences between fibrous and globular proteins.
- Describe the structures and functions of some proteins (α-keratin, collagen, silk fibroin, myoglobin, hemoglobin, and lysozyme).
- Describe mutations and how they affect living organisms.
- Describe protein denaturation by environmental changes.

The primary functions of the biomolecules that you have studied thus far—the carbohydrates and lipids—are to provide an organism with energy, with precursors to other biomolecules, and with molecules with which to construct cell membranes. The primary functions of proteins are the building and maintenance of the organism. Proteins are the most plentiful organic chemicals in the body, making up more than half of its dry weight. The protein family comprises about 100,000 different compounds or more—a far greater number than that of any other family of biomolecules. And proteins are responsible for the greatest range of functions. We can summarize the functions of proteins as follows:

▶▶ Enzymes are considered in Section 22.5.

- **Catalytic proteins,** or **enzymes,** catalyze the synthesis and utilization of proteins (including enzymes themselves), carbohydrates, lipids, nucleic acids, and almost all other biomolecules. The different enzymes that an organism produces are determined by heredity and are in fact what distinguishes one species from another and one individual member of a species from another. In Chapter 21, we shall see how deoxynucleic acids (DNAs) direct the synthesis of enzymes and all other proteins.
- **Transport proteins** bind and carry specific molecules or ions from place to place. Hemoglobin transports oxygen from the lungs to other tissues (Section 26.13). Integral membrane proteins transport molecules and ions across cell and organelle membranes (Section 19.10).

A PICTURE OF HEALTH

Examples of Proteins in the Body

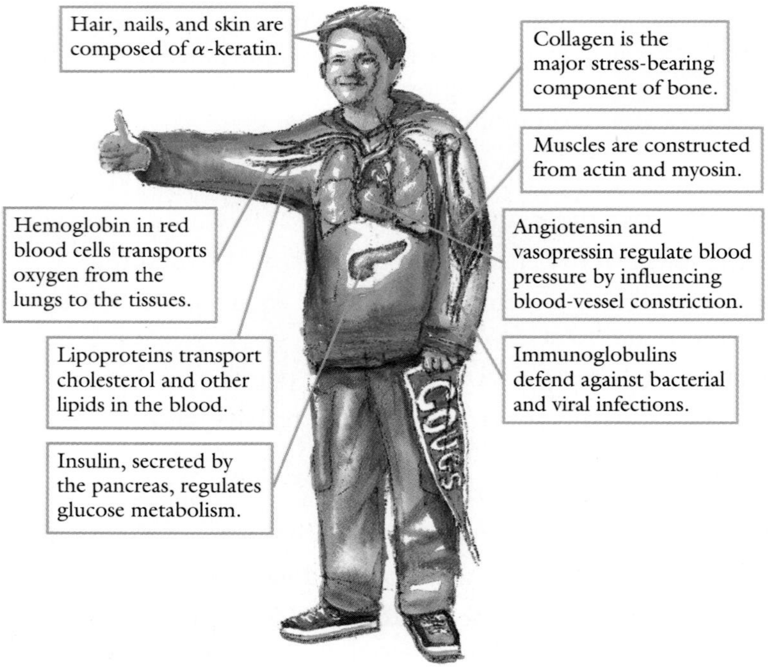
Hair, nails, and skin are composed of α-keratin.
Collagen is the major stress-bearing component of bone.
Muscles are constructed from actin and myosin.
Hemoglobin in red blood cells transports oxygen from the lungs to the tissues.
Angiotensin and vasopressin regulate blood pressure by influencing blood-vessel constriction.
Lipoproteins transport cholesterol and other lipids in the blood.
Immunoglobulins defend against bacterial and viral infections.
Insulin, secreted by the pancreas, regulates glucose metabolism.
COUGS

- **Regulatory proteins** control cellular activity. In addition to lipid hormones (considered in Section 19.7), there are protein hormones (for example, insulin, which regulates glucose metabolism; Section 26.11). Some neurotransmitters, which relay information within the nervous system, are proteins. Proteins called inducers and repressors control gene expression, the way in which information in DNA is translated into the synthesis of proteins.
- **Structural proteins** give the physical shape, and the strength to maintain it, to structures in animals—the role played by cellulose in plants. The protein collagen is the major component of bone, tendons, and cartilage; whereas hair, fingernails, feathers, and horns consist mostly of the protein α-keratin.
- **Contractile proteins** provide cells and organisms with the ability to change shape and move. Muscles contract and cells change shape through the actions of actin and myosin. A sperm cell is propelled by its flagellum, composed of the protein tubulin, and an ovum passes through a fallopian tube assisted by movements of cilia, also composed of tubulin.
- **Protective proteins** defend against invaders and prevent or minimize damage subsequent to injury. Immunoglobulins, or antibodies, recognize and inactivate invading bacteria, viruses, and other foreign

proteins. The blood-clotting proteins fibrinogen and thrombin prevent loss of blood when the vascular system is damaged. Toxins and venoms serve defensive or predatory functions for some plants and animals.

- **Storage proteins** provide a reservoir of nitrogen and other nutrients, especially when external sources are low or absent. Ovalbumin, the protein of egg white, supplies the developing bird embryo with nutrients during its isolation in the egg. Casein in milk serves the same function for mammalian infants. The seeds of many plants contain proteins needed for germination and early plant development. Ferritin serves as a storage form of iron.

Polypeptides are polymers constructed by the bonding together of α-amino acids. Proteins are naturally occurring polypeptides, usually aggregated with other polypeptides or with other types of molecules or ions or with both. In this chapter, we examine the structures and properties of the α-amino acids and the polypeptides formed from them. With this background, we turn to a consideration of the three-dimensional structures of proteins, which determine their biological functions; selected proteins serve as examples of the relation between structure and function. The chapter concludes with a discussion of what happens when protein structure is altered because of mutations in genes and ways in which protein function can be lost through environmental changes.

20.1 α-AMINO ACIDS

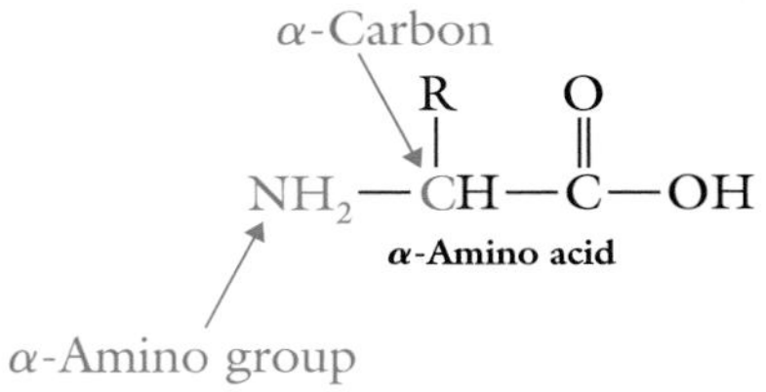

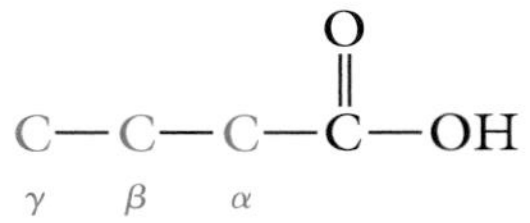

The building blocks for naturally occurring polypeptides are **α-amino acids,** compounds containing both a carboxyl group and an amino group attached to the same carbon. The amino acids are called α-amino acids because the first carbon attached to the carboxyl carbon is called the α-carbon. Other amino acids are known; for example, β- and γ-amino acids have the amino group located on the second and third carbons from the carboxyl carbon, respectively (see structures in margin). Nearly all polypeptides of all species of plants and animals, however, are constructed from the 20 α-amino acids shown in Table 20.1 on the next two pages. Whenever we use the term amino acid in discussing polypeptides and proteins, we mean α-amino acid, unless otherwise noted.

Mammals require all 20 α-amino acids for protein synthesis but are unable to synthesize half of them. These 10 **essential α-amino acids** must be obtained from the diet (Table 20.1). Diets containing animal proteins supply all of the essential α-amino acids. Vegetarians must exercise extra care to obtain sufficient amounts of all of the essential α-amino acids (Box 20.1 on page 592).

The α-amino acids are categorized according to the R group, or **side group,** on the α-carbon:

- **Nonpolar neutral** α-amino acids have neutral hydrophobic side groups.
- **Polar neutral** α-amino acids have neutral hydrophilic side groups.
- **Polar acidic** α-amino acids have acidic hydrophilic side groups.
- **Polar basic** α-amino acids have basic hydrophilic side groups.

The structures and functions of proteins strongly depend on the side groups of their constituent α-amino acids.

Proline differs from all other α-amino acids in having the nitrogen atom of its α-amino group as part of a ring structure. The α-amino group of proline is thus a secondary amine, whereas the α-amino groups of all other α-amino acids are primary amines. Note that, in Table 20.1, the structure shown for proline in the "side group (R)" column is the complete structure of proline; for all other α-amino acids, only the structure of the side group is given.

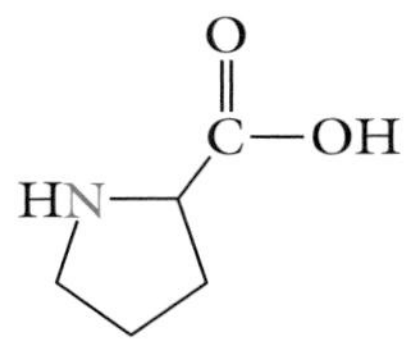

Proline

TABLE 20.1 α-Amino Acids $NH_2—\overset{\overset{R}{|}}{C}H—\overset{\overset{O}{\|}}{C}—OH$

Name	Abbreviations		Side Group (R)	Isoelectric Point (pI)
NONPOLAR NEUTRAL				
glycine	Gly	G	H—	5.97
alanine	Ala	A	CH_3—	6.01
valine[a]	Val	V	$(CH_3)_2CH$—	5.96
leucine[a]	Leu	L	$CH_3\overset{\overset{CH_3}{|}}{C}HCH_2$—	5.98
isoleucine[a]	Ile	I	$CH_3CH_2\overset{\overset{CH_3}{|}}{C}H$—	6.02
phenylalanine[a]	Phe	F	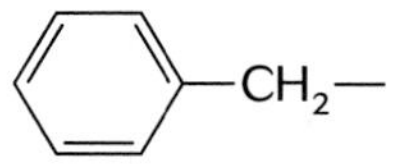	5.48
methionine[a]	Met	M	$CH_3SCH_2CH_2$—	5.74
proline[b]	Pro	P	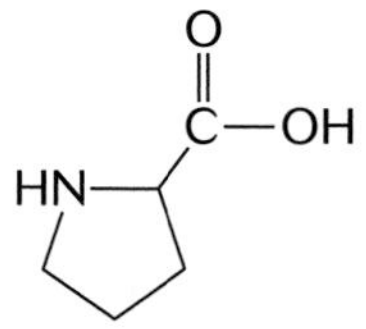	6.30
tryptophan[a]	Trp	W	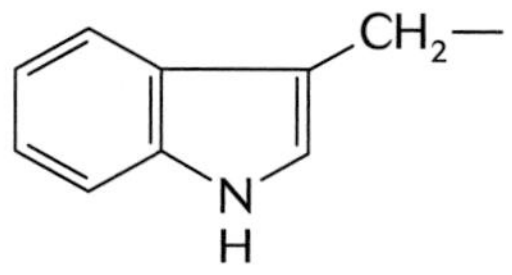	5.88
POLAR NEUTRAL				
cysteine	Cys	C	$HSCH_2$—	5.05
serine	Ser	S	$HOCH_2$—	5.68
threonine[a]	Thr	T	$CH_3\overset{\overset{OH}{|}}{C}H$—	5.60
asparagine	Asn	N	$H_2N\overset{\overset{O}{\|}}{C}CH_2$—	5.41
glutamine	Gln	Q	$H_2N\overset{\overset{O}{\|}}{C}CH_2CH_2$—	5.65
tyrosine	Tyr	Y	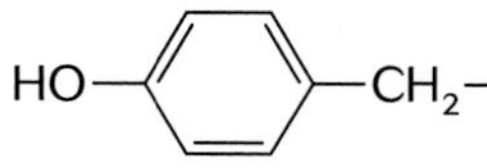	5.66

(Continued at the top of following page.)

The common names of the α-amino acids are also the accepted IUPAC names. Each α-amino acid has a standard three-letter and one-letter abbreviation as well. The three-letter abbreviations are the first three letters of the names of all except four α-amino acids. For those four α-amino acids, the first letter of the name is combined with two other letters in the name. The three-letter abbreviations are more widely used and easier to remember than the one-letter abbreviations and will be used in this text. Biochemists use the one-letter symbols mainly for denoting the often very long sequences of α-amino acids in polypeptides and proteins.

TABLE 20.1 Continued

Name	Abbreviations		Side Group (R)	Isoelectric Point (pI)
POLAR ACIDIC				
aspartic acid	Asp	D	$\mathrm{HO\overset{\overset{O}{\|}}{C}CH_2-}$	2.77
glutamic acid	Glu	E	$\mathrm{HO\overset{\overset{O}{\|}}{C}CH_2CH_2-}$	3.22
POLAR BASIC				
lysine[a]	Lys	K	$\mathrm{H_2NCH_2CH_2CH_2CH_2-}$	9.74
arginine[a]	Arg	R	$\mathrm{H_2N\overset{\overset{NH}{\|}}{C}NHCH_2CH_2CH_2-}$	10.76
histidine[a]	His	H	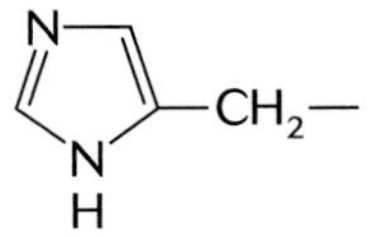	7.59

[a] Essential for mammals.

[b] Complete structure of proline is shown.

The α-carbon of all α-amino acids except glycine is a tetrahedral stereocenter: the four different groups attached to the tetrahedral stereocenter are the amino group, carboxyl group, H, and R. The α-carbon of glycine is not a tetrahedral stereocenter, because two of the attached groups are the same (hydrogen). Thus, all α-amino acids except glycine can exist as a pair of enantiomers, nonsuperimposable mirror-image molecules (Section 17.2). Figure 20.1 shows D-alanine and L-alanine; the D- and L-designations are based on the convention for D- and L-glyceraldehyde, as described in Section 17.4 and shown in Figure 20.1. The carbon substituents, COOH and R, are placed vertically, with the more highly oxidized substituent (COOH) placed at the top. In the D-enantiomers, the heteroatom substituent—NH_2 for an amino acid and OH for glyceraldehyde—is on the right; in the L-enantiomers, it is on the left.

With very rare exceptions, only the L-α-amino acids exist in the proteins of plants and animals. Some bacterial cell walls contain D-alanine and D-glutamic acid.

Unless indicated otherwise, the term amino acid in this book refers to the L-α-enantiomer.

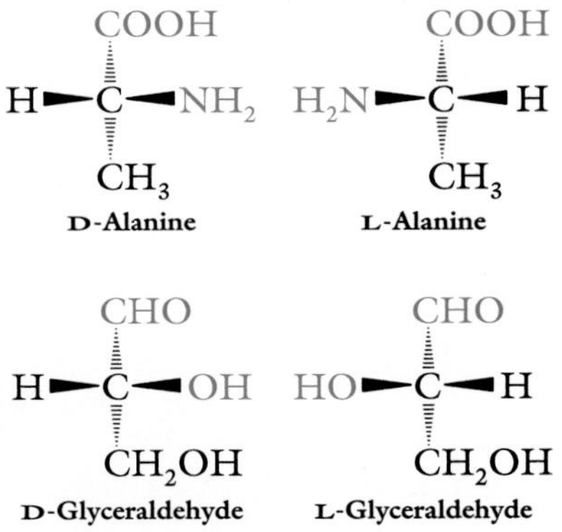

Figure 20.1 The D- and L-enantiomers of alanine and their relation to D- and L-glyceraldehyde.

20.2 ZWITTERIONIC STRUCTURE OF α-AMINO ACIDS

The structures representing α-amino acids to this point are not completely correct. An α-amino acid exists not as an uncharged molecule but as a **zwitterion,** or **dipolar ion,** which we can visualize to be formed by intramolecular proton transfer between the acidic —COOH and basic —NH_2 groups:

$$\mathrm{NH_2-\overset{\overset{R}{|}}{CH}-\overset{\overset{O}{\|}}{C}-OH \longrightarrow {}^{+}NH_3-\overset{\overset{R}{|}}{CH}-\overset{\overset{O}{\|}}{C}-O^-}$$

Uncharged (Nonexistent) — **Zwitterion** (Exists in both solid and solution)

20.1 Chemistry Within Us

Proteins in the Diet

Our bodies store significant amounts of lipids (mainly as triacylglycerols) and carbohydrates (as glycogen) to supply energy but only minimal reserves of α-amino acids for the production of proteins. This situation necessitates an almost daily intake of the essential α-amino acids, in the requisite amounts. Protein synthesis can proceed only when all 20 α-amino acids are present. It also calls for specific synthesis of nonessential α-amino acids from the essential α-amino acids or other precursors.

Dietary proteins are classified as **complete** or **incomplete proteins,** depending on their α-amino acid content. Complete proteins supply all of the essential α-amino acids in the amounts needed for protein synthesis; incomplete proteins do not. Animal proteins are complete proteins. For example, human milk protein contains all of the α-amino acids, both essential and nonessential, in the proportions needed by growing infants. Animal proteins such as beef, pork, poultry, seafood, cow's milk, and hen's eggs have very nearly the same nutritional value as human milk protein.

Plant proteins are incomplete proteins, varying in their nutritional value. All essential α-amino acids can be obtained from plant sources, but no plant provides all of them in sufficient amounts: different plant proteins are deficient in different α-amino acids. Proteins from grains such as corn, oats, rice, and wheat are low in lysine, and some are also low in tryptophan; proteins from legumes such as beans and peas are low in methionine. An all-grain or all-legume diet does not supply all essential α-amino acids in the required amounts.

A vegetarian diet must contain a complementary mixture of plant proteins so as to supply all essential α-amino acids. This mixture can be accomplished by combining a grain with a legume: the legume is low in methionine but high in lysine and tryptophan; the grain is high in methionine but low in lysine (and some are also low in tryptophan). Nuts and seeds such as almonds and walnuts have protein compositions similar to that of grains and can be used to complement legumes. However, nuts and seeds are high in lipids.

More of the world's population is vegetarian than meat eating. Some people follow a vegetarian diet for religious or ethical convictions; others do so for health reasons (to restrict cholesterol, saturated fat, and total fat intake); but most do so for economic reasons. Animal proteins are more expensive to produce than plant proteins. Many cultures have discovered specific dietary combinations of plant proteins that supply all the α-amino acids required for normal human growth: in Mexico, corn tortillas and refried beans; in Japan, rice and soybean curd (tofu); in the U.S. South, rice and black-eyed peas. Each of these diets provides a balance of the required α-amino acids.

The protein in steak is complete. Although the proteins in tofu and rice are individually incomplete, the combination of tofu and rice supplies all the essential α-amino acids.

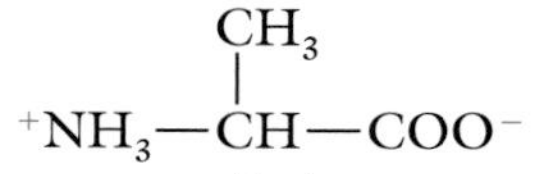

L-Alanine

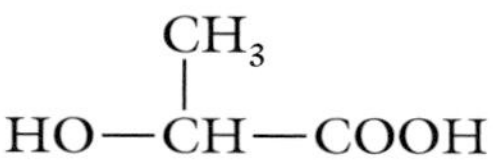

L-Lactic acid

A zwitterion contains one negative (carboxylate) and one positive (ammonium) charge center and has a net charge of zero. The zwitterion structure of α-amino acids is not unexpected: an acid-base reaction takes place whenever a carboxylic acid and an amine are mixed together (Section 16.5). In this case, the carboxyl and amino groups are part of the same molecule.

Strong secondary attractive forces between the negative and the positive charge centers of zwitterions result in high melting points. For example, the melting point of L-alanine (314°C) is much higher than that of L-lactic acid (53°C), a polar compound of similar molecular mass. Strong secondary attractive forces between zwitterion charge centers and water results in water solubility for α-amino acids.

Concept check

✔ The zwitterion structures of α-amino acids are responsible for their physical properties: high melting point and significant solubility in water.

α-Amino acids exist exclusively as zwitterions in the solid state. The structure in aqueous solution is more complicated. There is an equilibrium between three species—zwitterion, cation, and anion:

$$\underset{\text{Cation}}{{}^{+}NH_3-\underset{|}{\overset{R}{C}}H-COOH} \underset{HO^-}{\overset{H^+}{\rightleftharpoons}} \underset{\text{Zwitterion}}{{}^{+}NH_3-\underset{|}{\overset{R}{C}}H-COO^-} \underset{H^+}{\overset{HO^-}{\rightleftharpoons}} \underset{\text{Anion}}{NH_2-\underset{|}{\overset{R}{C}}H-COO^-}$$

High $[H^+]$ Low pH (<1)	pH = pI	Low $[H^+]$ High pH (>12)

The cation is formed from the zwitterion by protonation of the carboxylate group; the anion is formed by ionization (deprotonation) of the ammonium group. Compounds with this dual ability to act as both base (proton acceptor) and acid (proton donor) are called **amphoteric** compounds and are discussed in Section 9.6.

The equilibrium between the three species (zwitterion, cation, and anion) varies with pH. Each α-amino acid has a pH value, called the **isoelectric point** or **pI,** at which almost all molecules (>99.9%) are present as the zwitterion with no net electrical charge. Table 20.1 lists the pI values of the α-amino acids. As pH falls below pI, the concentration of cation increases and the concentration of zwitterion decreases. As pH rises above pI, the concentration of anion increases and the concentration of zwitterion decreases.

The acid-base behavior of neutral (nonpolar and polar) α-amino acids is quantitatively different from that of the acidic or basic α-amino acids. Neutral α-amino acids act as moderately strong buffers, resisting changes in ionization state with changes in pH. Calculations using the Henderson-Hasselbalch equation (Section 9.8) show that more than 97% of a neutral α-amino acid is still in its zwitterion form when the pH is within two pH units above or below pI. The cation and anion forms become the dominant forms (>90%) only in highly acidic ($pH < 1$) and highly basic ($pH > 12$) media, respectively. The pI values for the neutral α-amino acids fall in the relatively narrow range of 5.05 to 6.30. The carboxyl and amino groups of these α-amino acids are charged at physiological pH because physiological pH is near neutral (7.35 for blood and 6.8–7.1 for most cells), which is within two pH units of pI.

Acidic α-amino acids have pI values below (on the acidic side of) the pI of the neutral α-amino acids because of the acidic carboxyl group in the R side chain. The pI values of aspartic and glutamic acids are 2.77 and 3.22, respectively.

Basic α-amino acids have pI values above (on the basic side of) the pI of the neutral α-amino acids because of the basic amino side group in the R side chain. The pI values of lysine, arginine, and histidine are 9.74, 10.76, and 7.59, respectively.

Concept check

✔ All carboxyl and amino groups, including those on side groups, of all α-amino acids are charged at physiological pH. The charge on carboxyl groups is negative; that on amino groups is positive.

Whereas the neutral α-amino acids exist as zwitterions with zero overall charge at physiological pH, the acidic and basic α-amino acids have overall charges of $1-$ and $1+$, respectively:

$$
\begin{array}{cc}
\underset{\text{Aspartic acid (1-)}}{\overset{\displaystyle CH_2COO^-}{\underset{|}{^+NH_3-CH-COO^-}}} &
\underset{\text{Lysine (1+)}}{\overset{\displaystyle (CH_2)_4\overset{+}{N}H_3}{\underset{|}{^+NH_3-CH-COO^-}}}
\end{array}
$$

Example 20.1 — Determining the ionization state of α-amino acids in media of different pH

Show the predominant structure of valine at its isoelectric point (5.96), at physiological pH (7), at pH < 1, and at pH > 12.

Solution

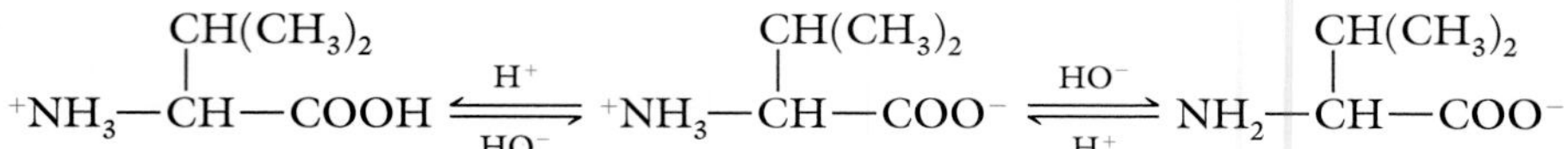

$$
\underset{\substack{2 \\ pH < 1}}{^+NH_3-\overset{\overset{\displaystyle CH(CH_3)_2}{|}}{CH}-COOH} \underset{HO^-}{\overset{H^+}{\rightleftharpoons}} \underset{\substack{1 \\ pH = 5.96 \text{ and } 7}}{^+NH_3-\overset{\overset{\displaystyle CH(CH_3)_2}{|}}{CH}-COO^-} \underset{H^+}{\overset{HO^-}{\rightleftharpoons}} \underset{\substack{3 \\ pH > 12}}{NH_2-\overset{\overset{\displaystyle CH(CH_3)_2}{|}}{CH}-COO^-}
$$

The zwitterion (structure 1) exists at pI. The zwitterion structure also predominates at physiological pH, which is within two pH units of pI. When conditions are much more acidic than pI, at pH < 1, the carboxylate group is protonated to form species 2, with an overall charge of $1+$. When conditions are much more basic than pI, at pH > 12, the ammonium group loses a proton to form species 3, with an overall charge of $1-$.

Problem 20.1 Show the predominant structure of phenylalanine at its isoelectric point (5.96), at physiological pH (7), at pH < 1, and at pH > 12.

The water solubility of α-amino acids varies with pH. Although α-amino acids are fairly soluble in water in general, they are least soluble at the isoelectric point. Zwitterions aggregate together; the positive end of one zwitterion associates with the negative end of another through strong secondary attractive forces, minimizing their ability to associate with water. At pH above or below pI, α-amino acids are cations or anions, and intermolecular associations between them are much weaker. Strong attractive interactions with water result in increased solubility.

Electrophoresis is useful for the analysis of mixtures of α-amino acids (and peptides, as described later). One type of electrophoresis, called **paper electrophoresis,** is shown in Figure 20.2. This technique identifies substances in an electric field by separation on the basis of their overall electric charge. Different α-amino acids show different migration behaviors in the electric field because of charge differences, which depend on structure and pH. An α-amino acid does not migrate in electrophoresis at pH equal to its pI, because it has zero charge. Electrophoresis at physiological pH distinguishes between neutral, acidic, and basic α-amino acids: a neutral α-amino acid has zero charge and does not migrate, an acidic α-amino acid is charged $1-$ and migrates to the anode (positive electrode), and a basic α-amino acid is charged $1+$ and migrates to the cathode (negative electrode).

20.3 PEPTIDES

Peptides are polyamides formed by α-amino acids reacting with each other. The details of the reaction are complex and will be considered in Section 21.7.

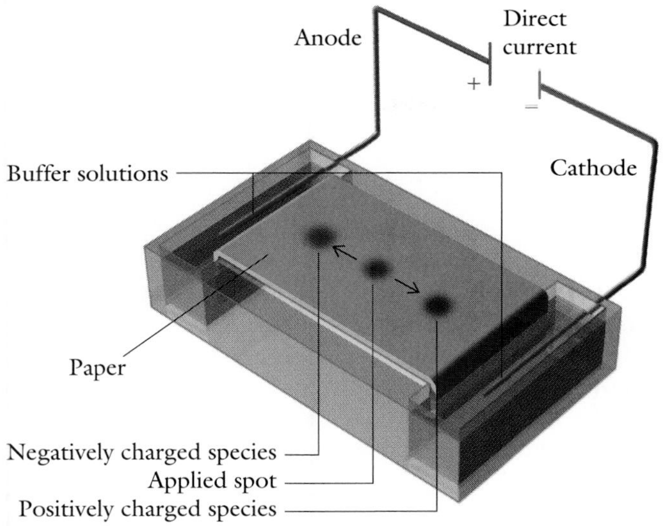

Figure 20.2 Electrophoresis. A strip of paper, saturated with a buffer solution, is positioned with its ends in buffer reservoirs that hold the electrodes. The researcher places a drop of sample solution in the middle of the paper and then turns on the electric current (direct current). After some period of time, the researcher turns off the current, removes and dries the paper strip, and stains it with a dye specific for the substance(s) under study.

Overall, the reaction is viewed, at least conceptually, as a dehydration between the carboxyl and amino groups of different α-amino acid molecules:

$$\overset{+}{H_3N}-\underset{\underset{R^1}{|}}{CH}-\overset{\overset{O}{\|}}{C}-O^- + \overset{+}{H_3N}-\underset{\underset{R^2}{|}}{CH}-\overset{\overset{O}{\|}}{C}-O^- \xrightarrow{-H_2O} \overset{+}{H_3N}-\underset{\underset{R^1}{|}}{CH}-\overset{\overset{O}{\|}}{C}-\underset{\underset{H}{|}}{N}-\underset{\underset{R^2}{|}}{CH}-\overset{\overset{O}{\|}}{C}-O^-$$

Peptide group

Dehydration takes place here

Peptide bond

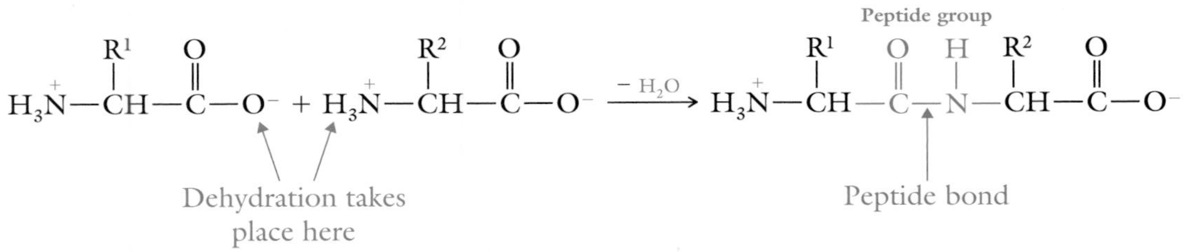

The α-amino acids in the peptide, called **amino acid residues** or **monomers,** are linked together by a **peptide bond,** the bond between the carbonyl carbon and the nitrogen. The functional group formed by this reaction is an amide group and is called the **peptide group** (—CO—NH— or —CONH—). In the laboratory, amides are formed by reacting carboxyl and amino groups at elevated temperatures (Section 16.8). In biological systems, peptide formation is catalyzed by an enzyme.

Peptides are called **dipeptides, tripeptides, tetrapeptides, pentapeptides, hexapeptides,** and so forth, depending on the number of amino acid residues. The name **oligopeptide** is loosely used to refer to peptides of fewer than from 10 to 20 amino acid residues, and **polypeptide** refers to larger peptides. Not all biochemists agree on the exact size of peptide at which an oligopeptide becomes a polypeptide. And biochemists often use the term "peptide" loosely to refer to peptides of any size.

Amino Acid Sequences and Constitutional Isomers

Different constitutional (structural) isomers are possible whenever α-amino acids combine to form peptides. For example, two different dipeptides are possible from the reaction of glycine and alanine, depending on which amino group reacts with which carboxyl group:

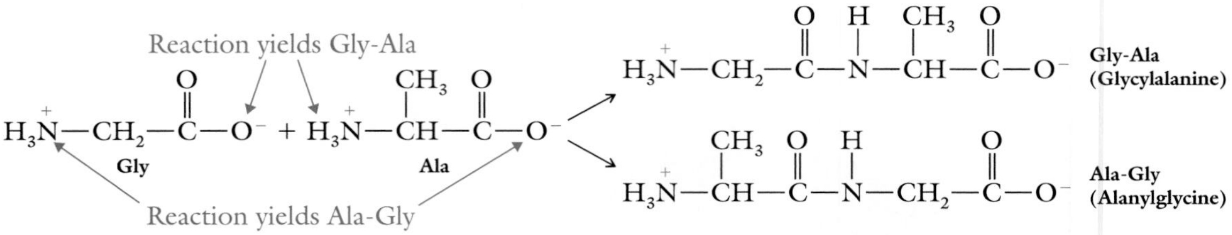

All peptides contain an α-amino group at one end and an α-carboxyl group at the other end. The α-amino acid residues containing the amino and carboxyl groups are the **N-terminal (amino-terminal)** and **C-terminal (carboxyl-terminal) residues,** respectively. The convention for drawing peptides is to place the N-terminal residue at the left and the C-terminal residue at the right. Peptides are named as follows:

Rules for naming peptides

- The C-terminal residue, located at the far right, keeps its amino acid name.
- For each of the other amino acid residues, the **-ine** or **-ic acid** ending of the amino acid name is replaced by **-yl,** except for tryptophan, for which **-yl** is added to the name.
- Naming begins at the N-terminal residue.

For example, the two dipeptides formed from glycine and alanine are glycylalanine and alanylglycine, as shown in the preceding reaction. More often, however, the three-letter abbreviations of the α-amino acid residues are used, such as Gly-Ala and Ala-Gly, rather than the full name.

Recall that all α-amino acids used in plants and animals are the L-enantiomers. The complete name for a peptide would include L-before the name of each residue—for example, L-Gly-L-Ala. The inclusion of L is usually not done, it being understood that only L-enantiomers are present (unless otherwise stated).

Example 20.2 Drawing and naming peptides

Show the formation of Val-Ser-Asp from the individual α-amino acids by using complete structural formulas. Give the full name for this tripeptide. Show all amino and carboxyl groups in their ionic forms.

Solution

Draw the α-amino acids in the order Val-Ser-Asp from left to right, each amino acid drawn with its α-amino group left and carboxyl group right. Carry out dehydrations between adjacent carboxyl and amino groups.

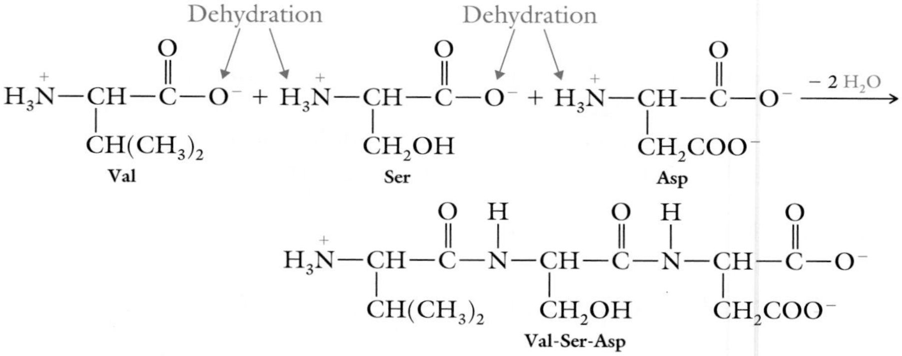

The full name of the peptide is valylserylaspartic acid.

Problem 20.2 Show the formation of Ala-Lys-Phe from the individual α-amino acids by using complete structural formulas. Give the full name for this tripeptide. Show all amino and carboxyl groups in their ionic forms.

The biological function of a peptide is determined largely by its **amino acid sequence,** the sequence of amino acid residues, which we list by convention from the N-terminal to the C-terminal residue. As we noted, peptides with the same amino acids but in different sequences are constitutional isomers: the number of isomers is the number of possible amino acid sequences. For a peptide with one each of *n* different amino acid residues, the number of constitutional isomers is *n*! (*n* factorial). For example, the number of constitutional isomers for a decapeptide containing ten different amino acid residues is

$$10! = 10 \times 9 \times 8 \times 7 \times 6 \times 5 \times 4 \times 3 \times 2 \times 1 = 3{,}628{,}800$$

The peptides that make up proteins are much larger than decapeptides, often containing hundreds and sometimes thousands of amino acid residues. Any polypeptide is but one of an astronomical number of possible constitutional isomers, another example of the high specificity of biological systems.

Example 20.3 Calculating the number of constitutional isomers of peptides

How many different constitutional isomers are possible for a tripeptide containing tyrosine, histidine, and proline? Give the abbreviated names for the different amino acid sequences.

Solution

There are three different amino acids and thus $3! = 3 \times 2 \times 1 = 6$ isomers. The isomers are simply all the possible sequences:

Tyr-His-Pro	His-Pro-Tyr	Pro-Tyr-His
Tyr-Pro-His	His-Tyr-Pro	Pro-His-Tyr

Problem 20.3 How many different constitutional isomers are possible for tripeptides containing glutamic acid, isoleucine, and lysine? Give the abbreviated names for the different amino acid sequences.

The Peptide Bond

The peptide C–N bond, although drawn as a single bond, does not have the properties of a single bond. There is no free rotation about this bond, and its length is shorter than expected for a single bond. The peptide bond, like all amide bonds (Section 16.11), has considerable double-bond character, resulting from resonance interaction between the π electrons of the carbonyl group and the nonbonded electron pair of nitrogen:

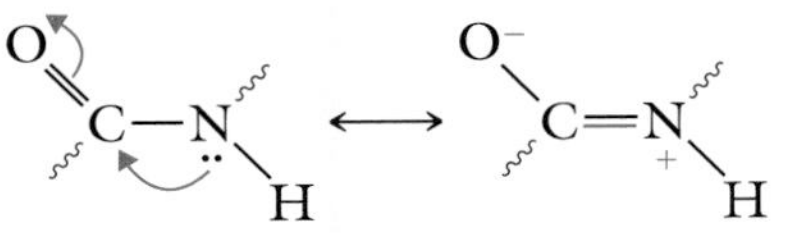

The peptide-chain segments, shown as wavy lines are trans to each other on the C=N bond. This configuration is more stable than the cis configuration, in which the peptide-chain segments would sterically (spatially) interfere with each other. The atoms of the double bond and the atoms directly attached to the double bond are coplanar—they lie in the same plane.

In simple amides, the double-bond character of the peptide bond is responsible for very high melting and boiling points and a lack of basicity (Section 16.11). In polypeptides, it plays a role in determining three-dimensional structure and function (Section 20.5).

Ionization of Peptides

A peptide, like the amino acids from which it is constructed, has an isoelectric point—the pH at which the peptide has an overall zero charge and does not migrate in electrophoresis.

- The pI value of a peptide containing only neutral α-amino acid residues or equal numbers of acidic and basic residues or both is in the range of pI values for neutral α-amino acids (pH = 5.05–6.30).
- The pI of a peptide containing acidic and basic α-amino acid residues is on the acidic side (lower than 5.05–6.30) if there is an excess of acidic residues and on the basic side (higher than 5.05–6.30) if there is an excess of basic residues.

The most important thing to keep in mind about the ionization state of peptides in biological systems is:

- All amino and carboxyl groups, including those on side groups of acidic and basic α-amino acid residues, are charged at physiological pH (Section 20.2).

Example 20.4 Writing the structures of peptides at different pH values

Show the structure of Asp-Lys-Gly at physiological pH. Does the peptide migrate in electrophoresis at physiological pH? If it migrates, to which electrode? Is the pI value for the peptide on the acidic or basic side of the pI values for polypeptides containing only neutral amino acid residues.

Solution

Asp-Lys-Gly has one carboxyl-containing side group and one amino-containing side group. All amino and carboxyl groups are charged at physiological pH and the resulting structure is:

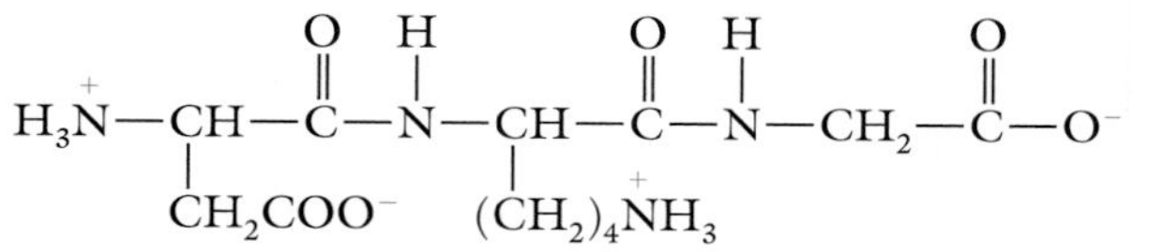

Asp-Lys-Gly has two positive charges and two negative charges, with an overall charge of zero, and does not migrate in electrophoresis at physiological pH.

The pI value is about the same as that for peptides containing only neutral amino acid residues because there are equal numbers of acidic and basic side groups.

Problem 20.4 Show the structure of the following peptides at physiological pH. (a) Ala-Lys-Ala; (b) Asp-Lys-Asp. Does each peptide migrate in electrophoresis at physiological pH? If it migrates, to which electrode? Is the pI value for each peptide on the acidic or basic side of the pI values for polypeptides containing only neutral amino acid residues.

Like α-amino acids, peptides have solubility and electrophoresis properties that are pH dependent (Section 20.2). A peptide shows minimum solubility at its pI; solubility increases at higher and lower pH values. Electrophoresis (Section 20.2) is useful for the identification and analysis of peptides. For example, the two peptides Asp-Lys-Gly and Asp-Lys-Asp have net charges of 0 and 1−, respectively, at physiological pH. The two tripeptides can be distinguished because, at physiological pH, Asp-Lys-Gly does not migrate in an electrophoresis experiment, whereas Asp-Lys-Asp migrates to the anode.

20.4 CHEMICAL REACTIONS OF PEPTIDES

Cysteine is the only α-amino acid that contains the **sulfhydryl** group, —SH (Section 13.11). Thus equipped, the cysteine residues of a peptide often have the function of linking together two peptides or different parts of the same peptide through formation of a **disulfide bridge.** Disulfide bridges are partly responsible for the three-dimensional structures (and resulting biological functions) of proteins (Section 20.5).

Disulfide bridges are formed by selective oxidation; that is, a loss of hydrogens from the sulfhydryl groups of a pair of cysteine residues:

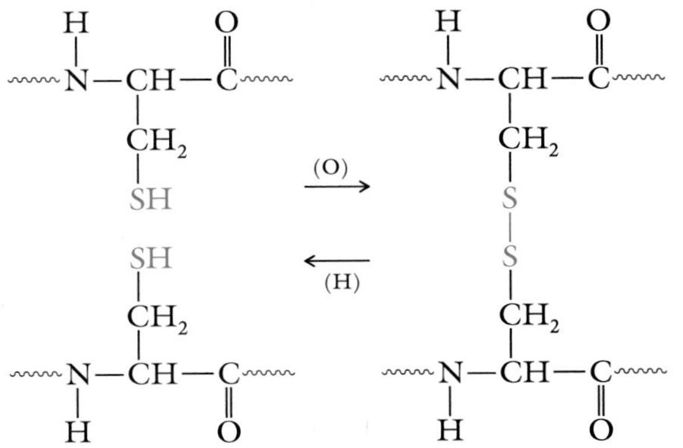

The reverse reaction, the selective reduction of disulfide bridges to form sulfhydryl groups, also takes place in biological systems. These reactions can also be carried out in the laboratory by selective oxidation and reduction.

In the digestion of a polypeptide, which takes place in the stomach and intestines, enzymes hydrolyze peptide bonds—the reverse of peptide formation. The digestive process is complex, but the end result is essentially the same as that in laboratory hydrolysis by acid or base (Section 16.12): α-amino acids. Disulfide bridges are not cleaved in digestion and hydrolysis; only reduction cleaves disulfides. Reduction of disulfide bridges takes place in the liver subsequent to digestion.

▶▶ The digestion of polypeptides is considered in Section 26.3.

Consider a hypothetical hexapeptide derived by disulfide-bridge formation between Ala-Cys-Ser peptides. Digestion (as well as laboratory hydrolysis) would break all peptide bonds but not the disulfide bridge:

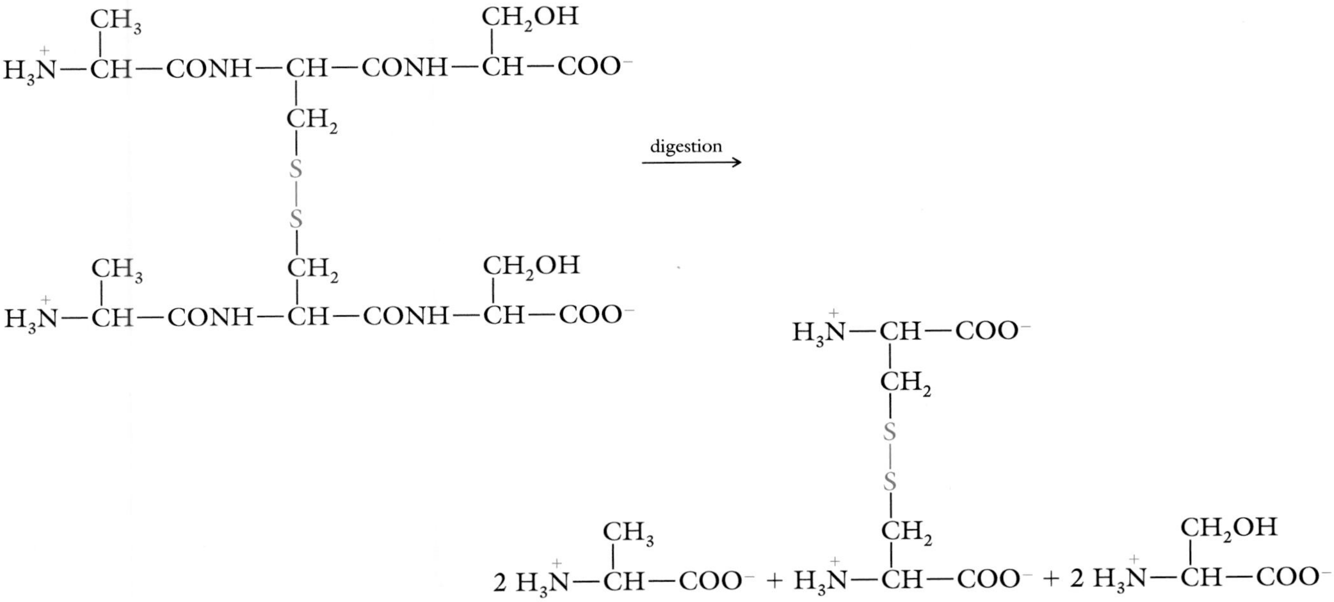

Cystine

The two cysteine residues joined by a disulfide bridge make up **cystine.** A reduction reaction cleaves cystine to form two cysteines. It is this type of cleavage that takes place in the liver:

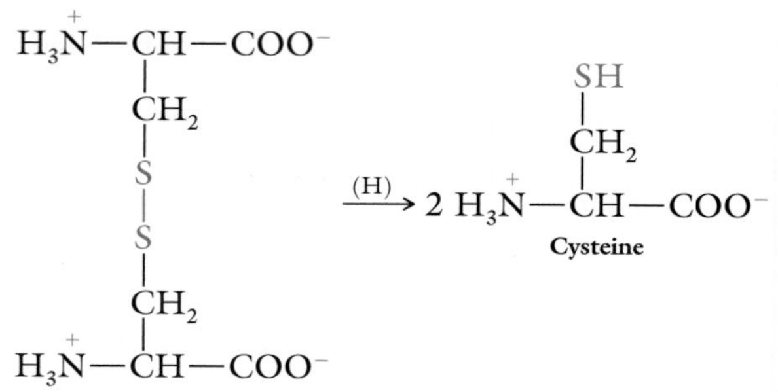

Cystine

Cysteine

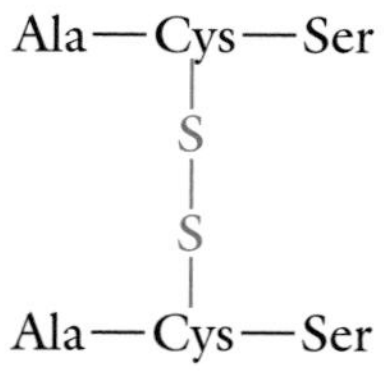

Peptide structures in which the three-letter abbreviations for amino acid residues are used can be written for peptides containing disulfide bridges by using —S—S— to bridge two cysteine residues, as shown in the margin.

20.5 THREE-DIMENSIONAL STRUCTURE OF PROTEINS

As mentioned earlier, biologically active peptides vary greatly in size, from a few α-amino acid residues to hundreds or thousands of residues. Not all are called proteins. Biochemists often distinguish between **peptides** and **proteins** on the basis of the number of α-amino acid residues. The term protein is generally not used for peptides containing fewer than 50 residues even when such peptides have biological functions similar to those of proteins (Table 20.2). They are referred to simply as peptides.

TABLE 20.2 Biological Functions of Some Peptides

Name	Number of α-Amino Acid Residues	Function
angiotensin	8	regulates blood pressure by constriction of arteries
enkephalin	5	reduces pain sensation by binding to brain receptors
gastrin	17	aids digestion by stimulating HCl and pepsinogen secretion in stomach
glutathione	3	maintains cysteine and iron in hemoglobin in reduced states by scavenging oxidizing agents
oxytocin	9	induces labor by contraction of uterine muscles
vasopressin	9	regulates blood pressure by stimulating excretion of water by kidneys

Polypeptides containing more than 50 residues are called proteins if they have biological function as individual polypeptide molecules. However, the typical protein is not a single polypeptide molecule; it is an aggregation of two or more identical or different polypeptides that forms a three-dimensional structure with a specific function. In many cases, the aggregation of polypeptides is associated with ions or molecules other than polypeptides. For example, hemoglobin, the carrier of oxygen in the blood, contains four polypeptide molecules (two each of two different polypeptides) and four heme molecules (Section 20.7).

Proteins containing only polypeptide molecules are called **simple proteins.** Those also containing nonpolypeptide molecules or ions are called **conjugated proteins:** the polypeptide part is the **apoprotein;** the nonpolypeptide molecules and ions, such as heme, are the **prosthetic groups.** Table 20.3 lists the major classes of conjugated proteins, with the prosthetic group of each class.

As stated earlier, there are more than 100,000 different proteins, each with a unique biological function. This enormous diversity in function begins with the diversity of α-amino acid sequences of polypeptide chains. Each sequence produces a unique three-dimensional structure, which in turn determines the unique function of the protein.

Amino acid sequence of polypeptide → Three-dimensional shape of protein → Biological function

The conformation of a polypeptide or protein determines its three-dimensional structure. The conformation of a protein containing more than one polypeptide is the sum of the conformations of each of its constituent polypeptides. Theoretically, a polypeptide can have many different conformations

TABLE 20.3 Classification of Conjugated Proteins

Class	Prosthetic Group	Example
glycoprotein	saccharide	immunoglobulin (antibody); interferon (antiviral agent); mucin (food lubricant in saliva)
hemoprotein	heme	hemoglobin (O_2 carrier in blood); myoglobin (O_2 storage in muscle)
lipoprotein	lipid	chylomicron, VLDL, LDL, HDL (lipid carriers)
metalloprotein	metal ion	Ca^{2+} in calmodulin (muscle contraction); Fe^{2+} in hemoglobin and myoglobin; Fe^{2+} in ferritin (Fe^{2+} storage); Zn^{2+} in carboxypeptidase (protein digestive enzyme)
nucleoprotein	nucleic acid	RNA-bound protein (protein synthesis in ribosome)
phosphoprotein	phosphate ester	casein (milk protein)

Abbreviations: VLDL, very low-density lipoprotein; LDL, low-density lipoprotein; HDL, high-density lipoprotein; RNA, ribonucleic acid.

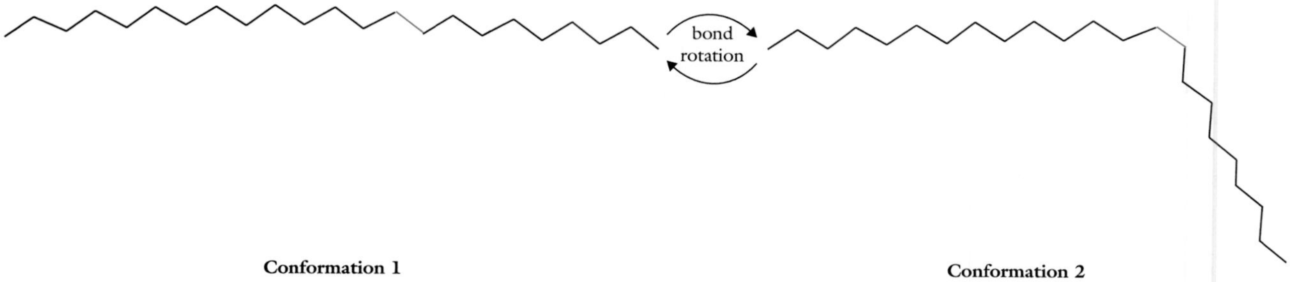

Figure 20.3 Conformations are interconvertible through rotation about single bonds in a polypeptide chain. The bond undergoing rotation in this hypothetical polypeptide is shown in red.

owing to rotations about the various single bonds in the polymer chain. The formation of polypeptide conformations by rotations about single bonds is referred to as the **folding** of the polypeptide. Figure 20.3 shows how rotation about one bond changes the conformation of a molecule. Rotations about many different bonds in a polypeptide result in an even greater number of possible conformations. Nevertheless, a polypeptide normally exists as a single specific conformation: the unique α-amino acid sequence of a polypeptide stabilizes one specific conformation more than any of the others. This one specific conformation is essential to the molecule's biological function under normal physiological conditions.

Protein structure is discussed at four levels:

- **Primary structure** is the α-amino acid sequence of a polypeptide.
- **Secondary structure** is the conformation in a local region of a polypeptide molecule. The conformations are the same in different regions of the molecule for some polypeptides but are different in different regions for other polypeptides.
- **Tertiary structure** exists when the polypeptide has different secondary structures in different local regions. Tertiary structure describes the three-dimensional relation among the different secondary structures in different regions.
- **Quaternary structure** exists only in proteins in which two or more polypeptide molecules aggregate together. It describes the three-dimensional relation among the different polypeptides.

Determinants of Protein Conformation

Two aspects of polypeptide structure determine the polypeptide's (and protein's) most stable conformation: (1) the bonds in the linear chain and (2) the interactions of the side groups. First, the allowed conformations are limited by the requirements that all bonds in the polymer chain (particularly the peptide bond with its double-bond nature; Section 16.11) maintain their normal bond angles and bond lengths and that all amino acid residues are the L-enantiomers. These structural features disallow a significant fraction of the possible conformations that would distort and destabilize the molecule. Then, of the very large number of possible conformations remaining, the most stable one is determined by the second aspect: the way in which the amino acid residues interact with each other and with the aqueous environment. One conformation is stable relative to other conformations because of:

1. **Shielding of nonpolar α-amino acid residues from water.** Most polypeptides fold into a conformation that buries the side groups of nonpolar α-amino acid residues in the interior of the polypeptide molecule or

in the interior of an aggregate of polypeptide molecules. This **hydrophobic effect** prevents the highly destabilizing contact between the nonpolar residues and the highly polar aqueous environment. (The same driving force is responsible for the lipid bilayer of cell and organelle membranes; Section 19.10.)

2. **Hydrogen bonding between peptide groups.** The peptide group (—CO—NH—) can hydrogen bond with water or with other peptide groups but prefers not to hydrogen bond with water, because such bonding does not minimize exposure of nonpolar α-amino acid residues to water. Hydrogen bonding between different peptide groups predominates because it produces conformations with maximum shielding of nonpolar residues from water, which has the predominant effect on the secondary structure of polypeptides.

3. **Attractive interactions between side groups of α-amino acid residues.** Folding proceeds with the placement of appropriate α-amino acid residues near each other to minimize repulsive interactions and maximize attractive interactions. Three types of attractive interactions take place (Figure 20.4) and have the predominant effect on the tertiary and quaternary structures of proteins:

   a. Nonpolar residues are placed near each other and participate in **hydrophobic attractions.** Hydrophobic attractions are not the same as

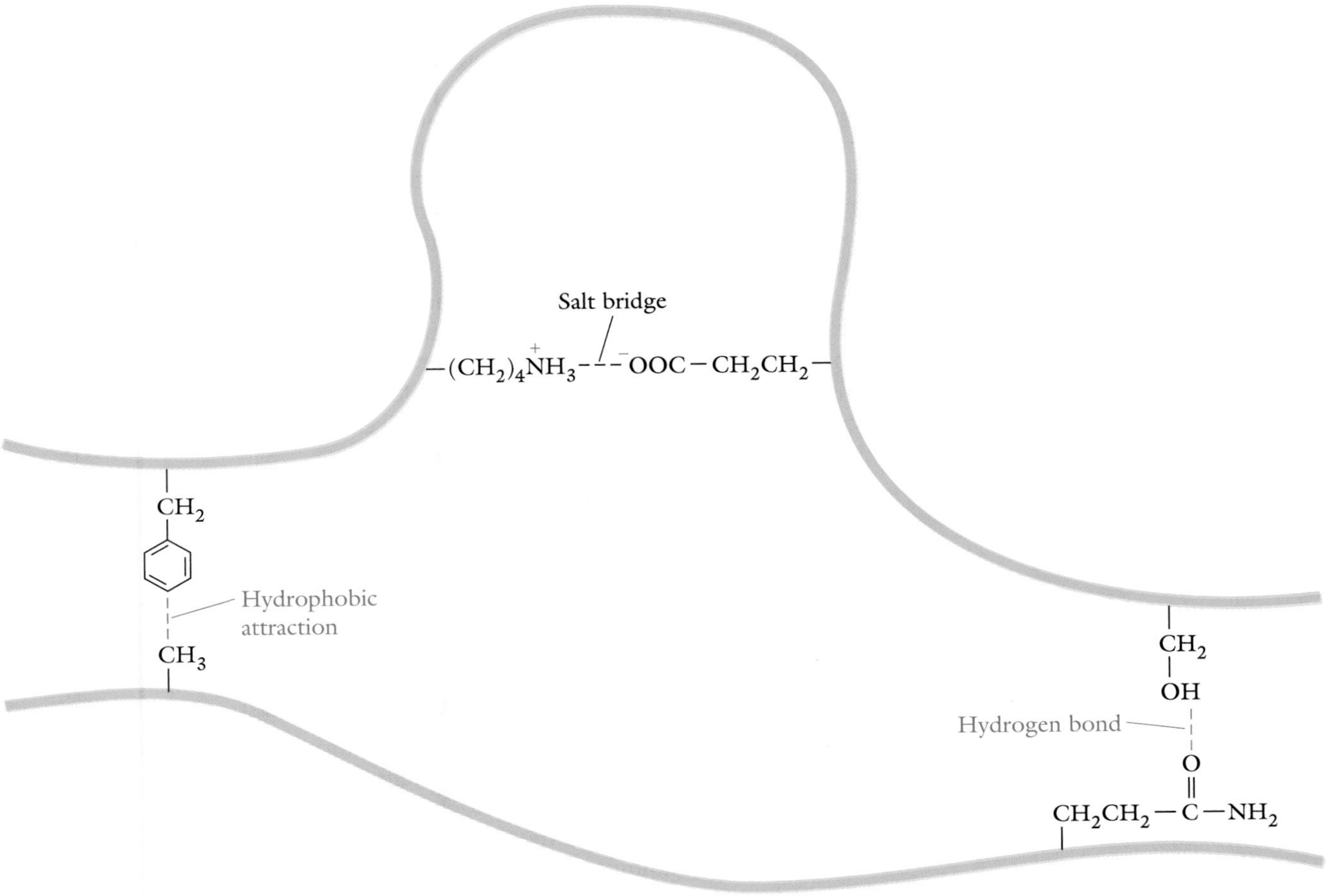

Figure 20.4 Noncovalent attractive interactions of amino acid side groups: hydrophobic attraction between Phe and Ala; salt bridge between Lys and Glu; hydrogen bonding between Ser and Gln.

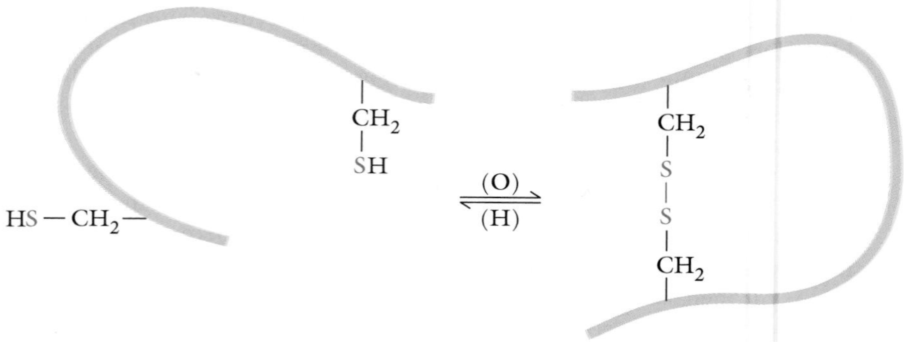

Figure 20.5 Formation of disulfide bridge between Cys residues.

the hydrophobic effect, but there is a relation between the two: the hydrophobic effect shields nonpolar residues from the aqueous environment and places them near each other, which results in hydrophobic attractions.

b. Polar, neutral residues are placed near each other and participate in **hydrogen-bonding attractions.**

c. Polar, acidic residues (Asp, Glu) are placed near polar, basic (Arg, His, Lys) residues and participate in **salt-bridge (ionic) attractions.** (Recall that all acidic and basic side groups are charged at physiological pH.)

4. Attractive interactions of side groups of polar α-amino acid residues with water. Some proteins fold so that polar residues, both charged and uncharged, are on the protein surface, where they undergo attractive interactions with water. **Globular proteins** (Section 20.7), such as hemoglobin, whose physiological functions require solublility in the aqueous environment, are folded in this way. The functions of **fibrous proteins** (Section 20.6), such as α-keratin of hair, nails, and skin, require water insolubility, and these molecules have far fewer polar residues on their surfaces.

5. Disulfide bridges formed in the folding process. Disulfide bridges are the only covalent bonds contributing to the stabilization of protein conformation through side-group interactions. The disulfide bridge in Figure 20.5 is an intramolecular disulfide bridge (forming between different parts of the same molecule). Intermolecular disulfide bridges, between different polypeptide molecules, also exist.

Concept checklist

- ✔ Hydrogen bonding between peptide groups is mostly responsible for the secondary structure of polypeptides.
- ✔ Attractive interactions between side groups is mostly responsible for the tertiary and quaternary structures of proteins.

Basic Patterns of Protein Conformation

The α-helix, β-pleated sheet, β-turn, and loop conformations are the most important secondary structures in naturally occurring polypeptides and proteins. The **α-helix** conformation has the overall shape of a coiled spring (Figure 20.6). This helical (spiral) folding of the polypeptide chain is due to stabilization by hydrogen bonding between the oxygen of each peptide group's C═O and the hydrogen of the peptide group's N—H of the fourth residue further along the polypeptide chain. The side groups of the residues protrude outward at right angles to the long axis (length) of the α-helix. The α-helix is a right-handed helix, a clockwise movement in either direction being required in a procession along the length of the helix.

In the ***β*-pleated sheet,** the polypeptide chains are not coiled but are in extended conformations with side-by-side alignment of adjacent chains, in either a parallel or antiparallel arrangement. The antiparallel arrangement (Figure 20.7) is the more common one: adjacent chains run in opposite directions; that is, the N-terminal to C-terminal-residue orientation of adjacent chains is in opposite directions. Hydrogen bonding between adjacent chains—that is, between the peptide-group oxygen of one chain and the peptide-group hydrogen of an adjacent chain—stabilizes the *β*-pleated sheet. Each polypeptide chain, except for residues at the edges of the sheet, is hydrogen bonded to adjacent chains on each side. The sheets are pleated, not flat, because of the planar peptide groups: side groups of *α*-amino acid residues alternately protrude above and below the plane of the sheets. The side-by-side alignment is not always between segments from different polypeptide molecules. In many polypeptides, the molecules turn back on themselves, in U-turn style; so different segments of the same chain are aligned side-by-side. Thus, in some polypeptides, the *β*-pleated-sheet conformation is intramolecular; in others, it is intermolecular.

The ***β*-turn,** also called the ***β*-bend,** is the part of a polypeptide chain where the chain abruptly changes direction. Such turns often connect adjacent intramolecular segments of *β*-pleated sheets in a polypeptide chain. *β*-Turns are less ordered than *α*-helices and *β*-sheets.

Loop conformations are segments of polypeptide chains that are less ordered than *β*-turns and much less ordered than *α*-helices and *β*-pleated sheets.

Specific conformations are favored or disfavored by certain *α*-amino acid residues:

- The *β*-pleated sheet is favored only for polypeptides with small side groups—specifically, for polypeptides with a high content of Gly, with most other residues being Ala and Ser. Residues with larger side groups are generally not accommodated by the *β*-pleated-sheet conformation. Large side groups spatially interfere with each other: polypeptide chains cannot pack sufficiently close to allow attractive interactions.
- The *α*-helix can better accommodate larger side groups than the *β*-pleated sheet, but certain residues disfavor the *α*-helix. For example, proline is called a **helix breaker** because its presence prevents formation of the *α*-helix: proline's rigid cyclic structure does not fit into the

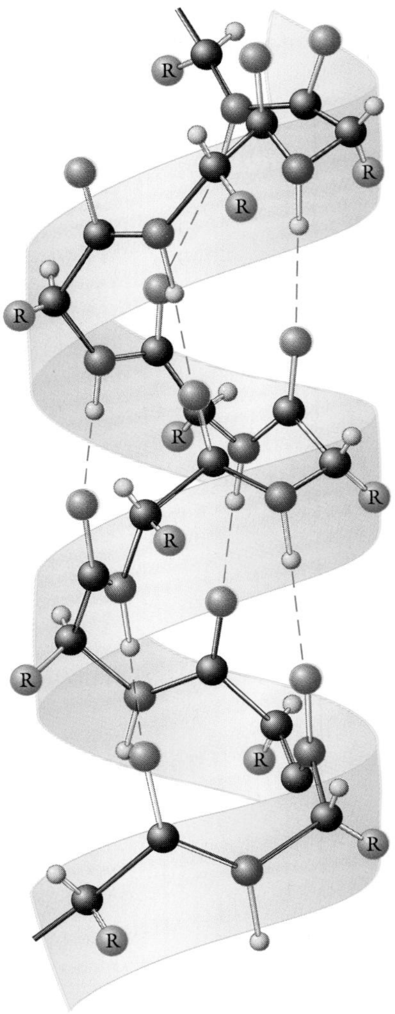

Figure 20.6 *α*-Helix conformation. (The C=O double bond is shown as a single bond to simplify the drawing.)

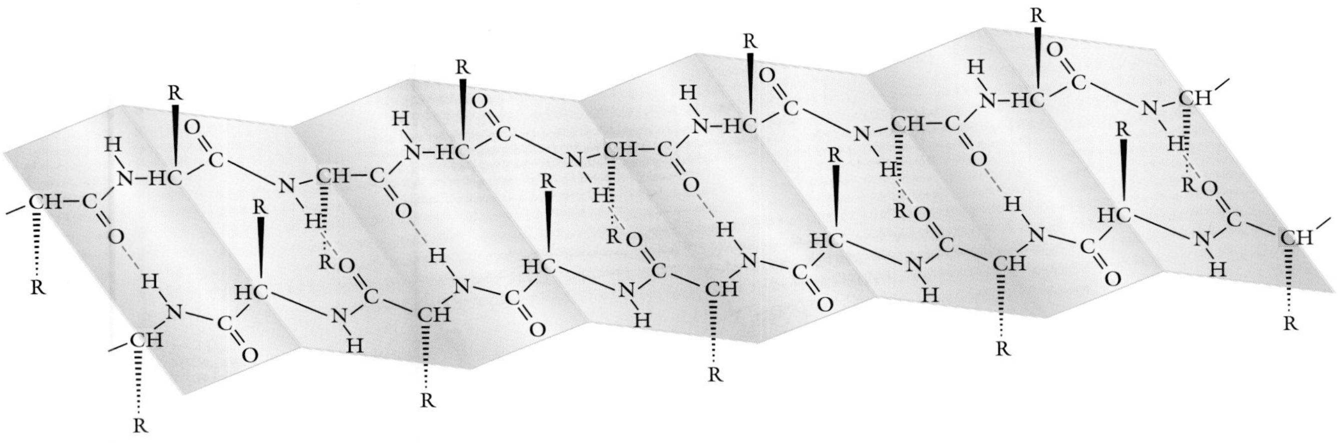

Figure 20.7 *β*-Pleated-sheet conformation.

helical conformation. Several consecutive residues with like-charged side groups or larger side groups also disfavor the α-helix because of repulsive interactions between side groups (electrostatic and spatial, respectively).

The effects of specific residues on conformation is a highly complex topic that we will not pursue further.

20.6 FIBROUS PROTEINS

Fibrous proteins—water-insoluble proteins with elongated shapes having one dimension much longer than the others—serve as structural and contractile proteins. A characteristic of the secondary structure of their polypeptides is the repetition of a single conformational pattern throughout all or almost all of the chain. For this reason, the elongated chains, whether α-helix or β-pleated sheet, aggregate tightly into fibers or sheets with strong secondary attractive forces, producing macroscopic structures with high strength.

Fibrous proteins are usually described as containing no tertiary structure, because the polypeptide chains have only one conformational pattern. On the other hand, fibrous proteins almost always have quaternary structure, because they are composed of two or more polypeptide molecules aggregated together into a specific conformational pattern.

α-Keratins

α-Keratins are the structural components of hair, horn, hoofs, nails, skin, and wool. Electron microscopy and X-ray diffraction studies of these materials show that they have hierarchical structures. An example is the structure of hair (Figure 20.8). Its individual polypeptide chains are almost entirely in the α-helix conformation, with very short disordered regions separating much longer α-helical regions. A pair of right-handed α-helices coil around each other in a left-handed manner to form a double-stranded helical coil (a **double helix**) called a **supercoil** or **superhelix.** A pair of supercoils coil around

INSIGHT INTO FUNCTION

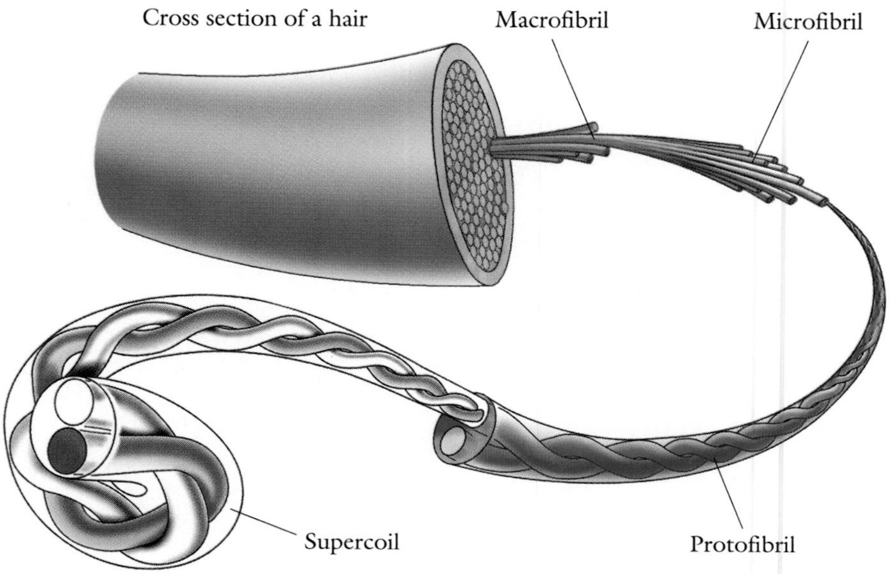

Figure 20.8 The hierarchical structure of a strand of hair explains many of the hair's properties.

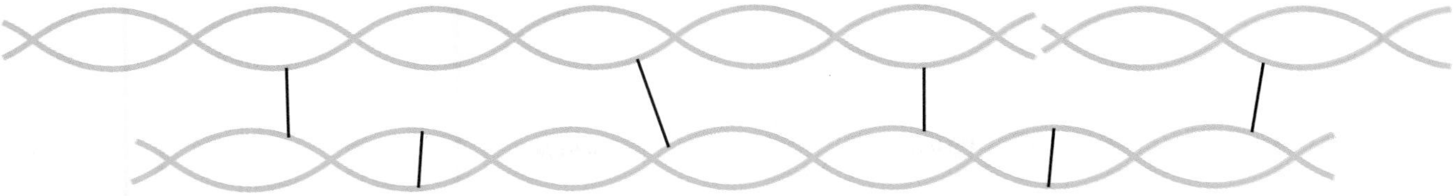

Figure 20.9 Disulfide cross-links between polypeptide chains in a protofibril. Cross-links are represented by straight lines between helices. The representation of disulfide cross-links is simplified by not showing the two superhelices coiled around each other.

each other to form a **protofibril.** Protofibrils coil together to form **microfibrils,** which in turn coil into **macrofibrils,** which are further assembled into a hair. Coiling at higher and higher levels is a mechanism for enhancing physical strength. At each level—superhelix, protofibril, microfibril, and macrofibril—structures are both packed side by side and staggered lengthwise to produce a final biological structure that is much longer and thicker than any of the individual elements. The springiness of hair and wool results from the tendency of the coiled structure to untwist when stretched and to recover its original conformation when released.

The horns of this bighorn ram are constructed of α-keratin proteins.

This packing into coiled conformations at higher and higher levels of organization is stabilized by secondary attractive forces and disulfide bridges between different polypeptide molecules. Disulfide bridges are more important than secondary interactions in imparting insolubility, strength, and resistance to stretching. Figure 20.9 shows disulfide bridges between the polypeptide chains in a protofibril: between chains of the same superhelix and between polypeptide chains of different superhelices. These disulfide bridges are intermolecular, which distinguishes them from the intramolecular disulfide bridges illustrated in Figure 20.5. Intermolecular bridges are also called **cross-links.**

α-Keratins are hard or soft, depending on the cysteine content and thus the number of disulfide bridges. Hard α-keratins (hair, horn, nail) have higher cysteine contents and are harder and less pliable than soft α-keratins (skin, callus). Permanent waving of hair is the deliberate rearrangement of the hair's disulfide bonds (Box 20.2 on the following page and Figure 20.10).

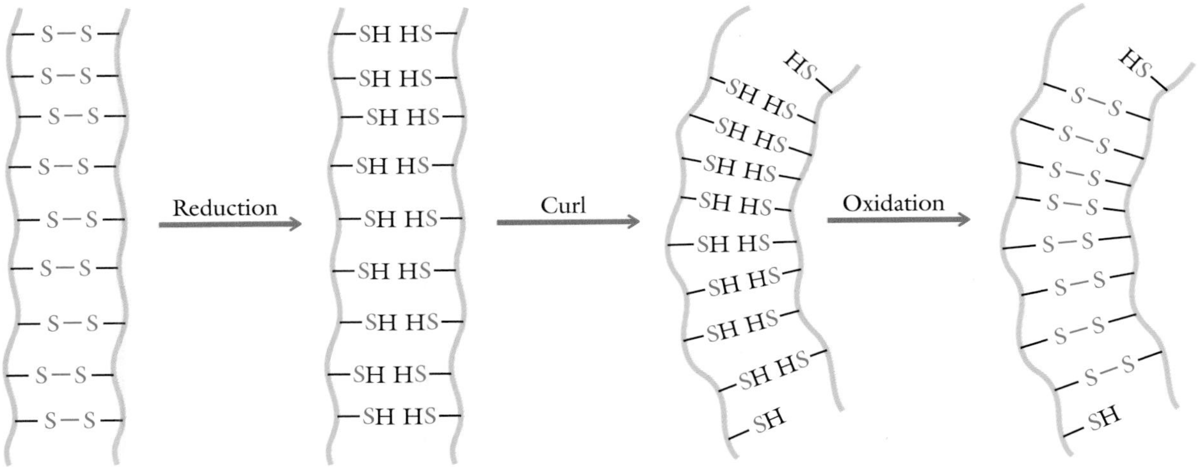

Figure 20.10 Permanent waving of hair.

20.2 Chemistry Within Us

Permanent Waving of Hair

Whether your hair is straight or curly and, if curly, whether loosely or tightly curled are hereditary traits. Heredity determines the primary structure of the α-keratin of your hair, and the amount and location of cysteine residues in that structure determine the pattern of disulfide bridges between polypeptide chains, which determines the shape of your hair. However, straight hair can be changed to curly hair or the size of the curl in curly hair can be changed by permanent waving.

First, the hair is treated with a reducing agent to break the disulfide bridges responsible for holding the hair in its present shape (see Figure 20.10). This treatment converts the disulfide bridges into sulfhydryl groups and imparts some conformational mobility to the α-helical polypeptide molecules of α-keratin. Next, a new shape is imposed on the hair through a process that adjusts the relative placement of the polypeptide chains: shaping the hair around curlers of the appropriate diameter and then setting the shape with an oxidizing agent to convert the sulfhydryl groups into a new pattern of disulfide bridges. When the oxidizing agent is rinsed out and the curlers removed, the hair keeps the new shape. A permanent wave is not truly permanent because new hair will grow in its genetically determined shape. The style can be maintained only by periodically repeating the permanent waving process.

In the last step of the permanent waving of hair, an oxidizing agent is applied to form disulfide bridges between cysteine residues in the proteins of the hair.

Collagen

The most abundant protein in vertebrates is **collagen,** the major stress-bearing component of connective tissues such as bone, cartilage, cornea, ligament, teeth, tendons, and the fibrous matrices of skin and blood vessels. Collagen, like the α-keratins, has a hierarchical structure based on helical polypeptide chains—but with some important differences. Collagen contains much more glycine and proline than does α-keratin, and a considerable proportion of the proline residues are converted into 4-hydroxyproline (Hyp) residues in the biosynthesis of collagen. Additionally, collagen contains very little cysteine. This difference in amino acid content and thus primary structure results in a different type of helical structure. The collagen polypeptide forms a left-handed helix, more elongated than the α-helix. Three collagen polypeptides wind around each other with a long right-handed twist to form a right-handed superhelix called a **triple helix** or **tropocollagen** (Figure 20.11). These triple helices are further organized into fibrils and higher-level structures. Coiling from one level to the next is in opposite directions. The packing at each level is very tight, and, along with the opposite-direction coiling, this packing produces a very strong structure. The same mechanism is used to produce the cables that support suspension bridges.

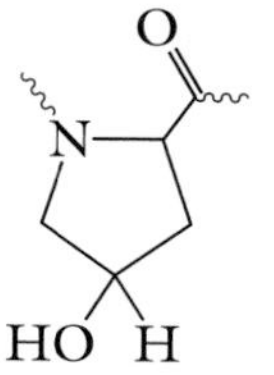

4-Hydroxyproline (Hyp) residue

Hydrogen bonding and cross-linking between polypeptide chains contribute considerably to the strength of collagen. Most of the hydrogen bonding between peptide groups is not within the same polypeptide chain but is between different chains—both within and between triple helices. The hydroxyl groups of 4-hydroxyproline residues also participate in hydrogen bonding. In collagen, cross-linking is not through disulfide bridges, because collagen is almost devoid of cysteine. Cross-linking is accomplished through a series of complex reactions between lysine and histidine residues.

In bone and teeth, collagen fibrils are embedded in **hydroxyapatite,** $Ca_5(PO_4)_3(OH)$, an inorganic calcium phosphate polymer. This composite structure has very high physical strength. The same strength-enhancing principle applies to the structure of wood (Box 18.3).

INSIGHT INTO FUNCTION

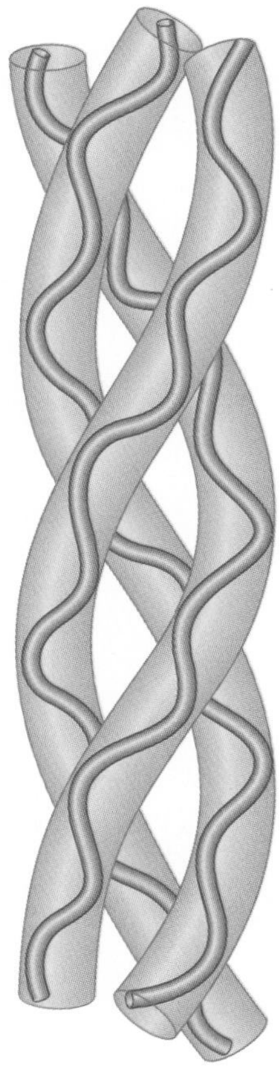

Figure 20.11 The triple helix of collagen explains that substance's flexibility and strength.

β-Keratins and Silk Fibroins

Among the relatively few proteins having conformations that are almost completely β-pleated sheets are the β-keratins, the proteins in bird feathers and reptile scales, and **silk fibroin,** the protein of the silk secreted by many insects to fabricate cocoons, webs, and nests. In most silks, a gummy protein called **serican** cements the silk fibroin fibers together.

20.7 GLOBULAR PROTEINS

Globular proteins, unlike fibrous proteins, do not aggregate into macroscopic structures. Whereas fibrous proteins form the structural elements of an organism, the globular proteins do most of the metabolic work—catalysis, transport, regulation, protection—which requires solubility in blood and the other aqueous media of cells and tissues. How are such large molecules made water soluble? In globular proteins, the primary structures are folded and organized into globular (globelike) conformations with a preponderance of hydrophilic α-amino acid residues on their outer surfaces. Globular proteins do not aggregate into macroscopic structures, because there are negligible secondary attractive forces between them. (Recall that highly branched molecules have lower boiling points than do unbranched molecules because their globular shape does not allow significant secondary attractive forces between molecules; Section 11.9.) Globular proteins are solubilized because the hydrophilic nature of their surfaces allows strong hydrogen-bond attractions with water. Unlike fibrous proteins, the typical globular protein possesses tertiary structure—different conformational structures in different segments of their polypeptide chains, with α-helix or β-pleated-sheet segments or both connected by β-turn or loop segments or both.

Myoglobin and Hemoglobin

The hemoglobin in red blood cells picks up oxygen in the lungs and transports it in the bloodstream to tissues throughout the body. Myoglobin, a similar protein present in muscle tissues, has a higher affinity for oxygen than does hemoglobin. It picks up oxygen from hemoglobin and stores it as a reserve for times when a working muscle's demand for oxygen is too high to be met by hemoglobin.

Myoglobin consists of one polypeptide containing 153 α-amino acid residues. Three-fourths of the residues are folded into eight α-helical sections

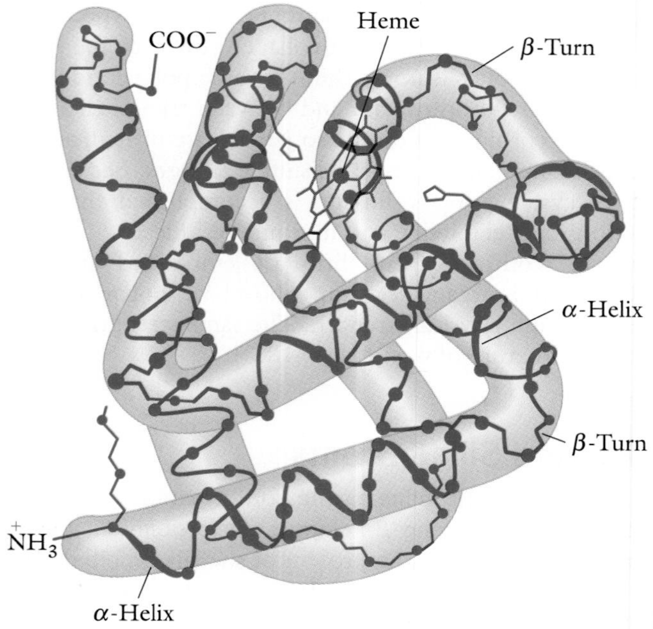

Figure 20.12 Myoglobin.

(shown as straight-tubular sections in Figure 20.12). The α-helical sections are connected by short β-turns containing mostly proline, the α-helix breaker (Section 20.5). The folding of the polypeptide chain places nonpolar residues together in the interior, where they are shielded from water. All polar residues except two histidines are located on the exterior surfaces, where they can hydrogen bond with water. The prosthetic group **heme,** an organic molecule with an Fe^{2+} at its center, is held in a cavity inside the polypeptide molecule by hydrophobic attractive forces. The Fe^{2+} is held in the center of heme by the free electron pairs of the four heme nitrogens and by ionic attractions with an interior histidine residue. The Fe^{2+} ion is the site where oxygen is bound when myoglobin takes on oxygen and stores it.

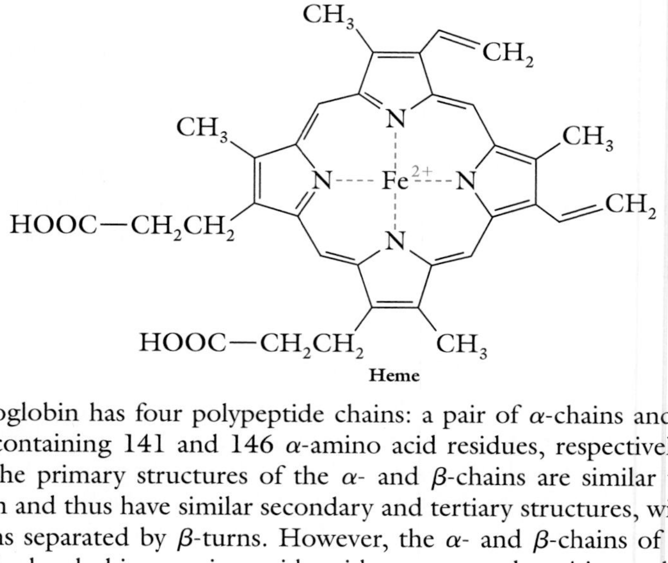

Heme

Hemoglobin has four polypeptide chains: a pair of α-chains and a pair of β-chains containing 141 and 146 α-amino acid residues, respectively (Figure 20.13). The primary structures of the α- and β-chains are similar to that of myoglobin and thus have similar secondary and tertiary structures, with α-helical sections separated by β-turns. However, the α- and β-chains of hemoglobin have hydrophobic α-amino acid residues at several positions where myoglobin has hydrophilic residues. These hydrophobic residues do not fold into the interior of the α- and β-chains but remain on their surfaces. Hydrophobic

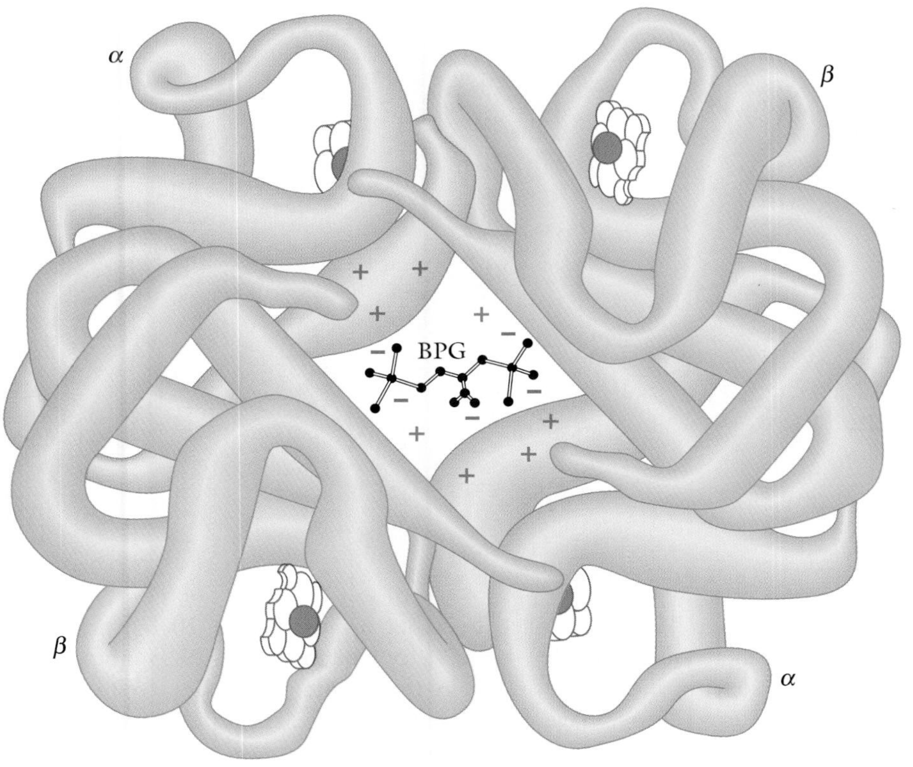

Figure 20.13 Hemoglobin's structure explains the ability of blood to transport oxygen. BPG, 2,3-bisphosphoglycerate.

interactions among these nonpolar residues, avoiding repulsive interactions with water, serve to hold the four polypeptide chains together in the quaternary structure.

Each polypeptide of hemoglobin carries a heme with its Fe^{2+} ion. In deoxygenated hemoglobin—that is, hemoglobin that is not carrying any oxygen—a 2,3-bisphosphoglycerate ion (BPG) is held in a central cavity between the four polypeptide chains by interaction with four positively charged groups on each of the two β-chains. BPG moves out of hemoglobin as hemoglobin picks up oxygen in the lungs; BPG is again picked up by hemoglobin after oxygen has been unloaded in peripheral tissues. BPG as well as environmental conditions (pH, CO_2 concentration) regulate the oxygen affinity of hemoglobin and are responsible for hemoglobin having a lower oxygen affinity than does myoglobin.

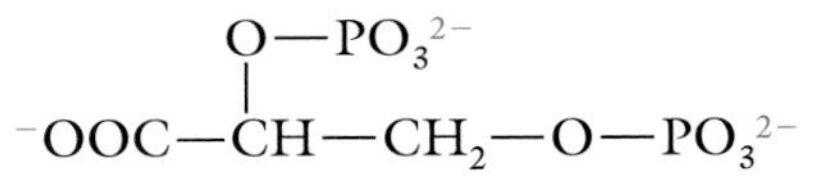

2,3-Bisphosphoglycerate (BPG)

▶▶ Section 26.13 describes the transport of oxygen by hemoglobin in more detail.

Carbon monoxide is a poison because it binds much more strongly than oxygen to the Fe^{2+} centers of heme, preventing oxygen from being picked up in the lungs and thus preventing the transport of oxygen to tissues, with consequent cessation of all metabolic processes. Death is the result unless oxygen is administered to reverse the binding of carbon monoxide to hemoglobin. Heavy smokers of tobacco and other plants have a significant fraction of their hemoglobin tied up by carbon monoxide (produced by incomplete combustion) at all times, causing their shortness of breath.

Lysozyme

Lysozyme, an enzyme in the cells and secretions of vertebrates, hydrolyzes bacterial cell walls. This enzyme does not kill bacteria but helps dispose of

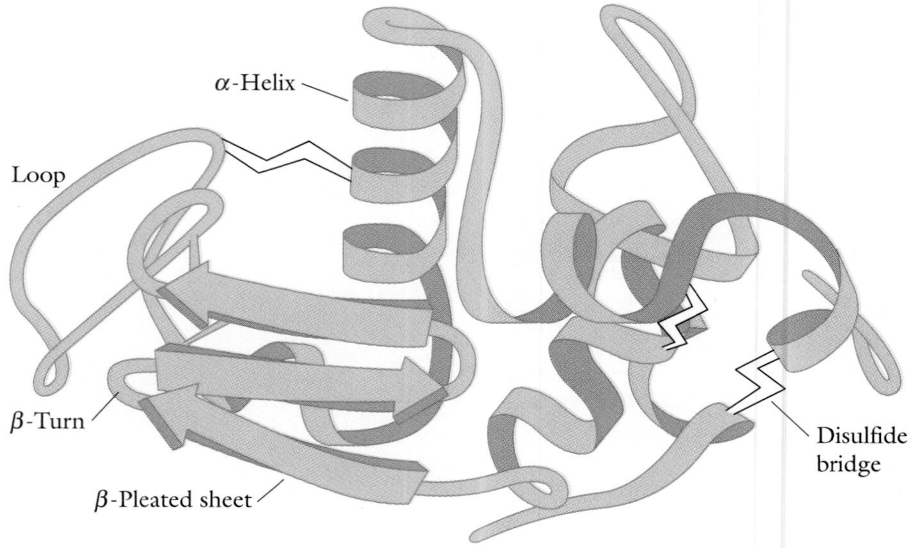

Figure 20.14 Hen egg lysozyme.

bacteria after they have been killed by other means. Figure 20.14 shows the structure of hen egg white lysozyme, a polypeptide of 129 α-amino acid residues. Lysozyme illustrates a protein that contains all four types of secondary structures that we have considered—α-helix, β-pleated sheet, β-turn, and loop. The β-pleated sheets are shown as flat arrows to indicate the direction of the polypeptide chain; the arrowhead points in the direction of the C-terminal residue. Lysozyme has intramolecular alignment of β-pleated sheets in contrast with the aforedescribed β-keratins and silk fibroin, which have intermolecular alignments of β-pleated sheets.

20.8 MUTATIONS: SICKLE-CELL HEMOGLOBIN

Protein function, as we have seen, is ultimately dependent on primary structure. Primary structure is, as we shall see in Section 21.7, determined by the structure of the DNA (deoxyribonucleic acid) of the gene(s) that direct the protein's synthesis. A **genetic mutation** is an alteration in the DNA structure of a gene that may in turn produce a change in the primary structure of a protein. Secondary, tertiary, and quaternary structures can change as a result, which could have consequences for the protein's function. However, not all alterations in the primary structure produce significantly altered function.

▶▶ Mutations are discussed further in Section 21.7.

- Substitution of an α-amino acid residue similar in size, charge, and polarity often has minimal effect on three-dimensional structure and may have no effect on function. Examples are substitution of Ile for Val, of Asp for Glu, and of Thr for Ser.
- Substitution of an α-amino acid residue very different in size, charge, and polarity may result in a large change in three-dimensional structure. Examples are Trp for Gly, Leu for Glu, and Ser for Ala. Substitutions of this type will most likely result in significant alterations in function unless the structural change is in a part of the protein without function.

Many mutant variations of human hemoglobin have been reported, of which a significant number have deleterious effects. **Sickle-cell hemoglobin,** the most harmful of mutant hemoglobins, causes misery and early death for many humans. Deoxygenated red blood cells containing sickle-cell hemoglobin take on an elongated sickle shape instead of the normal biconcave disk shape (Figure 20.15). The "sickled" cells aggregate together into long rodlike structures that do not move easily through the blood capillaries. The capillaries become inflamed, causing considerable pain. Blood circulation is impaired, leading to damage to many organs. Furthermore, the sickled cells are fragile; breakage leads to a decrease in functioning red blood cells—the anemia that leaves the person susceptible to infections and other diseases.

The severity of this hereditary, or genetic, disease depends on whether the person has received the gene for sickle-cell hemoglobin from both parents or only one parent. Those who inherit the mutation from both parents have **sickle-cell anemia:** all their deoxygenated red blood cells take the sickle shape, causing the aforedescribed problems. Many do not survive to adulthood, and those that do are seriously debilitated and have greatly reduced life spans. Persons who inherit the mutation from only one parent produce both normal and sickle-cell hemoglobins and often do not suffer severe symptoms except under conditions of severe oxygen deprivation, such as at high altitudes or after strenuous aerobic activity. This condition is called **sickle-cell trait.**

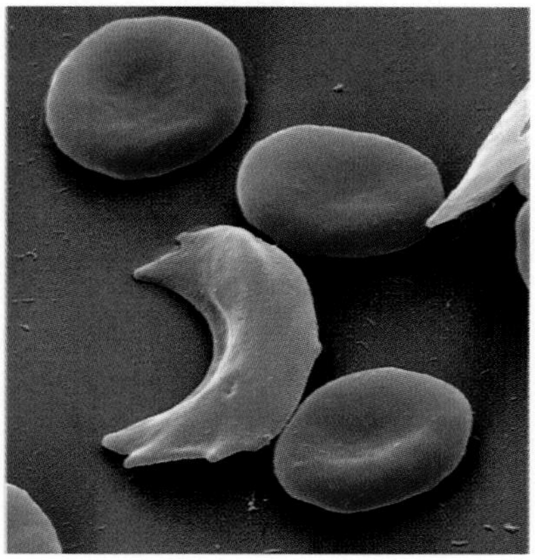

Figure 20.15 Electron micrograph of normal (circular) and sickled red blood cells.

Sickle-cell hemoglobin and normal hemoglobin differ in only one α-amino acid residue in each of the two β-polypeptide chains. Valine substitutes for the glutamic acid residue at position 6:

Normal hemoglobin	Val-His-Leu-Thr-Pro-Glu-Glu-Lys~~~~
Sickle-cell hemoglobin	Val-His-Leu-Thr-Pro-Val-Glu-Lys~~~~
	β 1 2 3 4 5 6 7 8

Although a hydrophobic residue replaces a charged residue, sickle-cell hemoglobin does not differ greatly from normal hemoglobin in its ability to pick up oxygen, because the regions critical to the picking up of oxygen (the heme and BPG cavities) are not altered. However, residue 6 of each of the β-polypeptide chains is at the surface of hemoglobin, and the substitution results in a disastrous interaction between different hemoglobins. A hydrophobic pocket is exposed at the surface of hemoglobin, both normal and sickle cell, when hemoglobin is deoxygenated. The valine residue at position 6 of one sickle-cell hemoglobin fits into and sticks (by hydrophobic attractions) to the hydrophobic pocket of the next hemoglobin (Figure 20.16). Sickling is the sequential aggregation of many such hemoglobins. It does not occur with normal hemoglobin, because there are no attractive interactions between the hydrophobic pocket and residue 6 when that residue is a charged glutamic acid residue. Moreover, sickling occurs only when there is a high concentration of deoxygenated hemoglobin. The hydrophobic pocket of hemoglobin is not exposed at the surface in oxygenated hemoglobin, because of conformational differences between oxygenated and deoxygenated hemoglobins.

Sickle-cell anemia is found almost exclusively in populations arising in tropical regions of the world, notably central and western Africa. Why should a genetic disease be restricted geographically? An answer is suggested by the

INSIGHT INTO FUNCTION

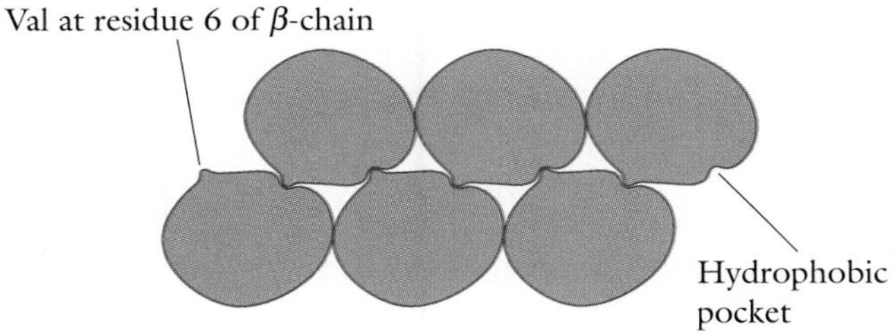

Figure 20.16 Formation of rodlike structures by aggregation of deoxygenated sickle-cell hemoglobin. This phenomenon explains the painful and dangerous symptoms of sickle-cell anemia.

high incidence of malaria, a tropical disease, found in the same geographical regions. The malarial parasite spends much of its life cycle in red blood cells, but sickled cells are a less hospitable environment for the malarial parasite than normal red blood cells. This built-in protection against malaria gives little advantage to persons with sickle-cell anemia, who already have limited life spans. But those with sickle-cell trait have a survival advantage against malaria, a higher likelihood of surviving to adulthood to pass on the trait to their offspring. Unfortunately, there is a 25% chance, based on the principles of genetics, that an offspring from two parents with sickle-cell trait will have sickle-cell anemia.

The dependence of biological function on protein structure is often subtle and not easy to predict. The substitution of valine for glutamic acid at position 6 of the β-polypeptide chain of hemoglobin results in a terrible genetic disease. However, replacement of the same glutamic acid residue by a different nonpolar residue, alanine, instead of valine, a mutation called hemoglobin G Makassar, has no significant effect on the function of hemoglobin. The methyl side group of alanine apparently does not interact strongly enough with the hydrophobic pocket of a neighboring hemoglobin. Other genetic diseases are described in Section 21.8.

The subtleness of the interrelation of primary, secondary, tertiary, and quaternary structures and biological functions is also evident in the use of bovine and pig insulins as substitutes for human insulin in the treatment of diabetes (Box 20.3).

20.9 DENATURATION

The **native conformation** of a protein is its conformation under normal physiological conditions; proteins in their native conformation are called **native proteins. Denaturation** is a loss of native conformation brought about by a change in environmental conditions, resulting in loss of biological function; the protein is then called a **denatured protein.**

Denaturation alters secondary, tertiary, and quaternary structures through conformational changes. Denaturation should not be confused with digestion or laboratory hydrolysis, which affect the primary structure of a protein. Denaturation occurs under milder conditions than does digestion or hydrolysis; the secondary, tertiary, and quaternary structures are held together mainly by noncovalent attractions, which are much weaker than covalent forces and more easily disrupted. Disulfide bridges impart considerable resistance to denaturation because they are stronger than the noncovalent interactions.

Globular proteins possess weaker secondary attractive forces and denature more readily than fibrous proteins. Denaturation of globular proteins usually results in an insoluble or much less soluble protein compared with the native protein as polypeptide chains unfold from the secondary, tertiary, and quaternary structures of the native protein and become tangled together in random conformations, with nonpolar residues no longer folded inside and protected from the aqueous environment. When we cook or whip egg white, the visible changes are due to denaturation of albumin.

The whipping of egg whites into a frothy solid for meringues and soufflés is achieved through a change in the conformational structure of the polypeptide chain.

A variety of denaturing conditions or agents lead to protein denaturation:

1. **Increased temperature** (increased thermal energy) increases the thermal motions of proteins, disrupting the noncovalent attractions responsible for the secondary, tertiary, and quaternary structures of the native protein. We take advantage of these effects to kill microorganisms during the sterilization of surgical and laboratory instruments and for food canning and food preparation equipment, and in the sealing and cauterization of small blood vessels in surgery. Cooking fish, meats, and

20.3 Chemistry Within Us

Diabetes Mellitus and Insulin

Insulin, a peptide hormone secreted by the pancreas, is essential to the regulation of glucose metabolism. Abnormalities in insulin action cause two types of the disorder known as diabetes mellitus. Insufficient production of insulin by the pancreas is the cause of **type I diabetes mellitus,** also called **insulin-dependent diabetes** or **juvenile-onset diabetes.** Insufficient insulin results in defects not only in carbohydrate metabolism, but also in lipid and protein metabolism because of the interrelations between metabolic pathways (Section 26.11). Insulin-dependent diabetes usually develops before the age of 40, often in adolescence. The consequences of untreated diabetes are severe: increased risk of athereosclerosis (Box 24.1); peripheral neuropathy (a nerve disorder); diabetic retinopathy, an eye disorder often resulting in blindness; and chronic kidney failure.

Insulin-dependent diabetes is treated by subcutaneous injection of replacement insulin; both bovine (beef) and porcine (pig) insulin have been used for this purpose. These insulins are not identical with human insulin, however. Insulin is composed of two polypeptide chains, A and B, covalently linked by disulfide bridges. Chain A has 21 and chain B has 30 α-amino acid residues. Human and pig insulins differ in only one residue, whereas human and beef insulins differ in three residues:

	Chain A			Chain B
Position of Residue	**8**	**9**	**10**	**30**
human insulin	Thr	Ser	Ile	Thr
pig insulin	Thr	Ser	Ile	Ala
beef insulin	Ala	Ser	Val	Ala

These differences in primary structure do not cause large differences in the physiological functions of the three insulins. Either the three-dimensional structures are the same or any differences that do exist are not in functionally important regions of the molecule. This is not necessarily what we would predict. Pig insulin has a nonpolar alanine residue instead of the human insulin's polar threonine residue at one position. Beef insulin has nonpolar alanine residues instead of human insulin's polar threonine residues at two positions and valine instead of isoleucine at a third position. Apparently, the polarity of the side group is not important. All of the variant residues have a common feature—the side groups are not too different in size.

Treatment with pig and beef insulins allows a reasonable quality of life for diabetics but does not completely solve the problem caused by the disease. On average, the life span is still decreased, perhaps because treatment does not maintain the ideal insulin concentration in the bloodstream at all times or because pig and beef insulins are good but not perfect replacements for human insulin. Many diabetics now test themselves for glucose blood levels throughout the day and adjust the amount of insulin that they use accordingly. The equivalent of human insulin is now available, synthesized by recombinant DNA technology (Section 21.11). This synthetic insulin may offer significant improvement over pig and beef insulins in the treatment of this form of diabetes.

About 90% of all diabetics have **type II diabetes mellitus,** called **non-insulin-dependent diabetes** or **adult-onset diabetes.** This condition generally develops in people over 40 who are obese, physically inactive, and have carbohydrate-rich diets. The pancreas secretes sufficient insulin, but cells throughout the body do not properly respond to it. Non-insulin-dependent diabetes is much less severe than insulin-dependent diabetes, with most patients being able to manage their glucose blood levels by diet and exercise alone.

poultry both kills microorganisms and makes the foods more digestible, by unraveling the native conformations. **Microwave radiation** also disrupts native conformations by increasing the thermal motions of proteins.

2. **Ultraviolet** and **ionizing radiations** disrupt native conformations by causing chemical reactions in the polypeptide chains.

3. **Mechanical energy** has similar effects. Violent mixing as in the whipping or shaking of egg white into a frothy solid for meringues and souffles causes polypeptide chains to unfold and become tangled up in a random fashion.

4. **Changes in pH** from normal physiological pH alter protein conformations by interfering with salt-bridge interactions. A fraction of the positively charged (basic) side groups become uncharged when the pH increases above 7, and a fraction of the negatively charged (acidic) side groups become uncharged when the pH decreases below 7. The extent and types of changes depend on the extent of the pH changes.

When bacteria cause milk to sour, the protein casein precipitates; this precipitation is the basis for cheese production. The pH of milk is normally about 6.5. Casein has an isoelectric point of 4.7, which means that it contains an excess of acidic side groups and has an overall negative charge at pH 6.5. Bacterial metabolism of milk produces lactic acid, lowering the pH. The carboxylate side groups of casein become protonated and the protein denatures.

Acid denaturation of proteins in the stomach aids in the digestive process. The low pH (1.5–2.0) of the stomach unfolds protein structures, making peptide bonds more accessible to enzymatic hydrolysis.

5. Organic chemicals such as soaps and detergents, alcohols, and urea (NH_2CONH_2) denature proteins by interrupting interactions between side groups of α-amino acid residues. This is the basis for using 70% alcohol as an antiseptic agent, for example, to sterilize skin before an injection. Alcohol passes through the bacterial cell walls and denatures bacterial proteins.

6. Salts of heavy metals such as Pb^{2+}, Hg^{2+}, and Ag^{+} react with sulfhydryl groups of cysteine residues to form metal disulfide bridges that prevent the formation of native conformations.

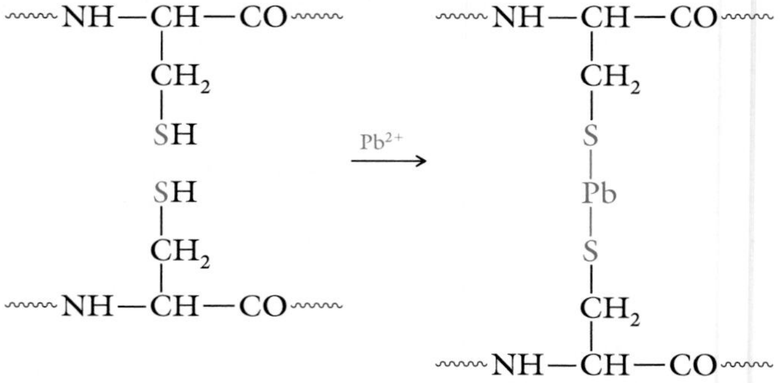

Young children who ingest flaked-off paint chips in very old buildings, where lead-based paint still remains on walls and woodwork, can suffer from lead poisoning. Lead poisoning can also result when acidic foods leach out lead pigments from the glazes used in some countries to manufacture ceramic dinnerware. Mercury poisoning results from ingesting too much fish taken from waters into which chemical wastes containing mercury salts have been dumped.

7. Oxidizing and reducing agents alter native conformations, the first by forming disulfide bridges and the second by cleaving them.

Summary

Proteins have a wider range of functions than any other family of biomolecules: enzymatic, transport, regulatory, structural, contractile, protective, and storage. Polypeptides are polymers constructed by the bonding together of α-amino acids. Proteins are naturally occurring polypeptides, usually aggregated with other polypeptides or with other types of molecules or ions or with both.

α-Amino Acids α-Amino acids contain a carboxyl group and an amino group attached to the same carbon. They are categorized as nonpolar neutral, polar neutral, polar acidic, and polar basic, depending on the side group attached to the α-carbon. Only L-α-amino acids are used in the polypeptides of plants and animals. Ten of the 20 α-amino acids needed for polypeptide synthesis are not synthesized by mammals; these essential amino acids must be obtained from the diet.

Zwitterionic Structure of α-Amino Acids α-Amino acids exist as zwitterions in equilibrium with cation and anion forms. The isoelectric point, pI, is the pH at which the α-amino acid is present almost exclusively in its zwitterion form. Neutral α-amino acids have pI values ranging from 5.05 to 6.30. Acidic and basic α-amino acids have pI values on the acidic and basic sides, respectively, of this range. All amino and carboxyl groups of α-amino acids are charged at physiological pH, which is also the case for the α-amino acid residues of peptides and proteins.

Peptides Peptides are formed by dehydration between carboxyl and amino groups of α-amino acids to form peptide bonds. Peptides are named by naming the α-amino acid residues in sequence, beginning at the end with the free α-amino group (the N-terminal residue) and proceeding to the end with the free carboxyl group (the C-terminal residue).

Chemical Reactions of Peptides Peptides undergo disulfide-bridge formation by oxidation of sulfhydryl groups on pairs of cysteine residues. Both intramolecular and intermolecular disulfide bridges exist in proteins. Digestion of peptides and proteins in the stomach and intestine consists of the hydrolysis of peptide bonds. Disulfide bonds are cleaved by reduction in the liver.

Three-Dimensional Structure of Proteins The primary structure of a peptide—the sequence of α-amino acid residues—determines its three-dimensional structure (secondary, tertiary, and quaternary), which in turn determines its biological function. Secondary structure is the conformation present in a local region of a polypeptide. The important secondary structures are the α-helix, β-pleated sheet, β-turn, and loop conformations. Tertiary structure exists when the polypeptide has different secondary structures in different local regions. Tertiary structure describes the three-dimensional relation among the different secondary structures in different regions. Quaternary structure exists only in proteins with two or more polypeptide molecules; it describes the three-dimensional relation among the different polypeptides of a protein.

Many factors determine the three-dimensional structure of a polypeptide or protein: shielding of nonpolar α-amino acid residues from water; hydrogen bonding between peptide groups; hydrophobic, hydrogen-bonding, and salt-bridge attractions between α-amino acid side groups; interaction of polar side groups with water; and disulfide bridges.

Fibrous and Globular Proteins Fibrous proteins are water-insoluble proteins with elongated shapes. The proteins aggregate tightly together into fibers or sheets with extensive intermolecular secondary forces to build macroscopic structures. These structures are the structural and contractile proteins of bone, hair, skin, and muscle.

Globular proteins do not aggregate into macroscopic structures. They perform a wide range of functions (enzymatic, transport, regulatory, protective) that usually require solubility in blood and the other aqueous media of cells and tissues. Solubility depends on primary structures that fold and organize into globular conformations with hydrophilic residues on their outer surfaces.

Mutations A genetic mutation—an alteration in DNA structure—can result in the production of a protein with an abnormal α-amino acid sequence. Some mutations result in proteins with drastically altered biological function. Sickle-cell anemia is one such mutation.

Denaturation Denaturation is loss of the native conformation of a protein brought about by a change in environmental conditions. Denaturing conditions or agents include heat; microwave, ultraviolet, and ionizing radiations; mechanical energy; pH changes; organic chemicals; heavy metal salts; and oxidizing and reducing agents.

Key Words

α-amino acid, p. 589
amino acid residue, p. 590
amino acid sequence, p. 595
β-bend, p. 605
denaturation, p. 614
disulfide bridge, p. 599
electrophoresis, p. 594
fibrous protein, p. 606
folding, p. 602
globular protein, p. 605
α-helix, p. 604
isoelectric point, p. 593
mutation, p. 612
peptide, p. 594
peptide group, p. 595
β-pleated sheet, p. 605
polypeptide, p. 595
primary structure, p. 602
protein, p. 600
quaternary structure, p. 602
secondary structure, p. 602
sickle-cell anemia, p. 612
sickle-cell trait, p. 612
side group, p. 589
tertiary structure, p. 602
zwitterion, p. 591

Exercises

α-Amino Acids

20.1 Classify the following as α-, β-, or γ-amino acids.

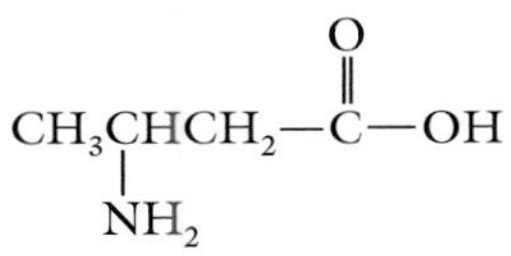
1

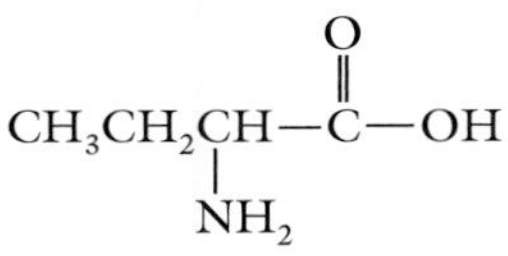
2

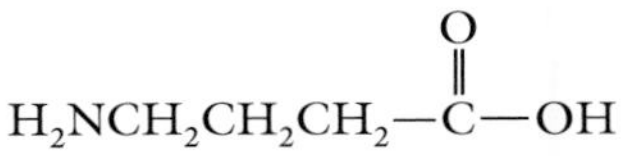
3

20.2 Classify the following amino acids as D or L.

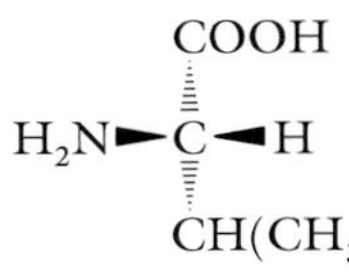
1

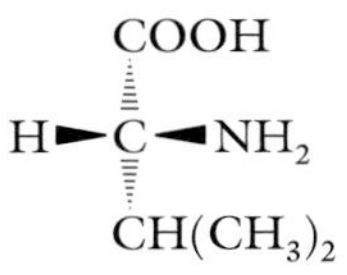
2

20.3 By reference to Table 20.1, use the three-letter abbreviations to name the α-amino acid(s) having: (a) the smallest side group; (b) the functional group of the alcohol family; (c) a benzene ring; (d) two tetrahedral stereocenters;

(e) a heterocyclic side group; (f) sulfur; (g) an acidic side group; (h) an α-amino group in a ring.

20.4 By reference to Table 20.1, use the three-letter abbreviations to name the α-amino acid(s) having: (a) the largest side group; (b) the functional group of the amide family; (c) a heterocyclic structure; (d) no tetrahedral stereocenter; (e) four nitrogen atoms; (f) a basic side group; (g) an —SH group; (h) a secondary α-amino group.

Zwitterionic Structure of α-Amino Acids

20.5 Why does an α-amino acid exist as a zwitterion, both in the solid state and when dissolved in water?

20.6 Define isoelectric point.

20.7 (a) Show the structures of alanine at its isoelectric point, at physiological (7) pH, at very low (< 1) pH, and at very high (> 12) pH.

(b) At what pH does alanine have its lowest solubility in water? its highest solubility?

(c) To which electrode does alanine migrate in electrophoresis at its pI? at physiological (7) pH?

20.8 (a) Show the structure of glutamic acid at physiological (7) pH.

(b) At what pH will glutamic acid have its lowest solubility in water?

(c) To which electrode does glutamic acid migrate in electrophoresis at pH = pI? at physiological (7) pH?

Peptides

20.9 Draw the structure of Ser-Met at physiological pH and give its full name.

20.10 Draw the structure of Cys-Phe at physiological pH and give its full name.

20.11 Draw the structure of Thr-Ala-Asp at physiological pH and give its full name.

20.12 Draw the structure of Leu-Val-Lys at physiological pH and give its full name.

20.13 Draw the structure of Gly-Pro-Ala at physiological pH and give its full name.

20.14 Draw the structure of Ser-Ala-Pro at physiological pH and give its full name.

20.15 How many constitutional isomers are possible for tripeptides containing threonine, cysteine, and leucine? Give abbreviated names for the different amino acid sequences.

20.16 How many different constitutional isomers are possible for tetrapeptides containing alanine, glutamic acid, tyrosine, and valine? Give abbreviated names for the different amino acid sequences for those tetrapeptides that have valine as the C-terminal residue.

20.17 There are six possible stereoisomers of the type based on tetrahedral stereocenters for Lys-Ala-Glu. For each stereoisomer, give an abbreviated name that indicates the configuration at each of the stereocenters. Which stereoisomer would be found in nature?

20.18 Cis-trans isomers of peptides are possible because of the double-bond character of the peptide group. The peptide groups of peptides found in nature possess the trans configuration. Draw Gly-Ala in the trans configuration.

20.19 Draw the structure of Asp-Ala-Asp at physiological pH. To which electrode will the tripeptide move in electrophoresis at physiological pH?

20.20 Draw the structure of Lys-Ala-Lys at physiological pH. To which electrode will the tripeptide move in electrophoresis at physiological pH?

Chemical Reactions of Peptides

20.21 Give the names of the products of digestion of Phe-Asp-Lys-Gly. What products are formed if the same peptide is subjected to acidic hydrolysis? basic hydrolysis?

20.22 Give the names of the products of digestion of Pro-Ala-Thr-Lys. What products are formed if the same peptide is subjected to acidic hydrolysis? basic hydrolysis?

20.23 Hydrolysis of a tripeptide yields equimolar amounts of Ala, Gly, and Lys. What is the amino acid sequence of the tripeptide?

20.24 Compare the products of digestion of the two pentapeptides Ala-Gly-Asp-Gly-Phe and Gly-Asp-Gly-Phe-Ala.

20.25 Consider the following hexapeptide:

$$\overset{+}{H_3N}-CH-CO-NH-CH-CO-NH-CH-COO^-$$
$$\quad\quad | \quad\quad\quad\quad\quad | \quad\quad\quad\quad\quad |$$
$$\quad\quad CH_3 \quad\quad\quad\quad CH_2 \quad\quad\quad CH_2C_6H_5$$
$$\quad\quad\quad\quad\quad\quad\quad\quad\quad | $$
$$\quad\quad\quad\quad\quad\quad\quad\quad\quad S$$
$$\quad\quad\quad\quad\quad\quad\quad\quad\quad | $$
$$\quad\quad\quad\quad\quad\quad\quad\quad\quad S$$
$$\quad\quad CH_3 \quad\quad\quad\quad CH_2 \quad\quad\quad CH_2C_6H_5$$
$$\quad\quad | \quad\quad\quad\quad\quad | \quad\quad\quad\quad\quad |$$
$$\overset{+}{H_3N}-CH-CO-NH-CH-CO-NH-CH-COO^-$$

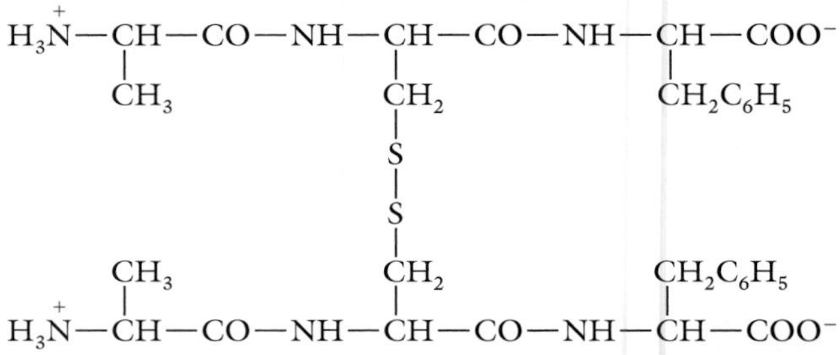

Write the structures of the products obtained, if any, when the tripeptide is (a) digested, (b) subjected to selective oxidizing conditions, (c) subjected to selective reducing conditions.

20.26 Consider the following tripeptide:

$$\quad\quad\quad\quad\quad\quad\quad\quad\quad\quad CH(CH_3)_2$$
$$\quad\quad\quad\quad\quad\quad\quad\quad\quad\quad | $$
$$\overset{+}{H_3N}-CH_2-CONH-CH-CONH-CH-COO^-$$
$$\quad\quad\quad\quad\quad\quad\quad | $$
$$\quad\quad\quad\quad\quad\quad\quad CH_2SH$$

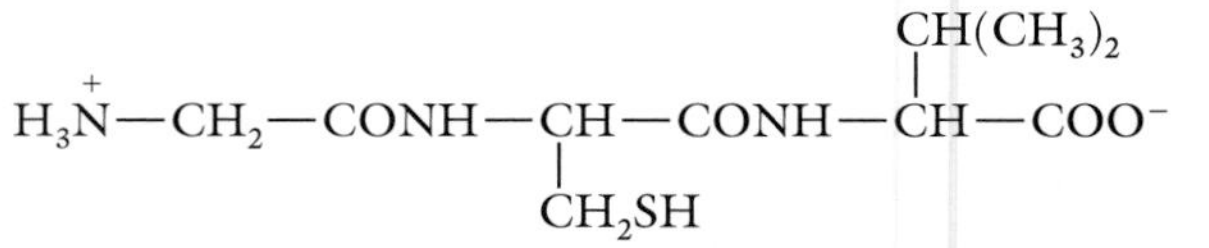

Write the structures of the products obtained, if any, when the tripeptide is (a) digested, (b) subjected to selective oxidizing conditions, (c) subjected to selective reducing conditions.

Three-Dimensional Structure of Proteins

20.27 Distinguish between conjugated and simple proteins.

20.28 Distinguish between an apoprotein and a prosthetic group.

20.29 What takes place in the conversion of one conformation of a molecule into a different conformation?

20.30 Which of the following structural levels of proteins are the result of conformational differences? (a) Primary; (b) secondary; (c) tertiary; (d) quaternary.

20.31 Which of the following interactions are most important in determining the secondary structure of polypeptides? (a) hydrogen bonding between peptide groups; (b) attractive interactions between side groups of residues.

20.32 Which of the following interactions are most important in determining the tertiary and quaternary structure of polypeptides? (a) hydrogen bonding between peptide groups; (b) attractive interactions between side groups of residues.

20.33 What attractive interaction (hydrogen bond, salt bridge, hydrophobic, or none) can take place between side groups for each of the following pairs of α-amino acid residues under physiological conditions? (a) Pro and His; (b) Ser and Tyr; (c) Pro and Phe; (d) Lys and Arg; (e) Lys and Glu; (f) Ser and Val.

20.34 What attractive interaction (hydrogen bond, salt bridge, hydrophobic, or none) can take place between side groups for each of the following pairs of α-amino acid residues under physiological conditions? (a) Met and Arg; (b) Met and Ile; (c) Thr and Tyr; (d) Asp and Glu; (e) Thr and Phe; (f) Arg and Asp.

20.35 What secondary forces other than those between side groups of pairs of α-amino acid residues are important in determining protein conformations?

20.36 What covalent bond is important in stabilizing protein conformations?

20.37 What structural feature dictates the secondary, tertiary, and quaternary structures of proteins?

20.38 Is intermolecular or intramolecular hydrogen bonding between peptide groups more important for the α-helix? for the β-pleated sheet?

Fibrous Proteins

20.39 Why are fibrous protein structures required for proteins that function as structural and contractile proteins?

20.40 Which of the following structural features contribute to the high physical strength of fibrous proteins? (a) Double helix (for example, α-keratin); (b) triple helix (for example, collagen); (c) heirarchy of increasing levels of coiling; (d) intermolecular disulfide bridges; (e) composite construction (for example, bone).

20.41 Briefly describe the difference between a double helix and a triple helix.

20.42 Does the hardness of α-keratins increase or decrease with increasing cysteine content?

20.43 Where are α-keratins found? What is their biological function?

20.44 Where are collagens found? What is their biological function?

Globular Proteins

20.45 Why are globular protein structures required for proteins that function as catalytic, transport, regulatory, and protective proteins?

20.46 Why does a description of hemoglobin include quaternary structure but that of myoglobin does not?

20.47 Why does a description of myoglobin include tertiary structure but that of an α-keratin does not?

20.48 Classify hemoglobin and myoglobin as conjugated or simple proteins.

20.49 Which is the apoprotein and which the prosthetic group in myoglobin?

20.50 How do the physiological functions of hemoglobin and myoglobin differ?

20.51 What is the function of heme in hemoglobin and myoglobin?

20.52 Draw the abbreviated structure when the following polypeptide forms an intramolecular disulfide bridge.

Gly-Ala-Cys-His-Phe-Asp-Cys-Met-Gly-Thr

20.53 Which of the following α-amino acid residues are most likely to be in the exterior of a globular protein: Asp, Ile, Glu, His, Lys, Phe, Trp, Val? Explain.

20.54 Which of the following peptides will be more soluble in water at physiological pH? (a) Ala-Glu-Phe-Gly-Leu-Val-Ala-Phe; (b) Ala-Glu-Phe-Glu-Leu-Val-Ala-Glu. Explain.

Mutations

20.55 Define mutation.

20.56 Distinguish between the sickle-cell trait and sickle-cell anemia.

20.57 A hemoglobin mutation known as hemoglobin Hammersmith has Ser instead of Phe at position 42 of the β-polypeptide chains. This substitution is in the pocket of the polypeptide chain that holds heme. The overall result is a decrease in the picking up of oxygen by hemoglobin because hemoglobin tends to lose heme. A replacement of Phe by Ile at the same position has much less effect on the ability of hemoglobin to pick up oxygen. Why?

20.58 What is the consequence of a mutation that changes a hydrophilic residue into a hydrophobic residue on the surface of a globular protein?

20.59 What is the consequence of a mutation that changes a hydrophobic residue into a hydrophilic residue in the interior of a globular protein?

20.60 Explain each of the following observations regarding the effect of mutations on biological function:
(a) replacement of Trp often has a large effect;
(b) replacement of Asp by Glu often has little effect;
(c) replacement of Ile by Leu often has little effect;
(d) replacement of Lys by Glu often has a large effect.

Denaturation

20.61 What is a native protein?

20.62 What is denaturation?

20.63 Compare protein digestion and denaturation, indicating the difference in the products.

20.64 Proteins with higher cysteine contents are more resistant to denaturation than those with lower cysteine contents. Why?

20.65 Drops of dilute silver nitrate solution placed in the eyes of newborn babies protect against gonorrhea by killing the gonorrhea organism. What mechanism is responsible for the protective effect?

20.66 Which of the following side-group attractive forces are most affected by pH changes? (a) Hydrophobic; (b) salt bridge; (c) hydrogen bond.

20.67 Polypeptides with low contents of acidic and basic residues are less sensitive to pH changes than those with high contents. Why?

20.68 What is the mechanism responsible for sterilization by heat?

Unclassified Exercises

20.69 Draw the structure of L-glutamic acid.

20.70 Place the following compounds in order of increasing melting point: glycine, butylamine, propanoic acid. Explain the order.

20.71 (a) Show the structure of methionine at its isoelectric point, at physiological pH, at very low (<1) pH and very high (>12) pH.

(b) At what pH will methionine have the lowest solubility in water? its highest solubility in water?

(c) To which electrode does methionine migrate in electrophoresis at its pI? at physiological pH? at pH = 1? at pH = 12?

20.72 Show the structure of lysine at physiological pH and indicate to which electrode it migrates?

20.73 How many different constitutional isomers are possible for pentapeptides containing one each of phenylalanine, lysine, threonine, proline, and valine? Give the sequence by using abbreviated names for the different amino acid sequences for those pentapeptides that have valine as the N-terminal residue, proline as the C-terminal residue, and lysine as the middle residue.

20.74 Draw Ser-Val in the trans configuration.

20.75 How many stereoisomers of the type based on tetrahedral carbons are possible for Lys-Gly-Ala? Give abbreviated names. Which is (are) probably present in nature?

20.76 Which would have higher solubility in aqueous solution—a peptide with a high lysine content or one with a high valine content?

20.77 Draw the structure of Met-Ser-Ile-Gly-Glu at physiological pH and give its full name.

20.78 Draw the structure of Pro-Gly-Pro at physiological pH and give its full name.

20.79 Show the structure of Asp-Lys-Lys at physiological pH and indicate to which electrode it migrates in electrophoresis at physiological pH. Is the pI for the peptide on the acidic or basic side of the pI for polypeptides containing only neutral amino acid residues?

20.80 To which electrode will Lys-Gly-Ile move in electrophoresis at physiological pH?

20.81 What is the net charge on Phe-Asp-Leu-Glu-Thr at physiological pH?

20.82 To which electrode will the following peptide migrate at physiological pH?

Phe-Lys-Leu-His-Thr-Glu-Met-Asp-Ser-Glu

20.83 Draw the abbreviated structure(s) of the product(s) formed when the following peptide is subjected to selective reduction.

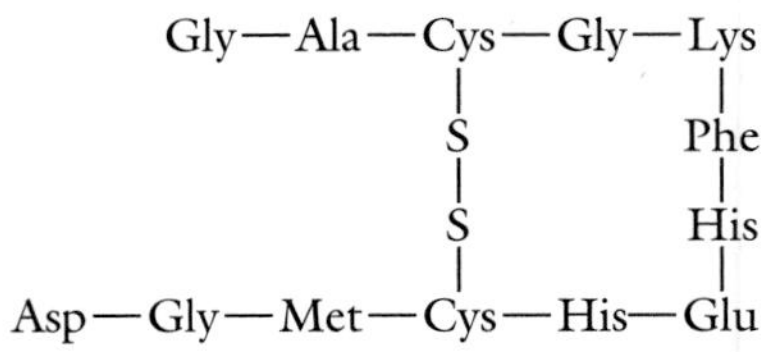

20.84 The isoelectric point of the protein lysozyme is 11.0. Which of the following α-amino acid residues are present in larger amounts: Asp, Glu, Gly, Lys, Ser, Trp, Val? Explain.

20.85 The isoelectric point of albumin is 4.9. (a) The water solubility of albumin will be minimum at which of the following pH values: <4.9, 4.9, or >4.9? (b) To which electrode will albumin migrate in electrophoresis at physiological pI?

20.86 An unknown protein sample is either albumin (pI = 4.9) or lysozyme (pI = 11.0). In electrophoresis at physiological pH, migration of protein is only to the cathode. Identify the unknown.

20.87 An unknown protein sample is either pepsin (pI = 1.0) or human growth hormone (pI = 6.9). In electrophoresis at physiological pH, migration of protein is only to the anode. Identify the unknown.

20.88 The primary structures of the α- and β-polypeptides of hemoglobin are very similar to the polypeptide of myoglobin except that several hydrophilic residues on the exterior surface of myoglobin are replaced by hydrophobic residues on the surfaces of the α- and β-polypeptides of hemoglobin. Reconcile this difference with the fact that a hemoglobin has four polypeptides and myoglobin has only one.

20.89 Which proteins have more polar residues on their surfaces—fibrous or globular proteins? Why?

20.90 Parts A and B of a polypeptide have different amino acid compositions. Part A contains mostly Gly, Ala, and Ser; part B contains only small amounts of Pro, Gly, Ala, Ser, Glu, Asp, Lys, Arg, and His. The secondary structure of one part is α-helix, and the other is β-pleated sheet. Which part is α-helix and which β-pleated sheet?

20.91 Part X of a polypeptide contains considerable amounts of Pro and Glu and very little Gly, Ala, and Ser. What is the secondary structure of part X? Explain.

20.92 Which of the following functions are performed by globular proteins and which by fibrous proteins? Catalytic; structural; transport; regulatory; contractile; protective.

20.93 Explain each of the following observations regarding the effect of mutations on function: (a) replacement of Gly by Trp often has a large effect; (b) replacement of Arg by Lys often has little effect; (c) replacement of Thr by Ser often has little effect; (d) replacement of Arg by Asp often has a large effect.

20.94 What mechanism is responsible for denaturation of a protein by each of the following changes: (a) decrease pH to 2.0; (b) microwave energy; (c) reducing agent; (d) mercury compounds.

Chemical Connections

20.95 The following amino acid residues are found in certain proteins: residues 1 and 2 in collagen and residue 3 in prothrombin (a protein in blood-clot formation). Each of these residues is formed by modification of one of the 20 amino acids after a peptide is synthesized. Which amino acid is modified to produce each residue?

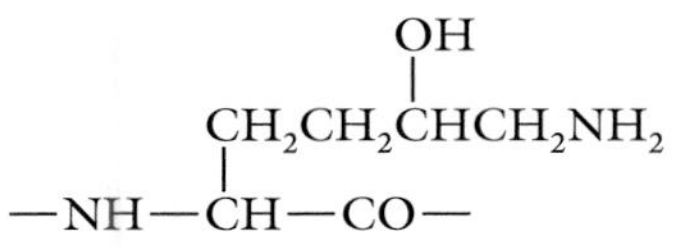

1

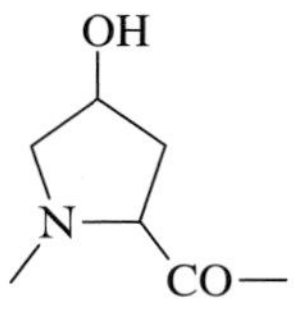

2

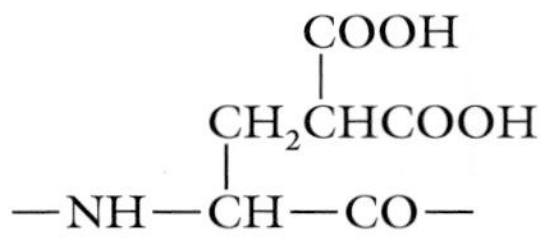

3

20.96 Glutathione (γ-glutamylcysteinylglycine) functions as a scavenger for oxidizing agents in red blood cells and other cells. It is an atypical peptide in that the γ-carboxyl group, not the α-carboxyl group, of glutamic acid is used in forming the peptide bond with the α-amino group of cysteine. Draw the structure of glutathione at physiological pH.

20.97 Consider an integral protein that spans the lipid bilayer of a cell membrane and protrudes on both the extracellular and intracellular sides. Of the α-amino acid residues Glu, Leu, Lys, Phe, Ser, and Val, predict which are in contact with (a) the interior of the bilayer, (b) the extracellular environment, and (c) the intracellular environment. Explain.

20.98 Describe the chemical reactions taking place in the permanent waving of hair (Box 20.2).

20.99 The defect in sickle-cell hemoglobin is caused by the replacement of only one α-amino acid residue: the glutamic acid residue at position 6 of the β-polypeptide chains is replaced by valine. On the other hand, beef insulin is useful for treating diabetics, even though it differs from human insulin in three residues (Box 20.3). Explain the difference.

20.100 α-Keratins are the structural components of hair and nails. Hair splits most easily along its longest dimension, whereas fingernails tend to split across the finger direction not along the finger direction. What differences most likely exist in the fibrous structures of hair and nails?

20.101 The lipids and proteins of cell membranes have lateral mobility. Explain why the lateral motion of proteins is much slower than that of lipids (Section 19.10).

CHAPTER 21

NUCLEIC ACIDS

CHEMISTRY IN YOUR FUTURE

You are a genetic counselor talking to a young married couple. The wife is pregnant, and you have the DNA test results from the amniocentesis that was performed on her fetus. These indicate that the baby will be born with sickle-cell anemia because the gene for the β-polypeptide chain of hemoglobin has a point mutation at one of its 438 nucleotide bases. The messenger RNA produced from the gene for the β-polypeptide chain has a total of 146 codons, and the point mutation results in the wrong base triplet at codon 6. What is a point mutation? What is messenger RNA? What is a codon? And why does a mutation at only one codon have such a large effect on a child's health? The answers are in Chapter 21.

LEARNING OBJECTIVES

- Describe and draw the structures of nucleotides.
- Describe and draw the structures of nucleic acids, and write equations for their synthesis from nucleotides.
- Describe the three-dimensional structures of DNA and RNA.
- Describe the replication of DNA.
- Describe the formation of rRNA, tRNA, and mRNA.
- Describe the genetic code for polypeptide synthesis.
- Describe the synthesis of polypeptides.
- Describe mutations and how they affect living organisms.
- Describe how antibiotics fight infections.
- Describe viruses and vaccines.
- Describe recombinant DNA technology and some of its uses.

We learned in Chapter 20 that each biological species is different from every other because of structural differences in their proteins. Humans, elephants, salmon, shrimp, string beans, and roses each have a different (though not necessarily an entirely different) set of proteins. Proteins also differ somewhat between individual members of the same species, but the differences are much smaller than the differences that distinguish species. The proteins of the offspring of the same parents are much more alike than those of unrelated members of the same species. Thus, Mary Kelley and her brother Tom have many more traits in common with each other and with one or the other of their parents than they do with their friend Jeffrey Rogers and his sister Gloria.

Chromosomes, located in the nuclei of cells, contain the hereditary information that directs the synthesis of the approximately 100,000 proteins unique to a human being. All human cells except germ cells (sperm and ova) are **diploid,** which means that they contain two copies of each chromosome: 23 pairs of chromosomes, for a total of 46. Germ cells are **haploid** cells, which means that they contain only one copy of each chromosome.

Every chromosome contains large numbers of **genes,** the fundamental units of heredity. Genes are responsible both for the traits common to a species and for the unique traits of an individual member of that species. Each gene carries the information for synthesizing one or more polypeptides, which are responsible for those hereditary traits.

At the molecular level, the growth and reproduction of organisms are directed and carried out by two types of **nucleic acids—ribonucleic acids (RNA)** and **deoxyribonucleic acids (DNA).** Chromosomes contain DNA molecules; each gene is just a part of a DNA molecule. DNA contains the hereditary information and directs its own reproduction and the synthesis of RNA. RNA molecules leave the cell nucleus and direct the synthesis of proteins in ribosomes, which are organelles in the cytosol. (The cytosol is the region of the cell outside the nucleus.)

21.1 NUCLEOTIDES

Nucleic acids are polymers, as are the carbohydrates and polypeptides that you studied in Chapters 19 and 20. Their building blocks (monomers) are called **nucleotides.** Unlike the building blocks for carbohydrates and polypeptides,

The DNA of all species is based on the same four nucleotides. The differences between this man and the rhinoceroses reside in the differences in the nucleotide sequences of their DNA.

Figure 21.1 Relation of components of nucleotides to nucleic acids.

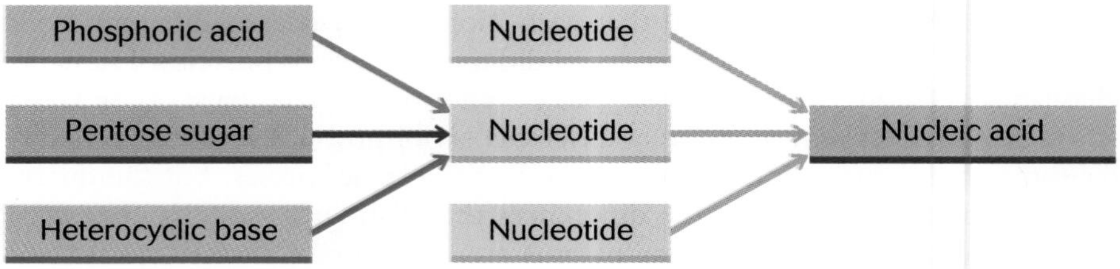

◀◀ The —OH group at C2 in D-ribose is replaced by —H in 2-deoxy-D-ribose, as described in Section 18.3.

however, nucleotides are themselves hydrolyzable into three components—phosphoric acid, a pentose sugar, and a heterocyclic nitrogen base (Figures 21.1 and 21.2). The pentose sugar found in RNA is D-ribose; the pentose sugar in DNA is 2-deoxy-D-ribose. The sugars are usually referred to simply as ribose and deoxyribose.

RNA and DNA also differ in the composition of their heterocyclic nitrogen bases. A total of five different heterocyclic bases are used: three pyrimidine and two purine bases. A pyrimidine base contains one ring, whereas a purine base has a bicyclic structure. Each of the five bases is symbolized by the first letter of its name—C, T, U, A, and G for cytosine, thymine, uracil, adenine, and guanine, respectively. A, G, and C are used in synthesizing both DNA and RNA nucleotides. T is used only for DNA, whereas U is used only for RNA.

Nucleotide synthesis is a complex process requiring metabolic energy, but the end result is easily understood by visualizing two dehydrations among the

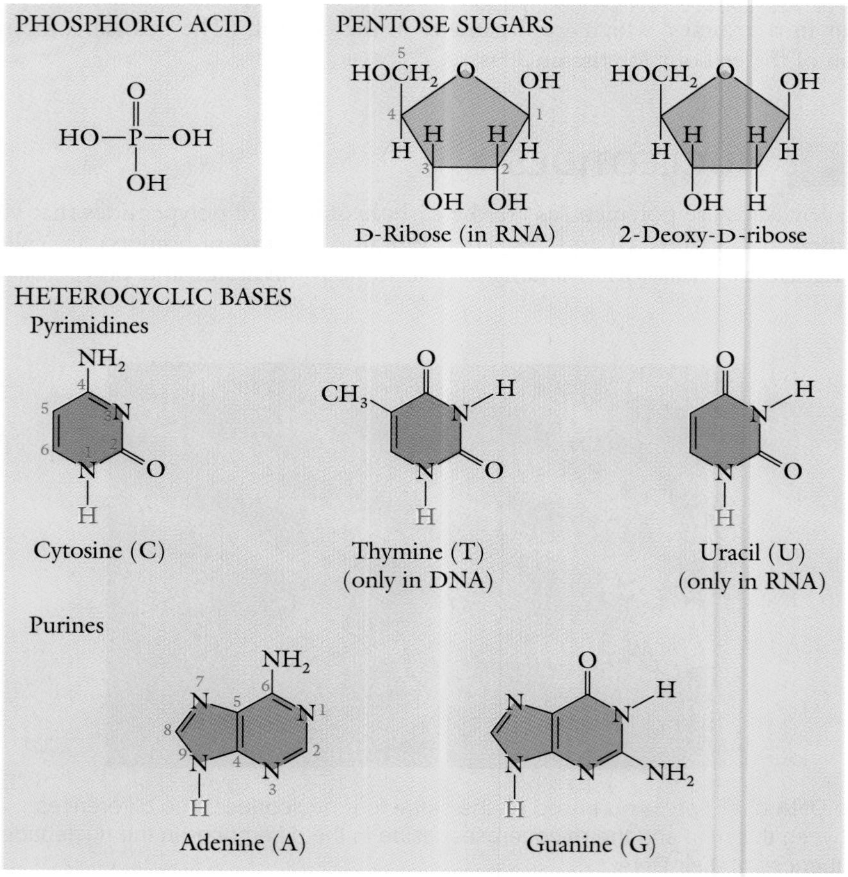

Figure 21.2 Individual components of nucleotides. The blue hydrogen is lost when a heterocyclic base is bonded to a sugar.

three components: one dehydration between phosphoric acid and the pentose sugar and the other between the pentose sugar and the heterocyclic base. The synthesis of a nucleotide from phosphoric acid, deoxyribose, and cytosine, for example, proceeds as follows:

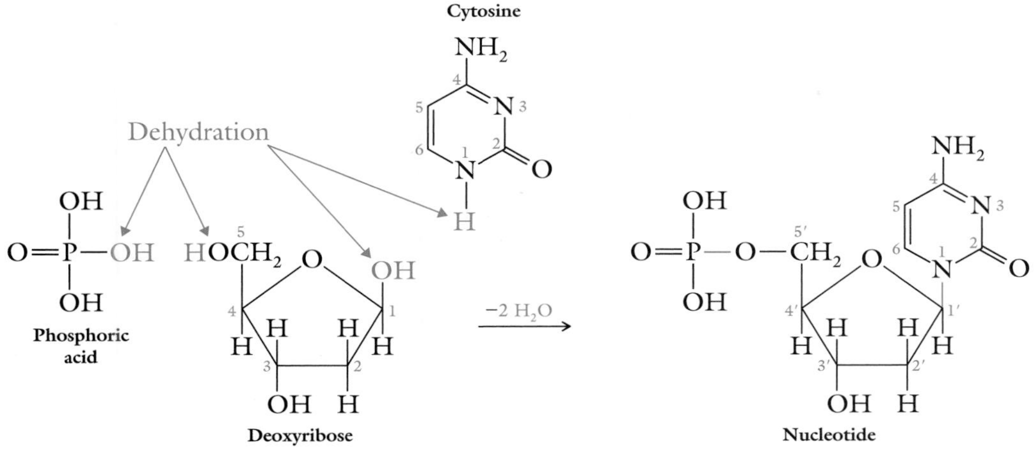

Dehydration between phosphoric acid and deoxyribose takes place at C5 of the pentose ring. Dehydration between deoxyribose and cytosine takes place at C1 of the pentose ring and at an N—H hydrogen of the heterocyclic base. In pyrimidine bases, it is the hydrogen at position 1; in purine bases, it is the hydrogen at position 9. Note the use of prime numbers in drawing the structure of the nucleotide, to distinguish ring positions in the sugar from ring positions in the heterocyclic base.

Nucleotides derived from ribose and deoxyribose are called **ribonucleotides** and **deoxyribonucleotides,** respectively.

Nucleotides are named as follows:

Rules for naming nucleotides

1. The first word of the name indicates the sugar and base components.
   a. The part of the molecule composed of purine base and sugar is named by replacing the ending **-ine** of the base by **-osine;** thus, **adenosine** = adenine + sugar and **guanosine** = guanine + sugar. The pyrimidine bases—cytosine, thymine, and uracil—combined with sugar are named **cytidine, thymidine,** and **uridine,** respectively.
   b. When deoxyribose is the sugar, the prefix **deoxy-** is used. Otherwise no prefix is used.
2. **5′-Monophosphate** is the second word of the name, indicating the phosphate group at C5′ of the sugar.

Nucleotide names are usually abbreviated. For example, the prefix deoxy- is shortened to **d-** and is followed by the one-letter symbol for the base **(C, T, U, A, G)** and **MP** for 5′-monophosphate. Thus, the nucleotide containing deoxyribose and cytosine, whose structure is shown near the top of this page, is named deoxycytidine 5′-monophosphate or dCMP.

Problem 21.1 Draw the structure of the nucleotide consisting of phosphoric acid, deoxyribose, and guanine. Give the full and abbreviated names.

Problem 21.2 What components make up the nucleotide UMP? Give the full name for UMP.

Concept checklist

- ✔ The bases adenine, guanine, and cytosine are used in both DNA and RNA nucleotides.
- ✔ Thymine is used only for DNA, whereas uracil is used only for RNA.

21.2 NUCLEIC ACID FORMATION FROM NUCLEOTIDES

As noted earlier, nucleic acids are polynucleotides. RNA is formed from ribonucleotides and DNA is formed from deoxyribonucleotides. Like that of nucleotides, the synthesis of nucleic acids is a complex process. The products of nucleic acid synthesis are those that would be produced by a simple dehydration reaction between the —OH of the phosphate group at C5′ of one nucleotide molecule and the —OH at C3′ of another nucleotide molecule. The group formed by this reaction, called a **phosphodiester group,** joins one nucleotide residue to another. For example, the dinucleotide UMP-CMP is produced by the reaction of UMP with CMP through dehydration between the —OH group at C3′ of UMP and the phosphate group at C5′ of CMP:

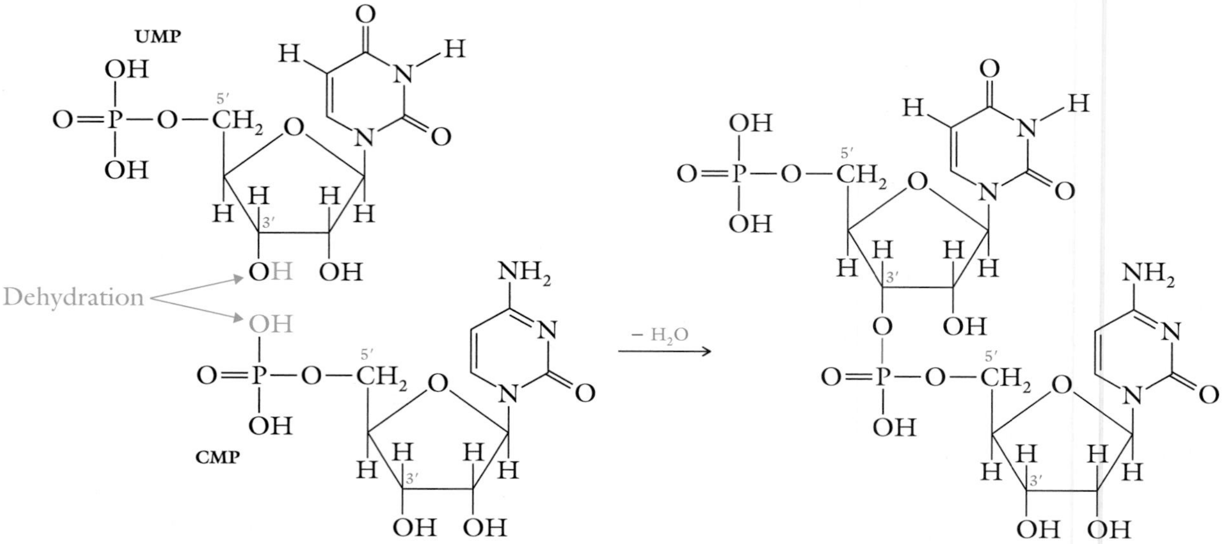

The formation of a three-nucleotide sequence, dCMP-dAMP-dTMP, is shown in Figure 21.3. By convention, nucleotide sequences are named in the 5′ ⟶ 3′ direction. A nucleic acid has one 5′-end and one 3′-end. The **5′-end** has a phosphate group at C5′ that is attached to only one pentose ring. All other phosphate groups are attached to two different pentose rings. The **3′-end** has a pentose ring with an unreacted —OH group at C3′. The nucleotide residues in a nucleic acid are named by proceeding from the 5′-end to the 3′-end.

When the ends of the nucleic acid are not given, an alternate method for describing the 5′ ⟶ 3′ convention is used: Proceed from the 5′-carbon of a pentose ring to the 3′-carbon of the same pentose ring and then move to succeeding pentose rings. Do not proceed from the 5′-carbon of one pentose ring to the 3′-carbon of the adjacent pentose ring.

The importance attached to the 5′ ⟶ 3′ convention stems from the fact that the three-nucleotide sequence dCMP-dAMP-dTMP is only one of a total of six constitutional isomers—corresponding to six different ways that three nucleotides can be joined together. The other five constitutional isomers are:

dTMP-dAMP-dCMP
dCMP-dTMP-dAMP
dAMP-dTMP-dCMP
dAMP-dCMP-dTMP
dTMP-dCMP-dAMP

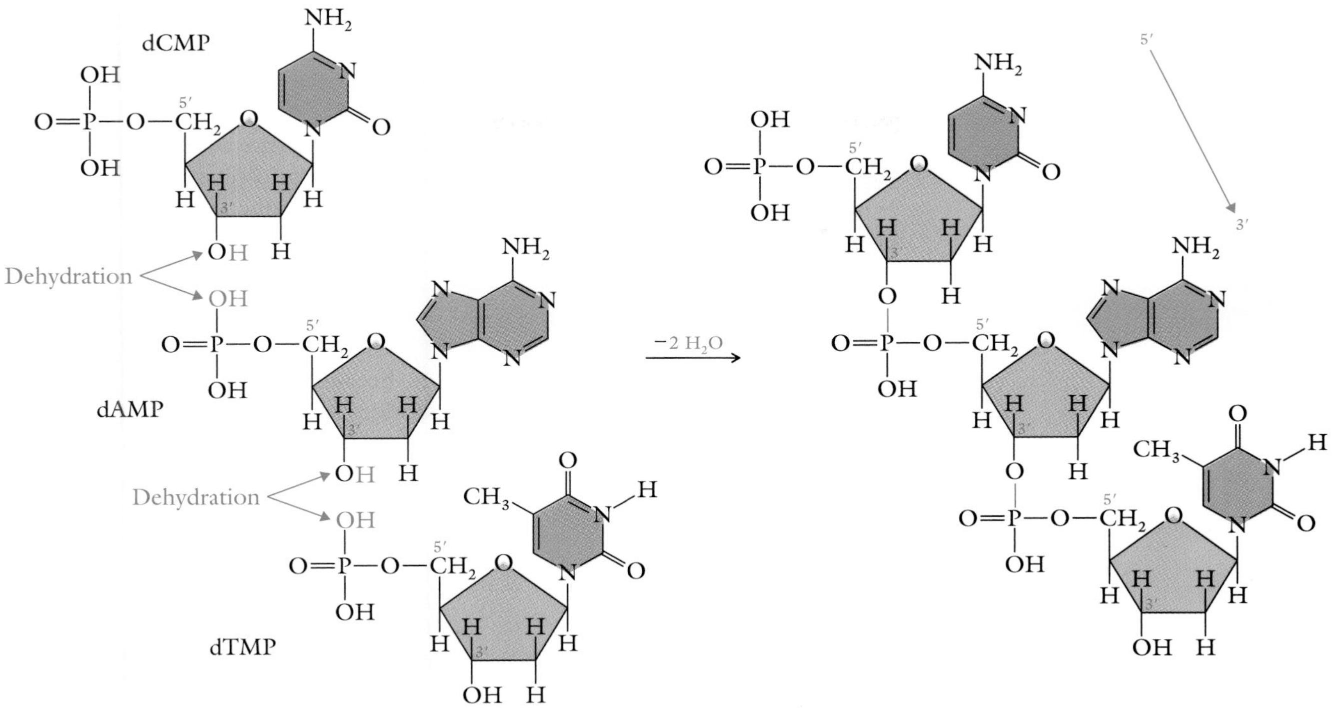

Figure 21.3 Synthesis of dCMP-dAMP-dTMP.

The nucleotide sequence in a specific nucleic acid is critical to its function. We will see in Section 21.7 that the nucleotide sequence in DNA is what determines the sequence of α-amino acid residues in a polypeptide.

Drawing the nucleic acid structures for sequences of more than two or three nucleotide residues becomes cumbersome. Figure 21.4 shows various abbreviated representations of a tetranucleotide sequence. Many structural details are left out of these representations so that, for example, instead of looking at the sugar to find out if a sequence represents DNA or RNA, we must look at other available information, such as whether T or U is present. By convention

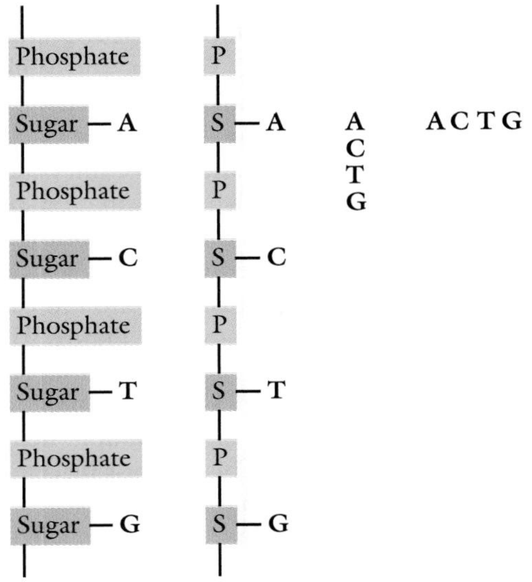

Figure 21.4 Different representations of deoxyribonucleic acid containing the nucleotide sequence dAMP-dCMP-dTMP-dGMP.

the $5' \longrightarrow 3'$ direction proceeds either from top to bottom or from left to right. For sequences of more than a dozen nucleotide residues, it is very inconvenient to use any representation other than the one on the far right in Figure 21.4, written left to right instead of top to bottom.

All of these representations emphasize the key feature of nucleic acids: the sequence of bases present as side groups on a sugar-phosphate polymer chain. Note the analogy to polypeptides, where there is a sequence of different side groups on a polyamide chain.

Concept check

✔ Nucleic acids are named in the $5' \longrightarrow 3'$ direction.

Nucleotides and nucleic acids are acidic because of the presence of P–OH groups. Although the structures of nucleotides and nucleic acids are shown with the P–OH in nonionized form, it should be understood that they are ionized to $P{-}O^-$ at physiological pH.

Problem 21.3 Draw the complete structure of GMP-UMP. Indicate the $5' \longrightarrow 3'$ direction.

21.3 THREE-DIMENSIONAL STRUCTURE OF NUCLEIC ACIDS

The primary structure of a nucleic acid—the sequence of bases along the sugar-phosphate polymer chain—determines its higher-level conformational structure (secondary, tertiary, quaternary). The situation is analogous to that of proteins except that there are only a few different types of three-dimensional structures for nucleic acids—corresponding to their limited (but enormously critical) range of functions.

Deoxyribonucleic Acids

The three-dimensional structure of DNA was deduced by James Watson and Francis Crick in 1953 on the basis of two key pieces of data. First, prior to 1953, Erwin Chargoff showed that, although the base compositions of the DNA in different species of plants and animals varied considerably, the molar amounts of A and T were equal and the molar amounts of G and C were equal for each species. Second, Maurice Wilkins and Rosalind Franklin obtained the X-ray diffraction structure of DNA in 1953. The 1962 Nobel Prize in Medicine or Physiology was awarded to Crick, Watson, and Wilkins for their work in elucidating the structure of DNA. Franklin deserved to be included in the award but lost out on two counts—the Nobel Prize regulations limit the number of co-awardees to three persons and, more important in this instance, the prize cannot be awarded posthumously (Franklin died in 1957).

The Watson-Crick model for DNA is the **DNA double helix.** DNA molecules do not exist as individual polynucleotide molecules. Instead, two paired DNA molecules (often called **DNA strands**) fold around each other to form a right-handed double helix **(duplex),** as shown in Figure 21.5. The two DNA strands of the DNA double helix run in opposite directions, one in the $5' \longrightarrow 3'$ direction and the other in the $3' \longrightarrow 5'$ direction. The phosphate groups are located on the outer surface of the double helix, and the heterocyclic bases are inside the double helix. This interior placement of heterocyclic bases is critical to the formation of the double helix.

The orientation of the heterocyclic base on each nucleotide residue is perpendicular to the axis of the double helix; that is, the bases are in the horizon-

INSIGHT INTO FUNCTION

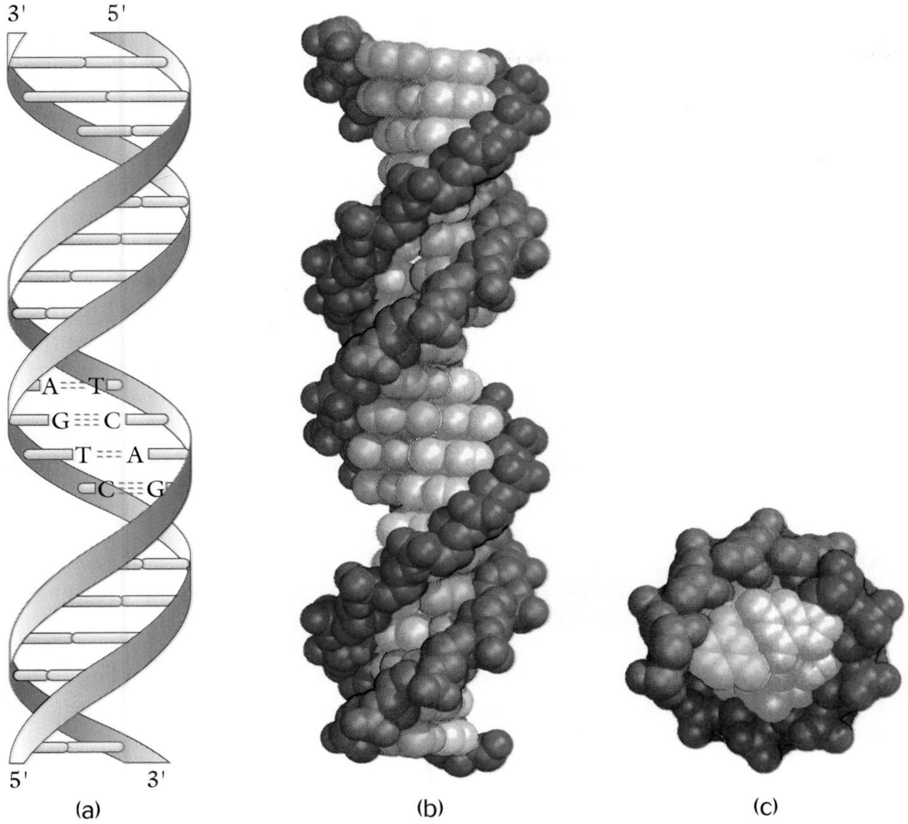

Figure 21.5 DNA double helix: (a) simple representation with hydrogen bonding shown between base pairs; (b and c) axial and radial views of space-filling models. One DNA strand is green and the other is red. The bases are shown in lighter shades of green and red. We shall see how this structure provides the basis for biological growth and development, reproduction, and evolution.

tal plane when the double helix is viewed in the vertical plane. Bases stack on top of each other in the vertical plane, like the steps in a spiral staircase.

There are strong hydrogen-bonding attractive forces between a base on one DNA strand and a base on the other DNA strand. This attraction is called **base-pairing,** and the bases pair in a **complementary** manner—adenine pairs only with thymine, and cytosine pairs only with guanine. Adenine and thymine pairing is accomplished through two hydrogen bonds; cytosine and guanine pairing is accomplished through three hydrogen bonds (Figure 21.6). The two DNA strands in the double helix are **complementary strands** in the sense that each heterocyclic base is located opposite its complementary base (A opposite T; G opposite C). Base-pairing to form a DNA double helix with maximum hydrogen bonding (and maximum stability) requires that the two strands run in opposite directions. If the strands ran in the same direction, the bases would

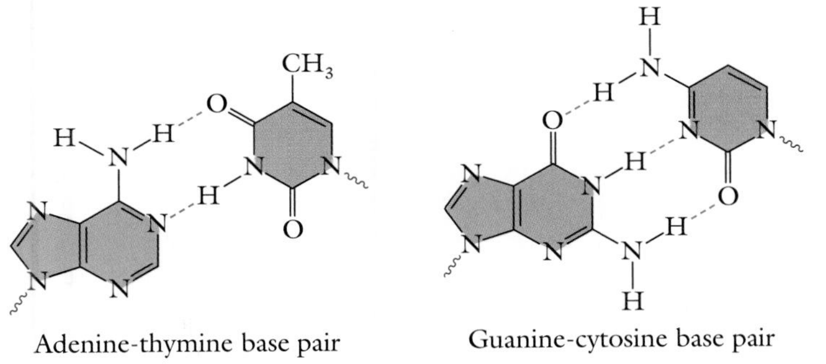

Figure 21.6 Complementary base-pairing in A-T and G-C.

not be in the correct proximity to each other to form the maximum number of hydrogen bonds.

Hydrophobic attractions between bases contribute—perhaps even more than hydrogen bonding—to the stabilization of the DNA duplex. Each heterocyclic base in a base pair has hydrophobic attractive interactions with the bases directly above and below it.

Concept check

✔ The double helix is stabilized by hydrogen bonding and hydrophobic attractions between heterocyclic bases.

Why is base-pairing only between A and T and between C and G but not between A and C, G and T, C and T, or A and G? As Watson and Crick were the first to realize, the A-T and C-G base pairs have the same dimension (see Figure 21.6), which results in a uniform cross-sectional dimension throughout the entire DNA double helix. Other combinations of bases have different dimensions and would decrease the stability of the DNA duplex by causing variations in the cross-sectional dimension and a consequent decrease in hydrogen bonding and hydrophobic attractions between bases.

Example 21.1 Determining the base sequence in a DNA strand from the sequence in its complementary strand

If the base sequence in a section of one DNA strand is ACGTAG, reading in the $5' \longrightarrow 3'$ direction, what is the sequence, reading in the $5' \longrightarrow 3'$ direction, for the corresponding section of the complementary DNA strand?

Solution

The complementary strand has A opposite T in the given strand, T opposite A, C opposite G, and G opposite C.

5′ ACGTAG 3′ **Given strand**

3′ TGCATC 5′ **Complementary strand**

Reading in the $5' \longrightarrow 3'$ direction, the sequence for the complementary strand is CTACGT.

Instead of writing a base sequence as "ACGTAG, reading in the $5' \longrightarrow 3'$ direction," we can write 5′-ACGTAG-3′ as an abbreviation or just simply ACGTAG because the $5' \longrightarrow 3'$ direction is the convention unless otherwise indicated.

Problem 21.4 If the base sequence in a section of one DNA strand is 5′-GGCTAT-3′, what is the sequence for the corresponding section of the complementary DNA strand?

DNA strands are enormously large molecules, with molecular masses estimated to be from a few billion to as high as 100 billion. Considering that the human cell nucleus has a diameter of only 10 micrometers ($1\ \mu m = 10^{-6}$ m), whereas the **total DNA** (called the **genome**) contained in its 46 chromosomes has about 3×10^9 base pairs (and a length of about 1 m if stretched end to end), the DNA duplex must be highly compacted to fit into the nucleus (Figure 21.7). This compaction is achieved by the double helix being folded around structures called **nucleosome cores.** Each nucleosome core is composed of two pairs each of four different proteins called **histones.** The DNA duplex wraps around the nucleosome cores to form a chain of **nucleosomes,**

INSIGHT INTO PROPERTIES

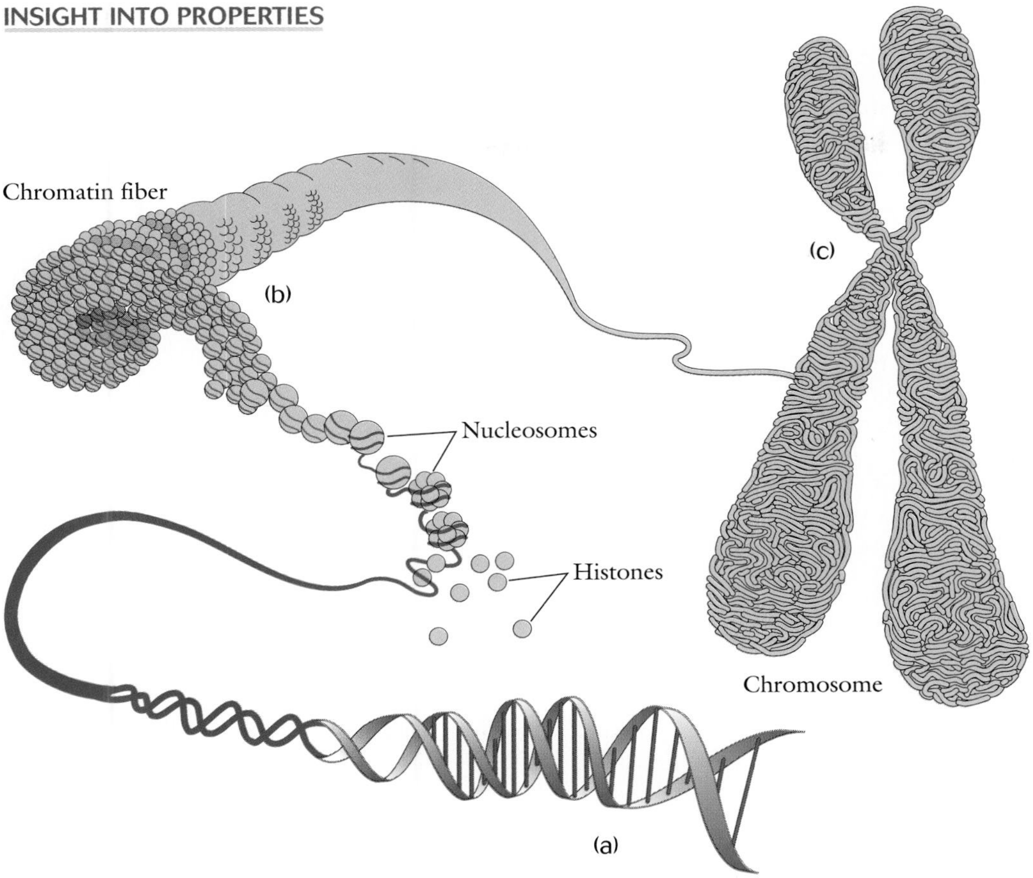

Figure 21.7 Different levels of DNA structure: (a) double helix (hydrogen bonding between base pairs is represented by straight lines between the two DNA strands); (b) highly supercoiled chromatin fiber; (c) chromosome. The compaction and supercoiling illustrated here are necessary to fit the enormously long DNA duplexes into the nuclei of human cells.

with about 150 to 200 base pairs per nucleosome. Large numbers of nucleosome cores are needed to compact each DNA duplex; chromosomes contain about equal masses of DNA and histones. The chain of nucleosomes is coiled to higher and higher levels to form a highly compact, highly supercoiled **chromatin fiber.**

Ribonucleic Acids

Ribonulceic acids exist as single-strand molecules, not as double helices as does DNA. The spatial bulk of the —OH group at C2′ of the ribose units in RNA (in place of the —H of deoxyribose in DNA) prevents the formation of long stretches of double-helix conformation. Several different types of RNA molecules exist, each with its own three-dimensional structure. Figure 21.8 on the following page shows one type of RNA molecules called **transfer RNA** or **tRNA.** The overall shape of tRNA resembles a cloverleaf, especially in the two-dimensional representation.

The secondary and tertiary structures of tRNA result from intramolecular base-pairing of nucleotide residues. The base-pairing scheme is the same as that for DNA except that uracil is used in RNA instead of thymine. Adenine pairs with uracil; cytosine pairs with guanine. Roughly half the nucleotide

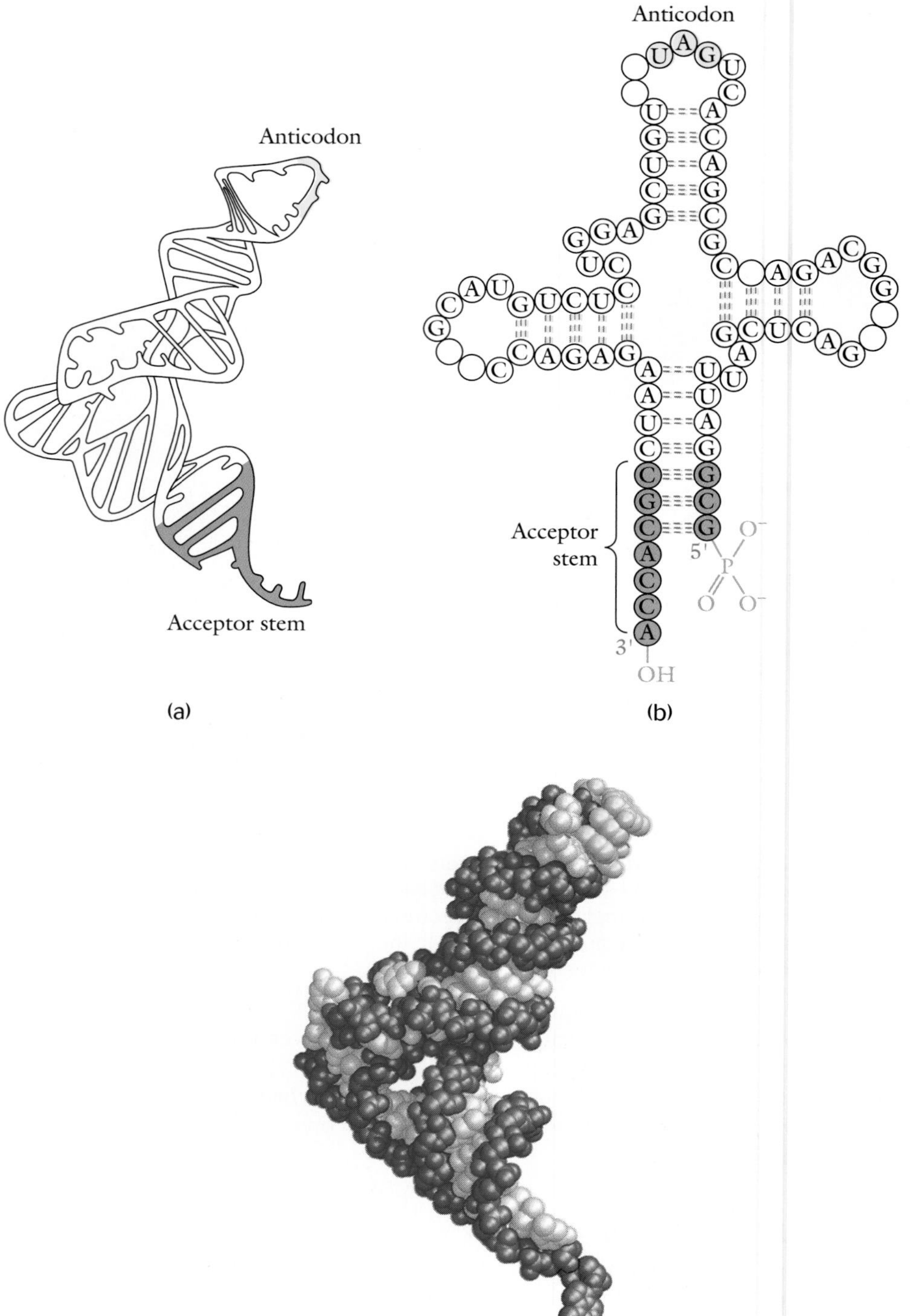

Figure 21.8 Structure of a tRNA molecule: (a) folded into its three-dimensional structure with hydrogen-bonded base pairs represented by straight bars connecting parts of tRNA; (b) two-dimensional structure with hydrogen-bonded base pairs represented by dashed lines. The unspecified nucleotides are modified analogues of A, C, G, or U. (c) A space-filling model of tRNA. The RNA polymer chain is shown in dark blue and the heterocyclic bases in silver. The anticodon is shown in yellow and the acceptor stem in red.

bases in a tRNA molecule participate in base-pairing. Base-pairing occurs not in one long continuous sequence but in short sequences. Sections of the tRNA molecule with approximately helical conformation are separated by nonhelical sections without base-pairing. The nonhelical sections, except for the linear section at the 3′-end of the tRNA chain, are curved and called **loops.** The linear section at the 3′-end is called the **acceptor stem,** and the loop farthest from it is called the **anticodon.** There is no base-pairing in the loop sections, because complementary bases are not opposite each other or, if opposite each other, are not within hydrogen-bonding distance. The acceptor stem and anticodon play key roles in polypeptide synthesis (Section 21.7).

21.4 INFORMATION FLOW FROM DNA TO RNA TO POLYPEPTIDE

Figure 21.9 gives an overview of molecular genetics—the storage, transmission, and use of genetic information at the molecular level. Genetic information is stored in DNA as a sequence of nucleotide residues. These sequences correspond to specific sequences of α-amino acid residues in polypeptides. DNA is the "master" molecule, directing its own replication as well as the formation of various RNA molecules. DNA remains in the nucleus, being too large to move through the nuclear membrane into the cytosol. The much smaller RNA molecules go out into the cytosol to carry out protein synthesis. The genetic process consists of three parts:

- **Replication** is the copying of DNA in the course of cell division.
- **Transcription** is the synthesis of RNA from DNA. Three types of RNA molecules are produced: **ribosomal RNA (rRNA), messenger RNA (mRNA),** and **transfer RNA (tRNA).**
- **Translation** is the synthesis of polypeptides through the combined efforts of rRNA, mRNA, and tRNA.

21.5 REPLICATION

Replication is an extremely complex process, requiring about two dozen enzymes. Copying the human genome consists of simultaneously copying 46 DNA duplexes, each duplex containing an average of about 70 million base pairs. Cell division of a typical human cell takes a little less than a day, with DNA replication accounting for about one-third of that time, which means that thousands of nucleotides are joined together each second during replication. All of this replication is carried out with an error level of less than 1 wrong nucleotide per 10 billion. This extreme level of accuracy preserves the integrity of the genome from generation to generation—not only from parents to offspring, but also from one generation of cells to the next generation within the same person. It is the reason that the replication process is so complex.

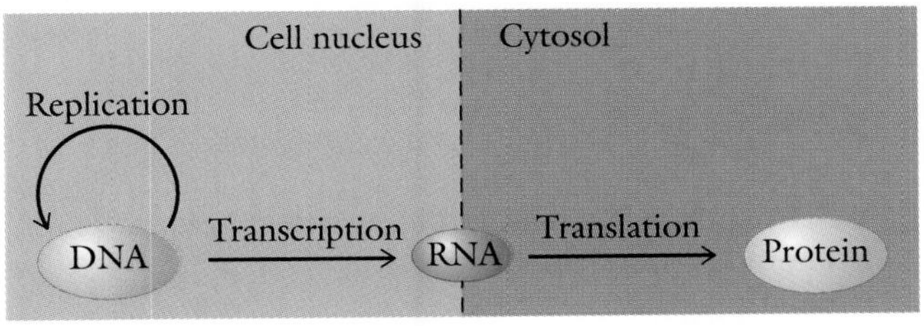

Figure 21.9 Overview of molecular genetics.

In replication, the strands of a DNA duplex unwind over a short span of 150 to 200 nucleotide residues, forming a **replication bubble** with a **replication fork** at each end (Figure 21.10a). Each strand acts as a **template strand** on which a complementary strand is synthesized. Enzymes catalyze the unwinding process by breaking the hydrogen bonding between base pairs; other proteins stabilize the unwound DNA strands by hydrogen bonding with the bases. The enzyme **DNA polymerase** catalyzes the addition of nucleotides to growing DNA strands.

The two new strands grow from replication forks at opposite sides of the bubble. They grow in opposite directions because new bases are always added in the $5' \longrightarrow 3'$ direction. As DNA synthesis proceeds, the DNA duplex un-

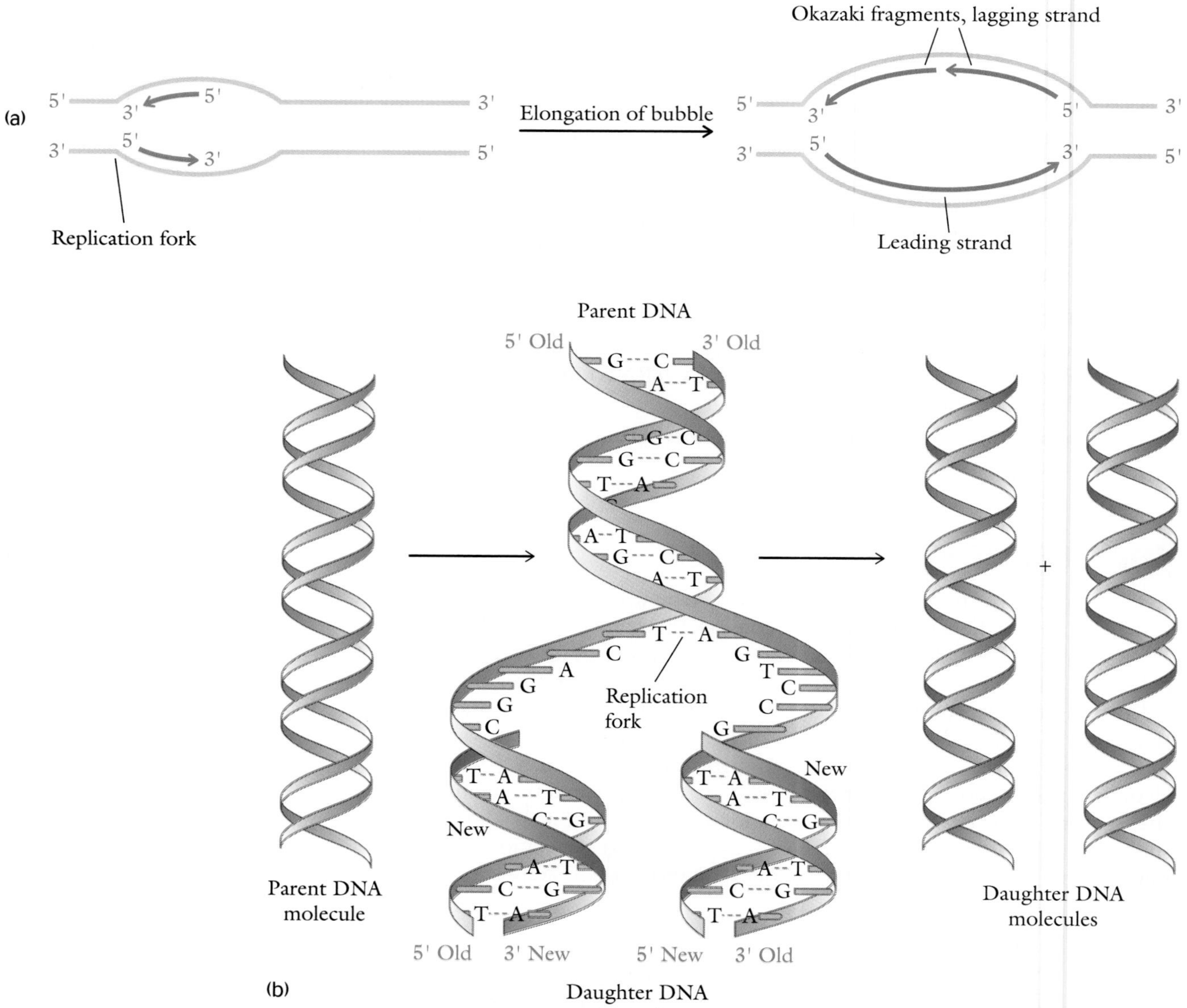

Figure 21.10 Replication of DNA duplex: (a) DNA synthesis at a pair of replication forks proceeds with elongation of the replication bubble (the figure is simplified by not showing the DNA strands in helical form); (b) semiconservative replication (old strands are red, new strands are green).

winds further, to elongate the replication bubble. One of the new DNA strands, the **leading strand,** is synthesized in a continuous manner. The other DNA strand, the **lagging strand,** is synthesized in a discontinuous manner as a series of smaller DNA fragments, called **Okazaki fragments.** These fragments are subsequently linked together by the enzyme **DNA ligase.**

Replication proceeds simultaneously at thousands of replication bubbles along each DNA strand, completing replication of this enormously large molecule in a reasonable amount of time. The DNA fragments synthesized in different replication bubbles are bonded together by DNA ligase to produce the final new DNA strands.

DNA polymerase, with a molecular mass of one-half million, directs the synthesis process by simultaneous interaction (through secondary forces) with the template DNA strand, the new growing DNA strand, and nucleotides. The pairing of bases and the action of DNA polymerase together ensure the accuracy of replication. Thus, if a template strand has T at a certain position, only a dAMP nucleotide will hydrogen bond and be incorporated into the new strand at that point. After addition of a nucleotide, DNA polymerase moves to the next nucleotide residue in the template strand and ensures that only the correct, complementary nucleotide will be added to the chain. DNA polymerase not only catalyzes the addition of nucleotides to growing DNA chains, but also has a role, along with other enzymes, in proofreading and correcting errors, thus ensuring that replication is carried out with an error level of less than 1 wrong nucleotide per 10 billion.

Through replication, a single **parent** DNA duplex is transformed into two **daughter** DNA duplexes, each containing one old and one new DNA strand. Replication is therefore described as **semiconservative** (one parental strand is conserved intact within each daughter DNA duplex). The new DNA strand of a daughter DNA duplex is identical with the old DNA strand of the other daughter duplex (Figure 21.10b).

Example 21.2 Determining the base sequences in parent and daughter DNA strands

The sequence 5′-GCGTAA-3′ is part of one strand of a parent DNA double helix. What is the corresponding sequence on the complementary strand of the parent DNA double helix? What is the corresponding sequence on the new daughter strand made from the parent strand during replication?

Solution

The complementary parent strand and the new daughter strand are identical (5′-TTACGC-3′) and complementary to the 5′-GCGTAA-3′ sequence of the parent DNA strand:

DNA parent complement	3′ CGCATT 5′ = 5′ TTACGC 3′
	↑
DNA parent	5′ GCGTAA 3′
	↓
DNA daughter complement	3′ CGCATT 5′ = 5′ TTACGC 3′

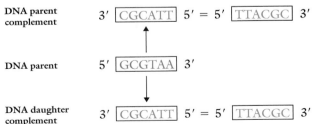

Problem 21.5 The sequence 5′-ACGTGC-3′ is part of one strand of a parent DNA double helix. What is the corresponding sequence on the complementary strand of the parent DNA double helix? What is the corresponding sequence on the new daughter strand made from this parent strand during replication?

21.6 TRANSCRIPTION

Transcription, the synthesis of rRNA, tRNA, and mRNA with the use of information from DNA, proceeds similarly to replication (Figure 21.11). The DNA double helix unwinds to form a **transcription bubble.** Only one of the two DNA strands in the bubble, called the **template strand,** is used as the template for RNA synthesis. The other DNA strand, called the **nontemplate strand,** is not used for RNA synthesis. Transcription proceeds under the control of **RNA polymerase,** which catalyzes the joining together of nucleotides. Synthesis begins at an **initiation site** whose base sequence is recognized by RNA polymerase as a start signal. Synthesis stops when RNA polymerase encounters a **termination site** whose base sequence is recognized as a stop signal. The RNA polymerase and the newly synthesized RNA are released, and the template and nontemplate DNA strands are rewound into the DNA duplex.

Unlike DNA polymerase, RNA polymerase has no proofreading function, with the result that the error level in RNA synthesis, less than 1 base in 10,000 to 100,000 bases, is not nearly as low as that in DNA synthesis. The higher error level in RNA synthesis compared with that in DNA synthesis is not a serious problem, because RNA synthesis does not have a role in preserving the integrity of the genome from generation to generation.

As in replication, base-pairing determines the base sequence of RNA synthesized from a DNA template. A, G, C, and T in the template strand of DNA result in the incorporation of U, C, G, and A, respectively, in the RNA strand. As mentioned earlier, uracil is used in place of thymine in RNA. Thus, an A in DNA has U as its complementary base in RNA.

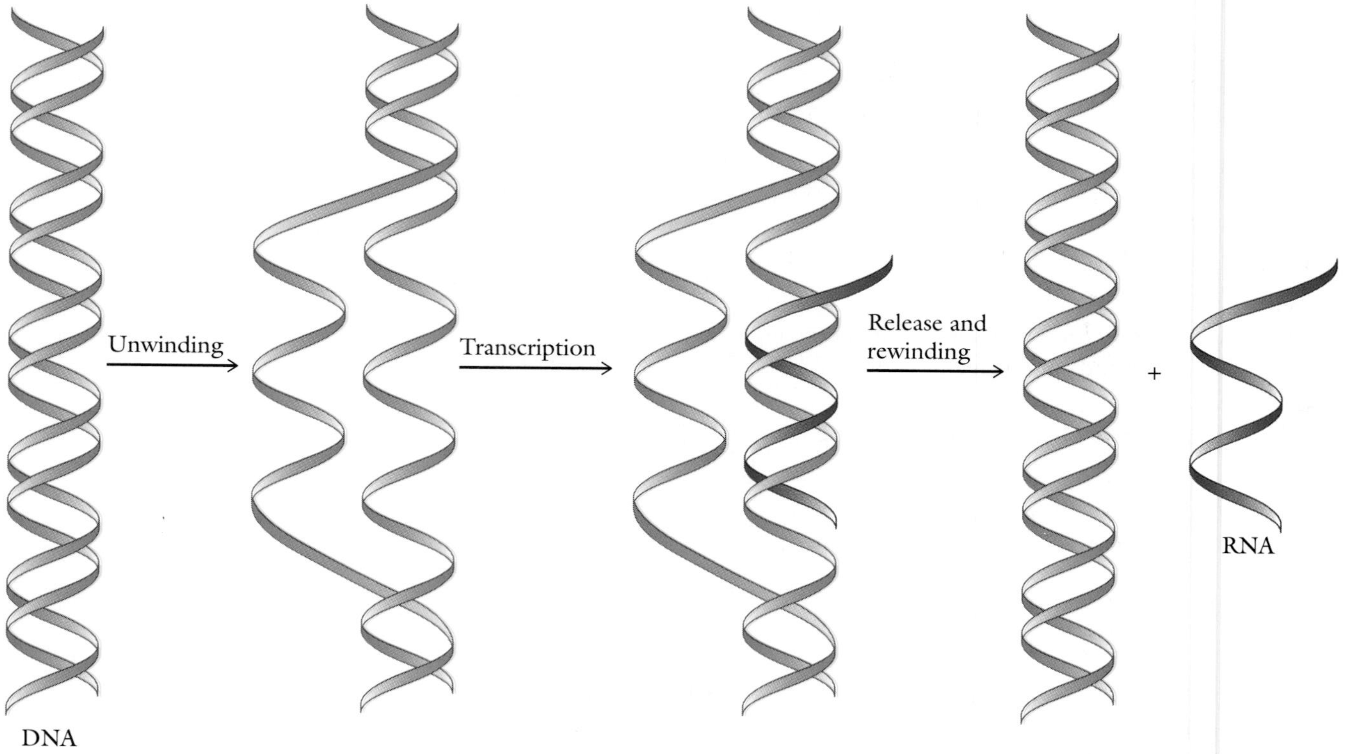

Figure 21.11 Transcription of DNA to form RNA.

Concept check

✔ The RNA synthesized from a DNA template strand has the same base sequence as the DNA nontemplate strand (except that RNA has U in place of T).

Example 21.3 Determining the base sequences in DNA nontemplate and RNA strands from the sequence in the DNA template strand

For the sequence 5′-ACATGC-3′ in a DNA template strand, what is the base sequence in the DNA nontemplate strand? What is the base sequence in the RNA synthesized from the DNA template strand?

Solution

The sequences in the DNA nontemplate and RNA strands are complementary to the sequence in the DNA template strand. The base sequence is 5′-GCATGT-3′ in the DNA nontemplate strand, and its equivalent, 5′-GCAUGU-3′, is in the RNA strand:

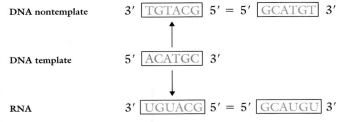

DNA nontemplate 3′ TGTACG 5′ = 5′ GCATGT 3′

↑

DNA template 5′ ACATGC 3′

↓

RNA 3′ UGUACG 5′ = 5′ GCAUGU 3′

Note that the direction of the RNA strand, like that of the DNA nontemplate strand, is opposite that of the DNA template strand. However, the convention for reading and writing base sequences of all nucleic acids is in the 5′ ⟶ 3′ direction unless otherwise stated.

Problem 21.6 For the sequence 5′-TAAGTCAAC-3′ in a DNA template strand, what is the base sequence for the DNA nontemplate strand? What is the base sequence for the synthesized RNA?

A ribonucleic acid is called a **primary transcript RNA (ptRNA)** when initially formed. ptRNA is modified by **posttranscriptional processing** to produce messenger, ribosomal, and transfer ribonucleic acids. Some of this processing takes place in the cell nucleus and some in the cytosol. The purpose of posttranscriptional processing is to stabilize and optimize a specific RNA molecule for its function in polypeptide synthesis. **End capping** adds certain nucleotide units to the 5′- and 3′-ends of the ribonucleic acid; **base modification** alters certain bases. **Splicing** deletes certain parts of the ribonucleic acid. The parts deleted are called **introns;** the parts retained are called **exons.** Figure 21.12 on the following page shows the splicing process for the conversion of a primary transcript into mRNA. The introns have no well-established biological functions. The exons are bonded together after excision of the introns to form the mRNA. The base sequence in these exons is what defines the amino acid sequence in the polypeptide whose synthesis they will subsequently direct.

The various RNAs are much smaller than DNA. Most rRNAs have molecular masses ranging from 0.5 million to 1 million. mRNAs have molecular masses ranging from 100,000 to 1 or 2 million. tRNAs are much smaller, with molecular masses below 50,000.

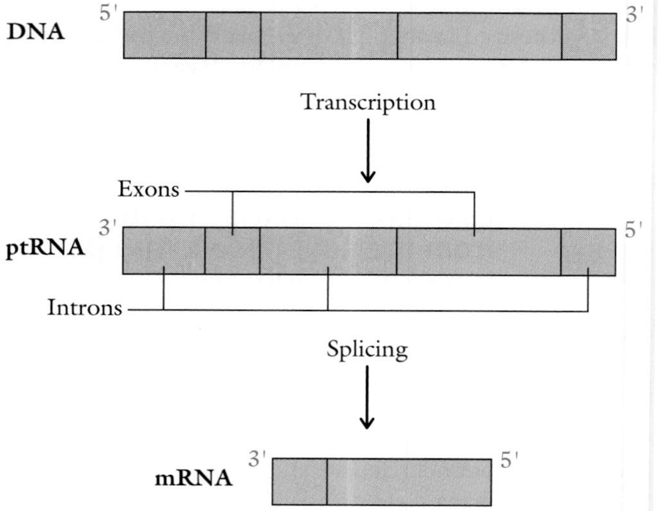

Figure 21.12 Posttranscriptional processing of ptRNA into mRNA by splicing. Introns are gray; exons are orange.

21.7 TRANSLATION

Translation is the process by which rRNA, mRNA, and tRNA work together outside the nucleus to carry out polypeptide synthesis. Polypeptides are produced at structures called **ribosomes** located in the cytosol. Ribosomes have two subunits, one large and one small, and are composed of rRNA (about 60%) and protein (most of which comprises the enzymes required for protein synthesis). mRNA, which carries the genetic message encoding the amino acid sequence of the polypeptide to be synthesized in the ribosome, binds to the ribosome and acts as the template for polypeptide synthesis. α-Amino acids are supplied for polypeptide synthesis when tRNAs transport them into the ribosome.

The **genetic message** carried by mRNA is its sequence of bases, which specifies the sequence of α-amino acids for the polypeptide to be synthesized. Each successive set of three consecutive bases **(base triplet)** in mRNA is called a **codon** because it codes for, or specifies, one specific tRNA that carries one specific α-amino acid. Mutual recognition between a codon of mRNA and its specified tRNA is accomplished through a base triplet, the **anticodon,** on the tRNA molecule (see Figure 21.8). By convention, the one-letter abbreviations for the bases are used in writing codons and anticodons: thus, UGC stands for uracil-guanine-cytosine.

The three-dimensional structures of tRNA and mRNA are such that a tRNA molecule hydrogen bonds (that is, base pairs) with an mRNA molecule only when the tRNA anticodon is complementary to the mRNA codon. For example, only the tRNA with anticodon 3′-ACG-5′ (5′-GCA-3′) base pairs to codon 5′-UGC-3′ of mRNA:

mRNA codon	5′-UGC-3′
tRNA anticodon	3′-ACG-5′

That particular tRNA carries the α-amino acid Cys.

The **genetic code** is the complete list of mRNA codons and the α-amino acids and other instructions that they specify (Table 21.1). Again, by convention, the codon sequences are always written in the 5′ ⟶ 3′ direction. Table

TABLE 21.1 Genetic Code: Codon Assignments

First Base (5′-end)	Second Base U	C	A	G	Third Base (3′-end)
U	Phe	Ser	Tyr	Cys	U
	Phe	Ser	Tyr	Cys	C
	Leu	Ser	Stop	Stop	A
	Leu	Ser	Stop	Trp	G
C	Leu	Pro	His	Arg	U
	Leu	Pro	His	Arg	C
	Leu	Pro	Gln	Arg	A
	Leu	Pro	Gln	Arg	G
A	Ile	Thr	Asn	Ser	U
	Ile	Thr	Asn	Ser	C
	Ile	Thr	Lys	Arg	A
	Met[a]	Thr	Lys	Arg	G
G	Val	Ala	Asp	Gly	U
	Val	Ala	Asp	Gly	C
	Val	Ala	Glu	Gly	A
	Val	Ala	Glu	Gly	G

[a] Codon AUG for Met also codes for the initiation of polypeptide synthesis.

21.1 is used as follows to find the α-amino acid or instruction specified by a codon:

- Find the first base of the codon in the far left vertical column.
- Move horizontally to the right of the first base to the vertical column headed by the second base of the codon. You are now in a section of the table with four entries.
- Of those four entries, the specified α-amino acid or instruction is the one on the same horizontal line as the third base, which is listed in the far right vertical column.

Problem 21.7 For each of the following codons, what is the complementary tRNA anticodon, and what α-amino acid is specified by the codon? (a) UCC; (b) CAG; (c) AGG; (d) GCU.

The genetic code is "degenerate"; that is, many of the 64 codons are redundant. Most of the α-amino acids are coded by two, three, or four codons, and Arg, Leu, and Ser are each coded by six codons. Only Met and Trp are coded by a single codon. This degeneracy acts as a protective mechanism against mutations (Section 21.8). The codon for Met, which is AUG, codes simultaneously for Met and the initiation of synthesis. This simultaneous coding means that the first codon transcribed in all mRNAs is AUG, and the first α-amino acid in all initially formed polypeptides is Met. Three of the codons (UAA, UAG, UGA) are "Stop" codons, coding for the termination of polypeptide synthesis, in that no tRNAs recognize and hydrogen bond with them.

The genetic code is nearly universal. It applies to all organisms, whether plant or animal, with very few exceptions.

Example 21.4 Determining the polypeptide structure coded by a template DNA strand

Sections *X*, *Y*, and *Z* of the following part of a template DNA strand are exons of a gene:

	5′				3′
DNA	TGT	GCG	GTG TAT	TGC AAC	CAT
	X		*Y*		*Z*

What are the structures of the primary transcript and mRNA made from this template DNA? What polypeptide sequence will be synthesized?

Solution

The base sequence in the primary transcript is the complement of that in template DNA, except that U is used in RNA instead of T. The direction of RNA synthesis is opposite the direction of the template DNA strand. In the splicing process, introns are removed and the result is mRNA:

	5′				3′
DNA	TGT	GCG	GTG TAT	TGC AAC	CAT
	X		*Y*		*Z*
			↓ transcription		
	3′				5′
Primary transcript	ACA	CGC	CAC AUA	ACG UUG	GUA
	X		*Y*		*Z*
			↓ splicing		

mRNA 3′ ACA CAC AUA GUA 5′ = 5′ AUG AUA CAC ACA 3′

The codons in mRNA must be read in the $5' \longrightarrow 3'$ direction to correctly use the genetic code (Table 21.1) to determine the sequence of α-amino acids in the polypeptide. The mRNA sequence AUG-AUA-CAC-ACA codes for the polypeptide sequence Met-Ile-His-Thr.

Problem 21.8 Sections *X*, *Y*, and *Z* of the following part of a template DNA strand are exons of a gene:

	5′				3′
DNA	CAC	CAC	ACC GTA	TGT GGA	CAT
	X		*Y*		*Z*

What are the structures of the primary transcript and mRNA made from this template DNA? What polypeptide sequence will be synthesized?

◀◀ The formation of an ester by reaction between —COOH and —OH groups was described in Section 15.8.

Each tRNA is designed to carry a specific α-amino acid, and the reaction that bonds the two together is catalyzed by the enzyme **aminoacyl-tRNA synthetase.** There is at least one synthetase for each α-amino acid. Each α-amino acid synthetase has recognition sites specific to a particular α-amino acid and its corresponding tRNA. The synthetase causes the α-amino acid to bond to the acceptor stem (see Figure 21.8) at the 3′-end of tRNA by formation of an ester linkage between the carboxyl group of the α-amino acid and the —OH group on C3′ (in some cases, the C2′) of ribose (Figure 21.13).

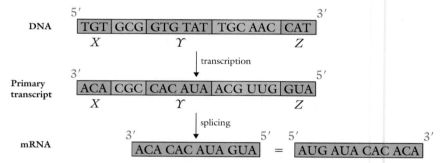

Polypeptide synthesis of the sequence Met-Gly-Ile is shown in Figure 21.14 on page 642. It takes place at **peptidyl (P)** and **aminoacyl (A) sites** in the ribosome, located on the left and right sides of the ribosome structures in Figure 21.14. The P site holds the growing polypeptide chain, and the A site holds the tRNA with its α-amino acid. Initiation starts with Met, followed by

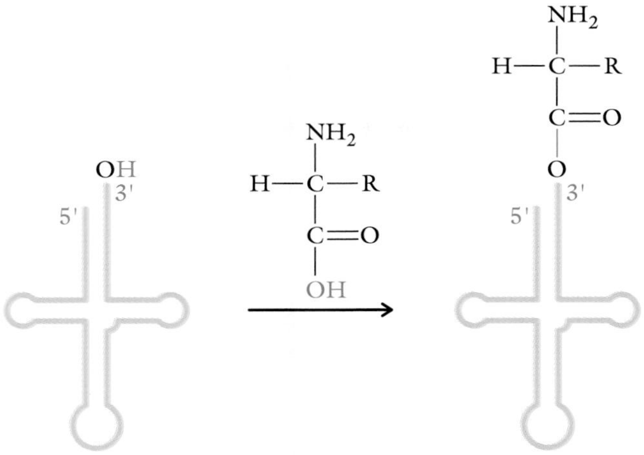

Figure 21.13 Picking up an α-amino acid at the 3′-end of tRNA.

the sequential addition of Gly and Ile. The process begins when the 5′-end of the mRNA enters the smaller subunit of the ribosome, which prepares it to receive the correct tRNAs.

In Figure 21.14a, the tRNAs for Met and Gly have already picked up Met and Gly from the cytosol, entered the ribosome, and lined up at the P and A sites, respectively (because of base-pairing between the mRNA codons and tRNA anticodons). The enzyme **peptidyl transferase,** contained in the large subunit of the ribosome, then catalyzes the transfer of Met from site P and the bonding of Met to Gly at site A. The reaction, called **acyl transfer,** yields a dipeptide at site A and an empty tRNA at site P (Figure 21.14b). The empty tRNA is then released from the ribosome (becoming available to transport another Met), and the ribosome moves along the mRNA in the $5' \longrightarrow 3'$ direction. The dipeptide is now at the P site (Figure 21.14c). A tRNA carrying Ile now base pairs at the empty A site (Figure 21.14d), after which the Met-Gly dipeptide is transferred and bonded to Ile to form the Met-Gly-Ile tripeptide at site A and an empty tRNA at site P (Figure 21.14e). Synthesis proceeds in this manner over and over again, with the ribosome moving toward the 3′-end of the mRNA.

Synthesis terminates when a Stop codon is read on mRNA, and the polypeptide is released from the ribosome. The polypeptide then undergoes **posttranslational processing,** which brings it to its final, functional form:

- The methionine residue that initiated polypeptide synthesis is cleaved from most polypeptides by a hydrolysis reaction.
- Polypeptides are folded into their active conformations.
- Some α-amino acid residues are modified by chemical reactions; examples are disulfide-bridge formation at cysteine residues (Section 20.4) and attachment of hydroxyl groups to proline residues (Section 20.6).
- For proteins with quaternary structure, different polypeptides and nonpolypeptide molecules or ions must be assembled together and folded into their active conformations.

Every step in the expression of the genetic message, from translation to posttranslation, requires the coordinated effort of large numbers of enzymes. Furthermore, superimposed on that overall process are higher-level, regulatory mechanisms that determine which of all possible genetic events take place and when they take place within each individual cell. This level of control is necessary because of the fact that the organism's entire genome—the complete

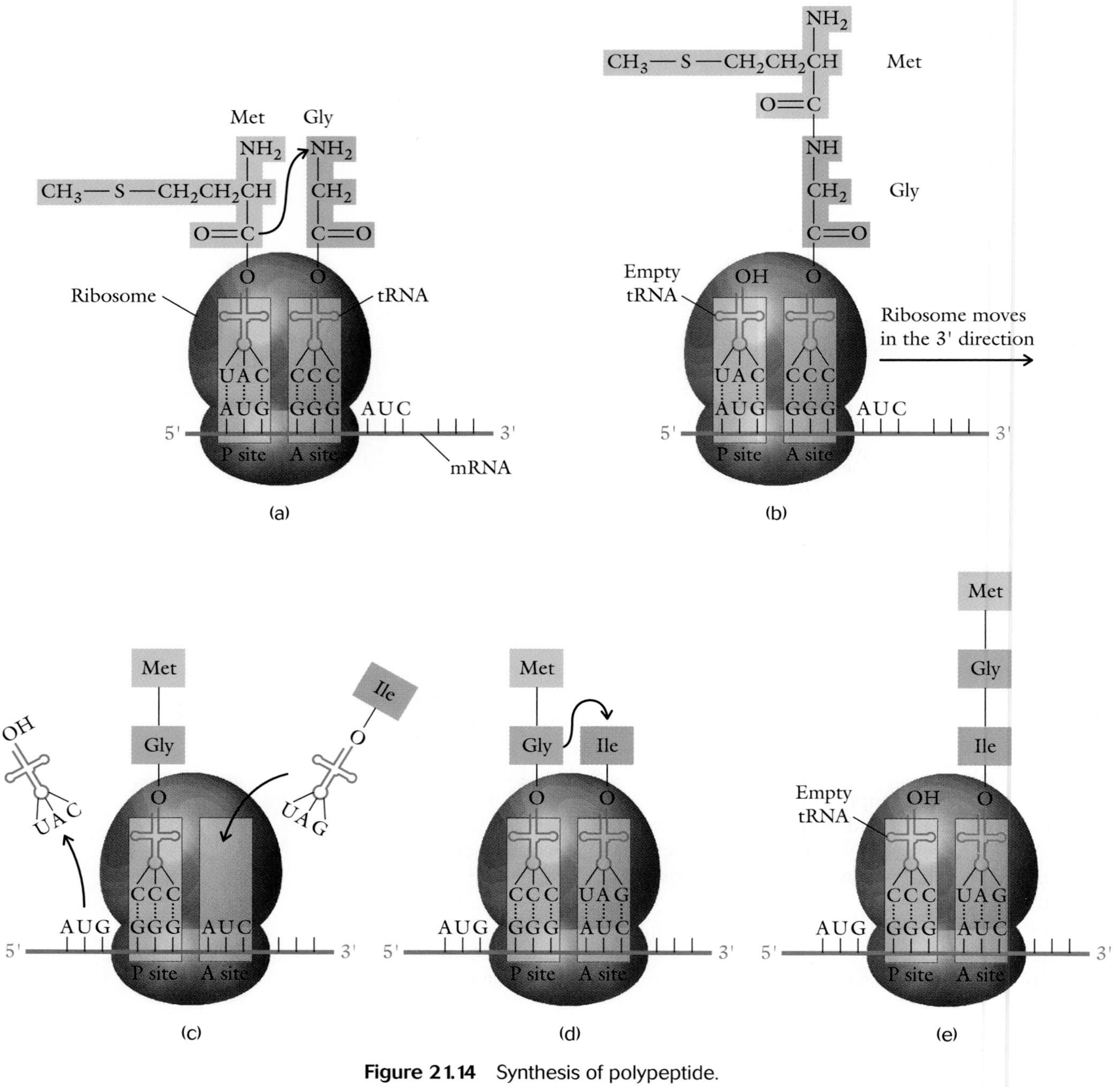

Figure 21.14 Synthesis of polypeptide.

blueprint for the production of all the proteins required by all cells—is found in every cell of the body. For example, heart cells contain the DNA not only for heart polypeptides and proteins, but also for bone, liver, skin, and other cells. Therefore, regulatory proteins are present in each cell to prevent the translation of most of the DNA messages contained in its chromosomes: only about 5 to 10% of all the DNA in any given cell is actually expressed. Heart cells do not need and must not synthesize the proteins found in liver, skin, and bone cells, and so forth. **Repressor proteins** turn off the synthesis of all proteins except those needed for the particular cell. Simultaneously, **inducer proteins** turn on the synthesis of proteins needed by the cell. The details of how these processes are coordinated and carried out are not well understood.

21.8 MUTATIONS

A **mutation** is an error in the base sequence of a gene. The end result can be alteration or cessation of a polypeptide's or protein's functioning because of a change in its α-amino acid sequence. There are two types of mutations:

- **Substitution (point) mutations,** in which one base substitutes for another in the normal base sequence: one purine for another, one pyrimidine for another, a purine for a pyrimidine, or a pyrimidine for a purine.
- **Frameshift mutations,** in which a base is inserted into or deleted from the normal base sequence.

A substitution mutation changes one codon in the DNA base sequence. A frameshift mutation, although rarer than a substitution mutation, has a much larger effect because it changes all codons that follow the insertion or deletion.

Mutations that arise from random errors in replication are called **spontaneous mutations.** Superimposed on spontaneous mutations are those caused by **mutagens,** agents that produce mutation. Various chemicals and types of radiation act as mutagens by modifiying the nucleotide bases. Sodium nitrite ($NaNO_2$), used as a preservative for frankfurters, bologna, and other processed meats, is a chemical mutagen. It converts cytosine, which base pairs with guanine, into uracil, which then base pairs with adenine; thus, it affects DNA structure by converting a C-G base pair into U-A. The overall danger is thought to be extremely low because of the very small amounts of nitrite used in meat preservation. However, it is probably prudent to limit the intake of such processed foods. Stopping the use of nitrite is not a practical alternative, because it would increase the incidence of botulism—food poisoning by the neurotoxins secreted by the airborne bacterium *Clostridium botulinum.* Viruses also may cause mutations (Section 21.10).

Frameshift mutations can occur when certain planar organic compounds such as benzopyrene and other fused-ring aromatic compounds (see Box 12.9) undergo reactions that subsequently result in their insertion between DNA base pairs. The insertion of these compounds distorts the DNA double helix and can result in deletion or insertion of a base pair. Benzopyrene and similar compounds are found in automobile exhaust, tobacco smoke, and burnt (especially barbecued) meats. (Here is one plausible mechanistic connection between smoking and cancer.)

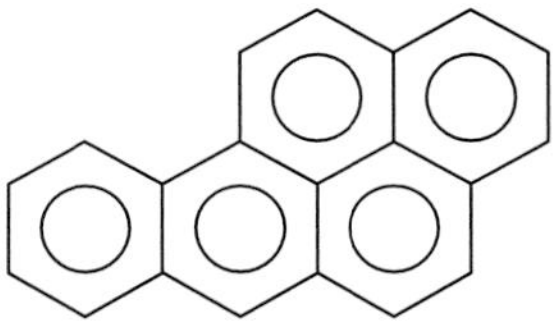

Benzopyrene

Chapter 10 described our exposure to various types of radiation: ultraviolet light and cosmic rays (α- and β-particles, neutrons) from outer space; α- and β-particles and neutrons from the radioisotopes in the ground beneath us; and X-rays and radioisotopic radiation from medical imaging equipment. When the energy of radiation is absorbed by DNA, bonds may be broken and bases and base sequences modified.

Radiation may also lead to oxidative damage, perhaps the most significant source of mutations. Excited-oxygen species such as hydroxyl ($HO\cdot$) and superoxide ($O_2^-\cdot$) free radicals (Section 10.2) are formed from the effects of radiation on living cells, as well as in the course of normal aerobic metabolism. Because of their unpaired electron structure, free radicals are extremely reactive species that can alter base structures and DNA base sequences. Fortunately, cells have elaborate antioxidative mechanisms for destroying these reactive species. Vitamins C and E are important antioxidants.

Preservation of the base sequence in an organism's genome is the key to the survival of the organism and the survival of its offspring. Two factors preserve the genome with great fidelity. First, replication errors are very rare because of the proofreading function of DNA polymerase. Second, several repair systems correct errors, both spontaneous mutations and mutations caused by mutagens.

Silent mutations are base-sequence errors that do not affect the functioning of an organism. Several factors are responsible for silent mutations:

- The redundancy of the genetic code makes the genome somewhat mutation resistant. Many base changes have no effect on the α-amino acid sequence in the polypeptide or protein synthesized from DNA (through RNA). Say, for example, that the codon CCU in mRNA has been mutated to CCC or CCA or CCG (see Table 21.1). The same α-amino acid, Pro, is encoded by all four codons. On the other hand, mutation of the first or second base in the CCU codon does result in a change in the encoded α-amino acid.
- Substitution of an α-amino acid residue for one similar in size, charge, and polarity often has no effect on function, because the three-dimensional structure of the polypeptide or protein is unchanged (Section 20.8).
- Substitution of an α-amino acid residue for a dissimilar residue may have little or no effect if the residue is located in a region of the polypeptide or protein that is unimportant for its function.
- Mutations, even frameshift mutations, in introns are constrained within the introns and have no effect on the base sequences in exons, because the introns are excised.
- Many genes are present in two or more copies. Unless all copies of a gene undergo mutation, an improbable occurrence, the organism still produces the required polypeptides and proteins.

Other mutations do have negative consequences, because they interfere with proper functioning:

- Substitution of an α-amino acid residue for a dissimilar residue in a region of the polypeptide or protein that is important for its function usually results in nonfunctional polypeptides and proteins.
- Changes that convert a codon for an an α-amino acid into a stop codon that terminates polypeptide synthesis usually result in nonfunctional polypeptides and proteins.
- Frameshift mutations in exons almost always result in nonfunctional polypeptides and proteins because the deletion or insertion changes all codons in the gene from that point on.
- The polypeptides or proteins produced by mutant DNA may be toxic.

Mutations in **somatic cells** (cells other than egg and sperm) affect the functioning of an individual organism. Mutations in **germ cells** (egg and sperm cells) have greater significance because they are passed onto offspring and subsequent generations. Thus, germ-cell mutations are the cause of **genetic (hereditary) diseases** that can be passed onto offspring. Some of the more common human genetic diseases are described in Table 21.2. Diabetes mellitus (Box 20.3 and Section 26.11), familial hypercholesterolemia (Box 24.1), galactosemia and lactose intolerance (Box 18.2), and sickle-cell anemia (Section 20.8) are described elsewhere.

Although the fidelity of replication is high and the repair systems are very efficient, mutations do accumulate over an individual organism's lifetime. There is probably a link between the accumulation of a lifetime of mutations and the changes seen in cancer (uncontrolled growth of cells) and aging. A mutagen giving rise to cancer is called a **carcinogen.** Many mutagens, but not all, are carcinogens. The ultimate effects of carcinogens depend on the quan-

TABLE 21.2 Hereditary Diseases

Disease	Affected Biological Function
albinism	Tyrosinase is deficient and melanin is not produced; melanin protects skin and the iris of the eye against UV damage and skin cancer.
cystic fibrosis	Viscous mucus in bronchial passages leads to pulmonary bacterial infections. Inadequate lipase and bile salts produced by the pancreas result in maladsorption of lipids and fat-soluble vitamins.
diabetes mellitus	Decreased insulin production results in poor glucose metabolism and the accumulation of glucose in the blood and urine, atherosclerosis, visual problems, and circulatory problems in legs and feet.
galactosemia	Defective transferase results in poor metabolism of galactose; accumulation of galactose in cells causes cataracts and mental retardation.
hemophilia	Defective blood-clotting factors result in poor clotting, excess bleeding, and internal hemorrhaging.
phenylketonuria (PKU)	Owing to lack of Phe hydroxylase for converting Phe into Tyr, Phe is converted into phenylpyruvic acid, which causes brain damage.
sickle cell	Defective hemoglobin; sickled red blood cells aggregate to cause anemia, plugged capillaries, low oxygen pressure in tissues.
Tay-Sachs	Defective hexosaminidase A causes accumulation of glycolipids in brain and eyes; mental retardation and loss of motor control.
thalassemia	Defective hemoglobin.

tity of accumulated mutations and the particular functions affected. Box 21.1 on the following page considers cancer and cancer therapy.

Not only are mutations inevitable, but they are the basis for the evolution of a species. Sometimes, but rarely, a mutation may occur that gives some members of a species an advantage over other members in the face of certain adverse environmental conditions. The population containing that mutation survives and reproduces in greater numbers, passing its altered genome along to subsequent generations.

Example 21.5 Determining the effect of a mutation on the α-amino acid sequence of a polypeptide

Imagine the following mutations in the template DNA strand in Example 21.4: (a) TAT of the *Υ* exon is mutated to GAT; (b) TAT of the *Υ* exon is mutated to TAA; (c) G is inserted in the middle of the second intron. What effect would each of these mutations have on the codon of the resulting mRNA? What effect would they have on the α-amino acid sequence of the resulting polypeptide?

21.1 Chemistry Within Us

Cancer and Cancer Therapy

We are made up of many different types of cells—brain, heart, lung, muscle, bone, skin, hair, and so forth—all dividing at different times and at different rates. (Some cells, such as brain and other nerve cells, do not divide at all after reaching mature development.) Cell division is a tightly regulated process in a healthy person, constrained at boundaries between different types of cells, tissues, and organs and controlled by regulatory proteins called **growth factors,** which circulate in the blood stream. Each type of cell is regulated by its own growth factor and a corresponding **receptor protein** on its membrane. The genes for the growth factors and receptors, responsible for normal cell function, are called **proto-oncogenes.**

The feature common to all cancers and distinct from other illnesses is the uncontrolled growth and division of cells to form large masses called **tumors.** A **benign tumor,** which is not a cancer, is a limited growth that remains in its initial location. Cancers are **malignant tumors**—uncontrolled growths that consume the body's resources and invade and crush nearby structures (such as when a tumor invades and compromises the lungs), destroying their functions. Even worse, cancers often shed cells that spread to and colonize new sites in the body—a process called **metastasis.**

Cancer begins when mutations convert proto-oncogenes into **oncogenes,** genes that no longer code for the correct regulatory proteins. Cell division goes out of control in the absence of the normal regulatory mechanism. Cancer-causing mutations occur through replication errors, the effects of chemical and radiation carcinogens, and some viruses (Sections 21.8 and 21.10). The United Nations World Health Organization estimates that environmental and life-style factors account for more than 80% of all cancers, with the remainder being caused by genetic factors and viruses. The environmental and life-style factors include cigarette smoking, air and water pollution, diet, and occupational hazards.

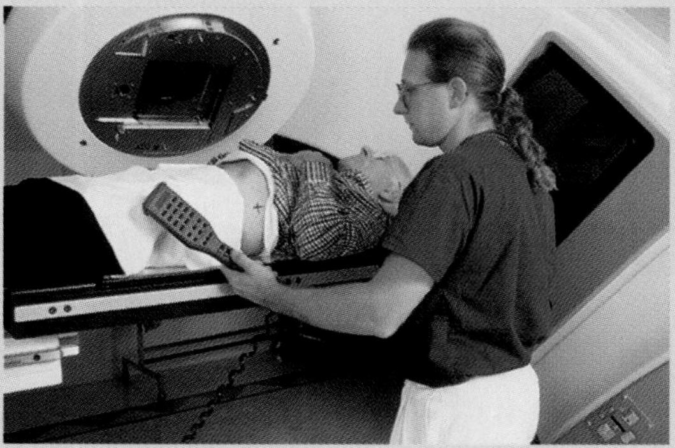

A patient is positioned before receiving radiation therapy to combat cancer.

The main strategies for treating cancers are surgery, radiotherapy, and chemotherapy. Surgical removal of cancerous tissues is an option only if the patient can survive without the affected tissues. It requires knowledge of the cancer's exact location because the cancer will rebound after surgery if all cancerous cells are not removed. The outlook after surgical treatment is often enhanced by some radiotherapy or chemotherapy to ensure that no cancerous cells remain.

Radiotherapy is the irradiation of cancers with ionizing radiations. γ-Rays from radioisotopic sources such as cobalt-60 and cesium-137 are those most often used. X-rays also are used. Irradiation stops the growth of cancer cells through mutation of its DNA. Hence, mutation, the mechanism that initiates cancer, may also be the means for stopping it. Radiotherapy damages healthy tissue along with the cancer cells, but, because cancer cells grow more rapidly than normal cells do, they are more sensitive to mutation than the healthy tissues are. The objective of radiotherapy, therefore, is to kill the cancerous cells while keeping the damage to healthy tissue at a minimum. The total radiation delivered to a patient is held to the lowest effective level, and collimated (narrowed and focused) radiation beams are used to confine the radiation as much as possible to the cancerous tissues.

Chemotherapy is the use of chemicals to interfere with the replication of cancer cells. Among the useful chemicals are 5-fluorouracil and cisplatin. Cisplatin binds with DNA and prevents its replication. 5-Fluorouracil deactivates one of the enzymes responsible for producing the nucleotide dUMP. Other chemicals used in chemotherapy interfere with replication by substituting for one of the heterocyclic bases, resulting in an incorrect base sequence. As with radiotherapy, the key to chemotherapy is that cancerous cells are more susceptible to disruption than are healthy cells. There is a continuing effort to find chemicals that are more selective in disrupting replication of cancerous cells than that of healthy cells.

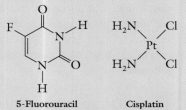

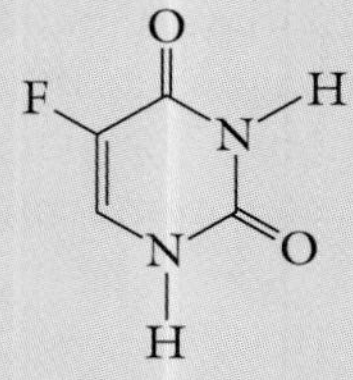

5-Fluorouracil

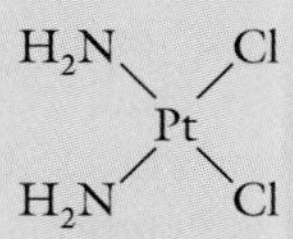

Cisplatin

A recent trend in cancer treatment is to combine different kinds of therapy—radiotherapy with chemotherapy, surgery with either radiotherapy or chemotherapy, or chemotherapy with a combination of different drugs. Fine tuning of combination therapies results in more complete destruction of cancerous cells with minimal damage to healthy cells.

Solution

(a) 5′-GAT-3′ in DNA results in 3′-CUA-5′ in mRNA. The codon is 5′-AUC-3′, which codes for Ile. There would be no change in the α-amino acid sequence, because Ile was the residue encoded by the codon before mutation.

(b) 5′-TAA-3′ in DNA results in 3′-AUU-5′ in mRNA. The codon is 5′-UUA-3′, which codes for Leu. The α-amino acid sequence would be altered, with Ile changed to Leu. The polypeptide would be Met-Leu-His-Thr instead of Met-Ile-His-Thr.

(c) There would be no effects on either the mRNA or the polypeptide. A mutation in an intron is constrained within the intron and has no effect on the base sequence in the exons, because the intron is excised.

Problem 21.9 Imagine the following mutations in the template DNA strand in Problem 21.8: (a) GTA of the *Y* exon is mutated to GAA; (b) GTA of the *Y* exon is mutated to TTA; (c) TGT of the second intron is mutated to AGT. What effect would each of these mutations have on the codon of the resulting mRNA? What effect would they have on the α-amino acid sequence of the resulting polypeptide?

21.9 ANTIBIOTICS

Antibiotics are chemicals, usually organic compounds, used to fight infections by microorganisms (bacterium, mold, yeast), especially bacterial infections. Bacteria enter the body but usually not the cells of the organisms that they infect; they harm the host through the toxicity of their metabolic products. Before the discovery and use of antibiotics in the 1930s, infectious diseases such as diphtheria, gonorrhea, pneumonia, and syphilis were a major cause of death. Although most of the original antibiotics were isolated from microorganisms, most present-day antibiotics are synthesized in the laboratory.

Antibiotics fight bacterial infections by interfering with the genetic apparatus in bacteria. Protein synthesis is blocked, usually by preventing translation, and this stops bacterial reproduction (Table 21.3). Large-scale use of antibiotics has been highly successful, but some bacteria have "fought back" by spontaneous mutation into antibiotic-resistant strains. Scientists attempt to keep ahead of the bacteria by synthesizing new antibiotics effective against the mutant strains.

To be useful, an antibiotic must be specific for bacteria, with minimal effect on protein synthesis in humans and other animals. There are probably no antibiotics that completely meet this criterion. The typical antibiotic affects the

TABLE 21.3 Antibiotic Inhibition of Protein Synthesis in Bacteria

Antibiotic	Mechanism for Inhibition
chloramphenicol	inhibits peptide bond formation by interfering with peptidyl transferase
erythromycin	prevents translocation of ribosome along mRNA
penicillin	inhibits formation of enzyme needed for cell-wall formation
puromycin	causes premature termination
streptomycin	inhibits initiation; causes misreading of mRNA
tetracyclin	prevents binding of tRNA to mRNA

host but less so than the bacteria. That is why antibiotics are prescribed for a short period only—say, 2 weeks—and then halted. Prolonged antibiotic use fosters the production of mutant strains of bacteria and may adversely affect the host. At the same time, patients are always instructed to complete the entire course of antibiotic treatment and not to stop before finishing should they feel better after a few days. If treatment is stopped too soon, the bacterial population may not have dropped sufficiently low for the patient's immune system to keep it under control, and the bacterial infection may rebound.

21.10 VIRUSES

Viruses are infectious, parasitic particles, usually smaller than bacteria, that consist of either DNA or RNA (but not both) encapsulated by a protein coat. **DNA viruses** contain only DNA; **RNA viruses** contain RNA. Many viruses have one or more additional protective coats, usually composed of lipids and proteins. Viruses lack some or most of the cellular apparatus (α-amino acids, nucleotides, enzymes) needed for replication, transcription, and translation. Thus, when they are outside host cells, they show no signs of life: they do not generate energy, reproduce, or synthesize proteins. To reproduce, viruses must enter a cell and take control of its metabolic machinery. Viruses are responsible for many diseases—AIDS, chicken pox, the common cold, hepatitis, herpes, influenza, leukemia, mononucleosis, poliomyelitis, rabies, shingles, smallpox, and various tumors (including some cancers).

Each kind of virus infects only specific types of cells in specific organisms. For example, there are large numbers of plant viruses, none of which invade animal cells. Infection of a host cell by a virus occurs when the virus adheres to the cell. Enzymes in the virus coat enable the virus to penetrate the cell membrane and insert itself or its nucleic acid into the host cell.

When inside, the virus diverts the host's cellular machinery (enzymes, α-amino acids, nucleotides) from its normal role and puts it to work replicating new daughter viruses (using the viral nucleic acid as a blueprint). Hundreds of progeny viruses result from the entry of one virus into a host cell. The progeny then escape from the host cell and attack other host cells. Some viruses exit the host cell by breaking open its membrane, killing the cell instantly. Other viruses escape through the cell membrane without breaking it open, but the host cell dies nevertheless, because its normal cell processes have been compromised.

Reproduction of RNA and DNA viruses follow different paths:

- DNA viruses enter the host nucleus and integrate themselves into the host genome. The modified DNA is treated by the host in the same manner as it treats its own (unmodified) DNA.
- RNA viruses need to replicate RNA from RNA. The host cells do not have the enzyme, **RNA replicase,** required for this process, but the RNA virus contains information for synthesizing it. RNA replicase is synthesized in the host cell, and it directs the reproduction of viral RNA.
- **Retroviruses** are RNA viruses that reproduce through the intermediate formation of DNA. The retrovirus contains the enzyme **reverse transcriptase,** which directs the synthesis of viral DNA from viral RNA, a direction contrary to the normal pathway. The viral DNA becomes integrated into the host genome, and then the host reproduces the viral RNA by using the viral DNA as a template. The retroviruses include HIV and cancer-causing viruses **(oncogenic viruses).** Box 21.2 on

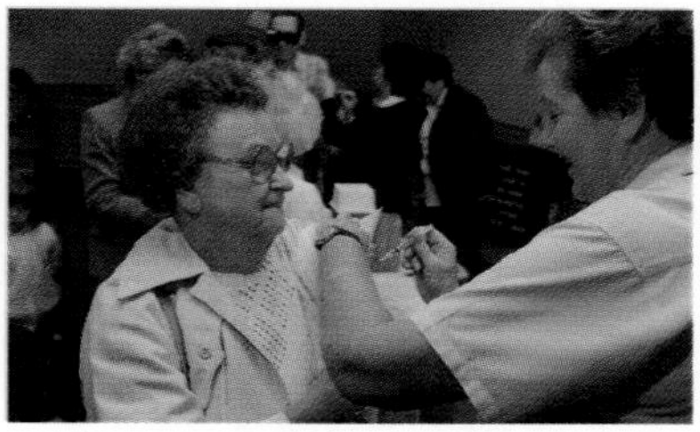

A nurse administers a vaccine to a senior citizen to protect her against the flu.

pages 650 to 651 describes AIDS and, together with Figure 21.15, the life cycle of the HIV virus.

Viral infections are generally more difficult to treat than bacterial infections because viruses, unlike bacteria, replicate inside host cells and spend little time outside of them. It is, therefore, much more difficult to design a drug that prevents the reproduction of the virus without compromising the host cell's normal activities.

The most successful approach to fighting viral infections is the preemptive use of **vaccines.** A vaccine is a preparation containing the weakened virus or its proteins. Its presence in the body stimulates the immune system to generate antibodies (immunoglobulins) that recognize the virus's antigens (Section 18.6). Subsequent entry of the virus into the host is thus quickly recognized

◀◀ Immunoglobulins are proteins that have a protective function, described in Chapter 20.

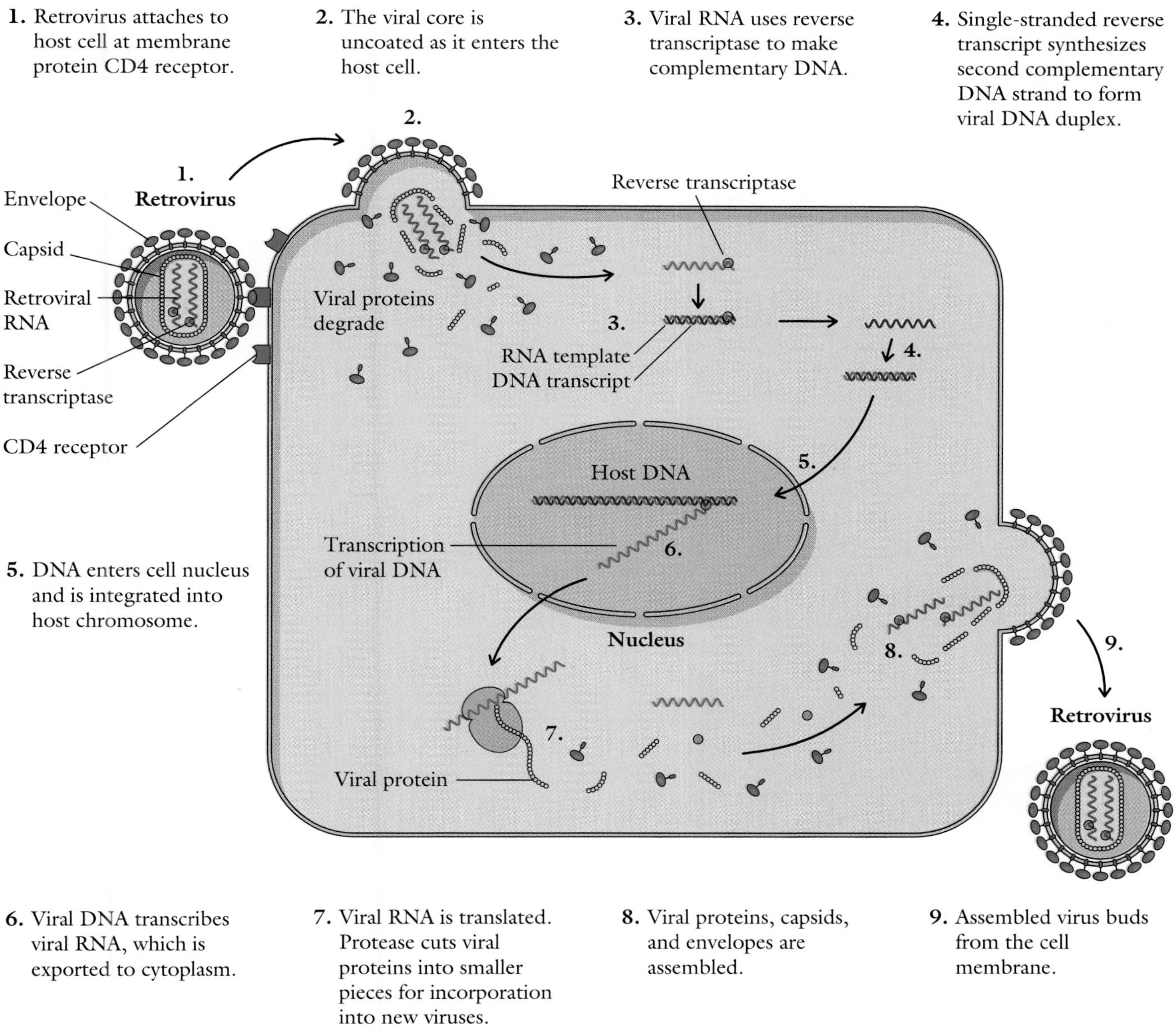
1. Retrovirus attaches to host cell at membrane protein CD4 receptor.
2. The viral core is uncoated as it enters the host cell.
3. Viral RNA uses reverse transcriptase to make complementary DNA.
4. Single-stranded reverse transcript synthesizes second complementary DNA strand to form viral DNA duplex.
1. Retrovirus
Envelope
Capsid
Retroviral RNA
Reverse transcriptase
CD4 receptor
2.
Reverse transcriptase
Viral proteins degrade
3.
RNA template
DNA transcript
4.
5.
Host DNA
Transcription of viral DNA
6.
Nucleus
5. DNA enters cell nucleus and is integrated into host chromosome.
7.
Viral protein
8.
9.
Retrovirus
6. Viral DNA transcribes viral RNA, which is exported to cytoplasm.
7. Viral RNA is translated. Protease cuts viral proteins into smaller pieces for incorporation into new viruses.
8. Viral proteins, capsids, and envelopes are assembled.
9. Assembled virus buds from the cell membrane.

Figure 21.15 Reproduction of HIV in a T cell.

by the host, which produces lymphocytes and antibodies to fight the invasion. Vaccines are also useful for protection against bacterial infections.

21.11 RECOMBINANT DNA TECHNOLOGY

Recombinant DNA technology (also called **genetic engineering**) alters the genome of an organism by transplanting DNA into it, either synthetic DNA or DNA from another organism (either the same or a different species). Research in recombinant DNA technology dates only to the mid-1970s, but we have already seen or are on the verge of seeing major benefits:

- Organisms with altered DNA producing therapeutic drugs
- Agricultural improvements in animal and plant stocks
- Gene therapy to alleviate genetic diseases

21.2 Chemistry Within Us

HIV and AIDS

The human immunodeficiency virus (HIV), the cause of acquired immune deficiency syndrome (AIDS), has a very specific target. It infects only the T cells of the human immune (lymphatic) system. Ironically, the normal function of these cells is to identify and destroy invaders in the body. The HIV count in the blood soars at the onset of the infection but usually drops to a very low number as the immune system fights back with an increased production of T cells. A struggle ensues between HIV and T cells, but, as long as the immune system stays ahead of the virus, the infected person remains symptom free.

Several months to several years after infection, the immune system usually begins to lose the struggle, and symptoms appear. These symptoms include swollen lymph nodes and fungal and herpes simplex infections. Most people with these symptoms progress to the stage at which their immune systems become compromised. People do not die of AIDS directly but become easy prey to opportunistic infections—such as pneumonia, toxoplasmosis, and various cancers—that do not normally affect healthy persons.

AIDS is primarily a sexually transmitted disease. The virus is spread from one person to another through blood, semen, vaginal fluid, and breast milk. It passes from the fluid of an infected person to the noninfected person's bloodstream through the thin linings of the noninfected person's anus, mouth, penis, vagina, or uterus. Torn or damaged tissues in these organs greatly facilitate entry of HIV. Intravenous drug users are at high risk of contracting AIDS because they often share needles. HIV-infected mothers transmit the virus to their infants during birth or subsequently through breast feeding. Before 1985, blood transfusions also resulted in some transmission of HIV, but blood banks now test donors' blood for HIV prior to accepting blood donations.

The patterns of AIDS infection differ throughout the world. In the developed Western countries, AIDS is most prevalent in homosexual males and intravenous drug users, although there are indications that the incidence may be rising in the heterosexual population. In the underdeveloped countries of Africa and Asia, AIDS is mostly a heterosexual disease equally divided between males and females. More than 300,000 people died from AIDS in the United States from 1983 to 1994. The United Nations' estimates for the number of persons infected with HIV as of the beginning of 1997 are 750,000 in the United States, 14 million in sub-Saharan Africa, 5 million in southern and Southeast Asia, and more than 22 million worldwide. The spread of HIV has slowed down in the United States, probably a result of education about the mechanism of transmission. However, an increasing proportion of new infections are found among persons younger than 22. The epidemic rages on in the underdeveloped countries, where the number of people infected with HIV continues to soar.

There is an enormous worldwide effort to find a cure for AIDS. Efforts to find a vaccine to prevent infection by HIV are not encouraging, because the virus is constantly mutating, and vaccines with promise against one strain offer no hope against new strains. However, there is consensus that ways can be found to control the reproduction of HIV within infected persons and delay the onset of AIDS by blocking one or more steps in the virus's life cycle (see Figure 21.15). New therapeutic drugs have been tested against almost all of the steps. To be useful, the drugs must slow or stop HIV reproduction without doing too much damage to the patient. No one drug has been found to be highly effective, although a number of drugs have shown promise. The problem in every instance has been that prolonged use of the drug results in resistant mutant

Production of Therapeutic Proteins

The first application of recombinant DNA technology was the use of an organism with altered DNA to produce human insulin for diabetics to use in place of bovine and ovine insulin. Both yeast and bacteria have been useful as the vehicle for this purpose. Figure 21.16 on the following page outlines the process for producing human insulin by altering the DNA of *Escherichia coli* bacteria. *E. coli* contains DNA in two forms, a single chromosome and a large number of circular, double-stranded DNAs called **plasmids.** The more plentiful plasmids are more useful than the single chromosome for genetic engineering.

The process begins after scientists have identified the gene encoding the synthesis of the desired polypeptide. The next critical step is to splice this gene, called **donor DNA,** into the plasmid **(vector DNA).** The vector DNA is the DNA that is to be altered—that is, altered by the introduction of the donor DNA. *E. coli* cells are broken up, and their plasmids are isolated by extraction and centrifugation. A **restriction enzyme** (*Eco*RI) is then added to the mix-

strains. However, there is considerable optimism that a combination of partly effective drugs may work well, because the virus is less likely to undergo several simultaneous mutations that would circumvent all of the drugs.

Initial results are encouraging for a "cocktail" of two **reverse transcriptase inhibitors,** AZT and 3TC, plus a **protease inhibitor** (such as Indinavir). The reverse transcriptase inhibitors block the transcription of viral RNA to viral DNA (step 3 in Figure 21.15). Reverse transcriptase cannot discriminate AZT and 3TC from the normal nucleotide containing deoxyribose and thymidine, but AZT and 3TC differ from the normal thymidine nucleotide in not having an —OH group at the 3′-position of deoxyribose. Because of the absence of the —OH group, DNA synthesis terminates whenever AZT or 3TC is incorporated into the growing DNA strands in the course of reverse transcription. Protease inhibitors block step 7, the cutting up of the proteins produced by translation of viral RNA, by inactivating the enzyme protease. The components needed to assemble the virus are then not available, and new HIVs are not produced.

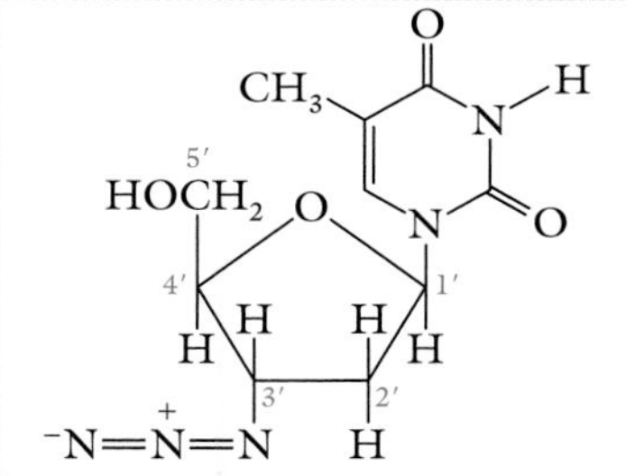

3′-Azido-2′,3′-dideoxythymidine (AZT)

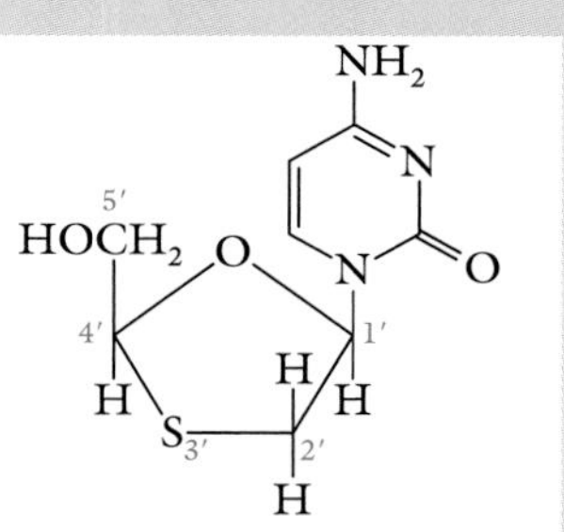

3′-Thia-2′,3′-dideoxycytidine (3TC)

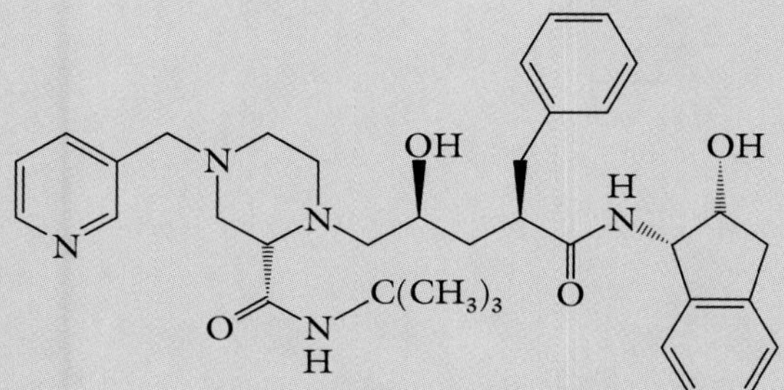

Protease inhibitor (Indinavir)

ture, and this enzyme cleaves the plasmid at one site containing a specific sequence of nucleotides. The cut is uneven, leaving overlapping, complementary, single strands at each end of the now linear DNA. For example, when *Eco*RI is used to cleave plasmids, one single-strand end may be TTAA and the other AATT.

The donor DNA must have uneven ends complementary to those on the vector DNA, because the two are to be spliced (bonded) together. The donor DNA is prepared either by using the same restriction enzyme on the DNA from an organism that contains the gene of interest or by in vitro DNA synthesis. In vitro synthesis of donor DNA requires knowledge of the primary structure of the DNA that codes for the desired polypeptide. To use recombinant DNA technology to make human insulin, the donor DNA must contain the base sequence that codes for human insulin.

DNA ligase joins together the vector and donor DNAs. Base-pairing of the complementary uneven ends ensures the proper alignment of vector and donor DNAs prior to covalent bonding by DNA ligase. The altered plasmid

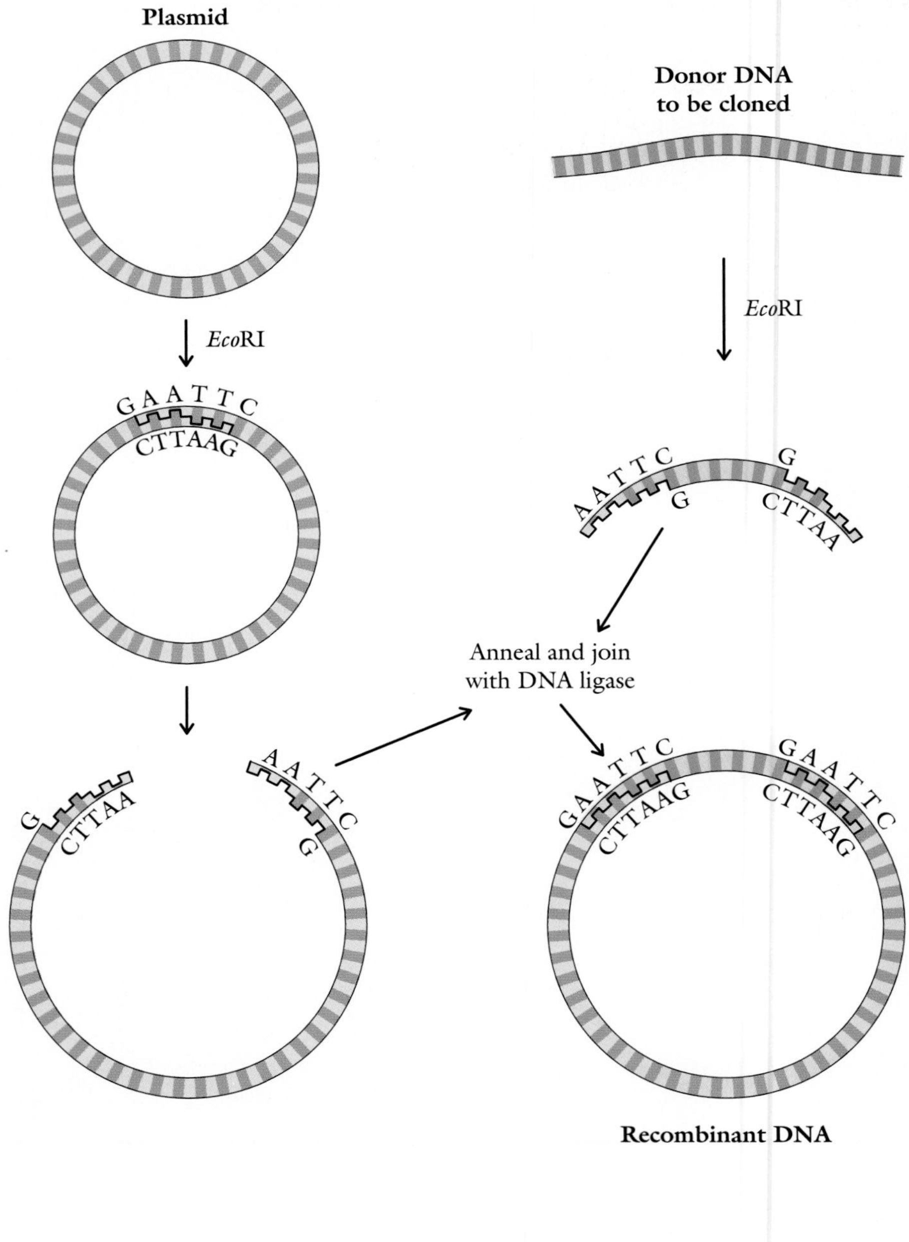

Figure 21.16 Formation of recombinant DNA.

DNA is called **recombinant DNA.** This altered DNA is introduced into *E. coli* by any of several treatments that make bacterial walls transiently permeable to plasmids. One such treatment is **chemical shock**—*E. coli* are placed in aqueous calcium chloride at low temperature and then incubated with the plasmids at higher temperature.

The altered *E. coli* now contain both normal and recombinant DNA. When they replicate, they reproduce many identical copies **(clones)** of the altered plasmid DNA, which is why recombinant DNA technology, or genetic engineering, is often also called **cloning.** When the altered plasmid DNAs undergo transcription and translation, they produce not only the normal *E. coli* polypeptides, but also the polypeptide coded by the donor DNA—human insulin in this case. The human insulin is then separated and purified.

Human insulin produced by genetic engineering was approved for use by the U.S. Food and Drug Administration in 1982. Many other substances produced by recombinant DNA technology are now in use or in various stages of development (Table 21.4). The applications are diverse and include use in animals and plants as well as in humans. For example, bovine growth hormone is used to increase milk production in cows.

Microinjection and viral vectors are techniques other than chemical shock that are useful for introducing foreign DNA into an organism's genome. **Microinjection** is the direct injection of DNA into the nucleus of a cell with the use of a very fine needle. This technique has a high success rate for the skilled practitioner but cannot be used to alter large number of cells, because each must be treated one at a time. It is well suited for the alteration of germ cells because large numbers of cells need not be altered. **Viral vectors** are altered viruses, often retroviruses, whose DNA has been engineered in the laboratory to include the desired DNA sequence and to lack the genes for viral reproduction. When they enter a host cell, they incorporate the desirable DNA into the host's genome. Because these viruses are altered to prevent their replication, they do no harm to the host cell.

Dolly is the world's first clone of an adult mammal.

TABLE 21.4 Products of Recombinant DNA Technology

Product	Use
blood-clotting factor	Promotes blood clotting for treating hemophilia; eliminates risk of treatment by blood transfusion.
erythropoietin	Treats anemia resulting from kidney disease; stimulates erythrocyte production.
human growth factor (somatotropin)	Treats dwarfism in children.
interferons	Prevent viral reproduction by inhibiting protein synthesis in infected cells.
interleukins	Stimulate production of T leukocytes in immune disorders.
monoclonal antibodies	Treat cancer by binding and transporting drug, toxin, or radioactive compound to tumor.
tissue plasmologen activator (TPA)	Activates plasmin (enzyme); dissolves clots in heart attack.
vaccines	Like the traditional weakened viruses, viral coat proteins act as vaccines but are safer.

Transgenic Breeding of Animals and Plants

Traditional practices for improving animal and plant stocks include selective breeding and selective feeding or fertilizing programs. **Transgenic breeding,** the selective production of organisms whose genes have been permanently altered by recombinant DNA technology, offers the potential for faster, more accurate, and more varied improvement in stocks. Accurate control of transplanted DNA allows desirable traits to be introduced into the transgenic animal or plant without the simultaneous introduction of too many other, possibly undesirable traits. The desirable traits can be introduced in one generation, compared with the several generations needed with traditional breeding. Furthermore, traits not normally available in a given species may be introduced into it through the transportation of DNA from a different species.

The objective of transgenic breeding is to produce stock that is larger and leaner, grows faster, uses feed more efficiently, and is more disease resistant. Many experimental efforts have reported success, although very few, if any, transgenic animals have yet reached the market. Transplantation of the growth hormone gene from the sockeye salmon into the coho salmon has accelerated the coho's growth so that it is ready for market much sooner than the standard farmed coho salmon, about 1 year instead of 3 years. Sheep that grow superior wool without needing the standard dietary supplements and faster-growing and leaner pigs and cattle are other examples of genetic engineering.

A number of transgenic plants have received regulatory approval, including herbicide-resistant cotton and soybean, insect-resistant corn and potato, virus-resistant squash, and delayed-ripening tomato. The delayed-ripening tomato is already being marketed. Tomatoes are normally picked in the hard, green state so that they can be shipped long distances without rotting. At their final destination, they are treated with ethylene, which triggers the ripening process. Transgenic tomatoes, designed to ripen much more slowly, can be harvested later in the ripening process (after they have turned red instead of while they are still green) and still reach market without losses caused by fast ripening during shipment.

Gene Therapy

The **Human Genome Project,** a research effort funded by the U.S. National Institutes of Health, has the objective of mapping the chromosomal locations and discovering the base sequences of the approximately 100,000 genes of the human genome. Data of this kind will lead to identification of the genes responsible for individual genetic diseases and eventually to better methods of treatment.

When the base sequence is known for a gene that is missing or defective in certain people, recombinant DNA technology offers two approaches to treatment. One approach is the production of therapeutic drugs, as described earlier, to be taken orally or by injection. However, most genetic diseases do not respond to this approach. The other approach is **gene therapy**—the use of recombinant DNA technology to modify a person's genome directly by inserting the gene that corrects the genetic disease. Research efforts are underway to furnish genes for missing or flawed genes to persons with cystic fibrosis, hemophilia, muscular distrophy, and other genetic diseases.

The critical step that must be solved before gene therapy is successful is delivery of recombinant DNA to the appropriate cells in the patient. The common cold virus, adenovirus, is the most-studied gene-delivery system. Researchers modify the adenovirus in two ways. First, certain viral genes are excised to prevent the virus from reproducing in cells and causing illness. Second, the excised genes are replaced with the DNA sequence whose absence is responsible for the genetic disease. A number of techniques are under study:

aerosol spraying into the air passages; direct injection into the bloodstream; and removal of cells from the patient, incubation with the modified viral DNA, and return of the treated cells to the patient. The results have shown much promise, and there is considerable hope of success in the near future.

Social, Ethical, and Other Considerations

Although major new technologies have many predictable social effects, history teaches that many consequences of progress will come as a surprise. What will be the effect of recombinant DNA technology? New diagnostic procedures allow the testing of people for genetic diseases and for predispositions to conditions such as alcoholism, Alzheimer disease, and hypercholesterolemia. The positive results include better and earlier treatments and genetic counseling for couples who are considering having children. On the other hand, the same information may be used to restrict people's access to health and life insurances and to employment.

Similarly, human gene therapy, which offers great potential for treating genetic diseases, poses equally great problems. Few would want to deny lifesaving treatment to anyone, but some treatments may be prohibitively expensive. If gene therapy can eliminate a genetic disease, can it also enhance human intelligence, physical strength, and athletic ability? What about treating germ cells to improve the genome of the offspring? Who will benefit, and who might suffer as a result? How will these questions be debated and resolved in different cultures and societies throughout the world? A fraction of the federal funds spent on the Human Genome Project are allocated for studies of these ethical issues.

Summary

Genes, the fundamental units of heredity, are responsible for the traits of species and of individual organisms. Each gene is a part of deoxyribonucleic acid and carries the information for synthesizing one or more polypeptides. Deoxyribonucleic (DNA) and ribonucleic (RNA) acids are responsible for carrying out the hereditary process.

Nucleotides Nucleotides—composed of phosphoric acid, a pentose sugar (either ribose or deoxyribose), and a heterocyclic nitrogen base—are the building blocks for nucleic acids. Adenine, guanine, cytosine, and thymine (A, G, C, and T) are the bases for DNA; uracil (U) replaces T for RNA.

Nucleic Acid Formation from Nucleotides Nucleic acids are polymers formed by dehydration between the —OH of the phosphate group at C5′ of one nucleotide residue and the —OH at C3′ of the next nucleotide residue. Nucleic acids are named in the $5' \longrightarrow 3'$ direction.

Three-Dimensional Structure of Nucleic Acids The three-dimensional structure of a nucleic acid is determined by its primary structure, the sequence of bases along the polymer chain. DNA molecules exist not as individual molecules but as complementary, paired DNA molecules that fold around each other to form a right-handed double helix (duplex). The double helix is stabilized by hydrogen bonding and hydrophobic attractions between heterocyclic bases. RNA molecules exist as individual molecules with three-dimensional structures determined by intramolecular hydrogen bonding between bases.

Replication, Transcription, and Translation Replication is the copying of DNA in the course of cell division and is directed by the DNA itself. Transcription is the synthesis of ribosomal RNA (rRNA), messenger RNA (mRNA), and transfer RNA (tRNA) from DNA. Translation is the synthesis of polypeptides through the combined efforts of rRNA, mRNA, and tRNA. The genetic code is the complete description of the α-amino acids specified by various base triplets (codons) on mRNA. The genetic code is degenerate and nearly universal.

Mutations A mutation is an error in the DNA base sequence of a gene. It can yield altered or no biological activity for a polypeptide or protein, because of a change in the α-amino acid sequence. A substitution mutation consists of the substitution of one base for another; a frameshift mutation consists of a base being inserted into or deleted from the normal base sequence. Spontaneous mutations are those resulting from random errors in replication. Other mutations are produced by radiation and chemical mutagens. Mutations in the DNA of germ cells may result in genetic (hereditary) diseases.

Antibiotics Antibiotics are organic compounds used to fight infections by bacteria, molds, and yeasts.

Viruses Viruses are infectious, parasitic particles, smaller than bacteria, composed of DNA or RNA. Viruses carry out life processes only within host cells, where they take over the host's cellular apparatus. The most successful approach to fighting viral infections is the preemptive use of vaccines. Vaccines contain the weakened virus or its proteins.

Recombinant DNA Technology Recombinant DNA technology (genetic engineering, or cloning) alters the genome of an organism by transplanting into it synthetic DNA or DNA from another organism (of the same or a different species). This technology allows the production of proteins and therapeutic drugs from altered organisms, agricultural improvements through transgenic breeding, and gene therapy to alleviate genetic diseases.

Key Words

antibiotic, p. 647
anticodon, p. 633
base-pairing, p. 629
base triplet, p. 638
cancer, p. 644, 646
cloning, p. 653
codon, p. 638
deoxyribonucleic acid, p. 625
DNA double helix, p. 628
gene therapy, p. 654
genetic code, p. 638
genetic engineering, p. 651
Human Genome Project, p. 654
mutation, p. 643
nucleic acid, p. 626
nucleotide, p. 624
primary transcript, p. 637
recombinant DNA technology, p. 651
replication, p. 633
ribonucleic acid, p. 625, 631
transcription, p. 636
translation, p. 638
virus, p. 648

Exercises

Nucleotides

21.1 What sugar is used to synthesize ribonucleotides? deoxyribonucleotides?

21.2 What bases are used to synthesize ribonucleotides? deoxyribonucleotides?

21.3 Draw the structure and give the abbreviated name of the deoxyribonucleotide whose components are phosphoric acid, deoxyribose, and thymine.

21.4 Draw the structure and give the full name of AMP.

21.5 What are the products of hydrolysis of dGMP?

21.6 What are the products of hydrolysis of UMP?

Nucleic Acid Formation from Nucleotides

21.7 Where in the cell is DNA found?

21.8 Where in the cell is RNA found?

21.9 Give the abbreviated name for the following dinucleotide:

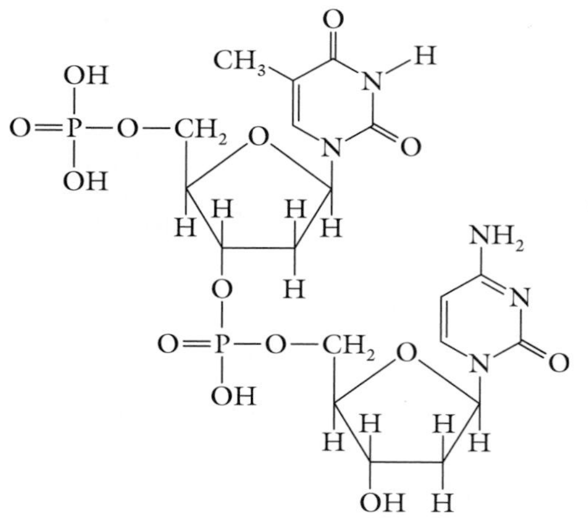

21.10 What is the relation between UMP-CMP and CMP-UMP?

21.11 Why does the sequence dUMP-dGMP not exist in nature?

21.12 Why does the sequence dGMP-CMP not exist in nature?

21.13 Draw the complete structure of dGMP-dTMP. Indicate the 5′ ⟶ 3′ direction.

21.14 Draw the complete structure of AMP-UMP. Indicate the 5′ ⟶ 3′ direction.

21.15 What nucleotides are obtained by hydrolysis of dAMP-dTMP-dTMP-dGMP-dCMP?

21.16 What bases are obtained by complete hydrolysis of dAMP-dTMP-dTMP-dGMP-dCMP? in what relative amounts?

Three-Dimensional Structure of Nucleic Acids

21.17 What are histones and what is their function?

21.18 What is a DNA duplex?

21.19 What attractive forces result in base-pairing?

21.20 Which are the complementary base pairs?

21.21 Which of the following is a variable part of a specific DNA double helix? (a) Ratio of moles of adenine to moles of thymine present; (b) ratio of moles of sugar to moles of bases present; (c) sequence of sugars attached to the basic units; (d) sequence of bases attached to the sugar units.

21.22 What is the difference, if any, between chromosome, DNA double helix, and gene?

21.23 The DNA contained in the 23 pairs of human chromosomes contains 30.4 mol-% adenine and 19.6 mol-% cytosine. What are the amounts of thymine and guanine?

21.24 One DNA strand in a DNA duplex contains 34 mol-% A, 28 mol-% T, 16 mol-% C, and 22 mol-% G. What is the composition of the complementary strand?

21.25 The base sequence in a section of one DNA strand is 5′-TAACCG-3′. What is the sequence for the corresponding section of the complementary DNA strand?

21.26 The base composition of one of the DNA chains of a DNA double helix contains 18 mol-% A, 35 mol-% T, 26 mol-% C, and 21 mol-% G.

(a) What is the base composition of the complementary DNA chain?
(b) Is the total amount of purine bases (A and G) equal to the total amount of pyrimidine bases (C and T) for the DNA double helix?

Information Flow from DNA to RNA to Polypeptide

21.27 Where is genetic information stored?

21.28 Distinguish among replication, transcription, and translation.

Replication

21.29 What is produced by replication?

21.30 Where does replication take place?

21.31 Define or describe the roles of the following elements in replication: (a) replication bubble; (b) leading and lagging strands; (c) Okazaki fragments.

21.32 What are the roles of the following enzymes in replication? (a) DNA polymerase; (b) DNA ligase.

21.33 Distinguish between parent and daughter DNA.

21.34 Why is DNA replication called semiconservative replication?

21.35 The sequence 5′-TTAGCG-3′ is contained in a part of the new DNA strand of a daughter DNA double helix.

(a) What is the corresponding sequence in the old DNA chain of the same daughter DNA double helix?
(b) What is the corresponding sequence in the old DNA chain in the other daughter DNA double helix?

21.36 The sequence 5′-CCTATC-3′ is contained in a part of one strand of a parent DNA double helix.

(a) What is the corresponding sequence in the complementary strand of the parent DNA double helix?
(b) What is the corresponding sequence in the new daughter strand made from this parent strand during replication?

Transcription

21.37 What is produced by transcription?

21.38 Where does transcription take place?

21.39 Which DNA strand is used in transcription, the template or nontemplate DNA strand?

21.40 What is the function of the enzyme RNA polymerase?

21.41 What is the purpose of posttranscriptional processing?

21.42 What chemical reactions take place in posttranscriptional processing?

21.43 Compare the molecular masses of DNA, rRNA, mRNA, and tRNA.

21.44 Define and distinguish between introns and exons. What are their functions?

21.45 The base composition of a DNA template strand for replication is 15 mol-% A, 25 mol-% C, 20 mol-% G, 40 mol-% T. What base composition is expected for the RNA synthesized from this template strand?

21.46 Consider the sequence 5′-ACAGGTTAC-3′ in a DNA template strand.

(a) What is the base sequence for the DNA nontemplate strand?
(b) What is the base sequence for the synthesized RNA?

Translation

21.47 What is produced by translation?

21.48 Where does translation take place?

21.49 What are the roles of rRNA, mRNA, and tRNA in translation?

21.50 Distinguish between codon and anticodon, indicating where each is located and their functions.

21.51 What constitutes the genetic message and what is it a message for?

21.52 What is the genetic code?

21.53 What is meant by the statement that the genetic code is nearly universal?

21.54 What is meant by the statement that the genetic code is degenerate?

21.55 What are the roles of aminoacyl-tRNA synthetase, the aminoacyl site, the peptidyl site, and acyl transfer in the translation process?

21.56 What are the functions of inducer and repressor proteins?

21.57 Define base triplet.

21.58 Consider each of the following codons. What is the anticodon on the tRNA complementary to the codon? What α-amino acid is specified by the codon? (a) UCC; (b) CAG; (c) AGG; (d) GCU.

21.59 Sections *X*, *Y*, and *Z* of the following part of a template DNA strand at its 3′ end are exons of a gene:

5′				3′
ACA	CAC	CAA ATG	TGT GGT	CAT
X		*Y*		*Z*

What are the structures of the primary transcript and mRNA made from this template DNA? What polypeptide sequence will be synthesized?

21.60 Sections *X*, *Y*, and *Z* of the following part of a template DNA strand at its 3′ end are exons of a gene:

5′				3′
TAC	AGC	GTA ACC	GAT CCT	CAT
X		*Y*		*Z*

What are the structures of the primary transcript and mRNA made from this template DNA? What polypeptide sequence will be synthesized?

Mutations

21.61 What is a mutation?

21.62 What is meant by the statement that the genetic code is very nearly mutation resistant?

21.63 Describe the difference between substitution and frameshift mutations.

21.64 Which has the greater potential for harm—insertion, deletion, or substitution mutations?

21.65 What is a spontaneous mutation?

21.66 What is a silent mutation?

21.67 Distinguish between germ and somatic cells.

21.68 What is a genetic (hereditary) disease?

21.69 What is cancer?

21.70 What is a mutagen? a carcinogen?

21.71 Which of the following substitution mutations is likely to be more harmful? Why? (a) Valine is substituted for glutamic acid; (b) aspartic acid is substituted for glutamic acid.

21.72 Which of the following substitution mutations is likely to be more harmful? Why? (a) Lysine is substituted for glutamic acid; (b) lysine is substituted for histidine.

21.73 The following diagram shows part of a template DNA strand at its 3′ end, with sections *Y* and *Z* being the exons of a gene:

5′			3′
CAC	TAA GTC	TGT GGT	CAT
	Y		*Z*

(a) What polypeptide sequence will be synthesized?
(b) What polypeptide sequence will be synthesized if the GTC triplet of exon *Y* is mutated to CTC?
(c) What polypeptide sequence will be synthesized if the GTC triplet of exon *Y* Is mutated to ATC?
(d) What polypeptide sequence will be synthesized if C is inserted between the two triplets of the intron between exons *Y* and *Z*?

21.74 The following diagram shows part of a template DNA strand at its 3′ end, with sections *Y* and *Z* being the exons of a gene:

5′			3′
GTG	GTA ACT	TGT GGT	CAT
	Y		*Z*

(a) What polypeptide sequence will be synthesized?
(b) What polypeptide sequence will be synthesized if the ACT triplet of exon *Y* is mutated to AAT?
(c) What polypeptide sequence will be synthesized if the ACT triplet of exon *Y* is mutated to GCT?
(d) What polypeptide sequence will be synthesized if C is inserted between the two triplets of exon *Y*?

Antibiotics

21.75 What is an antibiotic and what is its purpose?

21.76 What is the mechanism by which an antibiotic works?

Viruses

21.77 What is a virus?

21.78 How does a virus reproduce?

21.79 What is a vaccine and what is its purpose?

21.80 What is the mechanism by which a vaccine works?

Recombinant DNA Technology

21.81 What is the principle on which recombinant DNA technology is based?

21.82 What benefits are possible through recombinant DNA technology?

21.83 Distinguish among vector DNA, donor DNA, and recombinant DNA in recombinant DNA technology.

21.84 What is the role of *E. coli* plasmids in the production of human insulin?

Unclassified Exercises

21.85 What is the relation between nucleotides and nucleic acids?

21.86 What component besides a sugar and a base is contained in nucleotides?

21.87 Distinguish between the 5′-end and the 3′-end of a nucleic acid.

21.88 Why does the sequence CMP-TMP-AMP not exist in nature?

21.89 Why does the sequence dGMP-dCMP-dUMP not exist in nature?

21.90 Why is base-pairing only between adenine and thymine and between guanine and cytosine in DNA? Why is base-pairing not between adenine and guanine and between thymine and cytosine?

21.91 What nucleotides are obtained by hydrolysis of dGMP-dTMP-dAMP-dTMP-dTMP-dGMP-dCMP?

21.92 What are the major differences in the primary structures of DNA and RNA?

21.93 What is the major difference in the three-dimensional structures of DNA and RNA?

21.94 The total DNA in the chromosomes of carrots contains 36 mol-% adenine and 14 mol-% cytosine. What are the amounts of thymine and guanine?

21.95 Strand 1 of a DNA double helix contains 31 mol-% G and 14 mol-% T; strand 2 of the double helix contains 17 mol-% G and 38 mol-% T. Answer the following questions.

(a) What are the amounts of C and A in strand 2?
(b) What are the amounts of C and A in strand 1?

21.96 If one particular mRNA contains 26 mol-% U and 33 mol-% G, what are the amounts of A and C?

21.97 Sections *X*, *Y*, and *Z* of the following part of a template DNA strand at its 3′ end are exons of a gene:

5′				3′
ACT	CAC	AAA TTG	TGT GGT	CAT
X		*Y*		*Z*

What mRNA is made from this template DNA? What polypeptide sequence will be synthesized?

21.98 Which, if any, of the following statements is (are) true? (a) All mutagens are carcinogens; (b) All carcinogens are mutagens.

21.99 What is a transgenic plant or animal?

21.100 What is the objective of the Human Genome Project?

21.101 What is the objective of gene therapy?

21.102 What do transgenic breeding and gene therapy have in common?

Chemical Connections

21.103 At high temperatures, deoxynucleic acids become denatured—they unwind from double helices into disordered single strands. Account for the fact that the higher the content of guanine-cytosine base pairs relative to adenine-thymine base pairs, the higher the temperature required to denature a DNA double helix.

21.104 A mutant hemoglobin has aspartic acid at position 5 of the α-polypeptide chains instead of alanine. This mutation occurred by substitution of a single base in the normal codon for the alanine residue. Give the codons for the normal and mutant residues.

21.105 Explain why the ratio of guanine to cytosine is 1 : 1 in DNA, but it is usually not 1 : 1 in RNA.

21.106 Erwin Chargoff showed that the base compositions of the DNA in different species of plants and animals varies considerably. If two species are found with identical base compositions, does this finding necessarily indicate that the two species have identical DNA? Explain.

21.107 What is the minimum number of nucleotide bases in the gene for the β-polypeptides of human hemoglobin (146 amino acid residues)? Why is the number of nucleotide bases in the gene likely to be much larger?

21.108 The error level in DNA synthesis is much lower than that in RNA synthesis, less than 1 base in 10 billion compared with less than 1 base in 10,000 to 100,000. This difference is due to the fact that DNA polymerases can proofread and correct errors, but RNA polymerases cannot. Because an error of a single base in either replication or transcription can lead to an error in protein synthesis, what is the biological explanation for this difference in error levels between DNA and RNA syntheses?

CHAPTER 22

METABOLISM AND ENZYMES: AN OVERVIEW

CHEMISTRY IN YOUR FUTURE

You have applied for a job in a major hospital's inborn errors of metabolism program, and, before your first interview, you decide to brush up on what you learned in nursing school about inherited metabolic disorders. You locate an impressive book called *Inborn Errors of Metabolism* and, scanning its table of contents, find lengthy lists of disorders. Some you have heard of, and you know that they are rare and difficult to manage, although the high mortality rates of a generation ago are gradually going down. Each of these disorders is caused by the lack of some functional enzyme. What is so special about the role of enzymes in metabolism, anyway? This chapter throws some light on the question.

LEARNING OBJECTIVES

- Compare and contrast the characteristics of eukaryotic and prokaryotic cells.
- Describe the sense in which metabolism is a balance between catabolic and anabolic processes.
- Describe how adenosine triphosphate links catabolism to anabolism.
- Explain how enzymes and their modifiable activities make metabolism possible.
- Describe the role of high-energy compounds in metabolism, and give an example of how they work.

The preceding four chapters considered the structures and properties of the four principal families of organic biomolecules: the carbohydrates, lipids, proteins, and nucleic acids. In the body, these biomolecules combine to form structures (such as ribosomes, lysosomes, the Golgi apparatus, and mitochondria) whose conjoint task is to extract energy from the environment and use it to sustain life. Chapters 22 through 26 examine these molecular and subcellular structures for the purpose of understanding how they perform their biochemical functions—the functions known collectively as metabolism.

A key theme in any discussion of metabolism is the central importance of regulation and control. A living organism is an open system (Section 6.7), through which mass and energy flow at varying rates. In contrast with chemical systems in the laboratory which will eventually reach equilibrium, living systems never do. The life of each cell rests on the ability to achieve a balance between degradative processes that break tissues and molecules down and synthetic processes that build new ones up. This delicate balance is maintained by a variety of regulatory mechanisms that control the rates and concentrations of enzymes. We will not explore the details of every regulatory step in all the metabolic processes considered herein. But, because this theme of regulation and control is so important for understanding the phenomenon of metabolism, selected examples of regulation, demonstrating important aspects of molecular control mechanisms, will be presented.

◀◀ The principles underlying enzyme structure are covered in Chapter 20; the DNA-directed synthesis of enzymes is described in Chapter 21.

Metabolism requires many different enzymes and their catalyzed reactions but only a few central metabolic pathways. These central pathways, virtually identical in all forms of life, will be the principal topic of this chapter and the following three chapters.

◀◀ The basic concepts of biochemical catalysis are introduced in Section 8.4.

22.1 CELL STRUCTURE

In chemistry laboratories and in industrial manufacturing facilities, temperatures of hundreds of degrees, pressures of tens of atmospheres, organic solvents, and extremes of pH are commonly required to induce chemical compounds to react. Moreover, these reactions take place in vessels millions of times as large as living cells. Chemical reactions in cells, on the other hand, take place in aqueous media at neutral pH and at low and constant temperatures. The cell itself is an extraordinarily fragile vessel of extremely small dimensions. Let us start with a brief survey of its interior.

For our purposes, the living world can be divided into two types of organisms, called **eukaryotes** and **prokaryotes.** The biochemistry of the two groups is strikingly similar. Moreover, in both groups, all cells are enclosed within a complex membrane capable of exquisite control over the entry and exit of nutrients and waste products (Section 19.10). The differences between eukaryotes and prokaryotes lie in their internal organization and modes of reproduction. Eukaryotes include all cells of multicellular organisms, such as the vertebrates, and many single-celled organisms, such as the yeasts and the protists *Euglena* and *Paramecium.* Almost all prokaryotes are bacteria.

In eukaryotic cells, many of the cellular macromolecules are packaged into **organelles,** subcellular structures surrounded by their own membranes. Organelles form separate functional compartments within the cell. The internal landscape of prokaryotes, as we shall soon see, is quite different.

Figure 22.1 on the next page is a sketch of a typical animal cell, identifying its principal compartments and structures, and Table 22.1 on page 663 lists their major functions. The cell is surrounded by a **cell membrane** responsible for recognizing both small and large molecules and for the transport of molecules and ions into and out of the cell. The area outside of all the compartments in the cell, filled with a jellylike background substance, is called the **cytosol.**

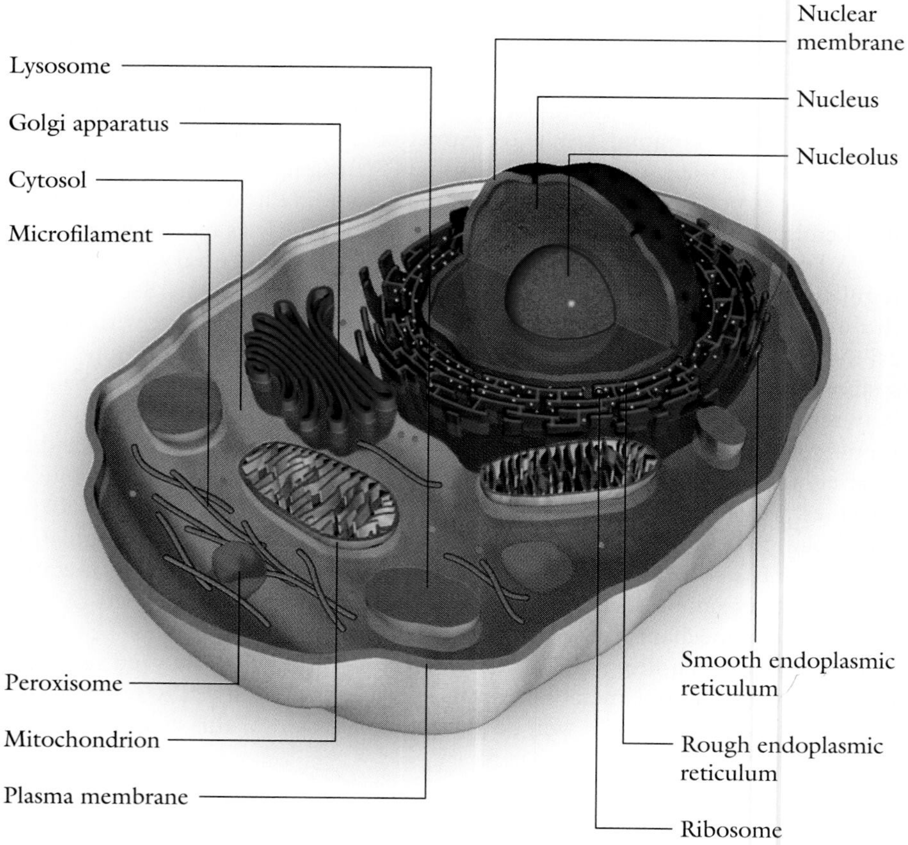

Figure 22.1 Drawing of a typical animal cell identifying its principal compartments and structures.

Biochemists have adopted the words aerobic and anaerobic from bacteriology, where they denote cell growth in the presence or absence of oxygen, respectively. Some anaerobic bacteria can only grow in the absence of oxygen and are called obligate anaerobes. Others can grow in the absence or presence of oxygen and are called facultative anaerobes. The biochemical use of the word **aerobic** describes a chemical reaction that requires oxygen. Biochemically, the word **anaerobic** describes a chemical reaction in which oxygen is not a reactant, but the reaction can take place in the presence of oxygen. In fact, oxygen is present in the cytosol, and anaerobic metabolic processes take place there. Aerobic metabolic processes, those requiring oxygen as a reactant, take place chiefly in **mitochondria.** Mitochondria are subcellular organelles that possess both an outer and an inner membrane specialized for specific transport of molecules and electrons.

Lysosomes are the cell's digestive system, responsible for the hydrolysis of carbohydates, proteins, lipids, and nucleic acids. In Section 24.7, we list some diseases that are the result of the failure of lysosomes to hydrolyze certain lipids. Protein synthesis takes place at RNA structures called **ribosomes** (Section 21.7). Ribosomes are attached to the **endoplasmic reticulum**—membranous structures that, in addition to protein and lipid synthesis, participate in the transport of macromolecules into and out of the cell. The **Golgi apparatus** is a network of flattened, smooth membranes surrounding vesicles in which proteins synthesized on the endoplasmic reticulum are modified and

TABLE 22.1 Animal Cell Compartments and Their Major Functions

Compartment	Functions
cell membrane	transport of ions and molecules; receptors for biomolecules
nucleus	DNA synthesis and repair; RNA synthesis
nucleolus	ribosome synthesis
endoplasmic reticulum	synthesis of proteins and the lipids used for manufacture of cell organelles; lipid synthesis
ribosome	protein synthesis; usually attached to the endoplasmic reticulum
Golgi apparatus	modification of proteins both for export and for incorporation into organelles
mitochondrion	cellular respiration, oxidation of lipids and carbohydrates; conservation of energy; urea synthesis
peroxisome	reactions with oxygen, decomposition of hydrogen peroxide
lysosome	hydrolysis of carbohydrates, proteins, lipids, and nucleic acids
microfilament	cell cytoskeleton; intracellular movements
cytosol	metabolism of carbohydrates, lipids, amino acids, nucleotides; protein synthesis

packaged for transport out of the cell. It is also an important site for the synthesis of new membrane material. Eukaryotic DNA is organized into chromosomes located within the **nucleus,** which is enclosed by the nuclear membrane. Within the nucleus is a structure called the **nucleolus,** where the synthesis of ribosomes takes place.

Eukaryotic cells also possess a cytoskeleton consisting of **microtubules** and **microfilaments.** They give cells a shape and provide a means for movement; for example, they impel the movement of chromosomes during mitosis (cell division) and meiosis (haploid germ-cell formation). At small structures called **peroxisomes,** direct reactions with oxygen and the decomposition of hydrogen peroxide take place. Reactions there are usually those that detoxify toxic substances.

In prokaryotes, subcellular macromolecular structures either are anchored directly to the inside of the cell membrane or move about freely within the cell. The DNA is organized into a single chromosome located in a **nuclear zone,** which is not surrounded by a membrane. Prokaryotic cells have little internal membranous structure, but their internal protein concentration is very high, about 20%. This leads to a very large internal osmotic pressure, perhaps from 2 to 3 atm. No lipid-based cell membrane could withstand such high internal pressure, but prokaryotic cells possess, in addition to the cell membrane, a strong, rigid polysaccharide framework called a cell wall. The cell wall is strong enough to resist the internal osmotic pressures typical of prokaryotes.

22.2 GENERAL FEATURES OF METABOLISM

Metabolism in animal cells has four functions: (1) to obtain energy in a chemical form by the degradation of nutrients; (2) to convert a wide variety of nutrient molecules into the few precursor molecules used to build proteins, carbohydrates, nucleic acids, lipids, and other cell molecules; (3) to synthesize cell molecules; and (4) to produce or modify the biomolecules necessary for specific functions in specialized cells or both.

Metabolism takes place through the interaction of two processes: catabolism and anabolism. **Catabolism** is the biochemical degradation of energy-containing compounds, leading to the capture of that energy in new chemical forms that the cell can use for biosynthesis, movement, and secretion. As a result of catabolic reactions, the many different types of energy-containing molecules are converted into a limited number of simpler molecules, which are then used for the synthesis of components vital to cell structure and function.

Anabolism is the biochemical synthesis of biomolecules. A few simple molecules serve as building blocks to create a large variety of macromolecules and complex cellular components. These simple, building-block molecules have other uses as well. Amino acids are also the precursors for hormones, alkaloids, and porphyrins, and serve as neurotransmitters. Nucleotides are also precursors of energy carriers and coenzymes. Anabolism includes the work necessary to transport ions and biomolecules across cell membranes.

Stoichiometrically, the oxidation of D-glucose in the body is identical with its combustion under commercial or laboratory conditions; that is

$$C_6H_{12}O_6 + 6\ O_2 \longrightarrow 6\ CO_2 + 6\ H_2O$$

In those combustions, the energy contained in the D-glucose is converted into heat. However, in cellular metabolism, some of the reaction energy is captured. The metabolic process consists of sequences of reactions in which nutrients are broken down to simpler molecules while, at the same time, substances that fuel biosynthetic processes are synthesized. These substances, called high-energy compounds, are discussed in Section 22.9. It is in this sense that metabolism extracts energy from the environment and uses it to sustain life.

22.3 STAGES OF CATABOLISM

Catabolism, the degradation of the major energy-yielding compounds ingested by cells, proceeds in the sequence of steps shown in Figure 22.2. These steps can be grouped into three major stages.

In the first stage, nutrient molecules are degraded to their lower-molecular-mass components: α-amino acids, monosaccharides, fatty acids, and glycerol.

The next stage converts these different products into one simple molecule—**acetyl-*S*-coenzyme A** (acetyl-*S*-CoA). The collected monosaccharides are first converted into a three-carbon intermediate, pyruvate, and then into the two-carbon acetyl-*S*-CoA (Section 22.9). The carbon chains of fatty acids and most of the amino acids also are converted into the same two-carbon compound. The second stage of catabolism ends with production of this one molecule from the myriad types entering the degradative process.

In the final stage of catabolism, the acetyl-*S*-CoA enters the **citric acid cycle.** In this part of the pathway, the acetyl-*S*-CoA is oxidized to carbon dioxide and water.

Concept check

✔ Catabolism is characterized by the conversion of the many different types of nutrient molecules into a final, common end product.

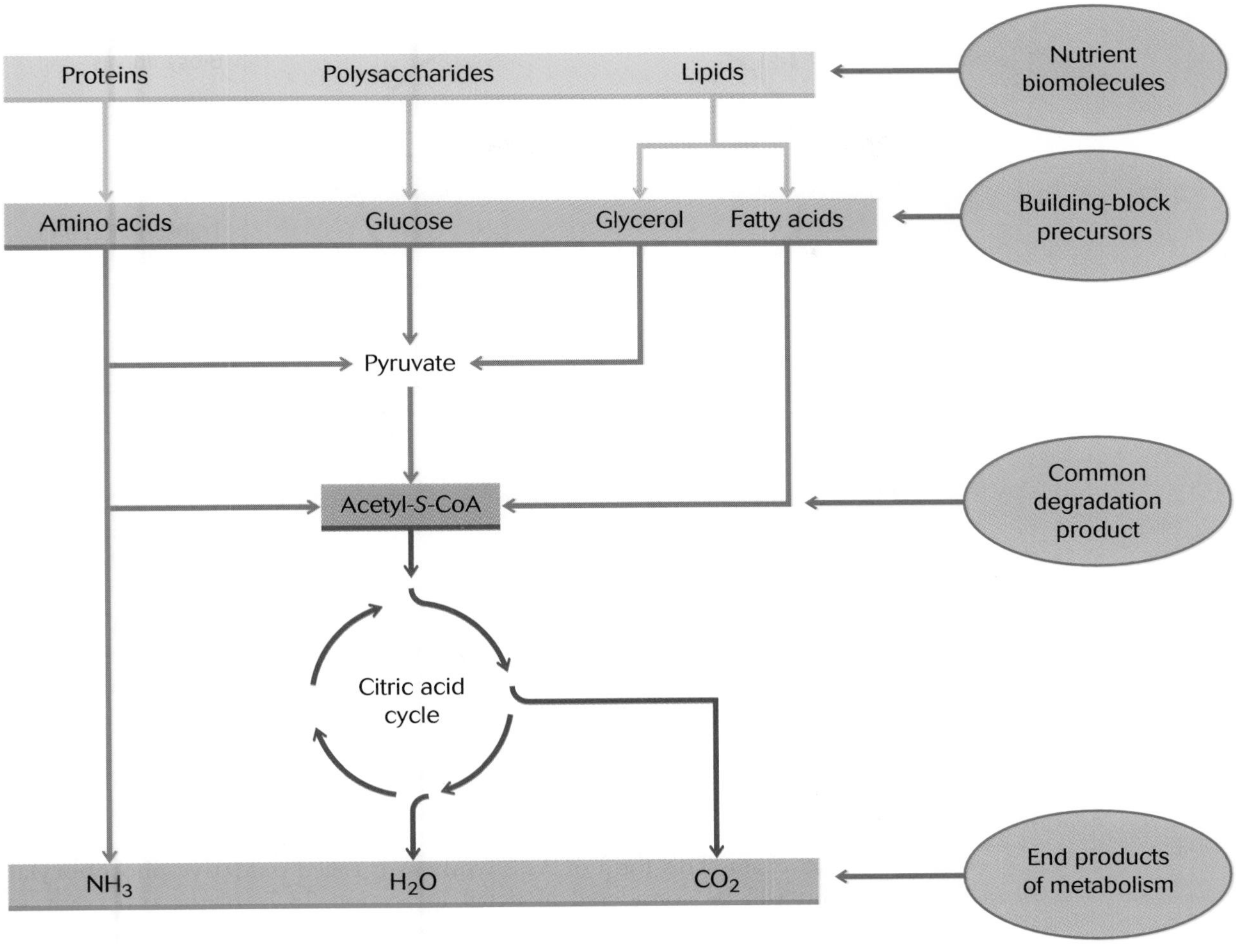

Figure 22.2 Catabolism of the major energy-yielding compounds of cells proceeds in a sequence of steps that consists of three major stages: (1) nutrient biomolecules are degraded to monosaccharides, amino acids, glycerol, and fatty acids; (2) these molecules are converted into a common degradation product, acetyl-*S*-CoA, which is (3) degraded into the end products carbon dioxide and water.

22.4 TRANSFORMATION OF NUTRIENT CHEMICAL ENERGY INTO NEW FORMS

Much of the energy of nutrients captured in the second and third stages of catabolism is used to synthesize three important compounds that fuel anabolism. The first, **adenosine triphosphate,** abbreviated **ATP,** functions as the carrier of energy to the energy-requiring processes of cells. Thus, it is the link between catabolism and anabolism. The second and third products, **NADH (nicotinamide adenine dinucleotide)** and **NADPH (nicotinamide adenine dinucleotide phosphate)** capture the reducing power that is a key requirement for cellular biosynthesis. NADH is primarily used for the synthesis of ATP. NADPH is used almost exclusively for reductive biosynthetic processes.

Figure 22.3 on the following page is a representation of how ATP links the processes of catabolism to those of anabolism. In general, as ATP yields its energy to drive an anabolic reaction, it is broken into its hydrolysis products, adenosine diphosphate (ADP) and inorganic phosphate (see Box 15.5). (Inorganic phosphate, consisting of all species of phosphate, HPO_4^{2-}, $H_2PO_4^{-}$, and

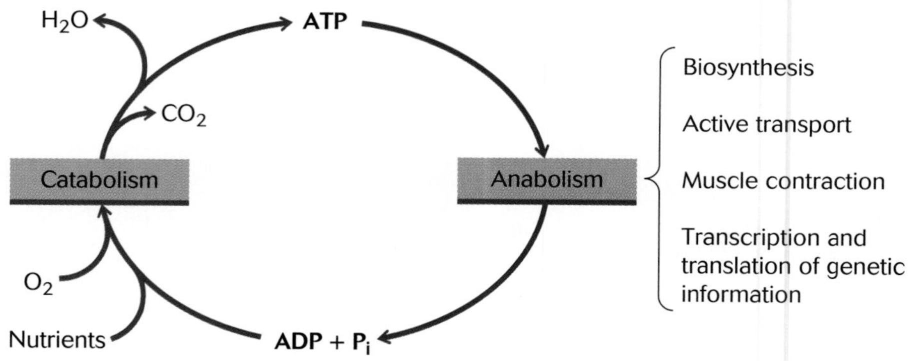

Figure 22.3 ATP is the principal link between the processes of catabolism and anabolism.

PO_4^{3-}, is represented by the symbol P_i for convenience in describing phosphorylation reactions in metabolism. Phosphorylation describes a reaction in which the hydroxyl group of an alcohol is converted into a phosphate ester group; see Section 15.13.)

The ATP molecule, seen in Figure 22.4, is composed of the nitrogen base adenine, ribose, and phosphates esterified to carbon 5 of the ribose group. Figure 22.4 indicates the cleavage point for the conversion of ATP into ADP. Note that, at cellular pH, the phosphate groups of both ATP and ADP are ionized.

ATP synthesis takes place in two ways. In one way, ADP reacts directly with P_i. This synthesis of ATP is associated with the simultaneous oxidation of NADH in a sequence of reactions within mitochondria called the electron-transport chain. This form of ATP synthesis is called **oxidative phosphorylation** and will be considered in detail in Section 23.7. In the other way in which ATP is synthesized, called **substrate-level phosphorylation,** a phosphate group is transferred to ADP directly from another phosphorylated compound. (These phosphorylated compounds, called high-energy compounds, are considered in more detail in Section 22.9.)

NAD^+ (Figure 22.5) is one of a class of substances that link metabolic oxidations to metabolic reductions. Metabolic oxidations, like any redox reaction (Section 4.7), consist of an oxidation half-reaction and a reduction half-

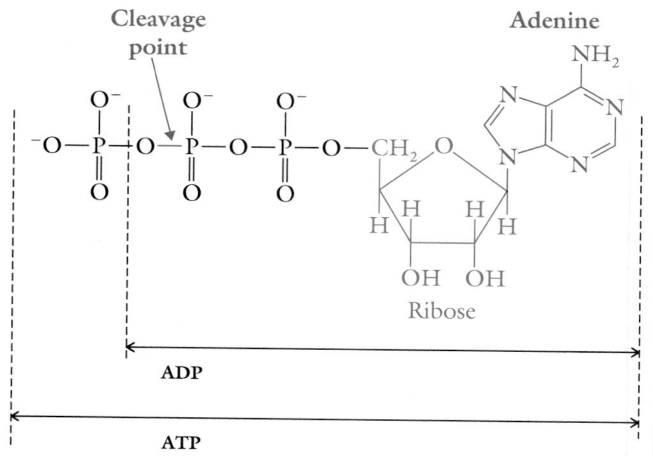

Figure 22.4 The structure of ATP is composed of the nitrogen base adenine, ribose, and phosphates esterified to carbon 5 of the ribose group. Note the cleavage point for the hydrolysis of ATP into ADP.

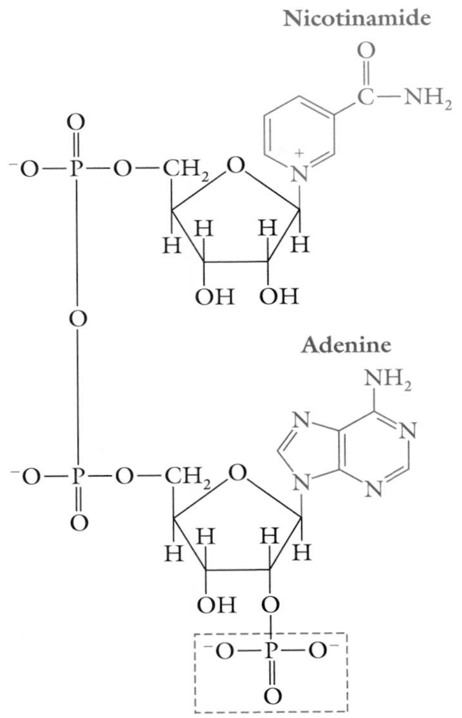

Figure 22.5 The structure of NAD^+ (and of $NADP^+$) shows that, in the oxidized state, the nicotinamide ring has a positive charge. In the reduced state of NAD(P)H, the positive charge is neutralized when the hydride ion is transferred to the nicotinamide ring. The phosphate group outlined by red dashed lines, esterified to the 2-OH of the ribose group of NAD^+, transforms the NAD^+ into $NADP^+$, nicotinamide adenine dinucleotide phosphate.

reaction. For the most part, metabolic oxidations are not direct reactions with oxygen but consist of transfers of hydrogen atoms and electrons and are better described as **dehydrogenations.** In a metabolic dehydrogenation, two hydrogen atoms and two electrons are transferred to an acceptor. Although comparatively few, direct reactions with oxygen do take place, and they are called oxygenations.

In catabolic dehydrogenations, NAD^+ receives electrons and a hydrogen atom in the form of a hydride ion ($H:^-$) to form NADH (Figure 14.8). It can then be oxidized in subsequent reactions through which hydrogen and electrons are eventually transferred to oxygen to form water. NADPH is chiefly responsible for providing the electrons and hydrogen atoms in anabolic syntheses—for example, the hydrogenation of double bonds in the formation of steroids.

The structures of NAD^+ and $NADP^+$ are identical except for the phosphate group, outlined by red dashed lines in Figure 22.5, esterified to the 2-OH of the ribose group of NAD^+. The addition of that phosphate group transforms the NAD^+ into **$NADP^+$** (nicotinamide adenine dinucleotide phosphate). The complex adenine nucleotide part of the molecule does not participate in redox reactions. Its function is to provide the structural properties that allow interaction with a specific enzyme.

Note that, in the oxidized state, the nicotinamide ring has a positive charge, and, in the reduced state (NADH), the positive charge is neutralized when the hydrogen and its electrons are transferred to the nicotinamide ring. Because NAD^+ and $NADP^+$ undergo the same reduction reaction, biochemists symbolize both of them by the formula $NAD(P)^+$. When $NAD(P)^+$ participates in the dehydrogenation of reduced nutrients, one hydrogen atom with two electrons is transferred as a hydride ion ($H:^-$) to the nicotinamide ring, whereas the second hydrogen enters the reaction medium as a proton (H^+). With AH_2 representing the reduced nutrient, the typical reaction is:

$$AH_2 + NAD(P)^+ \longrightarrow A + NAD(P)H + H^+$$

Figure 22.6 The structure of flavin adenine dinucleotide (FAD). The arrows point to the nitrogen atoms that become hydrogenated when FAD is reduced to form $FADH_2$.

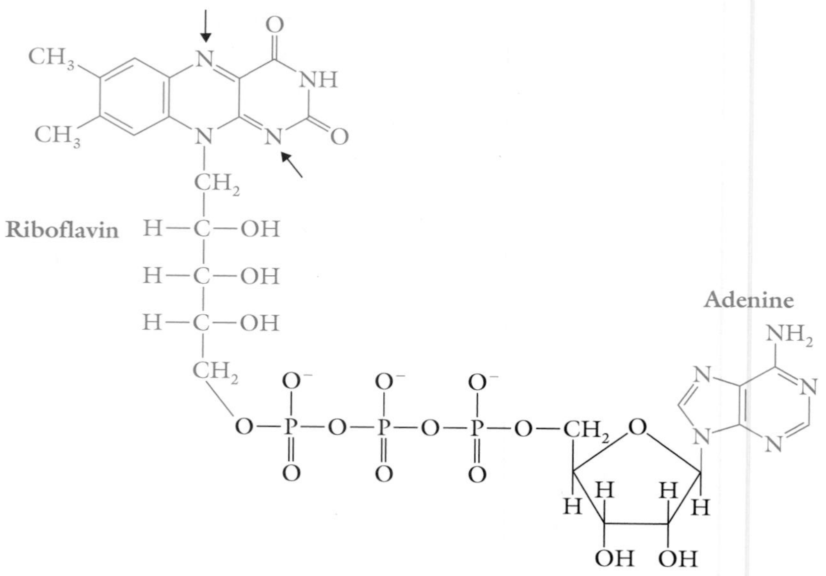

The other important substances that link metabolic oxidations to metabolic reductions are the flavin nucleotides, FAD (flavin adenine dinucleotide; Figure 22.6) and FMN (flavin mononucleotide; Figure 22.7). In contrast with the pyridine nucleotides [$NAD(P)^+$], both hydrogens are incorporated into a flavin nucleotide when it is reduced:

$$AH_2 + FAD \longrightarrow A + FADH_2$$

To function as hydrogen acceptors or donors, the pyridine and flavin nucleotides must be bound to enzymes specific for the particular reaction taking place. $NAD(P)^+$ is loosely bound to its enzyme, but FAD is so tightly bound that its combination with its enzyme is called a flavoprotein. As we shall see, they are cofactors (Section 22.5) in enzymatic dehydrogenations. The linking of the oxidizing reactions of catabolism and the reducing requirements of anabolism is illustrated in Figure 22.8.

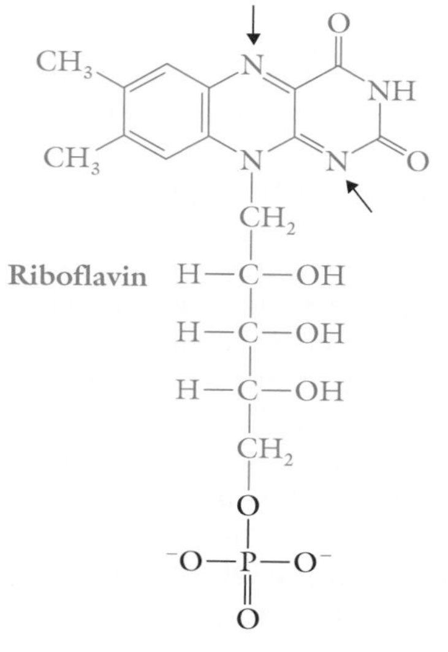

Figure 22.7 The structure of flavin mononucleotide (FMN). FMN is essentially phosphorylated riboflavin. Reduction (hydrogen addition) takes place at the nitrogen atoms identified by the arrows.

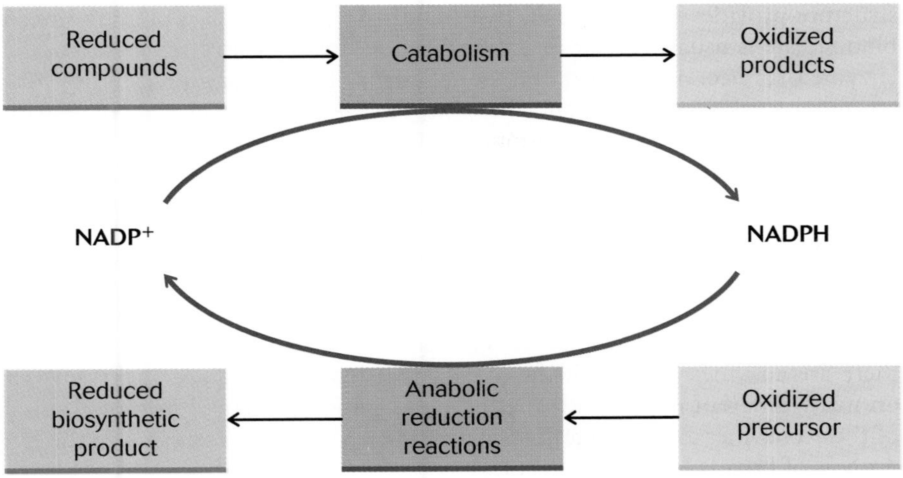

Figure 22.8 $NADP^+$ links the oxidation reactions of catabolism to the reduction reactions of anabolism. NADPH is primarily used for reductive biosynthetic reactions.

Concept checklist

- ✔ The chemical energy obtained from nutrients during the second and third stages of catabolism is used to build the major products, ATP and NAD(P)H.
- ✔ Adenosine triphosphate functions as the carrier of energy to the energy-requiring processes of cells.
- ✔ NAD(P)H provides the reducing power required for cellular biosynthesis.

22.5 ENZYMES

A single living cell is a far more efficient chemical factory than has ever been devised by chemical engineers. Multicellular organisms synthesize thousands of different types of proteins, as well as other molecules, all within a single cell having a volume of about 1×10^{-12} cm^3.

The cell is able to channel biomolecules into the construction of specific products in precisely the required amounts because its reactions are linked into specific sequences that can be controlled in a number of ways. The key to this control is the existence of unique catalysts called enzymes. Enzymes as catalysts were first considered in Section 8.3. The amounts and activities of the enzymes catalyzing each step of each reaction sequence can be modified, allowing the cell to carry out its metabolic functions in an integrated, harmonious way.

Enzymes Are Proteins

Although recent work has demonstrated modest catalytic activity in some nucleic acids, the overwhelming evidence is that enzymes are protein molecules possessing unique catalytic properties that depend on their structural integrity.

When an enzyme is hydrolyzed by the digestive enzyme trypsin, its catalytic activity disappears completely. This means that the primary structure (the backbone) of the enzyme is necessary for the enzyme to function. Catalytic activity will also be lost if the enzyme is subjected to conditions that denature proteins—for example, heat or low or high pH (Section 20.9). Clearly, the detailed folding of the enzyme is central to its catalytic efficiency.

Catalysis takes place at **active sites** on the enzyme's surface. These sites possess two independent structural characteristics: part of the active site's structure provides the catalytic activity, and part functions as a binding site. A binding site is usually a cleft or indentation occupying a very small part of the enzyme's surface. At that location, noncovalent or secondary forces bring **substrate** molecules—the substance the enzyme acts on—into close contact with the chemical groups responsible for catalytic activity.

Enzymes Are Specific

▶▶ Specific binding sites are also important in cell recognition (Section 18.6), lipid transport (Section 26.6), and nerve transmission (Section 26.4).

Many enzymes possess great specificity: only molecules of a specific structure will be accommodated at the binding site. This specificity permits great selectivity among the myriad molecules present in the cell and allows the efficient simultaneous operation of linked metabolic pathways. On the other hand, there are enzymes acting outside of cells that have a broad specificity and act on many different compounds that have a common structural feature. Typical of these enzymes are the digestive enzymes, each of which can catalyze the hydrolysis of broad classes of molecules.

The binding site accepts its substrate in much the way that a lock accepts its key, the substrate's structure complements the structure of the binding site. This **complementarity** of structure includes stereochemistry and shape (Section 17.5 and Figure 17.2) as well as electrical charge and hydrophobic-hydrophilic considerations. In other words, the shape and charge of a molecule must be complementary to the shape and charge of the binding site. Moreover, the affinity of a molecule for a binding site is enhanced if the incoming molecule has a hydrophobic character that is matched by the hydrophobic character of the binding site. In certain cases, the shape of an enzyme's binding site changes on binding of substrate. This phenomenon, called **inducible fit** because of the flexible nature of the protein, enhances the complementary character of the binding site as it binds substrate.

Enzyme Cofactors

Although some metabolic reactions can be catalyzed by proteins alone, most metabolic reactions are catalyzed by proteins that are combined with specialized small molecules called **cofactors.** A cofactor may be a metal ion (Table 22.2) or an organic molecule called a coenzyme (Table 22.3). A cofactor combines transiently with a noncatalytic protein, called an **apoenzyme,** to form the catalytic **holoenzyme.** Table 22.2 contains a list of enzymes that must combine with inorganic elements to have catalytic activity. Many of the dietary vitamins, such as the B vitamins, become cofactors and serve as temporary carriers of atoms or functional groups in redox and group-transfer reactions. Some of them are listed in Table 22.3. The pyridine and flavin nucleotide cofactors are derived from the vitamins nicotinic acid and riboflavin, respectively.

TABLE 22.2 Inorganic Cofactors and Their Enzymes

Cofactor	Enzyme
Cu^{2+}	cytochrome oxidase
Fe^{2+}, Fe^{3+}	catalase; cytochrome oxidase; peroxidase
K^{+}	pyruvate kinase (also needs Mg^{2+})
Mg^{2+}	hexokinase; glucose 6-phosphatase
Mn^{2+}	arginase
Ni^{2+}	urease
Se^{2+}	glutathione peroxidase

TABLE 22.3 Vitamins and Corresponding Coenzymes Serving as Group Transfer Carriers

Vitamin	Coenzyme	Group Transferred
biotin	biocytin	carbon dioxide
folic acid	tetrahydrofolate	other one-carbon groups
pantothenic acid	coenzyme A	acyl groups
cobalamin (vitamin B_{12})	5′-deoxyadenosylcobalamine	alkyl groups, hydrogen atoms
riboflavin	flavin adenine dinucleotide	hydrogen atoms
niacin (nicotinic acid)	nicotinamide adenine dinucleotide	hydride ion (H^-)
pyridoxine (vitamin B_6)	pyridoxal phosphate	amino groups
thiamine	thiamine pyrophosphate	aldehydes

22.6 ENZYME CLASSIFICATION

The early phases of every science are characterized by much naming but only rudimentary understanding. Thus, enzymes were named as they were discovered, before very much was known about their structure or the way in which they worked. In many cases, chemists simply added the suffix **-ase** to the name of the compound acted on by the enzyme, as in hexokinase, urease, cytochrome oxidase. In other cases, the names (trypsin or pepsin, for example) do not indicate the substrates. Although the historical names are still in common use, a new system of nomenclature has been developed that eliminates ambiguity in naming and classifying enzymes. This system places an enzyme in one of six major classes, listed in Table 22.4, with subclasses based on the reactions catalyzed.

Hexokinase is the common name for the enzyme catalyzing the reaction between adenosine triphosphate and hexoses such as D-glucose:

$$\text{D-Glucose} + \text{ATP} \rightleftharpoons \text{D-glucose-6-PO}_4 + \text{ADP}$$

The formal systematic name for the enzyme is ATP: hexose phosphotransferase. This name indicates that the enzyme catalyzes the transfer of a phosphate group from ATP to any hexose. The phosphotransferase specific for glucose is called glucokinase. Although the formal nomenclature eliminates ambiguity, its use is often quite cumbersome. Most writers and workers in the field use the simpler historically derived common or trivial names of enzymes unless the situation warrants formal nomenclature, and that will be our policy

TABLE 22.4 Enzyme Classification Based on Catalyzed Reactions

Class	Reaction Type
oxidoreductase	electron transfer
transferase	group-transfer reactions
hydrolase	hydrolysis reactions
lyase	addition or removal of groups to or from double bonds
isomerase	isomers produced as a result of group transfers within a molecule
ligase	formation of C–C, C–S, C–O, and C–N bonds by condensation reactions coupled to ATP cleavage

in this book. In Section 18.2, it was pointed out that the D form of monosaccharides predominates in nature. As a consequence, whenever the stereochemistry of a saccharide is not indicated, such as in "glucose" or "fructose," you should assume that the D-enantiomer is meant. The enantiomer will be identified in chemical equations.

22.7 ENZYME ACTIVITY

When an enzyme is discovered, one of the first aspects to be studied is its effect on the rate of its catalyzed reaction: how many moles of substrate are converted into product per unit time? To answer this question, the experimenter adds increasing concentrations of substrate to a fixed concentration of enzyme. For example, the experimenter might prepare a rack of test tubes each containing 0.01 m*M* of enzyme solution (Section 7.6) and then add a different amount of substrate to each one (Figure 22.9). After the initial rate of reaction in each tube has been measured and plotted as a function of substrate concentration, many enzymes (for example, pepsin or salivary amylase) produce curves resembling the one in Figure 22.7. The rate of the enzymatically catalyzed reaction is called the **enzyme activity.**

The graph in Figure 22.7 shows that the rate of this enzyme-catalyzed reaction reaches a maximum value when the substrate concentration exceeds the

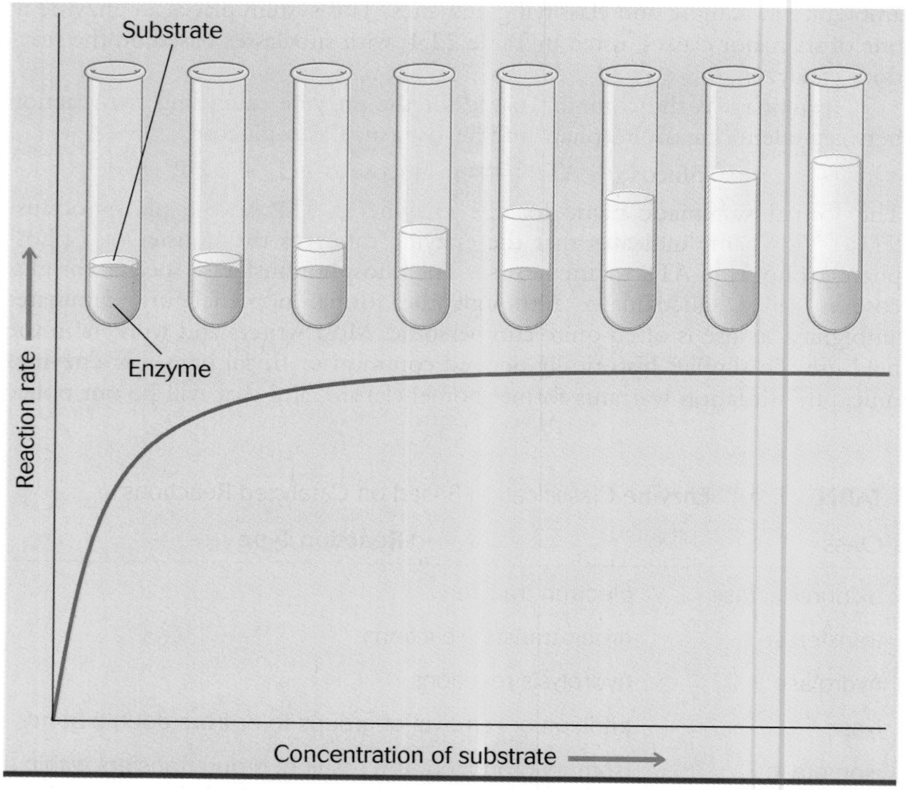

Figure 22.9 The rates of an enzyme-catalyzed reaction are measured in a series of test tubes in which the enzyme concentration is kept constant and the substrate concentration is continually increased. The rate of the enzyme-catalyzed reaction increases and reaches a limit as the substrate concentration increases.

enzyme concentration. At this point, the enzyme has reached its capacity to bind substrate and has become "saturated." A fixed concentration of enzyme means a fixed and limited number of active sites. The rate of reaction at each site is constant, so an increase in the substrate concentration past the enzyme saturation point cannot increase the rate of reaction. The added substrate must wait its turn to enter the active site, just as any passenger must do at a crowded bus stop.

When the reaction rate becomes constant, it is possible to define the enzyme activity in terms of the ratio of catalyzed rate to uncatalyzed rate. For example, the enzyme carbonic anhydrase catalyzes the formation of bicarbonate ion in the reaction of carbon dioxide with water. Found in high concentrations in the human erythrocyte, it is among the most active of all enzymes. The ratio of enzyme-catalyzed rate to the uncatalyzed rate of hydration of CO_2 is: $\text{rate}_{\text{catalyzed}}/\text{rate}_{\text{uncatalyzed}} = 1 \times 10^7$. In other words, the catalyzed reaction takes place 10 million times as fast as the uncatalyzed reaction.

22.8 CONTROL OF ENZYME ACTIVITY

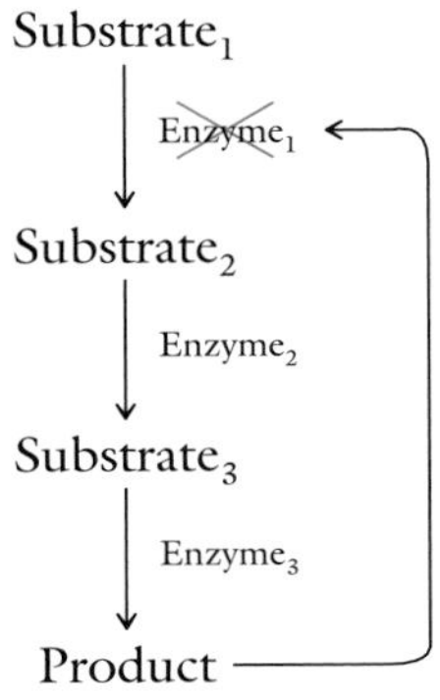

The activity of enzymes in the cell is regulated in a number of ways. For example, the enzymes catalyzing the synthesis and hydrolysis of glycogen can be activated by phosphorylation and deactivated by dephosphorylation (Section 23.10). This form of regulation is an example of **covalent modification.** Another example of covalent modification will be considered in Section 23.2, when we describe the enzymatic conversion of pyruvate into acetyl-*S*-CoA.

Amino acids are synthesized by sequences of as many as 13 separate enzymatically catalyzed reactions. In the hypothetical sequence depicted in the margin, when the end product (the final product of the synthetic sequence) is present in high concentrations, it combines specifically with the first enzyme in the synthetic sequence and prevents that enzyme from functioning. This is an example of **feedback inhibition.** Other kinds of feedback inhibition prevent the synthesis of the enzyme itself, by interfering with the transcription of its messenger RNA (Section 21.7).

The enzymes located within the cells of multicellular organisms, particularly of warm-blooded animals, are constrained to operate within very narrow limits of temperature and pH. Therefore metabolic activity does not respond to changing cellular conditions due to variations of temperature or pH; but rather it requires the action of intracellular enzymes called **regulatory** or **allosteric enzymes.** The activity of these enzymes is controlled by the binding of specific molecules called activators or inhibitors. A regulatory modifier of an allosteric enzyme combines not with the enzymes' active site, but with another specific binding site of the enzyme that is complementary to the modifier's structure. The binding of the modifier molecule at the nonenzymatic site in turn modifies the enzyme's conformational structure (Section 20.5), causing changes in its catalytic activity.

Regulatory enzymes control the balance between energy-rich and energy-poor states of the cell. When the cell is in an energy-rich state (high concentration of ATP), there is no need to sacrifice nutrients, and the result is that anabolism (biosynthesis) predominates over catabolism. Conversely, when the cell is in an energy-poor state (high concentration of ADP), synthesis of biomolecules will be curtailed, with the result that catabolism and the production of ATP will predominate over anabolism. In order for the cell's metabolic machinery to respond to these conditions, specific enzymes in the metabolic pathways must recognize those states and respond appropriately. For example, citrate is a major component of aerobic catabolism, so when the cell is in a high-energy state, citrate is present in high concentration in the cell. Under those conditions, citrate acts as a regulatory modifier and will combine with and activate the enzyme that begins the process for the synthesis of fatty acids. (Section 24.6.)

Concept check

✔ Regulatory enzyme activity is modified by specific, low-molecular-mass activator or inhibitor molecules that bind to an enzyme and alter its conformation.

Some enzymes are present in the cell as individual catalytic molecules. Others combine into structural complexes consisting of many different enzymes. The enzymes responsible for the synthesis of fatty acids in human cells, for example, are present in the cell cytosol in the form of a single multifunctional protein that catalyzes seven different reactions. The enzymes of aerobic respiration are arranged in close contact with each other within the membranes of mitochondria. The rates at which sequences of metabolic reactions take place within such complexes are markedly enhanced by the enzymes' organization into closely arranged, functionally integrated structures.

22.9 HIGH-ENERGY COMPOUNDS

Remember that catalysts cannot alter the equilibrium of a chemical reaction; they only increase the rate of approach to equilibrium. Many metabolic reactions have small equilibrium constants; that is, they are unfavorable and yield small quantities of product. They could not take place within cells without the presence of ATP and a small number of other **high-energy compounds** to drive them forward. The mechanism for achieving this impetus was detailed in Example 8.10, which showed how a reaction with a small equilibrium constant can be coupled to a favorable reaction through a common intermediate to produce a large overall equilibrium constant.

High-energy compounds tend to contain groups such as phosphate and acetate, whose transfer to water is characterized by very large equilibrium constants. These compounds turn up repeatedly in each metabolic pathway, serving as common intermediates in reactions whose equilibrium constants are small. To illustrate again how the strategy works, we will examine the first reaction of anaerobic metabolism, in which glucose is phosphorylated by ATP.

The reaction equation is:

$$\text{D-Glucose} + \text{ATP} \xrightleftharpoons{\text{hexokinase}} \text{D-Glucose-6-PO}_4 + \text{ADP}$$

It is critical to note that this reaction, like all others of metabolism, requires a specific enzyme to proceed at a rate appropriate to the time requirements of metabolism. The enzyme is specific in two key respects: (1) it transfers a phosphate group (all enzymes that catalyze the transfer of a phosphate group are called kinases) and (2) it is structurally adapted to hexose sugars.

In the enzyme-catalyzed reaction, ATP reacts directly with glucose. However, if we picture this single-step catalyzed reaction as taking place in two sequential steps with a common intermediate, the role of ATP in this and similar reactions can be made more clear. The two sequential steps will be (1) the reaction between glucose and inorganic phosphate and (2) the reaction of ATP with water. The first step, the direct reaction of glucose with inorganic phosphate, is characterized by a small equilibrium constant, whereas the second step, the hydrolysis of ATP, has a very large equilibrium constant:

1. $\text{D-Glucose} + \text{P}_i \rightleftharpoons \text{D-glucose-6-PO}_4 + \text{H}_2\text{O} \qquad K_1 = 3.8 \times 10^{-3}$
2. $\text{ATP} + \text{H}_2\text{O} \rightleftharpoons \text{ADP} + \text{P}_i \qquad K_2 = 2.26 \times 10^{6}$

When the two equations are added together, the P_i and H_2O on each side cancel, resulting in the equation describing the one-step process:

3. D-Glucose + ATP $\rightleftharpoons$ D-glucose-6-PO_4

The equilibrium constant for the one-step process is obtained by multiplying the equilibrium constants of reactions 1 and 2:

$$K_1 \times K_2 = K_3 = 8.6 \times 10^3$$

The overall equilibrium constant now favors the formation of D-glucose-6-PO_4.

◄◄ Section 8.7 presents an example of how to calculate the overall equilibrium constant for a sequence of reactions.

High-energy compounds are compounds that have large equilibrium constants in certain reactions, such as the transfer of phosphate or other groups to water (HOH) and other hydroxyl-bearing (ROH) compounds. However, keep in mind that a favorable equilibrium constant does not always mean that a product is formed quickly. It turns out that the activation energies for direct transfer to water are quite high, and the rates of hydrolysis for these compounds would be very slow if it were not for the presence of transferase enzymes such as hexokinase, which accelerate the reaction rate.

There are other high-energy compounds in metabolism that have even larger equilibrium constants for phosphate transfer than does ATP. As a consequence, the ATP-ADP system can operate as a phosphate shuttle between the high-energy compounds with large equilibrium constants for phosphate transfer and others with lower equilibrium constants for phosphate transfer. The high-energy compounds can transfer, for example, a phosphate group to ADP, which then can transfer that phosphate group to glucose. Reactions in which parts of molecules such as phosphate groups ($—PO_4$), amino groups ($—NH_2$), or methyl groups ($—CH_3$), are transferred from one molecule to another are called group-transfer reactions.

Two such compounds are **1,3-bisphosphoglycerate** and **phosphoenolpyruvate.** Both are synthesized during the initial stage of glucose catabolism and in turn are used to synthesize ATP by substrate-level phosphorylation; that is, direct phosphate-group transfer from a phosphorylated high-energy compound to a phosphate-accepting compound. Other high-energy compounds include the **thioesters,** such as **acetyl-*S*-CoA,** and **acid anhydrides,** such as **acetyl phosphate.**

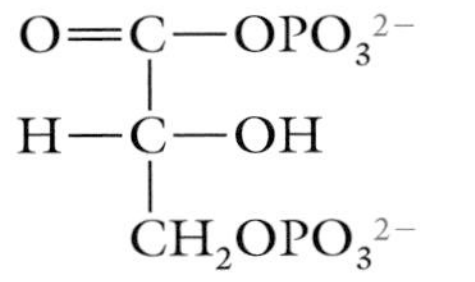

1,3-Bisphosphoglycerate

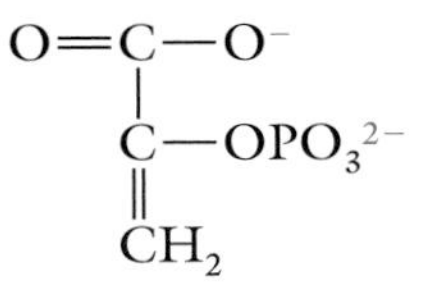

Phosphoenol pyruvate

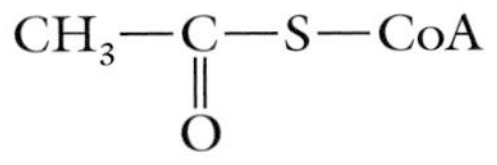

Acetyl-*S*-CoA

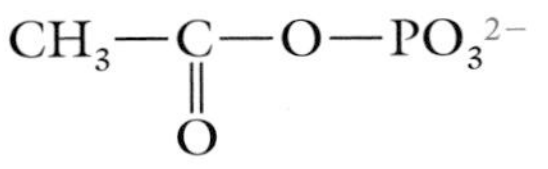

Acetyl phosphate

Concept check

✔ Reactions that include high-energy compounds enable the chemically unfavorable reactions of metabolism (that is, reactions characterized by small equilibrium constants) to produce adequate quantities of desired products.

Full details of the biochemical processes of catabolism and anabolism will be presented in subsequent chapters, beginning, in Chapter 23 with a discussion of carbohydrate metabolism.

Summary

General Features of Metabolism Metabolism consists of the physical and chemical processes with which an organism extracts energy from the environment and uses that energy for sustaining life. Metabolism can be described by individual equilibrium chemical processes, but living systems are never at equilibrium: they are open systems that are balanced between degradative and synthetic processes. The balance is maintained by a variety of controls over the rates and concentrations of enzymes. Metabolic processes are remarkably similar in all types of cells.

Metabolism consists of the interaction of two major processes, catabolism and anabolism. Catabolism is the chemistry of the degradation of energy-containing compounds. It leads to the capture of some of that energy in new chemical forms that the cell can use for biosynthesis, movement, and secretion—processes of anabolism.

Stages of Catabolism The first stage of catabolism is a collection step in which nutrient molecules are degraded to their building blocks: proteins to amino acids, polysaccharides to monosaccharides, and complex lipids to fatty acids and glycerol.

In the next stage, the collected monosaccharides are first converted into a three-carbon intermediate, pyruvate, and then into a unique two-carbon unit, acetyl-*S*-coenzyme A. The carbon chains of fatty acids and most of the amino acids are also converted into that same two-carbon fragment. In the final stage of catabolism, the acetyl-*S*-CoA enters the citric acid cycle and is oxidized to carbon dioxide.

Transformation of Nutrient Chemical Energy into New Forms The chemical energy of nutrients is converted principally into adenosine triphosphate, ATP, and into reducing power in the form of NADH, nicotine adenine dinucleotide, and NADPH, nicotine adenine dinucleotide phosphate.

Enzymes and Cofactors Enzymes—protein catalysts—are central to the integrated operation of all the metabolic pathways within cells. Enzymes permit chemical reactions to take place rapidly at low temperatures, in aqueous environments, and at neutral pH in cells. They are structurally specific for their substrates, and many of them require the presence of nonprotein molecules called cofactors to function. Enzyme activity can be regulated in a number of ways: the activity of allosteric regulatory enzymes is modified by certain molecules that are not their substrates; other types of control of enzyme activity are covalent modification and feedback inhibition.

High-Energy Compounds High-energy compounds act as common intermediates in group-transfer reactions and cause the overall equilibrium constants for reactions to be highly favorable. As a result, the chemically unfavorable reactions of metabolism are able to produce adequate quantities of desired products.

Key Words

acetyl-*S*-coenzyme A, p. 664
active site, p. 669
adenosine triphosphate (ATP), p. 665
anabolism, p. 664
catabolism, p. 664
cell structure, p. 661
citric acid cycle, p. 664
cofactor, p. 670
enzyme activity, p. 672
eukaryote, p. 661
high-energy compound, p. 674
metabolism, p. 664
NADH, p. 665
oxidative phosphorylation, p. 666
prokaryote, p. 661

Exercises

Cell Structure

22.1 In what way is biochemistry a new dimension of biology?

22.2 What are the two fundamental types of cells?

22.3 What is the chief difference between the two fundamental types of cells?

22.4 What is the cytosol?

22.5 Where do oxidative processes take place in animal cells?

22.6 Where is DNA located in animal cells?

22.7 Where does protein synthesis take place in animal cells?

22.8 What is the endoplasmic reticulum?

22.9 Where is DNA located in bacterial cells?

22.10 Why do bacteria possess cell walls in addition to cell membranes?

General Features of Metabolism

22.11 What are the functions of metabolism?

22.12 Define catabolism.

22.13 Define anabolism.

22.14 In what general way does the oxidation of glucose in the test tube differ from the oxidation of glucose in the body?

22.15 What is the function of the first stage of catabolism?

22.16 What occurs during the second stage of catabolism?

22.17 What occurs during the final stage of catabolism?

22.18 What are the two major products derived from the chemical potential energy of nutrients?

22.19 What is the most fundamental role played by ATP in metabolism?

22.20 What chemical process does ATP undergo when its energy is used in a metabolic process?

22.21 What are the two processes by which ATP is synthesized?

22.22 Write the reaction describing the fate of hydrogen when NAD^+ or $NADP^+$ takes part in a dehydrogenation.

22.23 What functions differentiate NAD^+ from $NADP^+$?

Enzymes, Cofactors, and High-Energy Compounds

22.24 To what family of biomolecules do enzymes belong, and what role do they play in the cell?

22.25 What is an enzyme's active site, and what functions does the active site serve?

22.26 What is complementarity?

22.27 What is meant by inducible fit?

22.28 What are cofactors, and why are they necessary?

22.29 What factors in our diets are an important source of enzyme cofactors?

22.30 What is the meaning of enzyme activity?

22.31 Define turnover number.

22.32 What is feedback inhibition?

22.33 What is a regulatory enzyme?

22.34 True or false: Enzymes always operate as individual molecules. Explain your answer.

22.35 What is the chemical role of high-energy compounds?

Chemical Connections

22.36 Are the biochemical processes of living cells at chemical equilibrium?

22.37 How is the metabolic machinery of cells able to handle the many different types of molecules used as nutrients?

22.38 Is the energy of nutrients used directly and without modification in the metabolic activity of cells?

22.39 Do all enzymes operate at full activity at all times?

22.40 How are biochemically unfavorable reactions able to produce adequate amounts of products?

CHAPTER 23

CARBOHYDRATE METABOLISM

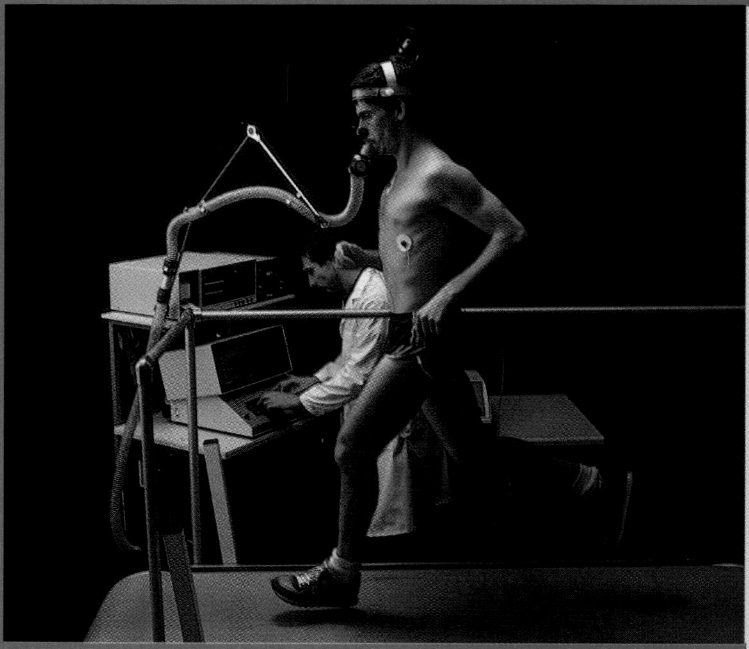

CHEMISTRY IN YOUR FUTURE

An increasingly busy sports physician has asked you, her assistant, to write a list of general guidelines for clients to follow whenever they are preparing for a demanding athletic event. She's provided some notes to get you started, the first of which reads, "Increase carbohydrate consumption, for efficient ATP synthesis." How can you reword this in a way that will motivate your clients? Chapter 23 should give you some ideas.

LEARNING OBJECTIVES

- Describe the first stage of carbohydrate catabolism, glycolysis, which takes place in the cell cytosol.
- Describe the second stage of carbohydrate catabolism, the citric acid cycle, which takes place in the cell's mitochondria.
- Describe gluconeogenesis, the synthesis of glucose from the product of glycolysis.
- Describe the role of hormones in the synthesis and degradation of glycogen.
- Describe how ATP is synthesized through the operation of the electron-transport chain.
- Describe the selective permeability of mitochondria.
- Explain the comparative efficiencies of glycolysis and the citric acid cycle.

In Chapter 22, you learned that catabolism has three phases. In the first, nutrients are degraded to monosaccharides, α-amino acids, and fatty acids. In the second stage, hexoses are converted into pyruvate, a three-carbon intermediate. Pyruvate, the carbon chains of fatty acids, and many of the α-amino acids are then converted into a unique two-carbon unit, acetyl-*S*-coenzyme A (acetyl-*S*-CoA). In the final stage of catabolism, acetyl-*S*-CoA enters a pathway in which it is oxidized to carbon dioxide and water. Two other end products—ammonia and urea—are produced subsequently by other pathways.

The biochemical pathway for the metabolism of carbohydrates includes the catabolic pathway through which α-amino acids and fatty acids also are oxidized. Our examination of carbohydrate metabolism here, in Chapter 23, prepares the way for a fruitful consideration of the metabolism of α-amino acids and fatty acids in Chapters 24 and 25.

23.1 GLYCOLYSIS

Glucose is the principal nutrient that fuels metabolism. Like most saccharides in biological systems, the glucose in our bodies consists of the D-enantiomer (Section 18.3). Whenever the stereochemistry of a biological saccharide or its derivatives is not indicated, such as in "glucose" or "fructose," you should assume that the D-enantiomer is meant.

It was pointed out in Section 11.10 that we do not regularly balance equations describing organic chemical reactions. The reasons were spelled out in that chapter and we will not repeat them here. What is important to note is that the equations describing biochemical reactions will often be treated the same way. We wish to bring attention to the transformations of one substance into another, rather than focusing on considerations of mass or charge balance. Where such considerations are of primary importance, they will be presented in detail, as in Section 23.15.

The first pathway that glucose encounters is called **glycolysis** and takes place in the cell cytosol. In glycolysis, the six-carbon glucose molecule is converted in a sequence of nine enzymatically catalyzed steps into two molecules of pyruvate, a three-carbon molecule. In the process, ATP and NADH are produced. The pyruvate produced in glycolysis proceeds to be oxidized, first, to acetyl-*S*-CoA and, then, to carbon dioxide and water in the next pathway of metabolism, the citric acid cycle (Section 23.5).

◀◀ Major cell structures are described in Section 22.1.

Glycolysis itself can be seen as consisting of two major stages. In the first stage, glucose and other hexoses, such as fructose and galactose, are converted into the three-carbon product glyceraldehyde-3-phosphate. In the second stage, glyceraldehyde-3-phosphate undergoes oxidation and then rearrangement into two high-energy phosphorylated compounds that are used to phosphorylate ADP. Although two ATPs are necessary to start the process, four ATPs are produced by it in phosphorylations of ADP during the process. Both NAD^+ and inorganic phosphate (P_i) are required.

The stoichiometry of the process, including priming by ATP, is:

$$\underset{\text{Glucose}}{C_6H_{12}O_6} + 2\ ATP + 2\ NAD^+ + 2\ ADP + 2\ P_i \longrightarrow$$
$$2\ \underset{\text{Pyruvate}}{C_3H_4O_3} + 4\ ATP + 2\ NADH + 2\ H^+ + 2\ H_2O$$

By subtracting the priming ATP from both sides, we find that the net result is:

$$C_6H_{12}O_6 + 2\ NAD^+ + 2\ ADP + 2\ P_i \longrightarrow$$
$$2\ C_3H_4O_3 + 2\ ATP + 2\ NADH + 2\ H^+ + 2\ H_2O$$

Figure 23.1 on the following page diagrams that glycolytic pathway.

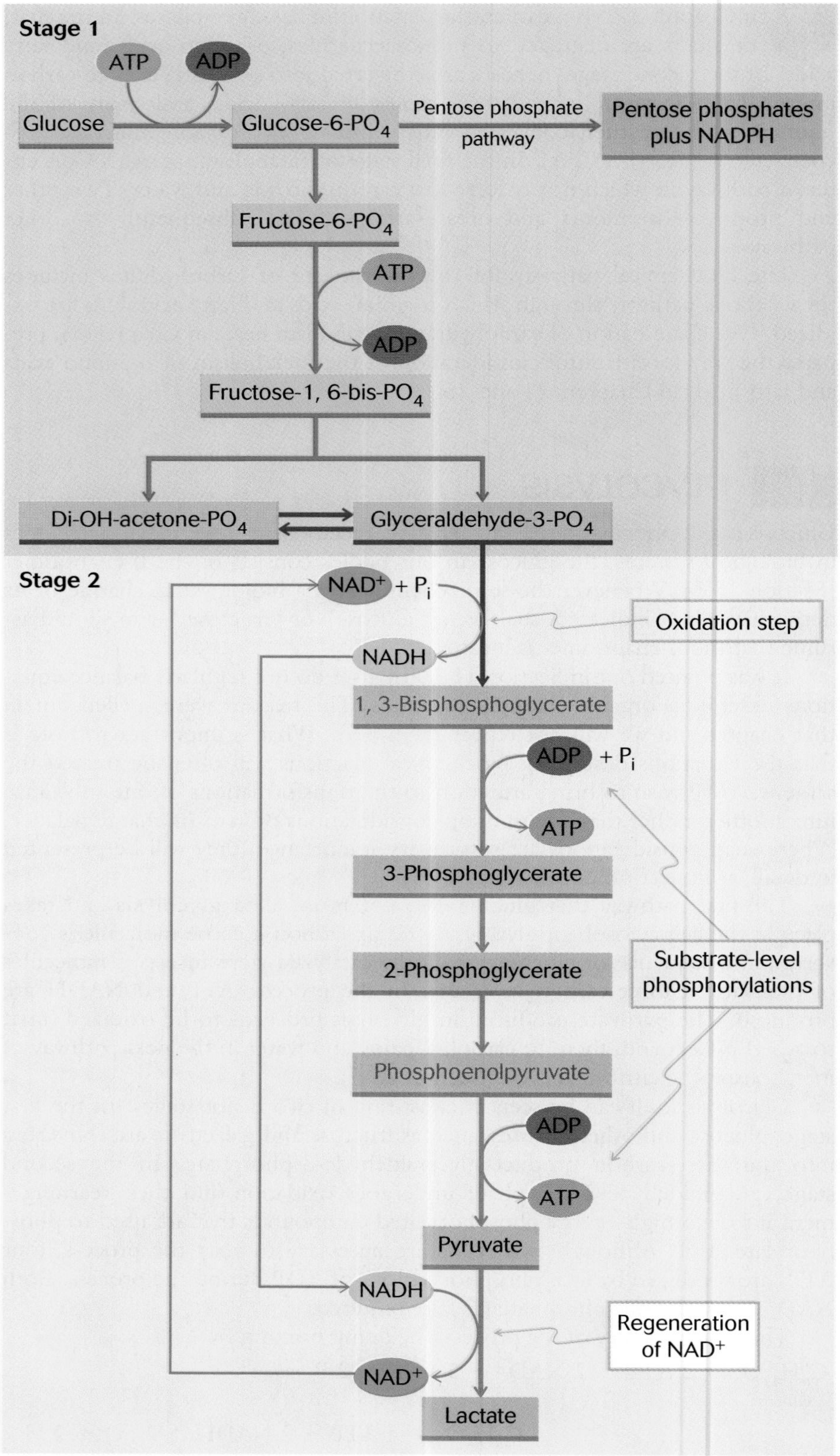

Figure 23.1 The glycolytic pathway. Note the steps for the production of the high-energy compounds 1,3-bisphosphoglycerate and phosphoenolpyruvate, the two substrate-level phosphorylations of ADP, and the regeneration of NAD^+. Keep in mind that 2 mol of glyceraldehyde-3-PO_4 are produced for each mole of glucose.

23.2 CHEMICAL TRANSFORMATIONS IN GLYCOLYSIS

The First Stage of Glycolysis

Step 1. Glucose is phosphorylated at carbon 6 by hexokinase.

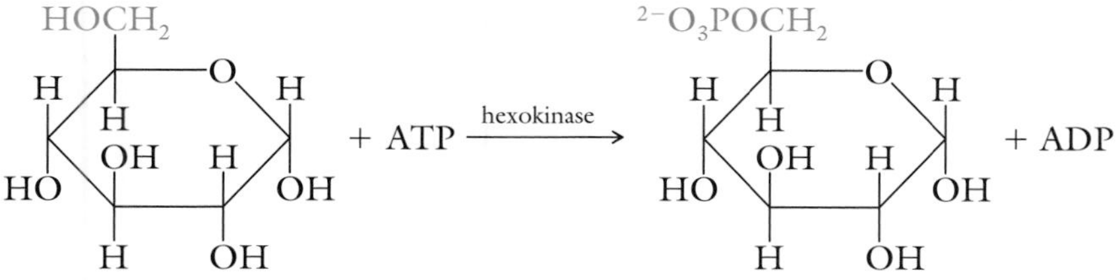

Glucose **Glucose-6-PO$_4$**

Step 2. Glucose-6-PO$_4$ is isomerized by glucose-6-phosphate isomerase to form fructose-6-PO$_4$.

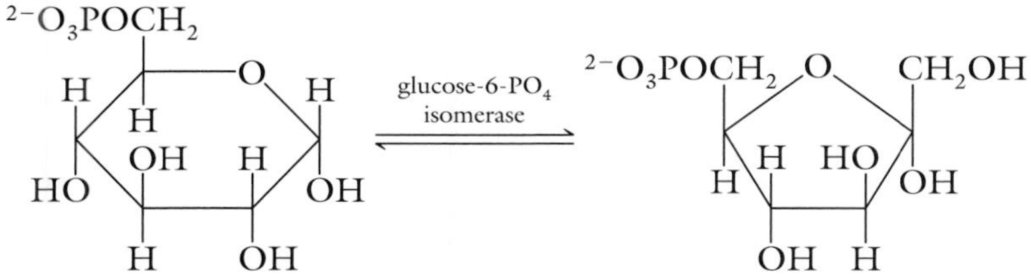

Glucose-6-PO$_4$ **Fructose-6-PO$_4$**

Step 3. Fructose-6-PO$_4$ is phosphorylated by ATP to form fructose-1, 6-bisphosphate (1,6-BPF). The reaction is catalyzed by **phosphofructokinase.**

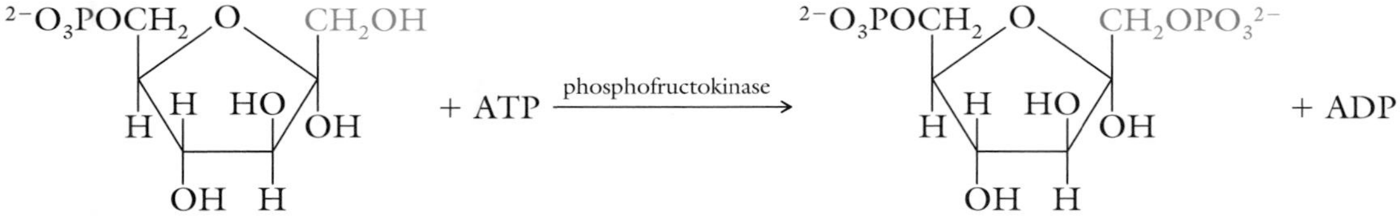

Fructose-6-PO$_4$ **Fructose-1,6-bis-PO$_4$**

Phosphofructokinase is a regulatory enzyme whose activity is decreased by high concentrations of ATP. When the energy state of the cell is high—that is, when the cell has a high ATP concentration—there is no need to put glucose through this pathway. Therefore this enzyme is a key control point in the glycolytic pathway. Unneeded glucose is stored as glycogen (Section 23.9) or triacylglycerols (Section 24.6).

Step 4. Aldolase catalyzes the cleavage of fructose-1,6-bisphosphate into an equilibrium mixture of dihydroxyacetone phosphate and glyceraldehyde-3-phosphate. Dihydroxyacetone phosphate is rapidly converted into glyceraldehyde-3-phosphate through the action of the enzyme triose phosphate isomerase. The equilibrium then rapidly shifts in the direction of the aldehyde because it is continually removed in the next step (step 5).

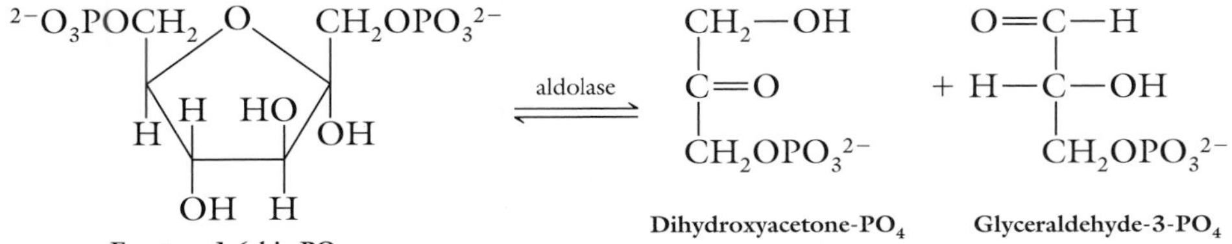

Fructose-1,6-bis-PO$_4$ **Dihydroxyacetone-PO$_4$** **Glyceraldehyde-3-PO$_4$**

In effect, 1 mol of glucose has been transformed into 2 mol of glyceraldehyde-3-PO_4. In the next steps, we will follow the fate of 1 mol of this three-carbon molecule, but keep in mind that two such molecules have been formed from 1 mol of glucose.

The Second Stage of Glycolysis

Step 5. In this step, the aldehyde group of glyceraldehyde-3-PO_4 (3-PGAL) is oxidized to a carboxyl group and esterified with inorganic phosphate to form the high-energy compound 1,3-bisphosphoglycerate (1,3-BPG). The reaction is catalyzed by glyceraldehyde-3-PO_4 dehydrogenase, during which NAD^+ is reduced to NADH.

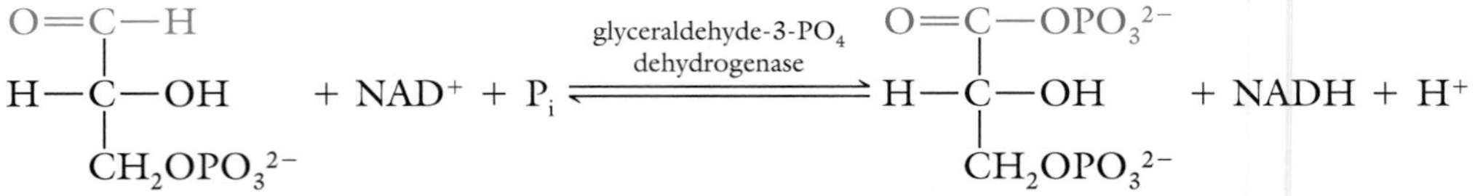

3-PGAL 1,3-Bisphosphoglycerate

Step 6. In this step, phosphoglycerate kinase catalyzes the phosphorylation of ADP by the high-energy compound 1,3-bisphosphoglycerate to form ATP and 3-phosphoglycerate.

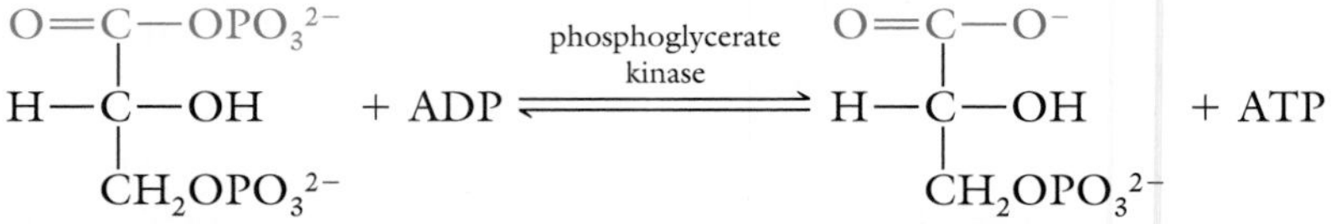

1,3-BPG 3-Phosphoglycerate

Step 7. In the next step, phosphoglyceromutase catalyzes the rearrangement of 3-phosphoglycerate into 2-phosphoglycerate.

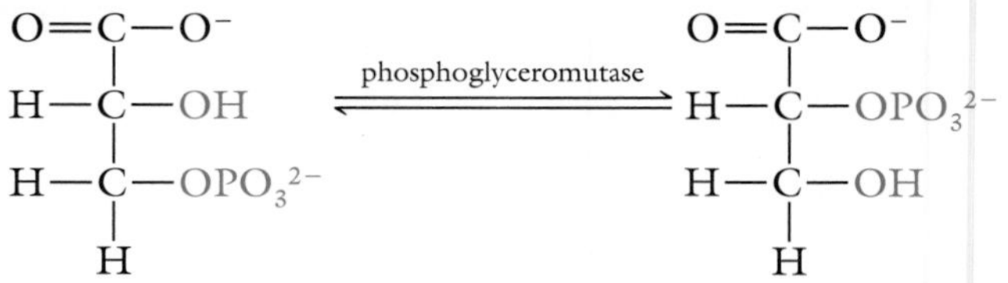

3-Phosphoglycerate 2-Phosphoglycerate

Step 8. Enolase then catalyzes the dehydration of 2-phosphoglycerate to yield the phosphorylated enol form of pyruvate, phosphoenolpyruvate (PEP), the second high-energy compound produced by the glycolytic process.

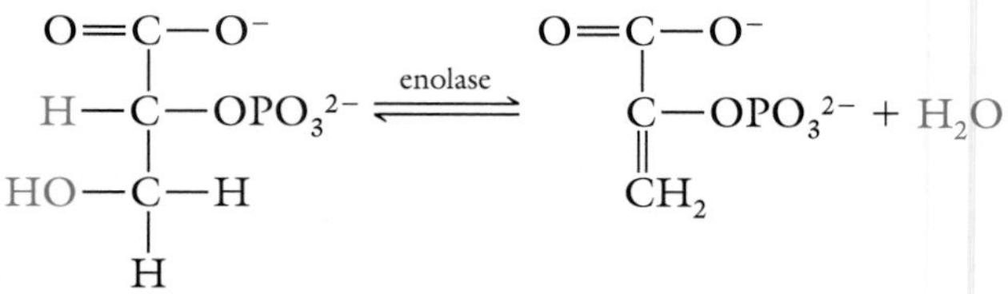

2-Phosphoglycerate Phosphoenolpyruvate

Step 9. ADP is then enzymatically phosphorylated by pyruvate kinase, and the enol form of pyruvate rapidly rearranges nonenzymatically to the keto form of pyruvate.

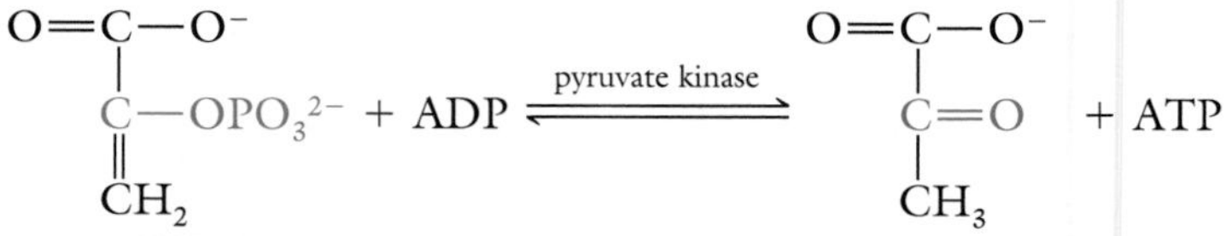

PEP Pyruvate

Glycolysis can continue only if NAD^+ can be regenerated from its reduced form, NADH, produced in step 5. In active skeletal muscle this regeneration is accomplished through the next step, step 10, which results in the production of large quantities of lactate. The lactate diffuses from the muscle cells, enters the circulation, and is transported to the liver, where it is converted back into glucose in a process called gluconeogenesis (Section 23.8).

Step 10. NAD^+ is regenerated by the reduction of pyruvate to lactate catalyzed by lactic dehydrogenase.

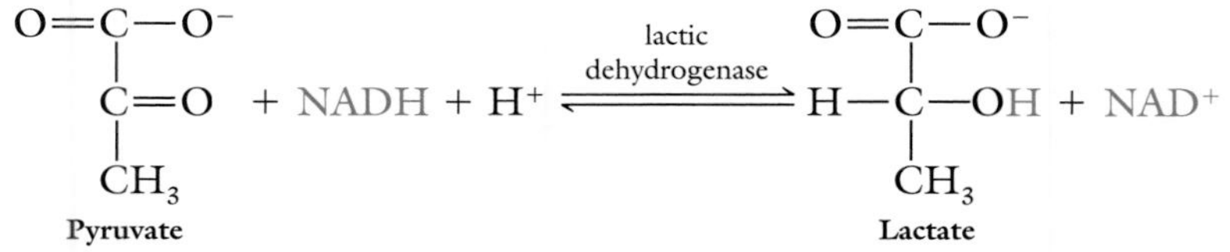

The overall result of steps 1 through 10 is:

$$\text{Glucose} + 2\ \text{ADP} + 2\ \text{P}_i \longrightarrow 2\ \text{lactate} + 2\ \text{ATP}$$

A much greater amount of ATP per mole of glucose is produced in the part of the carbohydrate catabolic pathway that follows glycolysis—the citric acid cycle. However, the rate at which ATP is produced in glycolysis is far greater than in the citric acid cycle. For this reason, glycolysis is the principal metabolic pathway for glucose in active skeletal muscle, where large and rapidly available quantities of ATP are required.

In other tissues, such as brain or kidney, where a high rate of ATP production is not needed, pyruvate becomes the final product of the glycolytic breakdown of glucose, and this process can be represented by steps 1 through 9:

$$C_6H_{12}O_6 + 2\ \text{NAD}^+ + 2\ \text{ADP} + 2\ \text{P}_i \longrightarrow 2\ \text{pyruvate} + 2\ \text{ATP} + 2\ \text{NADH} + 2\ \text{H}^+$$

The NADH generated by glycolysis is oxidized by transferring its reducing power from the cytosol to the electron-transport system in the mitochondria, where the electrons and hydrogen are eventually passed on to molecular oxygen to form water (Section 23.11):

$$2\ \text{NADH} + 2\ \text{H}^+ + O_2 \longrightarrow 2\ \text{NAD}^+ + 2\ H_2O$$

Some microorganisms such as yeast, *Saccharomyces cerevisiae,* regenerate NAD^+ by the conversion of pyruvate into ethanol and carbon dioxide. The reaction takes place in two steps: first, the decarboxylation of pyruvate to form acetaldehyde and, then, the reduction of the acetaldehyde by the use of NADH to form ethanol and the regenerated NAD^+:

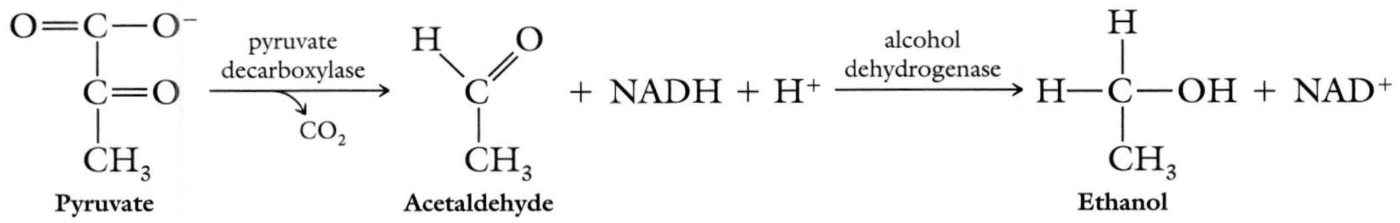

23.3 PENTOSE PHOSPHATE PATHWAY

Pentoses and NADPH (the reduced cofactor) are synthesized in a glucose-requiring metabolic sequence called the **pentose phosphate pathway.** In the initial step, 1 mol of glucose-6-PO_4 is oxidized, resulting in the formation of 2 mol of NADPH, 1 mol of ribose-5-PO_4 (Section 18.5), and 1 mol of CO_2:

$$\text{Glucose-6-PO}_4 + 2\ \text{NADP}^+ \longrightarrow \text{ribose-5-PO}_4 + 2\ \text{NADPH} + 2\ \text{H}^+ + CO_2$$

This step is an important source of reducing power in the biosynthesis of lipids (Section 24.5), and the pathway is quite active in tissue such as mammary glands. The pathway includes a set of enzymes that allow the interconversion of hexoses and pentoses so that, in combination with glycolysis, the pentose phosphate pathway can supply the cell with any of four alternatives: (1) both NADPH and pentoses, (2) only NADPH when no pentoses are needed, (3) pentoses (for nucleic acid or cofactor synthesis) when no NADPH is needed, and (4) both ATP (from glycolysis) and NADPH for situations in which both are needed.

23.4 FORMATION OF ACETYL-*S*-CoA

After glycolysis, the catabolism of glucose continues with the entry of pyruvate into the mitochondria. There the pyruvate is oxidatively transformed into acetyl-*S*-CoA, which is subsequently oxidized in the citric acid cycle (Section 23.5).

The overall reaction producing acetyl-*S*-CoA from pyruvate is:

$$\text{Pyruvate} + \text{NAD}^+ + \text{CoA-SH} \longrightarrow \text{acetyl-}S\text{-CoA} + \text{NADH} + \text{H}^+ + \text{CO}_2$$

(**Coenzyme A,** abbreviated **CoA-SH,** is a complex thioalcohol containing the vitamin pantothenic acid; see Figure 23.2.) The requisite dehydrogenation and decarboxylation of pyruvate is carried out in a series of steps by a complex of enzymes called the pyruvate dehydrogenase multienzyme complex, located in the mitochondria of eukaryotic cells. The details of this complicated process illustrate the economy and efficiency that result when related enzyme activities are organized into one macromolecular complex. They also demonstrate the role of vitamins as cofactors in enzyme reactions and they show how a regulatory enzyme responds both allosterically and by covalent modification to the energy needs of the cell.

▶▶ The body's daily vitamin requirements are presented in Section 26.3.

The Reaction Steps

The pyruvate dehydrogenase multienzyme complex consists of three different enzymes and five cofactors, assembled together in such a way that the reactions are carried out sequentially. All of these cofactors, except lipoic acid, are derived from vitamins required for human nutrition. The cofactors are:

1. **Thiamine** (vitamin B_1) in thiamine pyrophosphate (TPP). The combination of this cofactor with its apoenzyme is abbreviated as Enz_1-TPP.
2. **Lipoic acid,** a growth factor that can be synthesized by vertebrates. The combination of lipoic acid with its apoenzyme is abbreviated as shown below.

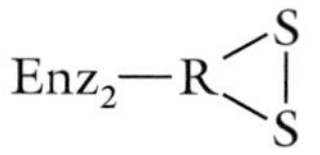

3. **Pantothenic acid** in coenzyme A.
4. **Riboflavin** in flavin adenine dinucleotide (FAD; see Figure 22.6). The combination of FAD with its apoenzyme is abbreviated as Enz_3-FAD.
5. **Nicotinic acid** in nicotinamide adenine dinucleotide (NAD; see Figure 22.5.)

The structures and names of the first three cofactors are shown in Figure 23.2. Table 23.1 lists the three enzymes and their cofactors that take part in the following sequence of steps of the overall reaction. Although the details of the reaction are complex, we will present the process in outline in five steps.

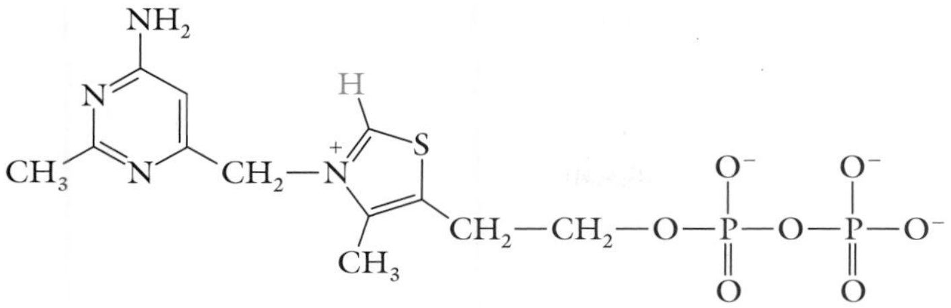

Thiamine pyrophosphate (TPP)

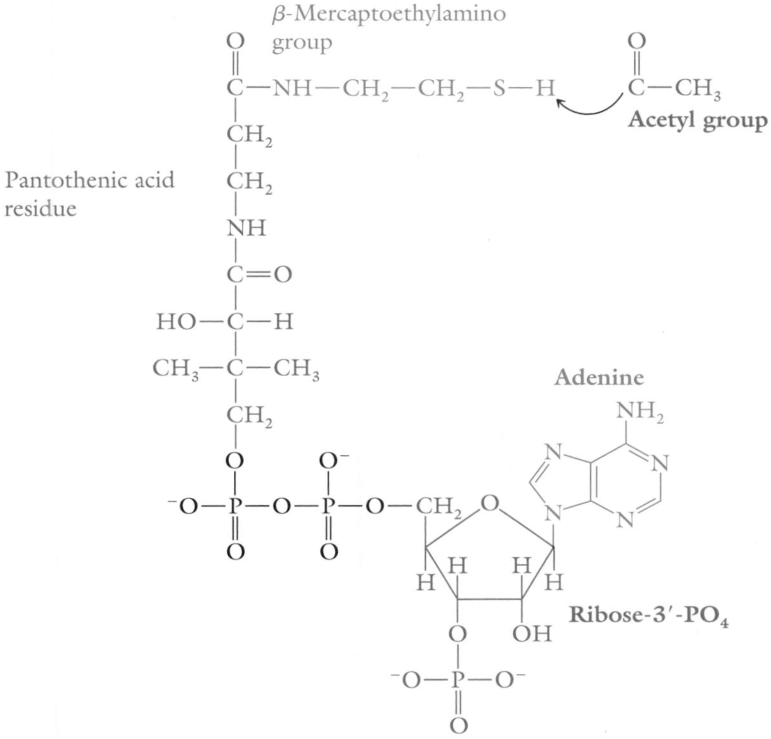

Coenzyme A (CoA-SH)

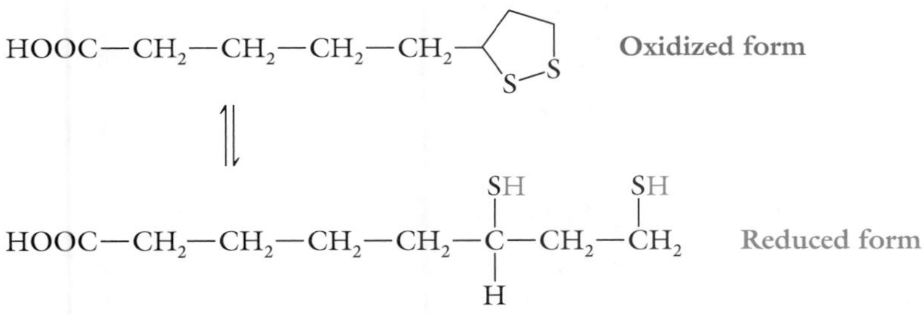

Lipoic acid

Figure 23.2 Three of the cofactors of pyruvate dehydrogenase required for the conversion of pyruvate into acetyl-*S*-CoA. Note that, in coenzyme A, the β-mercaptoethylamine group becomes acetylated at the SH group.

TABLE 23.1 Enzymes and Cofactors in the Conversion of Pyruvate into Acetyl-*S*-CoA

Enzyme	Cofactor
pyruvate dehydrogenase	thiamine pyrophosphate
dihydrolipoyl transacetylase	lipoic acid, CoA-SH
dihydrolipoyl dehydrogenase	flavin adenine dinucleotide, nicotinamide adenine dinucleotide

In the first step, pyruvate replaces the TPP hydrogen shown in blue in Figure 23.2, with the result that carbon dioxide is lost and thiamine pyrophosphate is converted into a hydroxyethyl derivative of the cofactor bound to enzyme 1.

Step 1. $Enz_1-TPP + \underset{O}{\overset{CH_3}{C}}-COO^- \longrightarrow Enz_1-TPP-\underset{OH}{\overset{CH_3}{C}}-H + CO_2$

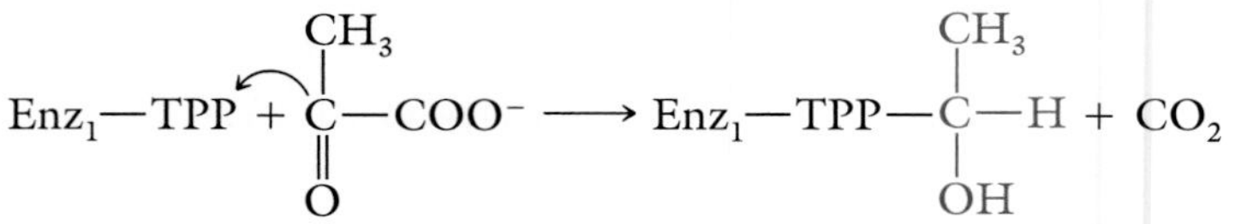

In step 2, the hydroxyethyl group is transferred enzymatically to the oxidized form of the lipoic acid cofactor of enzyme 2 to form an acetyl thioester along with a sulfhydryl group.

Step 2. $Enz_1-TPP-\underset{OH}{\overset{CH_3}{C}}-H + Enz_2-R\overset{S}{\underset{S}{\big|}} \rightleftharpoons$

$$Enz_1-TPP + Enz_2-R-\underset{SH}{S}-\underset{O}{C}-CH_3$$

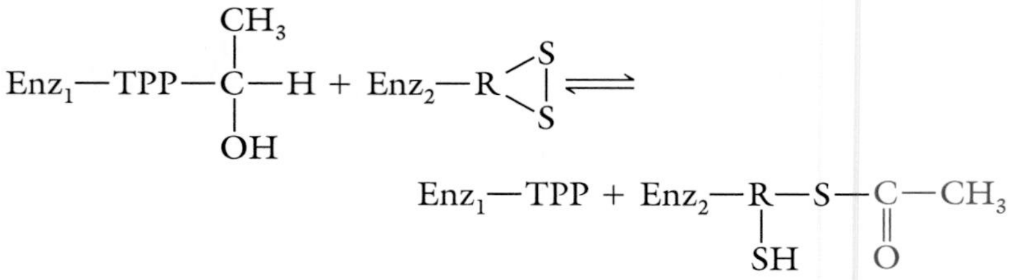

In step 3, acetyl-S-CoA is formed as the thioester is transferred to CoA-SH, and the lipoic acid becomes fully reduced.

Step 3. $Enz_2-R-\underset{SH}{S}-\underset{O}{C}-CH_3 + CoA\text{-}SH \rightleftharpoons$

$$Enz_2-R\overset{SH}{\underset{SH}{\diagdown}} + CoA-S-\underset{O}{C}-CH_3$$

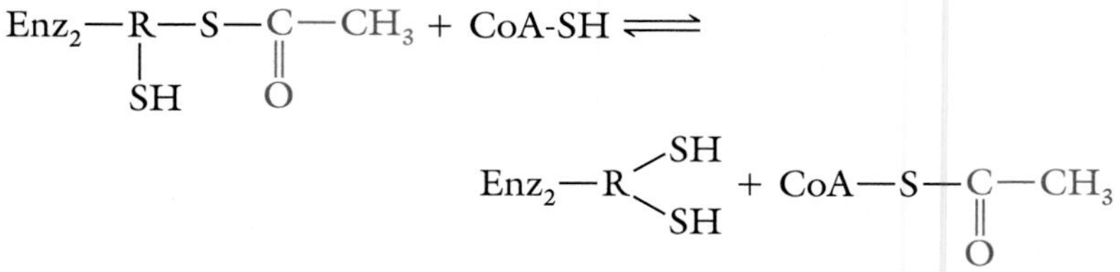

Steps 4 and 5 consist of the transfer of reducing power from reduced lipoic acid, first to FAD and then to NAD^+ to form NADH.

Step 4. $Enz_2-R\overset{SH}{\underset{SH}{\diagdown}} + Enz_3-FAD \rightleftharpoons$

$$Enz_2-R\overset{S}{\underset{S}{\big|}} + Enz_3-FADH_2$$

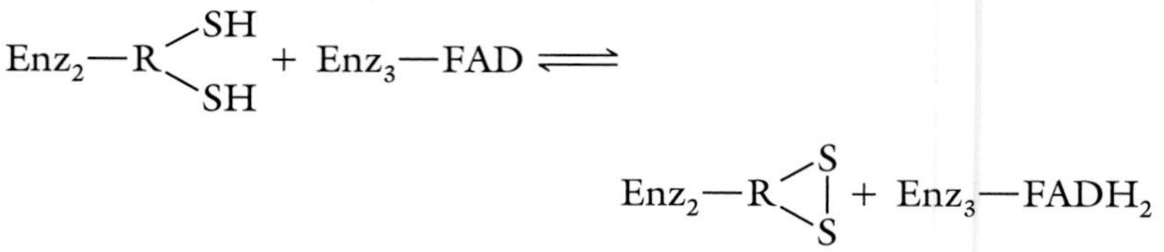

Step 5. $Enz_3-FADH_2 + NAD^+ \rightleftharpoons Enz_3-FAD + NADH + H^+$

The NADH then yields its electrons to the mitochondrial electronic-transport system.

Adding the five reactions and canceling common terms, we find the overall process to be:

$$Pyruvate + NAD^+ + CoA\text{-}SH \longrightarrow acetyl\text{-}S\text{-}CoA + NADH + H^+ + CO_2$$

Steps 2 through 5 are reversible; however, Step 1 is irreversible. Therefore, pyruvate cannot be synthesized by addition of carbon dioxide to acetyl-S-CoA.

Because the five cofactors and three enzymes are all located within the same complex: (1) diffusion distances are minimized, enhancing the rates of reaction and increasing efficiency; (2) side reactions are minimized, increasing economy; and (3) control mechanisms can be integrated and coordinated.

Concept check

✔ A multienzyme complex that carries out a sequence of metabolic reactions is a common structural theme of cellular metabolism.

The Control Mechanisms

The activity of the pyruvate dehydrogenase complex is controlled by both product inhibition and covalent modification.

TABLE 23.2 Allosteric Control of Pyruvate Dehydrogenase Kinase (PDK) and Pyruvate Dehydrogenase Phosphatase (PDP)

Enzyme	Activator	Inhibitor
PDK	acetyl-*S*-CoA	pyruvate
	NADH	ADP
		Ca^{2+}
PDP	Ca^{2+}	none
	Mg^{2+}	

Product inhibition is simply Le Chatelier's principle at work. When acetyl-*S*-CoA and NADH are present in high concentrations, steps 3 and 5 are driven in the reverse direction. Enzyme 1 cannot deliver its hydroxyethyl group to enzyme 2, and the decarboxylation of pyruvate is inhibited.

Within the multienzyme complex, two allosterically controlled enzymes—pyruvate dehydrogenase kinase and pyruvate dehydrogenase phosphatase (abbreviated PDK and PDP)—covalently modify enzyme 1, causing its activation or inactivation. PDK inactivates enzyme 1 by phosphorylation. PDK itself is allosterically activated by acetyl-*S*-CoA and NADH and inactivated by pyruvate, ADP, and Ca^{2+}. PDP removes the phosphate group by hydrolysis and reactivates enzyme 1. It has no inactivators and is active only in the presence of high concentrations of Ca^{2+} and Mg^{2+}. Table 23.2 summarizes the allosteric control of these two enzymes. There are many examples of this kind of control in cellular metabolism. We will see it again in glycogenolysis (Section 23.10).

Concept check

✔ Covalent modification of an enzyme by allosterically controlled phosphorylation and dephosphorylation is an important molecular theme of cellular metabolism.

23.5 CITRIC ACID CYCLE

In contrast with the glycolytic pathway, which is a linear sequence of enzymatically catalyzed reactions, the aerobic phase of glucose catabolism is cyclic. It begins with the formation of citric acid—hence its name, the **citric acid cycle.** It is also called the tricarboxylic acid cycle (TCA cycle) or the Krebs cycle (in honor of its discoverer, Hans Krebs).

The cyclic nature of the process is illustrated in Figure 23.3 on the following page. A turn of the cycle begins when acetyl-*S*-CoA donates an acetyl group to the four-carbon dicarboxylic acid oxaloacetate to form the six-carbon tricarboxylic acid citrate. Citrate is then converted into isocitrate, which, in turn is oxidatively decarboxylated with the loss of CO_2 and the production of NADH to form the five-carbon α-ketoglutarate. This compound undergoes an oxidative dehydrogenation and a series of transformations to yield a second molecule of CO_2 and NADH and emerges as the four-carbon succinate. The enzyme system that decarboxylates α-ketoglutarate is almost identical with the pyruvate complex except that it has no regulatory components. Succinate is then converted in three steps into the four-carbon oxaloacetate, which begins the cycle again.

The result is that two carbons enter the cycle in the form of acetyl-*S*-CoA, two carbons leave as CO_2, and oxaloacetate is regenerated to begin another

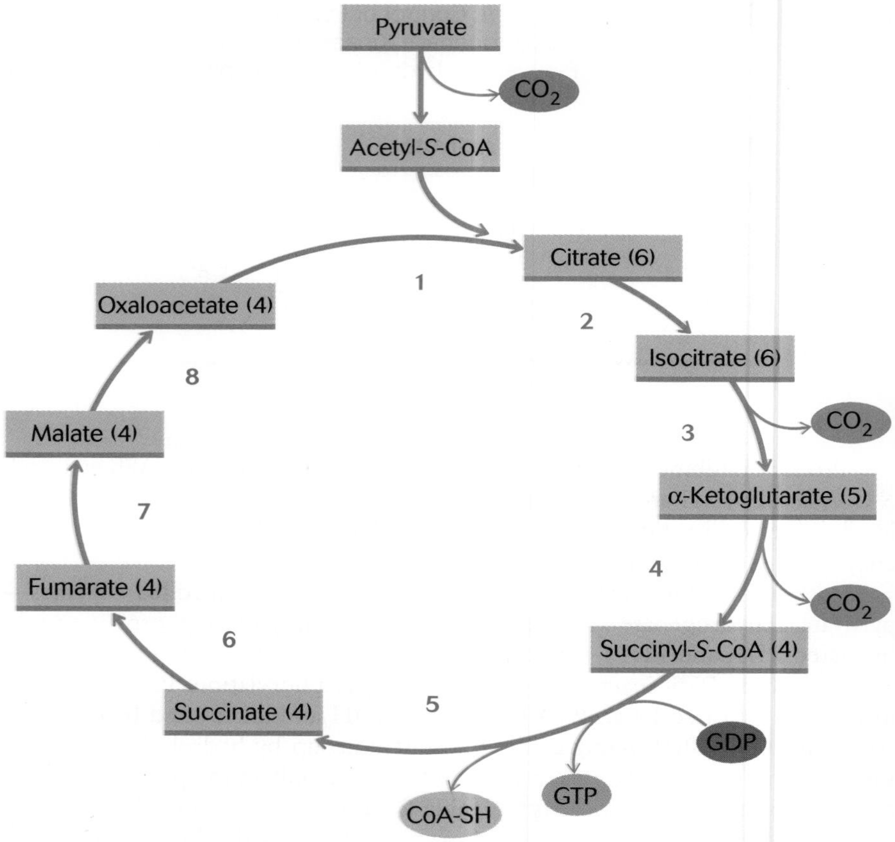

Figure 23.3 An outline of the citric acid cycle identifying components by name and number of carbon atoms (given in parentheses) in the intermediate.

turn of the cycle. No oxaloacetate is lost, and therefore one molecule of it can effect the oxidation of limitless numbers of acetyl groups.

All of the enzymes of the citric acid cycle, as well as those of the electron-transport system, are located in the mitochondria of animal cells. Their precise locations are listed in Table 23.3.

23.6 REACTIONS OF THE CITRIC ACID CYCLE

Let us now look more closely at the citric acid cycle's eight successive reactions. The first reaction, acetyl-*S*-CoA with oxaloacetate, is catalyzed by citrate synthase. The methyl carbon of acetyl-*S*-CoA reacts with the carbonyl group of

TABLE 23.3 Mitochondrial Location of Citric Acid Cycle Enzymes

Inner Membrane	Matrix Space
aconitase	citrate synthase
succinate dehydrogenase	isocitrate dehydrogenase
electron-transport chain	α-ketoglutarate dehydrogenase
succinyl-*S*-CoA synthetase	
fumarase	
malate dehydrogenase	
pyruvate dehydrogenase	

oxaloacetate, and, at the same time, the thioester bond is hydrolyzed to produce free CoA-SH.

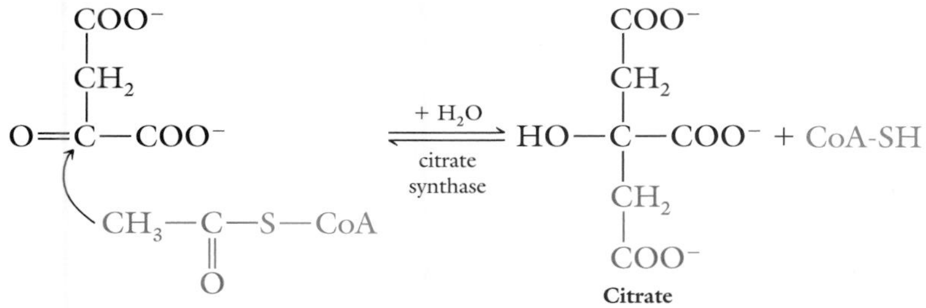

$$
\begin{array}{c}
\text{COO}^- \\
| \\
\text{CH}_2 \\
| \\
\text{O}=\text{C}-\text{COO}^- \\
\text{CH}_3-\overset{\displaystyle O}{\underset{\displaystyle \|}{\text{C}}}-\text{S}-\text{CoA}
\end{array}
\quad\underset{\substack{\text{citrate} \\ \text{synthase}}}{\overset{+\ \text{H}_2\text{O}}{\rightleftharpoons}}\quad
\begin{array}{c}
\text{COO}^- \\
| \\
\text{CH}_2 \\
| \\
\text{HO}-\text{C}-\text{COO}^- \ +\ \text{CoA-SH} \\
| \\
\text{CH}_2 \\
| \\
\text{COO}^- \\
\textbf{Citrate}
\end{array}
$$

CoA-SH is now free to participate in another round of the pyruvate dehydrogenation reaction. Citrate synthase cannot bind acetyl-S-CoA unless oxaloacetate is first bound to the enzyme.

In the next step of the cycle, the enzyme aconitase catalyzes the transformation of citrate into isocitrate. It does so through the reversible dehydration-hydration of the double bond of an enzyme-bound intermediate called *cis*-aconitate.

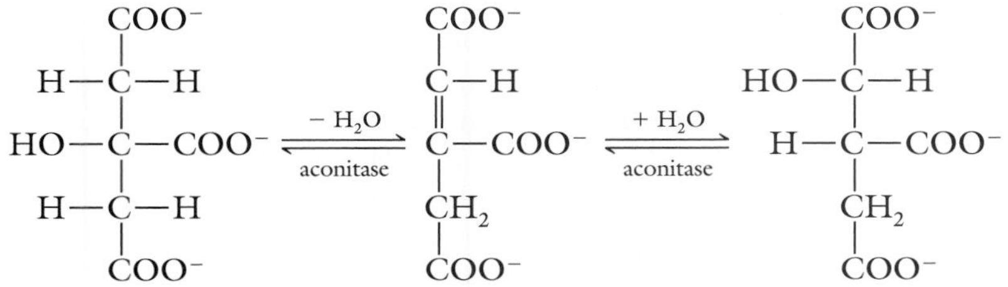

$$
\begin{array}{c}
\text{COO}^- \\
| \\
\text{H}-\text{C}-\text{H} \\
| \\
\text{HO}-\text{C}-\text{COO}^- \\
| \\
\text{H}-\text{C}-\text{H} \\
| \\
\text{COO}^- \\
\textbf{Citrate}
\end{array}
\quad\underset{\text{aconitase}}{\overset{-\ \text{H}_2\text{O}}{\rightleftharpoons}}\quad
\begin{array}{c}
\text{COO}^- \\
| \\
\text{C}-\text{H} \\
\| \\
\text{C}-\text{COO}^- \\
| \\
\text{CH}_2 \\
| \\
\text{COO}^- \\
\textit{cis}\text{-}\textbf{Aconitate} \\
\textbf{(enzyme bound)}
\end{array}
\quad\underset{\text{aconitase}}{\overset{+\ \text{H}_2\text{O}}{\rightleftharpoons}}\quad
\begin{array}{c}
\text{COO}^- \\
| \\
\text{HO}-\text{C}-\text{H} \\
| \\
\text{H}-\text{C}-\text{COO}^- \\
| \\
\text{CH}_2 \\
| \\
\text{COO}^- \\
\textbf{Isocitrate}
\end{array}
$$

In the next step, the six-carbon isocitrate is dehydrogenated and decarboxylated by the enzyme isocitrate dehydrogenase to form the five-carbon α-ketoglutarate. This reaction requires NAD^+.

$$
\begin{array}{c}
\text{COO}^- \\
| \\
\text{HO}-\text{C}-\text{H} \\
| \\
\text{H}-\text{C}-\text{COO}^- \ +\ \text{NAD}^+ \\
| \\
\text{CH}_2 \\
| \\
\text{COO}^- \\
\textbf{Isocitrate}
\end{array}
\quad\xrightarrow[\text{dehydrogenase}]{\text{isocitrate}}\quad
\begin{array}{c}
\text{COO}^- \\
| \\
\text{C}=\text{O} \\
| \\
\text{CH}_2 \qquad +\ \text{CO}_2\ +\ \text{NADH}\ +\ \text{H}^+ \\
| \\
\text{CH}_2 \\
| \\
\text{COO}^- \\
\boldsymbol{\alpha}\textbf{-Ketoglutarate}
\end{array}
$$

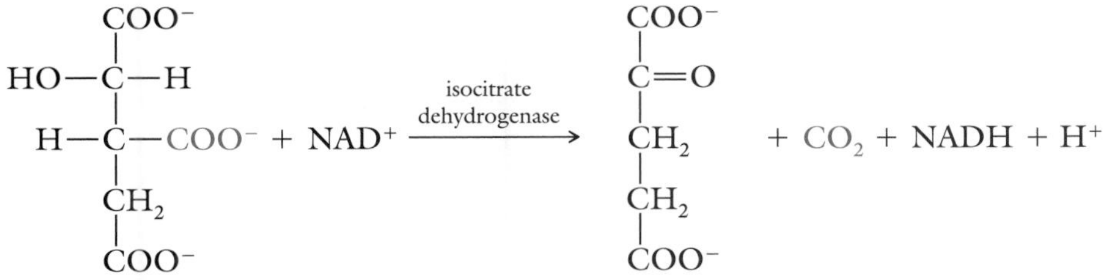

In the following step, α-ketoglutarate is acted on by the isocitrate dehydrogenase complex, which catalyzes the formation of succinyl-S-CoA with the loss of CO_2. The reaction is:

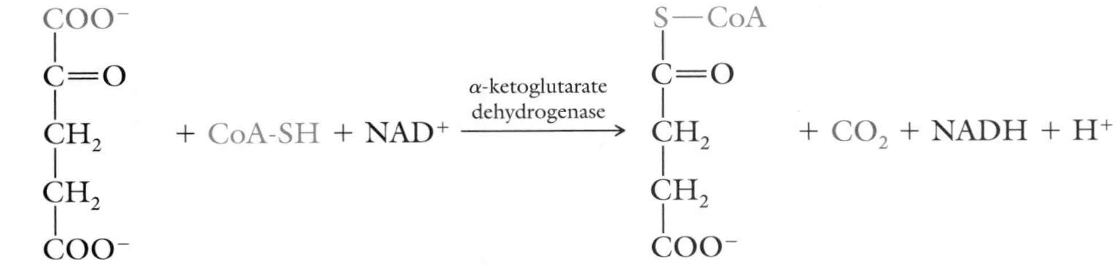

$$
\begin{array}{c}
\text{COO}^- \\
| \\
\text{C}=\text{O} \\
| \\
\text{CH}_2 \qquad +\ \text{CoA-SH}\ +\ \text{NAD}^+ \\
| \\
\text{CH}_2 \\
| \\
\text{COO}^- \\
\boldsymbol{\alpha}\textbf{-Ketoglutarate}
\end{array}
\quad\xrightarrow[\text{dehydrogenase}]{\alpha\text{-ketoglutarate}}\quad
\begin{array}{c}
\text{S}-\text{CoA} \\
| \\
\text{C}=\text{O} \\
| \\
\text{CH}_2 \qquad +\ \text{CO}_2\ +\ \text{NADH}\ +\ \text{H}^+ \\
| \\
\text{CH}_2 \\
| \\
\text{COO}^- \\
\textbf{Succinyl-}S\textbf{-CoA}
\end{array}
$$

The α-ketoglutarate complex is virtually identical in structure and function with the pyruvate dehydrogenase complex. However, the α-ketoglutarate complex does not possess the regulatory properties of the pyruvate dehydrogenase complex.

◀◀ High-energy compounds are the subject of Section 22.9.

Like acetyl-*S*-CoA, succinyl-*S*-CoA is a high-energy compound that undergoes an energy-conserving reaction in which the compound GDP is phosphorylated to form GTP. GDP and GTP have the same structures, respectively, as ADP and ATP except that adenine (A) is replaced by guanine (G; Section 21.1). Hence GDP and GTP are guanosine diphosphate and guanosine triphosphate, respectively. GTP can phosphorylate ADP and takes part in a variety of membrane processes. The reaction is:

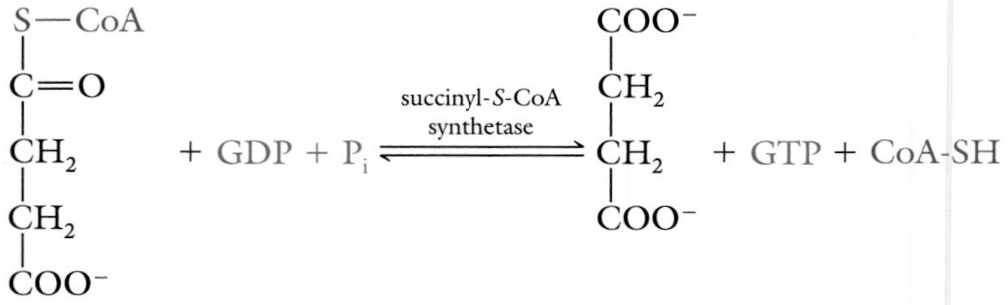

Succinyl-*S*-CoA **Succinate**

Succinyl-*S*-CoA is the main building block of porphyrins, compounds that are key components of the hemes of hemoglobin and myoglobin and the cytochromes of the electron-transport chain (Section 23.11).

In the next step, succinate is dehydrogenated by the enzyme succinic dehydrogenase to form the trans isomer, fumarate. Succinic dehydrogenase is a component of the inner mitochondrial membrane (Section 23.12). The hydrogen acceptor here is FAD (see Figure 23.2) represented as E-FAD. Dehydrogenases that contain FMN or FAD are called **flavoproteins** because these cofactors are so tightly bound that they are considered to be part of the protein molecule; E-FAD is therefore a flavoprotein.

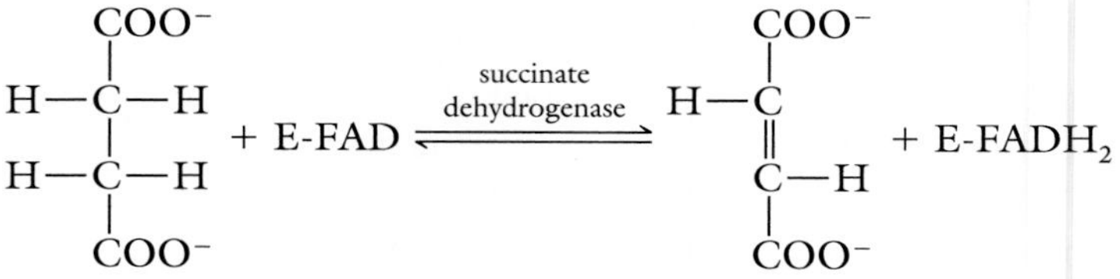

Succinate **Fumarate**

In the next reaction, fumarate is reversibly hydrated by the enzyme fumarate hydratase, also called fumarase. The product is the chiral L-malate. The hydration is specific for the trans dicarboxylic acid; fumarase will not hydrate maleate, the cis isomer of fumarate.

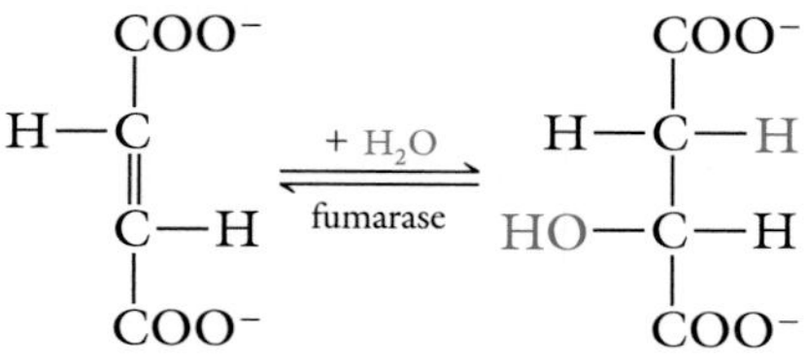

Fumarate **L-Malate**

The last step in the citric acid cycle is the dehydrogenation of L-malate to form oxaloacetate. The enzyme catalyzing this step is NAD^+-linked malate dehydrogenase.

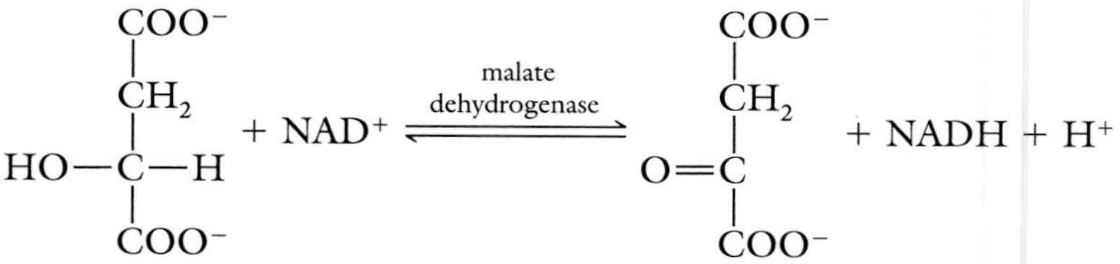

L-Malate **Oxaloacetate**

Oxaloacetate is again available to react with a new incoming acetyl-*S*-CoA to begin a new cycle. The equilibrium constant for this final reaction is 6.0×10^{-6}, which means that the cellular concentration of oxaloacetate is very low at all times. Therefore, slight changes in the concentration of oxaloacetate have a great effect on the overall rate of the cycle.

The net reaction of the citric acid cycle is:

$$\text{Acetyl-}S\text{-CoA} + 3\ NAD^+ + FAD + GDP + P_i \longrightarrow 2\ CO_2 + 3\ NADH + FADH_2 + GTP + \text{CoA-SH}$$

This equation should not be taken to convey that, after a single turn of the cycle, the two carbon atoms of acetyl-*S*-CoA are the same two carbon atoms that emerge as CO_2. They are not. They will emerge as CO_2 in the next cycle. One ATP in the equivalent form of GTP is generated. The hydrogens removed from four intermediates emerge as three molecules of NADH and one molecule of $FADH_2$. We shall examine their fates when we consider the electron-transport system in Section 23.11.

23.7 REPLENISHMENT OF CYCLE INTERMEDIATES

Some of the intermediates of the citric acid cycle are important components in other metabolic processes. For example, when α-ketoglutarate, succinate, and oxaloacetate are drained away to be converted into α-amino acids, their concentrations in the cycle must be replenished. Some of the intermediates can be replenished from the breakdown of α-amino acids and carbohydrates.

Oxaloacetate is replaced through the action of the enzyme pyruvate carboxylase, which catalyzes the following reversible reaction:

$$\text{Pyruvate} + CO_2 + ATP + H_2O \rightleftharpoons \text{oxaloacetate} + ADP + P_i + 2\ H^+$$

This complex reaction is described in detail in the next section on gluconeogenesis, because the concentration of oxaloacetate is also a key regulatory aspect of that process.

23.8 GLUCONEOGENESIS

Gluconeogenesis is a process that takes place in the liver; in this process, glucose is resynthesized from lactate, the chief product of glycolysis in active muscle. The pathway is largely but not quite the reverse of glycolysis: Three irreversible steps in glycolysis must be bypassed, which is accomplished by their replacement by different reactions and different enzymes. All the other, reversible steps in the synthesis utilize the reactions and enzymes of the glycolytic pathway. The irreversible steps that must be bypassed are indicated in Figure 23.4 on the following page.

The first bypass reaction is required because there is no enzymatic mechanism for the direct synthesis of the high-energy compound phosphoenolpyruvate from pyruvate. Phosphoenolpyruvate synthesis in gluconeogenesis begins in the mitochondria. There, oxaloacetate is synthesized from pyruvate by reaction with carbon dioxide:

$$\text{Pyruvate} + CO_2 + ATP \longrightarrow \text{oxaloacetate} + ADP$$

The enzyme catalyzing this reaction, pyruvate carboxylase, is a mitochondrial regulatory enzyme virtually inactive save in the presence of acetyl-*S*-CoA, which is a specific activator. Therefore, acetyl-*S*-CoA is the key to turning on this phase of carbohydrate synthesis, and it is present in significant quantities only when the cell is in an energy-rich phase.

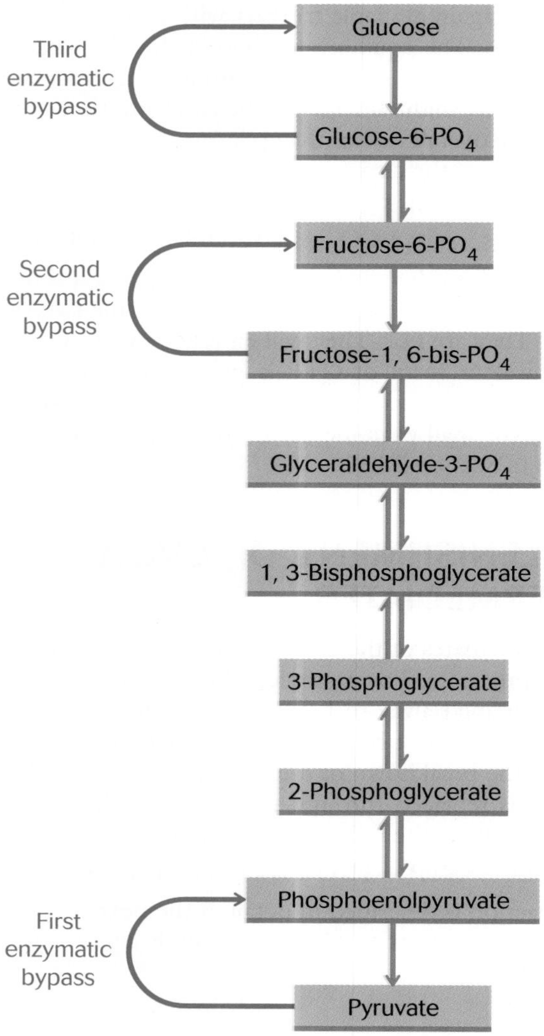

Figure 23.4 Gluconeogenesis is the reversal of glycolysis with the exception of three reactions that bypass the three irreversible reactions in the glycolytic pathway. These bypass reactions employ three new enzymes providing alternative reaction pathways.

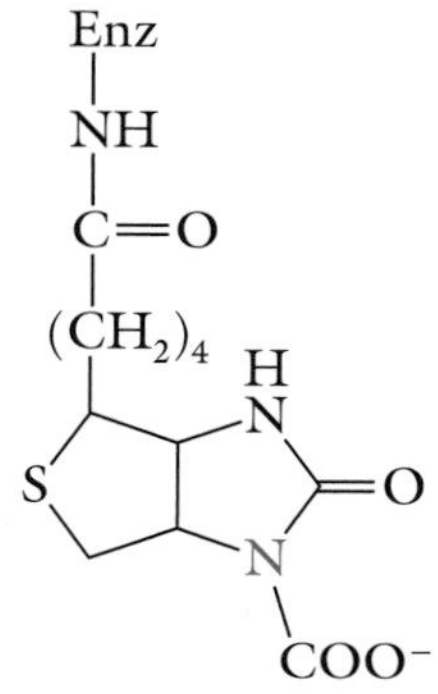

Carboxybiotinyl enzyme

The cofactor in carboxylations is biotin. It acts as a carrier of carboxyl ($—COO^-$) groups in enzymatically catalyzed carboxylation reactions in which ATP also is required. The structure in the margin shows that biotin is covalently attached to its carrier protein, indicated by Enz, and the transient carboxyl group is attached to the nitrogen atom shown in blue.

Remember that glycolysis takes place in the cytosol. Gluconeogenesis also must take place there. Oxaloacetate, which is synthesized in the mitochondria, cannot traverse the mitochondrial membrane into the cytosol, but malate can. So the next step is a reduction of mitochondrial oxaloacetate to malate by mitochondrial malate dehydrogenase with the use of NADH. This step is followed by the transport of malate into the cytosol, where it is reoxidized to oxaloacetate by cytosolic malate dehydrogenase.

Cytosolic phosphoenolpyruvate is synthesized from the extramitochondrial oxaloacetate in a reaction requiring GTP and catalyzed by the enzyme phosphoenolpyruvate carboxykinase. This is the first reaction bypassing the enzymes of the glycolytic sequence:

$$\text{Oxaloacetate} + \text{GTP} \longrightarrow \text{phosphoenolpyruvate} + CO_2 + \text{GDP}$$

Note that this reaction takes place only under conditions of high ATP and NADH concentrations, which also drive the next several reversible steps to produce fructose-1,6-bisphosphate.

The formation of fructose-1,6-bisphosphate from fructose-6-PO_4 in glycolysis is irreversible. Therefore, in a second bypass reaction, the gluconeogenesis pathway utilizes a new enzyme, fructose bisphosphatase, to irreversibly hydrolyze the bisphosphate to fructose-6-PO_4:

$$\text{Fructose-1,6-bis-}PO_4 + H_2O \longrightarrow \text{fructose-6-}PO_4 + P_i$$

The third bypass reaction accomplishes the hydrolysis of glucose-6-PO_4 to glucose, which then can leave the liver and enter the blood for distribution to other tissues. This reaction requires a third new enzyme, glucose-6-phosphatase:

$$\text{Glucose-6-}PO_4 + H_2O \longrightarrow \text{glucose} + P_i$$

In vertebrates, glucose-6-phosphatase is located only in the liver and is not found in any other tissue. Thus, only the liver can furnish free glucose to the blood, and therefore the liver is the control center that regulates the glucose concentration of the blood.

23.9 GLYCOGENESIS

Glucose can be converted into a polysaccharide called **glycogen** (Section 18.6), which consists of a chain of glucose molecules linked through $\alpha(1 \rightarrow 4)$ bonds (Section 18.7) and held in reserve as a future energy source. The human body stores about half its daily energy supply as glycogen. The process for the synthesis of glycogen, **glycogenesis,** requires a new set of enzymes that operate only under cellular conditions of high concentrations of ATP. Glycogen is synthesized only in the liver and in muscle cells and is stored as glycogen granules. The first step in its synthesis is an enzymatic isomerization in which the phosphate group attached to carbon 6 of glucose is transferred to carbon 1; that is, glucose-6-PO_4 is changed into glucose-1-PO_4.

Glucose-1-PO_4 must be activated to be incorporated into glycogen, a step accomplished by reaction with uridine triphosphate, UTP, to form uridine diphosphoglucose. This reaction is catalyzed by the specific enzyme UDP-glucose pyrophosphorylase:

$$\text{Glucose-1-}PO_4 + \text{UTP} \longrightarrow \text{UDP-glucose} + PP_i$$

PP_i stands for **inorganic pyrophosphate,** $P_2O_7^{4-}$, which undergoes an enzymatic hydrolysis characterized by a large equilibrium constant. That secondary process drives the overall activation reaction:

$$\begin{array}{ccc} & O^- \quad\quad O^- & \quad\quad\quad\quad O \\ & | \quad\quad\quad | & \quad\quad\quad\quad \| \\ ^-O-P-O-P-O^- + H_2O & \xrightarrow{\text{pyrophosphatase}} & 2\ HO-P-O^- \\ & \| \quad\quad\quad \| & \quad\quad\quad\quad | \\ & O \quad\quad\quad O & \quad\quad\quad\quad O^- \end{array}$$

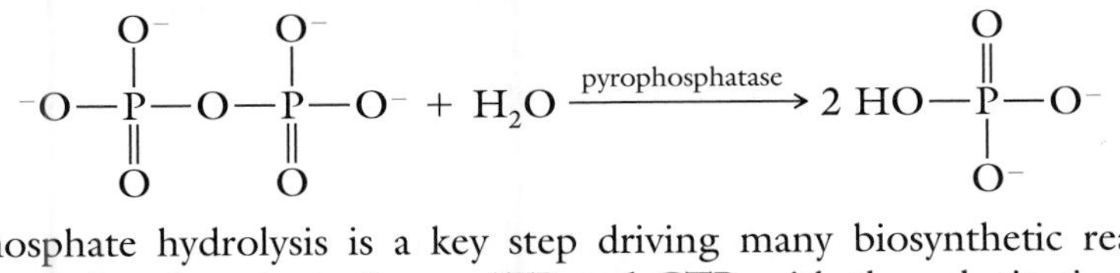

Pyrophosphate hydrolysis is a key step driving many biosynthetic reactions. Uridine triphosphate is similar to ATP and GTP, with the substitution of the nitrogen base uracil for either adenine or guanine (Section 21.1). The pyrophosphate derives from the two terminal phosphate groups that are cleaved from UTP when they are displaced by glucose-1-PO_4 to form UDP-glucose.

The activated glucose becomes the substrate for glycogen synthase, a regulatory enzyme that is activated by glucose-6-PO_4. This enzyme catalyzes the addition of the activated glucose to the nonreducing end (Section 18.6) of a growing chain of glycogen:

$$\text{UDP-glucose} + (\text{glucose})_n \longrightarrow (\text{glucose})_{n+1} + \text{UDP}$$

23.10 GLYCOGENOLYSIS

Glycogen in the liver is degraded to form glucose whenever the blood's concentration of glucose drops below normal levels, about 5 m*M*. This process, called **glycogenolysis,** proceeds by a route different from that of the synthesis of glycogen. The glycogen's chain-terminal glucose is removed through phosphorolysis by the enzyme phosphorylase in its active form, phosphorylase$_a$. Phosphorolysis accomplishes a result similar to that of hydrolysis except that the elements of phosphoric acid instead of water are used to cleave the acetal group between the glucose units.

$$(\text{Glucose})_n + \text{P}_i \xrightarrow{\text{phosphorylase}_a} (\text{glucose})_{n-1} + \text{glucose-1-PO}_4$$

Structurally, phosphate cleaves the $\alpha(1 \rightarrow 4)$ bond of glycogen at the nonreducing end of the polyglucose chain. In so doing, it (1) releases glucose-1-PO_4 and (2) shortens the polyglucose chain by one monomer. Schematically,

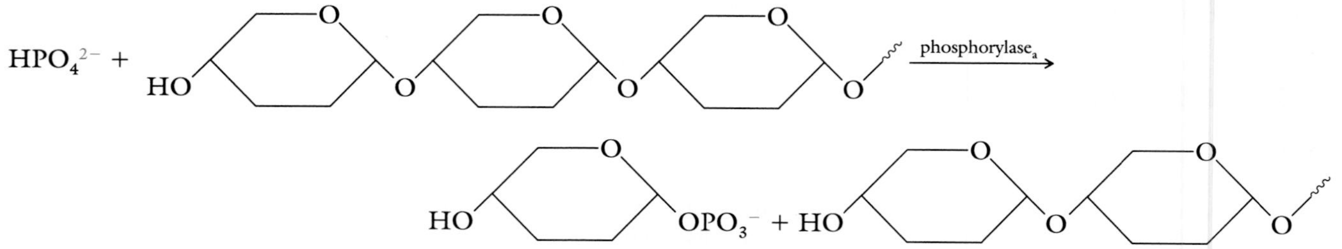

> ▶▶ The role of hormones in metabolism is considered in Section 26.2.

Phosphorylase is ordinarily present in its inactive form, phosphorylase$_b$. It must be activated through a series of steps initiated in the liver by the hormone glucagon and in both the liver and muscle by the hormone epinephrine. **Glucagon** is a polypeptide hormone synthesized by pancreatic α cells, and **epinephrine** (Box 16.3) is synthesized in the adrenal cortex from the α-amino acid tyrosine. The details of the process are presented in Figure 23.5.

The hormone epinephrine is released very quickly when any sudden environmental change is sensed by the central nervous system, a response known as the fight or flight reflex. It stimulates glycogenolysis primarily in muscle cells. Epinephrine is also under the control of cells in the hypothalamus of the brain that are sensitive to the blood concentration of glucose. Lowered glucose concentration will raise the epinephrine level. Epinephrine also increases blood pressure and heart rate.

The mechanism of action of the hormone glucagon is similar to that of epinephrine, but its role is to regulate the blood glucose concentration over the long term. Therefore, it is active during fasting periods and between meals. Its secretion into the blood is directly controlled by the blood-glucose concentration. When that concentration falls below from 80 to 100 mg/dL, the concentration optimum for brain function, the pancreatic α cells secrete glucagon. The hormone then stimulates glycogenolysis in both liver and muscle cells, which require glucose for energy, although, as you will recall, only liver cells, not muscle, can contribute glucose to the blood. Unlike epinephrine, glucagon has no effect on blood pressure or heart rate.

Hormones are classified as either water soluble or water insoluble. Water-insoluble hormones, such as steroids (Section 19.7), penetrate cell membranes, enter the cell, and directly affect their metabolic targets. Water-soluble hormones, such as epinephrine and glucagon, do not penetrate membranes but bind directly

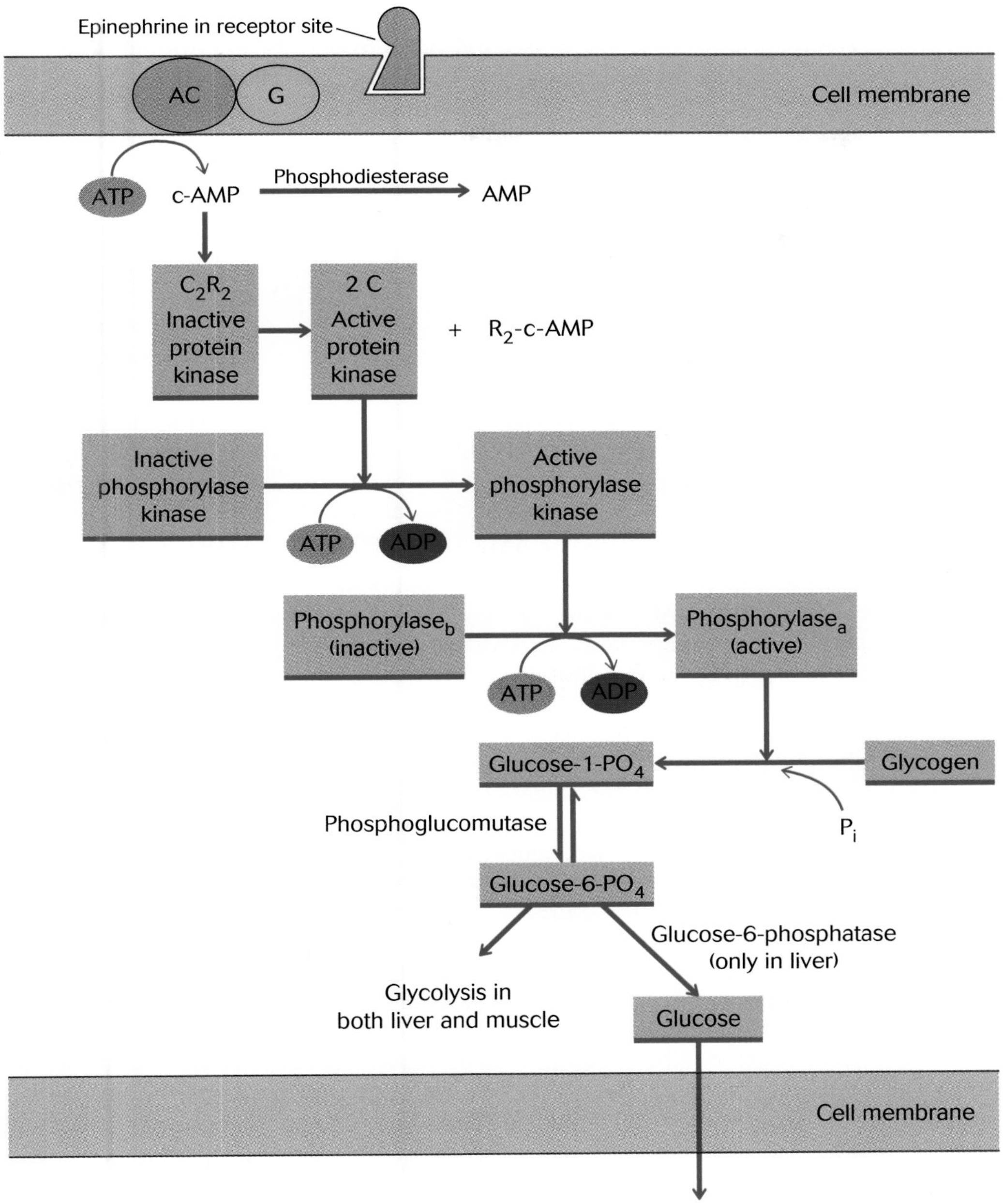

Figure 23.5 The enzymatic cascade in glycogenolysis is initiated by the action of the hormone epinephrine in the liver and in skeletal muscle cells. Abbreviations: G, GTP-binding protein; AC, active adenylyl cyclase. The pancreatic hormone glucagon causes the same result in the liver.

to specific receptors on the outer surface of the membrane. Each hormone that binds to the cell membrane has a unique receptor, but each receptor, when occupied by its hormone, ultimately activates the same membrane-bound enzyme, adenylyl cyclase. Activation takes place in three steps. First, hormone bound to its cell surface receptor causes a conformational change in the receptor protein. Second, this conformational change causes a second membrane protein, called a GTP-binding protein (abbreviated G protein) to bind GTP. In its inactive state, the G protein binds GDP. When the G protein is stimulated by the hormone-receptor complex, the GDP dissociates, GTP is bound in its place, and, in the last of the three steps, the GTP-protein then activates adenylyl cyclase.

Adenylyl cyclase catalyzes the formation of the cyclic form of AMP (c-AMP), which is called the **second messenger** (signaling within a cell), the hormone being the **first messenger** (signaling from cell to cell).

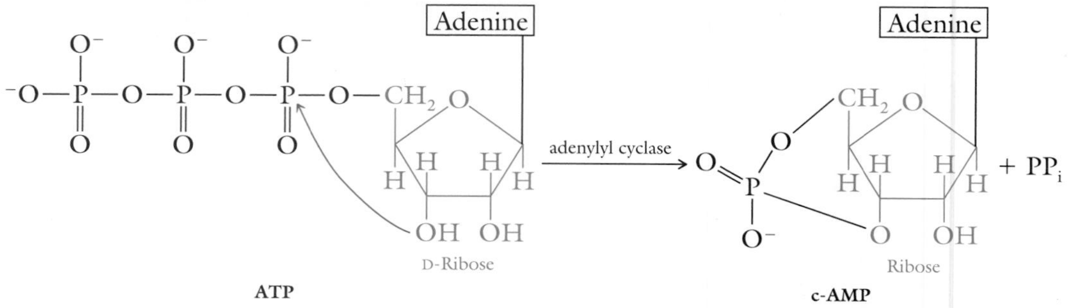

A second enzyme of the glycogenolysis pathway is ordinarily in its inactive form because it binds a regulatory inhibitor, R, to form an inactive C_2R_2 complex. When cyclic AMP binds to the R subunit, the C_2R_2 complex dissociates, thus freeing C, the active form of the second enzyme. The reaction is:

$$C_2R_2 + 2\text{ c-AMP} \longrightarrow 2\text{ C} + 2\text{ R-c-AMP}$$

In its active form, C catalyzes the phosphorylation and activation of a third enzyme. The third enzyme catalyzes the phosphorylation of inactive phosphorylase, phosphorylase$_b$, to yield the fourth enzyme, the active form of the phosphorylase, phosphorylase$_a$. Each phosphorylation requires ATP. Phosphorylase$_a$ in turn cleaves glucose from glycogen by phosphorolysis to yield glucose-1-PO_4. Glucose-1-PO_4 is converted into glucose-6-PO_4 by phosphoglucomutase.

Up to that point, the reactions in both liver and muscle cells are identical. However, glucose-6-phosphatase is present only in liver cells. Therefore, only liver cells can produce free glucose from glucose-6-PO_4, and, consequently, only liver cells can increase the blood-glucose concentration. The result of the process in muscle cells is a great increase in the rate of glycolysis, accompanied by increased availability of ATP.

In glycogenolysis, enzymes sequentially activate a series of other enzymes. The action of a single enzyme can be likened to opening a door through which other molecules may emerge. If each of the other molecules also is an enzyme, the number of catalytic units multiplies rapidly. Figure 23.6 depicts each enzyme as activating three other enzymes. In the cell, however, one enzyme may activate thousands of others. The original signal becomes greatly amplified, depending on how many activation steps take place, and, in this case, has been estimated to be as great as 2.5×10^7. Just a few molecules of epinephrine suffice to release many grams of glucose into the blood.

A fifth enzyme, phosphodiesterase, is active at all times in both muscle and liver cells. It catalyzes the hydrolysis of c-AMP to AMP and quickly puts an end to the enzyme cascade leading to glycogenolysis. This enzyme's presence ensures that glycogenolysis will take place only as long as either epinephrine or glucagon continue to be present at the cell membrane. After all, you can't fight or fly all the time. An interesting sidelight here is that phosphodiesterase is inhibited by caffeine and theobromine, the alkaloids in coffee and tea, respectively. Their effect is to prolong the active lifetime of c-AMP, as though epinephrine were constantly being delivered to cell membranes.

23.11 ELECTRON-TRANSPORT CHAIN

So far in our consideration of glucose metabolism, energy extracted from nutrients in the form of ATP has been obtained by substrate-level phosphoryla-

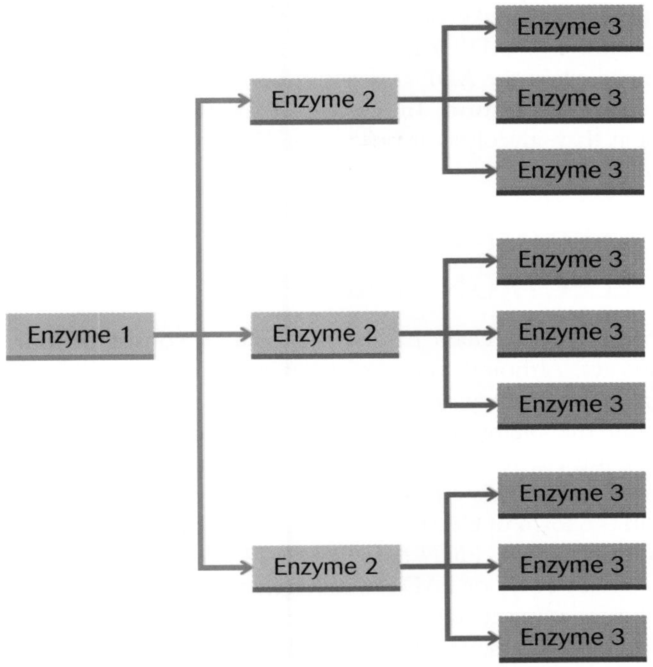

Figure 23.6 When an enzyme activates a series of enzymes of a pathway, the original signal that turned on enzyme 1 is amplified manyfold.

tions. In this section, we shall see how the oxidation of NADH also results in the formation of ATP. This NADH-dependent synthesis of ATP is called **oxidative phosphorylation** (Section 22.4). Stored electrons (in the form of NADH or $FADH_2$) pass through a highly organized sequence of enzymes and cofactors, called the **electron-transport chain,** in the inner mitochondrial membrane, to finally react with molecular oxygen to form water.

In Chapter 9, weak acids were arranged in an order based on their tendencies to lose a proton. The strongest lost protons more readily than the weakest, and the conjugate base of the weakest gained protons more easily than the strongest. An analogous ordering among partners in redox reactions is seen in the organization of the electron-transport chain: NADH tends to reduce components of its dehydrogenase, and that dehydrogenase tends to reduce components immediately adjacent to it, and so forth. Each component is alternately reduced and oxidized as electrons move along to the final electron acceptor, oxygen.

During electron transport, a proton gradient forms across the inner mitochondrial membrane: the pH outside that membrane is lower than the inside pH. ATP synthesis is driven by this proton gradient.

Oxidative phosphorylation completely depends on the structural and osmotic integrity of the mitochondria. A mitochondrion possesses two membranes, an outer and an inner membrane. The space enclosed by the inner membrane is called the mitochondrial matrix, and it contains the enzymes of the citric acid cycle. The components of the electron-transport chain are integral parts of the inner membrane, which is repeatedly folded so as to produce a very large surface area. The combination of the special organization of enzymes within the inner membrane and the large surface area enhances the efficiency of the electron-transport chain. Any breach in the inner mitochondrial membrane will destroy the hydronium ion gradient and stop the oxidative phosphorylation of ADP.

Under normal conditions, the movement of electrons and hydrogen along the electron-transport chain to oxygen is coupled to the phosphorylation of ADP: if the reduction of oxygen is halted, the synthesis of ATP also stops; at the same time, the uptake of oxygen depends on the availability of ADP. A high concentration of ADP leads to a rapid uptake of oxygen, and oxygen uptake is reduced when the ADP concentration is lowered. The uptake of oxygen by cells is called **respiration.** In Section 23.15, we shall see that respiration is more efficient than glycolysis at producing ATP from glucose.

23.12 ENZYMES OF THE ELECTRON-TRANSPORT CHAIN

The electron-transport chain has been shown to consist of four distinct enzyme complexes. Although the complexes are integral structures of the mitochondrial membrane, they do not seem to be fixed in position but are free to diffuse within the membrane. Figure 23.7 depicts the mitochondrial membrane containing the four enzyme complexes, with arrows indicating the flow of electrons. Complex I is essentially NADH dehydrogenase. Here, the reducing power in the form of electrons and hydrogen from NADH is transferred to flavine mononucleotide, FMN (see Figure 22.7). Remember that dehydrogenases that contain FMN or FAD are called flavoproteins. Also present is iron in the form of iron-sulfur centers incorporated into the protein. The iron in these centers undergoes reversible oxidation-reduction between Fe^{2+} and Fe^{3+} states during electron transport.

Electrons from complex I are transported to the next complex by a lipid-soluble molecule called ubiquinone (UQ) because of its ubiquitous nature, being found in virtually all forms of life. It is also called coenzyme Q (CoQ). UQ, shown in the margin on the facing page, is free to diffuse within the lipophilic environment of the inner mitochondrial membrane. It undergoes reduction in two steps, one electron at each step.

The oxidation of one NADH and the reduction of one UQ by complex I is accompanied by the net transport of protons from the matrix side of the mitochondrial membrane to the cytosolic side. The flow of electrons provides the energy to drive this proton transport, an example of active transport (Section 19.10). Approximately 3 mol of ATP are synthesized by the oxidation of

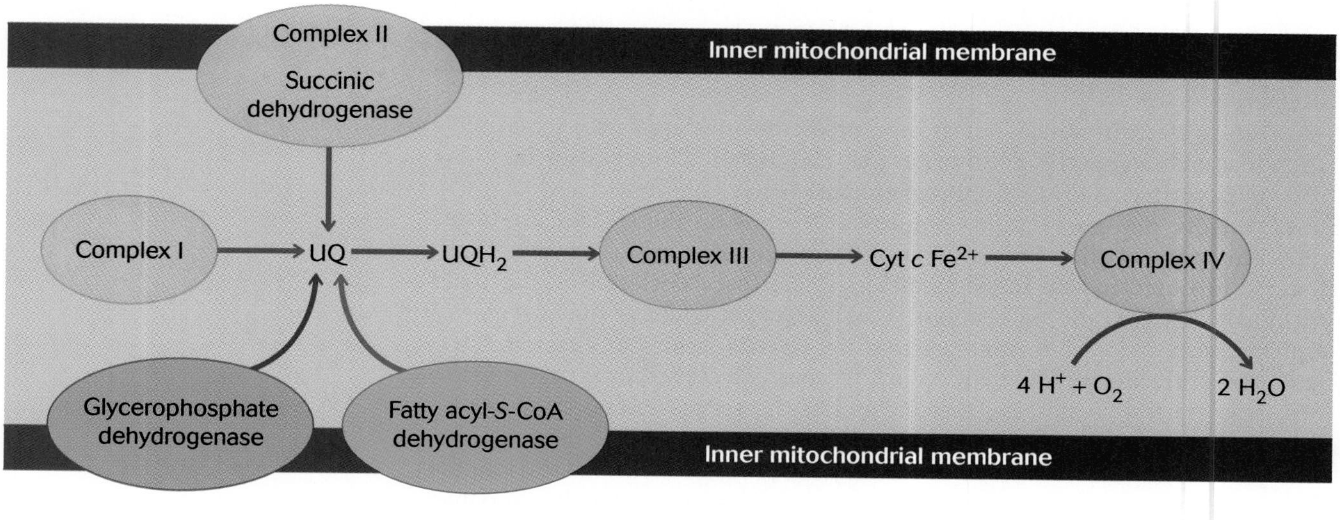

Figure 23.7 The enzyme complexes of the electron-transport chain in the mitochondrial membrane. Complex II, glycerophosphate dehydrogenase, and fatty acyl-*S*-CoA dehydrogenase are incorporated into the inner membrane. Abbreviations: UQ, ubiquinone; UQH_2, reduced ubiquinone; Cyt *c*, cytochrome *c*.

1 mol of NADH. Details about the stoichiometry of ATP production are presented in Section 23.15.

UQ receives electrons from three other sources. One of these sources is complex II. Complex II, the flavoprotein complex succinic dehydrogenase, is a member of the electron-transport chain in the mitochondrial membrane and, at the same time, is an integral member of the citric acid cycle. Succinic dehydrogenase contains an FAD covalently bound to its protein. The oxidation of succinate to fumarate results in the formation of $FADH_2$, which transfers its electrons to UQ. Oxidation of 1 mol of $FADH_2$ eventually results in the formation of 2 mol of ATP.

UQ is also reduced by electrons from the oxidation of fatty acids by an electron-transferring flavoprotein (Section 24.4) and from a reaction catalyzed by glycerophosphate dehydrogenase.

Reduced ubiquinone (UQH_2) delivers its reducing power to complex III. Complex III contains a family of iron-containing proteins called cytochromes—specifically cytochromes *b* and c_1—which undergo reversible oxidation-reductions. The cytochromes possess the same type of iron-containing heme as that found in hemoglobin, but, in the cytochromes, it combines with protein in such a way as to prevent its iron center from combining with molecular oxygen. Iron-sulfur centers also are present. The passage of electrons through complex III also is accompanied by the transport of protons to the cytosolic side of the inner membrane. Electrons from this complex are conducted to the next complex by a freely diffusing protein called cytochrome *c* (Cyt *c*).

Ubiquinone (UQ)
(R = hydrocarbon chain)

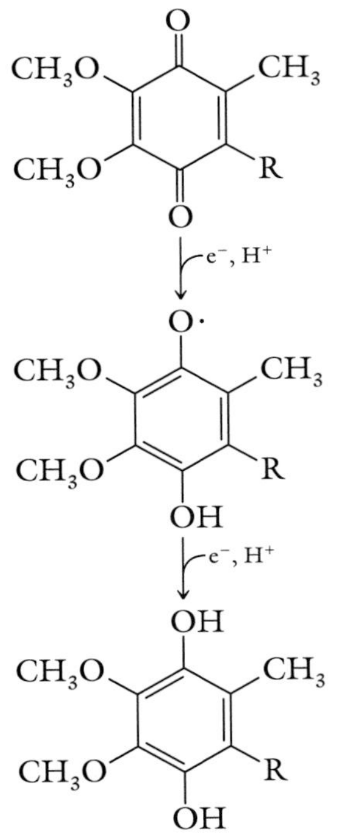

UQH_2

Complex IV is called cytochrome oxidase because it transfers electrons from the electron-carrier cytochrome *c* to molecular oxygen to form water. The complex contains cytochromes *a* and a_3 and two copper ions that shuttle between the Cu^{2+} and Cu^{+} states. The reaction of protons, electrons, and oxygen to form water is:

$$4\ \text{Cyt}\ c(Fe^{2+}) + 4\ H^+ + O_2 \xrightarrow{\text{complex IV}} 4\ \text{Cyt}\ c(Fe^{3+}) + 2\ H_2O$$

The reduction of oxygen in this reaction is accompanied by proton transport across the inner membrane. This is the final reaction in which electrons are derived from reduced nutrient molecules.

Carbon monoxide or cyanide poisoning prevents the combination of hydrogen and electrons with oxygen and shuts down the electron-transport chain at complex IV. Cyanide combines with Fe^{3+}, and carbon monoxide combines with Fe^{2+} of the cytochromes. Clinical treatment of cyanide poisoning includes the introduction of nitrites, which convert the Fe^{2+} of hemoglobin in the blood into the Fe^{3+} of methemoglobin. Blood methemoglobin can compete with cellular cytochrome Fe^{3+} for the cyanide and reverse the symptoms. Cyanide can then be transformed into the much less toxic cyanate by the therapeutic administration of thiosulfate. Carbon monoxide toxicity can be treated by placing the victim in a hyperbaric chamber, where oxygen, supplied at elevated pressures, can successfully compete with and displace carbon monoxide from the Fe^{2+} of the cytochromes.

23.13 PRODUCTION OF ATP

Reducing power is collected in the form of NADH from a wide variety of dehydrogenations. The reducing power of NADPH also can be transferred to NAD^+ by the enzyme pyridine nucleotide transhydrogenase. NADH cannot penetrate the mitochondrial membrane, however; so NADH generated by glycolysis or another cytosolic process must enter the mitochondria by an indirect route considered in Section 23.14.

Figure 23.8 catalogs the metabolic origins of reducing power and maps its entry into the electron-transport chain. Most reducing power is delivered directly to the flavoprotein at complex I. However, succinic dehydrogenase of the citric acid cycle is a flavoprotein and cannot reduce NADH dehydrogenase. This flavoprotein and others deliver reducing power directly to UQ and thence to complex III. Because it bypasses complex I, reducing power from the flavoprotein succinate dehydrogenase produces fewer moles of ATP. The origin of reducing power from fatty acids and α-amino acids will be taken up in Chapters 24 and 25.

As electrons move through the electron-transport chain during respiration, hydrogen ions are transported from the inner matrix to the intermembrane space of the mitochondrion. As a result, a large diffusion gradient of

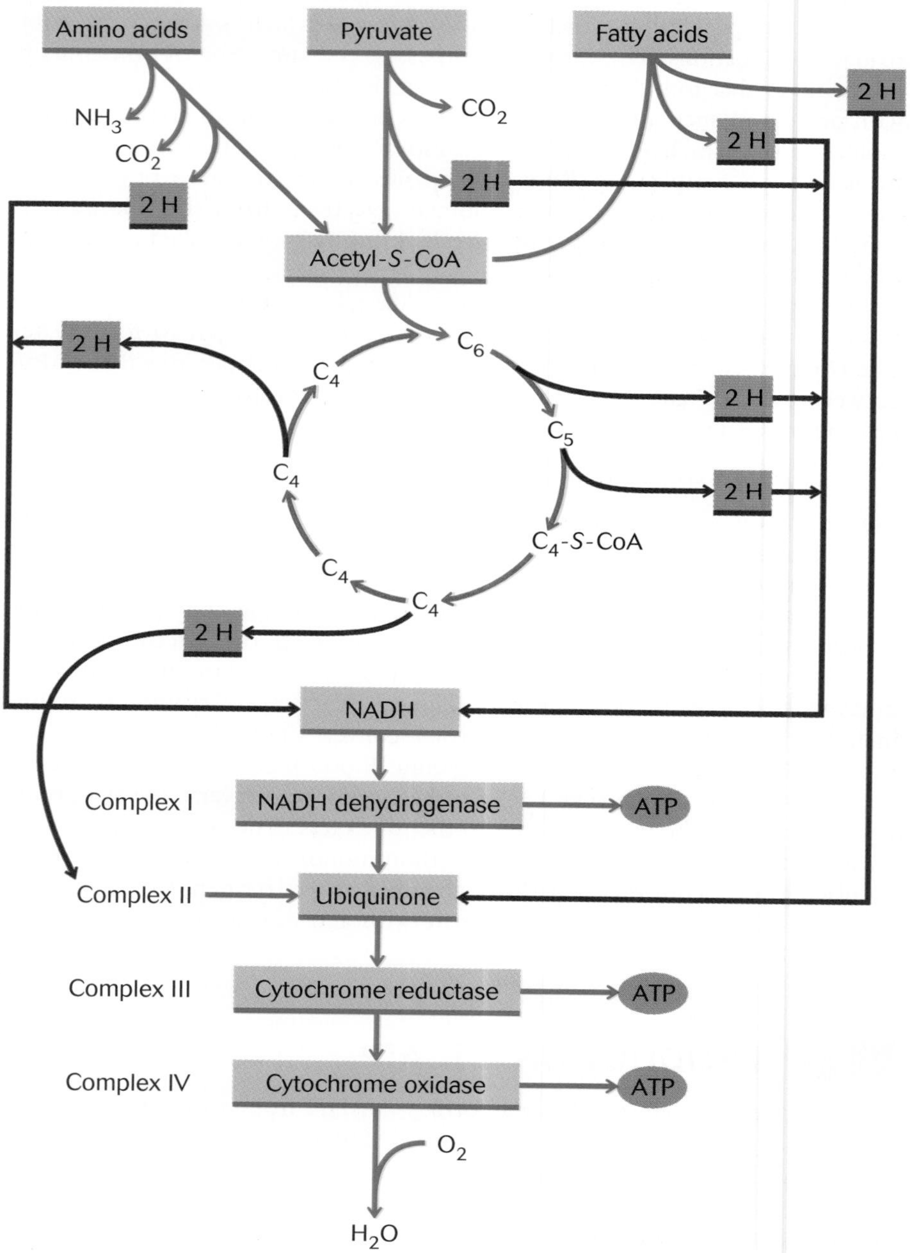

Figure 23.8 The origins and fate of reducing power extracted from nutrients in catabolism. Subscript numbers identify the components of the cycle: for example, C_6, citrate; C_5, α-ketoglutarate.

protons forms, which provides the driving force for the formation of ATP. This process is called the **chemiosmotic mechanism** because it effects chemical change through an osmotic gradient. The concept was developed by Peter Mitchell, who was awarded a Nobel Prize in 1969 for this powerful idea.

The inner membrane is the location of a complex enzyme that has been shown to possess two functions: a proton-transport channel, and an ATPase. This complex enzyme, called ATP synthase, has been extracted from mitochondria and dissociated into several structural components. The functional properties of the parts have been established and the parts have been reassembled into a working enzyme.

Structural studies with the use of electron microscopy have shown the enzyme to be an integral component of the inner mitochondrial membrane and to have a "lollypop" appearance with a stalk and head protruding into the mitochondrial matrix. Figure 23.9 contains one such image of a part of the inner mitochondrial membrane, showing the lollypop structure of ATP synthase. There can be as many as 100,000 of these ATP-synthesizing "factories" in a single mitochondrion.

As the transmembrane proton gradient grows, the electrical charge of the intermembrane space becomes increasingly positive. At some point, the transport of electrons along the electron-transport chain tends to stop because of this buildup of positive charge. The stoppage is relieved, however, when protons begin to flow inward through the proton transport channel. As a result of this inward flow, ATP synthase catalyzes the formation of a phosphoanhydride bond between ADP and P_i, which allows electron transport to resume. The process is diagrammed in Figure 23.10. Calculations show that the transmembrane proton gradient is great enough at three locations along the electron-transport chain to account for the synthesis of 1 mol of ATP at each location.

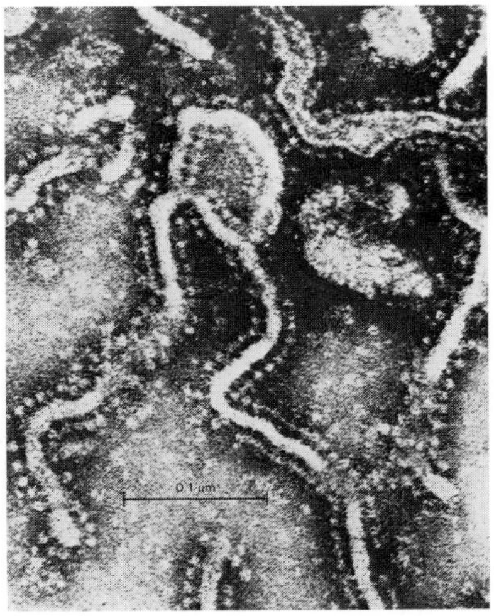

Figure 23.9 Electron micrograph of a part of the inner mitochondrial membrane showing the lollypop structure of ATP synthase.

23.14 MITOCHONDRIAL MEMBRANE SELECTIVITY

The outer mitochondrial membrane is freely permeable to almost all low-molecular-mass solutes. However, the inner mitochondrial membrane is permeable only to solutes corresponding to specific transport systems embedded

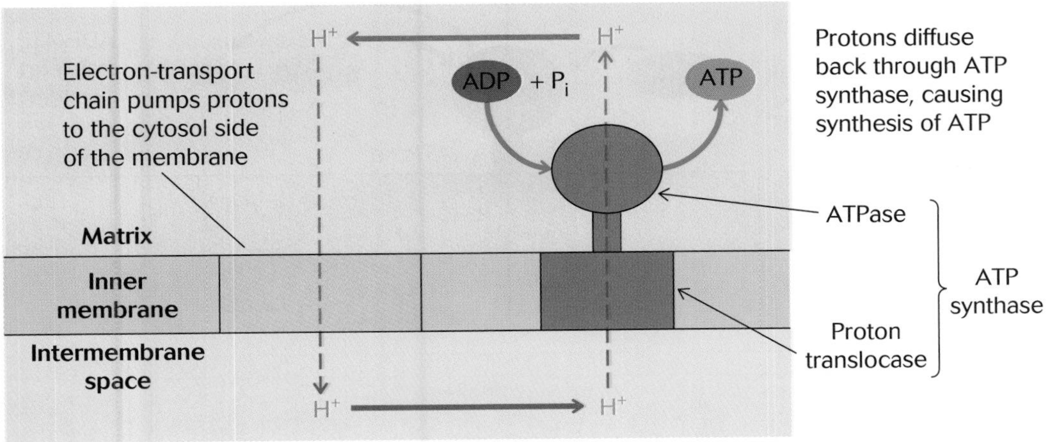

Figure 23.10 Formation of ATP as a result of the movement of protons along the proton gradient generated during electron transport in mitochondria, showing the "lollypop" structure of ATP synthase.

in that membrane. It is impermeable not only to H^+, but to OH^-, K^+, and Cl^- ions as well.

Adenine Nucleotide and Phosphate Translocases

Two of the most important inner-membrane-specific transport systems are:

1. **The adenine nucleotide translocase:** This transport system allows ADP and phosphate to enter the mitochondria and ATP to leave. It is specific for ADP and ATP and will not transport AMP or any other nucleotide such as GTP or GDP.
2. **The phosphate translocase:** This transport system promotes the simultaneous transport of $H_2PO_4^-$ and H^+ into the mitochondria.

The coordinated operation of these two transport systems brings ADP and P_i into the mitochondrion from the cytosol and sends the newly synthesized ATP out to the cytosol to energize the cell's activities. These coordinated systems are illustrated in Figure 23.11.

Other specific transport systems of the inner mitochondrial membrane are: a system for pyruvate, which is formed in the cytosol and must enter the mitochondrion to enter the citric acid cycle; a dicarboxylate transport system for malate and succinate; a tricarboxylate transport system for citrate and isocitrate; and two other systems that allow the reducing power of cytosolic NADH to enter the mitochondrion for oxidation.

Impermeability of the Mitochondrial Membrane to NAD^+ and NADH

NADH cannot pass through the inner mitochondrial membrane. It can deliver its reducing power to the electron-transport chain by one of two indirect routes. The first route, the malate-aspartate shuttle (shuttle A) is active in heart, liver, and kidney mitochondria, where it delivers its reducing equivalents to complex I, producing 3 mol of ATP per mole of NADH. The second route is the glycerol phosphate shuttle (shuttle B) active in brain and muscle mito-

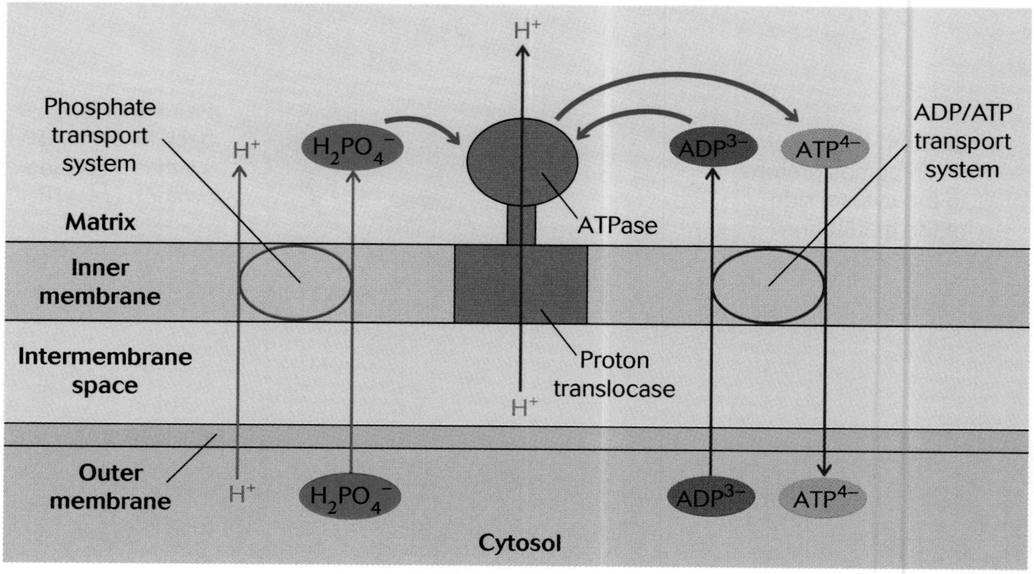

Figure 23.11 The coordinated transport of P_i and ADP into the mitochondria and ATP out of it.

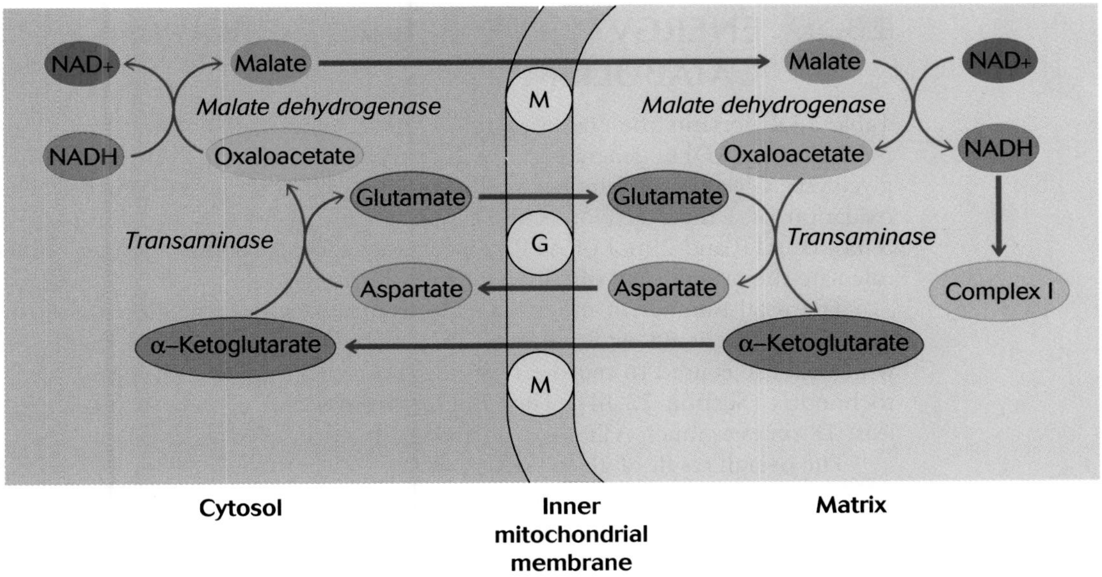

Figure 23.12 The malate-aspartate shuttle brings reducing equivalents generated in the cytosol into the mitochondria. Abbreviations: M, dicarboxylate acid transport system; G, α-amino acid transport system.

chondria, where it delivers its electrons directly to complex II to produce 2 mol of ATP per mole of NADH.

The malate-aspartate shuttle is illustrated in Figure 23.12. NADH from glycolysis is used to reduce cytosolic oxaloacetate to form malate. The reaction is catalyzed by cytosolic malate dehydrogenase. Malate is then transported into the mitochondrial matrix by the dicarboxylate transport system (M in Figure 23.12), where it is in turn oxidized by mitochondrial malate dehydrogenase to form mitochondrial oxaloacetate. The other product, mitochondrial NADH, delivers its reducing equivalents to complex I of the electron-transport chain. Cytosolic oxaloacetate must now be regenerated to continue the cycle. Mitochondrial oxaloacetate cannot pass through the membrane, but it is converted into aspartate by transamination with glutamate (see Section 25.2). Both aspartate and glutamate can penetrate the membrane through an α-amino acid transport system (G in Figure 23.12). The α-ketoglutarate generated in the transaminase reaction producing mitochondrial aspartate passes through the membrane by means of the same dicarboxylate transport system used by malate.

The glycerol phosphate shuttle begins in the cytosol with the reduction of dihydroxyacetone phosphate by NADH to form 3-phosphoglycerol:

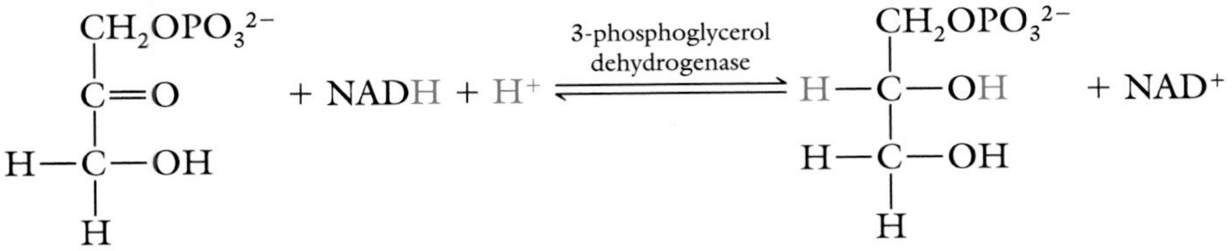

The product then enters the intermembrane space of the mitochondrion and interacts with a flavoprotein at the surface of the mitochondrial inner membrane. There it is oxidized back to dihydroxyacetone phosphate, and the reducing power, in the form of $FADH_2$, is passed on to ubiquinone and then to complex III of the electron-transport chain to result in the synthesis of 2 mol of ATP.

23.15 ENERGY YIELD FROM CARBOHYDRATE CATABOLISM

Table 23.4 presents the energy yield of carbohydrate metabolism in moles of NADH and $FADH_2$ generated at each step of the citric acid cycle and during glycolysis. Glycolysis produces a net of 2 mol of ATP per mole of glucose. The oxidation of 1 mol each of NADH and $FADH_2$ results in the synthesis of 3 mol of ATP and 2 mol of ATP, respectively. Table 23.4 uses these values to calculate the overall yield of ATP per mole of glucose.

The total number of moles of ATP produced by both glycolysis and the citric acid cycle is 38 or 36 mol of ATP per mole of glucose, depending on which shuttle is used to transfer reducing power from the cytosol into the mitochondria (Section 23.14). The citric acid cycle can be seen to produce at least 18 times as much ATP from glucose as that produced by glycolysis.

The overall result of glycolysis is:

$$\text{Glucose} + 2\ \text{ADP} + 2\ P_i + 2\ \text{NAD}^+ \longrightarrow 2\ \text{pyruvate} + 2\ \text{ATP} + 2\ \text{NADH} + 2\ \text{H}^+$$

TABLE 23.4 An Accounting of the ATP Yield from Carbohydrate Catabolism

Process	NADH	$FADH_2$	ATP
GLYCOLYSIS			
glucose → glucose-6-PO_4			−1
fructose-6-PO_4 → 1,6-BPF			−1
2 3-PGAL → 2 1,3-BPG	2		2
2 PEP → 2 pyruvate			2
CITRIC ACID CYCLE			
2 pyruvate → 2 acetyl-*S*-CoA + 2 CO_2	2		
2 isocitrate → 2 α-ketoglutarate + 2 CO_2	2		
2 α-ketoglutarate → 2 succinate + 2 CO_2	2		
2 succinyl-*S*-CoA + 2 GDP →			2
2 succinate → 2 fumarate		2	
2 malate → 2 oxaloacetate	2		
Sum of moles per mole of glucose	**10**	**2**	**4**

	Sum of ATP Yields	
	Shuttle A	**Shuttle B**
ATP from glycolysis	2	2
NADH from glycolysis	6	4
NADH from citric acid cycle	24	24
$FADH_2$ from citric acid cycle	4	4
ATP from GTP	2	2
Total ATP* per mole of glucose	**38**	**36**

* 3 ATP/NADH, 2 ATP/$FADH_2$

The net result of optimal oxidation of glucose is:

$$\text{Glucose} + 38\ \text{ADP} + 38\ \text{P}_i + 6\ \text{O}_2 \longrightarrow 6\ \text{CO}_2 + 38\ \text{ATP} + 44\ \text{H}_2\text{O}$$

Although mitochondrial oxidative phosphorylation produces more than 18 times as much ATP per mole of glucose as glycolysis produces, the rate of glycolysis is greater than that of the oxidative process. This greater rate is primarily because the concentration of the glycolytic enzymes is much greater than that of the enzymes in the citric acid cycle. This greater concentration allows anaerobic processes, primarily intense muscular activity, to take place at adaptively useful rates.

If all the potential energy contained in the glucose molecule could be harnessed in the form of ATP, it would generate about 94 mol of ATP per mole of glucose. Because 38 mol of ATP are optimally formed, the efficiency of the metabolic process can be calculated to be about 40%. This efficiency significantly exceeds the typical efficiencies of mechanical engines, about 10%, and is close to the efficiency of electrochemical cells that directly transform the oxidation of hydrocarbons into electricity.

Summary

Glycolysis Glucose is the principal reduced organic compound that fuels metabolism. The first catabolic pathway that glucose encounters is called glycolysis and is located in the cell cytosol. In that pathway, glucose is converted in a sequence of nine enzymatically catalyzed steps into two molecules of pyruvate, a three-carbon molecule. In the process, ATP and NADH are produced. Pyruvate enters the citric acid cycle to form first acetyl-*S*-CoA and then to be oxidized to carbon dioxide and water. In active muscle, pyruvate is reduced to lactate. This regenerates NAD^+, which allows glycolysis to take place at required rates.

Citric Acid Cycle In contrast with the glycolytic pathway, which is a linear sequence, the aerobic phase of glucose catabolism is cyclic; it is called the citric acid cycle. All of the enzymes of the citric acid cycle are located in the mitochondria of animal cells. A turn of the cycle begins with the condensation of an acetyl group of acetyl-*S*-CoA with the four-carbon dicarboxylic acid oxaloacetate to form the six-carbon tricarboxylic acid citrate. Citrate is then converted into isocitrate, which is oxidatively decarboxylated, with the loss of CO_2 and the production of NADH, to form the five-carbon α-ketoglutarate. This compound undergoes an oxidative dehydrogenation and a series of transformations to yield a second molecule of CO_2 and NADH and emerges as the four-carbon succinate. Succinate is then converted in three steps into the four-carbon oxaloacetate, which begins the cycle again. The NADH and $FADH_2$ produced as a result of the oxidations is used to generate ATP.

Gluconeogenesis Gluconeogenesis, the process in the liver in which glucose is resynthesized from the products of glycolysis, is not simply the reverse of glycolysis. There are three irreversible steps in glycolysis that must be bypassed and replaced by different reactions and different enzymes to synthesize glucose from pyruvate. All other steps in the synthesis utilize the reversible reactions and enzymes of the glycolytic pathway.

Glycogenesis Glucose is stored in the body as a polysaccharide called glycogen, whose synthesis requires a set of enzymes that operate only under cellular conditions of high concentrations of ATP. Glycogen is synthesized in the liver and muscle cells and is stored as glycogen granules.

Glycogenolysis The degradation of glycogen to release glucose proceeds by the phosphorolysis at the nonreducing chain end of glycogen by the enzyme phosphorylase in its active form. Phosphorylase is ordinarily present in its inactive form and must be activated through a series of steps initiated in the liver by the hormone glucagon and in both liver and muscle by the hormone epinephrine.

Electron-Transport Chain In oxidative phosphorylation, reduced coenzymes generated in catabolism are oxidized to form water, and that process is coupled to the simultaneous synthesis of ATP. ATP synthesis takes place in the inner mitochondrial membrane catalyzed by a sequence of enzymes called the electron-transport chain. The ATP synthesis is driven by a large diffusion gradient of protons established across the inner mitochondrial membrane. This mode of ATP synthesis is called the chemiosmotic mechanism of oxidative phosphorylation.

Specific transport systems in the inner mitochondrial membrane control the efficiency of oxidative phosphorylation by allowing the permeability only of solutes supporting that process. Specific proteins embedded in that membrane transport ADP, P_i, ATP, pyruvate, malate, succinate, citrate, isocitrate, and reducing equivalents into and out of the inner mitochondrial space.

Energy Yield of Carbohydrate Catabolism Glycolysis produces a net of 2 mol of ATP per mole of glucose. The total ATP produced by the citric acid cycle is from 36 to 38 mol per mole of glucose. Because it takes place at a much greater rate than does oxidative phosphorylation, glycolysis is the principal catabolic route in very active skeletal muscle, where ATP must be used at high rates.

Key Words

chemiosmotic mechanism, p. 701
citric acid cycle, p. 687
electron-transport chain, p. 696
flavoprotein, p. 690
gluconeogenesis, p. 691
glycolysis, p. 679
glycogenesis, p. 693
glycogenolysis, p. 694
hormone, p. 694
oxidative phosphorylation, p. 697

Exercises

Glycolysis

23.1 What nutrient is the principal source of energy for metabolism?

23.2 Does the first metabolic pathway require the presence of oxygen?

23.3 What are the most important products of glycolysis?

23.4 Are all the steps of the glycolytic pathway reversible? Explain your answer.

23.5 Are there any important regulatory control points in glycolysis?

23.6 Identify the oxidation step in glycolysis.

23.7 Identify the steps in glycolysis in which ATP is synthesized.

23.8 What is substrate-level phosphorylation?

23.9 How is NAD^+ regenerated in glycolysis in animal cells?

23.10 How is NAD^+ regenerated in glycolysis in yeast cells?

23.11 What would be the effect on glycolysis if inorganic phosphate were unavailable?

Formation of Acetyl-*S*-CoA

23.12 What reaction does pyruvate undergo when oxygen is available within a cell?

23.13 Where does the reaction in Exercise 23.12 take place?

23.14 What are the names of the four vitamins having roles in the oxidation and decarboxylation of pyruvate?

23.15 Is the pyruvate dehydrogenase complex under regulatory control? Explain your answer.

Citric Acid Cycle

23.16 What is the first reaction of the citric acid cycle?

23.17 In which of the reactions of the citric acid cycle is CO_2 generated?

23.18 Does a substrate-level phosphorylation take place in the citric acid cycle? Explain your answer.

23.19 Why is the citric acid cycle called a cycle rather than a pathway?

23.20 Is there an enzyme complex in the citric acid cycle similar to the pyruvate dehydrogenase complex? Explain your answer.

23.21 Explain why the citric acid cycle is sometimes referred to as a catalytic cycle.

23.22 Does the concentration of oxaloacetate have an effect on the overall rate of the citric acid cycle? Explain your answer.

23.23 How is oxaloacetate replaced after its concentration has been reduced by conversion into other biomolecules?

Gluconeogenesis

23.24 Is gluconeogenesis simply the reverse of glycolysis? Explain your answer.

23.25 How is phosphoenolpyruvate synthesized?

23.26 What is the cofactor in enzymatic carboxylation reactions?

23.27 How does oxaloacetate, synthesized in mitochondria, enter the cytosol?

23.28 Can muscle or brain supply free glucose to the blood? Explain your answer.

Glycogenesis

23.29 What is glycogen?

23.30 Can glycogen be synthesized by brain cells? Explain your answer.

23.31 What is the important metabolic function of pyrophosphate, $P_2O_7^{4-}$?

23.32 In what activated form is glucose added to a growing glycogen chain?

Glycogenolysis

23.33 Glucose is removed from a glycogen polymer by phosphorolysis. Compare that process with hydrolysis.

23.34 Is the enzyme that catalyzes glycogenolysis always present in its active form?

23.35 What are glucagon and epinephrine?

23.36 Explain the roles of glucagon and epinephrine in glycogenolysis.

23.37 Does either glucagon or epinephrine have any function other than in glycogenolysis?

23.38 What are the first and second messengers in glycogenolysis?

23.39 Why are liver cells the only cells that can produce free glucose from glucose-6-PO_4?

Electron-Transport Chain

23.40 What is meant by oxidative phosphorylation?

23.41 At how many points along the electron-transport chain can ATP be synthesized?

23.42 What are the cytochromes?

23.43 Can ATP synthesis take place in mitochondria whose inner membranes have been mechanically disrupted?

23.44 What fundamental process taking place in mitochondria provides the driving force for the synthesis of ATP?

23.45 It is possible for any cellular substance to diffuse into the inner mitochondrial space? Explain your answer.

23.46 ATP is used outside and is synthesized inside the mitochondria. Explain how this is accomplished.

Energy Yield from Carbohydrate Catabolism

23.47 What is the total amount of ATP produced under aerobic conditions per mole of glucose?

23.48 Compare the energy yields of glycolysis and the citric acid cycle.

23.49 Why is glycolysis the preferred carbohydrate catabolic pathway in active skeletal muscle cells?

Chemical Connections

23.50 Is it possible for a liver cell to synthesize glucose by reversing glycolysis?

23.51 What are the two critical conditions that must be fulfilled if muscle cells are to produce ATP during intense activity?

23.52 True or false: When a cell is in a high-energy state, pyruvate dehydrogenase becomes active. Explain your answer.

23.53 If NADH cannot penetrate the inner mitochondrial membrane, explain how it can deliver its reducing power to the electron-transport chain.

23.54 Why does the dehydrogenation of succinate lead to the synthesis of only 2 mol of ATP?

CHAPTER 24

FATTY ACID METABOLISM

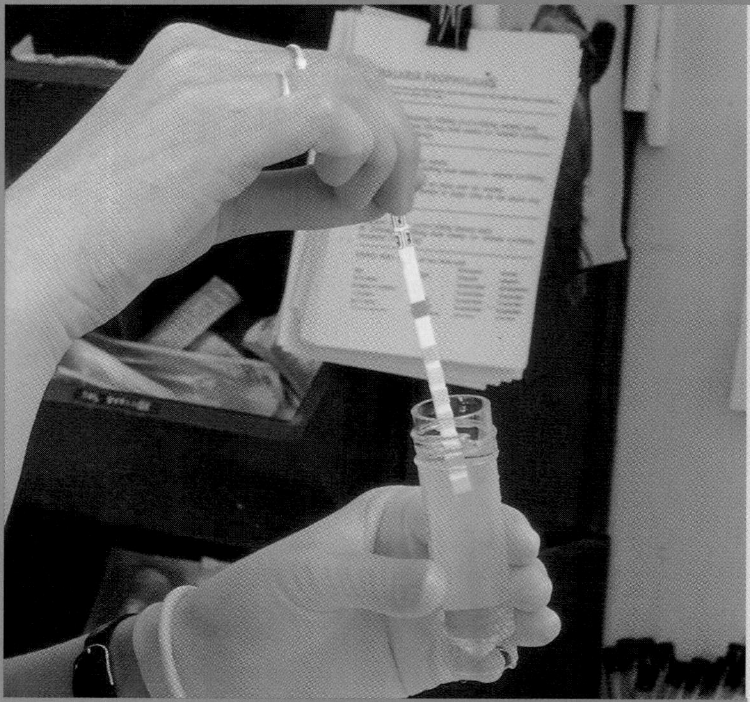

CHEMISTRY IN YOUR FUTURE

A patient admitted yesterday evening in a diabetic coma has recovered quite rapidly, and this morning his levels of blood glucose, urinary glucose, and urinary ketones, as well as his mental status, are almost normal. Nevertheless, you continue to monitor him carefully. Most of the urine and blood analyses are performed in the hospital lab, but you have been doing spot tests for ketonuria (ketones in the urine) by immersing a reagent test strip into a urine sample and observing the extent of the reagent's color change. In presenting the metabolism of fatty acids, Chapter 24 describes ketone bodies and explains why they are sometimes excreted in the urine.

LEARNING OBJECTIVES

- Describe how fatty acids are activated and transported from the cytosol into the mitochondria.
- Describe the fundamental steps by which a fatty acid chain is shortened by two carbons.
- Describe ketone bodies and how they are used in cellular metabolism.
- Describe the circumstances under which fatty acid biosynthesis is initiated.
- List the sequence of reactions by which a fatty acid chain is elongated by two carbons.
- Show how triacylglycerols are synthesized and describe their principal cellular function.
- Describe the biosynthesis of membrane lipids.

In this chapter, we will explore how fatty acids (Section 19.2) function as a major energy source for the cell, how they are synthesized, and how they serve as the building blocks for the biosynthesis of triacylglycerols, membrane lipids, and steroids. In addition, we will look at some health problems related to the metabolism of membrane lipids.

Both fatty acids, stored as triacylglycerols (Section 19.3) in adipose cells, and glucose, stored as glycogen (Section 18.6) in both the liver and in muscle, serve as sources of energy. In addition to serving as a source of energy, glycogen in the liver is used to maintain a constant blood concentration of glucose. The amount of glycogen stored in the liver and muscle of a 70-kg man is about 400 g. His body needs about 2400 kcal/day but, at 4 kcal/gram, 400 g of glycogen cannot supply all of that need. Both the need for energy and the need to spare glycogen for blood-glucose maintenance are satisfied by the use of stored triacylglycerols.

When we eat more carbohydrate than is required to satisfy our immediate energy needs, the liver and muscle cells convert the excess into glycogen. When the capacity to store glycogen is reached, any remaining carbohydrate is transformed into fatty acids and then stored as triacylglycerol. In contrast with the limitations of glycogen storage, there is no apparent limit to the amount of triacylglycerols that can be stored, and, as a result, the biosynthesis of triacylglycerols in animals, including humans, proceeds very actively.

To understand how good an energy source triacylglycerols are, consider that the average male human stores about 17% of body weight as triacylglycerols. Oxidation of triacylglycerols yields about 9 kcal of energy per gram; therefore the stored lipid provides a 6-week supply of energy for a 70-kg man who consumes about 2400 kcal/day. The need for triacylglycerols arises between meals, when the hormone glucagon is released in response to the lowered glucose concentration in the blood (Section 23.6). We begin our presentation of fatty acid metabolism with a consideration of the release of fatty acids from storage.

24.1 FATTY ACID MOBILIZATION

Triacylglycerols, commonly called triglycerides (Section 19.3), are the fats stored in fat cells. Mobilization of triacylglycerols from storage takes place principally through the action of the hormone glucagon. Between meals, glucagon is released in response to lowered blood-glucose concentration and initiates a sequence of enzymatic reactions that activate cellular triacylglycerol lipase. This enzyme hydrolyzes the triacylglycerols and frees fatty acids, along with glycerol, for oxidative processes.

The mechanism of action of glucagon in fatty acid mobilization is similar to that which results in the activation of phosphorylase in glycogenolysis

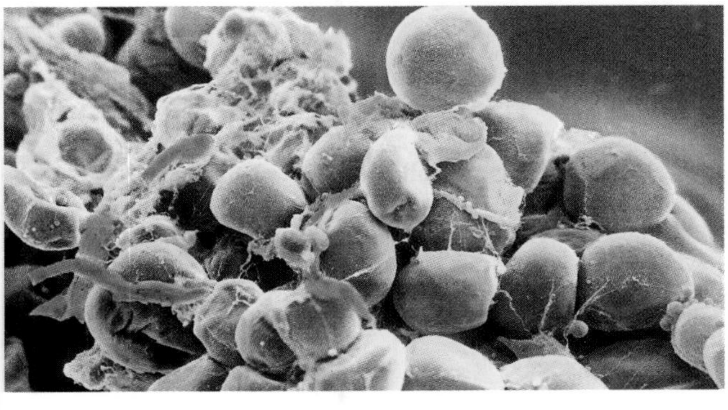

Adipose tissue. Colored scanning electron micrograph of adipocytes (yellow), which build up the adipose connective tissue. Connective tissue fibers at top left support the fat cells. Almost the entire volume of each adipocyte is occupied by a single lipid droplet of triacylglycerols. Fat to be stored arrives at these cells through small capillaries, seen here as blue tubules.

(Section 23.6). The free fatty acids leave the adipose-tissue cells and become bound to serum albumin. Water-soluble glycerol and the protein-bound fatty acids then circulate to other organs to be utilized as energy sources.

Glycerol is processed through the glycolytic pathway. It is first phosphorylated by ATP, then oxidized to dihydroxyacetone phosphate, and subsequently isomerized into glyceraldehyde-3-phosphate to form pyruvate. Pyruvate can then be oxidized through the citric acid cycle or converted into glucose through gluconeogenesis.

The fatty acids take a more complex route. Fatty acids can be oxidized only when in the form of an acyl-*S*-CoA thioester. This transformation, which takes place in the cytosol, requires the use of 2 mol of ATP. The fatty acid used in the following discussion is palmitic acid, a 16-carbon fatty acid. At the pH of the cell, the acid exists as palmitate and the activation reaction is:

$$\underset{\textbf{Palmitate}}{CH_3(CH_2)_{14}COO^-} + ATP + CoA{-}SH \longrightarrow \underset{\textbf{Palmitoyl-}\textit{S}\textbf{-CoA}}{CH_3(CH_2)_{14}CO{-}S{-}CoA} + AMP + PP_i$$

As mentioned in Chapter 23, PP_i represents pyrophosphate, $P_2O_7^{4-}$, derived from the hydrolysis of ATP. We have seen this type of process in the synthesis of glycogen (Section 23.9). The subsequent hydrolysis of pyrophosphate catalyzed by the enzyme pyrophosphatase is characterized by a very large equilibrium constant, thus providing the driving force for the overall reaction.

As described in the preceding equation, the activation process appears to require only 1 mol of ATP, when, in fact, 2 mol of ATP are consumed. The reason is that AMP (adenosine monophosphate) cannot be phosphorylated by either substrate-level or oxidative phosphorylation. It can, however, react with another mole of ATP, catalyzed by the enzyme adenylate kinase, to form 2 mol of ADP:

$$AMP + ATP \longrightarrow 2\ ADP$$

The ADP thus generated now can be phosphorylated by substrate-level or oxidative phosphorylation to form ATP. The result of adding the two equations is:

$$\underset{\textbf{Palmitate}}{CH_3(CH_2)_{14}COO^-} + 2\ ATP + CoA{-}SH \longrightarrow \underset{\textbf{Palmitoyl-}\textit{S}\textbf{-CoA}}{CH_3(CH_2)_{14}CO{-}S{-}CoA} + 2\ ADP + 2\ P_i$$

Oxidation of the fatty acid must take place in mitochondria, but the acyl-*S*-CoA derivative cannot penetrate the inner mitochondrial membrane. Therefore, in the next step, the CoA derivative exchanges its CoA part for another substance that confers membrane (lipid) solubility upon the activated fatty acid. The substance is carnitine. In a reaction catalyzed by the enzyme carnitine transferase, the thioacyl ester is exchanged for a carboxylic ester at the OH group of carnitine:

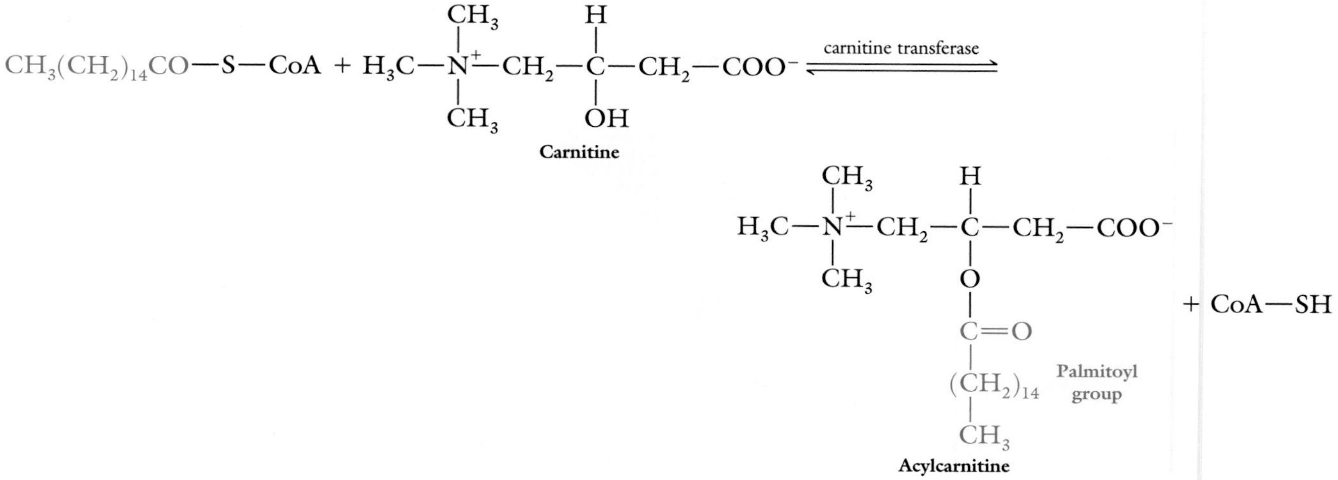

The acylcarnitine molecule passes through the inner mitochondrial membrane into the matrix. Once there, it reacts with mitochondrial CoA-SH, catalyzed by a mitochondrial form of carnitine acyltransferase, to reform the acyl-*S*-CoA derivative in the reverse of the preceding reaction. The oxidation process can now take place.

The pools of cytosolic and intramitochondrial CoA-SH are kept separated, as are those of ATP and NAD^+. Compartmentalization keeps regulatory control of anabolic processes separate from and independent of the regulatory controls of catabolic processes.

Concept checklist

- ✔ Fatty acids released from adipose tissue are transported in the blood to other tissues as fatty acid–serum albumin complexes.
- ✔ Before they can be oxidized, fatty acids must be converted into acetyl-*S*-CoA derivatives, and this process takes place in the cytosol.
- ✔ To leave the cytosol and enter mitochondria where oxidation takes place, the acetyl-*S*-CoA derivatives must be transformed into carnitine derivatives.

24.2 FATTY ACID OXIDATION

In stage 1 of fatty acid oxidation, which consist of four steps, the products of all fatty acids are acetyl-*S*-CoA and NADH and $FADH_2$. In stage 2, those products enter the citric acid cycle and the electron-transport chain for ultimate oxidation to CO_2 and H_2O.

Step 1. The first step in the oxidation of the 16-carbon fatty acid palmitate is a dehydrogenation between carbons 2 and 3 of the acyl-*S*-CoA derivative to produce the trans-unsaturated acyl-*S*-CoA derivative of palmitate. In the following reaction equation, the double bond between carbons 2 and 3 is indicated by the symbol Δ^2, in which Δ signifies a double bond and the superscript specifies that the double bond is located between carbons 2 and 3, counting from the carboxyl carbon. The coenzyme reduced in this reaction is FAD (Section 22.4 and Figure 22.6), and the enzyme is acyl-*S*-CoA dehydrogenase. With carbon atoms 2 and 3 identified, the reaction is:

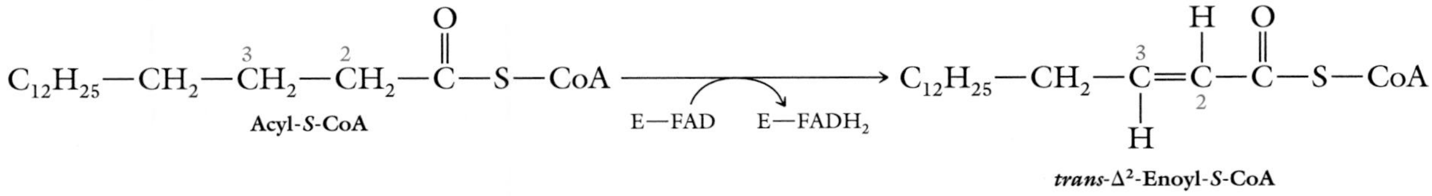

The E-$FADH_2$ delivers its reducing power to a flavoprotein, which passes it along to ubiquinone from which it moves on to complex II of the electron-transport chain (Section 23.8). This leads to the production of 2 mol of ATP.

Step 2. The next step is an enzymatic hydration of the trans double bond to form a 3-hydroxyacyl-*S*-CoA. The enzyme, enoyl-*S*-CoA hydratase, catalyzes the hydration of the double bond between carbon atoms 2 and 3 of this fatty acid derivative, leading to the L-hydroxy configuration. The reaction is:

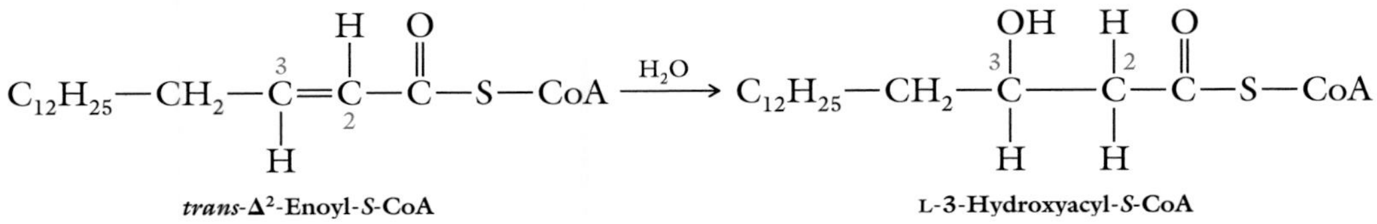

The synthesis of fatty acids takes place in the cytosol and is thus separated from their oxidation in the mitochondria. In the synthesis, the catabolic sequence is largely reversed (Section 24.4) Here it is important to note why we stress the stereochemical L-configuration of the 3-hydroxyacyl-*S*-CoA product of the oxidative pathway. The use of a different reductase in the cytosolic synthetic process that yields the D-hydroxy configuration also keeps control of and separates the synthetic and oxidative pathways.

Step 3. The next step is the dehydrogenation of L-3-hydroxyacyl-*S*-CoA by L-3-hydroxyacyl-*S*-CoA dehydrogenase, with NAD^+ as cofactor, to produce the 3-ketoacyl-*S*-CoA. The reaction is:

$$C_{12}H_{25}-CH_2-\underset{\underset{H}{|}}{\overset{\overset{OH}{|}}{C}}_3-CH_2-\overset{\overset{O}{\|}}{C}_2-S-CoA \xrightarrow[\text{NAD}^+ \quad \text{NADH}]{} C_{12}H_{25}-CH_2-\overset{\overset{O}{\|}}{C}_3-CH_2-\overset{\overset{O}{\|}}{C}_2-S-CoA$$

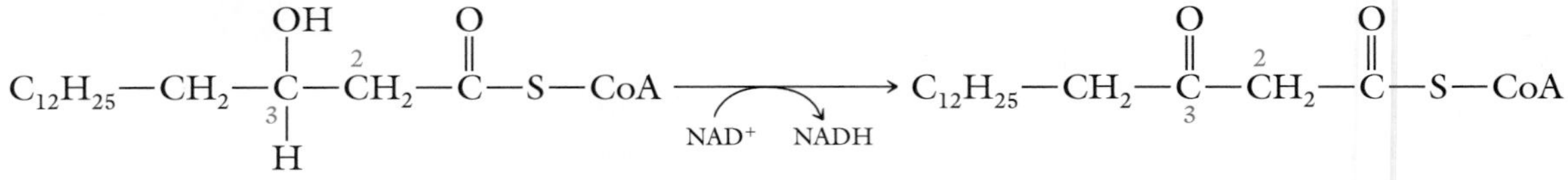

L-3-Hydroxyacyl-*S*-CoA   3-Ketoacyl-*S*-CoA

The dehydrogenase delivers its reducing power to complex I of the electron-transport chain (Section 23.12), thereby producing 3 mol of ATP per mole of NADH.

Step 4. At the fourth and last step of the oxidation cycle, an enzymatic cleavage takes place that requires a molecule of CoA-SH. The enzyme catalyzing this reaction is 3-ketothiolase. The result is the production of a molecule of acetyl-*S*-CoA derived from carbons 1 and 2 of the original fatty acid, along with another fatty acyl-*S*-CoA molecule having two fewer carbon atoms than the original acyl-*S*-CoA derivative. The reaction is:

$$C_{12}H_{25}-CH_2-\overset{\overset{O}{\|}}{C}_3-CH_2-\overset{\overset{O}{\|}}{C}_2-S-CoA \xrightarrow{\text{CoA—SH}} C_{12}H_{25}-CH_2-\overset{\overset{O}{\|}}{C}_3-S-CoA + CH_3-\overset{\overset{O}{\|}}{C}_2-S-CoA$$

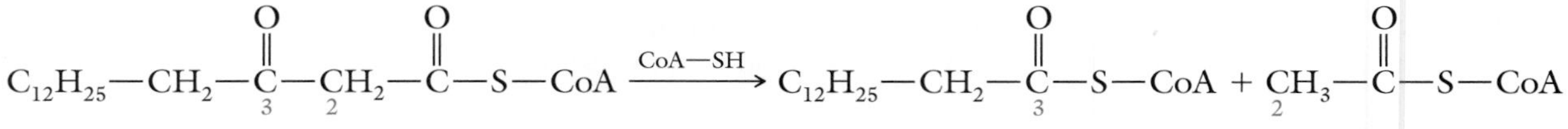

3-Keto(C_{16})acyl-*S*-CoA   (C_{14})Acyl-*S*-CoA   Acetyl-*S*-CoA

The new, shortened acyl-*S*-CoA undergoes the same sequence of reactions, ending again with the production of a second molecule of acetyl-*S*-CoA and another acyl-*S*-CoA derivative with two fewer carbon atoms in the chain. The reaction equation for one oxidation cycle consisting of steps 1 through 4 and beginning with palmitic acid (C_{16}) thioester is:

$$C_{16}\text{-}S\text{-CoA} + \text{CoA-SH} + \text{FAD} + \text{NAD}^+ + H_2O \longrightarrow$$
$$C_{14}\text{-}S\text{-CoA} + \text{acetyl-}S\text{-CoA} + \text{FADH}_2 + \text{NADH} + H^+$$

At each passage through this sequence of reactions, the fatty acid chain loses two carbons as acetyl-*S*-CoA; so, if we start with palmitoyl-*S*-CoA, a 16-carbon chain acid, seven of these sequences will take place. Furthermore, because the cycle begins with a CoA derivative, only seven more molecules of CoA-SH are required to produce overall eight molecules of acetyl-*S*-CoA, as indicated in the following sketch:

$$\overset{8}{C}-C \overset{7}{\vert} C-C \overset{6}{\vert} C-C \overset{5}{\vert} C-C \overset{4}{\vert} C-C \overset{3}{\vert} C-C \overset{2}{\vert} C-C \overset{1}{\vert} C-C-S-CoA$$

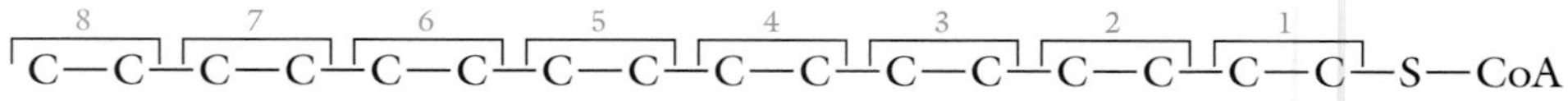

In addition, 14 pairs of hydrogen atoms are removed from palmitoyl-*S*-CoA, of which 7 emerge as $FADH_2$ and 7 as $NADH + H^+$. The equation for seven cycles that reduce the C_{16} acid to eight molecules of acetyl-*S*-CoA is:

$$\text{Palmitoyl-}S\text{-CoA} + 7 \text{ CoA-SH} + 7 \text{ FAD} + 7 \text{ NAD}^+ + 7 H_2O \longrightarrow$$
$$8 \text{ acetyl-}S\text{-CoA} + 7 \text{ FADH}_2 + 7 \text{ NADH} + 7 H^+$$

Each $FADH_2$ produces 2 ATP, and each NADH produces 3 ATP; so the dehydrogenations leading to the formation of acetyl-*S*-CoA, the first stage of fatty acid oxidation, result in the production of 35 ATP. The oxidation of unsaturated fatty acids skips step 1 and proceeds through steps 2 through 4.

The second stage of fatty acid oxidation is the disposal of the acetyl-*S*-CoA produced in the first stage. That acetyl-*S*-CoA is identical with the acetyl-*S*-CoA produced by the dehydrogenation of pyruvate (Section 23.4). Much of it enters the citric acid cycle and is oxidized to CO_2 and water (but see Section 24.3).

Table 23.4 shows that the oxidation of 1 mol of acetyl-*S*-CoA results in the production of 12 ATP. Therefore 8 mol of acetyl-*S*-CoA yield 96 ATP. The net production of ATP from the first and second stages of fatty acid oxidation is 35 ATP + 96 ATP − 2 ATP (activation step) = 129 ATP per mole of palmitate.

24.3 KETONE BODIES AND CHOLESTEROL

In the mitochondria of the human liver, acetyl-*S*-CoA is transformed in two ways: oxidation by the citric acid cycle as indicated in Section 24.2 and, in a second pathway, the formation of **ketone bodies,** a group of compounds consisting of acetoacetate, D-3-hydroxybutyrate, and acetone. The ketone bodies arise as an overflow from fatty acid oxidation. They are normally present and make up about 10% of the metabolic energy supply. However, they are present in high concentrations in untreated diabetes mellitus (Section 26.4, and Box 20.3). In diabetes mellitus, peripheral tissues fail to oxidize glucose, because it cannot penetrate into cells, and so fatty acids must be oxidized to compensate for the unavailable energy. Under those circumstances, the excess acetyl-*S*-CoA is converted into ketone bodies, a condition known as **ketosis.** Starvation is another condition that leads to ketosis. In starvation, oxaloacetate is necessarily very low in concentration. Therefore, the acetyl-*S*-CoA produced from fatty acid oxidation cannot react with oxaloacetate to form citric acid and so must go the route of ketone body formation.

The formation of acetoacetate takes place in liver mitochondria and consists of the reaction of 2 mol of acetyl-*S*-CoA to form acetoacetyl-*S*-CoA, catalyzed by the enzyme thiolase:

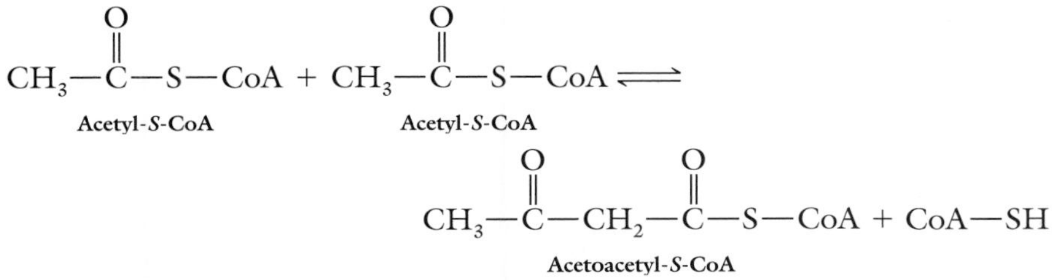

Next, to form acetoacetate, the acetoacetyl-*S*-CoA must lose its CoA component. This loss takes place in two steps, the first being the reaction of acetoacetyl-*S*-CoA with acetyl-*S*-CoA to form 3-hydroxy-3-methylglutaryl-*S*-CoA (HMG-*S*-CoA).

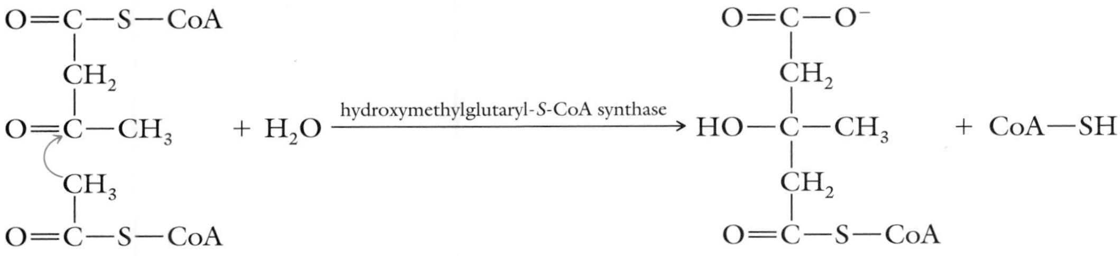

HMG-*S*-CoA

In the next step, acetoacetate is cleaved from HMG-*S*-CoA by hydroxymethylglutaryl-*S*-CoA lyase:

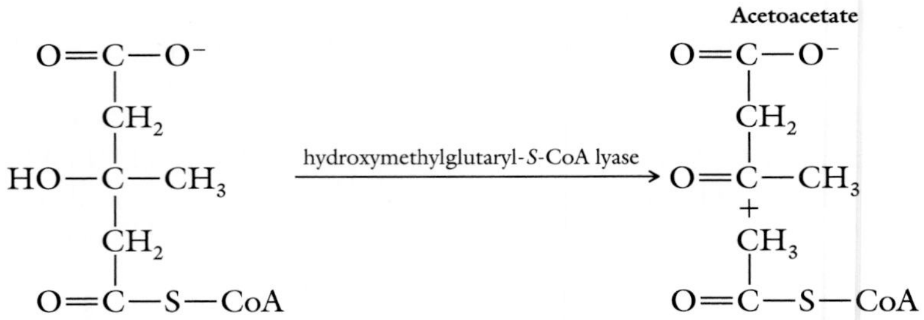

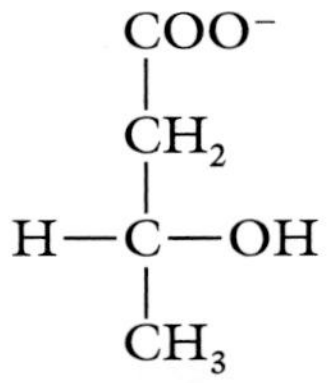

D-3-Hydroxybutyrate

Acetoacetate is reduced by D-3-hydroxybutyrate dehydrogenase to produce D-3-hydroxybutyrate, which is transported to peripheral tissues. There, by another complex series of reactions, D-3-hydroxybutyrate is converted into acetyl-*S*-CoA, which is oxidized in the citric acid cycle, supplying cardiac and skeletal muscle with as much as 10% of their daily energy requirements. In starvation, D-3-hydroxybutyrate can supply as much as 75% of the brain's energy requirement. **Acetoacetate** is unstable and can spontaneously or enzymatically lose its carboxyl group to form acetone, a volatile substance that can be detected in the breath of diabetics as a kind of sweet odor.

3-Hydroxy-3-methylglutaryl-*S*-CoA is also synthesized in the cytosol by a different group of enzymes. As the precursor molecule central to the synthesis of many important biomolecules such as cholesterol and vitamins A, E, and K, HMG-*S*-CoA plays a central role in lipid metabolism. Biosynthesis of cholesterol, for example, begins with the reduction of 3-hydroxy-3-methylglutaryl-*S*-CoA to form mevalonic acid catalyzed by HMG-*S*-CoA reductase. That reduction step controls the overall rate of cholesterol biosynthesis, because the reductase is a regulatory enzyme that exists in both phosphorylated (inactive) and dephosphorylated (active) forms. In cases of uncontrolled high concentrations of cholesterol, drugs such as lovastatin (Box 24.1) inhibit the reductase and block cholesterol's synthesis.

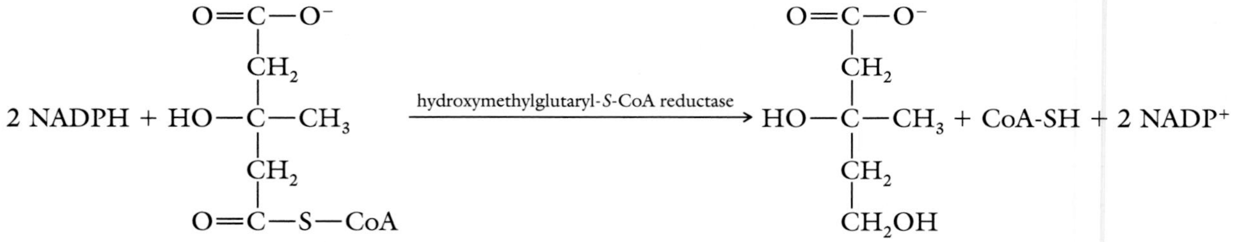

HMG-*S*-CoA Mevalonate

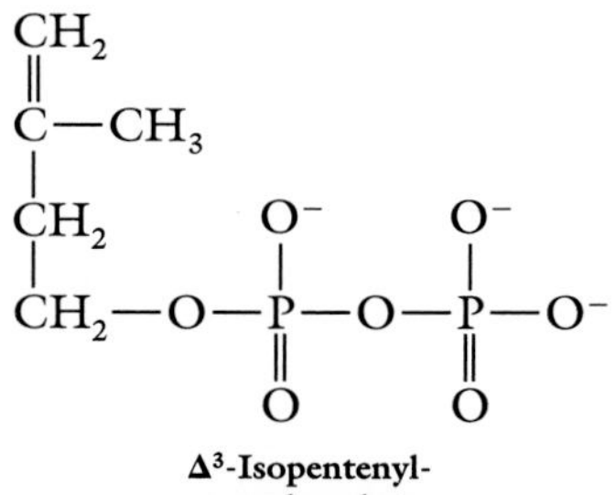

Δ^3-Isopentenyl-pyrophosphate

The mevalonate is converted by a complex series of reactions into Δ^3-isopentenylpyrophosphate. Six moles of this compound are polymerized and cyclized to form lanosterol, the precursor of cholesterol. Cholesterol (Section 19.7) is the precursor of the steroid hormones, bile salts, and vitamin D. Excessive cholesterol production leads to the formation of deposits, called plaque, on the internal surfaces of arteries and is thought to be a major factor in heart disease (see Box 24.1). Δ^3-Isopentenylpyrophosphate is the basic building block of vitamins A, E, and K and of natural rubber. In polymerized form, it forms the long hydrocarbon chain of ubiquinone (Section 23.7).

24.4 BIOSYNTHESIS OF FATTY ACIDS

The synthesis of fatty acids essentially utilizes the same chemical reactions of fatty acid catabolism. However, fatty acid synthesis takes place in the cytosol rather than in mitochondria and it requires a different activation mechanism and different enzymes and coenzymes. Fatty acids are synthesized when the cell is in an energy-rich state; that is, when there is an abundance of ATP, NADPH (Section 23.1), and acetyl-*S*-CoA.

The process begins with the entry of acetyl-*S*-CoA into the cytosol from the mitochondria. Acetyl-*S*-CoA cannot penetrate the mitochondrial membrane, but citrate can. Citrate leaves the mitochondrion through a tricarboxylate-transport system (Section 26.14) and, in the cytosol, is cleaved into acetyl-*S*-CoA and oxaloacetate by the enzyme citrate lyase:

$$\text{Citrate} + \text{ATP} + \text{CoA-SH} \longrightarrow \text{acetyl-}S\text{-CoA} + \text{oxaloacetate} + \text{ADP} + \text{P}_i$$

Citrate is also an allosteric signal for the activation of the next enzyme in the overall process, acetyl-*S*-CoA carboxylase.

Acetyl-*S*-CoA carboxylase catalyzes the rate-limiting step in fatty acid biosynthesis. In the presence of citrate, the enzyme forms a highly active, filamentous polymeric structure and catalyzes the formation of malonyl-*S*-CoA from acetyl-*S*-CoA by a carboxylation in which biotin is the cofactor. A similar carboxylation reaction initiates gluconeogenesis (Section 22.2).

$$\underset{\text{Acetyl-}S\text{-CoA}}{CH_3-\overset{\overset{\displaystyle O}{\|}}{C}-S-CoA} + CO_2 + ATP + H_2O \xrightarrow{\text{acetyl-}S\text{-CoA carboxylase}}$$

$$\underset{\text{Malonyl-}S\text{-CoA}}{{}^{-}OOC-CH_2-\overset{\overset{\displaystyle O}{\|}}{C}-S-CoA} + ADP + P_i + H^+$$

After its formation, malonyl-*S*-CoA then reacts with the fatty acid synthase complex.

In bacteria, the fatty acid synthase complex consists of seven different enzymes. In animal cells, however, these enzymatic activities take place at specialized regions, or catalytic domains, of a single multifunctional polypeptide chain.

◄◄ The advantage of a multifunctional enzyme complex is described in Section 22.8.

Acyl groups are esterified to CoA-SH in the oxidation of fatty acids, but, in fatty acid biosynthesis, acyl groups are esterified to —SH groups on the protein of the synthase complex. This protein is called the **acyl carrier protein,** abbreviated **ACP.** The special feature of the acyl carrier protein is the presence of two unique —SH groups that have different functions. We will use a simple method to differentiate these —SH groups. One of the —SH groups will be designated the **α-SH** group. It is located on a pantotheine residue as in CoA-SH (Figure 23.2) but is attached directly to the protein rather than to adenine. The second —SH group will be designated the **β-SH** group. It is a part of a specific cysteine residue (Section 20.4) of the synthase complex.

The chain-lengthening reaction begins with the reaction of acetyl-*S*-CoA with the α-SH group, catalyzed by acetyl-ACP transferase. In a companion reaction, malonyl-*S*-CoA reacts with the β-SH group, catalyzed by malonyl-ACP transferase.

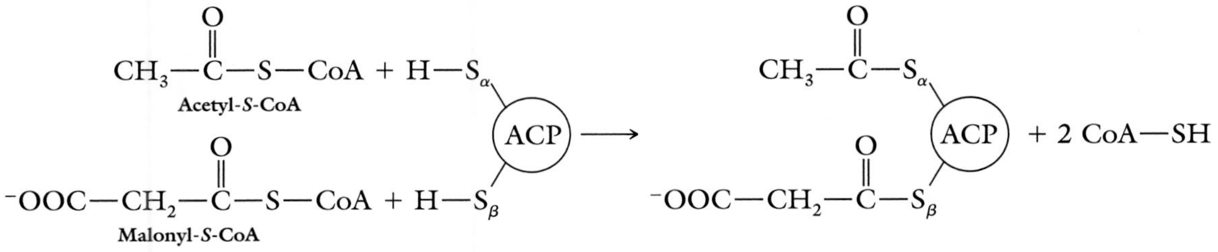

The next reaction is the first of four steps in the lengthening of the fatty acid chain. The acetyl group bound to the α-SH group is transferred to the

24.1 Chemistry Within Us

Atherosclerosis

Cardiovascular (blood-vessel) disease is the single largest killer in the developed Western world, being responsible for about half the deaths each year. Death occurs by heart attack or stroke, the end result of atherosclerosis (hardening of the arteries). Atherosclerosis is called the silent killer because it begins and progresses for many years before symptoms are present.

Atherosclerosis is a buildup of cholesterol-rich, fatty deposits called **plaque** on the interior endothelial lining of arteries. Cholesterol is cleared from the bloodstream when the lipid-transport system (Section 26.10) is working properly and the dietary intake of cholesterol is not excessive. Plaque develops in response to damaged arterial lining resulting from chronic high blood pressure **(hypertension),** stress, smoking, a high-fat and high-cholesterol diet, as well as genetic and other factors. White blood cells migrate into damaged endothelial tissue in an attempt to repair the lesion. Growth factors are released that cause proliferation of tissue at the site of damage, and this proliferation is followed by the deposition of lipids, mostly cholesteryl esters. Fibrous connective tissue including calcium deposits cap off the thickened area to yield narrowed and roughened arterial walls that are much less elastic.

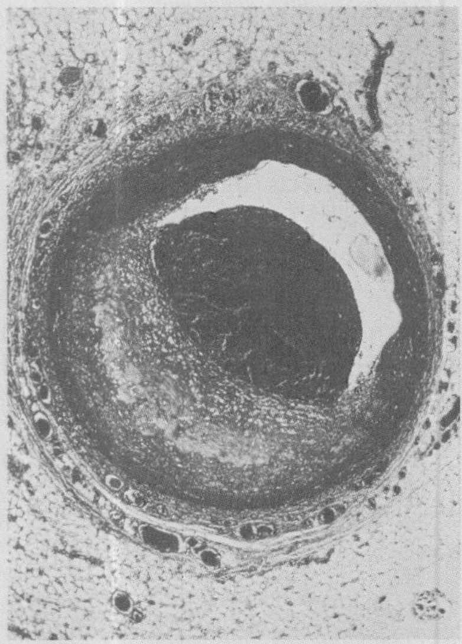
Light micrograph showing atherosclerosis in coronary artery. Most of the artery is blocked by plaque deposits.

Plaque buildup promotes the formation of a blood clot **(thrombus).** Blood flow becomes turbulent when blood flows over plaque. The roughed plaque damages blood platelets and acts as the nucleus for clot formation, which further narrows the artery. The coronary arteries—the arteries that feed the heart—are especially prone to narrowing and clot formation. A **heart attack (myocardial infarction)** with death of heart tissue results when a clot in a coronary artery **(coronary thrombosis)** stops blood flow. In a **stroke (cerebral infarction),** arteries in the brain are blocked (and often ruptured). Death of brain tissue results in muscle, sensory, or language impairment, depending on the part of the brain damaged. Strokes as well as damage to other tissues and organs can also occur when a piece of a coronary thrombus breaks loose, travels through the bloodstream, and lodges in a narrower artery. Plaque buildup also damages the cardiovascular system by further elevating the blood pressure. Hypertension weakens heart muscle (which thickens because of the strain) and decreases the efficiency of the heart's pumping action. Blood backs up into the heart and lungs, a condition called **congestive heart failure** that is often fatal.

Many epidemiological studies indicate a strong correlation between elevated cholesterol and lipoprotein levels in the blood and the incidence of cardiovascular disease. Various medical organizations and journals such as the American Heart Association and the *New England Journal of Medicine* have published guidelines for blood cholesterol and lipoproteins (Table 26.14). Experts believe that adults should maintain cholesterol levels below 200 mg/dL to minimize the risk of cardiovascular disease. **Low-density lipoprotein (LDL),** the main carrier of cholesterol in the blood, is sometimes called **"bad cholesterol."** LDL levels below 130 mg/dL are recommended. However, the relation between blood cholesterol and cardiovascular disease is very complex because **high-density lipoprotein (HDL),** sometimes called **good cholesterol,** protects against cardiovascular disease by returning cholesterol to the liver.

Heredity, diet, and life style have roles in the development of atherosclerosis. The importance of heredity is clearly evident given that there are people who consume a diet high in cholesterol and yet have blood-cholesterol levels below 200 mg/dL. Other people have high blood-cholesterol levels in spite of diets low in cholesterol. **Familial hypercholesterolemia** is a hereditary condition in which the blood-cholesterol level is grossly elevated (as high as 500 mg/dL and higher). Michael Brown and

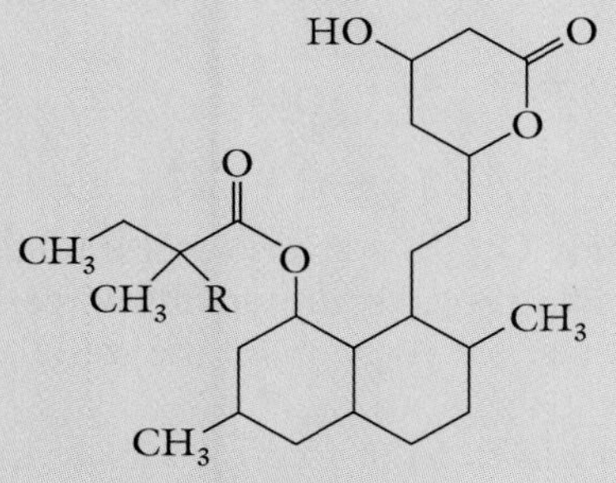

Lovastatin (R = H)
Simvastatin (R = CH_3)

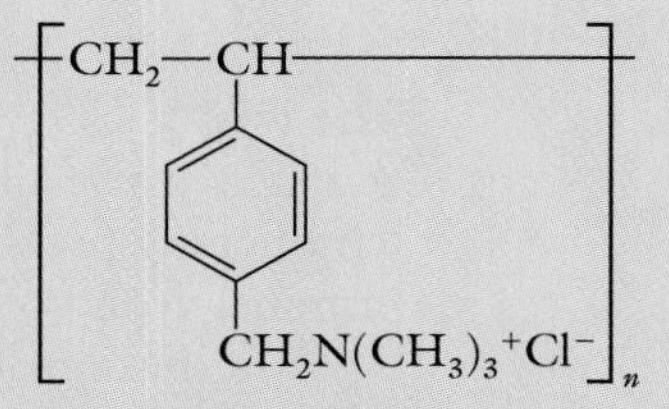

Cholestyramine resin

Joseph Goldstein were awarded the 1985 Nobel Prize in Physiology or Medicine for their study of this disorder. People with familial hypercholesterolemia have defective LDL receptors or an insufficient number of LDL receptors. Cholesterol is not taken up by cells, the feedback mechanism for shutting down endogenous synthesis of cholesterol does not function, and the cholesterol level in the blood becomes elevated. When dietary control is ineffective, the drugs lovastatin and simvastatin are useful for lowering cholesterol by inhibiting endogenous cholesterol synthesis through inhibition of the enzyme HMG-*S*-CoA reductase (Section 24.3). Also useful is cholestyramine resin (a synthetic polymer; Section 12.7), which binds bile salts in the intestine and thus promotes the conversion of more cholesterol into bile salts in the liver.

The extent to which blood cholesterol can be lowered by diet has limits, because cholesterol is essential for life and the body synthesizes cholesterol. However, diet clearly has an effect on the blood-cholesterol level. Prudent dietary recommendations are based on epidemiological correlations between diet and cardiovascular disease:

- Consume fewer animal products (meat and dairy products) relative to plant products (fruits, vegetables, and grains).
- Consume more chicken and fish relative to red meat.

These recommendations have two objectives. First, the total intake of lipids is lowered because meat, especially red meat, has a higher lipid content than do fruits, vegetables, and grains. Second, the amount of saturated fatty acids and cholesterol relative to unsaturated fatty acids is lowered because fruits and vegetables have no cholesterol and are lower in saturated fatty acids compared with meats. The amount of saturated fatty acids is also lowered because fish and chicken are lower in saturated fats relative to red meat.

Decreased lipid intake not only decreases the blood-cholesterol level, but also tends to reduce weight because lipids generate more than twice the energy per gram than does protein or carbohydrate. Lower lipid intake tends to translate into less excess (unmetabolized) lipid that is stored as fat (adipose) tissue. Decreased weight lowers the load on the heart and the rest of the cardiovascular system, decreasing the tendency for hypertension, lesions, and atherosclerosis. The specific effects of fatty acids are not simple. Saturated fatty acids increase cholesterol and LDL levels, whereas unsaturated fatty acids lower cholesterol and LDL levels. But not all unsaturated fatty acids have the same quantitative effect. ω-6-Polyunsaturated fatty acids (plant oils) are more effective than ω-6-polyunsaturated fatty acids (fish oils) in lowering cholesterol levels but less effective in lowering triacylglycerol levels. Monounsaturated fatty acids (olive oil) are more effective than polyunsaturated fatty acids in lowering LDL levels. Dietary fiber (Box 18.4) also lowers cholesterol levels.

Many other, poorly understood factors are important. Smoking and a stressful life style correlate with a higher incidence of atherosclerosis, perhaps by increasing arterial lesions. Aerobic exercise raises the level of HDL, the good cholesterol. Small amounts of alcohol, especially red wine, appear to decrease the tendency of cholesterol to adhere to lesions.

malonyl group at the β-SH site to form the acetoacetyl derivative. At the same time, CO_2 is displaced and lost as bicarbonate.

$$CH_3-\overset{O}{\overset{\|}{C}}-S_\alpha \quad \text{(ACP)} + H_2O \longrightarrow \quad H-S_\alpha \text{(ACP)} + HCO_3^-$$

$$^-OOC-CH_2-\overset{O}{\overset{\|}{C}}-S_\beta \qquad CH_3-\overset{O}{\overset{\|}{C}}-CH_2-\overset{O}{\overset{\|}{C}}-S_\beta$$

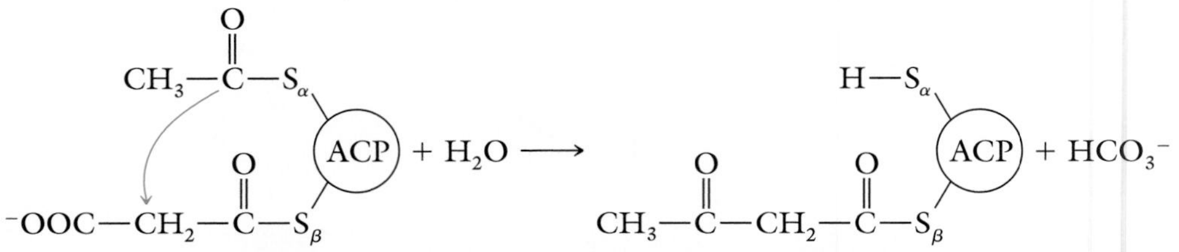

The loss of CO_2, as HCO_3^-, has a very large equilibrium constant, so the reaction is irreversible and provides the driving force for this step in the chain-lengthening process. The acetyl group on the α-SH group becomes the methyl terminal group of the growing chain.

The next reactions affect the four-carbon derivative covalently bound at the β-SH site. In the first of these reactions, the acetoacetyl-S_β-ACP-S_αH derivative is reduced to form the D-3-hydroxybutyryl derivative. The cofactor is NADPH, and the enzyme is 3-ketoacyl-ACP reductase. Note that the reductase uses NADPH rather than the NADH used in fatty acid oxidation. An additional contrasting feature is that the product is the D-derivative rather than the L-derivative seen in the fatty acid oxidation pathway.

$$CH_3-\overset{O}{\overset{\|}{C}}-CH_2-\overset{O}{\overset{\|}{C}}-S_\beta-ACP-S_\alpha H + NADPH + H^+ \longrightarrow$$

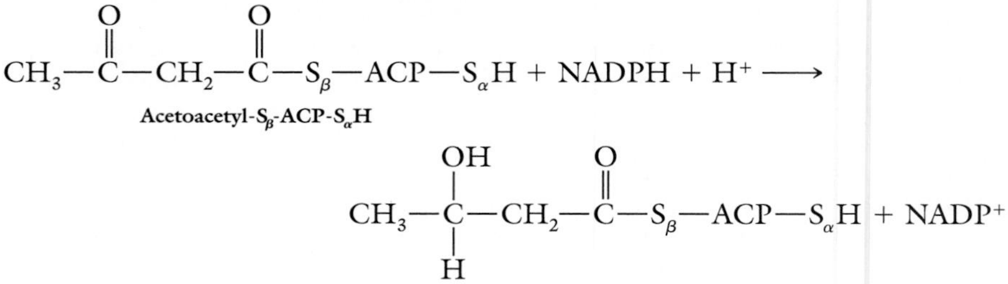

Acetoacetyl-S_β-ACP-S_αH

$$CH_3-\overset{OH}{\underset{H}{\overset{|}{\underset{|}{C}}}}-CH_2-\overset{O}{\overset{\|}{C}}-S_\beta-ACP-S_\alpha H + NADP^+$$

D-3-Hydroxyacyl-S_β-ACP-S_αH

The next step is a dehydration catalyzed by 3-hydroxyacyl-ACP-dehydratase, which produces the *trans-Δ^2-enoyl-S_β-ACP-S_αH* derivative.

$$CH_3-\overset{OH}{\underset{H}{\overset{|}{\underset{|}{C}}}}-CH_2-\overset{O}{\overset{\|}{C}}-S_\beta-ACP-S_\alpha H \xrightarrow{-H_2O} CH_3-\overset{H}{\overset{|}{C}}=\overset{}{\underset{H}{\overset{|}{C}}}-\overset{O}{\overset{\|}{C}}-S_\beta-ACP-S_\alpha H$$

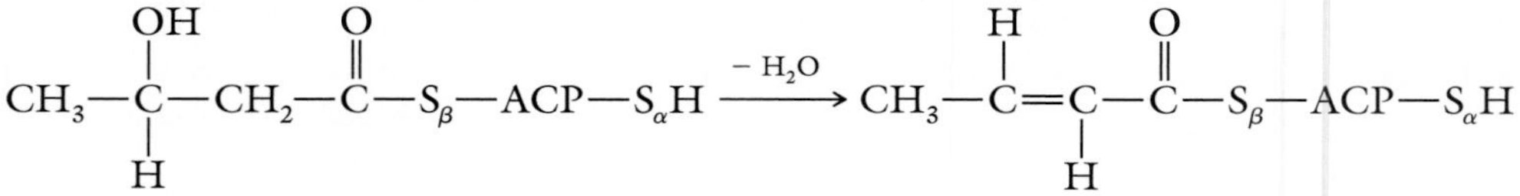

D-3-Hydroxyacyl-S_β-ACP-S_αH   ***trans-Δ^2-Enoyl-S_β-ACP-S_αH***

The dehydration reaction is followed by a reduction at the double bond to form the saturated acyl derivative. The enzyme is enoyl-S-ACP reductase, and the cofactor is NADPH.

$$CH_3-\overset{H}{\overset{|}{C}}=\overset{}{\underset{H}{\overset{|}{C}}}-\overset{O}{\overset{\|}{C}}-S_\beta-ACP-S_\alpha H + NADPH + H^+ \longrightarrow CH_3-\overset{H}{\underset{H}{\overset{|}{\underset{|}{C}}}}-\overset{H}{\underset{H}{\overset{|}{\underset{|}{C}}}}-\overset{O}{\overset{\|}{C}}-S_\beta-ACP-S_\alpha H + NADP^+$$

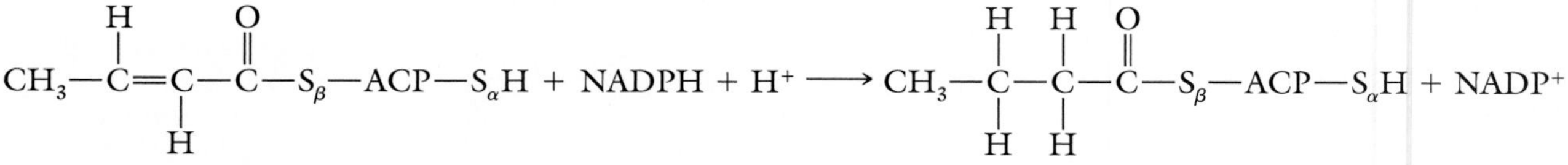

trans-Δ^2-Enoyl-S_β-ACP-S_αH   **Acyl-S_β-ACP-S_αH**

With the completion of this first round of reactions, the newly formed four-carbon saturated chain is transferred from the β-SH site to the α-SH site formerly occupied by acetyl-S-CoA.

$$H-S_\alpha \text{(ACP)} \qquad CH_3-CH_2-CH_2-\overset{O}{\overset{\|}{C}}-S_\alpha \text{(ACP)}$$

$$CH_3-CH_2-CH_2-\overset{O}{\overset{\|}{C}}-S_\beta \longrightarrow \qquad H-S_\beta$$

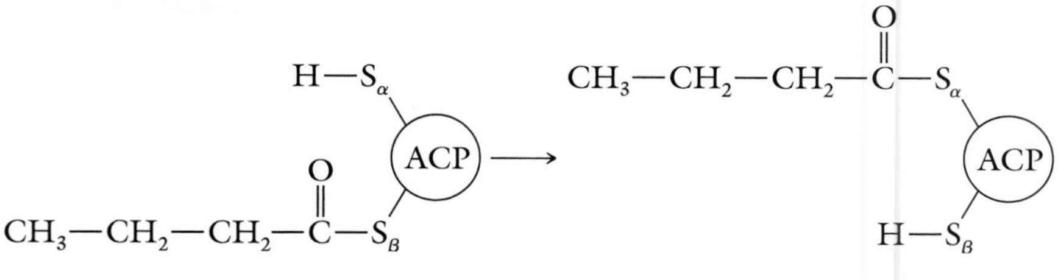

The next round of reactions begins with the addition of another malonyl-*S*-CoA to the now empty β-SH site. This reaction is followed by the transfer of the four-carbon chain at the α-SH site to the malonyl group, with the loss of CO_2 as before. We now have a six-carbon chain, and it will undergo the four reactions just detailed to yield a saturated chain lengthened by two carbons.

After seven rounds of addition of malonyl-*S*-CoA to the first molecule of acetyl-*S*-CoA, a molecule of palmitoyl-S_β-ACP-S_αH is produced. At this point, the chain-lengthening process ceases, and the free palmitate is released from the enzyme by the action of the hydrolytic enzyme palmitoyl thioesterase.

Because ionic bicarbonate appears as a product, the stoichiometry of palmitate synthesis, including charge balance, is:

$$\text{Acetyl-}S\text{-CoA} + 7\ \text{malonyl-}S\text{-CoA}^- + 14\ \text{NADPH} + 14\ \text{H}^+ \longrightarrow \text{palmitoyl-}S\text{-CoA} + 7\ \text{HCO}_3^- + 14\ \text{NADP}^+ + 7\ \text{CoA-SH}$$

When the ATPs required for the synthesis of seven molecules of malonyl-*S*-CoA derived from acetyl-*S*-CoA are taken into account,

$$7\ \text{Acetyl-}S\text{-CoA} + 7\ \text{HCO}_3^- + 7\ \text{ATP}^{4-} \longrightarrow 7\ \text{malonyl}^-\text{-}S\text{-CoA} + 7\ \text{ADP}^{3-} + 7\ \text{P}_i^{2-} + 7\ \text{H}^+$$

the overall stoichiometry for the synthesis of palmitoyl-*S*-CoA from acetyl-*S*-CoA is:

$$8\ \text{Acetyl-}S\text{-CoA} + 7\ \text{ATP}^{4-} + 14\ \text{NADPH} + 7\ \text{H}^+ \longrightarrow \text{palmitoyl-}S\text{-CoA} + 14\ \text{NADP}^+ + 7\ \text{CoA-SH} + 7\ \text{ADP}^{3-} + 7\ \text{P}_i^{2-}$$

Hormonal Regulation

The two hormones that regulate fatty acid biosynthesis are glucagon and insulin. Glucagon initiates a sequence of enzymatic reactions that phosphorylate and inactivate acetyl-*S*-CoA carboxylase. The mechanism is similar to the one that activates phosphorylase in glycogenolysis (Section 23.6). In its phosphorylated form, acetyl-*S*-CoA carboxylase is active only in the presence of high concentrations of citrate, so fatty acid biosynthesis virtually ceases. At the same time, triacylglycerol lipase is activated by phosphorylation and frees fatty acids for oxidative processes. Insulin, on the other hand, activates the phosphodiesterase that hydrolyzes cyclic AMP to noncyclic AMP, ending the cascade initiated by glucagon and allowing fatty acids to be synthesized and stored.

▶▶ **The roles of insulin and glucagon in the regulation of metabolic processes are summarized in Section 26.5.**

Further Processing of Palmitate

Chains longer than 16 carbon atoms—for example, the 18-carbon atom stearate—are made from palmitate by special elongation reactions in mitochondria or at the surface of the endoplasmic reticulum. Shorter chains are made when the growing fatty acid chain is released from the synthase complex before reaching its maximum 16-carbon length.

Eukaryotes can introduce a single cis double bond into stearate. It is a complex dehydrogenation requiring molecular oxygen and NADH, and it takes place at the surface of the endoplasmic reticulum with the formation of oleyl-*S*-CoA from stearoyl-*S*-CoA. Oleate is the only unsaturated fatty acid with the cis double bond as close to the methyl end (the 9,10 position of stearoyl-*S*-CoA) that animals can synthesize; however, animals require polyunsaturated fatty acids with double bonds even closer to the methyl end. These fatty acids are linoleic and linolenic acids, and they must be obtained through the diet. Such fatty acids are therefore called essential (Section 19.2). Mammals do, however, have the ability to introduce additional double bonds into the carbon chains of linoleic and linolenic acids between carbon 9 and the carboxyl carbon. In this way, mammals can synthesize arachidonic acid, a precursor of eicosanoids, a group of compounds derived from fatty acids that induce a wide variety of physiological responses (Section 19.8).

24.5 BIOSYNTHESIS OF TRIACYGLYCEROLS

The biosynthesis of cellular structures containing lipids begins with the fatty acids synthesized in the cytosol. The structures and properties of many of these lipids were considered in Sections 19.3, 19.4, and 19.6. In this section, we consider the biosynthesis of triacylglycerols.

Triacylglycerols, as we have seen, are the principal storage forms of fatty acids. Glycerol and fatty acids are the building blocks for synthesizing triacylglycerols. When biosynthesis and oxidation take place at about equal rates, the amount of fat in the body remains relatively constant. But, if the intake of carbohydrate, fat, or protein is in excess of required amounts, the excess calories are stored as triacylglycerols—in other words, as additional deposits of fat.

The rate of formation of triacylglycerols is strongly affected by the hormone insulin, which promotes the formation of triacylglycerols from carbohydrates (Section 24.4). Patients who suffer from diabetes mellitus cannot utilize glucose properly and are therefore unable to synthesize fatty acids and triacylglycerols from carbohydrates.

The precursors of triacylglycerols are fatty acyl-*S*-CoA and glycerol-3-phosphate (Section 23.7). The biosynthesis of triacylglycerols begins with acylation of the free hydroxy groups of glycerol-3-PO_4 with 2 mol of fatty acyl-*S*-CoA:

$$2 \text{ Fatty acyl-}S\text{-CoA} + \text{glycerol-3-PO}_4 \longrightarrow \text{diacylglycerol-3-PO}_4 + 2 \text{ CoA-SH}$$

The product diacylglycerol-3-PO_4, also known as **phosphatidic acid** (Section 19.6), exists as **phosphatidate** at cellular pH and is a key intermediate in membrane lipid biosynthesis.

In the next stage, phosphatidate is hydrolyzed to the diacylglycerol, which can then be acylated by a third fatty acyl-*S*-CoA to form the final product, triacylglycerol. Seven moles of ATP are required for this synthesis: 1 mol for phosphorylation of glycerol, and 2 mol for activation of each of three fatty acids to the fatty acyl-*S*-CoA form. The number of moles of ATP required for the total synthesis of tripalmitoylglycerol is:

$$\begin{aligned} 15 \text{ ATP/mole of palmitate} \times 3 \text{ mol of palmitate} &= 45 \text{ ATP} \\ \text{ATP/mole for tripalmitoylglycerol synthesis} &= \underline{\ 7 \text{ ATP}} \\ \text{Total moles of ATP per mole of tripalmitoylglycerol} &= 52 \text{ ATP} \end{aligned}$$

Although the synthesis of triacylglycerol entails a significant cost in ATP, the energy yield of triacylglycerol far outweighs the investment in its formation. To place the relative costs in perspective, let us calculate the ATP yield per mole of tripalmitoylglycerol oxidized. We can begin by noting that, as a result of oxidation, there are 8 mol of acetyl-*S*-CoA, 7 mol of NADH, and 7 mol of $FADH_2$ generated per mole of palmitate. The table below shows the ATP yield per mole of palmitate.

		ATP Yield
fatty acid activation	−2 ATP	−2
from electron transport	7 NADH	21
	7 $FADH_2$	14
from oxidation of 8 acetyl CoA	24 NADH	72
	8 $FADH_2$	16
from succinyl-CoA synthetase	8 GTP	8
Total ATP per mole of palmitate		129

Furthermore, the glycerol from triacylglycerol enters the glycolytic pathway after having been primed by ATP and therefore yields an additional 20 mol of ATP through the glycerol phosphate shuttle (or 22 mol ATP through the malate-aspartate shuttle). The total yield of ATP per mole of tripalmitoylglycerol oxidized is:

$$\begin{array}{lr} \text{Fatty acid} \quad 3 \times 129 = & 387 \\ \text{Glycerol} \qquad\qquad\quad = & \underline{20} \\ \text{Sum of ATP per mole} = & 407 \end{array}$$

Now subtract the 2 mol of ATP required for activation of each fatty acid to form the CoA derivative:

$$\begin{array}{rr} 3 \times 2 = & \underline{6} \\ \text{Total ATP yield per mole of tripalmitoylglycerol oxidized} = & 401 \end{array}$$

This yield is almost eight times the ATP cost of synthesizing the tripalmitoylglycerol.

It is interesting to compare our biochemical estimates of the energy available in lipids and carbohydrates with measured nutritional values (9.13 kcal/g for palmitic acid and 3.81 kcal/g for glucose). We can do so by comparing two ratios: (1) ATP yield per gram of tripalmitoylglycerol to ATP yield per gram of glucose and (2) the ratio of the measured nutritional energy of lipid to the measured nutritional energy of carbohydrate.

First, we calculate the ATP yield per gram of each compound.

Formula mass of tripalmitoylglycerol: 807.3 g/mol

$$407 \text{ mol ATP}/807.3 \text{ g} = 0.504 \text{ mol ATP/g}$$

Formula mass of glucose: 180.2 g/mol

$$38 \text{ mol ATP}/180.2 \text{ g} = 0.211 \text{ mol ATP/g}$$

Then we calculate the ratio of those two values:

$$\left(\frac{\dfrac{\text{mol ATP}}{\text{g TPG}}}{\dfrac{\text{mol ATP}}{\text{g glucose}}} \right) = \frac{0.504}{0.211} = 2.39$$

Finally, we calculate the ratio of nutritional energy values:

$$\frac{9.13 \text{ kcal/g palmitate}}{3.81 \text{ kcal/g glucose}} = 2.40$$

The closeness (less than 1% difference) between measured nutritional energy and that calculated from ATP yield suggests that metabolic energy can indeed be estimated by using the ATP yields from the stoichiometry of metabolic reactions.

24.6 BIOSYNTHESIS OF MEMBRANE LIPIDS

The structures of membrane lipids were presented in Section 19.6. They will not be described here except to repeat that it is useful to think of these lipids in the general sense illustrated in Figure 19.6; that is, as possessing a strongly polar or ionic head group, represented by a sphere, and large hydrophobic tail groups, represented by long rods. The long rods represent fatty acid chains with varying degrees of unsaturation. The polar head groups vary from the small ethanolamine head group in phosphatidylethanolamine to the complex carbohydrates in the sphingoglycolipids.

The biosynthesis of membrane lipids begins with diacylglycerol. The head groups are added in an activated form much as glucose, in the form of a

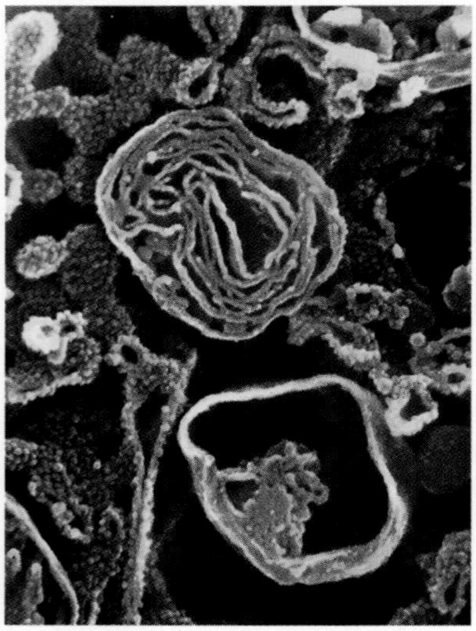

Lysosomes. Colored high-resolution scanning electron micrograph of two lysosomes in a pancreatic acinar cell. Lysosomes (rounded, center) are bound by a single membrane. Old or nonfunctioning organelles become enclosed in the lysosome and are digested by a concentrated mixture of digestive enzymes. Partially digested cell organelles can be seen in each lysosome.

uridine diphosphoglucose (UDP-glucose) derivative, is added to glycogen. In the formation of the membrane lipid phosphatidyl ethanolamine, the activated head group is in the form of **cytidine diphosphoethanolamine (CDP-ethanolamine).**

The formation of the activated head group begins with the phosphorylation of ethanolamine by ATP catalyzed by the enzyme ethanolamine kinase:

$$\text{Ethanolamine} + \text{ATP} \longrightarrow \text{phosphoethanolamine} + \text{ADP}$$

Phosphoethanolamine then reacts with cytidine triphosphate (CTP). The enzyme in this step is phosphoethanolamine cytidylyl transferase.

$$\text{Phosphoethanolamine} + \text{CTP} \longrightarrow \text{CDP-ethanolamine} + \text{PP}_i$$

This activation reaction is driven by the favorable hydrolysis of pyrophosphate. Finally, the activated head group is added to diacylglycerol by the enzyme ethanolaminephosphotransferase:

$$\text{Diacylglycerol} + \text{CDP-ethanolamine} \longrightarrow \text{phosphatidylethanolamine} + \text{CMP}$$

In similar reaction pathways, CDP derivatives react with diacylglycerol to produce other membrane phospholipids such as phosphatidylcholine, phosphatidylserine, phosphatidylinositol, and cardiolipin.

We saw earlier that the nucleotide adenine forms the core of the carrier of activated phosphate groups and the nucleotide uridine forms the core of the carrier for glucosyl groups. We can now add cytidine to our list of nucleotides that act as carriers of a specific chemical group in cellular metabolism.

Membrane lipids undergo very rapid turnover in the cell, but a critical balance between synthesis and degradation keeps their concentrations steady. Imbalances between the two processes are the causes of many early childhood diseases.

Membrane lipids as well as other biomolecules are degraded within lysosomes, membrane-bounded structures in the cytosol that contain a large array of hydrolytic enzymes of great specificity. Normally, membrane lipids find their way into the lysosomes and are degraded into soluble components that then leave the lysosome to be salvaged and incorporated into newly synthesized biomolecules. However, certain recessive human mutations result in the absence of specific lysosomal hydrolytic enzymes. In consequence, membrane lipids that enter lysosomes do not leave, and so the lysosomes swell, disrupting cell function. This causes mental retardation in many cases and death at an early age. There are no treatments for these lysosomal diseases at present, but genetic engineering offers hope for the future. In the interim, amniocentesis and genetic counseling have reduced the incidence of these diseases. Table 24.1 lists some of these lysosomal diseases and the enzyme that is defective or absent in each disease.

TABLE 24.1 Lysosomal Diseases Due to Faulty Degradation of Membrane Lipids

Disease	Defective Enzyme
Gaucher disease	glucocerebrosidase
Tay-Sachs disease	*N*-acetylhexosaminidase
Fabry's disease	trihexosylceramide galactosyl hydrolase
Hurler syndrome	α-L-iduronidase
Krabbe disease	galactosylceramide β-galactosylhydrolase
Niemann-Pick disease	sphingomyelinase

Summary

Fatty Acid Mobilization The glycogen storage capacity of the liver and muscle in humans is about 12 hours' worth of metabolic energy. The deficit is made up by stored triacylglycerols. Fatty acids derived from stored triacylglycerols are activated in the cytosol to acyl-*S*-CoA thioesters and are then transported into the mitochondria in the form of carnitine complexes.

Fatty Acid Oxidation Fatty acids are oxidized in four steps: (1) a dehydrogenation between carbons 2 and 3 of the acyl CoA derivative; (2) an enzymatic hydration of the trans double bond produced in step 1; (3) the dehydrogenation of the 3-hydroxyacyl-*S*-CoA produced in step 2; and finally (4) production of a molecule of acetyl-*S*-CoA from carbons 1 and 2 of the original fatty acid, along with another fatty acyl-*S*-CoA molecule having two fewer carbon atoms than the original acyl-*S*-CoA derivative. The acetyl-*S*-CoA formed in this process enters the citric acid cycle and is oxidized to CO_2 and water.

Ketone Bodies and Cholesterol Ketone bodies, formed in the mitochondria of the human liver, consist of acetoacetate, D-3-hydroxybutyrate, and acetone. They are an overflow from excessive fatty acid oxidation and are present in high concentrations in diabetes mellitus. A key intermediate in their biosynthesis is identical with 3-hydroxy-3-methylglutaryl-*S*-CoA, a precursor molecule synthesized in the cytosol that is central to the synthesis of many important biomolecules such as cholesterol, vitamins A, E, and K, and natural rubber.

Biosynthesis of Fatty Acids The biosynthesis of fatty acids takes place in the cytosol. It begins when citrate leaves the mitochondria and enters the cytosol, where it is enzymatically cleaved into acetyl-*S*-CoA and oxaloacetate. Citrate is also an allosteric signal for the activation of the rate-limiting step in fatty acid biosynthesis, the formation of malonyl-*S*-CoA by the carboxylation of acetyl-*S*-CoA.

The fatty acid synthase complex of animal cells is a multifunctional protein responsible for seven different enzymatic activities (each in a different region of the protein). The central feature of the complex is the presence of two unique—SH groups to which growing fatty acid chains are esterified.

In the chain-lengthening reaction, acetyl-*S*-CoA is transferred to one of the SH groups, and malonyl-*S*-CoA is transferred to the other. The acetyl group is then condensed with the malonyl group to form the acetoacetyl derivative. At the same time, CO_2 is displaced and lost. The acetyl group becomes the methyl-terminal group of the growing chain. The next three reactions are essentially the reverse of the oxidation steps: the β-keto group is reduced, dehydrated, and finally reduced to form the aliphatic chain increased by two carbons. The process is repeated until a 16-carbon chain is formed and released as palmitoyl-*S*-CoA.

Biosynthesis of Triacylglycerols The biosynthesis of triacylglycerols begins with the formation of diacylglycerol-3-PO_4, also known as phosphatidic acid. Phosphatidate is hydrolyzed to diacylglycerol, which is acylated by a third fatty acyl-*S*-CoA, to form the final product, triacylglycerol.

Biosynthesis of Membrane Lipids A membrane lipid is composed of a large hydrophobic tail group and strongly polar head groups. The biosynthesis of membrane lipids begins with diacylglycerol. The head groups are added in an activated form, an example being cytidine diphosphoethanolamine in the formation of phosphatidyl ethanolamine. Recessive human mutations result in the loss of lysosomal hydrolytic enzymes, creating an imbalance between synthesis and degradation of membrane lipids and causing childhood diseases such as Tay-Sachs disease.

Key Words

α-SH, p. 715
acetoacetate, p. 714
ACP (acyl carrier protein), p. 715
β-SH, p. 715
carnitine, p. 710
CDP-ethanolamine (cytidine diphosphoethanolamine), p. 722
ketone body, p. 713
ketosis, p. 713
lysosome, p. 722
phosphatidic acid, p. 720

Exercises

24.1 Can glycogen supply all of the body's daily energy requirements? Explain your answer.

24.2 What is the principal source of stored energy in humans?

24.3 Under what conditions is stored fat mobilized for cellular oxidation?

24.4 What is the first step in the mobilization of the fat stored in animals?

24.5 Show the fatty acid activation reaction that takes place in the cytosol.

24.6 How many moles of ATP are required for the activation of one mole of fatty acid? Explain.

Fatty Acid Transport into Mitochondria

24.7 Can activated fatty acids directly enter mitochondria? Explain your answer.

24.8 What is carnitine and what is its role in fatty acid oxidation?

Oxidation in Mitochondria

24.9 What is the first product of palmitate oxidation in mitochondria?

24.10 What is the coenzyme in the first oxidation step of palmitate, and what is the fate of its reducing power?

24.11 What is the second product of palmitate oxidation in the mitochondria?

24.12 What is unique about the structure of the product in Exercise 24.11?

24.13 What is the third product of palmitate oxidation in mitochondria?

24.14 What is the coenzyme in the third step of the mitochondrial oxidation of palmitate, and what is the fate of its reducing power?

24.15 What are the reactants in the fourth step of the oxidation of palmitate?

24.16 Describe the products of the fourth step of the oxidation of palmitate.

Ketone Bodies

24.17 Name the ketone bodies synthesized in the liver.

24.18 What is the word used to indicate that ketone bodies are in abnormal excess?

24.19 What are two physiological conditions leading to an abnormal excess of ketone bodies?

24.20 Do ketone bodies serve any normal metabolic role?

Fatty Acid Biosynthesis

24.21 What physical method is employed by animal cells to separate the fatty acid biosynthetic apparatus from the oxidative process?

24.22 What biochemical method is employed by animal cells to separate the fatty acid biosynthetic apparatus from the oxidative process?

24.23 What hormone is responsible for initiating fatty acid biosynthesis?

24.24 What hormone is responsible for inhibiting fatty acid biosynthesis?

24.25 What is the principal source of acetate in fatty acid biosynthesis?

24.26 What is the rate-limiting step in fatty acid biosynthesis?

24.27 Describe the steps that alter the fatty acid synthase molecule prior to chain lengthening.

24.28 Describe the process that leads to fatty acid chain lengthening.

24.29 What are the roles of CO_2 and biotin in the process that leads to fatty acid chain lengthening?

24.30 Describe a biochemical synthesis other than fatty acid chain lengthening in which CO_2 and biotin are required.

24.31 Describe conditions at the α-SH and β-SH sites of the fatty acid synthase at the start of chain lengthening.

24.32 Describe conditions at the α-SH and β-SH sites of the fatty acid synthase when chain lengthening is complete.

Biosynthesis of Triacylglycerols

24.33 What is the first step in the biosynthesis of triacylglycerols?

24.34 What is the role of phosphatidic acid in lipid metabolism?

24.35 Is the return—that is, the yield of stored energy—worth the investment of energy required to synthesize triacylglycerols? Explain.

24.36 Is it legitimate to express metabolic energy as quantities of ATP? Explain your answer.

Biosynthesis of Membrane Lipids

24.37 In what form is ethanolamine added to diacylglycerol?

24.38 Name the nucleotides that act as specific chemical-group carriers in cell metabolism.

24.39 What is the chemical basis of many lysosomal diseases?

24.40 What is the genetic basis of the lysosomal diseases?

Unclassified Exercises

24.41 What are some of the features differentiating the biosynthetic from the catabolic reactions of the fatty acids?

24.42 What is a central structural feature of the fatty acid synthase complex in animal cells?

24.43 What is the overall stoichiometry for the synthesis of palmitate from acetyl-*S*-CoA?

24.44 Criticize this statement: Ketone bodies are synthesized from cytosolic 3-hydroxy-3-methylglutaryl-*S*-CoA.

24.45 Can the oxidation of fatty acids generate ATP if the resultant acetyl-*S*-CoA is not oxidized? Explain.

Chemical Connections

24.46 What physiological condition arises when oxaloacetate is absent from animal cells?

24.47 True or false: The presence of high concentrations of citrate in cells triggers the biosynthesis of fatty acids. Explain your answer.

24.48 An unconscious older man was brought to the emergency room. The attending physician smelled the man's breath and immediately ordered intravenous infusion of glucose. What led him to order this treatment?

24.49 How do fatty acids arise in adipose tissue cells and move from those cells to where they are needed for energy?

24.50 True or false: The activation of fatty acids for oxidation requires 1 mol of ATP. Explain your answer.

CHAPTER 25

AMINO ACID METABOLISM

CHEMISTRY IN YOUR FUTURE

You have enjoyed the experience of interviewing Americans from all walks of life for the USDA's nationwide food-consumption survey, and it has made you more aware of your own eating habits. For example, because you worry about gaining weight, you have always tried to avoid foods that are high in fat. As a consequence, you almost never eat meats and dairy products. Now you realize that your diet includes few other sources of protein and that you are eating far less than the recommended daily amount. This chapter on amino acid metabolism will tell you why it is important to maintain a consistent level of protein intake.

LEARNING OBJECTIVES

- Describe transamination and oxidative deamination.
- Describe how amino groups are collected in the liver in the form of glutamate.
- Describe how ammonia is transported through the blood to the liver.
- Describe the ways in which ammonia is transformed into urea, and indicate where they take place in the cell.
- Describe the end products of the catabolism of an amino acid's carbon skeleton.
- Identify and describe some of the heritable genetic defects of amino acid metabolism.
- Learn why some amino acids are called "essential" and others "nonessential."

Amino acids, largely derived from dietary proteins, are chiefly used to synthesize proteins. A smaller proportion is used to synthesize specialized biomolecules such as the hormones epinephrine and norepinephrine, neurotransmitters, and the precursors of purines and pyrimidines. Whenever we use the term amino acid in discussing polypeptides and proteins, we always mean α-amino acids, whether or not the α prefix is shown. In addition, whether or not indicated and regardless of how drawn, all amino acids are L-amino acids unless otherwise noted (Section 20.1). The structures of the amino acids found in proteins are given in Table 20.1. Unlike carbohydrates and lipids, they cannot be stored in the body for later use; so any amino acids not required for immediate biosynthetic needs are either degraded, supplying energy in the process, or converted into acetyl-*S*-CoA and then into fatty acids (and thus triacylglycerol stores).

▶▶ This and other effects of diabetes are explained in Section 26.5.

When carbohydrates are not available—or cannot be used properly, as in diabetes mellitus—amino acids become a primary source of energy. Thus, when starvation, fasting, or the result of some other condition depletes all carbohydrate supplies, the body's own proteins, especially muscle proteins, are broken down and used for energy.

In this chapter, we shall explore the catabolic pathways by which amino acids are degraded and then consider some aspects of their biosynthesis. The discussion of catabolism will focus, first, on the production and fate of ammonia and, then, on the breakdown of the "carbon skeletons," the α-keto acids formed when the α-amino group is removed from the α-amino acids.

25.1 AN OVERVIEW OF AMINO ACID METABOLISM

Amino acids are produced from the breakdown of dietary protein and from the breakdown, through normal metabolic turnover, of the body's proteins. Enzymes and muscle proteins have a particularly rapid turnover.

The catabolism of amino acids takes place in three stages: (1) removal of the amino group, leaving the carbon skeleton of the amino acid; (2) breakdown of the carbon skeleton to an intermediate of the glycolytic pathway or citric acid cycle, or to acetyl-*S*-CoA; and (3) oxidation of these intermediates to carbon dioxide and water with production of ATP. This third stage takes place in the pathways considered in Chapter 23.

Unique to amino acid catabolism is the extraction of the amino group ($—NH_2$) from the rest of the amino acid molecule, a process that must be accomplished without producing toxic levels of ammonia (NH_3) in blood and tissues. At cellular pH, NH_3 exists as the NH_4^+ ion; so, when ammonia appears in cells, it is immediately converted into ammonium ion (see Table 9.5), and this is the form in which it participates in almost all biochemical reactions. The normal concentration of ammonium ion in the blood is in the range of 3.0×10^{-5} to 6.0×10^{-5} *M*. Above these concentrations (hyperammonemia), coma may result. The reasons for the toxicity of ammonium ion are not entirely clear. One possibility is the reaction of the ammonium ion with α-ketoglutarate to form glutamate (Section 25.3). α-Ketoglutarate is usually present at very low concentrations in cells, particularly brain cells, and any further reduction in concentration by conversion into glutamate could result in severe metabolic stress.

The pathway for the amino groups extracted from amino acids consists of:

- Conversion of amino groups from all amino acids into a single product, glutamate, by transamination

- Conversion of glutamate into α-ketoglutarate by oxidative deamination, releasing NH_4^+
- Conversion of NH_4^+ into the nontoxic compound urea, which the blood carries to the kidneys for excretion

These processes take place in the liver. Amino groups and ammonium ion that must be conveyed from other tissues to the liver are transported in the form of glutamine or alanine, as we shall see. The amino acids glutamate and glutamine play central roles in amino acid catabolism, as well as in biosynthesis.

◀◀ The structures and properties of the amino acids are described in Sections 20.1 and 20.2.

25.2 TRANSAMINATION AND OXIDATIVE DEAMINATION

After the digestion of a protein-rich meal, amino acids are absorbed from the intestine into the bloodstream, and those not used in biosynthetic reactions undergo degradation in the liver. The degradation is accomplished through two kinds of reaction: transamination and oxidative deamination.

In **transamination** reactions, the α-amino group of an amino acid is transferred to the carbon of the carbonyl group of an α-keto acid, which is then converted into the corresponding amino acid. The original amino acid is converted into its corresponding α-keto acid. That is:

$$\text{Amino acid}_1 + \alpha\text{-keto acid}_2 \longrightarrow \alpha\text{-keto acid}_1 + \text{amino acid}_2$$

(Recall that α-keto acids are components of the citric acid cycle, undergoing the catabolic degradation described in Section 23.5.) In essence, transamination is the exchange of an amino group for a carbonyl group. It can take place not only in liver cells but in all cells.

Transamination reactions are catalyzed by enzymes called amino transferases or transaminases. Most transaminases are specific for α-ketoglutarate but are less specific for the amino acid. For this reason, the principal product of transamination from a wide array of amino acids is glutamate. Therefore, the oxidative pathways for amino acids tend to converge to form a single product. An important exception is a group of transaminases in muscle cells that uses pyruvate as the amino group acceptor. Because of this, alanine is the principal product of amino acid catabolism in muscle cells. Both types of transaminase are of central importance in the transport systems considered in detail in Section 25.3.

The enzymes are named for the amino acid that donates the amino group: for example, alanine transaminase, aspartate transaminase, and leucine transaminase. The transamination reaction between aspartate and α-ketoglutarate, catalyzed by asparatate transaminase is:

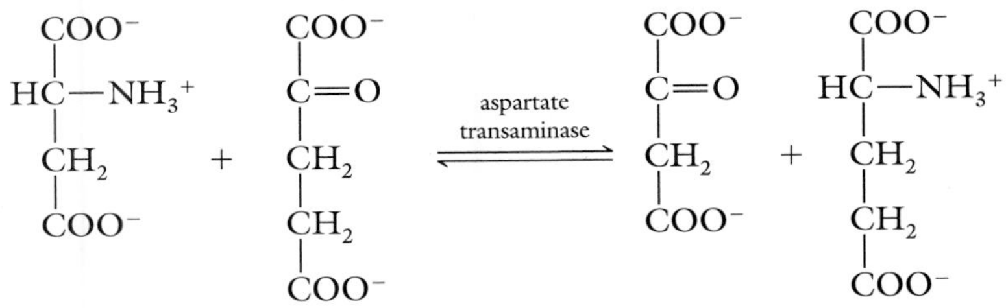

Aspartate **α-Ketoglutarate** **Oxaloacetate** **Glutamate**

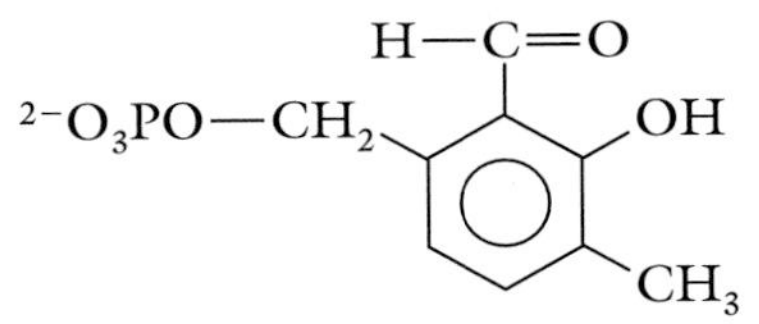

Pyridoxal phosphate

All transaminases utilize pyridoxine (vitamin B_6) in the form of **pyridoxal phosphate** as the coenzyme in the transamination reaction.

Measurements of the levels of two transaminases in the blood—alanine transaminase, also called **glutamate:pyruvate transaminase (GPT),** and aspartate transaminase, also called **glutamate:oxaloacetate transaminase (GOT)**—are used in the diagnosis of heart disease. A heart attack, or myocardial infarction, results in tissue damage, and, as a consequence, these transaminases, along with other enzymes, leak into the bloodstream (Box 24.1). The severity and stage of the heart attack can be monitored by measuring the concentrations of these enzymes in the blood serum, measurements known as the serum GPT and GOT or, more commonly, **SGPT** and **SGOT tests.**

Glutamate produced from transamination reactions can undergo **oxidative deamination,** releasing ammonium ion and producing α-ketoglutarate. The reaction is catalyzed by glutamate dehydrogenase, with NAD^+ as coenzyme.

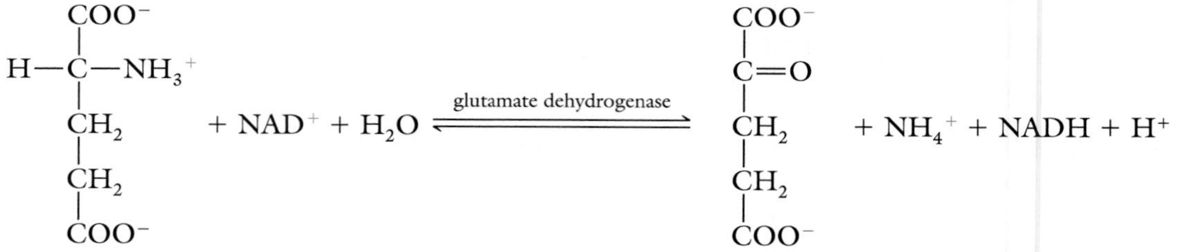

Glutamate α-Ketoglutarate

Because it is reversible, this reaction also affords a mechanism both for effectively assimilating ammonium ions into metabolic pathways in the human liver and kidney and for lowering toxic levels of ammonium ions in all cells.

In extrahepatic tissues, amino acids are produced from the metabolic turnover of proteins. These amino acids, if not reused, must be degraded. As in the liver, the amino acids are removed in transamination reactions to form glutamate. The glutamate cannot pass through the cell membrane; as shown in the next section, it must be converted into another compound for passage into the bloodstream.

25.3 AMINO GROUP AND AMMONIA TRANSPORT

The amino groups collected in the form of glutamate in extrahepatic tissues, as well as NH_4^+ formed in those tissues from the breakdown of amino acids and other nitrogen-containing biomolecules, are packaged in a nontoxic form that can leave the cell, enter the circulation, and travel to the liver. The conversion of NH_4^+ into nontoxic form takes place by two processes, resulting in two different transport forms:

- In most cell types, the production of glutamine
- In muscle cells, the production of alanine

The enzyme glutamine synthetase, in almost all cell types, catalyzes the formation of glutamine from glutamate.

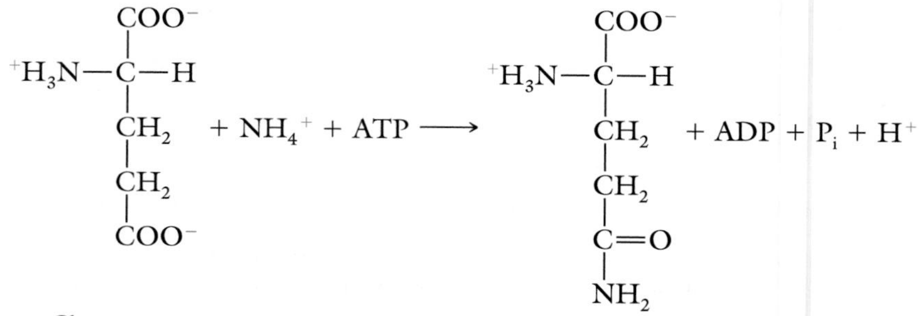

Glutamate Glutamine

The amination of the γ-glutamyl carboxyl group converts the negatively charged glutamate into glutamine, which has a net charge of zero and can pass through cell membranes into the blood.

The amide group of glutamine is the source of amino groups in many biosynthetic reactions, including amination of α-keto acids in amino acid synthesis (Section 26.7). The concentration of glutamine in blood is significantly higher than that of any other amino acid.

Concept check

✔ Glutamine is a nontoxic transport form of NH_4^+ and a temporary storage form of amino groups in the body.

On reaching the liver, glutamine can be deaminated in another reaction by the enzyme glutaminase:

$$\text{Glutamine} + H_2O \longrightarrow \text{glutamate} + NH_4^+$$

Some of the glutamate can undergo oxidative deamination, as described earlier, releasing another NH_4^+.

In active muscle cells, there is considerable turnover of amino acids and nucleotides, which gives rise to large quantities of ammonium ion. The ammonium ion is transported from muscle to the liver in the form of the amino acid alanine through the action of the **glucose-alanine cycle** (Figure 25.1).

The cycle begins in the mitochondria of muscle cells with a process called **reductive amination,** in which α-ketoglutarate reacts with NH_4^+ to form glutamate, a reaction catalyzed by glutamate dehydrogenase—in this case, an irreversible reaction in which NADPH is the cofactor:

$$NH_4^+ + \alpha\text{-ketoglutarate} + \text{NADPH} \longrightarrow \text{glutamate} + \text{NADP}^+ + H_2O$$

This reaction also constitutes another way in which ammonium ion can be incorporated to form an amino acid.

The glutamate then moves into the cytosol where it undergoes transamination with pyruvate catalyzed by alanine transaminase:

$$\text{Glutamate} + \text{pyruvate} \longrightarrow \text{alanine} + \alpha\text{-ketoglutarate}$$

Pyruvate is readily available in actively contracting muscle, where the glycolytic pathway is extremely active. Alanine has no net charge at physiological

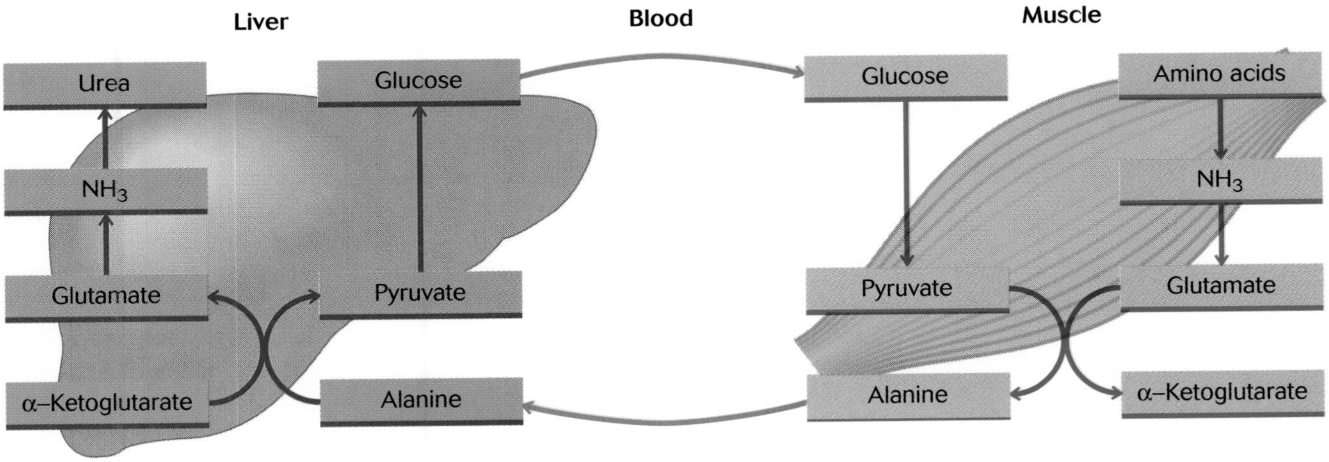

Figure 25.1 The glucose-alanine cycle by which ammonium ion is transported from muscle cells to the liver in the form of an amino acid.

pH and can pass through cell membranes to enter the circulation and thence the liver.

After alanine enters the liver, transamination in the cytosol transfers alanine's amino group to α-ketoglutarate, forming glutamate and pyruvate. The glutamate gives up its NH_4^+ by oxidative deamination catalyzed by glutamate dehydrogenase (Section 26.2) or, as we shall see in Section 25.4, by transamination with oxaloacetate to form aspartate. Pyruvate is converted into glucose by gluconeogenesis, and glucose returns through the circulation to muscle cells.

As we can see, the glucose-alanine cycle kills two birds with one stone: pyruvate produced in muscle cells is converted into glucose in the liver and returned to muscle; the waste product ammonium ion is removed from muscle cells and converted into urea in the liver. The urea is then excreted by the kidneys.

25.4 UREA CYCLE

Having considered the ways in which amino groups and NH_4^+ are transported to the liver and how NH_4^+ is released in liver cells, we now look at the way in which this toxic compound is converted into a nontoxic, excretable form. Humans as well as most other terrestrial animals (land-dwelling animals rather than those that live in water) transform ammonium ion into a water-soluble, nonionic compound called urea. Such organisms are called **urotelic.** Urea excretion is advantageous for two reasons. First, the water solubility of ammonium ion makes keeping it in the urine formed in the kidney problematic because the ammonium ion can easily pass back into the blood. Second, excretion of its water-soluble form, NH_4^+, requires the simultaneous excretion of an oppositely charged ion (a counterion). This requirement would lead to significant loss of metabolically important anionic counterions—for example, phosphate and bicarbonate. Conversion of ammonium ion into nontoxic, nonionic form nullifies these problems.

The biosynthesis of urea is a complex process that takes place only in the liver. It is a cyclic process, one that takes place partly in mitochondria and partly in the cytosol. Two intermediates of the cycle are amino acids not found in proteins: ornithine and citrulline. The other amino acid in the cycle is arginine, a common component of proteins. In the **urea cycle,** biosynthesis of 1 mol of urea fixes 2 mol of ammonium ion. The steps of this cycle are outlined in Figure 25.2.

Reactions in the Mitochondria

In liver mitochondria, NH_4^+ reacts with bicarbonate and two molecules of ATP to produce carbamoyl phosphate. The reaction is catalyzed by carbamoyl phosphate synthetase I (as distinct from the cytosolic carbamoyl phosphate synthetase II, an enzyme of nucleotide synthesis). This reaction is an activation reaction in which carbamoyl phosphate is an activated form of the carbamoyl group, which then enters the urea cycle.

$$HCO_3^- + NH_4^+ + 2\,ATP^{4-} \longrightarrow \underset{\textbf{Carbamoyl phosphate}}{H_2N-\overset{\overset{\displaystyle O}{\|}}{C}-OPO_3^{2-}} + 2\,ADP^{3-} + P_i^{2-} + 2\,H^+$$

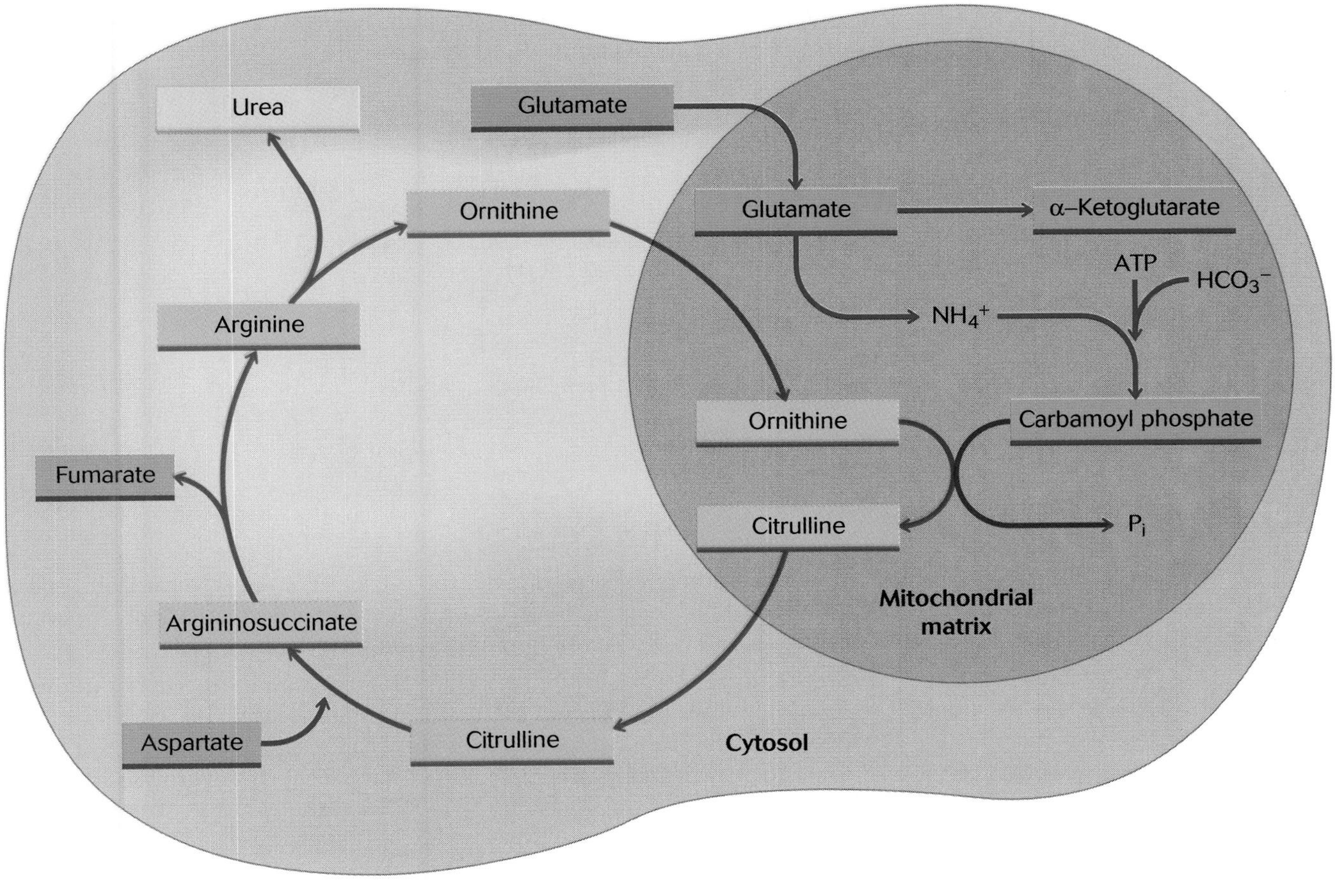

Figure 25.2 The urea cycle consists of reactions within the cytosol and reactions within the mitochondria.

Step 1. Carbamoyl phosphate reacts with ornithine to form citrulline, a reaction catalyzed by ornithine transcarbamoylase.

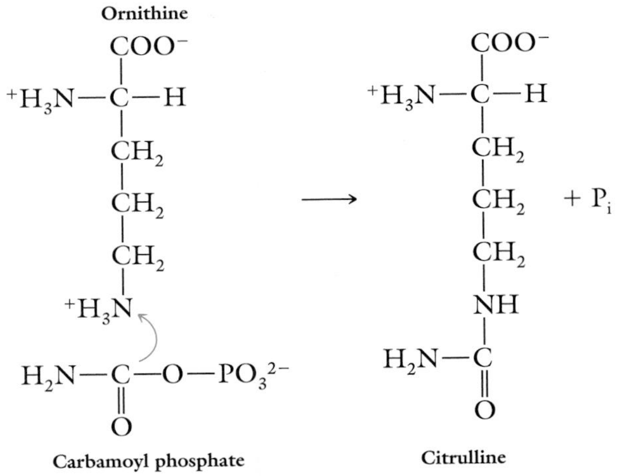

Citrulline then leaves the mitochondrion and enters the cytosol.

In subsequent reaction equations, the ornithine part of citrulline will be represented by Orn because it undergoes no fundamental structural changes and emerges unchanged at the end of the process to begin a new cycle. Thus,

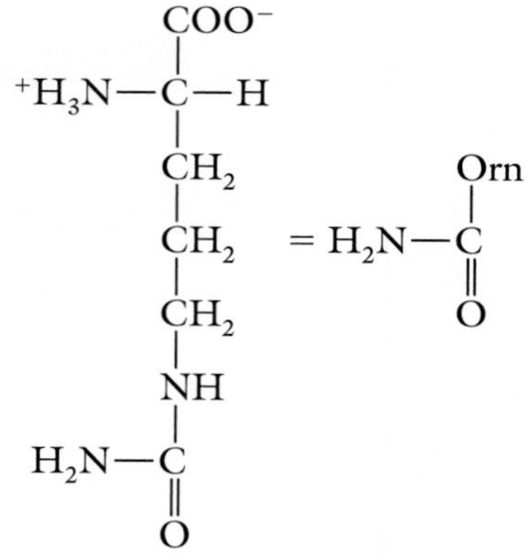

Citrulline

Reactions in the Cytosol

In the cytosol of liver cells, glutamate, the carrier of amino groups from amino acids, undergoes transamination with oxaloacetate to form aspartate. Aspartate carries the second molecule of ammonium ion into the urea cycle.

Step 2. Aspartate reacts with citrulline to form argininosuccinate. In this reaction, requiring ATP and catalyzed by argininosuccinate synthetase, the two terminal phosphates are split off as pyrophosphate ($P_2O_7^{4-}$), which is subsequently hydrolyzed to 2 P_i. As we have seen, this hydrolysis has a very large equilibrium constant (Section 23.9) and is the driving force for the synthesis of argininosuccinate.

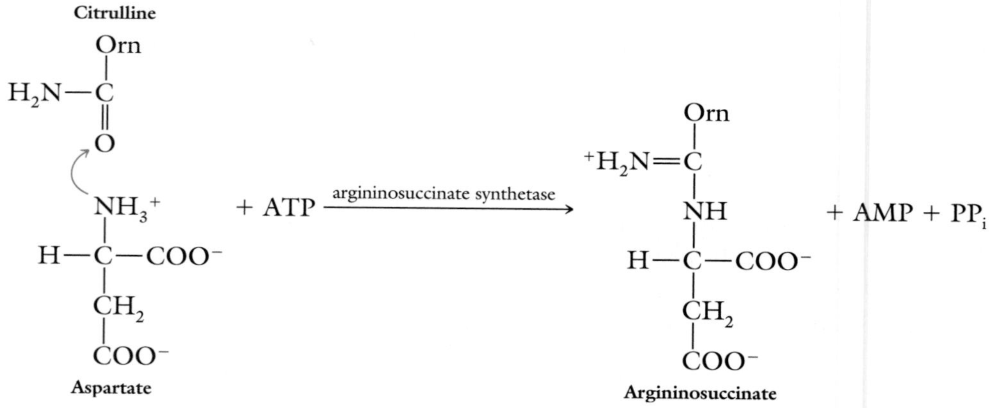

Step 3. In the next step, the argininosuccinate is cleaved into arginine and fumarate by the enzyme argininosuccinate lyase.

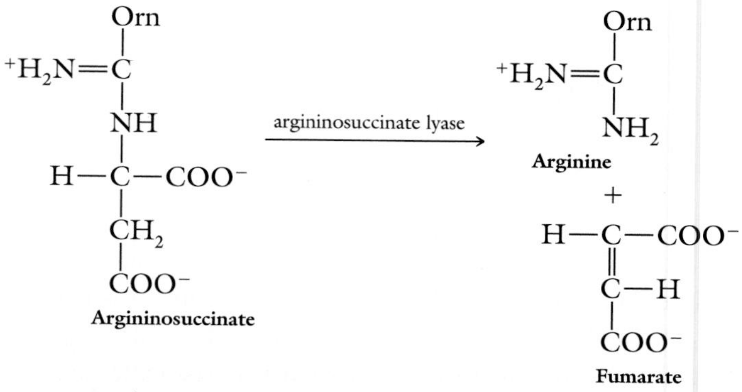

Fumarate enters the citric acid cycle.

Step 4. The enzyme arginase catalyzes the cleavage of arginine into urea. Birds, reptiles, and bony fishes do not possess this enzyme. It is found only in the liver of urotelic organisms.

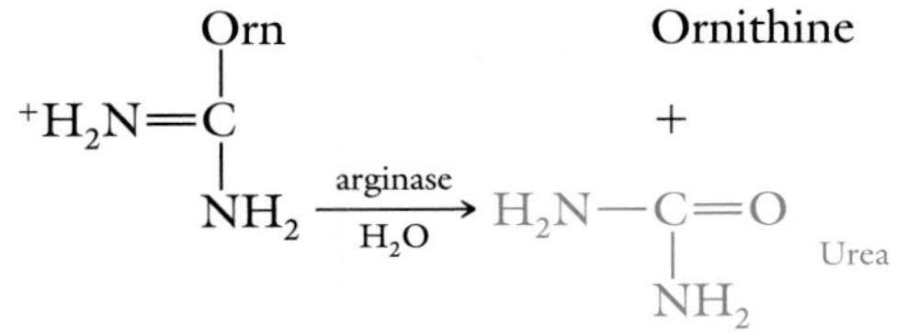

Ornithine is now available for another round of the urea cycle. Urea passes from liver cells into the blood and is excreted by the kidneys.

A significant amount of energy goes into urea synthesis: 2 ATP for the synthesis of carbamoyl phosphate, 2 ATP for the synthesis of argininosuccinate, 1 ATP converted into AMP in the synthetase reaction, and 1 ATP required for conversion of AMP into ADP (Section 24.1). Thus the total energy cost for the synthesis of a mole of urea is 4 mol of ATP. The price for detoxification of ammonium ion represents about 7% of the available energy in the amino acids oxidized.

Figure 25.3 summarizes the processes by which ammonium ion is removed from amino acids and converted into urea.

25.5 OXIDATION OF THE CARBON SKELETON

We now turn to the second stage of amino acid catabolism: the breakdown of the carbon skeletons of the deaminated α-keto acids.

The α-keto acids of many amino acids are the same as those found in the glycolytic pathway and the citric acid cycle. Therefore they undergo the oxidative degradation described in Chapter 23. These α-keto acids thus replenish the citric acid cycle intermediates, either directly or through formation of pyruvate and its carboxylation to oxaloacetate (Section 23.7).

Other α-keto acids produced from amino acids are broken down to acetyl-*S*-CoA or acetoacetyl-*S*-CoA or both and thus can enter the citric acid cycle or be converted into ketone bodies (Section 24.3). These α-keto acids—just like fatty acids, which produce acetyl-*S*-CoA—do not replenish citric acid cycle intermediates.

Every amino acid has a different degradative pathway; however, we will not consider these pathways in detail. All pathways lead to intermediates that

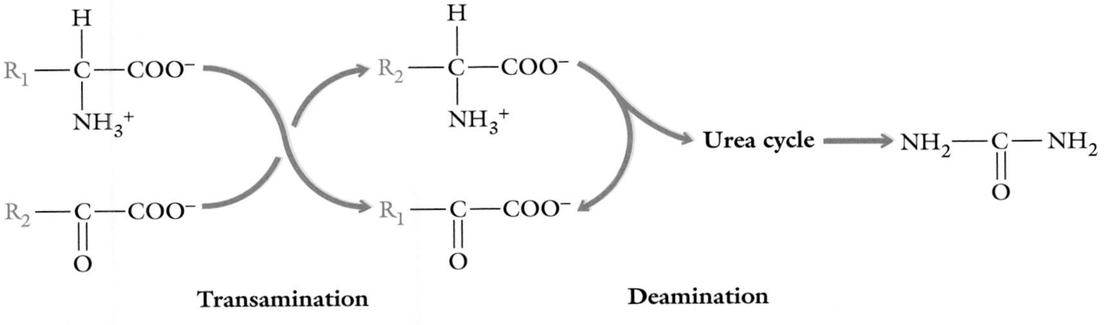

Figure 25.3 Outline of the processes through which amino groups of amino acids are converted into urea.

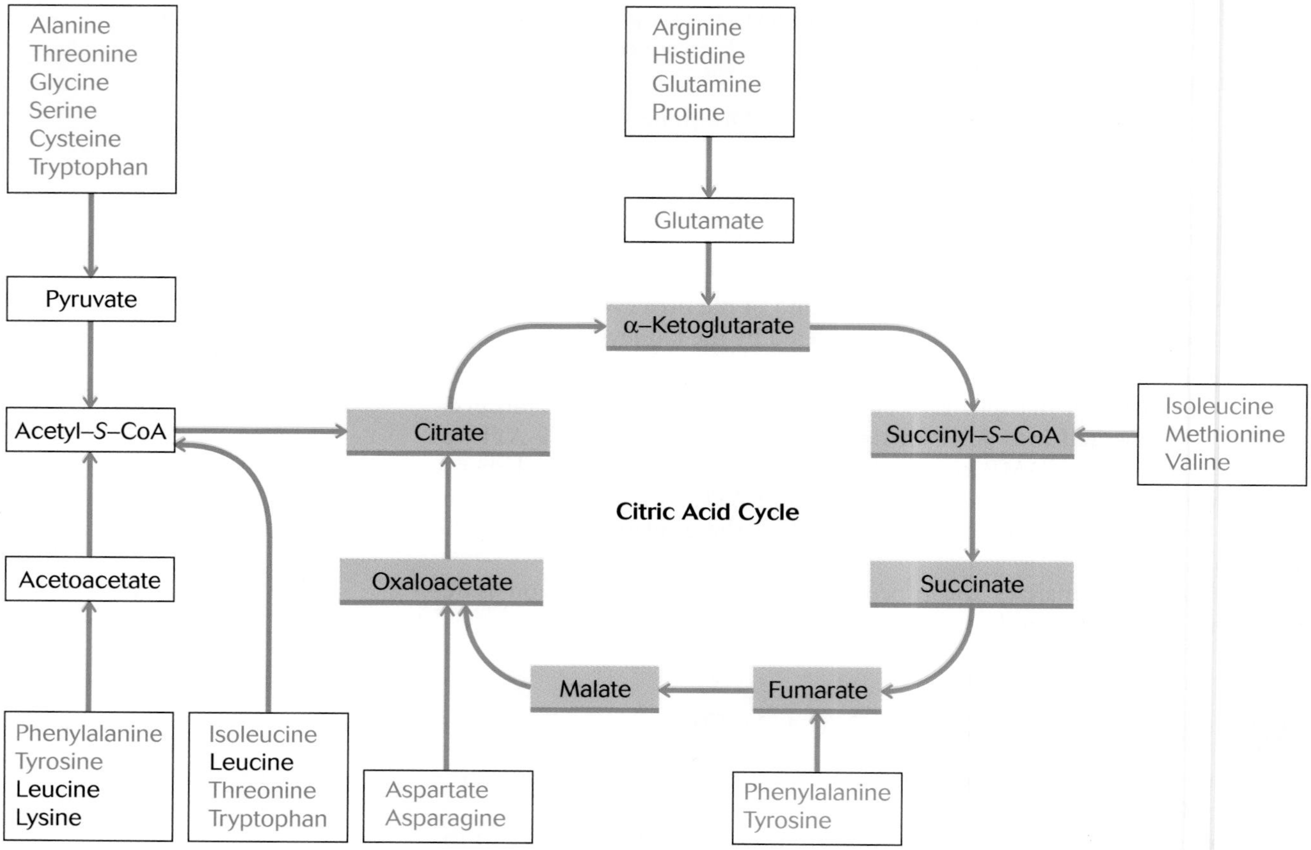

Figure 25.4 The relation between the catabolism of amino acids and the citric acid cycle. Glucogenic amino acids are in blue type; ketogenic amino acids are in black type; and amino acids that are both glucogenic and ketogenic are in red type.

find their way into the citric acid cycle for complete oxidation to CO_2 and H_2O. Six amino acids are degraded to pyruvate, seven to acetyl-*S*-CoA, four to α-ketoglutarate, three to succinyl-*S*-CoA, two to fumarate, and two to oxaloacetate (Figure 25.4).

The amino acids that can be converted into pyruvate, α-ketoglutarate, succinyl-*S*-CoA, fumarate, and oxaloacetate can give rise to glucose by gluconeogenesis (Section 23.8), and these amino acids are said to be **glucogenic.** Seven amino acids are converted into acetyl-*S*-CoA or acetoacetyl-*S*-CoA or both and can thus yield ketone bodies in the liver (Section 24.3); they are the **ketogenic** amino acids. Glucose cannot be produced from amino acids that form only acetyl-*S*-CoA or acetoacetyl-*S*-CoA.

Concept check

✔ Only amino acids that replenish oxaloacetate can give rise to glucose by gluconeogenesis.

Large amounts of ketone bodies are produced from amino acids, as well as from lipids (Section 24.3), in untreated diabetes mellitus. In starvation and fasting, as well as in diabetes, large amounts of glucose also are produced from amino acids. Phenylalanine, tyrosine, and isoleucine are both ketogenic and glucogenic: their degradation yields both glucose-forming and ketone body-forming precursors (Figure 25.4). Table 25.1 lists the glucogenic and ketogenic amino acids. Note that only leucine and lysine cannot act as precursors for glucose synthesis.

TABLE 25.1 Glucogenic and Ketogenic Amino Acids

Glucogenic Only	Glucogenic and Ketogenic	Ketogenic Only
alanine	phenylalanine	leucine
arginine	isoleucine	lysine
asparagine	tyrosine	
aspartate		
cysteine		
glutamate		
glutamine		
glycine		
histadine		
methionine		
proline		
serine		
threonine		
tryptophan		
valine		

25.6 HERITABLE DEFECTS IN AMINO ACID METABOLISM

A number of heritable diseases result from defects in enzymes catalyzing steps in amino acid catabolism. The defects arise from genetic mutations that cause errors in the amino acid sequence of the primary peptide chain with consequent loss in function of the enzyme (Section 20.5). Because there are usually several enzymatically catalyzed steps in a catabolic sequence, normal intermediate compounds are not processed; they therefore accumulate and often lead to negative physiological consequences. Defects in the degradative pathway for the amino acid phenylalanine are the cause of several heritable diseases.

The causes of genetic mutations are discussed in Section 21.8.

The first step in the normal degradation of phenylalanine is its oxidation to form the amino acid tyrosine. Aside from incorporation into proteins, tyrosine is the precursor of the hormone thyroxine. In the following structural diagram, this step is represented in the upper sequence. When that enzyme is defective (crossed-out arrow), the amino group of phenylalanine can be removed by transamination to form phenylpyruvate. In **phenylketonuria (PKU),** one of the first human defects of metabolism to be discovered, the enzyme catalyzing the oxidation step is defective, and phenylalanine accumulates. When concentrations of phenylalanine are high enough, it reacts (by transamination) with pyruvate to form phenylpyruvate.

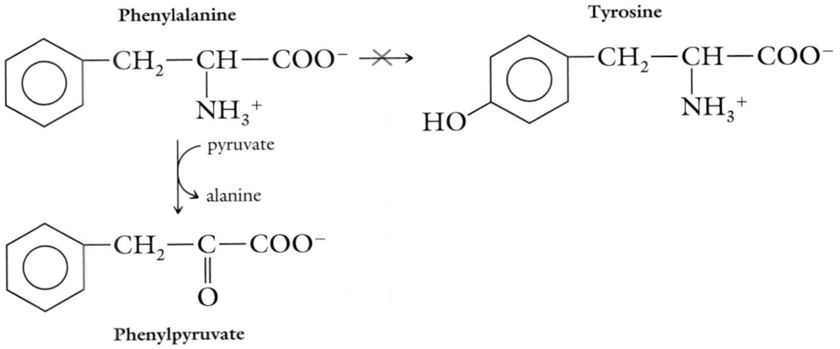

Phenylpyruvate is not further metabolized. It accumulates in tissue and blood, and, because it is soluble, it is excreted along with phenylalanine in the urine. Unfortunately, accumulation of phenylpyruvate in nerve cells in the early stages of life impairs normal development of the brain, causing severe mental retardation.

Phenylalanine and phenylpyruvate are easily detected in blood and urine, and, in the United States and many other countries, newborns are now routinely tested for PKU. When PKU is diagnosed early, mental retardation can be prevented by modifying the infant's diet. Because phenylalanine is an essential amino acid, it cannot be completely eliminated from the diet, but significant reductions can be made. For example, casein from milk can be hydrolyzed and most of the phenylalanine removed. This special low-phenylalanine diet is required only in childhood. When the nervous system is fully developed, such precautions are usually no longer necessary. However, people with PKU are cautioned not to drink soft drinks sweetened with aspartame (Section 17.6). Aspartame is the methyl ester of the dipeptide aspartylphenylalanine, and one of the products of its digestion is phenylalanine.

Another disorder of phenylalanine catabolism leads to the accumulation of the intermediate homogentisate, a nontoxic water-soluble compound excreted in the urine. Rapid air oxidation of this compound produces an intensely black pigment, which gave this disorder the name "black urine" disease, or **alkaptonuria.** The disease has no negative physiological consequences but, as might be imagined, led to some psychological and social problems before it was understood as a benign genetic defect.

A group of related conditions called maple syrup urine disease is caused by deficiencies in the catabolism of the branched-chain amino acids such as isoleucine. The disease is usually detected because it results in acidosis in newborns and young children. The most common metabolic defect lies in the inability to oxidize the α-keto acids resulting from transamination. All patients with this disease excrete α-keto acids and other side products. An unidentified product gives rise to the characteristic odor that lends its name to this group of diseases—maple syrup urine. Although some cases respond to dietary intervention, most cases result in mental retardation and early death.

A brief listing of several other genetic defects in amino acid metabolism includes a deficiency in mitochondrial ornithine transaminase. This deficiency results in a progressive loss of vision caused by atrophy of the retina. Plasma levels of ornithine, a critical component of the urea cycle, can be elevated as a result of a deficiency in an aminotransferase that catalyzes the first step in the conversion of ornithine into glutamate. Another disorder of ornithine metabolism has been found to be caused by defective transport of ornithine into the mitochondria. A deficiency or absence of cytosolic tyrosine transaminase results in a disease characterized by skin and eye lesions, often accompanied by mental retardation. The enzyme tyrosinase catalyzes the formation of a precursor of melanin through the oxidation of tyrosine. Melanin is a high-molecular-mass polymer that is insoluble and very dark in color. Its absence leads to the condition known as albinism—a lack of skin and hair color.

25.7 BIOSYNTHESIS OF AMINO ACIDS

Humans can synthesize 10 of the 20 amino acids required for protein synthesis. These amino acids are classified as nonessential—meaning nonessential in the diet (Section 20.1). Those amino acids that humans cannot synthesize are classified as essential and must be obtained by dietary intake. We consider here, very briefly, only the biosynthetic pathways for some nonessential amino acids.

Glutamate and glutamine play essential roles as amino-group donors in amino acid synthesis. Reductive amination, producing glutamate, takes place in all cells; the reaction is catalyzed by glutamate dehydrogenase, with NADPH as cofactor:

$$NH_4^+ + \alpha\text{-ketoglutarate} + NADPH \xrightarrow[\text{}]{\text{glutamate dehydrogenase}} \text{glutamate} + NADP^+ + H_2O$$

Glutamate is used to synthesize amino acids, by transamination, from appropriate precursors—primarily, α-keto acids of the citric acid cycle and glycolysis or α-keto acids derived from intermediates in these two processes. The amino acids synthesized directly from citric acid cycle intermediates by transamination or reductive amination are: alanine, aspartate, asparagine, glutamate, and glutamine. The biosynthetic pathways for the other amino acids are quite complicated and consist of many enzymatically catalyzed steps and many types of intermediate compounds.

Summary

Transamination and Oxidative Deamination Amino acids cannot be stored in the body; any that are not required for the synthesis of proteins or other nitrogen-containing biomolecules undergo catabolic degradation. The amino groups can be removed by transamination with α-ketoglutarate to form glutamate and the α-keto acid of the amino acid. The α-keto acids then undergo catabolic degradation. In the liver, the amino groups of all amino acids are collected in the form of glutamate. Oxidative deamination then removes ammonium ion from glutamate to form α-ketoglutarate. The ammonium ion is converted into a nontoxic form—urea—for excretion.

Amino Group and Ammonia Transport Two processes convert ammonium ion produced in extrahepatic tissues into a nontoxic, transportable form. The first process, synthesis of glutamine from glutamate, takes place in most tissues. Glutamine has no net charge, and so it can leave cells, enter the circulation, and travel to the liver, where the ammonium ion is released and converted into urea. The second process, the glucose-alanine cycle, takes place in muscle cells. Ammonium ion is carried from muscle to the liver in the form of the amino acid alanine. Alanine forms in muscle by transamination of pyruvate, a product of glycolysis. When alanine reaches the liver, its amine group is converted into urea, and the resulting pyruvate is converted into glucose by gluconeogenesis. Glucose circulates back to muscle to begin the cycle again.

Urea Cycle Humans and most other terrestrial animals package ammonium ion into water-soluble, nonionic, nontoxic urea, which is excreted by the kidneys. Biosynthesis of urea is complex and requires about 7% of the energy available in the amino acids. The urea cycle takes place only in liver cells. Its intermediates include two amino acids not found in proteins—ornithine and citrulline—and arginine. Urea is split from arginine by the enzyme arginase, unique to urotelic (urea-forming) organisms. The regenerated ornithine begins a new turn of the cycle.

Oxidation of the Carbon Skeleton All amino acid carbon skeletons are oxidized to CO_2 and H_2O in the citric acid cycle. Eighteen amino acids can be converted into glucose, and are called glucogenic. Seven amino acids are converted into acetyl-*S*-CoA. They yield ketone bodies in the liver and are called ketogenic. Phenylalanine, tyrosine, and isoleucine are both ketogenic and glucogenic. Their degradation yields both glucose-forming and ketone body-forming precursors.

Heritable Defects in Amino Acid Catabolism A number of diseases caused by heritable mutations are the result of incomplete catabolic degradation of amino acids. Normal intermediates are not processed past some point in the sequence; they accumulate and cause a variety of physiological abnormalities. A defect in the degradative pathway of phenylalanine is the cause of phenylketonuria (PKU). This heritable disease can lead to severe mental retardation, but its effects can be prevented by detection in newborns and strict adherence to a low-phenylalanine diet. Another defect in phenylalanine catabolism causes alkaptonuria.

Biosynthesis of Amino Acids Humans can synthesize 10 of the amino acids required for protein synthesis from intermediates of glycolysis and the citric acid cycle. Those amino acids that humans cannot synthesize must be obtained in the diet. Reductive amination produces glutamate, which acts as the amino-group donor in amino acid biosynthesis.

Key Words

Exercises

Amino Acid Metabolism

25.1 Describe the process called transamination. Give an example.

25.2 What vitamin forms the coenzyme used in transamination?

25.3 What happens to dietary amino acids in excess of those required for biosynthesis?

25.4 What happens to proteins of the body if carbohydrates are not available in the diet?

Transamination and Oxidative Deamination

25.5 Describe a reaction that is a first step in the catabolism of ingested amino acids.

25.6 Describe an alternative first step in the catabolism of ingested amino acids that takes place in the liver.

25.7 How is ammonium ion that is generated in the liver rendered nontoxic in humans?

25.8 How is ammonium ion that is generated in cells other than the liver rendered nontoxic in humans?

25.9 What is a urotelic animal?

25.10 What unique enzyme is possessed by urotelic animals?

25.11 What is the likely mechanism of ammonium ion toxicity in cells?

25.12 Describe the method used by all cells of the body to render ammonium ion harmless for transport through the blood to the liver.

25.13 What is the method used by muscle cells to render ammonium ion harmless for transport through the blood to the liver?

The Urea Cycle

25.14 In what organ is urea synthesized?

25.15 Do the reactions of the urea cycle take place in the cytosol or in mitochondria? Explain your answer.

25.16 Name and draw the structure of an amino acid that is not used to synthesize proteins but participates in urea's synthesis.

25.17 Name and draw the structure of an amino acid other than that named in Exercise 25.16 that also is not used to synthesize proteins but participates in urea's synthesis.

25.18 List the urea cycle reactions that require ATP.

25.19 List the urea cycle reactions that do not require ATP.

25.20 What is the total energy cost for the synthesis of 1 mol of urea in moles of ATP?

25.21 What percentage of the energy available in amino acids is used to synthesize urea?

Oxidation of the Carbon Skeleton

25.22 Name two glucogenic amino acids.

25.23 Why are the amino acids named in Exercise 25.22 called glucogenic?

25.24 Name two ketogenic amino acids.

25.25 Why are the amino acids named in Exercise 25.24 called ketogenic?

25.26 Name two products of the catabolism of tyrosine that cause that amino acid to be called both glucogenic and ketogenic (see Figure 25.4).

25.27 Name two products of the catabolism of phenylalanine that cause that amino acid to be called both glucogenic and ketogenic (see Figure 25.4).

Heritable Defects in Amino Acid Metabolism

25.28 What is the origin of heritable metabolic disorders of amino acid metabolism?

25.29 How are heritable metabolic defects in amino acid metabolism manifested?

25.30 What is the name of one of the heritable diseases caused by defects in phenylalanine catabolism?

25.31 What is the result of the disease named in Exercise 25.30 if untreated, and how can it be treated?

Biosynthesis of Amino Acids

25.32 What is meant by an essential amino acid?

25.33 What is meant by a nonessential amino acid?

Unclassified Exercises

25.34 True or false: The citric acid cycle is a source of amino acids. Explain your answer.

25.35 Aside from preventing its toxicity, is there any other advantage for converting ammonia into urea?

25.36 True or false: Dietary amino acids cannot be used to maintain normal blood-glucose concentrations.

25.37 Can arginase be extracted from the liver cells of a fish?

Chemical Connections

25.38 True or false: Ornithine is both a substrate and a product of the urea cycle. Explain your answer.

25.39 Aspartame, a nonnutritional sweetener, is L-aspartyl-L-phenylalanine methyl ester. Why do products containing it warn phenylketonurics not to use the product?

25.40 Explain how reductive amination allows a citric acid cycle intermediate to replenish supplies of amino acids.

25.41 True or false: Oxidative deamination of aspartate leads to the formation of pyruvate and ammonia. Explain your answer.

25.42 Humans can synthesize arginine but not in sufficient amounts. Is this amino acid considered essential or nonessential? Explain your answer.

CHAPTER 26

HORMONES AND THE CONTROL OF METABOLIC INTERRELATIONS

CHEMISTRY IN YOUR FUTURE

Every August, you work as resident nurse at a summer camp near your favorite mountain resort. The first days are busy, as the children wrestle with homesickness and as their bodies adjust to the moderately high altitude. The camp staff are well trained, however, and do not allow the children to overexert themselves. Soon the complaints of headache, nausea, sleeplessness, and malaise—whether due to homesickness or lack of oxygen—subside, leaving you time to enjoy your beautiful surroundings. The transport of oxygen through the blood (and the body's adjustment to breathing at high altitudes) is one of the topics in Chapter 26.

LEARNING OBJECTIVES

- Describe the cellular communication and mass-transport problems confronted by multicellular organisms.
- Describe the digestive process and the role of hormones in its regulation.
- Describe the chief metabolic requirements of the major organs.
- Describe the special role of the liver in metabolic regulation.
- Describe the biochemical and hormonal characteristics of the absorptive and postabsorptive states.
- Compare and contrast starvation and diabetes with the absorptive and postabsorptive states.
- Describe the transport of oxygen, carbon dioxide, cholesterol, and triacylglycerols in the blood

With minor variations, the processes that you have been studying in Chapters 22 through 25 are found in almost all living organisms. The cells of bacteria, mice, elephants, and human beings have some or all of the same central metabolic pathways and networks in common. Outside the cell, however, the functioning of single-cell and multicellular organisms differs greatly. Multicellular organisms have had to develop ways of coordinating and controlling the activities of their many specialized organs and tissues located in regions of the body remote from each other.

The goal in this chapter is to develop a coherent picture of the thousands of chemical reactions taking place simultaneously and continually in a familiar multicellular organism: a human being. To this purpose, we will compare and contrast physiological states in which the metabolic machinery is under stress and examine the body's efforts to reestablish homeostasis—the normal steady state of the body—in response to the different kinds of stress. Thus, we will look at the conditions in the body directly after a meal, about 4 h after a meal, during starvation, and in diabetes mellitus. Finally, we will explore some aspects of the transport of oxygen and carbon dioxide to and from actively metabolizing cells.

26.1 UNICELLULARITY VERSUS MULTICELLULARITY

◀◀ Membrane structure and permeability are explored in Section 19.10.

If a cell is to function properly, its internal environment must remain constant with respect to ion concentrations, pH, energy sources, cofactors, and so forth. Most of the responsibility for maintaining this stability falls to the cell's external membrane. This responsibility is a special challenge for single-celled organisms, such as bacteria or yeast, whose entire membrane faces a changeful external environment. No part of the membrane is exempt from protecting the cell's interior from fluctuations in the organism's surroundings. Such a membrane must be endowed with great biochemical versatility, and the cell must use a large part of its energy and chemical resources to execute that task. The payoff to such a creature, however, is that the exchange of nutrients and waste

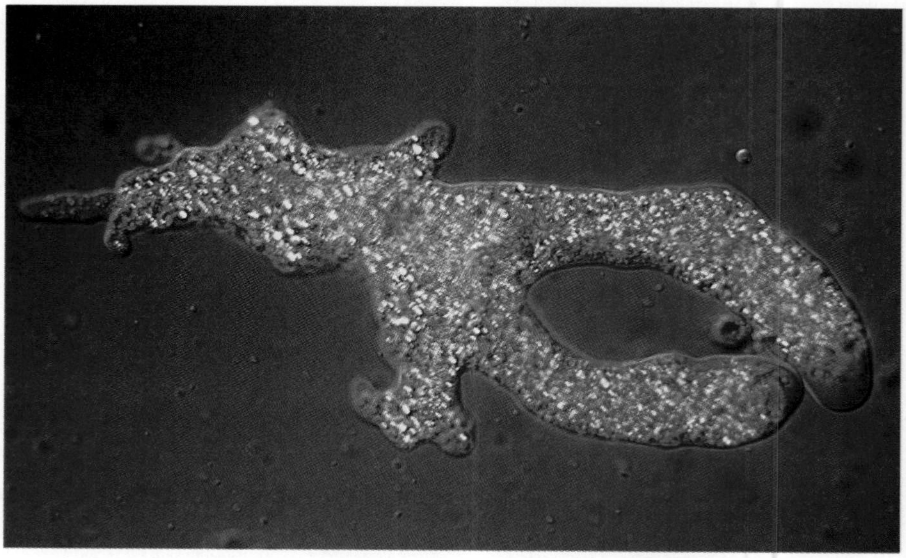

An amoeba prepares to engulf its prey. This single-celled organism has no need of complex transport systems to distribute nutrients.

products with the environment requires no special arrangements, because the distance between the outer and inner worlds is very short (the thickness of the membrane), and the exchange is rapidly accomplished by diffusion. The cells of multicellular organisms—living organisms that consist of many cells—face significantly fewer problems in maintaining their inner equilibrium than single-celled creatures do, but at the expense of the simplicity of a direct exchange of nutrients and waste products with the outside world.

In multicellular organisms, most cells are too remote from the external environment for direct exchange of matter and energy to be practicable. That problem is solved by transport systems such as the human vascular system, in which the blood rapidly carries nutrients to and waste products from all cells of the body. In addition to the diffusion problem, the coordinated control of the operation of tissues and organs that are remote from each other requires rapid communication between all parts of the body. All organs and tissues must be able to readjust their activities in response to the body's changing metabolic requirements. This readjustment is accomplished by the action of chemical and neuronal communication systems such as hormones, neurotransmitters (Section 26.5), and growth factors (Section 26.4).

◄◄ The limitations of diffusion in multicellular organisms are first laid out in Box 7.1.

26.2 HORMONES AND COMMUNICATION BETWEEN CELLS

The defining characteristics of the substances known as **hormones** are as follows: a hormone is secreted into the bloodstream in response to a neural or chemical signal and it acts at an organ or tissue other than its place of origin. Hormones can be low-molecular-mass substances (such as epinephrine), peptides (such as oxytocin), steroids (such as cortisol, Section 19.7 and Figure 19.4), or proteins (such as insulin). Many hormonal systems are governed by a part of the brain called the hypothalamus, but others, such as the digestive-system hormones, are subject to regulation within the target organs themselves.

When a hormone reaches its target organ, it alters some metabolic process in one of two general ways. One type of hormone—for example, water-soluble epinephrine (Section 22.6)—interacts with receptors on the surface of a cell. This interaction causes the production of a different molecule, a so-called second messenger, inside the cell. The second messenger modifies the activity of an enzyme or enzyme system within the cell. These hormones are rapidly inactivated, so their effects are short lived. The other type of hormone, such as the steroid hormones estradiol and testosterone, is carried through the bloodstream to its target cell, where, after entry, it forms a complex with a specific receptor protein. This intracellular hormone-receptor complex then interacts with the cell's DNA. The result is a modification of the level, or concentration, of a specific enzyme encoded by the DNA. The effects of this type of hormone are generally long lived. Both types act at very low concentrations, between micromolar and picomolar amounts.

Hormones (which were discussed at greater length in Section 23.6) are central to the coordinated control of metabolism, but even more critical are the concentrations of nutrients such as glucose, fatty acids, and amino acids in determining which catabolic and anabolic pathways the cell takes. These substances enter the human body in forms that are not amenable to transport across cell membranes. We therefore begin our consideration of metabolic interrelations with a description of the digestive processes that transform food into nutrients.

26.3 DIGESTIVE PROCESSES

In digestion, foods are enzymatically degraded to low-molecular-mass components to prepare them for absorption into the cells lining the gut (through which they pass into the bloodstream). This enzymatic degradation is necessary because the cells lining the intestine can absorb only small molecules, and most nutrients are ingested in the form of biopolymers—that is, proteins and carbohydrates.

The reduction of the major components of food—proteins, carbohydrates, and triacylglycerols—to low-molecular-mass components begins in the mouth, with the mechanical action of chewing and the secretion of amylase (a starch-degrading enzyme) in the saliva. The next stage takes place in the stomach, where secretion of the hormone gastrin is stimulated by the entry of protein into the stomach. A summary of the secretions of the human digestive system can be found in Table 26.1.

Gastrin stimulates the gastric glands in the stomach's lining to secrete pepsinogen and the parietal cells in the stomach's lining to secrete HCl (Figure 26.1). The HCl brings the pH of the stomach to between 1.5 and 2.0, a level of acidity that denatures, or unfolds, proteins, making their internal peptide bonds accessible to enzymatic hydrolysis.

Pepsinogen is a **zymogen,** or inactive enzyme precursor. The HCl in the stomach converts some of the pepsinogen into the active enzyme, pepsin, by removing a small terminal peptide. The pepsin thus activated becomes the catalyst for the rapid conversion of the remaining pepsinogen into pepsin. Pepsin, which attacks the peptide bonds of amino acids possessing hydrophobic side groups (leucine and phenylalanine, for example), reduces large proteins to mixtures of smaller peptides. Other enzymatic hydrolases secreted as zymogens (proenzymes) and activated by similar processing are found in the small intestine.

As the stomach contents pass into the small intestine, their low pH stimulates secretion of the hormone secretin by duodenal cells. When this hormone reaches the pancreas, it stimulates that organ to secrete bicarbonate into the gut. Because intestinal digestive enzymes function at or near neutral pH, the

TABLE 26.1 Secretions of the Human Digestive System

Location	Proenzyme (Zymogen)	Enzyme	Hormone	Other
mouth		amylase		
stomach	pepsinogen	pepsin	gastrin	HCl
intestine		enterokinase carboxypeptidase aminopeptidase	secretin	
within intestinal cells		nucleotidases disaccharidases		
pancreas	trypsinogen chymotrypsinogen prolipase	trypsin chymotrypsin lipase amylase		HCO_3^-
liver		colipase		bile salts

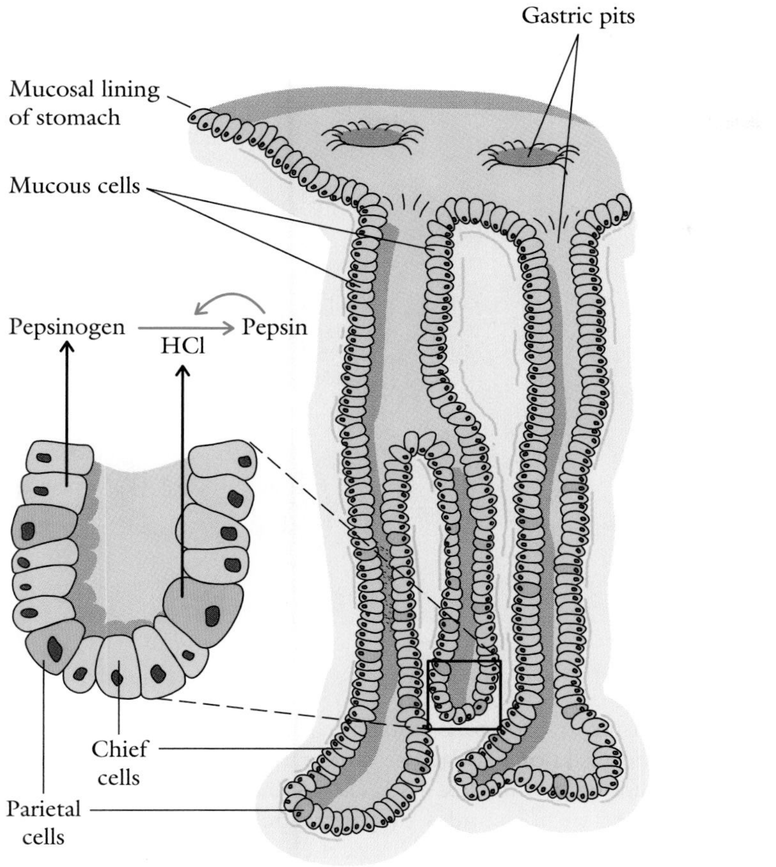

Figure 26.1 Gastric glands are located within gastric pits in the stomach lining. These glands contain both chief cells, which secrete pepsinogen, and parietal cells, which secrete HCl. (Adapted from Figure 37.12, page 802, Campbell, *Biology,* 3d ed. Benjamin Cummings: Redwood City, CA).

secretion of bicarbonate is necessary to neutralize the low pH of the entering stomach contents.

As amino acids enter the small intestine, they stimulate intestinal-cell secretion of the specialized enzyme enterokinase. This enzyme converts the zymogen trypsinogen, secreted by the pancreas, into the active proteolytic enzyme trypsin. Trypsin then converts pancreatic chymotrypsinogen into the active chymotrypsin, and the two enzymes proceed to reduce polypeptides to small peptides. The further hydrolysis of the small peptides to their constituent amino acids is accomplished by two other enzymes secreted by the pancreas and intestinal cells: carboxypeptidase, which attacks peptides at the free carboxyl end, and aminopeptidase, which attacks peptides at the free amino end. The mixture of amino acids is then transported across the intestinal cells and into the blood, which carries it directly to the liver.

The principal carbohydrates in food are starch and cellulose (from plants) and glycogen (from animals). Mammals lack enzymes capable of hydrolyzing the $\beta(1 \rightarrow 4)$ linkages of cellulose, which therefore passes through and out of the human gut in the form of roughage (Box 18.4). The hydrolysis of starch and glycogen at the $\alpha(1 \rightarrow 4)$ linkage, begun in the mouth, is completed in the small intestine chiefly by the action of pancreatic amylase. Disaccharides derived directly from dietary sources (such as sucrose from fruits and lactose from milk) are hydrolyzed to a mixture of monosaccharides by enzymes located within cells lining the small intestine; the monosaccharides are then absorbed into the blood and brought directly to the liver. However, many people lack the enzyme β-galactosidase, which hydrolyzes lactose (Box 18.2). In these persons, lactose remains in the gut, where it is partly fermented by intestinal

People with lactose intolerance can substitute a variety of soy-based nutritional supplements and drinks for lactose-containing milk.

microorganisms, a process resulting in diarrhea and the formation of gases. This condition is known as lactose intolerance.

Small amounts of nucleic acids (DNA and RNA) are present in food. They are hydrolyzed to nucleotides by pancreatic enzymes. Enzymes from the epithelial cells lining the intestine break the nucleotides down to free bases (Section 21.1) and monosaccharides.

The digestion of triacylglycerols begins in the small intestine. The pancreas secretes a zymogen called prolipase, which is converted into the active lipase by intestinal proteases. In the presence of a special protein called colipase and bile salts synthesized in the liver and stored in the gall bladder, lipase begins the hydrolysis of fatty acids from triacylglycerols. The bile salts are emulsifying agents that, in combination with the churning action of the intestine, produce a suspension of triacylglycerol droplets with sufficiently large total surface area to permit efficient enzymatic hydrolysis (Box 6.1 and Section 19.7).

Generally, only one or two fatty acid chains of a triacylglycerol are released by hydrolysis. The product is therefore a mixture of sodium and potassium salts of fatty acids along with monoacylglycerols. The mixture is absorbed by intestinal cells and reassembled within them into new triacylglycerols. These reassembled triacylglycerols combine with protein to form **chylomicrons** (Section 19.4), droplets of triacylglycerols surrounded by mono- and diacylglycerols covered by a surface layer of protein. The chylomicrons leave the intestinal cells by exocytosis (Section 19.10, Figure 19.8) to enter the interstitial space between the intestinal cells and the vascular system.

▶▶ Chylomicrons are considered further in Sections 26.5 and 26.10.

The chylomicrons do not penetrate the capillaries to be carried directly to the portal circulation of the liver, as do amino acids and monosaccharides. Instead, they pass into the lymphatic system by means of small lymph vessels called lacteals in the intestinal walls. The lymphatic system eventually connects with the vascular system at the thoracic duct, where the contents of the lymphatic system empty into the subclavian vein and enter the blood. Lipids absorbed into the blood combine with proteins produced by the liver to form **lipoproteins.** The details of lipoprotein structure, transport, and interaction with cells are described in Section 26.10.

26.4 NUTRITION

The end product of digestion is a complex mixture of biomolecules that must be sufficient to fulfill an organism's requirements for biosynthesis, motion (muscle contraction), ion transport, and secretion. To be adequate, it must contain five basic classes of nutrients: energy sources, essential amino acids, vitamins, minerals, and essential fatty acids. The components of each class are listed in Table 26.2.

The energy content of foodstuffs varies. The approximate caloric values of generic carbohydrates, fatty acids, and proteins listed in Table 26.3 on the following page are based on a varied diet in which all nutritional components are present. (Recall, for example, that humans cannot synthesize glucose from fatty acids. If carbohydrate is absent from the diet, fatty acid metabolism becomes inefficient, and the caloric value of fatty acids decreases.)

◀◀ Categories of amino acids are summarized in Table 20.1.

The energy requirement of a body at complete rest 12 h after eating is called the **basal metabolic rate** (Section 1.11), which is considered the energy required to maintain the body's basic "housekeeping" functions. The basal metabolic rates for men and women in their early twenties are 1800 and 1300 kcal/day, respectively. However, a person's daily activity controls his or her total caloric requirement. In general, variations depend on the extent of muscular

TABLE 26.2 Nutrients Required by Humans

ENERGY SOURCES		
carbohydrates		
fats		
proteins		
ESSENTIAL AMINO ACIDS		
arginine	lysine	threonine
histidine	methionine	tryptophan
isoleucine	phenylalanine	valine
leucine		
ESSENTIAL FATTY ACIDS		
linoleic acid		
linolenic acid		
VITAMINS		
thiamine	pantothenic acid	vitamin B_{12}
riboflavin	folic acid	ascorbic acid
niacin	biotin	vitamins A, D, E, K
pyridoxine		
MINERALS		
arsenic	iron	selenium
calcium	magnesium	silicon
chlorine	nickel	sodium
chromium	molybdenum	tin
copper	phosphorus	vanadium
fluorine	potassium	zinc
iodine		

TABLE 26.3 Energy Equivalents of Nutrients

Nutrient	Energy Content (kcal/g)*
carbohydrates	4.0
fats	9.2
proteins	4.2

*Per gram of dry weight.

activity, body weight, age, and sex. Table 26.4 lists the recommended caloric intake as a function of sex and age.

Proteins

A diet of only glucose would be adequate to fulfill all the body's carbohydrate requirements. However, the requirements for amino acids and fatty acids are more complicated, because certain amino acids and fatty acids are considered essential and others nonessential from a dietary point of view. Of the 20 amino acids required for protein synthesis, there are 10 that humans either cannot synthesize at all or cannot synthesize in sufficient quantities. Those ten essential amino acids must be obtained from the diet.

Proteins are required not for their caloric value but for their content of amino acids. The biosynthesis of specific proteins demands that each required amino acid be present at the synthesis site or synthesis will cease. If even one amino acid needed for the protein is not present, synthesis will stop, and the previously synthesized nascent polypeptide chain will be dismantled. Experi-

TABLE 26.4 Recommended Daily Energy Allowances

	Age (years)	Weight (kg)	Energy (kcal)
infants	0.0–0.5	6	650
	0.5–1.0	9	970
children	1–3	13	1300
	4–6	20	1700
	7–10	28	2400
females	11–14	46	2200
	15–18	55	2100
	19–22	55	2100
	23–50	55	2000
	50+	55	1800
males	11–14	45	2700
	15–18	66	2800
	19–22	70	2900
	23–50	70	2700
	50+	70	2400

mental animals fed a synthetic diet of amino acids do not grow if one essential amino acid is omitted from the diet. However, when that amino acid is added, growth begins within hours.

In evaluating the protein content of a meal, we must ask two questions: (1) does a food protein contain the correct types of amino acids and (2) how accessible are the amino acids—that is, how digestible is the food? These qualities of a dietary protein are expressed as its **biological value.** For example, if a given protein provides all the required amino acids in the proper proportions and all are released on digestion and absorbed, the protein is said to have a biological value of 100. When the biological value of a protein is high, only small daily amounts of that protein are required to keep a person in nitrogen balance. **Nitrogen balance** means that the body's intake of protein nitrogen is equal to the nitrogen excreted in urine and feces.

A protein's biological value will be less than 100 if (1) it is incompletely digestible, as is true of keratin; (2) it is a protein of plant origin surrounded by cellulosic husks, as is true of cereal grains; (3) it is deficient in one or more essential amino acids. In the last case, large quantities of the protein would have to be ingested to obtain enough of the essential amino acid, whereas the amino acids in abundance would be used calorically (see Box 20.1).

A somewhat different but experimentally useful protein classification is called the **chemical score,** obtained by completely hydrolyzing the protein and comparing its amino acid composition with that of human milk. The biological value and chemical score of some food proteins are listed in Table 26.5.

Fatty Acids

Fatty acids containing more than one unsaturated bond past carbon 9 of a saturated chain, counting from the carboxyl end, cannot be synthesized by humans. Therefore, two polyunsaturated fatty acids of plant origin, linoleic acid and linolenic acid, are essential to human nutrition. They are used by mammals to synthesize arachidonic acid, which in turn is used to synthesize leucotrienes, thromboxanes, and prostaglandins, a family of lipid-soluble organic acids that have hormonelike physiological functions (Section 19.8 and Figure 19.5). Deficiencies in these fatty acids are rare, because they are present in abundance in edible plants and in fowl and fish.

Although the caloric values of saturated and unsaturated fatty acids are comparable, the proportion of saturated to unsaturated fatty acids in the diet has significant physiological consequences. A great deal of evidence has accumulated over many years that correlates decreased concentrations of high-density lipoproteins, increased concentrations of low-density lipoproteins, and

TABLE 26.5 Chemical Scores and Biological Values of Some Food Proteins

Protein Source	Chemical Score	Biological Value
human milk	100	95
beefsteak	98	93
whole egg	100	87
cow's milk	95	81
corn	49	36
polished rice	67	63
whole wheat bread	47	30

British sailors came to be called "limeys" because they consumed limes in order to prevent scurvy.

total blood cholesterol with diets that are rich in saturated fatty acids. The studies also relate such diets to a predisposition to develop coronary artery disease. For this reason, experts are urging the people of developed countries (where foods tend to be high in saturated fatty acids) to increase the proportion of unsaturated fatty acids in the diet (Box 24.1). The typical compositions of various plant oils and animal fats are listed in Table 19.2.

Vitamins

Many vitamins—vitamin C (ascorbic acid), for example—were discovered when a disorder caused by their absence from the diet was cured by their addition to the diet. One of the earliest documentations of a vitamin deficiency appears in the journals of Jacques Cartier, who explored North America in 1535. He described a disease that came to be known as scurvy, which was manifested in his sailors as terrible skin disorders accompanied by tooth loss. Two hundred years later, a British physician found that he could cure scurvy by adding citrus fruits such as lemons and limes to the diet and, 200 years after that, in 1932, the antiscurvy vitamin was isolated and given the name vitamin C.

Another deficiency disease, known as beriberi, is characterized by neurological disorders. Originally thought to be an infectious disease, beriberi was unknown until the early nineteenth century, when rice-polishing machines were invented to remove the brown outer hull of the rice seed. The cure of beriberi was discovered when the addition of the outer hull of rice to a victim's diet completely reversed the symptoms. The critical dietary component found in rice hulls is thiamine, the coenzyme in decarboxylations. The blood of people with a thiamine deficiency contains elevated levels of pyruvate, which must be decarboxylated before it can enter the TCA cycle.

Other vitamins are known as growth factors, because experiments showed that test animals do not grow if certain substances other than carbohydrates, triacylglycerols, and proteins are omitted from the diet. Vitamins were chemically analyzed, and, in many cases, their metabolic roles as enzyme cofactors were identified. In other cases, the vitamin's precise biochemical function is yet to be made clear.

Vitamins, both water soluble and fat soluble (Section 19.9), can be divided into two groups that depend on the effects of their deficiencies. The first group includes thiamine, riboflavin, niacin, ascorbic acid, and folic acid. In affluent countries, marginal deficiencies of these vitamins are common, but in many parts of the world, the deficiencies are great enough to be life threatening. The other group of vitamins includes pyridoxine, pantothenic acid, biotin, vitamin B_{12}, and the fat-soluble vitamins A, D, E, and K. Deficiencies of these vitamins are rare. However, people suffering from fat-absorption disorders are deficient in the fat-soluble vitamins; vitamin A deficiency is the cause of xerophthalmia, or night blindness, owing to an insufficient synthesis of the visual pigment rhodopsin. Pyridoxine is needed for transaminations, so the body's requirement depends on the quantity of protein in the diet—the more dietary protein, the greater the need for pyridoxine. Biotin, pantothenic acid, and vitamin B_{12} are not ordinarily required in the diet, because they are usually synthesized in adequate amounts by intestinal bacteria. Nevertheless, a diet rich in egg-white protein can cause a serious biotin deficiency, because egg white contains the protein avidin, which binds very strongly to biotin to form an avidin-biotin complex that cannot be absorbed. Vitamin B_{12} deficiency, resulting in pernicious anemia, does occasionally occur, and its likely cause is the absence of **intrinsic factor,** a glycoprotein synthesized by the stomach. Vitamin B_{12} must be transported across the intestinal-cell membrane as a complex with intrinsic factor. In people who cannot synthesize this protein, the vitamin must be administered by injection directly into the bloodstream.

The fat-soluble vitamins are stored in body fat and therefore need not be ingested daily. However, the water-soluble vitamins are excreted or destroyed in the course of metabolic turnover and must be replaced by regular ingestion. Table 26.6 lists the known essential vitamins (both the fat soluble and the water soluble), their coenzyme forms if known, their biological functions, and the recommended daily allowance (RDA) for men between 23 and 50 years of age.

Minerals

Carbohydrates, proteins, triacylglycerols, and nucleic acids are composed of six elements: carbon, hydrogen, nitrogen, oxygen, phosphorus, and sulfur. In addition to these elements, many other minerals are required for experimental mammals and presumed to be required for humans. These minerals are divided

TABLE 26.6 Vitamin Needs of Men 23–50 Years of Age

Vitamin	Coenzyme Form, Where Known	Metabolic Role or Associated Deficiency Disease or Both, Where Known	Recommended Daily Allowance
thiamine	thiamine pyrophosphate	decarboxylation coenzyme; deficiency causes beriberi	1.5 mg
niacin	NAD^+	dehydrogenase coenzyme; deficiency causes pellagra	19 mg
ascorbic acid	unknown	unknown; deficiency causes scurvy	60 mg
riboflavin	FAD	dehydrogenase coenzyme	1.7 mg
pyridoxine	pyridoxal phosphate	transamination coenzyme	2.2 mg
folic acid	tetrahydrofolate	one-carbon-group transfer; deficiency causes anemia	400 μg
pantothenic acid	coenzyme A	fatty acid oxidation	5–10 mg
biotin	biocytin	CO_2-transferring enzymes	150 μg
vitamin B_{12}	deoxyadenosylcobalamine	odd-numbered fatty acid oxidation; deficiency causes pernicious anemia	3 μg
vitamin A_1	unknown	visual cycle intermediate; deficiency causes night blindness (xerophthalmia)	1 mg
vitamin D_3	1,25-dihydroxy-cholecalciferol	hormone controlling calcium and phosphate metabolism; deficiency causes rickets	10 μg
vitamin E	unknown	protects against damage to membranes by oxygen; deficiency causes liver degeneration	10 mg
vitamin K_1	unknown	activation of prothrombin; deficiency causes disorders in blood clotting	1 mg

TABLE 26.7 Minerals Required by Humans

BULK ELEMENTS*		
calcium	magnesium	potassium
chlorine	phosphorus	sodium
TRACE ELEMENTS†		
copper	molybdenum	nickel‡
fluorine	selenium	silicon‡
iodine	zinc	tin‡
iron	arsenic‡	vanadium‡
manganese	chromium‡	

* These minerals are required in doses higher than 100 mg/day.
† These minerals are required in doses of 1–3 mg/day.
‡ These minerals are known to be required in test animals and are likely to be required in humans.

into two groups—bulk and trace minerals—and are presented in Table 26.7. Table 26.8 lists some minerals whose functions are known or whose deficiencies result in well-recognized symptoms.

26.5 METABOLIC CHARACTERISTICS OF THE MAJOR ORGANS AND TISSUES

The cells of different organs have different metabolic requirements. Some use all three principal nutrients—glucose, fatty acids, and amino acids. Others may

TABLE 26.8 Minerals and Their Nutritional Functions

Element	Nutritional Function
calcium	bones, teeth
phosphorus	bones, teeth
magnesium	cofactor for many enzymes
potassium	water, electrolyte, acid-base balance; location is intracellular
sodium	water, electrolyte, acid-base balance; location is extracellular
iron	iron porphyrin proteins
copper	iron porphyrin synthesis and cytochrome oxidase
iodine	synthesis of thyroxin; lack leads to goiter
fluorine	forms fluoroapatite, strengthens bones and teeth
zinc	cofactor for many enzymes
tin	growth factor for mammals grown in ultraclean conditions*
nickel	growth factor for mammals grown in ultraclean conditions*
vanadium	growth factor for mammals grown in ultraclean conditions*
chromium	growth factor for mammals grown in ultraclean conditions*
silicon	growth factor for mammals grown in ultraclean conditions*
selenium	component of the enzyme glutathione peroxidase
molybdenum	component of the enzymes xanthine and aldehyde oxidases

*Probably required under normal conditions.

use only glucose or principally fatty acids. A good first step toward learning how metabolism is coordinated is to take a metabolic inventory of the major organ systems and tissues.

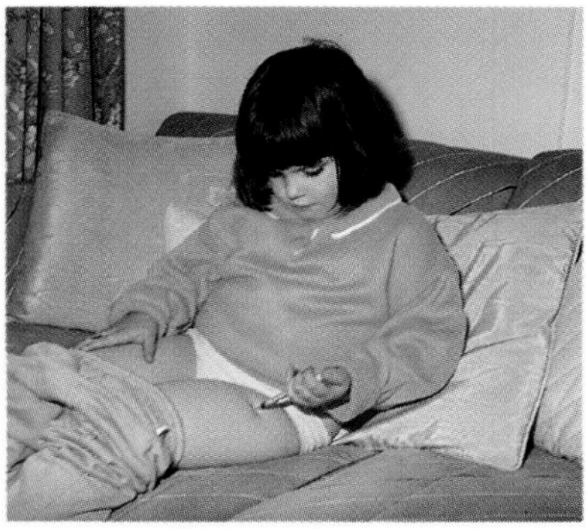

Young diabetic girl injecting herself with insulin. Diabetes mellitus is a disorder of carbohydrate metabolism due to a lack of the pancreatic hormone insulin. Treatment requires the use of daily insulin injections.

Brain and Nerve Tissue

The brain contains no stored energy sources and uses only glucose for its energy needs, which amount to about 60% of the total resting human glucose consumption. Its very active aerobic metabolism uses about 20% of the total oxygen consumed by the body at rest. Even more interesting, the actual amounts of oxygen consumed in liters per minute remain constant whether a person is asleep or actively thinking. Because the brain stores no glycogen or triacylglycerol, it depends critically on the constant availability of glucose from the circulating blood. This dependency means that our blood glucose must be maintained at a constant concentration at all times (see Table 26.12). Brain function can undergo significant and irreversible damage if blood glucose should fall below critical levels for even quite short periods of time. Although it cannot use fatty acids, the brain can adapt to use 3-hydroxybutyrate (Section 24.4) as an energy source under certain circumstances, such as starvation. The transport of glucose into brain cells is noninsulin dependent (Section 26.6); so, provided that glucose is above minimal blood levels, brain function in diabetic patients is unaffected.

Electrical activity of the brain The special function of the brain rests on its ability to generate a large electrical potential across its cell membranes and to transmit an electrical signal from cell to cell. The anatomical features of nerve cells necessary to this function are presented in Box 26.1 on the following page.

The electrical potential across nerve-cell membranes is the result of an asymmetric distribution of sodium and potassium ions across nerve-cell membranes. The concentration of potassium ions within nerve cells is about 125 m*M*, and the concentration outside is about 10 m*M*. The concentration of sodium ions outside is about 150 m*M*; inside it is about 15 m*M*. These concentration asymmetries are created by a unique membrane active transport enzyme system called the **sodium potassium ATPase** (**Na^+/K^+ ATPase,** sometimes called the Na^+/K^+ pump; Section 19.10). Utilizing a continuous supply of ATP, this enzyme system literally pumps three Na^+ ions out of nerve cells while simultaneously pumping two K^+ ions into nerve cells for every ATP hydrolyzed. The resulting asymmetric distribution generates an electrical potential of about -60 to -70 millivolts (mV) with respect to the outside of the membrane. (The inside of the cell is negatively charged with respect to the outside of the cell.) The Na^+/K^+ ATPase is found in many other cells as well, notably intestinal epithelium and the cells lining the kidney tubule.

When an incoming signal sufficiently reverses this polarization at a synaptic contact (described in Box 26.1), it causes membrane depolarization immediately adjacent to it, which is repeated again and again along the membrane. The depolarization travels down the axon membrane and is called an **action potential.** The axon therefore functions to transmit a signal rapidly, as fast as 90 m/s, over long distances. For example, the sciatic nerve extends from the lower end of your spinal cord to your foot.

When an action potential reaches a presynaptic terminal, neurotransmitters are released into the synaptic cleft, diffuse to the postsynaptic terminal, bind to receptor sites there, and affect the membrane potential. In this way, a nerve impulse is transmitted from one cell to another. The neurotransmitters must then be rapidly inactivated or the next incoming signal will not be detected. Inactivation of the neurotransmitter is accomplished either enzymatically or by its reabsorption into the presynaptic cell.

26.1 Chemistry Within Us

Nerve Anatomy

Nerve cells, or **neurons,** have a unique architecture. Short structures called dendrites emerge from one end of the neuronal cell body, and a single, long extension called the axon emerges from the other end. The axon may divide into many special branches called **synaptic terminals.**

There is no direct contact between neurons. They are separated by a space called the **synaptic cleft.** When a signal is transmitted from one neuron to the next, the end of the neuron that is sending the signal is called the **presynaptic ending,** and the part of the membrane of the adjacent neuron receiving the signal is called the **postsynaptic terminal.** All these functional components in the nerve junction constitute the **synapse.** The signal is transmitted across the synaptic cleft by chemicals called **neurotransmitters.** Neurotransmitters are synthesized in nerve-cell bodies and stored in secretory vesicles located in the presynaptic ending.

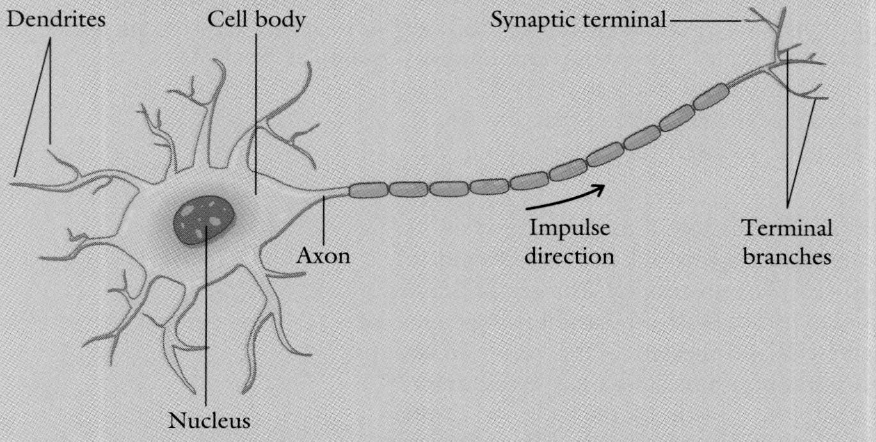

The structure of a typical vertebrate neuron. Two kinds of structures extend from the cell body: dendrites and an axon. The dendrites receive signals from other neurons, and the axon conveys signals away from the cell body. Synaptic terminals at the end of the axon make connections to other neurons.

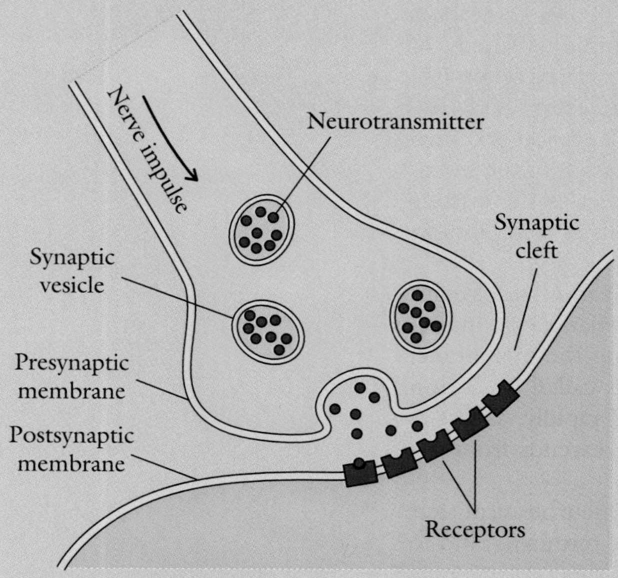

A synapse. An action potential causes synaptic vesicles in the presynaptic cell to fuse with the presynaptic membrane. Neurotransmitters are released into the synaptic cleft and diffuse to the postsynaptic membrane. There they bind to receptors that modify the permeability of the membrane, which in turn alters the membrane potential.

Sensory neurons, or **receptor cells,** are neurons that are specialized to detect particular environmental changes, reacting, for example, to temperature, touch, or particular chemicals. They respond to changes in a given variable by becoming depolarized to varying extents, depending on the intensity of the environmental change. Sensory neurons generate action potentials spontaneously and continuously. The different levels of depolarization cause the action potentials to be generated at different frequencies—for example 5/s, 15/s, or 35/s. Consequently, the intensity of environmental change is signaled by the frequency of action potentials, called the **nerve discharge.** The signal must be conveyed to some effector cell or organ in order for a response to be made to the receptor cell's output. Effector cells and organs are specialized to carry out functions appropriate to the incoming signal; for example, to cause muscle contraction or chemical secretion.

Many neurotransmitter and inhibitor substances function in the brain, between neurons outside the brain, and between neurons and effector organs such as muscles. Among them are aspartate, γ-aminobutyrate (GABA), glutamate, glutamine, glycine, and acetylcholine. Most of them and a number of other peptides and amino acid derivatives act specifically in particular regions of the brain.

Acetylcholine functions as a neurotransmitter both in the brain and at neuromuscular junctions; that is, at the junction between nerve and muscle. There it is inactivated by the enzyme acetylcholine esterase, which hydrolyzes acetylcholine to acetate and choline. These components must then be reabsorbed, resynthesized into acetylcholine, and stored in presynaptic vesicles. All these processes require considerable amounts of ATP and account for the large consumption of oxygen by nerve tissue.

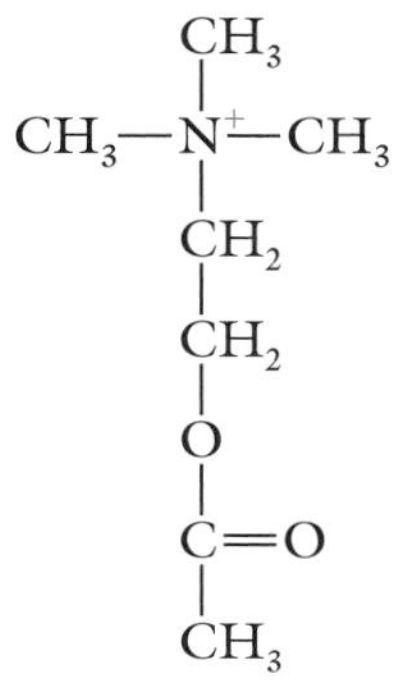
Acetylcholine

Hormones of the brain Among its other brain functions, the hypothalamus is the site of synthesis of a large number of hormones and hormone-releasing agents. These substances can affect organ systems directly or cause other endocrine glands to secrete their hormones. For example, the hypothalamus secretes a specific **hypothalamic releasing hormone (HRH)** that causes the anterior pituitary gland to secrete thyrotropic hormone, which, in turn, stimulates the thyroid gland to secrete thyroxin. Another HRH stimulates the anterior pituitary to secrete β-corticotropin, which stimulates the adrenal cortex to secrete cortisol (Figure 19.4). Yet another HRH causes the anterior pituitary to secrete prolactin, which acts directly on the mammary glands to stimulate milk production.

The hypothalamus releases hormones in response to input from the central nervous system. In higher animals, the central nervous system responds either to signals that are generated internally or to signals that originate in the outside environment, as illustrated in Figure 26.2 on the following page. After a signal has been transmitted either chemically or electrically to the hypothalamus, the signal is relayed to the pituitary and from there to a target gland, which secretes the final hormone that causes systemic events. Examples of target glands are the thyroid gland—which produces thyroxin, a hormone that influences cellular oxidations—and the adrenal gland—which produces cortisol, a steroid hormone that mobilizes the body's defenses against stress. As the signal progresses along the pathway, each subsequent factor or hormone is released in a larger amount. Therefore the original signal becomes amplified in intensity, and the amplification is often referred to as a hormonal cascade. The releasing hormones are usually secreted in nanogram amounts, the hormones released from the anterior pituitary may be in microgram amounts, and the concentration of the ultimate hormone released from the target—for example, cortisol from the adrenal gland—can be in milligram amounts.

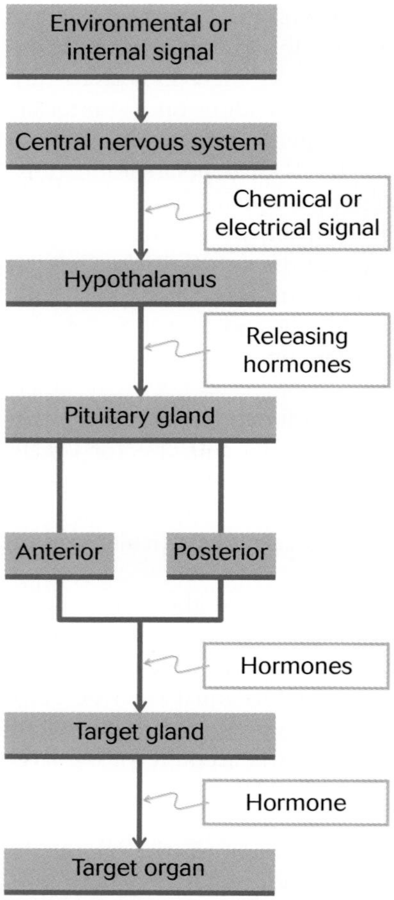

Figure 26.2 The signal pathway for hypothalamic releasing hormones in higher animals. Signals originate either internally or in the outside environment. The signal is then transmitted, either chemically or electrically (by specific neurons), first to the hypothalamus, then to the pituitary, and from there to a target gland, which secretes the final hormone that causes systemic events.

The posterior pituitary does not synthesize hormones but stores oxytocin and vasopressin and releases them in response to specific releasing hormones from the hypothalamus. Table 26.9 is a list of some hypothalamic releasing hormones and the anterior pituitary hormones that they affect.

TABLE 26.9 Some Hypothalamic Releasing Hormones

Releasing Hormone	Anterior Pituitary Hormone Effect
thyrotropic releasing hormone	thyrotropin released
gonadotropin-releasing hormone	lactogenic hormone and follicle-stimulating hormone (FSH) released
gonadotropin release inhibiting factor	lactogenic hormone and FSH release inhibited
corticotropin-releasing hormone (CRH)	adrenocorticotrophic hormone (ACTH) released
somatocrinin (growth-releasing hormone, GRH)	growth hormone (GH) released
somatostatin (growth-inhibiting hormone, GIH)	GH release inhibited

PROFESSIONAL CONNECTIONS

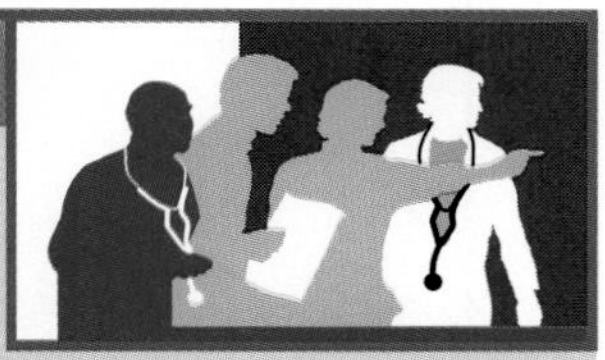

Gerald Bennett, Ph.D., Psychiatric Nurse and Health Science Educator

Gerald Bennett took chemistry as an undergraduate at the Medical College of Georgia.

Why did you choose the position that you are in today? What do you enjoy the most?

I'm an associate professor at a health science university, involved in three ongoing activities: teaching, research, and practicing psychiatric nursing, I learned what makes the best use of my talents: I not only love learning and books and the intellectual stimulation of a college environment, but also like to apply the knowledge in a direct way to help people. . . . I was exposed to several options when I was interested in a mental health career and saw psychiatric nursing as a balance of the social and biological sciences.

What positions did you have before this one?

In the past, I've been the head of this department, an instructor at another institution, and a staff nurse in a psychiatric unit in a large hospital.

What advances have been made in your field?

A major advance is in the neurosciences. Our understanding of the brain is increasing rapidly and, along with that, our knowledge of using drugs to treat mental illness. Drugs have been used for some time to treat mental illness, but they had only general effects and carried with them many side effects. The most recent research allows us to be specific and target precise behaviors with particular drugs.

In addition, the psychiatric nursing role is better recognized, particularly for advanced practitioners in psychiatric nursing, which requires a master's degree. This requirement varies from state to state, but generally they are able to provide psychiatric therapy and drugs to patients.

Have these advances changed your working life?

They have changed the scope of what you can do for patients. Psychopharmacology requires different types of reading and focus—the biological sciences have become more important. For decades, the behavioral sciences base was always dominant, but that's not so today.

Do you use chemistry in your job?

This has been called the Decade of the Brain in psychiatric health nursing. We didn't use chemistry on a daily basis that much until the past decade, but now the biochemistry of the brain and knowledge of . . . neurons, neurotransmitters, and the effect of drugs on them and how they help particular disorders are critical. We also need the chemistry knowledge to explain to patients how the drugs help them.

What advice do you have for students who are studying chemistry today?

This is a very exciting time for health science students. To keep connected with progress, a solid foundation in the sciences is critical. I hope that students will have a goal in their courses to get a grasp of basic concepts for the future, rather than just memorizing for the next exam. A solid foundation will allow them to develop a sophisticated way of thinking . . . about health and disease when they emerge from school. The connections between the basic concepts and their careers are important, and I would encourage students not to be shy about asking their professors about the relevance and practical issues of their course of study.

Heart

The heart contains almost no stored energy (glycogen or triacylglycerol) and, although it can use a variety of metabolic fuels (glucose, fatty acids, lactate, and ketone bodies), most of its energy is derived from the oxidation of fatty acids. As much as 50% of heart muscle cell volume is taken up by mitochondria carrying out oxidative processes that lead to the production of ATP. Because heart muscle, unlike skeletal muscle, cannot function anaerobically for a brief period, a lack of oxygen results immediately in cell death. When oxygenated blood is prevented from reaching the heart because of blockage in blood vessels, a process called **myocardial infarction** (heart attack) occurs (Box 24.1).

Muscle

Muscle can utilize a variety of fuels, but the choice depends on the muscle's degree of activity. The principal energy source of resting muscle is fatty acids. When exertion begins, glycogen reserves are mobilized to provide glucose in the form of glucose-6-phosphate. (About 75% of the body's glycogen supply is located in muscle, but, because muscle cells do not possess glucose-6-phosphatase, they cannot provide free glucose to other organs.) The rate of glycolysis, which produces lactate, is much greater than the rate of the citric acid cycle, which breaks it down; so lactate accumulates and is released into the blood. Alanine produced from pyruvate by transamination also is released (see Section 25.3), and both products are transported to the liver to be converted into glucose through gluconeogenesis. The glucose from liver is then transported back to the muscle as well as other tissues.

In addition to glycogen, muscle possesses a second energy storage depot, **creatine phosphate.** This high-energy compound can phosphorylate ADP to produce ATP and thus supply energy for a short time in periods of extreme exertion. Its breakdown product, creatinine, is a normal component of urine.

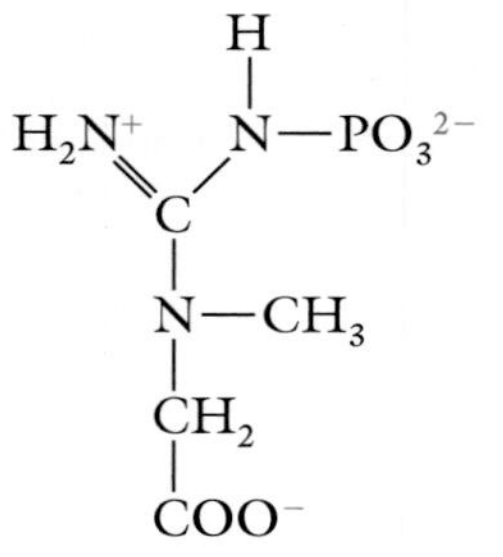

Creatine phosphate

Adipose Tissue

Adipose tissue consists of cells called **adipocytes.** About 17% of an average human male weighing 70 kg consists of triacylglycerol stored in adipose tissue. This stored triacylglycerol represents about 110,000 kcal (450,000 kJ) of stored energy—enough to sustain life for a few months.

When chylomicrons from intestinal absorption reach adipocytes, they are acted on by the lipoprotein lipase on the cell surface. The fatty acids freed by this action are either complexed with serum albumin for transport to other tissues for oxidation or absorbed into the adipocyte for storage.

The breakdown of triacylglycerols within adipocytes depends on the activity of a hormone-sensitive lipase. This internal lipase is activated by glucagon, which activates a phosphorylation cascade, as in glycogenolysis. The hormone insulin reverses this stimulation. As a result, when glucose levels are high, the rate of synthesis of triacylglycerols exceeds the rate of breakdown; however, when the glucose level falls and glucagon levels increase, triacylglycerol breakdown exceeds synthesis and fatty acids are released from the cell. These free fatty acids combine with the blood protein serum albumin to form soluble complexes that are thus able to travel through the blood to other tissues. The glycerol from the hydrolysis of triacylglycerols also enters the blood and travels to the liver, where it is used to produce glucose through gluconeogenesis.

The synthesis of triacylglycerols requires the presence of glycerol-3-phosphate as well as fatty acids. Glycerol-3-phosphate is produced by the reduction of dihydroxyacetone phosphate from glycolysis. The supply of glycerol-3-phosphate is controlled by the cell's concentration of glucose. Insulin stimulates the uptake of glucose in adipose tissue, and triacylglycerol synthesis will take place as long as the glucose supply is adequate. Lipid transport and the role of insulin are considered further in Section 26.6.

Kidney

About 80% of the oxygen consumed by the kidneys generates the ATP used to pump Na^+ into the interstitial space surrounding the kidney tubules. The high ion concentration thus created outside the tubules establishes a strong osmotic gradient that withdraws water from the ultrafiltrate flowing through the tubules and therefore concentrates the urine. At the same time, other important substances, such as glucose, are actively transported from the ultrafiltrate back into the blood.

Like the heart, the kidneys work continuously and have a very active aerobic metabolism. They can use glucose, fatty acids, ketone bodies, and amino acids as metabolic fuels. All of these substances are ultimately degraded through the citric acid cycle to produce by oxidative phosphorylation the required amounts of ATP.

The kidney is an important component of the body's pH control system. For example, in starvation and in diabetes, large amounts of organic acids are produced, causing the blood's pH to decrease significantly—perhaps to a greater degree than the blood's bicarbonate buffer system could handle alone. Fortunately, the kidney has its own buffer system, consisting principally of ammonia and ammonium ion. The ammonia is generated by deamination of amino acids in kidney cells and transported into the tubules. There it combines with excess hydronium ion to form ammonium ion. A significant amount of the ionic form of ammonia does not reenter the kidney from the tubules, and hydronium ion is therefore excreted. The body's reservoir of hydrogen ion also can be conserved. This conservation is accomplished by reducing the extent of deamination of amino acids within kidney cells, thereby reducing the concentration of ammonia in the kidney tubules.

◀◀ The blood's bicarbonate buffer system is described in Section 9.9.

Liver

With the exception of triacylglycerols (Section 26.3), all nutrients absorbed by the intestinal tract are transported directly to the liver. There they are processed and distributed to the other organs and tissues.

Glucose Hexokinase can phosphorylate cellular glucose at the concentrations of glucose normally found in the blood—about 5 m*M*. In contrast, after a meal, the concentrations of glucose in the portal circulation of the liver can rise to about two to three times the normal concentrations. At those concentrations, hexokinase cannot phosphorylate all the incoming glucose. However, liver cells possess a phosphorylating enzyme, **glucokinase,** that is not found in any other organ and is adapted to these high concentrations. Therefore all incoming dietary glucose is converted into glucose-6-phosphate by the liver.

Glucose-6-phosphate stands at the crossroads of the needs of all the body's organs. Liver cells contain glucose phosphatase—an enzyme not found in any other tissue. Therefore, when the concentration of blood glucose falls, the liver can dephosphorylate glucose-6-phosphate and supply free glucose to the blood. Glucose-6-phosphate not needed for maintenance of blood glucose is stored as glycogen (Section 22.5).

Glucose-6-phosphate in excess of these two needs is degraded to acetyl-*S*-CoA, which is converted into malonyl-*S*-CoA for the synthesis of fatty acids and cholesterol (Section 24.5). The fatty acids are used to synthesize triacylglycerols and phospholipids. Some of the triacylglycerols and phosphatides are

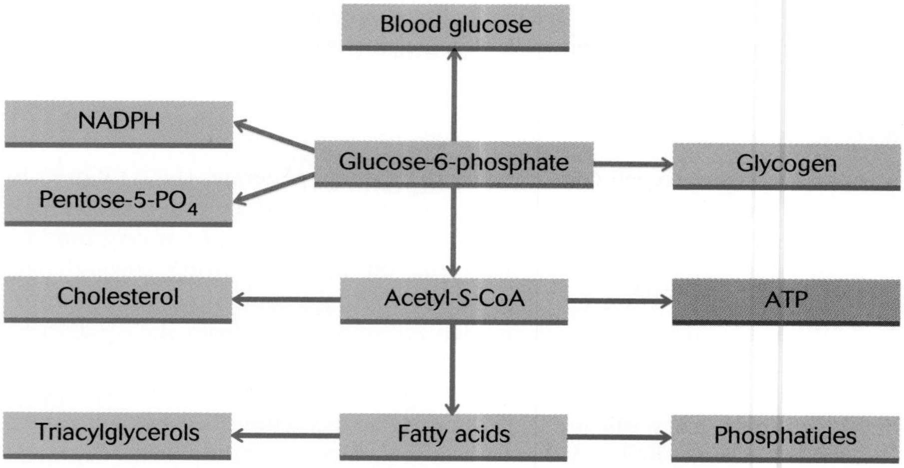

Figure 26.3 Metabolic pathways for glucose in the liver.

exported to other organs. The cholesterol is converted in part into bile salts stored in the gall bladder.

Some of the acetyl-*S*-CoA from glycolysis can be used to produce ATP through the citric acid cycle, but, normally, fatty acids are the principal oxidative fuel used by the liver. Finally, the reductive power used in fatty acid synthesis, NADPH, is produced in the liver by oxidation of glucose-6-phosphate through the pentose phosphate pathway (Section 23.1). These pathways are illustrated in Figure 26.3.

Amino acids Amino acids entering the liver after absorption in the intestines follow a number of metabolic pathways (Figure 26.4). The turnover rate of the liver's proteins is high. Therefore a significant portion of the entering amino acids is used to renew the liver's own proteins. Another portion enters the outgoing blood and travels to the other organs for biosynthesis into tissue proteins. Yet another portion is used by the liver to synthesize the plasma proteins of the blood (Section 26.7), with the important exception of the immunoglobulins.

Amino acids in excess of those needs are deaminated and degraded to acetyl-*S*-CoA and citric acid cycle intermediates. The ammonia from deamina-

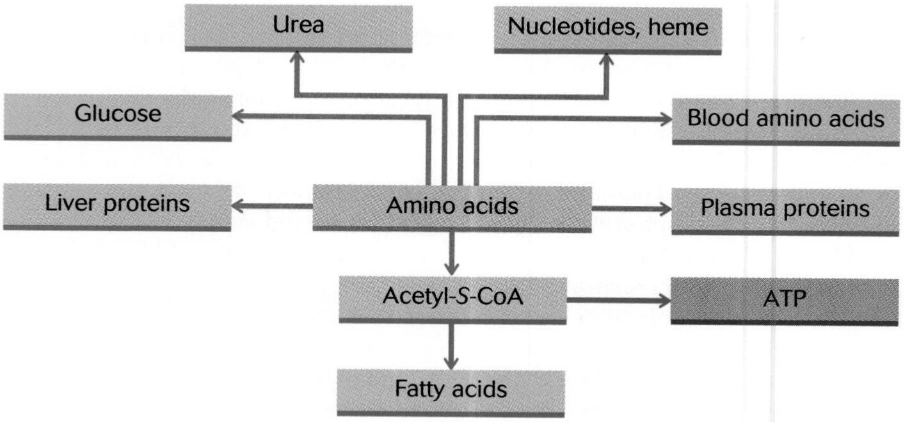

Figure 26.4 Metabolic pathways for amino acids in the liver.

tion is converted into urea (Section 25.4). The acetyl-*S*-CoA can be used to synthesize fatty acids or can be further oxidized through the citric acid cycle to form ATP. The citric acid cycle intermediates can be used to synthesize glucose through gluconeogenesis and glycogen through glycogenesis.

The liver is also a participant in the glucose-alanine cycle (Section 25.3), which is an aid in maintaining the blood-glucose concentration in periods between meals. Alanine arriving at the liver is ultimately a result of muscle protein breakdown. The deficit that this breakdown creates in amino acid concentration in muscle cells is corrected by nutrients from the next meal.

Amino acids are used as precursors of a variety of nitrogen-containing biomolecules, such as the heme of hemoglobin, peptide hormones, and the nucleotides used to synthesize ATP, RNA, and DNA. The degradation of the nitrogen bases (adenine, cytidine, and so forth) of nucleotides results in the production of uric acid.

◀◀ Nucleotide and nucleic acid structures are presented in Sections 21.1–21.3.

Fatty acids Fatty acids are the chief oxidative fuel of the liver (Figure 26.5). Their oxidation (Section 24.3) produces acetyl-*S*-CoA, which enters the citric acid cycle to yield ATP by oxidative phosphorylation. Acetyl-*S*-CoA in excess of that required for basic energy needs is converted into the ketone bodies acetoacetate and 3-hydroxybutyrate (Section 24.4). These ketone bodies enter the circulation to supply energy to peripheral tissues through the citric acid cycle. Ketone bodies can supply as much as one-third of the energy required by the heart. In starvation, the brain adapts to use 3-hydroxybutyrate in addition to glucose.

The bile salts necessary for fat digestion (Box 6.1 and Section 19.7) are synthesized from cholesterol, which is synthesized in the liver from acetyl-*S*-CoA derived from fatty acids and glucose. Fatty acids are also incorporated into the lipid parts of the plasma lipoproteins synthesized in the liver. The lipoproteins (Section 26.6) are critical in the transport of dietary lipids to adipose tissue for storage. Free fatty acids form complexes with serum albumin in the plasma in order to be transported to peripheral tissue for use as oxidative fuel.

Detoxification The liver possesses oxidative enzyme systems specialized to detoxify nonphysiological or foreign organic substances such as drugs (including caffeine), food additives, paint-thinner vapors, and so forth. The general result of the oxidation is to introduce hydroxyl groups into the foreign organic molecule. The hydroxyl group is then esterified to an amphipathic molecule such as an amino acid, significantly increasing the water solubility of the foreign substance and thus enhancing the ability of the kidneys to excrete it.

◀◀ Box 13.3 describes some consequences of the liver's attempt to detoxify alcohol.

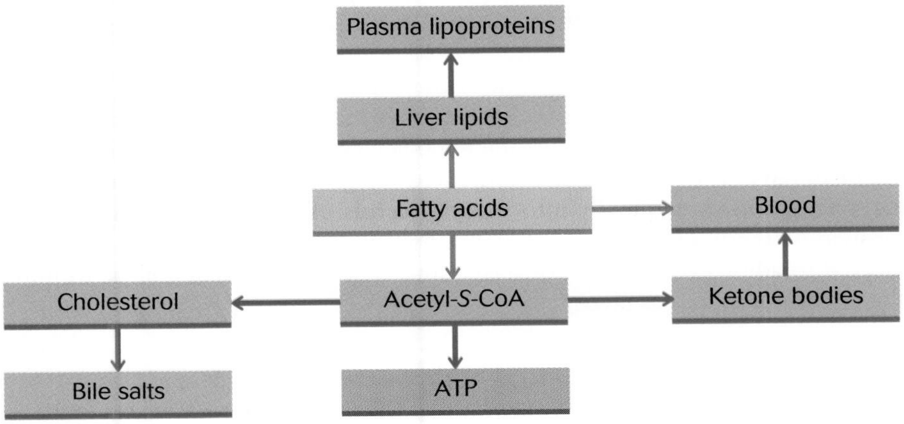

Figure 26.5 Metabolic pathways for fatty acids in the liver.

26.6 METABOLIC RESPONSES TO PHYSIOLOGICAL STRESS

When our bodies are subjected to physiological stress, they respond so as to counteract the stress and return to their normal steady state, or homeostasis. A study of these responses reveals the mechanisms underlying the coordinated control of metabolism. We will explore these mechanisms by first examining the conditions existing (1) just after a meal—the absorptive, or "fed," state—when there is a transitory increase in the blood concentrations of nutrients and (2) some time after a meal—the postabsorptive state—when the concentration of nutrients must be maintained above a certain minimal level. With that preparation, we will then examine the body's responses under two abnormal stressful conditions: starvation and diabetes mellitus.

The **absorptive state** describes conditions in the body when nutrients are entering the blood; that is, during the ingestion of a meal, which ordinarily consists of carbohydrate, protein, and lipid. The **postabsorptive state** describes conditions some time after a meal, when the gastrointestinal tract is empty and the body must subsist on stored forms of energy.

Absorptive and Postabsorptive States

Between meals, the glucose concentration of the blood is normally in the range of 70 to 90 mg/dL. After a meal, it may rise to a range between 120 and 150 mg/dL. Blood rich in glucose flows through the pancreas and stimulates the β cells to secrete the protein hormone insulin into the blood. The absorptive state is therefore characterized by high concentrations of insulin in the blood.

Insulin acts on all tissues of the body, with the exception of the brain and erythrocytes, and sends the metabolic machinery into an anabolic mode. It acts on a cell by adsorbing to receptors on the cell's surface. The mechanism of this membrane interaction is not clear—the second messenger (Section 23.6) has not been determined—but, in every case, the insulin alters either (1) membrane transport or (2) cellular enzyme activity. It enhances the facilitated transport (Section 19.10) of glucose and amino acids into most cells, promotes the storage of lipids and glycogen, and increases the biosynthesis of proteins and nucleic acids.

More specifically, insulin enhances the entry of glucose into muscle and adipose tissue, activates glycolysis in the liver (thus providing glycerol phosphate for triacylglycerol synthesis), increases the rate of fatty acid and triacylglycerol synthesis in both the liver and adipose tissue, inhibits gluconeogenesis in the liver, increases glycogenesis in the liver and in muscle, increases amino acid uptake in muscle, leading to protein synthesis, and, at the same time, inhibits muscle protein degradation.

In the absorptive state, glucose is the principal fuel used by all cells. Because it is usually present in excess of what is required for basic energy needs, most glucose is converted into storage molecules. The glycogen storage capacity of the liver is rapidly filled, and the leftover glucose there is converted into fatty acids and subsequently into triacylglycerols. Amino acids not needed for protein synthesis also are converted into lipid for storage. Glucose is also transported to adipose tissue and converted into triacylglycerols, as well as to muscle cells to be stored as glycogen.

Concept check

✔ The absorptive state is signaled by the presence of insulin and characterized by the preponderance of anabolic processes over catabolic processes.

About 4 to 5 h after a meal, the blood-glucose concentration falls to its normal level of about 70 to 90 mg/dL. It is prevented from dropping below this value by the response of another set of hormone-producing cells in the pancreas, the α cells. These cells, like the β cells, are sensitive to the blood-glucose concentration. When that concentration falls below about 80 mg/dL, the α cells secrete the protein hormone glucagon into the blood.

Glucagon essentially reverses the effects of insulin, shifting the metabolic machinery toward catabolism. Its primary target is the liver, and its chief physiological role is to keep the blood-glucose level constant by increasing the cyclic AMP levels within the cells. This enhances glycogenolysis in the liver and, through the action of liver glucose-6-phosphatase, releases glucose to the blood. Blood-glucose levels are also kept constant because, in this state, the insulin levels are significantly reduced. This means that glucose is primarily diverted to brain cells, and all other cells must switch to other metabolic fuels. Other fuel is available because glucagon also acts on adipose tissue. It increases the activity of the internal lipase of adipose tissue, thereby releasing fatty acids to the blood for transport to all tissues. The availability of fatty acids for oxidation spares the glucose concentration so that glucose can be directed principally to the brain.

The sympathetic nervous system also is capable of shifting the metabolic processes from an anabolic to a catabolic state. These shifts are short lived and are caused by the presence of the neurotransmitter **epinephrine.** Epinephrine is synthesized by the adrenal gland and released into the blood in response to anger or fear signaled by the sympathetic nervous system—the "flight or fight" reflex. There are no epinephrine receptors in the liver; instead, the neurotransmitter's effects are directed toward muscle and adipose tissue cells, where they cause rapid mobilization of stored energy resources—glycogenolysis in muscle and lipolysis in adipocytes.

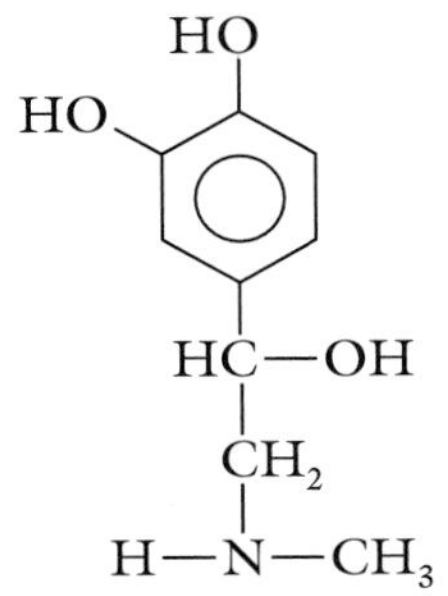

Epinephrine

Concept check

✔ The postabsorptive state is characterized by a shift to catabolic processes caused by the presence of the protein hormone glucagon.

Starvation

Starvation can be viewed as an unrelieved continuation of the postabsorptive state. The body must exist solely on endogenous supplies of energy. The question to be explored is: What happens when the postabsorptive state is extended past the usual fasting period that stretches from dinnertime to breakfast?

First, the mobilization of fat stores for their energy continues, but the problem of maintaining the required continuous glucose supply to the brain becomes more acute. Normally, glucose is supplied by glycogenolysis in the liver. But, after fasting, the liver glycogen supply becomes depleted and unavailable. Hydrolysis of triacylglycerols in both the liver and adipocytes will produce glycerol phosphate, which can be converted into glucose through gluconeogenesis; however, the amounts produced cannot equal those produced through glycogenolysis. In starvation, the glucose-alanine cycle in the liver that ordinarily provides some of the raw material for liver glucogenesis becomes a major route to gluconeogenesis. This, combined with the fact that muscle protein constitutes about 65% of the body's mass, allows the catabolism of glucogenic amino acids to proceed at a rate that maintains a constant blood-glucose concentration. As a result, protein breakdown speeds up in the initial stages of starvation. Aside from the shortage of water and vitamins, continuous catabolism of the body's protein mass is the cause of the most serious consequences of starvation.

Large amounts of ketone bodies are present in the blood during starvation, because the body's energy is coming from the oxidation of fat, and the required amounts of oxaloacetate are not available for acetyl-*S*-CoA to enter the citric acid cycle (Section 24.3). Within days of the initiation of starvation, the brain adapts to the use of ketone bodies. As a result, the rate of protein catabolism can be reduced, and the reservoir of the body's protein can be "spared." Within a week, the brain derives about a third of its energy needs from acetoacetate and its reduction product 3-hydroxybutyrate, and protein catabolism has slowed significantly. By the end of 6 weeks, the brain takes about two-thirds of its energy needs from the ketone-body pool. As the brain uses more and more of the ketone bodies as fuel, the catabolism of muscle protein decreases by a factor of four from the first week to the sixth week. Although our ability to respond to physiological stresses beyond those imposed by starvation are severely compromised, depending on the size of the adipose tissue depot, life can continue for long periods of time. However, to do so requires the presence of adequate amounts of water, without which death will occur within days, not weeks.

Concept check

✔ Starvation is a continuation of the postabsorptive state accompanied by significant loss of protein mass and the presence of ketone bodies in the blood and urine.

Diabetes Mellitus

Diabetes mellitus, the third leading cause of death in the United States, is due to a functional deficiency of insulin. The disease is seen in two forms: type 1, or juvenile-onset diabetes, which is insulin dependent (meaning insulin must be administered externally); and type 2, or adult-onset, diabetes, which is not insulin dependent. Type 1 may afflict people at any age but usually appears early in life, and type 2 usually affects people over the age of 40. The type 1 disease is characterized by an inability to produce insulin and is believed to be caused by the destruction of the β cells in the pancreas by either a viral infection or an autoimmune disorder. About 10% of all cases of diabetes are type 1. The other 90% of cases are type 2 disease, in which insulin is produced but is unable to effect the passage of glucose across cell membranes. The chief cause is believed to be a lack of insulin membrane receptors. It is often accompanied by obesity and can be controlled by diet and drugs.

The disease can have a genetic origin (Box 20.3). Just as some enzymes are synthesized as zymogens, the hormone insulin is synthesized as proinsulin. For proinsulin to be converted into insulin, a peptide of 33 amino acids must be excised from the proinsulin polypeptide. Mutations can cause defects preventing this conversion. Even if the conversion takes place properly, other mutations can render the insulin inactive. Other genetic defects affect cell surface receptors or the internal mechanisms through which the insulin manifests its activity or both.

The physiological situation created by diabetes mellitus is somewhat paradoxical. Because no insulin is available, glucose cannot enter any cells except brain tissue and erythrocytes, which means that most cells are starving in the midst of plenty. The cells of the body of a diabetic operate in a metabolic state somewhere between the postabsorptive state and starvation. The liver attempts to increase its glycogen supplies by speeding up gluconeogenesis. This process requires glucogenic amino acids, which must be supplied from muscle protein. That is why uncontrolled diabetes was called "the wasting disease." The catabolism of amino acids leads to greatly increased urea concentrations in the urine, as much as five times the normal amounts, indicative of the severe imbalance between the daily intake and the breakdown of protein.

As in starvation, adipose tissue lipids are mobilized through the action of glucagon, and fatty acid oxidation becomes the principal source of ATP. The result is an elevation of the concentrations of ketone bodies and organic acids (acetoacetate and β-hydroxybutyrate) in the blood. The normal pH of the blood is 7.4, but, in severe diabetics, the pH can drop as low as 6.8. In uncontrolled diabetes, this decrease in pH can cause unconsciousness. An unfortunate side effect of the low pH is that the unstable acetoacetate is decarboxylated nonenzymatically to form acetone. The sweet smell of acetone on the breath, accompanied by unconsciousness, is symptomatic of a life-threatening diabetic coma, yet all too often is mistaken for a drunken stupor.

In paradoxical contrast with the concurrent starvation of body cells, the blood-glucose concentration in diabetes can exceed 200 mg/dL. At levels above 180 mg/dL, the kidneys can no longer reabsorb the glucose in the tubule ultrafiltrate. The glucose then appears in the urine in large quantities (hence the name mellitus—from the Latin "like honey"). The high concentrations of glucose in the ultrafiltrate create a large osmotic effect that keeps water in the tubules and prevents it from being reabsorbed. Another typical symptom of this disease, therefore, is frequent and excessive urination, from 2 to 10 L/day, accompanied by extreme thirst. Initial diagnosis of diabetes is often based on these effects.

Type 1, or insulin dependent, diabetes, requiring daily injections of insulin, is the more difficult form to control. Insulin cannot be administered by mouth, because, as a protein, it is rapidly hydrolyzed in the gastrointestinal tract. The less-severe, adult-onset type of diabetes can often be controlled much more simply by diet and drugs. Until recently, the insulin administered to insulin-dependent diabetics was extracted from animals. Today, however, this insulin source has largely been replaced by recombinant DNA techniques (Section 21.11), which are used to induce *E. coli* bacteria to produce human insulin.

26.7 BLOOD: THE MASS-TRANSPORT SYSTEM OF THE BODY

All the cells of the human body depend on the blood to bring them oxygen and nutrients and to carry CO_2, the principal waste product of metabolism, away. As the chief transport system of the body, the blood is also responsible for the overall coordination of metabolic processes in the various organs and tissues.

Composition of Blood

The blood's composition is extremely complex, owing to its role in the coordinated operation and integration of all the body's organ systems. For this reason, routine blood tests provide many clues that enable physicians to diagnose pathological conditions.

Whole blood consists of about a 50:50 volume ratio of cells to a liquid fraction called plasma. The cells are mostly erythrocytes (red blood cells), leukocytes (white blood cells), and platelets (clotting cells). Although the plasma is not as complex as the blood's cellular contents, it contains a wide variety of dissolved inorganic components, organic metabolites, waste products, and so-called plasma proteins.

Plasma is 90% water and 10% solutes. The composition of the solute fraction is 70% plasma proteins, 20% organic metabolites, and 10% inorganic salts.

TABLE 26.10 Components of Human Blood Plasma

Inorganic Components	Organic Metabolites	Plasma Proteins
buffers { $NaHCO_3$, Na_2HPO_4 }	glucose	serum albumin
	amino acids	VLDLs
NaCl	lactate	LDLs
$CaCl_2$	pyruvate	HDLs
$MgCl_2$	ketone bodies	immunoglobulins
KCl	citrate	fibrinogen
Na_2SO_4	urea	prothrombin
	uric acid	specialized transport proteins
	creatinine	

Abbreviations: VLDL, very low density lipoprotein; LDL, low-density lipoprotein; HDL, high-density lipoprotein.

The major components of each of these categories are listed in Table 26.10. One class of plasma proteins, the immunoglobulins, is a group of structurally related proteins that may consist of hundreds of different types. Transferrin, a substance that transports iron, is an example of a number of specialized transport proteins.

The total plasma protein concentration is between 8.6 and 11.7 g/dL. The major protein components of plasma, their normal concentrations, and their chief functions are listed in Table 26.11, and the concentrations of the nonprotein components of plasma are listed in Table 26.12. Among the latter

TABLE 26.11 Principal Protein Fractions of Blood Plasma

Protein Component of Plasma	Normal Concentration Range (mg/dL)	Function
serum albumin	3500–4500	osmotic regulation of blood volume; transport of fatty acids
α_1-globulins	300–600	transport of lipids, thyroxine, and ACTH
α_2-globulins	400–900	transport of lipids and copper
β-globulins	600–1100	transport of lipids, iron, and hemes; antibody activity
γ-globulins	700–1500	almost all circulating antibodies
fibrinogen	3000	fibrin precursor for blood clotting
prothrombin	100	precursor of thrombin required for blood clotting

TABLE 26.12 Normal Concentrations of Organic Substances in Blood Plasma

Nonprotein Plasma Component	Normal Concentration Range (mg/dL)
CARBOHYDRATES	
glucose	70–90
fructose	6–8
ORGANIC ACIDS	
lactate	8–17
pyruvate	0.4–2.5
ketone bodies	1–4
citrate	1.5–3.0
NITROGENOUS COMPOUNDS	
amino acids	35–65
urea	20–30
uric acid	2–6
creatinine	1–2
LIPIDS*	
total lipids	300–700
triacylglycerols	80–240
cholesterol and esters	130–240
phospholipids	160–270

* All lipids in blood plasma are bound to proteins.

are urea, uric acid, and creatinine, the chief nitrogenous waste products of metabolism excreted by the kidneys.

The blood communicates with the external environment through the kidneys and the lungs. Nonvolatile substances are excreted through the kidneys, and volatile substances and gases are excreted through the lungs.

26.8 TRANSPORT OF OXYGEN

Oxygen is required by respiring cells, and carbon dioxide is generated by those cells. Simple diffusion would not allow tissues buried deep within a multicellular organism to obtain atmospheric oxygen or to dissipate carbon dioxide at a rate sufficient to sustain metabolism. Section 26.1 stated that this gross diffusion problem was solved by the development of circulatory systems and respiratory pigments such as hemoglobin. Now we will see how they work.

In terrestrial animals, oxygen enters the circulatory system through the respiratory membranes of the lungs. It diffuses across the membranes passively, under a gradient in concentration; that is, the concentration of oxygen in the alveolar spaces of the lung is greater than that in the blood plasma. The solubility of oxygen in plasma is too low for the plasma alone to absorb enough oxygen to fuel aerobic metabolism. However, sufficient oxygen is carried to the tissue level by another mechanism.

Oxygen is transported to the tissues not by plasma but by the erythrocytes of the blood—the red blood cells. The blood of an adult human has a volume of 5 to 6 L. About half of this volume is composed of erythrocytes, a mass equivalent to that of the liver. Erythrocytes are very small, degenerate cells containing no nuclei, mitochondria, or any other subcellular organelle. Their only metabolic fuel is glucose, and their metabolic engine consists only of glycolysis, which is their sole source of ATP. Their chief function is to transport oxygen from the lungs to the tissues and carbon dioxide from the tissues to the lungs. They contain large quantities of the iron-containing protein hemoglobin, which is approximately 90% of the erythrocyte's protein content.

Oxygen combines with hemoglobin. In doing so, the oxygen is not dissolving into the blood but instead is complexed reversibly with the Fe(II) of the hemoglobin protein's heme portion (Section 20.7). This complex increases the blood concentration of oxygen to such an extent that 100 mL of whole blood carries about 21 mL of oxygen, about 50 times the amount that dissolves in the plasma.

The oxygen-binding curves for hemoglobin at two values of pH are presented in Figure 26.6. The y-axis denotes the degree of hemoglobin's saturation with oxygen. If we designate hemoglobin as Hb and oxygenated hemoglobin as HbO_2, the percentage of saturation of hemoglobin with oxygen is defined as

$$\text{Percent saturation} = \left(\frac{[HbO_2]}{[Hb + HbO_2]}\right) \times 100\%$$

The x-axis shows the partial pressure of oxygen. The oxygen partial pressure at 50% saturation is known as the $\boldsymbol{P_{50}}$.

The curves in Figure 26.6 have a characteristic S-shape, known as a **sigmoid shape,** that has an important physiological significance. Each molecule of hemoglobin consists of four polypeptide subunits, and each subunit has an iron-bearing heme molecule that binds one molecule of oxygen (Section 20.7). The sigmoid shape indicates that the oxygen molecule bound to the first subunit increases the binding affinity of the remaining subunits for the next molecules of oxygen to be bound. In other words, the first oxygen bound increases the capacity of hemoglobin to bind more oxygen. This characteristic is called **cooperative binding.** The extent of binding increases as the partial pressure of oxygen increases until a maximum is reached and no more can be bound. That maximum partial pressure is about equal to the oxygen tension (Section 7.7) at the lungs.

Concept check

✔ The steepness of the sigmoid oxygen-binding curve shows that oxygen can be loaded (bound) and unloaded (dissociated) over a narrow range of oxygen tensions.

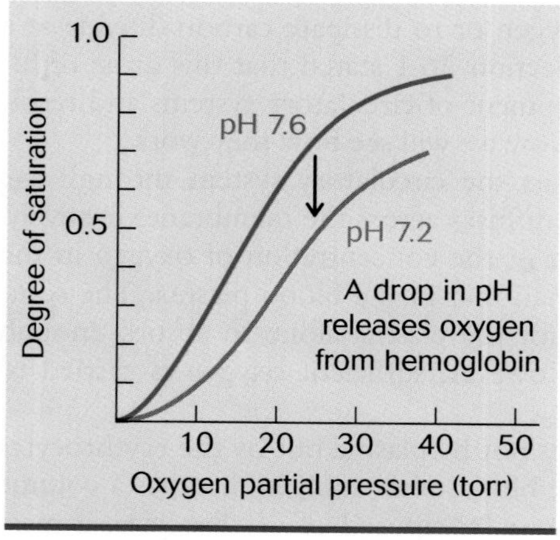

Figure 26.6 Oxygen saturation curves for hemoglobin at two values of pH.

The amount of oxygen that combines with hemoglobin depends not only on the partial pressure of oxygen present, but also on the pH, CO_2, and the presence of a compound, 2,3-bisphosphoglycerate, generated by the glycolytic apparatus of the erythrocyte. 2,3-Bisphosphoglycerate lowers the affinity of hemoglobin for oxygen (Section 20.7). In solution, the P_{50} for hemoglobin is 1 torr. In the presence of 2,3-bisphosphoglycerate, the P_{50} increases to 26 torr. This increase is helpful in situations where the external oxygen concentration is low. For example, at 15,000 feet, the altitude of Lake Titicaca in Peru, hemoglobin cannot be fully saturated. Therefore, at the partial pressure of oxygen at the tissue level, only a small part of the bound oxygen could be unloaded into the cells. The erythrocytes of people living at that altitude contain increased levels of 2,3-bisphosphoglycerate, lowering the oxygen affinity of hemoglobin enough so that a significant fraction of its oxygen can be unloaded at the tissue level. When people move from low to high altitudes, a hormone called erythropoietin is synthesized by the kidneys. This hormone initiates the synthesis of erythrocytes, thus increasing the oxygen-carrying capacity of the blood. The administration of erythropoietin is also used to alleviate clinical states of anemia.

The pH of the blood has an even more significant effect on the oxygen-carrying properties of hemoglobin. If we use the abbreviation **H**Hb to represent the deoxygenated form of hemoglobin, a Brønsted-Lowry acid, and HbO_2^- to represent the oxygenated form, a Brønsted-Lowry base—and if we print hydrogen ions in color—the oxygenation of hemoglobin can be represented by the following equilibrium:

$$HHb + O_2 \rightleftharpoons HbO_2^- + H^+$$

The reversibility of this system means that, if the pH is raised, more oxygen will be bound. If the pH is lowered, oxygenated hemoglobin will give up its oxygen. This effect is known as the **Bohr effect,** named after its discoverer, Christian Bohr, the father of Niels Bohr (Section 2.7). The effect of pH on the oxygen-binding curves of hemoglobin in Figure 26.6 can be understood if we follow the series of events taking place from the time of loading in the lungs to the unloading of oxygen at the tissue level.

Oxygen is bound, or loaded, by hemoglobin in the lungs at an oxygen tension of about 100 torr (Section 5.11). There the CO_2 concentration is low

This child is acclimatized to living at the high altitude of Lake Titicaca in Peru.

and, consequently, the pH is relatively high, about 7.6. At that pressure, the hemoglobin becomes almost completely saturated. The oxygen is then carried to the tissues, where the CO_2 concentration is high, and the pH is lowered to about 7.2 to 7.3. The oxygen concentration is also much lower in the tissues (an oxygen tension from about 25 to 40 torr) than in the lungs; so the unloading of oxygen (dissociation from hemoglobin) is favored. At the cell level, the partial pressure of oxygen is 40 torr, and the binding curve becomes quite steep. Therefore, over a narrow range of oxygen concentrations, hemoglobin cycles between about 60% and 95% saturation. Thus, the acid-base properties of hemoglobin allow a very efficient delivery of oxygen at the tissue level.

The fact that dissociation of hydrogen ion is caused by oxygenation of hemoglobin greatly affects events within the lungs. When CO_2 arrives at the lungs, it is in the form of HCO_3^-; the H^+ from the dissociation of H_2CO_3 in the erythrocytes, is bound to hemoglobin. As O_2 is loaded, H^+ ions are released from the newly oxygenated hemoglobin, and the following reaction takes place:

$$HCO_3^- + H^+ \rightleftharpoons H_2CO_3 \rightleftharpoons CO_2 + H_2O$$

Because the oxygenation of hemoglobin makes hydrogen ion available, the binding of oxygen at the lungs increases the efficiency of the release of CO_2 to the atmosphere.

The effect of CO_2 on the oxygenation equilibrium of hemoglobin will be considered next.

26.9 TRANSPORT OF CARBON DIOXIDE

The CO_2 generated in the tissues diffuses under a concentration gradient into the **interstitial space**—the space between cells—and then across the capillary walls into the blood, where it dissolves in the plasma. The bulk of the CO_2 produced by cellular metabolism arrives at the red blood cell in the dissolved state, CO_2 (*aq*). Erythrocytes possess the enzyme carbonic anhydrase, which rapidly catalyzes the formation of HCO_3^- from the incoming CO_2. Only about 0.5% of the CO_2 dissolved in plasma is in the form of H_2CO_3, and, of this amount, only a small proportion forms plasma HCO_3^- and H^+. The H^+ thus formed in the plasma is bound by the plasma proteins, so there is no shift in blood pH.

Carbonic anhydrase, utilizing zinc ion as a cofactor, catalyzes the reaction between CO_2 and an OH^- ion from water to form HCO_3^- and leave an H^+ ion:

$$CO_2(aq) + OH^-(aq) + H^+(aq) \rightleftharpoons HCO_3^-(aq) + H^+(aq)$$

We write the equation with H^+ ion on both sides to emphasize the fact that CO_2 does not react directly with water but with one of the dissociation products of water, OH^-, and leaves the other product, H^+, behind. The enzyme increases the rate of HCO_3^- formation by a factor of 1×10^7 (Section 22.3). At body temperature, the solubility of $NaHCO_3$ in water is about 109 g/L, compared with a solubility of only about 0.12 g/L for CO_2 at its tissue partial pressure of 40 torr. The body's strategy of converting most of the CO_2 in the blood into HCO_3^- and transporting it in that form greatly increases the amount of CO_2 that the blood is able to carry.

The bicarbonate ion concentration is greater in the red blood cell than in the plasma because of its continuous high rate of formation in red blood cells. Therefore bicarbonate tends to diffuse out of the red blood cell into the plasma. Under most circumstances, the HCO_3^- anion would diffuse across the membrane only if a cation accompanied it (to maintain the balance of charge on both sides of the membrane); however, the red blood cell membrane is not permeable to K^+, the cation found within these cells. An equivalent alternative mechanism, known as the **chloride shift,** is used instead, in which an HCO_3^- anion is exchanged across the cell membrane with a Cl^- anion that diffuses into the cell from the plasma.

Now let us turn our attention to the proton generated by the formation of HCO_3^- in the erythrocyte. After oxygenation in the lungs, hemoglobin is carried to the tissues in the form of HbO_2^-. Within the red blood cell, its negative charge is balanced by the K^+. When the red blood cells begin to absorb CO_2 at the tissues, the proton generated by the consequent formation of HCO_3^- reacts with the HbO_2^- formed at the lungs:

$$H^+ + HbO_2^- \rightleftharpoons HHb + O_2$$

In this way, the protons generated at the tissues enhance the unloading there of oxygen from the oxygenated hemoglobin formed in the lungs.

The carbon dioxide generated at the tissue level also has an effect on the unloading of oxygenated hemoglobin. A significant amount—about 20%—of the CO_2 in blood is carried in **carbamate** compounds formed by the reaction of CO_2 with amino groups of hemoglobin. The general reaction of CO_2 with any $—NH_2$ group can be represented as

$$R—NH_2 + CO_2 \rightleftharpoons \underset{\textbf{Carbamate}}{R—NH—COO^-} + H^+$$

When CO_2 reacts in this way with the amino groups of oxygenated hemoglobin, it causes the dissociation of the oxygen:

$$CO_2 + HbO_2^- \rightleftharpoons HbCO_2^- + O_2$$

Therefore, the direct loading of CO_2 as a carbamate compound on hemoglobin results in the additional unloading of O_2 at the tissue level.

Finally, the equilibrium between oxygenated and deoxygenated hemoglobin (Section 26.8) is sensitive to pH. When the pH decreases at the cell level because of the high concentration of CO_2, O_2 leaves its bound form and can diffuse into the tissue under its concentration gradient.

The processes that take place in the red blood cell at the tissue level are reversed at the lungs, as shown in the following reaction equations. In these sequences of equations, protons generated in the red blood cell at the tissue level and eliminated in the lungs are identified by color. Remember that the pH is about 7.2 at the tissue level and about 7.6 at the lungs.

$$\begin{aligned} \text{Tissue level:}\quad & CO_2 + H_2O \rightleftharpoons H^+ + HCO_3^- \\ & H^+ + HbO_2^- \rightleftharpoons HHb + O_2 \\ \text{Lungs:}\quad & HHb + O_2 \rightleftharpoons HbO_2^- + H^+ \\ & H^+ + HCO_3^- \rightleftharpoons CO_2 + H_2O \end{aligned}$$

The flow of CO_2 represented in these equations—from the tissues to the external environment—is entirely through diffusion. No active transport mechanism is required. The direction and rate of flow are completely dependent on the difference in concentrations of CO_2 between the internal and the external environment.

The HCO_3^- ion is one-half of the Brønsted-Lowry conjugate acid-base pair H_2CO_3-HCO_3^-. However, the red blood cell contains very little H_2CO_3, because of the action of carbonic anhydrase. As a consequence, the concentration of HCO_3^- in the plasma depends on dissolved CO_2, not H_2CO_3. The reactions of this conjugate pair at the tissue level and at the lungs constitute the bicarbonate buffer of the blood (Section 9.8):

$$\begin{aligned} \text{Tissue level:}\quad & CO_2 + H_2O \rightleftharpoons H^+ + HCO_3^- \\ \text{Lungs:}\quad & H^+ + HCO_3^- \rightleftharpoons CO_2 + H_2O \end{aligned}$$

Remember that both reactions are catalyzed by carbonic anhydrase. If they were not, then the $CO_2 + H_2O$ appearing in both equations could be replaced by H_2CO_3, and the more familiar Brønsted-Lowry conjugate acid-base pair would be present. Dissolved CO_2 takes the place of H_2CO_3 in the Brønsted-Lowry scheme (Section 9.8). The reactions in oxygen and carbon dioxide transport and the anatomical locations of these reactions are summarized in Figure 26.7.

An important point is that acid generated at the tissue level is eliminated by exhalation of CO_2. If CO_2 is not removed from the lungs rapidly enough, the system "backs up": hydrogen ion will not be removed by reaction with HCO_3^-, and the blood pH will fall, a condition known as **respiratory acidosis.** When other physiological events cause a lowering of blood pH, the condition is called **metabolic acidosis.** The healthy body's response to acidosis is to increase the rate of breathing (hyperventilation). Chemoreceptors in the carotid artery, sensing a rise in levels of CO_2 and H^+ ion, send a message to increase the activity of the respiratory center of the brain.

The reverse problem, in which blood pH is excessively high, is called alkalosis. Victims of emphysema run the risk that their rapid, shallow breathing will eliminate so much CO_2, and therefore H^+ ion, that they will begin to suffer from **respiratory alkalosis.** When other physiological conditions are responsible, the condition is called **metabolic alkalosis.** One example is excessive vomiting. The loss of H^+ from the stomach causes a flow of replacement H^+ from the plasma, with a consequent rise in blood pH. The healthy body's response to such a rise in blood pH is to retain as much CO_2 as possible by reducing the breathing rate, or hypoventilating. More on acidosis and alkalosis can be found in Section 9.9.

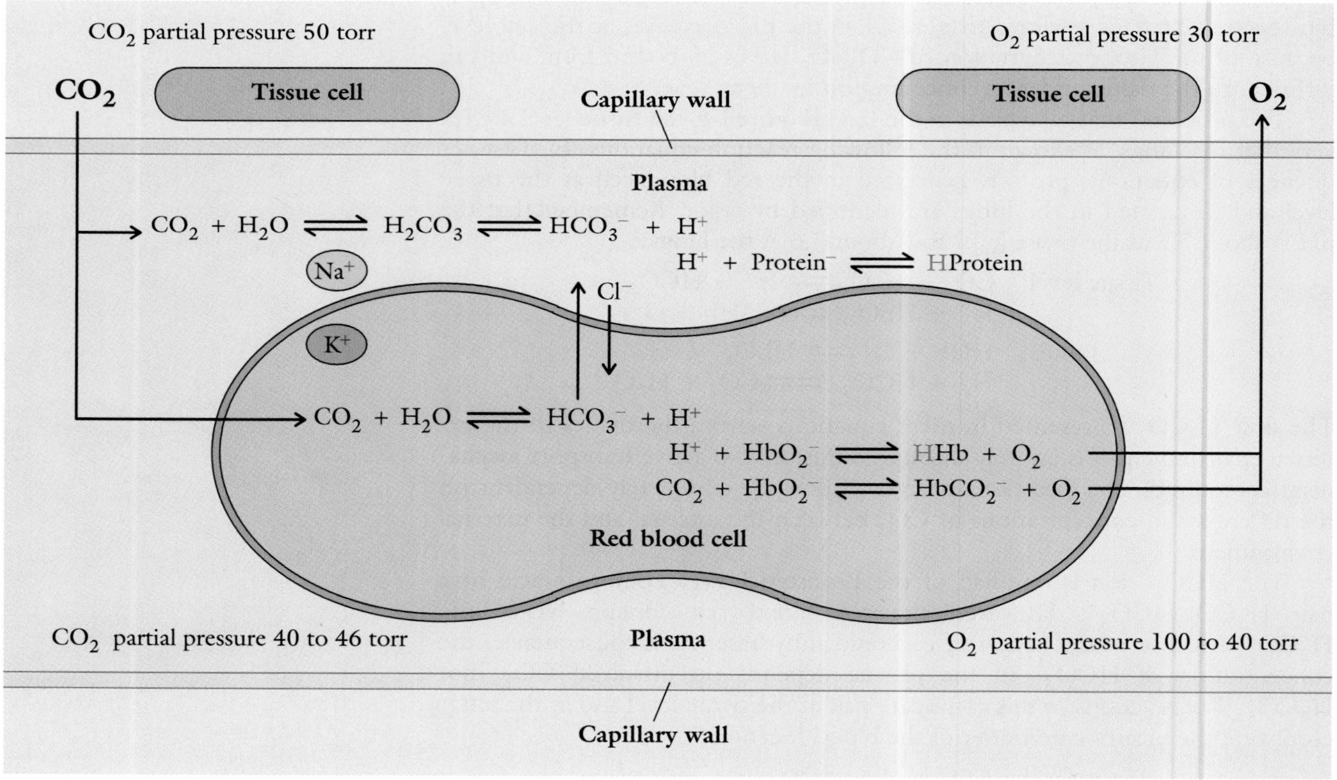

Figure 26.7 The reactions responsible for the transport of oxygen and carbon dioxide in the blood.

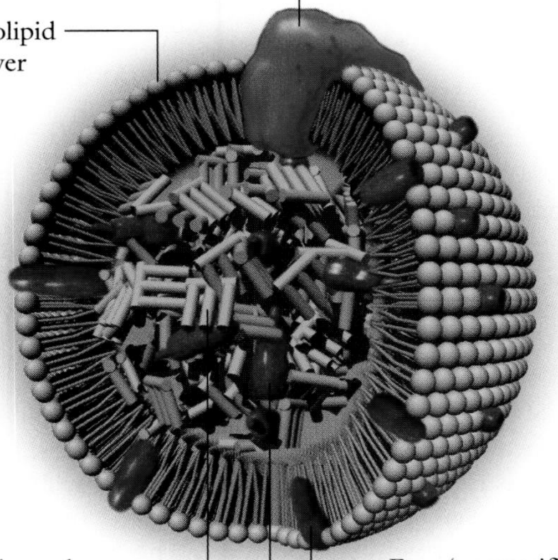

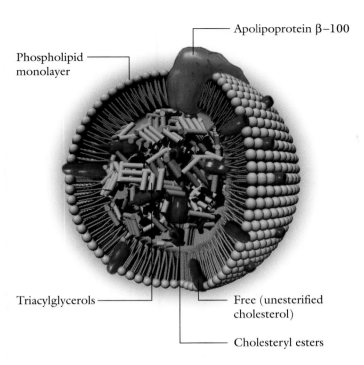

Figure 26.8 Structure of low-density lipoprotein (LDL). The protein is apolipoprotein β-100, which recognizes receptor sites on adipocytes.

26.10 TRANSPORT OF LIPIDS

The hydrophobic lipids (cholesterol and triacylglycerols) are insoluble in blood and must be packed inside of soluble, spherical **lipoprotein particles** to be transported from one tissue to another through the bloodstream. The lipoprotein particles' structure is reminiscent of the structure of the micelles formed when soap combines with hydrophobic substances (Box 7.5 and Section 15.7). Each lipoprotein particle consists of a core of hydrophobic lipids surrounded by a shell of amphipathic lipids (phospholipids and cholesterol, mostly in the form of cholesteryl esters) and proteins (Figure 26.8). The amphipathic lipids and proteins have their hydrophobic ends oriented inward toward the hydrophobic lipids and their hydrophilic ends oriented outward toward the aqueous plasma.

The lipoproteins differ in the relative amounts of lipid and protein that they contain, as well as the relative amounts of different lipids (Table 26.13). These differences result in differences in density, by which the lipoproteins are classified. The least-dense lipoproteins are the **chylomicrons,** followed by **very low density lipoproteins (VLDLs), low-density lipoproteins (LDLs),** and **high-density lipoproteins (HDLs).** A higher protein content results in

TABLE 26.13 Density and Composition of Lipoproteins

	Chylomicron	VLDL	LDL	HDL
Density (g/mL)	<0.95	0.95–1.006	1.006–1.063	1.063–1.210
Composition (wt-%)				
cholesterol	1	7	8	2
cholesteryl ester	3	13	38	15
triacylglycerol	85	51	10	4
phospholipid	9	19	20	24
protein	2	10	24	55

TABLE 26.14 Lipid Levels in Blood

	Concentration (mg/dL)		
	Desirable	**Borderline High**	**High**
cholesterol*	160	200–240	>240
low-density lipoprotein	< 130	130–160	> 160
high-density lipoprotein	>35		
triacylglycerol	<200	200–400	>400

* Includes cholesteryl esters.

higher density, because protein is more dense than lipid. Normal, borderline high, and high levels for blood lipids are listed in Table 26.14.

Each lipoprotein has a specific function based on its point of synthesis, lipid composition, and protein composition. The proteins in lipoproteins have several roles: solubilization of lipids, targeting of specific lipoproteins to specific tissues (by cell recognition between those proteins and the glycoproteins or glycolipids of cell membranes), and activation of enzymes that hydrolyze and unload lipids from lipoproteins. As lipoproteins circulate through the liver and other tissues in the body, cell recognition between proteins of the lipoproteins and receptors on the capillary membranes results in binding of the lipoproteins to the cells. Binding of lipoproteins is followed by transfer of lipid between lipoproteins and cells.

Chylomicrons, which transport dietary (exogenous) triacylglycerols from the intestine to adipose tissue and cholesterol to the liver, are assembled in intestinal cells and carried through the lymph system to the bloodstream. The enzyme lipoprotein lipase, located on the extracellular sides of the membranes of muscle and adipose cells, is activated by proteins on the chylomicrons. This enzyme hydrolyzes the triacylglycerols to fatty acids and glycerol, which are then taken up by the cells in the target tissues. (The fatty acids not oxidized for energy are reesterified for storage as triacylglycerols in adipose cells.) Devoid of triacylglycerols but still rich in cholesterol, the chylomicrons, now called **chylomicron remnants,** are taken up by the liver.

Very low density lipoproteins transport cholesterol and triacylglycerols synthesized in the liver to other tissues. Cholesterol delivered to the liver by chylomicron remnants also is packed into VLDLs. The VLDLs deliver triacylglycerols to extrahepatic tissue (other-than-liver tissue) by using the same mechanisms as chylomicrons do. As the lipoproteins unload their triacylglycerols, lipoprotein density increases, and the VLDLs become low-density lipoproteins. LDLs transport cholesterol from the liver to other cells for the construction of cell membranes and the synthesis of steroid hormones. The LDLs then return to the liver for reprocessing. Recognition between proteins on LDLs and LDL receptors on cell membranes is critical to the uptake of LDLs and the unloading of cholesterol in liver cells, as well as all other cells.

High-density lipoproteins, synthesized in the liver, remove cholesterol from dying cells and from membranes undergoing turnover in cells outside the liver. The cholesterol is returned to the liver, both directly by HDLs and through transfer to LDLs. Excess cholesterol is converted into bile acids, which are secreted into the intestine and then excreted in the feces.

The cholesterol level in the blood results from the interaction of the different lipoproteins with each other and with liver and other cells and from the regulation of biosynthesis of cholesterol in the liver. Cholesterol synthesis in cells is regulated through a feedback mechanism: cholesterol uptake from the bloodstream shuts down the synthesis of the enzyme HMG-*S*-CoA reductase required for cholesterol biosynthesis (Section 24.4).

Fatty acids liberated from adipose tissue cells by glucagon-sensitive cellular lipase are transported through the blood in the form of serum albumin–fatty acid complexes to other tissues for oxidation.

Summary

Hormones and Communication Between Cells The coordinated control of the operation of tissues and organs is accomplished by intercommunication through hormones. At the cell level, the concentration of substrates controls the choices of catabolic and anabolic pathways made by cells. Biopolymers are enzymatically degraded in the digestive tract to low-molecular-mass components to prepare them for absorption in the gut.

Metabolic Characteristics of the Major Organs and Tissues The cells of different organs have different metabolic requirements. The brain uses only glucose for its energy needs. The special function of the brain rests on the ability of neurons to generate a large electrical potential across their membranes and to transmit electrical signals from cell to cell. The latter is accomplished by chemical substances called neurotransmitters. The brain is also the center of synthesis of a large number of hormones and hormone-releasing agents.

Heart muscle utilizes fatty acids as its primary energy source and operates aerobically at all times. A lack of oxygen will result in myocardial infarction. The heart contains almost no stored energy; so metabolic fuels also must be continuously supplied.

Skeletal muscle can utilize a variety of fuels, depending on its degree of activity. The principal energy source of resting muscle is fatty acids. About 75% of the body's glycogen supply is located in muscle. When exertion begins, the glycogen reserves are mobilized to provide glucose.

Adipose tissue is the principal energy storage depot of the body. The normal adult possesses enough of this tissue to sustain life for a few months. The breakdown of triacylglycerols within adipocytes is activated by glucagon. Insulin reverses this stimulation.

About 80% of the oxygen consumed by the kidneys is used to establish an osmotic gradient for concentrating the urine. At the same time, other important substances, such as glucose, are actively transported from the ultrafiltrate back to the blood. The kidney is an important component of the pH control system of the body.

All nutrients absorbed by the intestinal tract, with the exception of triacylglycerols, are transported directly to the liver. There they are processed and distributed to the other organs and tissues.

Metabolic Responses to Physiological Stress The term absorptive state describes metabolic conditions during the ingestion of a meal. This state is signaled by the presence of insulin and characterized by the preponderance of anabolic processes over catabolic processes. The term postabsorptive state describes conditions some time after a meal, when the gastrointestinal tract is empty and the body must subsist on stored forms of energy. The postabsorptive state is characterized by a shift to catabolic processes caused by the presence of the protein hormone glucagon.

Starvation can be viewed as an unrelieved continuation of the postabsorptive state. The body must exist solely on endogenous supplies of energy.

Diabetes mellitus is caused by a functional deficiency of insulin, in consequence of which glucose cannot enter any cells except brain tissue and erythrocytes. All the cells of the body of a diabetic operate in a metabolic state somewhere between the postabsorptive state and starvation.

Blood: The Mass Transport System of the Body The transport of oxygen and carbon dioxide is controlled only by differences in their internal and external concentrations. Oxygen is transported to the tissues by the erythrocytes of the blood. They contain the protein hemoglobin, which combines reversibly with oxygen to increase O_2 concentration to about 21 mL/100 mL of blood. A decrease in pH and an increase in CO_2 enhance the unloading of hemoglobin at the tissue level. Most of the CO_2 carried by the blood is in the form of bicarbonate ion.

The hydrophobic lipids (cholesterol and triacylglycerols) are transported from one tissue to another through the bloodstream in the form of spherical lipoprotein particles. Lipoproteins differ in their relative amounts of lipid and protein as well as in relative amounts of different lipids.

Key Words

absorptive state, p. 760
acidosis, p. 770
action potential, p. 751
alkalosis, p. 770
chylomicron, p. 744
diabetes mellitus, p. 762
hormone, p. 741
lipoprotein, p. 744
neuron, p. 752
postabsorptive state, p. 760
zymogen, p. 742

Exercises

Unicellularity Versus Multicellularity

26.1 Describe major advantages and disadvantages of the single-cell way of life.

26.2 Describe major advantages and disadvantages of the multicellular way of life.

Digestive Processes

26.3 What is the outcome of the digestive process?

26.4 Why is the digestive process necessary?

26.5 Describe the way in which water-soluble hormones affect cellular metabolism.

26.6 Describe the way in which water-insoluble hormones affect cellular metabolism.

26.7 What is a zymogen?

26.8 Are all digestive enzymes secreted as zymogens? Explain.

26.9 How do the digestive products of carbohydrates and proteins reach the liver?

26.10 How do the digestive products of triacylglycerols reach the liver?

26.11 Describe the role of gastrin in digestion.

26.12 Describe the role of secretin in digestion.

Nutrition

26.13 What is the meaning of the biological value of a protein?

26.14 How is the chemical score of a protein determined?

26.15 Why are the polyunsaturated fatty acids linoleic acid and linolenic acid considered essential?

26.16 Why is methionine considered an essential amino acid?

26.17 What is meant by a dietary-deficiency disease?

26.18 What is an example of a dietary-deficiency disease and its cure by replacement.

Metabolic Characteristics of the Major Organs and Tissues

26.19 What are the nutrients used for energy production by brain cells?

26.20 What effect does blood composition have on brain functions?

26.21 What special physiological property is associated with brain cells?

26.22 What is the origin of the property referred to in Exercise 26.21?

26.23 Describe the quantitative distribution of sodium ions and potassium ions between the inside and the outside of nerve cells.

26.24 How is the distribution of sodium ions and potassium ions across nerve membranes achieved?

26.25 Describe the role of the brain in effecting changes in organ systems that are remote from the brain and its nerve connections.

26.26 Give a specific example of how the brain can affect the physiological behavior of an organ such as the kidney or the thyroid gland.

26.27 What are the forms of energy storage in heart cells?

26.28 What is the principal fuel of heart cells?

26.29 What is the biochemical consequence of a heart cell's preference for its principal energy source?

26.30 What is the physiological consequence of a heart cell's preference for its principal energy source?

26.31 What is the principal energy source of active muscle?

26.32 What is the role of muscle cells in the maintenance of the blood-glucose concentration?

26.33 How is the triacylglycerol content of chylomicrons in the blood transferred into adipocytes?

26.34 Is all of the triacylglycerol content of chylomicrons transferred into adipocytes? Explain.

26.35 How do adipocytes respond to the hormone glucagon?

26.36 How do adipocytes respond to the hormone insulin?

26.37 What becomes of the glycerol from the hydrolysis of triacylglycerols in adipocytes?

26.38 How does glycolysis contribute to the synthesis of triacylglycerols in adipocytes?

26.39 Most of the respiration of kidney cells is put to what purpose?

26.40 How do the kidneys concentrate urine?

26.41 Give an example of how the kidneys can help in excreting excessive hydrogen ion from the body.

26.42 Give an example of how the kidneys can correct the pH of the body when it rises above normal levels.

26.43 How are the high concentrations of glucose in the liver after a meal handled by that organ?

26.44 How does the liver take part in the maintenance of a supply of free glucose in the blood?

26.45 How is the ammonia derived from the catabolism of amino acids in the liver detoxified?

26.46 How is the ammonia derived from the catabolism of amino acids in muscle cells detoxified?

26.47 What are the ketone bodies?

26.48 What are the origin and source of most of the ketone bodies in circulating blood?

26.49 How does the liver contribute to the intestinal digestion of dietary lipids?

26.50 How does the liver contribute to the biosynthesis of the steroid hormones?

Metabolic Responses to Physiological Stress

26.51 Define the absorptive state.

26.52 Define the postabsorptive state.

26.53 What hormone is present in high concentrations in the absorptive state?

26.54 Describe the actions of the hormone given in answer to Exercise 26.53.

26.55 The presence of what hormone in the blood characterizes the postabsorptive state?

26.56 Describe the actions of the hormone given in answer to Exercise 26.55.

26.57 What is the major difference between the body's responses to the postabsorptive state and its responses to the initial stages of starvation?

26.58 What is the major difference between the body's initial response to starvation and its responses to the later stages?

26.59 What are the characteristics of juvenile-onset diabetes?

26.60 What are the characteristics of adult-onset diabetes?

26.61 How does the liver maintain its glycogen supplies in diabetes?

26.62 What is the explanation for the presence of high concentrations of urea in the urine of diabetics?

26.63 What is the explanation for the presence of high concentrations of ketone bodies in the blood of diabetics?

26.64 Explain the fact that the pH of the blood of diabetics is often lower than normal.

Blood: The Mass-Transport System of the Body

26.65 Describe the cellular composition of blood.

26.66 Describe the composition of blood plasma.

26.67 What characteristic of the blood increases its oxygen-carrying capacity 50-fold over the oxygen solubility of plasma alone?

26.68 What biochemical characteristic makes hemoglobin efficient in unloading oxygen at the low oxygen tension at respiring cells?

26.69 What is the function of 2,3-bisphosphoglycerate?

26.70 How does blood pH affect the unloading of oxygen at respiring cells and the lungs?

26.71 The major form in which CO_2 is carried in the blood is bicarbonate ion. How is it formed?

26.72 What is the chloride shift?

26.73 How does CO_2 affect oxygenated hemoglobin when hemoglobin reaches respiring cells?

26.74 How does the formation of carbamate hemoglobin affect the unloading of oxygen at respiring cells?

26.75 What is respiratory alkalosis?

26.76 What is respiratory acidosis?

26.77 Describe the structure of a typical lipoprotein.

26.78 List the various classes of lipoproteins and the chief property used in differentiating them.

26.79 What is the role of the protein components of lipoproteins?

26.80 What happens to a triacylglycerol at the surface of an adipocyte?

Unclassified Exercises

26.81 What is the meaning of nitrogen balance?

26.82 What are the consequences of a high proportion of saturated to unsaturated fatty acids in the diet?

26.83 Why is vitamin B_{12} not ordinarily required in the diet but must sometimes be supplemented?

26.84 What differentiates bulk elements from trace elements?

26.85 Describe how the liver treats glucose in excess of what is needed for maintenance of blood glucose and storage.

26.86 What is the source of reducing power for biosynthesis in the liver?

Chemical Connections

26.87 What is the biological logic behind the pancreatic synthesis of chymotrypsinogen—a zymogen—rather than chymotrypsin?

26.88 Can an injection of insulin improve the intellectual capabilities of a type-1 diabetic?

26.89 How does dietary lipid enter adipose tissue cells?

26.90 How would an abnormally low concentration of chloride ion in the blood affect the blood's pH?

26.91 How can a person suffer from protein deficiency when her diet consists of wheat products high in protein concentration?

Answers to Problems Following In-Chapter Worked Examples

Chapter 1

1.1 (a) 0.002 s (b) 0.05 m (c) 0.1 liters
1.2 7.068×10^{-4}
1.3 7.311×10^{-4}
1.4 27 or 2.7×10^{1}
1.5 4.8×10^{5}
1.6 6.1×10^{2} cm^3
1.7 40.35
1.8 87.3 m
1.9 0.0741 L
1.10 1.659 lb
1.11 6.048×10^{5}
1.12 2.87×10^{3} g
1.13 3.97×10^{3} mL
1.14 118.2 g
1.15 36°C
1.16 212°F
1.17 37.0°C and 310 K
1.18 0.20 J/g × °C
1.19 49 J
1.20 1412 kcal

Chapter 2

2.1 C = 58.54%, H = 4.071%, N = 11.38%, O = 26.02%
2.2 30 oranges, 10 cantaloupes
2.3 0
2.4 3−
2.5 Titanium, 48 amu
2.6 $^{14}_{7}N$ and $^{15}_{7}N$
2.7 (a) protons = 7, neutrons = 7
(b) protons = 7, neutrons = 8
2.8 24.3 amu
2.9 Four subshells, *s*, *p*, *d*, and *f*
2.10 $1s^2 2s^2 2p^6 3s^2 3p_x^{\,2} 3p_y^{\,1} 3p_z^{\,1}$
2.11

Element	Atomic Number	1*s*	2*s*	2*p*
Boron	5	↑↓	↑↓	↑

2.12

Element	Atomic Number	1*s*	2*s*	2p_x	2p_y	2p_z
Nitrogen	7	↑↓	↑↓	↑	↑	↑

2.13 [Ne]$3s^2 3p^4$
2.14 Na· ⟶ Na^+ + e
:Cl· + e ⟶ :Cl:$^-$

Chapter 3

3.1 (a) K^+; (b) Ca^{2+}; (c) In^{3+}; (d) P^{3-}; (e) S^{2-}; (f) Br^-
3.2 (a) KBr; (b) GaF_3; (c) Ca_3P_2
3.3 (a) KCl; (b) Mg_3P_2; (c) Ga_2O_3
3.4 (a) Titanic chloride; (b) mercuric chloride; (c) ferrous chloride; (d) plumbic oxide
3.5 (a) Ammonium hydrogen sulfite; (b) calcium hypochlorite; (c) magnesium cyanide; (d) potassium dichromate; (e) ammonium sulfate
3.6 (a) Nitrogen triiodide; (b) diphosphorus pentoxide; (c) disulfur dichloride; (d) sulfur trioxide
3.7 (a) CH_4; (b) PCl_3; (c) CO_2
3.8 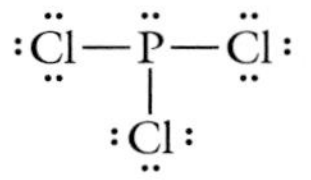
3.9 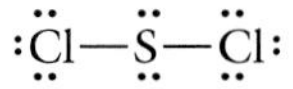
3.10 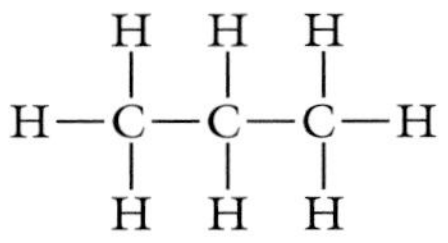
3.11 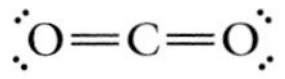
3.12 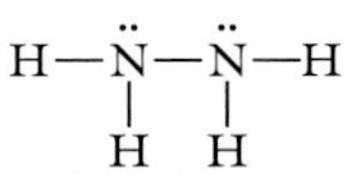
3.13 CCl_4 will have a tetrahedral shape.
3.14 NI_3 will have the shape of a trigonal pyramid.
3.15 OF_2 will have a bent shape.

Chapter 4

4.1 (a) 257.3 amu; (b) 231.6 amu; (c) 18.02 amu; (d) 98.08 amu; (e) 88.09 amu
4.2 (a) 142.1 g; (b) 119.0 g; (c) 40.31 g; (d) 86.17 g
4.3 1.251 mol
4.4 1.998×10^{-3} mol
4.5 (a) 1 mol each of Zn, H, and P atoms, 4 mol O atoms; (b) 2 mol H atoms, 1 mol S atoms, 4 mol O atoms; (c) 2 mol Al atoms, 3 mol O atoms; (d) 3 mol Ca atoms, 2 mol P atoms

4.6 Li, 1.153×10^{-23}g; N, 2.327×10^{-23} g; F, 3.156×10^{-23} g; Ca, 6.657×10^{-23} g; C, 1.995×10^{-23} g

4.7

Number of Calcium Atoms	Mass in grams
10	6.657×10^{-22}
100	6.657×10^{-21}
100000	6.657×10^{-18}
1.0×10^{15}	6.657×10^{-8}
6.022×10^{20}	0.04009
6.022×10^{23}	40.09

4.8 MgO
4.9 $MgSO_4$
4.10 P_2O_5
4.11 $3\ Ca(OH)_2 + 2\ H_3PO_4 \longrightarrow Ca_3(PO_4)_2 + 6\ H_2O$
4.12 $C_4H_{10} + \frac{13}{2}\ O_2 \longrightarrow 4\ CO_2 + 5\ H_2O$
$2\ C_4H_{10} + 13\ O_2 \longrightarrow 8\ CO_2 + 10\ H_2O$
4.13 $C_3H_6O_3 + 3\ O_2 \longrightarrow 3\ CO_2 + 3\ H_2O$
4.14 $\frac{3\text{ mol Mg(OH)}_2}{2\text{ mol FeCl}_3}$ $\frac{3\text{ mol Mg(OH)}_2}{2\text{ mol Fe(OH)}_3}$ $\frac{3\text{ mol Mg(OH)}_2}{3\text{ mol MgCl}_2}$
$\frac{2\text{ mol FeCl}_3}{2\text{ mol Fe(OH)}_3}$ $\frac{2\text{ mol FeCl}_3}{3\text{ mol MgCl}_2}$ $\frac{2\text{ mol Fe(OH)}_3}{3\text{ mol MgCl}_2}$
plus the reciprocal of each.
4.15 6.40 mol
4.16 1.0 mol
4.17 62.2 g $BaCl_2$, 60.2 g $Ba_3(PO_4)_2$

Chapter 5

5.1 (a) 0.741 atm; (b) 707 mmHg, or 707 torr
5.2 $V = 407$ mL
5.3 $P = 0.950$ atm
5.4 $V = 1.14 \times 10^3$ mL
5.5 $T = 619$ K, or 346°C
5.6 $P = 1.73$ atm
5.7 $T = 913$ K, or 640°C
5.8 $V = 1.11$ L
5.9 $CO_2 = 0.191$ g
5.10 $M = 32.2$ g/mol
5.11 $M = 40.05$ g/mol
5.12 $P_{hydrogen} = 494$ torr; $P_{oxygen} = 247$ torr
5.13 $V = 8.14$ L
5.14 0.0152 mL N_2/mL H_2O
5.15 0.0122 mL N_2/mL H_2O

Chapter 6

6.1 (c) H—C≡N
6.2 c and e.
6.3 d > c > b > a
6.4 No. Liquids having strong attractive forces will not mix with liquids possessing weak attractive forces.
6.5 (a) The secondary forces between CCl_4 molecules are weak London forces, and those between H_2O molecules are strong H-bonds. Solutions between substances possessing such very different secondary forces are not possible.
(b) The secondary forces between C_5H_{12} molecules and between CCl_4 molecules are London forces. Because they are the same, solutions of these two substances are possible.
6.6 Ethanol, 200 torr; hexane, 390 torr; chloroform, 500 torr.
6.7 Methyl alcohol, CH_3OH, in the liquid state is hydrogen bonded, and methyl iodide, CH_3I, is not.
6.8 Ammonia can form hydrogen bonds; phosphine cannot. Therefore ammonia has the higher boiling point.
6.9 b > a > d > c

Chapter 7

7.1 Dissolve 11.0 g of gelatin in 189 g of water.
7.2 0.0020 g/g of solution
7.3 438 g of solution
7.4 Add 4.8 mL of acetone to sufficient water so that the final volume of solution equals 100 mL.
7.5 6.6 g of sodium chloride
7.6 Dissolve 37.4 g of KH_2PO_4 in 1.0 L of solution.
7.7 0.918 mol
7.8 1.67 L
7.9 0.300 M
7.10 0.0200 mmol/mL
7.11 1.80×10^3 mL, or 1.80 L
7.12 503 mL, or 0.503 L
7.13 9.0×10^{-5} g/mL
7.14 (a)
$3\ Ca^{2+}(aq) + 6\ (Cl)^-(aq) + 6\ Na^+(aq) + 2\ PO_4^{3-}(aq) \longrightarrow Ca_3(PO_4)_2(s) + 6\ Cl^-(aq) + 6\ Na^+(aq)$
(b) $3\ Ca^{2+}(aq) + 2\ PO_4^{3-}(aq) \longrightarrow Ca_3\ (PO_4)_2(s)$

Chapter 8

8.1 $\Delta H_{rxn} = +6$ kJ. The reaction is endothermic.
8.2 $K_{eq} = \frac{[HI]^2}{[I_2][H_2]}$
8.3 $K_{eq} + \frac{[CO]^2}{[O_2]}$
8.4 [A] = [B]
8.5 $K_{eq} = 1.1 \times 10^{-3}$
8.6 $K_{eq} = 1 \times 10^2$
8.7 Shift of mass to the left (reactant side).
8.8 (a) Shift of mass to the right; (b) shift of mass to the left.

Chapter 9

9.1 0.050 M
9.2 0.050 M
9.3 0.060 M
9.4 2.00×10^{-12} M
9.5 2.00×10^{-12} M
9.6 $[H_3O^+] = 1.0 \times 10^{-12}$ M; pH = 12.00
9.7 pH = 2.43
9.8 pH = 10.00
9.9 pOH = 11.30
9.10 $[HI] = 3.47 \times 10^{-3}$
9.11 $K_{b(cyanate)} = K_w/K_{a(cyanic\ acid)}$
9.12 $pK_{b(cyanate)} = pK_w - pK_{a(cyanic\ acid)}$
9.13 $MgCl_2$ and KNO_3, neutral; $SnCl_2$ and NH_4NO_3, acidic.
9.14 pH = 4.89
9.15 26.3 mL
9.16 Dissolve 59 g of succinic acid in 1.0 L of water.
9.17 $[K^+] = 0.0050$ M; $[Mg^{2+}] = 0.0015$ M

Chapter 10

10.1 ${}^{223}_{88}\text{Ra} \longrightarrow {}^{219}_{86}\text{Rn} + {}^{4}_{2}\text{He}$

10.2 ${}^{230}_{88}\text{Ra} \longrightarrow {}^{230}_{89}\text{Ac} + {}^{0}_{-1}\text{e}$

10.3 ${}^{21}_{11}\text{Na} \longrightarrow {}^{21}_{10}\text{Ne} + {}^{0}_{+1}\text{e}$

10.4 1.56%

10.5 3.13%

10.6 7.18%

10.7 12 m

Chapter 11

11.1 CH_5N and C_2H_5Cl are correct molecular formulas.

11.2 (a) ketone; (b) ester; (c) amide; (d) amine.

11.3

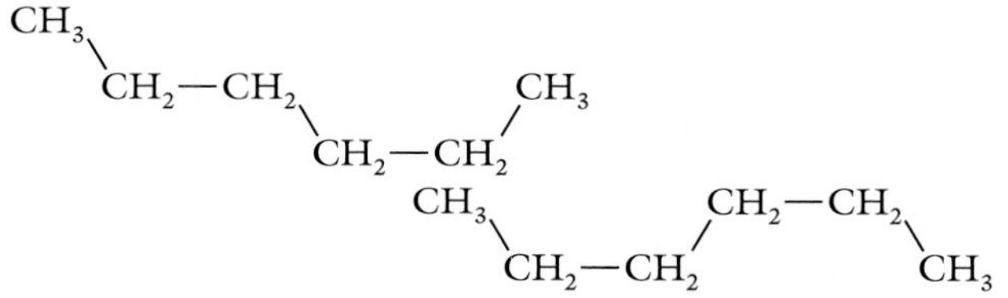

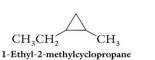

Different conformations

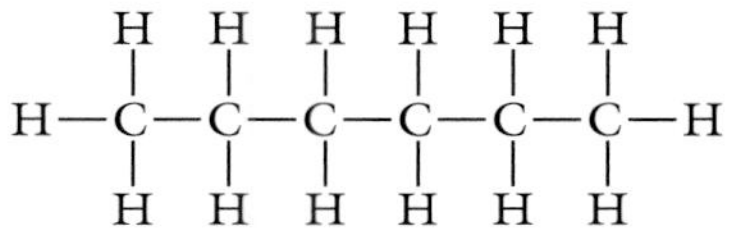

Expanded

$CH_3CH_2CH_2CH_2CH_2CH_3$

Condensed

C—C—C—C—C—C

Skeleton

Line

11.4 $CH_3—CH_2—CH_2—CH_2—CH_2—CH_3$

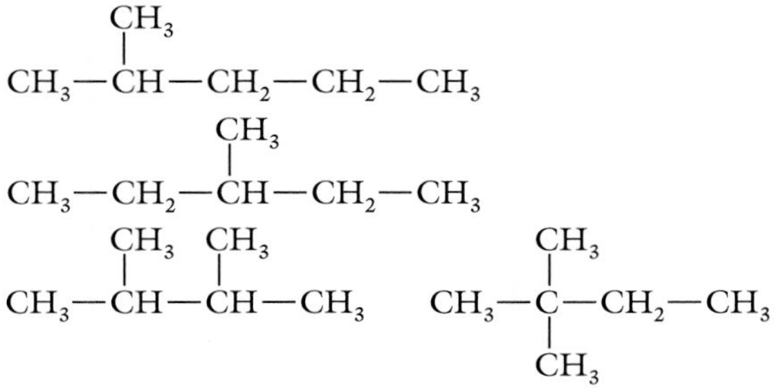

11.5

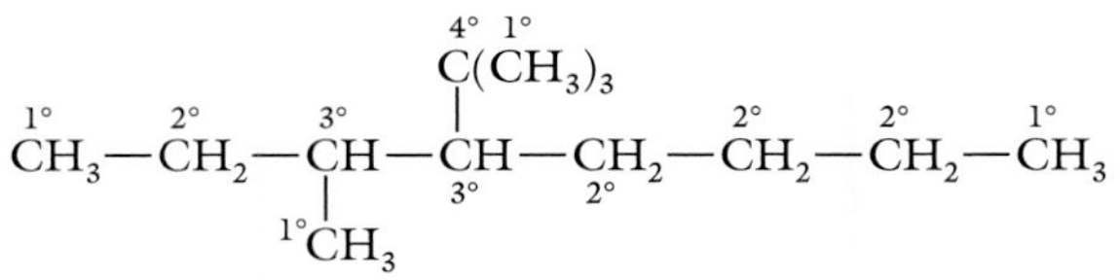

11.6 (a) 5-ethyl-2,3,3,8-tetramethylnonane

(b) 4-isopropyl-3-methylheptane

11.7

(a) 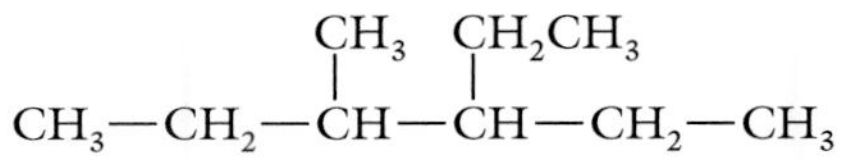

(b) 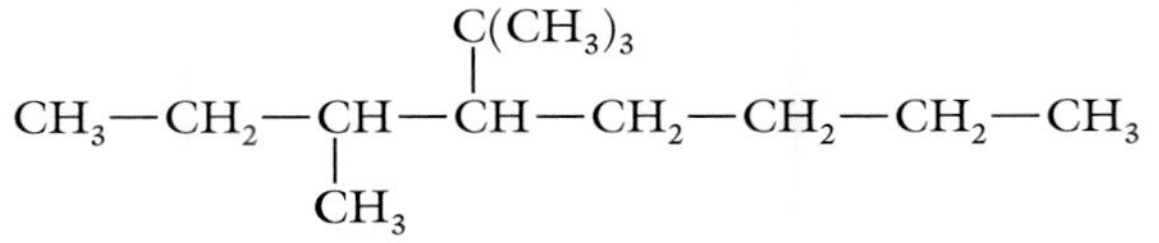

11.8 (a) 1-*t*-butyl-3-ethyl-5-isopropylcyclohexane

(b) 1-*s*-butyl-1,3-dimethylcyclobutane

Geometric stereoisomers (Section 11.9) are possible for both compounds.

11.9

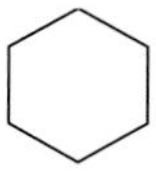

Cyclohexane

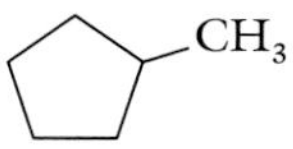

Methylcyclopentane

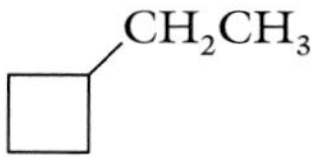

Ethylcyclobutane

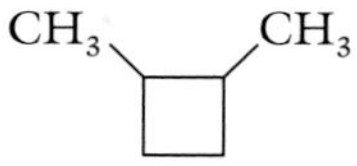

1,2-Dimethylcyclobutane

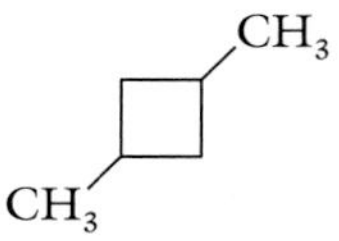

1,3-Dimethylcyclobutane

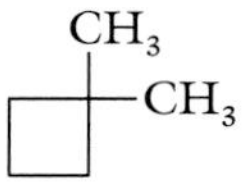

1,2-Dimethylcyclobutane

Propylcyclopropane

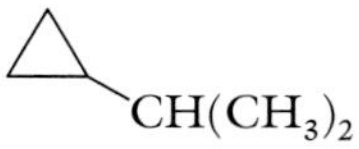

Isopropylcyclopropane

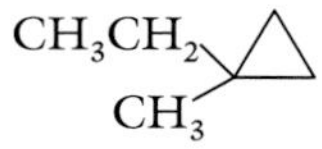

1-Ethyl-1-methylcyclopropane

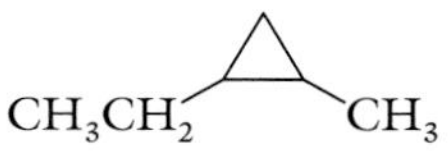

1-Ethyl-2-methylcyclopropane

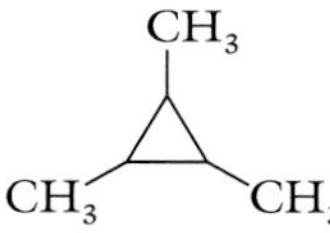

1,2,3-Trimethylcyclopropane

Geometric stereoisomers (Section 11.9) are possible for 1,2-dimethylcyclobutane, 1,3-dimethylcyclobutane, 1-ethyl-2-methylcyclopropane, and 1,2,3-trimethylcyclopropane.

11.10

(a)

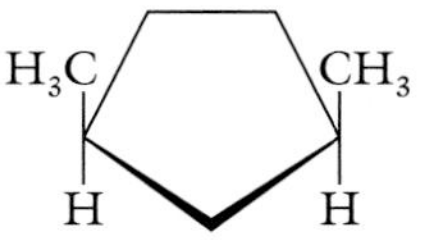

***cis*-1,3-Dimethylcyclopentane**

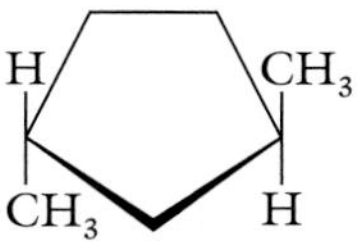

***trans*-1,3-Dimethylcyclopentane**

(b) Cis-trans are not possible.

(c)

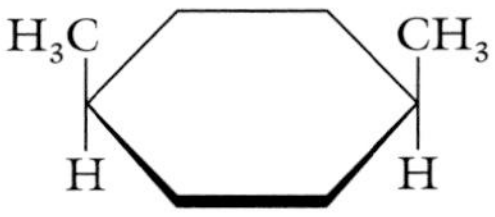

***cis*-1,4-Dimethylcyclohexane**

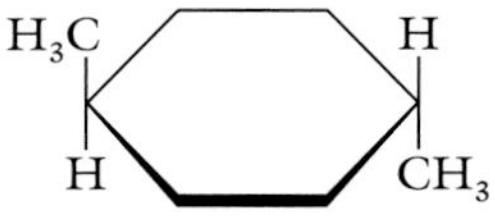

***trans*-1,4-Dimethylcyclohexane**

11.11 (a) Cyclopentane has the higher boiling point because its rigid structure allows tighter packing, which results in greater secondary forces. 2-Methylbutane, being branched and spherical in structure, has much smaller secondary forces.

(b) 3-Methylnonane has the higher boiling point. Both compounds have the same longest chain of nine carbons, but 3-methylnonane has an additional carbon (the 3-methyl branch).

(c) Cycloheptane has the higher boiling point because of its larger molecular mass.

11.12 No. NaCl is an ionic compound. The energy required to separate NaCl into Na^+ and Cl^- cannot be recouped by attraction between the ions and hexane, because hexane is nonpolar.

11.13 (a) $2\ C_5H_{10} + 15\ O_2 \longrightarrow 10\ CO_2 + 10\ H_2O$
(b) $C_9H_{20} + 14\ O_2 \longrightarrow 9\ CO_2 + 10\ H_2O$

11.14
$$CH_3-CH_3 \xrightarrow[\text{UV or heat}]{Cl_2} CH_3-CH_2Cl$$

11.15

$$CH_3-\underset{\underset{CH_3}{|}}{CH}-CH_3 \xrightarrow[\text{UV or heat}]{Br_2}$$

$$CH_3-\underset{\underset{CH_3}{|}}{CH}-CH_2-Br + CH_3-\underset{\underset{Br}{|}}{\overset{\overset{CH_3}{|}}{C}}-CH_3$$

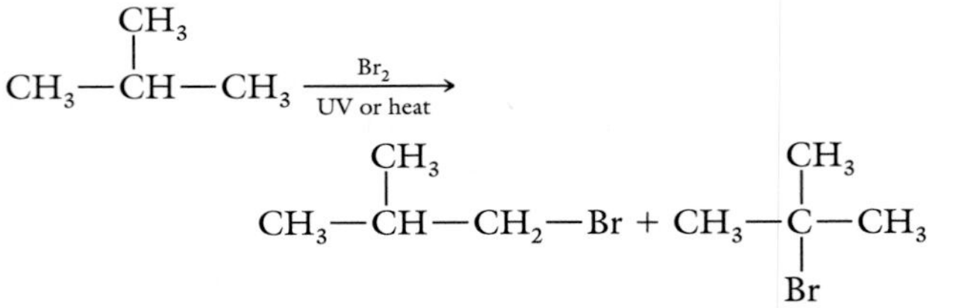

11.16 (a) IUPAC: 1-bromo-2-methylpropane; common: isobutyl bromide.   (b) IUPAC: The two possible names are *cis*-1-bromo-2-methylcyclopentane and *cis*-2-bromo-1-methylcyclopentane. Although both names have the same set of numbers (1, 2), the correct IUPAC name is *cis*-1-bromo-2-methylcyclopentane because the numbers are assigned in alphabetical order. There is no common name, because the alkyl group attached to Br is not a simple alkyl group.

Chapter 12

12.1 The unknown compound is an alkane because its molecular formula fits the general formula for an alkane, C_nH_{2n+2}, where $n = 7$.

12.2

$CH_2=CH-CH_2-CH_2-CH_2-CH_3$

$CH_3-CH=CH-CH_2-CH_2-CH_3$

$CH_3-CH_2-CH=CH-CH_2-CH_3$

$CH_2=\underset{\underset{CH_3}{|}}{C}-CH_2-CH_2-CH_3$

$CH_3-\underset{\underset{CH_3}{|}}{C}=CH-CH_2-CH_3$

$CH_3-\underset{\underset{CH_3}{|}}{CH}-CH=CH-CH_3$

$CH_3-\underset{\underset{CH_3}{|}}{CH}-CH_2-CH=CH_2$

$CH_2=CH-\underset{\underset{CH_3}{|}}{CH}-CH_2-CH_3$

$CH_3-CH=\underset{\underset{CH_3}{|}}{C}-CH_2-CH_3$

$CH_3-CH_2-\underset{\overset{\overset{CH_2}{||}}{}}{C}-CH_2-CH_3$ $CH_2=CH-\underset{\underset{CH_3}{|}}{\overset{\overset{CH_3}{|}}{C}}-CH_3$

$CH_3-\underset{\underset{CH_3}{|}}{C}=C-CH_3$ $CH_2=\underset{\underset{CH_3}{|}}{C}-\underset{\underset{CH_3}{|}}{CH}-CH_3$

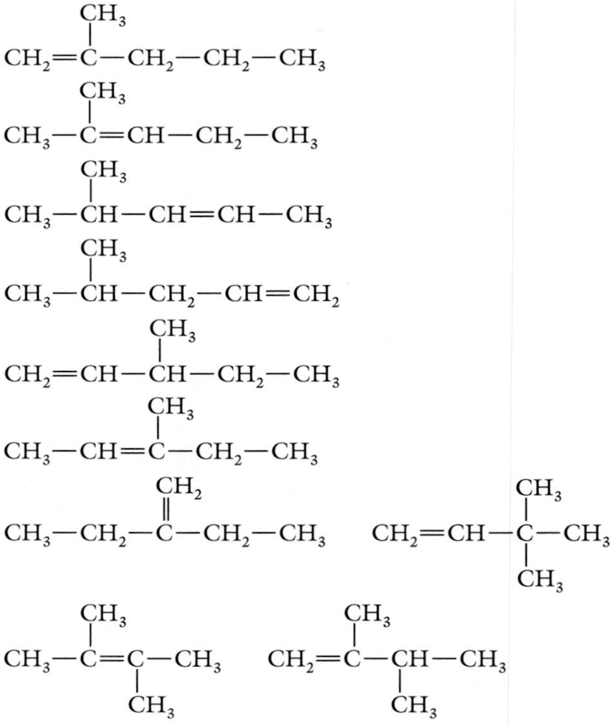

12.3 (a) 4-Methyl-1-hexene;   (b) 1-chloro-4-methyl-2-pentene;   (c) 3-*t*-butyl-2-isopropylcyclopentene;   (d) 2-methyl-1,4-pentadiene

12.4 (a) No cis-trans isomers

(b)

(c)

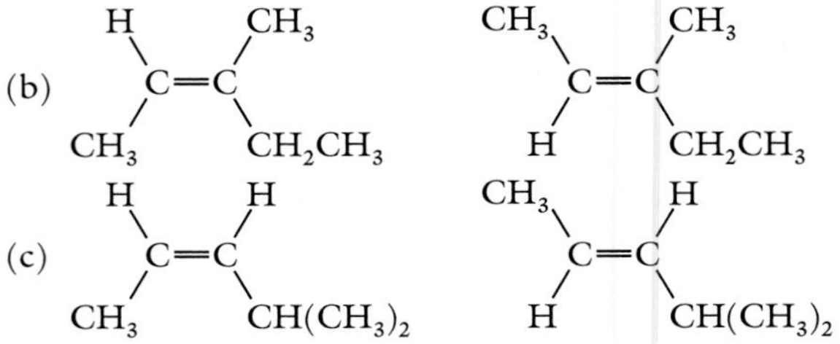

12.5 (a) $CH_3-CH=CH-CH_3 \xrightarrow[Pt]{H_2}$

$$CH_3-CH_2-CH_2-CH_3$$

(b) $CH_3-CH=CH_2 \xrightarrow{Br_2} CH_3-\underset{\underset{Br}{|}}{CH}-CH_2-Br$

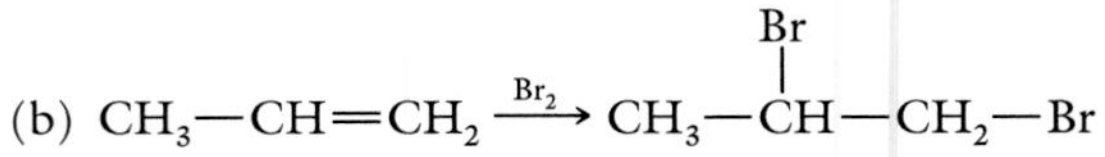

12.6 2-Pentene

12.7 (a) $CH_2=CH-CH_2-CH_3 \xrightarrow{Cl_2}$

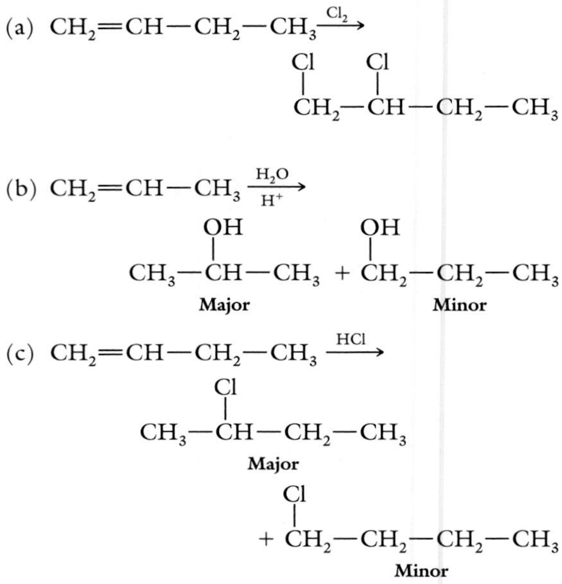

$$\underset{CH_2}{\overset{\overset{Cl}{|}}{}}-\underset{}{\overset{\overset{Cl}{|}}{CH}}-CH_2-CH_3$$

(b) $CH_2=CH-CH_3 \xrightarrow[H^+]{H_2O}$

$$CH_3-\underset{\overset{\overset{OH}{|}}{}}{CH}-CH_3 + \underset{\overset{\overset{OH}{|}}{}}{CH_2}-CH_2-CH_3$$
Major   **Minor**

(c) $CH_2=CH-CH_2-CH_3 \xrightarrow{HCl}$

$$CH_3-\underset{\overset{\overset{Cl}{|}}{}}{CH}-CH_2-CH_3$$
Major

$$+ \underset{\overset{\overset{Cl}{|}}{}}{CH_2}-CH_2-CH_2-CH_3$$
Minor

12.8 $n\ CH_2=\underset{\underset{CH_3}{|}}{CH} \xrightarrow{\text{polymerization catalyst}} \overset{}{-}(CH_2-\underset{\underset{CH_3}{|}}{CH})\overline{}_n$

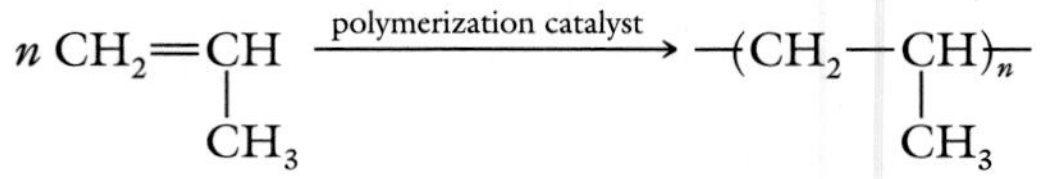

12.9 $C_8H_{16} + 12\ O_2 \longrightarrow 8\ CO_2 + 8\ H_2O$

12.10

$CH_3-C\equiv C-CH_3 \xrightarrow{HCl}$

$CH_3-\underset{\overset{\overset{Cl}{|}}{}}{C}=CH-CH_3 \xrightarrow{HCl} CH_3-\underset{\underset{Cl}{|}}{\overset{\overset{Cl}{|}}{C}}-CH_2-CH_3$

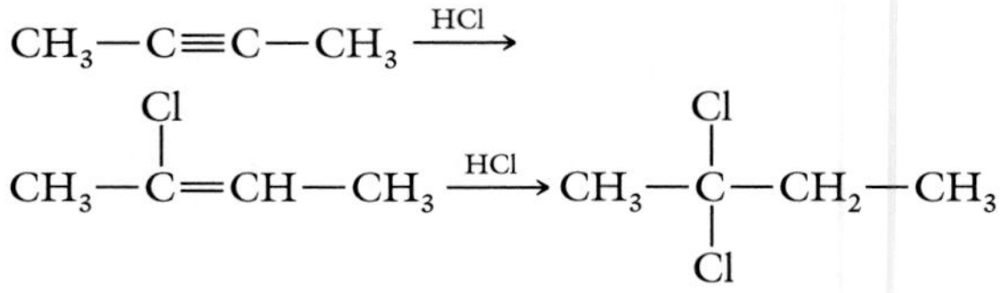

12.11

(a)

(b)

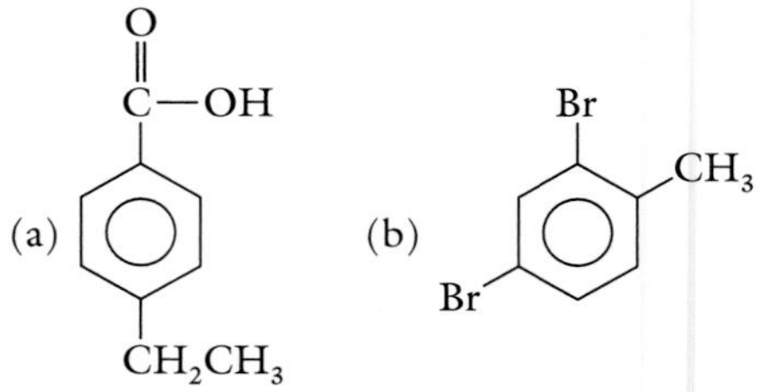

(c) 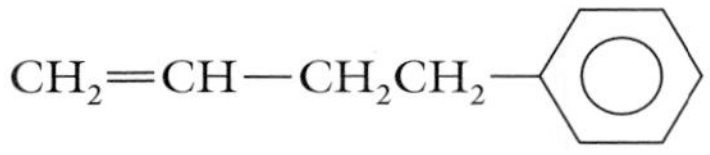

(d) 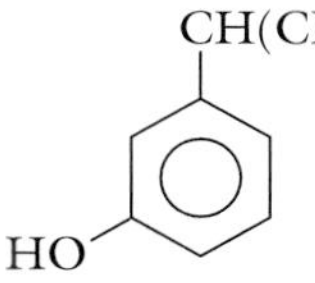 (e) 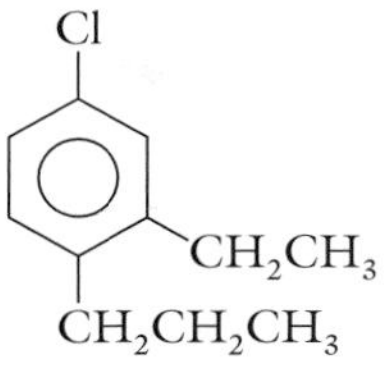

(f) $C_6H_5CH_2Cl$

12.12 (a) No reaction takes place, because alkylation requires a metal halide catalyst.

(b) 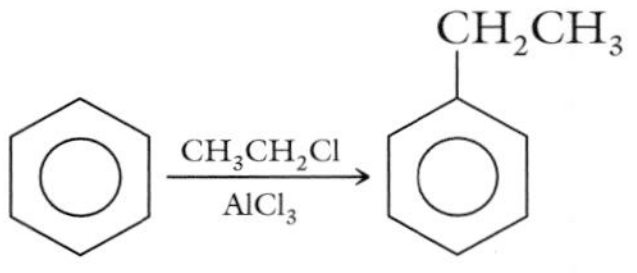

(c) 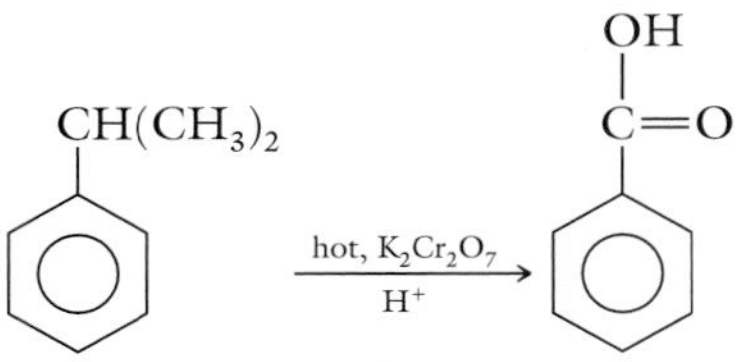

12.13 Benzene does not undergo bromine substitution unless a metal halide is present. Cyclohexane undergoes substitution in the presence of UV radiation (or high temperature):

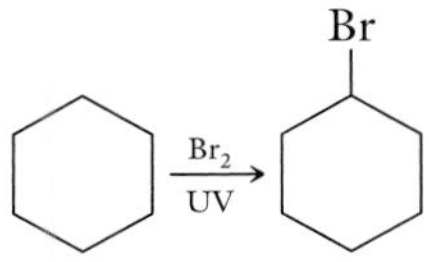

Cyclohexene undergoes addition irrespective of whether there is UV radiation or metal halide:

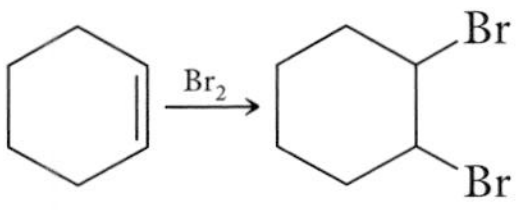

cis- and *trans-*

Chapter 13

13.1 Structure 1, alcohol; structure 2, ether; structure 3, phenol. Structure 4 is not an alcohol, phenol, or ether; it is an aldehyde.

13.2 Eight isomers:

$CH_3{-}CH_2{-}CH_2{-}CH_2{-}CH_2{-}OH$

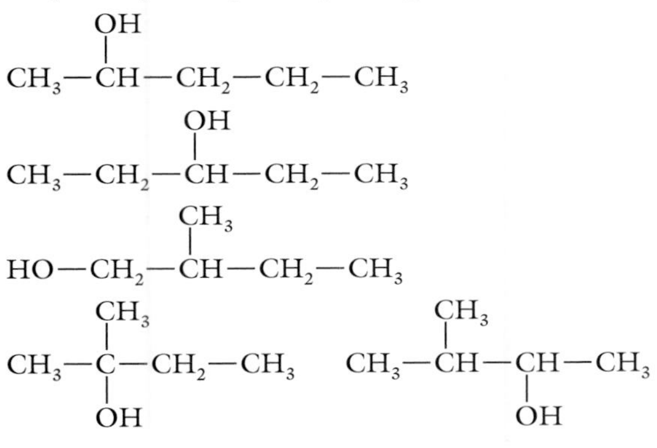

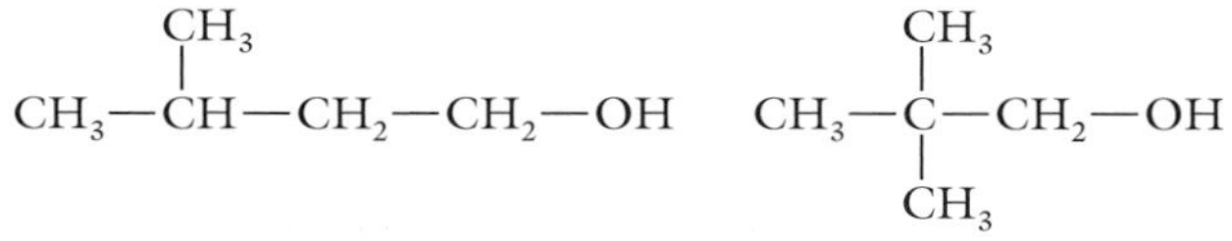

13.3 Alcohol 1 is $C_5H_{12}O$ with $n = 5$ and $2n + 2 = 12$. Alcohol 2 is C_5H_{10} with $n = 5$ and $2n = 10$.

13.4 (a) Tertiary; (b) primary; (c) secondary; (d) secondary

13.5 (a) 2,5-Dimethyl-3-hexanol; (b) *cis*-2-*t*-butylcyclohexanol; (c) 3-methyl-1,2-butanediol; (d) 4-isopropyl-1-heptanol

13.6 (a) 1,2-Ethanediol has a much higher boiling point (198°C versus 97°C) even though both compounds have nearly the same molecular mass. The presence of two OH groups results in more hydrogen bonding in 1,2-ethanediol. Higher temperatures are needed during boiling to overcome the resulting larger secondary attractive forces in 1,2-ethanediol.

(b) This trend is the same as that observed in all families. Secondary attractive forces and boiling points decrease with branching. Branching decreases the molecular surface area over which molecules attract each other (see Section 12.11).

(c) One compound dissolves in another only when both compounds have similar secondary attractions. Propanol participates in hydrogen bonding and cannot interact with nonpolar compounds. Secondary attractions exist between hexane and butane because both are nonpolar.

13.7

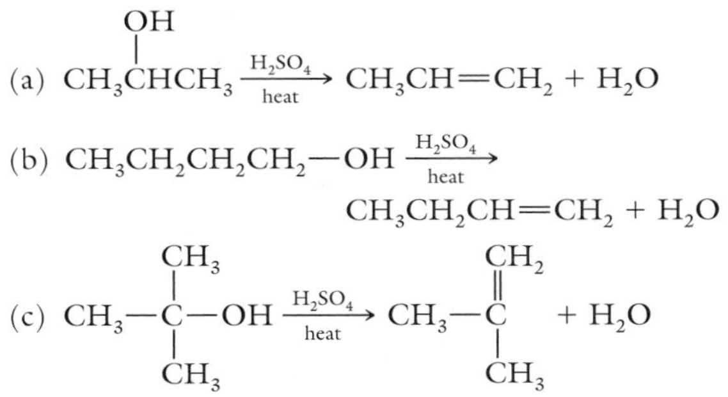

13.8

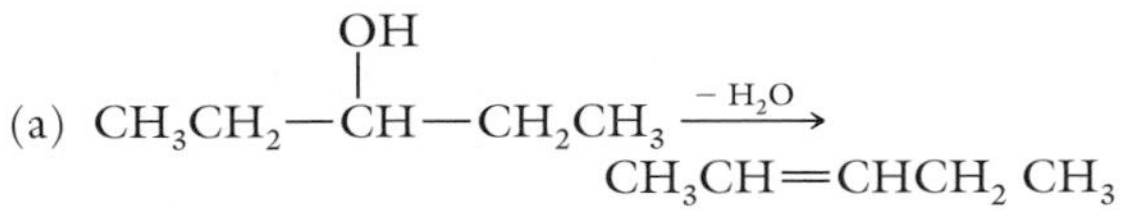

2-Pentene

trans-2-Pentene predominates over *cis*-2-pentene.

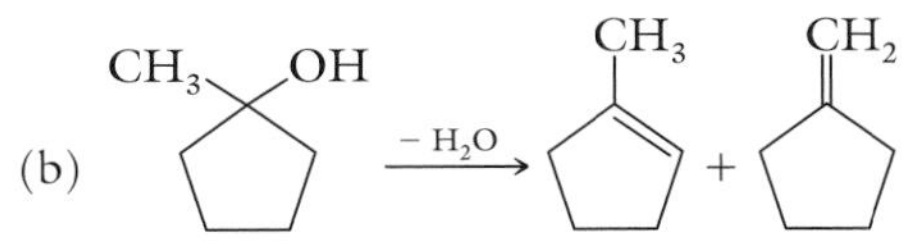

Major **Minor**

13.9

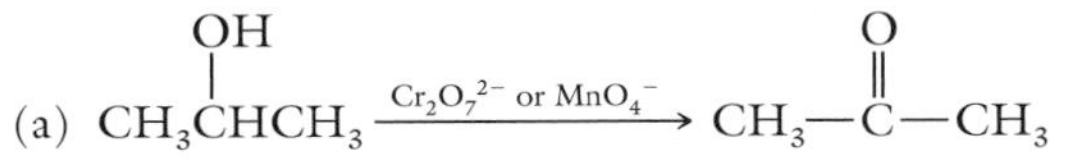

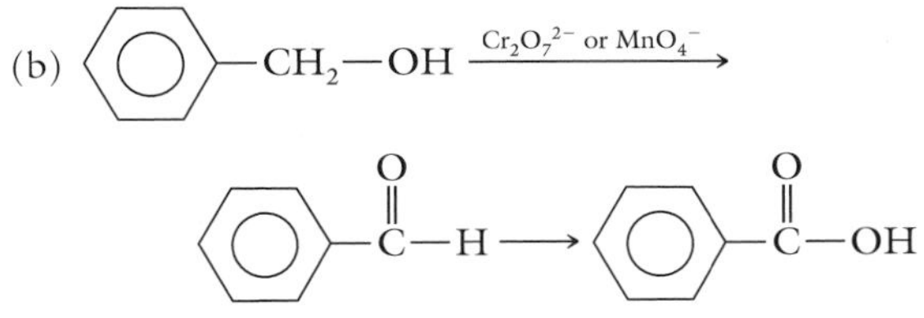

(c) $CH_3-\underset{\underset{\underset{\text{No H}}{CH_3}}{|}}{\overset{\overset{CH_3}{|}}{C}}-OH \xrightarrow{Cr_2O_7^{2-} \text{ or } MnO_4^-}$ no reaction 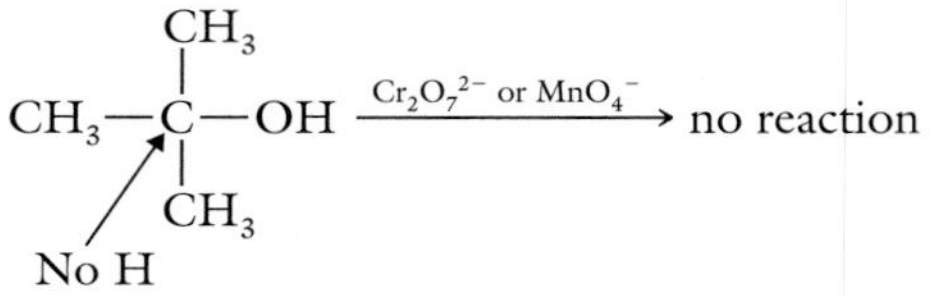

13.10 A positive test with dichromate does not differentiate between 1-butanol (primary alcohol) and 2-butanol (secondary alcohol), because both give positive tests with dichromate (and permanganate).

13.11 (a) 4-Butyl-3-t-butylphenol; (b) 3-bromo-2-methylphenol

13.12 Hydrogen bonding is described by representations 2 and 3.

13.13 (a) Butyl t-butyl ether; (b) 3-butoxy-3-methyl-1-butene; (c) s-butyl isopropyl ether

13.14 The four alcohols of Example 13.2 (butyl alcohol, s-butyl alcohol, t-butyl alcohol, isobutyl alcohol) are isomeric with the three ethers; all are C_4H_9O.

13.15 (a) p-Methylphenol has the higher boiling point because it has an —OH group and can hydrogen bond with itself. Methyl phenyl ether cannot hydrogen bond with itself.

(b) Dipropyl ether is more soluble than 1-hexanol in heptane. Like dissolves in like. The secondary forces in dipropyl ether are the weak London forces, like those in heptane. The secondary forces in 1-hexanol are the strong hydrogen-bonding attractions; there are only very weak attractions between 1-hexanol and the nonpolar heptane.

(c) Diethyl ether and 1-butanol have similar water solubilities. 1-Butanol is slightly more soluble than diethyl ether in water.

13.16

$CH_3CH_2-O-H \xrightarrow[-H_2O]{\substack{180°C, \\ H_2SO_4}}$

$\underset{\text{Minor}}{CH_3CH_2-O-CH_2CH_3} + \underset{\text{Major}}{CH_2=CH_2}$

13.17

(a) $CH_3CH_2CH_2SH + NaOH \longrightarrow$
$CH_3CH_2CH_2S^- + Na^+ + H_2O$

(b) $CH_3CH_2-S-S-CH_2CH_3 \xrightarrow{(H)}$
$2\ CH_3CH_2-SH$

(c) $CH_3CH_2CH_2SH + Hg^{2+} \longrightarrow$
$CH_3CH_2CH_2S-Hg-SCH_2CH_2CH_3 + 2\ H^+$

Chapter 14

14.1 Compound 1, ester; 2, aldehyde; 3, ketone; 4, amide

14.2 There are four aldehydes and three ketones for $C_5H_{10}O$.

$CH_3CH_2CH_2CH_2-\overset{\overset{O}{||}}{C}-H$

$CH_3CH_2\underset{\underset{CH_3}{|}}{CH}-\overset{\overset{O}{||}}{C}-H \qquad CH_3\underset{\underset{CH_3}{|}}{CH}CH_2-\overset{\overset{O}{||}}{C}-H$

$(CH_3)_3C-\overset{\overset{O}{||}}{C}-H \qquad CH_3-\overset{\overset{O}{||}}{C}-CH_2CH_2CH_3$

$CH_3CH_2-\overset{\overset{O}{||}}{C}-CH_2CH_3 \qquad CH_3-\overset{\overset{O}{||}}{C}-CH(CH_3)_2$

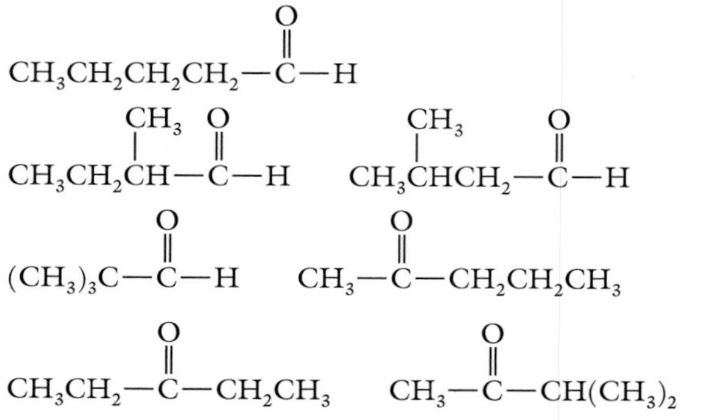

14.3 (a) 2,5-Dimethyl-3-hexanone; (b) 4,4-dimethylpentanal; (c) 3-chlorocyclopentanone; (d) 2-bromo-4-methylbenzaldehyde

14.4 (a) t-Butyl cyclohexyl ketone; (b) 3-chlorophenyl propyl ketone

14.5 (a) Hexanal has the higher boiling point because of its higher molecular mass. Secondary attractive forces increase with molecular mass within a family of organic compounds.

(b) 1-Propanal has the higher solubility in water. Increasing molecular mass decreases the fraction of the molecule that is able to hydrogen bond with water.

(c) 2-Propanol has the higher boiling point because it hydrogen bonds with itself. Propanal does not hydrogen bond with itself. The dipole–dipole secondary attractive forces in propanal are weaker than the hydrogen-bonding forces in 2-propanol.

(d) Butanal and 1-butanol have nearly the same solubility in water because both can hydrogen bond with water.

14.6 (a) No oxidation

(b) $CH_3CH_2CH_2CH_2\overset{\overset{O}{||}}{C}H \xrightarrow{MnO_4^- \text{ or } Cr_2O_7^-}$ 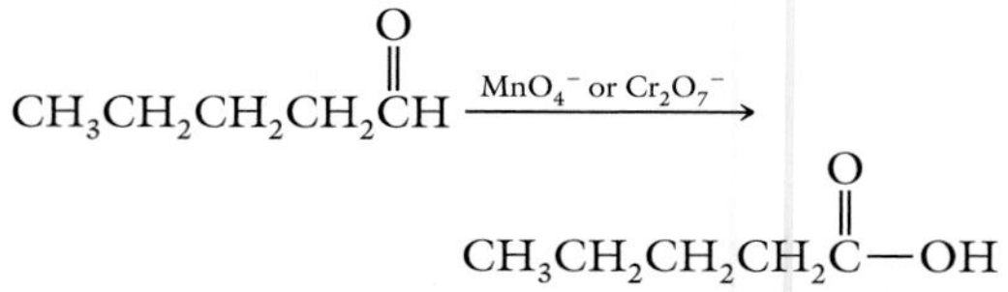

$CH_3CH_2CH_2CH_2\overset{\overset{O}{||}}{C}-OH$

14.7 (a) Compounds 2 and 4 are aldehydes and give positive tests with Tollens's reagent. Compounds 1 and 3 are ketones and give negative tests with Tollens's reagent.

(b) Compounds 1, an α-hydroxy ketone, and 4, an α-hydroxy aldehyde, give positive tests with Benedict's reagent. Compound 2, a simple aldehyde, gives a negative test. Compound 3, a simple ketone, gives a negative test.

14.8

(a) $\xrightarrow[\text{2. }H_2O]{\text{1. }LiAlH_4}$

(b) $\xrightarrow[\text{Pt}]{H_2}$

(c) $CH_3\overset{\overset{O}{||}}{C}CH_2CH_2CH_3 \xrightarrow[\text{2. }H_2O]{\text{1. NaDH}}$

$CH_3\underset{\underset{H}{|}}{\overset{\overset{OH}{|}}{C}}CH_2CH_2CH_3$

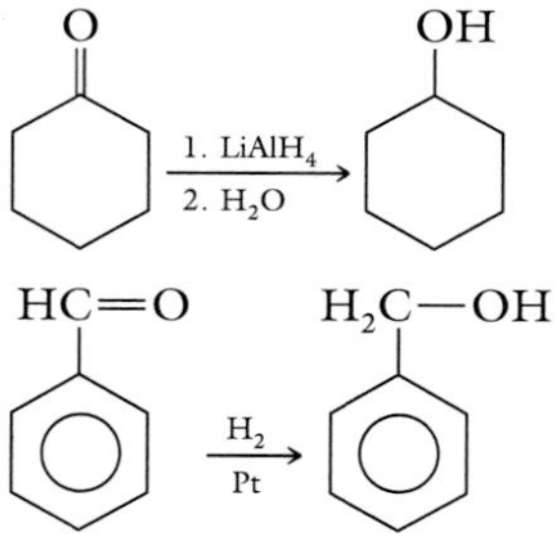

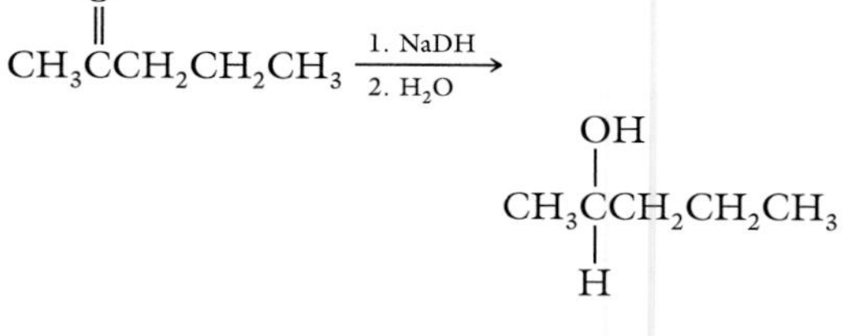

14.9

$CH_3CH_2CH_2\overset{\overset{O}{||}}{C}H \xrightarrow[H^+]{CH_3OH} CH_3CH_2CH_2\underset{\underset{OH}{|}}{\overset{\overset{OCH_3}{|}}{C}}H \xrightarrow[H^+]{CH_3OH}$

$CH_3CH_2CH_2\underset{\underset{OCH_3}{|}}{\overset{\overset{OCH_3}{|}}{C}}H$

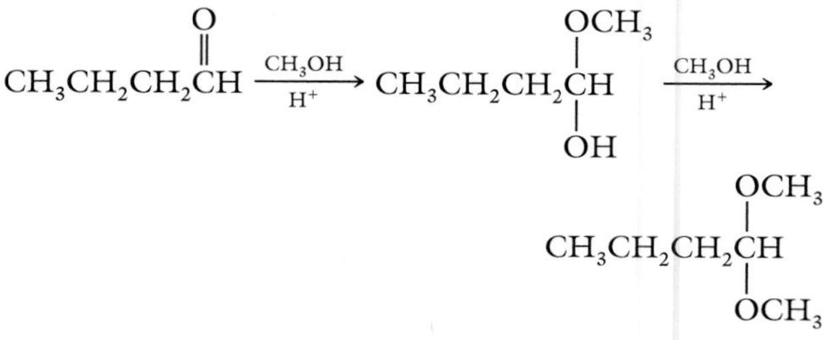

14.10

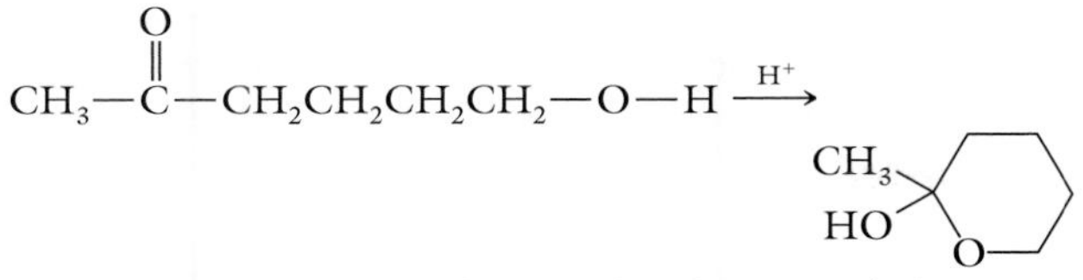

14.11 Compounds 1 and 4, acetal; 2, hemiacetal; 3, something different, a diether.

14.12

(a) $CH_3CH_2O{-}CH(CH_2C_6H_5){-}OCH_2CH_3 \xrightarrow{H^+} 2\ CH_3CH_2OH + HC(=O){-}CH_2C_6H_5$

(b)

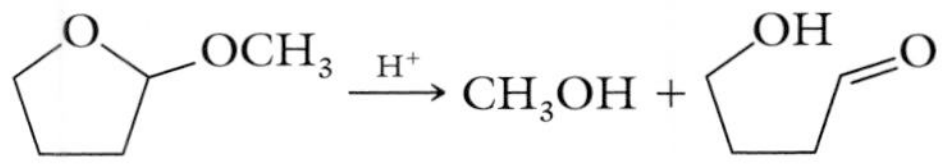

Chapter 15

15.1 Compound 1, ester; 2, aldehyde; 3, amide; 4, acid halide; 5, acid anhydride; 6, carboxylic acid

15.2

(a)

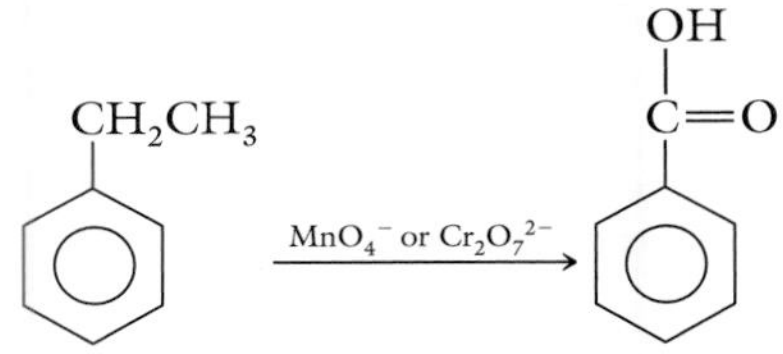

(b) $CH_3CH_2{-}OH \xrightarrow{MnO_4^- \text{ or } Cr_2O_7^{2-}} CH_3C(OH){=}O$

(c) $CH_3C(H){=}OH \xrightarrow{MnO_4^- \text{ or } Cr_2O_7^{2-}} CH_3C(OH){=}O$

(d) No reaction takes place, because ketones do not undergo selective oxidation.

15.3 (a) 2, 5-Dimethylheptanoic acid; (b) 4-*t*-butylbenzoic acid or *p*-*t*-butylbenzoic acid; (c) 2-methyl-1,5-pentanedioic acid

15.4

(a) $CH_3CH_2C(=O){-}O{-}H + H_2O \rightleftharpoons CH_3CH_2C(=O){-}O^- + H_3O^+$

(b) $CH_3CH_2C(=O){-}O{-}H + NaOH \longrightarrow CH_3CH_2C(=O){-}O^-Na^+ + H_2O$

(c) $CH_3CH_2C(=O){-}O{-}H + NaHCO_3 \longrightarrow CH_3CH_2C(=O){-}O^-Na^+ + H_2O + CO_2$

(d) no reaction

15.5 (a) Sodium 3-methylbutanoate; (b) aluminum ethanoate (common name: aluminum acetate)

15.6 $CH_3(CH_2)_{14}COOK$ is water soluble, whereas $CH_3(CH_2)_{14}COOH$ and $CH_3(CH_2)_{14}OH$ are water insoluble.

15.7

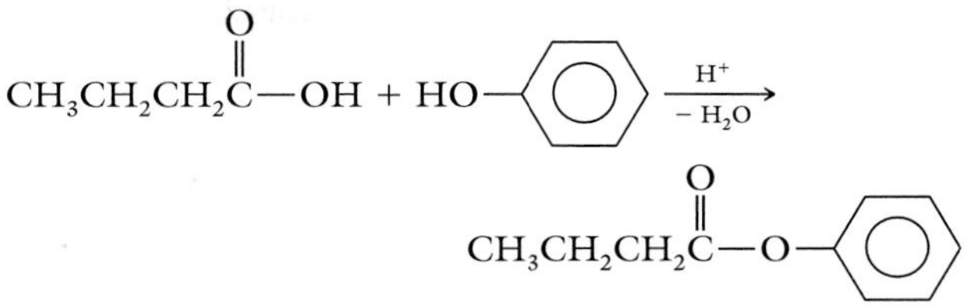

15.8 (a) *p*-Ethylphenyl 2-methylbutanoate or 4-ethylphenyl 2-methylbutanoate; (b) pentyl methanoate (common name: pentyl formate); (c) methyl hexanoate

15.9 Polymer is produced only when both reactants are bifunctional. Only the pair in part c meets this requirement. The pairs in parts a and b each have one bifunctional reactant and one monofunctional reactant.

$$HO{-}C(=O)CH_2CH_2CH_2CH_2C(=O){-}OH + HO{-}CH_2CH_2{-}OH \xrightarrow[-H_2O]{H^+} \left[{-}C(=O)CH_2CH_2CH_2CH_2C(=O){-}O{-}CH_2CH_2{-}O{-}\right]_n$$

15.10

(a)

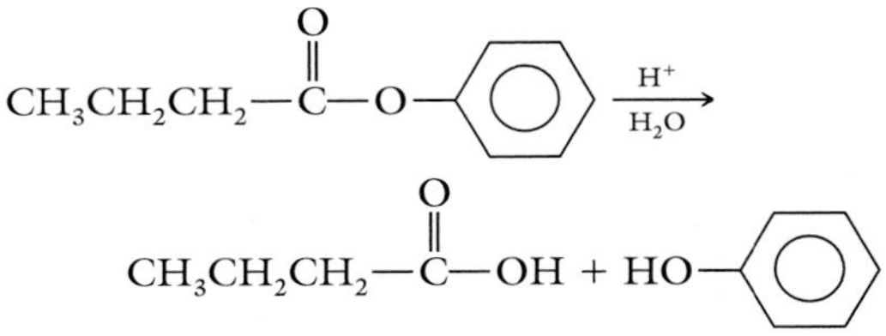

(b) $CH_3CH_2{-}C(=O){-}OCH_2CH_2CH_3 \xrightarrow[KOH]{H_2O} CH_3CH_2{-}C(=O){-}OK + HOCH_2CH_2CH_3$

15.11

(a) $CH_3CH_2CH_2{-}C(=O){-}Br$

(b) $CH_3{-}C(=O){-}O{-}C(=O){-}CH_3$

15.12

$$C_6H_5{-}C(=O){-}Cl + CH_3OH \longrightarrow C_6H_5{-}C(=O){-}OCH_3 + HCl$$

15.13

$$HO{-}P(=O)(OH){-}O{-}P(=O)(OH){-}OH \xrightarrow{KOH} KO{-}P(=O)(OK){-}O{-}P(=O)(OK){-}OK$$

15.14

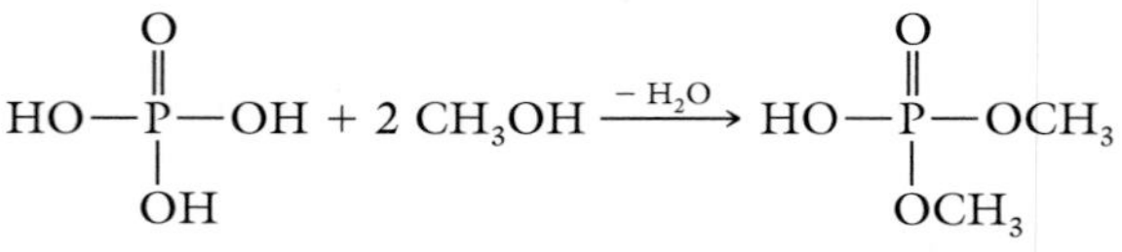

Dimethyl phosphate

Chapter 16

16.1 (a) All are amines except compound 5, which is an amide; compound 6 is an amine, not an amide, because the carbonyl carbon is not bonded to nitrogen. (b) Compounds 1, 3, and 7 are primary amines, 4 is a secondary amine, 2, 6, and 8 are tertiary amines. (c) Only compound 7 is an aromatic amine; the other amines are aliphatic amines.

16.2

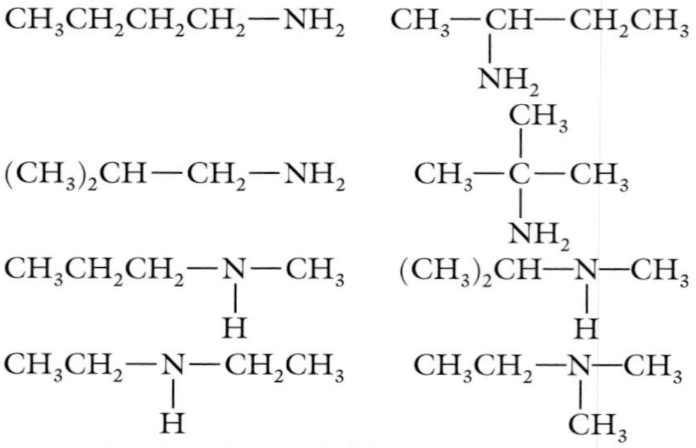

16.3 (a) Common name: s-butylethylmethylamine; (b) *CA*: 1,2-butanediamine; (c) *CA: trans*-3-ethyl-*N,N*-dimethylcyclopentanamine; (d) common: *N*-ethyl-3-isopropyl-4-methylaniline; (e) *CA*: 4-ethyl-2,6-dimethyl-2-heptanamine

16.4 (a) $2\ C_6H_5NH_2 + H_2SO_4 \longrightarrow [C_6H_5NH_3^+]_2\ SO_4^{2-}$
(b) $(CH_3CH_2)_2NH + H_2O \rightleftharpoons (CH_3CH_2)_2\overset{+}{N}H_2 + OH^-$

16.5

(a)

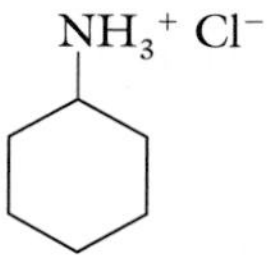

(b)

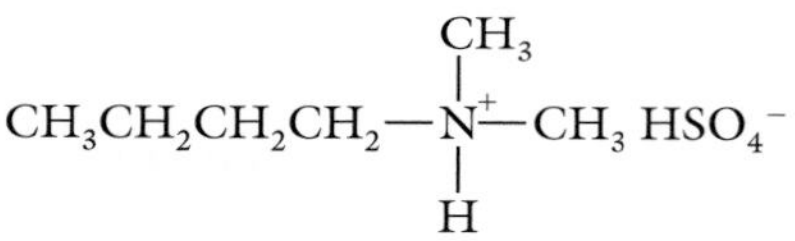

(c)

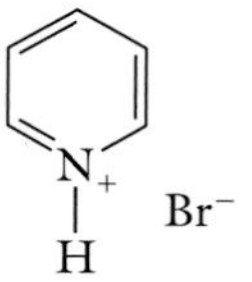

16.6 The order is $CH_3(CH_2)_{14}NH_2 < CH_3(CH_2)_6NH_2 \ll CH_3(CH_2)_{14}NH_3^+Cl^-$. The ionic amine salt is much more soluble than the covalent amines because ions are attracted to water more strongly than are covalent molecules. The lower-molecular-mass amine is more soluble than the higher-molecular-mass amine because the amine group, which hydrogen bonds with water, constitutes a larger part of the molecule.

16.7 Compound 1, tertiary amide; 2, secondary amide; 3 is not an amide—it is an amine and an aldehyde; 4, tertiary amide.

16.8

(a)

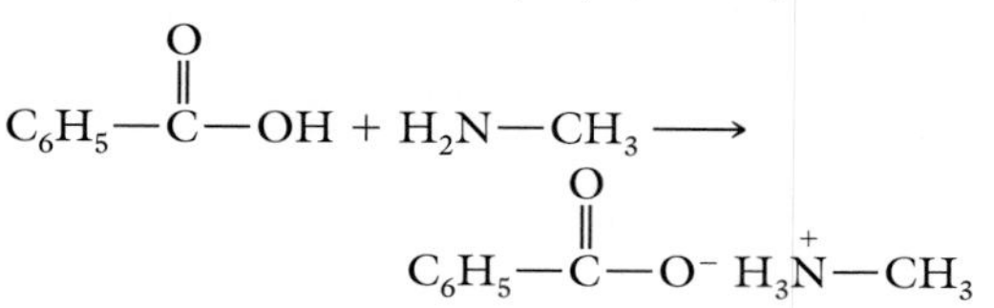

(b) The reaction temperature is that given for amide formation, but amide formation does not take place when tertiary amines are used, because there is no hydrogen attached to the nitrogen of the amine. The only possible reaction is acid-base reaction:

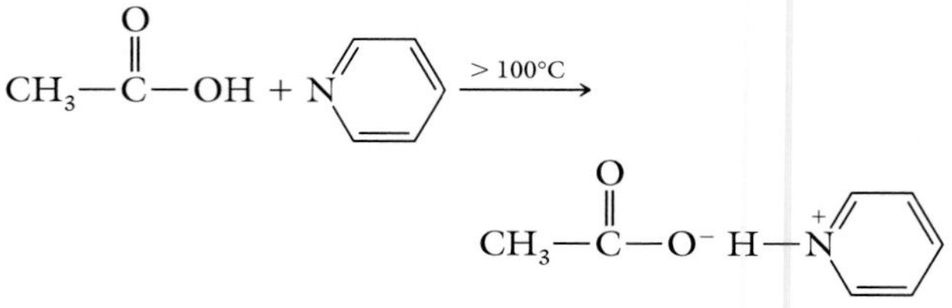

(c)

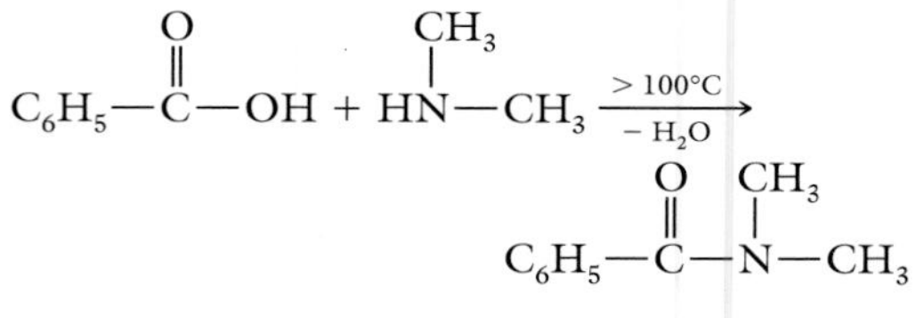

16.9

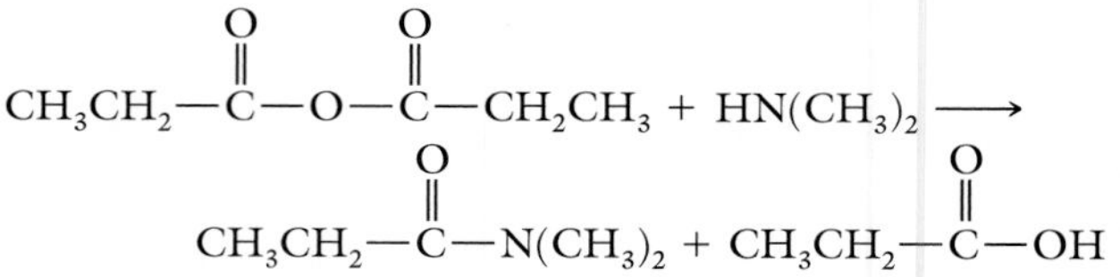

16.10 (a) *N*-Butyl-*N*-methylpropanamide or *N*-butyl-*N*-methylpropionamide; (b) *N*-phenylbenzamide; (c) 2-aminopropanal or 2-aminopropionaldehyde; (d) *N*-benzyl-*N*-methyl-2-methylpropanamide

16.11

(a)

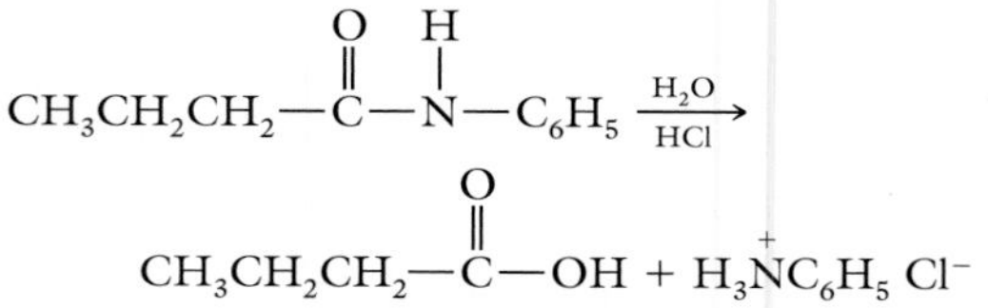

(b)

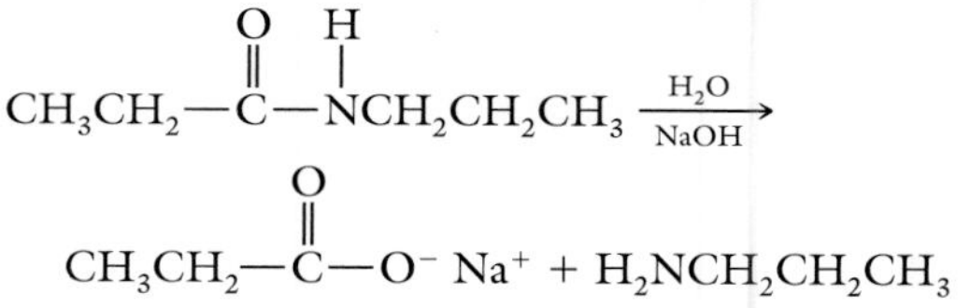

Chapter 17

17.1 (a) There is only one 3-hydroxypropanal because none of the carbon atoms have four different substituents.

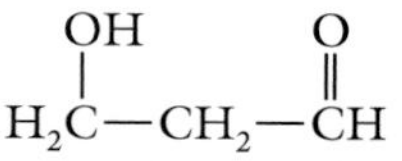

(b) A pair of enantiomers exists because C2 is a tetrahedral stereocenter.

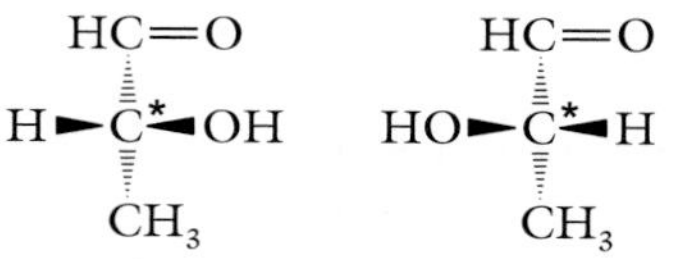

1 2

17.2 (a) Step 1: Structures 1 and 2 are not superimposable. Step 2: Structures 1 and 2 are not mirror images. Step 3: Rotation of structure 1 by 180° in the plane of the paper yields structure 1′, which is the nonsuperimposable mirror image of 2. Structures 1 and 2 are a pair of enantiomers.

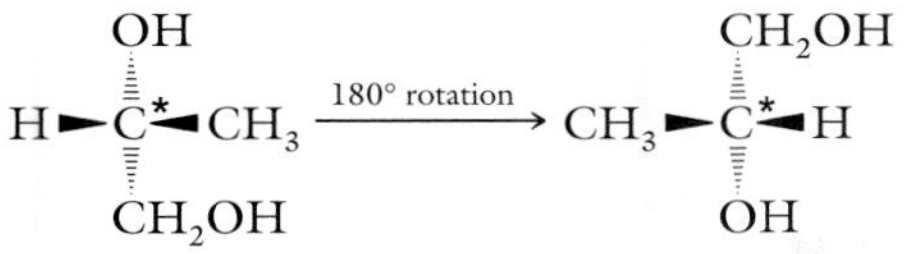

1 1′

(b) Step 1: Structures 3 and 4 are not superimposable. Step 2: Structures 3 and 4 are not mirror images. Step 3: Rotation of structure 3 by 180° in the plane of the paper yields structure 3′, which is the same as structure 4. Structures 3 and 4 are the same.

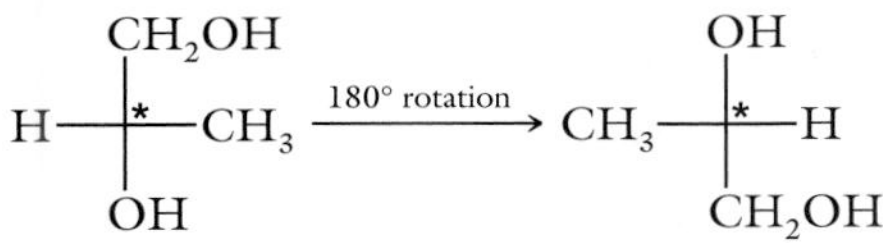

3 3′

17.3 The two carbon substituents are vertical, with the more-substituted substituent (COOH) upward. This enantiomer is the L-enantiomer because the hetero substituent (NH_2) is on the left side of the tetrahedral stereocenter.

17.4 $\alpha = [\alpha]CL = +53° \times 0.030 \text{ g/mL} \times 1.5 \text{ dm} = +2.4°$

17.5 There are two tetrahedral stereocenters. Only three stereoisomers are possible because the tetrahedral stereocenters possess the same four different substituents:

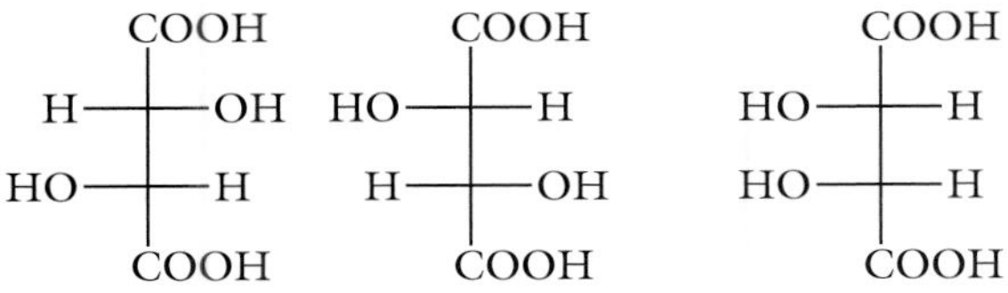

1 2 3

Stereoisomers 1 and 2 are optically active. Stereoisomer 3 is optically inactive because it is a meso compound. Stereoisomers 1 and 2 are enantiomers of each other, and each is a diastereomer of 3. Stereoisomer 3 is diastereomer of both 1 and 2, and vice versa.

17.6 (a) There are no stereoisomers, because the two halves of the ring at C1 are the same.

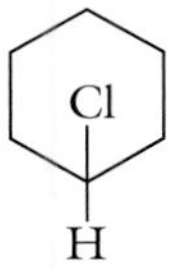

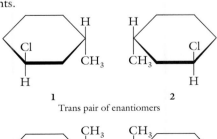

(b) There are four stereoisomers because the two tetrahedral stereocenters do not have the same four different substituents.

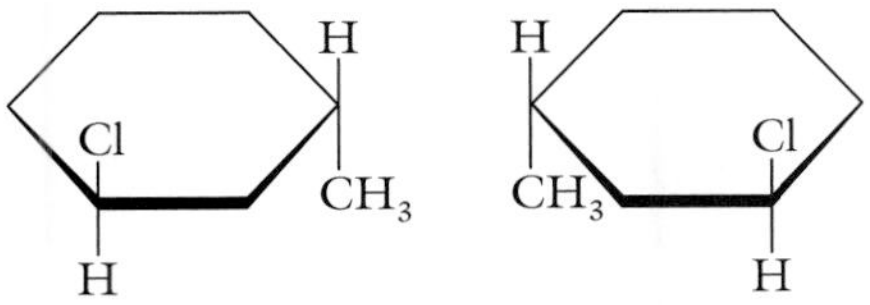

1 2

Trans pair of enantiomers

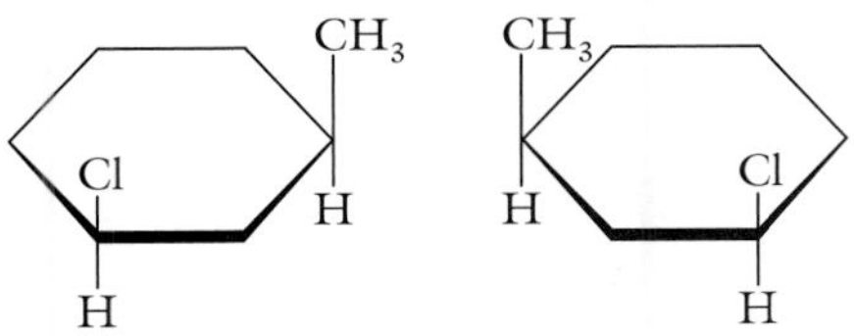

3 4

Cis pair of enantiomers

All four compounds are optically active. Each stereoisomer is simultaneously both an enantiomer and a diastereomer. Stereoisomers 1 and 2 are enantiomers of each other, and each is a diastereomer of both 3 and 4. Stereoisomers 3 and 4 are enantiomers of each other, and each is a diastereomer of both 1 and 2.

Chapter 18

18.1 1, aldotriose; 2, ketohexose; 3, aldopentose; 4, aldotetrose

18.2 (a) Diastereomers; they are stereoisomers that are not enantiomers. (b) Not isomers; D-ribose is a pentose and D-sorbose is a hexose. (c) Enantiomers; they are nonsuperimposable mirror-image stereoisomers. (d) Consititutional isomers; D-erythrose is an aldotetrose and L-erythrulose is a ketotetrose. (e) Diastereomers; they are stereoisomers that are not enantiomers.

18.3 Figure 18.2 shows that D-sorbose differs from D-fructose at the configurations at C3 and C4. Draw the cyclic structure of β-D-fructose and then reverse the configurations at C3 and C4.

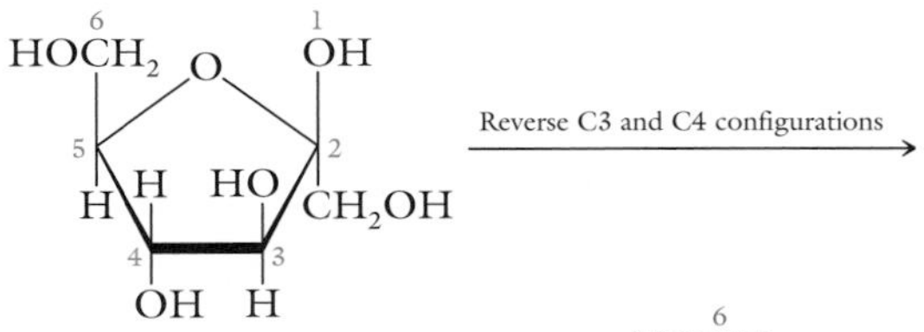

β-D-Fructose

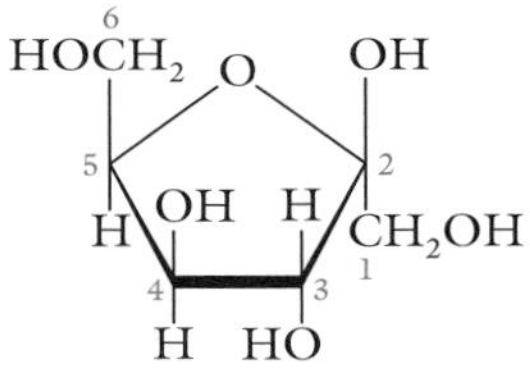

β-D-Sorbose

18.4

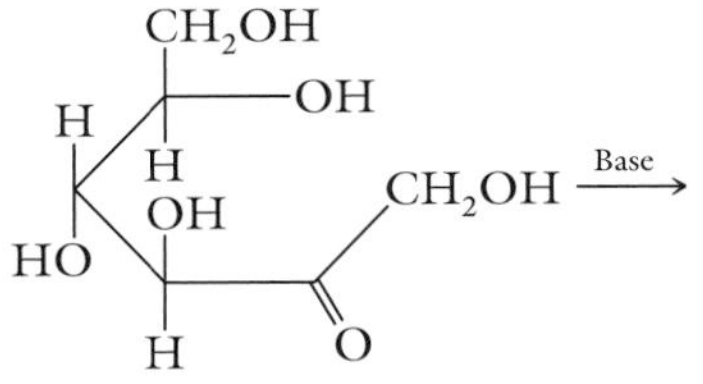

Ketose

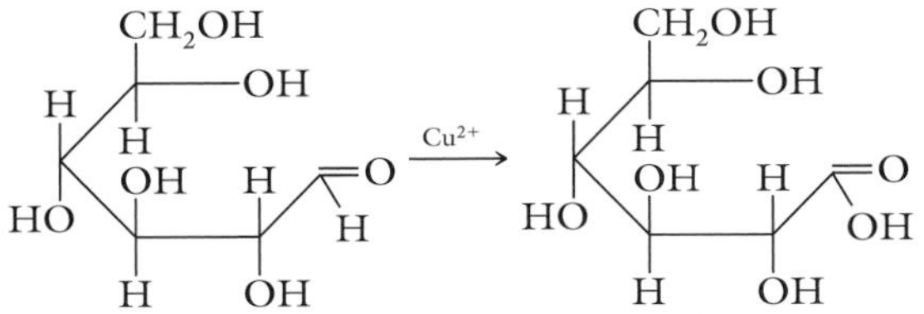

Aldose

18.5 The monosaccharides are D-mannose and D-fructose. D-Mannose differs from D-glucose only at the configuration of C2 (Figure 18.1). The linkage between disaccharide residues is an $\alpha(1 \rightarrow 6)$-glycosidic linkage from D-mannose to D-fructose. The disaccharide is a reducing sugar and undergoes mutarotation because the hemiacetal linkage of D-fructose is intact.

Chapter 19

19.1 Acids 1 and 3 are found in lipids because both are unbranched and contain an even number of carbons, and the double bond in acid 3 has a cis configuration. Acid 2 is not usually found in lipids, because it is branched.

19.2

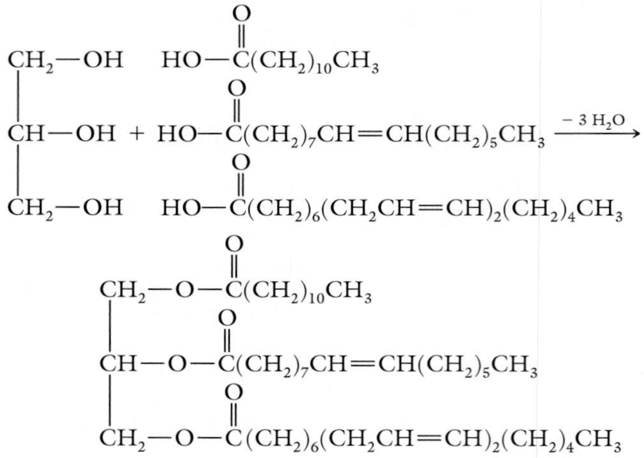

19.3

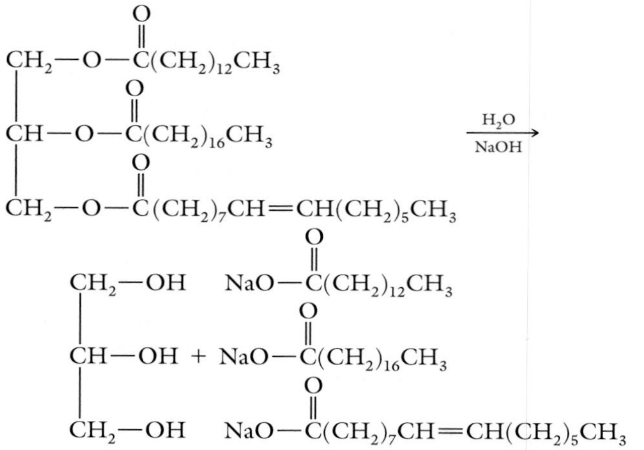

19.4

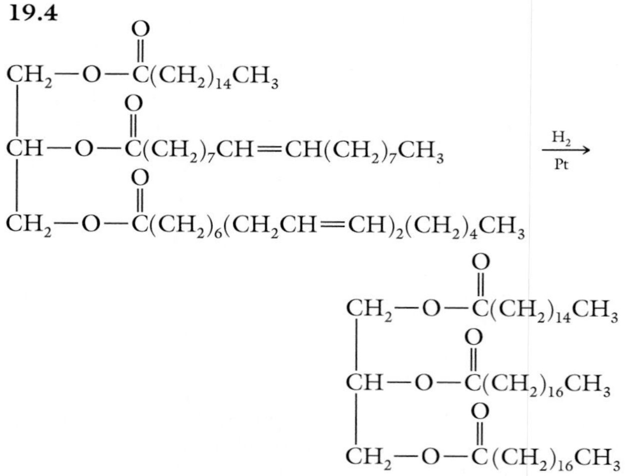

19.5

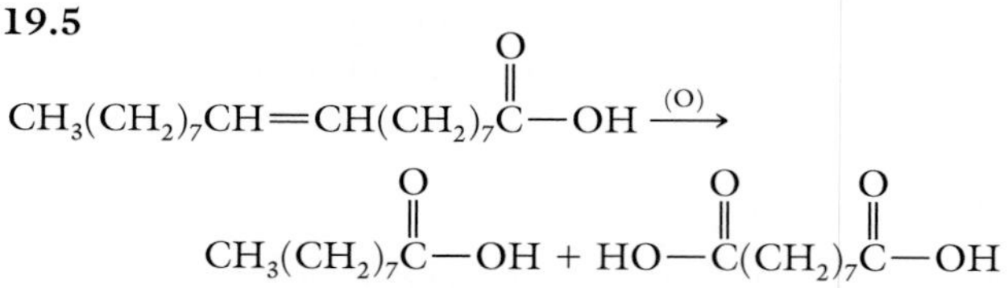

19.6

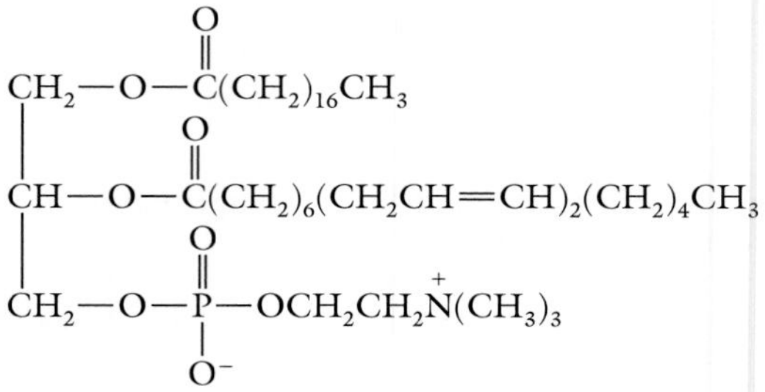

19.7

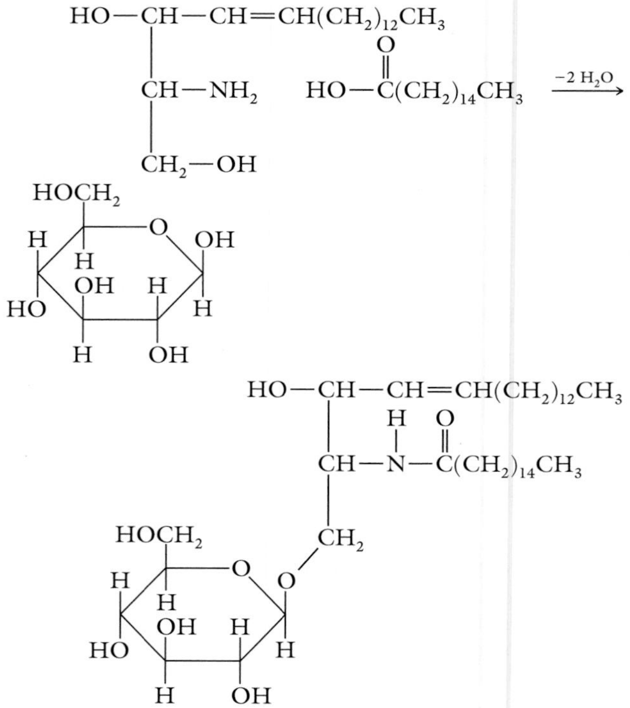

Chapter 20

20.1

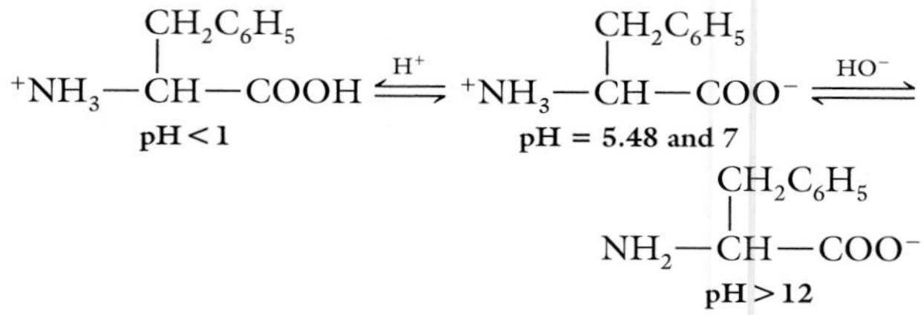

20.2

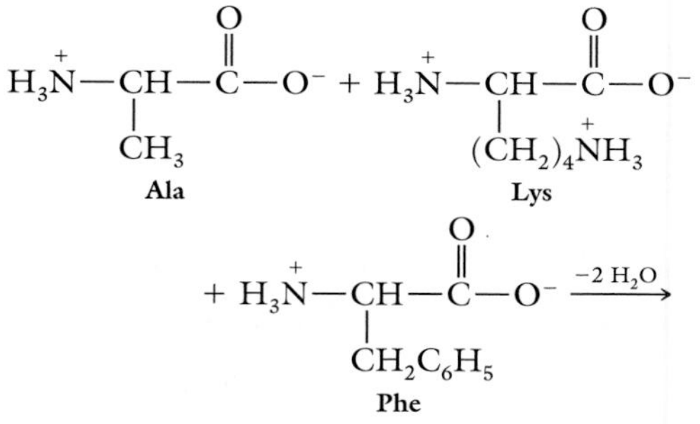

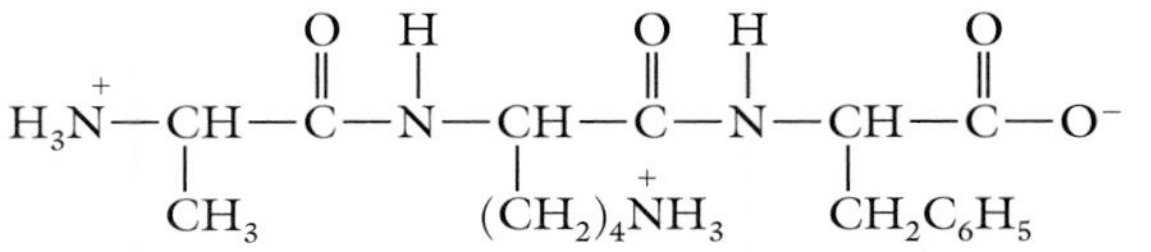

$$H_3\overset{+}{N}-\underset{\underset{CH_3}{|}}{CH}-\underset{\underset{}{\overset{\overset{O}{\|}}{C}}}-\underset{\overset{H}{|}}{N}-\underset{\underset{(CH_2)_4\overset{+}{NH_3}}{|}}{CH}-\underset{\underset{}{\overset{\overset{O}{\|}}{C}}}-\underset{\overset{H}{|}}{N}-\underset{\underset{CH_2C_6H_5}{|}}{CH}-\overset{\overset{O}{\|}}{C}-O^-$$

Ala-Lys-Phe
Alanyllysylphenylalanine

20.3 There are 3! = 6 constitutional isomers: Glu-Ile-Lys, Glu-Lys-Ile, Ile-Glu-Lys, Ile-Lys-Glu, Lys-Glu-Ile, Lys-Ile-Glu.

20.4 (a)

$$H_3\overset{+}{N}-\underset{\underset{CH_3}{|}}{CH}-\overset{\overset{O}{\|}}{C}-\underset{\overset{H}{|}}{N}-\underset{\underset{(CH_2)_4\overset{+}{NH_3}}{|}}{CH}-\overset{\overset{O}{\|}}{C}-\underset{\overset{H}{|}}{N}-\underset{\underset{CH_3}{|}}{CH}-\overset{\overset{O}{\|}}{C}-O^-$$

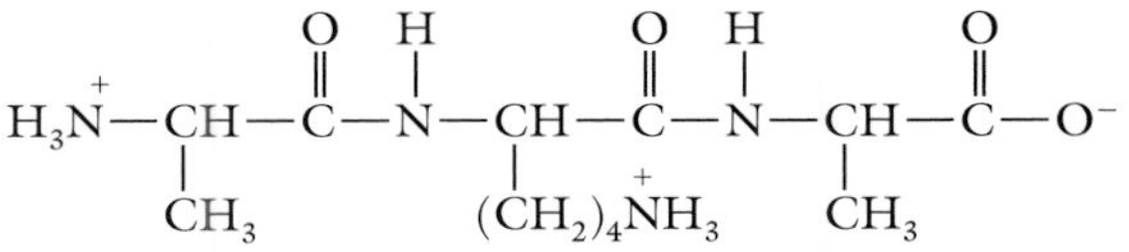

Ala-Lys-Ala has a charge of 1+ and migrates to the cathode. The pI is on the basic side of that of a peptide containing only neutral residues.

(b)

$$H_3\overset{+}{N}-\underset{\underset{CH_2COO^-}{|}}{CH}-\overset{\overset{O}{\|}}{C}-\underset{\overset{H}{|}}{N}-\underset{\underset{(CH_2)_4\overset{+}{NH_3}}{|}}{CH}-\overset{\overset{O}{\|}}{C}-\underset{\overset{H}{|}}{N}-\underset{\underset{CH_2COO^-}{|}}{CH}-\overset{\overset{O}{\|}}{C}-O^-$$

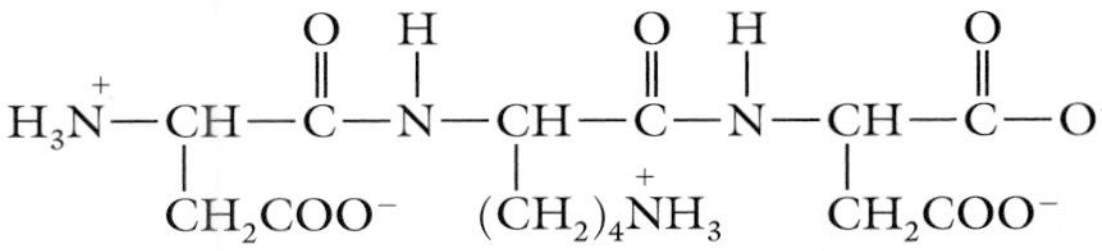

Asp-Lys-Asp has a charge of 1− and migrates to the anode. The pI is on the acid side of that of peptides with only neutral amino acid residues.

Chapter 21

21.1 The name is deoxyguanosine 5′-monophosphate, or dGMP.

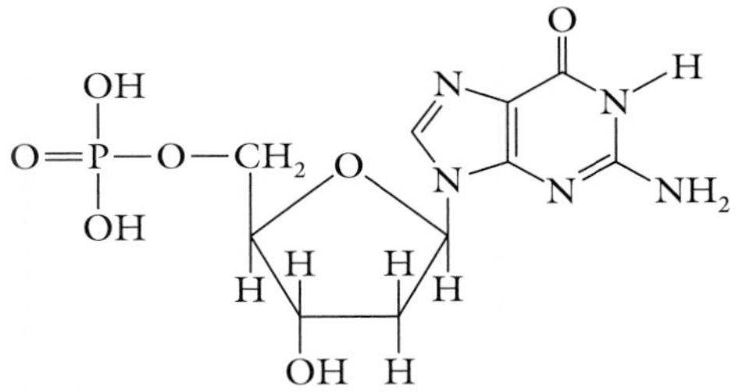

21.2 The components are phosphoric acid, ribose, and uracil. The name is uridine 5′-monophosphate.

21.3

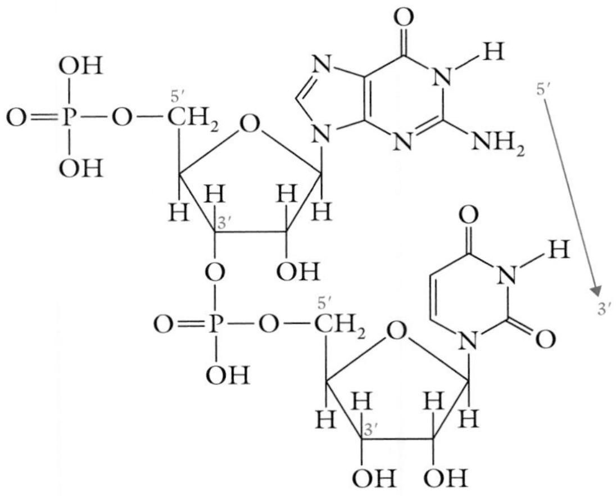

21.4

5′ ⬚GGCTAT⬚ 3′ **Given strand**

3′ ⬚CCGATA⬚ 5′ **Complementary strand**

   The sequence for the complementary strand is 5′-ATAGCC-3′.

21.5 The complementary parent strand and the new daughter strand are identical and complementary to the ACGTGC sequence. The sequence is 3′-TGCACG-5′ (5′-GCACGT-3′).

21.6 DNA nontemplate strand: GTTGACTTA; RNA: GUUGACUUA

21.7 Anticodon (written 5′ ⟶ 3′): (a) GGA;   (b) CUG; (c) CCU;   (d) AGC. α-Amino acid: (a) Ser;   (b) Gln; (c) Arg;   (d) Ala.

21.8

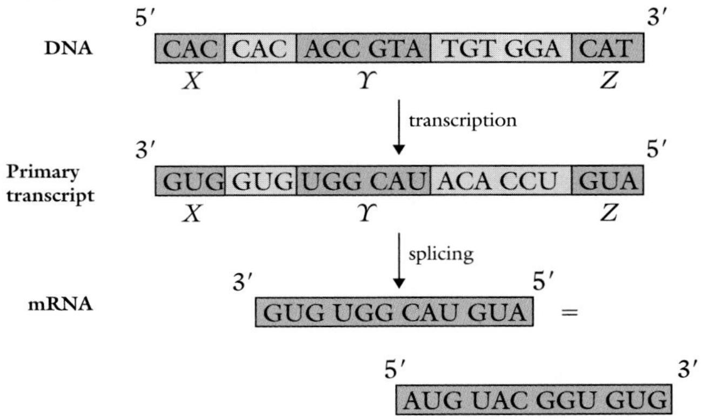

The mRNA written in the 5′ ⟶ 3′ direction is AUG-UAC-GGU-GUG, which codes (Table 21.1) for the polypeptide sequence Met-Tyr-Gly-Val.

21.9 (a) 5′-GAA-3′ in DNA results in 3′-CUU-5′ in mRNA. The codon is 5′-UUC-3′ and codes for Phe. The α-amino acid sequence would be changed: Tyr would be replaced by Phe.

   (b) 5′-TTA-3′ in DNA results in 3′-AAU-5′ in mRNA. The codon is 5′-UAA-3′ and codes for Stop. Polypeptide synthesis would be terminated prematurely.

   (c) There would be no effects on either the mRNA or the polypeptide. A mutation in an intron is constrained within the intron and has no effect on the base sequence in the exons, because the intron is excised.

Answers to Odd-Numbered Exercises

Chapter 1

1.1 (a) chemical; (b) physical; (c) physical; (d) chemical

1.3 (a) mixture; (b) pure substance; (c) pure substance; (d) pure substance

1.5 length, meter; mass, kilogram; temperature, kelvin; time, second; amount of substance, mole

1.7 (a) 0.630 km; (b) 1.440 s; (c) 65 mg; (d) 1.3 mg

1.9 (a) 3; (b) 4; (c) 3; (d) 3

1.11 (a) 6; (b) 3; (c) 1; (d) 4; (e) 3

1.13 (a) 8.39×10^{-3}; (b) 8.3264×10^{4}; (c) 3.72×10^{2}; (d) 2.08×10^{-5}

1.15 (a) 9.368×10^{5}; (b) 1.638×10^{3}; (c) 5.68×10^{-5}; (d) 9.17×10^{-3}

1.17 (a) 48.4 cm^2; (b) 8.4

1.19 (a) 14.02; (b) 28.615

1.21 (a) 1.29 km; (b) 1.29×10^{3} m; (c) 1.29×10^{5} cm; (d) 1.29×10^{6} mm

1.23 1.42×10^{4} mL

1.25 6.214 miles

1.27 28.35 g/oz

1.29 (a) 0.27 kg; (b) 2.7×10^{2} g; (c) 2.7×10^{5} mg

1.31 10 cm

1.33 79 g

1.35 20°C

1.37 300 K

1.39 0.069

1.41 1.130

1.43 1.1 lb

1.45 1.04 g/mL

1.47 125 cm^3

1.49 (a) chemical; (b) physical; (c) physical; (d) chemical

1.51 (a) 9.62×10^{6} kg; (b) 5.487×10^{4} days; (c) 2.53×10^{-1} s; (d) 2.74×10^{-4} km

1.53 1.62×10^{-3} oz

1.55 (a) 3.83×10^{4}; (b) 5.48×10^{-2}; (c) 7.33×10^{-3}

1.57 1×10^{3} cm^3

1.59 1.5×10^{5} kJ/24 h

1.61 The unspecified solution contains two components, NaCl and water. You could obtain pure NaCl by evaporating all the water from the mixture. Five grams of the pure NaCl could then be weighed on a laboratory balance, and the specified mixture of salt and water could then be prepared.

1.63 You must modify your original hypothesis and retest it.

Chapter 2

2.1 No mass is lost as a result of a chemical reaction.

2.3 C = 40.00%, H = 6.671%, O = 53.33%

2.5 Atomic mass O = 16.0000 amu

2.7 Atomic mass N = 14.0 amu

2.9 Proton: mass, 1.00728 amu; charge, +1
Neutron: mass, 1.00894 amu; charge, 0
Electron: mass, 0.0005414 amu; charge, −1

2.11 Atomic mass includes the mass of neutrons.

2.13 It would be a cation; charge, 3+

2.15 Atomic mass, Sr = 87.62 amu

2.17 $^{16}_{8}O$, $^{17}_{8}O$; $^{24}_{12}Mg$, $^{25}_{12}Mg$; $^{28}_{14}Si$, $^{29}_{14}Si$

2.19 (a) Lithium, (b) calcium, (c) sodium, (d) phosphorus, (e) chlorine

2.21

Element	Group	Period
Li	I	2
Na	I	3
K	I	4
Rb	I	5
Cs	I	6

2.23 Li, Na, K, Rb, and Cs are metals.

2.25 Main-group elements

2.27 Group VII

2.29 No. The ground state is stable; no energy can be lost.

2.31 Shells identified by a principal quantum number. Subshells within shells identified by the letters *s, p, d,* etc. Orbitals within subshells.

2.33 An orbital defines the probability of finding an electron in a region of space around an atomic nucleus.

2.35 One orbital in an *s* subshell.

2.37 No, an *s* subshell is spherically symmetrical.

2.39 (a) magnesium; (b) sulfur; (c) potassium

2.41 Sodium ion

2.43 Group I. All possess one outer electron.

2.45 Group II. All possess two outer electrons.

2.47 True

2.49 Its energy increases.

2.51 They have virtually identical chemical properties.

2.53 Metal

2.55 $^{16}_{8}O$, $^{18}_{8}O$

2.57 0.022%

2.59 Blue light

2.61 The octet rule

2.63 You need to know the natural abundances of each of the isotopes.

2.65 Dalton's atom was indestructible, but the modern atom can be decomposed into subatomic particles. In addition, the discovery of isotopes showed that the masses of the atoms of an element are not identical.

Chapter 3

3.1 Chemical-bond formation is the result of a process that allows the reacting elements to achieve stability.

3.3 By losing, gaining, or sharing electrons.

3.5 (a) potassium sulfate; (b) manganese(II) hydroxide; (c) iron(II) nitrate; (d) potassium dihydrogen phosphate; (e) calcium acetate; (f) sodium carbonate

3.7 (a) ammonium; (b) nitrate; (c) sulfate; (d) phosphate (e) acetate

3.9 (a) $AlCl_3$; (b) Al_2S_3; (c) AlN

3.11 (a) Li_2O; (b) $CaBr_2$; (c) Al_2O_3; (d) Na_2S

3.13 1 (Group I):1 (Group VII)

3.15 1 (Group II):2 (Group VII)

3.17 (a) CaO; (b) CaS; (c) CaSe

3.19 (a) Al_2O_3; (b) Al_2S_3; (c) Al_2Se_3

3.21 (a) AlF_3; (b) $AlCl_3$; (c) $AlBr_3$; (d) AlI_3

3.23

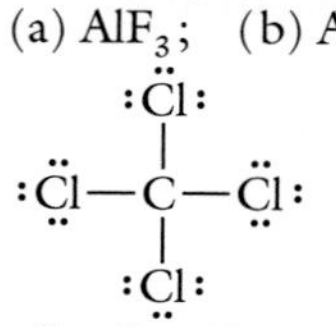

3.25 F—O—F

3.27 O=C=O

3.29 The four fluorine atoms are situated at the corners of a tetrahedron with the boron atom at its center.

3.31 (a) a tetrahedron, no dipole moment; (b) a tetrahedron, no dipole moment; (c) linear, no dipole moment

3.33 A chemical bond that consists of two electrons with paired spins.

3.35 (a) Silver nitrate; (b) mercury(II) chloride; (c) sodium carbonate; (d) calcium phosphate

3.37 (a) Magnesium chloride; (b) magnesium sulfide; (c) magnesium nitride

3.39 (a) Barium oxide; (b) barium sulfide; (c) barium selenide

3.41 (a) Potassium nitride; (b) potassium phosphide; (c) potassium arsenide

3.43 The atoms in $TeBr_2$ are oriented toward the corners of a tetrahedron. The molecule is bent.

3.45 (a) Covalent; (b) covalent; (c) covalent; (d) ionic; (e) covalent

3.47 LiF; LiCl; LiBr; LiI

3.49 Yes. Hydrogen requires only two electrons, and beryllium and boron require four and six electrons, respectively, to complete their valence shells.

3.51 Test its ability to conduct electricity when dissolved in water. An electrolyte is an ionic substance.

3.53 H—O—O—H

3.55

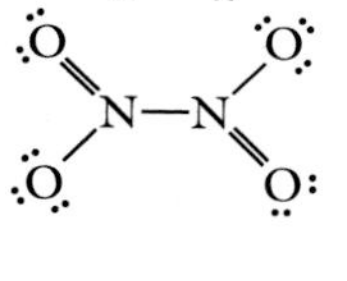

Chapter 4

4.1 (a) 156.1; (b) 58.32; (c) 189.7; (d) 262.0; (e) 60.10

4.3 (a) 474.0; (b) 204.1; (c) 271.5; (d) 254.5; (e) 391.5

4.5 0.4221 mol

4.7 101 g

4.9 1.20×10^{23}

4.11 280.8 g

4.13 $ZnCrO_4$

4.15 (a) $2\ N_2 + 3\ Br_2 \rightarrow 2\ NBr_3$
(b) $2\ HNO_3 + Ba(OH)_2 \rightarrow Ba(NO_3)_2 + 2\ H_2O$
(c) $HgCl_2 + H_2S \rightarrow HgS + 2\ HCl$

4.17 (a) $4\ P + 5\ O_2 \rightarrow 2\ P_2O_5$
(b) $FeCl_2 + K_2SO_4 \rightarrow FeSO_4 + 2\ KCl$
(c) $HgCl + 2\ NaOH + 2\ NH_4Cl \rightarrow Hg(NH_3)_2Cl + 2\ NaCl + 2\ H_2O$

4.19 $C_6H_6 + 7.5\ O_2 \rightarrow 6\ CO_2 + 3\ H_2O$
$2\ C_6H_6 + 15\ O_2 \rightarrow 12\ CO_2 + 6\ H_2O$

4.21 $C_5H_{12} + 8\ O_2 \rightarrow 5\ CO_2 + 6\ H_2O$
$2\ C_8H_{18} + 25\ O_2 \rightarrow 16\ CO_2 + 18\ H_2O$
$C_{10}H_{20} + 15\ O_2 \rightarrow 10\ CO_2 + 10\ H_2O$

4.23 $CH_4 + 2\ O_2 \rightarrow CO_2 + 2\ H_2O$

4.25 $\frac{2\ Al(OH)_3}{3\ H_2SO_4}$ $\frac{1\ Al(OH)_3}{3\ H_2O}$ $\frac{1\ H_2SO_4}{2\ H_2O}$
$\frac{2\ Al(OH)_3}{1\ Al_2(SO_4)_3}$ $\frac{3\ H_2SO_4}{1\ Al_2(SO_4)_3}$ $\frac{1\ Al_2(SO_4)_3}{6\ H_2O}$
and the inverse of each of these unit-conversion factors.

4.27 0.83 mol

4.29 3.3 mol

4.31 38.47 g

4.33 (a) 0.44 mol; (b) 2.91 mol; (c) 9.38 g; (d) 8.49 g

4.35 (a) $Ca_3(PO_4)_2 + 4\ H_3PO_4 \rightarrow 3\ Ca(H_2PO_4)_2$
(b) $FeCl_2 + (NH_4)_2S \rightarrow FeS + 2\ NH_4Cl$
(c) $2\ KClO_3 \rightarrow 2\ KCl + 3\ O_2$
(d) $3\ O_2 \rightarrow 2\ O_3$
(e) $2\ C_5H_6 + 13\ O_2 \rightarrow 10\ CO_2 + 6\ H_2O$

4.37 (a) 0.5199 mol; (b) 0.7508 mol; (c) 1.500 mol; (d) 0.800 mol; (e) 0.36 mol

4.39 1.80

4.41 1.1×10^{24}

4.43 AgCl

4.45 C_2H_6O

4.47 (a) $AgNO_3 + KCl \rightarrow AgCl + KNO_3$
(b) $Ba(NO_3)_2 + Na_2SO_4 \rightarrow BaSO_4 + 2\ NaNO_3$
(c) $2\ (NH_4)_3PO_4 + 3\ Ca(NO_3)_2 \rightarrow Ca_3(PO_4)_2 + 6\ NH_4NO_3$
(d) $3\ Mg(OH)_2 + 2\ H_3PO_4 \rightarrow Mg_3(PO_4)_2 + 6\ H_2O$

4.49 684.3 g

4.51 Carbon: 67.33%; hydrogen: 6.931%; nitrogen: 4.620%; oxygen: 21.12%

4.53 (a) 45 g of H_2SO_4; (b) 133 g of $Ca(OH)_2$; (c) 35 g of C_2H_6O; (d) 110 g of C_4H_{10}; (e) 210 g of $FeCl_2$

4.55 P_4O_{10}

Chapter 5

5.1 (a) 364 torr; (b) 483 torr; (c) 675 torr; (d) 735 torr

5.3 (a) 735 torr; (b) 0.967 atm
5.5 (a) 912 torr; (b) 342 mm Hg; (c) 1.12 atm
5.7 T and n
5.9 V and n
5.11 2.5 L
5.13 1.00 L
5.15 348°C, or 621 K
5.17 3.25 atm
5.19 2.25 L
5.21 1.52 L
5.23 6.24×10^4 (torr × mL)/(K × mol)
5.25 3.59 L
5.27 5.60 L
5.29 1.00 g of H_2, 4.25 g of NH_3
5.31 0.920 atm
5.33 0.243 mol
5.35 All volumes are the same.
5.37 200 torr
5.39 40.4 L
5.41 153°C
5.43 2.0 L
5.45 0.900 L
5.47 0.0906 g/L
5.49 4.49×10^3 mL
5.51 (a) 570 torr; (b) 935 torr; (c) 722 torr; (d) 1.330×10^3 torr
5.53 (a) 27.7 L; (b) 23.1 atm; (c) 329 K; (d) 1.00 mol
5.55 2.03 mol
5.57 0.031 mL CO_2/mL arterial blood

Chapter 6

6.1 All matter would exist only in the gaseous state. No matter would be present in either liquid or solid form.

6.3 The melting of a solid is the transition from a structure consisting of a highly ordered array of molecules to a disordered structure with the molecules still in contact but now free to move about. This transition is effected by adding energy to the system to overcome the strong attractive forces in the solid.

6.5 True. The vapor pressure of a liquid is a measure of the escaping tendency of its molecules. Increasing the temperature increases the kinetic energy of its molecules, thus increasing their escaping tendency.

6.7 Because the molecules are in contact.

6.9 The molar heat of vaporization is always greater than the molar heat of fusion.

6.11 True. The transformation between the solid state and the liquid state occurs at the same temperature. If the transformation is approached by raising the temperature of the solid until the liquid forms, this temperature is called the melting point. If the transformation is approached by lowering the temperature of the liquid until the solid forms, it is called the freezing point.

6.13 9.7 kcal of heat will be released.

6.15 Through the weak London forces (temporary dipoles).

6.17 Only b.

6.19 The greater the number of electrons (atomic number) in a molecule, the larger the temporary dipole (London force). This means that, in general, the greater the molecular mass, the greater the strength of London forces exerted between molecules.

6.21 The four C–Cl polar bonds are arranged symmetrically (tetrahedrally) with carbon at the center of the tetrahedron. The symmetric arrangement of dipoles results in a net dipole moment of zero.

6.23

	London Force	Dipole–Dipole	Hydrogen Bond
CH_4	Yes	No	No
$CHCl_3$	Yes	Yes	No
NH_3	Yes	Yes	Yes

6.25 Compound (a) has the greater heat of vaporization, because it can form hydrogen bonds between molecules, whereas compound (b) cannot.

6.27 The molecules of a liquid have acquired sufficient kinetic energy to partly overcome the intermolecular forces holding them in fixed positions. Although they are free to assume new positions under an applied force (for example, gravity), they still remain in contact.

6.29 In an open system, vapor molecules will continuously escape into the surroundings. Therefore the rate of evaporation will always exceed the rate of condensation, and evaporation will occur continuously.

6.31 When a solid melts, the ordered solid-state structure becomes disordered, but the molecules in the liquid state remain in close contact. Therefore the volume of the liquid is close to that of the solid.

6.33 The word vapor is used to describe the gaseous state of a substance whose liquid and gaseous states are present at the same time. Under normal conditions of room temperature, the liquid states of oxygen or nitrogen are not possible.

6.35 $c < a < b$

6.37 No. Vapor pressure is a balance between evaporation and condensation. In an open container, the rate of evaporation must be greater than the rate of condensation, and equilibrium cannot be established.

6.39 Equilibrium describes a situation that does not change over time.

6.41 The temperature at which vapor bubbles form throughout a liquid at an external pressure of 1 atmosphere.

6.43 Both molecules are polar, but, in addition, water can form hydrogen bonds, so the secondary attractive forces in water are greater than those in chloroform. Therefore water will have the greater surface tension of the two.

6.45 Acetic acid molecules form hydrogen bonds both among themselves and with water. Pentane molecules can interact only by London forces. The attractive forces between acetic acid and water are similar, and therefore acetic acid will dissolve in water. The attractive forces between pentane molecules and water are very different and will therefore not form solutions.

6.47 A crystalline solid has a definite geometric shape, has a definite melting point, and can shatter into small pieces that resemble the original crystal.

6.49 $H_2S < H_2O < KCl$

6.51 52%

6.53 Hydrogen, H_2; methane, CH_4; nitrogen, N_2.

6.55 At Denver's altitude, atmospheric pressure is less than at sea level. Therefore the boiling point of water is lower in Denver and it takes longer to boil an egg there than in San Francisco.

6.57 Rubbing alcohol is a volatile liquid. Its evaporation from an open system (skin) is rapid and requires heat. The skin provides the heat and, in the process, is cooled.

6.59 This substance undergoes sublimation at room temperature and atmospheric pressure. Therefore its vapor is present at concentrations that can be detected by the human nose because of its distinctive odor.

Chapter 7

7.1 (a) 1.07 g; (b) 8.19 g

7.3 1.50 g

7.5 87.1 mL of solution

7.7 1.25 g

7.9 448 mL

7.11 1.21 mg %

7.13 0.0042% w/v

7.15 1.0 mol

7.17 (a) 2.0 *M*; (b) 3.38 *M*

7.19 (a) 2.78 mol; (b) 0.384 mol

7.21 (a) 1.50 L; (b) 3.83 L

7.23 9.09×10^{-4} *M*

7.25 3.75×10^3 mg

7.27 5.40×10^2 mL

7.29 0.020 L

7.31 (a) 1.50 L; (b) 325 mL

7.33 9.0% w/v

7.35 0.500 *M*

7.37 No. The data show that the solubility of strontium acetate decreases with increase in temperature. If the solution had been saturated, strontium acetate would have crystallized out.

7.39 6.71 *M*

7.41 (a) 0.40 osmol/L; (b) 0.20 osmol/L

7.43 The dextrose solution is hypertonic; erythrocytes will lose water and shrink.

7.45 (a) 8.90% w/w; (b) 7.04% w/w; (c) 4.0% w/w

7.47 (a) 2.20 g; (b) 13.1 g

7.49 (a) 9.70% w/v; (b) 10.4% w/v; (c) 4.00% w/v

7.51 (a) 0.0030 mol; (b) 0.078 mol; (c) 0.124 mol; (d) 1.09 mol

7.53 (a) 0.559 mol; (b) 2.52 mol; (c) 0.960 mol; (d) 0.679 mol

7.55 Add water to 13.3 g of KI to a final volume of 500.0 mL of solution.

7.57 (a) 5.66% w/w; (b) 4.83% w/w; (c) 1.77% w/w

7.59 (a) 69.5 g NaCl; (b) 69.5 g $Mg(NO_3)_2$; (c) 98.7 g K_2CO_3

7.61 15.0 L

7.63 9.0% w/v

7.65 (a) Precipitate forms: $Ag^+(aq) + Cl^-(aq) \rightarrow AgCl(s)$
(b) Precipitate forms: $Mg^{2+}(aq) + 2\,OH^-(aq) \rightarrow Mg(OH)_2(s)$
(c) Precipitate forms: $Pb^{2+}(aq) + 2\,Cl^-(aq) \rightarrow PbCl_2(s)$
(d) Precipitate forms: $3\,Ca^{2+}(aq) + PO_4^{3-}(aq) \rightarrow Ca_3(PO_4)_2(s)$

7.67 Propylene glycol is the solvent, and water is the solute.

7.69 Water moves out of the arterial blood.

7.71 0.30 osmol/L (see Box 7.3)

Chapter 8

8.1 2.25 mol/min

8.3 The proportion of molecules undergoing reactive collisions doubled.

8.5 Reaction rate a is greater than reaction rate b.

8.7 $E_{back} = +58$ kJ

8.9 No. The reaction takes place in an open system, the CO_2 escapes into the atmosphere, and the system can never come to equilibrium.

8.11 Forward reaction: $CaCO_3(s) \rightarrow CaO(s) + CO_2(g)$
Back reaction: $CaO(s) + CO_2(g) \rightarrow CaCO_3(s)$

8.13 $K_{eq} = [CO_2]$

8.15 $K_{eq} = 0.099$

8.17 Equilibrium shifts to the right (to product).

8.19 (a) Color shift to pink; (b) color shift to blue.

8.21 The square brackets around chemical components in an equilibrium constant expression denote their molar concentrations at equilibrium.

8.23 (a) $K_{eq} = \dfrac{[CO][H_2]^3}{[CH_4][H_2O]}$; (b) $K_{eq} = \dfrac{[CO_2]^6[H_2O]^6}{[O_2]^6}$

8.25 (a) $K_{eq} = [CO_2][NH_3]^2$; (b) $K_{eq} = \dfrac{1}{[O_2]^{0.5}}$

8.27 (a) $COCl_2(g) \rightleftharpoons CO(g) + Cl_2(g)$
(b) $CH_4(g) + H_2O(g) \rightleftharpoons CO(g) + 3\,H_2(g)$

8.29 (a) $K_{eq} = \dfrac{[NO_2][NO_3]}{[N_2O_5]}$; (b) $K_{eq} = \dfrac{[N_2]^2[H_2O]^6}{[NH_3]^4[O_2]^3}$

8.31 (a) 1000/1; (b) 1/100; (c) 7.1/1; (d) 12.0/1

8.33 Some of the solid would dissolve.

8.35 $E_{forward} = +72$ kJ

8.37 E_a of the back reaction was larger.

8.39 If a system in an equilibrium state is disturbed, the system will adjust to neutralize the disturbance and restore the system to equilibrium.

Chapter 9

9.1 $K_W = 1.00 \times 10^{-14}$

9.3 A substance that can act as an acid or a base.

9.5 A strong base is 100% dissociated in aqueous solution. Examples: NaOH and KOH.

9.7 $[H_3O^+] = [Cl^-] = 0.30$ *M*

9.9 $[OH^-] = [Tl^+] = 0.25$ *M*

9.11 $[OH^-] = 2.86 \times 10^{-14}$ *M*

9.13 $[H_3O^+] = 2.5 \times 10^{-13}$ *M*

9.15 The pH of pure water at 25°C is 7.00.

9.17 pH = 2.0

9.19 pH = 12.18; pOH = 1.82

9.21 (a) 2.46; (b) 1.60; (c) 3.14; (d) 1.13

9.23 A weak acid is one that is incompletely dissociated in aqueous solution. Examples: acetic and phosphoric acids.

9.25 Neutral

9.27

	Acid	Base
(a)	HNO_2	NO_2^-
	H_3O^+	H_2O
(b)	$H_2PO_4^-$	HPO_4^{2-}
	H_3O^+	H_2O

9.29 $K_b = 5.75 \times 10^{-10}$

9.31 $pK_a = 4.76$; $pK_b = 9.24$

9.33 $pK_a + pK_b = 14.00$

9.35 $H_2CO_3(aq) + H_2O(l) \rightleftharpoons H_3O^+(aq) + HCO_3^-(aq)$
$HCO_3^-(aq) + H_2O(l) \rightleftharpoons H_3O^+(aq) + CO_3^{2-}(aq)$

9.37 $K_{a1} = \dfrac{[H_3O^+][HCO_3^-]}{[H_2CO_3]} = 4.45 \times 10^{-7}$

$K_{a2} = \dfrac{[H_3O^+][CO_3^{2-}]}{[HCO_3^-]} = 4.72 \times 10^{-11}$

9.39 Minor significance

9.41 (a) Neutral; (b) basic; (c) acidic; (d) acidic; (e) neutral

9.43 (a) Neutral; (b) basic; (c) basic; (d) basic; (e) basic

9.45 The conjugate acid prevents shifts to basicity, and the basic form prevents shifts to acidity.

9.47 pH = 4.89

9.49 pH = 7.38

9.51 [HCl] = 0.017 *M*

9.53 $[H_2SO_4] = 0.0170$ *M*

9.55 Dissolve 16.3 g of H_3PO_4 in sufficient water to make 1.00 L of solution.

9.57 0.072 *N*

9.59 (a) $Na^+ = 0.125$ *M*; (b) $K^+ = 0.035$ *M*; (c) $Ca^{2+} = 0.036$ *M*; (d) $Mg^{2+} = 0.014$ *M*

9.61 (a) 2.50; (b) 0.00; (c) 2.16; (d) 1.72; (e) 1.64

9.63 (a) 2.21 (b) 2.09; (c) 1.25; (d) 0.00; (e) 12.0

9.65 (a) Acidic; (b) neutral; (c) basic; (d) acidic; (e) basic

9.67 Molar ratio of $[HPO_4^{2-}/H_2PO_4^-] = 10/1$

9.69 0.0620 *N*

9.71 (a) pOH = 11.53, $[OH^-] = 3.0 \times 10^{-12}$ *M*; (b) pOH = 12.40, $[OH^-] = 4.0 \times 10^{-13}$ *M*; (c) pOH = 10.86, $[OH^-] = 1.4 \times 10^{-11}$ *M*; (d) pOH = 12.87, $[OH^-] = 1.4 \times 10^{-13}$ *M*

9.73 Ammonium acetate in water will form a neutral solution.

9.75 A buffered solution contains approximately equal concentrations of a conjugate acid-base pair.

9.77 A conjugate acid-base pair consists of a weak acid and the basic anion resulting from its dissociation.

9.79 5.31 g

Chapter 10

10.1 Alpha-radiation consists of helium nuclei, charge 2+, mass, 4 amu.

10.3 Gamma-radiation has no mass or charge and is highly penetrating.

10.5 ^4_2He

10.7 $^{\ 0}_{-1}e$

10.9 Radioactivity can be induced by bombardment of nonradioactive elements with high-energy subatomic particles.

10.11 A radioactive decay series is a series of nuclear reactions that begins with an unstable nucleus and ends with the formation of a stable one.

10.13 12,033 years old

10.15 Beta-radiation

10.17 It becomes attenuated.

10.19 The ionization of water and the production of hydrated electrons

10.21 The secondary chemical processes

10.23 Radiation sickness is caused by nonlethal exposure to radiation and is characterized by nausea and a severe drop in white blood cell count.

10.25 She must double the distance from the source to reduce the intensity by 1/4—move from 4 feet away to 8 feet away from the radiation source.

10.27 The detection of ions caused by radiation

10.29 A rad is the amount of energy absorbed in matter when exposed to radiation without regard to the effects caused.

10.31 No. The RBE for α-rays is 10, and that for γ-rays is 1.0.

10.33 No. The radiation will be absorbed by tissue before it can be detected externally.

10.35 Yes. Cobalt-60 is an energetic source of γ-radiation that can penetrate deeply into tissue to cause radiation damage.

10.37 No. Positrons can be detected only when they interact with an electron to produce γ-rays, which are in turn detected indirectly by the presence of ions that they produce as they pass through matter.

10.39 No. Beta-emission would be absorbed by tissue and could not be used for detection of disorders within the body.

10.41 A CAT scan produces images of internal structures of the body with the use of X-rays.

10.43 31.25 mg

10.45 2.33×10^8 disintegrations per second

10.47 22.1%

10.49 Radioactivity is the name of the process in which atomic nuclei spontaneously decompose.

10.51 No. Gamma-radiation does not affect atomic mass or number.

10.53 Rapidly dividing tissue—for example, intestinal epithelium

10.55 Both processes release enormous amounts of energy. In nuclear fission, unstable nuclei of high atomic mass are split by interaction with a high-energy particle into lower-atomic-mass nuclei. In nuclear fusion, low-atomic-mass nuclei fuse to form new nuclei of higher atomic mass.

Chapter 11

11.1 The number of covalent bonds formed by the atom of an element in forming a compound.

11.3 a

11.5 b

11.7 (a) C—C and C—H single bonds

(b) 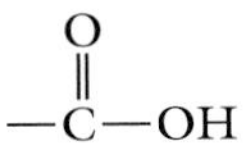 (c) 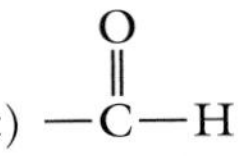

11.9 (a) alcohol; (b) alkene; (c) ketone; (d) aromatic; (e) ketone; (f) ether; (g) carboxylic acid; (h) amide

11.11 sp^3 orbitals are proposed because carbon is tetravalent with four equivalent bonds that have 109.5° bond angles.

11.13

(a) 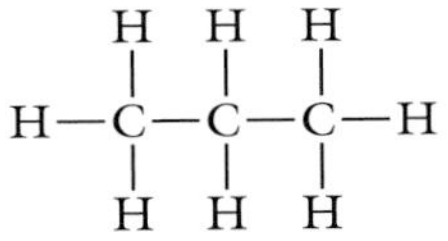

(b) 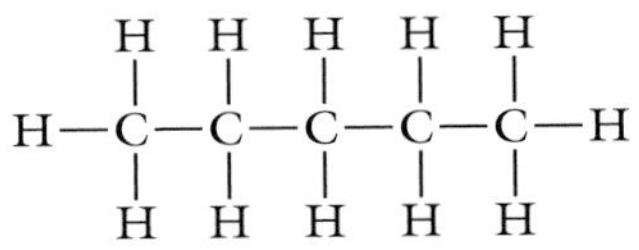

(c) 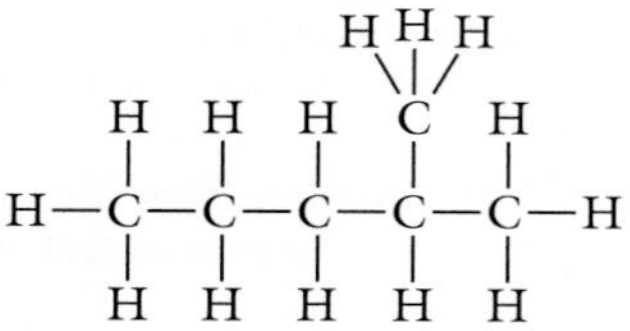

11.15 (a) 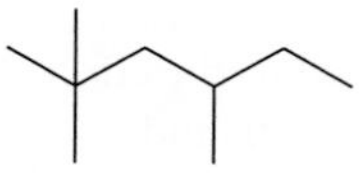

(b) 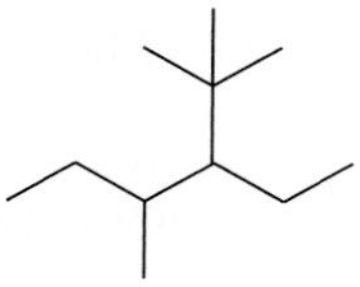

(c) 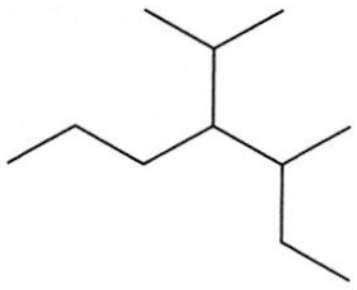

(d) 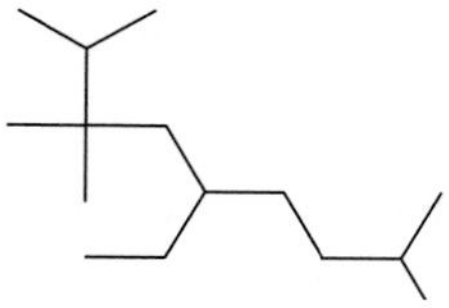

11.17 (a) 1; (b) 2; (c) 2

11.19 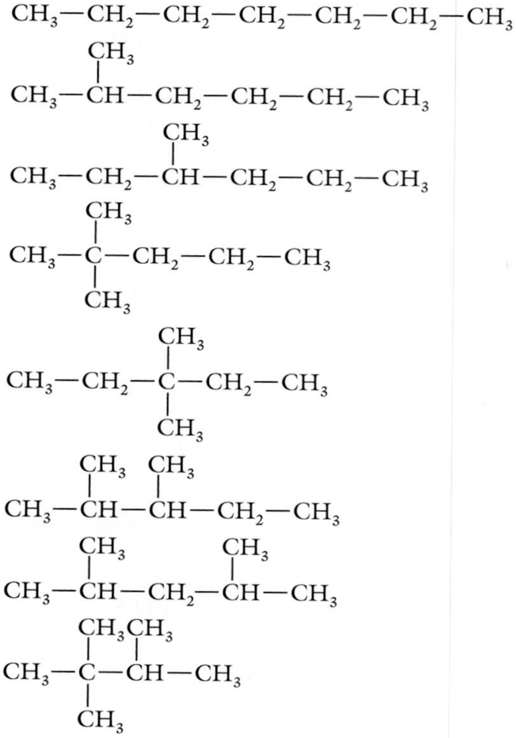

11.21 (a) 2-methylbutane
(b) 3-methylpentane
(c) 2,2,4-trimethylhexane
(d) 3-isopropyl-4-methylhexane

11.23 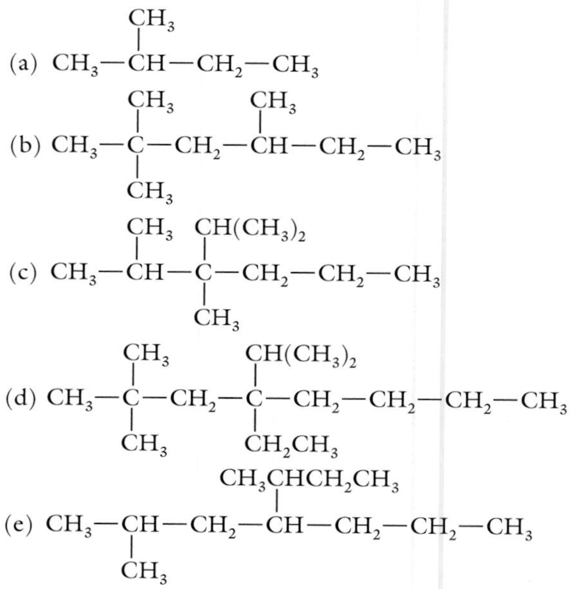

11.25 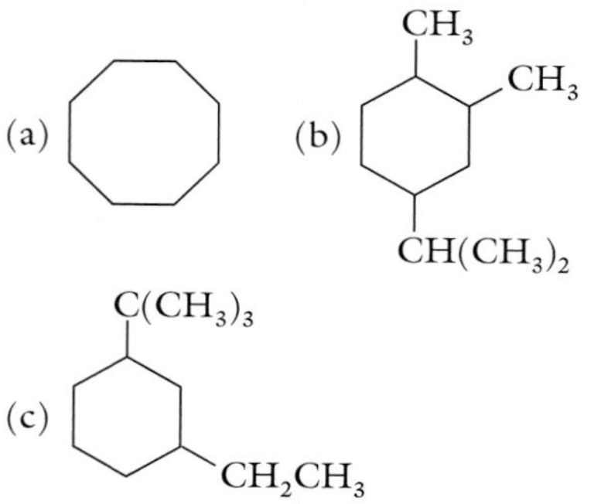

Geometric stereoisomers (Section 11.9) are possible for b and c.

11.27 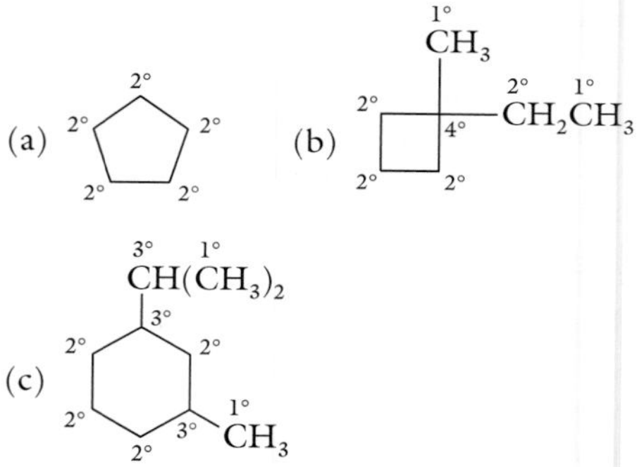

11.29 (a) 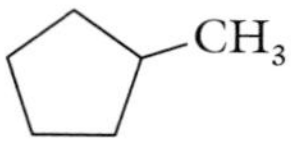 Methylcyclopentane

(b) 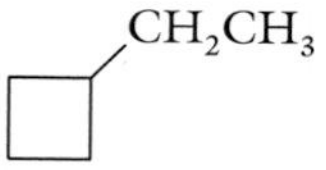 Ethylcyclobutane

11.31 (a) No cis and trans isomers.

(b) 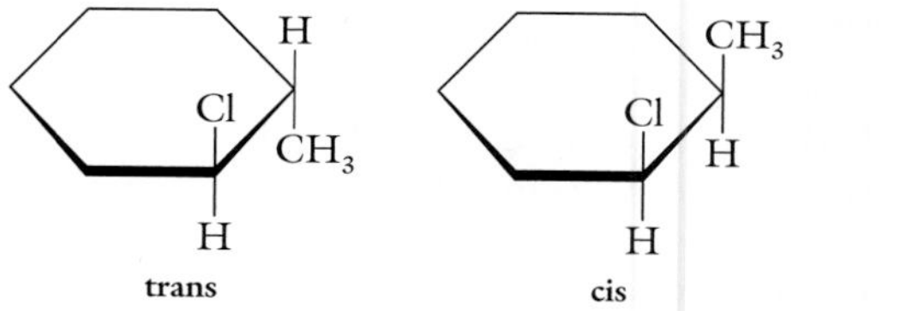

(c) 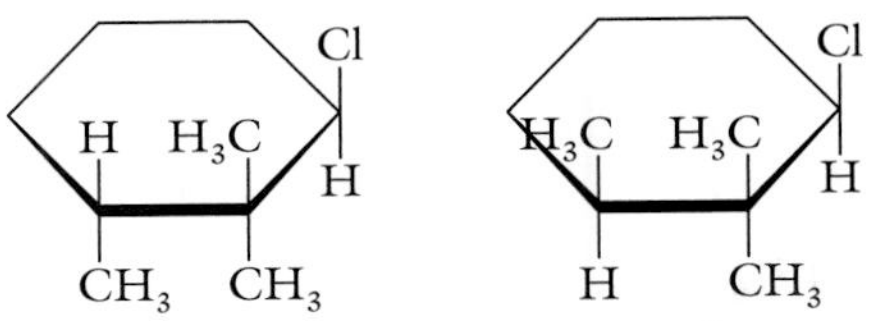

trans cis

11.33 (a) Hexane has the higher boiling point because it has the higher molecular mass and, therefore, greater secondary forces.

(b) Cyclohexane has the higher boiling point because it has the higher molecular mass and, therefore, greater secondary forces.

(c) Methylcyclohexane has the higher boiling point because it has the higher molecular mass and, therefore, greater secondary forces.

(d) Hexane has the higher boiling point because it is not branched and therefore has greater secondary forces.

(e) Cyclopentane has the higher boiling point because it is cyclic, packs tightly, and therefore has greater secondary forces.

11.35 No reaction for a, b, c, and d.

(e) $CH_3—CH_2—CH_2—CH_2—Cl$

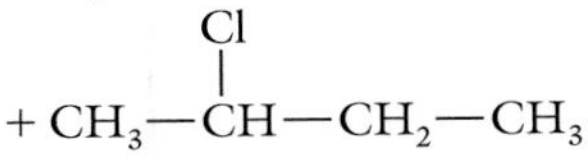

11.37 (a, b) The same equation describes both reactions because both compounds have the same molecular formula:

$$C_5H_{12} + 8\ O_2 \longrightarrow 5\ CO_2 + 6\ H_2O$$

(c, d) The same equation describes both reactions because both compounds have the same molecular formula:

$$C_6H_{12} + 9\ O_2 \longrightarrow 6\ CO_2 + 6\ H_2O$$

11.39 Carbon atoms bond to one another in many different patterns.

11.41 $CH_3—CH_2—CH_2—CH_2Cl$

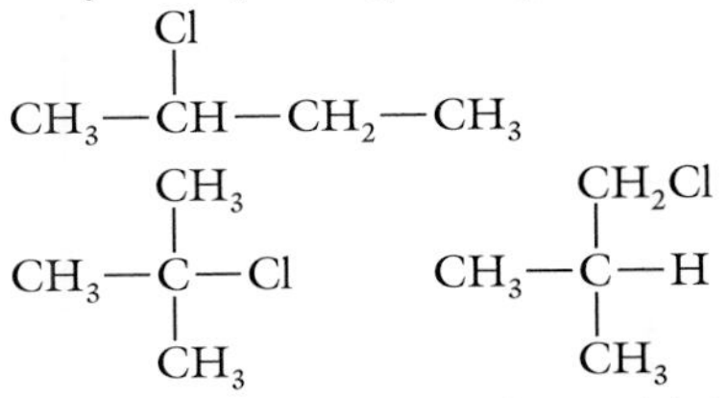

11.43 (a) 4; (b) 4; (c) 2; (d) 2; (e) 4; (f) 2; (g) 2; (h) 3; (i) 2; (j) 1

11.45 Cyclic. (Chapter 12 will show that another possibility is an acyclic alkene.)

11.47 (a) C_6H_{14}; (b) C_8H_{18}; (c) $C_6H_{11}Cl$; (d) $C_6H_{11}Cl$; (e) C_9H_{20}; (f) C_6H_{12}

11.49 (a) $2\ C_6H_{14} + 19\ O_2 \longrightarrow 12\ CO_2 + 14\ H_2O$

(b) 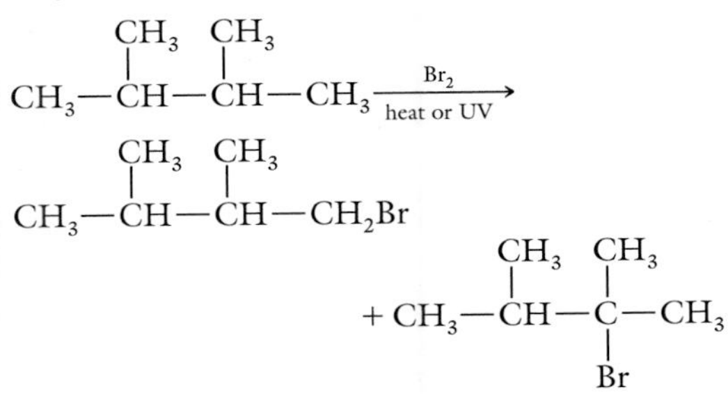

11.51 $CH_3—CH_2—CH_2—CHCl_2$
$+ CH_3—CH_2—CCl_2—CH_3$
$+ CH_2Cl—CHCl—CH_2—CH_3$
$+ CH_2Cl—CH_2—CHCl—CH_3$
$+ CH_2Cl—CHCl—CH_2—CH_2Cl$

11.53 The alkanes in oil possess only the weak London attractive forces, whereas water contains the much stronger hydrogen-bonding forces. Alkane molecules are much more weakly attracted to each other than are water molecules and this results in a lower density for oil.

11.55 Chemical reactions are used to convert alkanes into members of other families.

11.57 The molecules of lipstick and petroleum jelly attract each other because they are both hydrocarbons with the same type of secondary force (London).

11.59 8030 kJ

Chapter 12

12.1 Alkane

12.3 (a) Incorrect; (b) correct; (c) correct

12.5 (a) C_5H_{10}; (b) C_7H_{12}

12.7 Carbons 3, 4, 8, 9

12.9 $CH_2{=}CH—CH_2—CH_3$ $CH_3—CH{=}CH—CH_3$ (*cis-* and *trans-*)

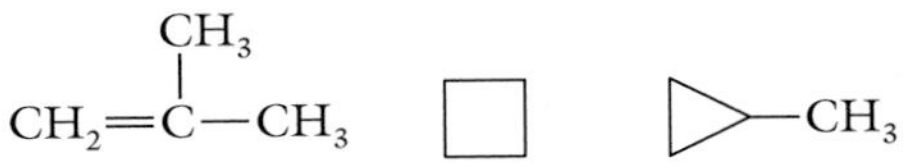

12.11 (a) $CH_2{=}CH—CH_2—CH_2—CH_3$ and $CH_3—CH{=}CH—CH_2—CH_3$ (*cis-* and *trans-*)

(b) 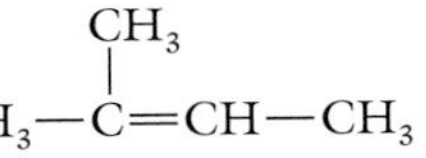

(c) 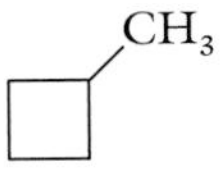

(d) 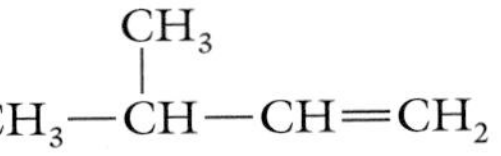

12.13 (a) 3; (b) 3; (c) 2

12.15

(a) 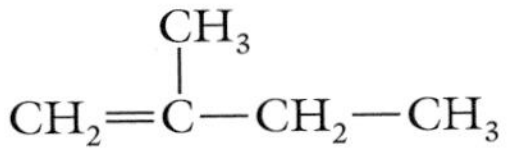

(b) 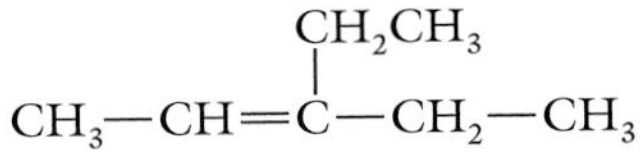

(c) 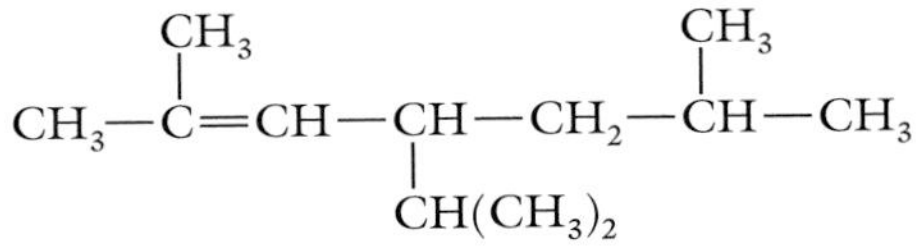

(d) 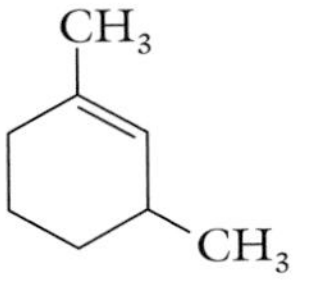

(e) $CH_2{=}\underset{\underset{\displaystyle C(CH_3)_3}{|}}{\overset{\overset{\displaystyle CH_3}{|}}{C}}{-}\underset{}{CH}{-}\underset{\overset{|}{\displaystyle CH_3}}{CH}{-}CH_3$ 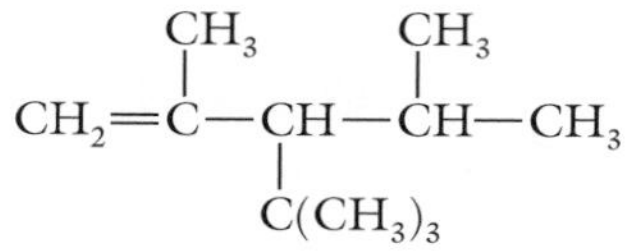

(f) $CH_2{=}CH{-}CH_2{-}CH{=}\underset{}{\overset{\overset{\displaystyle CH_3}{|}}{C}}{-}CH_3$ 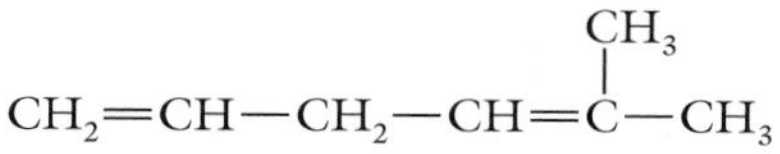

12.17 (a) No cis-trans isomers

(b) 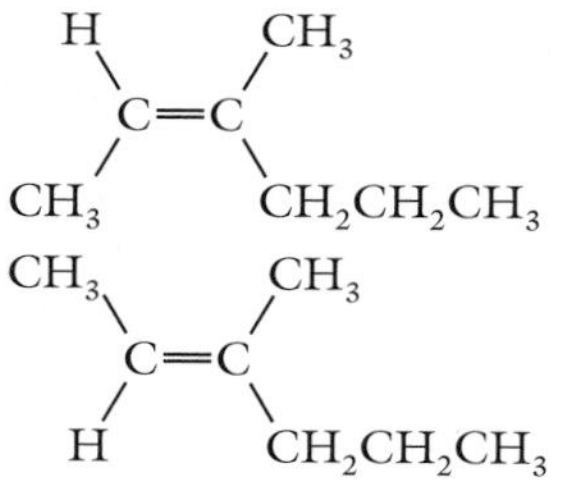

$\underset{CH_3}{\overset{H}{}}C{=}C\underset{CH_2CH_2CH_3}{\overset{CH_3}{}}$

$\underset{H}{\overset{CH_3}{}}C{=}C\underset{CH_2CH_2CH_3}{\overset{CH_3}{}}$

(c) 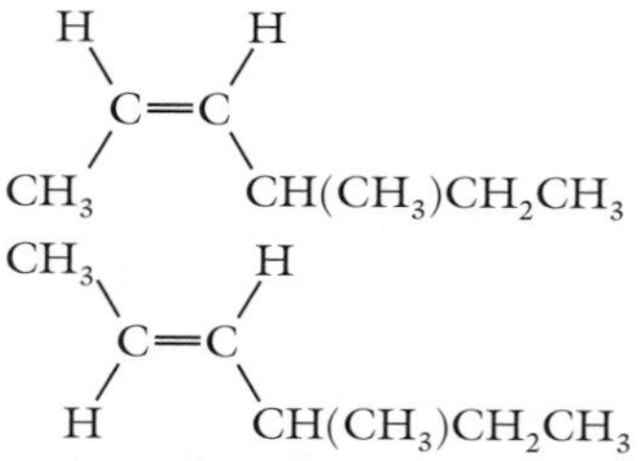

$\underset{CH_3}{\overset{H}{}}C{=}C\underset{CH(CH_3)CH_2CH_3}{\overset{H}{}}$

$\underset{H}{\overset{CH_3}{}}C{=}C\underset{CH(CH_3)CH_2CH_3}{\overset{H}{}}$

(d) no cis-trans isomers
(e) no cis-trans isomers

(f) 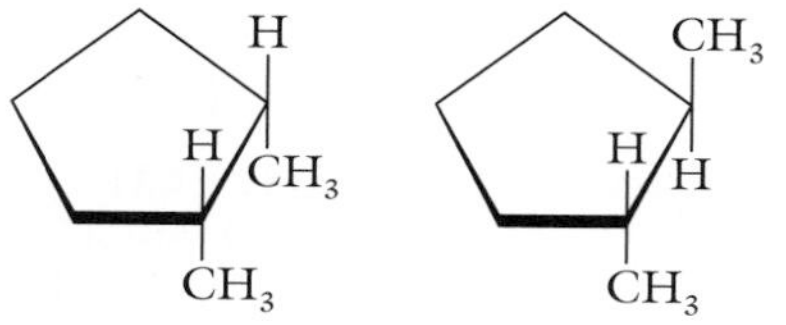

12.19 (a) 3; (b) 1; (c) 2; (d) 4

12.21 (a)

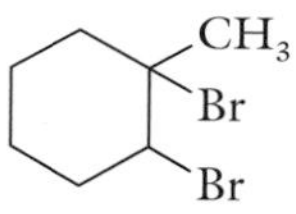

(b) 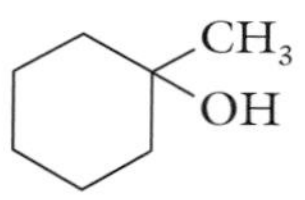

cis and *trans*

(c) $(CH_3)_3CH$ (d) $(CH_3)_3CCl$

(e)

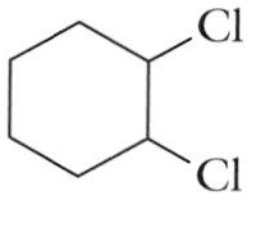

(f) $CH_3{-}CH_2{-}\underset{\overset{|}{\displaystyle Cl}}{CH}{-}CH_3$ 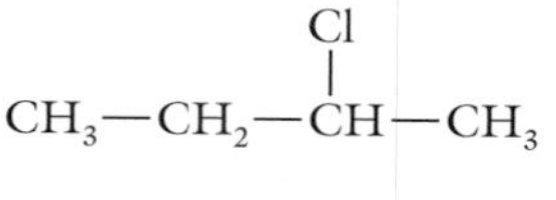

(g) $CH_3{-}CH_2{-}\underset{\overset{|}{\displaystyle Br}}{CH}{-}CH_2CH_3$ 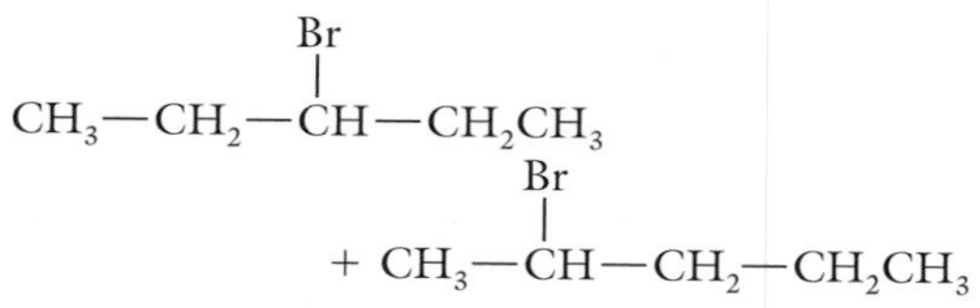
$+\ CH_3{-}\underset{\overset{|}{\displaystyle Br}}{CH}{-}CH_2{-}CH_2CH_3$

(h) $CH_3{-}\underset{\overset{|}{\displaystyle Br}}{CH}{-}\underset{\overset{|}{\displaystyle Br}}{CH}{-}CH_2CH_3$ 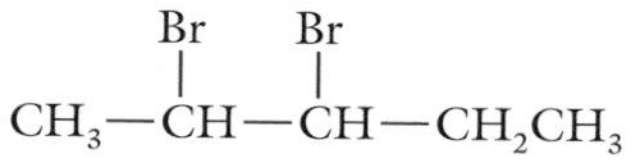

(i) $CH_3{-}\underset{\overset{|}{\displaystyle Cl}}{CH}{-}\underset{\overset{|}{\displaystyle Cl}}{CH}{-}CH_3$ 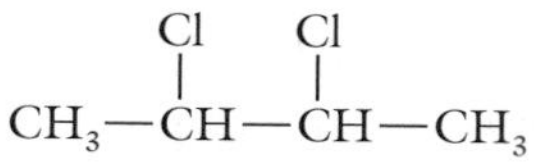

(j) $CH_3{-}CH_2{-}\underset{\overset{|}{\displaystyle OH}}{CH}{-}CH_2CH_3$ 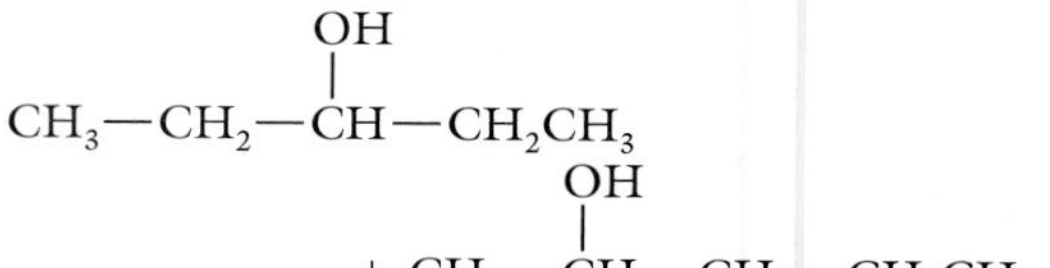
$+\ CH_3{-}\underset{\overset{|}{\displaystyle OH}}{CH}{-}CH_2{-}CH_2CH_3$

(k) $CH_3{-}CH_2{-}CH_2{-}CH_2CH_3$

12.23 Hexane or methylcyclopentane

12.25 $n\ CH_2{=}\underset{\overset{|}{\displaystyle CN}}{CH}\ \xrightarrow{\text{polymerization catalyst}}\ {-}(CH_2{-}\underset{\overset{|}{\displaystyle CN}}{CH}){-}_n$ 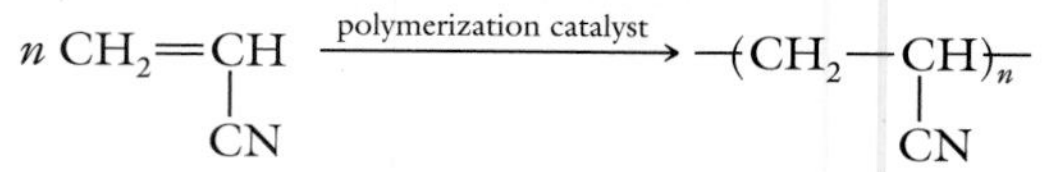

12.27 $C_6H_{12} + 9\ O_2 \rightarrow 6\ CO_2 + 6\ H_2O$

12.29 Pentane or cyclopentane

12.31 $HC{\equiv}C{-}\underset{\overset{|}{\displaystyle CH_3}}{CH}{-}CH(CH_3)_2$ 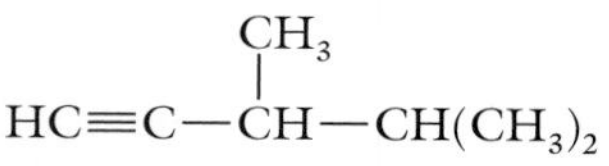

12.33 Aromatic compounds have three double bonds alternating with three single bonds in a six-membered ring structure. Aromatic compounds have high stability, with a very low tendency to undergo addition reactions. Substitution reactions take place without loss of the aromatic structure.

12.35 (a) 1; (b) 2; (c) 1

12.37 (a) 3-Chlorotoluene or *m*-chlorotoluene; (b) 4-bromotoluene or *p*-bromotoluene; (c) 1-benzyl-3-chlorobenzene or *m*-benzylchlorobenzene; (d) 2-bromophenol or *o*-bromophenol; (e) 4-ethylaniline or *p*-ethylaniline; (f) *cis*-1-chloro-2-phenylcyclohexane

12.39 (a) No reaction (b) no reaction

(c)

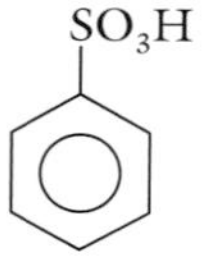

(d) 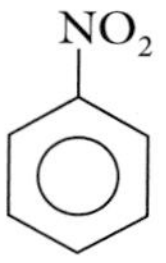

(e) $2\ C_6H_6 + 15\ O_2 \rightarrow 12\ CO_2 + 6\ H_2O$

12.41 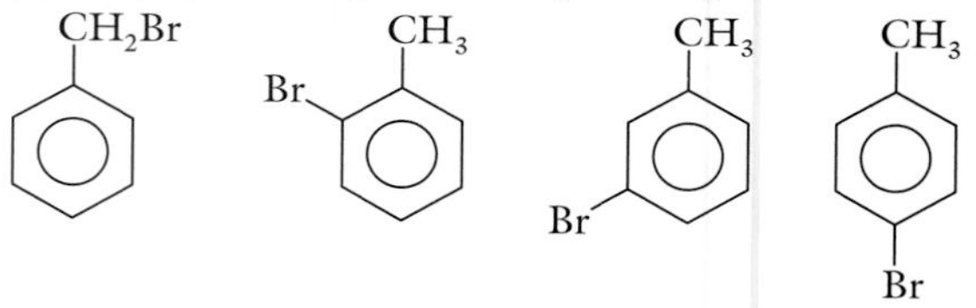

12.43

(a) 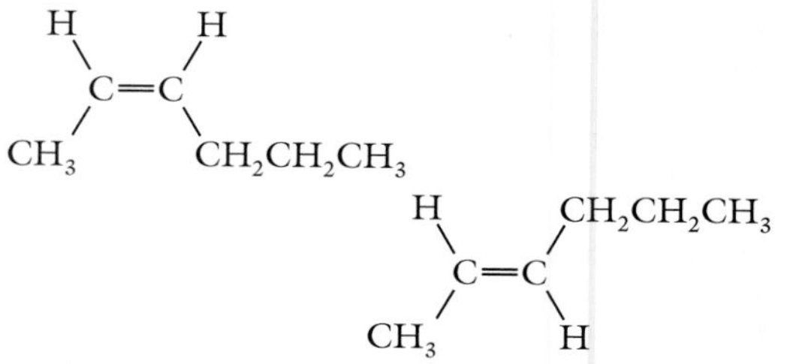

$\underset{CH_3}{\overset{H}{}}C{=}C\underset{CH_2CH_2CH_3}{\overset{H}{}}$

$\underset{CH_3}{\overset{H}{}}C{=}C\underset{H}{\overset{CH_2CH_2CH_3}{}}$

(b) no cis-trans isomers

(c) 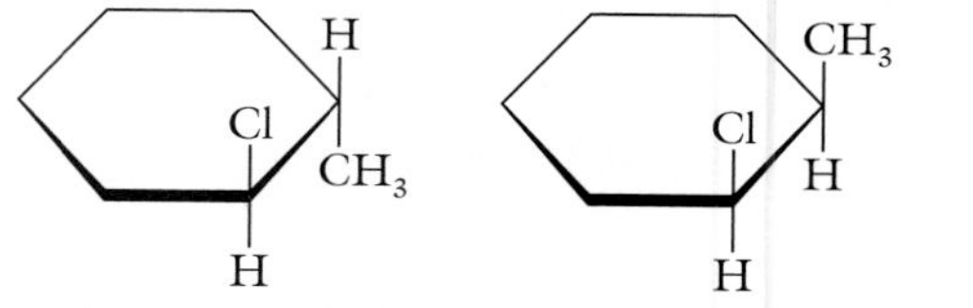

(d) no cis-trans isomers; (e) no cis-trans isomers;
(f) no cis-trans isomers

12.45 (a) 2; (b) 2; (c) 4; (d) 2; (e) 2; (f) 4

12.47 All three compounds are nonpolar with very similar boiling points and neglible solubility in water.

12.49 (a) All three compounds burn in air; (b) none of the compounds react; (c) only cyclohexene reacts; (d) only cyclohexene reacts; (e) only cyclohexene reacts.

12.51 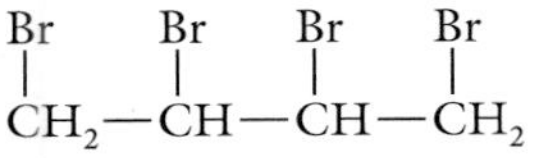

12.53 (a) C_6H_8; (b) C_9H_{12}; (c) C_5H_8; (d) $C_8H_7BrClNO_2$

12.55 Sample A is 1-hexene; sample B is 1-hexyne.

12.57

(a) 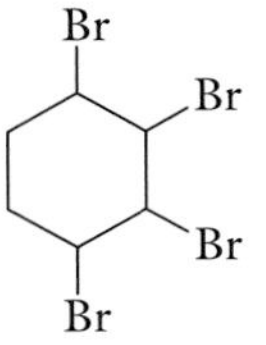 (b) no reaction

(c) 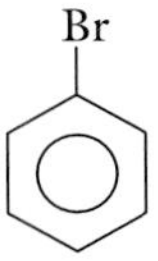 (d) no reaction

(e) no reaction (f) 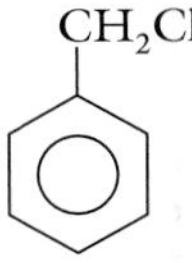

(g) 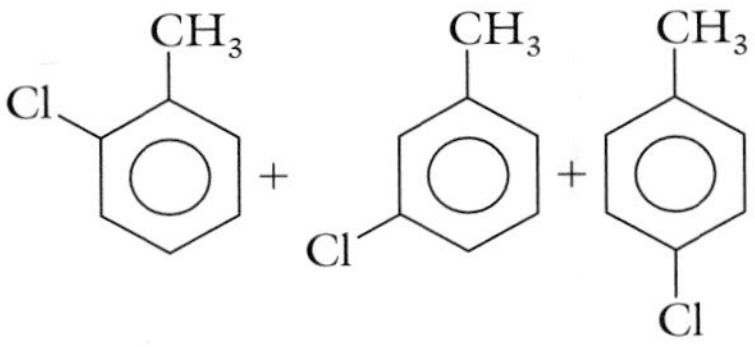

(h) $CO_2 + H_2O$ (i) $CO_2 + H_2O$

(j) no reaction (k) 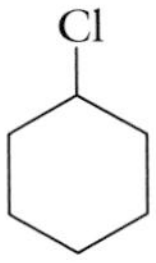

12.59 1-Hexene

12.61 Mary Smith is correct. Addition of water to propene will proceed according to Markovnikov's rule to produce 2-propanol not 1-propanol, as the major product.

12.63 Perform the bromine test in a quantitative manner. Weigh out an amount of the shipped compound. Calculate the number of moles, assuming the compound is geraniol. Determine the amount of bromine that reacts with the compound. Calculate the number of moles of bromine used. If the compound is geraniol, the number of moles of bromine will be twice that of the compound. If more than twice the moles of bromine are used, the shipped compound is most likely myrcene. This likelihood can be verified by performing the calculation assuming the compound is myrcene. Myrcene will react with three times the number of moles of bromine.

12.65 Unsaturated fats contain carbon–carbon double bonds; saturated fats contain only single bonds.

12.67 The reaction rate increases with increasing stability of the carbocation formed when H^+ adds to the double bond of each alkene. More-stable carbocations are formed faster than less-stable carbocations. 2-Methylpropene forms a tertiary carbocation, propene forms a secondary carbocation, and ethene forms a primary carbocation. The order of carbocation stability is 3° > 2° > 1° and this results in the observed order of reactivity of alkenes.

Chapter 13

13.1 An alcohol has an —OH bonded to a saturated (sp^3) carbon. A phenol has the —OH bonded to a carbon of a benzene ring. An ether has an oxygen directly bonded to two different carbons.

13.3

(a) 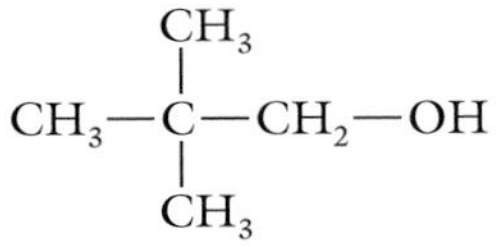

(b) $CH_3—CH_2—CH_2—CH_2—OH$

13.5 Compound 1, $C_nH_{2n+2}O$; compound 2, $C_nH_{2n}O$

13.7 (a) 2-Methyl-1-butanol, primary; (b) 3-methyl-2-butanol, secondary; (c) 2-methyl-1-butanol, primary; (d) 2-methyl-2-propanol, tertiary

13.9 1,4-Butanediol > 1-pentanol > 2-methyl-2-butanol > hexane. The alcohols have higher boiling points than that of hexane because the secondary forces are stronger—hydrogen bonding compared with London forces. 1-Pentanol's boiling point is higher than that of 2-methyl-2-butanol because the secondary forces are stronger for unbranched than for branched molecules. 1,4-Butanediol has stronger secondary forces compared with the other two alcohols because it has two OH groups compared with one.

13.11 There is no change, because ethanol is not sufficiently acidic to affect blue litmus paper.

13.13 $CH_3CH_2CH_2CH_2—OH + HCl \rightleftharpoons CH_3CH_2CH_2CH_2—\overset{+}{O}H_2 + Cl^-$

13.15 (a) Acid-base reaction takes place, but dehydration does not, because there is no heat:

$CH_3CH_2CH_2CH_2—OH + H_2SO_4 \rightleftharpoons CH_3CH_2CH_2CH_2—\overset{+}{O}H_2 + HSO_4^-$

(b) no reaction; (c) $CH_2{=}CHCH_2CH_3$

13.17 (a) The primary alcohol is first oxidized to an aldehyde, and then the aldehyde is oxidized to a carboxylic acid:

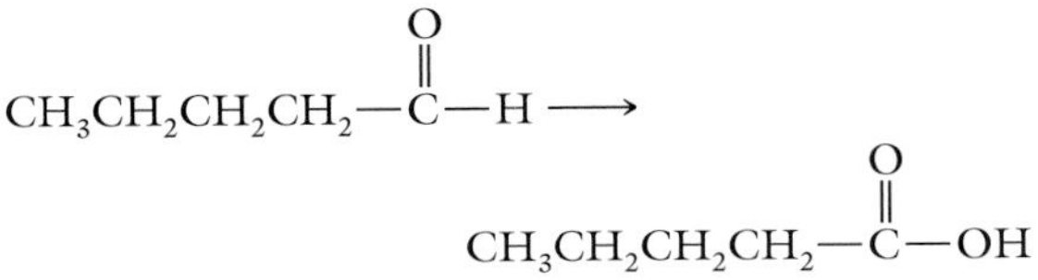

(b) no reaction

13.19 The test does not distinguish between the two possibilities, because both 1-butene and 1-butanol react with permangante solution.

13.21 2-Isopropyl-5-methylphenol

13.23 $p\text{-}CH_3\text{-}C_6H_5—O—H + H_2O \rightleftharpoons p\text{-}CH_3\text{-}C_6H_5—O^- + H_3O^+$

13.25 (a) $C_6H_6O + 7\ O_2 \rightarrow 6\ CO_2 + 3\ H_2O$

(b) $C_6H_5—O—H + KOH \rightarrow C_6H_5—O^- + K^+ + H_2O$

13.27

(a) 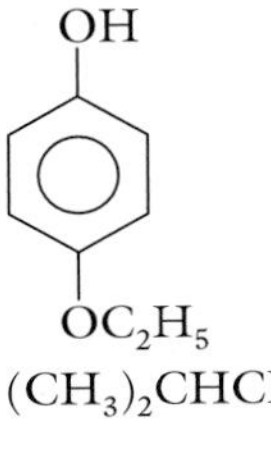

(b) $(CH_3)_2CHCH_2—O—CH_2CH_2CH_3$

(c) 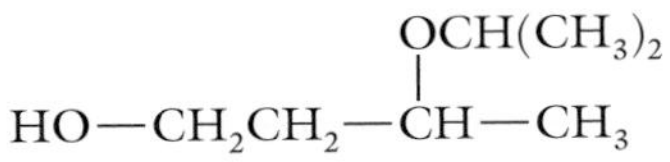

13.29 Ethanol > dimethyl ether > propane

13.31 1-Methylcyclopentene is formed at both temperatures. Although the higher temperature favors ether formation for primary alcohols, ethers are not formed with tertiary alcohols.

13.33 (a) $CH_3{-}\underset{\displaystyle SH}{\underset{|}{CH}}{-}CH_2CH_3$

(b) 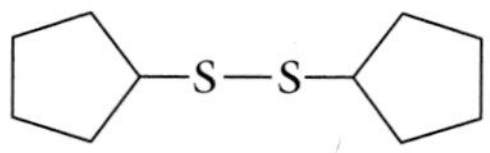

13.35 A, phenol; B, ether; C, alcohol; D, thiol; E, ether

13.37 Compound 1, disulfide; compound 2, alcohol; compound 3, thiol; compound 4, ether

13.39 The formula fits either an alcohol or an ether with either a double bond or a ring.

13.41 (a) $CH_3CH_2CH_2{-}O{-}CH_2CH(CH_3)_2$

(b) 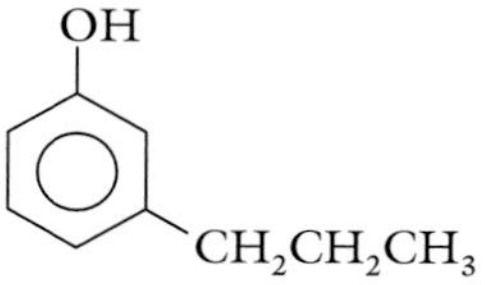

(c) 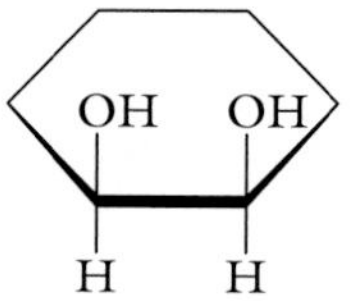

(d) $CH_3{-}\underset{\displaystyle SH}{\underset{|}{CH}}{-}CH_2CH_2CH_3$

(e) $CH_2{=}CH{-}\overset{\displaystyle OH}{\overset{|}{CH}}{-}CH_2{-}CH_3$

13.43 (a) 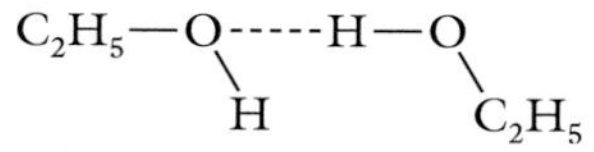

(b) 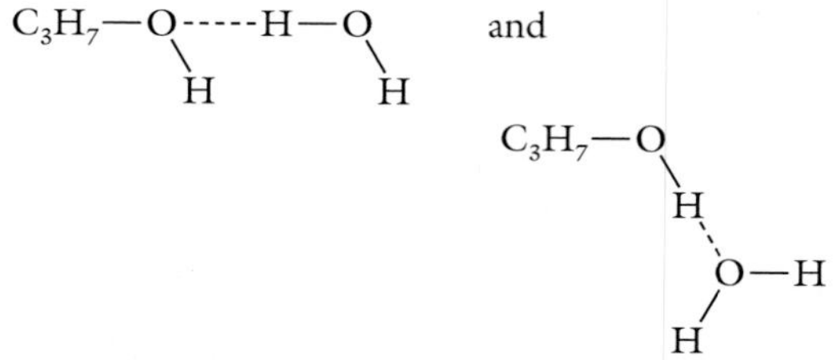

(c) no hydrogen bonding

(d) 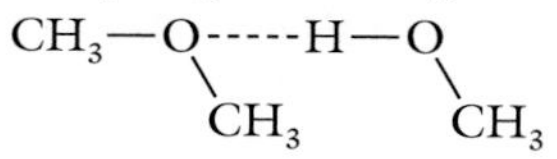

(e) 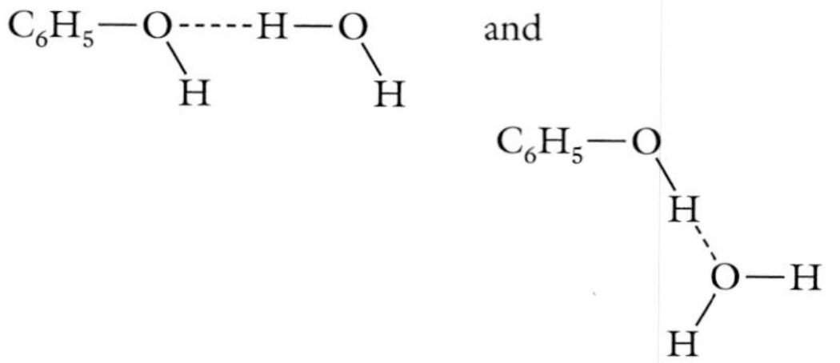

13.45 HCl > phenol > cyclohexanethiol > cyclohexanol

13.47 (a) 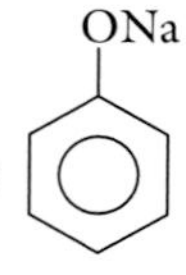

(b) no reaction

(c) 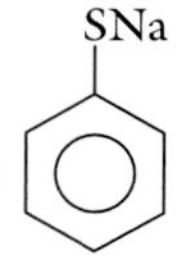

(d) no reaction

(e) 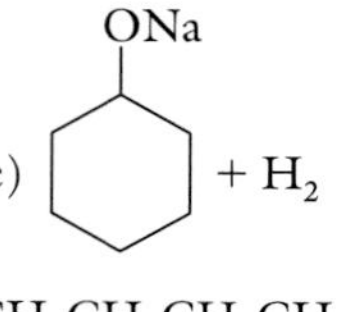

(f) $CH_3CH_2CH_2CH_2{-}S{-}Pb{-}S{-}CH_2CH_2CH_2CH_3$

(g) $CH_3CH_2CH_2CH_2{-}S{-}S{-}CH_2CH_2CH_2CH_3$

(h) 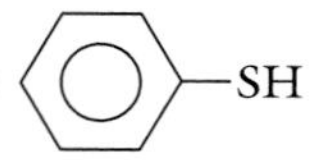

13.49 (a) 1-propanol; (b) HCl; (c) same; (d) phenol; (e) 1-propanethiol

13.51 Only 1-butanol is eliminated with the permangante test. None of the other possibilities react with permanganate. The unknown is either *t*-butyl alcohol, ethyl methyl ether, or pentane.

13.53 Phenol. Neither toluene nor cyclohexanol turns blue litmus paper red.

13.55 A = C = D = E = 104.5°; B = 120°

13.57 Hexylresorcinol is a phenol and sufficiently acidic that it turns blue litmus paper red. 3-Hexyl-1,2-cyclohexanediol is an alcohol and is neutral to litmus paper.

13.59 1-Propanol evaporates rapidly from the skin surface, removing heat and creating a cooling effect. 1-Decanol and 2-decanol are much less volatile because of their higher molecular masses, which results in less heat removed from the skin surface.

13.61 Carbocations are intermediates in the reaction. The carbocation formed from 2-propanol is secondary and more stable than the primary carbocation formed from 1-propanol. The more stable carbocation is formed faster and this results in a faster overall rate for the dehydration reaction.

13.63 Ethylene oxide is relatively unstable due to the distortion of its bond angles from the normal tetrahedral bond angle. Reaction to form an acyclic product with tetrahedral bond angles relieves the instability. The six-membered ring of dioxane possesses tetrahedral bond angles and is as stable as an acyclic compound. Opening of the cyclic structure does not increase stability.

Chapter 14

14.1 Alcohol or ether

14.3

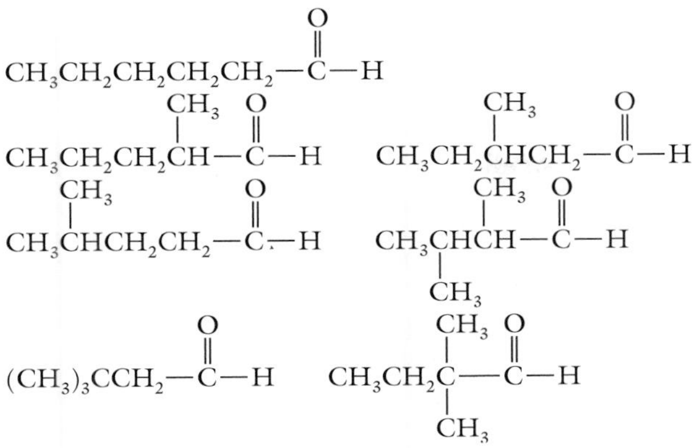

14.5 Compound 1, aldehyde; 2, ester; 3, ketone; 4, amide

14.7 a = d = e = sp^3; b = c = sp^2

14.9

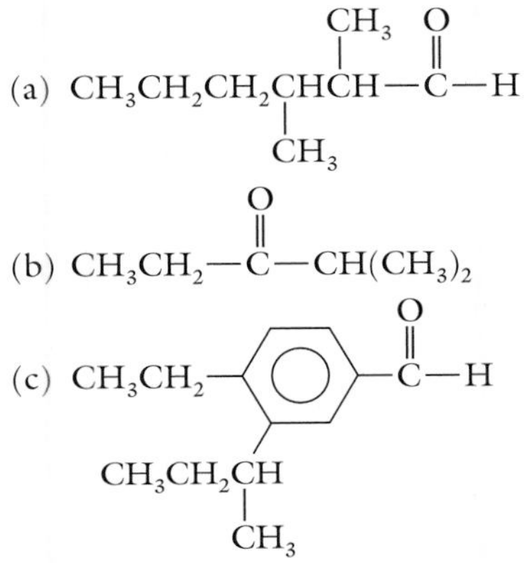

14.11 (a) 2-Isopropylbutanal; (b) *t*-butyl phenyl ketone; (c) 2-methylcyclopentanone; (d) 3-chloro-4,4-dimethylpentanal

14.13 (a) 3-Pentanone. Both compounds are ketones, but 3-pentanone has a higher molecular mass.

(b) Aldehydes and ketones of the same molecular mass have about the same boiling point.

(c) Aldehydes and ketones of the same molecular mass have about the same water solubility.

(d) Butanal. Aldehydes have much stronger secondary attractive forces, dipole–dipole, compared with the weaker London forces of hydrocarbons.

(e) Butanal. Aldehydes hydrogen bond with water, but there are no attractive forces between pentane and water.

(f) 2-Butanone. A larger fraction of the lower-molecular-mass compound is able to hydrogen bond with water.

(g) 2-Butanol. 2-Butanol has the strongest secondary attractive forces with itself (that is, hydrogen bonding), whereas 2-butanone has the weaker dipole–dipole forces.

(h) The water solubilities are very nearly the same because both ketones and alcohols hydrogen bond with water.

14.15 Compound 1 is not oxidized. Compound 2 is oxidized to CH_3CH_2COOH. Compound 3 is oxidized to the aldehyde, which is further oxidized to the carboxylic acid:

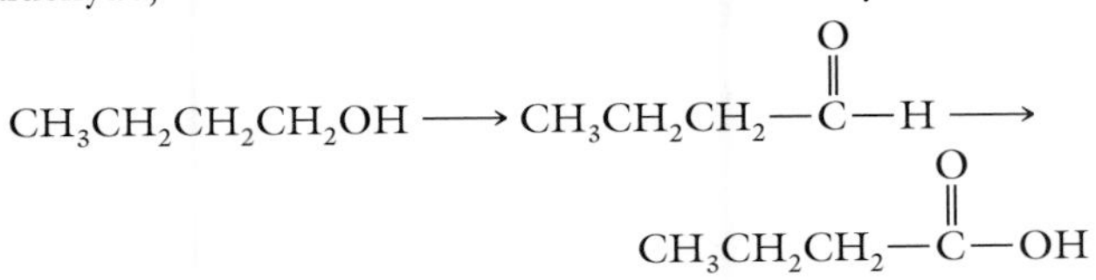

14.17 (a) Compounds 3 and 4 are both aldehydes and give positive tests with Tollens's reagent. A positive Tollens's test is indicated by the formation of a black precipitate or shiny mirror. Compounds 1 and 2 are not aldehydes and give negative tests.

(b) Compounds 1, an α-hydroxy ketone, and 4, an α-hydroxy aldehyde, give positive tests with Benedict's reagent. A positive test is indicated by a red precipitate. Compound 2, a hydroxy carboxylic acid, gives a negative test. Compound 3, a simple aldehyde, gives a negative test.

14.19 Permanganate does not distinguish between 1-pentanol and pentanal, because both give a positive test.

14.21

(a) 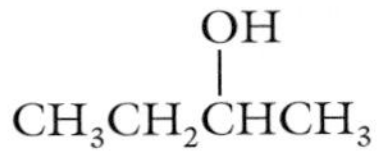

(b) $CH_3CH_2CH_2OH$

(c) no reduction

14.23 $H:^- + H_2O \rightarrow H_2 + HO^-$

H^- is a base because it accepts a proton in the reaction.

14.25

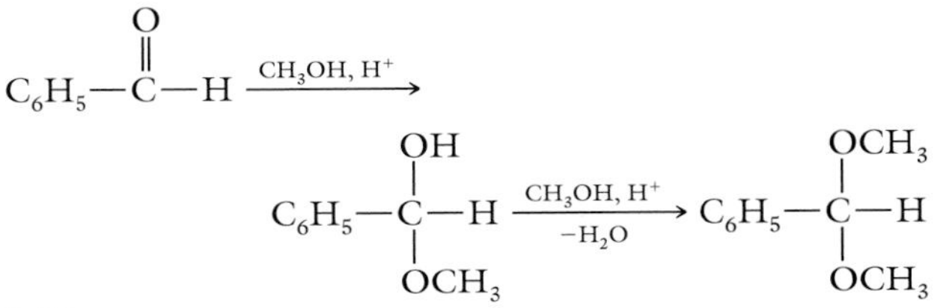

14.27

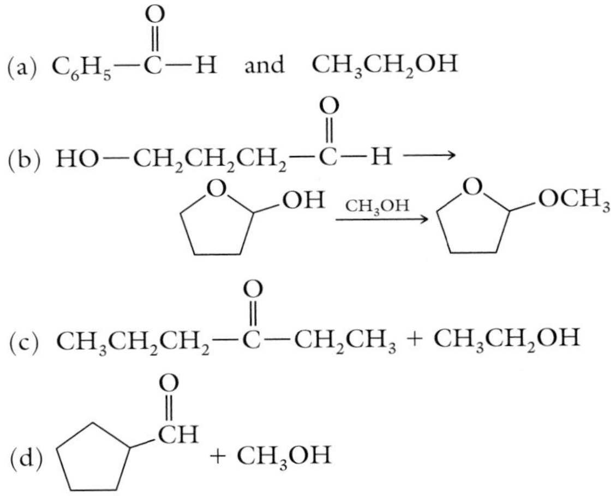

14.29 (a) The carbon atoms of the C=C double bond as well as the carbon and oxygen atoms of the C=O double bond are sp^2 hybridized; (b) all bond angles are 120°; (c) the C=C is nonpolar, whereas the C=O is polar; (d) both double bonds undergo a variety of addition reactions, although not with exactly all the same reagents.

14.31

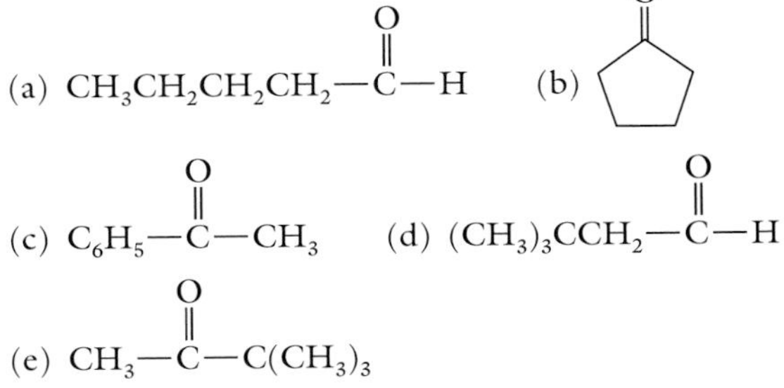

14.33 (a) 3-Methylcyclopentanone; (b) 2-ethylpentanal; (c) pentanal; (d) 3-chloro-2-pentanone

14.35 (a) 1-Butanol. The alcohol has stronger secondary attractive forces (that is, hydrogen bonding) compared with dipole–dipole forces in the aldehyde.

(b) The water solubilities are very nearly the same because both alcohols and ketones hydrogen bond with water.

(c) Butanal. Aldehydes have stronger secondary attractive forces because the carbonyl group in aldehydes is much more polar than the C–O single bond in ethers.

(d) The water solubilities are very nearly the same because both aldehydes and ethers hydrogen bond with water.

14.37 Diethyl ether ≅ butanal ≅ 1-butanol > pentane. Any compound containing oxygen atoms can hydrogen bond with water, although alcohols do so slightly better than aldehydes, which do so slightly better than ethers.

14.39 Tollens's reagent does not distinguish the two compounds, because both ketones and tertiary alcohols are unreactive with Tollens's reagent.

14.41 Yes

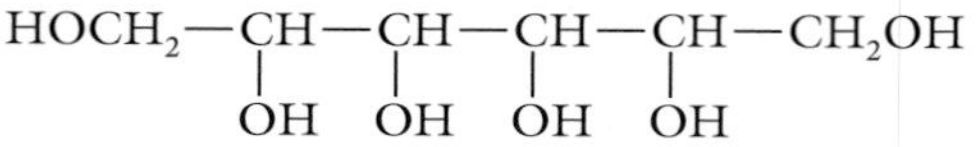

14.43

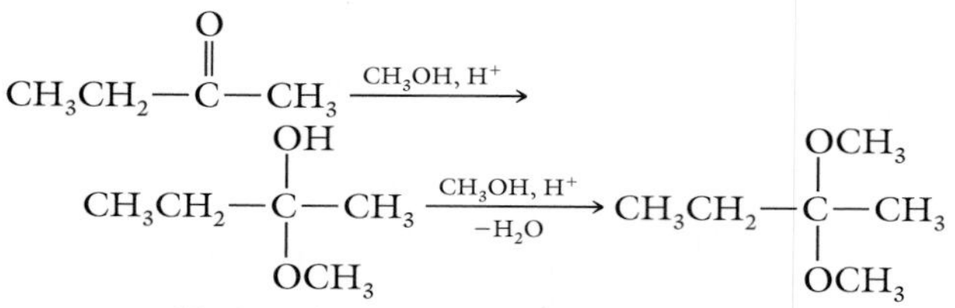

Hemiacetal Acetal

14.45 (a) 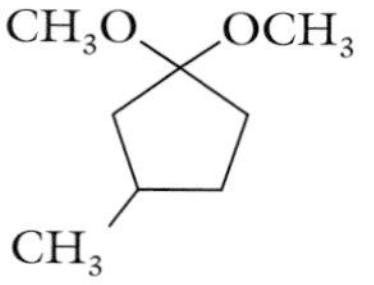 (b) 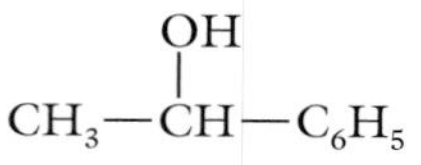

(c) no reaction (d) 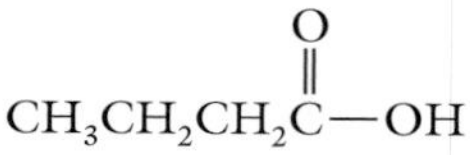

(e) no reaction (f) $HOCH_2CH_2C_6H_5$

(g) 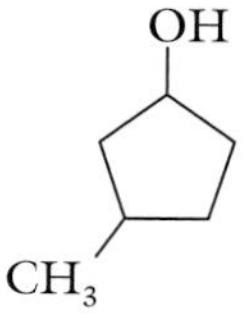 (cis and trans)

(h) 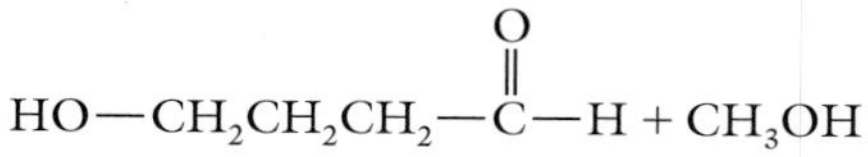

14.47

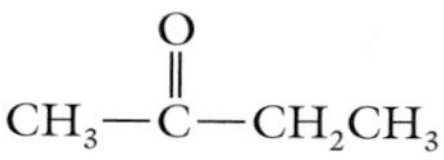

14.49 (a) None (b) 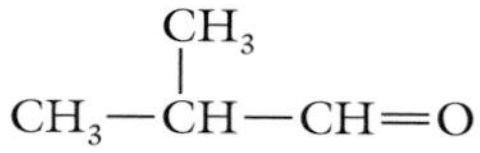

(c) 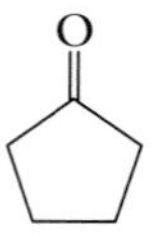 (d) none

14.51 Vanillin turns blue litmus paper red because it is a phenol. Vanillin undergoes oxidation with Tollens's reagent because it is an aldehyde. Menthone is neutral to litmus paper. It gives a negative Tollens's test because it is a ketone, not an aldehyde.

14.53 Hydride reduction reduces only the C=O group; catalytic reduction reduces both the C=O and the C=C groups.

14.55 The cyclic structures are hemiacetals formed from the acyclic structure by reaction between the —OH group at C5 and the carbonyl group (C1). There are two cyclic structures, α-D-glucose and β-D-glucose, formed because there are two positions (up and down) available for the hemiacetal —OH group at C1.

Chapter 15

15.1 Compound 1, something else (ketone and ether); 2, ester; 3, acid halide; 4, something else (ketone and alcohol); 5, carboxylic acid; 6, amide; 7, carboxylic acid and ether; 8, anhydride

15.3

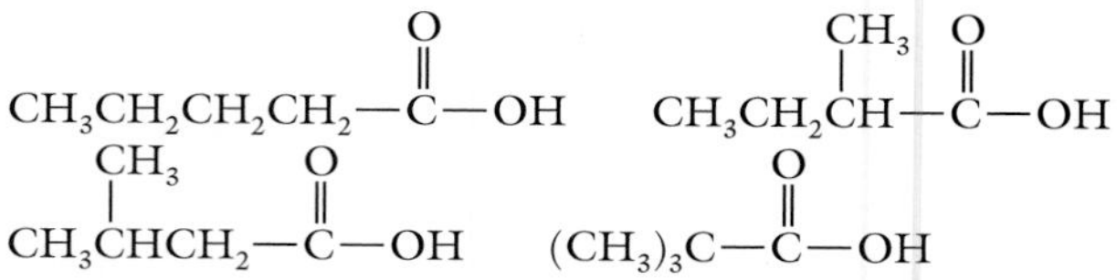

15.5

(a) 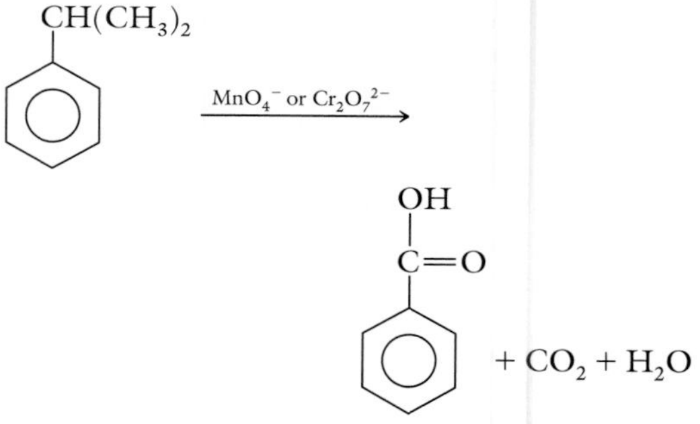

(b) 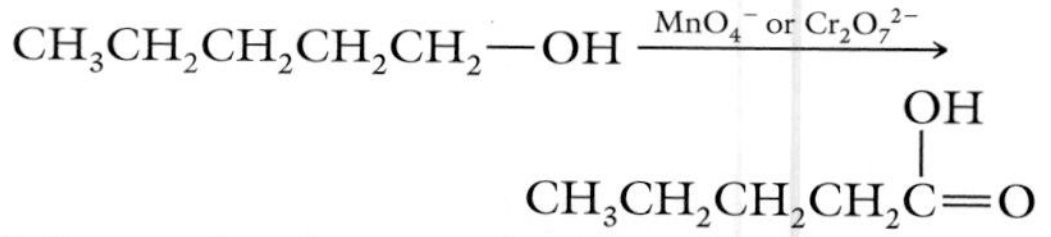

(c) 2-Pentanol undergoes selective oxidation to a ketone (2-pentanone) but the ketone is not oxidized to a carboxylic acid.

15.7 (a) 2-Isopropyl-2,5-dimethylhexanoic acid; (b) 3-chloro-4,4-dimethylpentanoic acid; (c) 3-phenylbutanoic acid; (d) 1,2-butanedioic acid

15.9

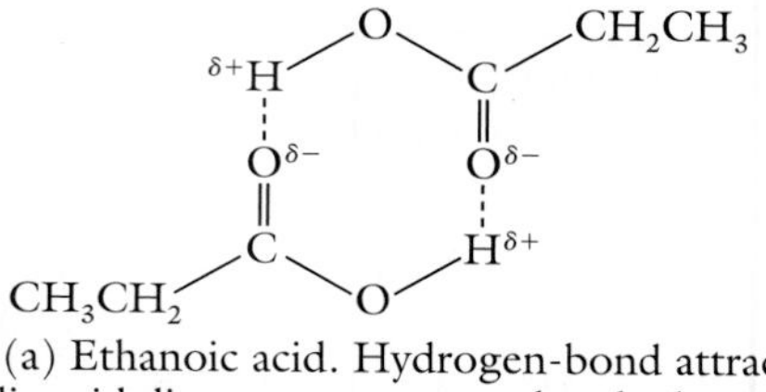

15.11 (a) Ethanoic acid. Hydrogen-bond attractions in carboxylic acid dimers are stronger than hydrogen-bond

attractions in alcohols. (b) The solubilities are about the same because both have strong hydrogen-bond attractions with water.

15.13 Compound 1 is more acidic than compound 2 because compound 1 is a carboxylic acid, whereas compound 2 is an alcohol and ketone.

15.15

(a) $CH_3\overset{O}{\overset{\|}{C}}-O-H + H_2O \rightleftharpoons CH_3\overset{O}{\overset{\|}{C}}-O^- + H_3O^+$

(b) $CH_3\overset{O}{\overset{\|}{C}}-O-H + NaOH \longrightarrow CH_3\overset{O}{\overset{\|}{C}}-O^-Na^+ + H_2O$

(c) $CH_3\overset{O}{\overset{\|}{C}}-O-H + NaHCO_3 \longrightarrow CH_3\overset{O}{\overset{\|}{C}}-O^-Na^+ + CO_2 + H_2O$

(d) $C_6H_5-O-H + H_2O \rightleftharpoons C_6H_5-O^- + H_3O^+$

(e) $C_6H_5-O-H + NaOH \longrightarrow C_6H_5-O^- + Na^+ + H_2O$

15.17 Propanoic acid predominates in water because the extent of ionization is less than about 1 to 2%. Propanoate ion is the predominant species at pH = 7 because the buffered solution is much more basic compared with the acidity of a carboxylic acid.

15.19 Propanoic acid

15.21

(a) $CH_3CH_2CH_2CH_2CH_2\overset{O}{\overset{\|}{C}}-ONa$

(b) $(C_6H_5COO)_2Ca$

15.23 (a) Sodium propanoate has the higher melting point because it is an ionic compound, whereas propanoic acid is a covalent compound. (b) Sodium hexanoate. The attractive forces between the ions of the carboxylate salt and water are stronger than the hydrogen-bond attraction between the carboxylic acid and water. (c) Octanoic acid. The carboxylic acid is a polar covalent compound and interacts more strongly with the nonpolar hexane than does the ionic carboxylate salt.

15.25 Sodium butanoate

15.27 Only compound 3 is a soap. A soap must have both an ionic group and a long hydrocarbon chain.

15.29

(a) $CH_3CH_3CH_3CH_3-\overset{O}{\overset{\|}{C}}-OH + HOCH_2CH_3 \xrightarrow[-H_2O]{H^+} CH_3CH_2CH_2CH_2-\overset{O}{\overset{\|}{C}}-OCH_2CH_3$

(b) $C_6H_5-\overset{O}{\overset{\|}{C}}-OH + HOCH_2CH_2CH_2CH_3 \xrightarrow[-H_2O]{H^+} C_6H_5-\overset{O}{\overset{\|}{C}}-OCH_2CH_2CH_2CH_3$

15.31 (a) 4,4-Dimethylpentanoic acid and 2-propanol; (b) *p*-methylbenzoic acid and 2-butanol; (c) butanoic acid and *p*-methylphenol

15.33

(a) $CH_3(CH_2)_2\overset{O}{\overset{\|}{C}}-OC(CH_3)_3$

(b) 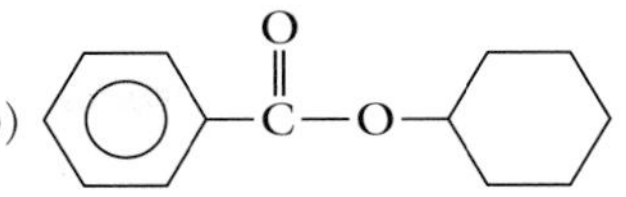

15.35 (a) Pentanal's boiling point is higher than that of methyl propanoate because of higher secondary attractive forces. (b) The solubilities in water are similar because both hydrogen bond equally well with water. (c) 1-Pentanol. The hydrogen-bond secondary attractive forces in alcohols are stronger than the dipole–dipole forces in esters. (d) Hexanoic acid is more soluble in water because it hydrogen bonds more extensively with water than does methyl pentanoate. (e) Butanoic acid. The hydrogen-bond secondary attractions in carboxylic acid dimers are much stronger than the dipole–dipole attractions in esters.

15.37 1,6-Hexanedioic acid and 1,3-propanediol

15.39

(a) $CH_3CH_2-\overset{O}{\overset{\|}{C}}-OCH_2CH_2CH_3 \xrightarrow[H_2O]{H^+} CH_3CH_2-\overset{O}{\overset{\|}{C}}-OH + HOCH_2CH_2CH_3$

(b) 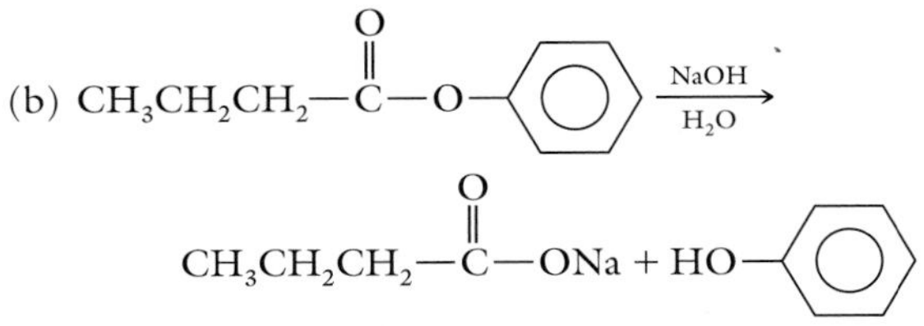

15.41

(a) $(CH_3)_2CH-\overset{O}{\overset{\|}{C}}-Cl$

(b) $CH_3-\overset{O}{\overset{\|}{C}}-O-\overset{O}{\overset{\|}{C}}-CH_3$

15.43

$C_6H_5-\overset{O}{\overset{\|}{C}}-Cl \xrightarrow[-HCl]{H_2O} C_6H_5-\overset{O}{\overset{\|}{C}}-OH$

15.45

$HO-\underset{OH}{\overset{O}{\overset{\|}{P}}}-O-\underset{OH}{\overset{O}{\overset{\|}{P}}}-OH + CH_3CH_2CH_2CH_2OH \xrightarrow{-H_2O} HO-\underset{OH}{\overset{O}{\overset{\|}{P}}}-O-\underset{OH}{\overset{O}{\overset{\|}{P}}}-OCH_2CH_2CH_2CH_3$

Butyl diphosphate

15.47

$^-O-\underset{O^-}{\overset{O}{\overset{\|}{P}}}-O-\underset{O^-}{\overset{O}{\overset{\|}{P}}}-O^-$

15.49

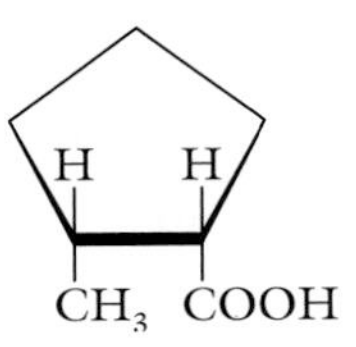

15.51 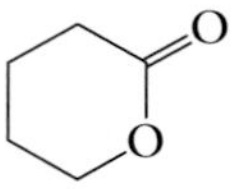

15.53

(a) 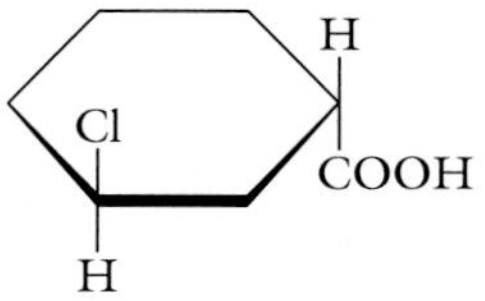

(b) $(CH_3COO)_2Ca$

(c) $C_6H_5-\overset{O}{\overset{||}{C}}-OCH(CH_3)_2$

(d) $CH_3CH_2CH_2CH_2-\overset{O}{\overset{||}{C}}-O-\overset{O}{\overset{||}{C}}-CH_2CH_2CH_2CH_3$

(e) $CH_3CH_2-\overset{O}{\overset{||}{C}}-Cl$

(f) $HO-\overset{O}{\overset{||}{\underset{OH}{P}}}-O-\overset{O}{\overset{||}{\underset{OH}{P}}}-OCH_3$

(g) $HO-\overset{O}{\overset{||}{\underset{OCH_3}{P}}}-OCH_3$

15.55 $a = c = sp^3$; $b = d = sp^2$; $A = C = 120°$; $B = 104.5°$

15.57 Phenol or propanoic acid

15.59

$$CH_3\overset{O}{\overset{||}{C}}-OH + HSCH_2CH_2CH_2CH_3 \xrightarrow[-H_2O]{H^+}$$

$$CH_3\overset{O}{\overset{||}{C}}-SCH_2CH_2CH_2CH_3$$

15.61 (a) No reaction

(b) $CH_3CH_2\overset{O}{\overset{||}{C}}-OH + HOCH_2C_6H_5$

(c) $CH_3CH_2\overset{O}{\overset{||}{C}}-ONa + HOCH_2C_6H_5$

(d) $HOCH_2CH_2\overset{O}{\underset{CH_3}{\overset{||}{C}}}HC-ONa$

(e) $HO-\overset{O}{\overset{||}{\underset{OH}{P}}}-O-\overset{O}{\overset{||}{\underset{OH}{P}}}-O-\overset{O}{\overset{||}{\underset{OH}{P}}}-OCH_2CH_2CH_3$

(f) $CH_3CH_2CH_2\overset{O}{\overset{||}{C}}-ONa + CO_2 + H_2O$

(g) $(CH_3)_2CHCH_2CH_2\overset{O}{\overset{||}{C}}-OC_6H_5$

(h) $CH_3CH_2-\overset{O}{\overset{||}{C}}-OH$

(i) $CH_3CH_2\overset{O}{\overset{||}{C}}-OH$

(j) $CH_3CH_2\overset{O}{\overset{||}{C}}-OH + HSCH_2C_6H_5$

15.63 Add a drop of the unknown to blue litmus paper. The unknown is the carboxylic acid if the litmus paper changes color to red. If there is no color change, the unknown is the ester.

15.65

$$\begin{matrix} CH_2-OH \\ | \\ CH-OH \\ | \\ CH_2-OH \end{matrix} + 3\,HO-NO_2 \xrightarrow{-3\,H_2O} \begin{matrix} CH_2-O-NO_2 \\ | \\ CH-O-NO_2 \\ | \\ CH_2-O-NO_2 \end{matrix}$$

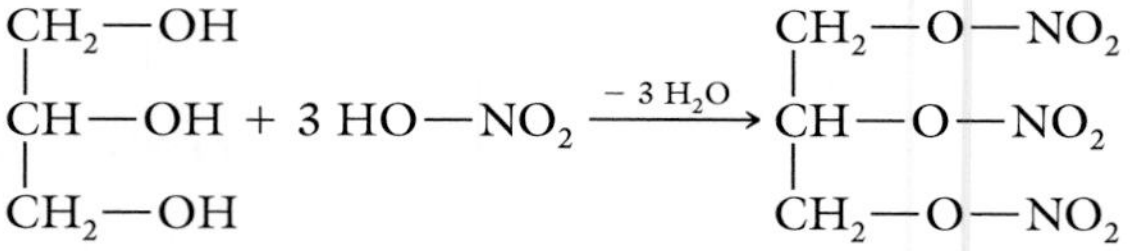

15.67 $[H_3O^+] = 4.17 \times 10^{-3}$; pH = 2.38

Chapter 16

16.1

$$CH_3CH_2-\underset{\underset{CH_3}{|}}{N}-CH_2CH_3 \qquad CH_3CH_2CH_2-\underset{\underset{CH_3}{|}}{N}-CH_3$$

$$(CH_3)_2CH-\underset{\underset{CH_3}{|}}{N}-CH_3$$

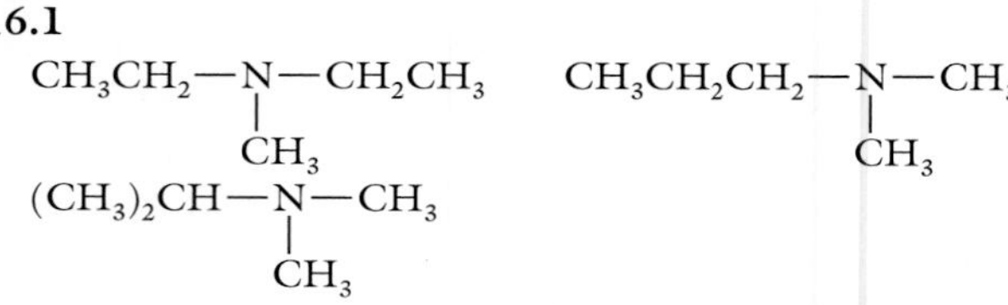

16.3 1, aliphatic, secondary amine; 2, ketone and aliphatic, primary amine; 3, aromatic amine; both amine groups are tertiary; 4 and 5, amides; 6, neither an amine nor an amide — it is nitrobenzene.

16.5 (a) $CH_3CH_2CH_2CH_2CH_2-\underset{\underset{H}{|}}{N}-CH_3$

(b) $(CH_3)_2CH-\underset{\underset{H}{|}}{N}-CH_3$

(c) 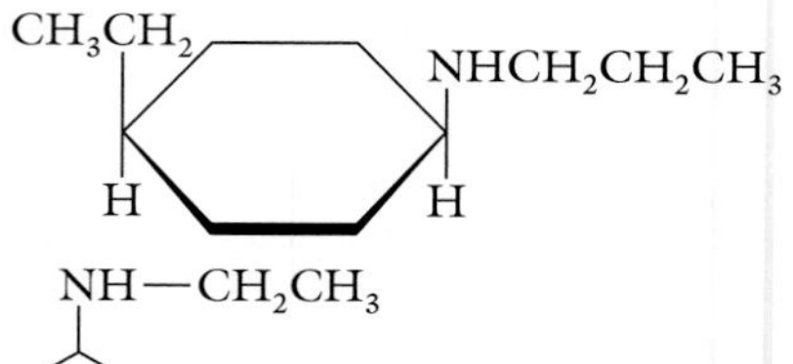

(d) 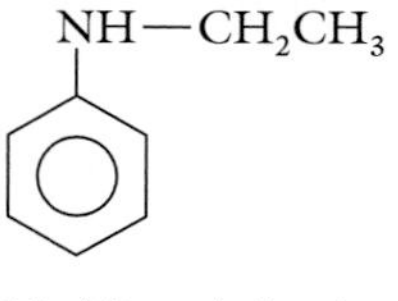

16.7 (a) Ethyldimethylamine; (b) *trans*-2-methylcyclohexanamine; (c) *N*-ethyl-*N*,4-dimethyl-2-pentanamine; (d) *N*-ethyl-*N*-methylaniline

16.9 $C_2H_5-\underset{\underset{H}{|}}{\overset{H}{\overset{/}{N}}}-----H-\underset{\underset{H}{|}}{N}-C_2H_5$ 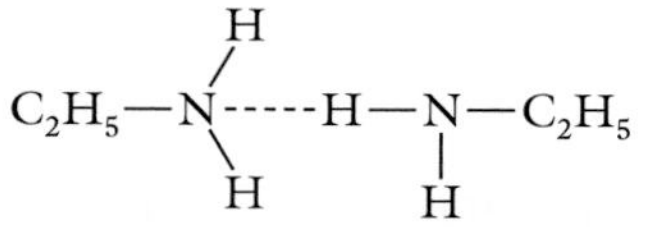

16.11 (a) 1-Butanol. The O—H bond is more polar than the N—H bond, because oxygen is more electronegative than nitrogen, and this polarity results in stronger hydrogen bonding in alcohols than in primary or secondary amines.

(b) The solubilities of butylamine and butanal in water are about the same. The hydrogen bonding of water with the N–H of butylamine is no stronger than that of water with the carbonyl oxygen of butanal.

(c) Butylamine. The hydrogen-bond attractive forces in the primary amine (butylamine) are stronger than the dipole–dipole attractive forces in the tertiary amine (ethyldimethylamine).

(d) Primary and secondary amines have about the same solubility in water. Both have N–H bonds that hydrogen bond with water.

16.13 $CH_3CH_2CH_2NH_2 + HCl \rightarrow CH_3CH_2CH_2NH_3^+ \ Cl^-$

16.15 Water = 1-propanol < aniline < propylamine < KOH. Basicity measures the ability of a compound to bond to a proton. Amines are more basic than water and 1-propanol because nitrogen's nonbonded pair of electrons is more available for bonding to a proton, the result of nitrogen being less electronegative than oxygen. Aniline is less basic than propylamine because the benzene ring, through its electron-withdrawing effect, decreases the availability of nitrogen's nonbonded pair of electrons. KOH is the strongest base because the actual base is OH^-. The negative charge on oxygen in this anion greatly increases its ability to accept a proton.

16.17 Propylamine

16.19 $H_2NCH_2CH_2CH_2CH_2CH_2CH_2NH_2 + 2\ HCl \rightarrow$
$Cl^- \ H_3\overset{+}{N}CH_2CH_2CH_2CH_2CH_2CH_2\overset{+}{N}H_3 \ Cl^-$

16.21 $(CH_3)_2NH^+ \ Cl^-$

(a) 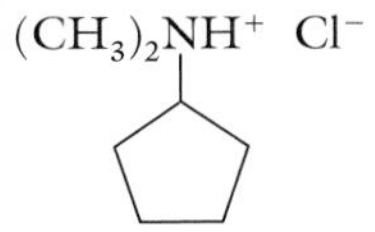

(b) $\left(\underset{CH_3 \quad CH_2CH_3}{\overset{}{N^+}} \right)_2 SO_4^{2-}$ 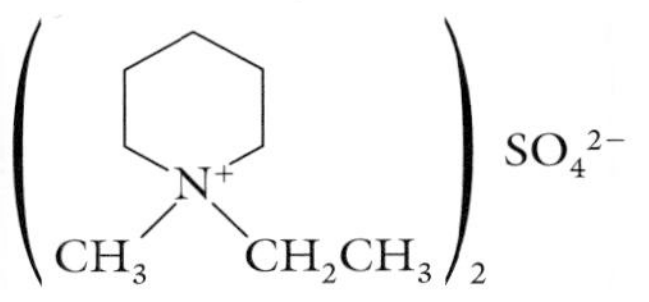

(c) $CH_3CH_2 - \underset{CH_3CH_2CH_2}{\overset{CH_3}{\underset{|}{\overset{|}{\underset{}{N^+}}}} - H} \ Br^-$ 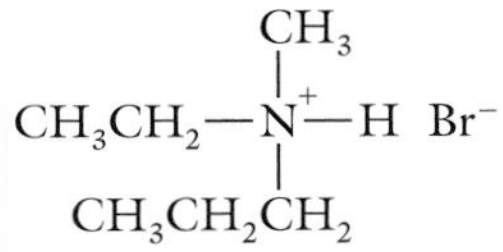

16.23 (a) Trimethylammonium chloride has the higher boiling point because the ionic attractive forces of amine salts are stronger than the hydrogen-bond secondary attractive forces of amines.

(b) Hexylammonium chloride has the higher solubility because amine salts interact more strongly with water than does an amine.

16.25 Butylamine

16.27 1, tertiary amide; 2 is not an amide—it is a ketone and amine; 3, primary amide and primary amine; 4, secondary amide

16.29

(a) $C_6H_5-\overset{O}{\overset{||}{C}}-OH + H_2NCH_2CH_2CH_2CH_3 \xrightarrow{25°C}$ 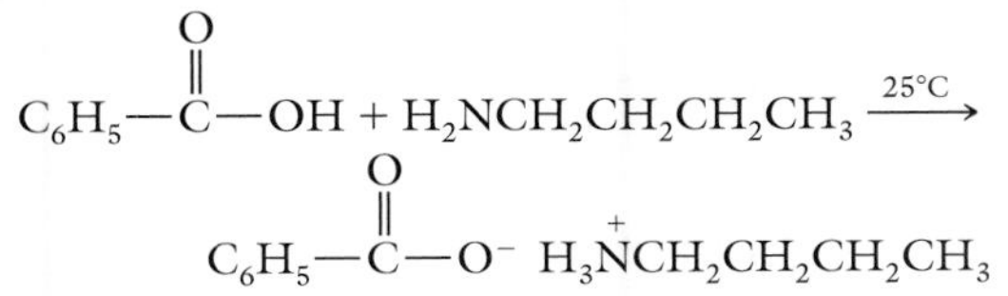
$C_6H_5-\overset{O}{\overset{||}{C}}-O^- \ H_3\overset{+}{N}CH_2CH_2CH_2CH_3$

(b) $C_6H_5-\overset{O}{\overset{||}{C}}-OH + H_2NCH_2CH_2CH_2CH_3 \xrightarrow[-H_2O]{>100°C}$ 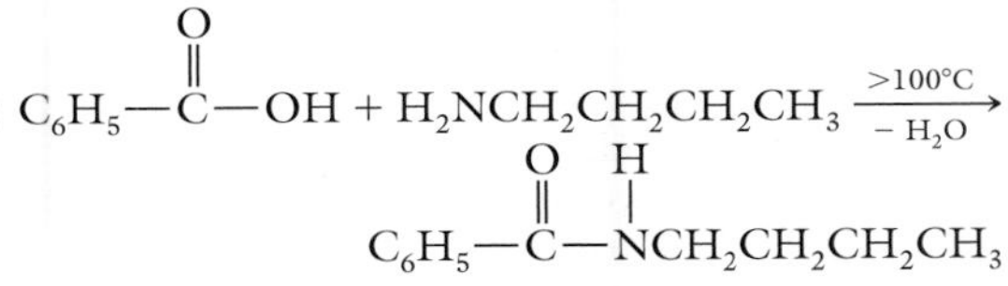
$C_6H_5-\overset{O}{\overset{||}{C}}-\overset{H}{\overset{|}{N}}CH_2CH_2CH_2CH_3$

(c) $C_6H_5-\overset{O}{\overset{||}{C}}-Cl + H_2NCH_2CH_3 \xrightarrow{-HCl}$ 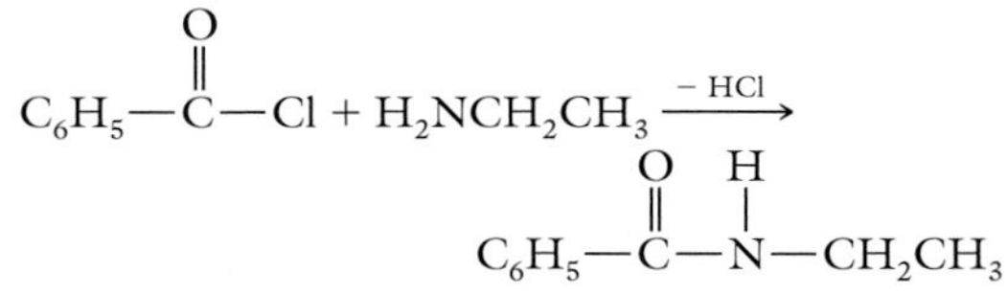
$C_6H_5-\overset{O}{\overset{||}{C}}-\overset{H}{\overset{|}{N}}-CH_2CH_3$

(d) $CH_3-\overset{O}{\overset{||}{C}}-O-\overset{O}{\overset{||}{C}}-CH_3 + HN\overset{CH_3}{\underset{|}{}}-C_6H_5 \longrightarrow$ 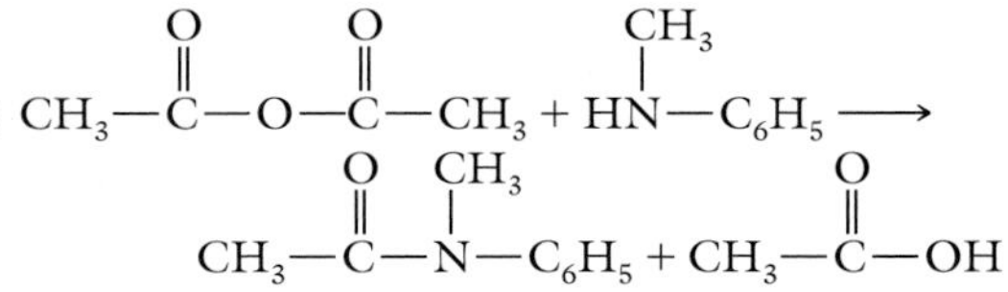
$CH_3-\overset{O}{\overset{||}{C}}-\overset{CH_3}{\underset{|}{N}}-C_6H_5 + CH_3-\overset{O}{\overset{||}{C}}-OH$

16.31

(a) $(CH_3)_3CCH_2CH_2-\overset{O}{\overset{||}{C}}-OH + H_2NCH(CH_3)_2$ 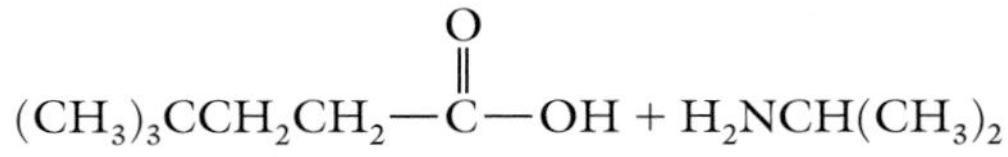

(b) $2\ CH_3CH_2\overset{O}{\overset{||}{C}}-OH + H_2NCH_2CH_2CH_2NH_2$ 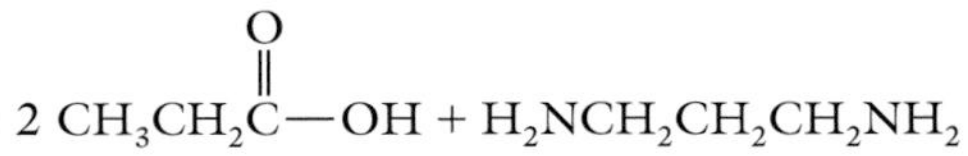

16.33

$HO-\overset{O}{\overset{||}{C}}-CH_2CH_2CH_2CH_2-\overset{O}{\overset{||}{C}}-OH$
$+ H_2NCH_2CH_2CH_2NH_2$

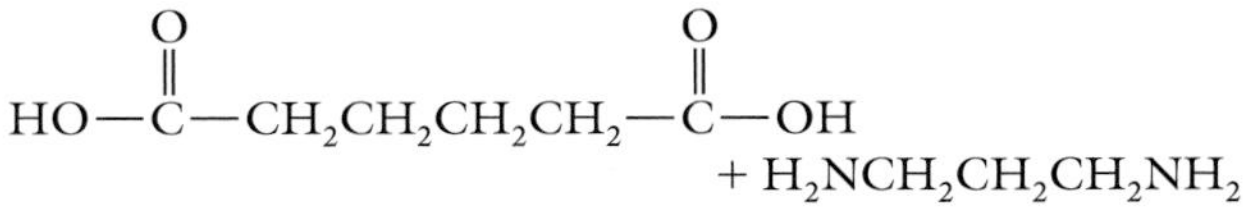

16.35

(a) $CH_3CH_2CH_2-\overset{O}{\overset{||}{C}}-NHC(CH_3)_3$ 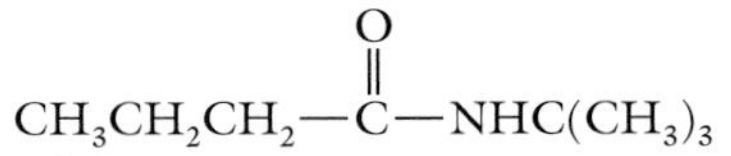

(b) $CH_3CH_2CH_2\overset{CH_3}{\underset{|}{C}}HCH_2-\overset{O}{\overset{||}{C}}-NHC_6H_5$ 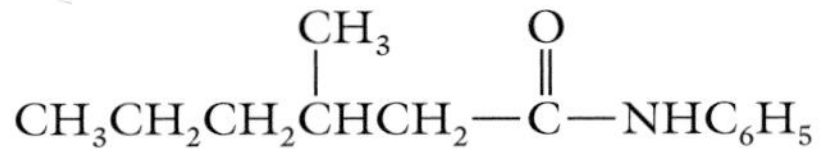

16.37 (a) *N*-Isopropylpentanamide;
(b) *N*-phenylpropanamide; (c) *N,N*-diethylbenzamide

16.39 (a) Pentanamide has the higher boiling point because it possesses hydrogen bonding, whereas *N,N*-dimethylpropanamide does not.

(b) The water solubilities are close for the two compounds because both can hydrogen bond with water. Pentanamide may be slightly more soluble because of its N–H bonds, which allow for more hydrogen-bond possibilities with water.

(c) Butanamide has the higher boiling point because of its dipolar ion structure ($^+N=C-O^-$), in addition to the hydrogen bonding that both compounds possess, which results in the strongest of secondary attractive forces.

(d) The water solubilities are close. Hexanamide may have slightly higher solubility because of the strong interactions of the dipolar structure with water.

(e) Butanamide has the higher boiling point because the secondary forces resulting from its hydrogen bonding and dipolar structure are stronger than the hydrogen-bonding forces in 1-pentanol.

16.41 Water = 1-propanol = ethanamide < propanamine < NaOH. Basicity measures the ability of a compound to bond to a proton. Amines are more basic than water and 1-propanol because nitrogen's nonbonded pair of electrons is more available for bonding to a proton, the result of nitrogen being less electronegative than oxygen. Amides are no more basic than water or alcohols because the carbonyl group, through its electron-withdrawing effect, decreases the availability of nitrogen's nonbonded pair of electrons. NaOH is the strongest base because the actual base is OH^-. The negative charge on oxygen in this anion greatly increases its ability to accept a proton.

16.43

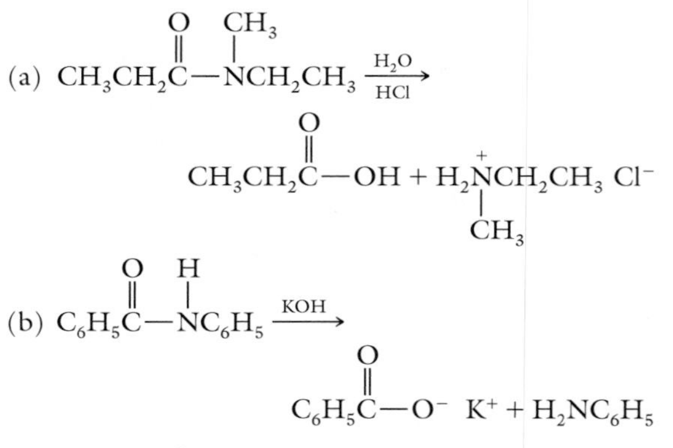

16.45

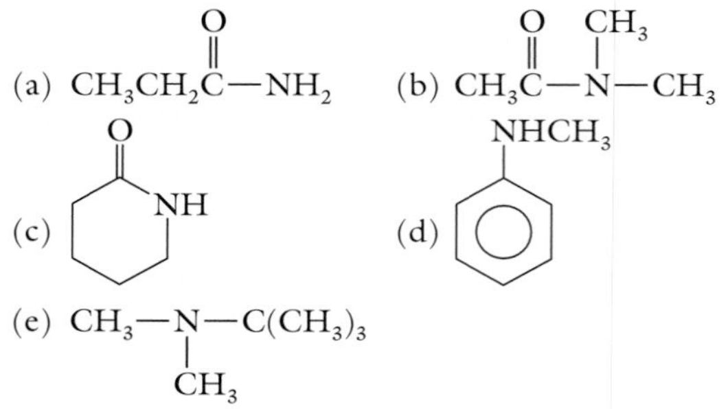

16.47

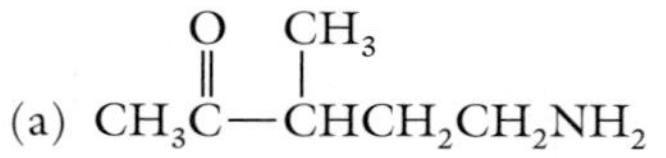

(b) $H_2NCH_2CH_2CH_2CH{=}CH_2$

16.49 (a) a = b = c = sp^3; A = B = 109.5°.
(b) a = b = c = sp^3; A = B = 109.5°.

16.51 (a) Butanamide. Amides possess, in addition to hydrogen bonding, very strong secondary attractive forces because of the dipolar structure of the amide group ($^+N{=}C{=}O^-$).

(b) Butanamide is slightly more water soluble because of the strong interaction of $^+N{=}C{=}O^-$ with water, although both compounds can hydrogen bond with water.

(c) Pentanamide. Amides possess the strongest of all secondary attractive forces because of the dipolar structure of the amide group ($^+N{=}C{=}O^-$), although both compounds possess hydrogen bonding.

(d) Pentanamide is slightly more water soluble because of the strong interaction of $^+N{=}C{=}O^-$ with water, although both compounds can hydrogen bond with water.

(e) Trimethylammonium chloride has the higher boiling point because it is a salt whose ionic attractive forces are stronger than those of a covalent compound.

(f) Sodium butanoate has the higher boiling point because it is a salt whose ionic attractive forces are stronger than those of a covalent compound.

(g) The two compounds have the same solubility because there is one NH_2 group per four carbons in each compound.

16.53 Add a drop of the unknown to red litmus paper. If there is a change from red to blue, the unknown is an amine. If there is no color change, the unknown is an amide.

16.55 Propanol or ethanamide

16.57 Polymer is formed only when both reactants are bifunctional. Only c fulfills this requirement:

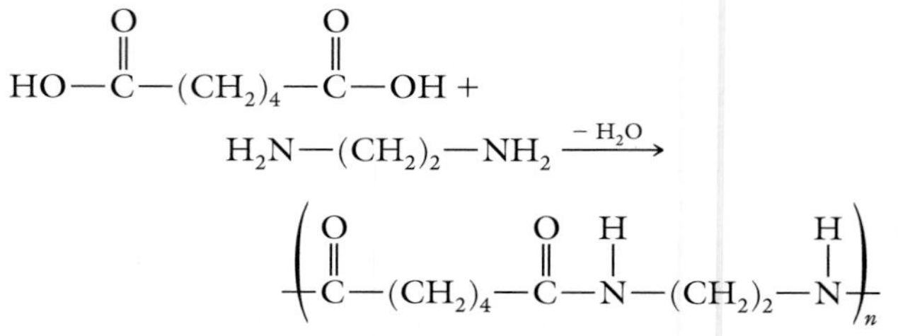

16.59 (a) 4; (b) 2; (c) 2; (d) 2

16.61 $[H_3O^+] = 2.6 \times 10^{-13}$; pH = 12.6

16.63 Propanoic acid

16.65 Ammonium chloride is not volatile, because it is an ionic compound.

Chapter 17

17.1 $CH_3CH_2CH_2CH_3$; $(CH_3)_3CH$

17.3 $CH_3CH_2OCH_2CH_3$; $CH_3OCH_2CH_2CH_3$; $CH_3OCH(CH_3)_2$

17.5

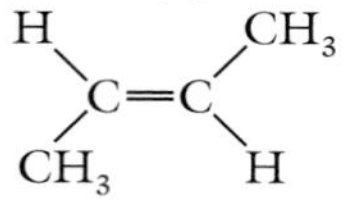

17.7 (a) Chiral; (b) chiral; (c) achiral; (d) chiral

17.9 Structures a, b, and d cannot exist as enantiomers; c and e can.

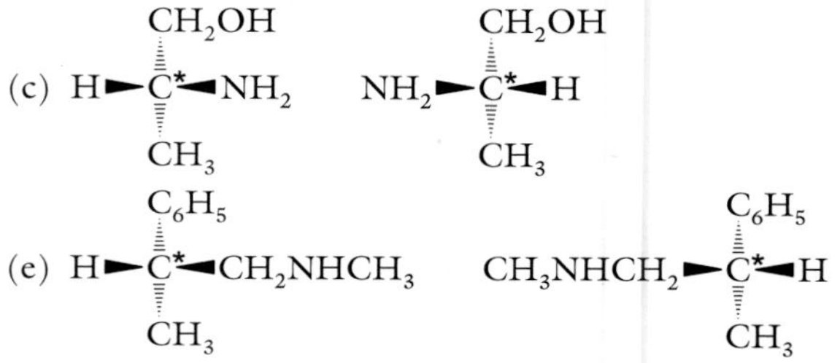

17.11 (a) 3; (b) 1; (c) 1; (d) 3; (e) 1

17.13 (a) D; (b) L; (c) D; (d) L

17.15 Only b and c rotate plane-polarized light. (+)-Phenylalanine rotates in the clockwise direction. The direction for D-glucose is not known from the D configuration; it must be experimentally determined.

17.17 +31.5°

17.19

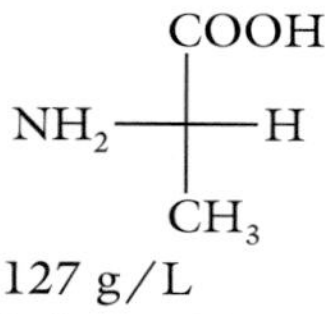

17.21 127 g/L

17.23 $[\alpha] = -90°$

17.25 $\alpha = +20.1°$

17.27 (a) One pair of enantiomers, both of which are optically active.

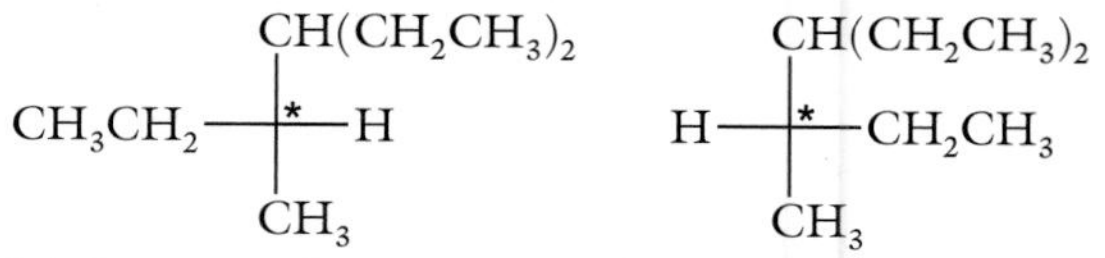

(b) One pair of enantiomers (1 and 2) and one meso compound (3). Structures 1 and 2 are diastereomers of 3, and vice versa. Structures 1 and 2 are optically active; 3 is not.

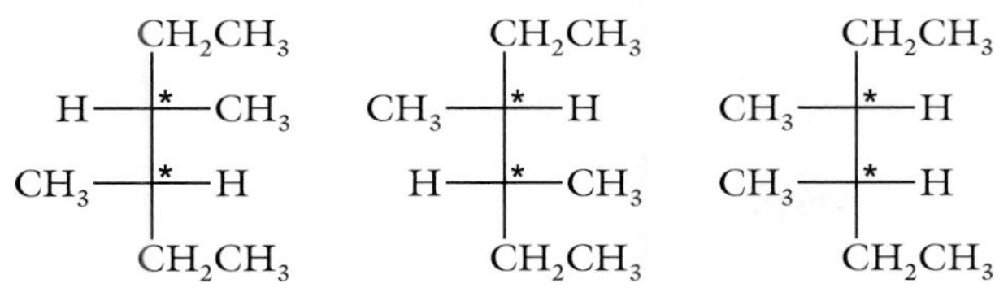

1 2 3

(c) Two pairs of enantiomers (5 and 6, 7 and 8). Enantiomers 5 and 6 are diastereomers of 7 and 8, and vice versa. All stereoisomers are optically active.

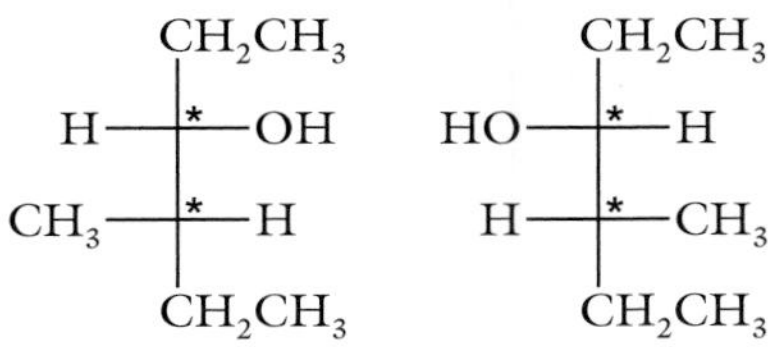

5 6

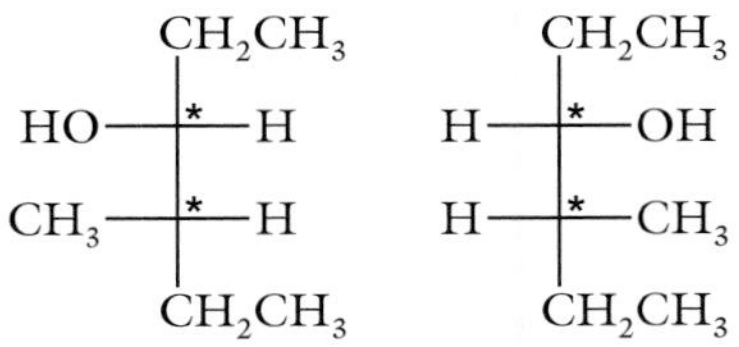

7 8

17.29 Stereoisomer 3 is the enantiomer of 1; 4 is the enantiomer of 2. Stereoisomer 3 has a specific rotation of − 15° and a boiling point of 180–182°C. Stereoisomer 4 has a specific rotation of − 26° and a boiling point of 163–165°C.

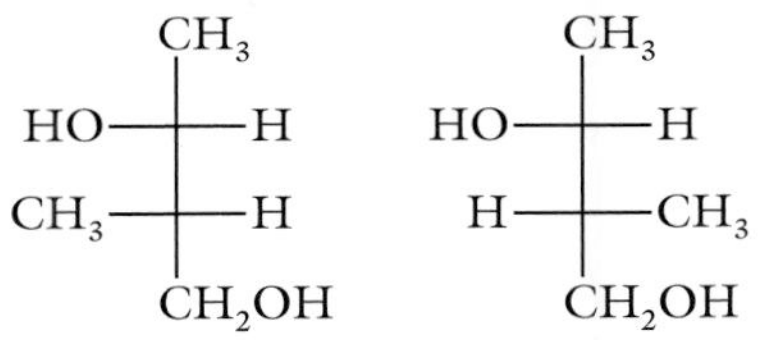

3 4

17.31 (a) 16 stereoisomers

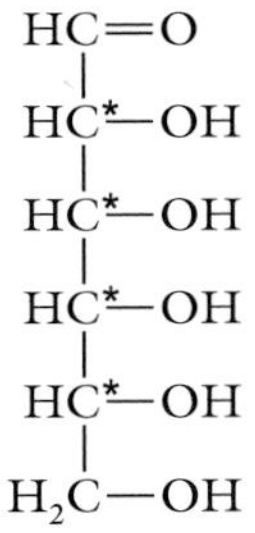

(b) 8 stereoisomers

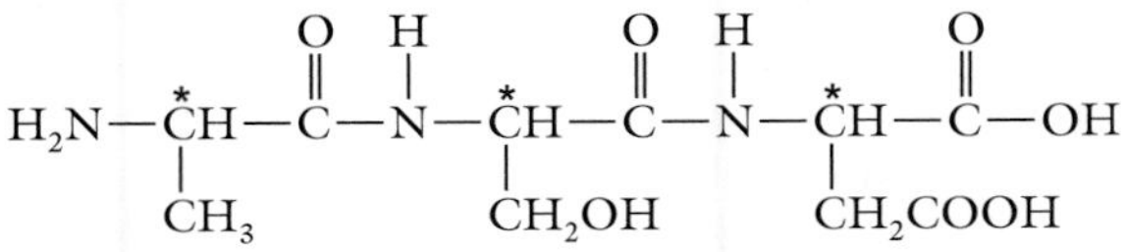

17.33 (a) One pair of optically active enantiomers.

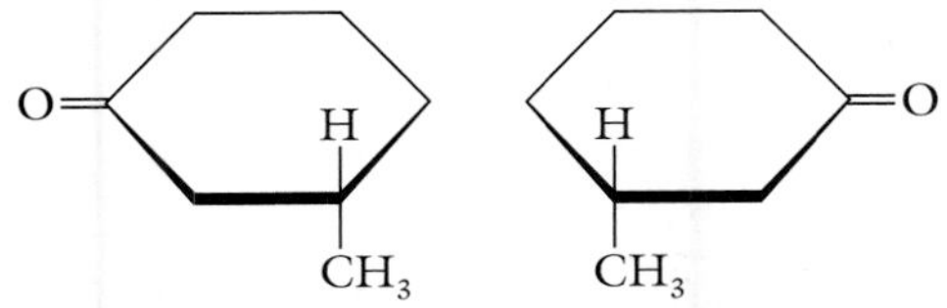

(b) No stereoisomers. (c) No stereoisomers. (d) One pair of trans enantiomers (1 and 2) and one pair of cis enantiomers (3 and 4). All stereoisomers are optically active. Stereoisomers 1 and 2 are each diastereomers of 3 and 4, and vice versa.

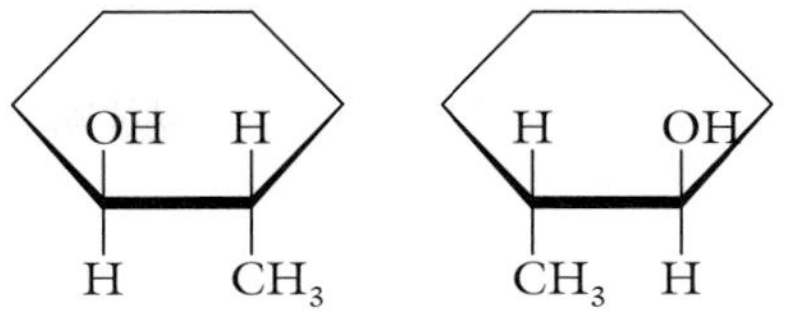

1 2

Trans pair of enantiomers

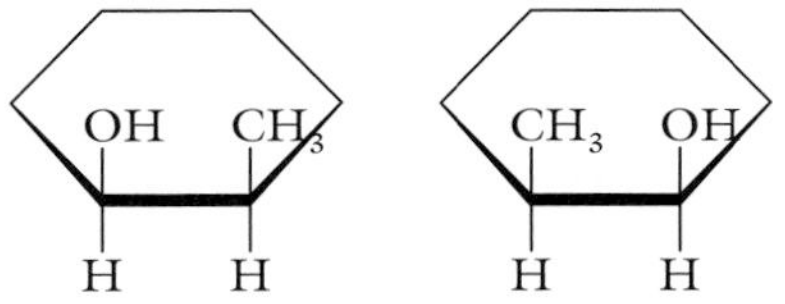

3 4

Cis pair of enantiomers

(e) One pair of optically active enantiomers (5 and 6) and one optically inactive meso compound (7). Stereoisomers 5 and 6 are diastereomers of 7, and vice versa.

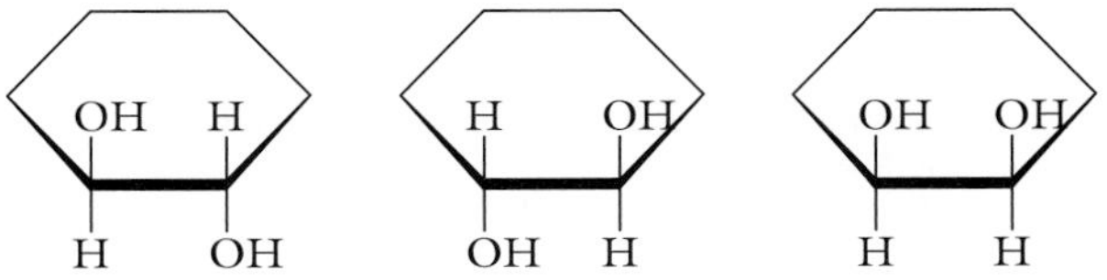

5 6 7

Trans pair of enantiomers **Cis meso**

17.35 (a) Four tetrahedral stereocenters and 16 stereoisomers.

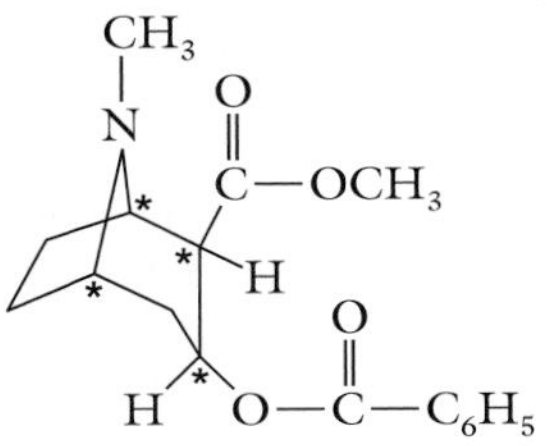

(b) One tetrahedral stereocenter and two stereoisomers.

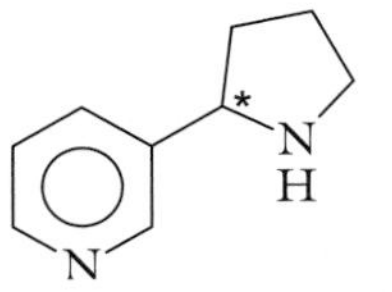

(c) Six tetrahedral stereocenters and 64 stereoisomers.

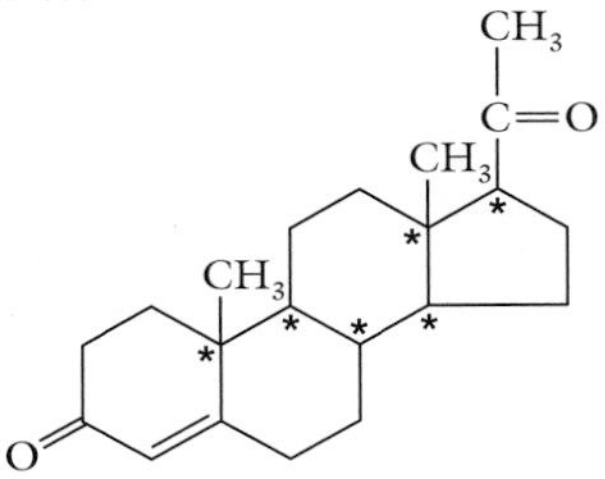

17.37 (a) 2; (b) 1; (c) 3; (d) 2; (e) 1; (f) 4; (g) 1; (h) 3; (i) 1; (j) 1; (k) 4; (l) 3; (m) 2; (n) 4; (o) 5

17.39 One pair of cis enantiomers and one pair of trans enantiomers.

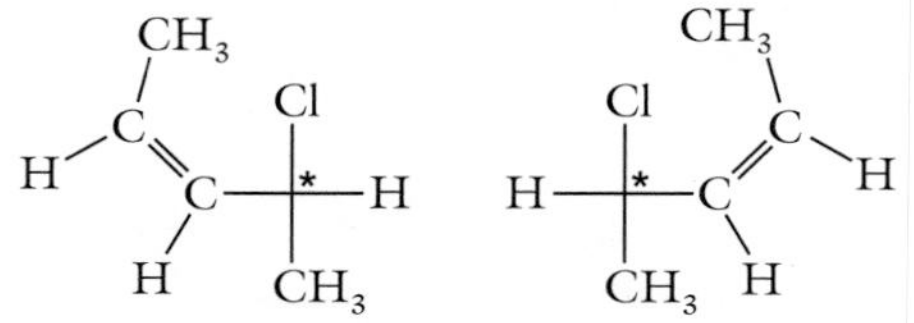

Cis pair of enantiomers

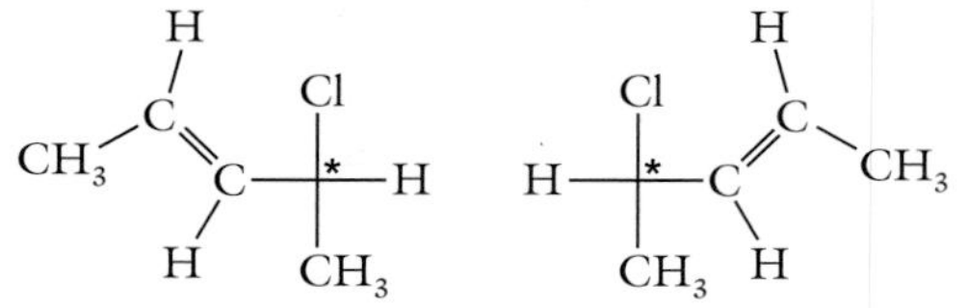

Trans pair of enantiomers

17.41 The product is the racemic mixture of the two enantiomers of 2-chlorobutane. There is equal probability of formation of the two enantiomers by addition of Cl to C2 from either side of the double bond.

17.43 Chiral recognition is not responsible, because neither isomer is chiral. Biological discrimination based on geometric isomerism is responsible.

Chapter 18

18.1 (a) An aldose is a polyhydroxyaldehyde.
(b) A hexose is a six-carbon polyhydroxyaldehyde or polyhydroxyketone.
(c) A ketopentose is a five-carbon polyhydroxyketone.
(d) An aldotetrose is a four-carbon polyhydroxyaldehyde.
(e) All saccharides in the D family have the D configuration at the tetrahedral stereocenter farthest from the carbonyl carbon—the —OH group is on the right side of this carbon atom in the Fischer projection.

18.3 Eight aldopentoses; four L-aldopentoses

18.5 Aldohexoses have four tetrahedral stereocenters and 16 stereoisomers of which 8 are D-aldohexoses; ketohexoses have three tetrahedral stereocenters and 8 stereoisomers, of which 4 are D-ketohexoses.

18.7 (a) 1; (b) 3; (c) 2; (d) 2; (e) 4

18.9 Haworth projections are used to represent cyclic structures.

18.11 A pyranose ring is a six-membered cyclic hemiacetal ring.

18.13 In α-D-glucose, the —OH group at C1 and the $-CH_2OH$ group at C5 have a trans relation; in β-D-glucose, the two groups have a cis relation.

18.15

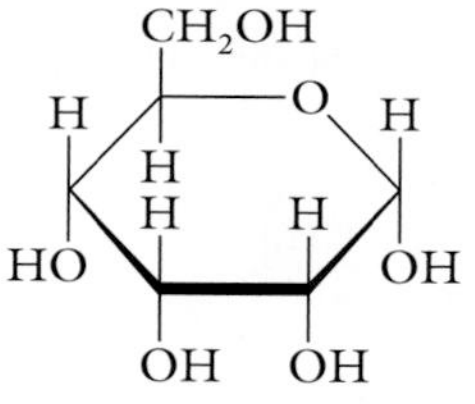

18.17

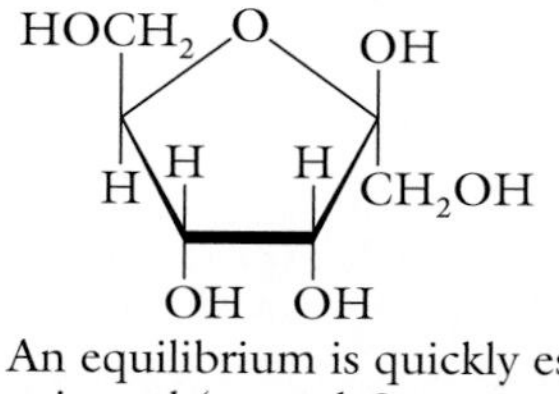

18.19 An equilibrium is quickly established between the cyclic hemiacetal (α- and β-anomers) and the open-chain aldehyde structures. It is the aldehyde structure that gives the positive Benedict's test. As reaction proceeds, the hemiacetals open up to become the aldehyde, and eventually all of the D-galactose reacts with Benedict's reagent. A positive test is indicated by the disappearance of the blue color of Benedict's reagent and the formation of a red precipitate of Cu_2O.

18.21 (a) 1; (b) 3; (c) 4; (d) 2

18.23

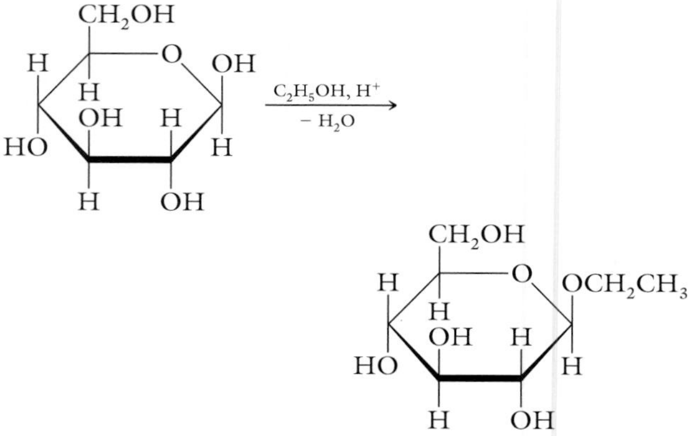

18.25

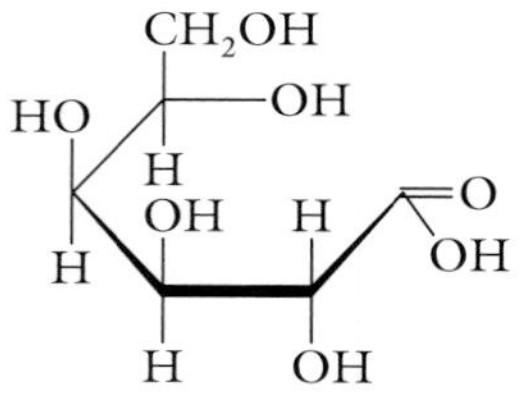

18.27 All except b are reducing sugars and undergo mutarotation.

18.29 One aldohexose is linked through the α-anomeric position (C1) of its hemiacetal to the C4 position of the other aldohexose.

18.31 (a) Glucose; (b) glucose, galactose; (c) glucose, fructose; (d) cellobiose cannot be digested by humans

18.33 Figure 18.8 shows the structure of sucrose. Sucrose does not have a hemiacetal group.

18.35

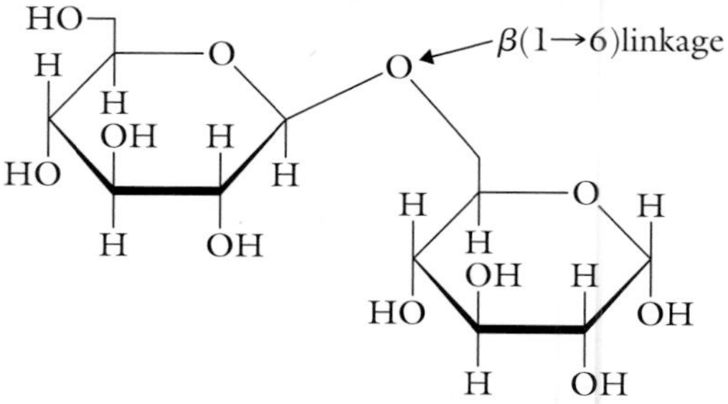

Gentiobiose is a reducing sugar and undergoes mutarotation.

18.37 Benedict's test is negative; the unknown is sucrose. Sucrose does not give a positive Benedict's test, but lactose does.

18.39 Both are polymers of D-glucose linked together through $\alpha(1 \rightarrow 4)$-glycosidic linkages, but, in addition, amylopectin branches repeatedly through $\alpha(1 \rightarrow 6)$-glycosidic linkages. Starch, a mixture of amylose and amylopectin, is the storage form of D-glucose in plants. Starch is a nutritional carbohydrate for animals.

18.41 Digestion of amylose, amylopectin, and glycogen by humans yields D-glucose; humans cannot digest cellulose. Digestion of amylose, amylopectin, and cellulose by grazing animals yields D-glucose. Because they are herbivores (plant eaters) and glycogen is present only in animals, grazing animals do not ingest glycogen.

18.43 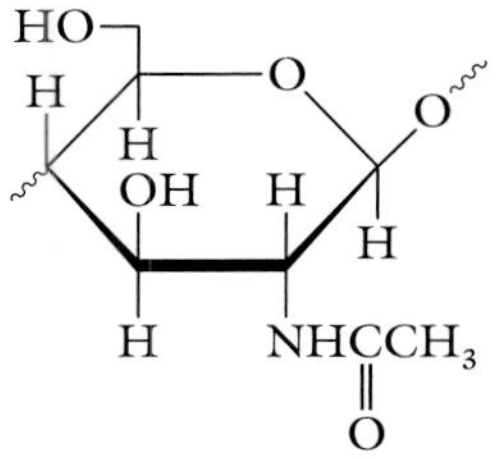

18.45 Plants are the immediate source; photosynthesis is the ultimate source.

18.47 A carbohydrate is a polyhydroxyaldehyde or polyhydroxyketone.

18.49 Monosaccharides and disaccharides only

18.51 The —OH group on the tetrahedral stereocenter (C5) farthest from the carbonyl group is on the right side in the Fischer projection. The $-CH_2OH$ group at C5 extends upward from the plane of the ring in the Haworth projection.

18.53 (a) Disaccharide; (b) monosaccharide; (c) disaccharide; (d) polysaccharide; (e) polysaccharide; (f) monosaccharide; (g) polysaccharide

18.55 (a) 3; (b) 1; (c) 2; (d) 1; (e) 4; (f) 4; (g) 1

18.57 Storage polysaccharides are the storage forms of molecules used to generate energy or synthesize other needed molecules. Structural polysaccharides are used to construct plant-cell walls and the structures responsible for the macroscopic shape of a plant.

18.59 Grazing animals digest cellulose because microorganisms in their digestive tracts possess cellulase and cellobiase, the enzymes for cleaving the $\beta(1 \rightarrow 4)$-glycosidic linkages of cellulose. Grazing animals themselves possess amylase and maltase, the enzymes for cleaving the $\alpha(1 \rightarrow 4)$-glycosidic linkages of starch.

18.61 Tetrasaccharide

18.63 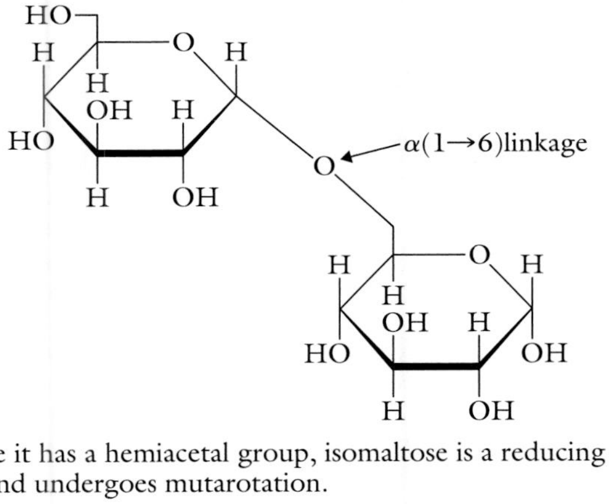

Because it has a hemiacetal group, isomaltose is a reducing sugar and undergoes mutarotation.

18.65

$$6\ CO_2 + 6\ H_2O + \text{sunlight} \xrightarrow{\text{chlorophyll}} (CH_2O)_6 + 6\ O_2$$

18.67 Plants produce oxygen as a by-product of photosynthesis. Plants and animals use oxygen in metabolism to generate energy.

18.69 The glucose residue, $C_6H_{10}O_5$, in amylose and cellulose has a molecular mass of 162 amu. The number of residues is the polysaccharide molecular mass divided by the residue molecular mass. There are 1000 residues for both the amylose and the cellulose of molecular mass 162,000 amu.

18.71 This is an example of chiral recognition whereby a stereospecific enzyme allows the two monosaccharides to link together only in the specified manner—α-D-glucose and β-D-fructose.

18.73 The abdominal distention and cramping, nausea, pain, and diarrhea of lactose intolerance can be minimized by taking lactase with ingestion of milk and milk products and limiting the amount of such products. There are no long-term irreversible effects of lactose intolerance. Galactosemia has long-term adverse effects, including mental retardation, impaired liver function, cataracts, and even death, if not detected early in infancy. The only treatment is a lactose-free diet.

Chapter 19

19.1 1

19.3 Saturated fatty acids have no C=C double bonds. Unsaturated fatty acids contain cis C=C double bonds; monosaturated contain one double bond, polyunsaturated contain more than one.

19.5 Compound 2 has the higher melting point. Molecules of compound 2 pack tightly together because of their extended saturated chains, which results in stronger secondary attractive forces. Molecules of compound 1 do not pack tightly and have weaker secondary attractive forces, because the cis double bonds cause the chains to kink.

19.7 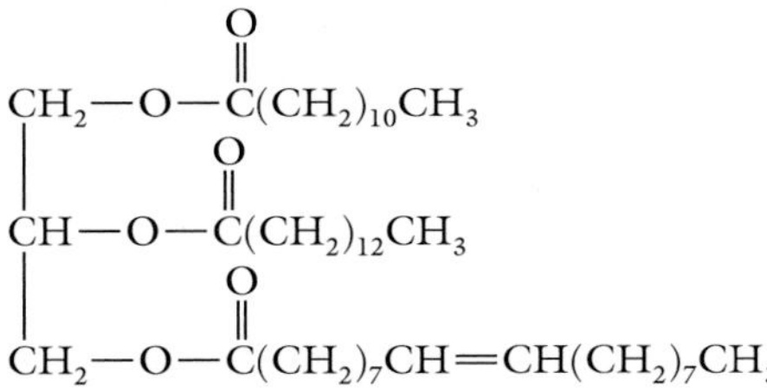

19.9 a > c > b. Melting point decreases with increasing unsaturated fatty acid content.

19.11

(a) 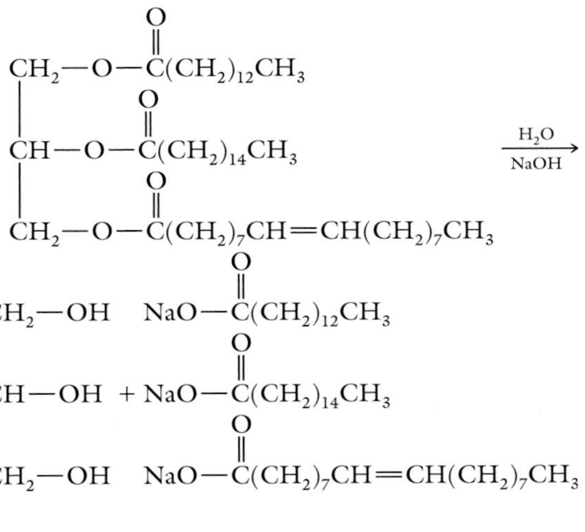

(b)

$$CH_2-O-\overset{\overset{O}{\|}}{C}(CH_2)_{12}CH_3$$
$$CH-O-\overset{\overset{O}{\|}}{C}(CH_2)_{14}CH_3$$
$$CH_2-O-\overset{\overset{O}{\|}}{C}(CH_2)_7CH=CH(CH_2)_7CH_3$$

$\xrightarrow[\text{Pt}]{\text{H}_2}$

$$CH_2-O-\overset{\overset{O}{\|}}{C}(CH_2)_{12}CH_3$$
$$CH-O-\overset{\overset{O}{\|}}{C}(CH_2)_{14}CH_3$$
$$CH_2-O-\overset{\overset{O}{\|}}{C}(CH_2)_{16}CH_3$$

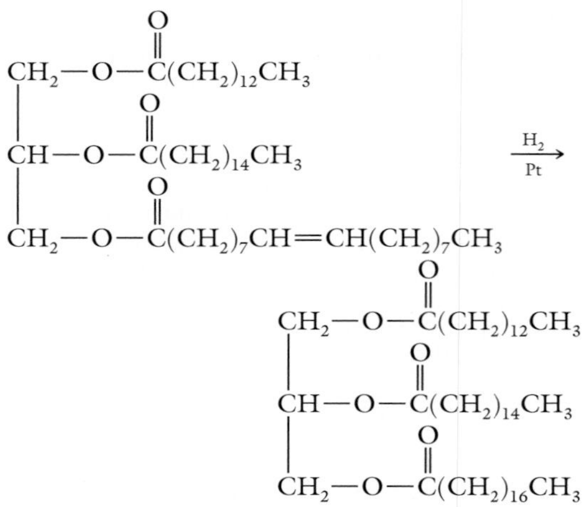

19.13 Butanoic acid is one of the hydrolysis products. Its low molecular mass makes it volatile, and this volatility is responsible for the foul odor.

$$CH_2-O-\overset{\overset{O}{\|}}{C}(CH_2)_2CH_3$$
$$CH-O-\overset{\overset{O}{\|}}{C}(CH_2)_{14}CH_3$$
$$CH_2-O-\overset{\overset{O}{\|}}{C}(CH_2)_7CH=CH(CH_2)_5CH_3$$

$\xrightarrow{\text{H}_2\text{O}}$

$$CH_2-OH \quad HO-\overset{\overset{O}{\|}}{C}(CH_2)_2CH_3$$
$$CH-OH \; + \; HO-\overset{\overset{O}{\|}}{C}(CH_2)_{14}CH_3$$
$$CH_2-OH \quad HO-\overset{\overset{O}{\|}}{C}(CH_2)_7CH=CH(CH_2)_5CH_3$$

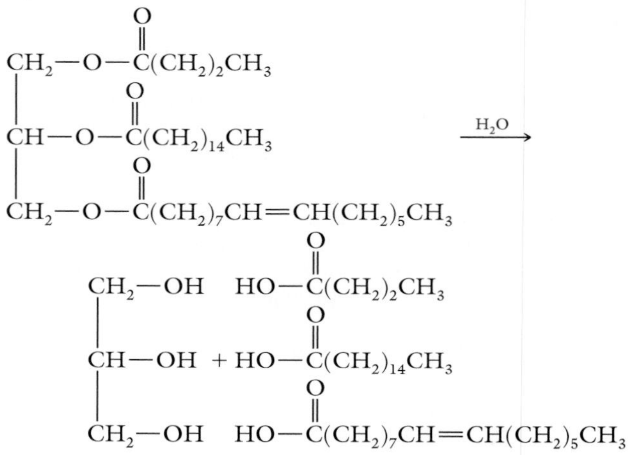

19.15 Reduction of C=C double bonds with H_2 in the presence of a metal catalyst such as Ni or Pt

19.17 2

19.19 Like triacylglycerols, glycerophospholipids have fatty acid ester groups at two of the carbons of glycerol; but, at the third carbon, a glycerophospholipid has a phosphodiester group.

19.21

$$CH_2-O-\overset{\overset{O}{\|}}{C}(CH_2)_{12}CH_3$$
$$CH-O-\overset{\overset{O}{\|}}{C}(CH_2)_7CH=CH(CH_2)_7CH_3$$
$$CH_2-O-\overset{\underset{O^-}{\overset{O}{\|}}}{P}-OCH_2\overset{\overset{+}{NH_3}}{\underset{COO^-}{CH}}$$

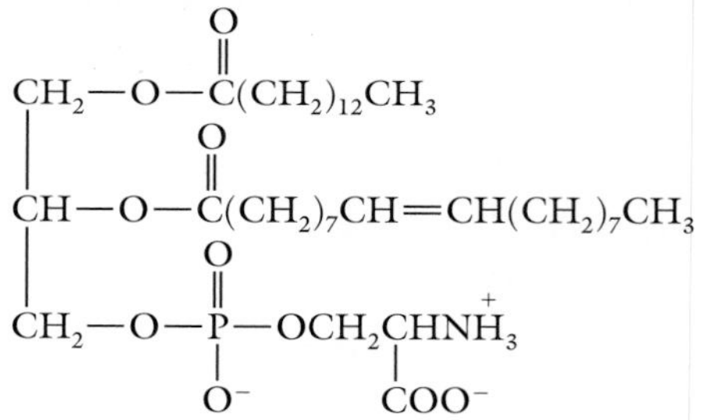

19.23

$$CH_2-OH \quad HO-\overset{\overset{O}{\|}}{C}(CH_2)_{14}CH_3$$
$$CH-OH \quad HO-\overset{\overset{O}{\|}}{C}(CH_2)_{16}CH_3$$
$$CH_2-OH \quad HO-\overset{\underset{O^-}{\overset{O}{\|}}}{P}-OH \quad HOCH_2CH_2\overset{+}{N}(CH_3)_3$$

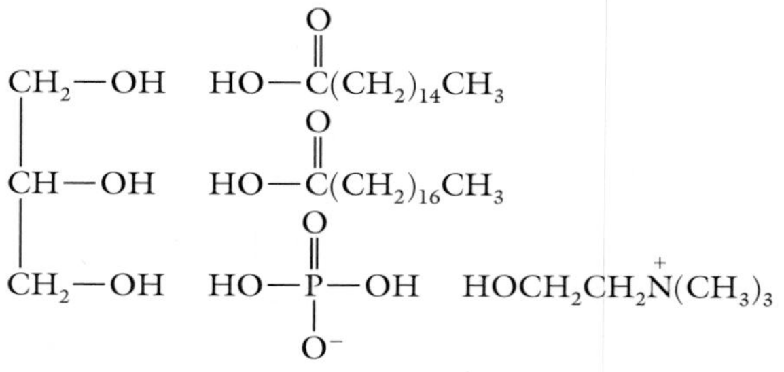

19.25

$$HO-CH-CH=CH(CH_2)_{12}CH_3$$
$$CH-\overset{\overset{H}{|}}{N}-\overset{\overset{O}{\|}}{C}(CH_2)_6(CH_2CH=CH)_2(CH_2)_4CH_3$$
$$CH_2-O-\overset{\underset{O^-}{\overset{O}{\|}}}{P}-OCH_2CH_2\overset{+}{N}(CH_3)_3$$

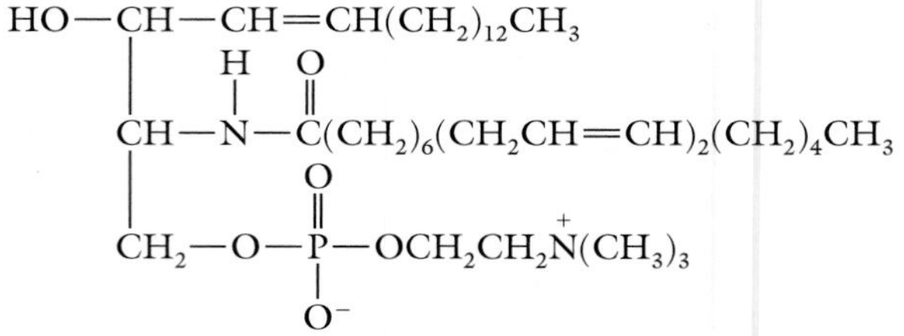

19.27

$$HO-CH-CH=CH(CH_2)_{12}CH_3$$
$$CH-NH_2 \quad HO-\overset{\overset{O}{\|}}{C}(CH_2)_7CH=CH(CH_2)_7CH_3$$
$$CH_2-OH \quad HO-\overset{\underset{OH}{\overset{O}{\|}}}{P}-OH \quad HOCH_2CH_2\overset{+}{N}(CH_3)_3$$

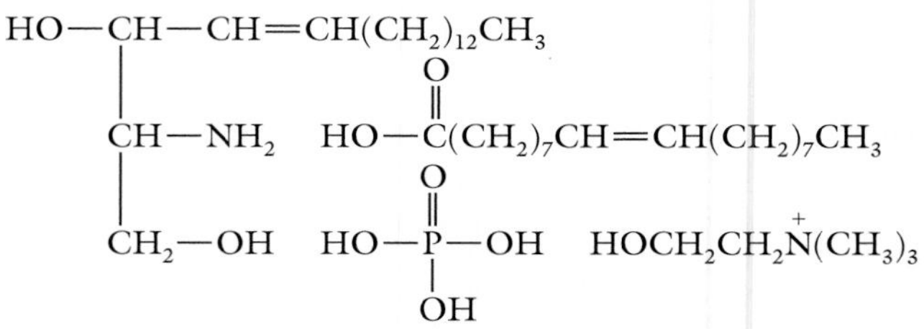

19.29 There is no functional group such as ester or amide whose hydrolysis cleaves the lipid into smaller molecules.

19.31 $2^6 = 64$ stereoisomers

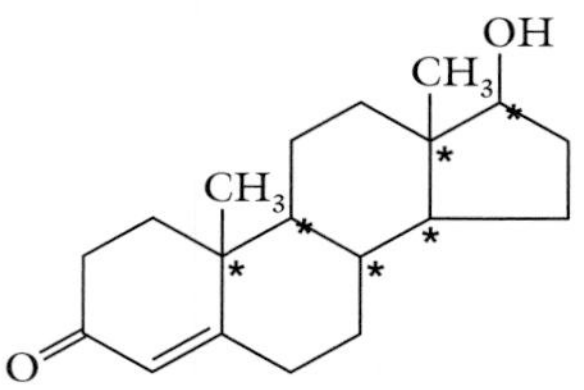

19.33 The sex hormones regulate the development of the sex organs, production of sperm and ova, and development of secondary sex characteristics: lack of facial hair, increased breast size, and high voice in women; facial hair, increased musculature, and deep voice in men.

19.35 Cholesterol

19.37 There is a carboxylate salt group attached to the steroid ring structure.

19.39 Hydrolysis of the amide group does not cleave a large part of the molecule; only a small part is cleaved.

19.41 In leukotrienes, the 20-carbon chain of arachidonic acid and its carboxyl group are intact. Prostaglandins are similar to leukotrienes but have a cyclopentane ring formed by bond formation between C8 and C12 of the 20-carbon chain.

19.43 Hormones are substances synthesized in and secreted by endocrine glands and then transported in the bloodstream to target tissues, where they regulate the functions of cells. They are effective at very low concentrations. Eicosanoids are called local hormones because they are not transported in the bloodstream from their sites of synthesis to their sites of action. A local hormone acts in the same tissue in which it is synthesized.

19.45 A vitamin is an organic compound required in trace amounts for normal metabolism, but it is not synthesized by the organism that requires it. Vitamins must be included in the diet.

19.47 Fat-soluble vitamins are fat soluble and water insoluble; water-soluble vitamins are water soluble and fat insoluble.

19.49 See Table 19.4.

19.51 Membranes surround all cells and all organelles.

19.53 Amphipathic lipids (glycerophospholipids, sphingolipids, cholesterol) form the membrane by assembly into the lipid bilayer.

19.55 Unsaturated fatty acid chains are kinked because of the cis double bonds; they do not pack tightly in the lipid bilayer, secondary attractive forces are weaker, and flexibility is increased.

19.57 Hydrophilic molecules and ions are repelled by the hydrophobic regions of the lipid bilayer.

19.59 Energy is required to transport solute against a concentration gradient, from the low-concentration side to the high-concentration side.

19.61 Active transport

19.63 CH_3OH

19.65 (a) No ω number, because there is no C=C; (b) ω-7; (c) ω-6

19.67 Waxes have a variety of protective functions in plants and animals. They protect against parasites and mechanical damage, prevent excessive water loss, and waterproof waterfowl.

19.69 Digestion does not proceed with cleavage of all hydrolyzable groups. A mixture of products is obtained; for example, triacylglycerols yield mostly monoacylglycerol and fatty acids. Hydrolysis with base forms the carboxylate salts of fatty acids instead of the fatty acids.

19.71 a, c, f, g

19.73 a, c, f, g

19.75

$$
\begin{array}{l}
CH_2-O-\overset{O}{\overset{\|}{C}}(CH_2)_{12}CH_3 \\
CH-O-\overset{O}{\overset{\|}{C}}(CH_2)_{14}CH_3 \\
CH_2-O-\overset{O}{\overset{\|}{C}}(CH_2)_{16}CH_3
\end{array}
\qquad
\begin{array}{l}
CH_2-O-\overset{O}{\overset{\|}{C}}(CH_2)_{12}CH_3 \\
CH-O-\overset{O}{\overset{\|}{C}}(CH_2)_{16}CH_3 \\
CH_2-O-\overset{O}{\overset{\|}{C}}(CH_2)_{14}CH_3
\end{array}
$$

$$
\begin{array}{l}
CH_2-O-\overset{O}{\overset{\|}{C}}(CH_2)_{14}CH_3 \\
CH-O-\overset{O}{\overset{\|}{C}}(CH_2)_{12}CH_3 \\
CH_2-O-\overset{O}{\overset{\|}{C}}(CH_2)_{16}CH_3
\end{array}
$$

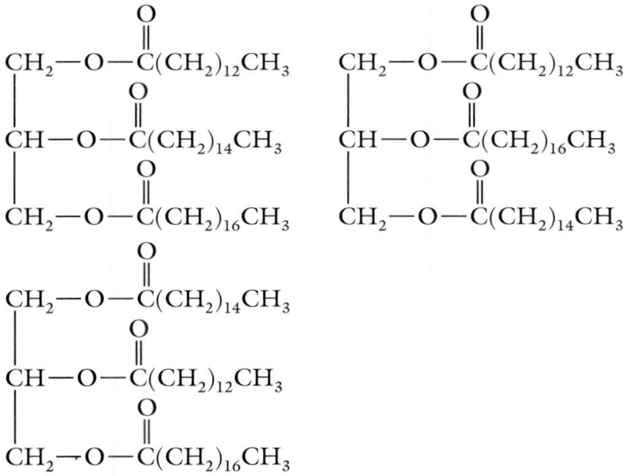

19.77 The unsaturated fatty acid component (oleic) of triacylglycerol B yields a carboxylic acid on air oxidation. The carboxylic acid is malodorous because it is sufficiently low in molecular mass to be volatile.

19.79 $2^6 = 64$ stereoisomers

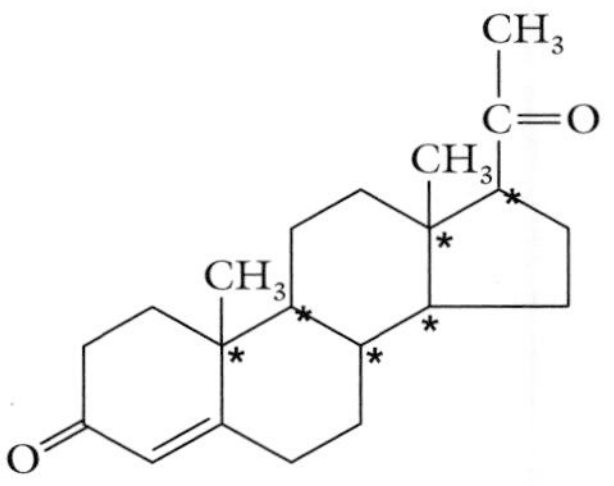

19.81 Triacylglycerols are completely hydrophobic; there is no hydrophilic part.

19.83 Active transport

19.85 Melting point is increased with increasing saturated fatty acid composition; it is decreased by a decreasing number of carbons in the fatty acid components. Coconut oil has a high saturated fatty acid composition, but a high percentage of the saturated fatty acids have fewer than 14 carbons. The lower molecular mass lowers the melting point sufficiently that coconut triacylglycerols are liquids at ambient temperatures.

19.87

$$CH_3(CH_2)_4(CH=CHCH_2)_2(CH_2)_6\overset{O}{\overset{\|}{C}}-O-$$

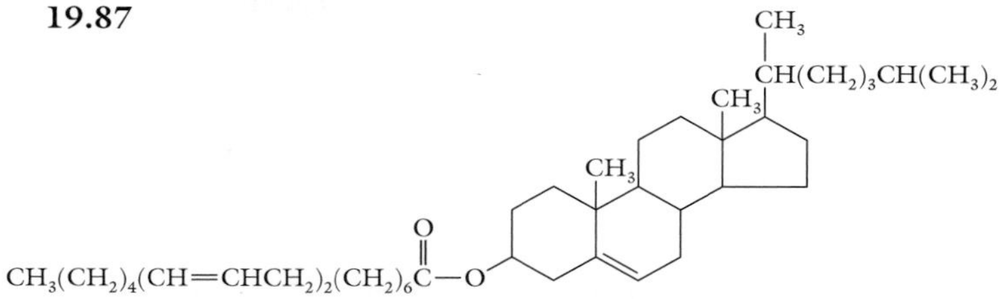

19.89 NaOH hydrolyzes the triacylglycerols (fats) to glycerol and sodium carboxylate salts, which are much more soluble in water than are the triacylglycerols.

19.91 Reverse bilayers are used. Hydrophilic heads are in the middle of the membrane; nonpolar tails are at the two surfaces of the membrane and in contact with the nonpolar heptane.

Chapter 20

20.1 1, β; 2, α; 3, γ

20.3 (a) Gly; (b) Ser, Thr; (c) Phe, Tyr, Trp; (d) Ile, Thr; (e) Trp, His; (f) Met, Cys; (g) Asp, Glu; (h) Pro

20.5 An acid-base reaction takes place between the carboxyl and amino groups.

20.7 (a)

$$
\overset{CH_3}{\underset{}{{}^+NH_3-CH-COOH}} \qquad \overset{CH_3}{\underset{}{{}^+NH_3-CH-COO^-}}
$$
$$\text{pH} < 1 \qquad\qquad \text{pH} = \text{pI and } 7$$

$$\overset{CH_3}{\underset{}{NH_2-CH-COO^-}}$$
$$\text{pH} > 12$$

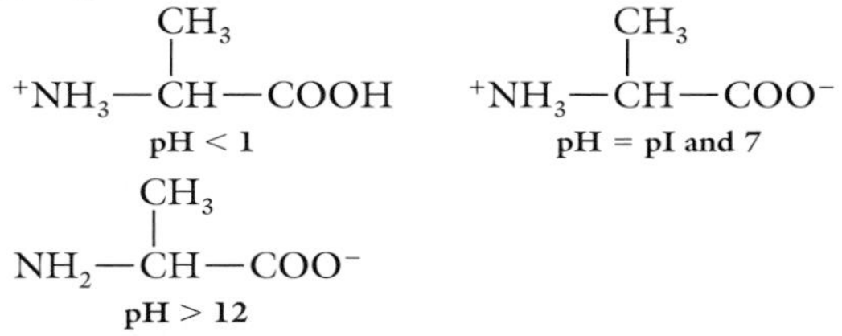

(b) Solubility is minimum at pI (6.01), increasing at both higher and lower pH values.

(c) No migration at pI and physiological pH; migration to anode at pH > 12, to cathode at pH < 1.

20.9 Serylmethionine

$$
\overset{+}{H_3}N-CH-\overset{O}{\overset{\|}{C}}-\overset{H}{\overset{|}{N}}-CH-\overset{O}{\overset{\|}{C}}-O^-
$$
$$
\underset{CH_2OH}{} \qquad\qquad \underset{CH_2CH_2SCH_3}{}
$$

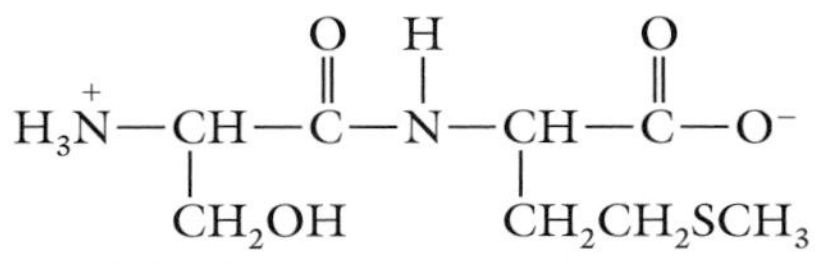

20.11 Threonylalanylaspartic acid

$$
\overset{+}{H_3}N-CH-\overset{O}{\overset{\|}{C}}-\overset{H}{\overset{|}{N}}-CH-\overset{O}{\overset{\|}{C}}-\overset{H}{\overset{|}{N}}-CH-\overset{O}{\overset{\|}{C}}-O^-
$$
$$
\underset{\underset{OH}{HCCH_3}}{} \qquad \underset{CH_3}{} \qquad\qquad \underset{CH_2COO^-}{}
$$

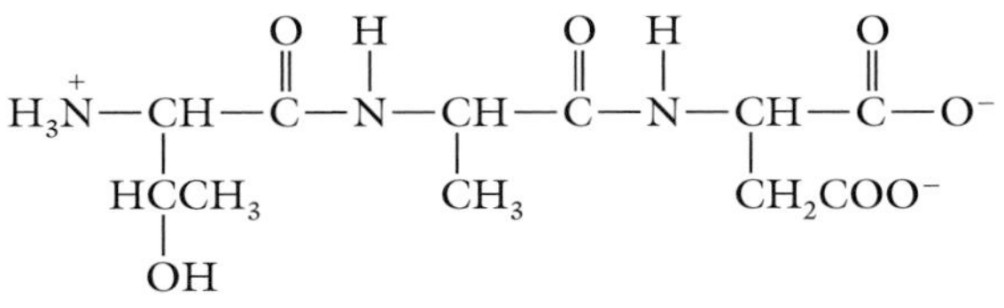

20.13 Glycylprolylalanine

$$
\overset{+}{H_3}N-CH_2-\overset{O}{\overset{\|}{C}}\diagdown_{N}\diagup\overset{O}{\overset{\|}{C}}-\overset{H}{\overset{|}{N}}-CH-\overset{O}{\overset{\|}{C}}-O^-
$$
$$\underset{CH_3}{}$$

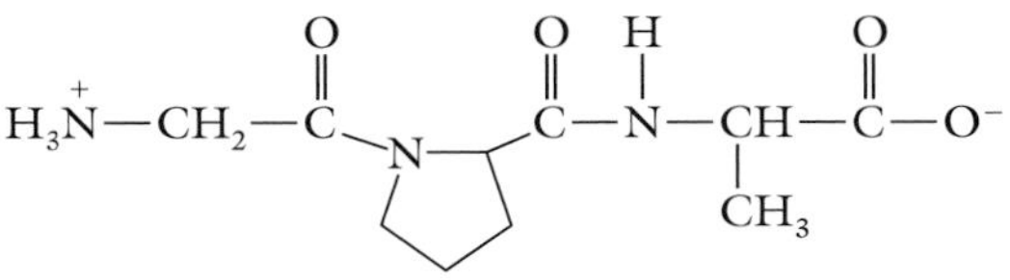

20.15 Six constitutional isomers: Thr-Cys-Leu, Thr-Leu-Cys, Cys-Leu-Thr, Cys-Thr-Leu, Leu-Cys-Thr, Leu-Thr-Cys.

20.17 L-Lys-L-Ala-L-Glu, D-Lys-D-Ala-D-Glu, L-Lys-L-Ala-D-Glu, D-Lys-D-Ala-L-Glu, L-Lys-D-Ala-D-Glu, D-Lys-L-Ala-L-Glu. The stereoisomer with all L residues would be found in nature.

20.19 The peptide has a charge of 2− and migrates to the anode.

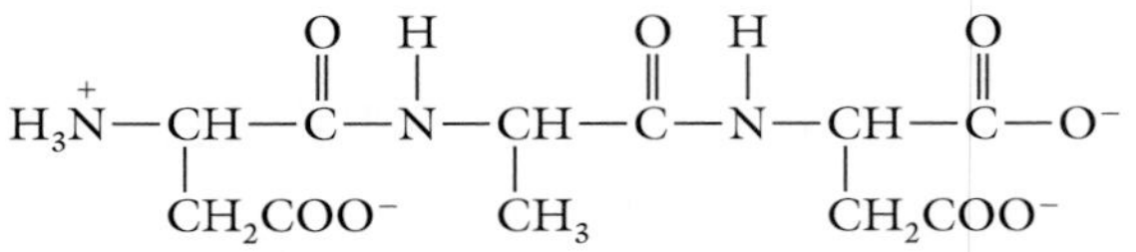

20.21 The products are the same—the individual amino acids (Phe, Asp, Lys, and Gly)—in digestion and both acidic and basic hydrolysis. There is a difference between acidic and basic hydrolysis. In acidic hydrolysis, amino groups are charged but carboxyl groups are uncharged. In basic hydrolysis, carboxyl groups are charged but amino groups are uncharged.

20.23 Hydrolysis yields the overall composition of the tripeptide but not the amino acid sequence. There are six possible sequences with the same composition: Ala-Gly-Lys, Ala-Lys-Gly, Gly-Ala-Lys, Gly-Lys-Ala, Lys-Gly-Ala, Lys-Ala-Gly.

20.25 (a)

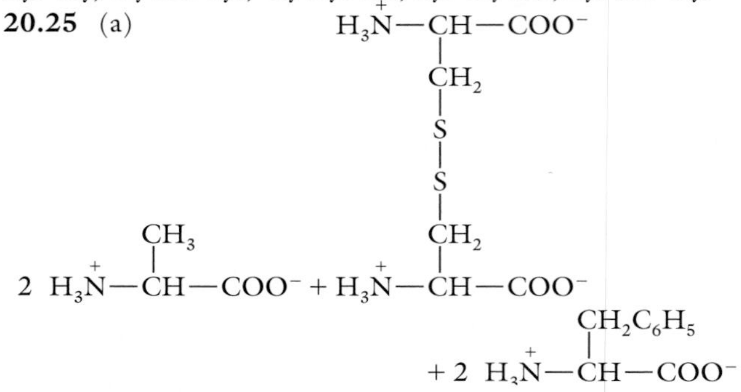

(b) no reaction

(c)

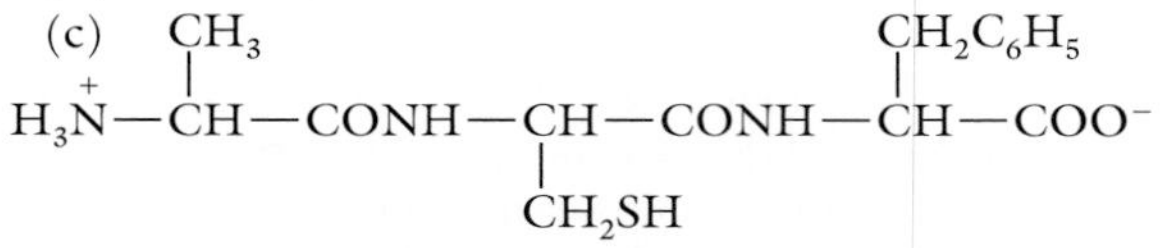

20.27 Simple proteins contain only peptide molecules. Conjugated proteins contain nonpeptide molecules or ions together with peptide molecules.

20.29 Rotation around a single bond

20.31 a

20.33 (a) None; (b) hydrogen bond; (c) hydrophobic; (d) none; (e) salt bridge; (f) none

20.35 Shielding of nonpolar residues from water; hydrogen bonding between peptide groups; interaction of polar residues with water

20.37 Primary structure

20.39 The functions of structural and contractile proteins require the formation of strong, macroscopic structures. These structures are possible only with fibrous proteins whose shape allows their aggregation by strong secondary attractive forces.

20.41 A double helix has two polypeptide chains coiled around each other; a triple helix has three chains coiled around each other.

20.43 α-Keratins are the structural components of cilia, hair, horn, hoof, skin, and wool.

20.45 The functions of catalytic, transport, regulatory, and protective proteins require that proteins not aggregate together to form macroscopic structures. The globular shape is not conducive to aggregation of proteins, because there are minimal secondary attractive forces between them. At the same time, globular proteins are solubilized because the hydrophilic nature of their surfaces allows strong hydrogen-bond attractions with water.

20.47 Tertiary structure describes the relation between the different conformational patterns in different local regions of a polypeptide. α-Keratins have the same conformational pattern throughout the polypeptide chain; myoglobin has different conformational patterns in different local regions of the polypeptide.

20.49 The polypeptide is the apoprotein; the heme is the prosthetic group.

20.51 Oxygen is picked up and held at the Fe^{2+} ion of hemoglobin.

20.53 Asp, Glu, His, Lys. The protein is solubilized by attractive interactions of these side groups with the aqueous environment.

20.55 A mutation is an alteration in the DNA structure of a gene that may in turn produce a change in the primary structure of a protein, which in turn may alter its function.

20.57 Side-group polarity is important to the three-dimensional structure and function of hemoglobin. In the detrimental mutation, a polar residue replaces a nonpolar residue. In the mutation with little effect, one nonpolar residue replaces another.

20.59 There will be a tendency for a change in conformation to place that residue on the exterior surface where attractive interaction with the aqueous environment can take place.

20.61 A native protein has the conformation that exists when the protein functions under physiological conditions.

20.63 Digestion destroys primary structure and produces α-amino acids by hydrolysis of peptide bonds. Denaturation alters secondary, tertiary, and quaternary structures but not the primary structure of proteins.

20.65 Ag^+ reacts with sulfhydryl groups of cysteine residues to form metal disulfide bridges that disrupt the protein's native conformation.

20.67 The conformations of polypeptides with lower contents of acidic and basic residues are less dependent on salt-bridge attractions.

20.69

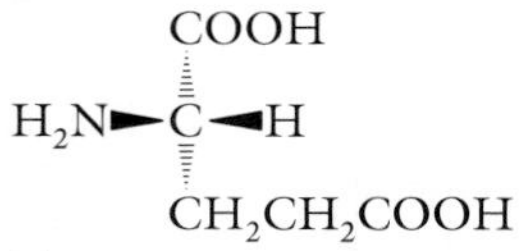

20.71 (a)

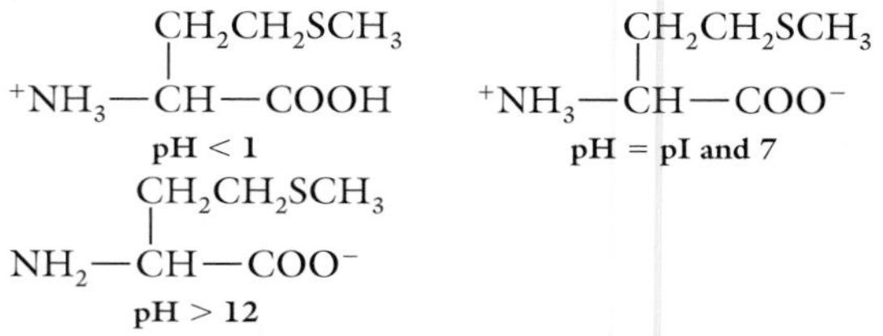

(b) Solubility is minimum at pI (5.74), increasing at both higher and lower pH values.

(c) No migration at pI and physiological pH; migration to anode at pH > 12, to cathode at pH < 1.

20.73 5! = 120: Val-Phe-Lys-Thr-Pro, Val-Thr-Lys-Phe-Pro

20.75 Four: L-Lys-Gly-L-Ala, D-Lys-Gly-D-Ala, L-Lys-Gly-D-Ala, D-Lys-Gly-L-Ala. L-Lys-Gly-L-Ala would be found in nature.

20.77 Methionylserylisoleucylglycylglutamic acid

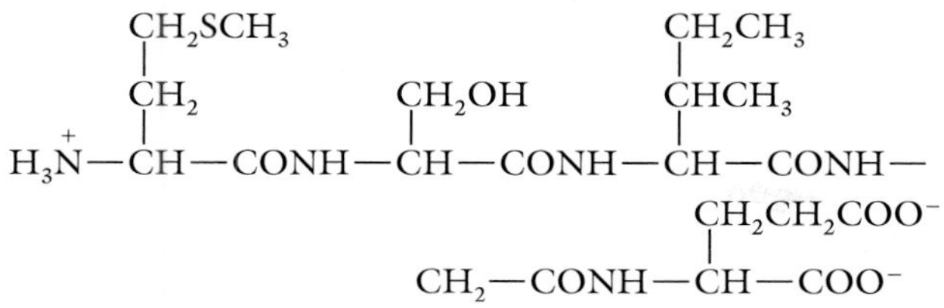

20.79

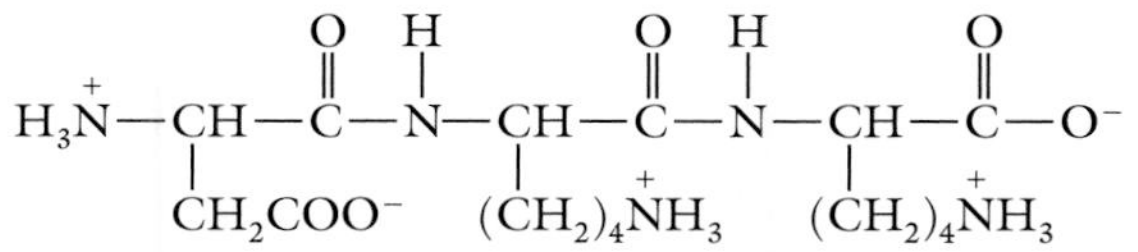

It has a charge of 1+ and migrates to the cathode. The pI is on the basic side of that of a peptide containing only neutral residues.

20.81 2−

20.83 Gly-Ala-Cys-Gly-Lys-Phe-His-Glu-His-Cys-Met-Gly-Asp

20.85 (a) 4.9; (b) anode

20.87 Pepsin

20.89 Globular proteins. Their function requires water solubility.

20.91 X is not a β-pleated sheet, because it has a low content of Gly, Ala, and Ser. It is not an α-helix, because of the high content of Pro and Glu. X is either a β-turn or a loop.

20.93 Protein conformation and function are generally least affected by mutations in which replacements are by residues of the same size, charge, and polarity: (a) a very small residue is replaced by a very large residue; (b) one basic residue (positive charge at physiological pH) is replaced by another; (c) one neutral polar residue is replaced by another; (d) a basic residue (positive charge at physiological pH) is replaced by an acidic residue (negative charge at physiological pH).

20.95 Residue 1 from lysine; 2 from proline; 3 from glutamic acid.

20.97 (a) Hydrophobic residues (Leu, Phe, Val) are in the interior of the bilayer because it is hydrophobic; (b, c) polar residues (Glu, Lys, Ser) are in the extracellular and intracellular environments because those environments are aqueous.

20.99 The alteration in sickle-cell anemia hemoglobin must be in regions of the polypeptide chains that greatly affect conformation and function. The difference in residues between beef and human insulins must be in regions of the polypeptide chains that do not significantly alter conformation and function.

20.101 The proteins are much larger molecules.

Chapter 21

21.1 Ribose for RNA; deoxyribose for DNA

21.3

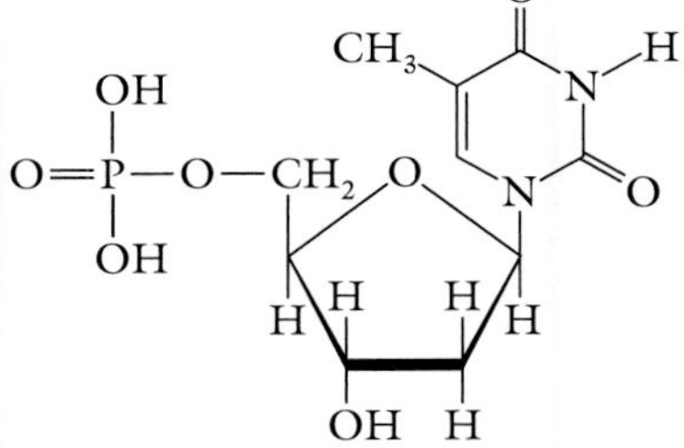

21.5 Deoxyribose, guanine, phosphoric acid

21.7 Nucleus

21.9 dTMP-dCMP

21.11 DNA does not contain U.

21.13

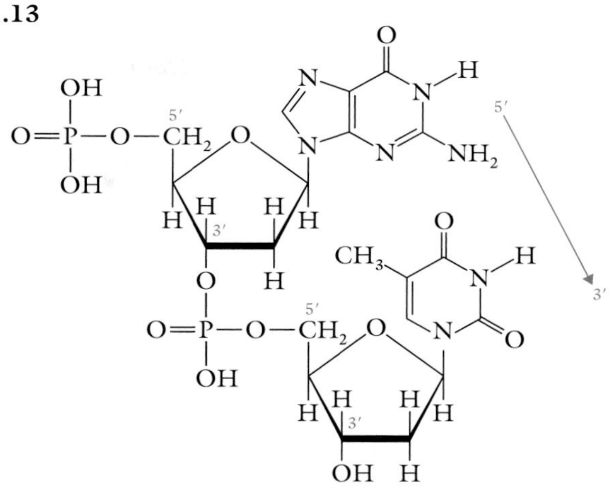

21.15 dAMP, 2 dTMP, dGMP, dCMP

21.17 Histones are proteins that compact the high-molecular-mass DNA double helix into the small volume of the cell nucleus.

21.19 Hydrogen bonding

21.21 d

21.23 30.4 mol-% T, 19.6 mol-% G

21.25 5′-CGGTTA-3′

21.27 DNA

21.29 A chromosome

21.31 (a) Replication bubbles are unwound parts of the DNA duplex where replication takes place; (b) the leading strand is the DNA strand that is synthesized in a continuous manner and the lagging strand is synthesized in a discontinuous manner; (c) Okazaki fragments are the discontinuous DNA fragments of the lagging strand.

21.33 Replication results in one parent DNA being converted into a pair of daughter DNAs.

21.35 (a) 5′-CGCTAA-3′; (b) 5′-TTAGCG-3′

21.37 RNA

21.39 The template DNA strand is used to produce the primary transcript.

21.41 Various primary transcripts are modified by different chemical reactions to produce rRNA, mRNA, and tRNA.

21.43 DNA: many billions, as high as 100 billion. rRNA: 500,000 to 1 million. mRNA: 100,000 to 1 or 2 million. tRNA: $<$ 50,000.

21.45 15 mol-% U, 25 mol-% G, 20 mol-% C, 40 mol-% A

21.47 Polypeptide

21.49 rRNA together with proteins form the ribosomes, in which polypeptide synthesis takes place. mRNA carries the genetic message encoding the polypeptide's amino acid sequence to be synthesized in the ribosome; binds to the ribosome and acts as the template for polypeptide synthesis. α-Amino acids are supplied for polypeptide synthesis when tRNAs transport them into the ribosome.

21.51 The genetic message carried by mRNA is its sequence of bases, which specifies the sequence of α-amino acids for the polypeptide to be synthesized.

21.53 With very rare exceptions, all species—plant and animal—follow the same genetic code.

21.55 Aminoacyl-tRNA synthetase is the enzyme that catalyzes the joining together of amino acids at aminoacyl and peptidyl sites on mRNA. The reaction is called acyl transfer.

21.57 Three consecutive bases on DNA or RNA.

21.59 Primary transcript = 3′-UGU-GUG-GUU-UAC-ACA-CCA-GUA-5′; mRNA = 5′-AUG-CAU-UUG-UGU-3′; polypeptide = Met-His-Leu-Cys

21.61 A mutation is an error in the base sequence of a gene.

21.63 In a substitution mutation, one base substitutes for another in the normal base sequence of DNA. In a frameshift mutation, a base is inserted into or deleted from the normal base sequence.

21.65 A spontaneous mutation is a mutation resulting from random errors in replication.

21.67 Germ cells are egg and sperm cells; all other cells are somatic cells.

21.69 A cancer is an uncontrolled growth of cells.

21.71 In mutation a, a negatively charged amino acid residue is replaced by an uncharged residue and is thus more harmful; in mutation b, one negatively charged residue is replaced by another. Biological function is most often altered when a residue is replaced by another that is different in size, polarity, or charge.

21.73 (a) Met-Asp-Leu; (b) Met-Glu-Leu; (c) Met-Asp-Leu; (d) Met-Asp-Leu (no effect of mutation in intron)

21.75 An antibiotic is a chemical, usually an organic compound, that kills microorganisms (bacterium, mold, yeast).

21.77 A virus is an infectious, parasitic particle, usually smaller than bacteria, that consists of either DNA or RNA (but not both) encapsulated by a protein coat.

21.79 A vaccine contains a weakened virus or its proteins and is used to prevent viral infections.

21.81 The genome of an organism is altered by transplanting DNA into it from another organism (either the same or a different species).

21.83 Vector DNA is some organism's DNA that is to be altered by the introduction of donor DNA (from some other organism or laboratory synthesis). The altered DNA is the recombinant DNA.

21.85 Nucleotides are building blocks for nucleic acids.

21.87 The 3′-end has an unreacted OH at C3′ of either ribose or deoxyribose; the 5′-end has only one of the oxygens of the phosphate group at C5′ attached to carbon.

21.89 Deoxyribonucleic acids do not contain U.

21.91 GMP, TMP, AMP, and CMP in the ratio 2 : 3 : 1 : 1.

21.93 DNA is double-stranded; RNA is single-stranded.

21.95 (a) 31 mol-% C, 14 mol-% A; (b) 17 mol-% C, 38 mol-% A

21.97 mRNA = 5′-AUG-CAA-UUU-AGU-3′; polypeptide = Met-Glu-Phe-Ser

21.99 A plant or animal conceived from a germ cell containing recombinant DNA

21.101 To introduce recombinant DNA into an organism to correct a hereditary disease

21.103 G-C base pairs have three hydrogen bonds, whereas A-T base pairs have two. Stability increases as the number of hydrogen bonds increases.

21.105 DNA consists of pairs of DNA molecules held together by the specific base-pairings G-C and A-T. Such base-pairing requires G = C and A = T for the pair (DNA duplex). These ratios are not needed for RNA, because RNA exists as single molecules, not pairs of double-helix molecules.

21.107 The minimum number of nucleotide bases is 438 because three nucleodide bases specify one amino acid residue. There are usually many more nucleotide bases because of the introns that are present. The 438 nucleotide bases are contained only in the exons of the gene.

Chapter 22

22.1 Biochemistry is the extension of the relation of structure to function to the molecular and subcellular levels of organization.

22.3 In contrast with eukaryotic cells, prokaryotic cells have no subcellular membrane-bounded organelles or structures.

22.5 Oxidative processes take place in the mitochondria of animal cells.

22.7 Protein synthesis takes place at ribosomes.

22.9 DNA in bacterial cells is not bounded by a membrane and is located in a microscopically visible nuclear zone.

22.11 (1) To obtain energy in a chemical form by the degradation of nutrients, (2) to convert nutrient molecules into precursor molecules used to build cell macromolecules, (3) to synthesize cell macromolecules, and (4) to produce or modify the biomolecules necessary for specific functions in specialized cells or both

22.13 Anabolism comprises those processes taking part in the synthesis of biomolecules.

22.15 To reduce nutrient biomolecules to monosaccharides, amino acids, and fatty acids

22.17 Acetyl-*S*-CoA is oxidized to CO_2 and H_2O.

22.19 ATP functions as the carrier of energy to the energy-requiring processes of cells and is the link between catabolism and anabolism.

22.21 Substrate phosphorylation and oxidative phosphorylation

22.23 NAD^+ functions in catabolic reactions, and the reduced form of $NADP^+$ (NADPH) is used in reductive biosynthetic reactions.

22.25 An active site is a region on an enzyme where catalysis takes place. It is a binding site and it has catalytic function.

22.27 Inducible fit describes cases in which a substrate molecule can influence the complementarity of a binding site.

22.29 The vitamins

22.31 The number of moles of substrate that react per unit time per mole of enzyme

22.33 An enzyme whose activity can be modified by combination with specific activators or inhibitors.

22.35 High-energy compounds are responsible for driving forward essentially unfavorable chemical reactions.

22.37 Catabolism consists of several distinct stages, the first of which is a collection step in which nutrient biomolecules are degraded to their building blocks, proteins to amino acids, polysaccharides to monosaccharides, and hydrolyzable lipids to fatty acids and glycerol. After that, hexoses, the carbon chains of fatty acids, and most of the amino acids are converted into acetyl-*S*-CoA, which is eventually oxidized to carbon dioxide and water.

22.39 No. The activities of cellular enzymes are regulated so that they can respond appropriately to the immediate metabolic needs of the cell. The activities of allosteric regulatory enzymes are modified by the binding of molecules that are not their substrates; other types of control are by covalent modification or feedback inhibition.

Chapter 23

23.1 Glucose is the principal source of energy for metabolism.

23.3 ATP, NADH, lactate (pyruvate under aerobic conditions)

23.5 The enzyme phosphofructokinase is the key regulatory control point in glycolysis.

23.7 In glycolysis, ATP is synthesized in two reactions:

ADP + 1,3-bisphosphoglycerate $\longrightarrow$
ATP + 3-phosphoglycerate

ADP + phosphoenolpyruvate $\longrightarrow$ ATP + pyruvate

23.9 NADH + H^+ + pyruvate $\rightarrow$ NAD^+ + lactate

23.11 Glycolysis would cease if inorganic phosphate were unavailable.

23.13 In the mitochondria

23.15 Yes. It is deactivated by phosphorylation and activated by dephosphorylation.

23.17 Isocitrate $\rightarrow$ α-ketoglutarate + CO_2
α-Ketoglutarate + CoA-SH $\longrightarrow$ succinyl-S-CoA + CO_2

23.19 Because the product of the reaction sequence, oxaloacetate, is the first reactant of the same sequence.

23.21 Because oxaloacetate is regenerated at the conclusion of each cycle, one molecule of it can effect the oxidation of limitless numbers of acetyl groups.

23.23 The enzyme pyruvate carboxylase is activated by acetyl-S-CoA and, in the presence of ATP, catalyzes the formation of oxaloacetate from pyruvate and CO_2.

23.25 Not directly. Oxaloacetate is first synthesized in the mitochondria and reduced to malate, which then enters the cytosol and is reoxidized to oxaloacetate. There, the oxaloacetate reacts with GTP under the influence of the enzyme phosphoenolpyruvate carboxykinase to form phosphoenolpyruvate, CO_2, and GDP.

23.27 Oxaloacetate is first reduced to malate, which can pass through the mitochondrial membrane and enter the cytosol.

23.29 Glycogen is the polymeric storage form of glucose in animal tissue.

23.31 Its hydrolysis has a very large equilibrium constant, is coupled to the overall process, and is the driving force for the fomation of the activated glucose molecule.

23.33 Phosphorolysis results in the formation of glucose-1-PO_4, whereas hydrolysis would yield glucose.

23.35 Glucagon and epinephrine are hormones.

23.37 They both regulate the blood-glucose concentration, but epinephrine also affects blood pressure and heart rate.

23.39 Only liver cells possess the enzyme glucose-6-phosphate phosphatase, which hydrolyzes glucose-6-PO_4 to form free glucose.

23.41 There are three locations along the electron-transport chain where there is sufficient energy for the synthesis of ATP.

23.43 No. The membrane must be intact.

23.45 No. The inner membrane possesses transport proteins specific for only a few substances.

23.47 The maximum total amount of ATP produced under aerobic conditions is 38 mol of ATP per mole of glucose.

23.49 In active muscle cells, the rate of glycolysis is much greater than the rate of the citric acid cycle.

23.51 Active muscle cells produce ATP primarily through glycolysis. For glycolysis to continue at a maximal rate, NAD^+ must be regenerated by oxidation of NADH and inorganic phosphate must be available for the formation of 1,3-bisphosphoglycerate.

23.53 NADH delivers its reducing power to the electron-transport chain by reducing oxidized cytosolic substances to their reduced counterparts, which can then penetrate the mitochondrial membrane.

Chapter 24

24.1 No. The upper limit for the mass of glycogen stored in the liver and muscles is less than the human daily caloric requirement.

24.3 When the blood-glucose concentration reaches its lowest levels between meals, and glucagon is released.

24.5 R-COO$^-$ + ATP + CoA-SH $\rightarrow$
RCO-S-CoA + AMP + 2P_i

24.7 No. There is no specific transport system for fatty acyl-S-CoA derivatives in the mitochondrial membrane.

24.9 The first product is:

$$C_{12}H_{25}-CH_2-CH=CH-\overset{\displaystyle O}{\overset{\displaystyle \|}{C}}-S-CoA$$

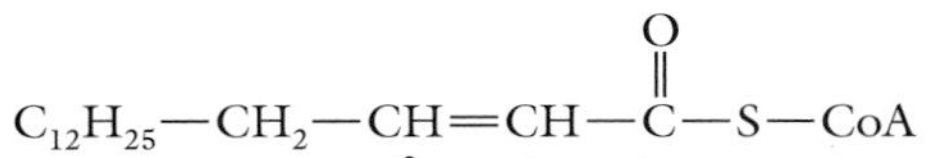

trans-Δ^2-Enoyl-S-CoA

24.11 The second product is:

$$C_{12}H_{25}-CH_2-\overset{\displaystyle OH}{\underset{\displaystyle H}{\overset{\displaystyle |}{\underset{\displaystyle |}{C}}}}-CH_2-\overset{\displaystyle O}{\overset{\displaystyle \|}{C}}-S-CoA$$

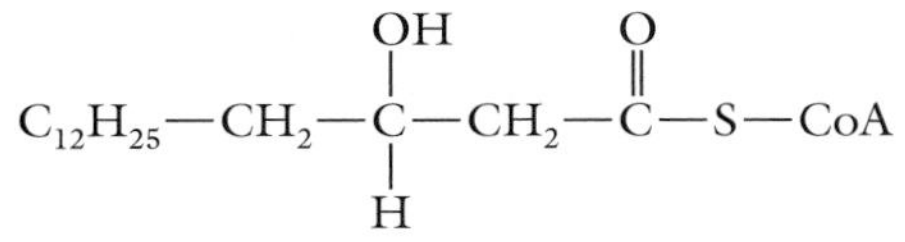

L-3-hydroxyacyl-S-CoA

24.13 The third product is:

$$C_{12}H_{25}-CH_2-\overset{\displaystyle O}{\overset{\displaystyle \|}{C}}-CH_2-\overset{\displaystyle O}{\overset{\displaystyle \|}{C}}-S-CoA$$

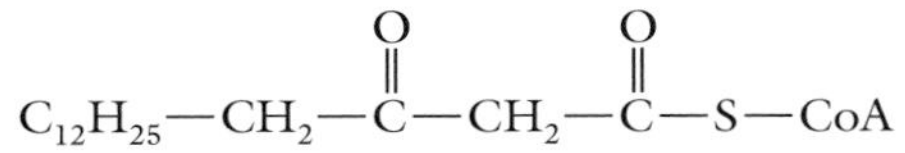

3-ketoacyl-S-CoA

24.15 The reactants in the fourth oxidation step are 3-keto(C_{16})acyl-S-CoA and CoA-SH.

24.17 Acetoacetate, D-3-hydroxybutyrate, and acetone

24.19 Ketosis arises during starvation and in diabetes mellitus.

24.21 Fatty acid anabolism takes place in the cytosol, and catabolism takes place in the mitochondria.

24.23 Insulin

24.25 Citrate from the mitochondria

24.27 Acetyl-S-CoA is transferred to the α-SH site, and malonyl-S-CoA is transferred to the β-SH site.

24.29 CO_2 is lost as HCO_3^-, so the reaction is irreversible and provides the driving force for the reaction. Biotin is the cofactor required for the formation of malonyl-S-CoA.

24.31 Prior to chain lengthening, the α-SH site is occupied by an acetyl group, and a carboxylated acyl derivative is at the β-site.

24.33 The formation of phosphatidate

24.35 The cost of the synthesis of a triacylglycerol is about 15% of the ATP generated in its oxidation.

24.37 In the form of cytidine diphosphoethanolamine

24.39 The hydrolysis of complex cellular glycolipids that normally takes place in the lysosomes does not take place.

24.41 In catabolism, the acyl carrier is CoA-SH, but, in anabolism, the acyl carrier is an —SH protein. Reduction in catabolism employs NADH, but, in anabolism, NADPH is used.

24.43 8 Acetyl-S-CoA + 7 ATP^{4-} + 14 NADPH +
7 H^+ $\rightarrow$ palmitoyl-S-CoA +
14 $NADP^+$ + 7 CoA-SH + 7 ADP^{3-} + 7 P_i^{2-}

24.45 Yes. The oxidation of the fatty acid chain produces NADH and $FADH_2$, both of which provide reducing power to the electron-transport chain for the synthesis of ATP through oxidative phosphorylation.

24.47 True. The presence of high concentrations of citrate indicates that the cells are in a high-energy state (ATP in high concentration). Equally important is the fact that citrate is a specific allosteric activator of acetyl-*S*-CoA carboxylase, which catalyzes the rate-limiting step in the fatty acid synthase system.

24.49 Fatty acids arise in adipose tissue by enzymatic hydrolysis of the stored triacylglycerols. The hydrolysis is catalyzed by a lipase that is activated by glucagon. The fatty acids leave adipose cells, become solubilized by being bound to serum albumin, and in that form travel throughout the circulatory system.

Chapter 25

25.1 Transamination is a process in which the amino group of an amino acid is interchanged with the carbonyl group of an α-keto acid.

25.3 They are catabolized and used as energy sources.

25.5 The amino groups of ingested amino acids are transferred to α-ketoglutarate by transamination.

25.7 By conversion into urea

25.9 A urotelic animal excretes ammonium ion in the form of urea.

25.11 The excessive lowering of α-ketoglutarate concentrations

25.13 The glucose-alanine cycle

25.15 The reactions constituting the urea cycle begin in the cytosol, continue in the mitochondria, and end in the cytosol.

25.17 Citrulline:

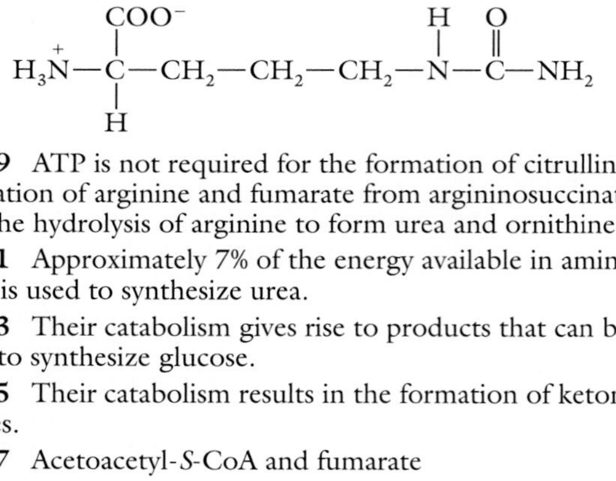

25.19 ATP is not required for the formation of citrulline, the formation of arginine and fumarate from argininosuccinate, and the hydrolysis of arginine to form urea and ornithine.

25.21 Approximately 7% of the energy available in amino acids is used to synthesize urea.

25.23 Their catabolism gives rise to products that can be used to synthesize glucose.

25.25 Their catabolism results in the formation of ketone bodies.

25.27 Acetoacetyl-*S*-CoA and fumarate

25.29 Intermediates of catabolic or anabolic sequences accumulate in cells.

25.31 It leads to severe mental retardation. Treatment requires that phenylalanine be eliminated or severely restricted from the diet of newborns.

25.33 A nonessential amino acid can be synthesized by humans.

25.35 The nonionic water-soluble compound can be excreted without the loss of important anions such as phosphate.

25.37 No. Arginase is an enzyme found only in the livers of terrestrial animals.

25.39 Phenylketonuria, which causes defects in the central nervous system, is the result of a defect in the enzyme that oxidizes phenylalanine to form tyrosine. The hydrolysis of aspartame in the intestine will produce phenylalanine, which phenylketonurics must avoid.

25.41 False. Oxidative deamination of aspartate leads to the formation of pyruvate and ammonium ion. Ammonia cannot exist at physiological pH.

Chapter 26

26.1 The exchange of nutrients and waste products with the external environment is easily accomplished by simple diffusion, but there is a high energy cost for maintaining a constant internal environment in the face of changing external environmental conditions.

26.3 Foods are enzymatically degraded to low-molecular-mass components to prepare them for absorption in the gut.

26.5 They bind to protein receptors on cell membranes and cause a second messenger to be activated in the cytosol, which initiates a variety of enzymatic cascades.

26.7 A zymogen is an inactive enzyme precursor.

26.9 They are transported across intestinal cells into the bloodstream and directly into the portal circulation.

26.11 Gastrin is secreted in the stomach in response to the entry of protein and stimulates the secretion of pepsinogen and HCl.

26.13 If a given protein provides all the required amino acids in the proper proportions and all are released upon digestion and absorbed, the protein is said to have a biological value of 100.

26.15 Fatty acids containing more than one unsaturated bond past carbon 9 of a saturated chain, counting from the carboxyl end, cannot be synthesized by humans. Eicosanoids, a family of lipid-soluble organic acids that are regulators of hormones, are synthesized by mammals from arachidonic acid—a polyunsaturated fatty acid that mammals can synthesize with the use of dietary polyunsaturated fatty acids of plant origin as precursors.

26.17 A dietary deficiency disease is caused by a deficiency in a factor essential to cellular function which can only be obtained through the diet.

26.19 The brain uses mostly glucose and some 3-hydroxybutyrate for energy needs.

26.21 A large electrical potential across its cell membranes and the ability to transmit an electrical signal from cell to cell

26.23 $Na^+_{in} = 5\ mM$, $Na^+_{out} = 150\ mM$; $K^+_{in} = 125\ mM$, $K^+_{out} = 10\ mM$

26.25 The brain synthesizes a large number of hormones and hormone-releasing agents that affect distant organs and tissues.

26.27 Virtually none

26.29 It must depend on the citric acid cycle for its ATP.

26.31 Glycogen

26.33 Lipases on adipocyte cell surfaces hydrolyze the triacylglycerols of the chylomicrons, allowing the resulting fatty acids to enter the cells.

26.35 Glucagon stimulates lipases within adipocytes.

26.37 It eventually reaches the liver to contribute to gluconeogenesis.

26.39 To produce the ATP necessary to effect the assymmetric distribution of Na^+ ions around the kidney tubules

26.41 Excess amounts of hydrogen ion can be eliminated by increasing the ammonia concentration of the urine, thereby increasing the amount of ammonium ion excreted.

26.43 The liver alone possesses glucokinase, an enzyme able to phosphorylate glucose at the high concentrations present after a meal.

26.45 It is converted into urea.

26.47 Acetoacetate and 3-hydroxybutyrate

26.49 It synthesizes the bile acids from cholesterol.

26.51 The absorptive state is the condition of the body immediately after a meal, when the gastrointestinal tract is full.

26.53 Insulin

26.55 Glucagon

26.57 In the postabsorptive state, blood glucose is supplied by glycogenolysis. When starvation begins, glycogen is unavailable, and gluconeogenesis with the use of glucogenic amino acids from muscle provides the glucose.

26.59 It typically appears early in life and can be controlled by insulin replacement.

26.61 Gluconeogenesis with the use of glucogenic amino acids provides the glucose.

26.63 Because glucose is not available for energy production, fatty acid oxidation becomes the main source of ATP.

26.65 The cells present are erythrocytes, leukocytes, and platelets.

26.67 The presence of hemoglobin in erythrocytes

26.69 It lowers the affinity of hemoglobin for oxygen.

26.71 Bicarbonate ion is formed in erythrocytes from CO_2, under the influence of the enzyme carbonic anhydrase.

26.73 A high concentration of CO_2 lowers the pH, and the Bohr effect enhances oxyhemoglobin dissociation.

26.75 The blood pH rises to higher than normal levels because CO_2 is being eliminated faster than it is being formed by respiring cells.

26.77 A lipoprotein consists of a core of hydrophobic lipids surrounded by a shell of amphipathic lipids and proteins.

26.79 The proteins in lipoproteins solubilize lipids, direct specific lipoproteins to particular tissues, and activate enzymes that hydrolyze and unload lipids from lipoproteins.

26.81 Nitrogen balance is achieved when the intake of protein nitrogen is equal to the loss of nitrogen in the urine and feces.

26.83 This vitamin is not ordinarily essential, because it is synthesized in adequate amounts by intestinal bacterial flora. However, vitamin B_{12} is transported across the intestinal cell membrane as a complex with intrinsic factor; in persons who cannot synthesize this protein, the vitamin cannot enter the bloodstream through intestinal absorption.

26.85 The liver uses excess glucose for the synthesis of fatty acids and cholesterol.

26.87 If it were synthesized as the active proteolytic enzyme chymotrypsin, it would digest the pancreas itself.

26.89 Dietary lipid in the form of chylomicrons is hydrolyzed at the surface of capillary membranes within muscle and adipose tissue. The free fatty acids then penetrate the cells of the tissue to be stored as triacylglycerol or oxidized for energy.

26.91 The protein contained in the wheat is difficult to extract because it is located within an indigestible husk.

Glossary

accuracy *See* **error.**
acetal An organic compound that contains two —OR or —OAr groups attached to the same carbon.
achiral Refers to a molecule or object that is not chiral; that is, it is superimposable on its mirror image.
acid *See* **Brønsted-Lowry acid.**
acid anhydride *See* **carboxylic acid anhydride.**
acid derivative *See* **carboxylic acid derivative.**
acid dissociation constant An equilibrium constant for the dissociation of a weak acid.
action potential After stimulation, an electrical depolarization that moves along a neuron's membrane.
activated complex A transitory molecular structure formed by the collision of two reacting molecules.
active site A location on an enzyme's surface where catalysis takes place.
acyclic compound An organic compound that does not contain a cyclic structure.
acyl group The —CO—R or —CO—Ar group that is found in carboxylic acids and their derivatives.
acyl transfer reaction A reaction that transfers an acyl group from one molecule to another.
addition polymerization The synthesis of a polymer by the self-addition of large numbers of alkene molecules.
addition reaction A reaction in which all the elements of a reactant add to the double or triple bond of a compound.
adipocyte *See* **adipose cell.**
adipose cell The cell in which triacylglycerols are stored.
alcohol An organic compound that contains a hydroxyl group (—OH) attached to an sp^3-hybridized carbon atom.
aldehyde An organic compound that contains a hydrogen and either an alkyl or an aryl group attached to a carbonyl group.
aldose A saccharide that contains an aldehyde group.
aliphatic hydrocarbon A nonaromatic hydrocarbon; that is, either an alkane, alkene, or alkyne.
alkali metal An element in Group I of the periodic table.
alkaline earth metal An element in Group II of the periodic table.
alkane A hydrocarbon that contains only carbon–hydrogen and carbon–carbon single bonds.
alkene An unsaturated hydrocarbon that contains a carbon–carbon double bond.
alkyl group A saturated hydrocarbon group that is attached to another group or to an atom in a molecule.
alkyne An unsaturated hydrocarbon that contains a carbon–carbon triple bond.
alpha particle The nucleus of a helium atom emitted by a radioactive substance.
amide An organic compound that contains a nitrogen atom attached to a carbonyl group.
amide bond The bond between the carbonyl carbon and the nitrogen in an amide group (—CO—N).
amine An organic compound that contains a nitrogen atom that is not attached to a carbonyl group.
amine salt The product of an amine and a strong acid such as HCl.
α-amino acid An amino acid containing both a carboxyl group and an amino group attached to the same carbon.
amino acid residue An α-amino acid that has been incorporated into a peptide.
amino acid sequence The sequence of amino acid residues in a peptide, listed from the N-terminal to the C-terminal residue.
amphipathic molecule A molecule that contains both hydrophilic and hydrophobic groups.
amphoteric compound A compound that is both an acid and a base.
anabolism The reactions of metabolism in which energy is conserved and biomolecules are synthesized.
anhydrous substance A substance that does not contain water.
anion An ion with a negative charge.
anode The positive electrode in a battery or an electrophoretic or other apparatus.
anomers Saccharides that are diastereomers differing only in the configuration at a hemiacetal or acetal carbon.
anticodon A base triplet on tRNA that is complementary to a codon on mRNA.
antioxidant A substance that protects other substances against damage from oxidation.
apoprotein The polypeptide part of a conjugated protein.
aqueous solution A solution with water as the solvent.
aromatic compound An organic compound with high stability due to the presence of six π electrons in a cyclic structure.
aryl group An aromatic group attached through one of its sp^2 carbons to another group or to an atom in a molecule.
atmosphere *See* **standard atmosphere.**
atomic symbol A one- or two-letter symbol for an element or an element's atoms.

basal metabolic rate The measurement of basal metabolic activity.
basal metabolism The minimal metabolic activity of a human at rest whose gastrointestinal tract is empty.
base *See* **Brønsted-Lowry base.**
base (of nucleic acids) A heterocyclic nitrogen base that is a component of a nucleotide.
base dissociation constant An equilibrium constant for the dissociation of a weak base.
base pairing The strong hydrogen-bonding attractive forces between a base on one DNA strand of a DNA double helix and a base on the other DNA strand.
base triplet A set of three consecutive bases in DNA or RNA.
base unit A fundamental unit of measurement for one of the base quantities in the SI system.
β-bend *See* **β-turn**

beta particle An electron emitted from the nucleus of a radioactive substance.
bifunctional reactant An organic compound with two functional groups per molecule.
binary compound A compound consisting of two elements.
binding site The structural component of an enzyme's catalytic site where substrate is bound by secondary forces.
biochemistry The study of the structures and functions of living organisms at the molecular level.
biological membrane A membrane that surrounds a cell or organelle.
biomolecules The molecules of which living organisms are composed.
boiling point The temperature at which a substance boils when the atmospheric pressure is 760 torr.
bond angle The angle between two bonds that share a common atom.
branched chain Refers to an organic compound in which not all carbon atoms in the molecule are connected one after the other in a continuous chain.
Brønsted-Lowry acid Any substance that can donate a proton.
Brønsted-Lowry base Any substance that can accept a proton.
buffer system An aqueous solution containing a Brønsted-Lowry acid with its conjugate base.

calorie The heat absorbed when the temperature of 1.0 g of water rises 1 Celsius degree between 14.5°C and 15.5°C.
carbocation An organic species carrying a positive charge on a carbon atom.
carbohydrate A polyhydroxyaldehyde or a polyhydroxyketone and its derivatives and polymers.
carbonyl group The carbon–oxygen double bond C═O, present in aldehydes and ketones.
carboxyl group The —COOH group present in carboxylic acids.
carboxylic acid An organic compound that contains a carboxyl group.
carboxylic acid anhydride An organic compound that contains the —CO—O—CO— group.
carboxylic acid derivative A compound that can be synthesized from or converted into a carboxylic acid.
carboxylic acid halide An organic compound that contains a halogen attached to a carbonyl group.
carboxylic ester An organic compound that contains an —OR or —OAr group attached to a carbonyl group.
carboxyl-terminal residue The end of a peptide that contains the carboxyl group.
catabolism Metabolic reactions in which molecules are degraded.
catalysis A process in which the rate of a chemical reaction is increased by the presence of a catalyst.
catalyst A substance that takes part in and accelerates the rate of a chemical reaction but emerges unchanged at the reaction's conclusion.
cathode The negative electrode in a battery or an electrophoretic or other apparatus.
cation An ion with a positive charge.
cellular respiration Metabolic reactions in which oxygen is used and carbon dioxide is produced. *See also* **respiration.**
Celsius scale A temperature scale, in degrees, that defines the freezing point of water at 0°C and the boiling point at 100°C.
centimeter A length equal to 1/100 of a meter.
chair conformation The puckered, nonplanar shape of a six-membered ring.
chemical bond An electrical force of attraction strong enough to hold atoms together to form compounds.
chemical change A process through which substances lose their chemical identities and form new substances with new properties.
chemical equation A shorthand representation of a chemical reaction that uses formulas for reactants and products and numbers before components to represent their mole proportions.
chemical equilibrium A state in which the rate at which products form is equal to the rate at which reactants form.
chemical kinetics The study of the rate of a chemical reaction.
chemical property The ability of a pure substance to undergo chemical change.
chemiosmotic mechanism Describes the formation of ATP by a hydronium ion gradient across the inner mitochondrial membrane caused by electron flow through the electron-transport chain.
chiral Refers to a molecule or object that is not superimposable on its mirror image.
chloride shift The movement of chloride ions from plasma into a red blood cell in response to the movement of bicarbonate ions out of the red blood cells.
chromosome A double helix of DNA located in the nucleus of a cell and containing the hereditary information that directs the synthesis of proteins.
chylomicron A small droplet consisting of about 90% triacylglycerol and small amounts of phospholipid, cholesterol, free fatty acids, and protein, formed during absorption of dietary fat and released into the extracellular space surrounding intestinal cells.
cis isomer The diastereomer of an alkene or cyclic compound in which similar substituents are on the same side of the double bond or ring.
class of organic compounds *See* **family of organic compounds.**
codon A base triplet on mRNA that codes for a specific tRNA carrying a specific α-amino acid.
coenzyme An organic molecule, often derived from a vitamin, that is essential to the functioning of an enzyme.
cofactor Any molecule or ion essential to the functioning of an enzyme.
colloidal particle A particle smaller than 1×10^{-4} cm.
combining power The number of bonds formed by an atom when the atom is present in a covalent compound.
combustion The burning of an element or a compound in air.
complementarity principle A principle that accounts for the specificity of enzymes. The structure of the substrate must complement the structure of the enzyme's binding site. *See also* **lock-and-key theory.**
complementary Refers to the base pairing of adenine with thymine in DNA, adenine with uracil in RNA, and guanine with cytosine in both DNA and RNA.
complementary strands The two DNA strands in a DNA double helix.
compound A pure substance composed of atoms of two or more elements present in a fixed and definite ratio.
concentration The quantity of a component of a mixture in a unit of mass or a unit of volume of the mixture.
condensation The conversion of the gaseous state into the liquid state.
condensation polymerization The synthesis of a polymer by the reaction between two bifunctional compounds, with the formation of a small-molecule (usually water) byproduct.

condensed structural formula An abbreviated structural formula.
configuration Describes the relative orientations in space of the atoms of a stereoisomer, independent of changes due to rotation about single bonds.
conformation Describes the different orientations of the atoms of a molecule that result only from rotations about its single bonds.
connectivity The order of attachment of the atoms in a molecule.
constitutional isomers Different compounds possessing the same molecular formula but differing in connectivity.
conversion factor *See* **unit-conversion factor.**
cosmic radiation Ionizing radiation emanating from the sun and outer space and consisting mostly of protons.
covalent bond The attractive force holding two atoms together resulting from the sharing of a pair of electrons.
cyclic compound An organic compound that contains a cyclic structure in which the first and last atoms of a chain are connected to each other.

deamination The removal of an amino group from an amino acid.
degree Celsius *See* **Celsius scale.**
degree Fahrenheit *See* **Fahrenheit scale.**
dehydration reaction A reaction that proceeds with the loss of water from within a molecule or between a pair of molecules.
dehydrogenation A reaction that proceeds with the loss of two hydrogen atoms from a molecule.
denaturation A loss of a protein's native conformation brought about by a change in environmental conditions, resulting in loss of biological function.
density A derived unit defined as mass per unit volume.
deoxyribonucleic acid (DNA) A polynucleotide formed from deoxyribonucleotides.
deoxyribonucleotide A nucleotide that contains deoxyribose, phosphoric acid, and a heterocyclic base bonded together.
derived quantity A unit that is a mathematical relation between two or more base quantities, such as centimeters squared (cm^2) or centimeters per second (cm/s).
derived unit of measurement *See* **derived quantity.**
dextrorotatory compound A compound that rotates plane-polarized light in the clockwise direction.
diastereomers Stereoisomers that are not mirror images of each other.
diatomic molecule A molecule consisting of two atoms, such as O_2.
diffusion A reduction in a concentration gradient resulting from the random motion of particles.
digestion The hydrolysis of the amide, ester, and other hydrolyzable groups of various foodstuffs in the digestive tract of an organism.
dipolar ion A molecule with an overall zero charge but containing atoms bearing opposite charges.
dipole Any molecule that is electrically neutral overall but electrically asymmetric—that is, containing separated partial and opposite electrical charges.
dipole–dipole force An attraction between dipolar molecules.
dipole moment A measure of the size of a dipole.
diprotic acid An acid containing two dissociable hydrogen atoms.
disaccharide A saccharide composed of two monosaccharide units bonded together.
dissolution The process of dissolving or of preparing a mixture that will form a solution.
disulfide An organic compound that contains the —S—S— group.
D/L system A system for naming enantiomers.
DNA *See* **deoxyribonucleic acid.**
DNA double helix Two paired DNA molecules wound around each other to form a right-handed double helix.
DNA strand A DNA molecule.
double bond Two bonds between a pair of adjacent atoms, either two carbons or one carbon and one oxygen.
double helix Two polymer molecules wound around each other in a helical manner.
dynamic equilibrium An equilibrium that is the result of two opposing processes both taking place at the same rate.

eicosanoid A nonhydrolyzable lipid derived from arachidonic acid, a polyunsaturated C_{20} fatty acid.
electrode An electrically conductive solid suspended in a conductive medium through which electricity can flow.
electrolyte Any substance that, when dissolved in water, will allow the solution to conduct electricity.
electromagnetic radiation Radiation, such as heat and light, that has its origin in the oscillation of charged particles.
electron configuration The complete description of the organization of the electrons of an atom.
electronegativity The ability of an atom covalently bonded to another atom to draw the bonding electrons toward itself.
electron shell An organization level of atomic electrons defined by a principal quantum number.
electron volt (eV) An energy unit used for radiation (1 eV = 96.5 kJ/mol).
element A substance in which all the atoms have the same atomic number and electron configuration.
elemental symbol A symbolic representation of the name of an element, such as He for helium.
empirical formula *See* **formula, empirical.**
emulsifying agent A substance that stabilizes a suspension of colloidal droplets of one liquid in a continuous phase of another liquid.
emulsion A stable suspension of colloidal droplets of one liquid in a continuous phase of another liquid.
enantiomers Stereoisomers that are nonsuperimposable mirror images of each other.
endothermic reaction A reaction in which heat is absorbed.
endpoint In a titration, the experimentally determined point at which the unknown acid or base is completely neutralized.
energy The capability to cause a change that can be measured as work.
energy level An atomic energy state defined by a principal quantum number.
enzyme A molecule, usually a protein, that catalyzes a biochemical reaction.
enzyme-substrate complex A temporary combination of enzyme with its substrate prior to catalysis.
equilibrium *See* **chemical equilibrium.**
equilibrium concentration The concentration of a product or a reactant of a chemical reaction at equilibrium.
equilibrium constant A constant that is calculated from the relation between molar concentrations of products and reactants of a chemical reaction at equilibrium.
equilibrium expression The relation between molar concentrations of products and reactants of a chemical reaction at equilibrium.
equivalence point In a titration, the theoretically expected point at which the unknown acid or base should be completely neutralized.

equivalent mass The equivalent mass of an acid is the formula mass of the acid divided by the number of reacting H^+ ions per mole of acid.
error The difference between the value considered true or correct and the measured value.
erythrocyte A red blood cell.
ester *See* **carboxylic ester.**
esterification The formulation of an ester by the reaction of a carboxylic acid with an alcohol or phenol.
ester linkage The functional group, —CO—O—, present in esters.
ether An organic compound that contains an oxygen bonded directly to two different carbons.
eukaryote An organism whose DNA is contained inside the nuclei of the organism's cells.
evaporation The vaporization of a liquid into the atmosphere.
exact number A number that can be considered to have an infinite number of significant figures.
exothermic reaction A chemical reaction that evolves heat.
expanded structural formula A structural formula that shows the bonds of a molecule.
extensive property A property, such as mass or volume, that is directly proportional to the size of the sample.
extracellular fluid The fluid that is outside a cell.

Fahrenheit scale A temperature scale, in degrees, on which the freezing point of water is 32°F and the boiling point is 212°F.
family of organic compounds A large number of different organic compounds that have a characteristic functional group and pattern of behavior in common.
fatty acid A long-chain aliphatic carboxylic acid.
feedback inhibition The inhibition of the first enzyme in a biosynthetic sequence by the last product of the sequence.
fibrous protein A water-insoluble protein whose molecules have an elongated shape with one dimension much longer than the others and with a tendency to aggregate together to form macroscopic structures.
Fischer projection A representation of the bonds at a carbon atom, in which horizontal lines represent bonds extending in front of the plane of the paper and vertical lines represent bonds extending behind the plane of the paper.
flavoprotein A protein combined with a coenzyme containing riboflavin.
fluid-mosaic model The conceptual model that describes the structure and functioning of cell and organelle membranes.
folding The formation of polypeptide conformation by rotations about single bonds.
formula, chemical A representation of how many atoms of each element are in a fundamental unit of a compound.
formula, empirical A representation that gives the smallest whole-number ratio of the atoms of the elements of a compound.
formula, molecular The formula that shows the numbers of the atoms of the elements in a molecule of a compound.
formula, structural The formula that shows how the various atoms in a molecule are bonded together.
formula mass The sum of the atomic masses of the atoms in a formula of a chemical compound.
formula unit The smallest particle that has the composition of the chemical formula of the compound.
formula weight *See* **formula mass.**
free radical *See* **radical.**
freezing point The temperature at which a substance undergoes the transition from a liquid to a solid.
functional group A specific atom or bond or a specific group of atoms in a specific bonding arrangement.
fused-ring compound An organic compound that contains adjacent rings sharing two or more ring atoms.

gamma ray Radiation of energy higher than that of an X-ray.
gene A part of a chromosome.
genetic code The complete list of mRNA codons and the α-amino acids and other instructions that the codons specify.
genetic disease A disease, caused by a mutation in a germ cell, that is passed from parent to offspring.
genetic message The sequence of codons in mRNA that specifies the sequence of α-amino acids for a polypeptide to be synthesized in a ribosome.
genome The total DNA contained in all the chromosomes of an organism.
geometric isomers *See* **cis isomer; trans isomer.**
germ cell An egg or sperm cell.
globular protein A water-soluble protein whose molecules have a globelike shape and do not aggregate together to form macroscopic structures.
glycogenesis The process by which glycogen is synthesized from glucose.
glycogenolysis The process by which glucose is obtained by the breakdown of glycogen.
glycoprotein A compound having a saccharide bonded to a protein.
glycoside A saccharide that is an acetal.
gradient The change in value of a physical quantity with distance.
gram A mass equal to 1/1000 of a kilogram.
group, organic *See* **substituent.**
group or family of the periodic table The elements contained in one of the vertical columns of the periodic table.
growth factor A substance, such as a protein or trace element, that regulates cell division.

half-life The time required for a substance to lose one-half of its physical or chemical activity.
Haworth structure (projection) The cyclic structure of a saccharide.
heat A form of energy that moves between two objects in contact that are at different temperatures.
heat of fusion *See* **molar heat of fusion.**
heat of reaction *See* **molar heat of reaction.**
heat of vaporization *See* **molar heat of vaporization.**
α-helix A polypeptide or other polymer molecule whose conformation is in the shape of a coiled spring.
hemiacetal An organic compound that contains one each of an —OH group and an —OR or —OAr group attached to the same carbon.
heterocyclic compound A cyclic organic compound that contains an oxygen or nitrogen atom in the ring.
heterogeneous mixture A mixture in which there are visual discontinuities in composition.
homeostasis The steady-state physiological condition of the body.
homogeneous mixture A mixture in which there are no visual discontinuities in composition.
hormone A substance synthesized in and secreted by an endocrine gland and then transported in the bloodstream to a target tissue where it regulates a cell function.
hydrate A compound in which water is present in a fixed molar proportion of the other constituents.
hydration A reaction that proceeds with addition of water to a double or triple bond. In aqueous solutions, the association of water with ions by secondary forces.

hydrocarbon An organic compound that contains only carbon and hydrogen.
hydrogen bond The intramolecular or intermolecular attractive interaction between the partly positive H of an O—H, N—H, or H—F bond and the partly negative O, N, or F of another O—H, N—H, or H—F bond.
hydrolysis reaction The reaction of an organic compound with water resulting in cleavage of the compound into two organic fragments each of which combines with a fragment (H^+ or OH^-) from water.
hydrophilic Refers to a molecule or a part of a molecule that is attracted to water.
hydrophobic Refers to a molecule or a part of a molecule that is repelled by water.
hydroxyl group An —OH group attached to a carbon atom.

inorganic chemistry The study of compounds that contain elements other than carbon.
inorganic compound A compound that contains elements other than carbon.
intermolecular forces *See* **secondary force.**
intermolecular process A process (physical or chemical) that takes place between molecules.
International System of Units *See* **SI units.**
interstitial fluid Fluid outside cells and not in the blood.
intracellular fluid Fluid that is inside a cell.
intramolecular process A process (physical or chemical) that takes place within a molecule.
in vitro Refers to a substance or a process that is outside an organism.
in vivo Refers to a substance or a process that is inside an organism.
ionizing radiation Radiation that enters a medium and creates ions from the molecules therein.
isomers Different compounds that have the same molecular formula.

joule The SI unit of energy (4.184 J = 1 cal).

kelvin The SI unit of temperature; its size is equal to 1/100 of the temperature interval between the freezing point and the boiling point of water.
ketone An organic compound that contains two alkyl or two aryl groups or one alkyl and one aryl group attached to a carbonyl group.
kilocalorie The quantity of heat equal to 1000 calories.
kilogram The SI unit of mass.
kilojoule The quantity of energy equal to 1000 joules.
kinetic energy The energy of a moving body.
kinetics *See* **chemical kinetics.**

leukocyte A white blood cell.
levorotatory compound A compound that rotates plane-polarized light in the counterclockwise direction.
like-dissolves-like rule A solute is soluble in a solvent only if the secondary attractive forces between molecules of solute are similar to the secondary attractive forces between molecules of solvent.
linear chain Refers to an organic compound in which all carbon atoms in the molecule are connected one after the other in a continuous chain.
lipid A naturally occurring compound that is soluble in nonpolar or low-polarity solvents.
lipid bilayer Two lipid layers are arranged with the hydrophobic sides in contact with each other and the hydrophilic sides forming the inner and outer surfaces of the membrane, which are in contact with the internal and external aqueous environments.
lipoprotein A protein that has a lipid part either covalently bonded to the protein or held in place by secondary forces.
liter A volume equal to 1000 cm^3, or 1000 mL.
lock-and-key theory A theory to account for the specificity of enzymes for their substrates. The substrate must fit into a binding site just as a key fits into a lock.
longest continuous chain The number of carbon atoms in a molecule that are connected in a successive manner.
loop A region of polypeptide conformation that is less ordered than β-turns and much less ordered than α-helices and β-sheets.

macromolecule *See* **polymer.**
mass A measure of the quantity of matter relative to a reference standard.
mass number The sum of protons and neutrons in an isotope of an element.
matter Anything that has mass and occupies space.
measurement An instrumental determination of a physical quantity.
mechanism of reaction The molecular-level details of how reactants change into products.
melting point The temperature at which a solid is transformed into a liquid.
meso compound A diastereomer that contains tetrahedral stereocenters but is achiral and optically inactive.
metabolism All the chemical reactions that take place in an organism.
metal An element or combination of elements that is shiny, conducts electricity, and, if solid, is malleable.
metalloid An element that has some of the properties of metals and nonmetals.
meter The SI unit of length (1 m = 1000 cm).
metric system A decimal system of weights and measures in which base units are converted into smaller or larger multiples by movement of a decimal point; superceded by the International System of Units (SI system).
microgram A mass equal to 1/1000 of a milligram.
microliter A volume equal to 1/1000 of a milliliter.
milliliter A volume equal to 1/1000 of a liter.
millimeter A unit of length equal to 1/1000 of a meter.
mixture Matter consisting of two or more pure substances in varying proportions.
molar concentration A solution's concentration in units of moles of solute per liter of solution; molarity.
molar heat of fusion The amount of heat required to melt 1 mole of a pure solid.
molar heat of reaction The amount of heat generated or absorbed per mole of reactant when a chemical reaction takes place.
molar heat of vaporization The amount of heat required to vaporize 1 mole of a pure liquid.
molarity *See* **molar concentration.**
mole An Avogadro number of a substance's formula units.
molecular compound A compound whose atoms are joined by covalent bonds.
molecular formula *See* **formula, molecular.**
molecular mass The formula mass of a substance.
molecular weight *See* **molecular mass.**
molecule The smallest particle of a molecular compound.
monomer A reactant from which polymers are synthesized.
monoprotic acid An acid containing one dissociable hydrogen atom.

monosaccharide The simplest saccharide; monosaccharides cannot be hydrolyzed into smaller saccharides.
multiple bond A double or triple bond.
mutarotation The change in optical rotation that takes place when a pure α- or β-saccharide with a hemiacetal structure is dissolved in water.
mutation An alteration in the base sequence of a gene that may in turn produce a change in the primary structure of a peptide.

native protein The conformation of a protein under normal physiological conditions.
neuron Any of a number of specialized cells constituting the body's nervous system.
neutralization reaction A reaction between an acid and a base.
noble gas An element in Group VIII of the periodic table.
nomenclature The naming of compounds.
nonbonding electron An electron located in the outer shell of an atom but not participating in bonding to other atoms.
nonelectrolyte Any substance that, when dissolved in water, will not allow the solution to conduct electricity.
nonmetal An element that is not shiny, cannot conduct electricity, and is not malleable.
nucleic acid A polynucleotide.
nucleotide A building block (monomer) for nucleic acids.
nucleus of a cell A membrane-enclosed organelle in which the cell's genetic information is stored and nucleic acids are synthesized.

open chain *See* **acyclic compound.**
optical activity The ability of a compound to rotate the plane of plane-polarized light.
optical rotation The rotation of the plane of plane-polarized light.
orbital The region of space in which an electron resides.
orbital hybridization The excitation of an element's ground state to produce a different electronic configuration.
organelle A specialized, self-contained structure surrounded by a membrane and found inside a cell.
organic chemistry The study of compounds that contain carbon.
organic compound A compound that contains carbon.
oxidation of an inorganic compound An increase in the number of oxygen atoms or an increase in the positive charge of the metallic element or both.
oxidation of an organic compound An increase in the number of oxygen atoms or a decrease in the number of hydrogen atoms (or both) bonded to one or more of the carbon atoms of the compound.
oxidative phosphorylation The formation of ATP coupled to the flow of hydronium ions through an ATP synthase in the inner mitochondrial membrane.
oxidizing agent A substance that oxidizes another substance.

peptide A polyamide formed by the bonding together of α-amino acids.
phenol An organic compound that contains a hydroxyl group (—OH) attached to a carbon atom of a benzene ring.
phosphate ester An ester of a phosphoric acid, synthesized in the reaction of a phosphoric acid with an alcohol or phenol.
phospholipid An amphipathic lipid that contains a phosphodiester group, either a glycerophospholipid or sphingophospholipid.
photon A unit of light energy.
physical change A change in a physical property.
physical property Any observable physical characteristic of a substance other than a chemical property.
physical quantity A property that is described by a quantity and a unit.
physical state The state of being either a gas, a liquid, or a solid.
physiological function A function of a living organism or of an individual cell, tissue, or organ of which the organism is composed.
physiological saline solution A solution of sodium chloride with the same osmolarity as that of blood.
pi (π) bond The weaker bond in a double bond.
plasma The liquid part of blood, in which blood cells and other substances are dissolved or suspended.
β-pleated sheet A polypeptide or other polymer molecule whose conformation has side-by-side alignment of adjacent extended chains, in either a parallel or antiparallel arrangement.
polar covalent bond A covalent bond in which the electron pair resides closer to one of the bond partners than the other.
polar molecule A molecule that has a dipole moment.
polyatomic ion An ion, such as OH^- or CO_3^{2-}, made up of two or more atoms.
polyatomic molecule A molecule, such as $C_6H_{12}O_6$ or H_2O, made up of two or more atoms.
polyfunctional molecule A molecule that contains two or more functional groups.
polymer High-molecular-mass molecule produced by bonding together large numbers of smaller molecules.
polymerization The synthesis of polymer from monomer.
polypeptide A peptide that contains large numbers of α-amino acid residues.
polyprotic acid An acid that has two or more dissociable protons.
polysaccharide A polymer that contains large numbers of monosaccharide units.
precipitate A solid that forms in a solution as the result of a chemical reaction.
precipitation The formation of a precipitate.
precision A measure of how close a series of measurements agree with each other.
pressure Force per unit area.
primary carbon A carbon directly bonded to one other carbon.
primary protein structure The α-amino acid sequence of a polypeptide.
product A substance that is produced by a chemical reaction.
proenzyme A protein that will have enzymatic activity subsequent to some form of activation, such as hydrolysis, of a part of its polypeptide chain.
prokaryote An organism (such as a bacterium) whose DNA is not contained inside a nucleus.
protein A polypeptide that contains more than 50 α-amino acid residues and has biological function.
protein turnover The net result of the breakdown and the synthesis of proteins.
pure substance *See* **substance, pure.**

quantum A quantity of energy contained by a photon of light.
quaternary carbon A carbon directly bonded to four other carbons.
quaternary protein structure The three-dimensional relation among the different polypeptide molecules of a protein.

racemic mixture An equimolar (1:1) mixture of the two enantiomers of a pair of enantiomers.

radical A species, such as $\cdot OH$, with an unpaired electron.
reactant A substance that undergoes chemical change in a reaction.
reaction, chemical *See* **chemical change.**
reducing agent A substance that reduces another substance.
reduction of an inorganic compound A decrease in the number of oxygen atoms or a decrease in the positive charge of the metallic element or both.
reduction of an organic compound A decrease in the number of oxygen atoms or an increase in the number of hydrogen atoms (or both) bonded to one or more of the carbon atoms of the compound.
renal threshold A blood concentration of a substance above which it appears in the urine.
representative element Any element that appears in an A group of the periodic table.
respiration The uptake of oxygen and release of carbon dioxide by either cells or the body.
reversible reaction A chemical reaction that can reach equilibrium by starting with either reactants or products.
ribonucleic acid (RNA) A polynucleotide formed from ribonucleotides.
ribonucleotide A nucleotide that contains ribose, phosphoric acid, and a heterocyclic base bonded together.
ring compound *See* **cyclic compound.**

salt Any crystalline compound that consists of oppositely charged ions, with the exception of H^+, OH^-, and O^{2-}.
saponification The hydrolysis of an amide or ester with a strong base such as NaOH.
saturated fatty acid A fatty acid with no C=C double bonds.
saturated hydrocarbon A hydrocarbon with only single bonds and no multiple bonds.
scientific notation A method of writing a number as a product of a number between 1 and 10 multiplied by 10^x where x can be either a positive or a negative number.
secondary carbon A carbon directly bonded to two other carbons.
secondary force An attractive force between identical molecules (such as H_2O and H_2O) or between different molecules (such as formaldehyde and water) or between different parts of the same molecule (such as a polypeptide).
secondary protein structure The conformation in a local region of a polypeptide molecule.
sigma (σ) bond The single bonds and the stronger bond of double and triple bonds in organic compounds.
significant figure The number of digits in a numerical measurement or calculation that are known with certainty plus one additional digit.
single bond One bond between two atoms.
SI units These units have the same names as those in the older, metric system but with new reference standards for the base units.
solute The substance present in smaller amount in a solution.
solution A homogeneous mixture of two or more substances that is visually uniform throughout.
solvent The substance present in larger amount in a solution.
somatic cell A cell other than an egg or sperm cell.
specific gravity The ratio of the density of a test liquid to the density of a reference liquid.
specific heat The heat absorbed or lost per Celsius degree change in temperature per gram of substance.
sphingolipid An amphipathic hydrolyzable lipid containing sphingosine.
standard, reference A base unit of measurement such as the meter or the kilogram.
standard atmosphere The pressure that supports a column of mercury 760 mm high at zero degrees Celsius.
standard conditions of temperature and pressure (STP) A temperature of zero degrees Celsius and one atmosphere of pressure.
standard solution A solution for which its concentration is known with accuracy.
state of matter A condition in which matter can exist: solid, liquid, or gas.
stereocenter A carbon atom in a molecule at which exchange of two substituents converts one stereoisomer into the other.
stereoisomers Different compounds that have the same connectivity but differ in configuration.
stoichiometry The calculation of the quantities of the elements or compounds taking part in a chemical reaction.
STP *See* **standard conditions of temperature and pressure.**
straight chain *See* **linear chain.**
strong acid An acid that is completely dissociated in aqueous solution.
strong base A base that is completely dissociated in aqueous solution.
structural formula *See* **formula, structural.**
structural isomers *See* **constitutional isomers.**
substance, pure An element or a compound; not a mixture.
substituent An atom or group of atoms, such as —Cl or $—CH_3$, in an organic molecule.
substrate The substance that is the object of an enzyme's catalysis.
sugar A monosaccharide or disaccharide.
surface tension A force at the surface of a liquid that reduces the area of the surface.
suspension A mixture in which the solute particles are larger than colloidal in size.

temperature A measure of the hotness or coldness of an object.
tertiary carbon A carbon directly bonded to three other carbons.
tertiary protein structure The three-dimensional relation among the different secondary structures in different regions of a polypeptide.
tetrahedral bond angle The 109.5° bond angle at sp^3-hybridized carbons.
tetrahedral stereocenter A carbon atom with four different substituents.
tetrahedron A geometric shape with four triangular faces of the same size.
theory A fundamental assumption that explains a large number of observations, facts, or hypotheses.
thermal expansion The increase of volume of a substance in response to an increase in its temperature.
thermal property The manner in which a substance responds to changes in its temperature.
trans isomer The diastereomer of an alkene or cyclic compound in which similar substituents are on opposite sides of the double bond or ring.
transition element The elements between Group IIA and Group IIIA and in the actinide and lanthanide families.
transport (active, facilitated, simple) The passage of a solute across a cell or organelle membrane.
triacylglycerol A triester of glycerol.
triatomic molecule A molecule containing three atoms, such as H_2O or CO_2.
trigonal bond angle The 120° bond angle at the carbons of a double bond.

triprotic acid An acid containing three dissociable hydrogen atoms.

β-turn The conformation of a polypeptide chain in a region where the chain abruptly changes direction.

turnover number A quantitative measure of the replacement rate of a cellular component.

unbranched chain *See* **linear chain.**

uncertainty The estimate of the last number in a measurement.

unit-conversion factor A fraction, such as 2.54 cm/in., that allows the conversion of one unit into another unit.

unsaturated fatty acid A fatty acid with one or more C=C double bonds.

unsaturated hydrocarbon A hydrocarbon with a multiple bond.

vacuum An enclosed space containing no matter.

valence electron An electron in the outermost electron shell of an atom.

valence shell The outermost electron shell of an atom.

valence-shell electron-pair repulsion theory (VSEPR theory) A theory that accounts for the symmetric distribution of atoms around a central atom in a covalent compound.

vaporization The conversion of a liquid into a gas.

vapor pressure The pressure of a vapor in equilibrium with its liquid.

vitamin An organic compound that is required in trace amounts for normal metabolism but is not synthesized by an organism and must be included in the diet.

volatile liquid A liquid with a relatively high vapor pressure.

volume The capacity of an object to occupy space.

water of hydration Water contained in a pure solid in a specific ratio of moles of water to moles of substance.

weak acid An acid that undergoes incomplete dissociation in aqueous solution; one that has a small dissociation constant.

weak base A base that undergoes incomplete dissociation in aqueous solution; one that has a small dissociation constant.

wedge-bond representation Similar to a Fischer projection except that horizontal bonds are shown as solid wedge bonds and vertical bonds are shown as dashed wedge bonds.

weight The gravitational force on an object relative to some reference standard.

zymogen *See* **proenzyme.**

Illustration Credits

CHAPTER 1

2: Mary K. Denny/ Photo Edit. **5:** Figure 1.2, Chip Clark. **6:** Figure 1.4 a & b, Chip Clark. **7:** Figure 1.5, Chip Clark. **7:** Figure 1.6 a, b, c, Richard Megna/ Fundamentals Photographs. **8:** Figure 1.7, Chip Clark; Figure 1.8, Richard Megna/ Fundamentals Photographs; Figure 1.9, W. H. Freeman photo by Ken Karp. **22:** Figure 1.11, Chip Clark. **23:** Figure 1.12, Richard Megna/ Fundamentals Photographs. **26:** Box 1.4, G. L. Kooyman/ Animals Animals. **33:** St Bartholomew's Hospital/Science Photo Library/ Photo Researchers.

CHAPTER 2

37: David J. Sams/ Tony Stone. **40:** Figure 2.2, Ernest O. Lawrence/ Berkeley National Laboratory. **46:** Figure 2.4, Alexander Boden. **47:** Figure 2.5, Richard Megna/ Fundamentals Photographs; Figure 2.6, Chip Clark. **49:** Figure 2.7, Photo Edit. **50:** Figure 2.8 a, b, c, W. H. Freeman photos by Ken Karp. **51:** Figure 2.10 a & b, Dept. of Physics, Imperial College/Science Photo Library/ Photo Researchers. **51:** Box 2.1, Courtesy Perkin Elmer.

CHAPTER 3

65: Bill Aron/ PhotoEdit. **68:** Box. 3.1, E.R. Degginger, Convent Station, New Jersey. **86:** Figure 3.7 a & b, W. H. Freeman photos by Ken Karp. **88:** Figure 3.10, Håkon Hope.

CHAPTER 4

94: Donald C. Johnson/ The Stock Market. **96:** Figure 4.1, W. H. Freeman photo by Ken Karp. **111:** Richard Hutchings/ PhotoEdit.

CHAPTER 5

120: Michael DeMocker/ Visuals Unlimited. **124:** Figure 5.3, Mary Kate Denny/ PhotoEdit. **125:** Box 5.2, Jose L. Pelaez/ The Stock Market. **126:** Science Photo Library/ Photo Researchers. **133:** Box 5.4, Ulrich Sapountsis/Okapia/ Photo Researchers. **143:** Box 5.5, SIU/ Visuals Unlimited.

CHAPTER 6

148: Spencer Grant/ PhotoEdit. **157:** Figure 6.7, Chip Clark. Figure 6.8, Hermann Eisenbeiss/ Photo Researchers. **159:** Box 6.2, Peter Ryan/Science Photo Library/ Photo Researchers. **163:** Box 6.3, Richard Megna/ Fundamental Photographs. **166:** Julian Baum/ Bruce Coleman. **168:** Figure 6.13, Chip Clark; Figure 6.14, a, Steven Smale; b, W. H. Freeman photo by Ken Karp.

CHAPTER 7

173: Steve Weinrebe/ Picture Cube. **180:** Figure 7.4 a, b, c, Chip Clark. **187:** Figure 7.6, Chip Clark; Figure 7.7, Chip Clark. **193:** Figure 7.11, Courtesy Ionics, Inc. **194:** Figure 7.12 a, b, c, From " Biological Membranes and Bilayer Couples," by M. Sheetz, R. Painter, and S. Singer, 1976, *Journal of Cell Biology,* 70:193. **196:** Figure 7.13, E.R. Degginger, Convent Station, New Jersey.

CHAPTER 8

202: Pete Saloutos/ The Stock Market. **208:** Spencer Grant/ Photo Researchers. **219:** Box 8.3, Richard R. Hansen/ Photo Researchers.

CHAPTER 9

223: Michael Newman/ PhotoEdit. **225:** Figure 9.1, Chip Clark. **227:** Figure 9.2, both, W. H. Freeman photos by Ken Karp. **229:** Figure 9.3 a & b, W. H. Freeman photos by Ken Karp. **242:** Box 9.2, John D. Gordon/ Visuals Unlimited. **243:** Figure 9.7, Chip Clark. **251:** Figure 9.10, Chip Clark. **252:** Figure 9.11, Chip Clark.

CHAPTER 10

258: Richard Megna/ Fundamental Photographs. **270:** Figure 10.2, Mulvehill/ The Image Works. **270:** Top left, PhotoEdit; bottom left, D. Wray/ The Image Works. **271:** Michael Dalton/ Fundamental Photographs. **272:** Figure 10.3, Yoav Levy/ Phototake; Figure 10.4, Courtesy BICRON, a division of Saint-Gobian Industrial Ceramics, Inc. 276: Box 10.4, Blair Seitz/ Photo Researchers. **278:** Figure 10.5, Reprinted with permission of Bruce McEwen of the Rockefeller Institute. **279:** Figure 10.6, N.I.H./Science Source/ Photo Researchers.

CHAPTER 11

286: David R. White, Richmond, Virginia. **297:** Box 11.1, Gasoline, motor oil, and kerosene, Tony Freeman/ PhotoEdit; petroleum jelly, Young-Wolff/ PhotoEdit; mineral oil and plastic, Tony Freeman/ PhotoEdit; textiles, Frank Siteman/ Picture Cube; drugs, J.C. Carton/ Bruce Coleman; detergent and dyes, Tony Freeman/ PhotoEdit. **314:** Najlah Feanny/ Stock Boston. **317:** Box 11.3, Michael Rosenfeld/ Tony Stone. **318:** Poulides/Thatcher/ Tony Stone. **321:** Box 11.5, Goddard Space Flight Center, Greenbelt, Maryland.

CHAPTER 12

329: Lance Nelson/ The Stock Market. **332:** Box 12.1, Joy Spurr/ Bruce Coleman. **339:** Box 12.2, RubberBall Productions/ PNI. **340:** Box 12.3, R.A. Steinbrecht. **343:** Figure 12.8, Richard Megna/ Fundamental Photographs. **347:** Figure 12.9, top left, and center, Tony Freeman/ PhotoEdit; top right, Felicia Martinez/

PhotoEdit; bottom left, center, and right, Tony Freeman/ PhotoEdit. **360:** Box 12.9, Network Production/ The Image Works.

CHAPTER 13

369: David R. White, Richmond, Virginia. **379:** Box 13.2, Paul Harris/ Tony Stone. **380:** Box 13.3, L. Kolvoord/ The Image Works. **381:** Figure 13.4, Richard Megna/ Fundamental Photographs. **392:** Box 13.6, SIU/ Photo Researchers, **396:** Karl H. Switak/ Photo Researchers.

CHAPTER 14

403: Michael Rosenfeld/ Tony Stone. **406:** Box. 14.1, left, Inga Spence/ Picture Cube; center, Ed Bock/ The Stock Market; right, Wally Eberhart/ Visuals Unlimited. **408:** Box 14.2 , Richard Megna/ Fundamental Photographs. **415:** Figure 14.5, Richard Megna/ Fundamental Photographs.

CHAPTER 15

433: David Joel/ Tony Stone. **440:** Richard Megna/ Fundamental Photographs. **441:** Box 15.1, Richard Megna/ Fundamental Photographs. **444:** Box 15.2, Bill Aron/ PhotoEdit. **446:** David Parker/SPL/ Photo Researchers. **450:** SIU/ Visuals Unlimited.

CHAPTER 16

463: Bill Aron/ PhotoEdit. **471:** Box 16.1, Rick Strange/ Picture Cube. **474:** Box 16.3, Dion Ogust/ Image Works. **479:** Box 16.4, Lawrence Migdale/ Photo Researchers. **480:** Charles Gupton/The Stock Market.

CHAPTER 17

495: David Hanover/ Tony Stone. **498:** Figure 17.3, Martin Bough/ Fundamental Photographs. **498:** Bottom left, A. Kerstitou/ Bruce Coleman. **511:** Box 17.1, SIU/ Photo Researchers.

CHAPTER 18

524: Lynn M. Stone/ Picture Cube. **530:** Scott Camazine/ Photo Researchers. **540:** Ronnie Kaufman/ The Stock Market. **546:** N/A /Jasmin/ PNI. **549:** David M. Phillips/ Photo Researchers.

CHAPTER 19

556: Simon Fraser/ Science Photo Library/ Photo Researchers. **564:** Top, Kim Taylor/ Bruce Coleman; bottom, Richard Megna/ Fundamental Photographs. **565:** Box 19.1, Richard Megna/ Fundamental Photographs. **567:** Ron Sanford/ The Stock Market. **574:** Box 19.3, Daemmrich/ The Image Works

CHAPTER 20

587: Mark Richards/ PhotoEdit. **592:** Left, Roy Morsch/ The Stock Market; center, Michael A. Keller/ The Stock Market; right, Don Mason/ The Stock Market. **607:** Leonard Lee Rue III/ Photo Researchers. **608:** Box 20.2, Brent T. Madison, Shizuoka.Ken, Japan. **613:** Figure 20.15, Meckes/Ottawa/ Photo Researchers. **614:** David Hundley/ The Stock Market.

CHAPTER 21

622: Charles Gupton/ Stock Boston. **623:** David Weintraub/ Photo Researchers. **632:** Figure 21.8c, From *Biochemistry,* 4th ed., by Lubert Stryer. © 1975, 1981, 1988, 1995 by Lubert Stryer. **646:** L. Steinmark/ Custom Medical Stock Photo. **648:** Kenneth Murray/ Photo Researchers. **653:** AP/ Wide World Photos.

CHAPTER 22

660: John Coletti, Boston.

CHAPTER 23

678: Tom Tracy/ The Stock Market. **701:** Figure 23.9, Courtesy Dr. Donald F. Parsons.

CHAPTER 24

708: Hattie Young/Science Photo Library/ Photo Researchers. **709:** P. Motta/Dept. of Anatomy/University "LaSapienza," Rome/Science Photo Library/ Photo Researchers. **716:** Box 24.1, Biophoto Associates/ Photo Researchers. **722:** P. Motta & T. Naguro/Science Photo Library/ Photo Researchers.

CHAPTER 25

725: Bruce Ayres/ Tony Stone.

CHAPTER 26

739: Charles Gupton/ The Stock Market. **740:** Michael Abbey/ Visuals Unlimited. **744:** Keith/ Custom Medical Stock. **748:** Culver Pictures. **751:** Science Photo Library/ Photo Researchers. **767:** Kenneth Murray/ Photo Researchers.

Index

Relations Between Units

Property	Common Unit	SI Unit
mass	2.205 pounds (lb)	1.000 kilogram (kg)
	1.000 lb	453.6 grams (g)
	1.000 ounce (oz)	28.35 g
	1 tonne (metric ton)	10^3 kg
length	1.094 yards (yd)	1.000 meter (m)
	0.3937 inch (in.)	1.000 centimeter (cm)
	0.6214 mile (mi.)	1.000 kilometer (km)
	1 in.	**2.54 cm**
	1 foot (ft)	**30.48 cm**
	1 angstrom (Å)	**10^{-10} m**
volume	**1 liter (L)**	**10^3 cm^3**
	1.000 gallon (gal)	3.785 L
	1.00 quart (qt)	0.946 L
pressure	1.000 torr	1.000 mm Hg
	1 atmosphere (atm)	**760 torr**
time	**1 minute (min)**	**60 seconds (s)**
	1 hour (h)	**3,600 s**
	1 day	**86,400 s**
energy	**1 calorie (cal)**	**1/4.184 joules (J)**
	1 electron-volt (eV)	96.485 kilojoules/mole (kJ/mol)
temperature conversions	(Fahrenheit temperature)/°F = 1.8 × (Celsius temperature)/°C + 32 (Celsius temperature)/°C = {[(Fahrenheit temperature)/°F] − 32}/1.8 (Kelvin temperature)/K = (Celsius temperature)/°C + 273.15	

Note: Entries in boldface type are exact. The numbers in the temperature conversion formulas also are exact.

Fundamental Constants

Name	Value
atomic mass unit	1.66054×10^{-24} g
Avogadro constant	6.02214×10^{23}/mol
electron charge	1.60218×10^{-19} C
gas constant	8.20578×10^{-2} L·atm·K^{-1}·mol^{-1}
mass of electron	9.10939×10^{-28} g
mass of neutron	1.67493×10^{-24} g
mass of proton	1.67262×10^{-24} g

SI Prefixes, Symbols, and Values

Prefix	Symbol	Value	Prefix	Symbol	Value
femto	f	10^{-15}	deka	da	10
pico	p	10^{-12}	hecto	h	10^2
nano	n	10^{-9}	kilo	k	10^3
micro	μ	10^{-6}	mega	M	10^6
milli	m	10^{-3}	giga	G	10^9
centi	c	10^{-2}	tera	T	10^{12}
deci	d	10^{-1}			